# SOIL MECHANICS AND FOUNDATION ENGINEERING

## SECOND EDITION

# SOIL MECHANICS AND FOUNDATION ENGINEERING

## SECOND EDITION

**P. Purushothama Raj**

Director
Adhiparasakthi Engineering College, Melmaruvathur
Kancheepuram District, Tamil Nadu

ISBN 9788131790816
**First Impression, 2013**
**Fourteenth Impression, 2024**
**Fifteenth Impression, 2025**



Published by Pearson India Education Services Pvt. Ltd, CIN: U72200TN2005PTC057128.

Head Office: 1st Floor, Berger Tower, Plot No. C-001A/2, Sector 16B, Noida - 201 301, Uttar Pradesh, India.
Registered Office: Featherlite, 'The Address' 5th Floor, Survey No 203/10B, 200 Ft MMRD Road, Zamin Pallavaram, Chennai – 600 044.
Website: in.pearson.com; Email: companysecretary.india@pearson.com

*Printed in India at: Manipal Technologies Limited, Manipal*

# Brief Contents

# Contents

# Preface

Popularity of this book amongst the undergraduate students and practising engineers has made the author to revive the book with updated materials.

The second edition of the book comprises 24 chapters dealing with four components, viz. Basic Principles of Soil Mechanics, Laboratory Determination of Soil Parameters Under Different Field Conditions, Earth Pressure Problems Including Foundations and Advanced Topics on Soil Mechanics Applications.

Chapters 1 9 deal with Basic Principles of Soil Mechanics. Chapter 1 deals with soil formation and composition, highlighting the types of weathering, soil formation in nature and major soil deposits of India. Chapters 2 and 3 explain the methods of identification and classification of soil including Bureau of Indian Standards, with the background knowledge of the index properties of soils. The basic properties of soils are compaction, permeability, consolidation and shear strength. Chapters 4–9 detail these properties. Principles of compaction and field compactions are explained in Chapter 4. Flow through porous medium and its applications are dealt within Chapters 5 and 6. Stresses in nature and applied stresses cause consolidation and failure due to shear. These aspects are dealt with at length in Chapters 7, 8 and 9. Chapter 7 details the different types of loading and the methods of computing stresses. Chapter 8 explains the basic theory of consolidation followed by computation of settlements while Chapter 9 explains the methods of determining shear strength of different types of soils under different loading and drainage conditions.

Chapter 10 discusses the basic techniques of testing of soils as per the Bureau of Indian Standards. Further, it includes methods of material collection, data presentation, computation and presentation of results and discussion. Necessary diagrams and standard values are included in the chapter.

Chapters 11–19 explain the principles of earth pressure and its applications in the design of earth-retaining structures, stability of slopes and foundations. Principles of earth pressure theories, in particular, the classical earth pressure theories of Coulomb and Rankine, and other modern theories are dealt with in Chapter 11. Chapter 12 gives the design concepts of retaining walls, including sheet piles and cuts, for both active and passive cases with different backfill conditions. Stability of slopes for different soil conditions, seepage conditions and pure pressure conditions are dealt with in detail in Chapter 13 with different methods of analysis.

Chapter 14 deals with the bearing capacity of soils and the connected theories for various ground conditions. Determination of safe bearing capacity and allowable soil pressure for different loading conditions are presented in Chapter 14. Chapters 15–17 cover the subjects on foundation engineering viz. shallow foundations, pile foundations drilled and caisson foundations. The subject matter has been dealt with in depth so as to introduce the student to the field of foundation engineering such that he/she will gain the ability to investigate and select the most suitable type of foundation. A foundation engineer or a student can select the best foundation, provided he/she knows the subsoil conditions and methods of ground improvements. These aspects are dealt with in Chapters 18 and 19. The latest methods of ground investigation and soil improvement techniques are explained in these chapters.

Advanced topics on Soil Mechanics Applications viz. Embankment Dams, Dynamic Loading of Soils, Environmental Geotechnology and Introductory Rock Mechanics are explained in Chapters 20–23. Chapter 24 on Highway Pavements, an application of soil mechanics, has been added in this second edition. Chapter 20 on Embankment Dams covers both homogeneous and non-homogeneous dams, including rock-fill dams. Theory of vibrations, theory of machine and machine foundations and design requirements are explained in Chapter 21. Chapter 22 gives a brief account of Environmental Geotechnology. An introduction on rock mechanics has been given in Chapter 23 which explains the index properties of rocks, classification and in situ stresses.

The second edition would not have been possible but for the excellent encouragement given by M/S Pearson Education for which the author expresses his gratitude.

**Pondicherry**
**January 2013** **P. Purushothama Raj**

# 1

# Soil Formation and Composition

## CHAPTER HIGHLIGHTS

Origin of soils – Types of rocks – Effects of weathering – Soil formation – Major soil deposits of India – Components of soils – Size and shape of soil particles – Inter-particle forces – Soil minerals – Soil–water system – Soil structure

## 1.1 INTRODUCTION

Soil is an unconsolidated material that has resulted from the disintegration of rocks. It includes sediments and deposits beneath rivers and seas and on land along with all organic and inorganic materials overlying the bedrock. It, thus, constitutes the earth's surface both on land and beneath water.

The type and characteristics of soil depend largely on its origin. Transportation causes the sizes and shapes of the particles to alter and sort into sizes. Cementation due to carbonates, oxides, or organic matter provides additional particle binding. Thus, the engineering properties, viz., permeability, consolidation, and shear strength, of a soil deposit are governed by the mode of formation, stress history, groundwater condition, and physico-chemical characteristics of the parent material.

Soil deposits constitute an assemblage of solid particles resulting in the formation of certain voids or pore spaces. These voids are in turn filled with a gas or liquid or both. These components, viz., solid particle, gas, and liquid, play a significant role in the fundamental behaviour of soil. Thus, we can visualize the soil deposit as a particulate system comprising three phases, viz., the solid phase, the liquid phase, and the gaseous phase.

This chapter deals with the geological aspects of the formation of several types of soil deposits. The composition of such soil deposits has been treated as a three-phase system. The factors contributing to the behaviour of each phase and to the soil structure formation are dealt with in this chapter.

## 1.2 ORIGIN OF SOILS

The earth's crust consists of both rock and weathered rock (as soil) up to an approximate depth of 20 km. Soils (weathered rocks) originated from the rocks and minerals of the earth's crust. The principal minerals subjected to weathering to produce soil at or near the earth's surface and available in the order of abundance are quartz, feldspar, pyroxene, amphibole, etc. Continuous slow weathering processes, aided by crustal deformities in the past, are believed to have decomposed the solid rock to fragments, creating soils. The type of soil developed depends on the rock type, its mineral constituents, and the climatic regime of the area. Rock types are grouped into three major classes: igneous, sedimentary, and metamorphic.

Cooling and hardening of molten magma resulted in the formation of igneous rocks. Slow cooling of molten magma yields large crystals, while rapid cooling results in small crystals. Granite, syenites, diorites, and gabbros have large crystals, while basalts, rhyolites, and andesites have small crystals.

Rocks containing quartz or orthoclase minerals with high silica content (e.g., granite and rhyolite) mostly decompose into sands* or gravelly** soils with a little clay.*** On the other hand, rocks (e.g., gabbros and basalt) whose minerals contain iron, magnesium, calcium, or sodium with a little silica decompose to yield fine-textured silty+ and clayey soils. Clays are not fragments of primary minerals from the parent rock but secondary minerals formed by the decomposition of primary minerals. Thus, the behaviour of clay is different from that of gravel and sand as the latter are composed of primary minerals.

Transportation agents such as wind, water, and ice may move the loose weathered rock materials and deposit them in layers called sediments. Such sediments, with the cementing properties of fragments, when subjected under heavy pressure to compaction and cementation, result in sedimentary rocks. Sedimentary type of rocks are classified as chemical (e.g., limestones and dolomites), clastic (e.g., shale and sandstone), and biochemical or organic (e.g., fossil limestone, chalk, coral, and coal in the form of peat, lignite, bitumen, or anthracite).

Sedimentary rocks, and to a lesser extent igneous rocks, when subjected to metamorphism (changes brought about by combinations of heat, pressure, and plastic flow), undergo changes in their texture, structure, and mineral composition, resulting in rocks called metamorphic rocks. Metamorphism is of two types, thermal and dynamic. Thermal metamorphism occurs primarily due to temperature increase and high hydrostatic pressure, whereas dynamic metamorphism is due to differential pressure. Metamorphism changes limestone to marble, sandstone to quartzite, and shale to slate.

Metamorphic rocks may be categorized as foliated or non-foliated. Foliation occurs during the process of metamorphism when some metamorphic rocks reduce back to sedimentary rocks. Gneiss and schist decompose into silt–sand mixtures with mica, slates and phyllites to clays, marble to limestone, and quartzite to sands and gravels.

The cyclic process of transforming rock to soil and vice versa is a continuous process occurring over millions of years through complex chemical and physical processes. This phenomenon, referred to as the geological cycle, is schematically shown in Fig. 1.1.

---

Particle sizes: *sand, 0.075 to 2 mm; **gravel, 2 to 4.75 mm; ***clay, < 0.002 mm; +silt, 0.002 to 0.075 mm.

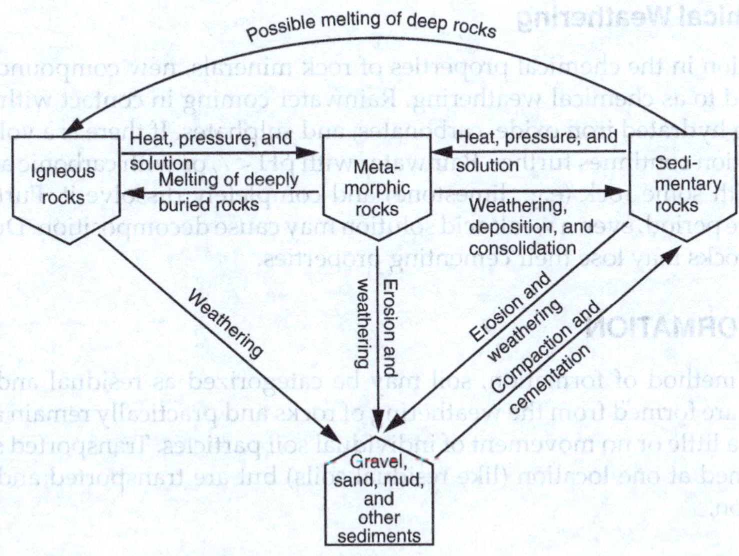


Fig. 1.1 Geological cycle (*Source*: Bowles, 1984)

## 1.3 TYPES OF WEATHERING

Rock disintegration, also called weathering, is one of the important geological processes. This disintegration of rock produces and deposits unconsolidated sediments as soils for plant and animal life. Weathering may be either physical (or mechanical) or chemical.

### 1.3.1 Physical Weathering

The process by which rock disintegrates into smaller fragments due to factors like stress changes, climatic changes, etc., without involving any change in its properties is called physical or mechanical weathering.

The principal factor causing physical weathering is climatic change. Adverse temperature changes coupled with different thermal coefficients of rock minerals produce rock fragments. The effect will be greater when temperature changes cause a freeze–thaw cycle. Similarly, heavy rainfall also brings about physical weathering.

Stress readjustments during regional uplift, accompanied by water runoff, cause the outer shell to separate from the main rock. This process is called exfoliation and may also be caused by sudden temperature changes.

On a rugged topography, heavy wind and rain may cause erosion of the rock surface and move disintegrated fragments. This is a continuing event and depends on the type of topography and the velocity of wind and water. Abrasion of rock is caused by ice under pressure or by the pushing of large unconsolidated materials. Mechanical weathering may also be caused by organic activity, such as cracking forces exerted by plants growing in the crevasses of rocks and the moving of fragments towards the surface by animals or insects.

### 1.3.2 Chemical Weathering

Due to alteration in the chemical properties of rock minerals, new compounds are formed. This is referred to as chemical weathering. Rainwater coming in contact with rock surfaces reacts forming hydrated iron oxide, carbonates, and sulphates. If there is a volume increase, the disintegration continues further. Rainwater with pH < 7, or with carbonic acid, may react chemically with some rock (e.g., limestone) and completely dissolve it. Further, during a geological time period, even a weak acid solution may cause decomposition. Due to leaching, sedimentary rocks may lose their cementing properties.

## 1.4 SOIL FORMATION

Based on the method of formation, soil may be categorized as residual and transported. Residual soils are formed from the weathering of rocks and practically remain at the location of origin with a little or no movement of individual soil particles. Transported soils are those that have formed at one location (like residual soils) but are transported and deposited at another location.

### 1.4.1 Residual Soils

Weathering (due to climate effects) and leaching of water-soluble materials in the rock are the geological processes in the formation of these soils. The rate of rock decomposition is greater than the rate of erosion or transportation of weathering material and results in the accumulation of residual soils. As the leaching action decreases with depth, there is a progressively lesser degree of rock weathering from the surface downwards, resulting in reduced soil formation, until one finally encounters unaltered rock (Fig. 1.2). Residual soils generally comprise a wide range of particle sizes, shapes, and composition.

In general, the rate of weathering is greater in warm, humid regions than in cold, dry regions. Humid, warm regions are favourable to chemical weathering. Also, because of the

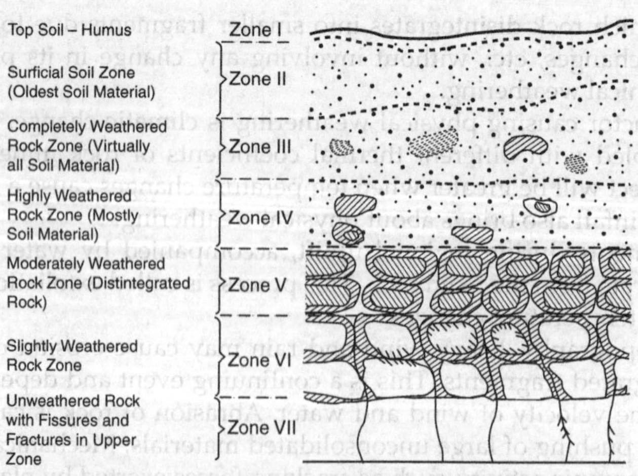


Fig. 1.2 Stages of formation of residual soil (*Source*: McCarthy, 1982)

presence of vegetation, there is less possibility of transportation of the decomposed materials as sediments. Residual soils exist in different parts of the world, viz., Asia, Africa, south-eastern North America, Central America, and South America. Sowers (1963) reported that the depth of residual soils varies from 6 to 25 m in general and from 7.5 to 15 m in South India.

### 1.4.2 Transported Soils

Weathered materials have been moved from their original location to new locations by one or more of the transportation agencies, viz., water, glacier, wind, and gravity, and deposited to form transported soils. Such deposits are further classified depending on the mode of transportation causing the deposit.

**Water-transported Soils**. Swift-running water is capable of moving a considerable volume of soil. Soil may be transported in the form of suspended particles or by rolling and sliding along the bottom of the stream. The size of the particle that can be in suspension is related to the square of the velocity of the flowing water. Particles transported by water range in size from boulders to clay.

Coarser particles are dropped when a decrease in water velocity occurs as the stream or river deepens, widens, or changes direction. Fine particles still remain in suspension and get deposited in quieter waters downstream. This is a typical case of a stream moving downhill, passing over a valley, and ultimately reaching a large body of water. Soils that are carried and deposited by rivers are called *alluvial deposits*. River deltas are formed in this manner. Soils carried by rivers, while entering a lake, deposit all the coarse particles because of a sudden decrease in velocity. Such coarse soil deposits are called *lake deltas*. But the fine-grained particles move to the centre of the lake and settle when the water becomes quiet. Alternate layers are formed with season, and such lake deposits are called *lacustrine deposits*. These deposits are weak and compressible and pose problems for foundations.

If coarse and fine-grained deposits are formed in sea water areas, then they are called *marine deposits*. Marine sediments are made up of terrestrial and marine contributions. The terrestrial contribution consists of particulate material eroded from the shore, as well as mineral matter, in true or colloidal solution, and this contribution decreases both in proportion and in grain size with increasing distance from the shoreline. The marine contribution is represented by the organic and inorganic remnants of dead marine life, and this normally increases with time (Iyer, 1975).

In marine deposits, marine life and environment play a more significant role than the salt concentration of the water. The clay particles absorb certain chemical elements from the organisms, which in turn can extract mineral substances from sea water. Some acids produced by the digestive tracts of marine organisms can alter the composition of the clay minerals (Iyer, 1975).

Marine clay deposits (excluding the deep deposits which have been subjected to many further changes and are overlain by other deposits) are generally weak, compressible, and problematic for foundations. If such a deposit is exposed above sea level and experiences leaching of sodium due to percolation of fresh water, it becomes very sensitive to disturbance.

Glacial soils transported by rivers from melting glacial water create deposits of stratified glacial drift and are referred to as *glacio fluvial deposit* or *stratified drift*.

**Glacial Deposits**. Compaction and re-crystallization of snow leads to the formation of glaciers. Glacier growth and movement depend on the formation of ice. Glacial deposits form a very large group of transported soils. A glacier moves extremely slowly but deforms and scours the surface and the bedrock over which it passes. Melting of a glacier causes deposition of all the materials, and such a deposit is referred to as *till*. The land form or topographic surface after a glacier has receded is called a *ground moraine* or *till plain*. Till deposits which have been overrun by glaciers contain coarser particles and form good construction material.

Soils deposited by the surface and sub-surface glacial rivers that remain in the form of long-winding ridges are called *eskers*. They may vary from about 10 to 30 m in height and about 0.5 km to several kilometres in length. Isolated mounds of glacial debris varying from about 10 to 70 m in height and 200 to 800 m in length are called *drumlins*. Large boulders picked up by a glacier, transported to a new location, and dropped are called *erratics*.

Glacial deposits provide a poor to excellent foundation. In many locations it is often found that the material is dense and contains considerable sand and gravel. It is believed that glaciers covered a large portion of the land during the ice age. Northern USA, Northern Europe, and Canada were subjected to continent glaciers. Now glaciers cover approximately 10% of the earth's surface. Almost all glaciers are now concentrated in Greenland and Antarctica.

**Wind-transported Soils**. Like water, wind can erode, transport, and deposit fine-grained soils. Soils carried by wind are subsequently deposited as *aeolian deposits*. *Dunes* are formed due to the accumulation of such wind-deposited sands. Dunes are a rather common occurrence in the desert areas of Africa, Asia, and the USA. Sands from dunes may be used to a limited extent for construction purposes.

Fine-grained soils such as silts and clays can be transported by wind in arid regions. Wind-blown silts and clays deposited with some cementing minerals in a loose, stable condition are classified as *loess*. Loess deposits have low density, high compressibility, and poor bearing resistance when wet. Loess is a clastic sediment comprising a uniformly sorted mixture of silt, fine sand, and clay-size particles. The structure of a loess deposit is susceptible to collapse on saturation.

**Gravity Deposits**. Gravity can transport materials only for a short distance. As the movement is limited, there is no appreciable change in the materials moved. Gravity deposits are termed *talus*. They include the material at the base of cliff and landslide deposits. The talus material at the cliff is formed due to the disintegration and subsequent failure of the cliff face. These fragments are generally loose and porous.

**Swamp and Marsh Deposits**. In water-stagnated areas where the water table is fluctuating and vegetational growth is possible, swamp and marsh deposits develop. Soils transported and deposited under this environment are soft, high in organic content, and unpleasant in odour. Accumulation of partially or fully decomposed aquatic plants in swamps or marshes is termed *muck* or *peat*. Muck is a fully decomposed material, spongy, light in weight, highly compressible, and not suitable for construction purposes.

### 1.4.3 Desiccated Soils

If a fine-grained soil is exposed to atmosphere, water is drawn from the interior to the surface. From the surface, the water gets evaporated. This sort of drainage is referred to as drainage by desiccation. During this process the soil becomes stiffer and ultimately becomes hard. The point at which evaporation ceases depends on the relative humidity of the air around.

Under field conditions, desiccation may take place whenever the surface of the soil is not permanently flooded. Due to periodic desiccation, even fine-grained silty sands show apparent cohesion.* As the apparent cohesion is very large, even rains of long duration cannot completely remove the cohesion. This phenomenon of desiccation is very much pronounced in soils of semi-arid and arid regions. Such soils are quite often mistaken for soft rocks. In the case of soft clays, the desiccation proceeds very slowly from the exposed surface and forms a thick crust, and the thickness grows with age.

## 1.5 MAJOR SOIL DEPOSITS OF INDIA

Among different types of soils spread over the Indian Peninsula, only five major deposits have been identified (Katti et al., 1975), viz., marine deposits, black cotton soils, laterite and lateritic soils, alluvial deposits, and desert soils. Figure 1.3 shows the regions covered by these soil deposits.

### 1.5.1 Marine Deposits

The marine deposits all along the Indian coast are generally derived from terrestrial sources. These deposits cover a narrow belt of tidal flats all along the coast from Porbandar in the west to Puri in the east. However, they are present over wide areas in places such as Rann of Kutch. These tidal flats experience high tide inundation. The deposits are very soft to soft clays, and the thickness varies from 5 to 20 m. The clay is medium sensitive and inorganic in nature. These deposits generally need a pre-treatment before application of any external load (Iyer, 1975). In order to prevent failures during construction, controlled loading should be planned (Katti et al., 1975).

### 1.5.2 Black Cotton Soils

Black cotton soil is one of the major soil deposits of India and is spread over a wide area of 3,00,000 km$^2$. The primary bed rock is basalt or trap, and in some locations, quartzites, schists, and sedimentary rocks are also reported (Katti et al., 1975). The black cotton soil is expansive in nature due to the presence of montmorillonite and illite clay minerals. The top black subsoil varies in thickness up to a maximum of 20 m. Based on the pedological conditions, crack depth and pattern vary. The soil surface is hard during summer and becomes slushy during the rainy season. The effect of seasonal moisture change brings in volume changes up to a maximum depth of 1.5 m. Because of the swelling and shrinking nature of the soil, there is a necessity for treatment of the soil, and special foundations need to be adopted in such soils to prevent failure of structures.

### 1.5.3 Laterites and Lateritic Soils

In tropical regions of high moisture and temperature, weathering activity is so intense that a tremendously thick soil (exceeding 30 m) may be formed from the parent rock through processes collectively termed *laterization*. Laterization is mainly due to the decomposition of

*A temporary shear strength gained by weathering.

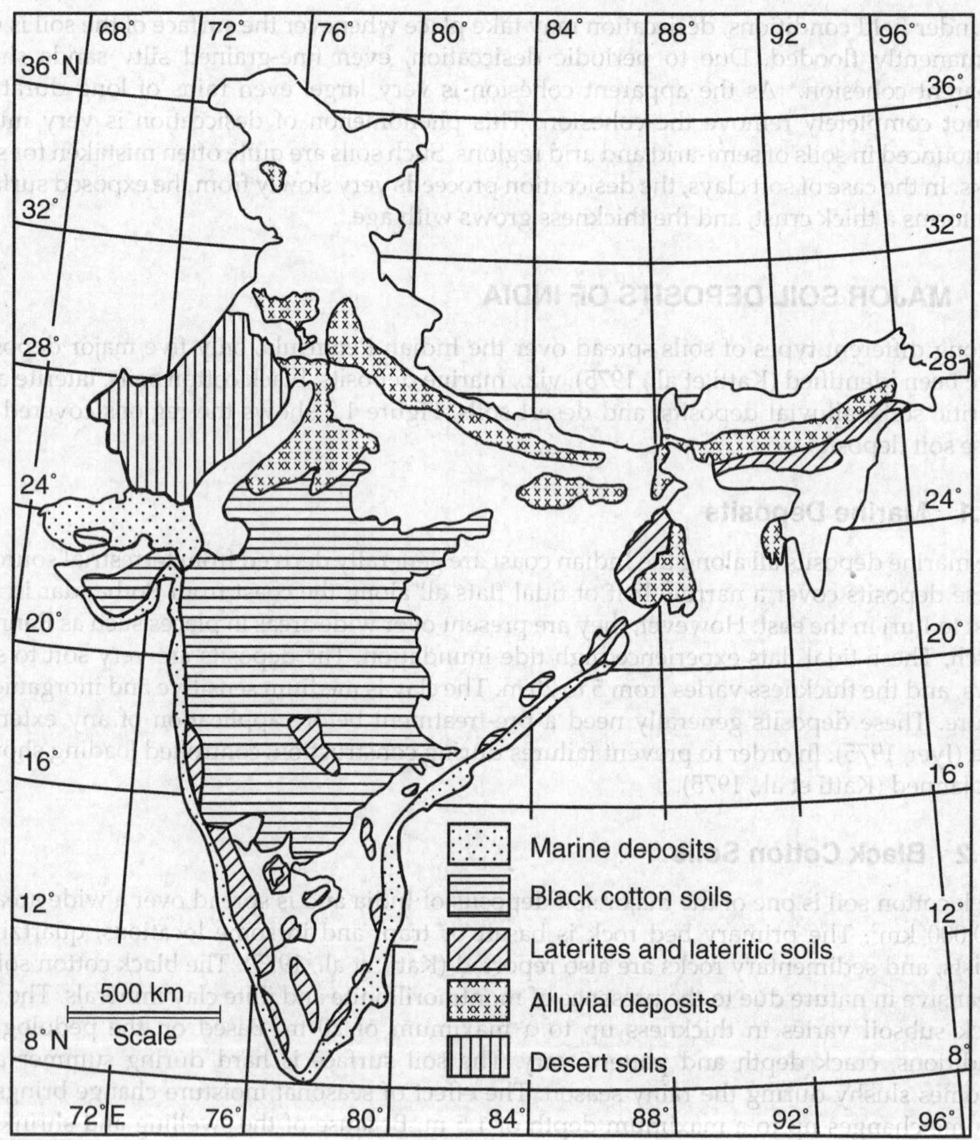


**Fig. 1.3** Map of India showing approximate extent of major regional deposits (*Source:* Katti et al., 1975)

rock, removal of silica and bases, and accumulation of aluminium and iron sesquioxides. The red, pink, or brown colour of laterites is essentially due to the presence of iron oxide. If about 90% of the material contains coarse grains, then this is called *laterite*; instead, if relatively fine grains are present, it is referred to as *lateritic soil*. In India, lateritic soils spread over an area of 100,000 km$^2$. Indian laterites are mostly residual soils.

The characteristic property of this type of soil is high strength when it is cut and dried in the sun. The specific reason for such a behaviour has been attributed to the dehydration of iron oxides and the presence of halloysite type of clay mineral. Some of the laterites show extremely high strength comparable to that of burnt bricks. After hardening, the strength gained is not affected when it comes in contact with water. Rao and Raymahashay (1981) studied the mineralogy of Calicut and Rajahmundry laterites. Calicut laterites were found to be rich in halloysite and crystalline goethite, whereas Rajahmundry laterites showed the presence of crystalline kaolinite and metahalloysite. The reason for the difference has been attributed to the geological environment of the areas.

The formation of sesquioxides in the top layers during laterization and weathering of the bottom layers present serious problems for civil engineers in the assessment of lateral stresses in lateritic profiles (Iyer and Pillai, 1972). Further road cuts in such deposits pose a serious stability problem.

### 1.5.4 Alluvial Deposits

The well-known alluvial deposits of India are in the Indo-Gangetic and Brahmaputra flood plains. Alluvial deposits exist up to a depth of 100 m. The north of the Vindhya Satpura range is covered with river alluvium, and other alluvial deposits of deltaic type are also present. Alluvial deposits exhibit alternate layers of sandy silt and clay, and in some locations organic layers are also encountered. The Bengal basin is another important alluvial deposit. The subsoil of the upper strata (which is of immediate relevance to civil engineers) is of recent origin and is believed to have been deposited by the Ganga river system. The soil around the Calcutta (now Kolkata) region, usually referred to as the Calcutta deposit, consists primarily of desiccated brownish grey silty clay up to a depth of about 15 m. Another deposit of the Bengal basin is the river channel deposits, consisting of sandy silts to silty sands up to a depth of 30 m (Som, 1975).

Alluvial sands are used as fine aggregates in concrete, whereas alluvial clay is used for manufacturing of bricks.

### 1.5.5 Desert Soils

The Thar Desert covers most of the area of the continent which forms the desert soil of India. These are wind-blown deposits generally present in the form of sand dunes. These deposits are formed under arid conditions and are predominantly of fine or silty sands. Scarcity of water is a serious problem for any construction activity.

## 1.6 COMPONENTS OF SOILS

The composition of natural soils may include diverse components which may be classified into three groups (Fig. 1.4): (i) solid phase (minerals, cementation, and organic material), (ii) liquid phase (water with dissolved salts), and (iii) gaseous phase (air or some other gas with water vapour). These are the components of a soil which affect its engineering properties.

### 1.6.1 Solid Phase

This consists of primary rock minerals, clay minerals, and cementing and organic materials. One or all may be present in a soil.

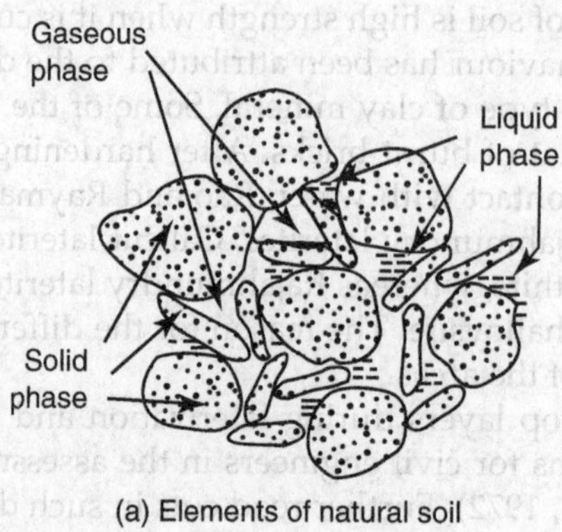


(a) Elements of natural soil

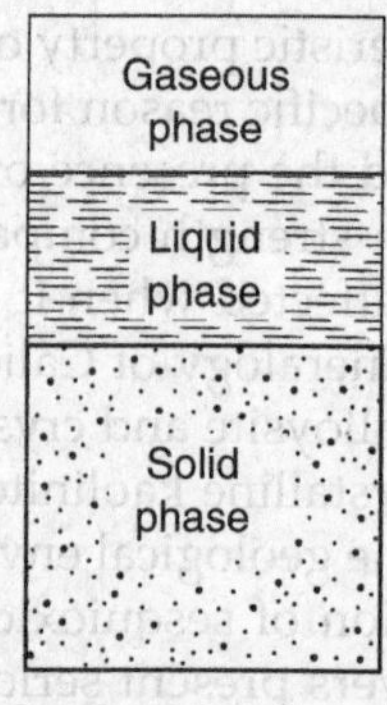


(b) Representation of soil elements

**Fig. 1.4** Components of natural soil

**Primary Rock Minerals**. These are rock fragments from the parent rock, formed due to weathering. In general, they are relatively large in size and rounded or angular in shape. When such particles form a major part of the soil minerals (as in gravels and sands), the engineering properties will be governed by the gradation and packing of the grains. The shape and texture of such particles (discussed in the next section) may have some bearing on the properties.

**Clay Minerals**. These are secondary minerals formed by chemical weathering, and the particle size is less than 2 μm. The particles commonly occur in the form of flat plates and are flaky in shape. The main characteristic of such particles is their large surface areas. A detailed treatment of clay minerals is presented in the next section.

**Cementing and Organic Materials**. Due to the decomposition of minerals by leaching or due to the presence of dissolved salts, certain cementing materials (such as calcite, iron oxide, or silica) may be deposited on the surface of the soil particles. Such materials improve the engineering properties of soils.

Organic matter in the soil has originated from plant or animal remains. It generally occurs in the top soil up to a depth of 0.5 m. Muck or peat deposits are primarily organic in nature and occur at considerable depths. Organic matter absorbs more water, compresses considerably under a load, fails due to low bearing resistance, and affects the setting of foundation concrete. Thus, organic materials have many undesirable properties harmful for engineering structures.

### 1.6.2 Liquid Phase

In soils of interest to the civil engineers, the only liquid phase is water. In geotechnical engineering, water is the prime factor which governs the engineering properties of soils. It is an incompressible fluid capable of taking normal stresses but not shear stresses.

Water can dissolve and transport, in solutions, various salts and compounds, some of which may seriously affect the soil behaviour. Calcium sulphate occurs in many clays but is only slightly soluble. Sulphate ion solutions in water have adverse effects on the properties of concrete structures.

### 1.6.3 Gaseous Phase

Air is the gaseous phase found in soil. It is extremely difficult to get a perfectly dry soil or fully saturated soil. The so-called saturated soils contain about 2% of air voids. Similarly, in dry soil, water vapour may exist. In partially saturated soils, because of vapour pressure and a continuous air path, there may be migration of water in the form of water vapour.

## 1.7 PARTICLE SIZES AND SHAPES

Naturally occurring soil deposits comprise soil particles of varying sizes and shapes. Size and, to a lesser extent, shape are factors that affect the material behaviour of a soil. However, most engineering properties are not controlled by particle size and shape but depend on soil mineral composition, interaction with water, and soil structure.

### 1.7.1 Particle Size

A soil particle does not have a specific size and shape so that a unique linear dimension can be assigned (as in a solid of regular geometrical shape). Thus, a representative size for the particle has to be fixed, based on a certain analysis (as defined in sieve or hydrometer analysis, discussed in the next chapter).

Soil may have particle sizes as big as several centimetres (pebbles) or as small as $10^{-6}$ mm (colloid). It is reasonable to assign a name to a certain size range. Such names and their size ranges are given below.

Cobbles or pebbles – rock fragments, size range 150 to 300 mm
Gravel – rock particles, size range 4.75 to 150 mm
Sand – rock particles, size range 0.075 to 4.75 mm
Silt – rock particles, size range 0.002 to 0.075 mm
Clay – mineral particles, size <0.002 mm (i.e., <2 μm)

The lowest particle size which could be identified without magnification is sand, and hence particle sizes (sand and above) bigger than 0.075 mm are grouped under *coarse-grained soil*. Particle sizes less than 0.075 mm are grouped under *fine-grained soil* because they are too small to be seen without magnification.

### 1.7.2 Particle Shape

The shape of individual particles depends on the parent rocks, their age, degree of weathering, and transporting processes. In general, all soil particles except clay-size particles which are plate-like in shape are approximately equidimensional. Other shapes common to clay-size particles are rods and laths.

Generally, the edges of a new coarse-grained particle are angular and rough surfaced, but with time they become round and smooth. The sharpness of the edges and corners of a particle is referred to as the degree of roundness. The degree of roundness in increasing order may be given as angular, sub-angular, sub-rounded, rounded, and well rounded. Figure 1.5 shows five levels of degree of roundness.

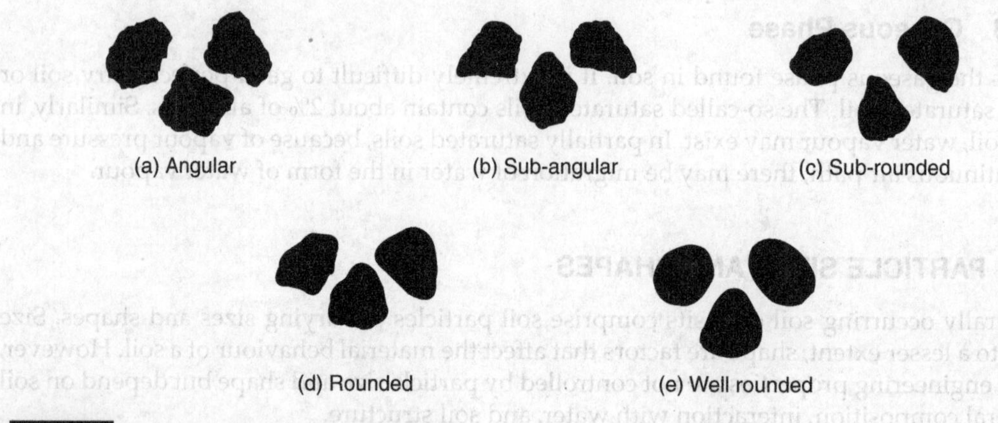


Fig. 1.5 Degree of particle roundness

The minor features of a particle surface are defined as the *surface texture* of a soil. This is independent of the size, shape, or degree of roundness of a particle. Dull, polished, smooth, rough, striated, frosted, etched, or pitted are terms used to define the surface texture of a particle.

## 1.8 INTER-PARTICLE FORCES

The size of a particle is an indirect measure of the magnitude of forces between particles. Inter-particle forces are broadly classified as gravitational forces or mass-derived forces and surface-bonding forces or surface-derived forces. Gravitational forces are proportional to mass or approximately to the volume of the soil particles. Thus, there is the same gravitational interaction between particles of equal volume. In coarse-grained soils, mass-derived forces are predominant. Surface-bonding forces between particles are dependent on the surface area, its character, and its environment. Accordingly, the magnitude of bonding forces need not be similar for particles of equal surface area. Surface-derived forces are predominant in fine-grained soils.

For a complete understanding of the formation of soil minerals and the connected inter-particle forces, a basic knowledge of atomic and molecular bonds is needed. There are two main types of bonds of interest in the discussion of soil minerals: (i) primary or high-energy bonds and (ii) secondary valence or weak bonds.

Primary or high-energy bonds are also referred to as electrostatic bonds and are of two types, viz., ionic and covalent bonds. An ionic bond is formed when one atom joins with another by adding electrons to its outer shell or shells or by losing them to arrive at a stable configuration. This is the strongest and simplest of the bonds that hold atoms together. For example, aluminium and oxygen combine to form aluminium oxide, and sodium and chlorine combine to form sodium chloride. Sometimes, one electron can be shared by two atoms and thereby create a bond called covalent bond (e.g., carbon). Primary valence bonds are also termed *intra-molecular bonds*.

Secondary valence bonds are also known as *inter-molecular bonds*. They are of two types, viz., hydrogen bonds and van der Waals forces. The hydrogen atom possesses only one electron in its shell. It can take one electron from another element to form an anion or can lose its electron to another element to form a proton. Thus, under certain conditions, if a hydrogen atom is attracted by two atoms instead of one, then the bond formed is called a hydrogen bond. The hydrogen bond is weaker than the ionic bond, and only two atoms can be bonded by the hydrogen ion. Fluorine, oxygen, nitrogen, and chlorine form hydrogen bonds.

Electrical or electromagnetic attraction between molecules is referred to as van der Waals forces. In a molecule, the centre of gravity of the positive and negative charges may not coincide, thus developing an electric moment. This system is referred to as polar. Thus, in such a system van der Waals forces develop depending on the orientation of the particle.

The relative magnitudes of these bonds are given below (Leonards, 1962):

| | |
|---|---|
| van der Waals forces | 1 to 10 |
| Hydrogen bonds | 10 to 20 |
| Primary valence bonds | 40 to 400 |

## 1.9 SOIL MINERALS

The composition of a soil particle plays an important role in the fundamental understanding of soil behaviour. Based on their origin, soils may be either organic or inorganic. Apart from the fact that organic soils are mixtures of decayed plant life, skeletons or shells of small organisms, and soil particles, not much is known about the composition of organic soils.

Inorganic soils are formed due to mechanical or chemical weathering. An inorganic soil particle may be a rock fragment or a mineral. In the present context, a mineral is a chemical element or a compound formed in nature during a geological process, and a rock fragment is representative of the parent rock, which may have a combination of one or more minerals.

Based on the nature of atoms, minerals are classified as carbonates, phosphates, oxides, and silicates. Of these, silicate minerals are important to civil engineers. Different arrangements of atoms in the silicate minerals give rise to different silicate structures. The most abundant soil minerals are the common rock-forming minerals, viz., sheet and framework silicate minerals. These are discussed below.

### 1.9.1 Clay Minerals

Soil minerals are formed from two basic structural units: tetrahedral and octahedral. The tetrahedral unit comprises a central silicon atom surrounded by four oxygen atoms positioned at the vertices of the tetrahedron (the basic unit and its symbolic representation are shown in Fig. 1.6). The tetra-hedrons are combined into hexagonal units called silica sheets. The octahedral unit comprises a central ion of either aluminium or magnesium surrounded by six hydroxyl ions (the basic unit and its symbolic representation are given in Fig. 1.6). Combining aluminium–oxygen octahedrons gives a *gibbsite sheet*, and combining magnesium–oxygen octahedrons gives a *brucite sheet*. Considering the valencies of atoms in the tetrahedral and two octahedral units, it is clear that the units are not electrically neutral and do not exist as single units. However, gibbsite and brucite sheets exist as they are electrically neutral.

The sheets may combine to form two-layer or three-layer sheet minerals or clay minerals. The basic kaolinite unit (two-layer unit) is formed by stacking a gibbsite sheet on a silica

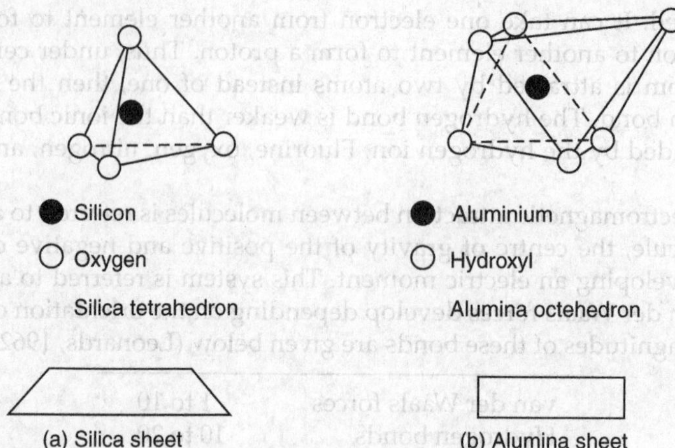


**Fig. 1.6** Basic units and symbolic blocks of clay minerals

sheet. These units may be stacked one on top of the other to form a *lattice* of the mineral (Fig. 1.7a). The bondage between the basic kaolinite units is due to hydrogen bonding and secondary valence forces. Kaolinite minerals are thus very stable and do not permit water to enter the lattice and hence do not expand when saturated. Kaolinite is the most abundant constituent of residual clay deposits. *Isomorphous substitution* (replacement of one kind of atom with another) is quite common during the formation of sheet silicate minerals. In kaolinite minerals, the amount of isomorphous substitution is negligible. Other clay minerals with two-layer sheets are serpentine (brucite silica sheet combination, Fig. 1.7b) and hallosite (similar to kaolinite except for the presence of water between the sheets).

Among three-layer sheet minerals, montmorillonite and illite clay minerals are the most common. Basic three-layer sheet units are formed by keeping one silica sheet on the top and one at the bottom of a gibbsite sheet. Figure 1.8a shows the unit and lattice of the

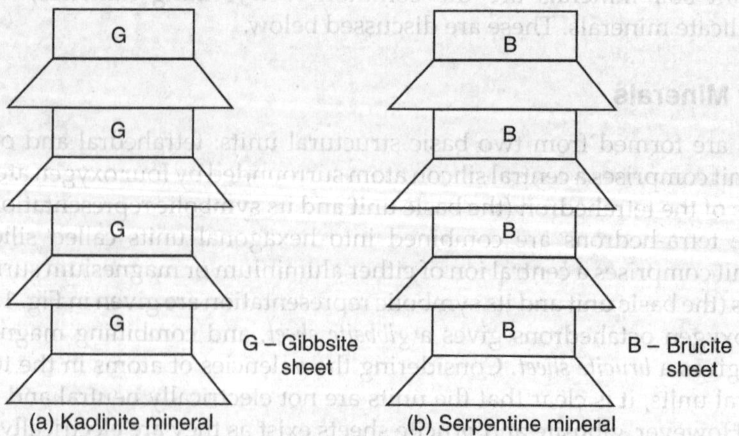


**Fig. 1.7** Symbolic structure of two-layer sheet minerals

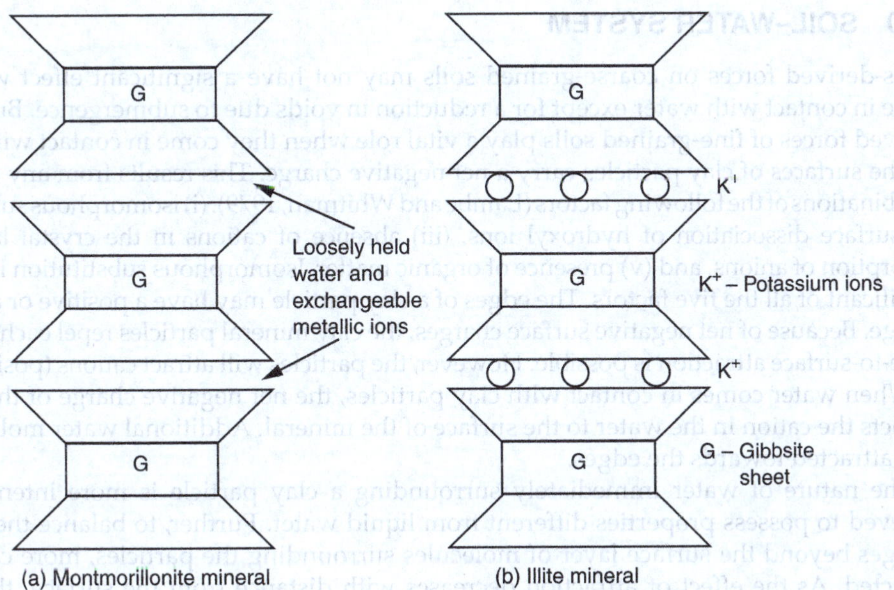


**Fig. 1.8** Symbolic structure of three-layer sheet minerals

montmorillonite mineral. Isomorphous substitution of magnesium or iron for the aluminium in the gibbsite sheet is common. The bonding between the three-layer units is by secondary valence forces and exchangeable ion linkage. The bonding of these sheets is very weak, and water may enter between the units. The mineral has significant affinity for water and can imbibe a large quantity of water, causing swelling.

Accordingly, there will be shrinkage during dry weather. Most of the Indian black cotton soils contain this mineral, and sufficient care has to be exercised in designing structures on such soils.

Illite consists of the basic montmorillonite units but is bonded by secondary valence bonds and potassium ions ($K^+$), as shown in Fig. 1.8b. There is about 20% replacement of aluminium with silicon in the gibbsite sheet due to isomorphous substitution. The mineral is a very stable one and does not swell or shrink unlike montmorillonite.

As the basic units of clay particles are sheeted structures, the particle formed by stacking such basic units ends up with a plate shape. The surface area of clay particles per unit mass is generally referred to as the *specific surface*. Clays with the smallest particles have the largest surface area. The surface area of kaolinite is quite less compared to that of montmorillonite. The specific surfaces of kaolinite, illite, and montmorillonite are about 15, 90, and 800 $m^2/g$, respectively.

### 1.9.2 Framework Silicate Minerals

Quartz and feldspar are very common rock-forming minerals, and their frameworks are abundant in soils. Particles of framework minerals are approximately equidimensional because of the nature of their structure. Quartz is one of the minerals which can resist weathering the most. Feldspars can be weathered to form clay minerals. Thus, these minerals sometimes find place in clay and silt-size particles.

## 1.10 SOIL–WATER SYSTEM

Mass-derived forces on coarse-grained soils may not have a significant effect when they come in contact with water except for a reduction in voids due to submergence. But surface-derived forces of fine-grained soils play a vital role when they come in contact with water.

The surfaces of clay particles carry a net negative charge. This results from any one of the combinations of the following factors (Lambe and Whitman, 1979): (i) isomorphous substitution, (ii) surface dissociation of hydroxyl ions, (iii) absence of cations in the crystal lattice, (iv) adsorption of anions, and (v) presence of organic matter. Isomorphous substitution is the most significant of all the five factors. The edges of a clay particle may have a positive or a negative charge. Because of net negative surface charges, the clay mineral particles repel each other, but edge-to-surface attraction is possible. However, the particles will attract cations (positive ions).

When water comes in contact with clay particles, the net negative charge of the mineral attracts the cation in the water to the surface of the mineral. Additional water molecules are also attracted towards the edges.

The nature of water immediately surrounding a clay particle is more intense and is believed to possess properties different from liquid water. Further, to balance the negative charges beyond the surface layer of molecules surrounding the particles, more cations are attracted. As the effect of attraction decreases with distance from the surface, there is no attraction and the pore water is normal. The distance from the surface of the particle to the limit of attraction is termed the *diffuse double layer* (Fig. 1.9). The water contained in the diffuse double layer and adjacent to the soil particle is termed the *rigid layer*.

The cations in the diffuse double layer are not permanently attached to the surface of clay particles and can be replaced by other cations. For example, if a soil with sodium ions is leached with potassium chloride solution, most of the sodium ions will be replaced by potassium ions. This transformation is known as cation or *base exchange*, and the ions involved in the replacement are called *exchangeable ions*. Such a transformation has a significant effect on the behaviour of clayey soils.

## 1.11 PHYSICO-CHEMICAL BEHAVIOUR OF CLAYS

Physico-chemical contributions of soils and the intrinsic forces acting in the soil–water system are of utmost importance for a better understanding of the deformation and strength characteristics of the soil. This subject of physico-chemical properties is a common topic of discussion in colloidal chemistry, mineralogy, soil physics, and soil mechanics.

### 1.11.1 Characteristics of Electrolytes

Substances that form ions in water solutions are called electrolytes. Substances which ionize completely are called strong electrolytes, and others which ionize slightly are called weak electrolytes. Substances which do not conduct electricity are called non-electrolytes. Most of the salts and all soluble metal hydroxides are strong electrolytes. Bases are weak electrolytes.

Dielectric constant is a measure of a material's ability to perform as an insulator. The higher the dielectric constant of a material, the larger its insulating capacity. Compared to other liquids, water has a high dielectric constant. This is the reason for water having great solvent power for strong electrolytes.

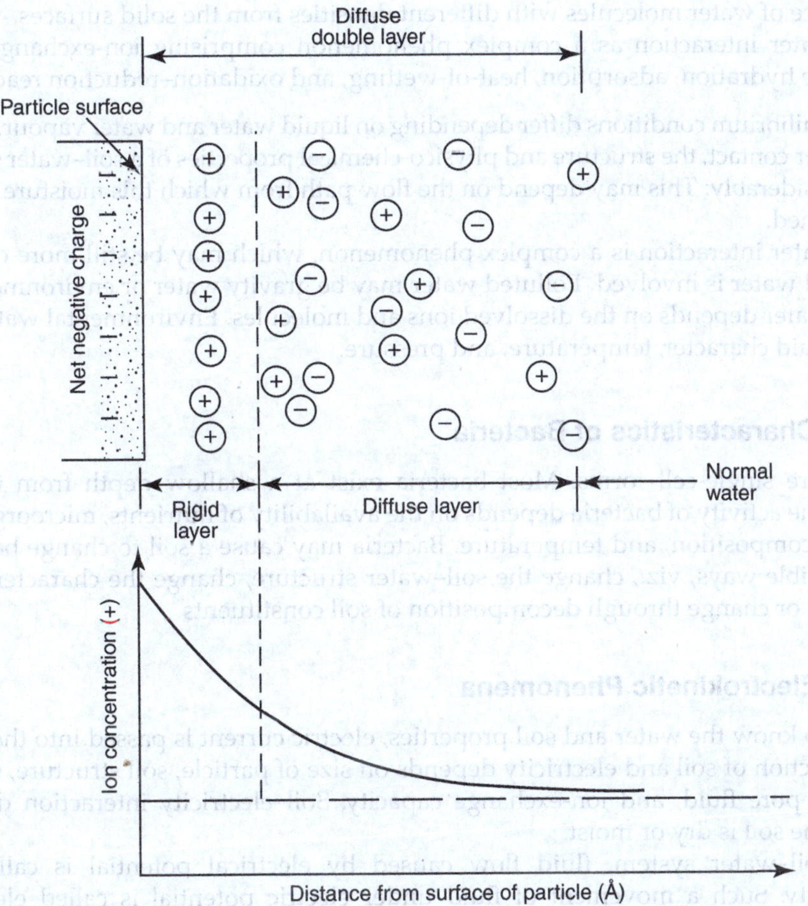


Fig. 1.9 Electric double layer

Zeta potential is the electric potential developed at the solid–liquid interface in response to movement of colloidal particles. The magnitude of the zeta potential depends on the thickness of the double layer. Zeta potential depends on ion types, temperature, adsorption, particle size, shape of particles, and pH value.

### 1.11.2 Electrochemical Characteristics

Interaction with the pore fluid occurs on the solid's surface. The surfaces of the soil solids may affect the water structure due to the presence or absence of interaction. The essential components of a soil–water system are as follows:

1. Surfaces of the soil mineral present with their characteristic geometry and electric patterns.
2. Location of cations so as to maintain equilibrium between the kinetic dispersive forces and the Coulombic attractive forces.

3. Presence of water molecules with different densities from the solid surfaces.
4. Soil–water interaction as a complex phenomenon comprising ion-exchange, leaching, osmotic hydration, adsorption, heat-of-wetting, and oxidation–reduction reactions.

Soil equilibrium conditions differ depending on liquid water and water vapour. Even at the same water contact, the structure and physico-chemical properties of a soil–water system may differ considerably. This may depend on the flow path from which this moisture content has been attained.

Soil–water interaction is a complex phenomenon, which may be still more complicated if polluted water is involved. Polluted water may be gravity water or environmental water. Gravity water depends on the dissolved ions and molecules. Environmental water depends on pore fluid character, temperature, and pressure.

### 1.11.3 Characteristics of Bacteria

Bacteria are single-cell forms. Most bacteria exist at a shallow depth from the ground surface. The activity of bacteria depends on the availability of nutrients, microorganic types, substrate composition, and temperature. Bacteria may cause a soil to change behaviour in three possible ways, viz., change the soil–water structure, change the characteristic of the pore fluid, or change through decomposition of soil constituents.

### 1.11.4 Electrokinetic Phenomena

In order to know the water and soil properties, electric current is passed into the soil mass. The interaction of soil and electricity depends on size of particle, soil structure, soil surface condition, pore fluid, and ion-exchange capacity. Soil–electricity interaction depends on whether the soil is dry or moist.

In a soil–water system, fluid flow caused by electrical potential is called electric conductivity. Such a movement of fluid under electric potential is called electrokinetic phenomena. These phenomena may be categorized under two groups:

1. *Phenomena between soil particles*: In this case, the fluid flows from one electrode to another electrode through the voids of the soil and the soil particles remain stagnant. This phenomenon is referred to as *electro-osmosis* (dealt in detail elsewhere).
2. *Phenomena in clay suspensions*: In this case, soil particles in a clay suspension are attracted or repelled from one another; e.g., negatively charged particles move toward an anode in a soil suspension. This phenomenon is referred to a *electrophoresis*.

Both these methods are effectively used in de-watering, soil densification, etc., in the field of ground improvement.

### 1.11.5 Characteristics of Heat

Heat is a form of energy which may be induced into the soil mass through man-made activities or due to natural causes. The man-made sources are heat released from steam pipes, electrical cable lines, and heat pump systems and heat caused due to environmental wastes. Natural sources include solar and geothermal heat.

Thermal conductivity is synonymous to hydraulic conductivity. Thermal conductivity of soil is defined as the quantity of heat flowing due to unit thermal gradient in unit time. The reciprocal of thermal conductivity is called thermal resistivity, which is a measure of a material's ability to resist heat flow.

In a saturated soil–water system, increase in temperature decreases the bound water and increases the free water and decreases the viscosity of the free water. This may be true for a limited range of temperature in a clay–water system. Under high temperature conditions in a clay–water system, dispersion or flocculation may occur depending on exchangeable ions and type of clay mineral.

## 1.12 SOIL STRUCTURE

Soil structure is a more generalized term, applicable to all types of soils. This includes gradation, compactness, geometric and skeletal arrangements of particles, inter-particle forces, and bonding agents. The term *soil fabric* has been explicitly used for the structure of clay-size particles.

### 1.12.1 Structure of Granular Soils

In granular soils, the ratio of the volume to the surface area is large, so that mass-derived forces are dominant and surface-derived electrical forces are negligible. Single-grained structures are formed when the soil grains settle out independently due to mass-derived forces. The individual granular particle is strong enough to withstand the normal pressures encountered in practice. Thus, the aggregate physical properties in coarse-grained soils are a function of relative density and particle shape.

Packing is the term used for the arrangement of individual soil particles in granular soils. But gradation and particle shape affect packing. For an idealized spherical particle, the minimum and maximum void ratios* are 0.35 and 0.91, respectively. But such a condition does not arise. However, ideal particle size distributions never exist in real soils. Possible packings representing upper (well graded) and lower limits (poorly graded) are shown in Fig. 1.10. Optimum packing gives rise to more particle contacts, more shear strength, less settlement, and more stability.

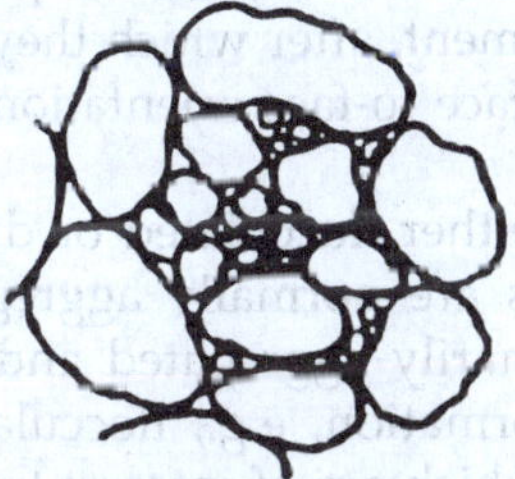

(a) Upper limit
well-graded particles

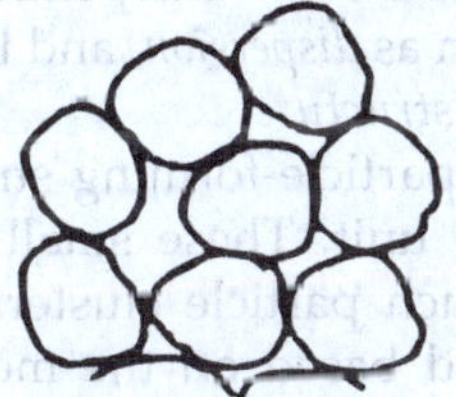

(b) Lower limit
poorly graded particles

**Fig. 1.10** Possible limits of packing

*Ratio of volume of voids to volume of soil solids.

### 1.12.2 Structure of Cohesive Soils

The structure of a fine-grained cohesive soil can be described fully with the understanding of inter-particle forces and the geometrical arrangement, or fabric, of the particles. In a cohesive soil deposit, the structural arrangement of individual particles or a group of particles at a microsocopic scale is defined as the *microstructure* or *microfabric* of the deposit. Miscrostructure is fundamentally important in the general understanding of soil behaviour. The larger, usually visible, structural features of cohesive deposits such as joints, fissures, rootlets, varves, silt and sand seams, and other discontinuities define the *macrostructure* or *macrofabric* of the deposit. Macrostructure of cohesive deposits has an important practical engineering influence on the entire soil behaviour.

The concepts of clay structure were proposed earlier by Terzaghi (1925) and Casagrande (1932b). Terzaghi postulated that during the process of sedimentation, adsorbed water surrounding the clay particles is squeezed out, thus making particles attain high attractive forces. Application of external pressure breaks the contacts between the solid water shells, leading to displacement and complete collapse of structure. Casagrande's concept of clay structure was that of a coarse-grained skeleton cemented together by a highly compressed "bond clay." The interstices are filled with partly consolidated high-water-content "matrix clay." Remoulding causes the mixing of matrix and bond clays and, thus, destroys the primary load-carrying structure.

It was Lambe (1953, 1958) who fundamentally delineated structures into two extreme groups – flocculated and dispersed. They are discussed below.

If two clay particles approach each other in a suspension, attractive forces due to van der Waals forces and repulsive forces due to positively ionized adsorbed layers come into action. At lesser separation, van der Waals forces dominate, and so particles adhere. However, there is a decrease in van der Waals forces with an increase in separation. If the adsorbed layer is thin, the repulsive force will be absent or negligible, and the random movement of particles will bring them into contact. Further, the net repulsive force will be greatest when particle faces approach closer. As a result, the flocculating particles form an edge-to-face arrangement. A group of such particles settles down together rapidly. This process is called *flocculation*, and the structure so formed is termed *flocculated structure*. Clays deposited under sea water have a high cation concentration with thin adsorbed water, thus resulting in a flocculated structure.

On the other hand, if clay particles have a thick adsorbed layer, the repulsion is greater and the particles are led to a slow independent settlement, after which they remain dispersed. This process is known as *dispersion*, and it causes face-to-face orientation of particles, which is known as *dispersed structure*.

The smallest clay-particle-forming structure, either flocculated or dispersed, is termed the first-order fabric unit. These small particles are normally aggregated. Van Olphen (1963) categorized such particle clusters as primarily aggregated and ultimately flocculated or deflocculated based on the mode of formation, e.g., flocculated–aggregated or deflocculated–aggregated. Ideal clay structures which are of interest to civil engineers are shown in Fig. 1.11.

The structural arrangement in nature can be highly complicated because of the presence of bulky particles along with clay particles. Interaction between single clay mineral particles is rare. Individual clay particles aggregate or flocculate together to form submicrosopic fabric units called *domains* (Fig. 1.12). Domains, in turn, group together to form submicrosopic groups called *clusters*. These submicroscopic groupings are attributed to

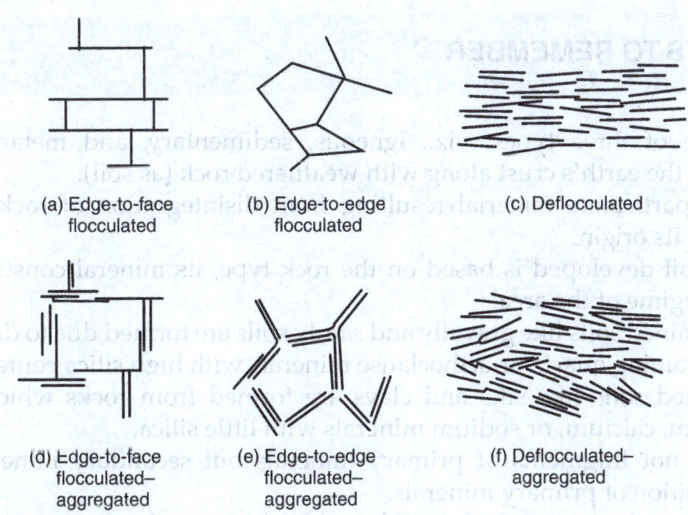


**Fig. 1.11** Idealized clay structures

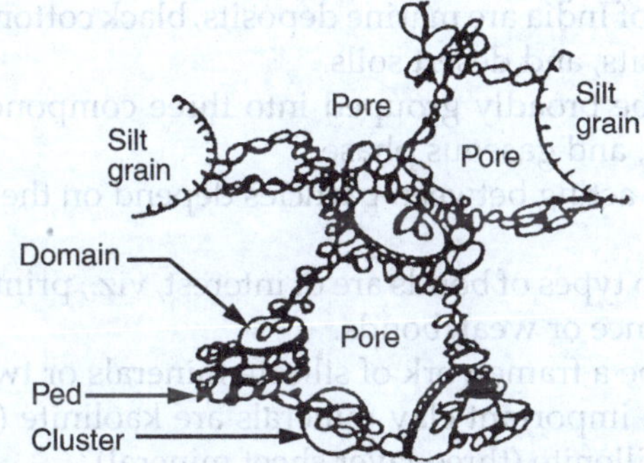


**Fig. 1.12** Structure of natural clay (*Source:* Bowles, 1984)

inter-particle forces acting between small basic units. Clusters group together to form *peds*, which are macroscopic.

Clay deposits which exhibit no visible variation in macrostructure are said to be *uniform* or *intact*. Many clays are layered or stratified, with predominant bedding planes. Clay deposits formed in glacial melt-water lakes exhibit thin layers of silt, fine sand and clay, and such deposits are referred to as *varved clays*. Stiff clays exist in thin natural states with a network of hair cracks, joints, or fissures; such clay deposits are called *fissured*. The fissures are formed due to stress release of highly stressed clays, past earth movements, or volume changes caused by desiccation.

A well-defined macrostructue significantly influences the engineering behaviour of clay deposits. Presence of fissures in a clay indicate planes of weakness and hence low strength along the planes, compared to a clay deposit which is intact. Apart from low strength, fissured clays or laminated clays show higher seepage than an intact clay, due to the preferential drainage path formed along the silt-filled fissures or bedding planes.

## POINTS TO REMEMBER

1.1 Rocks are of three types, viz., igneous, sedimentary, and metamorphic, which constitute the earth's crust along with weathered rock (as soil).
1.2 Soil is a particulate material resulting from disintegration of rocks and depends largely on its origin.
1.3 Type of soil developed is based on the rock type, its mineral constituents, and the climatic regime of the area.
1.4 Coarse-grained soils like gravelly and sandy soils are formed due to decomposition of rocks containing quartz or orthoclause minerals with high silica content.
1.5 Fine-grained soils like silts and clays are formed from rocks which contain iron, magnesium, calcium, or sodium minerals with little silica.
1.6 Clays are not fragments of primary minerals but secondary minerals formed by decomposition of primary minerals.
1.7 Physical weathering and chemical weathering are the two types which cause disintegration of rock.
1.8 Major formations of soils are residual soils, transported soils, and desiccated soils.
1.9 Major soil deposits of India are marine deposits, black cotton soils, laterite and lateritic soils, alluvial deposits, and desert soils.
1.10 Natural soils may be broadly grouped into three components or phases, viz., solid phase, liquid phase, and gaseous phase.
1.11 Inter-particle forces acting between particles depend on the surface area, its character and environment.
1.12 In soil minerals, two types of bonds are of interest, viz., primary or high-energy bonds and secondary valence or weak bonds.
1.13 Soil minerals may be a framework of silicate minerals or two-layer/three-layer sheet minerals. The three important clay minerals are kaolinite (two-layer sheet mineral), illite, and montmorillonite (three-layer sheet mineral).
1.14 Mass-derived forces are responsible for the formation of soil structure in coarse-grained soils. Surface-derived forces contribute to the formation of soil structure in fine-grained soils.
1.15 Water in clay leads to the attractive and repulsive forces on the particles, resulting in the formation of the electric double layer. Electric double layer (or diffused double layer) contributes to the formation of a particular type of structure.
1.16 Flocculated structure and dispersed structure are the two basic structures in clays.
1.17 Natural clays may be uniform or varied with cracks, joints, or fissures.

## QUESTIONS

### Objective Questions

1.1 Glaciers are formed by
(a) Compaction and recrystallization of snow
(b) Continuous freezing of water

(c) A sudden drop in temperature below 0°C
(d) None of the above processes

1.2 When the products of rock weathering are not transported as sediments but remain in place, the soil is known as
(a) Alluvial soil (b) Glacial soil
(c) Residual soil (d) Aeolian soil

1.3 Among the clay minerals, the one having the maximum swelling tendency is
(a) Kaolinite (b) Illite
(c) Montmorillonite (d) Halloysite

1.4 Cations from soil moisture are attracted to the surface of clay minerals to
(a) Balance the unsatisfied valence bonds
(b) Balance the negative electrical charge
(c) Form a diffuse double layer
(d) Replace the low-valence bonds

1.5 Identify the true statements from the following:
(a) A soil transported by gravitational force is called talus.
(b) Lateritic soil is a category of organic soil.
(c) Water held firmly to the clay particles has the same properties as ordinary water.
(d) A clay deposit which exhibits no evidence of fissuring is described as intact.
(e) Most clay crystals consists of atomic sheets, principally of two types: silica and alumina.

1.6 Chemical weathering may include the following processes:
(1) Oxidation (2) Solution (3) Leaching (4) Hydrolysis
Of these statements
(a) All are correct (b) 1, 2, and 3 are correct
(c) 2, 3, and 4 are correct (d) 3, 4, and 1 are correct

1.7 Clays which are deposited in marine water and later leached by groundwater are called
(a) Quick clays (b) Fat clays (c) Stiff clays (d) Fissure clays

1.8 Match list I with list II, and select the correct answer using the codes given below the lists:

| List I (soil deposit) | List II (soil name) |
|---|---|
| A. Gravity | 1. Stratified drift |
| B. Lake | 2. Talus |
| C. Glacier | 3. Loess |
| D. Wind | 4. Lacustrine |

*Codes:*

| | A | B | C | D |
|---|---|---|---|---|
| (a) | 1 | 3 | 2 | 4 |
| (b) | 2 | 4 | 1 | 3 |
| (c) | 3 | 4 | 1 | 2 |
| (d) | 4 | 3 | 2 | 1 |

1.9 Indian marine deposits are
(a) Very soft to soft clays (b) Soft to medium stiff clays
(c) Stiff to very stiff clays (d) Hard clays

1.10 Calicut laterites are rich in
(a) Halloysite (b) Montmorillonite
(c) Crystalline kaolinite (d) Metalalloysite

## Descriptive Questions

1.11 What are the different types of land forms associated with igneous rocks?
1.12 Distinguish between mechanical weathering and chemical weathering.
1.13 Which sedimentary deposits are of concern to civil engineers? Give examples.
1.14 What are loess soils, and what is the potential danger to loss of stability in loess soils?
1.15 Describe briefly the origin of soils and bring out the factors which control their formation.
1.16 Describe briefly the physical properties of residual and transported soils.
1.17 Name the parent rocks from which the following soils are derived: clay, sand, laterite, black cotton soil, and loess.
1.18 Explain the significance of the structure of clay minerals as constituents of engineering soils.
1.19 Explain why electrical surface forces are important for fine-grained soils but have little effect on coarse-grained soils.
1.20 What effect does the structure of clay have on the physical properties of soils?
1.21 From an engineering point of view, explain the properties of clay minerals.
1.22 Explain some of the stronger influences that are continually breaking rock into grains of soil.
1.23 What is degradation? Explain.
1.24 What influences can produce chemical changes in rock rather than mechanical breakage?
1.25 How does clay absorb water? Explain.
1.26 What soil types would be expected in a river or stream delta?
1.27 Explain the potential danger to stability in areas where the land is formed from marine clay.
1.28 Distinguish between the shape and size of clay particles with silts and sands.
1.29 What is a hydrogen bond? Explain its role in the physico-chemical behaviour of clays.
1.30 How does the modern concept of clay structure contrast with that given by Casagrande?

# 2

# Index Properties of Soils

## CHAPTER HIGHLIGHTS

Three-phase system Definition and applications of void ratio, porosity, degree of saturation, air void ratio, specific gravity of soil solids, moisture content, density, and density index – Particle-size analysis – Grain-size distribution curves – Consistency limits and indices of soils – Activity of clays

## 2.1 INTRODUCTION

Bulk soil, as it exists in nature, is a more or less random assembly of soil particles, water, and air. The properties of soils are complex and variable. Every civil engineering work involves the determination of soil type and its associated engineering application; certain properties are more significant than others. The common problems faced by civil engineers are related to the bearing capacity and compressibility of soil and the seepage through the soil. The possible solution to these problems is arrived at based on a study of the physical and index properties of the soil. A knowledge of these properties with value-oriented judgement gives a complete solution to such problems. The phase concept for soils introduced in the previous chapter is of great assistance in understanding these properties.

## 2.2 THREE-PHASE SYSTEM

Soil deposits are particulate systems containing three distinct phases. For the purpose of engineering analysis and design, it is essential to express the relationships among them in numerical terms.

As the relative proportions of these three phases, viz., soil solids, water, and air, are intermixed, it is customary to consider a soil model which will represent these phases distinctly and properly quantify the amount of each phase. A schematic diagram of the three-phase system is shown in Fig. 2.1.

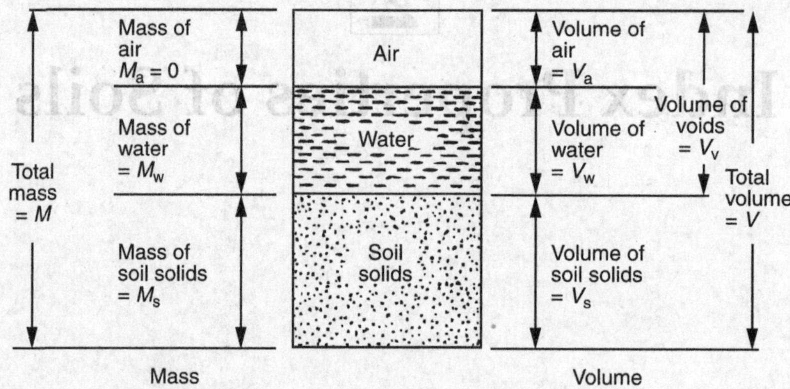


Fig. 2.1 Three-phase system

In dry and saturated soils, the three-phase system reduces to two phases, viz., soil solids and air, and soil solids and water, respectively (Fig. 2.2). Thus, soil exists in either the two-phase or three-phase state. The following are the basic relations.

### 2.2.1 Void Ratio (*e*)

Void ratio is the ratio of the volume of voids ($V_v$) to the volume of soil solids ($V_s$); that is,

$$e = \frac{V_v}{V_s} \tag{2.1}$$

It is expressed as a decimal and may take values greater than 1.0. Typical values of the void ratio in soils may range from 0.50 to 1.50.

### 2.2.2 Porosity (*n*)

Porosity is the ratio of the volume of voids to the total volume of soil (*V*); that is,

$$n = \frac{V_v}{V} \tag{2.2}$$

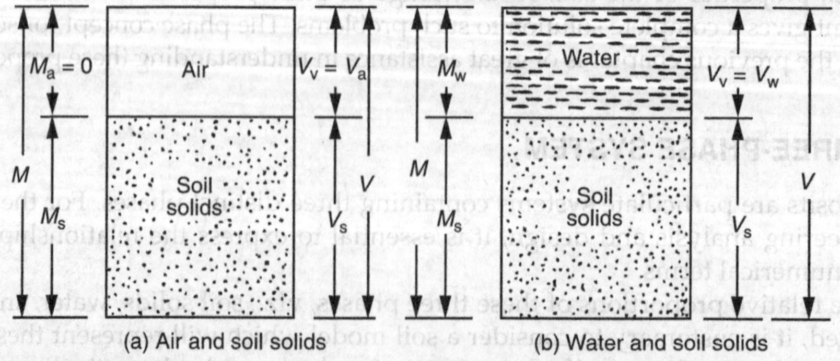


Fig. 2.2 Two-phase systems

It is expressed as a percentage. The range of porosity is $0\% \leq n \leq 100\%$. Void ratio and porosity are related as

$$e = \frac{n}{1-n} \tag{2.3}$$

and

$$n = \frac{e}{1+e} \tag{2.4}$$

### 2.2.3 Degree of Saturation ($S_r$)

Degree of saturation is the ratio of the volume of water ($V_w$) to the volume of voids; that is,

$$S_r = \frac{V_w}{V_v} \tag{2.5}$$

This is expressed as a percentage. Thus for dry soil, $S_r = 0\%$; for saturated soil, $S_r = 100\%$; and for partially saturated soil, $0\% < S_r < 100\%$.

### 2.2.4 Air Void Ratio ($A_v$) or Air Content

Air void ratio is the ratio of the volume of air ($V_a$) to the total volume of the soil; that is,

$$A_v = V_a/V$$

$$= \frac{V_v - V_w}{V_v + V_s} = \frac{V_v/V_v - V_w/V_v}{V_v/V_v + V_s/V_v} = \frac{1-S_r}{1+1/e}$$

$$= \frac{e}{1+e}(1-S_r) \tag{2.6}$$

$$A_v = n(1-S_r) \tag{2.7}$$

### 2.2.5 Specific Gravity of Soil Solids (G)

Specific gravity is the ratio of the mass of soil solids ($M_s$) to the mass of an equal volume of water, given by

$$G = \frac{M_s}{V_s \rho_w} \tag{2.8}$$

$$G = \frac{\rho_s}{\rho_w}$$

where $\rho_s = M_s/V_s$ is the density of soil solids and $\rho_w$ is the density of water (1.0 Mg/m$^3$ = 1,000 kg/m$^3$).

$G$ can be determined as follows (Fig. 2.3). An oven-dried sample of soil is placed in an empty pycnometer of mass $M_1$, and the combined mass $M_2$ is found. Water is added to the soil and agitated to remove the air. The mass of the pycnometer with soil and water, $M_b$, and the mass of the pycnometer with water, $M_a$, are found. Then,

$$G = \frac{M_s}{M_s - (M_b - M_a)}$$

where $M_s = M_2 - M_1$.

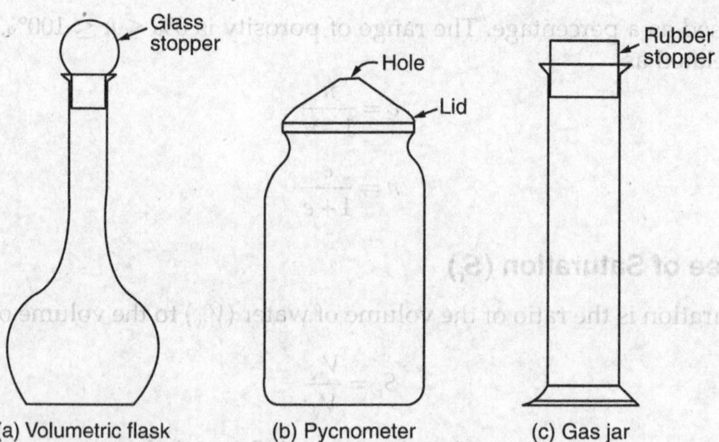


**Fig. 2.3** Apparatus for specific gravity determination

The detailed procedure for determining specific gravity is given in Chapter 10. Typical values of specific gravity are given in Table 2.1(a).

Specific gravity is relatively important as far as the qualitative behaviour of the soil is concerned. It is determined in a majority of testing programmes. The specific gravity of most soils falls within a narrow range and thus it is not used in the identification or classification of soils as it is in the identification of minerals.

### 2.2.6 Water Content (*w*) or Moisture Content

Water content is the ratio of the mass of water ($M_w$) to the mass of soil solids; that is,

$$w = \frac{M_w}{M_s} = \frac{V_w \rho_w}{V_s \rho_s}$$

**Table 2.1(a)** Typical values of specific gravity

| Type of soil | Specific gravity, *G* |
|---|---|
| Sand | 2.65–2.67 |
| Silty sand | 2.67–2.70 |
| Inorganic clay | 2.70–2.80 |
| Soil with mica or iron | 2.75–3.00 |
| Organic soils | Variable but may be under 2.00 |

*Source:* Bowles (1978).

$$= \frac{(V_w/V_v)\rho_w}{(V_s/V_v)\rho_s} \quad (2.9)$$

$$= \frac{S_r\rho_w}{(1/e)\rho_w G}$$

$$w = \frac{S_r e}{G} \quad (2.10)$$

This is represented as a percentage. The water content for a dry soil is zero. The natural water content for most soils is around 60%, but it can be as high as 400%.

The water content is obtained by weighing a moist sample of the soil, drying it in a temperature-controlled oven at 105°C for 24 hours, and then re-weighing. The detailed procedure is given in Chapter 10.

The dry soil mass is a function of the absolute magnitude of certain soil parameters, such as field moisture content and consistency limits (discussed elsewhere), and is used in the determination of void ratio and the degree of saturation.

The dry mass of a soil is not a definite quantity. It can be arbitrarily, but conveniently, defined as that mass which is obtained at 105°C (Lambe, 1949). It is believed that drying at 105°C for 24 hours completely removes the free water and some of the adsorbed water which forms part of the effective soil grain. For certain organic soils (like peat), for soils containing an appreciable amount of gypsum or some other mineral, and for some tropical soils, the conventional oven temperature of 105°C may be high. At this high temperature, the loosely bound water of hydration or molecular water may be lost, resulting in a change in the soil characteristics, such as the consistency limit. For such soils, the Indian Standards (IS: 2720 – Part 2, 1973) recommends a low oven temperature of 60 to 80°C.

For certain field controls (such as compaction), moisture content results are needed immediately. The conventional oven-drying method of measuring moisture content is so slow that results will not be available in time to make corrections in the field where deficiencies exist. Hence, rapid methods of testing are resorted to (this is discussed under compaction control in Chapter 4).

### 2.2.7 Soil Densities

**Dry Density ($\rho_d$).** It is the ratio of the total (dry soil) mass to the total volume (Fig. 2.2a); that is,

$$\rho_d = \frac{M}{V} = \frac{M_s + M_a}{V}$$

$$= \frac{M_s}{V} \quad \text{(mass of air } M_a = 0\text{)}$$

$$= \frac{M_s}{V_s + V_v} = \frac{M_s/V_s}{1 + V_v/V_s} = \frac{\rho_s}{1+e}$$

$$\rho_d = \frac{G\rho_w}{1+e} \quad (2.11)$$

**Saturated Density ($\rho_{sat}$).** It is the ratio of the total saturated soil mass to the total volume (Fig. 2.2b); that is,

$$\rho_{sat} = \frac{M}{V} = \frac{M_s + M_w}{V_s + V_v}$$

$$= \frac{M_s/V_s + M_w/V_s}{1 + V_v/V_s} = \frac{\rho_s + (V_w \rho_w)/V_s}{1+e}$$

$$\rho_{sat} = \left(G\rho_w + \frac{V_v}{V_s}\rho_w\right) \Big/ (1+e) \qquad \text{(Since } V_w = V_v\text{)}$$

$$\rho_{sat} = \frac{(G+e)}{1+e}\rho_w \tag{2.12}$$

**Submerged Density or Bouyant Density ($\rho'$).** It is the effective mass per unit volume when submerged. When a unit volume of soil is submerged in water, it displaces an equal volume of water, and the density is only ($\rho_{sat} - \rho_w$). Thus,

$$\rho' = \rho_{sat} - \rho_w = \left[\frac{G+e}{1+e}\right]\rho_w - \rho_w$$

$$\rho' = \left[\frac{G-1}{1+e}\right]\rho_w \tag{2.13}$$

**Bulk Density or Total Density.** It is the ratio of the total soil mass to the total volume:

$$\rho = \frac{M}{V} = \frac{M_s + M_w + M_a}{V_s + V_v}$$

$$= \frac{M_s/V_s + M_w/V_s}{1 + V_v/V_s} \qquad \text{(Since } M_a = 0\text{)}$$

$$\rho = \frac{\rho_s + (V_w \rho_w/V_s)}{1+e} = \frac{G\rho_w + (V_w/V_s) \times (V_v/V_v)\rho_w}{1+e}$$

$$= \frac{G\rho_w + eS_r\rho_w}{1+e}$$

$$\rho = \frac{G + eS_r}{1+e}\rho_w \tag{2.14}$$

That is,

$$\rho = \frac{G\rho_w}{1+e} + \frac{eS_r\rho_w}{1+e}$$

$$= \rho_d + \frac{(wG/S_r)S_r\rho_w}{1+e}$$

$$\rho = \rho_d(1+w) \tag{2.15}$$

For dry soil:

$$S_r = 0, \quad \rho = \frac{G\rho_w}{1+e} = \rho_d$$

For saturated soil:

$$S_r = 1, \quad \rho = \frac{G+e}{1+e}\rho_w = \rho_{sat}$$

Dry density of the *in situ* soil is usually measured using the core cutter and sand replacement methods.

The core cutter method consists of a cylindrical cutter with dolly and a rammer to drive the cutter into the soil. The cutter containing soil is then carefully dug out and excess soil on the ends trimmed off. The mass and volume (from the dimensions of the core) of the soil within are found and the wet density calculated. Dry density of the soil is found after determining the water content.

In the sand replacement method, a hole about 100 mm in diameter is dug with suitable tools at the selected location. The mass of the soil removed is weighed and a specimen sample taken for determination of water content. The mass of sand required to fill the conical portion, and hence of the sand filling the hole, is determined. The volume of sand required to fill the hole is obtained from the mass of sand filling the hole and the bulk density of the sand. The bulk density of sand is calibrated by filling a can of known volume under similar conditions. From the mass of the soil and volume of the hole, the wet density and, hence, the dry density are determined. Details of both procedures are given in Chapter 10.

Knowing the dry density and specific gravity of soil solids, the *in situ* void ratio is obtained from the expression

$$e = \frac{G\rho_w}{\rho_d} - 1 \tag{2.16}$$

### 2.2.8 Unit Weights

The unit weight ($\gamma$) of a soil is the ratio of the weight of the soil (force) to its total volume. The units of unit weight are force per unit volume (kN/m$^3$), whereas the units of density are mass per unit volume (Mg/m$^3$ or g/cm$^3$).

1. Dry unit weight $\gamma_d = \rho_d\, g$ (2.17)
2. Saturated unit weight $\gamma_{sat} = \rho_{sat}\, g$ (2.18)
3. Submerged unit weight $\gamma' = \rho'\, g$ (2.19)
4. Bulk unit weight $\gamma = \rho\, g$ (2.20)
5. Unit weight of water $\gamma_w = \rho_w\, g$ (2.21)

The value of gravitational acceleration, g, to be considered in Eqs. 2.17 to 2.21 is 9.807 m/s$^2$, and the unit weight of water at 4°C is 9.807 kN/m$^3$. Table 2.1(b) shows how $\gamma$, $\gamma_d$ and $\gamma_{sat}$ are related through various known parameters.

**Table 2.1(b)** Relationship for $\gamma$, $\gamma_d$, and $\gamma_{sat}$ in various forms

| Sl. No. | Bulk unit weight ($\gamma$) | | Sl. No. | Dry unit weight ($\gamma_d$) | | Sl. No. | Saturated unit weight ($\gamma_{sat}$) | |
|---|---|---|---|---|---|---|---|---|
| | **Known** | **Relationship** | | **Known** | **Relationship** | | **Known** | **Relationship** |
| 1. | $w, G, e$ | $\frac{(1+w)G\gamma_w}{1+e}$ | 1. | $\gamma, w$ | $\frac{\gamma}{1+w}$ | 1. | $G, e$ | $\frac{(G+e)\gamma_w}{1+e}$ |
| 2. | $S, G, e$ | $\frac{(G+Se)\gamma_w}{1+e}$ | 2. | $G, e$ | $\frac{G\gamma_w}{1+e}$ | 2. | $G, n$ | $[(1-n)G+n]\gamma_w$ |
| 3. | $w, G, s$ | $\frac{(1+w)G\gamma_w}{1+\frac{wG}{S}}$ | 3. | $G, n$ | $G\gamma_w(1-n)$ | 3. | $G, w_{sat}$ | $\left(\frac{1+w_{sat}}{1+w_{sat}G}\right)G\gamma_w$ |
| 4. | $w, G, n$ | $G\gamma_w(1-n)(1+w)$ | 4. | $G, w, s$ | $\frac{G\gamma_w}{1+\left(\frac{wG}{s}\right)}$ | 4. | $e, w_{sat}$ | $\left(\frac{e}{w_{sat}}\right)\left(\frac{1+w_{sat}}{1+e}\right)\gamma_w$ |
| 5. | $S, G, n$ | $G\gamma_w(1-n)+nS\gamma_w$ | 5. | $e, w, s$ | $\frac{es\gamma_w}{(1+e)w}$ | 5. | $n, w_{sat}$ | $n\left(\frac{1+w_{sat}}{w_{sat}}\right)\gamma_w$ |
| | | | 6. | $\gamma_{sat}, e$ | $\gamma_{sat}-\frac{e\gamma_w}{1+e}$ | 6. | $\gamma_d, e$ | $\gamma_d+\left(\frac{e}{1+e}\right)\gamma_w$ |
| | | | 7. | $\gamma_{sat}, n$ | $\gamma_{sat}-n\gamma_w$ | 7. | $\gamma_d, n$ | $\gamma_d+n\gamma_w$ |
| | | | 8. | $\gamma_{sat}, G$ | $\frac{(\gamma_{sat}-\gamma_w)G}{(G-t)}$ | 8. | $\gamma_d, s$ | $\left(1-\frac{1}{G}\right)\gamma_d+\gamma_w$ |
| | | | | | | 9. | $\gamma_d, w_{sat}$ | $\gamma_d(1+w_{sat})$ |

*Source:* Das (2002)

### 2.2.9 Density Index ($I_D$)

Density index is the ratio of the difference between the void ratio of a cohesionless soil in the loosest state and any given void ratio to the difference between its void ratios in the loosest and the densest states (IS: 2720 – Part 14, 1983). That is,

$$I_D = \frac{(e_{max} - e)}{(e_{max} - e_{min})} \times 100 \tag{2.22}$$

or in terms of dry densities,

$$I_D = \frac{\rho_{d\,max}}{\rho_d} \times \frac{\rho_d - \rho_{d\,min}}{\rho_{d\,max} - \rho_{d\,min}} \times 100 \tag{2.23}$$

where $e_{max}$ is the void ratio of soil in the loosest state, $e_{min}$ is the void ratio of soil in the densest state, $e$ is the void ratio of soil in the *in situ* state, $\rho_{d\,max}$ the dry density of soil in the densest state, $\rho_{d\,min}$ the dry density of soil in the loosest state, and $\rho_d$ the *in situ* dry density of the deposit.

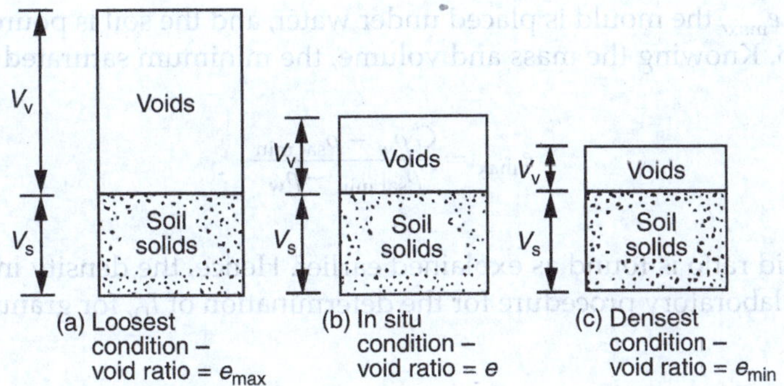


**Fig. 2.4** Relative density conditions

Figure 2.4 shows the three density conditions. Density index is a measure of the degree of compaction, and the stability of a stratum is indirectly reflected in the compactness. Table 2.2 presents designations associated with the density index for granular soils.

Simple practical methods may be adopted for determination of the maximum and minimum void ratios of granular soils (Whitlow, 1983). Minimum void ratio is determined by placing a soil in three layers in a mould of known volume, the mould being placed under water. Each layer is compacted thoroughly using a vibratory hammer or tamping rod. The collar of the mould is removed, the soil is levelled, and the mass of the soil, $M$, in the mould is found. Then, maximum saturated density is given as

$$\rho_{\text{sat max}} = \frac{M}{V} = \frac{G + e_{\min}}{1 + e_{\min}} \rho_w$$

Hence,

$$e_{\min} = \frac{G\rho_w - \rho_{\text{sat max}}}{\rho_{\text{sat max}} - \rho_w}$$

**Table 2.2** Density index of granular soils

| Designation | $I_D$ (%) |
|---|---|
| Very loose | 0–15 |
| Loose | 15–35 |
| Medium dense | 35–65 |
| Dense | 65–85 |
| Very dense | 85–100 |

*Source:* Lambe and Whitman (1979).

To determine $e_{max}$, the mould is placed under water, and the soil is poured quickly into it just from the top. Knowing the mass and volume, the minimum saturated density is found and, hence,

$$e_{max} = \frac{G\rho_w - \rho_{sat\,min}}{\rho_{sat\,min} - \rho_w}$$

The *in situ* void ratio is found as explained earlier. Hence, the density index can be computed. A simple laboratory procedure for the determination of $I_D$, for granular soils is given in Chapter 10.

## 2.3 PARTICLE-SIZE ANALYSIS

In view of the great diversity of soil grain sizes, civil engineers have long recognized a need for particle- or grain-size analysis. Particle-size analysis of soils determines the range of size of particles and the percentage of particles in each of the sizes between the maximum and the minimum. The designations given to different grain size ranges in soils by the Indian soil classification system (IS: 1498, 1970) are gravel $>4.75$ mm, sand $>0.075$ mm, silt $>0.002$ mm, and clay $<0.002$ mm.

The coarse particles (gravel and sand) may be separated by sieving, while a sedimentation procedure is used for analysing fine-grained soils (silt and clay). These are explained in the subsequent paragraphs.

### 2.3.1 Sieve Analysis

This is the most direct method for determining particle sizes but has a lower limit with respect to sieve opening. This lower limit corresponds to the fine-sand-size particles. In this method, the soil sample is passed through a stack of standard sieves having successively smaller mesh sizes. The percentage of weight retained, the cumulative percentage of weight retained, and the percentage passing (by weight) in each sieve are calculated.

The resulting data are presented with grain size along the $x$-axis (long scale) and percentage passing or finer along the $y$-axis (arithmetic scale). All the points are connected by a smooth curve which is referred to as a grain or particle size distribution curve. A detailed procedure for conducting sieve analysis is given in Chapter 10.

In the case of clayey soils, the fine fraction cannot be easily passed through a 75 μm sieve in the dry condition. In such cases, the materials are washed through with water (preferably mixed with 2 gm of sodium hexametaphosphate per litre), until the wash water is clean. The washed material is allowed to dry and then weighed. This is referred to as *wet sieve analysis*.

### 2.3.2 Sedimentation Analysis

The procedure commonly used for obtaining the particle-size distribution of fine-grained soil or the fine-grained fraction of a coarse-grained soil is the sedimentation method. The procedure is based on Stokes' law, which states that in a suspension the velocity of a spherical particle is governed by the diameter of the particle and the properties of the suspension. Thus, the terminal velocity $v$ (m/s) is given by

$$v = \frac{D^2(\gamma_s\gamma_w)}{18\eta_w} \tag{2.24}$$

where $D$ is the diameter of the particle (m), $\gamma_s$ the unit weight of grains of particles (kN/m$^3$), $\gamma_w$ the unit weight of suspension fluid (usually water; kN/m$^3$), and $\eta_w$ the viscosity of the suspension fluid (kN s/m$^2$). Then,

$$D = \sqrt{\frac{18\eta_w}{(\gamma_s - \gamma_w)}}\sqrt{\frac{H_e}{t}} \tag{2.25}$$

where $H_e$ is the height of the distance fallen (in metres) by the particles in time (seconds). Equation 2.25 is valid for particles larger than 0.002 mm because smaller-sized particles will be influenced by Brownian movement. The size of the particle is taken to be that of an equivalent sphere which will have the same settling velocity as that of the particle.

In this method, the soil is placed as a suspension in distilled water. To ensure independent settling of particles, a deflocculating agent is added to the suspension. The soil particles in the suspension are allowed to settle out. A sample at a depth $H_e$ below the surface, after allowing the suspension to settle for time $t$, will contain no particles of size larger than $D$. All particles smaller than $D$ will be present in the sample in the same proportion as at the beginning of the test. Thus, the effect is the same as if the sample had been separated on a sieve of mesh size $D$.

The concentration of particles remaining in the suspension at any level at any time may be determined by adopting any one of the following methods.

**Pipette Method**. The sample of suspension is drawn off with a pipette (Fig. 2.5) at the specified depth from the surface. The sample will contain only particles smaller than the

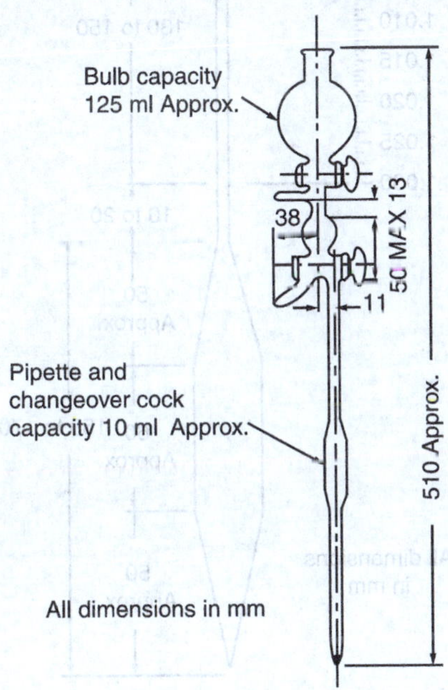


**Fig. 2.5** Pipette

size $D$. Similarly, samples are taken at the specified depth at times corresponding to other chosen particle sizes. The samples are dried, and the weight of the solid residue is recorded. If $M_b$ is the mass of the soil sample taken for sedimentation analysis after pre-treatment and $M'_i$ is the mass of material (of specific sizes of particles) in the entire suspension from corresponding samplings, then the percentage of finer particles $N$ is given as

$$N = \frac{M'_i}{M_b} \times 100$$

Now considering $D$ (in mm), $H$ (in cm), $t$ (in min), and $\eta_w$ (in poise) (1 poise $= 10^{-4}$ kN s/m$^2$) and substituting the respective units in Eq. 2.25 we get

$$D = \sqrt{\frac{30\eta_w}{980(\rho_s - \rho_w)}}\sqrt{\frac{H_e}{t}} \tag{2.26}$$

The diameter of the particle at every specified depth is obtained from Eq. 2.26. Then, the grain-size distribution is obtained.

**Hydrometer Method**. This method measures the specific gravity of the suspension using a special hydrometer (Fig. 2.6). The specific gravity of the solution decreases as the

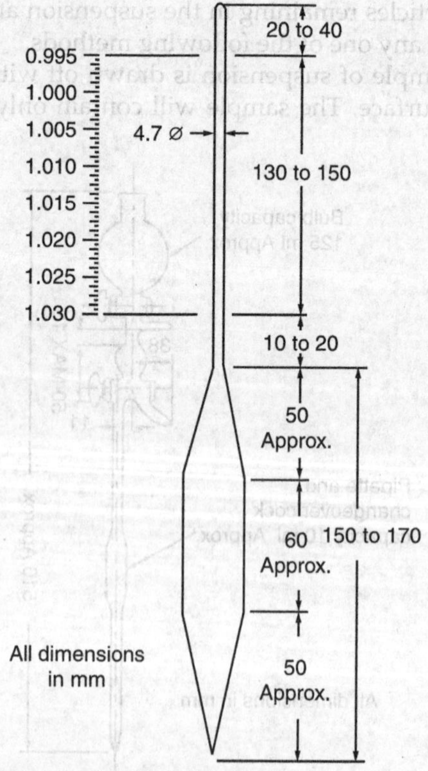


**Fig. 2.6** Hydrometer

settling starts. Specific gravity readings from the hydrometer at different time intervals provide information about the size of the particles that have settled down and the mass of soil remaining in solution.

A number of corrections are made for the hydrometer reading: (i) the meniscus correction (cm), which is necessary as the suspension is opaque and the lower meniscus cannot be seen clearly; (ii) the correction for the expansion of the hydrometer bulb due to increase in temperature (°C); and (iii) the correction due to the addition of a dispersing agent. The first two factors lead to a lower reading (note the hydrometer is graduated in increasing order from top to bottom), and hence the corrections are positive. The third factor increases the density of the suspension, and hence the correction is negative. The methods of determining these corrections are explained in Chapter 10. A calibration curve is drawn between the hydrometer reading corrected for meniscus correction ($R_b$) and the height of fall of the particle ($H$). The calibration procedure is given in Chapter 10. Thus, the diameter of the particle at time $t$ after the starting of sedimentation is obtained from Eq. 2.26.

To obtain the percentage of finer particles corresponding to each hydrometer reading, the density of suspension at that particular depth is required. Let $M$ be the mass of the pre-treated soil used in suspension of volume $V$. Before the start of the test; that is, at time $t = 0$, the density of the suspension is uniform and the mass of solids in a unit volume of the suspension is $M_b/V$.

Thus, the volume of solids $V_s$ in unit volume of suspension at time $t = 0$ and at any depth $H_e$ is

$$V_s = \frac{M_b}{V\rho_s} = \frac{M_b}{VG\rho_w} \qquad \left(\text{Since } G = \frac{\rho_s}{\rho_w}\right)$$

Therefore, the volume of water in unit volume of suspension (at $t = 0$) is

$$V_w = (1 - V_s) = \left[1 - \frac{M_b}{VG\rho_w}\right]$$

The initial density of the suspension is

$$\rho_i = \frac{M_b}{V} + \left[1 - \frac{M_b}{VG\rho_w}\right]\rho_{wT}$$

where $\rho_w$ is the density of water at 4°C and is 1 g/cm$^3$, and $\rho_{wT}$ is the density of water at test temperature, $T$°C.

The density of suspension after time $t$ and at temperature $T$°C can be written in a form similar to the above expression for $\rho_i$ as

$$\rho_f = \frac{M_D}{V} + \left(1 - \frac{M_D}{VG\rho_w}\right)\rho_{wT}$$

where $M_D/V$ is the mass of particles of diameter smaller than $D$ in unit volume of suspension at depth $H_e$, after time $t$. Substituing $\rho_{wT} = G_{wT}\,\rho_w$, we have

$$\rho_f = \frac{M_D}{V} + \left(1 - \frac{M_D}{VG\rho_w}\right)G_{wT}\rho_w$$

Dividing both sides by $\rho_w$, we have

$$\frac{\rho_f}{\rho_w} = \frac{M_D}{V\rho_w} + \left(1 - \frac{M_D}{VG\rho_w}\right)G_{wT}$$

$\rho_f/\rho_w$ is nothing but the specific gravity of suspension obtained after making the necessary corrections of meniscus, temperature, and dispersion agent. Let the hydrometer reading be $\bar{r}_h$. Then,

$$\bar{r}_h = \frac{M_D}{V\rho_w} + \left(1 - \frac{M_D}{VG\rho_w}\right)G_{wT}$$

$$\bar{r}_h = \frac{M_D}{V\rho_w}\left(1 - \frac{G_{wT}}{G}\right) + G_{wT}$$

or

$$(\bar{r}_h - G_{wT}) = \frac{M_D}{V\rho_w}\left(\frac{G - G_{wT}}{G}\right)$$

$$M_D = V\rho_w(\bar{r}_h - G_{wT})\left(\frac{G}{G - G_{wT}}\right)$$

$$N = \frac{M_D}{M_b}$$

$$N\% = \frac{V\rho_w}{M_b}\left(\frac{G}{G - G_{wT}}\right)(\bar{r}_h - G_{wT}) \times 100$$

Taking $V = 1{,}000$ ml, $\rho_w = 1\ \text{g/cm}^3$, and $G_{wT} = 1$, we get

$$N\% = \frac{1000}{M_b}\left(\frac{G}{G-1}\right)(\bar{r}_h - 1) \times 100$$

Let $\bar{r}_h/1000 = \bar{r}_h - 1$. Therefore,

$$N\% = \frac{100G}{M_b(G-1)}\bar{r}_h \tag{2.27}$$

From the data for $D$ and the corresponding percentage of finer particles after each instant of time, the grain-size distribution curve is obtained.

The sedimentation method is not absolutely correct as this is based on Stokes' assumption that (i) the particles are spherical, (ii) the flow around the particles is laminar, and (iii) the particles are much larger than the molecular size. Assumptions (i) and (iii) are not valid for fine-grained soils. Departure from spherical shape and molecular influence cause the particles to settle slowly. The dispersion of particles may be incomplete, and viscosity is not constant but varies due to changes in temperature.

However, this method has been in wide use and is more applicable to silts than to clays. The gradation curve is not used for evaluation of the engineering properties of fine-grained soils, and hence, a slight variation is insignificant.

Procedures for conducting pipette and hydrometer analyses are described in Chapter 10.

### 2.3.3 Grain-Size Distribution Curves

Grain-size distribution curves, as obtained from wet and dry methods, can be combined to form one complete grain-size distribution curve. It has to be remembered that the particle sizes defined in the two methods are different. In sieve analysis, a cylindrical particle of diameter $D$ and a spherical particle of the same diameter $D$ would fit through the same sieve opening. Though these two particles have different shapes, the sieve analysis identifies them as having the same size. Further, particle size as measured in the sedimentation method assumes an equivalent diameter of a spherical particle which would settle at the same rate. Thus, the accuracy of the gradation curve is questionable. In spite of serious limitations, particle-size curves of sands and silts have some practical value in the design of filters and in the assessment of permeability, capillarity, and frost susceptability, based on certain representative sizes of the particle.

However, very relevant and useful information may be obtained from a grain-sized curve, such as (i) the total percentage of larger or finer particles than a given size (to identify gravel-, sand-, silt-, and clay-size percentages) and (ii) the uniformity or the range in grain-size distribution.

The range of particle sizes present in a soil is reflected in the flatness of the curve. The flatter the curve, larger the range of size of particles and steeper the curve, smaller the range. The effective particle size has been defined by Hazen (1892) as that for which 10% of the material by weight is smaller than that size. Other particle sizes are also often used in describing or classifying soils; e.g., $D_{50}$ of a soil is used to represent the medium particle size, while $D_{85}$ and $D_{15}$ sizes are used to decide certain filter criteria.

A soil is said to be well graded or non-uniform if there is a distribution of particles over a relatively wide range (Fig. 2.7). A soil is said to be poorly graded if the sample has a

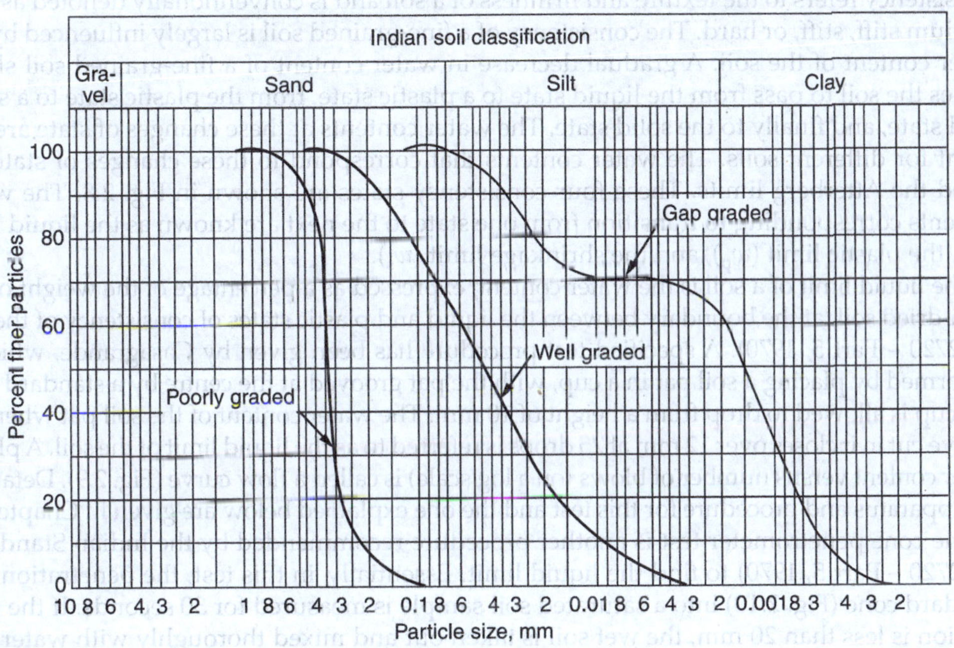


Fig. 2.7 Typical particle-size distribution curves

very narrow range of particles (also called uniform soil) or the sample is deficient in certain intermediate grain sizes (also called gap graded).

We can obtain a numerical measure of the gradation by defining the uniformity coefficient $C_u$ and the curvature coefficient $C_z$:

Uniformity coefficient:

$$C_u = \frac{D_{60}}{D_{10}} \tag{2.28}$$

Curvature coefficient:

$$C_z = \frac{D_{30}^2}{D_{60} \times D_{10}} \tag{2.29}$$

Soils with $C_u < 4$ are said to be uniform, and soils with $C_u > 4$ (6 for sands) are well graded as long as the grain-size distribution curve is smooth and symmetrical.

The curvature coefficient $C_z$ is a measure of the symmetry and shape of the gradation curve. For a well-graded soil, $C_z$ will be around 1. For $C_z$ much smaller or much greater than 1, the soil is viewed as poorly graded.

The uniformity coefficient and curvature coefficient are used as part of the Unified and Indian soil classification systems.

## 2.4 CONSISTENCY OF SOILS

### 2.4.1 Atterberg Limits

Consistency refers to the texture and firmness of a soil and is conventionally denoted as soft, medium stiff, stiff, or hard. The consistency of a fine-grained soil is largely influenced by the water content of the soil. A gradual decrease in water content of a fine-grained soil slurry causes the soil to pass from the liquid state to a plastic state, from the plastic state to a semi-solid state, and finally to the solid state. The water contents at these changes of state are different for different soils. The water contents that correspond to these changes of state are called the Atterberg limits. These four consistency states are shown in Fig. 2.8. The water contents corresponding to transition from one state to the next are known as the liquid limit ($w_L$), the plastic limit ($w_P$), and the shrinkage limit ($w_S$).

The liquid limit of a soil is the water content, expressed as a percentage of the weight of the oven-dried soil, at the boundary between the liquid and plastic states of consistency of the soil (IS: 2720 – Part 5, 1970). A specified test procedure has been given by Casagrande, which is performed by placing a soil pat in a cup, with the pat grooved at the centre by a standard tool. The cup is allowed to drop from a height of 10 mm. The water content of the soil pat when the groove cut in it closes over 12 mm at 25 drops is referred to as the liquid limit of the soil. A plot of water content versus number of blows (on a log scale) is called a flow curve (Fig. 2.9). Details of the apparatus and procedure for this test and the one explained below are given in Chapter 10.

The cone penetrometer test is another procedure recommended by the Indian Standards (IS: 2720 – Part 5, 1970) to find the liquid limit. Essentially, in this test, the penetration of a standard cone (Fig. 2.10) into a saturated soil sample is measured for 30 seconds. If the penetration is less than 20 mm, the wet soil is taken out and mixed thoroughly with water and the test is repeated till the penetration is between 20 and 30 mm. The exact penetration value

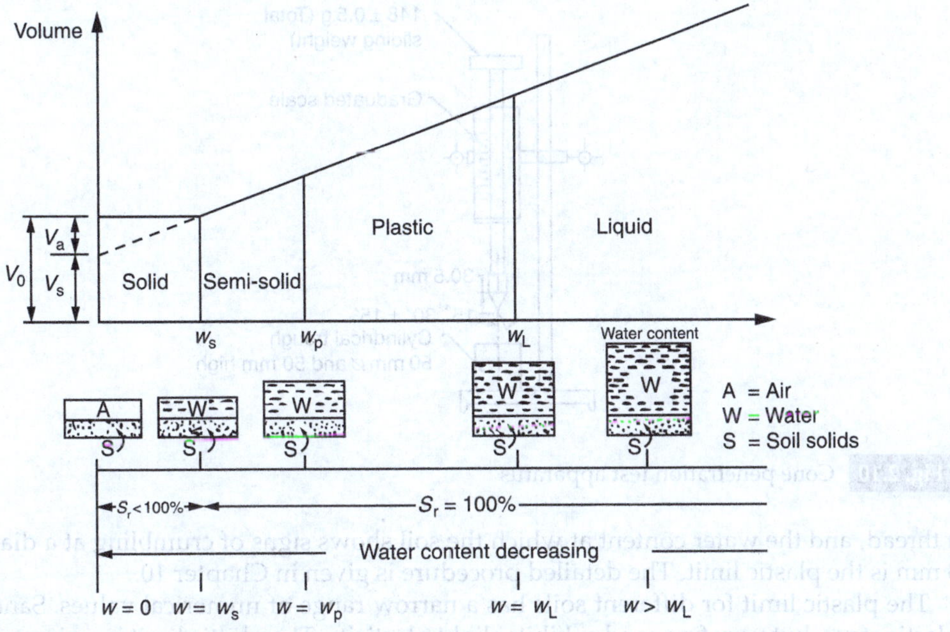


**Fig. 2.8** Consistency relationships

is noted down and the corresponding water content determined. The test is repeated for a variety of water contents, and the water content corresponding to a penetration of 25 mm is taken as the liquid limit of the soil. The test is quicker, and the results are accurate and reproducible. This has several advantages over the mechanical method. Therefore, it has been recognized as a standard method by Indian Standards.

The plastic limit of a soil is the water content, expressed as a percentage of the weight of oven-dried soil, at the boundary between the plastic and semi-solid states of consistency of the soil (IS: 2720 – Part 5, 1970). The plastic limit is determined by rolling a pat of soil into

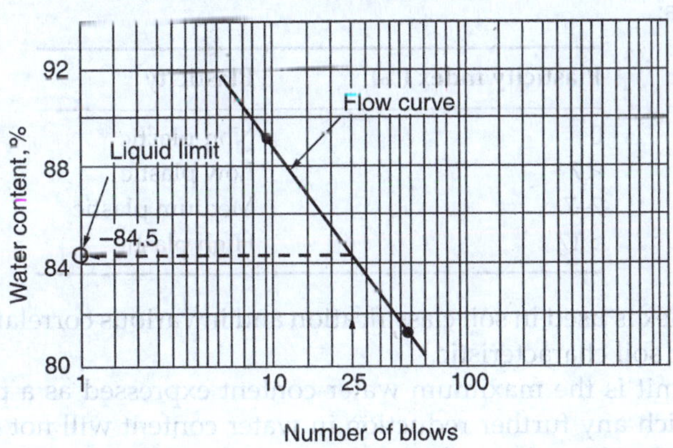


**Fig. 2.9** Flow curve

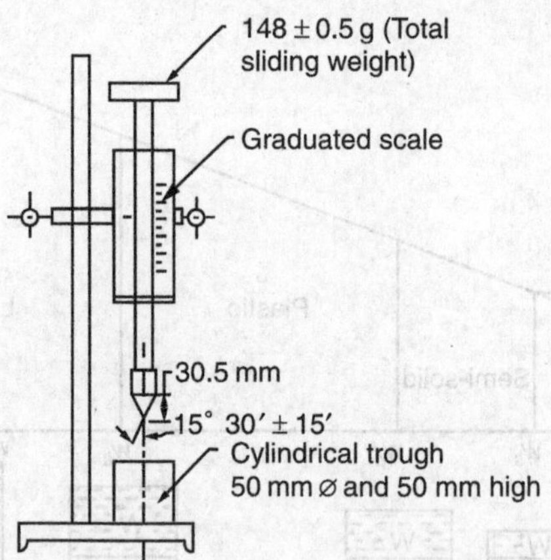


**Fig. 2.10** Cone penetration test apparatus

a thread, and the water content at which the soil shows signs of crumbling at a diameter of 3 mm is the plastic limit. The detailed procedure is given in Chapter 10.

The plastic limit for different soils has a narrow range of numerical values. Sand has no plastic stage, but very fine sand exhibits slight plasticity. The plastic limit is an important soil property. Earth roads are easily usable at this water content. Excavation work and agricultural cultivation can be carried out with the least effort with soils at the plastic limit.

Soil is said to be in the plastic range when it possesses water content in the range between $w_L$ and $w_p$. The range of the plastic state is given by the difference between $w_L$ and $w_p$ and is defined as the plasticity index. That is,

$$I_p = w_L - w_p \tag{2.30}$$

The plasticity index represents the range of water content over which a soil is plastic. The greater the plasticity index, higher will be the attraction between the particles of the soil and greater the plasticity of the soil. Based on the plasticity index, the soils are classified by Atterberg as follows:

| Plasticity index (%) | Plasticity |
|---|---|
| 0 | Non-plastic |
| <7 | Low plastic |
| 7–7 | Medium plastic |
| >17 | High plastic |

The plasticity index is used in soil classification and in various correlations with other soil properties as a basic soil characteristic.

The shrinkage limit is the maximum water content expressed as a percentage of oven-dried weight at which any further reduction in water content will not cause a decrease in volume of the soil mass, the soil mass being prepared initially from remoulded soil (IS: 2720 – Part 6, 1972). Based on the above definition, the shrinkage limit is determined by completely

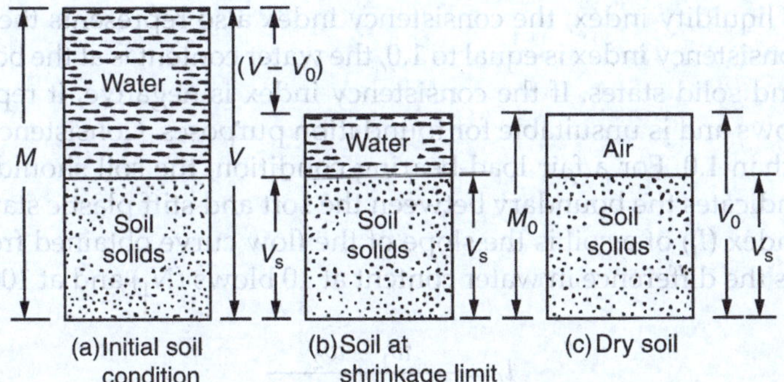


**Fig. 2.11** Phase diagrams representing shrinkage limit conditions

drying out a lump of soil and measuring its final volume and mass. Thus, referring to the phase diagram in Fig. 2.11, the shrinkage limit is given as

$$w_s = \frac{(M - M_0) - (V - V_0)\rho_w}{M_0} \times 100 \tag{2.31}$$

where $M$ is the initial wet mass of soil, $M_0$ the final dry mass of soil, $V$ the initial volume of soil, and $V_0$ the final volume of dry soil mass.

The shrinkage limit test can also be performed on undisturbed soil; in that case, the notation $w_{su}$ is used.

The finer the particles of the soil, the greater is the amount of shrinkage. Soils that contain montmorillonite clay mineral shrink more. Such soils shrink heterogeneously during summer, as a result of which cracks develop on the surface. Further, these soils imbibe more and more water during the monsoon and swell. Soils that shrink and swell are categorized as expansive soils. Indian black cotton soils belong to this group.

A detailed test procedure for determination of the shrinkage limit is given in Chapter 10.

The relationship between different limits of consistency and natural or *in situ* water content is given below (IS: 2720 – Part 5, 1970):

1. The liquidity index or water plasticity ratio ($I_L$) is the ratio expressed as a percentage of the natural water content ($w_n$) of a soil minus its plastic limit to its plasticity index. That is,

$$I_L = \frac{w_n - w_P}{I_P} \tag{2.32}$$

The *in situ* state of a soil is represented by the liquidity index: when $I_L < 0$, the soil is in the semi-solid state; when $I_L = 0$, the soil is in the stiff state; when $0 > I_L < 1$, the soil is in the plastic state, when $I_L = 1$, the soil is in a very soft state; and when $I_L > 1$, the soil is in the liquid state.

2. The consistency index ($I_c$) of a soil is the ratio of the liquid limit minus the natural water content to its plasticity index. That is,

$$I_c = \frac{w_L - w_n}{I_p} \tag{2.33}$$

Similar to the liquidity index, the consistency index also represents the *in situ* state of a soil. When the consistency index is equal to 1.0, the water content is at the boundary between the semi-solid and solid states. If the consistency index is negative, it represents the state where the soil flows and is unsuitable for foundation purposes. Consistency index can have a value greater than 1.0. For a fair load-bearing condition, the soil should have a value of $I_c = 0.5$, which indicates the boundary between the soft and stiff plastic states.

3. The flow index ($I_f$) of a soil is the slope of the flow curve obtained from a liquid limit test, expressed as the difference in water content at 10 blows ($N_1$) and at 100 blows ($N_2$):

$$I_f = \frac{w_1 - w_2}{\log_{10}(N_2/N_1)} \tag{2.34}$$

where $w_1$ and $w_2$ are the water contents corresponding to $N_1$ and $N_2$ drops, respectively.

Slopes of flow curves distinguish between the degree of cohesiveness and the shear strength of various soils. Two soils with the same plasticity index but different liquid limits will have different flow indices. The one with a steeper flow curve indicates soil of low shear strength.

4. The toughness index ($I_T$) of a soil is the ratio of the plasticity index to the flow index

$$I_T = \frac{I_p}{I_f} \tag{2.35}$$

The shear strength of a fine-grained soil at a water content close to the plastic limit is a measure of its toughness. The toughness of two fine-grained soils with the same plasticity index is inversely proportional to the flow indices. For clay, the toughness index is generally less than 3.

The concept of analysing various states in a soil based on water content is a sound one. The limits have been fixed arbitrarily and cannot be accepted as fundamental properties. Thus, not much significance should be attached to their accurate values.

### 2.4.2 Activity of Clays

As the particle size decreases, the surface area of the particle and the amount of water attracted to the soil surface increase. Thus, the amount of water attracted will depend considerably on the number of clay-size particles present in the soil. On the basis of this reasoning, Skempton (1953) proposed a relationship between the plasticity index and the percentage of particle sizes finer than 2 μm and called the quantity, the activity of clay, *A*.

$$A = \frac{\text{Plasticity index}}{\text{Percentage by weight of particles finer than } 2\,\mu\text{m}}$$

The activity of clay *A* gives a qualitative measure of the behaviour of the soil as active, normal, or inactive. Skempton (1953) has classified the clay as inactive if $A < 0.75$, normal if *A* is 0.75–1.25, and active if $A > 1.25$. Table 2.3 shows typical values of liquid limit, plastic limit, and activity.

**Table 2.3** Typical values of $\omega_L$, $\omega_P$, and $A$ of some minerals

| Mineral | Liquid limit ($\omega_L$) | Plastic limit ($\omega_P$) | Activity ($A$) |
|---|---|---|---|
| Kaolinite | 35–100 | 20–40 | 0.3–0.5 |
| Halloysite | 40–55 | 30–45 | 0.4–0.6 |
| Illite | 60–120 | 35–60 | 0.5–1.2 |
| Moutmorillonite | 100–900 | 50–100 | 1.5–7.0 |

*Source:* Das (2002).

## WORKED EXAMPLES

**Example 2.1** Adopting a routine laboratory procedure, the specific gravity of a river sand was determined. The mass of dry sand was 198.6 g. The mass of the calibrated flask filled with water was 1508.2 g. The masses of the flask, water, and sand were 1632.6 g. Determine the particle specific gravity of the soil.

If the true specific gravity was 2.72 and an error was made in recording the mass of dry sand, what is the correct mass of dry soil? The other two observations are correct.

***Solution***

Mass of dry soil = 198.6 g

Mass of an equal volume of water = (1508.2 – 1632.6) + 198.6 = 74.2 g

$$\text{Specific gravity of soil solids, } G = \frac{\text{Mass of dry soil}}{\text{Mass of an equal volume of water}}$$

$$= \frac{198.6}{74.2} = 2.68$$

Because of wrong recording of the mass of dry sand, both the numerator and denominator are affected. Let $M_s$ be the true mass of dry sand. Therefore,

$$2.72 = \frac{M_s}{(1508.2 - 1632.6) + M_s}$$

Solving for $M_s$, the true mass of dry sand is 196.73 g.

**Example 2.2** An attempt was made to determine the water content of a given moist soil of known specific gravity, using a pycnometer. The usual laboratory procedure for specific gravity determination of dry soil is used for the wet soil.

The following are the observations:

Mass of pycnometer ($M_1$) = 545 g

Mass of pycnometer with moist soil ($M_2$) = 790 g

Mass of pycnometer with soil and water ($M_3$) = 1,540 g

Mass of pycnometer and water ($M_4$) = 1,415 g
Specific gravity of soil grains = 2.67

Determine the water content of the soil from first principles.

***Solution***

Consider the two-phase diagrams shown in Fig. 2.12 representing the observations. Thus,

$$M_3 - M_4 = M_s - (\text{mass of an equal volume of water})$$

$$= M_s - \left(\frac{M_s}{G\rho_w}\right)\rho_w \qquad \left(\text{since } V_s = \frac{M_s}{\rho_s} = V_w\right)$$

$$= M_s\left(\frac{G-1}{G}\right)$$

That is, $M_s = (M_3 - M_4)\left(\dfrac{G}{G-1}\right)$

Substituting the respective values,

$$M_s = (1540 - 1415)\frac{2.67}{2.67-1} = 199.85\,\text{g}$$

Therefore,

$$w = \frac{(M_2 - M_1) - M_s}{M_s} \times 100$$

Substituting,

$$w = \frac{(790 - 545) - 199.85}{199.85} \times 100 = 22.6\%$$

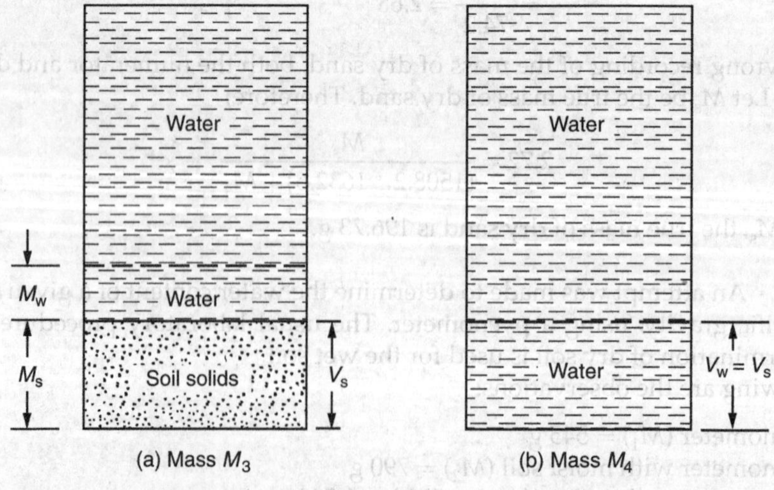


Fig. 2.12

**Example 2.3** A sample of soil, extracted in its natural state using a sampling tube of volume 0.001 m$^3$, was found to have a mass of 1,730 g, the degree of saturation being 61.6%. The oven-dried mass of soil was 1,440 g. Determine (i) the natural water content, (ii) the specific gravity of soil solids, (iii) the void ratio, (iv) the bulk density, (v) the saturated density, and (vi) the submerged density.

***Solution***

1. Natural water content $w = \dfrac{1730-1440}{1440} \times 100 = 20.14\%$

Dry density $\rho_d = \dfrac{1440}{0.001 \times 100^3} = 1.44\,\text{g/cm}^3$

Also,

$$\rho_d = \frac{G\rho_w}{1+e} = \frac{G\rho_w}{1+(wG/S_r)} \qquad \left(\text{since } e = \frac{wG}{S_r}\right)$$

$$1.44 = \frac{G \times 1}{1+(20.14/61.6)G} = \frac{G}{1+0.3279}$$

Rearranging,

2. $G = 2.72$

3. $e = \dfrac{wG}{S_r} = \dfrac{20.14 \times 2.72}{100 \times (61.6/100)} \times 2.72 = 0.89$

4. Bulk density $\rho_t = \rho_d\,(1+w) = 1.44\left(1+\dfrac{20.14}{100}\right) = 1.73\,\text{g/cm}^3$

5. Saturated density $\rho_{sat} = \rho_d\left(1+\dfrac{e}{G}\right)$ $\qquad (S_r = 100\%)$

$$= 1.44\left(1+\frac{0.89}{2.72}\right) = 1.91\,\text{g/cm}^3$$

6. Submerged density $\rho' = \rho_{sat} - \rho_w = 1.91 - 1.0 - 0.91\,\text{g/cm}^3$

**Example 2.4** A saturated specimen of undisturbed clay has a volume of 22.5 m$^3$ and mass of 35 g. After oven drying, the mass reduces to 20 g. Find its moisture content, specific gravity of solids, void ratio, and dry density.

***Solution***

Moisture content $w = \dfrac{M_w}{M_s} = \dfrac{35-20}{20} \times 100 = 75\%$

Dry density $\rho_d = \dfrac{M_s}{V} = \dfrac{20}{22.5} = 0.898\text{ g/cm}^3$

Also, $\rho_d = G\rho_w/(1+e)$, and for a saturated soil, $e = wG$. Therefore,

$$\rho_d = \frac{G\rho_w}{1+wG}$$

Rearranging,

$$G = \frac{\rho_d}{1-\rho_d w} = \frac{0.89}{1-0.89\times 0.75} = 2.697 \qquad \text{(since } \rho_w = 1.0\text{)}$$

Therefore, the void ratio $e = wG = \frac{75}{100}\times 2.697 = 2.02$

**Example 2.5** A wet soil sample weighs 3.46 N. After drying at 5°C, its weight is 2.84 N. The bulk unit weight of the soil is 18.6 kN/m$^3$. The specific gravity of the solid particles is 2.7. Determine (i) the water content, (ii) the void ratio, (iii) the degree of saturation, and (iv) the porosity.

***Solution***

1. Water content $w = \frac{3.46-2.84}{2.84}\times 100 = 21.83\%$

   Dry unit weight $\gamma_d = \frac{\gamma_t}{1+w} = \frac{18.6}{1+(21.83/100)} = 15.27\,\text{kN/m}^3$

2. Void ratio $e = \frac{G\gamma_w}{\gamma_d} - 1 = \frac{2.7\times 9.81}{15.27} - 1 = 0.735$

3. Degree of saturation $S_r = \frac{wG}{e} = \frac{21.83}{100}\times\frac{2.7}{0.735}\times 100 = 80.19\%$

4. Porosity $n = \frac{e}{1+e} = \frac{0.735}{1+0.75}\times 100 = 42.36\%$

**Example 2.6** The bulk unit weight of a soil is 19.10 kN/m$^3$, the water content is 12.5%, and the specific gravity of soil solids is 2.67. Determine the dry unit weight, void ratio, porosity, and degree of saturation.

***Solution***

1. $\gamma_d = \frac{\gamma}{1+w} = \frac{19.1}{1+(12.5/100)} = 16.98\,\text{kN/m}^3$

   $$\gamma_d = \frac{G\gamma_w}{1+e}$$

   Therefore,

2. $e = \frac{G\gamma_w}{\gamma_d} - 1 = \frac{2.67\times 9.81}{16.98} - 1 = 0.54$

Therefore,

3. $n = \frac{e}{1+e} \times 100 = \frac{0.54}{1+0.54} \times 100 = 35.07\%$

4. $S_r = \frac{wG}{e} \times 100 = \frac{12.5}{100} \times \frac{2.67}{0.54} \times 100 = 61.8\%$

**Example 2.7** A soil sample has a porosity of 40%. The specific gravity of solids is 2.7. Calculate the (i) void ratio, (ii) dry density, (iii) unit weight if the soil is 50% saturated, and (iv) unit weight if the soil is completely saturated.

***Solution***

1. Void ratio $e = \frac{n}{1-n} = \frac{0.40}{1-0.40} = 0.67$

2. Dry density $\rho_d = \frac{G\rho_w}{1+e} = \frac{2.7 \times 1}{1+0.67} = 1.62\,g/cm^3$

3. Wet unit weight $\gamma_t = \frac{G+eS_r}{1+e}\gamma_w = \frac{2.7+0.67\times 0.5}{1+0.67} \times 9.81$

$$= 17.82\,kN/m^3$$

4. Saturated unit weight $\gamma_{sat} = \frac{G+e}{1+e}\gamma_w = \frac{2.7+0.67}{1+0.67} \times 9.81$

$$= 19.80\,kN/m^3$$

**Example 2.8** How many cubic metres of fill can be constructed at a void ratio of 0.65 from 2,21,000 $m^3$ of borrow material that has a void ratio of 1.25?

***Solution***

Let $e_b$ and $e_f$ be the void ratios of the borrow material and the fill, respectively. Also, let $V_{vb}$ and $V_{vf}$ be the volume of voids in the borrow and the fill, respectively. The volume of soil solids is the same in both the cases. From Fig. 2.13,

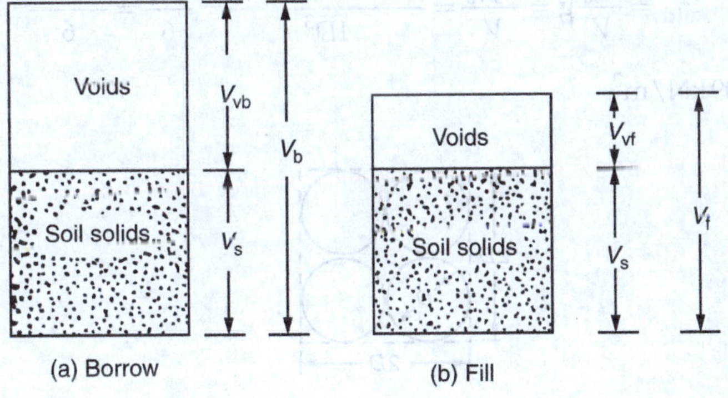


Fig. 2.13

$e_b = \frac{V_{vb}}{V_s}$ and $e_f = \frac{V_{vf}}{V_s}$

Therefore,

$V_{vb} = e_b V_s$ and $V_{vf} = e_f V_s$

Total volume of soil in the borrow is $V_b = V_{vb} + V_s$.

That is, $V_b = e_b V_s + V_s = (1+e_b)V_s$

Therefore, $V_s = \frac{V_b}{1+e_b}$

Total volume of soil in the fill $= V_f = (1+e_f)V_s$.

$$V_f = (1+e_f)\frac{V_b}{1+e_b} = \frac{1+0.65}{1+1.25} \times 221000 = 162066.7\,\text{m}^3$$

**Example 2.9** For a stable packing of regular spheres at the minimum density, find the void ratio and the dry unit weight. Unit weight of soil solids is 25 kN/m$^3$.

***Solution***

Refer to Fig. 2.14.

Let $D$ be the diameter of each sphere.

Volume of each sphere $= \frac{\pi D^3}{6}$

For the arrangement in Fig. 2.14, the density will be minimum.

Volume $= 2D \times 2D \times D = 4D^3$

Therefore, $e = \frac{4D^3 - 4\times \pi D^3/6}{4\times \pi D^3/6} = \frac{1-\pi/6}{\pi/6} = \frac{6-\pi}{\pi} = 0.91$

Also,

$$\gamma_d = \frac{M_s}{V}g = \frac{V_s\gamma_s}{V} = \frac{4\times \pi D^3/6 \times \gamma_s}{4D^3} = \frac{\pi\gamma_s}{6} = \frac{\pi \times 25}{6}$$

That is, $\gamma_d = 13.09\,\text{kN/m}^3$

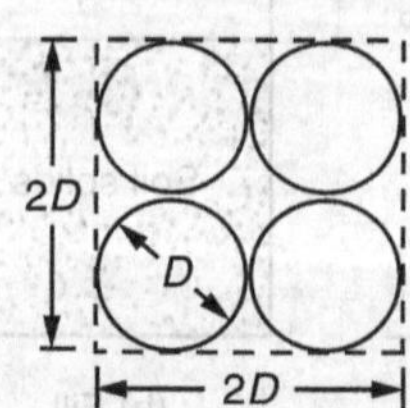


Fig. 2.14

**Example 2.10** A fully saturated soil sample was extracted during an oil well drilling. The wet mass of the sample was 3.15 kg, and the volume of the sampling tube was 0.001664 m$^3$. After analysis, the soil sample was found to contain 28.2% of the liquid as kerosene and had a dry mass of 2.67 kg. The specific gravity of soil grains was 2.68. Determine the bulk density, void ratio and water content of the sample.

***Solution***

$$\text{Bulk density} = \frac{3.15}{0.001664} = 1893\,\text{kg/m}^3 = 1.89\,\text{Mg/m}^3$$

$$\text{Volume of soil grains} = \frac{2.67}{2.68 \times 1000} = 0.000996\,\text{m}^3$$

$$\text{Volume of voids} = 0.001664 - 0.000996 = 0.000668\,\text{m}^3$$

$$e = \frac{0.000668}{0.000996} = 0.67$$

As the soil was fully saturated,

$$\text{Volume of liquid} = \text{volume of voids} = 0.000668\,\text{m}^3$$

$$\text{Volume of water} = (1 - 0.282) \times 0.000668 = 0.00048\,\text{m}^3$$

$$\text{Mass of water} = 0.48\,\text{kg}$$

$$\text{Water content} = \frac{0.48}{2.67} \times 100 = 17.89\%$$

**Example 2.11** A mass of soil is coated with a thin layer of paraffin wax. The paraffin wax weighs 6.906 g, and the soil alone weighs 443 g. When the sample is immersed in water, it displaces 346 ml of water. The specific gravity of the soil solids is 2.67, and that of wax is 0.89. Find the void ratio and degree of saturation, if the water content is 17.2%.

***Solution***

$$\text{Volume of paraffin wax } V_p = \frac{M_p}{G_p \rho_w} = \frac{6.906}{0.89 \times 1} = 7.76\,\text{cm}^3$$

$$\text{Volume of soil solids } V_s = \frac{M_s}{G_s \rho_w} = \frac{443}{2.65 \times 1} = 167.17\,\text{cm}^3$$

$$\text{Volume of soil } V = 346 - 7.76 = 338.24\,\text{cm}^3$$

$$\text{Volume of voids } V_v = 338.24 - 167.17 = 171.07\,\text{cm}^3$$

$$\text{Void ratio } e = \frac{V_v}{V_s} = \frac{171.07}{167.17} = 1.02$$

$$\text{Degree of saturation } S_r = \frac{wG}{e} = \frac{17.2}{100} \times \frac{2.65}{1.02} \times 100$$

$$= 44.69\%$$

**Example 2.12** A compacted cylindrical specimen 50 mm in diameter and 100 mm long is to be prepared from dry soil. If the specimen is required to have a water content of 15%, find the percentage of air voids required in the preparation of the soil when the specific gravity is 2.69.

***Solution***

Volume of cylinder = Volume of soil $= \frac{\pi}{4} \times 5^2 \times 10 = 196.3 \text{ cm}^3$

Air void ratio $A_v = \frac{e}{1+e}\left(1 - \frac{wG}{e}\right) = \left(\frac{e - wG}{1+e}\right)$

$$(1+e)A_v = e - wG$$

Rearranging,

$$e = \frac{wG + A_v}{1 - A_v} = \frac{(15/100) \times 2.69 + (20/100)}{1 - (20/100)}$$

$$= \frac{0.15 \times 2.69 + 0.20}{0.80} = \frac{0.2035}{0.80} = 0.75$$

Also,

$$e = \frac{V_v}{V_s} = \frac{V - V_s}{V_s} = \frac{V}{V_s} - 1$$

That is, $0.75 = \frac{196.3}{V_s} - 1$

$$V_s = \frac{196.3}{1.75} = 112.2 \text{ cm}^3$$

Weight of soil $M_s = \rho_s V_s = G\rho_w V_s = 2.69 \times 1 \times 112.2$

Weight of water $M_w = w M_s = \frac{15}{100} \times 301.8 = 45.27 \text{ g}$

**Example 2.13** A test of the density of the soil in place was performed by digging a small hole in the soil, weighing the extracted soil, and measuring the volume of the hole. The soil (moist) weighed 8.95 N; the volume of the hole was 426 cm$^3$. After drying, the sample weighed 7.78 N. Of the dried soil, 4 N was poured into a vessel in a very loose state. Its volume was subsequently determined to be 276 cm$^3$. That same 4 N was then vibrated and tamped to a volume of 212 cm$^3$. The specific gravity of the solid particles is 2.7. Find the relative density of the soil.

***Solution***

Natural density $\rho_d = \frac{7.78}{426} = 0.0183 \text{ N/cm}^3$

Loose density $\rho_{d\,min} = \frac{4}{276} = 0.0145 \text{ N/cm}^3$

Maximum density $\rho_{d\,max} = \frac{4}{212} = 0.01887 \text{ N/cm}^3$

From Eq. 2.23,

$$I_D = \frac{\rho_{d\,max}}{\rho_d} \times \frac{0.0183 - 0.0145}{0.01887 - 0.0145} \times 100$$

$$= \frac{0.01887}{0.0183} \times \frac{0.0038}{0.00437} \times 100 = 92.29\%$$

**Example 2.14** In order to determine the in-place density of a highway sub-grade, a sand bottle method was adopted. The mass of soil extracted from a hole at the surface was 4.87 kg. The hole was then filled with sand from the sand bottle and found to have a mass of 3.86 kg. While calibrating the sand bottle, to fill a container of volume 0.0048 $m^3$, a mass of 6.82 kg of sand was needed. In a moisture content determination, 28.26 g of the moist soil weighed 22.2 g after oven drying. If the specific gravity of the soil was 2.67, determine the bulk and dry densities and the degree of saturation of the soil.

***Solution***

$$\text{Density of sand in the sand bottle} = \frac{6.82}{0.0048} = 1420.8\,\text{kg/m}^3$$

$$= 1.42\,\text{Mg/m}^3$$

$$\text{Volume of the hole} = \frac{3.86}{1420.8} = 0.00272\,\text{m}^3$$

$$\text{Bulk density} = \frac{4.87}{0.00272} = 1790.4\,\text{kg/m}^3 = 1.79\,\text{Mg/m}^3$$

$$\text{Water content} = \frac{28.26 - 22.2}{22.2} \times 100 = 27.3\%$$

$$\text{Dry density} = \frac{1790.4}{1 + (27.3/100)} = 1406.4\,\text{kg/m}^3 = 1.41\,\text{Mg/m}^3$$

$$e = \frac{G\rho_w}{\rho_d} - 1 = \frac{2.67 \times 1000}{1406.4} - 1 = 0.899$$

$$S_r = \frac{wG}{e} \times 100 = \frac{27.3}{100} \times \frac{2.67}{0.899} \times 100 = 81.08\%$$

**Example 2.15** A relative density test conducted on a sandy soil yielded the following results: maximum void ratio = 1.23, minimum void ratio = 0.48, relative density = 42%, and $G = 2.67$.

Find the dry density of the soil in the present state. If a 3 m thickness of this stratum is densified to a relative density of 62%, how much will the soil reduce in thickness? What will be the new density in dry and saturated conditions?

***Solution***

$$I_D = \frac{e_{max} - e}{e_{max} - e_{min}} \times 100$$

or

$$0.42 = \frac{1.23 - e}{1.23 - 0.48} \times 100$$

or

$$(1.23 - e) = 0.42(1.23 - 0.48)$$

or

$$e = 0.915$$

Now,

$$\rho_d = \frac{G\rho_w}{1+e} = \frac{2.67 \times 1}{1 + 0.915} = 1.37\ \text{g/cm}^3$$

and

$$e = \frac{V_v}{V_s} = \frac{V - V_s}{V_s}$$

Therefore,

$$0.915 = \frac{3 - V_s}{V_s}$$

Therefore,

$$V_s = 1.57\ \text{m}^3$$

For 62% relative density, the void ratio to which the soil has to be compacted is obtained from

$$0.62 = \frac{1.23 - e}{1.23 - 0.48} \times 100$$

or

$$e = 0.765$$

Therefore,

$$0.765 = \frac{V - 1.57}{1.57}$$

or

$$V = 0.765 \times 1.57 + 1.57 = 2.77$$

Therefore, the reduction in thickness is 0.23 m.

**Example 2.16** From the results of a sieve analysis given below, plot a grain-size distribution curve and then determine (i) the effective size, (ii) the uniformity coefficient, and (iii) the coefficient of gradation. Mass of soil taken for sieve analysis was 500 g.

| IS sieve no. | Mass of soil retained in each sieve (g) |
|---|---|
| 480 | 3.8 |
| 240 | 32.2 |
| 120 | 52.8 |
| 60 | 38.7 |
| 30 | 122.5 |
| 15 | 15 |
| 8 | 26.4 |

***Solution***

| Sieve no. | Sieve opening (mm) | Mass retained (g) | Percent retained | Cumulative percent retained | Percent finer |
|---|---|---|---|---|---|
| 480 | 4.76 | 3.8 | 0.76 | 0.76 | 99.24 |
| 240 | 2.40 | 32.2 | 6.44 | 7.20 | 92.80 |
| 120 | 1.20 | 52.8 | 10.56 | 17.76 | 82.24 |
| 60 | 0.60 | 38.7 | 7.74 | 25.50 | 74.50 |
| 30 | 0.30 | 122.5 | 24.50 | 50.00 | 50.00 |
| 15 | 0.15 | 159.9 | 31.98 | 81.98 | 18.02 |
| 8 | 0.075 | 26.4 | 5.28 | 87.26 | 12.74 |

The grain-size distribution curve is plotted as given in Fig. 2.15.

Effective size $D_{10} = 0.07$ mm

Uniformity coefficient $C_u = \dfrac{D_{60}}{D_{10}} = \dfrac{0.43}{0.07} = 6.14$

Coefficient of gradation $C_z = \dfrac{{D_{30}}^2}{D_{60} \times D_{10}} = \dfrac{0.21^2}{0.43 \times 0.07} = 1.47$

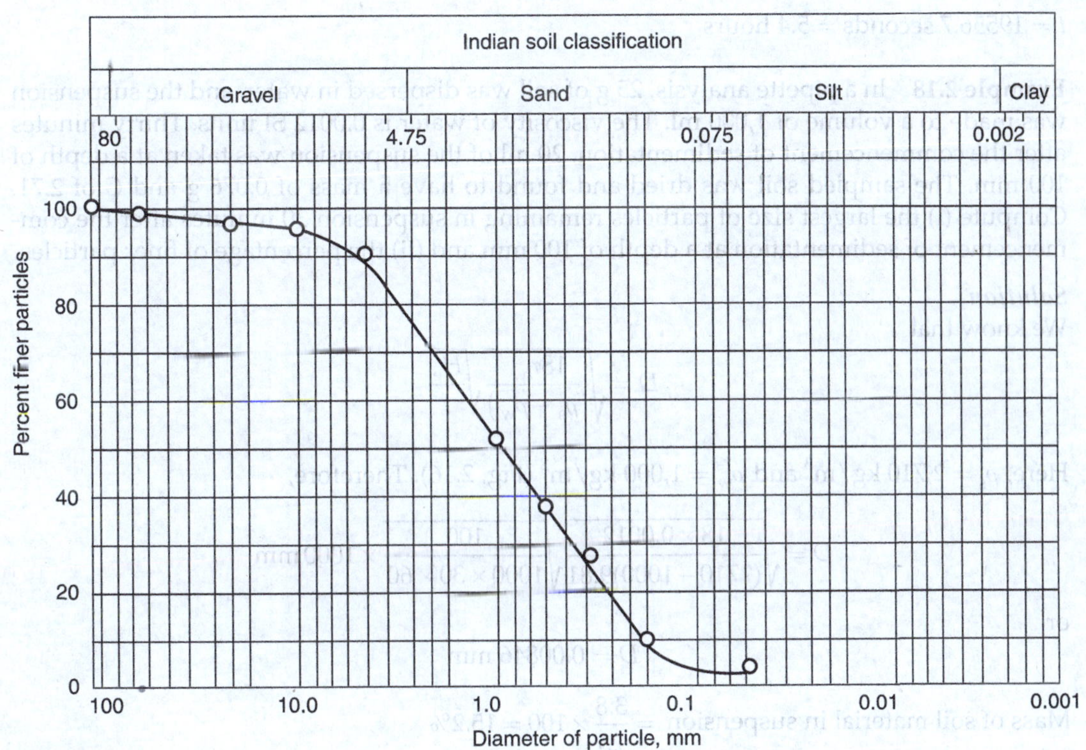


Fig. 2.15

**Example 2.17** In a sedimentation analysis 48 g of soil passing, 75 μm is dispersed in 1,000 ml of water. In order to estimate the percentage of particles of size less than 0.003 mm, how long after the commencement of sedimentation is the hydrometer reading to be taken? The centre of the hydrometer is 165 mm below the surface of the water. The specific gravity of soil grains is 2.72, and viscosity of water is 0.001 N-s/m$^2$.

***Solution***
From Eq. 2.24,

$$v = \frac{D^2(\gamma_s - \gamma_w)}{18\eta}$$

$$v = \frac{(0.003/1000)^2(2.72-1.0)9.81}{18 \times (0.001/1000)} = 8.437 \times 10^{-6}\,\text{m/s}$$

Also,

$$v = \frac{H_e}{t}$$

That is,

$$8.437 = \frac{165}{1000t}$$

Rearranging,

$t = 19556.7$ seconds $= 5.4$ hours

**Example 2.18** In a pipette analysis, 25 g of soil was dispersed in water, and the suspension was made to a volume of 1,000 ml. The viscosity of water is 0.0012 SI units. Thirty minutes after the commencement of sedimentation, 20 ml of the suspension was taken at a depth of 100 mm. The sampled soil was dried and found to have a mass of 0.076 g and $G$ of 2.71. Compute (i) the largest size of particles remaining in suspension 30 minutes after the commencement of sedimentation at a depth of 100 mm and (ii) the percentage of finer particles.

***Solution***
We know that

$$D = \sqrt{\frac{18\eta_w}{(\rho_s - \rho_w)}}\sqrt{\frac{H_e}{t}}$$

Here, $\rho_s = 2{,}710$ kg/m$^3$ and $\rho_w = 1{,}000$ kg/m$^3$ (Fig. 2.16). Therefore,

$$D = \sqrt{\frac{18 \times 0.0012}{(2710-1000)9.81}}\sqrt{\frac{100}{1000 \times 30 \times 60}} \times 1000\,\text{mm}$$

or

$$D = 0.00846\ \text{mm}$$

Mass of soil material in suspension $= \frac{3.8}{25} \times 100 = 15.2\%$

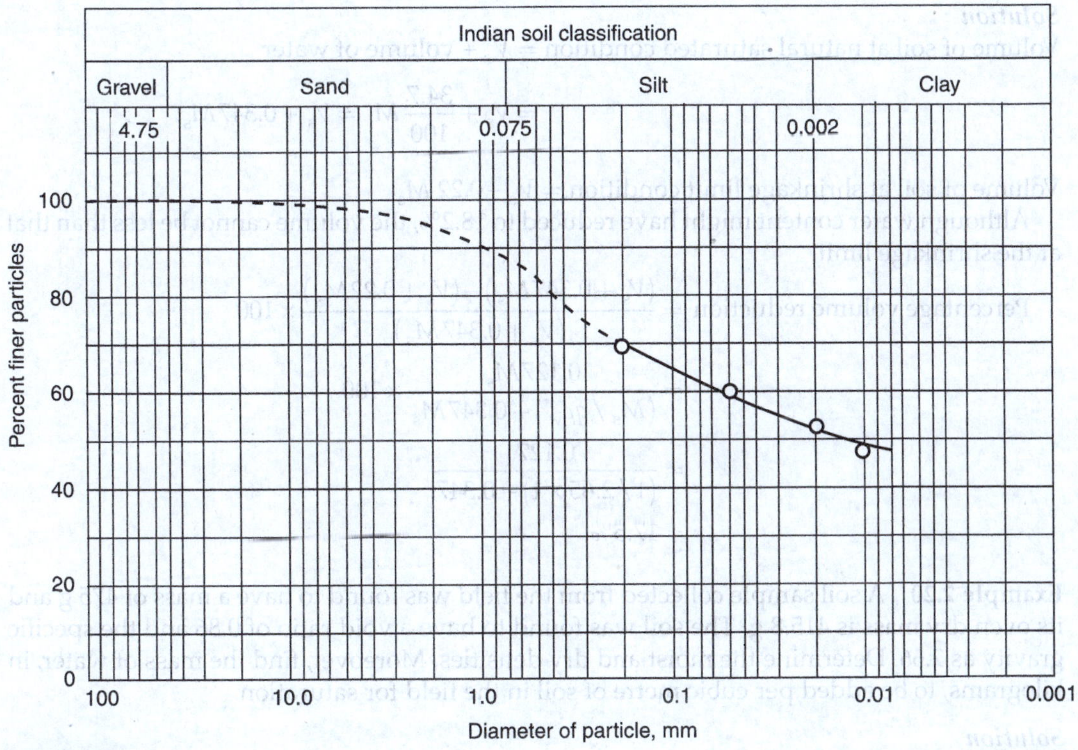


Fig. 2.16

**Example 2.19** The liquid limit and plastic limit of a soil are 65% and 31%, respectively. The natural water content is 25%. Find the liquidity index and activity number. Comment on the consistency of the soil.

***Solution***

Plasticity index $I_P = w_L - w_P = 65 - 31 = 34\%$

$$\text{Liquidity index } I_L = \frac{w_n - w_P}{I_P} = \frac{25-31}{34} = -0.176$$

$$\text{Activity number } A = \frac{I_P}{\%\text{ Particle less than } 2\ \mu\text{m}} = \frac{34}{24} = 1.42$$

The consistency of the soil is very stiff as the liquidity index is negative. The soil is highly plastic as the plasticity index is greater than 17%. The soil is active as the activity number is greater than 1.25.

**Example 2.20** The shrinkage limit of a clay is 22%, its natural moisture content 34.7%, and its specific gravity 2.65. Calculate the percentage decrease to be expected in a unit volume of clay if the moisture content is reduced by evaporation to 18.2%.

***Solution***

Volume of soil at natural saturated condition = $V_s$ + volume of water

$$= V_s + \frac{34.7}{100} M_s = V_s + 0.347 M_s$$

Volume of soil at shrinkage limit condition = $V_s + 0.22\, M_s$

Although water content might have reduced to 18.2%, the volume cannot be less than that at the shrinkage limit.

$$\text{Percentage volume reduction} = \frac{(V_s + 0.347 M_s) - (V_s + 0.22 M_s)}{(V_s + 0.347 M_s)} \times 100$$

$$= \frac{0.127 M_s}{(M_s / q\rho_w) + 0.347 M_s} \times 100$$

$$= \frac{0.127}{(1/2.65 \times 1) + 0.347}$$

$$= 17.5\%$$

**Example 2.20** A soil sample collected from the field was found to have a mass of 475 g and its oven dry mass is 415.8 g. The soil was found to have a void ratio of 0.86 and the specific gravity as 2.66. Determine the moist and dry densities. Moreover, find the mass of water, in kilograms, to be added per cubic metre of soil in the field for saturation.

***Solution***

$$\omega = \frac{\mu_\omega}{\mu_\lambda} = \frac{475 - 415.8}{475} \times 100 = 12.46\%$$

$$\rho = \frac{G\rho_\omega(1+\omega)}{1+e} = \frac{2.66 \times 1{,}000(1+0.1246)}{1+0.86} = 1{,}608.3 \text{ kg/m}^3$$

$$\rho_d = \frac{G\rho_\omega}{1+e} = \frac{2.66 \times 1{,}000}{1.86} = 1{,}430.11 \text{ kg/m}^3$$

$$\left.\text{Mass of water to be added for saturation}\right\} = \rho_{sat} = \rho$$

$$\rho_{sat} = \frac{(G+e)\rho_\omega}{1+e} = \frac{(2.66+0.86)1{,}000}{1.86} = 1{,}892.47 \text{ kg/m}^3$$

$$\left.\begin{array}{l}\text{Therefore, the mass of water to be}\\ \text{added per cubic metre}\end{array}\right\} = 1{,}892.47 - 1{,}608.3$$

$$= 284.17 \text{ kg.}$$

**Example 2.21** A fully saturated clay sample has specific gravity of 1.98 at 25% water contact. After oven-drying, the mass specific gravity reduces to 1.63. Find the specific gravity of the clay and the shrinkage limit.

***Solution***

As the clay is fully saturated void ratio, $e\} = \omega_G = 0.25\,G$

We know $$\rho_{\text{sat}} = \left(\frac{G+e}{1+e}\right)\rho_\omega$$

i.e., $$\frac{\rho_{\text{sat}}}{\rho_\omega} = \frac{G+e}{1+e}$$

i.e., $$1.98 = \frac{G+0.25G}{1+0.25G} = \frac{1.25G}{1+0.25G}.$$

By rearranging, we get

$$1.98 + 1.98 \times 0.25G = 1.25G$$

$$1.98 + 0.495G = 1.25G$$

$$\therefore \quad G = 2.62$$

As the soil is allowed to dry gradually in an oven, the dry mass specific gravity is at the shrinkage limit stage. Then the water contact at this stage is the shrinkage limit, $\omega_s$

i.e., $$\omega_s = \frac{\rho_\omega}{\rho_d} - \frac{1}{G} = \left(\frac{1}{1.63} - \frac{1}{2.62}\right) \times 100$$

$$= 23.2\%.$$

Hence shrinkage limit = 23.2%.

## POINTS TO REMEMBER

2.1 Soil deposits are particulate systems of three distinct phases, viz., soil solids, water, and air. This is referred to as the three-phase system. Dry soil (absence of water phase) and fully saturated soil (absence of air phase) constitute the two-phase systems.

2.2 Void ratio is an important parameter which governs the permeability, settlement, and stability problems of soil. Values of void ratio may range from 0.50 to 1.50 in soils.

2.3 Water content and degree of saturation represent the amount of water present in a soil. The behaviour of dry and saturated soils is easy to assess compared to partially saturated soils. Control of compaction is governed by moisture content.

2.4 Specific gravity of soil solids has a narrow range of variation (2.65–3.00), and the presence of organic material reduces the specific gravity.

2.5 The density of a soil (dry, saturated, or submerged) is a function of void ratio and moisture content and has a major role to play in all stability problems.

2.6 Grain-size distribution curves (obtained from sieve and sedimentation analyses) reflect the range of particle sizes present. The flatter the curve, the larger is the range of size of particles, and the steeper the curve, the smaller the range.

2.7 A numerical measure of the gradation of a soil is obtained by defining the uniformity coefficient and curvature coefficient and is also used in soil classification.
2.8 Selective particle sizes are used in the classification ($D_{60}$ and $D_{10}$) design of filters ($D_{50}$, $D_{85}$, and $D_{15}$).
2.9 Consistency of a soil refers to the texture and firmness of a soil and is denoted as soft, medium stiff, and stiff.
2.10 Consistency limits or Atterberg limits, viz., liquid limit, plastic limit, and shrinkage limit, are the water contents at the changes of states from liquid to plastic, plastic to semi-solid, and semi-solid to solid, respectively.
2.11 Liquidity index and consistency index represent the *in situ* firmness condition of a soil.

## QUESTIONS

### Objective Questions

2.1 Choose the correct statement from the following:
(i) The porosity of a soil can be greater than 100%.
(ii) The water content of a soil cannot be greater than 100%.
(iii) The natural water content of a soil cannot exceed the liquid limit.
(iv) The consistency index of a soil can be negative.

2.2 Void ratios of a micaceous sand sample in the densest and the loosest conditions are 0.4 and 1.2, respectively. The relative density of the soil for the in-place void ratio of 0.6 will be
(a) 60% (b) 75% (c) 65% (d) 80%

2.3 Consistency, in general, is that property of a soil which is manifested by its resistance to
(a) Impact (b) Rolling (c) Flow (d) None of the above

2.4 A clay is identified as a normal clay if the activity range is between
(a) 0.25 and 0.75 (b) 0.75 and 1.25 (c) 1.25 and 3.00 (d) 0.15 and 0.25

2.5 Swelling of clayey soil directly depends on the
(a) Percentage of clay fraction (b) Plasticity index of the soil
(c) Type of clay mineral (d) Liquid limit of the soil

2.6 Sand-bath method of determining water content is not suitable for
(a) Inorganic silts (b) Fine sands
(c) Soils with a high percentage of organic matter
(d) Soils with particle size ranging from 0.02 to 0.075 mm

2.7 For soils containing soluble salts, the specific gravity is determined using
(a) Salt water (b) De-aired water
(b) White spirit (d) Benzene

2.8 For a fine-grained soil with a plasticity index of 15 to 40%, the degree of plasticity is referred to as
(a) Non-plastic (b) Moderately plastic
(c) Plastic (d) Highly plastic

2.9 Identify the incorrect statcment. A semi-log plot is used for grain-size distribution so that
(a) A wide range of grain size can be accommodated
(b) Equal emphasis can be given to all grain sizes
(c) Comparison can be made between two or more soils
(d) An S-type curve can be obtained

2.10 *Assertion A*: Uniformity of a soil is reflected by the grain-size distribution curve.
*Reason R*: Uniformity coefficients indicate gradations of grain sizes in a soil sample.
*Select the correct code.*
(a) Both A and R are true, and R is the correct explanation of A.
(b) Both A and R are true, and R is not the correct explanation of A.
(c) A is true, but R is false.
(d) A is false, but R is true.

## Descriptive Questions

2.11 What is a unit phase diagram? Explain with examples.
2.12 Two soils of similar mineralogy have extreme percentage of clay content. How might the plasticity indices of the soil vary?
2.13 Give reasons for determining the grain-size distribution of a soil mass. Why is the study generally confined to coarse-grained soils?
2.14 Two clayey silty sands have identical particle sizes with 20% fines. When exposed to air, one dries out easily while the other does not. Why? Explain.
2.15 It is said that consistency index of a fine-grained soil and density index of a coarse-grained soil are synonymous. Explain.

## EXERCISE PROBLEMS

2.1 The following data are obtained from a pycnometer test of a soil sample:

Mass of pycnometer full of water = 2770.6 g.
Mass of pycnometer with soil and water = 2948.8 g.
Mass of moist soil = 315.5 g.
Specific gravity of soil solids = 2.67.

Find the water content of the soil.

2.2 A fully saturated clay has a moisture content of 42.4% and specific mass gravity (or bulk density) of 1.78 g/cm$^3$. Determine from first principles the void ratio and specific gravity of the soil grains.

2.3 Derive an expression for water content from first principles in terms of the unit weight of dry soil, the unit weight of water, the degree of saturation, and the specific gravity of soil solids.

2.4 The volume of soil taken from a field is 450 $cm^3$. The weight of soil mass is 760 g, and the dry weight is 620 g. Taking $G = 2.7$, find the (i) water content, (ii) void ratio, (iii) porosity, (iv) degree of saturation, and (v) mass specific gravity.

2.5 In a research project on synthetic soils, a soil with dry unit weight $\gamma_d$ is mixed with organic matter of unit weight $\gamma_0$ to have varied organic content $O_c$. Organic content is defined as the ratio of the dry organic matter to the total dry weight of the sample. Derive an expression for the unit weight of the synthetic soil in terms of $\gamma_d$, $\gamma_0$, and $O_c$.

2.6 An undisturbed sample was extracted using a sampling tube of volume 1,200 $cm^3$. The mass of the clay sample and tube was 5.00 kg, and the same sample after oven drying was 4.31 kg. The mass of the empty tube was 2.12 kg. Determine the water content, wet density, and dry density of the sample. The specific gravity of the soil solids was found to be 2.69. Find the void ratio and degree of saturation of the clay.

2.7 The porosity of a soil sample is 35%, and specific gravity of its particle is 2.70. Calculate its void ratio, dry density, saturated density, and submerged density.

2.8 A clayey soil has moisture content of 15.8%. The specific gravity is 2.72, and the saturation percentage is 70.8%. The soil is allowed to absorb water. After some time, the saturation increased to 90.8%. Find the water content in the latter case.

2.9 A dry soil sample of volume 280 $cm^3$ weighs 450 g. Determine the water content at 100% saturation without any change in volume. What will be the water content when the volume is allowed to increase by 12% of the original dry volume?

2.10 A 1,000-$cm^3$ container was filled with a sand first in its loosest possible state and then in its densest possible state, and the weight of the sand was 1,520 g and 1,830 g, respectively. The sand, *in situ*, had a void ratio of 0.64. If the specific gravity of the sand particles is 2.65, determine the limiting void ratios and the relative density *in situ*.

2.11 A soil sample has 80% of particles (by weight) finer than 0.1 mm, 7.5% finer than 0.01 mm, and 4% finer than 0.001 mm. Draw the grain-size curve and determine the percentage of total weight in each of the various size ranges, the effective size, and the uniformity coefficient of the soil.

2.12 Draw the grain-size distribution curve for the soil with the following data:

| | | | | |
|---|---|---|---|---|
| Aperture size (mm) | 4.76 | 2.38 | 2.0 | 0.85 |
| Percentage passing | 100 | 97 | 92 | 87 |
| Aperture size (mm) | 0.425 | 0.25 | 0.15 | 0.075 |
| Percentage passing | 53 | 42 | 15 | 8 |

Find the uniformity and curvature coefficients.

2.13 In a hydrometer analysis of a fine-grained soil, the initial reading was found to be 1.05. The corrected hydrometer reading after 70 minutes was 1.03, which corresponds to an effective depth of 11.5 cm. The suspension volume was 1,000 $cm^3$. Calculate the initial weight of the soil, the particle size corresponding to the 12 minute reading, and the percentage of particles finer than this size. Take $G = 2.68$ and $\eta = 0.1$ poise.

2.14 For a particle of diameter 0.005 mm, how many hours are required to settle to a depth of 3 m from the surface in a tank? The specific gravity of the particle is 2.70 and the coefficient of viscosity is 0.001 M-s/$m^2$.

2.15 In a liquid limit test, a soil sample showed water contents of 48%, 40%, 38.8%, and 37.1% against 12, 26, 28, and 31 blows, respectively. The plastic limit of the clay is 18.2%, and the natural water content is 34.5%. Find the liquid limit, plasticity index, liquidity index, relative consistency, flow index, and toughness index of the soil.

2.16 The liquid limit of a soil is 86%, and its plastic limit is 34%. If the natural water content is 48%, what is the state of consistency of the soil? What is the shrinkage limit of the soil if the void ratio at the shrinkage limit state is 0.89? Take G=2.68.

2.17 A saturated specimen was immersed in mercury, and its displaced volume was 20.8 $cm^3$. The weight of the sample was 0.312 N. After oven drying for 48 hours, the weight reduced to 0.196 N, while the volume came down to 10.2 $cm^3$. Find the shrinkage limit, void ratio, specific gravity, and shrinkage ratio of the soil.

2.18 In a big project, the Atterberg limits and natural water contents of three soils are determined as given below:

| Soil | $w_L$ (%) | $w_p$ (%) | $w_n$ (%) |
|---|---|---|---|
| 1 | 126 | 42 | 165 |
| 2 | 63 | 32 | 42 |
| 3 | 86 | 36 | 78 |

Determine the consistency of the natural soil and the liquidity indices.

2.19 An undisturbed saturated specimen of clay has a volume of 18.9 $cm^3$ and a mass of 30.2 g. In oven drying, the mass reduces to 18.0 g. Assuming the volume of dry specimen to be 9.9 $cm^3$, determine the shrinkage limit, shrinkage ratio, and volumetric shrinkage.

2.20 The Atterberg limits for a clay soil used for an earth dam are liquid limit 60%, plastic limit 40%, and shrinkage limit 25%. If a specimen of the soil of volume 10 $cm^3$ at the liquid limit has a volume of 6.5 $cm^3$ when dried, what would be the specific gravity of the soil particles?

# 3

# Identification and Classification of Soils

**CHAPTER HIGHLIGHTS**

Field identification of soils – Engineering classification of soils: Purpose of classification, Unified soil classification system, Indian soil classification system, AASHTO soil classification system, Textural soil classification system

## 3.1 INTRODUCTION

It is necessary to have a standard language for a careful description and classification of a soil. In principle, soil description is different from soil classification.

A soil description should include the material characteristics (viz., primary characteristics: particle size distribution and plasticity; secondary characteristics: colour, shape, texture, and composition) and the *in situ* soil mass (viz., firmness or strength, bedding planes, discontinuities, weathering, etc.). On the other hand, soil classification is the arrangement of soils into various groups or sub-groups so as to express briefly the primary material characteristics (viz., particle size distribution and plasticity) without detailed descriptions.

Generally, soils have various constituents in different proportions. The soil is denoted by the major constituent, and the minor constituents are indicated by adjectives. Further, the colour and density or stiffness and moisture conditions are added to fully describe the field condition of a soil, e.g., brownish red loose to medium dense silty sand.

In this chapter, field identification tests for soils and different engineering classification of soils are presented.

## 3.2 FIELD IDENTIFICATION OF SOILS

Soils can be broadly grouped as coarse-grained or non-cohesive and fine-grained or cohesive soils.

### 3.2.1 Coarse-grained Soils

Coarse-grained soils are mineral fragments which are easily identified by the naked eye on the basis of grain size. The major coarse-grained materials are gravel and sand. Based on experience, one can identify the sand or gravel as fine, medium, or coarse. Further, the grains may be rounded, sub-rounded, angular, or sub-angular. Sometimes other mineral grains such as mica or shale may also be present, which can be identified with the help of a magnifying glass. In addition to the major classification as sand or gravel, the lesser significant percentage of material should be identified. For example, a gravel with a significant percentage of sand has to be categorized as sandy gravel.

### 3.2.2 Fine-grained Soils

Fine-grained soils are silts and clays. Depending on the significant absence or presence of organic material, they are categorized as inorganic soils or organic soils, respectively.

**Inorganic Soils**. Field identification of these soils can be made by conducting the following tests:

1. Dry strength test
2. Dilatancy test
3. Plasticity test
4. Dispersion test

*Dry Strength Test.* The strength of a soil in a dry state is an indication of the presence of cohesion. A pat of soil about 6 mm thick is dried under natural conditions or in an oven. The dry strength can be estimated by breaking and crushing between the fingers. Dry inorganic clay shows high strength and can be broken only with a great effort. On the other hand, inorganic silts have little or no dry strength and crumble easily between the fingers.

*Dilatancy Test.* A pat of soil is made with water so that it is soft and not sticky. The pat is placed in the open palm in a horizontal position. Several times, the side of the hand is struck against the other hand. The appearance of a shiny film of water on the surface of the pat signifies silt. As clay is less permeable, no significant change on the surface of the pat can be seen after shaking.

*Plasticity Test.* A small quantity of soil is rolled into a thread form on a flat surface or on the palm. If the soil can be rolled into a long thread of about 3 mm diameter, it signifies that it contains a large quantity of clay, but silts cannot be rolled into a long thread of 3 mm diameter without severe cracking.

*Dispersion Test.* A small quantity of soil is put into a jar of water, allowing the particles to settle. Coarse-grained particles settle initially, followed by fine-grained particles. In a 10 cm depth of water, sand particles settle within 30 seconds, whereas silt particles may take 15 to 20 minutes, but clay particles remain in suspension for several hours or even days provided flocculation does not take place.

**Organic Soils**. Organic soils contain a significant proportion of dispersed vegetable matter. The organic matter in soil is due to disintegrated plant roots and other vegetable matter, such as muck or more fibrous materials. Organic soils have a distinctive odour and often are dark brown, dark grey, or bluish grey in colour.

Organic silts are less plastic, containing silt-size particles and finer particles of organic material and shell fragments.

Organic clay contains clay-size particles and finely divided organic material. Highly organic soils such as peat consists predominantly of plant remains, usually dark brown or black in colour and with a distinctive odour.

## 3.3 ENGINEERING CLASSIFICATION OF SOILS

### 3.3.1 Purpose of Classification Systems

The simple way of classifying a soil as non-cohesive or cohesive is inadequate as it does not specify other properties, such as gradation, grain sizes involved, plasticity, activity, and other relevant properties which help to identify the soil for a specific construction purpose. Thus, the aim of a classification system is to establish a set of conditions which will allow useful comparisons to be made between different soils. Soils classified to have a preference for one set of conditions may not be preferred for another set of conditions. Thus, a number of classification systems are available to cater to a particular purpose. Hence, to be of general use, a system must be simple, lucid, and directly involved with the engineering properties of the soil.

### 3.3.2 Unified Soil Classification System

Unified soil classification system is the most popular soil classification system among civil engineers. As in many of the systems, the grain-size characteristic has been used as the basis for grouping the soil particles into gravel, silt, or clay, i.e.,

Gravel > 4.75 to 80 mm
Sand > 0.075 to 4.75 mm
Silt > 0.002 to 0.075 mm
Clay < 0.002 mm

Further, Atterberg limits are used as an additional criterion for identifying the compressibility or plasticity of fine-grained soils. This system was first developed by A. Casagrande (1948) as the Airfield Classification system. After minor modifications, it was adopted by the US Bureau of Reclamation and US Corps of Engineers and later (1969) accepted as a standard classification system by the American Society for Testing Materials (Table 3.1).

Unified soil classification system divides soils into two major groups, viz., coarse-grained soils and fine grained soils, and is defined by a set of two letters, a prefix and a suffix. Coarse-grained soils are those for which more than 50% of the material has particle sizes greater than 0.075 mm. They are basically divided into gravels (G) and sands (S) and are further grouped according to gradation and the presence of silt and clay-size fraction.

They are (i) well graded (W), (ii) poorly graded (P), (iii) containing silt fines (M), and (iv) containing clay fines (C). For example, the symbol SP refers to poorly graded sand with no fines.

Fine-grained soils are those for which more than 50% of material has particle sizes < 0.0075 mm. They are divided into inorganic silts and very fine sand (M), inorganic clays (C), and organic silts and clays (O). They are further classified based on the liquid limit of the soil as low plasticity, L ($w_L < 50\%$) and high plasticity, H ($w_L > 50\%$). Highly organic soils (peat) are visually identified. The fine-grained soils are presented in a chart form called the plasticity chart (Fig. 3.1) based on their liquid limit and plasticity index. The "A" line (after A. Casagrande) separates the inorganic clays from the silts and organic soils.

**Table 3.1** Unified soil classification including identification and description

| Major divisions | | | Field identification procedures (excluding particles larger than 3 sq. inches and basing fractions on estimated weights) | | | Group symbols[a] | Typical names |
|---|---|---|---|---|---|---|---|
| (The 0.075 mm size sieve is about the smallest particle visible to the naked eye) | Gravels (more than half of coarse fraction is larger than 4.75 mm) (For visual classifications, the 1/4" size may be used as equivalent to the No. 4 sieve size) | Clean gravels (little or no fines) | Wide range in grain size and substantial amounts of all intermediate particle sizes | | | GW | Well-graded gravels, gravel-sand mixtures, little or no fines |
| | | | Predominantly one size or an range of sizes with some intermediate sizes missing | | | GP | Poorly graded gravels, gravel-sand mixtures, little or no fines |
| | | Gravels with fines (appreciaable amount of fines) | Non-plastic fines (for indentification procedures, see ML below) | | | GM | Silty gravels, poorly graded gravel-sand-silt mixtures |
| | | | Plastic fines (for indentification procedures, see CL below) | | | GC | Clayey gravels, poorly graded gravel-sand-clay mixtures |
| | Sands (more than half of coarse fraction is smaller than 4.75 mm) | Clean sands (little or no fines) | Wide range in grain sizes and substantial amounts of all intermediate particle sizes | | | SW | Well-graded sands, gravelly, little or no fines |
| | | | Predominantly one size or a range of sizes with some intermediate sizes missing | | | SP | Poorly graded sands, gravelly sands, little or no fines |
| | | Sands with fines (appreciable amount of fines) | Non-plastic fines (for indentification procedures, see ML below) | | | SM | Silty sands, poorly graded sand-silt mixtures |
| | | | Plastic fines (for indentification procedures, see CL below) | | | SC | Clayey sands, poorly graded sand-clay mixtures |
| | Identification procedures of fraction smaller than 0.425 mm | | | | | | |
| | | | Dry strength (crushing characteristics) | Dilatancy (reaction to shaking) | Toughness (consistency near plastic limit) | | |
| | Silts and clays (liquid limit less than 50) | | None to slight | Quick to slow | None | ML | Inorganic silts and very fine sands, rock flour, silty or clayey fine sands with slight plasticity |
| | | | Medium to high | None to very slow | Medium | CL | Inorganic clays of low to medium plasticity, clays, sandy clays, siltyclays, lean clays |
| | | | Slight to medium | Slow | Slight | OL | Organic silt sand, organic silt clays of low plasticity |
| | Silts and clays (liquid limit greater than 50) | | Slight to medium | Slow to none | Slight to medium | MH | Inorganic silts, micaceous or diatomaceous fine sandy or silty soils, elastic silts |
| | | | High to very high | None | High | CH | Inorganic clay of plasticity, fat clays |
| | | | Medium to high | None to very slow | Slight to medium | OH | Organic clays of medium to high plasticity |
| High organic soils | | | Readily identified by colour, odour, spongy feel, and frequently by fibrous texture | | | Pt | Peat and other highly organic soils |

[a]Boundary classifications: Soils possessing characteristics of two groups are designated by combinations of group symbols,

| Information required for describing soils | | Laboratory classification criteria | | |
|---|---|---|---|---|
| Give typical name; indicate approximate percentages of sand and gravel, maximum size, angularity, surface condition, and hardness of the coarse grains; local or geological name and other pertinent descriptive information; and symbol in parentheses | Use grain-size curve in identifying the fractions as given under field indentification | Determine percentages of gravel and sand from grain-size curve depending on percentage of fines (fraction smaller than No. 200 sieve size). Coarse-grained soils are classified as follows: Less than 5% – GW, GP, SW, SP; More than 12% – GM, GC, SM, SC; 5% to 12% – Borderline cases requiring use of dual symbols | $C_u = \frac{D_{60}}{D_{10}}$ greater than 4<br>$C_c = \frac{(D_{30})^2}{D_{10} - D_{60}}$ between 1 and 3 | |
| | | | Not meeting all gradation requirements for GW | |
| | | | Atterberg limits below "A" line or $w_p$ greater than 4 | Above "A" line with $w_p$ between 4 and 7 are borderline cases requiring use of dual symbols |
| For undisturbed soils, add information on stratification degree of compactness, cementation, moisture conditions, and drainage characteristics | | | Atterberg limits above "A" line *with* $w_p$ greater than 7 | |
| | | | $C_u = \frac{D_{60}}{D_{10}}$ greater than 6<br>$C_c = \frac{(D_{30})^2}{D_{10} - D_{60}}$ between 1 and 3 | |
| *Example:*<br>Silty sand, gravelly, about 20% hard angular gravel particles 1/2-inch maximum size, rounded and sub-angular sand grains, coarse to fine, about 15% non-plastic fines with low dry strength, well compacted and moist in places, alluvial sands (SM) | | | Not meeting all gradation requirements for SW. | |
| | | | Atterberg limits below "A" line or $w_p$ less than 4 | Above "A" line with $w_p$ between 4 and 7 are borderline cases requiring use of dual symbols |
| | | | Atterberg limits above "A" line or $w_p$ greater than 7 | |
| Give typical name; indicate degree and character of plasticity, amount and maximum size of coarse grains, colour in wet condition, odour if any; local or geological name and other pertinent descriptive information, and symbol in parentheses<br><br>For undisturbed soils, add information on structure, stratification, consistency in undistured and remoulded states, moisture and drainage conditions<br><br>*Example.*<br>Clayey silt, brown, slightly plastic, small percentage of fine sand; numerous vertical root holes; firm and dry in places; loess (ML) | | Plasticity chart (Plasticity index 0–60 vs. Liquid limit 0–100): Comparing soils at equal liquid. Toughness and dry strength increase with increasing plasticity index; "A" line; CH; OH; CL; CL; OL; CL-ML; NL | | |

Plasticity chart for laboratory classification of fine soils

for example, GW-GC, well-graded gravel–sand mixture with clay binder.

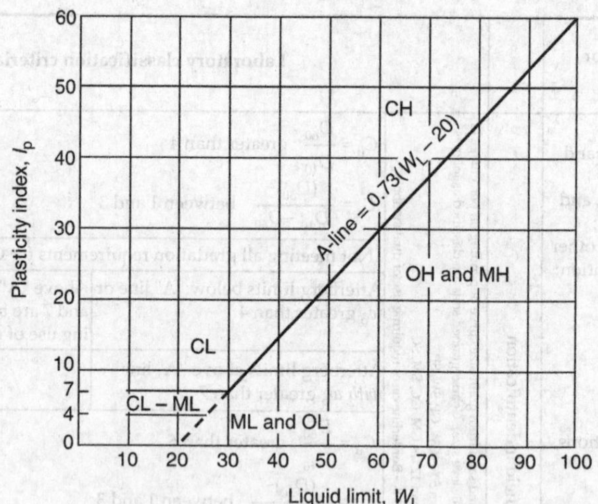


Fig. 3.1 Plasticity chart (Unified soil classification) (*Source*: A. Casagrande, 1948)

### 3.3.3 Indian Soil Classification System

The Indian soil classification (IS: 1498, 1970) is basically the same as that of the Unified soil classification system but for a slight modification in the plasticity chart. In this system, the fine-grained soils, viz., inorganic silts, inorganic clays, and organic silts and clays, are further divided into three groups based on the liquid limit of the soil as low compressibility, L ($w_L <$ 35%), medium compressibility, I (35% $< w_L <$ 50%), and high compressibility, H ($w_L >$ 50%). Figure 3.2 represents the plasticity chart as adopted by the Indian soil classification system. Highly organic soils (e.g., peat) are classified as Pt. Table 3.2 gives the details of the Indian soil classification system.

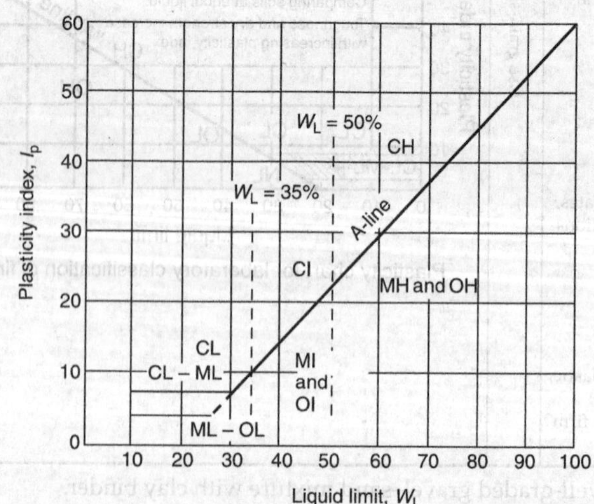


Fig. 3.2 Plasticity chart (Indian soil classification) (*Source*: IS: 1498, 1970)

**Table 3.2** Indian soil classification system (including field identification and description) (IS: 1498, 1970)

| Division | Sub-division | | Group letter symbol | Hatching | Mapping colour | Typical names | Field identification procedures (excluding particles larger than 80 mm and using fractions of estimated weights) | Information required for describing soils |
|---|---|---|---|---|---|---|---|---|
| 1 | 2 | | 3 | 4 | 5 | 6 | 7 | 8 |
| Coarse-grained soil More than half of material is larger than 75 μm IS sieve size; the smallest particle is visible to the naked eye — Gravels | More than half of coarse fraction is larger than 4.75 mm IS sieve size (For visual classification the 5 mm size may be used as equivalent to 4.75 mm IS sieve size) | Clean gravels (little or no fines) | GW | | Red | Well-graded gravels, gravel–sand mixtures, little or no fines | Wide range in grain sizes and substantial amounts of all intermediate particle sizes | For undistributed soils, add information on stratification, degree of compactness, cementation, moisture conditions, and drainage characteristics |
| | | | GP | | Red | Poorly graded gravels or gravel–sand mixtures, little or no fines | Predominantly one size or a range of sizes with some intermediate sizes missing | |
| | | Gravels with fines (appreciable amount of fines) | GM | | Yellow | Silty gravels, poorly graded gravel, sand–silt mixtures. | Non-plastic fines or fines with low plasticity (for identification procedures, see ML, and MI below) | Give typical name; indicate approximate percentage of sand and gravel, maximum size, angularity, surface condition, and hardness of the coarse grains, local or geological name, and other pertinent descriptive information; and symbol in parentheses |
| | | | GC | | Yellow | Clayey gravels, poorly graded gravel, sand–clay mixtures | Plastic fines (for identification procedures, see CL and CI below) | |
| Sands | More than half of coarse fraction is smaller than 4.75 mm IS sieve size | Clean sands (little or no fines) | SW | | Red | Well-graded sands gravelly sands, little or no fines | Wide range in grain size and substantial amounts of all intermediate particle sizes | |
| | | | SP | | Red | Poorly graded sands or gravelly sands, little or no fines | Predominantly one size or a range of sizes with some intermediate sizes missing | *Example:* |
| | | Sands with fines (appreciable amount of fines) | SM | | Yellow | Silty sands, poorly graded sand–silt mixures | Non-plastic fines or fines with low plasticity (for intermediate procedures, see ML and MI below) | Silty sand, gravelly, about 20% hard angular gravel particles 10-mm maximum size; rounded and sub-angular sand grains; about 15% non-plastic fines with low dry strength; well compacted and moist; compact alluvial sand (SM) |
| | | | SC | | Yellow | Clayey sands, poorly graded sand–clay mixtures | Plastic fines (for identification procedures, see CL and CI below) | |
| | | | | | | | Identification procedures (on fraction smaller than 425 μm IS sieve size) | |

*Table 3.2 Contd.*

**Table 3.2** Contd.

| Division | Sub-division | Group letter symbol | Hatching | Mapping colour | Typical names | Field identification procedures (excluding particles larger than 80 mm and using fractions of estimated weights) | | | Information required for describing soils |
|---|---|---|---|---|---|---|---|---|---|
| 1 | 2 | 3 | 4 | 5 | 6 | Dry strength | Dilatancy | Tough-ness | 8 |
| Fine-grained soil More than half the material is larger than 75 μm IS sieve size; the smallest particle visible to the naked eye | Silts and clays with low compressibility and liquid limit less than 35 | ML | | Blue | Inorganic silts and very fine sands, rock flour, silty or clayey fine sands, or clayey silts with none to low plasticity | None to low | Quick | None | For undistributed soils, add information on structure, stratification, consistency in undisturbed and remoulded states, moisture and draingae conditions<br>Give typical name; indicate degree and character of plasticity, amount and maximum size of coarse grains, colour in wet condition, odour, if any, local or geological name, and other pertinent descriptive information and symbol in parentheses<br>*Example:*<br>Clayey silt, brown; slightly plastic; small percentage of fine sand; numerous vertical root holes; firm and dry in places; loess (ML) |
| | | CL | | Green | Inorganic clays, gravelly clays, sandy clays, silty clays, lean clays of low plasticity | Medium | None to very slow | Medium | |
| | | OL | | Brown | Organic silts and organic silty clays of low plasticity | Low | Slow | Low | |
| | Silts and clays with medium compressibility and liquid limit greater than 35 and less than 50 | MI | | Blue | Inorganic silts, silty or clayey fine sands, or clayey silts of medium plasticity | Low | Quick to slow | None | |
| | | CI | | Green | Inorganic clays, gravelly clays, sandy clays, silty clays, lean clays of medium plasticity | Medium to high | None | Medium | |
| | | OI | | Brown | Organic silts and organic silty clays of medium plasticity | Low to medium | Slow | Low | |
| | Silts and clays with high compressibility and liquid limit greater than 50 | MH | | Blue | Inorganic silts of high compressibility, micaceous or diatomaceous fine sandy or silty soils, elastic silts | Low to medium | Slow to none | Low to medium | |
| | | CH | | Green | Inorganic clays of high plasticity, fat clays | High to very high | None | High | |
| | | OH | | Brown | Organic clays of medium to high plasticity | Medium to high | None to very slow | Low to medium | |
| Highly organic Soils | | Pt | | Orange | Peat and other highly organic soils with very high compressibility | Readily identified by colour, odour, spongy feel and frequently by fibrous texture | | | |

### 3.3.4 AASHTO Soil Classification System

The American Association of State Highway and Transport Officials (AASHTO) system was developed by the US Bureau of Public Roads (now referred to as the Federal Highway Administration) primarily based on the Public Road Administration classification system, 1978. The system is based on both the particle size and the plasticity characteristics.

According to the system, the soils are classified into eight groups, viz., A-1 to A-7 with an additional group A-8 for peat or muck. Several sub-groups are included in the system. The details of the sub-groups are presented in Table 3.3. Soils within each group are evaluated according to the *group index* (GI) obtained from the empirical formula (Eq. 3.1)

$$GI = 0.2a + 0.005ac + 0.01bd \tag{3.1}$$

where $a$ is that part of the percentage of soil particles passing the 75 μm sieve greater than 35 and not exceeding 75, expressed as a positive whole number (1 to 40); $b$ is that part of the percentage of soil particles passing the 75 μm sieve greater than 15 and not exceeding 55, expressed as a positive whole number (1 to 40); $c$ is that part of the liquid limit of the soil greater than 40 and not greater than 60, expressed as a positive whole number (1 to 20); and $d$ is that part of the plasticity index greater than 10 and not exceeding 30, expressed as a positive whole number (1 to 20).

For using the above equation and Table 3.3, the grain-size distribution, liquid limit, and plasiticity index values of the soil are to be determined.

If the specific index value for a soil falls below the minimum limit corresponding to $a$, $b$, $c$, or $d$, the value of the respective term is taken as 0 and the term is dropped out while calculating the GI. Similarly, when the value of $a$, $b$, $c$, or $d$ is more than the prescribed maximum value, then the respective value of 20 or 40 is assigned.

The classification is carried out by proceeding from left to right in the Table 3.3, and the first group which fits the test data is selected. The GI value shows if a soil is fit as a sub-grade material or not.

A group index of 0 indicates a good sub-grade material, while a group index of 20 corresponds to a very poor sub-grade material.

### 3.3.5 Textural Soil Classification System

In this system, soil fractions as per the US Bureau of Soils and Chemistry System are used. Accordingly, the following is the grain-size classification:

Gravel > 1.00 mm
Sand 1.00 to 0.05 mm
Silt 0.05 to 0.005 mm
Clay < 0.005 mm

A triangular chart has been developed by the Bureau using grain-size limits. In addition to gravel, sand, silt, and clay, the system uses another term, *loam*. A loam is a mixture of sand, silt, and clay particles in varying proportions. The term *loam* has originated from agriculturists and is also adopted by highway engineers as they too deal with surface layers.

As a first step, the grain-size distribution of the soil is found and the percent soil fractions are determined. With the known percentages of sand, silt, and clay, a point is located in the triangular chart, as shown in Fig. 3.3. The specified term designated in the chart for the area where the point falls is taken as the classification of the soil.

**Table 3.3** AASHTO soil classification system

| General classification | Granular materials (35% or less passing 0.075 mm) | | | | | | | Silt–clay materials (>35% passing 0.075 mm) | | | |
|---|---|---|---|---|---|---|---|---|---|---|---|
| Group classification | A-1 | | | A-2 | | | | | | | A-7 |
| | A-1a | A-1b | A-3 | A-2-4 | A-2-5 | A-2-6 | A-2-7 | A-4 | A-5 | A-6 | A-7-5, A-7-6 |
| Sieve analysis | | | | | | | | | | | |
| % Passing | 50 max | | | | | | | | | | |
| No. 10 (2 mm) | 30 max | 50 max | 51 max | | | | | | | | |
| No. 40 (0.425 mm) | 15 max | 25 max | 10 max | 35 max | 35 max | 35 min | 35 min | 36 min | 36 min | 36 min | 36 min |
| No. 200 (0.075 mm) | | | | | | | | | | | |
| Characteristics of fraction passing No. 40 (0.425 mm) | | | | | | | | | | | |
| Liquid limit | | | | 40 max | 41 max | 40 max | 41 max | 40 max | 41 min | 40 max | 41 min |
| Plasticity index | 6 max | | NP | 10 max | 10 max | 11 max | 11 max | 10 max | 10 max | 11 max | 11 min |
| Group index | 0 | | 0 | 0 | 0 | 4 max | 4 max | 8 max | 12 max | 16 max | 20 max |
| Usual types of significant constituent materials | Stone fragments, gravel, and sand | | Fine sand | Silty or clayey gravel and sand | | | | Silty soils | | Clayey soils | |
| General rating as sub-grade | Excellent to good | | Fair to poor | | | | | | | | |

*Note*: A-8, peat or muck is by visual classification and is not shown in the table. NP, non-plastic.

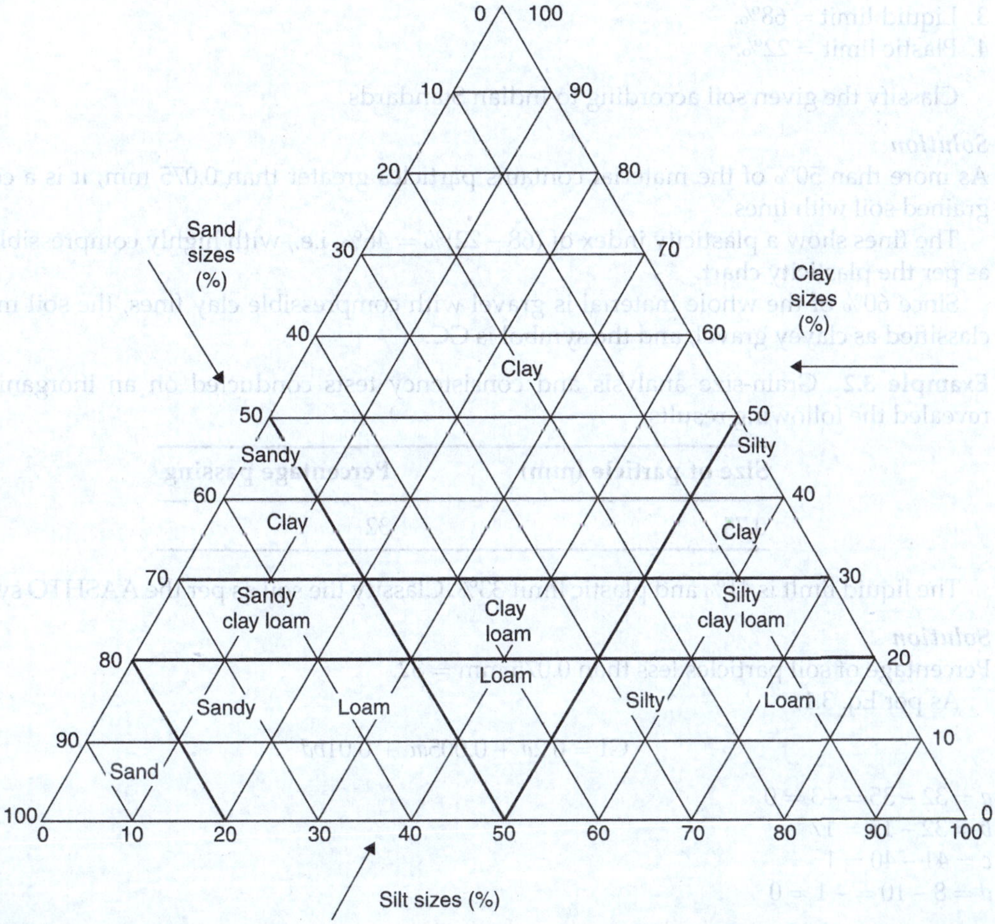

**Fig. 3.3** Textural soil classification system

This is a simple classification system widely used in the fields of agriculture and highway engineering. This classification depends on the grain-size distribution and does not reveal any other property of the soil.

## WORKED EXAMPLES

**Example 3.1** A soil has the following characteristics:

1. Percentage of soil passing 75 μm sieve = 55.
2. Percentage of coarse fraction passing 4.75 mm sieve = 60.

3. Liquid limit = 68%.
4. Plastic limit = 22%.

Classify the given soil according to Indian Standards.

***Solution***

As more than 50% of the material contains particles greater than 0.075 mm, it is a coarse-grained soil with fines.

The fines show a plasticity index of (68 – 22)% = 46%, i.e., with highly compressible clay as per the plasticity chart.

Since 60% of the whole material is gravel with compressible clay fines, the soil may be classified as clayey gravel, and the symbol is GC.

**Example 3.2** Grain-size analysis and consistency tests conducted on an inorganic soil revealed the following results:

| Size of particle (mm) | Percentage passing |
|---|---|
| 0.75 | 32 |

The liquid limit is 41% and plastic limit 33%. Classify the soil as per the AASHTO system.

***Solution***

Percentage of soil particles less than 0.075 mm = 32.

As per Eq. 3.1,

$$GI = 0.2a + 0.005ac + 0.01bd$$

$a = 32 - 35 = -3 = 0$
$b = 32 - 15 = 17$
$c = 41 - 40 = 1$
$d = 8 - 10 = -1 = 0$

$$\begin{aligned} GI &= 0.02 \times 0 + 0.005 \times 0 \times 1 + 0.01 \times 17 \times 0 \\ &= 0 \end{aligned}$$

From Table 3.3, on the basis of percentage of fine-grained soil, liquid limit, and plasticity index values, the soil is classified as A–2–5 (0).

**Example 3.3** The sieve analysis of a sample of a soil gave the following details. Classify the soil as per the Textural soil classification system.

Sand = 36%
Silt = 42%
Clay = 22%

***Solution***

Using the above values, the triangular chart is entered, and the soil is fit to be classified as clay loam.

## POINTS TO REMEMBER

3.1 Coarse-grained soils are mineral fragments which are easily identified in the field by the naked eye on the basis of grain size. The major coarse-grained materials are gravel and sand.

3.2 Fine-grained soils are silts and clays which are classified as inorganic or organic soils depending on the amount of organic material present.

3.3 Field identification tests for fine-grained inorganic soils are dry strength test, dilatancy test, plasticity test, and dispersion test.

3.4 Field identification of fine-grained organic soils can be made by the presence of disintegrated plant roots and other vegetable matter, a distinctive odour, and often dark brown, dark grey, or bluish grey.

3.5 The aim of a classification system is to establish a set of conditions which will allow useful comparisons to be made between soils.

3.6 In the Unified soil classification system, the grain-size characteristics have been used as the basis for grouping the soil particles into gravel, silt, or clay. Further, Atterberg limits are used as an additional criterion for identifying the compressibility or plasticity of fine-grained soils.

3.7 The Indian soil classification system is basically the same as the Unified soil classification system but for a slight modification in the plasticity chart.

3.8 The AASHTO soil classification system was developed by the US Bureau of Public Roads. The system is based on both the particle size and the plasticity characteristics. According to the system, the soils are classified into eight groups, viz., A-1 to A-7, with an additional group A-8, for peat or muck.

3.9 Group index (GI) is obtained from the empirical formula $GI = 0.2a + 0.005ac + 0.01bd$, where $a$ and $b$ depend on a certain percentage of particles and $c$ and $d$ depend on the liquid limit and plasticity index. GI is adopted in the AASHTO soil classification system.

3.10 In the Textural soil classification system, soil fractions as per the US Bureau of Soils and Chemistry are adopted. This system provides a triangular chart to classify the soil as sand, silt, clay, or loam. This system is widely followed by agriculturists and highway engineers.

## QUESTIONS

### *Objective Questions*

3.1 Choose the correct statements

1. A gravel with a significant percentage of clay has to be categorized as gravelly clay.
2. In a dilatancy test, the appearance of a shiny film of water on the surface of the soil part signifies silt.

3. Plasticity chart used in the Unified soil classification system is the same as in the Indian soil classification system.
4. The AASHTO soil classification system is based on particle size, shape, and roughness only.

3.2 The strength of a soil in the dry state is an indication of a high amount of ____:
(a) Sand (b) Silt (c) Clay (d) Gravel

3.3 Silt particle size as per Unified soil classification system is
(a) 0.075 to 4.75 mm (b) 0.002 to 0.075 mm
(c) >4.75 mm (d) <0.002 mm

3.4 As per the plasticity chart of Indian soil classification system, the soil is of medium compressibility when the liquid limit is
(a) < 35% (b) 35% to 50 %
(c) >50% (d) 25% to 34 %

3.5 A loam is a mixture of
(a) Gravel and sand (b) Sand and silt
(c) Sand, silt, and clay (d) Sand and clay

3.6 In the AASHTO soil classification system, the group classification A-4 to A-7 signifies the sub-grade as
(a) Excellent to good (b) Good to fair
(c) Fair to poor (d) Poor to very poor

### *Descriptive Questions*

3.7 Explain briefly the object of classifying soils for engineering purposes.
3.8 Discuss the physical properties and factors which are considered in any particular system of soil classification.
3.9 Explain the tests to be conducted to identify the soils in the field.
3.10 List different systems of engineering classification of soils. Discuss the merits and demerits of each system.
3.11 How is the plasticity chart useful for classifying fine-grained soils?
3.12 Explain the Indian soil classification system.
3.13 What are the advantages in using a triangular chart?
3.14 How is suitability of sub-grade soils assessed by the AASHTO soil classification system?

## EXERCISE PROBLEMS

3.1 The following data were obtained from a laboratory test on a soil:
Percentage of particles finer than 4.75 mm = 100
Percentage of particles finer than 75 μm = 96.9
Coefficient of uniformity = 1.40

Coefficient of curvature = 1.03
Plasticity index = 6.23

Classify the soil as per Unified soil classification system.

3.2 Liquid and plastic limits determination on a soil gave the following results: (Casagrande's apparatus was used for the liquid limit test).

| **Test no.** | **Mass of cup (g)** | **Mass of cup + wet soil (g)** | **Mass of cup + dry soil (g)** | **Number of blows** |
|---|---|---|---|---|
| *Liquid limit test* | | | | |
| 1 | 23.68 | 40.86 | 34.68 | 13 |
| 2 | 22.93 | 42.82 | 35.78 | 20 |
| 3 | 26.27 | 38.02 | 34.27 | 47 |
| *Plastic limit test* | | | | |
| 1 | 25.34 | 32.17 | 31.01 | — |
| 2 | 24.83 | 30.48 | 29.51 | — |

Determine the plasticity index and classify the soil as per BIS plasticity chart.

3.3 The sieve analysis of a soil revealed that 58% of the particles are finer than 75 μm. The liquid limit and plastic limit of the soil were 61% and 27%, respectively. Classify the soil as per the AASHTO system.

3.4 The following results were obtained from a laboratory test on three soil samples. Classify the soil as per Indian soil classification system.

| **Sieve size** | **Percentage passing** | | |
|---|---|---|---|
| | **Soil A** | **Soil B** | **Soil C** |
| 4.75 mm | 68 | 98 | 100 |
| 2.00 mm | 55 | 96 | 100 |
| 1.00 mm | 43 | 90 | 95 |
| 450 μm | 30 | 90 | 84 |
| 200 μm | 25 | 88 | 78 |
| 150 μm | 16 | 86 | 75 |
| 75 μm | 10 | 84 | 70 |
| Liquid limit | NP | 22.4% | 32.8 |
| Plastic limit | NP | 15.2% | 24.3 |

3.5 The following test results were obtained on a soil sample:

Percentage passing 4.75 mm IS sieve = 98.5%
Percentage passing 75 μm IS sieve = 42.0%
Coefficient of uniformity = 6.7
Coefficient of curvature = 1.2
Plasticity index = 2.2

Classify the soil by a suitable classification system.

# 4 Compaction of Soils

**CHAPTER HIGHLIGHTS**

Principles of compaction – Compactive effort – Laboratory compaction: Standard Proctor test, Modified Proctor test, BIS Light Compaction Test, BIS Heavy Compaction Test – Field compaction and equipment – Compaction specification and control – Factors affecting compaction – Compaction of sand

## 4.1 INTRODUCTION

Compaction may be defined as the process of increasing the density of a soil using force to pack the particles closer together, with a reduction in air voids without any significant change in the volume of water in the soil.

The reduction in air voids is deliberately brought about by some mechanical means during a construction process in the field or in the preparation of a sample in the laboratory. The higher the compaction, the lower will be the compressibility of the soil and higher the shear strength.

Typical examples are the construction of fills, embankments, and earth dams and strengthening of sub-grades of highways and runways.

## 4.2 PRINCIPLES OF COMPACTION

Soil compaction is the process whereby soil particles are forced to pack more closely together by reducing air voids. This is attained by applying some mechanical force (static or dynamic loads) on the soil. The purpose of compaction is to produce a soil having certain physical properties suitable for a given project. The state of compaction of a soil is measured by the dry density and the associated moulding water content.

The increase in the dry density of the soil produced by compaction basically depends on the water content of the soil and the applied energy. For each soil, there exists a moisture content termed the *optimum moisture content* at which a *maximum dry density* is attained for a given amount of compaction. Addition of water to dry soil results in adsorbed water around particles. When the water content is low, the soil is stiff and difficult to compress, and this results in a low dry density. As the adsorbed water films increase in thickness and act as a lubricant and bring the particles more closely together, they increase the dry density by reducing the air content. After a certain point, the effect of lubrication stops and the adsorbed water pushes the particles away, and any further increase in moisture decreases the dry density. Thus the maximum dry density occurs at an optimum moisture content.

## 4.3 COMPACTIVE EFFORT

The concept of compactive effort is used in both field and laboratory compactions. For laboratory conditions, compactive effort is defined as the application of a given amount of energy per unit volume of compacted soil. For field conditions, compactive effort is defined as the compaction obtained by allowing a piece of equipment to pass a given number of times on a given thickness of lift. Compactive effort can be varied in the laboratory tests by changing the weight of the compacting hammer, height of fall, number of blows per layer, and number of layers. If a vibratory method is used, the compactive effort can be changed by changing the frequency, amplitude, and time of vibration. In the field, the compactive effort can be increased by increasing the number of passes of a roller. For all soils, both in the laboratory and in field compaction, an increase in compactive effort results in an increase in the dry density and decrease in the optimum moisture content.

## 4.4 LABORATORY COMPACTION

Laboratory compaction tests are designed to estimate the dry density of soils. Two such tests were developed by Proctor (1933): (i) the Standard Proctor test (IS: 2720 – Part 7, 1974), which causes adequate compaction for most applications, such as backfills, highway fills, and earth dams, and (ii) the Modified Proctor test (IS: 2720 – Part 8, 1983), which is used for heavier-load applications, such as airport and highway base courses. Procedures for conducting laboratory tests are explained in Chapter 10.

The tests are performed by compacting a wet soil sample in a mould in a specified number of layers. Each layer is compacted at a stipulated compactive effort. The compactive effort is measured in terms of the energy per unit volume of compacted soil. The required compactive effort is attained by controlling the weight of the hammer, height of drop, number of layers, and number of blows for each layer. After the final layer is compacted, the bulk density of the soil and its moisture content are determined. Tests are repeated on fresh samples with increasing moisture content. From the values of the bulk density and the moisture content obtained, the dry density is calculated. Thus,

$$\rho_d = \rho/(1 + w)$$

A graph of dry density versus moisture content is plotted (Fig. 4.1) and the maximum dry density and optimum moisture content are read from the graph.

It is not feasible to expel air completely by compaction and obtain the maximum dry density. The maximum theoretical value of the dry density is referred to as the zero air void–dry density or the saturation dry density and can be computed from

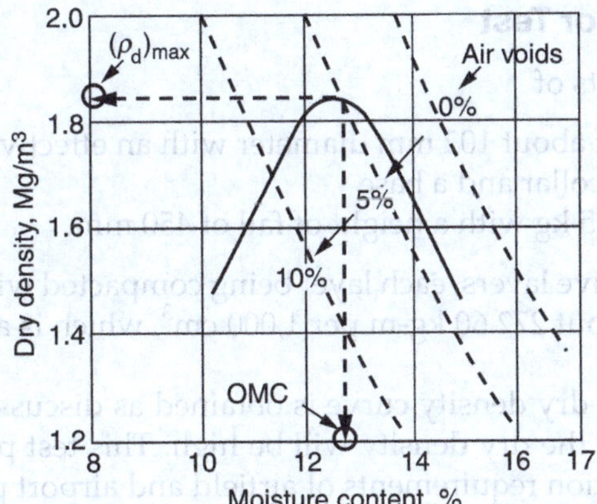


**Fig. 4.1** Dry density–moisture content relationship

$$\rho_d = \frac{G\rho_w}{1+wG}$$

and the dry density for any particular degree of saturation can be computed from

$$\rho_d = \frac{G(1-A_v)\rho_w}{1+wG}$$

Theoretical curves for 0%, 5%, and 10% air voids are shown in Fig. 4.1.

### 4.4.1 Standard Proctor Test

In connection with the construction of earth dams, a standard compaction test was developed by Proctor (1933).

The test apparatus consists of

1. A cylindrical mould of internal diameter 102 mm and an effective height of 177 mm, with a volume of 0.945 litres
2. A detachable collar extension of the mould (used during compaction)
3. A detachable base plate
4. A 50 mm diameter metal rammer of weight 2.5 kg with a height of 300 mm moving in a metallic sleeve

The soil is compacted in three equal layers with 25 blows on each layer, and the energy transmitted to the soil, i.e., the compactive effort, is about 60.50 kg-m per 1,000 $cm^3$ of soil.

About 3 kg of air-dried soil is used. An initial water content of 4% is added for coarse-grained soils, and 10% for fine-grained soils. Other details of the test procedure are given in Chapter 10.

This test is also adopted as the standard test by the American Association of State Highway Officials (AASHO) and is usually called the AASHO test. The moisture content–density relationship is obtained as discussed earlier.

### 4.4.2 Modified Proctor Test

The test apparatus consists of

1. A cylindrical mould of about 102 mm diameter with an effective height of 177 mm
2. A suitable detachable collar and a base
3. A rammer of weight 4.5 kg with a height of fall of 450 mm.

The mould is filled in five layers, each layer being compacted with 25 blows. The compactive effort imparted is about 272.60 kg-m per 1,000 cm$^3$, which is about 4.5 times that of the Standard Proctor test.

The moisture content–dry density curve is obtained as discussed earlier. Because of the higher compactive effort, the dry density will be high. This test procedure has emerged to meet the heavier compaction requirements of airfield and airport pavements.

### 4.4.3 Indian Standard Compaction Tests

The Indian Standard equivalent of the Standard Proctor test is called the *light compaction test* (IS: 2720 – Part 7, 1974), and the Indian Standard equivalent of the Modified Proctor test is called the *heavy compaction test* (IS: 2720 – Part 8, 1974).

Light and heavy compaction details are given below:

| Type of Compaction | Weight of rammer (kg) | Height of fall (mm) | No. of layers | No. of blows on each layer |
|---|---|---|---|---|
| Light | 2.6 | 310 | 3 | 25 |
| Heavy | 4.89 | 450 | 5 | 25 |

The other test procedure is the same, and the moisture content–dry density relationship is obtained as discussed earlier.

It may be observed that the Indian compaction test is not significantly different from the Proctor tests.

## 4.5 FIELD COMPACTION AND EQUIPMENT

Classification of soil layers using compacting equipment is referred to as field compaction. The field compaction process involves one or more of the following (Lambe and Whitman, 1979): (i) selection of soil from borrow areas, (ii) transfer of soil from a borrow area to a construction site, (iii) spreading of the soil to a suitable thickness, (iv) alteration of the moisture content of the soil, (v) breaking of lumps and making the soil uniform, and (vi) rolling the soil by adopting a specified procedure to attain the property required. Different types of equipment are available commercially for following the above steps. The choice of equipment will depend on the type of soil and economic considerations.

Compaction equipment consists of excavating and hauling equipment, rock separation equipment, spreading equipment such as bulldozers and graders, discs, harrows and watering equipment, rollers, and special compacting equipment. Out of the above, the main compaction equipment is the rollers, which are discussed below.

### 4.5.1 Smooth Wheel Rollers

Smooth wheel rollers consist of hollow steel drums. The mass of the main roller can be increased using water or sand ballast. These are suitable for proof rolling of sub-grades and for finishing the construction of fills with sandy or clayey soils. They can be successfully used in places where a mixing or kneading action is not required. They provide excellent coverage, and the contact pressure can be as high as 300 to 400 kN/m$^2$. Smooth wheel rollers can be either towed or self-propelled.

### 4.5.2 Pneumatic-Tyred Rollers

In these rollers, wheels are placed close together on two axles and placed such that the rear set of wheels overlap the lines of the front set to ensure complete coverage of the soil surface. Wide tyres with flat treads are provided so that the soil is not displaced laterally. The action produced by these rollers is somewhat better than that of smooth wheel rollers in that they produce a combination of pressure and kneading on the soil. Rubber-tyred rollers are effective for a wide range of soils from clean sand to silty clay.

### 4.5.3 Sheep's Foot Rollers

These rollers consist basically of drums with numerous club-shaped tapered projections. The mass of the drum can be varied by adding ballast. The area of each projection may be 4,000 to 6,500 mm$^2$. The projections or feet penetrate into the layer during the rolling operations. During compaction, the initial passes compact the lower portion of a lift. In successive passes, compaction is obtained in the middle and the top sections of the layer. For effective rolling, the lift thickness should be small and the contact pressure under the projections very high, of the order of 1,500 to 7,500 kN/m$^2$. These rollers are most suitable for plastic and non-plastic fine-grained soils. Although not suitable for clean granular soils, they may be used in such soils too if more than 20% fines are present. Due to the excellent bonding caused by the kneading effect of the sheep's foot, these rollers are generally recommended for water-retaining earthworks.

### 4.5.4 Vibratory Rollers

Both smooth wheel and rubber-tyred rollers can be modified so that they impart an impacting motion to the soil being compacted. A power-driven vibration mechanism is provided to reasonably match the resonant frequency of the soil type and layer thickness. These rollers are by far the predominant type used in the compaction of granular soils.

### 4.5.5 Grid Rollers

These are intermediate between smooth wheel and sheep's foot rollers, with their rotating wheels, made of a network of steel bars, forming a grid with square holes. These rollers provide less kneading action but high contact pressures. These are most suitable for coarse-grained soils.

### 4.5.6 Tampers and Rammers

Hand-held tampers or rammers operated by compressed air or gasoline power are commonly used to compact small areas to which access is difficult. They are also used for compacting backfills in trenches.

### 4.5.7 Vibrating Plates

Manually controlled vibrating plates operate efficiently on clean granular soils. Satisfactory compaction can also be achieved in other types of soils. Vibrating plates are also gang-mounted on machines for use in less-restricted areas.

Sand or sand–gravel mixtures with silt (without clay) show good compaction characteristics, and it is generally recommended to use vibratory drum rollers, vibratory rubber tyres, or pneumatic tyre equipments for compaction of such soils. If some clay fraction is present in sand or sand–gravel mixtures, vibratory sheep's foot–type equipment can be used. Fine-grained silt and clay show a varied compaction performance depending on the plasticity of the material, and such soils can be compacted more effectively by using pneumatic tyres, vibratory rubber tyres, or vibratory sheep's foot–type equipment. Organic soils are not recommended for structural earthfills.

Field compaction depends on several factors, such as the soil type, moisture content, lift thickness, number of passes, and speed and type of compactor. The control of field compaction is discussed in the next section.

Table 4.1 gives the compaction performance and recommended compaction equipment (McCarthy, 1982).

**Table 4.1** Compaction performance and recommended equipment

| General soil description | Unified soil classification | Compaction characteristics | Recommended compaction equipment |
|---|---|---|---|
| Saud or sand–gravel mixtures (no silt or clay) | SW, SP, GW, GP | Good | Vibratory drum roller, vibratory rubber tyre, pneumatic tyre |
| Sand or sand–gravel with silt | SM, GM | Good | Vibratory drum roller, vibratory rubber tyre, pneumatic tyre |
| Sand or sand–gravel with clay | SC, GC | Good to fair | Pneumatic tyre, vibratory rubber tyre, vibratory sheep's foot |
| Silt | ML | Good to poor | Pneumatic tyre, vibratory rubber tyre, vibratory sheep's foot |
| | MH | Fair to poor | Pneumatic tyre, vibratory rubber tyre, vibratory sheep's foot, sheep's foot |
| Clay | CL | Good to fair | Pneumatic tyre, sheep's foot, vibratory |
| | CH | Fair to poor | Sheep's foot and rubber tyre |
| Organic soil | OL, OH, PT | Not recommended for structural earthfills | |

*Source:* McCarthy (1982).

## 4.6 COMPACTION SPECIFICATION AND CONTROL

### 4.6.1 Specification

Two methods have been used (Foster, 1962) to specify the compaction requirements in the various compancted layers. In one type called the *performance-type specification*, the compaction requirement is stated in terms of the physical properties of the compacted layer. A typical specification of this type is the percentage of maximum dry density obtained in a standard compaction test. Sometimes both the dry density and the moisture content are specified. For example, a specification of 95% maximum dry density is frequently designated. Specification in terms of percentage of the maximum dry density in a standard test is preferable to a stated value of the dry density because it ensures that a compactive effort comparable to that of the standard laboratory test is applied in the field. This type of specification is applicable for cohesive soils. Other physical properties that are used are the percentage of air voids at a specific moisture content, void ratio, and relative density. The last two are used primarily for non-cohesive materials. In performance-type specification, the contractor is given a wide scope in the selection of equipment and lift thickness.

In the other type, generally called the *work-type specification*, the type of equipment, the lift thickness, the moisture content, and the amount of work required to obtain the necessary density are specified.

Performance-type specification has found widespread usage in highway and airfield pavement work, whereas work-type specification has been used more for dam and levee work.

### 4.6.2 Field Control

It is the responsibility of the field engineer to check the density and water content during the process of rolling each layer. The only way this can be accomplished is by taking soil samples for moisture and density determination. Density readings are taken, usually a designated number, for every lift or for a specified volume of fill placed. It is common to take such readings for each layer for every 500 to 1,000 $m^2$, depending on the importance of the site. During the initial stages of the work itself, there should be enough technicians and equipment to conduct the test.

The most important aspect of construction control is the speed with which the moisture and density are determined when the soil is under rolling for rectification. Conventional methods of measuring the moisture content are slow. Rapid methods of determining the moisture content involve the use of a frying pan, a hot plate, a forced-draft, radiant-heat oven, a Proctor needle, and a nuclear moisture gauge. The latter two methods are important and are described below.

A Proctor needle (Fig. 4.2) consists of a spring-loaded plunger with a calibrated stem and a needle. The calibration is made in $kg/cm^2$ so that the penetration resistance can be read directly. The needle is provided with various bearing points to measure a wide range of penetration resistance. A laboratory "penetration resistance curve" of the field soil is needed to use the Proctor needle. A routine laboratory compaction test is performed on the soil. Before determining the wet density for each moisture content, the penetrometer (with a known bearing area) is pushed into the wet soil with a uniform push, up to a depth of 7.5 cm, and the penetration resistance versus moisture content along with dry density is plotted (Fig. 4.3). To

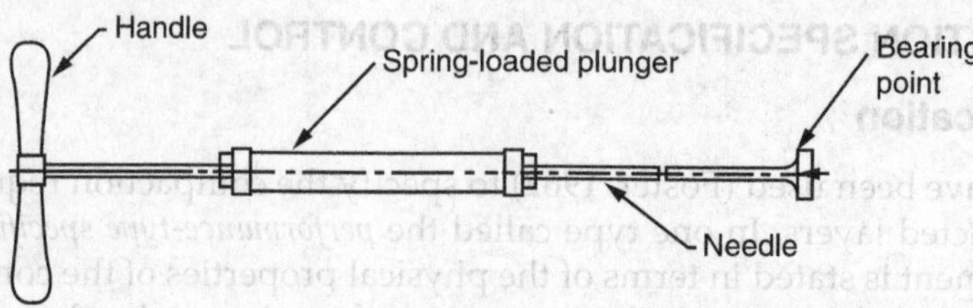


Fig. 4.2 Proctor needle

determine the moisture content in the field, the wet soil from the rolling yard is compacted into a mould under conditions similar to laboratory conditions. For this compacted soil, using the Proctor needle, the penetration resistance is read off. Then, from the laboratory curve, the moisture content corresponding to this penetration resistance is obtained. This method is sufficiently fast and accurate for fine-grained soils.

The nuclear moisture gauge is a modern instrument which is rapid and gives precise results. When using this, a source of fast neutrons is placed in the soil, and the neutrons move randomly and collide with atoms in the soil and rebound as slow neutrons. A counter is provided to record the counts of these slow neutrons. The quantity of hydrogen atoms in a soil is due to the presence of water, and hence such a count may be used to indicate the amount of water at the location of the source of fast neutrons. A radium–beryllium mixture is commonly used as the source of fast neutrons.

The wet density of the compacted soil in the field is determined using a core-cutter method or sand replacement method.

The Bureau of Indian Standards (IS: 10379, 1982) recommends three methods for non-gravelly soils and one method for soils containing gravels and rockfills. As per the first method for non-gravelly soils, the in-place density is determined using conventional methods such as the sand replacement method, core-cutter method, or rubber balloon method (IS: 2720 – Part 34, 1972). The moisture content is obtained using any of the rapid methods of

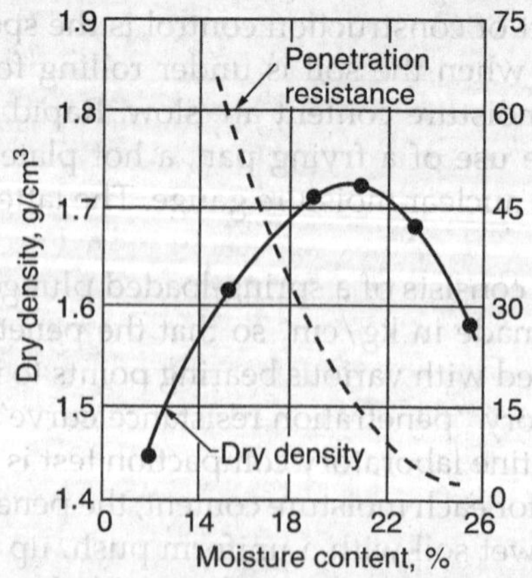


Fig. 4.3 Penetration resistance and dry denstiy/moisture content curves

water content determination such as the calcium carbide method, torsion balance method, sand bath method, or alcohol method (IS: 2720 – Part 2, 1973). Further, it is recommended that control tests be performed after removing the top 5 cm layer of earth. In the second method, Hilf's method for compaction control may be adopted (IS: 2720 – Part 38, 1977). This method suggests a relationship between the field moisture content, dry density, and laboratory optimum conditions determined without measurement of the water content. The third method is intended for certain weathered soils. For such soils, a test embankment under identical field conditions is used to determine the field moisture content and dry density.

In soils containing gravels and rockfills, the total density of the soil increases and the moisture content decreases with increasing gravel per cent up to 60% to 75%, beyond which the density again decreases. For soils with 30% gravel, the use of conventional light or heavy compaction methods (IS: 2720 – Parts 7 and 8, 1974, 1983) on the soil fraction passing a 40 mm IS sieve is recommended. The field density may preferably be determined using the ring and water replacement method (IS: 2720 – Part 33, 1971) or alternatively using the sand replacement method.

## 4.7 FACTORS AFFECTING COMPACTION

The degree of compaction of a soil is measured in terms of the dry density, which is the mass of soil solids per unit volume of the soil. The degree of compaction contributes to the shear strength, permeability, compressibility, and sustainability for repeated loads.

The major factors which affect compaction are (i) the moisture content, (ii) the compactive effect, (iii) the type of soil, and (iv) the method of compaction.

### 4.7.1 Effect of Moisture Content

As explained earlier, at lower levels of moisture content, the soil particles offer more resistance to compaction and the soil behaves like a stiff material. Increasing the moisture content helps the particles to move closer because of the lubrication effect. On further increasing the moisture content beyond a certain limit, the water starts to replace the soil particles. Thus, the dry density increases up to a limiting moisture content (optimum moisture content), beyond which an increase in the moisture content decreases the dry density. The effect of the formation of a structure with increasing moisture content is another meaning given for the increase in the dry density and the subsequent decrease beyond a certain limit (Lambe, 1958).

### 4.7.2 Effect of Compactive Effort

The maximum dry density and the optimum moisture content are both affected by a change in the compactive effort. An increase in the compactive effort increases the maximum dry density but decreases the optimum moisture content. However, the air void ratio at the peak density remains approximately the same (reflected by the constant air content line passing through the peak points). Further, it can be seen (Fig. 4.4) that there is only a marginal increase in the density with an increase in the compactive effort. It reflects the fact that only a very small improvement in the dry density results from the use of heavier equipment. Heavy equipment is generally preferred for economic reasons as it can produce the required compactive effort more cheaply.

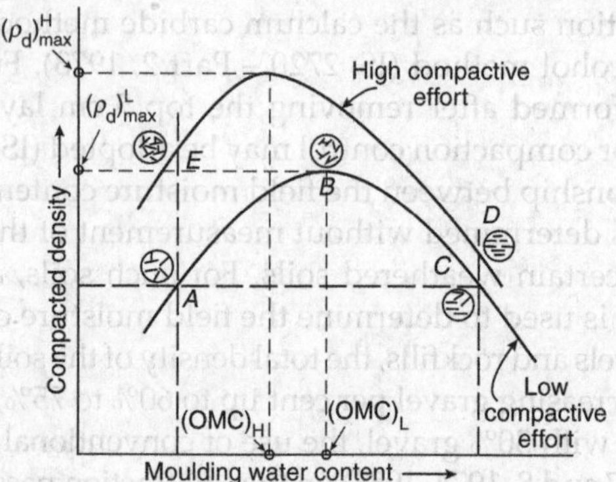


Fig. 4.4 Effect of compactive effort (Source: Lambe, 1962)

### 4.7.3 Effect of Type of Soil

Well-graded coarse-grained soils with smooth rounded particles show a high dry density, whereas uniform sands have a low maximum dry density (Fig. 4.5). Clayey soils have lower dry densities and higher optimum moisture contents than do sands. The effect of increasing the compactive effort is also more in clayey soils.

Figure 4.6 shows typical curves for different soils at the same compactive effort.

### 4.7.4 Effect of Method of Compaction

It is ideal to develop a laboratory test which could produce a reasonable moisture–density curve so as to assess the maximum dry density and optimum moisture content. As the processes of imparting energy to the soil are different in the field and laboratory, there may be different degrees of compaction depending on the method of compaction. Field compaction is essentially a rolling or kneading type of compaction, whereas the laboratory compaction is

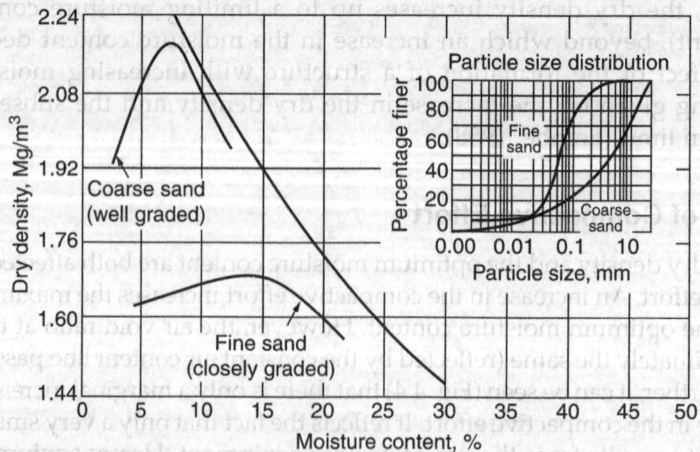


Fig. 4.5 Dry density versus moisture content for two grades of sands

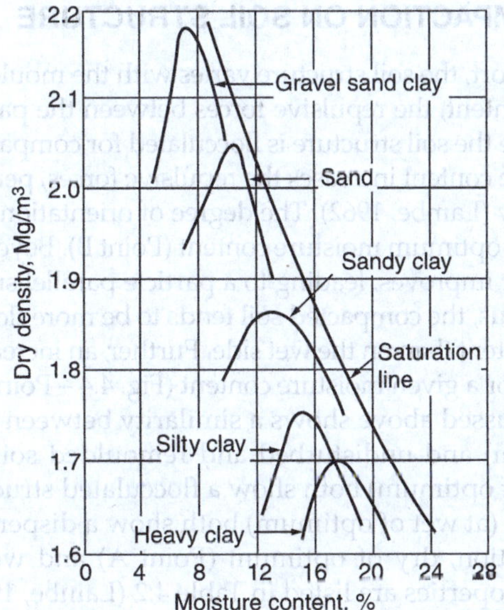


Fig. 4.6 Typical curves for different soils at the same compactive effort

of the dynamic-impact type. Dry unit weight–moisture curves (Turnbull, 1950) for the same soil subjected to different methods of compaction are given in Fig. 4.7. It can be seen that Curves 3 to 6 approximately give the same maximum dry density although the methods of compaction are different. Further, it may be reasoned out that the Standard AASHO test is the best fit to simulate the field moisture–unit weight relationship.

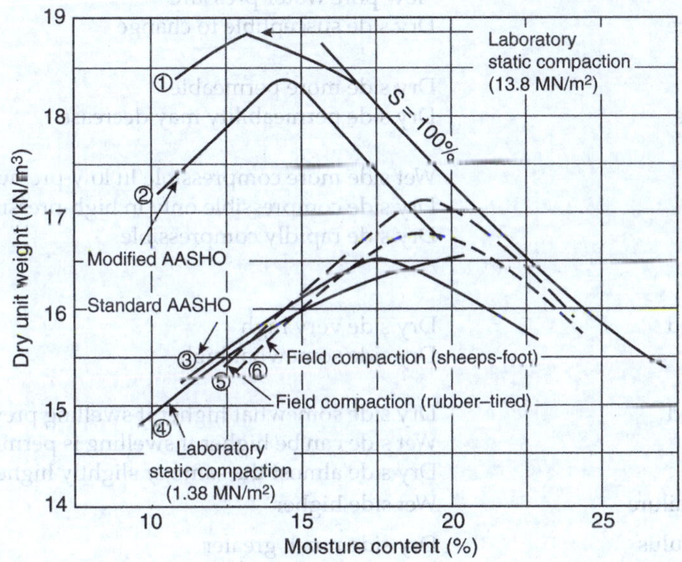


Fig. 4.7 Dry unit weight–moisture content curves for different methods of compaction (Source: Lambe and Whitman, 1979)

## 4.8 EFFECT OF COMPACTION ON SOIL STRUCTURE

For a given compactive effort, the soil structure varies with the moulding moisture content (Fig. 4.4). At a low moisture content, the repulsive forces between the particles are weaker than the attractive forces, and hence the soil structure is flocculated for compaction on the dry side (Point A). Increasing the moisture content increases the repulsive forces, permitting the particles to orient in a more orderly array (Lambe, 1962). The degree of orientation becomes such that the dry density is maximum at the optimum moisture content (Point B). Beyond this point, the degree of particle orientation further improves, leading to a particle parallelism, resulting in a dispersive soil structure (Point C). Thus, the compacted soil tends to be more flocculated on the dry side of the optimum moisture content than on the wet side. Further, an increase in the compactive effort tends to disperse the soil for a given moisture content (Fig. 4.4 – Points A and E and C and D).

The soil structure discussed above shows a similarity between compacted soils (dry side and wet side of optimum) and undisturbed and remoulded soils. Undisturbed soils and compacted soils (at dry of optimum) both show a flocculated structure, whereas remoulded soils and compacted soils (at wet of optimum) both show a dispersed structure.

The effects of compaction, dry of optimum (Point A) and wet of optimum (Point C), on several engineering properties are listed in Table 4.2 (Lambe, 1958).

**Table 4.2** Comparison of properties of cohesive soil on dry and wet sides of OMC

| Property | Comparison |
|---|---|
| Structure | |
| (i) Particle arrangement | Dry side more randomly oriented (flocculated) |
| (ii) Water deficiency | Dry side more deficient, takes more water, swells more; low pore water pressure |
| (iii) Permanence | Dry side susceptible to change |
| Permeability | |
| (i) Quantity | Dry side more permeable |
| (ii) Permanence | Dry side permeability may decrease |
| Compressibility | |
| (i) Quantity | Wet side more compressible in low-pressure range<br>Dry side compressible only in high-pressure range |
| (ii) Rate | Dry side rapidly compressible |
| Strength | |
| As moulded | |
| (i) Undrained | Dry side very high |
| (ii) Drained | Dry side somewhat high |
| At saturation | |
| (i) Undrained | Dry side somewhat higher if swelling prevented<br>Wet side can be higher if swelling is permitted |
| (ii) Drained | Dry side almost the same or slightly higher |
| Pore pressure at failure | Wet side higher |
| Stress–strain modulus | Dry side much greater |
| Sensitivity | Dry side more sensitive |

*Source:* Lambe (1958).

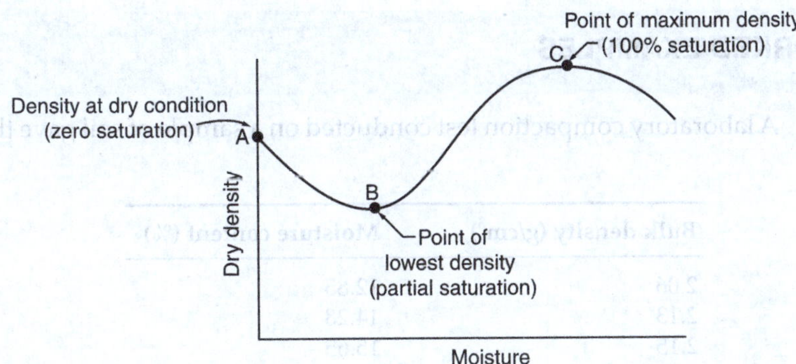


Fig. 4.8 Moisture–density relationship for sand

## 4.9 COMPACTION BEHAVIOUR OF SAND

The moisture–density curve (inverted V shaped) discussed so far is applicable to soils possessing some value of plasticity. In non-plastic soils like sands, the moisture–density curve is different.

The moisture–density relationship, as obtained from a laboratory test, on a cohesionless sand is shown in Fig. 4.8. The curve shows that a thin film of water around the grains keeps the particles away due to surface tension and causes a decrease in density in the initial stages. This phenomenon is referred to as *bulking of sand,* which occurs at a moisture content of 4% to 5%.

With increase in moisture content, the menisci are broken and the particles come closer, leading to an increase in the dry density. This density increases till the saturation is 100%, beyond which the density decreases due to occupation by water of the positions of grains.

An increase in the compactive effort in sand causes no significant change compared to cohesive soils. The above moisture–dry density relationship is of less practical importance.

Usually, cohesionless soils are densified by vibration in the dry or fully saturated condition or by simply flooding the area to be compacted. The degree of compaction is measured by the relative density or void ratio.

## 4.10 CALIFORNIA BEARING RATIO TEST

The California Bearing Ratio (CBR) test for the design of flexible pavements was developed by the California Division of Highways. The basic procedure of this test was adopted by the Corps of Engineers of the US Army. Certain modifications were made in the test procedure, and now the modified method is adopted by the Corps of Engineers and regarded as the standard method of determining the CBR. The Bureau of Indian Standards (IS: 2720 – Part 16, 1987) has also adopted the modified procedure.

The Corps of Engineers have developed design curves using CBR values for determining the required thickness of flexible pavements for airport runways and taxiways.

The detailed laboratory test procedure is explained in Chapter 10 for a remoulded soil. However, this penetration test can also be performed on undisturbed samples. A field CBR test is also available (IS: 2720 – Part 31, 1990) for finding the CBR of existing sub-grades.

## WORKED EXAMPLES

**Example 4.1** A laboratory compaction test conducted on a sample of soil gave the following results:

| Bulk density (g/cm³) | Moisture content (%) |
|---|---|
| 2.06 | 12.85 |
| 2.13 | 14.28 |
| 2.15 | 15.65 |
| 2.16 | 16.86 |
| 2.14 | 17.89 |

1. Find the maximum dry density and optimum moisture content
2. Plot the zero-air void line and 5% air void line. The specific gravity of soil is 2.72.

***Solution***

The dry density $\rho_d = \rho/(1+w)$,

Dry density values are calculated from the above formula as

$$\frac{2.06}{1+0.129} = 1.82\,\text{g/cm}^3; \quad \frac{2.13}{1+0.143} = 1.85\,\text{g/cm}^3; \quad \frac{2.15}{1+0.157} = 1.86\,\text{g/cm}^3;$$

$$\frac{2.16}{1+0.169} = 1.84\,\text{g/cm}^3; \quad \frac{2.14}{1+0.179} = 1.81\,\text{g/cm}^3$$

The moisture–dry density curve is plotted as given in Fig. 4.9. The maximum dry density and optimum moisture content (OMC) are read as

Max. $\rho_d = 1.87\,\text{g/cm}^3$

OMC $= 14.9\%$

The dry density in terms of air voids is given as

$$\rho_d = \frac{G\rho_w(1-V_a)}{1+mG}$$

Using this expression, the $\rho_d$ values for zero-air void (i.e., $V_a = 0$) and 5% air void (i.e., $V_a = 0.05$) are obtained as

| $w$ (%) | 14 | 15 | 16 | 17 | 18 |
|---|---|---|---|---|---|
| $V_a = 0$ | 1.98 | 1.94 | 1.90 | 1.87 | 1.83 |
| $V_a = 5\%$ | 1.88 | 1.84 | 1.80 | 1.78 | 1.74 |

These lines are plotted as shown in Fig. 4.9.

**Example 4.2** As per the compaction specification, a highway fill has to be compacted to 95% of Standard Proctor Compaction test density. A borrow area available near the project site has a dry density of 1.65 g/cm³ at 100% compaction and a natural void ratio of 0.61. The specific gravity of the soil solids is 2.65. Compute the volume of borrow material needed

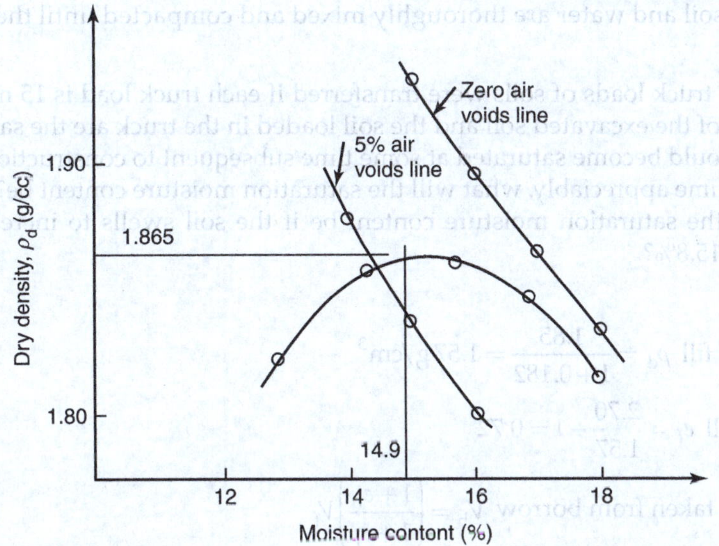


**Fig. 4.9** Compaction curve and zero and 5% air void lines.

to construct a highway fill of heigh 5 m and length 1 km with side slopes of 1:1.5. The top width of the fill is 8 m.

*Solution*

Top width of highway fill $= 8$ m

Base width of highway fill $= 2\,(1.5 \times 5) + 8 = 23$ m

Volume of highway fill $= ½(8 + 23) \times 5 \times 1 \times 1000$

$= 77{,}500\ \text{m}^3$

Dry density of soil at borrow $= 1.65\ \text{g/cm}^3$

Void ratio at borrow $e_b$ $= 0.61$

Dry density of highway fill $= 0.95 \times 1.65 = 1.57\ \text{g/cm}^3$

Void ratio at fill $e_f$ $= \dfrac{2.65}{1.57} - 1 = 0.69$

Volume of soil from borrow $V_b = V_f \dfrac{(1+e_b)}{(1+e_f)}$

$$= 77{,}500\left(\frac{1+0.61}{1+0.69}\right) = 73{,}832\ \text{m}^3$$

**Example 4.3** The undisturbed soil at a given borrow pit is found to have a water content of 16.8%, a void ratio of 0.62 and $G$ of 2.70. The soil from the borrow pit is to be used to construct a rolled fill having a finished volume of 4,800 m$^3$. The soil is excavated and dumped in trucks. In the construction process, the trucks dump their loads on the fill, and the material is spread and broken up after which water is added until the water content is 18.2%. The soil and water are thoroughly mixed and compacted until the wet density is 1.85 Mg/m$^3$.

1. How many truck loads of soils were transferred if each truck load is 15 m$^3$? Assume the void ratios of the excavated soil and the soil loaded in the truck are the same.
2. If the fill should become saturated at some time subsequent to construction and does not change volume appreciably, what will the saturation moisture content be?
3. What will the saturation moisture content be if the soil swells to increase its original volume by 15.8%?

***Solution***

Dry density of fill $\rho_d = \dfrac{1.85}{1+0.182} = 1.57\,\text{g/cm}^3$

Void ratio in fill $e_f = \dfrac{2.70}{1.57} - 1 = 0.72$

Volume of soil taken from borrow $V_b = \left(\dfrac{1+e_b}{1+e_f}\right) V_f$

$$= \frac{1+0.62}{1+0.72} \times 48000 = 45,209\,\text{m}^3$$

1. Number of truck loads required $= \dfrac{45209}{15} = 3,014$
2. Saturation moisture content, $w = \dfrac{eS_r}{G} = \dfrac{0.72\times 100}{2.70} = 26.67\%$
3. Increased volume = 48000 (1 + 0.158) = 55,584 m$^3$

   Void ratio in the swelled condition $= \dfrac{V_f}{V_b}(1+e_b) - 1$

   $$= \frac{55584}{45209}(1+0.62) - 1 = 0.99$$

4. Saturation moisture content, $w = \dfrac{eS_r}{G} = \dfrac{0.99\times 100}{2.70} = 36.73\%$

**Example 4.4** A sub grade soil of $G = 2.67$ and dry density 1.53 Mg/m$^3$ is available. With this soil an aggregate of the same specific gravity in a proportion of 75% of aggregate to 25% soil is mixed. The mixture is then compacted to a dry density of 1.84 Mg/m$^3$. At 100% saturation, the aggregate has a moisture content of 3%. What is the saturation moisture content for the soil in the compacted mixture?

***Solution***

Consider 1 m$^3$ of soil mixture.
Of this 25% is soil, i.e., 0.25 × 1.53 = 0.383 Mg
The balance is aggregate = 1.84 – 0.383 = 1.457 Mg

$$V_s \text{ of soil grains } = \frac{0.383}{2.67\times 1} = 0.143\,\text{m}^3$$

$$V_s \text{ of aggregate } = \frac{1.457}{2.67\times 1} = 0.546\,\text{m}^3$$

Soil grains fill the voids formed by the aggregate, and the balance volume of the voids is filled by soil grains = 1 – 0.546 – 0.143 = 0.311 m³.

$$\text{Void ratio } e = \frac{0.111}{0.546+0.143} = 0.45$$

At saturation of the compacted mixture, $w = \frac{0.45\times 1}{2.67}\times 100 = 16.91\%$

Of the water, 3% is for aggregate saturation.

Therefore, the total water content at 100% saturation of mixture = 19.91%

**Example 4.5** Some soil has been dumped loosely from a scraper. It has a unit weight of 16 kN/m³, a water content of 10.5%, and a specific gravity of solids of 2.68. Find the void ratio, porosity, density, and unit weight of the soil in the loose condition.

To make the compaction process more workable, an optimum water content of 15% is necessary. How much of water should be added in litres per cubic metre of soil to raise the water content to the optimum?

The soil is compacted with the optimum water content until it is 95% saturated. Find the new void ratio, porosity, dry density, and dry unit weight.

***Solution***

1. *In loose condition*

$$\text{Dry unit weight } \gamma_d = \frac{16}{1+0.105} = 14.40\,\text{kN/m}^3$$

$$\text{Void ratio } e = \frac{G\gamma_w}{\gamma_d} - 1 = \frac{2.68\times 9.81}{14.40} - 1 = 0.826$$

$$\text{Dry density } \rho_d = \frac{\gamma_d}{g} = \frac{14.40}{9.81} = 1.468\,\text{g/cm}^3$$

$$\text{Porosity } n = \frac{e}{1+e}\times 100 = \frac{0.826\times 100}{1+0.826} = 45.2\%$$

2. *Water to be added*

Volume of soil in fill $V_f = (1 + e_f)\,V_s$

Hence in an unit volume of fill,

$$V_s = \frac{V_f}{1+e_f} = \frac{1}{1+0.826} = 0.548\,\text{m}^3$$

Water to be added per cubic metre of fill is given as

$$\frac{15-10.5}{100}=\frac{M_w}{M_s}=\frac{M_w}{\rho_s V_s}=\frac{M_w}{2.68\times 1000\times 5.48}$$

That is, $M_w=\frac{3.5}{100}\times 2.68\times 1000\times 0.548=51.4\,\text{kg}$.

Therefore, water to be added per cubic metre of fill = 51.4 litres.

**Example 4.6** In the construction of a levee, the compaction specification required was 95% of Proctor maximum dry density at a field moisture content within 2% of the optimum moisture content.

The maximum dry density and optimum moisture content obtained in the laboratory from the Proctor test were 1.94 Mg/m$^3$ and 13.5%, respectively. A field supervisor conducted sand-cone tests at two locations and obtained the results presented below. The sand in the sand bottle was found to have a density of 1.87 Mg/m$^3$. Check whether the specification was satisfied.

| Location no. | Mass of soil removed (g) | | Mass of sand used (g) |
|---|---|---|---|
| | Wet | Dry | |
| 1 | 43.86 | 38.46 | 39.51 |
| 2 | 37.38 | 32.21 | 32.39 |

***Solution***

95% of $\rho_{d\,max}=0.95\times 1.94=1.843\ \text{g/cm}^3$

2% of OMC = 13.5 ± 2% = 15.5% to 11.5%

*Location 1*

$$\text{Volume of pit}=\frac{3951}{1.87}=2112.8\,\text{cm}^3$$

$$\text{Wet density}=\frac{4386}{2112.8}=2.08\,\text{g/cm}^3$$

$$\text{Water content}=\frac{4386-3846}{3846}\times 100=14.04\%$$

$$\text{Dry density }\rho_d=\frac{2.08}{1+0.1404}=1.82\,\text{g/cm}^3$$

Therefore, the moisture content requirements are satisfied, but the density requirement is not satisfied.

*Location 2*

$$\text{Volume of pit}=\frac{3239}{1.87}=1732.1\,\text{cm}^3$$

$$\text{Wet density}=\frac{3738}{1732.1}=2.16\,\text{g/cm}^3$$

$$\text{Water content} = \frac{3738-3221}{3221}\times 100 = 16.05\%$$

$$\text{Dry density } \rho_d = \frac{2.16}{1+0.1605} = 1.86\,\text{g/cm}^3$$

Therefore, the density requirement is satisfied, but the moisture content requirement is not satisfied.

**Example 4.7** An airfield sub-grade is compacted with a thickness of 350 mm. The rammer used for compaction has a foot area 0.06 $m^2$ and imparts an energy of 50 kg m. Find the number of passes required to develop a compactive energy of 30,000 kg fm/$m^3$.

***Solution***

$$\text{Compactive energy imparted by rammer per cubic metre of the soil} = \frac{50}{0.06\times 350\times 10^{-3}}$$

$$= 2380.95\ \text{kg fm/m}^3$$

$$\text{No. of passes required} = \frac{30000}{2380.95} = 12.6, \text{say } 13$$

## POINTS TO REMEMBER

4.1 Soil compaction is the process of increasing the density of the soil by applying some mechanical energy and thereby reducing air voids.

4.2 The compactive effort is the energy input to a soil for compacting it. The compactive effort can be varied in the laboratory and in the field. Increasing the compactive effort increases the dry density and decreases the optimum moisture content.

4.3 For most soils, the Standard Proctor test (BIS Light Compaction Test) is applicable. In situations where a heavy load is anticipated, the BIS Heavy Compaction Test is adopted. The compactive effort used in the Modified Proctor test is 4.5 times that of the Standard Proctor test.

4.4 Field compaction equipment consists of excavating and handling equipment, rock separation equipment, spreading equipment, discs, harrows, watering equipment, rollers, vibrators, and other special compacting equipment.

4.5 Smooth wheel rollers are suitable for rolling of sub grades and for finishing construction fills of sandy or clayey soils. Rubber-tyred rollers are effective for a wide range of soils from clean sand to silty clay. Sheep's foot rollers are effective for a wide range of soils from clean sand to silty clay.

4.6 Field control of compaction is the process of checking the density and moisture content during compaction by rollers or other compacting equipment. The most important aspect of construction control is the speed with which the moisture and density are determined and rectified if needed.

4.7 Factors affecting the compaction of a soil are the moisture content, compactive effort, and method of compaction.

4.8 For compaction on the dry side of optimum, the soil structure is flocculated and dispersed in the wet side of optimum.

4.9 The permeability, swelling, undrained strength, and sensitivity are high on the soil compacted dry of optimum.

4.10 The moisture-density curves for non-plastic soils, like sand, are not the same as those for plastic soils. The curve for a non-plastic soil shows a trough at a low moisture content and a peak at a high moisture content. Increasing the compactive effort in sand causes no significant change in the maximum dry density and the optimum moisture content.

## QUESTIONS

### *Objective Questions*

4.1 Choose the correct statements from the following:

1. The compactive effort in the laboratory can be varied only by varying the height of fall.
2. It is not feasible to expel air completely by compaction.
3. In field compaction, the effect of compaction is more with a small lift thickness.
4. The conventional method of measuring the moisture content is used in field control of moisture content.

4.2 In the light compaction test, the number of blows used per layer is
(a) 15 (b) 25 (c) 30 (d) 35

4.3 In the nuclear moisture gauge a source of ______ is used.
(a) Slow protons (b) Slow neutrons
(c) Fast protons (d) Fast neutrons

4.4 An increase in the compactive effort in laboratory compaction causes the OMC to
(a) Remain the same (b) Decrease
(c) Increase by 5% (d) Decrease by 5%

4.5 The soil structure at the dry side of optimum is
(a) Partially flocculated (b) Fully flocculated
(c) Fully dispersed (d) None of the above

4.6 *Assertion A:* The process of compaction is accompanied by the expulsion of air.
*Reason R:* The degree of compaction of a soil is charactenzed by its dry density.
*Select the correct code:*
(a) Both A and R are correct, and R is the correct explanation of A.
(b) Both A and R are correct, and R is not the correct explanation of A.
(c) A is true, but R is false.
(d) A is false, but R is true.

4.7 A zero-air void density can
(a) Be obtained with a high compactive effort
(b) Be obtained with a low compactive effort
(c) Be obtained using static compaction
(d) Not be obtained in practice

4.8 The use of sheep's foot rollers to compact cohesion-less soils in
(a) Very effective (b) Moderately effective
(c) Effective (d) Ineffective

4.9 The relative compaction corresponding to zero relative density is
(a) 80% (b) 70% (c) 60% (d) 50%

4.10 The swelling is greater and shrinkage is less for clay compacted
(a) at OMC (b) at OMC ± 2% (c) on the dry side of optimum

### Descriptive Questions

4.11 Distinguish between the Standard Proctor and Modified Proctor tests.

4.12 How do you differentiate between the compactive effort imparted in the laboratory and that in the field.

4.13 How is the required compactive effort for a particular soil to attain a desired dry density assessed in the field?

4.14 What effect does increased compaction have on the properties of a granular soil?

4.15 What is the ratio of the compactive energy of the IS heavy compaction test and the IS light compaction test?

4.16 With illustrative compaction curves, discuss various factors which influence the compaction of a cohesive soil of high compressibility.

4.17 Explain the methods of finding the placement density of a compacted fill. Name the method which is suitable for all types of soils.

4.18 How can the Standard Proctor test be modified to suit the compacting machinery used at the site for compacting a cohesive soil?

## EXERCISE PROBLEMS

4.1 The following data refer to a compaction test as per Indian Standards using light compaction:

| Water content (%) | 8.5 | 12.2 | 13.75 | 15.5 | 18.2 | 20.2 |
|---|---|---|---|---|---|---|
| Mass of wet sample (N) | 18.0 | 19.4 | 20.0 | 20.5 | 20.3 | 19.8 |

Plot the compaction curve and obtain the maximum dry unit weight and optimum moisture content. Also plot the 80% saturation line. Take $G = 2.7$ and the volume of the mould = 1,000 cm$^3$.

4.2 A laboratory compaction test conducted on a 900 ml volume of mould yielded the following results:

| Mass of dry soil (g) | 160.1 | 154.80 | 155.75 | 165.35 | 182.74 |
|---|---|---|---|---|---|
| Moisture content (%) | 9.2 | 11.56 | 13.45 | 15.20 | 17.34 |

Plot the moisture–dry density curve and find the maximum dry density and optimum moisture content. Find the void ratio at OMC and at 5% of OMC. Take $G = 2.65$.

4.3 In an embankment filling, the field density of the dry soil is 19.2 kN/m$^3$ and the maximum dry density (Proctor's density) of the soil is 20.0 kN/m$^3$. Calculate the percentage compaction.

4.4 The bulk unit weight and the moisture content of a borrow area are 16.85 kN/m$^3$ and 12.8%, respectively. It is intended to construct an embankment of 5 m height and, 10 m top width with 1:1.5 side slopes and 2 km length with a finished dry unit weight of 19.50 kN/m$^3$. Specific gravity of soil = 2.67.

1. Determine the quantity of soil required from the borrow pit for construction of 1 m of the embankment.
2. If the construction is to be made with a moisture content of 15.2%, estimate the amount of water to be added.

4.5 The details of two borrow areas identified for the construction of an embankment are given below:

| Borrow area | Bulk density (g/cc) | Moisture content (%) | Specific gravity |
|---|---|---|---|
| A | 1.65 | 14.5 | 2.65 |
| B | 1.30 | 15.6 | 2.67 |

The borrow areas are approximately at the same distance. The embankment is of length 1 km, top width 8 m, height 6 m, and side slopes 1:1.6.

Which of the above two borrow areas would you recommend? Reason out your choice.

4.6 A sand replacement test was conducted on a compacted field soil. The following are the observations made:

Bulk density of sand used for the test = 1.5 g/cm$^3$
Mass of soil excavated from the pit = 980 g
Mass of sand filling the pit = 720 g
Moisture content of compacted fill = 14.8%
Specific gravity of soil solids = 2.68%

Compute the wet density, dry density, void ratio, and degree of saturation of the compacted fill.

4.7 A core cutter of 10 cm diameter and 18 cm height is used in an in-place density determination of a compacted fill. The following are the other observations made:

Mass of empty core cutter = 2,330 g
Mass of soil + core cutter = 5,020 g
Mass of wet soil sample for water content determination = 54.8 g
Mass of oven-dried sample = 50.2 g
Specific gravity of soil solids = 2.68

Compute the field dry density, void ratio, and degree of saturation.

# 5

# Permeability and Capillarity

**CHAPTER HIGHLIGHTS**

Water flow – Darcy's law – Validity of Darcy's law – Laboratory and field permeability tests – Permeability of stratified soils – Values of permeability – Factors affecting permeability – Capillary phenomenon in soils – Shrinkage and swelling of soils

## 5.1 INTRODUCTION

The amount, distribution, and movement of water in soil have an important bearing on the properties and behaviour of soil. The engineer should know the principles of fluid flow, as groundwater conditions are frequently encountered on construction projects. Water pressure is always measured relative to atmospheric pressure, and water table is the level at which the pressure is atmospheric. Soil mass is divided into two zones with respect to the water table: (i) below the water table (a saturated zone with 100% degree of saturation) and (ii) just above the water table (called the capillary zone with degree of saturation $\leq$100%). Below the water table, the pore water may be static or seeping through the soil under hydraulic potential. This chapter and the next have been devoted to give an accurate and complete knowledge of the water condition in the soil.

## 5.2 WATER FLOW

Soil is a particulate material and has pores that provide a passage for water. Such passages vary in size and are tortuous and interconnected. A sufficiently large number of such paths of flow are grouped to act together, and the average rate of flow is considered to represent a property of the soil. This property is termed permeability of the soil and may be defined as the capacity of a soil to permit water to pass through its interconnected void spaces. As in any other porous medium, water transmission takes place between two points due to the difference in pressure heads or total heads ($h$).

As per Bernoulli's equation, the total head consists of three components, viz., position or elevation head ($z$), pressure head due to water pressure, $u_w$ ($h_w = u_w / \gamma_w$), and velocity head ($h_v = v^2/2g$, where $v$ is the velocity). Seepage velocities in soils are normally small, and hence the velocity head is ignored. Thus, the total head causing the flow of water in soil is

$$h = \frac{u_w}{\gamma_w} + z \tag{5.1}$$

The movement of water through a pipe or pore may take on either of the two characteristic states of motion, viz., laminar or turbulent. Laminar flow indicates that each water particle follows a definite path and never crosses the path of another particle. This is an orderly and steady flow with no mixing. Turbulent flow indicates a random path of irregular and twisted movement. This is a disorderly and unsteady flow with more mixing. The velocity of flow depends directly on the nature of motion. Because of small pores in most soils, the flow of water is steady and laminar except in a few cases, such as flow in very coarse-grained soils and high velocities causing internal soil erosion (Taylor, 1948). In general, for flow in the laminar range, energy losses are proportional to the first power of velocity.

## 5.3 DARCY'S LAW

Considering one-dimensional flow in a saturated medium obeying laminar flow, Darcy (1856) demonstrated experimentally (a schematic set-up of Darcy's sand filtration experiment is shown in Fig. 5.1) that the flow velocity is proportional to the hydraulic gradient ($i$), i.e.,

$$v \propto i \tag{5.2}$$

or

$$v = ki \tag{5.3}$$

where $v$ is the flow velocity (mm/s or m/s), $k$ the coefficient of permeability (mm/s or m/s), $i$ the hydraulic gradient $= h/L$, $h$ the difference in pressure heads $= (h_1 - h_2)$, where $h_1$ and

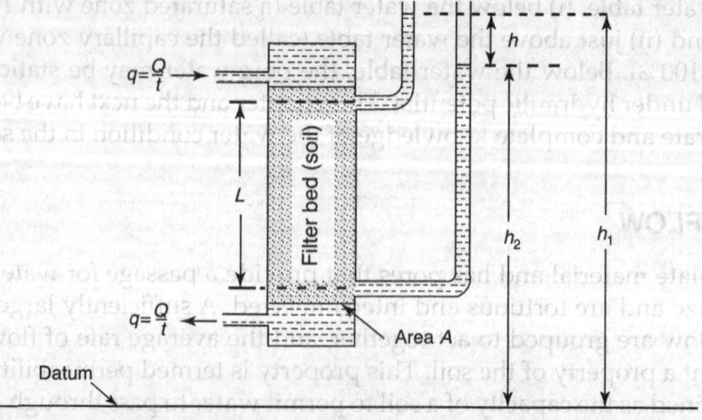


Fig. 5.1 Schematic set-up of Darcy's experiment

$h_2$ are the pressure heads (m), $L$ = length of specimen (mm), and the rate of flow, $q$ ($m^3/s$) is given as

$$q = kiA \tag{5.4}$$

where $A$ is the cross-sectional area.

The velocity $v$ is the overall velocity, also called discharge velocity, Darcian velocity, or superficial velocity. This velocity is different from the velocity inside the soil pores, which is known as the seepage velocity, $v_s$. As the flow is continuous, $q$ must be the same throughout the system. Thus,

$$q = Av = A_v v_s$$

where $A_v$ is the cross-sectional area of the voids.

$$v_s = \left(\frac{A}{A_v}\right)v = \frac{v}{n} = k\,i/n = \left(\frac{k}{n}\right)i - k_p i$$

where $k_p$ is the coefficient of percolation, or

$$k_p = k/n \tag{5.5a}$$

and

$$v = v_s n = \left(\frac{1+e}{e}\right)v_s \tag{5.5b}$$

This shows that the seepage velocity is greater than the superficial velocity, since $A_v < A$ and $v_s > v$, to keep the flow constant. In soil flow problems, it is more convenient to use the total cross-sectional area of flow rather than the area of voids.

Microscopically, the flow follows a tortuous path, but macroscopically it may be presumed to be orderly and in a straight line.

## 5.4 RANGE OF VALIDITY OF DARCIAN FLOW

Darcy's type of flow is stable in character as long as the four basic conditions are always satisfied, viz., (i) the steady-state laminar flow, (ii) 100% saturation (no compressible air present), (iii) flow fulfilling continuity conditions, and (iv) no volume changes (compression or swelling) during or as a result of flow. The validity of Darcy's flow may be analysed with respect to particle size, velocity, and hydraulic gradient.

Similar to flow through pipes, for flow through soils, Reynold's number $R_n$ may be expressed as

$$R_n = \frac{vD\gamma_w}{\eta_w g} \tag{5.6}$$

where $D$ is the average diameter of the soil particle.

It has been accepted that Darcy's law is valid and flow will be laminar as long as Reynold's number is equal to one. For $R_n = 1$, the corresponding value of $D = 0.5$ mm, which is in the coarse-sand range. This appears to be the upper limit of particle size beyond which the flow

**Table 5.1** Realm of validity for Darcy's flow of water in granular soils

| $D_{10}$ (mm) | 76.2 | 25.4 | 9.52 | 2.0 | 0.59 | 0.25 | 0.074 | 0.02 |
|---|---|---|---|---|---|---|---|---|
| | Gravel | | | Sand | | | Silt | |
| | Coarse | Medium | Fine | Coarse | Medium | Fine | Coarse | Fine |
| Realm of flow of water | Practically always turbulent flow | | | Darcy's laminar flow only for $i$ less than about 0.2 to 0.3 for the loose state and 0.3 to 0.5 for the dense state | | | Always laminar flow for the range of $i$ found in nature | |

*Source:* Burmister (1954).

may be turbulent. It has been shown by several authors that the flow is laminar in fine-grained soils for the range of gradients found in nature.

Darcian linear relationship between velocity and gradient deviates faster in very fine sands than in coarser sands with respect to gradient. In coarse materials, the pores are wider, and therefore, the turbulence may begin at lower values of gradient than in fine sands. Burmister (1954), based on experiments in granular soils, fixed a certain range for gradients, which is presented in Table 5.1.

In dense clays and heavy loams, in which the water is of molecularly bound nature, seepage starts only when the gradient exceeds a certain value, $i_0$, called the initial or threshold gradient. This gradient represents the gradient required to overcome the maximum binding energy of mobile pore water. For dense clays, $i_0$ may attain values from 20 to 30.

However, for a wide range of soils (silts through medium sands) for which the range of gradients usually met with in nature, Darcy's law stands valid. This is also true for clays under steady state of flow.

## 5.5 LABORATORY PERMEABILITY TESTS

Basically, there are two laboratory experiments for the determination of the coefficient of permeability, viz., the constant head and falling or variable head permeameters.

### 5.5.1 Constant Head Permeameter

This test is preferred for coarse-grained soils, such as gravels and sands, for permeabilities $>10^{-4}$ m/s. A schematic diagram of the apparatus is shown in Fig. 5.2 (details of the apparatus and procedure are given in Chapter 10).

The soil specimen is placed at an appropriate density in the permeameter. A steady vertical flow of water under a constant total head is maintained. After saturation, a certain quantity of water passing through the soil for a given time is collected and $q$ is calculated. Thus, from Darcy's law,

$$q = Ak\frac{h}{L}$$

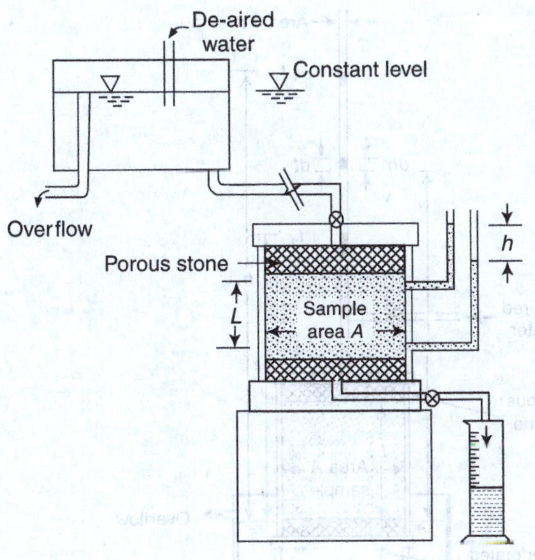


Fig. 5.2 Constant head permeameter

or

$$k = \frac{qL}{Ah} \tag{5.7}$$

### 5.5.2 Falling Head Permeameter

For fine-grained soils, such as silt and clay (with $k$ between $10^{-4}$ and $10^{-7}$ m/s), this is generally used. The experimental set-up is shown schematically in Fig. 5.3. After saturation, the standpipe (with area of cross-section $a$) is filled with water, and time $t_1$ corresponding to $h_1$ is noted down. Water is allowed to fall to $h_2$, and time $t_2$ is noted. The coefficient of permeability, $k$, is calculated from Eq. 5.7.

At any intermediate time $t$, let the water level be $h$ and its rate of change be $-dh/dt$. At time $t$, the head difference is $h$, hence $i = h/L$.

$$-a\frac{dh}{dt} = Ak\frac{h}{L}$$

$$-a\int_{h_1}^{h_2} \frac{dh}{h} = \frac{Ak}{L}\int_{t_1}^{t_2} dt$$

or

$$a\int_{h_2}^{h_1} \frac{dh}{h} = \frac{Ak}{L}\int_{t_1}^{t_2} dt$$

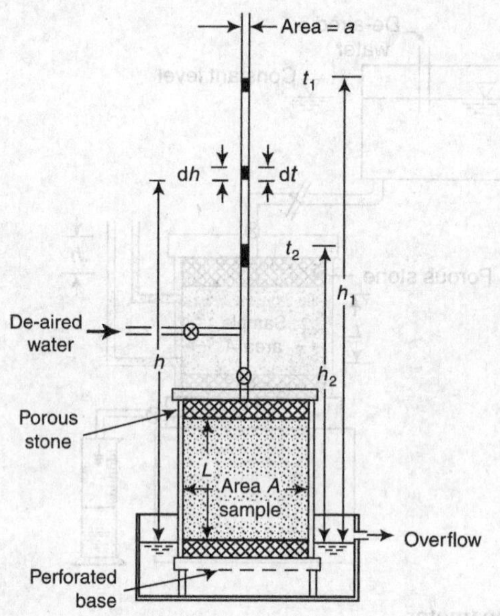


**Fig. 5.3** Falling head permeameter

or

$$k = \frac{aL}{A(t_2 - t_1)} \log_e \frac{h_1}{h_2}$$

Let $(t_2 - t_1)$ be $t$, then

$$k = 2.303 \frac{aL}{At} \log_{10} \frac{h_1}{h_2} \tag{5.8}$$

Tests should be repeated using different values of $h_1$ and $h_2$. The permeability of fine-grained soils can also be found from the consolidation test (Chapter 8).

## 5.6 FIELD PERMEABILITY TESTS

Data from field permeability tests are needed in the design of various civil engineering works, such as cut-off wall design of earth dams, to ascertain the pumping capacity for dewatering excavations and to obtain aquifer constants. The *in situ* tests, although expensive, take into account the effects of stress release, the direction of flow, and boundary conditions. However, field measurements are not sometimes precise because of the uncertainty of soil and water conditions at the location (Lambe and Whitman, 1979).

Under field conditions, the rate of flow of water is measured by a quantity, *coefficient of transmissivity* ($T$). It is defined as the rate of flow of water through a vertical strip of aquifer of unit width under a unit hydraulic gradient (IS: 5529 – Part 1, 1985). This coefficient depicts the ability of an aquifer to transmit water. This is related to the coefficient of permeability as

$$T = kh_t \tag{5.9}$$

where $h_t$ is the aquifer thickness.

Generally, the tests are carried out in boreholes where sub-surface explorations are carried out. These tests can be done effectively up to a depth of 30 m and give the most reliable results. The tests may be either pumping in or pumping out type. Pumping in test can be conducted irrespective of the position of the water table in a stratum, while pumping out test is suited for tests below the water table. The water table (or phreatic surface) is the level to which undergroundwater will rise in a soil and will be at atmospheric pressure.

The pumping in test is suitable for low permeability and thin strata where adequate yield may not be available for pump out test. By this test, permeability of the soil at the bottom of the borehole is obtained. Thus, this is recommended for permeability determination of stratified deposits and, hence, to check the effectiveness of grouting in such deposits. This test is economical since it does not require an elaborate test arrangement as in pump out tests. The types of pumping in tests as recommended in IS: 5529 – Part 1 (1985) are constant head method, falling head method, and slug method. The constant head method is recommended in highly permeable strata. The falling head method is more suitable for tests below the water table. Further, this method is applicable for strata with low permeability and where the soil below the casing pipe can stand without collapsing. The slug method is conducted in artesian aquifers with small to moderate permeabilities. The reader may refer to IS: 5529 – Part 1 (1985) for details of these methods.

The pumping out test is a more general and accurate method for permeability determination below the water table. This method is most suitable for all groundwater problems. There are basically two conditions of flow; accordingly, the pump out tests may be grouped as unconfined flow (or gravity well) test or confined flow (or artesian well) test. In IS: 5529 – Part 1 (1985), three methods are given, viz., unsteady-state, steady-state, and Bailer methods. Of these methods, the steady-state method, also known as Theim's steady state or equilibrium method, is the most accurate. The steady-state method for two flow conditions is explained below.

### 5.6.1 Unconfined Flow Pumping Out Test

Figure 5.4 represents a permeable layer underlain by an impermeable stratum and the arrangement of wells. The soil is assumed to be homogeneous and coarse grained. The initial

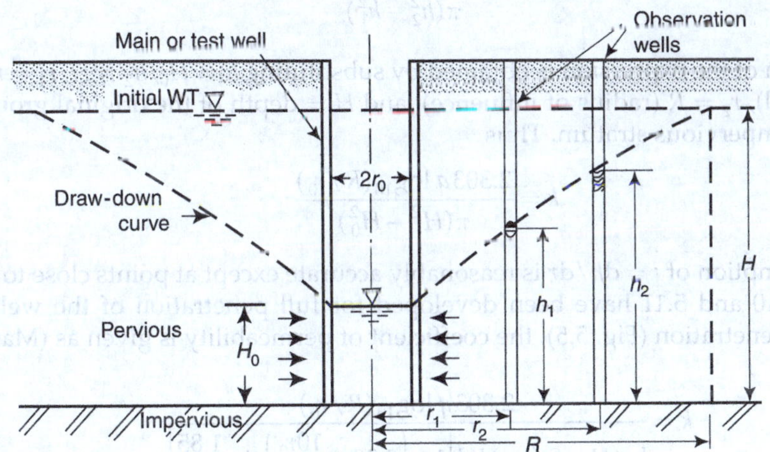


Fig. 5.4 Pumping test from unconfined aquifer

water table position is considered horizontal, and the hydraulic gradient is assumed to be constant at any given radius.

One large-diameter perforated casing is sunk up to the impervious stratum or to a considerable depth, and this well is used as the main or test well. Two additional small-diameter perforated casings are sunk at some distance from the test well. Water is pumped from the main well at a constant rate. The draw-down of the water table takes place, and the steady-state water table in each of the nearby observation wells is recorded.

The steady state is established when the water level in the main well and the observation wells becomes constant.

Assume that the water is flowing into the well in a horizontal and radial direction. Consider an elementary cylinder of soil having radius $r$, thickness d$r$, and height $h$. Let the water level fall in the observation wells at the rate of d$h$. At the steady state, the rate of discharge, $q$, due to pumping is

$$q = kiA$$

where

$$i \approx \frac{dh}{dr} \quad \text{(this is referred to as Dupit's assumption)}$$

and

$$A = 2\pi rh$$

Therefore,

$$q = k\frac{dh}{dr}2\pi rh$$

Rearranging and integrating

$$\int_{r_1}^{r_2} \frac{dr}{r} = \frac{2\pi k}{q}\int_{h_1}^{h_2} h\,dh$$

Therefore,

$$k = \frac{2.303\,q\log_{10}(r_2/r_1)}{\pi(h_2^2 - h_1^2)} \tag{5.10}$$

Another form of the expression is obtained by substituting $h_1 = H_0$, $h_2 = H$, $r_1 = r_0$ (radius of the main well), $r_2 = R$ (radius of influence), and $H$ = depth of the original groundwater table from the impervious stratum. Thus,

$$k = \frac{2.303\,q\log_{10}(R/r_0)}{\pi(H^2 - H_0^2)} \tag{5.11}$$

Dupit's assumption of $i = dh/dr$ is reasonably accurate except at points close to the well.

Equations 5.10 and 5.11 have been developed for full penetration of the well, and in case of partial penetration (Fig. 5.5), the coefficient of permeability is given as (Mansur and Kaufman, 1962)

$$k = \frac{2.303\,q\log_{10}(R/r_0)}{[\pi(H-S)^2 - H_0^2]\left[1+\left(0.30+\frac{10r_0}{H}\right)\sin\frac{1.85}{H}\right]} \tag{5.12}$$

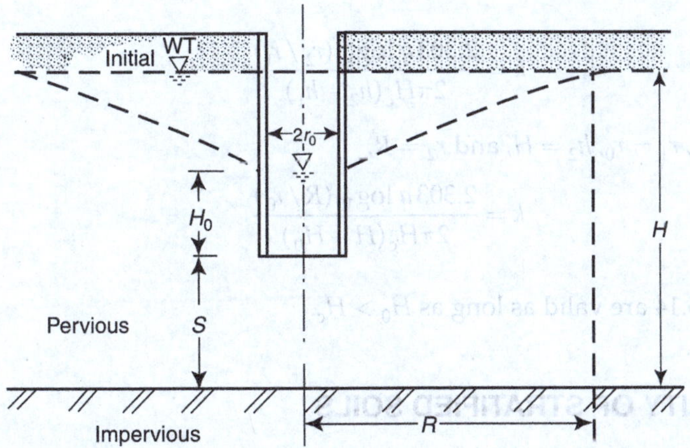


Fig. 5.5 Partial penetration of well in unconfined aquifer

### 5.6.2 Confined Flow Pumping Out Test

An artesian well penetrating the full depth of the aquifer is shown in Fig. 5.6. Herein, a permeable layer is sandwiched between two impermeable layers. Because of the artesian effect, the piezometric surface locates itself above the upper surface of the aquifer. In the steady-state condition, consider an elementary cylinder of radius $r$, thickness d$r$, and height $h$. The rate of flow is given as

$$q = kiA = k\frac{\mathrm{d}h}{\mathrm{d}r}2\pi rH_c$$

where $H_c$ is the depth of the confined aquifer. Thus,

$$\int_{r_1}^{r_2}\frac{\mathrm{d}r}{r} = \int_{h_1}^{h_2}\frac{2\pi kH_c\,\mathrm{d}h}{q}$$

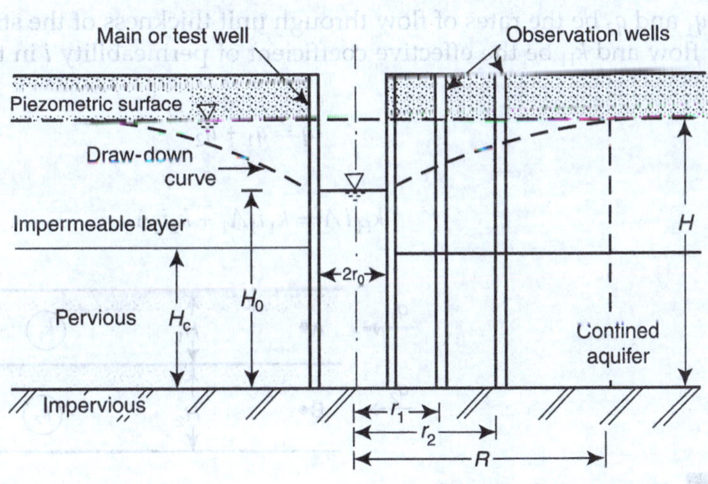


Fig. 5.6 Pumping test from confined aquifer

or

$$k = \frac{2.303\, q \log_{10}(r_2 / r_1)}{2\pi H_c (h_2 - h_1)} \tag{5.13}$$

Substituting $h_2 = H_0$, $r_1 = r_0$, $h_2 = H$, and $r_2 = R$,

$$k = \frac{2.303\, q \log_{10}(R / r_0)}{2\pi H_c (H - H_0)} \tag{5.14}$$

Equations 5.13 and 5.14 are valid as long as $H_0 > H_c$.

## 5.7 PERMEABILITY OF STRATIFIED SOILS

In nature, soils are usually deposited in successive layers, and the permeabilities of the layers may differ. It is not justifiable to find the numerical average of the coefficient of permeabilities of different layers. The stratifications can be considered as horizontal and continuous, and the effective or overall coefficient of permeability for flow in horizontal and vertical directions can be estimated.

### 5.7.1 Horizontal Flow

Consider the soil profile, shown in Fig. 5.7, consisting of two layers with isotropic permeabilities $k_1$ and $k_2$ in the first and second layers, respectively.

Assume the total head along the line AB to be constant. Similarly, the total head along CD may also be taken as constant, but the value will be less than that along AB. Thus, the overall gradient ($i$) and the individual gradients in each layer ($i_1$ and $i_2$) are equal, that is,

$$i = i_1 = i_2 \tag{5.15}$$

Let $q_1$ and $q_2$ be the rates of flow through unit thickness of the stratum. Let $q$ be the total rate of flow and $k_H$ be the effective coefficient of permeability $i$ in the horizontal direction. Thus,

$$q = q_1 + q_2$$

or

$$k_H i A = k_1 i_1 A_1 + k_2 i_2 A_2$$



Fig. 5.7 Flow parallel to stratification

or

$$k_H A = k_1 A_1 + k_2 A_2 \qquad \text{(Since } i = i_1 = i_2\text{)}$$

or

$$k_H (H_t \times 1) = k_1 (H_1 \times 1) + k_2 (H_2 \times 1)$$

or

$$k_H = \frac{k_1 H_1 + k_2 H_2}{H_t} \tag{5.16}$$

or

$$k_H H_t = k_1 H_1 + k_2 H_2 \tag{5.17}$$

If there are $m$ layers, then

$$k_H = \frac{\sum_{j=1}^{m} k_j H_j}{\sum_{j=1}^{m} H_j} \tag{5.18}$$

or

$$k_H \sum_{j=1}^{m} H_j = \sum_{j=1}^{m} k_j H_j \tag{5.19}$$

### 5.7.2 Vertical Flow

In this condition, the flow takes place in the direction perpendicular to the stratification (Fig. 5.8). To satisfy the continuity condition,

$$q = q_1 = q_2$$

Let $h_1$ and $h_2$ be the loss in heads in the first and second layers, respectively. Let $k_V$ be the effective coefficient of permeability in the vertical direction. Now,

$$i_1 = \frac{h_1}{H_1}, \quad i_2 = \frac{h_2}{H_2}, \quad \text{and } i = \frac{h}{H_t}$$

$$k_v i A = k_1 i_1 A = k_2 i_2 A$$

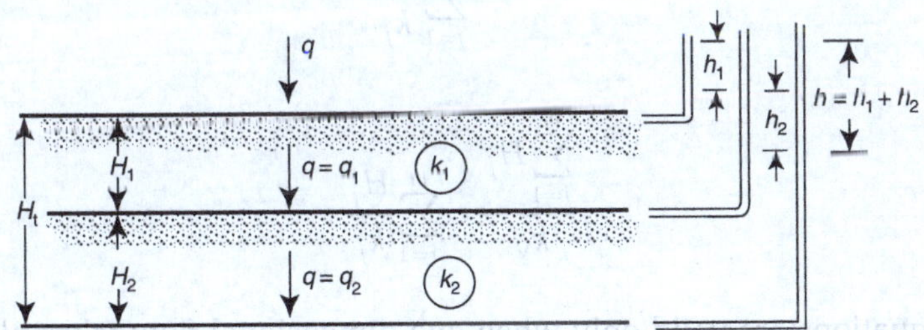


Fig. 5.8 Flow perpendicular to stratification

or

$$k_V \frac{h}{H_t} = k_1 i_1$$

or

$$k_V \left( \frac{h_1 + h_2}{H_t} \right) = k_1 i_1$$

or

$$k_V \left( \frac{i_1 H_1 + i_2 H_2}{H_t} \right) = k_1 i_1$$

or

$$k_V = \frac{H_t k_1 i_1}{i_1 H_1 + i_2 H_2}$$

Dividing by $k_1 i_1$,

$$k_V = \frac{H_t}{i_1 H_1 / k_1 i_1 + i_2 H_2 / k_2 i_2} \quad \text{(Since } k_1 i_1 = k_2 i_2\text{)}$$

or

$$k_V = \frac{H_t}{H_1 / k_1 + H_2 / k_2} \tag{5.20}$$

or

$$\frac{H_t}{k_V} = \frac{H_1}{k_1} + \frac{H_2}{k_2} \tag{5.21}$$

For $m$ layers,

$$k_V = \frac{\sum_{j=1}^{m} H_j}{\sum_{j=1}^{m} \frac{H_j}{k_j}} \tag{5.22}$$

or

$$\frac{\sum_{j=1}^{m} H_j}{k_V} = \sum_{j=1}^{m} \frac{H_j}{k_j} \tag{5.23}$$

The above equations are valid only when one-dimensional flow takes place in the horizontal or vertical direction.

## 5.8 VALUES OF PERMEABILITY

Table 5.2 represents a classification of soil based on permeability values (Terzaghi and Peck, 1967).

Typical permeability values for different soils are as follows:

| | |
|---|---|
| Uniform sand to fine sand | $0.4 \times 10^{-3}$ to $10^{-7}$ m/s |
| Silty sand to fine silt | $10^{-7}$ to $0.5 \times 10^{-7}$ m/s |
| Clay | $0.5 \times 10^{-7}$ to $0.01 \times 10^{-7}$ m/s |
| Colloidal clay | $\leq 10^{-9}$ m/s |

**Table 5.2** Permeability values

| Degree of permeability | k (m/s) |
|---|---|
| High | $>10^{-3}$ |
| Medium | $10^{-3}$ to $10^{-5}$ |
| Low | $10^{-5}$ to $10^{-7}$ |
| Very low | $10^{-7}$ to $10^{-9}$ |
| Practically impermeable | $<10^{-9}$ |

*Source:* Terzaghi and Peck (1967).

## 5.9 FACTORS AFFECTING PERMEABILITY

The coefficient of permeability of a soil depends basically on the characteristics of both the soil medium and the pore fluid. Lambe and Whitman (1979) have grouped particle size, void ratio, composition, fabric, and degree of saturation as major soil characteristics, and viscosity, unit weight, and polarity as major pore fluid characteristics. For a civil engineer dealing with soils, the permeant is water, whose variation in property may be presumed to be very less. Thus, soil characteristics may have to be given more importance.

### 5.9.1 Soil Characteristics

Based on Poiseuille's law for flow through a bundle of capillary tubes, Taylor (1948) has given a theoretical expression for flow through soil medium as

$$k = D_s^2 \frac{\gamma_w}{\eta_w} \frac{e^3}{1+e} C_s \tag{5.24}$$

where $D_s$ is the effective particle diameter and $C_s$ the composite shape factor.

**Particle Size.** Equation 5.24 considers only particle size and void ratio among the soil characteristics. This shows that permeability may be empirically related to the square of some representative particle diameter. Such estimations may be true only in coarse-grained soils, like silts and sands. Hazen (1911) proposed an expression for $k$ for filter sands as

$$k = C D_{10}^2 (\text{mm/s}) \tag{5.25}$$

**Table 5.3** Values for Hazen's coefficient

| Sand (one or more applies) | $C_s\left(\frac{1}{\text{mm}}\right)$ |
|---|---|
| (i) Very fine, well graded or with appreciable fines (≤75-μm size) | 4–8 |
| (ii) Medium, coarse, very poorly graded, clean, coarse but well graded | 8–12 |
| (iii) Very coarse, very poorly graded, gravelly, clean | 12–15 |

*Source:* Bowles (1984).

where $D_{10}$ is the effective size (mm) and $C$ the experimental coefficient dependent on the nature of the soil. Values of Hazen's coefficient are given in Table 5.3.

Several correlations have been reported in literature utilizing some characteristic grain size.

**Void Ratio**. Different linear relationships have been attempted, relating void ratio and permeability, viz., $k \propto e^3(1 + e)$, $k \propto e^2$, and $\log k \propto e$. These relationships have been found to indicate a straight-line relationship in non-cohesive soils but not in fine-grained soils. But $e$ versus $\log k$ is always close to a straight line for nearly all soils (Lambe and Whitman, 1979; Taylor, 1948).

**Composition**. The effect of soil composition is more predominant in clayey soils than in silts and sands. Depending on the type of clay mineral and the exchangeable cation present in the clay, the permeability varies from $10^{-6}$ to $10^{-10}$ m/s, and accordingly the variation of void ratio is from 16 to 1. The effect of exchangeable ion on permeability is less for low ion exchange capacity of a soil. An increase in the thickness of the diffuse double layer (effect by cation exchange capacity and cation valency) decreases its permeability, as the pore path is reduced by the thickness of water held onto the soil particles.

**Fabric**. The permeability of a soil deposit is greatly influenced by the in-place micro- and macrostructure. Clays are very significantly affected by the fabric component of a structure. At similar void ratios in a clay, the permeability has been shown to be many times greater in a flocculated state than in a dispersed state (Lambe, 1955). The above discussion is confined to microstructural changes. The significance of macrostructure is extremely important, e.g., the effect of stratification. Variation in the permeabilities of layered soils contributes more to the effective coefficient of permeability parallel to stratification than to the coefficient of permeability perpendicular to stratification.

### 5.9.2 Pore Fluid Characteristics

**Pore Fluid**. Water may be considered as the pore fluid generally met with, and accordingly, the discussion may be confined to water as the pore fluid. Equation 5.24 indicates that permeability is directly proportional to unit weight $\gamma_w$ and inversely proportional to viscosity $\eta_w$. Of these two parameters, viscosity is considerably affected by change in temperature. A viscosity decrease caused by increase in temperature results in high permeability. A correction for temperature has to be effected, and $k$ at 27°C has to be reported as per Indian Standards (IS: 2720 – Part 17, 1986), i.e.,

$$k_{27^\circ\text{C}} = k_T \frac{\eta_{wT}}{\eta_{w\,27^\circ\text{C}}} \tag{5.26}$$

The variables of water, i.e., $\gamma_w$ and $\eta_w$, may be eliminated by defining another permeability term as *specific* or *absolute permeability*, i.e.,

$$K = \frac{k\eta_w}{\gamma_w} \tag{5.27}$$

and $K$ will have units of length square. Michaels and Lin (1954) conducted permeability studies on kaolinite with permeants of different polarity and observed marked variation in permeability. Thus, apart from viscosity and unit weight of the permeant, a factor representing polarity should be included in Eq. 5.24 (Lambe and Whitman, 1979).

**Degree of Saturation**. The degree of saturation has an important bearing on permeability. In general, the higher the degree of saturation, the higher is the permeability. As the degree of saturation increases, there is an increase in the flow channels for water and, hence, high permeability.

## 5.10 SURFACE TENSION

If the water in soil pores is interconnected and subjected only to gravity, the soil mass above the water table would be dry. But in nature, the soil pores are saturated, fully above the water table and partially at large distances from it. The water in the soil voids located above the water table is referred to as soil moisture. The phenomenon of water rise in the pores of soils above the water table against the gravitational pull is called the capillary rise.

The principle of capillary rise of water is basically related to the rise of water in a glass capillary tube. Water, like any other liquid, behaves as though the surface is tightly stretched due to the intermolecular attraction of forces. This phenomenon is termed *surface tension*. When a clean capillary tube is brought in contact with a source of water, the water rises up in the tube and remains there. The rise is attributed to the combination of surface tension with the attraction between the glass and water molecules. The shape of air–water is concave in the downward, and the curved liquid surface is termed the *meniscus*. The rise of water in the tube reaches a maximum height ($h_c$) when the equlibrium condition is reached (Fig. 5.9). At this stage, the downward pull of gravity on the capillary column of water is balanced by the surface of water due to surface tension ($T_s$) effects.

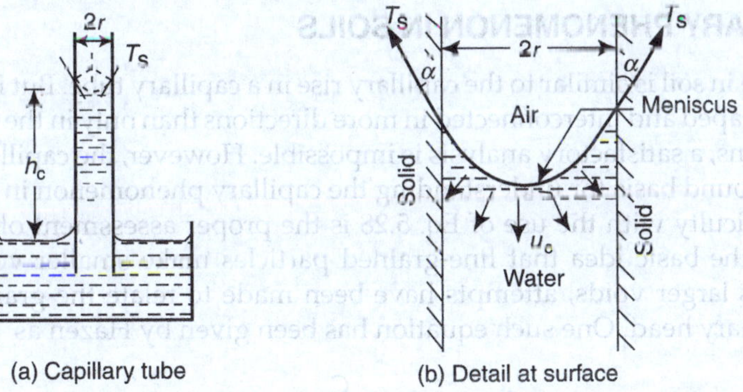


Fig. 5.9 Rise of water in a capillary tube

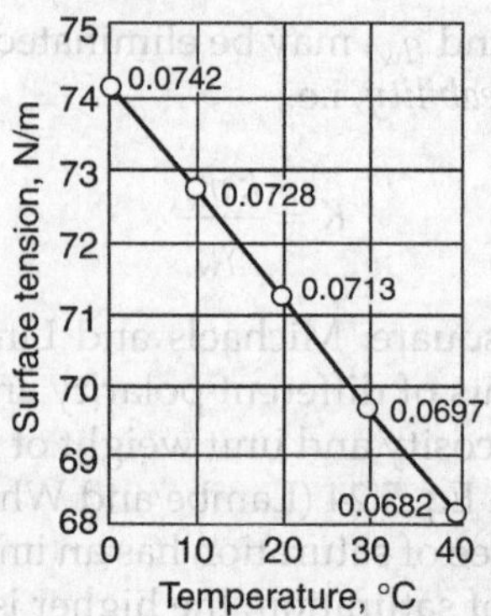


**Fig. 5.10** Surface tension of water as a function of temperature

Upward force due to surface tension acting around the periphery = $(2\pi r) \times T_s \cos \alpha$
Downward pull due to gravity on the column of water = $(\pi r^2) \times h_c \times \gamma_w$

where $r$ is the radius of the capillary tube (m), $T_s$ the surface tension of water (N/m), $\alpha$ the contact angle (pore water makes zero contact angle with glass), and $h_c$ the capillary rise (m).

$$2\pi r T_s \cos \alpha = \pi r^2 h_c \gamma_w$$

$$h_c = \frac{2T_s}{r\gamma_w} = \frac{4T_s}{d\gamma_w} \tag{5.28}$$

where $d$ is the diameter of the capillary tube.

The surface tension $T_s$ is temperature dependent (Fig. 5.10). The height of the capillary rise is not affected by the variations in the shape and size of tubes at levels below the meniscus. The capillary pore water pressure (negative) is given as

$$u_c = h_c\, r_w \tag{5.29}$$

This is a measure of the suction exerted on the pore water by the soil.

## 5.11 CAPILLARY PHENOMENON IN SOILS

The capillary rise in soil is similar to the capillary rise in a capillary tube. But in soils, the pores are irregularly shaped and interconnected in more directions than only in the vertical. Because of these limitations, a satisfactory analysis is impossible. However, the capillary tube concept does serve as a sound basis for understanding the capillary phenomenon in soils.

The main difficulty with the use of Eq. 5.28 is the proper assessment of the diameter of pores. Keeping the basic idea that fine-grained particles make smaller voids and coarse-grained particles larger voids, attempts have been made to relate the grain size and void ratio to the capillary head. One such equation has been given by Hazen as

$$h_{cr} = \frac{C}{e\,D_{10}} \tag{5.30}$$

where $C = 0.01 \times 10^{-3}$ to $0.05 \times 10^{-3}$ (m$^2$) and $h_{cr}$ = maximum capillary rise (m).

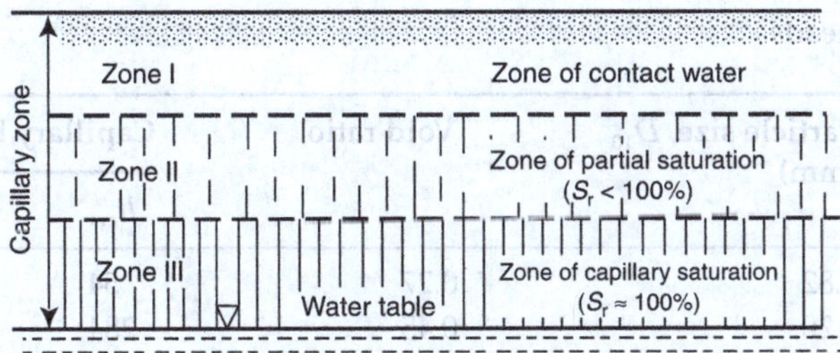


**Fig. 5.11** Capillary zones

This expression shows that decrease in the effective grain size causes a decrease in the void ratio and an increase in the capillary rise in soils.

Equation 5.28 indicates that the effect of molecular attraction will be near the water table, and hence, irrespective of the void space, all the pores will be filled. At a distance from the water table, only smaller voids would be filled with water. Thus, the capillary zone may be divided into three zones of arbitrary boundaries. The zone just above the water table is called the *zone of capillary saturation* (with almost 100% saturation). Above the zone of capillary saturation is the *zone of partial saturation* wherein only small pores are filled with water and the large ones with air; evidently, here the degree of saturation is less than 100%. The third zone near the ground surface contains water surrounding the particles at contact points, but there is no continuity. This zone is referred to as the *zone of contact water* (Fig. 5.11). Water may also reach this zone from the ground surface by percolation and may be held in suspension by the capillary forces.

Capillarity of a dry soil is its capacity to draw up water to elevations above the phreatic line and also to retain the water above the phreatic line in a draining soil. The height of water that a soil can support is generally called the *capillary head*. So far, only the first aspect is considered.

Figure 5.12a represents a column of cohesionless soil. The maximum capillary rise, $h_{cr}$, and the minimum capillary head, $h_{cn}$ (for the maximum degree of saturation) are identified and represented in Fig. 5.12b.

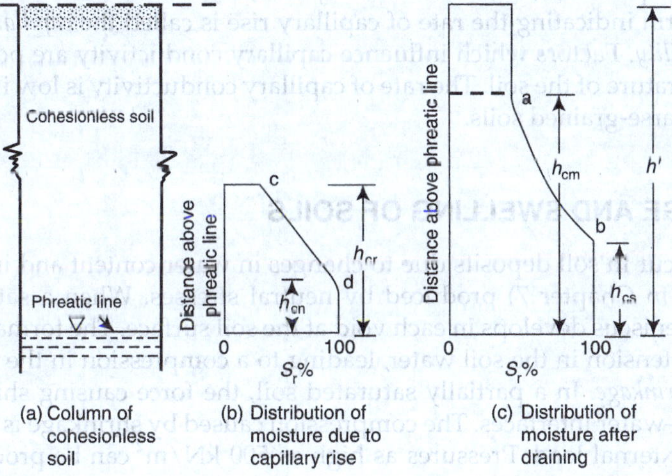


**Fig. 5.12** Capillary heads in soil

**Table 5.4** Capillary heads

| Soil | Particle size, $D_{10}$ (mm) | Void ratio | Capillary head (mm) | |
|---|---|---|---|---|
| | | | $h_{cr}$ | $h_{cs}$ |
| Coarse gravel | 0.82 | 0.27 | 54 | 60 |
| Sandy gravel | 0.20 | 0.45 | 284 | 200 |
| Fine gravel | 0.30 | 0.29 | 195 | 200 |
| Silty gravel | 0.06 | 0.45 | 1,060 | 680 |
| Coarse sand | 0.11 | 0.27 | 820 | 600 |
| Medium sand | 0.02 | 0.48–0.66 | 2,396 | 1,200 |
| Fine sand | 0.03 | 0.36 | 1,655 | 1,120 |
| Silt | 0.006 | 0.95–0.93 | 3,592 | 1,800 |

*Source:* Lane and Washburn (1946).

Let us consider a situation where the same soil has been saturated up to a height $h$ above the water table and allowed to drain. Then, the moisture will be as shown in Fig. 5.12c. Point "a" represents the highest elevation up to which there exists a continuous water path above the phreatic line. This distance is referred to as the *maximum capillary head*, $h_{cm}$. Point "b" shows the point up to which the soil is fully saturated, and this height is called the *saturation capillary head*, $h_{cs}$.

Thus, any capillary head associated with drainage has a maximum value of $h_{cm}$, and that with capillary rise has a maximum value of $h_{cr}$. There is more possiblity of bridging effect of surface water on large voids during draining than the pulling effect during rising. Accordingly, it is reasonable to expect that $h_{cs} > h_{cn}$ and $h_{cm} > h_{cr}$. Many capillary heads might exist between the two extremes $h_{cm}$ and $h_{cn}$. Table 5.4 shows the range of capillary heads for cohesive soils (Lane and Washburn, 1946).

In certain practical problems, the time necessary for the attainment of maximum capillary rise is more. The term indicating the rate of capillary rise is called the *capillary conductivity* or *capillary permeability*. Factors which influence capillary conductivity are pore size, water content, and temperature of the soil. The rate of capillary conductivity is low in fine-grained soils and high in coarse-grained soils.

## 5.12 SHRINKAGE AND SWELLING OF SOILS

Volume changes occur in soil deposits due to changes in water content and in the effective stresses (discussed in Chapter 7) produced by neutral stresses. When a saturated soil is allowed to dry, a meniscus develops in each void at the soil surface. The formation of such a meniscus brings in tension in the soil water, leading to a compression in the soil structure, which is termed *shrinkage*. In a partially saturated soil, the force causing shrinkage arises from the curved air–water interfaces. The compression caused by shrinkage is as effective as that produced by external load. Pressures as high as 500 kN/m$^2$ can be produced in fine-grained soils due to shrinkage.

**Table 5.5** Volume change potential

| Volume change potential | Plasticity index (%) | | Shrinkage limit (%) |
|---|---|---|---|
| | Arid area | Humid area | |
| Low | 0–15 | 0–30 | >12 |
| Moderate | 15–30 | 30–50 | 10–12 |
| High | >30 | >50 | <10 |

*Source:* Holtz and Gibbs (1956).

The degree of shrinking depends on the initial water content, the type and amount of clay content, and the mode and environment of geological deposition. The presence of sand and silt-size particles in a clay deposit reduces the total shrinkage. Shrinkage occurs horizontally as well as vertically, causing vertical shrinkage cracks. In highly compressible clays, the cracks may be as high as 0.5 m wide and 5 m deep.

Some soils (expansive soils like the black cotton soils of India) not only shrink on drying but also swell when they come in contact with water. Swelling is caused mainly by repulsive forces which separate the clay particles causing volume increase. The mechanism causing swelling is more complex, and the factors which contribute to it are

1. the elastic rebound of soil grains,
2. the clay mineral's affinity for water,
3. the cation exchange capacity and electrical repulsive forces, and
4. the expansion of entrapped air.

Free swell of a soil is defined as the increase in the volume of a soil, without any external constraints, on submergence in water (IS: 2720, 1977). In general, the free swell ceases when the moisture reaches the plastic limit, i.e., at a water plasticity ratio of about 0.25.

Clay soils with high montmorillonite clay mineral show an almost reversible swelling and shrinking on rewetting and redrying. Clay soils with illite or kaolinite show a large initial volume decrease on drying, with only a limited swelling on rewetting. Holtz and Gibbs (1956) have given an indication of the potential volume change based on shrinkage limit and plasticity index (Table 5.5).

Volume change is a serious problem in shrinkage-susceptible soils present in arid or semi-arid areas.

## WORKED EXAMPLES

**Example 5.1** A sand sample of 25 cm length was subjected to a constant head permeability in a permeameter having an area of 30 m$^2$. A discharge of 100 cm$^3$ was obtained in a period of 1 minute under a head of 39 cm. Mass of dry sand in the sample was 1,350 g. The specific gravity of the sand particles was 2.67. Determine (i) the coefficient of permeability, (ii) the superficial velocity, and (iii) the seepage velocity.

***Solution***

From Eq. 5.7,

(i) $$k=\frac{qL}{Ah}=\left(\frac{100}{60}\right)\frac{25}{30\times39}=0.0356\text{ cm/s.}$$

Dry density $\rho_d = G\,\rho_w/(1+e)$. That is,

$$e=\frac{G\rho_w}{\rho_d}-1$$

$$=\frac{2.67\times1}{1350/(30\times25)}-1$$

$$=1.483-1$$

$$=0.483$$

$$\text{Porosity } n=\frac{e}{1+e}=\frac{0.483}{1+0.483}=\frac{0.483}{1.483}=0.326.$$

(ii) Superficial velocity $v_s=\dfrac{q}{A}=\dfrac{100}{60\times30}=0.056\text{ cm/s.}$

(iii) Seepage velocity $v=v_s n=0.056\times0.326$

$$=0.0183\text{ cm/s.}$$

**Example 5.2** A soil sample has a length of 3.5 m and a cross-sectional area of 2 $m^2$. If water flows through such a soil sample and the fluid energy lost is 1,650 N-m for every cubic metre flow of water, estimate Darcy's velocity and permeability. The time of flow for 1 $m^3$ of water is 26 hours. Find also the seepage velocity if the void ratio of the sample is 0.58.

***Solution***

$$q=\frac{Q}{t}=\frac{1}{26\times60\times60}=1.068\times10^{-5}\text{ m}^3\text{/s.}$$

$$\text{Darcy's velocity } v=\frac{q}{A}=\frac{1.068}{2}\times10^{-5}=5.34\times10^{-6}\text{ m/s.}$$

$$\text{Loss of head } h=\frac{\text{Total energy loss}}{\text{Energy loss per unit length}}$$

$$=\frac{\text{Total energy loss}}{\text{Volume}\times\text{Unit weight of water}}$$

$$=\frac{1650}{1\times9810}=0.168\text{ m}$$

Now,

$$i=\frac{h}{L}=\frac{0.168}{3.5}=0.048$$

$$k = \frac{v}{i} = \frac{5.34}{0.048} \times 10^{-6} = 1.11 \times 10^{-4} \text{ m/s}$$

$$\text{Seepage velocity } v_s = \frac{v(1+e)}{e} = 5.34 \times 10^{-6} \frac{(1+0.58)}{0.58}$$

$$= 1.46 \times 10^{-5} \text{ m/s.}$$

**Example 5.3** In a falling head permeability test, the time taken for the head to fall from $h_1$ to $h_2$ is $t$. If the test is repeated with the same initial head, what would be the final head in a time interval of t/2?

***Solution***

Let $h'$ be the head after the lapse of time $t/2$.

Now, the time taken for the head to fall from $h_1$ to $h'$ and from $h'$ to $h_2$ is the same. Thus, for a given soil sample,

$$k_1 = \left(\frac{2.303\,aL}{At/2}\right) \log_{10} \frac{h_1}{h'}$$

Again,

$$k_2 = \left(\frac{2.303\,aL}{At/2}\right) \log_{10} \frac{h'}{h_2}$$

As the test is performed on the same sample on the same permeameter, $k_1 = k_2$. Therefore,

$$\log_{10} \frac{h_1}{h'} = \log_{10} \frac{h'}{h_2}$$

That is,

$$(h')^2 = h_1 h_2$$

$$h' = \sqrt{h_1 / h_2}$$

**Example 5.4** In a falling head permeability test, the head causing flow was initially 500 mm, and it drops 20 mm in 5 minutes. How much time is required for the head to fall to 250 mm?

***Solution***

From Eq. 5.8, the coefficient of permeability

$$k = 2.303 \frac{aL}{At} \log_{10} \frac{h_1}{h_2}$$

Therefore,

$$t = \frac{2.303\,aL}{kA} \log_{10} \frac{h_1}{h_2}$$

Therefore, the values in the bracket are constant for a given permeameter and the soil at a particular void ratio.

Hence,

$$t_1 = (c)\left[\log_{10}\frac{h_1}{h_2}\right]_1 = (c)\left[\log_{10}\frac{500}{20}\right]$$

$$t_2 = (c)\left[\log_{10}\frac{h_1}{h_2}\right]_2 = (c)\left[\log_{10}\frac{500}{250}\right]$$

$$\frac{t_1}{t_2} = \left[\log_{10}\frac{500}{20}\right] \Big/ \left[\log_{10}\frac{500}{250}\right]$$

$$= 5\,[\log_{10}{}^{2}]/[\log_{10}{}^{25}]$$

$$= 5\times\frac{0.301}{0.699}$$

$$= 2.15 \text{ minutes}$$

**Example 5.5** A falling head permeameter contains a soil sample 8 cm high and 60 $cm^3$ in cross-sectional area. The permeability of the sample is expected to be $1 \times 10^{-4}$ cm/s. If it is desired that the head in the standpipe should fall from 30 to 10 cm in 40 minutes, determine the size of the standpipe which should be used.

***Solution***

From Eq. 5.8,

$$k = 2.303\frac{aL}{At}\log_{10}\frac{h_1}{h_2}$$

Rearranging,

$$a = \frac{kAt}{2.303\times L\log_{10}h_1/h_2}$$

$$a = \frac{1\times10^{-4}\times60\times40\times60}{2.303\times8\times\log_{10}30/10}$$

$$= \frac{1\times10^{-4}\times60\times40\times60}{2.303\times8\times0.477}$$

$$= 1.64\text{ cm}^2$$

**Example 5.6** A permeameter of 80 mm diameter with a sample length of 300 mm has been used for constant head and falling head tests. While conducting a constant head test, the loss of head was 1,150 mm for a length of 250 mm and the rate of flow was 2,700 $mm^3$/s. Find the coefficient of permeability (in mm/s).

If a falling head test were then performed on the same sample at the same void ratio, find the time taken for the head to fall from 900 to 450 mm. The diameter of the standpipe in the falling head test was 25 mm.

***Solution***

Constant head test:

$$\text{Gradient } i = \frac{h}{L} = \frac{1150}{250} = 4.6$$

$$\text{Area of sample } A = \pi\left(\frac{80^2}{4}\right) = 5026.6\,\text{mm}^2$$

Rearranging Eq. 5.4,

$$k = \frac{q}{iA} = \frac{2700}{4.6 \times 5026.6} = 0.117\,\text{m/s}.$$

Falling head test:

$$\text{Area of standpipe } a = \pi\left(\frac{25^2}{4}\right) = 490.9\ \text{mm}^2$$

Rearranging Eq. 5.8,

$$t = \frac{2.303\,aL}{kA}\log_{10}\frac{h_1}{h_2}$$

$$t = \frac{2.303 \times 490.9 \times 300}{5026.6 \times 0.117}\log_{10}\frac{900}{450}$$

$$= 173.6 \text{ seconds}$$

**Example 5.7** Prove that for stratified deposits of soils, the average permeability in the horizontal direction is greater than the average permeability in the vertical direction.

***Solution***

Consider two layers of thicknesses $H_1$ and $H_2$ with coefficients of permeability $k_1$ and $k_2$ in the first and second layers, respectively.

From, Eqs. 5.16 and 5.20,

$$k_{\text{H}} = \frac{k_1 H_1 + k_2 H_2}{H_{\text{t}}}$$

and

$$k_{\text{V}} = \frac{H_{\text{t}}}{(H_1/k_1) + (H_2/k_2)}$$

Therefore,

$$\frac{k_{\text{H}}}{k_{\text{V}}} = \frac{(k_1 H_1 + k_2 H_2)/H_{\text{t}}}{H_{\text{t}}/[(H_1/k_1) + (H_2/k_2)]}$$

$$= \frac{H_1^2 + H_2^2 + [(k_1^2 + k_2^2)/k_1k_2]H_1H_2}{H_t^2}$$

$$= \frac{H_1^2 + H_2^2 + [(k_1^2 + k_2^2)/k_1k_2]H_1H_2}{H_1^2 + H_2^2 + 2H_1H_2}$$

Since $k_1$ and $k_2$ are always positive,

$$(k_1 - k_2)^2 \geq 0$$

or

$$k_1^2 + k_2^2 \geq 2k_1k_2$$

or

$$\frac{k_1^2 + k_2^2}{k_1k_2} \geq 2$$

Only when $k_1 = k_2$ the ratio is equal to 2, otherwise it is greater than 2. Here, $k_1 \neq k_2$. Therefore, in $k_H/k_V$, the numerator is greater than the denominator.

Therefore, $$k_H > k_V$$

Hence, the average coefficient of permeability in the horizontal direction is greater than the average coefficient of permeability in the vertical direction.

**Example 5.8** A horizontal stratified soil deposit consists of three layers, each uniform in itself. The permeabilities of the layers are $4 \times 10^{-4}$, $25 \times 10^{-4}$ and $7.5 \times 10^{-4}$ mm/s; their thicknesses are 6, 3, and 12 m, respectively. Find the effective average permeability of the deposits in the horizontal and vertical directions.

***Solution***

From Eq. 5.19,

$$k_H = \frac{\sum_{j=1}^{m} k_j H_j}{\sum_{j=1}^{m} H_j}$$

$$k_H = \frac{(4\times6 + 25\times3 + 7.5\times12)1000\times10^{-4}}{(6+3+12)1000}$$

$$= \frac{189\times10^{-4}}{21} = 9\times10^{-4} \text{ mm/s}$$

From Eq. 5.20,

$$k_V = \frac{\sum_{j=1}^{m} k_j H_j}{\sum_{j=1}^{m} H_j / k_j}$$

$$k_V = \frac{(6+3+12)1000 \times 10^{-4}}{[(6/4)+(3/25)+(12/7.5)] \times 1000} = \frac{21}{3.22}$$

$$k_V = 6.5 \times 10^{-4} \text{ mm/s}$$

**Example 5.9** In a falling head permeameter, the sample was 18 cm long with cross-sectional area of 22 cm$^2$. Calculate the time required for the drop of head from 25 to 10 cm if the cross-sectional area of the standpipe was 2 cm$^2$. The sample of soil was heterogeneous, with a coefficient of permeability of $3 \times 10^{-4}$ cm/s for the first 6 cm, $4 \times 10^{-4}$ cm/s for the second 6 cm, and $6 \times 10^{-4}$ cm/s for the last 6 cm of thickness. Assume the flow taking place perpendicular to the bedding planes.

***Solution***

From Eq. 5.20, the effective vertical coefficient of permeability

$$k_V = \frac{\sum_{j=1}^{m} k_j H_j}{\sum_{j=1}^{m} H_j / k_j}$$

$$k_V = \frac{(6+6+6)}{[(6/3)+(6/4)+(6/6)]} \times 10^{-4} = \frac{18}{4.5} \times 10^{-4}$$

$$k_V = 4 \times 10^{-4} \text{ cm/s}$$

Rearranging Eq. 5.8,

$$t = \frac{2.303\, a L}{kA} \log_{10} \frac{h_1}{h_2}$$

$$t = \frac{2.303 \times 2 \times 18}{4 \times 10^{-4} \times 22} \times \frac{1}{60} \log_{10} \frac{25}{10}$$

$$t = \frac{829080}{5280} \times 0.398 = 62.5 \text{ minutes}$$

**Example 5.10** In a falling head permeability test on a soil of length $l_1$, the head of water in the stand pipe takes 5 seconds to fall from 900 to 135 mm above the tail water level. When another soil of length 60 mm is placed on the first soil, the time taken for the head to fall between the same limits is 150 seconds. The permeameter has a cross-sectional area of 4,560 mm$^2$ and a standpipe area of 130 mm$^2$. Calculate the permeability of the second soil.

***Solution***

Refer to Fig. 5.13 for a two-layer system.
From Eq. 5.21, we have

$$\frac{l}{k_V} = \frac{l_1}{k_1} + \frac{l_2}{k_2}$$

From the permeability test on Sample 1 only, we have

$$k_1 = \frac{2.303\,al_1}{At_1}\log_{10}\frac{h_1}{h_2}$$

or

$$\frac{l_1}{k_1} = \frac{At_1}{2.303\,a\log_{10}(h_1/h_2)} = \frac{4560\times 50}{2.303\times 130\times \log_{10}(900/135)}$$

or

$$\frac{l_1}{k_1} = 924.31 \text{ seconds}$$

For permeability tests on both the soils,

$$k_V = \frac{2.303\,al}{At_2}\log_{10}\frac{h_1}{h_2}$$

or

$$\frac{l}{k_V} = \frac{At_2}{2.303\,a\log_{10}(h_1/h_2)} = \frac{4560\times 150}{2.303\times 130\times \log_{10}(900/135)}$$

$$= 2772.93 \text{ seconds}$$

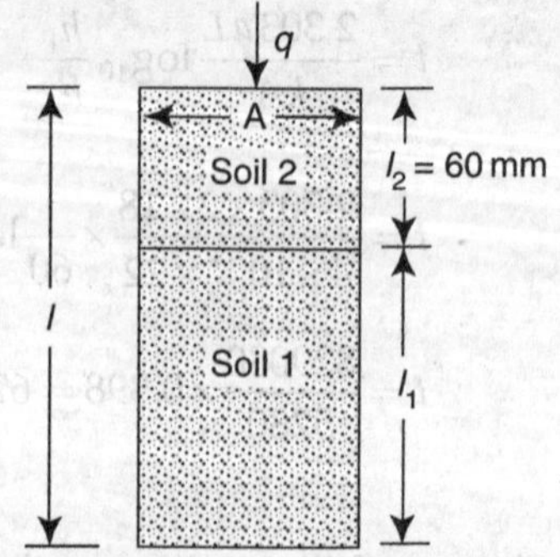


Fig. 5.13

Therefore,

$$\frac{l_2}{k_2} = \frac{l}{k_V} - \frac{l_1}{k_1} = 2772.93 - 924.31$$

or

$$\frac{l_2}{k_2} = 1848.62 \text{ seconds}$$

Therefore,

$$k_2 = \frac{l_2}{1848.62} = \frac{60}{1848.62} = 0.0325 \text{ mm/s}$$

**Example 5.11** A sand deposit of 12 m thickness overlies a clay layer. The water table is 3 m below the ground surface. In a field permeability pump out test, the water is pumped out at a rate of 540 litres/min when steady-state conditions are reached. Two observation wells are located at 18 m and 36 m from the centre of the test well. The depths of the draw-down curve are 1.8 m and 1.5 m, respectively, for these two wells. Determine the coefficient of permeability.

***Solution***

This is an unconfined aquifer, hence $k$ is given by Eq. 5.10. That is,

$$k = \frac{2.303 q \log_{10}(r_2 / r_1)}{\pi(h_2^2 - h_1^2)}$$

Here, $q = 540$ litres/min $= 540 \times 1000$ cm$^3$/min

$r_1 = 18$ m and $r_2 = 36$ m

$h_1 = 9 - 1.8 = 7.2$ m $= 720$ cm

$h_2 = 9 - 1.5 = 7.5$ m $= 750$ cm

Therefore,

$$k = \frac{2.303 \times 540 \times 1000(36/18)}{\pi[(750)^2 - (720)^2]}$$

$$k = \frac{2.303 \times 540 \times 1000 \times 2}{\pi \times 44100} = 17.96 \text{ cm/min}$$

**Example 5.12** A pumping test is conducted in an unconfined aquifer with a partially penetrated well. The following are the details:

Diameter of well = 1 m

Height of water level from the impermeable layer before pumping = 30 m

Depth of water in the well = 8 m

Depth of the bottom of the well from the impervious layer = 10 m

Radius of influence = 50 m

Discharge at steady state of pumping = 0.45 m$^3$/min

Find the coefficient of permeability of the soil.

***Solution***

Here, $R = 50$ m, $r = 0.5$ m, $s = 10$ m, $H_0 = 8$ m

$q = 0.45\ \text{m}^3/\text{min}$

From Eq. (5.12),

$$k = \frac{2.303q\log_{10}(R/r_0)}{\pi[(H-S)^2 - H_0^2]\left\{1+[0.3+(10r_c/H)]\sin\frac{1.8S}{H}\right\}}$$

Substituting the above values,

$$k = \frac{2.303\times 0.45\times \log_{10}(50/0.50)}{\pi[(30-10)^2 - 8^2]\{1+[0.3+(10\times 0.5/30)]\sin(8\times 10/30)\}}$$

Reducing,

$$k = 1.95\times 10^{-3}\ \text{m/min} = 0.0326\ \text{mm/s}$$

**Example 5.13** A glass capillary tube is 0.2 mm in diameter. What is the theoretical maximum height of capillary rise for a tube of this size? The surface tension is 0.0735 N/m. Estimate the pressure in the capillary water just under the meniscus.

***Solution***

From Eq. 5.28,

$$h_c = \frac{4Ts}{d\gamma_w} = \frac{4\times 0.0735}{(0.20/1000)\times 9.81\times 10^3}$$

$$h_c = 0.15\ \text{m}$$

From Eq. 5.29

$$u_c = h_c\gamma_w = 0.15\times 9.81 = 1.47\ \text{kN/m}^2$$

**Example 5.14** A silt deposit has a series of clay seams of about 5 mm thick at an average vertical spacing of about 1.65 m. Permeability of silt deposit is about 120 times as that of clay seam. Find the ratio of equivalent horizontal permeability to vertical permeability.

***Solution***

Naturally, alternate layers are found in silt and clay; therefore it is enough to find the ratio only for the two adjacent layers.

Height of silt layer, $H_1 = 1.65$ m.

Height of clay seam, $H_2 = 0.005$ m.

Permeability of silt, $k_1 = 120$ times $k_2$, where $k_2$ is permeability of clay.

Hence,

$$\text{Horizontal flow, } k_h = \frac{1.65k_1 + 0.005(k_2)}{1.655}$$

$$= 119.64\,k_2$$

$$\text{Vertical flow, } k_v = \frac{1.655}{\left(\frac{1.65}{k_1}\right)+\left(\frac{0.005}{k_2}\right)}$$

$$= \frac{1.655}{\frac{(0.0825+0.005)}{k_2}}$$

$$= 19.94\, k_2$$

$$\frac{k_h}{k_v} = \frac{119.64 k_2}{19.94 k_2} = 6.0.$$

## POINTS TO REMEMBER

5.1 Permeability of a soil is its capacity to permit water to pass through its inter-connected void spaces.

5.2 Potential or total head causing flow of water through soil consists of position or elevation head, pressure head due to water pressure, and velocity head. As seepage velocities are small in soil, the velocity head is ignored.

5.3 Darcy's law (velocity proportional to gradient) is valid as long as the flow is laminar, the soil is fully saturated, no volume change occurs during flow, and the continuity condition is present. Further, Darcy's law is valid for Reynold's number equal to unity.

5.4 Constant head permeameter is preferred for coarse-grained soils, whereas falling head permeameter is suitable for fine-grained soils.

5.5 In a layered soil, the average coefficient of permeability in the horizontal direction is greater than the average coefficient of permeability in the vertical direction.

5.6 Pumping out test is suitable for tests below the water table, whereas pumping in test can be conducted irrespective of the position of the water table for finding field k.

5.7 Factors affecting permeability are soil characteristics, viz., particle size, void ratio, composition, and fabric, and pore fluid characteristics, viz., pore fluid and degree of saturation.

5.8 Water table is the level of water in a soil stratum at which the pressure is atmospheric. Soil below the water table is called saturated zone ($s_r$ = 100%), while soil above is called capillary zone ($s_r \leq 100\%$)

5.9 In capillaries, the water surface is tightly stretched due to intermolecular attraction of forces, which is referred to as surface tension.

5.10 Capillary rise is the phenomenon of water rise in the pores of soils above the water table against the gravitational pull.

5.11 Shrinkage of soil depends on the initial water content, the type and amount of clay content, and the mode and environment of geological deposition. The presence of sand and silt-size particles in clays reduces the shrinkage.

5.12 Clayey soils with montmorillonite clay mineral show reversible swelling and shrinking, whereas kaolinite and illite show less swelling than shrinking.

## QUESTIONS

### *Objective Questions*

5.1 State whether the following are true or false:
  (i) Coefficient of permeability is greater for coarse-grained soils than for fine-grained soils.
  (ii) The constant head permeability test is the most reliable and accurate for clayey soils.
  (iii) Moisture rises above the groundwater table as a result of capillary tension.
  (iv) In sandy soils, the seepage velocity is equal to Darcy's flow velocity.
  (v) The capillary pressure in a soil may be more than 5 m head of water.

5.2 Coefficient of permeability of a fine-grained soil increases with
  (a) Increase in temperature of the pore fluid
  (b) Increase in viscosity of the pore fluid
  (c) Increase in density of soil
  (d) None of the above

5.3 Select the correct range of permeability (m/s) of a soil whose degree of permeability is low:
  (a) $10^{-8}$ to $10^{-9}$ (b) $10^{-3}$ to $10^{-5}$ (c) $10^{-5}$ to $10^{-7}$ (d) $10^{-1}$ to $10^{-3}$

5.4 Artesian conditions are said to exist when the piezometric surface lies
  (a) Below ground level
  (b) Between ground level and the aquifer
  (c) Above ground level
  (d) Below groundwater level

5.5 The pressure on a phreatic surface is
  (a) Less than atmospheric pressure
  (b) Equal to atmospheric pressure
  (c) Greater than atmospheric pressure
  (d) Not related to atmospheric pressure

5.6 Compacted well-graded, gravelly sands with little or no fines will be
  (a) Impervious (b) Semi-pervious to pervious
  (c) Semi-pervious (d) Pervious

5.7 Identify the wrong factor. The following three major characteristics influence permeability of clays:
  (a) Fabric (b) Composition
  (c) Degree of saturation (d) Particle shape

5.8 In a sedimentary soil deposit, the permeability
  (a) Is uniform in all directions
  (b) Is greater in the horizontal direction than in the vertical direction
  (c) Is lesser in the horizontal direction than in the vertical direction
  (d) Is double in the vertical direction of that in the horizontal direction

5.9 Which of the following pairs are correctly matched?
(1) Aquifer — Source for obtaining surface water
(2) Unconfined aquifer — Porous soil constitutes the surface stratum
(3) Confined aquifer — Porous soil bounded above and below aquitards
(4) Artesian aquifer — Confined aquifer with high pressure
Select the correct answer using the codes given below:
(a) 1, 2, and 3 (b) 2, 3, and 4 (c) 3, 4, and 1 (d) 4, 1, and 2

5.10 The mechanical effect of cohesion in sand due to contact moisture depends on the
(a) Shape of sand grains (b) Surface roughness of sand grains
(c) Relative density of sand (d) Gradation of sand

## Descriptive Questions

5.11 Two fluids with extreme viscosities are to be passed through a porous material. Reason out the condition for which the coefficient of permeability will be the greatest.

5.12 In fine-grained soils, what effects does the presence of adsorbed water have on the coefficient of permeability?

5.13 In what condition is the capillary system if the angle of wetting is zero?

5.14 Explain the ways by which the capillary water and the effect of capillarity can be removed from soil.

5.15 Capillary rise is greater for fine-grained soils than for coarse-grained soils. Substantiate this statement.

5.16 What are the applications of the capillary tube theory to soil engineering?

## EXERCISE PROBLEMS

5.1 A constant head permeability test was conducted on a sand sample of 250 mm length and 2,000 mm$^2$ area. The head loss was 500 mm, and the discharge was found to be 260 ml in 130 seconds. Determine the coefficient of permeability of the sand sample. Find the superficial and seepage velocities if the dry unit weight and specific gravity of the samples were 17.98 kN/m$^3$ and 2.62, respectively.

5.2 Three soil samples, x, y, z, with coefficients of permeability $1 \times 10^{-1}$, $2 \times 10^{-2}$ and $5 \times 10^{-3}$ m/s are placed in a tube of cross-section 100 mm $\times$ 100 mm, as shown in Fig. 5.14. Water is supplied through the apparatus such that the head difference is maintained at 300 mm. Find the rate of supply in litres per hour.

5.3 The changes caused by a rise in temperature in viscosity and unit weight of a pore fluid are 82.5% and 97.8%, respectively. Compute the percent change in the coefficient of permeability assuming other factors to remain constant.

5.4 A variable head permeability test is conducted on a 100 mm long specimen. The diameter of the standpipe is 1/10th that of the specimen. The test took 900 seconds to fall from a height of 300 to 100 mm. Determine the permeability of the specimen.

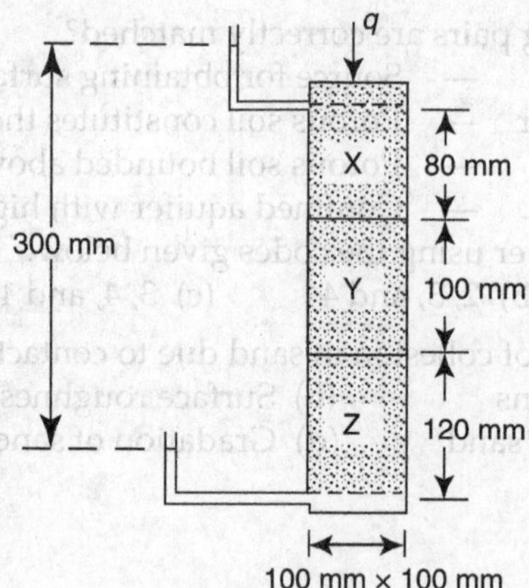


Fig. 5.14

5.5 A field pumping test was performed for a horizontal stratum of sandy soil 4 m thick, sandwiched between two impermeable strata. After the steady-state flow equilibrium, the rate of flow was 90 litres/hour. The elevation of water level in a borehole 3 m away from the test well was 2.1 m, and in a borehole 6 m away, it was 2.7 m above the top of the lower impermeable stratum. Estimate the coefficient of permeability of the soil.

5.6 In a falling head permeameter test, the initial head at $t = 0$ was 600 mm. The head dropped 30 mm in a time of 40 seconds. Find the time required to run the test to a final head of 200 mm.

5.7 A sand sample at a void ratio of 0.52 has a permeability of $0.4 \times 10^{-3}$ m/s. Assuming a reasonable relationship between void ratio and permeability for this soil, estimate the permeability at a void ratio of 0.65.

5.8 During a falling head permeability test, the sample on close investigation was found to be in two layers 60 and 40 mm thick. The routine falling head test on this sample yielded the following results: diameter of standpipe, 4 mm; sample diameter, 80 mm; length of sample 100 mm; initial head, 1,100 mm; final head, 420 mm, and time for fall in head, 6 minutes. After the test, independent tests were made on each soil; the permeabilities were found to be $5 \times 10^{-4}$ and $17 \times 10^{-4}$ mm/s, respectively. Check the average permeability through the sample in the laboratory test with the estimated value considering the layer effect. Also, estimate the average permeability in a direction at right angles to sampling. Comment on the results.

5.9 A graded filter has to be constructed with four soils of different layer thicknesses. The layer thicknesses are 350, 250, 200, and 100 mm and are to be placed at different compacted densities such that the permeabilities are $2 \times 10^{-2}, 3 \times 10^{-1}, 6.8 \times 10^{-1}$, and 1.5 mm/s, respectively. Calculate the average coefficients of permeability in directions parallel and orthogonal to the layers.

5.10 Three layers of soil represent the soil profile beneath a reservoir. The depth of water in the reservoir is 15 m and the area of spread is 4,500 $m^2$. The permeability and thickness of each layer are given below:

| **Layer** | **Thickness (m)** | ***k* (mm/s)** |
|---|---|---|
| 1 | 2.5 | $3.6 \times 10^{-5}$ |
| 2 | 1.5 | $2.8 \times 10^{-5}$ |
| 3 | 2.8 | $1.8 \times 10^{-5}$ |

A sand layer lies below this profile. The sand has horizontal drainage. Assuming vertical flow through the layers, compute the water loss in a period of 30 days from the reservoir.

5.11 The rise in a capillary tube is 520 mm above the free water surface. Determine the surface tension if the radius of the tube is 0.03 mm.

5.12 A sand sample has a porosity of 32.4%, and Hazen's effective grain size is 0.056 mm. Estimate the capillary rise in the soil sample.

# 6

# Seepage

## CHAPTER HIGHLIGHTS

Seepage forces – General flow equation – Significance of Laplace's equation – Properties and applications of flow nets – Construction of flow nets: boundary conditions, construction methods, flow nets for sheet piles and dams – Anisotropic soil conditions – Non-homogeneous soil conditions – Piping –Design of filters

## 6.1 INTRODUCTION

Groundwater is frequently encountered in construction projects. The movement of water through soil is referred to as seepage, and such movement leads to several groups of problems in civil engineering. Seepage of water has a bearing on three major types of problems, viz., (i) loss of stored water through an earth dam or foundation, (ii) instability of slopes and foundations of hydraulic structures due to the force exerted by the percolating water, and (iii) settlement of structures founded on or above compressible layers due to explusion of water from the voids caused by load applications.

Theoretical solutions based on simple assumptions in problems related to stability and settlement have been successful. But hydraulic problems do not lead to simple solutions because of adverse field conditions. This chapter discusses some of the techniques used for analysing seepage flow.

## 6.2 SEEPAGE FORCES

When water flows through soil, the water head is dissipated in viscous friction. During energy dissipation, a drag force is exerted on the soil particles in the direction of flow.

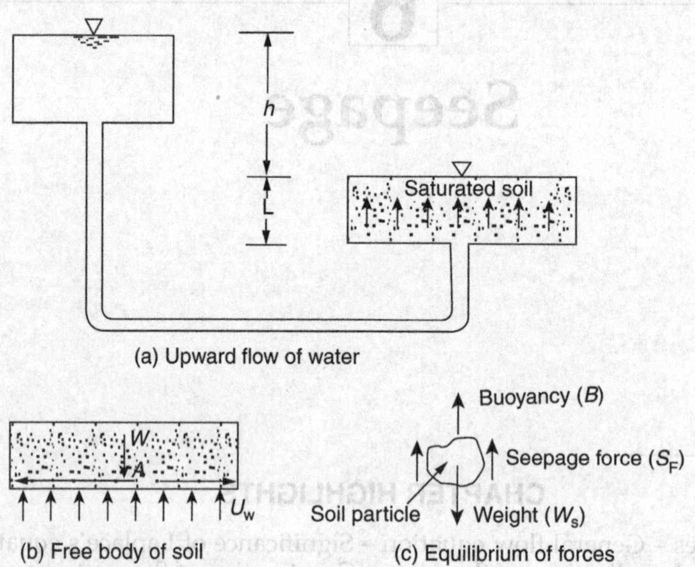


**Fig. 6.1** Flow showing seepage and other forces

Consider the column of soil shown in Fig. 6.1a. If the height $h$ of the water surface in the reservoir is raised, the water pressure at the bottom of the soil sample is increased and the drag force on the soil particle becomes greater. The drag force and the buoyant weight of the particles are in balance at a critical height $h = h_c$, and an increase in height will cause the soil particles to be washed out of the container. At this critical condition, the force acting on the bottom of the soil sample will just equal the weight of the soil and water mass in the container.

Now, consider the upward and the downward forces at the bottom of the soil mass (Fig. 6.1b).

$$\text{Downward force, } W = \left(\frac{G+1}{1+e}\right)\gamma_w AL$$

Upward force due to water pressure $U = (b + L)\,\gamma_w A$.

Assuming no friction on the sides of the container, and considering the critical condition (i.e., $h = h_c$),

$$(h_c + L)\gamma_w A = \frac{G+e}{1+e}\gamma_w AL$$

that is,

$$i_c = \frac{h_c}{L} = \frac{G-1}{1+e} \tag{6.1}$$

where $i_c$ is the *critical hydraulic gradient*.

This condition also occurs when individual soil particles are freely suspended in flowing water. The equilibrium of forces is shown in Fig. 6.1c for such a condition.

$$W_s = \frac{G\gamma_w}{1+e}AL = \text{weight of soil particles acting downwards}$$

$B = \gamma_w V_s$ = force due to buoyancy acting upwards

$S_F$ = seepage force on particles

For equilibrium,

$$W_s = S_F + B$$

$$\text{or } S_F = W_s - B = \frac{G\gamma_w AL}{1+e} - \gamma_w V_s = \frac{G\gamma_w AL}{1+e} - \gamma_w(1-n)AL$$

$$\text{or } S_F = \left(\frac{G\gamma_w}{1+e} - \frac{\gamma_w}{1+e}\right)AL$$

$$\text{or } S_F = i_c \gamma_w V \tag{6.2}$$

The seepage force expressed per unit volume is referred to as the *seepage pressure*. If $h$ is less than $h_c$, the seepage force is proportionately less than $i_c\gamma_w V$.

This critical condition described above is responsible for the phenomenon of boil in soils, usually referred to as *quicksand*. The quicksand condition is likely to occur at hydraulic gradients of about 1.0 in non-cohesive soils. Contrary to common belief, quicksand is not a type of sand but a phenomenon caused due to the flow condition. In cohesive soils, the cohesive strength of soil must be overcome before soil particles are washed out of the soil mass.

## 6.3 GENERAL FLOW EQUATION

In formulating the general flow equation for soils, the following assumptions are made:

1. The soil medium is saturated, incompressible, homogeneous, and isotropic with respect to permeability.
2. The flow is laminar and follows Darcy's law.
3. Water is incompressible.

Consider an element of soil of dimensions d$x$, d$y$, and d$z$ with velocities $v_x$, $v_y$, and $v_z$ in the $x$, $y$, and $z$ directions, respectively. This represents a generalized flow condition in three dimensions for a homogeneous isotropic medium (Fig. 6.2a).

There are many seepage problems in which the flow is essentially two-dimensional, for example, seepage under long sheet pile walls, dams, and water-retaining structures and through embankments and earth dams. Thus, ignoring the flow in the $y$ direction (i.e., $v_y = 0$), a two-dimensional flow condition may be considered (Fig. 6.2b).

Let the hydraulic gradients be $i_x$ and $i_z$ in the $x$ and $z$ directions and the permeability be $k$.

The quantity of water entering the element in unit time is $v_x \mathrm{d}y\,\mathrm{d}x + v_z \mathrm{d}x\,\mathrm{d}y$ and that leaving is $(v_x + \partial v_x/\partial x\,\mathrm{d}x)\mathrm{d}y\,\mathrm{d}z + (v_z + \partial v_z/\partial z\,\mathrm{d}z)\mathrm{d}x\,\mathrm{d}y$.

As the element is undergoing no volume change and the water is incompressible, the quantities of water entering and leaving should be equal, and thus

$$v_x\,\mathrm{d}y\,\mathrm{d}z + v_z\,\mathrm{d}x\,\mathrm{d}y = \left(v_x + \frac{\partial v_z}{\partial x}\mathrm{d}x\right)\mathrm{d}y\,\mathrm{d}z + \left(v_z + \frac{\partial v_z}{\partial z}\mathrm{d}z\right)\mathrm{d}x\,\mathrm{d}y$$

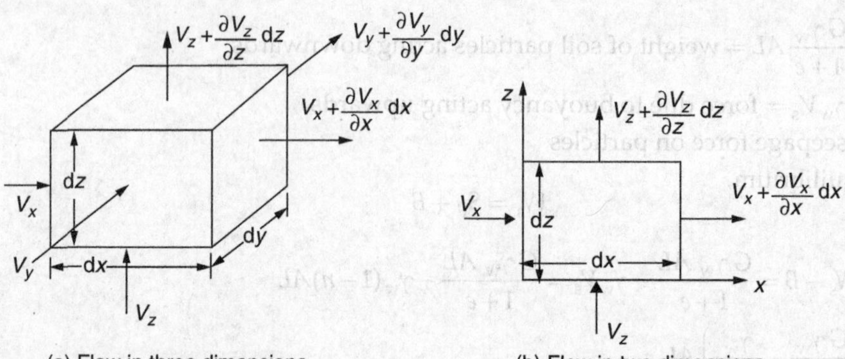


Fig. 6.2 Generalized flow condition

or

$$\left(\frac{\partial v_x}{\partial x}+\frac{\partial v_z}{\partial z}\right)\mathrm{d}x\ \mathrm{d}y\ \mathrm{d}z=0 \tag{6.3}$$

Since $\mathrm{d}x\ \mathrm{d}y\ \mathrm{d}z \neq 0$, Eq. 6.3 becomes

$$\frac{\partial v_x}{\partial x}+\frac{\partial v_z}{\partial z}=0 \tag{6.4}$$

Equation 6.4 is referred to as the equation of continuity in two dimensions.* Now, based on Darcy's law,

$$\left.\begin{aligned} v_x &= ki_x = k\frac{\partial h}{\partial x} \\ v_z &= ki_z = k\frac{\partial h}{\partial z} \end{aligned}\right\} \tag{6.5}$$

The partial derivatives in Eq. 6.5 suggest a *potential function* of the form $\phi(x, z)$, such that

$$v_x = \frac{\partial \phi}{\partial x} \text{ and } v_z = \frac{\partial \phi}{\partial z} \tag{6.6}$$

Substituting Eq. 6.6 in Eq. 6.4, we obtain

$$\frac{\partial^2 \phi}{\partial x^2}+\frac{\partial^2 \phi}{\partial z^2}=0 \tag{6.7}$$

Equation 6.7 is the Laplace equation which presents the two-dimensional steady flow of an incompressible fluid through an incompressible isotropic porous medium. In simple terms, it represents the balancing of gradient changes in the $x$ and $z$ directions when the volume is constant.

* For an element which experiences volume change, the continuity equation becomes $[(\partial v_x / \partial x) + (\partial v_z / \partial z)]\mathrm{d}x\ \mathrm{d}y\ \mathrm{d}z = (\mathrm{d}v / \mathrm{d}t)$, the volume change per unit time (refer to Chapter 8).

The existence of a potential function requires an *irrotational flow* such that

$$-\frac{\partial v_z}{\partial x}+\frac{\partial v_x}{\partial z}=0 \tag{6.8}$$

If a *flow function* ($\psi$) is defined such that

$$v_x=\frac{\partial \psi}{\partial z} \quad \text{and} \quad v_z=\frac{\partial \psi}{\partial x}$$

then

$$\frac{\partial^2 \psi}{\partial x^2}+\frac{\partial^2 \psi}{\partial z^2}=0 \tag{6.9}$$

Equation 6.9 also satisfies the Laplace condition.

Since $\phi = \phi(x, z)$ and $v_x = \partial\phi/\partial x$ and $v_z = \partial\phi/\partial z$

$$\begin{aligned} d\phi &= \frac{\partial \phi}{\partial x}dx+\frac{\partial \phi}{\partial z}dz \\ &= v_x\,dx+v_z\,dz \end{aligned}$$

For $\phi$ = constant, $d\phi = 0$, then

$$\frac{dz}{dx}=-\frac{v_x}{v_z} \tag{6.10}$$

Similarly, since $\psi = (x, z)$ and $v_x = (\partial\psi/\partial z)$ and $v_z = -(\partial\psi/\partial x)$

$$\begin{aligned} d\psi &= \frac{\partial \psi}{\partial x}dx+\frac{\partial \psi}{\partial z}dz \\ &= -v_z\,dx+v_x\,dz \end{aligned}$$

For $\psi$ = constant, $d\psi = 0$, then

$$\frac{dz}{dx}=\frac{v_z}{v_x} \tag{6.11}$$

Thus, the curves of constant $\phi$ are normal to curves of constant $\psi$ since the product of their gradients is –1. The form of the curve depends on the boundary conditions of the problem.

## 6.4 SIGNIFICANCE OF LAPLACE EQUATION

The solution of two Laplace equations for the potential and flow functions takes the form of two families of orthogonal curves.

One set of curves (constant-$\psi$ lines) represent the trajectories of seepage and are termed *flow lines* (Fig. 6.3). The space between two adjacent flow lines may be imagined to be a flow channel with an impervious boundary such that water does not cross the flow lines.

The other set of curves (constant-$\phi$ lines with $\phi = kh$) represent lines of equal head and are termed *equipotential lines*. The head loss caused by water crossing two adjacent equipotential lines is termed the *potential drop*.

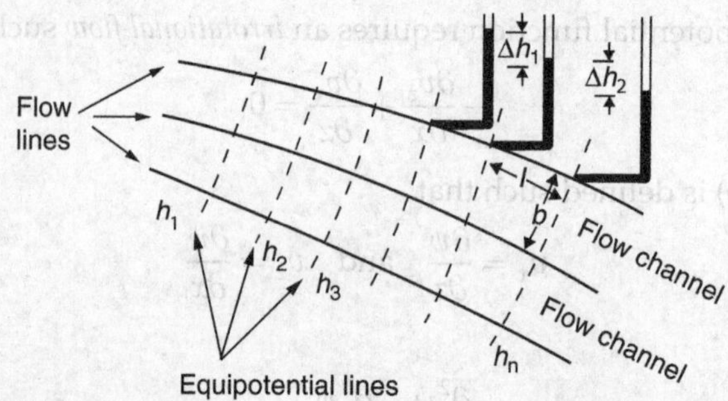


Fig. 6.3 Flow net definitions

The entire pattern of flow lines and equipotential lines is referred to as a *flow net*. Thus, a flow net is a graphical representation of the head and direction of seepage at every point. Seepage losses and their related flow pattern, the uplift pressure, and pore pressures are frequently estimated using flow nets.

## 6.5 PROPERTIES AND APPLICATIONS OF FLOW NETS

### 6.5.1 Properties

Let us consider the example of the flow net shown in Fig. 6.4. Each flow line starting at the upstream boundary with head $h_1$, dissipates the head in viscous friction and attains head $h_2$ when terminating at the downstream boundary. All such lines are shown by continuous lines. Along each flow line there must be a point where the total head may have a specific value. A line connecting all such points of equal head represents an equipotential line, and such lines are shown by broken lines.

An infinite number of flow lines and equipotential lines could be drawn for any given condition. But an important consideration to be borne in mind is that the geometric figures formed by the equipotential and flow lines should approach a square shape.

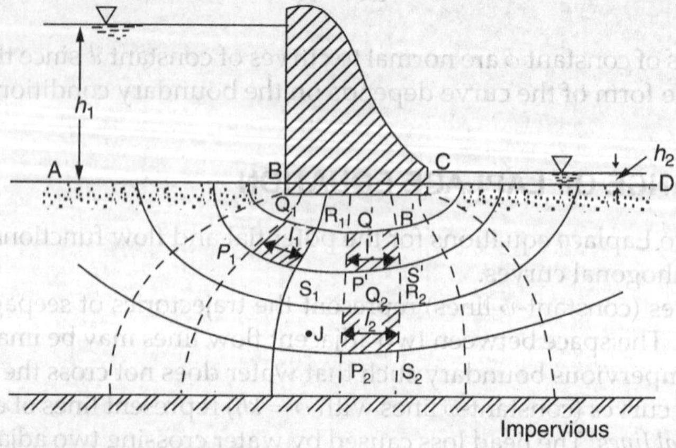


Fig. 6.4 Flow under dam

Choose $P_1Q_1R_1S_1$ and $P_2Q_2R_2S_2$, the figures formed between two pairs of flow and equipotential lines in two different channels, and P′Q′R′S′, an auxiliary figure bounded by the same pair of flow lines of the first figure and by the same pair of potential lines of the second figure.

Flow through any one channel may be given as

$$\Delta q = ki(h \times 1) = kih = k\frac{\Delta h}{l}h = k\Delta h\frac{h}{l}$$

Considering the three figures, the discharge equation for each case is given as

$$\left.\begin{aligned} \Delta q_1 &= k\,\Delta h_1 \frac{h_1}{l_1} \quad (\text{Fig. } P_1Q_1R_1S_1) \\ \Delta q_2 &= k\,\Delta h_2 \frac{h_2}{l_2} \quad (\text{Fig. } P_2Q_2R_2S_2) \\ \Delta q' &= k\,\Delta h' \frac{h'}{l'} \quad (\text{Fig. } P'Q'R'S') \end{aligned}\right\} \tag{6.12}$$

where $\Delta h_1$, $\Delta h_2$ and $\Delta h'$ are the potential drops considering two successive equipotential lines.

If all the figures in the flow net are drawn as squares, then

$$h_1 = l_1, \quad h_2 = l_2, \quad \text{and} \quad h' = l'$$

The auxiliary square and the first square have the same flow boundaries; thus $\Delta q' = \Delta q_1$, and they have the same equipotential boundaries as the second square, and thus $\Delta h' = \Delta h_2$. Hence,

$$\Delta q_1 = \Delta q_2 \quad \text{and} \quad \Delta h_1 = \Delta h_2 \tag{6.13}$$

Thus, it is shown that when all the figures are squares, there must be the same quantity of flow in each channel and the same potential drop in crossing each figure. To have these conditions, it is just sufficient if the ratio $h/l$ is maintained the same, but drawing square figures is far more convenient than drawing rectangular figures.

From the above discussion for a flow net with square figures, the properties of flow nets can be summarized as

1. Flow lines and equipotential lines intersect or meet orthogonally.
2. The quantity of water flowing through each channel is the same.
3. The potential drop between any two successive equipotential lines is the same.
4. The velocity of flow is more (because of high gradients) in figures of small dimensions so that the discharge remains the same.
5. Flow lines and equipotential lines are smooth continuous curves, being either elliptical or parabolic in shape.

### 6.5.2 Applications

**Seepage Quantity**. We have shown that in a flow net with square figures the flow through one channel is

$$\Delta q = k\Delta h \tag{6.14}$$

Let $H$ be the total head loss, i.e., $H = h_1 - h_2$.

Let $N_d$ be the number of potential drops

Therefore,

$$\Delta h = \frac{H}{N_d} \tag{6.15}$$

Therefore,

$$\Delta q = k\frac{H}{N_d} \tag{6.16}$$

Let $N_f$ be the number of flow channels.

The total discharge through the complete flow net per unit length is given as

$$q = N_f \Delta q = k\frac{H}{N_d}N_f$$

that is,

$$q = kH\frac{N_f}{N_d} \tag{6.17}$$

The ratio $N_f/N_d$ is independent of $k$ and $H$ and is characteristic of the flow net. This is called the *shape factor* of the flow net.

**Seepage Pressure**. Let $n_d$ be the number of potential drops (each of vale $\Delta h$) lost by a water particle before reaching point J, the point where the seepage pressure is needed. Let $h_1$ be the net potential at point J, that is,

$$h_1 = H - n_d \Delta h \tag{6.18}$$

Hence, the seepage pressure $p_s = h_1 \gamma_w$

or

$$p_s = (H - n_d\,\Delta h)\gamma_w \tag{6.19}$$

This pressure acts in the direction of the flow.

**Uplift Pressure**. The uplift pressure $u_w$ (also called hydrostatic pressure) at any point within the soil mass is given by

$$u_w = h_w \gamma_w \tag{6.20}$$

where $h_w$ is the piezometric head $= h_1 - z$, where $z$ is the position head of the point.

The downstream water level is usually considered as the datum, and all points above the datum are considered as positive.

**Exit Gradient**. The maximum hydraulic gradient at the downstream end of the flow lines is termed the *exit gradient*. This is given as

$$i_e = \frac{\Delta h}{l} \tag{6.21}$$

where $l$ is the length of the smallest square in the last field.

## 6.6 CONSTRUCTION OF FLOW NET

### 6.6.1 Boundary Conditions

Prior to the construction of a flow net, it is essential to study the hydraulic boundary conditions associated with that particular problem and ascertain the characteristics of flow lines and equipotential lines.

Two practical problems associated with their boundary conditions are shown in Fig. 6.5. The boundaries are as detailed below:

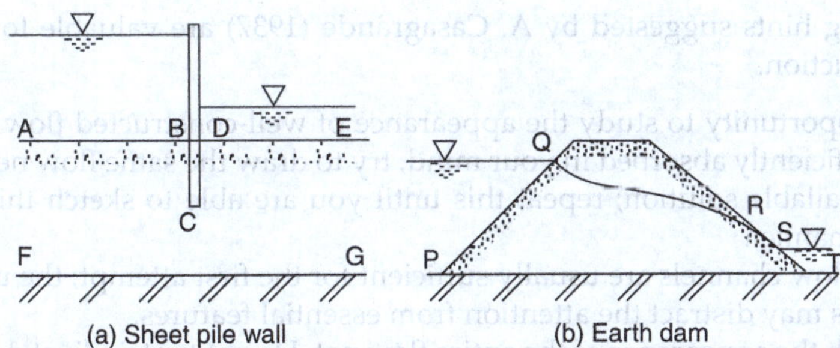


**Fig. 6.5** Boundary conditions

1. A–B, D–E, P–Q, and S–T are permeable boundaries. These surfaces have a constant head and hence are equipotential lines.
2. F–G, B–C–D, and P–T are impermeable boundaries. Further, there is no flow across these boundaries; i.e., $\partial h_t / \partial z = 0$ and $\psi$ is constant. Thus, these are flow lines.
3. R–S is a seepage surface, and along this surface the pore pressure is zero, and if $\Delta\phi$ is constant the equipontential lines meet the seepage surface at constant vertical intervals.
4. Q–R is the piezometric surface or a free surface, and the pore pressure is zero and since $\psi$ is constant it is a flow line (Whitlow, 1983).

### 6.6.2 Construction Methods

A solution of the Laplace equation for the boundary condition of the given seepage problem may yield data for plotting the flow net; that is, for such a solution the functions $\phi(x, z)$ and $\psi(x, z)$ have to be determined for the relevant boundary conditions.

The construction of a flow net can be done by adopting any one of the following methods.

**Graphical Method**. The commonest procedure for obtaining flow nets is a graphical, trial and error sketching method for seepage problems with well-defined boundary conditions. For this method, both practice and a natural aptitude are needed. Reasonable good flow nets can be obtained by practice by adhering to the correct boundary conditions and by the use of square figures. A square figure may be defined as the one in which the median lengths of the flow lines and equipotential lines are equal and have right angle corner intersections.

The following procedure may be adopted to obtain a reasonably good flow net:

1. Make a scale drawing showing the structure, soil mass, the pervious boundaries (through which water enters and leaves the soil), and the impervious boundaries (which confine the flow).
2. Keeping the properties of a flow net in mind, sketch two or more flow lines entering and leaving at right angles to the pervious boundaries and approximately parallel to the impervious boundaries.
3. Then draw equipotential lines at right angles to the flow lines such that the median lengths of the flow lines and equipotential lines are equal. This cannot be achieved in the first trial as the positions of the flow lines were approximate ones.
4. Readjust the flow lines and equipotential lines such that the condition stipulated in Step 3 is attained.
5. Introduce more flow lines and equipotential lines.

The following hints suggested by A. Casagrande (1937) are valuable for a beginner in flow net construction.

1. Use every opportunity to study the appearance of well-constructed flow nets; when the picture is sufficiently absorbed in your mind, try to draw the same flow net without looking at the available solution; repeat this until you are able to sketch this flow net in a satisfactory manner.
2. Four or five flow channels are usually sufficient for the first attempt; the use of too many flow channels may distract the attention from essential features.
3. Always watch the appearance of the entire flow net. Do not try to adjust details before the entire flow net is approximately correct.
4. The beginner usually makes the mistake of drawing overly sharp transitions between the straight and curved sections of flow lines or equipotential lines. Keep in mind that all transitions are smooth, of elliptical or parabolic shape. The size of the squares in each channel will change gradually.

Typical flow nets for flows below sheet pile walls and dams are shown in Figs. 6.6 and 6.7.

**Electrical Analogy Method**. Laplace equation not only governs a steady-state flow of ground-water but is also encountered in a steady flow of electric current through a conductor and the flow of heat through a plate. The correspondence between water and current flows is reflected from the following comparison:

| Steady-State Seepage | Electric Current |
|---|---|
| Total head $h_t$ | Voltage $V$ |
| Coefficient of permeability $k$ | Coefficient of electric conductivity $k_E$ |
| Discharge velocity $v$ | Current $I$ |
| Darcy's law $v \propto$ hydraulic gradient | Ohm's law $I \propto$ voltage gradient |
| $\nabla^2 h_t = 0$ | $\nabla^2 V = 0$ |
| Equipotential lines $= h_t =$ constant | Equipotential lines $= V =$ constant |
| Impervious boundary $(\partial h_t/\partial x) = 0$ | Insulated boundary $(\partial V/\partial x) = 0$ |

Thus, the flow domain of a porous medium has to be transferred into an electrical conductor field with similar configuration and boundary conditions (Fig. 6.8). The methodology is to obtain the locus of the lines of equal voltage drop which is in correspondence to the location of equipotential lines for the given flow domain.

A typical electrical analogy set-up is shown in Fig. 6.9. The flow domain is simulated with different conducting materials such as various metal sheets, heavy paper coated with graphite, dilute copper sulphate solution, salt water, and gelatins. The inflow face is at a potential $V_1$ and the outflow surface at a potential $V_2$. Points of equal voltage drop are found using the probe. Alternatively, the voltages at different pre-defined nodal points may be found. Then the contours of equal voltage may be sketched by hand after transferring the voltage values of nodal points on a separate sheet of paper with pre-drawn boundaries.

It should be noted that the electrical analogue simply provides only the equipotential lines. To get the complete flow net, flow lines are drawn orthogonal to the equipotential lines conforming to the boundary conditions. However, direct determination of flow lines is possible by interchanging the locations of metal bars and insulators.

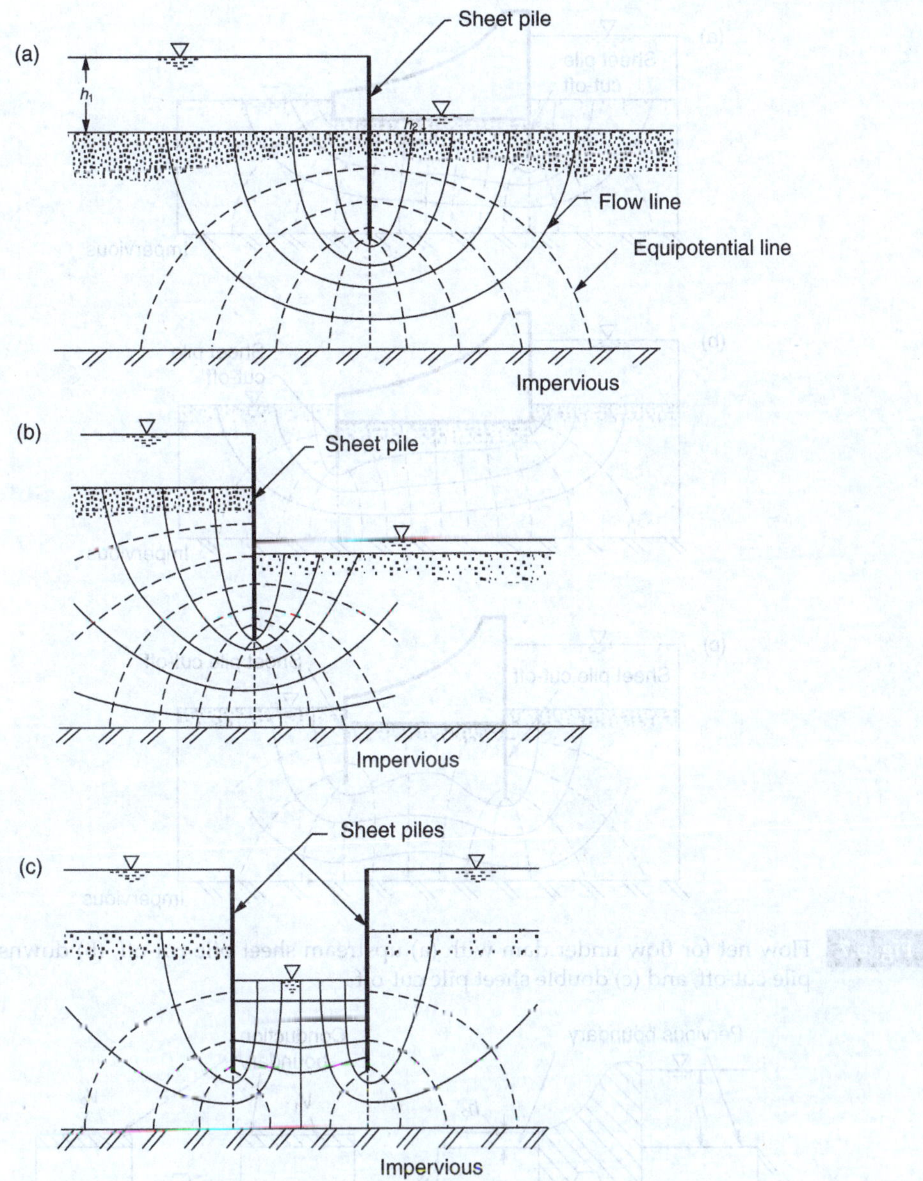


Fig. 6.6 Flow net for (a) sheet pile with level ground surface, (b) sheet pile with varied ground surface, and (c) double sheet piles

**Other Methods**. Flow tanks and viscous flow models have also been in use for constructing flow nets. In the flow tank model, the scale model of the prototype is used with sand as the porous medium. In the viscous flow model, a viscous fluid like glycerine is used for the medium. In both the cases, coloured dyes are injected at the upstream boundary which traces the path of flow lines. The equipotential lines are estimated later.

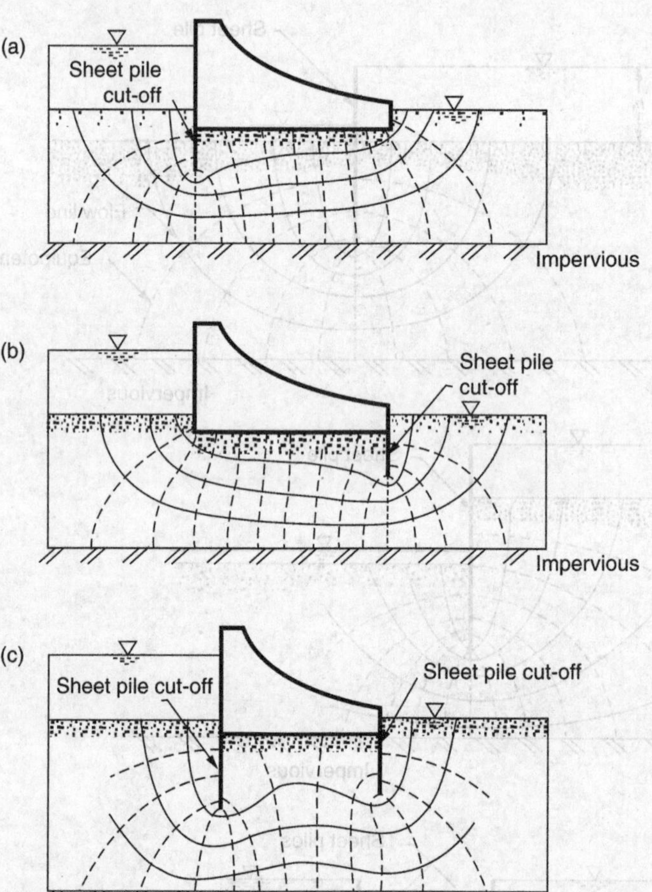


**Fig. 6.7** Flow net for flow under dam with (a) upstream sheet pile cut-off, (b) downstream sheet pile cut-off, and (c) double sheet pile cut-off

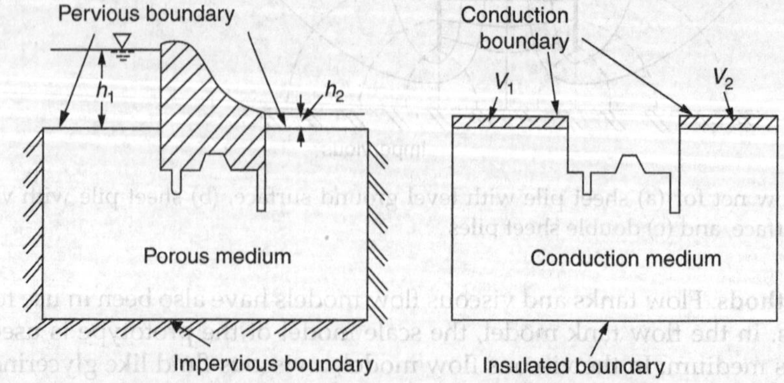


**Fig. 6.8** Seepage and current flow analogy

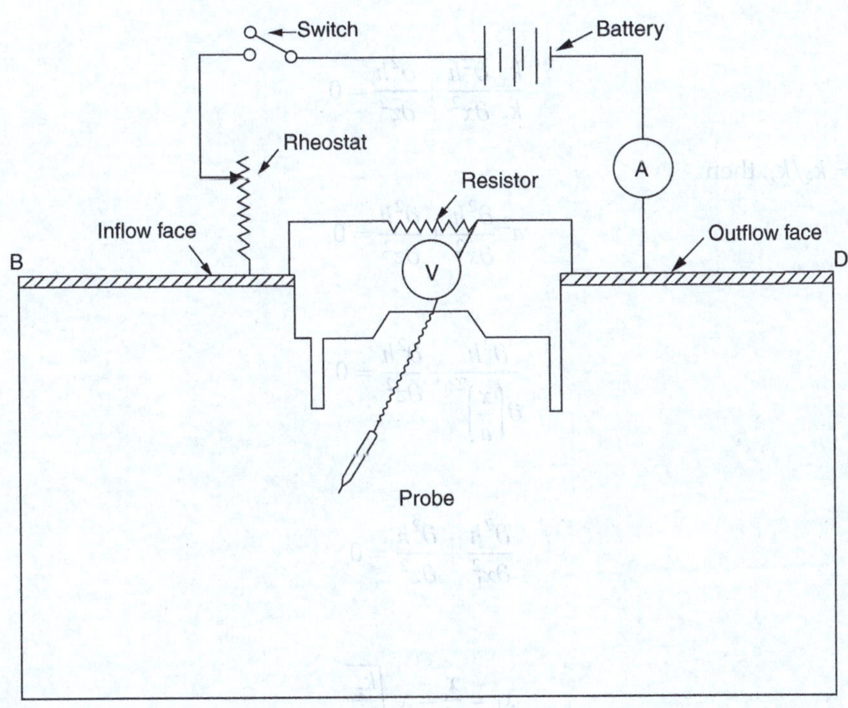


Fig. 6.9 Typical electrical analogy set-up

The solution to the Laplace's equation may also be obtained with the help of computers using analytical methods such as the finite difference method, finite element methods, and complex variable methods.

Construction of flow nets for flow through earth dams is discussed in Chapter 20.

## 6.7 ANISOTROPIC SOIL CONDITIONS

Most natural soil deposits depict anisotropy in permeability, with a higher permeability coefficient in the horizontal direction ($x$ direction) than in the vertical direction ($z$ direction), i.e., with $k_x > k_z$. Thus, Eq. 6.5 can be re-written as

$$\left.\begin{aligned} v_x &= k_x\, i_x = k_x \frac{\partial h}{\partial x} \\ v_z &= k_z\, i_z = k_z \frac{\partial h}{\partial z} \end{aligned}\right\} \tag{6.22}$$

and substituting Eq. 6.22 in the continuity equation, we have

$$k_x \frac{\partial^2 h}{\partial x^2} + k_z \frac{\partial^2 h}{\partial z^2} = 0 \tag{6.23}$$

or

$$\frac{k_x}{k_z}\frac{\partial^2 h}{\partial x^2}+\frac{\partial^2 h}{\partial z^2}=0$$

Let $a^2 = k_x/k_z$; then

$$a^2\frac{\partial^2 h}{\partial x^2}+\frac{\partial^2 h}{\partial z^2}=0$$

or

$$\frac{\partial^2 h}{\partial\left(\frac{x}{a}\right)^2}+\frac{\partial^2 h}{\partial z^2}=0$$

or

$$\frac{\partial^2 h}{\partial x_1^2}+\frac{\partial^2 h}{\partial z^2}=0 \tag{6.24}$$

where

$$x_1=\frac{x}{a}=x\sqrt{\frac{k_z}{k_x}}$$

Equation 6.24 satifies Laplace's condition for an isotropic soil in an $x_1$–$z$ plane. Transform an anisotropic flow region into a fictitious isotropic flow region by transferring all $x$ dimensions as $x_1$ (Fig. 6.10b). Now construct an artificial flow net by the usual method which will satisfy all the requirements of seepage. Redraw this flow net on the true scale by multiplying each $x$ dimension (measured from some arbitrary baseline, such as the centre line of the dam) by $a$, while keeping the $z$ dimension the same. This flow net on the true scale may consist of parallelograms and rectangles but not squares (Fig. 6.10a).

Figure 6.11 shows one flow field in natural and transformed scales. The quantity of flow $\Delta q_N$ and $\Delta q_T$ through these sections may be expressed as

$$\Delta q_N = k_x\frac{\Delta h\, b}{b\sqrt{k_x/k_z}}$$

and

$$\Delta q_T = k_e\frac{\Delta h}{b}b \tag{6.25}$$

where $k_e$ is the effective coefficient of permeability.

But

$$\Delta q_T = \Delta q_N$$

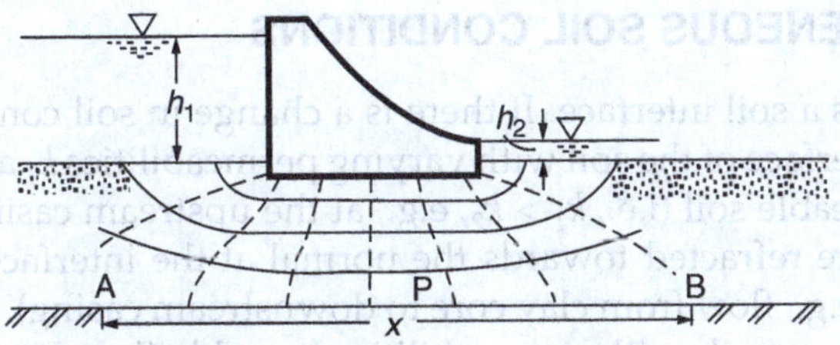

(a) Flow net for actual section

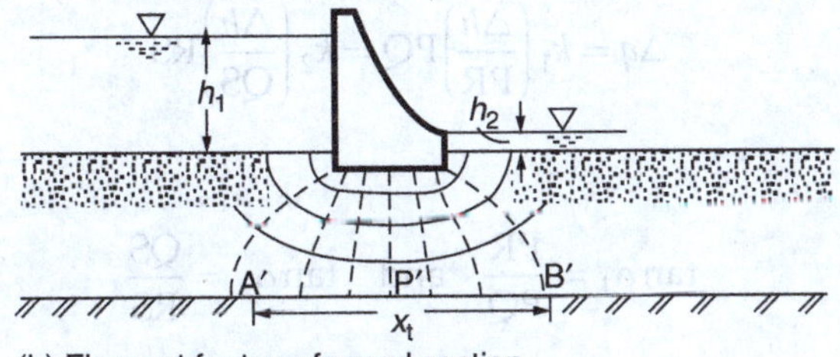

(b) Flow net for transformed section

**Fig. 6.10** Transformed flow net for anisotropic soil

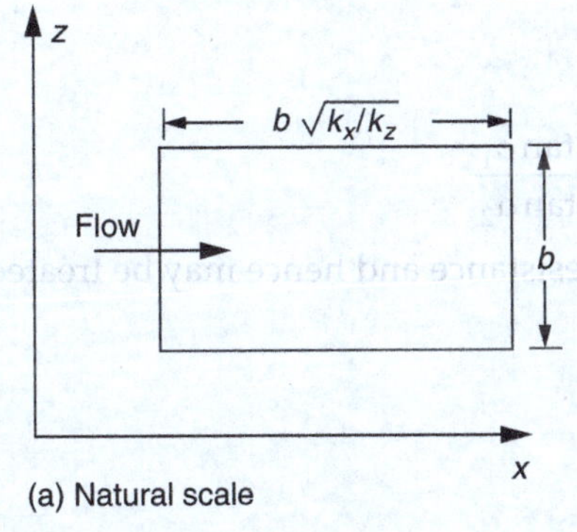

(a) Natural scale

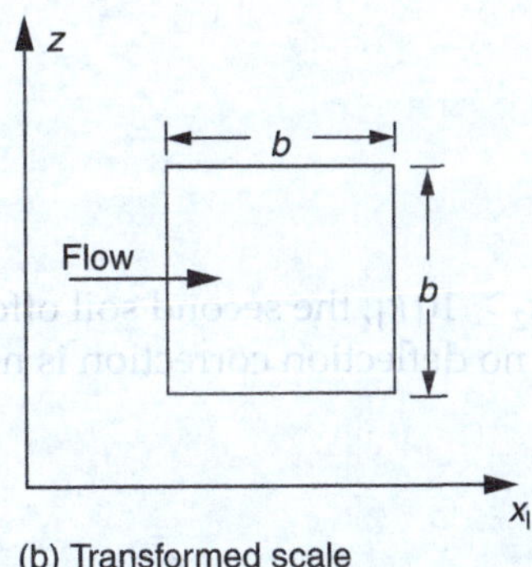

(b) Transformed scale

**Fig. 6.11** Flow fields

Therefore,

$$k_e \Delta h = k_x \frac{\Delta h}{\sqrt{k_x / k_z}}$$

that is

$$k_e = \sqrt{k_x k_z} \tag{6.26}$$

Thus,

$$q = k_e H \frac{N_f}{N_d} \tag{6.27}$$

## 6.8 NON-HOMOGENEOUS SOIL CONDITIONS

Consider a flow across a soil interface. If there is a change in soil conditions, the flow lines are deflected at the interface of the soil with varying permeabilities $k_1$ and $k_2$. If the flow takes place into a less permeable soil (i.e., $k_1 > k_2$, e.g., at the upstream casing and clay core junction), the flow lines are refracted towards the normal at the interface, and away from the normal when $k_1 < k_2$ (e.g., flow from clay core to downstream casing). Figure 6.12 shows the interface condition to two soils with permeabilities $k_1$ and $k_2$ ($k_1 > k_2$).

Let the potential drop from point P to Q and from R to S be $\Delta h$; then

$$\Delta q = k_1 \left(\frac{\Delta h}{\text{PR}}\right)\text{PQ} = k_2 \left(\frac{\Delta h}{\text{QS}}\right)\text{RS}$$

But

$$\tan\alpha_1 = \frac{\text{PR}}{\text{PQ}} \quad \text{and} \quad \tan\alpha_2 = \frac{\text{QS}}{\text{RS}}$$

Hence,

$$\frac{k_1}{\tan\alpha_1} = \frac{k_2}{\tan\alpha_2}$$

or

$$\frac{k_1}{k_2} = \frac{\tan\alpha_1}{\tan\alpha_2} \tag{6.28}$$

When $k_2 \geq 10\,k_1$, the second soil offers no resistance and hence may be treated as an open drain and no deflection correction is needed.

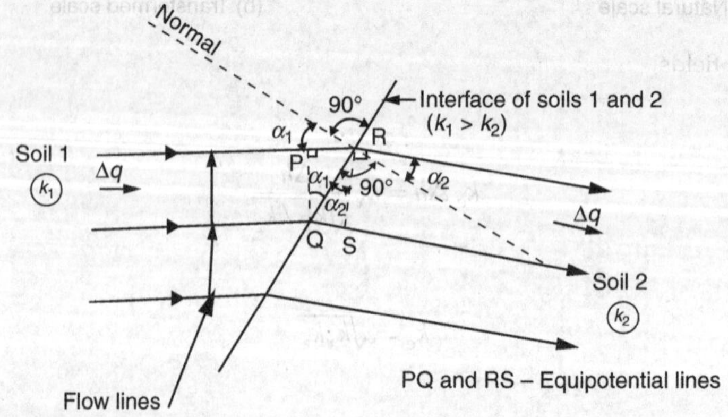


Fig. 6.12 Interface condition of two soils

## 6.9 PIPING

Because of local instability caused by a high hydraulic gradient at the exit face of a percolating soil mass, soil grains are dislodged and eroded. Such erosions gradually cause a pipe-shaped discharge channel. The width of the channel and the hydraulic gradient will increase with time and lead to a failure of the structure constructed on or with the soil. Such a mode of failure is called failure by piping.

Failures by piping may be due to scour or sub-surface erosion starting downstream and propagating inwards, causing an ultimate failure. Such a piping failure is called failure by sub-surface erosion, and no theoretical approach is possible. Piping failure is also initiated when the upward seepage pressure at the toe becomes greater than the effective weight of the soil (i.e., due to a quick condition). Such a piping failure is referred to as failure by heave. The mechanics of failure by piping due to heave is discussed below.

It has been found that the failure due to piping takes place within a distance of $D/2$ from the sheet pile, where $D$ is the depth of the sheet pile (Fig. 6.13).

Consider prism of soil ABB′A′ with width A′B′ = ½AA′ at the exit end of the structure shown in Fig. 6.13. The effective vertical pressure at the time of failure on any horizontal section through the prism is approximately equal to zero. Thus piping occurs when the seepage force on the base of the prism ($U$) becomes equal to the effective weight of the overlying sand ($W$). Let the hydraulic potential at A and B be $h_A$ and $h_B$, Then,

$$U = \tfrac{1}{2} D \gamma_w (h_A + h_B)$$

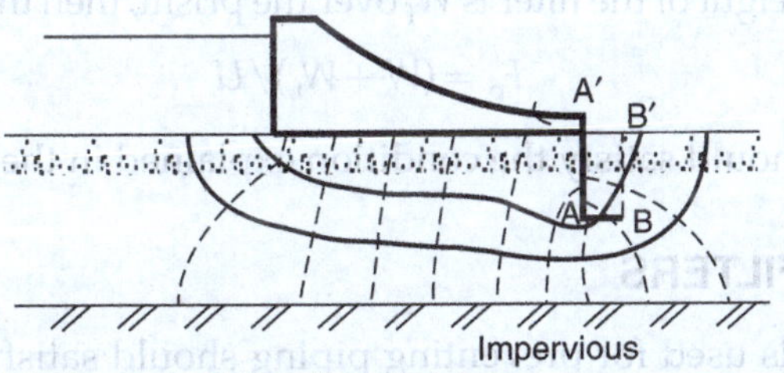

(a) Flow net with location of piping

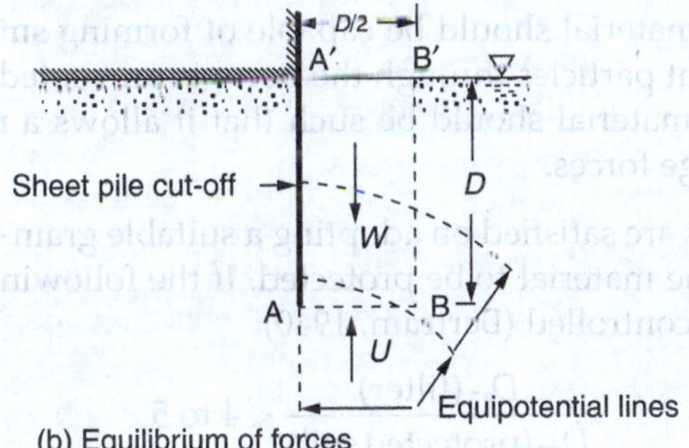

(b) Equilibrium of forces

**Fig. 6.13** Effect of piping

and

$$W = \tfrac{1}{2} D^2 \gamma'$$

Let

$$h_a = \frac{h_A + h_B}{2}$$

Therefore, factor of safety with respect to piping,

$$F_P = \frac{W}{U} = \frac{D\gamma'}{h_a \gamma_w}$$

and

$$i_{av} = \frac{(h_A + h_B)/2 - (h'_A + h'_B)/2}{D} \tag{6.29}$$

where $h'_A$ and $h'_B$ are the hydraulic potentials at A′ and B′.

Then,

$$F_P = i_c / i_w \tag{6.30}$$

where $F_p$ = factor of safety against failure by piping, and $i_c = (G - 1)/(1 + e)$ and is normally greater that 3 or 4. In cohesive soils, because of cohesion this method gives conservative values.

If the factor of safety against failure by piping is small, this may be increased by providing inverted filters. If the weight of the filter is $W_f$ over the prism, then the increased factor of safety

$$F'_P = (W + W_f)/U \tag{6.31}$$

The filter material should satisfy the condition explained in the next section.

## 6.10 DESIGN OF FILTERS

Filter or drain materials used for preventing piping should satisfy two requirements apart from adding weight as follows:

1. The gradation of filter material should be capable of forming small-size pores such that the migration of adjacent particles through the pores is prevented.
2. The gradation of filter material should be such that it allows a rapid drainage without developing large seepage forces.

The above requirements are satisfied on adopting a suitable grain-size distribution for the filter material, based on the material to be protected. If the following filter criterion is met, piping will be adequately controlled (Bertram, 1940):

$$\frac{D_{15}(\text{filter})}{D_{85}(\text{protected soil})} < 4 \text{ to } 5 \tag{6.32}$$

This criterion emphasizes that the $D_{15}$ size of the filter soil should not be more than four or five times the $D_{85}$ size of the protected soil. The second criterion is

$$\frac{D_{15}(\text{filter})}{D_{15}(\text{protected soil})} > 4 \text{ to } 5 \tag{6.33}$$

The second requirement is that the $D_{15}$ size of the filter soil should not be more than four or five times the $D_{15}$ size of the protected soil.

The US Corps of Engineers have recommended that

$$\frac{D_{50}(\text{filter})}{D_{50}(\text{protected soil})} \leq 25 \tag{6.34}$$

Based on this criterion the $D_{50}$ size of the filter should be less than or equal to 25 times the $D_{50}$ size of the protected soil.

Generally, the filter is not of one material but of different materials placed in layers. Each of these layers satisfies the requirements with respect to the preceding layer. Further, as a rough guideline, the grain-size distribution curves of the fine- and coarse-grained soils should be roughly parallel.

## WORKED EXAMPLES

**Example 6.1** At the toe of a dam, the foundation soil has a void ratio of 0.72. The specific gravity of the soil solids is 2.65. To ensure safety against piping, the upward gradient must not exceed 30% of the critical gradient at which quicksand conditions occur. Estimate the maximum permissible upward gradient.

***Solution***

Critical gradient $i_c = \dfrac{G-1}{1+e} = \dfrac{2.65-1}{1+0.72} = 0.959$

The permissible upward gradient is 30% of the critical gradient.

$$\text{Maximum permissible upward gradient} = 0.959 \times \frac{30}{100} = 0.288$$

**Example 6.2** A concrete gravity dam, 150 m long and 90 m wide, lies on a permeable soil with a coefficient of permeability of $30 \times 10^{-3}$ mm/s. The head of water is maintained at 30 mm upstream and zero at the tail-end. The soil is underlain by an impervious stratum. The depth from the base of the dam to the impervious stratum is 40 m. A flow net constructed for this condition yielded 7 flow channels and 16 equipotential drops. What is the seepage loss per day under the dam, considering a two-dimensional flow. Estimate also the approximate seepage loss under the dam using Darcy's law directly.

***Solution***

For a two-dimensional flow,

$$q = kH\frac{N_f}{N_d}$$

or

$$q = \frac{3\times10^{-3}}{1000}\times30\times\frac{7}{16}\times60\times60\times24 = 3.40\,\text{m}^3\,/\,\text{day}\,/\,\text{m}$$

For the entire length of the dam, $q = 3.402 \times 150 = 510.3$ m$^3$/day.
Using Darcy's law (one-dimensional flow) directly,

$$q = kiA$$

$$= \frac{3\times10^{-3}}{1000}\times\frac{30}{90}\times(40\times150)\times60\times60\times24$$

$$= 518.4\,\text{m}^3\,/\,\text{day}$$

**Example. 6.3** For the dam of Fig. 6.14, draw the flow net and determine the following:

1. the quantity of flow,
2. the seepage pressure in the middle of square B,
3. the uplift pressure at point B, and
4. the exit gradient at point A.

The coefficient of permeability is $4.0 \times 10^{-2}$ mm/s.

***Solution***

The flow net is drawn as in Fig. 6.14. Number of flow channels, $N_f = 5$

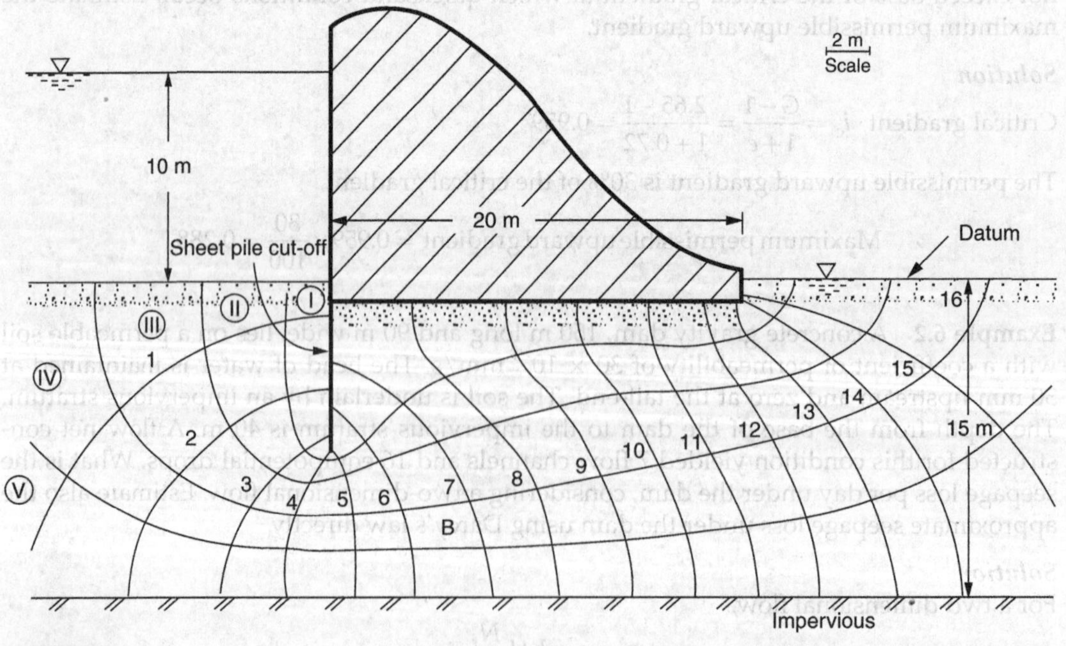


Fig. 6.14

Number of potential drops, $N_d = 16$
Head loss, $H = 10$ m
Potential drop

$$\Delta h = \frac{H}{N_d} = \frac{10}{16} = 0.625$$

1. The quantity of flow

$$q = kH\frac{N_f}{N_d}$$

$$q = \frac{4.0\times10^{-2}}{1000}\times10\times\frac{5}{16} = 1.25\times10^{-4}\ \text{m}^3/\text{s}/\text{m length}$$

2. The potential head at B is

$$h_t = H - n_d\Delta h$$
$$= 10 - 6.5\times0.625 = 5.94\ \text{m}$$

$$\text{Seepage pressure } p_s = h_t\,\gamma_w = 5.94\times9.81 = 58.27\ \text{kN/m}^2$$

3. The uplift pressure head $h_w = h_t - z$. Consider the downstream water level as the datum

$$h_w = 5.94 + 11.6 = 17.54\ \text{m}$$

and uplift pressure head $u_w = 17.54\times9.81 = 172.07\ \text{kN/m}^2$

4. Exit gradient $i_e = \dfrac{\Delta h}{l} = \dfrac{0.625}{0.60} = 1.04$

**Example 6.4** A masonry dam 50 m long and overlying an impermeable soil is founded on a soil with anisotropy in permeability. The upstream water level is 9.6 m, and the tail water level is 0.6 m above the ground level. The vertical permeability of the soil is 1.39 m/day and the horizontal permeability is six times the vertical permeability. The flow net drawn on a transformed section yields five flow channels and eight equipotential lines. Determine the seepage flow per day.

***Solution***

The horizontal permeability $= 6\times139 = 8.34$ m/day

Effective coefficient of permeability $k_e = \sqrt{k_x\,k_z}$
or

$$k_e = \sqrt{1.39\times8.34} = 3.4\ \text{m/day}$$

$$\text{Seepage flow per day} = k_e\times H\times\frac{N_f}{N_d}\times(\text{length of dam})$$
$$= 3.4\times9.0\times\frac{5}{8}\times50 = 956.25\ \text{m}^3/\text{day}$$

**Example 6.5** A section through a dam is shown in Fig. 6.15. Plot the distribution of the uplift pressure on the base of the dam.

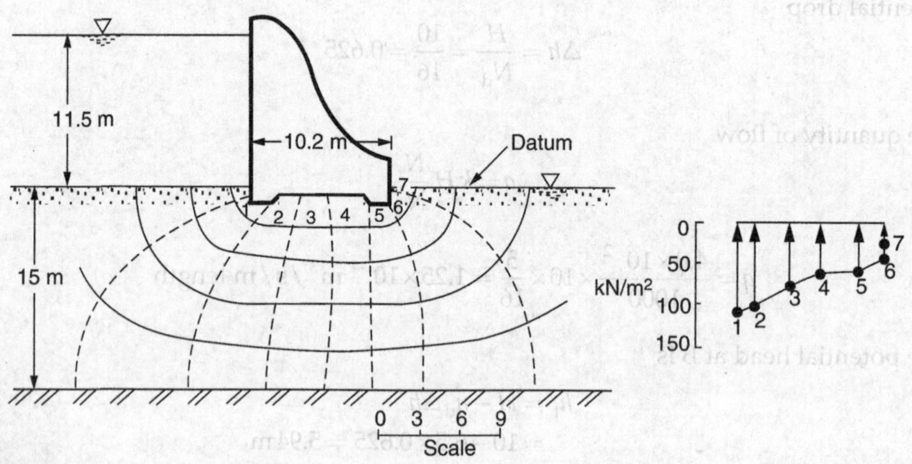


Fig. 6.15

*Solution*

The flow net is drawn as shown in Fig. 6.15.

Potential drop $\Delta h = \dfrac{11.5}{8} = 1.438 \text{ m}$

For point 1, the potential head $h_t = H - n_d \Delta h$

or

$$h_t = 11.5 - 1 \times 1.438 = 10.06 \text{ m}$$

The datum head $z = -0.9$ m

Therefore, uplift pressure head $h_w = h_t - z = 10.06 + 0.9 = 10.96$ m

and the uplift pressure at point $1 = 10.96 \times 9.81 = 107.5 \text{ kN/m}^2$

Similarly, the uplift pressures at other points are calculated and tabulated as follows. The uplift pressure diagram is shown in Fig. 6.15.

| Point | $n_d$ | $h_t$ (m) | $z$ (m) | $h_w$ (m) | $u$ (kN/m$^2$) |
|---|---|---|---|---|---|
| 1 | 1 | 10.06 | –0.9 | 10.96 | 107.5 |
| 2 | 2 | 8.62 | –1.8 | 10.42 | 102.2 |
| 3 | 3 | 7.18 | –0.9 | 8.08 | 79.27 |
| 4 | 4 | 5.75 | –0.9 | 6.65 | 65.24 |
| 5 | 5 | 4.31 | –1.8 | 6.11 | 59.94 |
| 6 | 6 | 2.87 | –1.8 | 4.67 | 45.81 |
| 7 | 7 | 1.43 | –0.8 | 2.23 | 21.88 |
| 8 | 8 | 0 | 0 | 0 | 0 |

**Example 6.6** Find the factor of safety against piping for the sheet pile wall shown in Fig. 6.16. The saturated density of the sand is 20.1 kN/m$^3$.

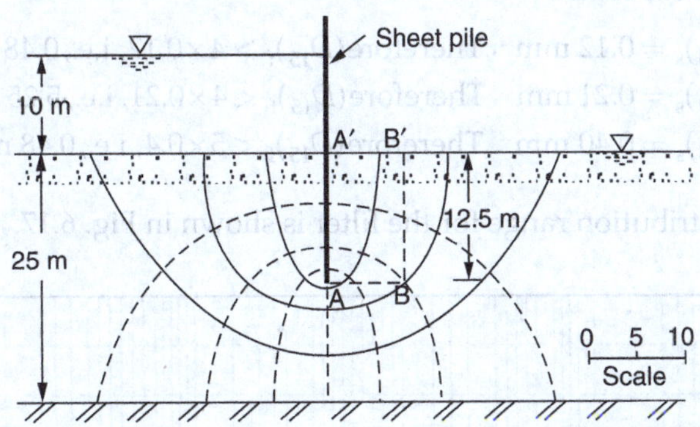


Fig. 6.16

***Solution***

The flow net is drawn as shown in Fig. 6.16.

Potential drop

$$\Delta h = \frac{10}{8} = 1.25 \text{ m}$$

Pressure head at A

$$h_A = 10 - 4 \times 1.25 = 5.0 \text{ m}$$

Pressure head at B

$$h_B = 10 - 5.9 \times 1.25 = 2.63 \text{ m}$$

Therefore,

$$h_a = \frac{5 + 2.63}{2} = 3.82 \text{ m}$$

Now, depth of penetration $D = 12.5$ m.

Therefore, factor of safety with respect to piping

$$F_p = \frac{D\gamma'}{h_a \gamma_w}$$

$$= \frac{12.5(20.1 - 9.81)}{3.82 \times 9.81} = 3.43$$

**Example 6.7** A filter is required to be provided at the downstream side of a weir. A sieve analysis conducted on the soil to be protected is as follows.

| **Sieve no.** | **1.2 (mm)** | **600 μm** | **300 μm** | **150 μm** | **75 μm** |
|---|---|---|---|---|---|
| Per cent finer | 96 | 88 | 83 | 23 | 2 |

Suggest a suitable grain-size distribution range for the filter.

***Solution***

The grain-size distribution curve of the soil is plotted as shown in Fig. 6.17. From the plot the following grain sizes are taken.

$$(D_{15})_s = 0.12 \text{ mm} \quad \text{Therefore} (D_{15})_f > 4 \times 0.12, \text{ i.e.}, 0.48 \text{ mm}$$
$$(D_{50})_s = 0.21 \text{ mm} \quad \text{Therefore} (D_{15})_f < 4 \times 0.21, \text{ i.e.}, 5.25 \text{ mm}$$
$$(D_{85})_s = 0.40 \text{ mm} \quad \text{Therefore} (D_{15})_f < 5 \times 0.4, \text{ i.e.}, 0.48 \text{ mm}$$

The grain-size distribution range for the filter is shown in Fig. 6.17.

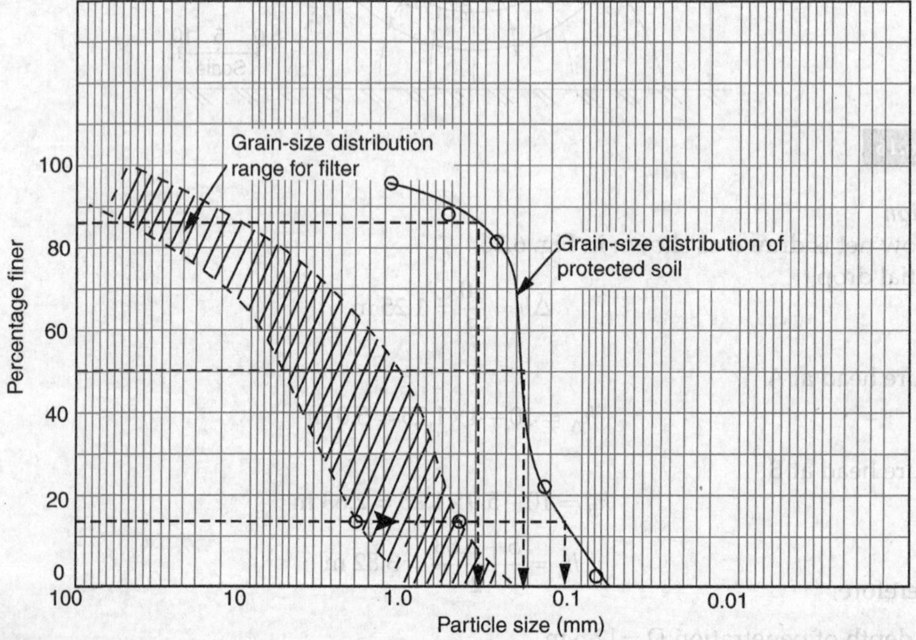


Fig. 6.17

## POINTS TO REMEMBER

6.1 Quicksand is not a type of sand but a phenomenon caused due to a flow condition. Quicksand condition is likely to occur at hydraulic gradients of about 1.0.

6.2 The general flow equation for soils is based on the assumptions that the soil medium is saturated, incompressible, and homogeneous; has isotropic permeability; the flow is laminar; and the fluid is incompressible.

6.3 Solution of the Laplace equations yields two sets of curves: flow lines which represent the trajectories of seepage and equipotential lines which represent the lines of equal head.

6.4 The entire pattern of flow lines and equipotential lines is referred to as the flow net. Thus, a flow net is a graphical representation of the head and direction of seepage at every point.

6.5 The properties of a flow net are as follows: (i) the flow lines and equipotential lines meet orthogonally, (ii) the quantity of flow through each channel is the same, and (iii) the head loss (potential drop) between any two successive equipotential lines is the same.

6.6 The uplift pressure $u_w$, also called the hydrostatic pressure, at any point within a soil mass is the pressure caused by the piezometric head at that point (i.e., the total head minus the position head).

6.7 Piping is caused by a high hydraulic gradient at the exit face of the percolating soil mass. Failures by piping may be due to scour or sub-surface erosion starting downstream and propogating inwards, causing an ultimate failure.

6.8 Filter or drain materials are used for preventing piping. Apart from providing weight, the filters should satisfy two grain-size requirements.

## QUESTIONS

### *Objective Questions*

6.1 State whether the following statements are true or false:
(1) In practically all seepage problems, velocity heads are disregarded.
(2) In a flow through a porous medium, lines connecting points of equal total energy head are termed equipotential lines.
(3) The uplift pressure at any point within a soil mass is independent of the position of the point.
(4) The flow of water through a soil specimen in a laboratory constant head permeability test is under two-dimensional flow conditions.
(5) The seepage loss through an anisotropic soil medium is less than in an isotropic medium.

6.2 For a flow under a concrete dam founded on a homogeneous isotropic porous medium, will the flow net alter (answer yes or no)
(a) If the horizontal permeability is altered?
(b) If the difference in head is changed?
(c) If the shape factor of the net is increased?
(d) If the width of the dam is reduced?

6.3 Piping in soils occur when
(a) The effective pressure becomes zero
(b) A sudden change of permeability takes place
(c) The soil is fissured and cracked
(d) The soil is highly porous

6.4 The seepage taking place beneath a long masonry dam founded on pervious soils is often considered as a ______ flow.
(a) Three-dimensional
(b) One-dimensional
(c) Two-dimensional
(d) Radial

6.5 Seepage flow in a porous medium is determined by the absolute value of
(a) $N_f$
(b) $N_d$
(c) $N_f / N_d$
(d) $N_d / N_f$

6.6 The velocity potential defined in the Laplace equation is a
(a) Scalar function of space
(b) Vector function of space
(c) Scalar function of space and time
(d) Vector function of space and time

6.7 The quantity of seepage depends on
(1) The coefficient of permeability
(2) The length of the flow path
(3) The differential head across the flow path
(4) The number of flow paths
Of these statements,
(a) 1, 2, and 3 are correct
(c) 2, 3, and 4 are correct
(b) 3, 4, and 1 are correct
(d) All are correct

6.8 Identify the incorrect flow net property
(a) Flow lines and equipotential lines intersect orthogonally.
(b) The quantity of water flowing through each channel is the same.
(c) The potential drop between any two successive equipotential lines is different.
(d) Flow lines and equipotential lines are smooth curves.

6.9 In order to prevent piping, the exit gradient should be
(a) Equal to the critical gradient
(c) Greater than the critical gradient
(b) Much less than the critical gradient
(d) Not a function of the critical gradient

6.10 Which of the following pairs are correctly matched?

| | |
|---|---|
| (1) Piping | A progressive failure |
| (2) Piping ratio | A filter criterion |
| (3) Graded filter | Material provided to prevent seepage |
| (4) Quicksand condition | When the pore pressure equals the total pressure |

Select the correct answer using the codes given below:
(a) 1, 2, and 3
(b) 2, 4, and 1
(c) 2 and 4
(d) 3 and 1

## *Descriptive Questions*

6.11 What do you understand by the mechanism of piping? Explain the methods that are adopted to increase the factor of safety against piping.

6.12 What constitutes a flow net? State any four methods of obtaining flow net in any given case.

6.13 State reasons for the quantity of seepage between two successive flow lines being equal.
6.14 Give reasons for limiting the size of particles used in constructing drainage filters.
6.15 What soil conditions in the foundations are vulnerable to the problem of piping danger?
6.16 Explain how weighted filters are useful in seepage problems for improving the stability.
6.17 What methods do you suggest to reduce the exit gradient in the case of flows under concrete dams?
6.18 Distinguish between seepage pressure and uplift pressure. Which one should be considered in the design of a masonry weir? Why?
6.19 Discuss the effects of anisotropy and non-homogeneity of a soil on the seepage loss.

## EXERCISE PROBLEMS

6.1 A sandy soil collected from an excavation showed void ratios of 0.48 and 0.97 in its densest and loosest states, respectively. The range of critical hydraulic gradients at which quicksand conditions might occur is needed to decide the depth of excavation. Take $G = 2.65$ and estimate the range.

6.2 Explain the phenomenon of quicksand. What hydraulic head is required to create a quicksand condition in a non-cohesive soil sample of length = 6 m, void ratio = 0.65, $G = 2.65$?

6.3 In a vertically upward flow of groundwater (an artesian condition) the hydraulic gradient in a sand mass is 0.95. Check whether a condition for quicksand or erosion could develop.

6.4 From the flow net shown in Fig. 6.18 find
1. the flow rate through the soil,
2. the water pressure in the middle of square X,
3. the seepage force per unit volume at X, and
4. the factor of safety against piping.

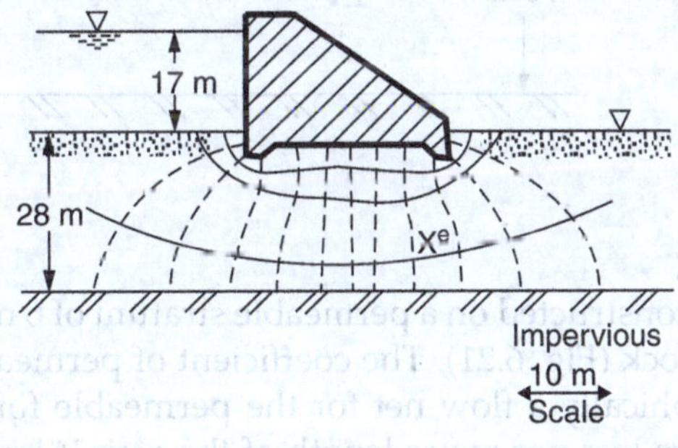


Fig. 6.18

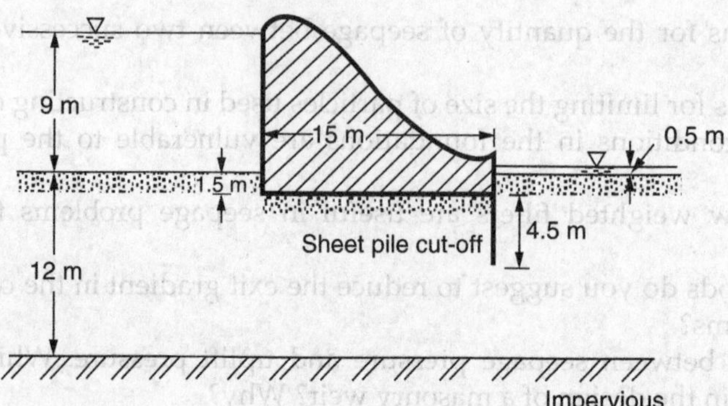


Fig. 6.19

6.5 It is proposed to construct a dam 100 m long on a permeable soil of permeability 0.00152 mm/s. The cross section of the dam is shown in Fig. 6.19. Estimate the quantity of water that will be lost per day by seepage. Also calculate the percentage reduction in the rate of flow if a 6 m upstream impervious apron is provided.

6.6 It is proposed to design a dam of 150 m length with a pool level of 40 m (Fig. 6.20). Three designs, all with impervious cut-off walls to decrease seepage, are being considered. Calculate the seepage per day for each design. Use flow net solutions. The value of $k = 2.5 \times 10^{-8}$ m/s.

| Design | Depth of cut-off wall (m) |
|---|---|
| 1 | 20 |
| 2 | 30 |
| 3 | 45 |

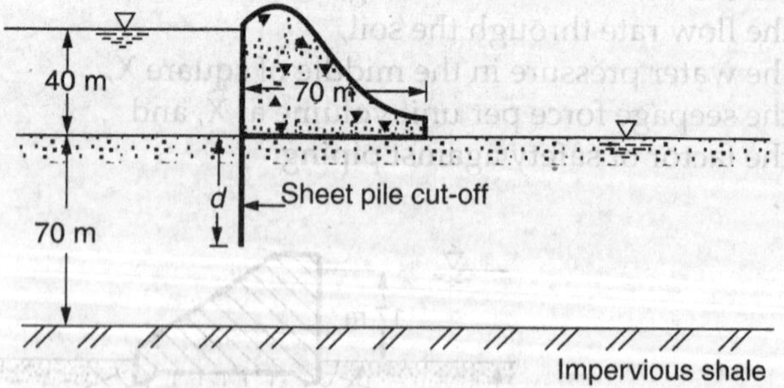


Fig. 6.20

6.7 A masonry weir is constructed on a permeable stratum of 6 m thickness and underlain by an impervious rock (Fig. 6.21). The coefficient of permeability of the soil is $0.54 \times 10^{-4}$ m/s. Plot graphically a flow net for the permeable foundation of the weir and estimate the seepage loss per metre length of the weir. Also compute the hydrostatic uplift pressure at 1, 1.5, 2, and 4 m from the upstream edge of the floor.

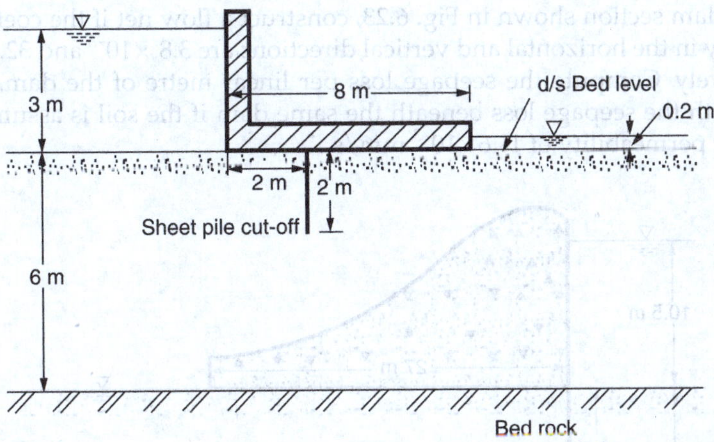


Fig. 6.21

6.8 The cross section of a dam is shown in Fig. 6.22. Make a flow net and determine the quantity of seepage under the weir. Also plot the distribution of the uplift pressure on the base of the dam. The coefficient of permeability of the soil is $2.8 \times 10^{-5}$ m/s.

6.9 A single-row vertical sheet piling penetrates 6 m into a soil of 15 m thickness overlying an impermeable rock. The coefficients of permeability of the soil in the vertical and horizontal directions are $2 \times 10^{-2}$ and $4 \times 10^{-2}$ mm/s, respectively. The depth of water on one side of the piling is 9 m and on the other side 2 m. Draw a neat sketch of a flow net and estimate the quantity of seepage in $m^3$/day/m run of piling. Re-plot the flow net from the transformed section to the natural section.

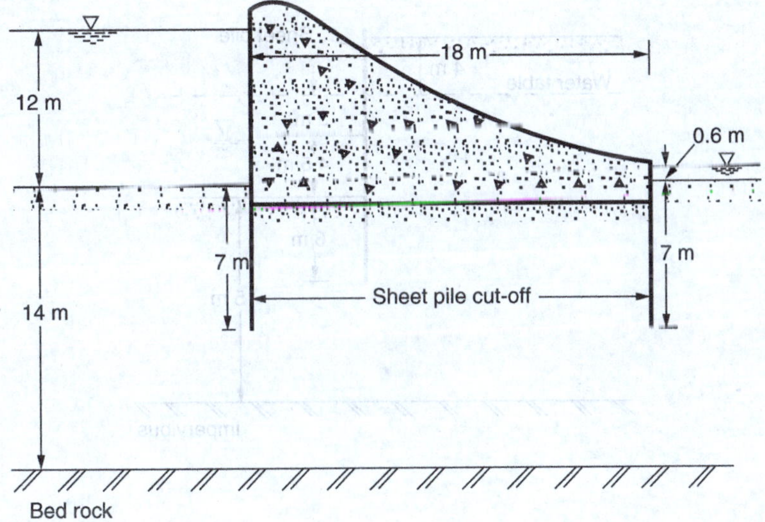


Fig. 6.22

6.10 For the dam section shown in Fig. 6.23, construct a flow net if the coefficients of permeability in the horizontal and vertical directions are $3.8 \times 10^{-3}$ and $32.4 \times 10^{-3}$ mm/s, respectively. Compute the seepage loss per linear metre of the dam. Compare this value with the seepage loss beneath the same dam if the soil is assumed to have an isotropic permeability of $11.6 \times 10^{-3}$ mm/s.

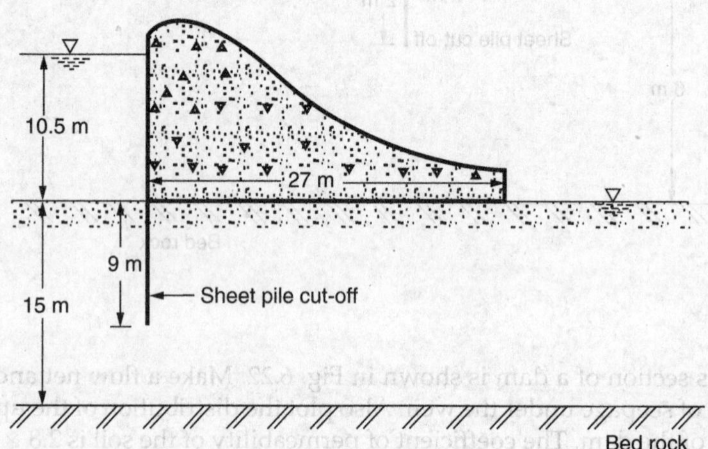


Fig. 6.23

6.11 In a tidal estuary, during low tide, the depth of water in front of a sheet pile wall is 5 m and the water table behind the wall lags 3 m behind the tidal level (Fig. 6.24). Plot the net distribution of water pressure on the piling.

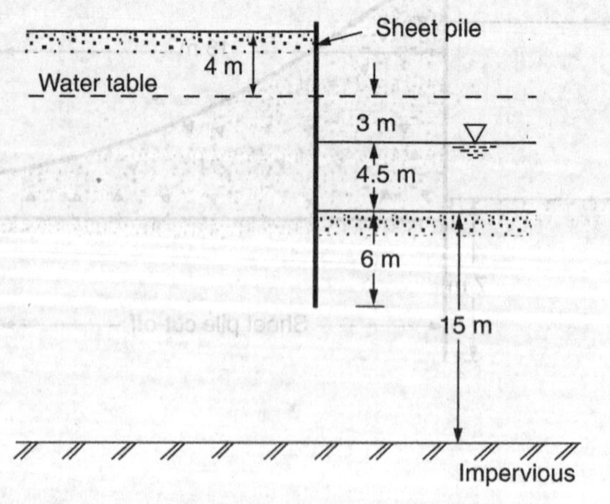


Fig. 6.24

# 7

# Stress and Stress Distribution in Soil

## CHAPTER HIGHLIGHTS

Stresses at a point – Mohr's circle – Stress paths – Stress concepts: total stress, neutral stress, effective stress – Geostatic stresses – Different positions of water table – Stresses due to surface loads: elastic half space, Boussinesq's theory, pressure distribution diagrams, Westergaard equation, types of surface loads, Newmark's influence chart, approximate solutions – Contact pressure

## 7.1 INTRODUCTION

Internal stress develops in a soil mass by the weight of the overburden and due to external loadings caused by the construction of structures. It is impossible to keep track of forces acting at different points of a soil mass because of its heterogenic nature, and thus stress developed over a zone is used. A stress-induced soil is associated with deformation (may be settlement or heave). Depending on the method of application of a load and the mode of distribution of stresses, the stress developed might strengthen the soil by expelling pore water pressure or induce a soil mass failure by actuating the stresses. This chapter and Chapters 8 and 9 discuss the predication of stress and the associated volume change and strength of soil.

## 7.2 STRESSES AT A POINT

In most branches of engineering, materials are regarded as a continuum and stresses and strains are evaluated considering an infinitesimal element having the same properties as the whole mass. This is generally so in geotechnical engineering when dealing with soil and rock. The stress at a point within a soil mass has to be viewed as a large point with representative materials of the whole mass.

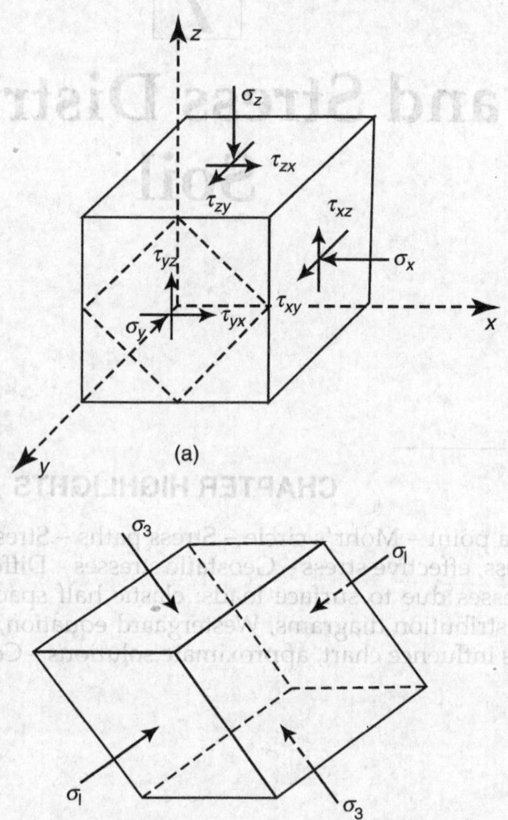


(b) Major and minor principal planes and stresses

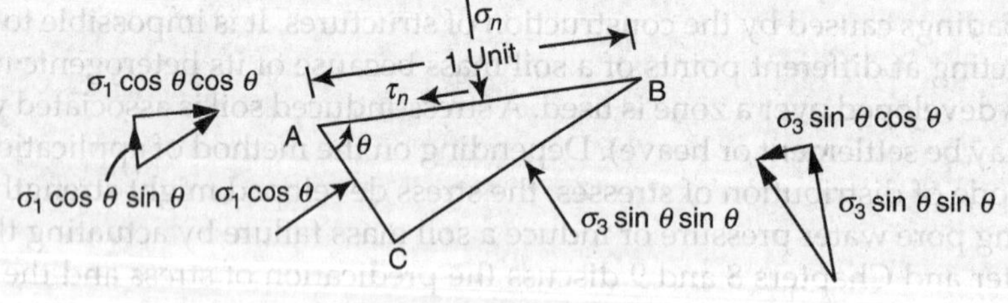


(c) Cut element with forces

**Fig. 7.1** State of stresses

Consider an incremental element and the stresses acting on the planes to represent the stress conditions at a point as shown in Fig. 7.1a. Here, $\sigma_x$, $\sigma_y$, and $\sigma_z$ are the normal stresses and $\tau_{xy}$, $\tau_{yz}$, and $\tau_{zx}$ are the shear stresses. To satisfy the rotational movement equilibrium condition, the shear stress acting on orthogonal planes should be zero (i.e., $\tau_{xy} = \tau_{yx}$, $\tau_{yz} = \tau_{zy}$ and $\tau_{xz} = \tau_{zx}$).

Now consider all the planes passing through the point and locate the planes on which there are no shear stresses. Such planes also mutually maintain orthogonality and are represented in Fig. 7.1b.

These normal stresses are called *principal stresses* and the planes, *principal planes*. These three principal stresses are termed the major principal stress, $\sigma_1$ (the largest stress), the intermediate principal stress, $\sigma_2$, and minor principal stress, $\sigma_3$ (the smallest stress). The respective strains in these three directions may be taken as $\varepsilon_1$, $\varepsilon_2$, and $\varepsilon_3$. In many practical geotechnical problems in soils, the principal stresses act vertically and horizontally, e.g., the stress condition below a horizontal ground surface.

It is convenient to assume one of the principal stresses or principal strains to be zero and convert the problem to one in a two-dimensional state. These two-dimensional states are called (i) plane stress (ignoring strain, $\varepsilon_2$) or (ii) plane strain (ignoring stress, $\sigma_2$). Many geotechnical problems are plane strain problems. Further, compression is considered as positive and tension as negative, and a shear stress causing a counter-clockwise torque about the centre of a free body is considered positive.

Now consider the cut element ABC as shown in Fig. 7.1c. Let the plane AB be 1 unit in length; then AC = cos $\theta$ and CB = sin $\theta$. The normal and shear stresses on plane AB can be determined by resolving the forces parallel and normal to plane AB as

$$\sigma_n = \sigma_1 \cos^2\theta + \sigma_3 \sin^2\theta$$

where $\sigma_n$ is the normal stress acting on plane AB, that is,

$$\sigma_n = \frac{\sigma_1 + \sigma_3}{2} + \frac{\sigma_1 - \sigma_3}{2} \cos 2\theta \qquad (7.1)$$

and

$$\tau_n = (\sigma_1 - \sigma_3)\sin\theta\cos\theta \qquad (7.2)$$

where $\tau_n$ is the shear stress acting on plane AB.

This shows that the normal and shear stresses on any plane orthogonal to the intermediate principal plane may be determined from Eqs. 7.1 and 7.2. It may be observed that these two equations have not included any material properties, but are based merely on the principles of mechanics.

From Eq. 7.2 it may be seen that the maximum shear stress value is $(\sigma_1 - \sigma_3)/2$ for 45°, and on this plane the normal stress is always $(\sigma_1 + \sigma_3)/2$.

## 7.3 MOHR'S CIRCLE

The normal and shear stresses given by Eqs. 7.1 and 7.2 for different values of $\theta$ may be presented graphically on a coordinate system and the locus of these points represents a circle (Fig. 7.2). This graphical representation of the state of stress in a lucid form is known as a Mohr circle after Mohr (1882). The graphical representation is a convenient aid in solving problems.

The circle has a radius of $(\sigma_1 - \sigma_3)/2$, with the centre on the $x$- axis with coordinates $[(\sigma_1 + \sigma_3)/2, 0]$. Any point on the circle represents $\sigma_n$ and $\tau_n$ on some plane. Point D represents

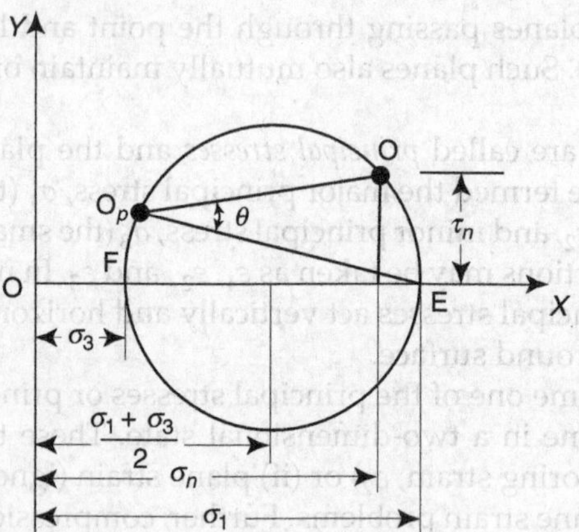


Fig. 7.2 Mohr's stress circle

the state of stress on the plane inclined at $\theta$ with the major principal plane. Points E and F represent the major and minor principal stresses, respectively.

The Mohr diagram is an excellent visualization of the orientations of various planes. If, through the coordinates of $\sigma_n$ and $\tau_n$ on the Mohr circle, a line is drawn parallel to the plane on which these stresses act, this line intersects the Mohr circle at a unique point. If parallels are drawn from E($\sigma_1$, 0) and F($\sigma_3$, 0) to the respective planes, these planes pass through the same unique point. This point is referred to as the *origin of planes* or *pole, $O_p$*. Thus, any line drawn from the pole, parallel to a plane (on which the stresses are needed), intersects the circle at a point, the coordinates of which represent the normal and shear stresses acting on that plane.

## 7.4 STRESS PATHS

Progressive changes in the state of a particular load application can be represented by a series of Mohr circles. For example, Fig. 7.3a represents successive states as $\sigma_1$ is increased with $\sigma_3$ constant. Such a diagram with several complete stress circles can appear cluttered.

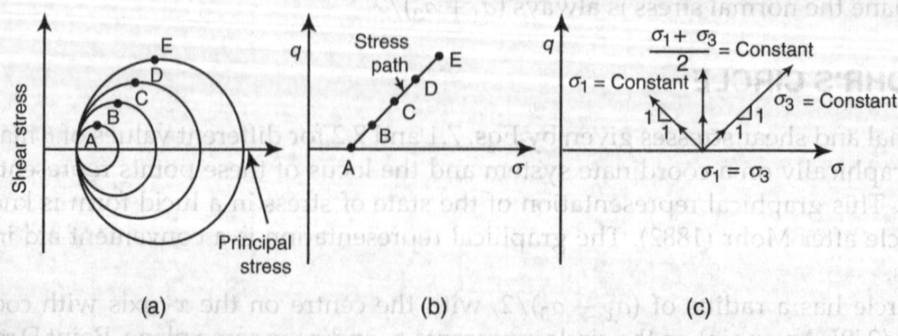


Fig. 7.3 Stress paths

It is convenient to plot only the point of maximum shear stress, and if needed the complete circle can be reconstructed using such a point. Thus, the locus of points (Fig. 7.3b) on the Mohr diagram whose coordinates represent the maximum shear stress and the associated principal stress for the entire stress history is defined as a *stress path* (Lambe, 1967). For given principal stresses $\sigma_1$ and $\sigma_3$, the coordinates of a point on the stress path are

$$p = \frac{\sigma_1 + \sigma_3}{2} \quad \text{and} \quad q = \frac{\sigma_1 - \sigma_3}{2}$$

Such a plot is referred to as a *p–q* diagram.

The stress path for $\sigma_3$ = constant and $\sigma_1$ increasing is a 45° line (Fig. 7.3b). Figure 7.3 shows stress paths for different variations of $\sigma_1$ and $\sigma_3$. A stress path diagram may be constructed for total or effective stress conditions.

## 7.5 EFFECTIVE STRESS CONCEPT

In a natural soil stratum or in man-made earth structures, three conditions based on moisture content may be visualized, viz., dry, saturated, or partially saturated. All earth structures or structural foundations may experience one or all of these conditions during their life span. Thus, the stress conditions present during these stages have to be completely understood.

### 7.5.1 Dry Soil

In a dry soil system any stress has to be visualized as the force in the mineral skeleton per unit area of the soil. Accordingly, a normal stress can be defined as the sum of the normal components of the forces ($\Sigma N$) over a plane divided by the area of the plane ($A$). Let us consider a dry soil medium of unit width (Fig. 7.4a) and unit length (normal to the plane of the paper) as the surface over which the normal stress is to be computed. The soil at the section would have attained equilibrium due to the overburden pressure, $\sigma_1$, and this overburden pressure would have changed the mechanical properties of the soil. Thus, this pressure may be termed the *effective pressure* or *effective stress*, $\sigma'$.

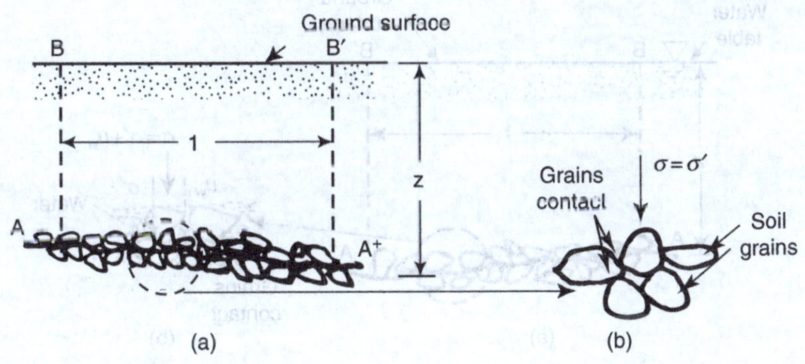


Fig. 7.4 Vertical stress in dry soil

Weight of prism of soil = $\gamma_d (z \times 1 \times 1)$
Area along section AA′ = 1
Therefore, the total stress is

$$\sigma = \frac{\gamma_d z}{1} \tag{7.3}$$

Thus the effective stress $\sigma' = \sigma$. (7.4)

Figure 7.4b shows a close-up view of the particles on plane AA′.

### 7.5.2 Saturated Soil

Consider a saturated soil condition as shown in Fig. 7.5. This condition is similar to the previous case, but the voids are completely filled with water (Fig. 7.5a). As before, the overburden pressure, $\sigma$, is given as

$$\sigma = \gamma_{sat} z \tag{7.5}$$

This total normal stress acting on section AA′ has two components, one of which acts on the pore water and the other on the soil skeleton. The component on the water acts equally in all directions and does not cause any change in the mechanical properties of the soil and is known as the *neutral stress* or *pore water pressure*, $u_w$.

The remaining part

$$\sigma' = \sigma - u_w \tag{7.6}$$

is that component of the total stress which rests entirely on the soil skeleton of the soil. Thus, only this component of the total stress will cause a change in the mechanical properties and, hence, is known as the effective stress.

This classical equation was put forth by Terzaghi (1925, 1943).

$$\sigma' = \gamma_{sat} z - u_w$$
$$\sigma' = \gamma_{sat} z - z\gamma_w$$
$$= z(\gamma_{sat} - \gamma_w)$$
$$\sigma' = z\gamma' \tag{7.7}$$

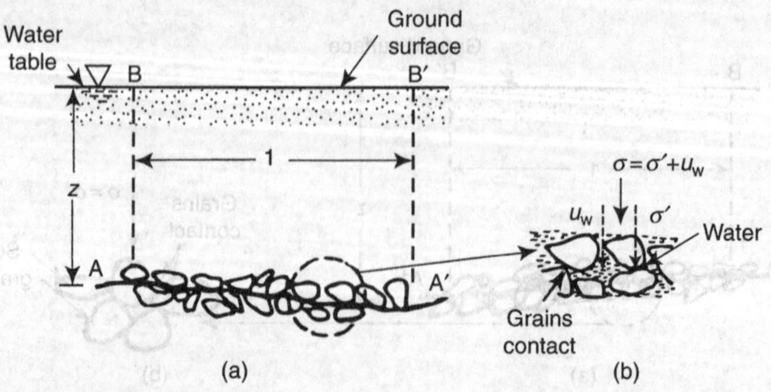


Fig. 7.5 Vertical stress in saturated soil

Let $A$ be the total area (Fig. 7.5 b), $A_w$ the area of water in contact with the total area minus the mineral contact area), and $A_c$ the mineral contact area. Then

$$A_w + A_c = A$$

$$\frac{A_w}{A} + \frac{A_c}{A} = 1$$

$$a_w + a_c = 1 \tag{7.8}$$

where $a_w = A_w / A$ is the ratio of area of water contact to the total area and $a_c = A_c/A$ the ratio of area of mineral contact to the total area.

Actually, the pore water pressure acts only on $a_w$, rather than on the complete area. Hence,

$$\sigma' = \sigma - a_w u_w$$

or

$$\sigma' = \sigma - (1 - a_c) u_w$$

It has been widely accepted that $a_c$ is negligible (Lambe and Whitman, 1979; Rcosenqvist, 1959). Hence

$$a_c = 0 \quad \text{and} \quad \sigma' = \sigma - u_w$$

The above relationship is generally valid. This expression indirectly assumes that no other stresses except the external applied stress, $\sigma$, and the pore water pressure, $u_w$, exist within the system. Further, the attractive and repulsive forces between particles are not accounted for in this expression.

### 7.5.3 Partially Saturated Soil

Consider the situation shown in Fig. 7.6a, which represents a partially saturated soil. Again, the total stress

$$\sigma = \gamma z \tag{7.9}$$

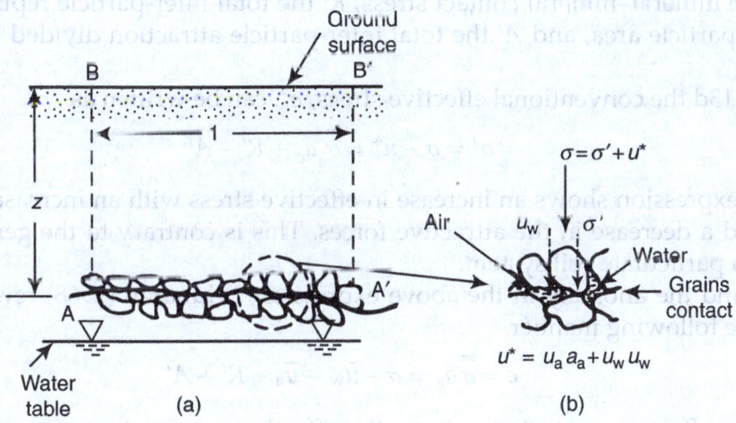


Fig. 7.6 Vertical stress in partially saturated soil

Here the contact areas may be considered as (Fig. 7.6b)

$$a_c + a_a + a_w = 1 \tag{7.10}$$

where $a_a$ is the ratio of area of air contact to the total area.

Again, considering $a_c = 0$ and $a_a + a_w = 1$, $\sigma'$ can be defined as

$$\sigma' = \sigma - u^* \tag{7.11}$$

where $u^* = u_a\, a_a + u_w\, a_w$ and $u_a$ is the pressure in the gas and vapour phase; that is,

$$u^* = u_a + a_w(u_w - u_a) \tag{7.12}$$

$$\sigma' = \sigma - [u_a - a_w(u_a - u_w)] \tag{7.13a}$$

Bishop (1959) based on his intuition replaced $a_w$ in the above expression by an empirical parameter $\chi$, and thus

$$\sigma' = \sigma - [u_a - \chi(u_a - u_w)] \tag{7.13b}$$

The parameter $\chi$ has to be determined experimentally (Bishop et al., 1960). It is believed that the parameter depends on the degree of saturation, i.e., it has a value of 1.0 for saturated soil and 0 for dry soil. $\chi$ may have different values at a given degree of saturation in relation to the shear strength and volume change. Further, in certain situations where the stress history plays a more important role than does the degree of saturation, the factor $\chi$ has been found to take negative values and values greater than unity. Thus, the factor $\chi$ is purely an empirical one and may to some extent depend on the degree of saturation.

There was another school of thought which related the total external pressure to the internal stress in the particulate soil system (Lambe, 1960).

$$\sigma' = \bar{\bar{\sigma}}\, a_c + u_a\, a_a + u_w\, a_w + R' - A' \tag{7.13c}$$

That is,

$$\sigma' = \bar{\bar{\sigma}}\, a_c + u^* + R' - A' \tag{7.13d}$$

where $\bar{\bar{\sigma}}$ is the mineral–mineral contact stress, $R'$ the total inter-particle repulsion divided by total inter-particle area, and $A'$ the total inter-particle attraction divided by total inter-particle area.

From Eq. 7.13d the conventional effective stress, $\sigma'$, can be written as

$$\sigma' = \sigma - u^* = \bar{\bar{\sigma}}\, a_c + R' - A' \tag{7.13e}$$

The above expression shows an increase in effective stress with an increase in the repulsive forces and a decrease in the attractive forces. This is contrary to the general physical behaviour in a particulate soil system.

Having found the anomaly in the above expression, Sridharan (1968) rewrote Lambe's equation in the following manner.

$$\bar{c} = \bar{\bar{\sigma}}\, a_c = \sigma - \bar{u}_w - \bar{u}_a - R' + A' \tag{7.13f}$$

where $\bar{c}$ is the effective contact stress, $\bar{u}_w$ the effective pore water pressure, and $\bar{u}_a$ the effective pore air pressure.

This can be represented as a combination of two effective stresses as

$$\bar{c} = \sigma' + \sigma'' \quad (7.13g)$$

where

$$\sigma' = \text{conventional effective stress}$$
$$= \sigma - u^* = \sigma - \bar{u}_a - \bar{u}_w$$
$$= \sigma - u_w \text{ (for saturated soil)}$$
$$\sigma'' = \text{intrinsic effective stress}$$
$$= A' - R'$$

The effective contact stress, $\bar{c}$, has been defined as the *modified effective stress*. Equation 7.13f agrees with the general behaviour of a soil system showing an increase in the effective stress with an increase in the attractive forces and a decrease in the repulsive forces. The application of this expression to predicting the volume change and shear strength behaviour of clays has been hypothesized by Sridharan and Venkatappa Rao (1973, 1979). In a soil system with low $A' - R'$ forces, such as in granular soils, the expression for $\bar{c}$ tends to $\sigma'$, the conventional effective stress.

## 7.6 GEOSTATIC STRESSES

Stresses within a soil mass are caused by the self-weight of the soil and the external applied load. The stress patterns due to these effects are complicated. The magnitude of the subsurface stresses is affected by the presence of groundwater. Stresses induced by the overburden pressure are called *geostatic stresses*. This situation gives rise to simple stress calculations when the ground surface is horizontal, and there is no marked variation of the soil properties in the horizontal direction.

Hence, the vertical stress caused by the soil at a point below the surface is equal to the weight of the soil lying directly above the point. Considering the unit weight, $\gamma$, to be constant with depth, the vertical stress, $\sigma_v$ due to overburden at a depth $z$ from the ground surface is given as (Fig. 7.7.a)

$$\sigma_v = \gamma z \quad (7.14)$$

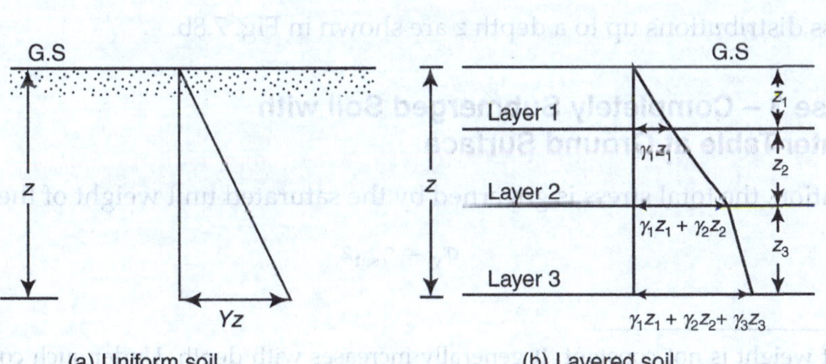


Fig. 7.7 Vertical stress distribution in uniform and layered soils

As the ground surface is horizontal, there are no shear stresses upon the horizontal or vertical planes. The vertical stress increases with depth.* If the soil stratum is layered with different total unit weights, then the vertical stress at a depth $z$ will be equal to the total weight of the individual soil layers (Fig. 7.7b); that is,

$$\sigma_v = \gamma_1 z_1 + \gamma_2 z_2 + \gamma_3 z_3 = \Sigma\gamma z \tag{7.15}$$

The situation will be different depending on the groundwater position. In general, five situations may be recognized for the static water condition, and they are explained below.

### 7.6.1 Case 1 – Soil Entirely Dry

Consider a level ground with the water table at a lower depth (Fig. 7.8a). The total stress, $\sigma_v$, at any depth $z$ is given as

$$\sigma_v = \gamma_d z \tag{7.16}$$

$$u_w = 0 \tag{7.17}$$

$$\sigma'_v = \sigma_v \tag{7.18}$$

where $\sigma'_v$ is the effective vertical stress due to overburden. The total, neutral, and effective stress variations up to a depth $z$ are shown in Fig. 7.8a.

### 7.6.2 Case 2 – Moist Soil

This is a situation of partially saturated soil wherein it is difficult to predict the neutral pressure distribution. Thus, this condition is treated as in Case 1 but with $\gamma$ instead of $\gamma_d$. Hence,

$$\sigma_v = \gamma z \tag{7.19}$$

$$u_w = 0 \tag{7.20}$$

$$\sigma'_v = \sigma_v \tag{7.21}$$

The stress distributions up to a depth $z$ are shown in Fig. 7.8b.

### 7.6.3 Case 3 – Completely Submerged Soil with Water Table at Ground Surface

In this situation, the total stress is governed by the saturated unit weight of the soil. Thus,

$$\sigma_v = \gamma_{sat} z \tag{7.22}$$

*But the unit weight is not constant. It generally increases with depth. Under such conditions, $\sigma_v$ is given as $\sigma_v = \int_0^z \gamma\, dz$.

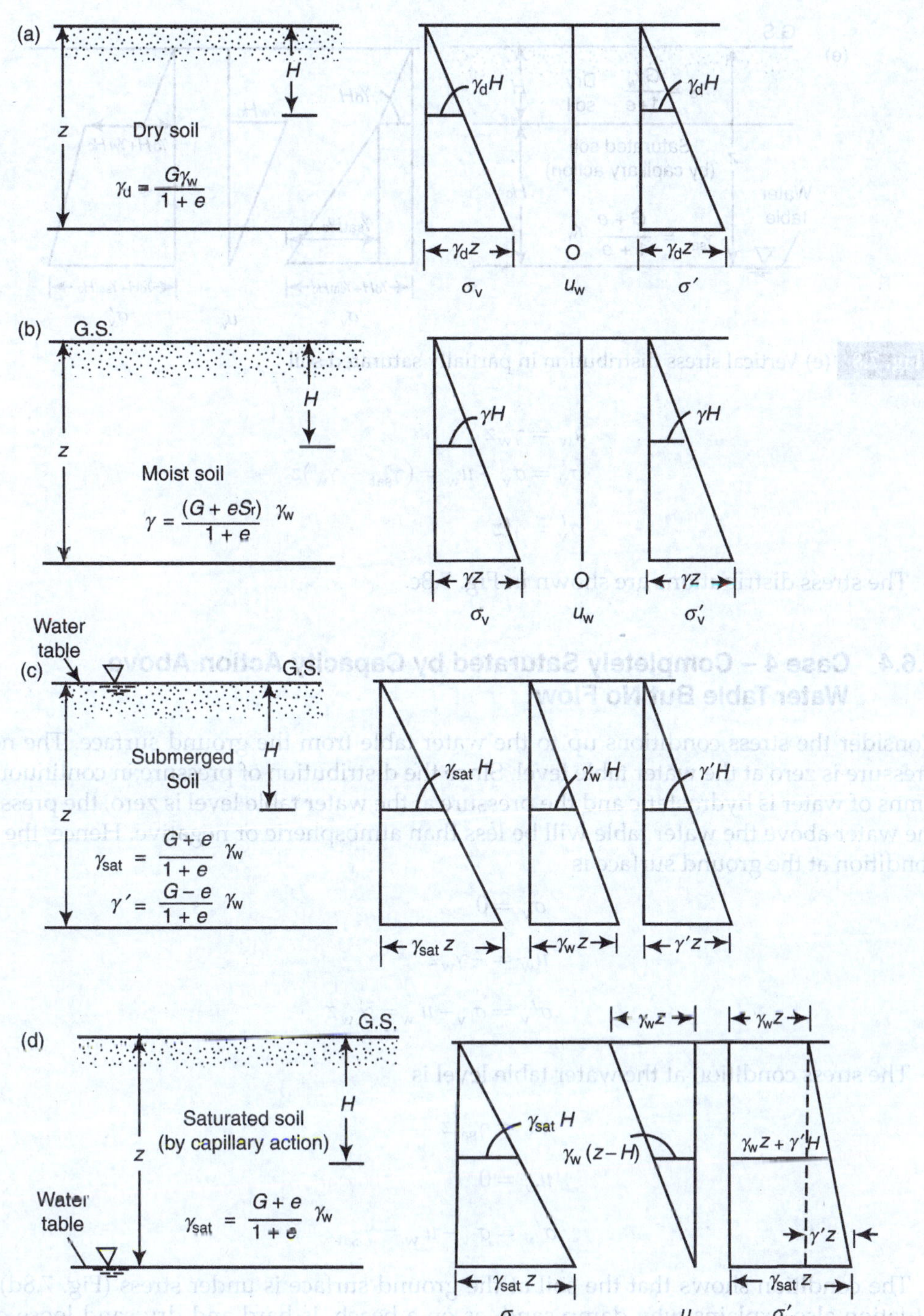


**Fig. 7.8** Vertical stress distribution in (a) dry soil, (b) moist soil, (c) submerged soil, and (d) saturated soil

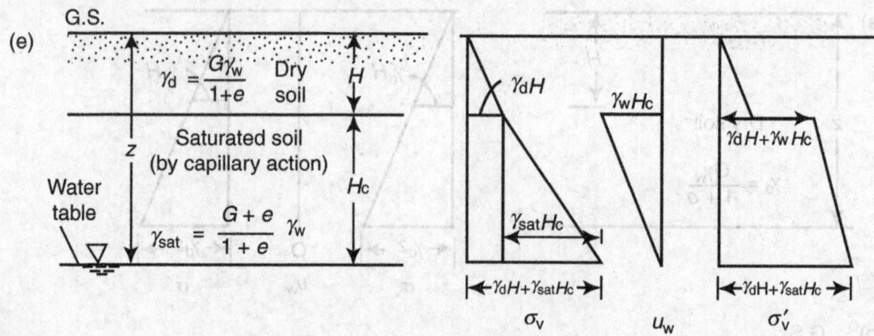


Fig. 7.8 (e) Vertical stress distribution in partially saturated soil

$$u_w = \gamma_w z \tag{7.23}$$

$$\sigma'_v = \sigma_v - u_w = (\gamma_{sat} - \gamma_w)z$$

$$\sigma'_v = \gamma' z \tag{7.24}$$

The stress distributions are shown in Fig. 7.8c.

### 7.6.4 Case 4 – Completely Saturated by Capacity Action Above Water Table But No Flow

Consider the stress conditions up to the water table from the ground surface. The neutral pressure is zero at the water table level. Since the distribution of pressure in continuous columns of water is hydrostatic and the pressure at the water table level is zero, the pressure in the water above the water table will be less than atmospheric or negative. Hence, the stress condition at the ground surface is

$$\sigma_v = 0 \tag{7.25}$$

$$u_w = -\gamma_w z \tag{7.26}$$

$$\sigma'_v = \sigma_v - u_w = \gamma_w z \tag{7.27}$$

The stress condition at the water table level is

$$\sigma_v = \gamma_{sat} z \tag{7.28}$$

$$u_w = 0 \tag{7.29}$$

$$\sigma'_v = \sigma_v - u_w = \gamma_{sat} z \tag{7.30}$$

The condition shows that the soil at the ground surface is under stress (Fig. 7.8d). This situation also explains why damp sand, as on a beach, is hard and dry sand loose on the surface.

### 7.6.5 Case 5 – Same as Condition 4 Except that the Height of Capillary Rise Is Less than *z*

Consider a situation in which the capillary rise is less than *z* and equals $(z-H) = H_c$ and soil above this level is dry up to the ground surface. The stresses at the ground surface are zero.

The stresses at a depth *H* from the ground surface are given as

$$\sigma_v = \gamma_d H \tag{7.31}$$

$$u_w = -\gamma_w H_c \tag{7.32}$$

$$\sigma'_v = \sigma_v - u_w = \gamma_d H - (-\gamma_w H_c) \tag{7.33}$$

$$\sigma'_v = \gamma_d H + \gamma_w H_c$$

That is, the stresses at a depth *z* from the ground surface are given as

$$\sigma_v = \gamma_d H + \gamma_{sat} H_c \tag{7.34}$$

$$u_w = 0 \tag{7.35}$$

$$\sigma'_v = \sigma_v - u_w = \gamma_d H + \gamma_{sat} H_c \tag{7.36}$$

The stress distributions for the conditions explained above are given in Fig. 7.8e.

For calculation of the stresses below the water table in Cases 4 and 5, Case 3 is combined with Case 1.

In the design of structures, such as retaining walls, sheeting, and pile foundations, the geostatic stresses acting in the horizontal direction are needed. The horizontal stress, $\sigma_h$, is a function of the vertical stress at the point under consideration. The ratio of the horizontal or lateral stress to the vertical stress is represented by a coefficient *K*, termed the *coefficient of lateral pressure*, that is,

$$K = \frac{\sigma_h}{\sigma_v} \tag{7.37}$$

$$\sigma_h = K\sigma_v \tag{7.38}$$

Depending on the stress history of the soil medium, *K* has a wide range of values. A detailed discussion of the parameter *K* is given in Chapter 11.

## 7.7 STRESSES DUE TO SURFACE LOADS

### 7.7.1 Elastic Half-Space

A soil medium with a horizontal ground surface extending laterally to infinite length and downwards from the horizontal is called a *semi-infinite medium* or *semi-infinite half-space*. If such a medium is assumed to be homogeneous, isotropic, and elastic, then it is called an *elastic half-space*. The theoretical treatment for determining the stresses in such a medium involves the theory of elasticity.

### 7.7.2 Elastic Properties of Soil

The relationship between the deformation or strain with stress is important in understanding the behaviour of any material. For some building materials, Hooke's law provides an useful approximation between the stress and strain. But this law does not necessarily hold good for soils, as their behaviour in general is non-linear and not perfectly elastic. It is observed that the entire stress–strain graph is a curve, unlike in steel, where the initial portion is predominantly a straight line. The slope of the initial portion of the curve is defined as the *stress–strain modulus, E*.

The determination of the stress–strain modulus is a much more critical problem. The most common method of computing the stress–strain modulus is to use the initial tangent value or the slope of the stress–strain curve at the origin (Fig. 7.9). This is referred to as the *initial tangent modulus* ($E_i$). The stress–strain modulus is also taken as the *initial secant modulus* ($E_s$), which is obtained using the origin and the secant line intercept at a stress level of 1/3 to 1/2 of the ultimate or failure stress. In another procedure, a cyclic loading test is done and the tangent modulus after the fifth or sixth cycle is taken. The tangent modulus for the re-loading curves is called the *re-load modulus* ($E_r$).

The initial tangent modulus is quite often used to represent the stress–strain modulus of a soil. The two main reasons for this choice are that the soil is elastic only near the origin and the region near the origin is nearly the same for different test plots. However, it has been recommended by many researchers that the re-load modulus is a better choice. The re-load modulus is generally higher than the initial tangent modulus of the first cycle due to the effect of strain hardening.

The other two important elastic properties of soil are the *shear modulus, G,* and *Poisson's ratio, ν*. The shear modulus is used in soil dynamic problems to estimate amplitudes of vibrations. The stress–strain modulus and Poisson's ratio are used in the evaluation of stresses and settlements.

Typical values of the stress–strain modulus and Poisson's ratio are given in Tables 7.1 and 7.2, respectively (Bowles, 1982). Table 7.2 shows that $\nu$ has a very narrow range of variation. An accurate prediction of $\nu$ is neither possible nor necessary. Fortunately, the value of $\nu$ usually has a relatively small effect on engineering predictions.

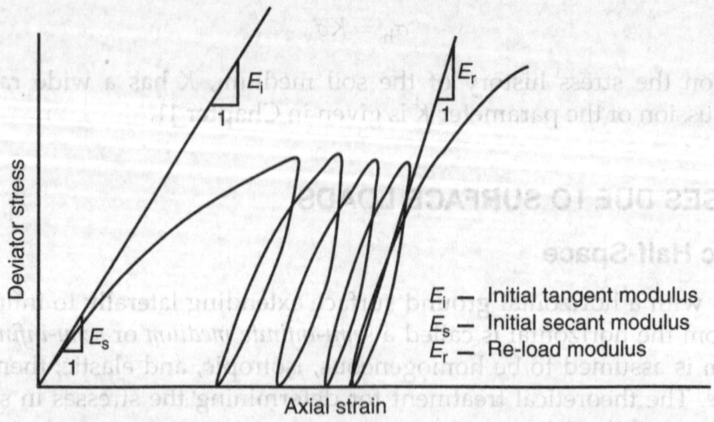


Fig. 7.9 Definitions of modulus

**Table 7.1** Typical range of values for stress–strain modulus, $E_s$ for soils

| Soil | $E_s$ (kN/m$^2$ $\times 10^3$) |
|---|---|
| Clay | |
| Very soft | 2–15 |
| Soft | 5–25 |
| Medium | 15–50 |
| Hard | 50–100 |
| Sandy | 25–250 |
| Glacial till | |
| Loose | 10–153 |
| Dense | 144–720 |
| Very dense | 478–1,440 |
| Loess sand | |
| Silty | 7–21 |
| Loose | 10–24 |
| Dense | 48–81 |
| Sand and gravel | |
| Loose | 48–144 |
| Dense | 96–192 |
| Shale | 144–14,400 |
| Silt | 2–20 |

*Source:* Bowles (1982).

**Table 7.2** Typical range of values for Poisson's ratio

| Soil | $\nu$ |
|---|---|
| Clay, saturated | 0.4–0.5 |
| Clay, unsaturated | 0.1–0.3 |
| Sandy clay | 0.2–0.3 |
| Silt | 0.3–0.35 |
| Sand | 0.15–0.40 |
| Loess | 0.10–0.30 |

*Source:* Bowles (1982).

Soil does not completely fulfil the basic assumptions of homogeneity and isotropy made in the elastic half-space concept. However, the civil engineer has to apply the results of this theory with judgement.

### 7.7.3 Boussinesq's Theory

One of the methods for computing stresses based on the theory of elasticity was given by Boussinesq (1885). Boussinesq assumed a weightless, elastic half-space, and given components of stresses caused by a vertical-point-surface load (Fig. 7.10). The stress components

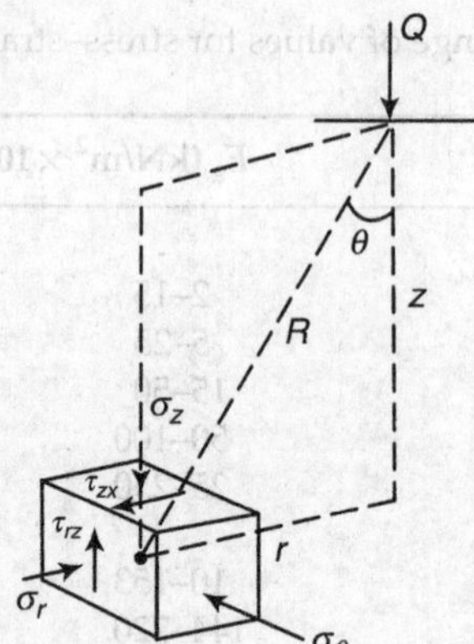


Fig. 7.10 Stresses at a point due to point load

due to the surface load, viz., the vertical stress, $\sigma_z$, radial stress, $\sigma_r$, circumferential stress, $\sigma_\theta$, and shear stress, $\tau_{rz}$, are given as (using polar coordinates, $r$, $\theta$, and $z$),

$$\sigma_z = \frac{3Q}{2\pi z^2}\left[\frac{1}{1+(r/z)^2}\right]^{5/2} \tag{7.39}$$

$$\sigma_r = \frac{Q}{2\pi}\left[\frac{3r^2 z}{(r^2/z^2)^{5/2}} - \frac{1-2\nu}{r^2+z^2+z(r^2+z^2)^{1/2}}\right] \tag{7.40}$$

$$-\sigma_\theta = \frac{Q}{2\pi}(1-2\nu)\left[\frac{z}{(r^2/z^2)^{3/2}} - \frac{1}{r^2+z^2+z\sqrt{(r^2+z^2)}}\right] \tag{7.41}$$

and

$$\tau_{rz} = \frac{3Q}{2\pi}\left[\frac{rz^2}{1+(r/z)^{5/2}}\right] \tag{7.42}$$

Equation 7.39 is most frequently used in practice. This equation represents a high stress beneath the point of load application ($z = 0$) and a decrease in stress with increase in depth. Further, the stress decreases with increasing distance from the point of load application. It should also be observed that it does not depend on the elastic or other properties of the soil, i.e., it is independent of the material content of the medium (i.e., clay or sand). This equation can be written in terms of an influence factor, called Boussinesq's vertical stress coefficient, $N_B$, when

$$N_B = \frac{3Q}{2\pi}\left[\frac{1}{1+(r^2/z^2)}\right]^{5/2} \tag{7.43}$$

Then

$$\sigma_z = \frac{Q}{z^2}N_B \tag{7.44}$$

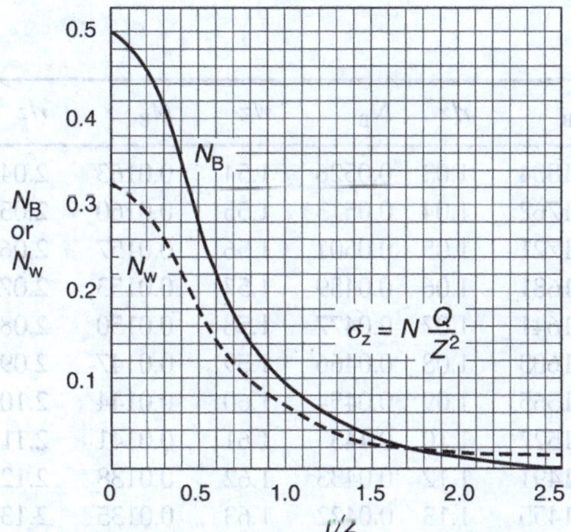


**Fig. 7.11** Boussinesq and Westergaard coefficients for a concentrated load

Equation 7.44 shows that the vertical stress is (i) directly proportional to the load, (ii) inversely proportional to the depth squared, and (iii) proportional to some function of the ratio $r/z$.

The solid line in Fig. 7.11 shows a variation of $N_B$ with the ratio $r/z$. Table 7.3 presents Boussinesq's vertical stress coefficients.

**Table 7.3** Boussinesq coefficients

| r/z | $N_B$ | r/z | $N_B$ | r/z | $N_B$ | r/z | $N_B$ | r/z | $N_B$ | r/z | $N_B$ |
|---|---|---|---|---|---|---|---|---|---|---|---|
| 0.00 | 0.4775 | 0.34 | 0.2733 | 84 | 0.0844 | 1.35 | 0.0251 | 1.85 | 0.0085 | 2.35 | 0.0034 |
| 0.01 | 0.4773 | 0.35 | 0.2679 | 85 | 0.0823 | 1.36 | 0.0245 | 1.86 | 0.0084 | 2.36 | 0.0033 |
| 0.03 | 0.4764 | 0.37 | 0.2571 | 87 | 0.0783 | 1.38 | 0.0234 | 1.88 | 0.0081 | 2.38 | 0.0032 |
| 0.04 | 0.4756 | 0.38 | 0.2518 | 88 | 0.0764 | 1.39 | 0.0229 | 1.89 | 0.0079 | 2.39 | 0.0032 |
| 0.05 | 0.4756 | 0.39 | 0.2466 | 89 | 0.0744 | 1.40 | 0.0224 | 1.90 | 0.0078 | 2.40 | 0.0031 |
| 0.06 | 0.4723 | 0.40 | 0.2414 | 90 | 0.0727 | 1.41 | 0.0219 | 1.91 | 0.0076 | 2.41 | 0.0031 |
| 0.07 | 0.4717 | 0.41 | 0.2363 | 91 | 0.0709 | 1.42 | 0.0214 | 1.92 | 0.0075 | 2.42 | 0.0030 |
| 0.08 | 0.4699 | 0.42 | 0.2313 | 92 | 0.0691 | 1.43 | 0.0209 | 1.93 | 0.0073 | 2.43 | 0.0030 |
| 0.09 | 0.4679 | 0.43 | 0.2263 | 93 | 0.0671 | 1.44 | 0.0204 | 1.94 | 0.0072 | 2.44 | 0.0029 |
| 0.10 | 0.4657 | 0.44 | 0.2214 | 94 | 0.0658 | 1.45 | 0.0200 | 1.95 | 0.0070 | 2.45 | 0.0029 |
| 0.11 | 0.4633 | 0.45 | 0.2165 | 95 | 0.0641 | 1.46 | 0.0195 | 1.96 | 0.0069 | 2.46 | 0.0028 |
| 0.12 | 0.4607 | 0.46 | 0.2117 | 96 | 0.0626 | 1.47 | 0.0191 | 1.97 | 0.0068 | 2.47 | 0.0028 |
| 0.13 | 0.4579 | 0.47 | 0.2070 | 97 | 0.0610 | 1.48 | 0.0187 | 1.98 | 0.0066 | 2.48 | 0.0027 |
| 0.14 | 0.4548 | 0.48 | 0.2040 | 98 | 0.0595 | 1.49 | 0.0183 | 1.99 | 0.0065 | 2.49 | 0.0027 |
| 0.15 | 0.4516 | 0.49 | 0.1978 | 99 | 0.0581 | 1.50 | 0.0171 | 2.00 | 0.0064 | 2.50 | 0.0026 |
| 0.16 | 0.4482 | 0.50 | 0.1934 | 1.00 | 0.0567 | 1.51 | 0.0171 | 2.01 | 0.0063 | 2.51 | 0.0026 |
| 0.17 | 0.446 | 0.51 | 0.1889 | 1.01 | 0.0553 | 1.52 | 0.0171 | 2.02 | 0.0062 | 2.52 | 0.0025 |
| 0.18 | 0.4409 | 0.52 | 0.1846 | 1.02 | 0.0539 | 1.53 | 0.0167 | 2.03 | 0.0060 | 2.53 | 0.0025 |

*Table 7.3 Contd.*

**Table 7.3** Contd.

| $r/z$ | $N_B$ | $r/z$ | $N_B$ | $r/z$ | $N_B$ | $r/z$ | $N_B$ | $r/z$ | $N_B$ | $r/z$ | $N_B$ |
|---|---|---|---|---|---|---|---|---|---|---|---|
| 0.19 | 0.4370 | 0.53 | 0.1804 | 1.03 | 0.0526 | 1.54 | 0.0163 | 2.04 | 0.0059 | 2.54 | 0.0025 |
| 0.20 | 0.4329 | 0.54 | 0.1762 | 1.04 | 0.0513 | 1.55 | 0.0160 | 2.05 | 0.0058 | 2.55 | 0.0024 |
| 0.21 | 0.4286 | 0.55 | 0.1721 | 1.05 | 0.0501 | 1.56 | 0.0157 | 2.06 | 0.0057 | ... | ... |
| 0.22 | 0.4242 | 0.56 | 0.1681 | 1.06 | 0.0489 | 1.57 | 0.0153 | 2.07 | 0.0056 | 2.57 | 0.0023 |
| 0.23 | 0.4197 | 0.57 | 0.1641 | 1.07 | 0.0477 | 1.58 | 0.0150 | 2.08 | 0.0055 | ... | ... |
| 0.24 | 0.4151 | 0.58 | 0.1603 | 1.08 | 0.0466 | 1.59 | 0.0147 | 2.09 | 0.0054 | 2.59 | 0.0023 |
| 0.25 | 0.4103 | 0.59 | 0.1565 | 1.09 | 0.0454 | 1.60 | 0.0144 | 2.10 | 0.0053 | ... | ... |
| 0.26 | 0.4054 | 0.60 | 0.1527 | 1.10 | 0.443 | 1.61 | 0.0141 | 2.11 | 0.0052 | 2.61 | 0.0022 |
| 0.27 | 0.4004 | 0.61 | 0.1491 | 1.12 | 0.0433 | 1.62 | 0.0138 | 2.12 | 0.0051 | ... | ... |
| 0.28 | 0.3954 | 0.62 | 0.1455 | 1.13 | 0.0422 | 1.63 | 0.0135 | 2.13 | 0.0050 | 2.63 | 0.0021 |
| 0.29 | 0.3902 | 0.63 | 0.1420 | 1.14 | 0.0412 | 1.64 | 0.0132 | 2.14 | 0.0049 | ... | ... |
| 0.30 | 0.3849 | 0.64 | 0.1386 | 1.15 | 0.0402 | 1.65 | 0.0129 | 2.15 | 0.0048 | 2.65 | 0.0021 |
| 0.31 | 0.3796 | 0.65 | 0.1353 | 1.16 | 0.0393 | 1.66 | 0.0126 | 2.16 | 0.0047 | ... | ... |
| 0.32 | 0.3742 | 0.66 | 0.1320 | 1.17 | 0.0384 | 1.67 | 0.0124 | 2.17 | 0.0047 | 2.67 | 0.0019 |
| 0.33 | 0.3687 | 0.67 | 0.1288 | 1.18 | 0.0374 | 1.68 | 0.0121 | 2.18 | 0.0046 | ... | ... |
| 0.34 | 0.3632 | 0.68 | 0.1257 | 1.19 | 0.0365 | 1.69 | 0.0119 | 2.19 | 0.0045 | 2.69 | 0.0017 |
| 0.35 | 0.3577 | 0.69 | 0.1226 | 1.20 | 0.0357 | 1.70 | 0.0116 | 2.20 | 0.0044 | ... | ... |
| 0.36 | 0.3521 | 0.70 | 0.1196 | 1.21 | 0.0348 | 1.71 | 0.0114 | 2.21 | 0.0043 | 2.71 | 0.0015 |
| 0.37 | 0.3465 | 0.71 | 0.1166 | 1.22 | 0.0340 | 1.72 | 0.0112 | 2.22 | 0.0043 | ... | ... |
| 0.38 | 0.3408 | 0.72 | 0.1138 | 1.23 | 0.0332 | 1.73 | 0.0109 | 2.23 | 0.0042 | 2.73 | 0.0013 |
| 0.39 | 0.3351 | 0.73 | 0.1110 | 1.24 | 0.0324 | 1.74 | 0.0107 | 2.24 | 0.0041 | ... | ... |
| 0.40 | 0.3294 | 0.74 | 0.1083 | 1.25 | 0.0317 | 1.75 | 0.0105 | 2.25 | 0.0040 | 2.75 | 0.0011 |
| 0.41 | 0.3238 | 0.75 | 0.1057 | 1.26 | 0.0309 | 1.76 | 0.0103 | 2.26 | 0.0040 | ... | ... |
| 0.42 | 0.3181 | 0.76 | 0.1031 | 1.27 | 0.0302 | 1.77 | 0.0101 | 2.27 | 0.0039 | 2.77 | 0.0009 |
| 0.43 | 0.3124 | 0.77 | 0.1005 | 1.28 | 0.0295 | 1.78 | 0.0099 | 2.28 | 0.0038 | ... | ... |
| 0.44 | 0.3068 | 0.78 | 0.0981 | 1.29 | 0.0288 | 1.79 | 0.0097 | 2.29 | 0.0038 | 2.79 | 0.0007 |
| 0.45 | 0.3011 | 0.79 | 0.0956 | 1.30 | 0.0282 | 1.80 | 0.0095 | 2.30 | 0.0037 | ... | ... |
| 0.46 | 0.2955 | 0.80 | 0.0933 | 1.31 | 0.0275 | 1.81 | 0.0093 | 2.31 | 0.0036 | 2.81 | 0.0005 |
| 0.47 | 0.2899 | 0.81 | 0.0910 | 1.32 | 0.0269 | 1.82 | 0.0091 | 2.32 | 0.0036 | ... | ... |
| 0.48 | 0.2843 | 0.82 | 0.0887 | 1.33 | 0.0263 | 1.83 | 0.0089 | 2.33 | 0.0035 | 2.83 | 0.0003 |
| 0.49 | 0.2788 | 0.83 | 0.0865 | 1.34 | 0.0257 | 1.84 | 0.0087 | 2.34 | 0.0034 | ... | ... |
| | | | | | | | | | | 2.85 | 0.0001 |
| | | | | | | | | | | 6.15 | 0.0001 |

### 7.7.4 Pressure Distribution Diagrams

Boussinesq's vertical stress equation may be used to draw three types of pressure distribution diagrams (Fig. 7.12). They are

1. the stress isobar,
2. the vertical stress distribution on a horizontal plane at a depth of $z$ below the ground surface, and
3. the vertical stress distribution on a vertical plane at a distance of $r$ from the load point.

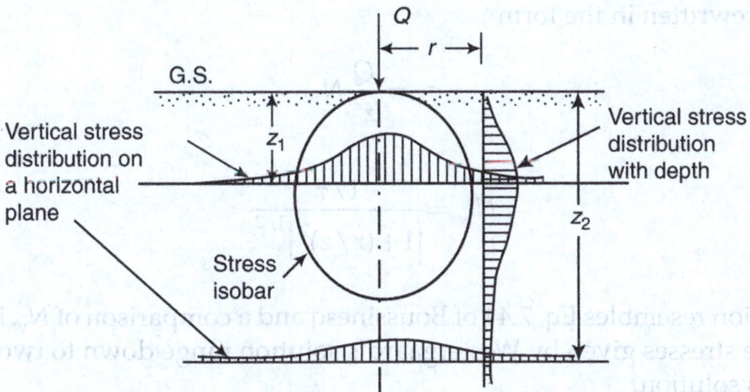


Fig. 7.12 Stresses at a point due to point load

The stress isobar is a stress contour connecting all points of equal stress below the ground surface. There are many isobars for a given load system. The stress isobar is also referred to as bulb of pressure or pressure bulb. The soil mass bounded within a pressure bulb furnishes the support power of a footing.

The vertical stress distribution on a horizontal plane at a depth $z_1$ from the ground surface is obtained by varying $r$. The magnitude of the vertical stress along the load-line decreases with an increase in depth and this is reflected in the distribution diagram at a depth $z_2$ (Fig. 7.12).

The vertical stress distribution on a vertical plane at a distance $r$ from the load points is obtained by varying $z$. The diagram represents a maximum value at a depth nearer to the ground surface, which decreases with depth. The magnitude of the maximum value will decrease with increasing distance from the load point (Fig. 7.12).

### 7.7.5 Westergaard Equation

Some fine-grained soils are interspersed with thin lenses of coarse-grained material that partially prevent lateral deformation of the soil. Such a situation represents the non-homogeneous condition. Westergaard (1938) suggested a solution to such a material by considering an elastic medium in which the lateral strain was assumed to be zero. As some of the soils, e.g., sedimentary soils, are of this type, Westergaard's solution may be taken as a better approximation for such soils than compared with that proposed by Boussinesq for homogeneous soils. Westergaard's expression for the vertical stress is given as

$$\sigma_z = \frac{Q}{z^2} 1/2\pi \frac{\sqrt{(1-2\nu)}}{(2-2\nu)/\left[(1-2\nu)/(2-2\nu)+(r/z)^2\right]^{3/2}} \tag{7.45}$$

Westergaard further considered obtaining a maximum by letting $\nu = 0$; hence, Eq. 7.45 reduces to

$$\sigma_z = \frac{Q}{z^2} \frac{1/\pi}{\left[1+2(r/z)^2\right]^{3/2}} \tag{7.46}$$

This can be rewritten in the form

$$\sigma_z = \frac{Q}{z^2} N_w \tag{7.47}$$

where

$$N_w = \frac{1/\pi}{\left[1+(r/z)^2\right]^{3/2}}$$

This expression resembles Eq. 7.44 of Boussinesq and a comparison of $N_w$ is made with $N_B$ in Fig. 7.11. The stresses given by Westergaard's solution range down to two-thirds of those of Boussinesq's solution.

### 7.7.6 Types of Surface Loads

In practice, the loads are applied over finite areas and never as point loads. Boussinesq's point load solution may be conveniently integrated to obtain stresses due to surface loads distributed over a particular area.

**Line Loads of Infinite Length**. Stresses at a point A due to a line load $q$ per unit length on the surface are given as (Fig. 7.13)

$$\sigma_z = \frac{2q}{\pi} \frac{z^3}{(x^2+z^2)^2} \tag{7.48}$$

and

$$\tau_{rz} = \frac{2q}{\pi} \frac{xz^2}{(x^2+z^2)^2} \tag{7.49}$$

The lateral pressure on earth-retaining structures caused by a line load (e.g., a railway) on the surface of the backfill may be computed using Eq. 7.48.

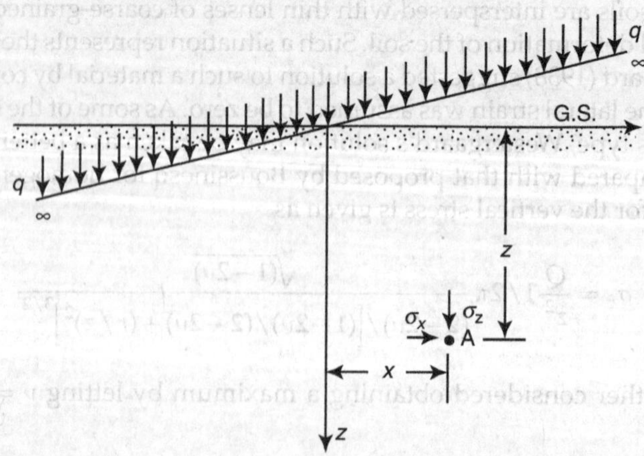


Fig. 7.13 Uniformly distributed infinite linear load

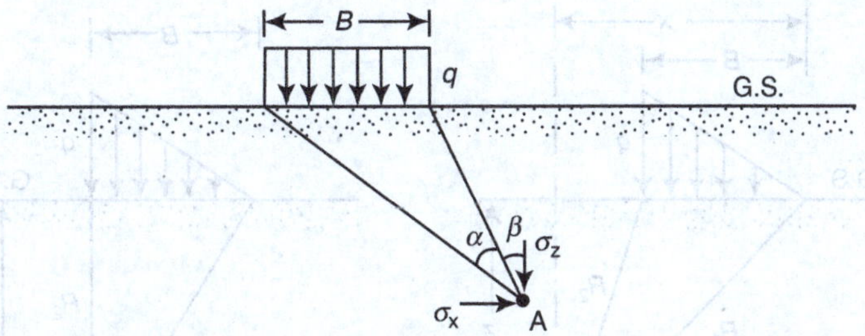


Fig. 7.14 Uniformly distributed infinite strip load

**Strip Area Carrying Uniform Pressure**. A strip of width $B$ and infinite length, loaded with a uniform pressure, is shown in Fig. 7.14 (similar to the pressure of a wall footing). The stresses at point A are given as

$$\sigma_z = \frac{q}{\pi}[\alpha + \sin\alpha\cos(\alpha + 2\beta)] \tag{7.50}$$

and

$$\tau_{rz} = \frac{q}{\pi}[\sin\alpha\sin(\alpha + 2\beta)] \tag{7.51}$$

A plot of contours of equal vertical stresses is shown in Fig. 7.15 for different stress ratios. As explained earlier, this enables one to fix the depth of the stress influence. The distribution of stress beneath an uniform strip load is important in estimating settlements.

**Strip Area of Triangular Shape**. A triangular strip area carrying a linearly increasing pressure over width $B$ and of infinite length is shown in Fig. 7.16a. The vertical stress at a point A due to such a surface load is

$$\sigma_z = \frac{q}{\pi}\left(\frac{x}{B}\alpha - \frac{1}{2}\sin 2\beta\right) \tag{7.52}$$

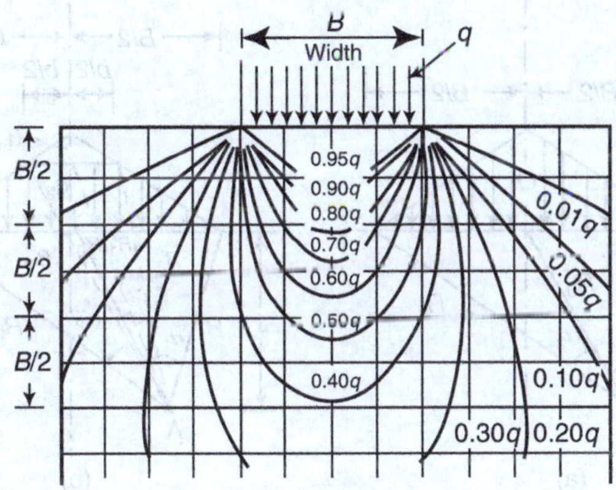


Fig. 7.15 Bulbs of vertical pressure under uniform strip load

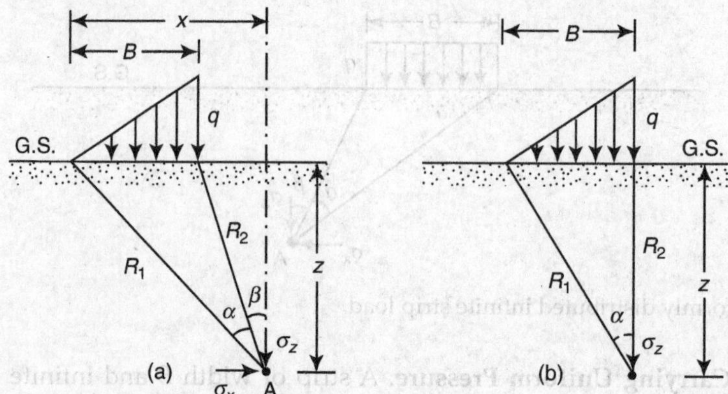


**Fig. 7.16** Uniformly distributed infinite triangular load

The vertical stress beneath the vertical face (Fig. 7.16b) is obtained by making $\beta = 0$ and $x = B$; thus,

$$\sigma_z = \frac{q}{\pi}\alpha \tag{7.53}$$

The shear stress is given as

$$\tau_{xz} = \frac{q}{\pi}\left(1 + \cos 2\beta - 2\frac{z}{B}\alpha\right) \tag{7.54}$$

For a symmertrically distributed triangular load (Fig. 7.17a), the stresses are

$$\sigma_z = \frac{2q}{\pi B}\left[\frac{B}{2}(\alpha_1 + \alpha_2) + x(\alpha_1 - \alpha_2) - 2z\ \log_e \frac{R_1\ R_2}{R_0^2}\right] \tag{7.55}$$

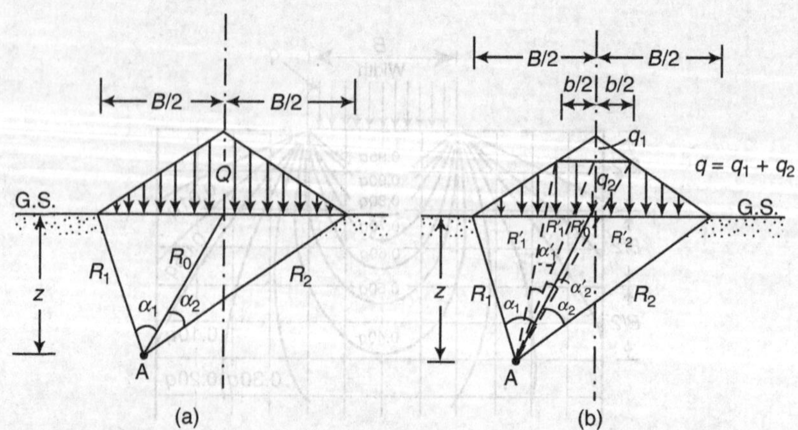


**Fig. 7.17** Symmetrically distributed triangular load

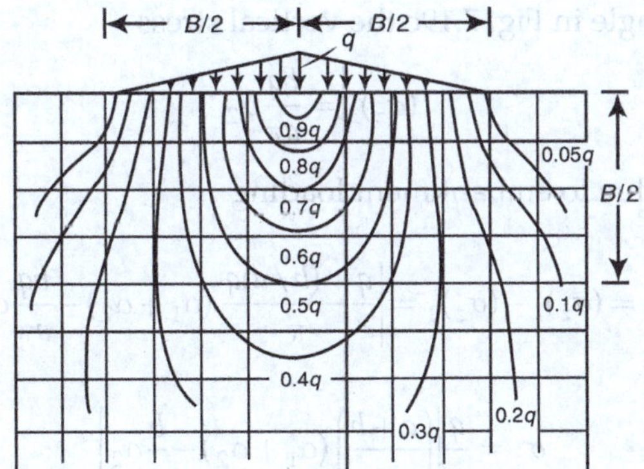


Fig. 7.18 Bulbs of vertical pressure under symmetrical triangular load

and

$$\tau_{rz} = \frac{2qz}{\pi B}(\alpha_1 - \alpha_2) \tag{7.56}$$

Equation 7.55 is of use in the estimation of settlement of embankments. An embankment section may be considered as the difference between two triangles of equal angles but of unequal base width (Fig. 7.17b). The stress beneath such an embankment may be obtained by subtracting the stresses due to the small triangle from those due to the large triangle. The pressure bulbs under a triangular strip load are shown in Fig. 7.18.

**Strip Area Loaded with Embankment Loading.** An increase in vertical stress in a soil mass due to embankment type loading may be handled using the method of superposition.

One half of the embankment is shown in Fig. 7.19a as a half-sectional elevation. The vertical stress at A due to this loading is equal to the stress caused at A by the large triangle (Fig. 7.19b) minus the stress caused at A by the small triangle (Fig. 7.19c).

Applying Eq. 7.53 to Fig. 7.19b, the vertical stress

$$(\sigma_z)_1 = \frac{q + (b/a)q}{\pi}(\alpha_1 + \alpha_2) \tag{7.57}$$

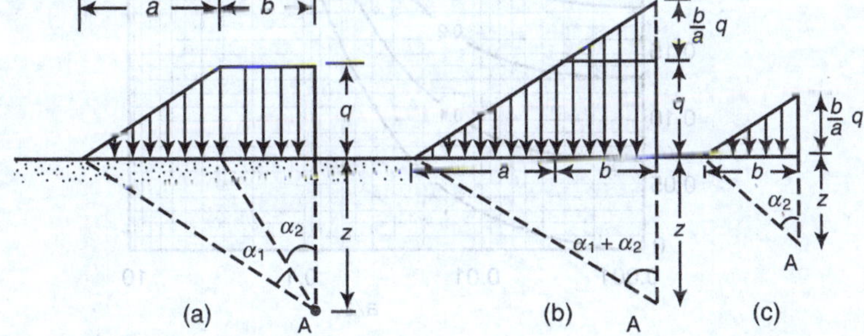


Fig. 7.19 Vertical stress due to embankment loading

Similarly, for the triangle in Fig. 7.19c the vertical stress

$$(\sigma_z)_2 = \frac{bq}{a\pi}\alpha_2 \tag{7.58}$$

Therefore, the stress due to embankment loading

$$\sigma_z = (\sigma_z)_1 - (\sigma_z)_2 = \left[\frac{q+(b/a)q}{\pi}(\alpha_1+\alpha_2) - \frac{bq}{a\pi}\alpha_2\right] \tag{7.59}$$

Therefore,

$$\sigma_z = \frac{q}{\pi}\left[\left(\frac{a+b}{a}\right)(\alpha_1+\alpha_2) - \frac{b}{a}\alpha_2\right] \tag{7.60}$$

or $\sigma_z = N_E q$, where

$$N_E = \frac{1}{\pi}\left[\left(\frac{a+b}{a}\right)(\alpha_1+\alpha_2) - \frac{b}{a}\alpha_2\right] \tag{7.61}$$

$$= \frac{1}{\pi} f\left(\frac{a}{z}, \frac{b}{z}\right) \tag{7.62}$$

The values of the influence factor for various $a/z$ and $b/z$ are given in Fig. 7.20 (Osterberg, 1957).

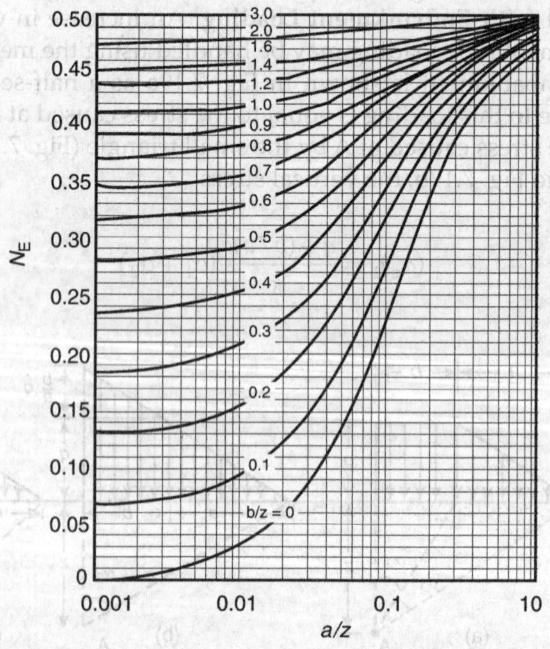


Fig. 7.20 Influence chart of embankment loading (*Source:* Osterberg, 1957)

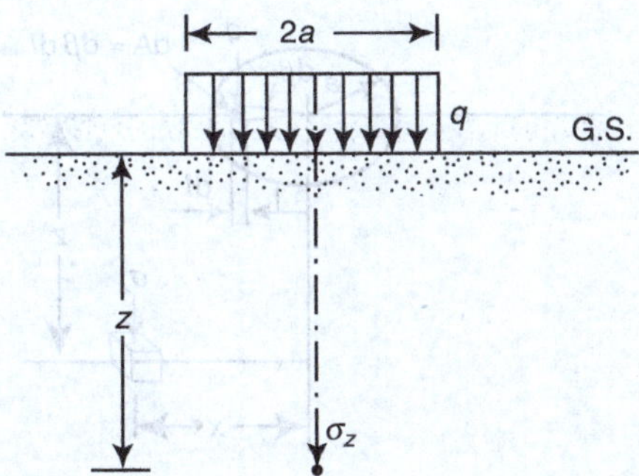


**Fig. 7.21** Vertical stress at centre of circular loaded area

**Circular Area Carrying Uniform Pressure**. Two cases of stresses due to a uniform pressure on a circular area are available, viz., (i) stresses under the centre of the circular area and (ii) stresses at any point on the soil.

The vertical stress at a depth $z$ under the centre of a circular area of diameter $2a$ is (Fig. 7.21)

$$\sigma_z = q\left[1-\left\{\frac{1}{(1+a/z)^2}\right\}^{3/2}\right] \tag{7.63}$$

$$= qN_{cc} \tag{7.64}$$

where

$$N_{cc} = \left[1-\left\{\frac{1}{(1+a/z)^2}\right\}^{3/2}\right] \tag{7.65}$$

The value of the influence factor $N_{cc}$ is given in Fig. 7.22. Equation 7.63 is valid only for the stress along the centre line. Foster and Ahlvin (1954) have given a chart for finding $\sigma_z$ at any point lying under as well as outside the loaded area (Fig. 7.23).

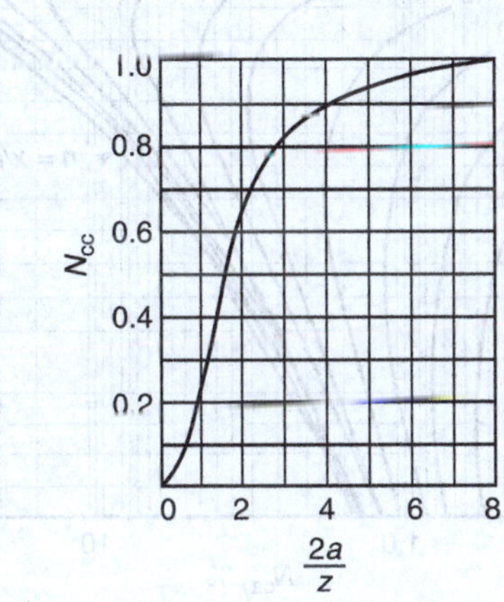


**Fig. 7.22** Influence chart for vertical stress under centre of circular area

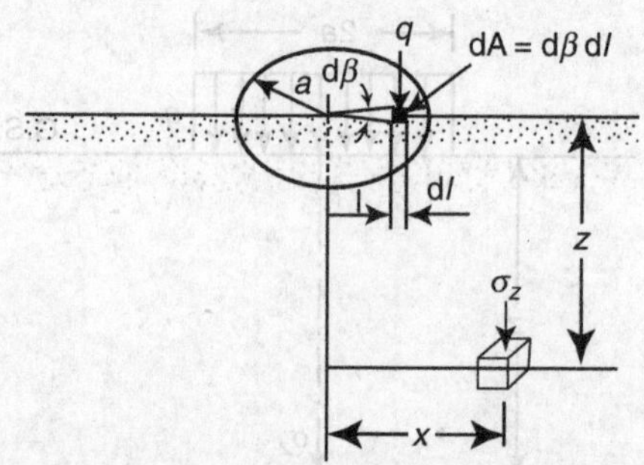


**Fig. 7.23** Vertical stress at any point due to uniformly loaded circular area

The expression for $\sigma_z$ is of the form

$$\sigma_z = \left(\frac{3qa^3}{2\pi}\right)\int_0^{2\pi}\int_0^{a} \frac{l\,\mathrm{d}\beta\,\mathrm{d}l}{(l^2+z^2+x^2-2xl\cos\beta)^{5/2}} \tag{7.66}$$

or

$$\sigma_z = q\,N_{CA}(m,n) \tag{7.67}$$

where $N_{CA}$ is a shape function of dimensionless variables,

$$m = \frac{z}{a}, \quad n = \frac{x}{a}$$

This chart (Fig. 7.24) is based on the assumption that Poission's ratio $\nu = 0.5$. This is applicable to points under the centre as well as at all points away from the centre. The pressure bulbs for a uniform circular load are given in Fig. 7.25.

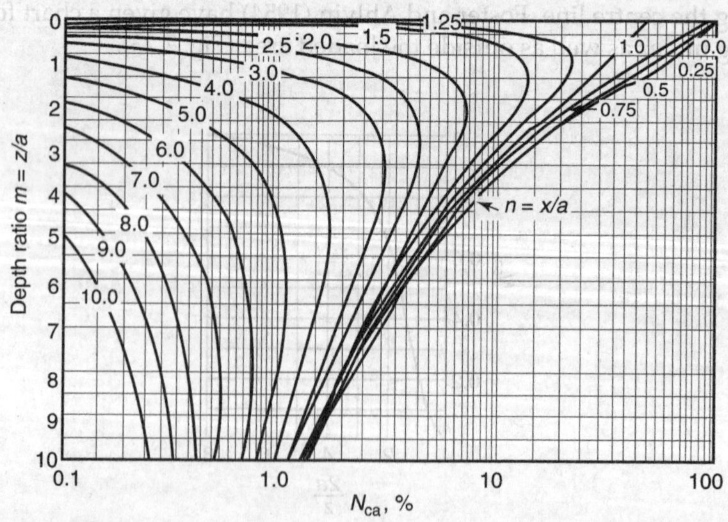


**Fig. 7.24** Influence chart for vertical stress at any point due to uniformly loaded circular area (*Source:* Foster and Ahlvin, 1954)

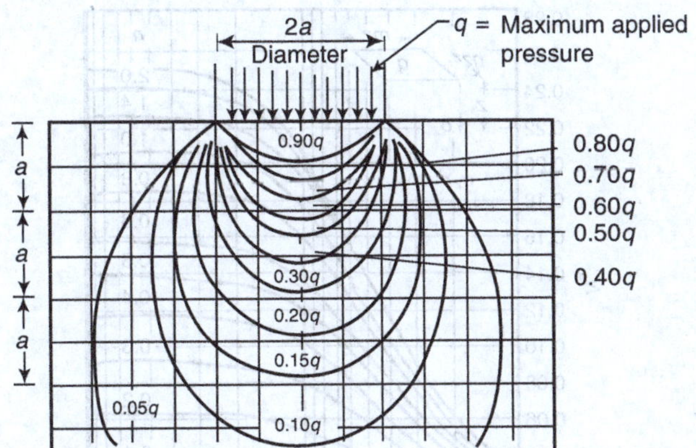


Fig. 7.25 Bulbs of vertical pressure under uniform circular load

**Rectangular Area Carrying Uniform Pressure.** The vertical stress beneath the corner of a uniformly loaded rectangular area can be expressed as (Fig. 7.26)

$$\sigma_z = \frac{q}{4\pi}\left[\frac{2mn(m^2+n^2+1)^{1/2}}{m^2+n^2+m^2n^2+1}\cdot\frac{m^2+n^2+2}{m^2+n^2+1}+\tan^{-1}\frac{2mn(m^2+n^2+1)^{1/2}}{m^2+n^2+1-m^2n^2}\right] \tag{7.68}$$

Here the width, $B$, and length, $L$, of the rectangle are given as $mz$ and $nz$, where $z$ is the depth under consideration. Equation 7.68 can be written as

$$\sigma_z = qN_R \tag{7.69}$$

where

$$N_R = \frac{1}{4\pi}\left[\frac{2mn(m^2+n^2+1)^{1/2}}{m^2+n^2+m^2n^2+1}\cdot\frac{m^2+n^2+2}{m^2+n^2+1}+\tan^{-1}\frac{2mn(m^2+n^2+1)^{1/2}}{m^2+n^2+1-m^2n^2}\right] \tag{7.70}$$

Figure 7.27 shows the variation of $N_R$ with $m$ and $n$ (Fadum, 1948). The factors $m$ and $n$ in the chart are interchangeable. This chart can be adopted for any area based on rectangles under any point within or outside the area to be obtained by the method of superposition.

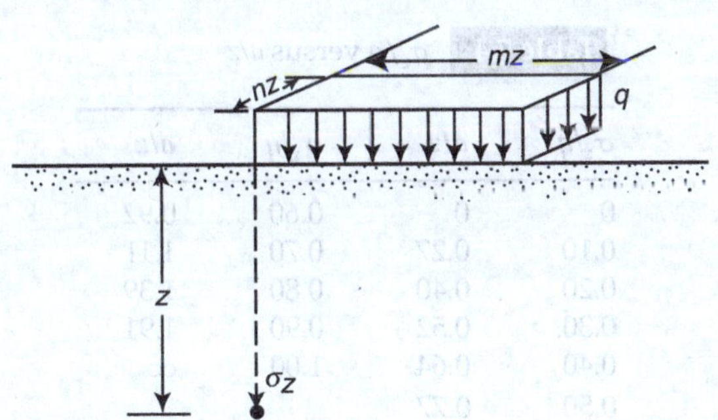


Fig. 7.26 Vertical stress under corner of uniformly loaded rectangular area

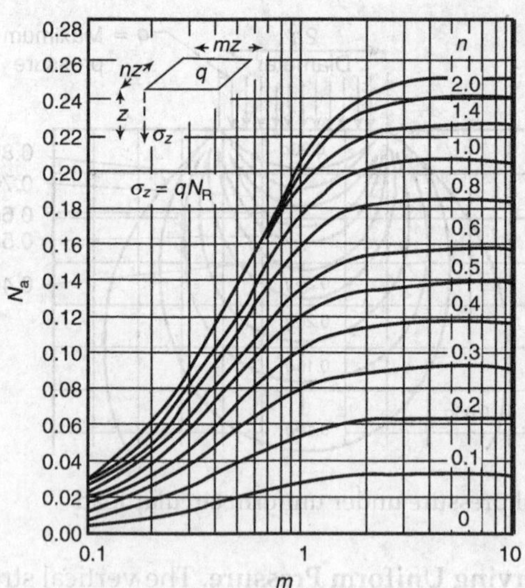


**Fig. 7.27** Influence chart for vertical stress under corner of uniformly loaded rectangular area (*Source:* Fadum, 1948)

**Newmark's Influence Chart**. The preceding sections outline the stress distributions due to loaded areas of regular geometry and cannot be applied without error to irregularly shaped areas. Newmark (1942) devised a graphical procedure based on the expression for the vertical stress under the centre of a loaded circular area (Fig. 7.21).

Equation 7.63 for a circular loaded area can be rewritten in the form

$$\frac{\sigma_z}{q} = \left[1 - \left\{\frac{1}{1+(a/z)^2}\right\}^{3/2}\right]$$

For different values $\sigma_z/q$, the values of the ratio are calculated. The $a/z$ values for $\sigma_z/q$ varying from 0 to 1.0 are given in Table 7.4.

A suitable depth scale is chosen and all the radii for nine circles are calculated and drawn (Fig. 7.28); e.g., the first circle, with $\sigma_z/q = 0.10$, will have a radius $a = 0.27z$. If the depth $z$ (in metres) is represented by the length AB (in mm), then the radius $a$ (in mm) $= 0.27$AB.

**Table 7.4** $\sigma_z/q$ versus $a/z$

| $\sigma_z/q$ | $a/z$ | $\sigma_z/q$ | $a/z$ |
|---|---|---|---|
| 0 | 0 | 0.60 | 0.92 |
| 0.10 | 0.27 | 0.70 | 1.11 |
| 0.20 | 0.40 | 0.80 | 1.39 |
| 0.30 | 0.52 | 0.90 | 1.91 |
| 0.40 | 0.64 | 1.00 | ∞ |
| 0.50 | 0.77 | | |

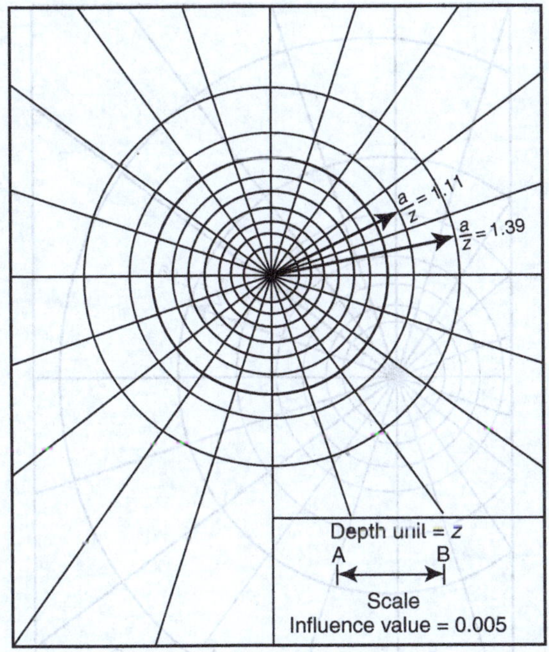


**Fig. 7.28** Newmark influence chart for vertical stress at any depth z = AB (*Source:* Newmark, 1942)

Now, divide each circular ring into 20 convenient parts (i.e., 200 influence units). The stress transferred by one annular ring is 0.10, which is divided into 20 parts. Thus, the influence value for one block (irrespective of size) is

$$N_N = \frac{0.10}{20} = 0.005 \tag{7.71}$$

The stress at a depth z for a specifc point is

$$\sigma_z - q \times N_N \times (\text{number of influence blocks}) \tag{7.72}$$

To use this chart, the loaded surface is drawn to a scale such that the distance AB equals the depth of the point in question. The point beneath the loaded area for which the vertical stress is sought is then located over the centre of the chart. The plotted area covers a number of influence blocks, and the number of influence units are counted. Thus, the vertical stress is found from Eq. 7.72.

Figure 7.29 shows Newmark's influence chart for the vertical stress based on Westergaard's theory.

**Approximate Soultion of Vertical Stress**. Approximate estimates of vertical stress at a depth z due to a uniformly loaded circular or rectangular area can be obtained by the 60° distribution (Fig. 7.30) or a 2:1 distribution (about 63°) assumption (Fig. 7.30b). This method

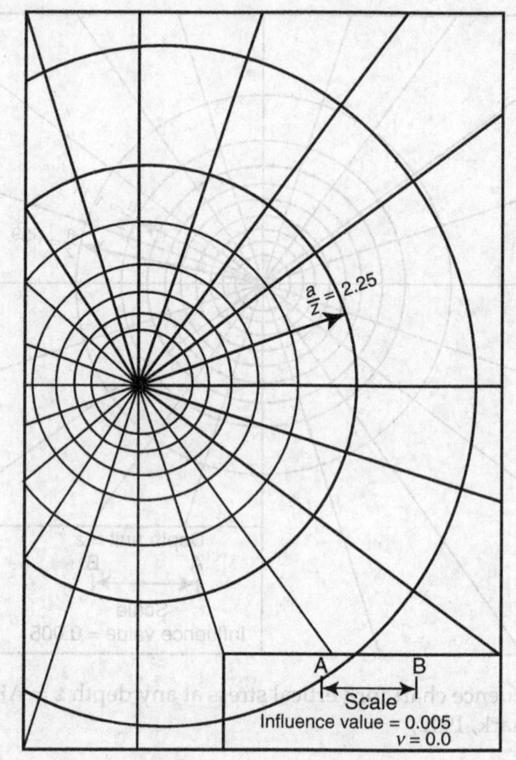


Fig. 7.29 Influence chart for vertical stress based on Westergaard theory (*Source:* Bowles, 1982)

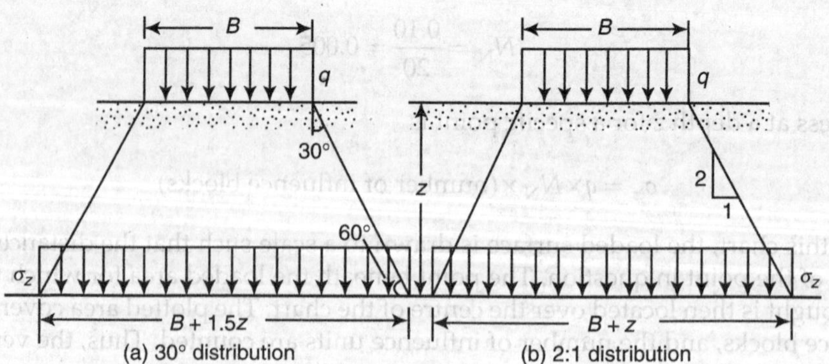


Fig. 7.30 Approximate solutions to vertical stress

predicts values nearer to those obtained from elastic solutions when $z/B$ is in the range $1.5 < z/B < 5$. For smaller depths this approach yields lower values (i.e., $z/B < 1.5$) and higher values at greater depth (i.e., $z/B > 5.0$). However, this method may be used for the determination of $\sigma_z$ for a preliminary analysis.

For a retangular area of dimensions $B \times L$, the vertical stress is given below. For a 30° slope (or distribution),

$$\sigma_z = q\left[\frac{BL}{(B+1.5z)(L+1.5z)}\right] \tag{7.73}$$

For a 2:1 slope,

$$\sigma_z = q\left[\frac{BL}{(B+z)(L+z)}\right] \tag{7.74}$$

### 7.7.7 Contact Pressure

The pressure transmitted from the base of a foundation to the soil is termed the contact pressure. This depends on the rigidity of the foundation structure and the nature of the soil.

The presence of a thick compressible layer, like soft clay, beneath a flexible foundation presents a bowl-shaped settlement profile with more settlement at the centre and almost zero at the edge. But the pressure distribution is uniform. This is the conventional distribution pattern used in the calculation of stressed settlements (Fig. 7.31a).

An extremely rigid footing on the same clay will settle an uniform amount across its breadth. Thus, the compressible cohesive soil under a rigid footing has to transmit a higher contact pressure near the edges than at the centre so as to maintain a uniform settlement. The contact pressure distribution is shown in Fig. 7.31b.

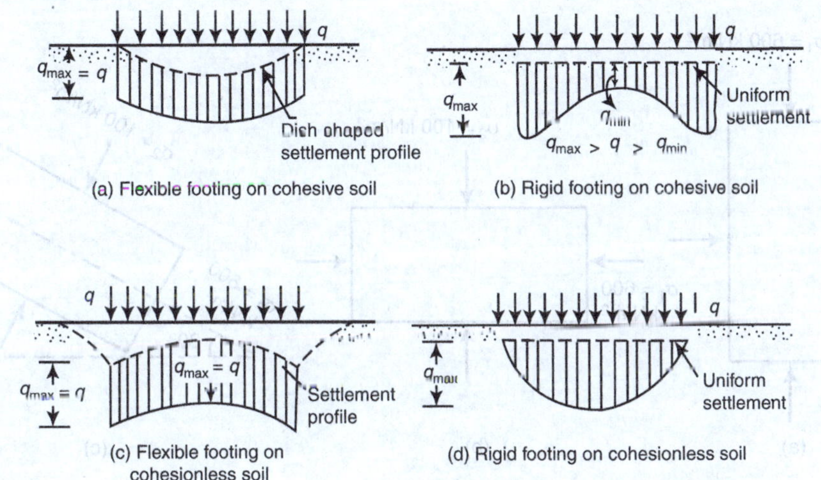


**Fig. 7.31** Effect of rigidity of footing on contact pressure

For a flexible foundation resting on a non-cohesive soil, the distribution of contact pressure is uniform, but the edges of the foundation experience a large settlement. Because of the lack of confining pressure at the edges, the foundation settles more (Fig. 7.31c).

The settlement of a rigid footing on a sand layer is uniform and the contact pressure increases from zero at the edge to a maximum at the centre (Fig. 7.31d).

In actual practice, no foundation is perfectly flexible or infinitely rigid, and hence the actual distribution of the contact pressure is somewhere between the extreme values. Sufficient accuracy in the calculation of stresses and displacements can be obtained by assuming a uniform distribution of the contact pressure.

### 7.7.8 Validity of Elastic Theory Application

The stresses obtained from the application of elastic theory can be accepted only when favourable field results are available. Only a limited number of field measurements are reported in the literature (e.g., Taylor, 1945; Turnbull et al., 1961). A great number of such observations are needed.

Based on a few excellent comparisons, it has been reported that a good agreement was found in the case of vertical stresses. For a lack of better knowledge, the civil engineer is compelled to adopt elastic theory for the computation of stresses. In any case, an error of ±25% between the field observations and elastic theory may be expected.

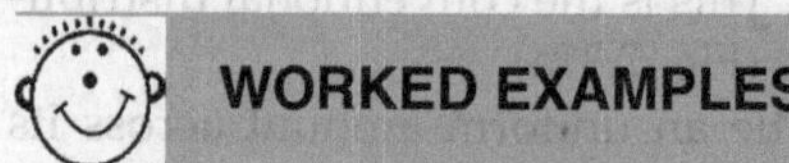

## WORKED EXAMPLES

**Example 7.1** Three soil samples are tested with the state of stresses shown in Fig. 7.32.

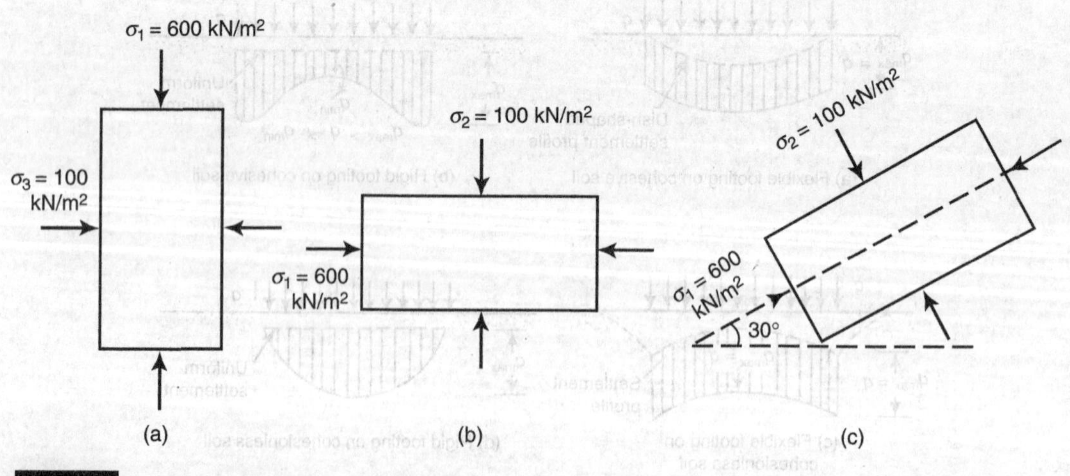


Fig. 7.32

1. Draw the Mohr's circle for each case.
2. Locate the pole for each case.
3. Locate the state of stress acting on a plane at 30° with the major principal plane for each case.

***Solution***

Mohr's circles are drawn as shown in Fig. 7.33. Also, the pole is marked in each case.

For each case, the state of stress acting on a plane at 30° with the major principal plane is marked in Fig. 7.33 and given below as

1. $\sigma_n = 500 \text{ kN/m}^2$
   $\tau_n = 220 \text{ kN/m}^2$
2. $\sigma_n = 500 \text{ kN/m}^2$
   $\tau_n = 220 \text{ kN/m}^2$
3. $\sigma_n = 500 \text{ kN/m}^2$
   $\tau_n = 220 \text{ kN/m}^2$

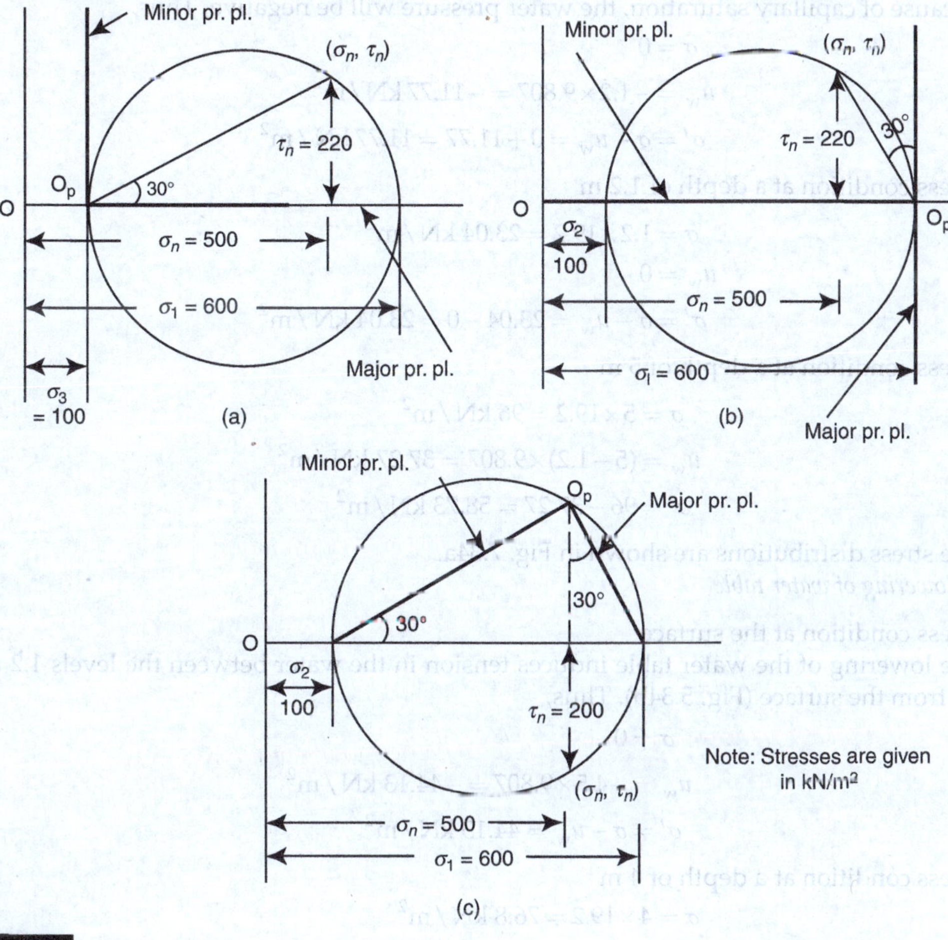

Fig. 7.33

**Example 7.2** In the process of an excavation for a wall footing, the water table level was lowered from a depth of 1.2 m to a depth of 4.5 m in a clayey sand deposit. Considering that the soil above the water table remains saturated at a water content of 28%, compute the following:

1. The effective stress at a depth of 4 m after the lowering of the water table. Take G = 2.68.
2. The increase in effective stress at a depth of 5 m.

***Solution***

$$\gamma_{sat} = \frac{G+e}{1+e}\gamma_w = \frac{G+wG}{1+wG}\gamma_w$$

$$= \frac{2.638+0.28\times 2.68}{1+0.28\times 2.68}\times 9.807 = 19.2 \text{ kN/m}^2$$

*Before lowering of water table*

1. Stress condition at the surface
   Because of capillary saturation, the water pressure will be negative, Thus,

$$\sigma = 0$$

$$u_w = -1.2\times 9.807 = -11.77 \text{ kN/m}^2$$

$$\sigma' = \sigma - u_w = 0 + 11.77 = 11.77 \text{ kN/m}^2$$

2. Stress condition at a depth of 1.2 m

$$\sigma = 1.2\times 19.2 = 23.04 \text{ kN/m}^2$$

$$u_w = 0$$

$$\sigma' = \sigma - u_w = 23.04 - 0 = 23.04 \text{ kN/m}^2$$

3. Stress condition at a depth of 5 m

$$\sigma = 5\times 19.2 = 96 \text{ kN/m}^2$$

$$u_w = (5-1.2)\times 9.807 = 37.27 \text{ kN/m}^2$$

$$\sigma' = 96 - 37.27 = 58.73 \text{ kN/m}^2$$

The stress distributions are shown in Fig. 7.34a.

*After lowering of water table*

1. Stress condition at the surface
   The lowering of the water table induces tension in the water between the levels 1.2 and 4.5 m from the surface (Fig. 5.34b). Thus,

$$\sigma = 0$$

$$u_w = -4.5\times 9.807 = -44.13 \text{ kN/m}^2$$

$$\sigma' = \sigma - u_w = 44.13 \text{ kN/m}^2$$

2. Stress condition at a depth of 4 m

$$\sigma = 4\times 19.2 = 76.8 \text{ kN/m}^2$$

$$u_w = -(4.5-4.0)\times 9.807 = -4.9 \text{ kN/m}^2$$

$$\sigma' = \sigma - u_w = 76.8 + 4.9 = 81.7 \text{ kN/m}^2$$

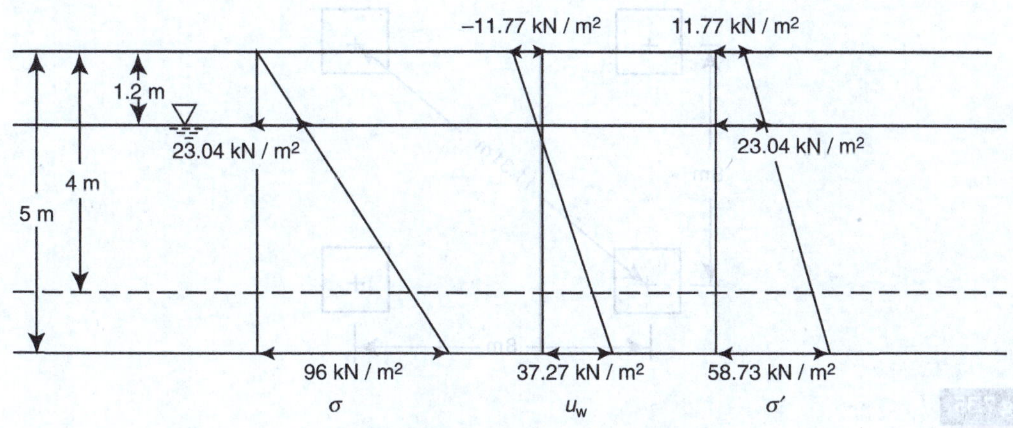

(a) Before lowering of water table

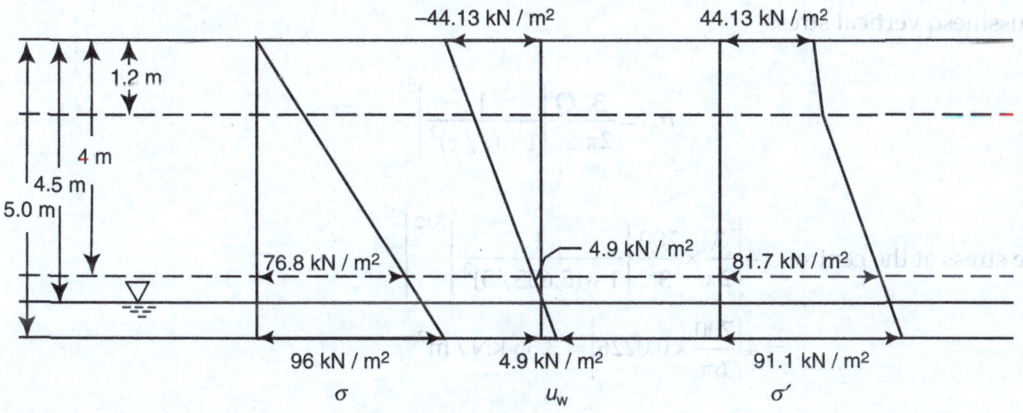

(b) After lowering of water table

Fig. 7.34

3. Stress condition at a depth of 5 m

$$\sigma = 5 \times 19.2 = 96 \text{ kN/m}^2$$

$$u_w = (5 - 4.5) \times 9.807 = 4.9 \text{ kN/m}^2$$

$$\sigma' = \sigma - u_w = 96 - 4.9 = 91.1 \text{ kN/m}^2$$

Therefore, change in effective stress at 5 m depth = 91.1 − 58.73 = 32.37 kN/m$^2$

**Example 7.3** An overhead water tank is supported at a depth of 3 m by four isolated square footing of sides 2 m each placed in a square pattern with a centre-to-centre spacing of 8 m (Fig. 7.35). Compute the vertical stress at the foundation level (i) at the centre of the four footings and (ii) at the centre of one footing. Adopt Boussinesq's point load approximation. The load on each footing is 700 kN.

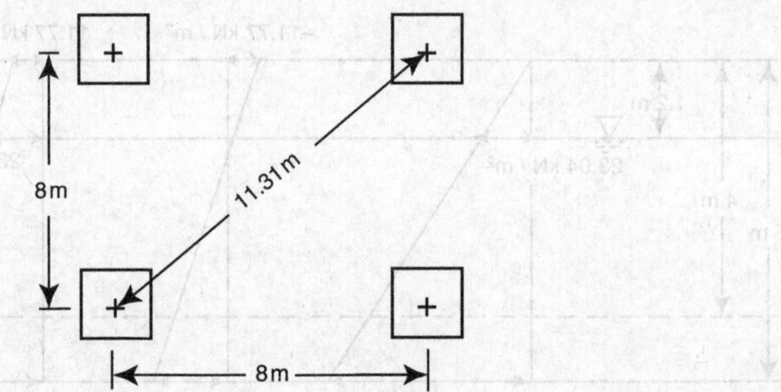


Fig 7.35

***Solution***

Boussinesq vertical stress

$$\sigma_z = \frac{3}{2\pi}\frac{Q}{z^2}\left[\frac{1}{1+(r/z)^2}\right]^{5/2}$$

$$\text{The stress at the centre} = 4\left[\frac{3}{2\pi}\times\frac{700}{3^2}\left\{\frac{1}{1+(5.655/3)^2}\right\}^{5/2}\right]$$

$$= 4\left[\frac{700}{6\pi}\times 0.0226\right] = 3.36\ \text{kN/m}^2$$

The stress at the centre of any corner footing

$$= \frac{3}{2\pi}\frac{Q}{z^2}\left[2\times\left\{\frac{1}{1+(8/3)^2}\right\}^{5/2} + \left\{\frac{1}{1+(0/3)^2}\right\}^{5/2} + \left\{\frac{1}{1+(11.31/3)^2}\right\}^{5/2}\right]$$

$$= \frac{3}{2\pi}\times\frac{700}{9}[0.0107 + 1 + 0.001108]$$

$$= 37.58\ \text{kN/m}^2$$

**Example 7.4** Two railway wagon lines in a harbour yard are located at 6 m centre-to-centre. The average loads per metre run in the lines are 100 and 80 kN/m. Find the vertical stress induced by this loading at a depth of 2 m beneath each load and halfway between them. If a 100 kN crane is installed exactly midway between the lines, what additional stress is caused below the crane at the same depth.

***Solution***

Consider the railway wagon load as a line load of infinite extent. The vertical stress is given as

$$\sigma_z = \frac{2qz^3}{\pi(x^2+z^2)^2}$$

The stress below the 100 kN/m load

$$= \frac{2\times100}{\pi}\left[\frac{2^3}{(0^2+2^2)^2}\right] + \frac{2\times80}{\pi}\left[\frac{2^3}{(6^2+2^2)^2}\right]$$

$$= 31.83 + 0.26 = 32.09 \text{ kN/m}^2$$

The stress below the 80 kN/m load

$$= \frac{2\times80}{\pi}\left[\frac{2^3}{(0^2+2^2)^2}\right] + \frac{2\times100}{\pi}\left[\frac{2^3}{(6^2+2^2)^2}\right]$$

$$= 25.78 \text{ kN/m}^2$$

The stress midway between the two loadings

$$= \frac{2\times100}{\pi}\left[\frac{2^3}{(3^2+2^2)^2}\right] + \frac{2\times80}{\pi}\left[\frac{2^3}{(3^2+2^2)^2}\right]$$

$$= 5.42 \text{ kN/m}^2$$

The additional stress below the crane, considering the crane load as a vertical concentrated load, is given as

$$\sigma_z = \frac{3Q}{2\pi z^2}\left[\frac{1}{1+(r/z)^2}\right]^{5/2}$$

$$= \frac{3\times100}{2\times\pi\times2^2}\left[\frac{1}{1+(0/2)^2}\right]^{5/2} = 11.94 \text{ kN/m}^2$$

**Example 7.5** An embankment is to be constructed with the following dimensions:

Top width – 8 m; height = 4 m; side slopes = 1:1½.

The unit weight of the soil is 21 kN/m$^3$. Compute the vertical pressure at a depth of 6 m below the ground surface at the following locations:

1. On the central longitudinal plane of the embankment
2. Below the toes of the embankment

If a surcharge load of 50 kN/m$^2$ is acting on the road surface, what is the increase in stress at the same central point. Assume the surcharge load is distributed vertically downwards.

***Solution***

Refer to Fig. 7.36.

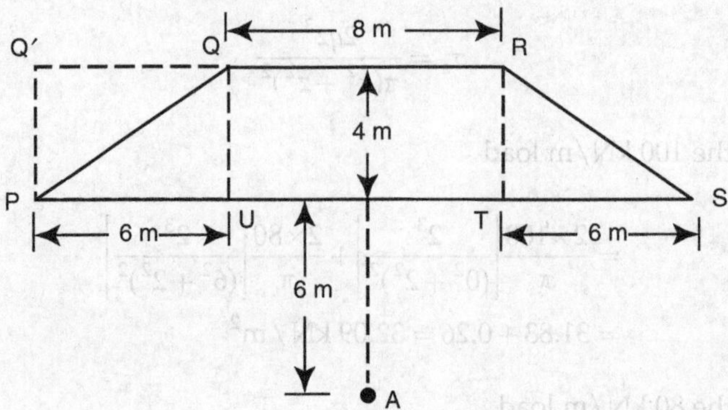


Fig. 7.36

The stress at the centre of the embankment

$$\frac{a}{z}=\frac{6}{6}=1, \quad \frac{b}{z}=\frac{4}{6}=0.67$$

$$q=\gamma z=21\times 4=84\,\text{kN/m}^2$$

$$(\sigma_z)_A = 2\times q\,N_E = 2\times 84\times 0.425 = 71.4\ \text{kN/m}^2$$

The stress below the toe = stress due to PQ′QRS – stress due to PQ′Q.
For the stress due to PQ′QRS,

$$\frac{a}{z}=\frac{6}{6}=1, \quad \frac{b}{z}=\frac{14}{6}=2.33$$

For the stress due to PQ′Q,

$$\frac{a}{z}=\frac{6}{6}=1, \quad \frac{b}{z}=0$$

The respective $N_E$ values from Fig. 7.20 are 0.49 and 0.24.
Hence, $(\sigma_z)_B$ = 84 (0.49 – 0.24) = 21 kN/m².
The additional stress due to the surcharge load can be obtained using Fig. 7.20, assuming a very low value of $a/z$, say $a/z = 0.1$. Here, $b/z = 8/10 = 0.80$. For $a/z = 0.1$ and $b/z = 0.80$, the value of $N_E = 0.37$. Therefore, the additional stress at A = 50 × 0.37 = 18.5 kN/m².

**Example 7.6** Calculate the stress in a soil mass below the centre of a uniformly loaded circular area of radius 1.5 m with a pressure of 60 kN/m² and thus obtain the exact depth at which the stress reduces to 10% of the applied stress.

***Solution***
The vertical stress at a depth $z$ under the centre of a circular area of diameter $2a$ is given as

$$\sigma_z = q\left[1-\left\{\frac{1}{1+(a/z)^2}\right\}^{3/2}\right]$$

Here $\sigma_z = 0.10q$. Therefore,

$$0.10q = q\left[1 - \left\{\frac{1}{1+(1.5/z)^2}\right\}^{3/2}\right]$$

or

$$\frac{1}{\{1+(1.5/z)^2\}^{3/2}} = 0.90$$

or $z^2 = 2.25/0.075 = 30$, or $z = 5.48$ m. Therefore, the depth at which the stress is 10% of the applied stress is 5.48 m.

**Example 7.7** A total load of 900 kN is uniformly distributed over a rectangular footing of size 2 m × 3 m. Find the vertical stress at a depth of 2.5 m below the footing at point C, under one corner, and D, under the centre. If another footing of size 1 m × 3 m with a total load of 450 kN is constructed adjoining the previous footing, what is the additional vertical stress at the point C at the same depth due to the construction of the second footing.

***Solution***
Refer to Fig. 7.37.

For the first footing,

$$q = \frac{900}{2\times3} = 150 \text{ kN/m}^2$$

For the stress under corner C,

$$m = \frac{3}{2.5} = 1.2 \quad \text{and} \quad n = \frac{2}{2.5} = 0.8$$

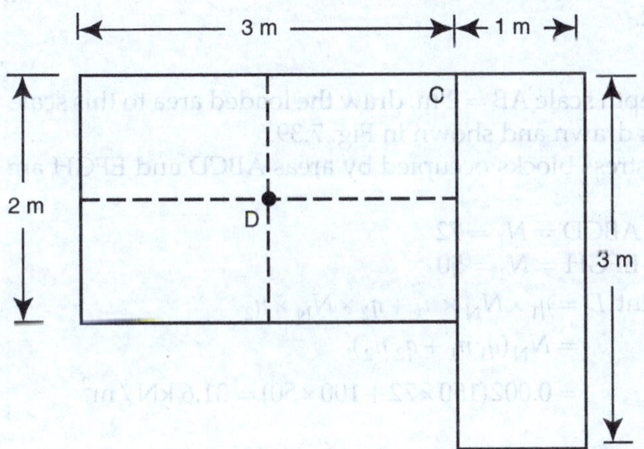


Fig. 7.37

From Fig. 7.27, $N_R = 0.168$. Therefore,

$$(\sigma_z)_C = qN_R = 150 \times 0.168 = 25.2\,\text{kN/m}^2$$

For the stress under centre D,

$$m = \frac{1.5}{2.5} = 0.6 \quad \text{and} \quad n = \frac{1}{2.5} = 0.4$$

Again, from Fig. 7.27, $N_R = 0.08$. Therefore,

$$(\sigma_z)_D = 4(qN_R) = 4 \times 150 \times 0.08 = 45\ \text{kN/m}^3$$

For the second footing,

$$q = \frac{450}{1 \times 3} = 150\ \text{kN/m}^2$$

For the additional stress under corner C,

$$m = \frac{3}{2.5} = 1.2 \quad \text{and} \quad n = \frac{1}{2.5} = 0.4$$

From Fig. 7.27, $N_R = 0.105$. The additional stress at corner C, due to the construction of the second footing = $150 \times 0.105 = 15.8$ kN/m$^2$.

**Example 7.8** A foundation is constructed to take a stress of 150 kN/m$^2$ and is flush with another existing foundation (Fig. 7.38) taking a load of 100 kN/m$^2$. Find the vertical stress at a depth of 2 m below the point D. Use the Newmark's chart given in Fig. 7.39 with an influence value of 0.002.

***Solution***

Considering the depth scale AB = 2 m, draw the loaded area to this scale with D at the centre of the chart. This is drawn and shown in Fig. 7.39.

The number of stress blocks occupied by areas ABCD and EFGH are separately counted and given as

No. of blocks in ABCD = $N_1 = 72$

No. of blocks in EFGH = $N_2 = 50$

$$\begin{aligned} \text{Stress under point } D &= q_1 \times N_N \times n_1 + q_2 \times N_N \times n_2 \\ &= N_N(q_1 n_1 + q_2 n_2) \\ &= 0.002(150 \times 72 + 100 \times 50) = 31.6\ \text{kN/m}^2 \end{aligned}$$

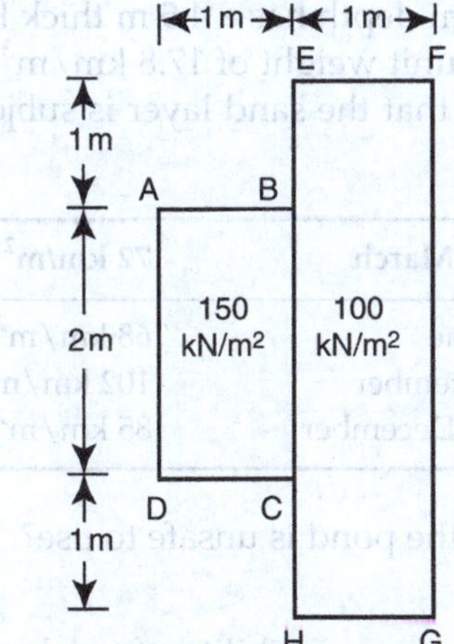


Fig. 7.38

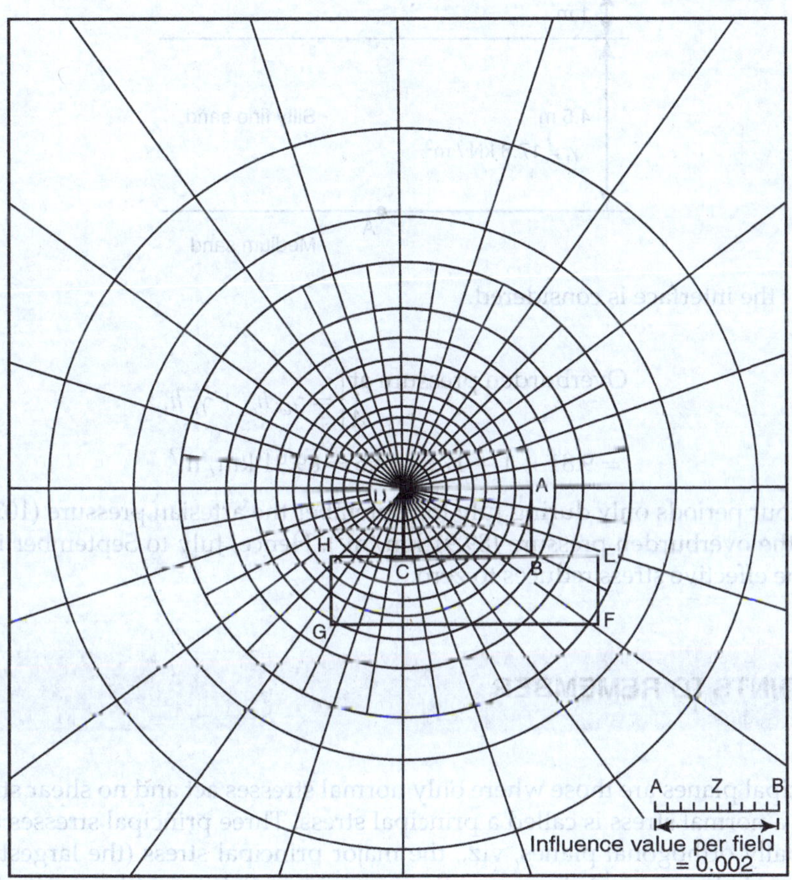


Fig. 7.39

**Example 7.9** A shallow pond of 1 m depth has a 4.5 m thick layer of silty fine sand below the bottom of the pond with a total unit weight of 17.8 km/m$^3$. This is underlain by a layer of medium sand. It has been found that the sand layer is subjected to an artesian pressure throughout the year as given below.

| **January to March** | **72 km/m$^2$** |
|---|---|
| April to June | 68 km/m$^2$ |
| July to September | 102 km/m$^2$ |
| October to December | 85 km/m$^2$ |

During which period the bottom of the pond is unsafe to use?

***Solution***

As there is an artesian pressure there is a possibility of quick sand formation at the bottom of the pond. The resisting pressure is the overburden pressure above the sand layer.

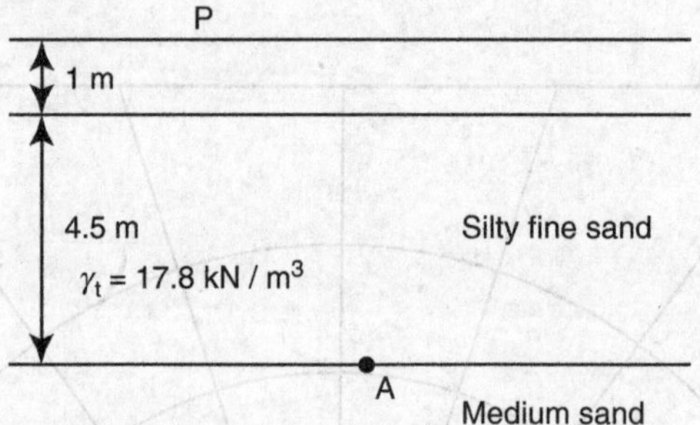


A point $A$ at the interface is considered.

$$\left.\begin{matrix}\text{Overburden pressure at} \\ A\end{matrix}\right\} = \gamma_\omega h_\omega + \gamma_t h_1$$

$$= 9.81 \times 1 + 17.8 \times 4.5 = 89.91 \text{ km/m}^2$$

Out of the four periods only during July to September the artesian pressure (102 km/m$^2$) is more than the overburden pressure (89.91 km/m$^2$). Hence, July to September is an unsafe period as the effective stress reduces to zero.

## POINTS TO REMEMBER

7.1 Principal planes are those where only normal stresses act and no shear stresses exist. Such a normal stress is called a principal stress. Three principal stresses act on three mutually orthogonal planes, viz., the major principal stress (the largest stress), the intermediate principal stress, and the minor principal stress (the smallest stress).

7.2 The graphical representation of the state of the normal and shear stresses acting on different planes in a lucid form is known as the Mohr's circle.

7.3 The locus of points on the Mohr diagram whose coordinates represent the maximum shear stress and the associated principal stress for the entire stress history is defined as a stress path.

7.4 In a saturated soil, the total stress ($\sigma$) has two stress components. The stress component on the water, which does not cause any change in the mechanical properties of the soil, is known as the neutral stress or pore water pressure ($u_w$). The other component of total stress which rests entirely on the soil skeleton of the soil is responsible for changes in the properties of the soil.

7.5 The elastic half-space is an idealized soil medium which is homogeneous, isotropic, and elastic. The behaviour of the medium is governed by the stress–strain modulus, $E$, and Poisson's ratio, $\nu$.

7.6 Boussinesq's theory assumes a weightless, elastic half-space, and gives components of stresses caused by a vertical point surface load.

7.7 The stress isobar or pressure bulb is a stress contour connecting all points of equal stress below the ground surface. The soil bounded within a pressure bulb furnishes the support power of a footing.

7.8 Westergaard assumed an elastic half-space medium in which there is no lateral strain and the medium suits the condition of a sedimentary soil.

7.9 Newmark's chart is a graphical procedure for determining the vertical stress due to a surface load of any shape. The chart is based on the expression for the vertical stress under the centre of a loaded circular area.

7.10 The contact pressure is the pressure transmitted from the base of a foundation of the soil and depends on the rigidity of the foundation structure and the nature of the soil.

## QUESTIONS

### Objective Questions

7.1 State which of the following statements are true or false:

1. In several situations the effective stress will be greater than the total applied stress.
2. The effective stress in a soil mass is always the actual grain-to-grain contact stress.
3. Application of Boussinesq's vertical stress overestimates the settlement.
4. The effective stress of a soil is not affected by the type of pore fluid.
5. Westergaard's expression for the vertical stress considers the weight of the soil medium.

7.2 The contact pressure distribution under a rigid footing on saturated clay and dense sand is ________ and ________, respectively.

(a) Uniform
(b) Concave parabolic
(c) Convex parabolic

7.3 For a vertical concentrated load acting on the surface of a semi-infinite elastic soil mass, the vertical normal stress at depth $z$ is
(a) Directly proportional to $z$ (b) Inversely proportional to $z$
(c) Directly proportional to $z^2$ (d) Inversely proportional to $z^2$

7.4 The approximate depth at which the effective vertical pressure is equal to 100 kN/m$^2$ in a typical deposit of submerged soil is
(a) 5 m (b) 10 m
(c) 20 m (d) 100 m

7.5 If the entire semi-infinite mass is loaded with a load intensity of $q$ at the surface, the vertical stress at any depth is
(a) 0.2$q$ (b) $q$
(c) Zero (d) Infinity

7.6 A rise in the groundwater table up to the capillary zone results in
(a) A decrease in the degree of saturation (b) An increase in the effective stress
(c) A decrease in the effective stress (d) No change in the pore water pressure

7.7 *Assertion A*: The effective stress is that part of the load which is transmitted by the particles divided by the gross area.
*Reason R*: The effective stress is not a stress that can be measured directly but a computed value.
*Select the correct code.*
(a) Both A and R are true, and R is the correct explanation of A.
(b) Both A and R are true, and R is not the correct explanation of A.
(c) A is true, but R is false.
(d) A is false, but R is true.

7.8 The diameter of the Mohr's circle for plane stress conditions is the
(a) Octahedral stress (b) Shear stress
(c) Deviator stress (d) Principal stress ratio

7.9 Which of the following pairs is correctly matched?
1. Westergaard's theory — For soils with a thin layer of coarse materials only
2. Newmark's chart — Graphical procedure based on circular loaded area
3. Boussinesq's theory — For vertical stress only
Select the correct answer using the codes given below.
*Codes:*
(a) 1 and 2 are correct (b) 1 and 3 are correct
(c) 2 and 3 are correct (d) 2 alone is correct

7.10 As the depth of the stress isobar increases, the intensity of stress
(a) Increases (b) Decreases
(c) Remains constant (d) Initially decreases and then increases

### Descriptive Questions

7.11 Discuss the essential differences between Boussinesq's and Westergaard's theories. For which condition do both these theories yield approximately the same value of vertical stress?

7.12 How far is it justifiable to adopt Boussinesq's theory for predicting the vertical stress in sand deposits?
7.13 What is the basic principle involved in the development of Newmark's chart?
7.14 Explain why the effective stress evaluation in a partially saturated clay is very complex.
7.15 Discuss the limitations of Lambe's effective stress expression.

## EXERCISE PROBLEMS

7.1 A sample is subjected to the following principal stresses: $\sigma_1 = 700\,\text{N}/\text{m}^2$ and $\sigma_3 = -200\,\text{N}/\text{m}^2$. Draw the Mohr's circle and locate the origin of planes. Find $\sigma_n$ and $\tau_n$ on a plane making an angle 50° with the major principal plane. Also find the maximum shear stress.

7.2 The normal stresses acting on two mutually perpendicular planes are 150 and 60 kN/m$^2$ and the shear stress on each plane is 110 kN/m$^2$. Draw Mohr's circle and find
1. The principal stresses and the planes
2. The possibility of tension occurring on any plane for this stress condition
3. The shear and normal stresses acting on a plane making an angle of 70° with the major principal stress

7.3 An impervious, saturated clay layer of 12.5 m thickness lies over a sand aquifer. Piezometers inserted into the sand layer show an artesian pressure condition with the piezometric surface 3.5 m above the surface of the clay. Determine the effective stress at the top of the sand layer. The parameters of the clay layer are $e = 1.26$ and $G = 2.72$. How deep an excavation can be made in the clay layer without the danger of a bottom heave?

7.4 The soil conditions shown in Fig. 7.40 are revealed during a boring operation. Represent by diagrams the variation with depth of the total vertical overburden pressure, the pore water pressure, and the effective overburden pressure. Assume the top 1 m is dry. Compute the total and effective overburden pressure at the bottom of the clay layer immediately after the lowering of the water table, if the water level in the gravel layer is suddenly lowered by 2.1 m.

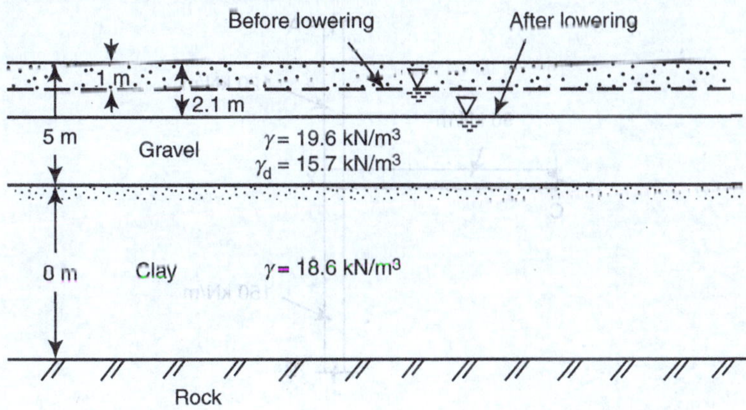


Fig. 7.40

7.5 In a fine sand deposit, the water table is located at 4 m from the ground surface. Due to capillary action only 1 m depth of sand above the water table is saturated and the remaining is dry. The dry and saturated unit weights of sand are, respectively, 19.2 and 21.6 kN/m$^3$. Estimate the effective vertical stress in the sand at a depth of 10 m below the surface.

7.6 A boring log indicates the presence of 5 m of silty sand from the ground surface followed by 3 m of coarse sand, which in turn rests on a deep deposit of gravel. The groundwater table is located at the top of the sand layer. The soil characteristics are as given below:

| | Specific gravity | Void ratio |
|---|---|---|
| Silty sand | 2.67 | 0.90 |
| Sand | 2.65 | 0.60 |

Compute the total, neutral, and effective stresses and draw the stress diagrams from the ground surface to the top of the gravel layer. Assume there is no capillary rise of water.

7.7 A point load of 1,200 kN acts on the surface of a deep clay layer. Compute the vertical stresses in horizontal layers spaced at 1 m increments of depth up to 5 m and take a radial distance up to 4 m on either side of the point load with 1 m increments. From these results plot
1. the vertical pressure bulb for 20 kN/m$^2$
2. the distribution of stress directly beneath the load,
3. the distribution of stress on horizontal place at 3 m depth.

Adopt Boussinesq's theory.

7.8 Two concentrated loads $Q_1 = 900$ kN and $Q_2 = 1{,}200$ kN are spaced 4 m apart. Draw the 5 kN/m isobar of the system. Adopt Boussinesq's theory for a single, vertical concentrated load.

7.9 A 200 kN load is transferred through a steel stanchion. Compute the vertical stresses beneath the stanchion at depths of 1, 3, and 9 m. Estimate the depth at which the load is 25% of the applied load. Adopt Westergaard's theory ($\nu = 0$).

7.10 Three load-bearing walls meet at a point X, as shown in Fig. 7.41. Compute the vertical stress at a point 3 m under X. The loads transferred by the walls may be approximated as line loads. (Hint: Take half of the stress caused by an infinite line load.)

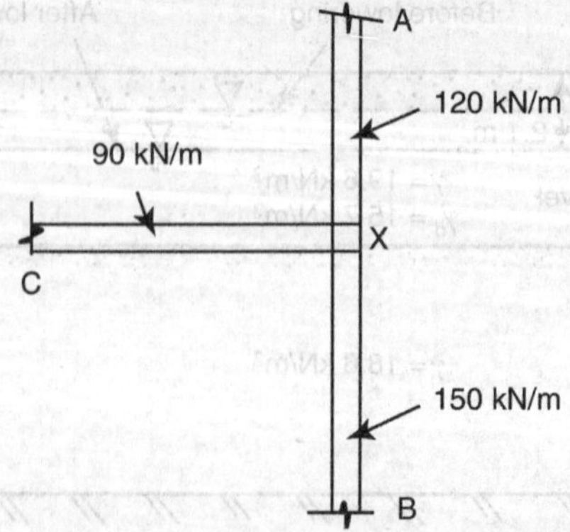


Fig. 7.41

7.11 A uniformly distributed load of infinite extent in both lateral directions, when applied at the surface of a natural soil formation, produces an increase of 75 kN/m$^2$ in the vertical stress at a depth of 3 m. Find the stress increment at a depth of 5 m.

7.12 Show that the vertical stress, $\sigma_z'$, at depth $z$, for a long strip area with a triangular distributed load (Fig. 7.42) is

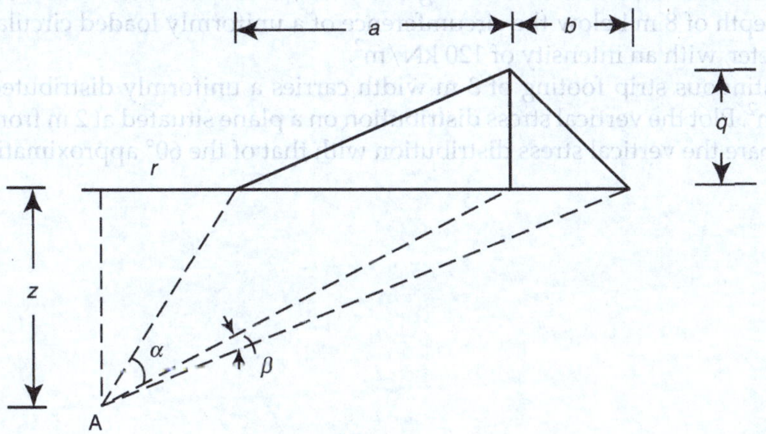


Fig. 7.42

$$\sigma_z = \frac{q}{\pi}\left[\frac{a+b+r}{b}\beta - \frac{r}{a}\alpha\right]$$

7.13 A very long embankment is to be built with a top width of 10 m and side slopes of 1:1½. The height of the embankment is 10 m. Compute the vertical stresses at a depth of 5 m from the base at the following points: (i) below the toe, (ii) below the central line, and (iii) below a point midway on the slope. Assume $\gamma = 21$ kN/m$^3$.

7.14 A circular area is loaded with a uniform load intensity of 100 kN/m$^2$ at the ground surface. Calculate the vertical pressure at point A, so situated on the vertical line through the center of the loaded area that the area subtends an angle of 90° at it.

7.15 A circular ring footing for an overhead water tank carries a load of 1,000 kN whose outer diameter is 3 m and inner diameter is 1.5 m. Assume the surface pressure to be uniform over the area. Determine the vertical stress at depths of 2 m and at radial distances of 2 and 4 m from the center. Use Newmark's chart with an influence value of 0.005.

7.16 A square footing of 3 m×3 m carries a uniformly distributed load of 200 kN/m$^2$. Find the vertical stress at 3 m below the footing and under a point (i) 1.2 m away from the corner and in line with the side, (ii) 1.2 m inside the corner and in line with the side.

7.17 A wheel load of 1,200 kN is applied at the surface of a road. What will be the total load on the crest of a culvert situated at 2.5 m below the surface? Assume that an area of 2 m×3 m at the crest level is transferring the load. Use Boussinesq's stress coefficients for a uniformly loaded rectangular area.

7.18 The ground surface of a building is lowered 2 m below the existing surface. A 1.2 m square footing carrying a load of 200 kN/m$^2$ is then constructed at the level of the new surface. Estimate the net increase in stress in the soil mass 1.2 m below the center of the foundation. Take the unit weight of soil to be 18 kN/m$^3$.

7.19 Construct a Newmark's chart for vertical stresses based on Boussinesq's theory with an influence value of 0.0025. Using this chart, determine the vertical stress induced at a depth of 8 m below the circumference of a uniformly loaded circular area of 6 m diameter, with an intensity of 120 kN/m$^2$.

7.20 A continuous strip footing of 3 m width carries a uniformly distributed load of 110 kN/m$^2$. Plot the vertical stress distribution on a plane situated at 2 m from the surface. Compare the vertical stress distribution with that of the 60° approximation.

# 8

# Consolidation and Consolidation Settlement

**CHAPTER HIGHLIGHTS**

Rheological models of soil – Compressibility of soils – One-dimensional consolidation – Consolidation test – Compressibility characteristics – Types of clay deposits – Prediction of pre-consolidation pressure – Terzaghi's theory of one-dimensional consolidation, time factor, coefficient of consolidation, fitting methods – Secondary compression – Consolidation settlement and its rates – Acceleration of consolidation – Compressibility of sands

## 8.1 INTRODUCTION

When a soil layer is subjected to compressive stress due to construction activities, or otherwise, it undergoes compression. The compression may be caused by rearrangement of particles, seepage of water, crushing of particles, and elastic distortions. The compression may be progressive and cumulative, dependent on the type, magnitude, and duration of load and on the properties of the materials. Although the stresses induced may not cause a failure, the civil engineer is concerned with them as the magnitude of the compression may be detrimental for some special structures or block the normal function of conventional structures. Compressibility is one of the three fundamental principles of geotechnical engineering to be understood by a civil engineer.

Although the mass is heterogeneous and does not have simple predictable characteristics, the engineer is compelled to provide a safe and economical design. Thus, the stress change – compression behaviour – may be dealt with by idealizing the soil material as elastic for certain conditions and treating the soil as a mathematical model.

Settlement of a structure has to be analysed for three reasons: appearance of the structure, utility of the structure, and damage to the structure. The aesthetic view of a structure may be spoiled due to the presence of cracks or tilt of the structure caused by settlement. Settlement caused to a structure may damage some of the utilities like cranes, drains, pumps, electrical lines, etc. Further settlement can cause a structure to fail structurally and collapse.

Settlement is the combination of time-independent (e.g., immediate compression) and time-dependent compression (called consolidation). The engineer is interested in assessing the magnitude and rate of compression, as well as the total and relative or differential settlement of a structure.

This chapter deals with the process of consolidation and the methods of evaluating consolidation settlement for different field conditions.

## 8.2 RHEOLOGICAL MODELS OF SOILS

The study of the behaviour of a material in a fluid state is referred to as rheology. Soil is a particulate system in which the soil skeleton undergoes a sort of statistical flow caused by particle rolling, sliding, and slipping, resulting in a void ratio reduction, and such a behaviour can be considered as a problem in rheology.

Three basic rheological models (Fig. 8.1a–c), have been identified to represent soil behaviour. They are the spring element (spring constant, $k_s$) or Hookean model, which represents the elasticity of the soil; the dashpot element or Newtonian element (constant $C$), which relates to the permeability of the soil; and the yield stress model, which depicts the permanent reduction in void ratio of the soil. The basic spring element is combined with one or both of the other elements.

The combined Hookean and yield stress model is used as a simple rheological model to represent immediate soil settlement (Fig. 8.1d). The yield or slip here occurs at a particular stress level $f$, and in soil, this may stop after a certain strain level. If pore water is involved, the Newtonian model may be combined with other elements. The combination of the Hookean and the Newtonian models in parallel is referred to as the *kelvin model* (Fig. 8.1e). This can also be represented as in Fig. 8.1f (piston–spring combination), where the dashpot effect is produced by the pore water pressure and the subsequent drainage through the valve. This piston–spring combination is used to represent the compressibility and consolidation of soils.

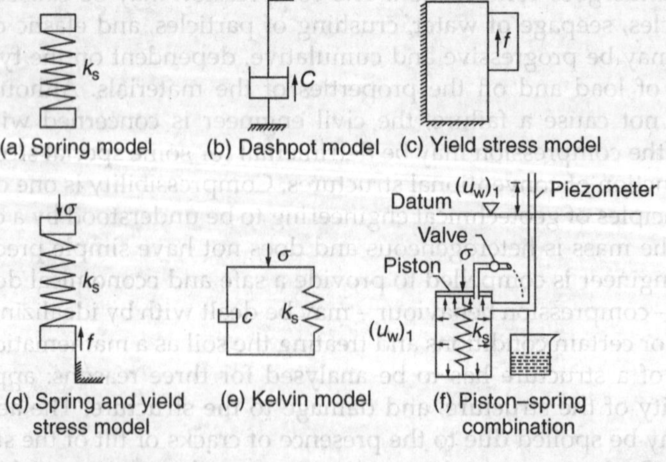


Fig. 8.1 Rheological soil models

## 8.3 COMPRESSIBILITY OF SOILS

A soil mass when subjected to a change in the stress system results in a change in volume of the mass. Consider the condition that a change in stress causes a volume decrease. The reduction of volume causes compression or settlement of the boundaries of the mass.

All soils, dry or partially saturated, undergo elastic distortion almost immediately after the load application. Similarly, in a saturated soil under no drainage condition, shear stresses induce elastic shear strain, which also occur more or less simultaneously with the application of load (Leonards, 1962). Compression caused due to these processes is termed *initial* or *immediate compression*.

The time-dependent volume decrease may be attributed to (i) a compression of soil matter, (ii) a compression of water and air within the voids, and (iii) escape of water and air from the voids. Under the stresses normally encountered in civil engineering problems, the soil solids and pore water (relatively incompressible fluid) undergo a negligible percentage of the total compression. Thus, all the compression is mainly due to the reduction in void volume within the soil. In a saturated system, the reduction in void volume is due to escape of water. But in partially saturated soils, because of the compressible nature of air, there may be an appreciable reduction in volume, even though there may be no seepage of water. The time-dependent compression in a partially saturated soil is beyond the scope of current knowledge. The time-dependent compression in a saturated soil is explained below.

When a pressure is applied to a saturated soil–water system, the applied pressure is immediately transferred as an excess pressure in pore water. The resulting hydraulic gradient initiates a flow of water and the soil mass begins to compress, and the portion of applied stress is transferred to the soil skeleton. This causes a reduction of void volume and dissipation of excess pore water pressure because of seepage of water from the voids. This process of gradual load transfer from pore water to soil skeleton and the corresponding gradual compression is called *consolidation*. That part of consolidation which is completely controlled by the resistance to flow of water under the induced hydraulic gradient is called *primary consolidation*. The other part, called *secondary consolidation* (creep), is due to the plastic deformation of the soil at zero excess pore water pressure. The primary consolidation is normally more than the secondary consolidation. The primary consolidation is generally referred to as consolidation settlement, and the same is followed in this book.

The immediate compression (including for undrained condition) is computed by assuming the soil mass as an elastic medium, whereas the time-dependent compression is computed assuming one-dimensional consolidation (explained in the subsequent paragraphs).

## 8.4 ONE-DIMENSIONAL CONSOLIDATION

Consider a fine-grained soil layer (say clay) of thickness $H$, sandwiched between two permeable sand layers and below the water table (Fig. 8.2a). If a pressure intensity, $\Delta\sigma$, is applied on the ground surface, the immediate increase of the pore water pressure will be $\Delta u_w$, which will be equal to the applied total pressure $\Delta\sigma$. Thus, immediately after application of pressure ($t \approx 0$),

$$(\Delta u_{\mathrm{w}})_0 = \Delta\sigma \tag{8.1}$$

and

$$\Delta\sigma' = 0 \tag{8.2}$$

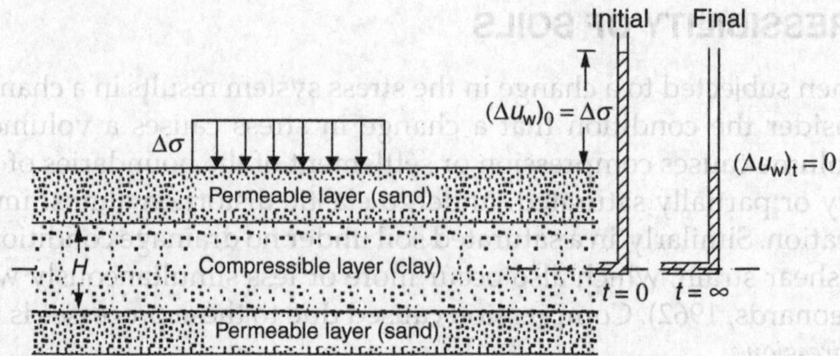


(a) One-dimensional compression-field condition

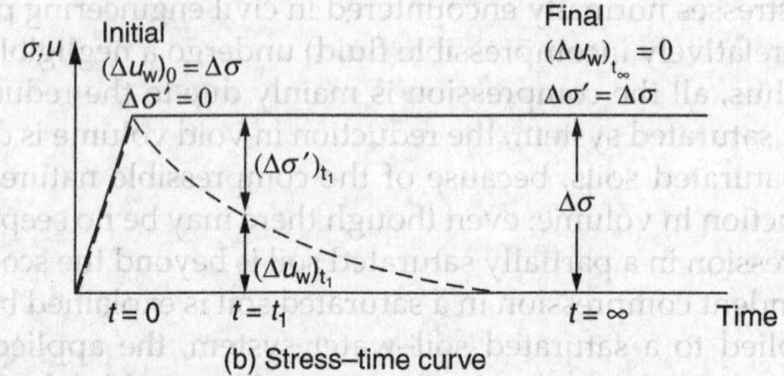


(b) Stress–time curve

**Fig. 8.2** Process of one-dimensional consolidation

Once the consolidation process starts by gradual squeezing of water from the soil pores, the excess pore water pressure decreases, and the effective stress also increases $[(\Delta\sigma')_{t_1}]$ by the same amount such that the total stress always remains equal to $\Delta\sigma$ at $t = t_1$. That is,

$$\Delta\sigma = (\Delta u_w)_{t_1} + (\Delta\sigma')_{t_1} \tag{8.3}$$

where

$$(\Delta u_w)_{t_1} = \Delta\sigma - (\Delta\sigma')_{t_1} \tag{8.4}$$

This fact is represented in Fig. 8.2b. At $t = \infty$, the excess pore water pressure at all depths of the clay layer will be dissipated completely such that

$$\Delta\sigma = (\Delta\sigma')_{t_\infty} = \Delta\sigma' \tag{8.5}$$

where

$$\Delta u_w = (\Delta u_w)_{t_\infty} = 0$$

This is the stage at which consolidation is said to be completed. In fact, only the primary consolidation is over, and secondary consolidation may be on.

Terzaghi (1925, 1943) postulated a rigorous mathematical solution to the process of consolidation, with a piston-spring rheological model (Fig. 8.1f) to explain the load transfer

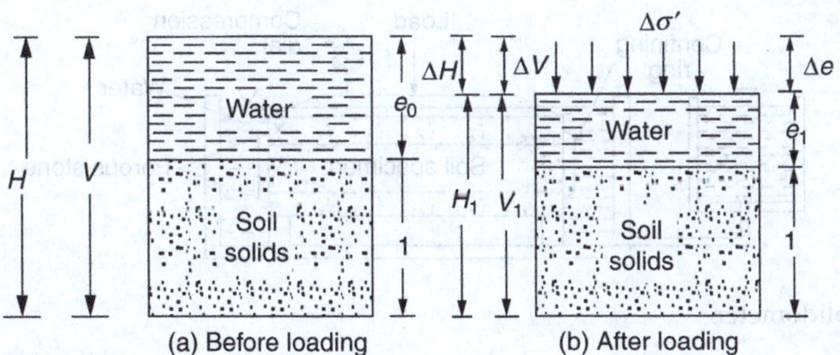


**Fig. 8.3** Interpretation of compressibility

technique. He considered the one-dimensional vertical consolidation, as this is the general field situation corresponding to the state of the soil during deposition and compression under its own weight of overburden. The lateral extent of the stratum is very large in comparison with the thickness of the layer, and it is justified to assume the lateral strains to be small.

The one-dimensional consolidation is based on the following considerations: (i) all displacements are vertical so that there are no lateral strains, (ii) all the flow of water from the soil layer is in a vertical direction only, and (iii) the change in void ratio is a direct function of the vertical component of effective stress.

As the consolidation is one-dimensional, the change in volume, $\Delta V$, per unit of original volume, $V$, may be taken equal to the change in height, $\Delta H$, per unit of original height, $H$ (Fig. 8.3), i.e.,

$$\frac{\Delta H}{H} = \frac{\Delta V}{V} \tag{8.6}$$

It is convenient to represent $V$ and $\Delta V$ in terms of void ratio as

$$\frac{\Delta H}{H} = \frac{\Delta e}{1+e_0}$$

where $\Delta e$ is the change in void ratio and $e_0$ is the original void ratio. Rearranging,

$$\Delta H = H\frac{\Delta e}{1+e_0} \tag{8.7}$$

This relationship is very general in nature and independent of the degree of saturation of soil and the mechanism causing volume change.

## 8.5 CONSOLIDATION TEST

The consolidation or oedometer test is used to determine the compressibility characteristics of a saturated undisturbed or remoulded soil. A trimmed sample is fitted in a cylindrical container, and a seating pressure of about 12 kPa is applied (Fig. 8.4). Water moves from the saturated soil both in upward and downward directions towards the porous stones.

When equilibrium is attained under the seating load, an additional increment of load is applied and allowed to consolidate. In the standard test (explained in detail in Chapter 10), the pressure is doubled every time until the maximum anticipated pressure in the field is attained.

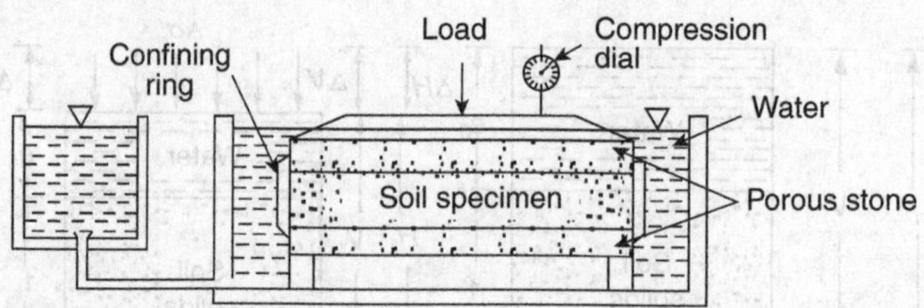


Fig. 8.4 Consolidometer

Each pressure is normally maintained for a period of 24 hours, compression readings being observed at suitable intervals during this period. The effective stress in the specimen is equal to the applied pressure at the end of the load increment period. The expansion of the specimen due to the successive decrease in applied pressure may be measured.

The above test procedure is referred to as the conventional procedure. Based on a detailed study of this procedure, Leonards and Ramiah (1959) have reported that the void ratio and effective stress relationship was not significantly affected by

1. moderate variations in room temperature,
2. variations in specimen size (the diameter to height ratio of about 2.75 or more), and
3. variations in the duration of load increment (provided primary consolidation is complete and secondary compression is not important).

The void ratio at the end of each increment is obtained from the difference in dial gauge readings and dry weight of the specimen at the end of the test. The method of calculation is as follows (Fig. 8.5):

Mass of sample measured at the end of test $= M_s$

Thickness at the end of any increment period $= H_1$

Area of specimen $= A$

Equivalent thickness of soil solids, $H_s$ $= \dfrac{M_s}{AG\rho_w}$

Void ratio $e_1$ corresponding to pressure $p_1$ is calculated as follows:

$$e_1 = \frac{H_1 - H_s}{H_s} = \frac{H_1}{H_s} - 1 \tag{8.8}$$

In the same way, void ratios at the end of each increment period are calculated.

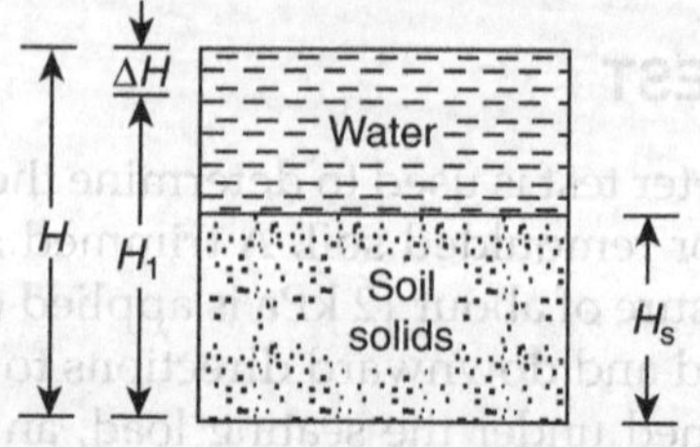


Fig. 8.5 Phase diagram for specimen

## 8.6 COMPRESSIBILITY CHARACTERISTICS

### 8.6.1 Pressure–Void Ratio Curves

Figure 8.6 represents the results of a typical laboratory one-dimensional pressure–void ratio curve. In Fig. 8.6a, the pressure is taken on an arithmetic scale, whereas in Fig. 8.6b, it is taken on a logarithmic scale. The shape of the *e–p* curve is dependent on the consolidation history of the soil.

A flat and somewhat straight curve up to a certain pressure (AB, Fig. 8.6b) is followed by a steep and fairly straight line (CD) with a smooth transition from the first limb. If the pressure is released, the soil rebounds and does not reach back to the previous void ratio, depicting permanent deformation. The recompression curve is somewhat parallel to the first and gradually blends into the straight line (CDE) (Fig. 8.6b). The point of transition, B, in the first loading curve corresponds to a state of pressure known as pre-consolidation pressure. More discussion on the *e–p* curve is presented in the next section.

### 8.6.2 Compression Index

The compression index ($C_c$) is the slope of the linear portion of the pressure–void ratio curve on a semi-log plot, with pressure on the log scale (IS: 8009 – Part 1, 1976). This is a dimensionless parameter. For any two points on the linear portion of the plot,

$$C_c = \frac{e_0 - e_1}{\log(p_1 / p_0)} = \frac{\Delta e}{\log(p_1 / p_0)} \tag{8.9}$$

where $p_0$ and $p_1$ are pressures corresponding to $e_0$ and $e_1$.

The slope of the expansion or decompression part of the *e*–log *p* plot (approximated to a straight line) is referred to as the expansion index, $C_e$.

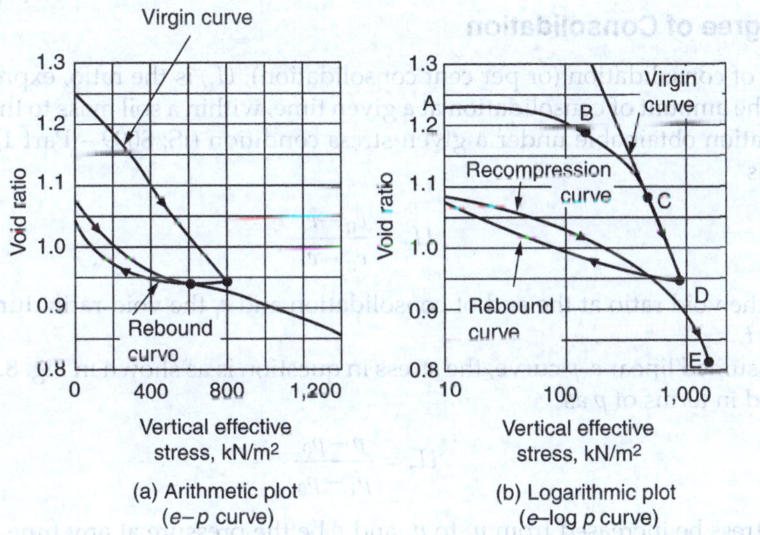


**Fig. 8.6** Pressure–void ratio curves

### 8.6.3 Coefficient of Compressibility

The coefficient of compressibility, $a_v$ ($m^2/N$), is the secant slope, for a given pressure increment, of the effective pressure–void ratio curve (IS: 8009 – Part 1, 1976); that is,

$$a_v = \frac{\Delta e}{\Delta p} \tag{8.10}$$

### 8.6.4 Coefficient of Volume Compressibility

The coefficient of volume compressibility, $m_v$ ($m^2/N$), is the compression of a soil layer per unit of original thickness due to a given unit increase in pressure (IS: 8009 – Part 1, 1976). If for an increase in effective pressure from $p_0$ to $p_1$ the void ratio decreases from $e_0$ to $e_1$, then

$$m_v = \frac{1}{1+e_0}\left(\frac{e_0 - e_1}{p_1 - p_0}\right) = \frac{1}{1+e_0}\left(\frac{\Delta e}{\Delta p}\right) \tag{8.11}$$

or

$$m_v = \frac{1}{H}\left(\frac{H - H_1}{p_1 - p_0}\right) = \frac{1}{H}\left(\frac{\Delta H}{\Delta p}\right) \tag{8.12a}$$

The coefficient of volume compressibility is numerically related to the coefficient of compressibility as

$$m_v = \frac{a_v}{1+e_0} \tag{8.12b}$$

The value of $m_v$ for a particular soil is not constant but depends on the stress range considered.

### 8.6.5 Degree of Consolidation

The degree of consolidation (or per cent consolidation), $U_z$, is the ratio, expressed as a percentage of the amount of consolidation at a given time, within a soil mass to the total amount of consolidation obtainable under a given stress condition (IS: 8009 – Part 1, 1976). This is expressed as

$$U_z = \frac{e_0 - e_t}{e_0 - e_f} \tag{8.13}$$

where $e_f$ is the void ratio at the end of consolidation and $e_t$ the void ratio during consolidation at time $t$.

For an assumed linear $e$–$p$ curve, the stress in question is as shown in Fig. 8.7; then, $U_z$ can be expressed in terms of $p$ as

$$U_z = \frac{p - p_0}{p_1 - p_0} \tag{8.14}$$

Let the stress be increased from $p_0$ to $p_1$ and $p$ be the pressure at any time. Also, let $(u_w)_0$ be the pore water pressure before the increase in total stress, $(u_w)_i$ the increase in pore water

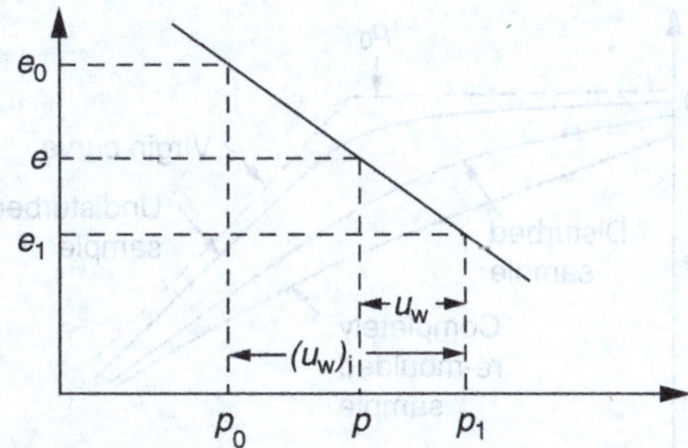


Fig. 8.7 Assumed linear *e–p* curve

pressure above $(u_w)_0$ due to increase in pressure, and $u_w$ the pore water pressure at any time in excess of $(u_w)_0$ then,

$$p_1 = p_0 + (u_w)_i = p + u_w$$

$$U_z = \frac{(u_w)_i - u_w}{(u_w)_t} = 1 - \frac{u_w}{(u_w)_t} \tag{8.15}$$

## 8.7 TYPES OF CLAY DEPOSITS

In the natural process of deposition, fine-grained soils, like silt and clay, undergo the process of consolidation under their own weight of overburden pressure. A state of equilibrium is reached after a lapse of several years, and the compression ceases. This process continues, season after season, and sometimes erosion or removal of overburden takes place, and sometimes the process of consolidation may be continuously taking place due to frequent deposition. So it is evident that clay soil deposits exist in the field under different conditions, and their stress history should be known.

### 8.7.1 Normally Consolidated Clay

If the present effective overburden pressure in the deposit is the maximum pressure to which the deposit has ever been consolidated at any time in the past, such a deposit is called a normally consolidated clay deposit. There is no reliable procedure available to predict the *in situ* effective stress–void ratio relationship.

A field *e–p* relationship has to be obtained only from a carefully obtained undisturbed soil sample. Whatever the care with which the sampling operation is performed, there is bound to be some disturbance due to stress removal. Accordingly, the shape of the *e*–log *p* (or *e–p*) curve is strongly influenced depending on the degree of disturbance. Increase in the degree of disturbance flattens the curve considerably, but the straight line portion (in the *e*–log *p* curve) converges at a low void ratio (Fig. 8.8). Schmertmann (1955) assumed that the straight line portion of the laboratory and *in situ* curves coincide at 0.42 $e_0$. The *in situ* void ratio may be taken equal to the initial void ratio at the start of the test without appreciable loss of accuracy. Further, for normally consolidated clays the

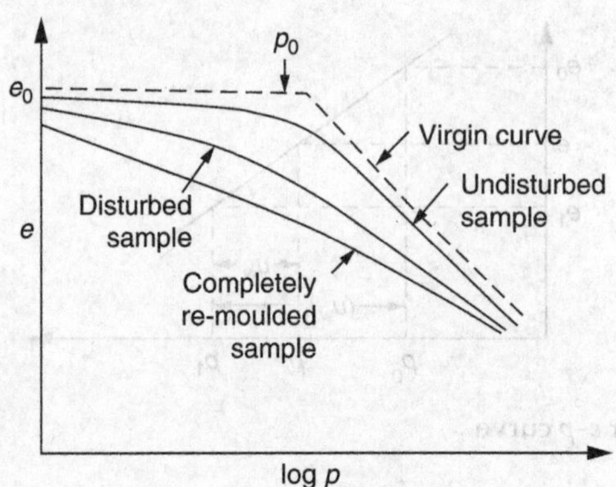


**Fig. 8.8** Effects of sample disturbance (*Source:* Leonards, 1962)

pre-consolidation pressure ($p_c$) is equal to the effective overburden pressure ($p_0$). This is not the case in the deposits which are subjected to thixotropic, secondary compression, or cementation effects (Leonards, 1962). Thus, the *in situ* (or unsampled) compression curve, called virgin curve, may be taken as line CD in Fig. 8.6b. Values of $m_v$ or $C_c$ have to be obtained from the virgin curve and used in the computation of the settlement. Terzaghi and Peck (1967), based on Skempton's earlier concept, proposed an empirical equation for $C_c$, as $C_c = 0.009(w_L - 10\%)$. Indian Standard (IS: 8009 – Part 1, 1976) suggests $C_c = 0.30(e_0 - 0.27)$.

### 8.7.2 Over-consolidated Clay

A clay soil deposit that has been fully consolidated under a pressure $p_c$ in the past, larger than the present overburden pressure $p_0$, is called an over-consolidated (pre-consolidated or pre-compressed) clay deposit. The ratio $(p_c - p_0)/p_0$ is called the over-consolidation ratio (OCR). Over-consolidation of clay may be caused by any or a combination of the following loads (Leonards, 1962):

1. Pressures due to overburden which have been removed (e.g., due to erosion or due to removal of an old structure)
2. Glacial ice sheets which have since disappeared
3. Sustained seepage forces
4. Tectonic forces caused due to movements in the earth's crust
5. Fluctuation of the water table

The *in situ e–p* relationship is radically changed by over-consolidation. Factors other than pressure which may affect the *in situ e–p* relationship are weathering, deposition of cementation materials, and leaching of ions from the pore water. The *in situ e*–log *p* curve is obtained following the procedure given below (Leonards, 1962) (as shown in Fig. 8.9):

1. Consolidation tests are performed on undisturbed samples and loading; unloading and reloading are continued till a reasonable straight line portion of the *e*–log *p* curve is obtained.

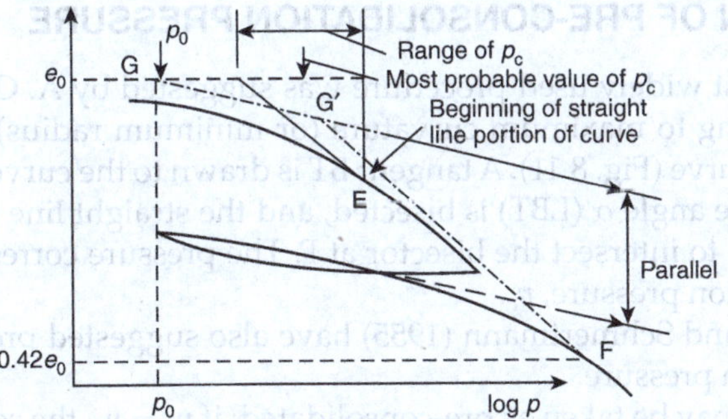


Fig. 8.9 *In situ e*–log *p* curve (*Source:* Leonards, 1962)

2. The pre-consolidation pressure, $p_c$, is estimated (discussed in the next section).
3. A line from the point G ($e_0$, $p_0$) is drawn parallel to the mean slope of the rebound curve.
4. Select a point G′ on this line corresponding to $p_c$. From this point, a line is drawn to meet the point (F) of intersection of the laboratory curve at 0.42 $e_0$. (Schmertmann, 1955, suggested a range of 0.35 $e_0$ to 0.45 $e_0$.)
5. The dot–dashed lines GG′ and G′F are then used to obtain $C_c$ for calculating the settlement.

### 8.7.3 Under-consolidated Clay

Rapid natural deposition or deposits under recent fillings may not be fully consolidated under the present overburden pressure; such clay deposits are called under-consolidated clays (Fig. 8.10). In such cases, $p_c < p_0$, and structures constructed on this deposit will cause additional compression. No specific procedure is available to get the *in situ e–p* curve. However, for all practical purposes, this may be treated as normally consolidated for the purpose of calculating the settlement.

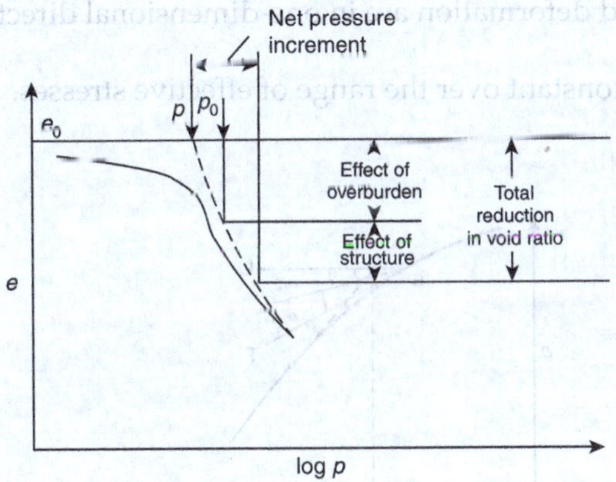


Fig. 8.10 Under-consolidated clay deposit

## 8.8 PREDICTION OF PRE-CONSOLIDATION PRESSURE

The earliest and most widely used procedure was suggested by A. Casagrande (1936). The point B corresponding to maximum curvature (or minimum radius) is chosen on the first laboratory loading curve (Fig. 8.11). A tangent BT is drawn to the curve at B, and a horizontal BL is also drawn. The angle $\alpha$ (LBT) is bisected, and the straight line portion of the curve is projected backwards to intersect the bisector at P. The pressure corresponding to this point is the pre-consolidation pressure, $p_c$.

Burmister (1951) and Schmertmann (1955) have also suggested procedures to determine the pre-consolidation pressure.

If $p_c > p_0$, the soil may be taken as pre-consolidated; if $p_c = p_0$, the soil is normally consolidated; and if $p_c < p_0$, the soil is probably under-consolidated (if pressure difference is large) or nearing normal consolidation condition.

## 8.9 RATE OF CONSOLIDATION

### 8.9.1 Terzaghi's Theory of One-Dimensional Consolidation

A consolidation theory in three dimensions involving stress and strain condition would be highly complicated because of the heterogeneous nature of the soil. The solutions to three-dimensional consolidations have to be attended through numerical methods. But the one-dimensional consolidation theory, as proposed by Terzaghi (1925, 1943), simplifies the problem smoothly and, at the same time, satisfies the relevant factors connected with settlement.

In analysing the rate of one-dimensional consolidation, Terzaghi (1925, 1943) made the following assumptions:

1. The soil mass is homogeneous.
2. The void spaces are completely filled with water.
3. The soil solids and water are incompressible.
4. Darcy's law is valid.
5. The seepage flow and deformation are in one-dimensional direction.
6. Strains are small.
7. The permeability is constant over the range of effective stresses.

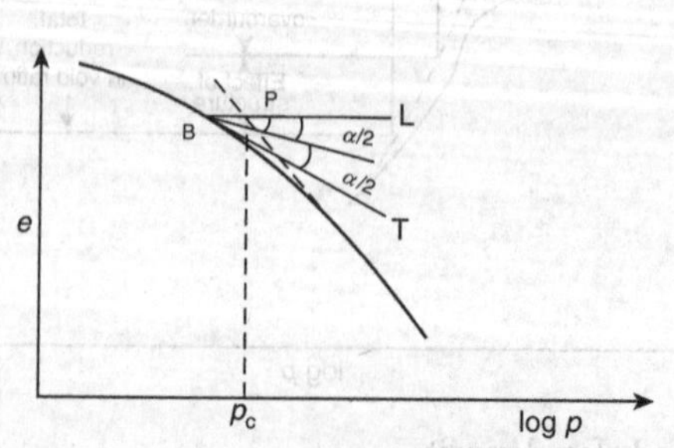


**Fig. 8.11** Determination of $p_c$

8. There is an unique pressure–void ratio relationship, i.e., the coefficient of compressibility is constant.
9. The time lag in consolidation is entirely due to the low permeability of the soil.

Consider an element of soil of constant area $dx \times dy$ and thickness dz within a clay layer of 2 $d$ depth free to drain in the $z$ direction only (Fig. 8.12). Let an increment of vertical stress $\Delta\sigma$ be instantaneously applied and maintained constant.

The volume of water flowing into the element in unit time is

$$q_{\text{in}} = v_z \, dx \, dy \tag{8.16a}$$

and the volume of water flowing out of the element

$$q_{\text{out}} = \left( v_z + \frac{\partial v_z}{\partial z} dz \right) dx \, dy \tag{8.16b}$$

As the soil is fully saturated and solid particles and water are incompressible, the law of conservation of matter requires that $q_{\text{out}} - q_{\text{in}}$ = change in volume of the element per unit time. That is,

$$\Delta q = \frac{\partial V}{\partial t} \tag{8.17}$$

or

$$\frac{\partial V}{\partial t} = \frac{\partial v_z}{\partial z} dx \, dy \, dz \tag{8.18}$$

and

$$V = V_s(1+e_0) = dx \, dy \, dz \tag{8.19}$$

or

$$\frac{\partial V}{\partial t} = V_s \frac{\partial e}{\partial t} \tag{8.20}$$

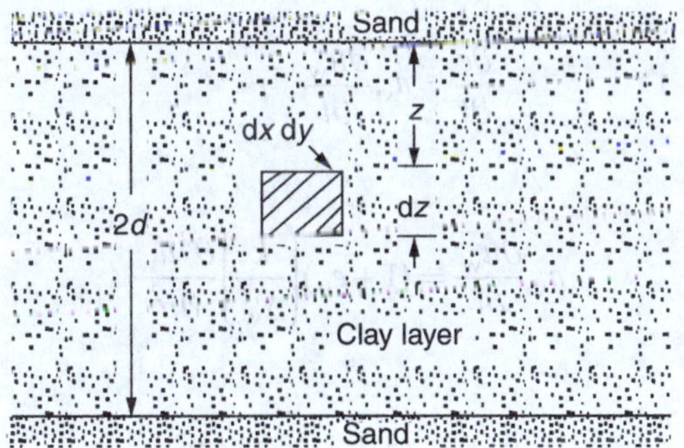


Fig. 8.12 Element within clay layer

Substituting for

$$V_s = \frac{dx\, dy\, dz}{1+e_0}$$

$$\frac{\partial V}{\partial t} = \frac{dx\, dy\, dz}{1+e_0}\frac{\partial e}{\partial t} \tag{8.21}$$

or

$$\frac{dx\, dy\, dz}{1+e_0}\frac{\partial e}{\partial t} = \frac{\partial v_t}{\partial t} dx\, dy\, dz$$

$$\frac{\partial e}{\partial t} = (1+e_0)\frac{\partial v_z}{\partial z} \tag{8.22}$$

The hydraulic gradient $i = \partial h/\partial z$

or

$$i = \frac{1}{\gamma_w}\frac{\partial u_w}{\partial z} \qquad \left(\because,\ h = \frac{u_w}{\gamma_w}\right)$$

or

$$v_z = ki = \frac{k}{\gamma_w}\frac{\partial u_w}{\partial z}$$

or

$$\frac{\partial v_z}{\partial z} = \frac{k}{\gamma_w}\frac{\partial^2 u_w}{\partial z^2}$$

$$\frac{\partial e}{\partial t} = \frac{(1+e_0)\,k}{\gamma_w}\frac{\partial^2 u_w}{\partial z^2}$$

Since $\partial e = a_v \partial u_w$

$$\frac{\partial e}{\partial t} = a_v \frac{\partial u_w}{\partial t}$$

or

$$a_v \frac{\partial u_w}{\partial t} = (1+e_0)\left(\frac{k}{\gamma_w}\right)\frac{\partial^2 u_w}{\partial z^2}$$

or

$$\frac{\partial u_w}{\partial t} = \left(\frac{1+e_0}{a_v}\right)\left(\frac{k}{\gamma_w}\right)\frac{\partial^2 u_w}{\partial z^2}$$

or

$$\frac{\partial u_w}{\partial t} = c_v \frac{\partial^2 u_w}{\partial z^2} \tag{8.23}$$

where

$$c_v = \frac{k}{\gamma_w}\left(\frac{1+e_0}{a_v}\right) \tag{8.24}$$

Equation 8.23 is the Terzaghi's one-dimensional consolidation equation, where $c_v$ is called the coefficient of consolidation (m$^2$/year).

As consolidation proceeds, $k$, $e$, and $a_v$ all decrease with time, but the ratio $c_v$ remains approximately constant. The main limitations of Terzaghi's theory (apart from its one-dimensional nature) is the non-linearity of the void ratio–pressure relationship.

### 8.9.2 Time Factor

The solution of Eq. 8.23 for a constant initial pore pressure $[(u_w)_i = (u_w)_0]$ satisfies the following conditions:

at $z = 0$ $u_w = 0$

at $z = 2d$ $u_w = 0$

at $t = 0$ $u_w = (u_w)_0$

where

$$u_w = \sum_{m=0}^{\infty} \frac{2(u_w)_0}{[\sin(Mz/d)]} \exp(-M^2T_v) \tag{8.25}$$

where $M = (\pi/2)(2m+1)$, $m$ = any integer (0, 1, 2, 3), and $d$ = length of the longest drainage path.

Now, time factor

$$T_v = \frac{c_v t}{d^2} \tag{8.26}$$

This is a dimensionless factor containing the physical constants of a soil stratum influencing its time rate of consolidation.

But

$$U_z = \left\{1 - \frac{u_w}{(u_w)_i}\right\} \times 100 \tag{8.27}$$

Combining Eqs. 8.25 and 8.26, a basic $U_z$ versus $T_v$ relationship can be developed; thus,

$$U_z = 1 - \sum_{m=0}^{\infty} \frac{2}{M}\left(\sin\frac{M_z}{d}\right)\exp(-M^2T_v) \tag{8.28}$$

A numerical relationship between $U_z$ and $T_v$ may be obtained by substituting values for $m$ from 0 to $\infty$ (Fig. 8.13).

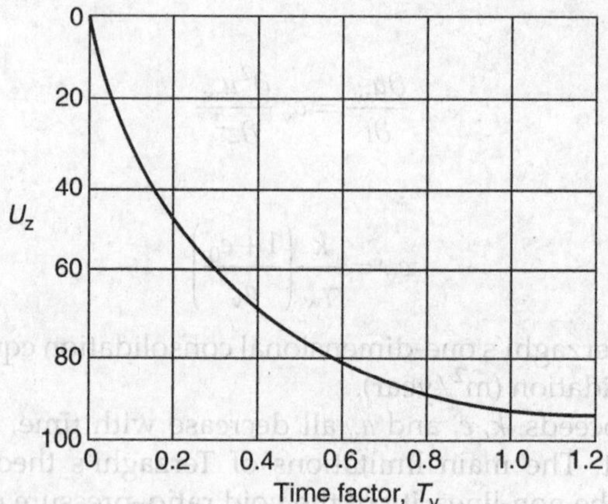


Fig. 8.13 Average percent consolidation versus time factor

The above numerical relationship is not valid for other types of distributions (e.g., rectangular, sinusoidal, and trapezoidal) except constant initial excess hydrostatic pressure. However, since the variation is very small, Fig. 8.13 may be used with reasonable accuracy (Taylor, 1948).

Two empirical equations instead of Eq. 8.28 are in use, viz.,

$$T_v = \frac{\pi}{4}(U_z/100)^2 \quad \text{when } U_z \le 60\% \tag{8.29}$$

and

$$T_v = -0.933\log_{10}\left(1-\frac{U_z}{100}\right) - 0.0851 \quad \text{when } U_z > 60\% \tag{8.30}$$

### 8.9.3 Determination of Coefficient of Consolidation

A comparison of laboratory compression versus time and the theoretical $U_z$ versus $T_v$ has shown similar shapes. This observation resulted in two types of transformation plots, one using the square root of time (Taylor, 1948) and the other the logarithm of time (Casagrande, 1936b), which are used to determine the coefficient of consolidation and are called fitting methods.

**The Square Root of Time Fitting Method**. The theoretical curve $U_z$ versus $T_v$ is a straight line up to 60% consolidation (Fig. 8.14a), and the abscissa of curve at 90% consolidation is 1.15 times the abscissa of an extension of the straight line. The characteristics of the theoretical curve have been used by Taylor (1948) to determine a point of 90% consolidation on the laboratory time curve.

A plot of compression dial reading versus $\sqrt{\text{time}}$ is made (Fig. 8.14b). In the early portion of the laboratory curve, a straight line is drawn through the observed points. A second line is drawn coinciding with the first line at $t = 0$, such that the abscissa for the new line is 1.15

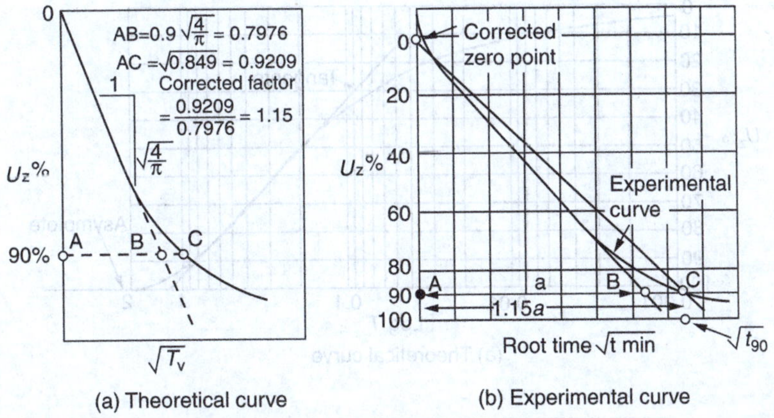


(a) Theoretical curve (b) Experimental curve

**Fig. 8.14** Square root of time method

times the abscissa of the previous line at a given dial reading. The observed zero reading and the points of coincidence of the two straight lines will not be the same, but the latter will usually lie below. This point is called the corrected zero point. The point of intersection of the second straight line and the laboratory curve corresponds to 90% consolidation, and the time is designated as $t_{90}$. Therefore,

$$(T_v)_{90} = \frac{c_v t_{90}}{d^2} \tag{8.31}$$

Finding $(T_v)_{90}$ and deciding the drainage path, $d$, which is equal to half the thickness of the layer for double drainage and full thickness for single drainage, we have

$$c_v = \frac{0.848 d^2}{t_{90}} \tag{8.32}$$

**The Logarithm of Time Fitting Method**. The intersection of the tangent and the asymptote to the theoretical consolidation curve is at 100% consolidation (Fig. 8.15a). As the shapes of the theoretical curve ($U_z$ versus log $T_v$) and the laboratory compression dial versus log $t$ curve resemble one another, Casagrande (1936b) suggested that the 100% consolidation point be obtained by drawing tangents to the straight line portions of the primary consolidation curve.

The early portion of the curve approximates a parabola. The corrected zero point may be located by taking the difference in ordinates ($z_0$) between two points corresponding to time $t$ and 4 $t$ on the early portion of the curve and laying off this value ($z_0$) above the curve at point $t$ (Fig. 8.15b).

As the dial reading corresponding to zero and 100% primary consolidation is known, the time for 50% consolidation ($t_{50}$) can be obtained; thus,

$$(T_v)_{50} = \frac{c_v t_{50}}{d^2} \tag{8.33}$$

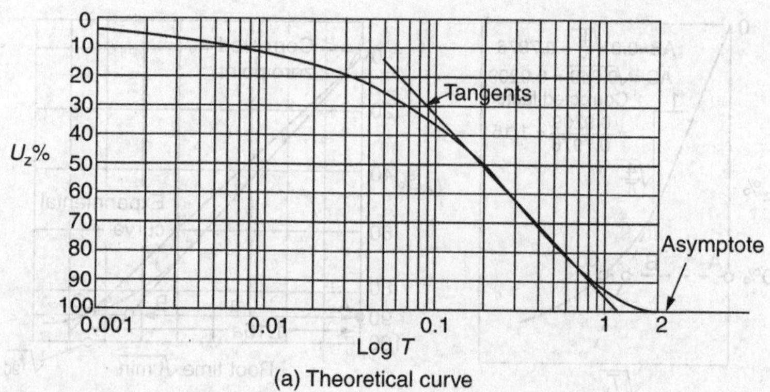

(a) Theoretical curve

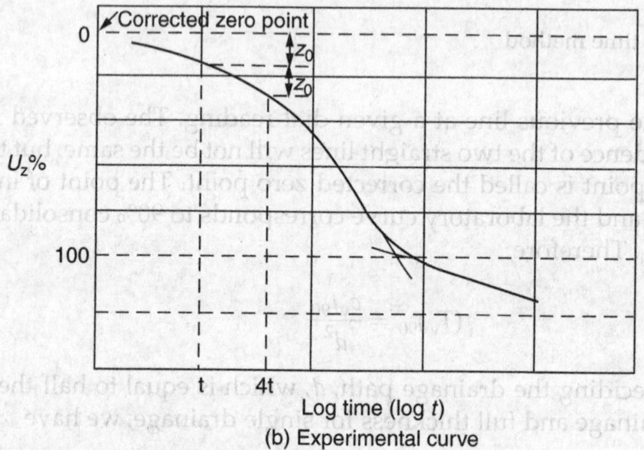

(b) Experimental curve

**Fig. 8.15** Logarithm of time method

or

$$c_v = \frac{0.196d^2}{t_{50}} \tag{8.34}$$

In general, both fitting methods show good agreement.

**Hyperbola Method.** The following procedure is adopted for the determination of $C_v$:

1. From laboratory consolidation test, the time ($t$) and the specimen deformation ($\Delta H$) are obtained.
2. A graph of $t/\Delta H$ against $t$ is drawn in (Fig. 8.16).
3. The straight line portion $bc$ is projected back to point $d$. The intercept $D$ is determined.
4. The slope $m$ of the line $bc$ is determined.
5. Then coefficient of consolidation, $C_v$ is calculated as

$$C_v = 0.3\left(\frac{mH^2}{D}\right).$$

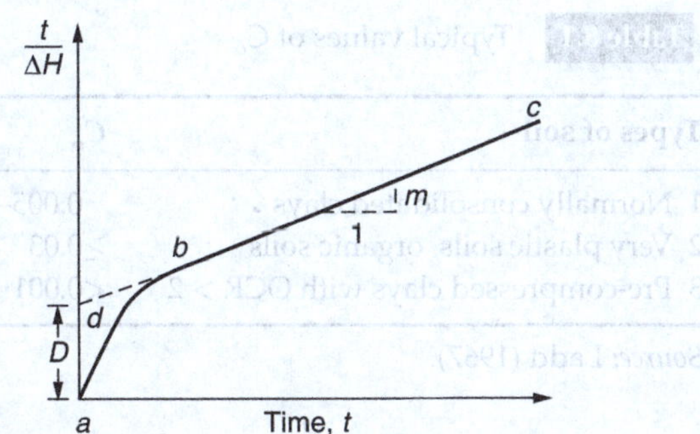


Fig. 8.16 Hyperbola method of determination of $C_v$

As the unit of $D$ is time/length and the unit of $m$ is (time/length)/time – 1/length, the unit of $C_v$ is $(\text{length})^2$/time.

The hyperbola method is fairly simple to use, and it gives good results for $U = 60\%$–$90\%$.

## 8.10 SECONDARY COMPRESSION

A soil mass is said (theoretically) to be fully consolidated under a given pressure when the excess pore water pressure is zero, depicting the termination of primary consolidation. Actually, consolidation does not cease but continues slowly even after the excess pore water pressure dissipation, and this continued time-dependent compression is referred to as the secondary compression. In natural soil deposits where the ground surface is inclined, creeps and, perhaps, volume changes due to dissipation of non-measurable excess pore water pressure will be taking place (Fig. 8.17). Compression occuring during creep due to readjustment of the soil skeleton is so low that it may continue for a very long time. Practical evidence shows that creep ceases or becomes so small that it is not measurable.

Evaluation of secondary compression is difficult. However, by maintaining a constant pressure on a clay long enough past the point of primary consolidation, a relationship between secondary compression and time may be obtained, provided temperature control and equipment corrosion are taken care of. The magnitude of secondary compression is

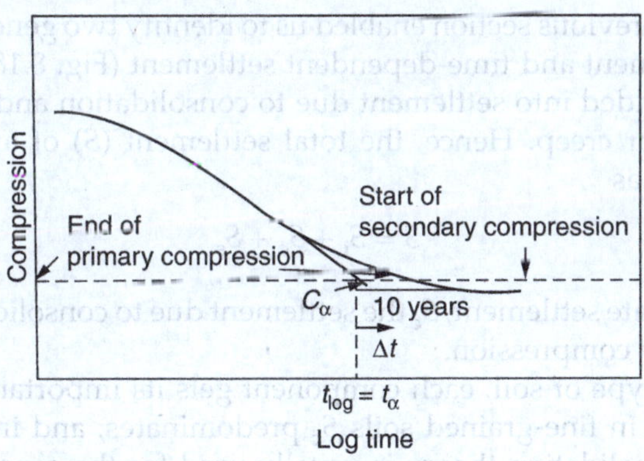


Fig. 8.17 Secondary compression and definition of its rate

**Table 8.1** Typical values of $C_\alpha$

| Types of soil | $C_\alpha$ |
|---|---|
| 1 Normally consolidated clays | 0.005–0.02 |
| 2 Very plastic soils, organic soils | $\geq 0.03$ |
| 3 Pre-compressed clays with OCR > 2 | <0.001 |

*Source:* Ladd (1967).

referred to as a coefficient of secondary compression and expressed as the slope $C_\alpha$ of the final portion of the time compression curve (Fig. 8.17).

That is,

$$C_\alpha = \frac{\Delta H/H}{\log(t_p + \Delta t/t_p)} \tag{8.35}$$

where $t_p$ is the time when the primary consolidation is complete and $\Delta t$ the time increment producing $\Delta H$.

Therefore, the secondary compression or creep *in situ* for the time increment $\Delta t$ for a stratum of thickness $H$ is

$$\Delta H = HC_\alpha \log \frac{t_p + \Delta t}{t_p} \tag{8.36}$$

Secondary compression may be insignificant for inorganic soils but may be predominant in highly organic soils. Based on different relationships between $C\alpha$ and $C_c$, Bowles (1984) suggested for inorganic soils a value of $C\alpha/C_c \leq 0.05$ and for organic soils a range of 0.07 to 0.10. Typical values for the coefficient of secondary compression for some soils are given in Table 8.1 (Ladd, 1967).

## 8.11 CONSOLIDATION SETTLEMENT AND ITS RATES

The discussion in the previous section enabled us to identify two general types of settlement, viz., immediate settlement and time-dependent settlement (Fig. 8.18). The time-dependent settlement may be divided into settlement due to consolidation and settlement due to secondary compression or creep. Hence, the total settlement ($S$) of any soil mass under an applied stress is given as

$$S = S_i + S_c + S_s \tag{8.37}$$

where $S_i$ is the immediate settlement, $S_c$ the settlement due to consolidation, and $S_s$ the settlement due to secondary compression.

Depending on the type of soil, each component gets its importance. For coarse-grained soils $S_i$ predominates, in fine-grained soils $S_c$ predominates, and in organic soils $S_s$ has a vital role to play. Consolidation theory is usually used for the consolidation settlement in saturated fine-grained soils.

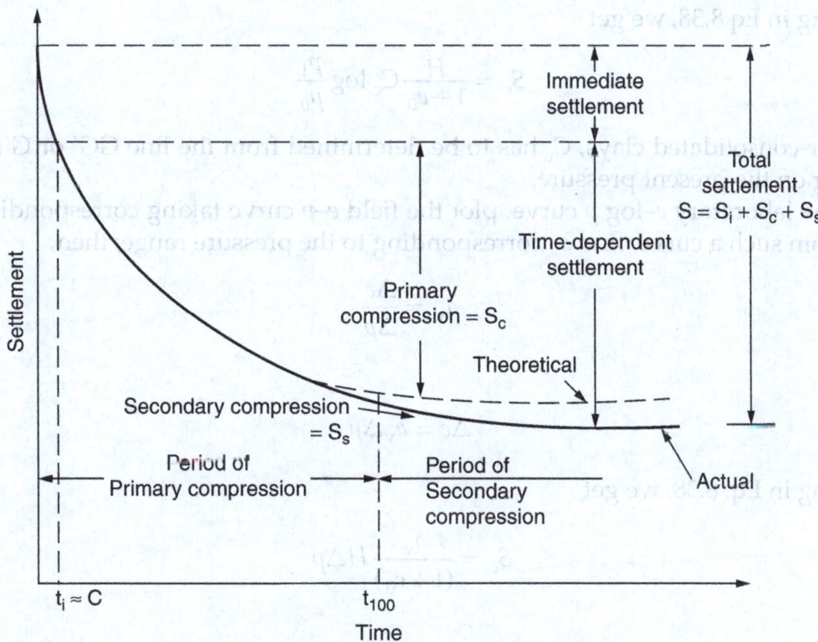


Fig. 8.18 Components of total settlement

In this section, only the consolidation settlement is discussed, and immediate settlement component is considered in Chapter 15.

### 8.11.1 Consolidation Settlement

Consolidation settlement is calculated knowing the following:

1. Thickness of compressible layer, $H$
2. Initial void ratio, $e_0$
3. Field $e$–log $p$ curve or $C_c$, $m_v$, or $a_v$ corresponding to the pressure range

Considering Eq. 8.7 and letting $\Delta H = S_c$, we get the generalized equation for consolidation settlement; that is,

$$S_c = H\frac{\Delta e}{(1+e_0)} \tag{8.38}$$

Compression index from the field $e$–log $p$ curve is given in Eq. 8.9 as

$$C_c = \frac{\Delta e}{\log(p_1/p_0)}$$

$$\Delta e = C_c \log\frac{p_1}{p_0}$$

Substituting in Eq.8.38, we get

$$S_c = \frac{H}{1+e_0} C_c \log \frac{p_1}{p_0} \tag{8.39}$$

For over-consolidated clays, $C_c$ has to be determined from the line GG′ or G′F (Fig. 8.9) depending on the present pressure.

From the laboratory $e$–log $p$ curve, plot the field $e$–$p$ curve taking corresponding $e$ and $p$ values. From such a curve, find $a_v$ corresponding to the pressure range; then,

$$a_v = \frac{\Delta e}{\Delta p}$$

or

$$\Delta e = a_v \Delta p$$

Substituting in Eq. 8.38, we get

$$S_c = \frac{a_v}{(1+e_0)} H \Delta p \tag{8.40}$$

But

$$m_v = \frac{a_v}{1+e_0'}$$

Thus,

$$S_c = m_v H \Delta p \tag{8.41}$$

Equations 8.38 to 8.41 can be used for the computation of settlement.

Equation 8.41 can be used only if the thickness of the compressible layer is less and the assumption of constant $m_v$ and $\Delta p$ with depth is valid. Settlement computation for thick layers is discussed below.

As the lateral strain is zero, the decrease in volume per unit volume is equal to the decrease in thickness per unit thickness. This is nothing but the settlement per unit depth, that is,

$$\frac{\Delta V}{V} = \frac{e_0 - e_1}{1+e_0}$$

or

$$\frac{dS_c}{dz} = \frac{e_0 - e_1}{1+e_0}$$

or

$$dS_c = \left(\frac{e_0 - e_1}{p_1 - p_0}\right)\left(\frac{p_1 - p_0}{1+e_0}\right) dz = \frac{a_v}{1+e_0} \Delta p \, dz$$

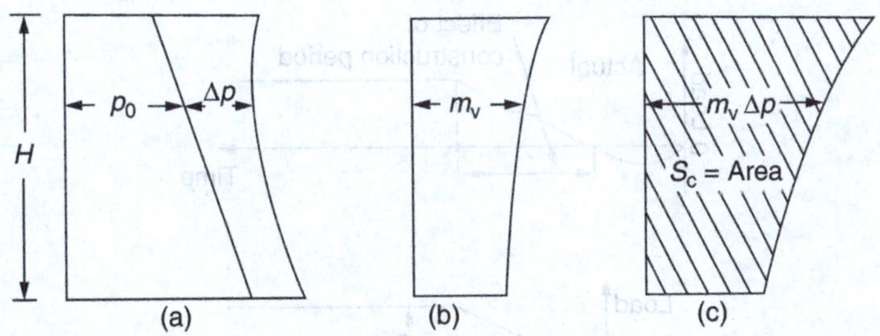


Fig. 8.19 Graphical procedure for consolidation settlement

or

$$dS_c = m_v \Delta p \, dz \tag{8.42}$$

$$S_c = \int_0^H m_v \Delta p \, dz \tag{8.43}$$

If $m_v$ and $\Delta p$ are constants, Eq. 8.43 reduces to Eq. 8.41.

The variations of initial effective vertical stress ($p_0$) and effective vertical stress increment ($\Delta p$) over the depth of the layer are shown in Fig. 8.19. The variation of $m_v$ is shown in Fig. 8.19b. Figure 8.19c represents the variation of non-dimensional factor $m_v \Delta p$ with depth. The area under this curve is the settlement of the layer.

### 8.11.2 Construction Period Correction

In practice, real loads are added to the soil over a period of time, perhaps after many months or years, and not instantaneously. Due to excavation, there is a reduction in the net load, resulting in swelling of the clay. Settlement begins only when the applied load exceeds the weight of the excavated soil. The corrected settlement may be obtained adopting Terzaghi's approximate method (also adopted by IS: 8009 – Part 1, 1976). This is based on the following assumptions: (i) the actual settlement at the end of the construction time is the same as that resulting from the total load acting for half of the loading time and (ii) the load–time relationship is linear. Thus, the settlement at the end of the loading period is equal to the settlement on the instantaneous load curve corresponding to one-half of the total loading period.

The instantaneous settlement–time curve is plotted from point O (Fig. 8.20). As per the assumption, the amount of consolidation settlement at time $t_c$ is the same as that obtained from the instantaneous curve at $1/2t_c$. A point P is located on the instantaneous curve at time $t_c/2$, and a point Q is established by drawing a parallel to the time axis from P to meet the ordinate at $t_c$. From points P and Q onwards, the abscissa of the corrected curve will be $1/2t_c$ greater than for the instantaneous curve. To get points during the construction period, choose a point R at time $1/2t$, and point S is found by projecting horizontally onto the ordinate at $t$ with the line OS. That is, the settlement at $t$ (i.e., ordinate BT), is equal to (OB/OA) AS, i.e., $t/t_c A_s$. This process is repeated until additional points are obtained for construction of the corrected curve.

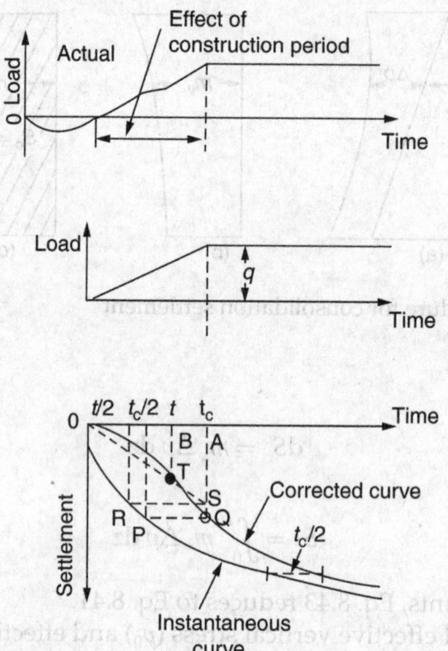


Fig. 8.20 Correction for construction period

### 8.11.3 Settlement Due to Secondary Compression

Considering Eq. 8.35 and letting $\Delta H = S_s$, the settlement due to secondary compression is

$$S_s = HC_\alpha \log \frac{t_p + \Delta t}{t_p} \tag{8.44}$$

### 8.11.4 Predicting the Rate of Settlement

The settlement $S_c$ is computed from Eqs. 8.38 to 8.41. An approximate value of $c_v$ is selected corresponding to the average pressure during consolidation. The time rate of settlement is calculated using the equation given below:

$$t = \frac{T_v (d)^2}{c_v} \tag{8.45}$$

where $T_v$ is calculated corresponding to the desired degree of consolidation and $d$ is the drainage path in the field.

Now,

$$\varepsilon_{av} = U_z \varepsilon_f \tag{8.46}$$

$$\varepsilon_f = \frac{S_c}{H} \tag{8.47}$$

where $\varepsilon_{av}$ is the average strain in the clay layer, $S_c$ the total calculated consolidation settlement, $H$ the height of the field clay layer, and

$$\varepsilon_{av} = \frac{S_t}{H} \tag{8.48}$$

where $S_t$ is the settlement of field layer at time $t$.

The degree of consolidation for the full clay layer is

$$U_z = \frac{S_t}{S_c} \tag{8.49}$$

or

$$S_t = U_z S_c \tag{8.50}$$

Settlement of clay layer, $S_t$, and the time for consolidation are computed for $U_z$ varying from 10% to 100%. The time values are calculated from Eq. 8.45 and the settlement values from Eq. 8.50.

For very thick clay layers, to determine $c_v$, several tests on undisturbed samples have to be made. If variation in $c_v$ for each layer is not wide, an arithmetical average may be taken.

## 8.12 ACCELERATION OF CONSOLIDATION BY SAND DRAINS

The normally slow process of consolidation of fine-grained soils may be accelerated by shortening the length of the drainage path of the pore water escaping from the consolidating soil. This may be accomplished by installing a system of vertical sand drains as shown in Fig. 8.21a. A sand drain essentially consists of a vertical borehole put down through the saturated fine-grained soil and extending to a relatively firm bottom. By this provision, the excess pore water pressures dissipate by both vertical and radial flow. However, the resulting settlement will still be in the vertical direction.

The basic theory of sand drains was presented by Rendulic (1935) and Barron (1948). A review of the theories of sand drains was made by Richart (1957). In the analysis of theories, two conditions prevail:

1. *Free-strain condition*: Application of a flexible surcharge load will cause an uneven settlement at the surface, and this condition is referred to as free-strain condition.
2. *Equal-strain condition*: Application of a rigid surcharge load will cause an equal settlement at the surface, and this condition is referred to as equal-strain condition.

Considering an element of soil A with vertical and radial flows (Fig. 8.21b) and developing the governing differential equations similar to one-dimensional theory (for detailed derivation, refer Leonards, 1962), we have

$$c_v \frac{\partial^2 u_w}{\partial z^2} + c_h \left[\frac{\partial^2 u_w}{\partial r^2} + \frac{1}{r}\frac{\partial u_w}{\partial r}\right] = \frac{\partial u_w}{\partial t} \tag{8.51}$$

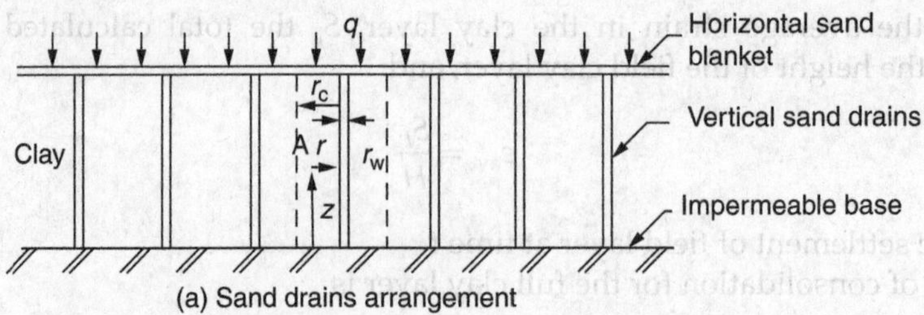


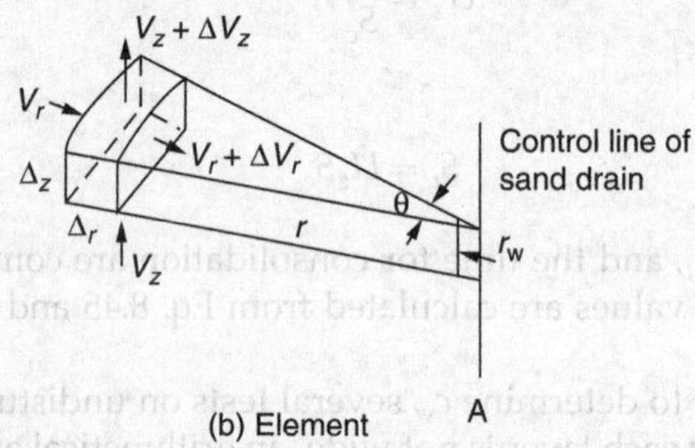


Fig. 8.21 Three-dimensional consolidation

where

$$c_v = \frac{k_v(1+e_0)}{\gamma_w a_v}$$

is the coefficient of vertical consolidation (as previously defined, Eq. 8.24) and

$$c_h = \frac{k_h(1+e_0)}{\gamma_w a_v} \tag{8.52}$$

is the coefficient of radial and horizontal consolidation, where $k_v$ and $k_h$ are referred to as vertical and horizontal permeabilities, respectively.

Equation 8.51 is the governing differential equation for three-dimensional consolidation and may be considered to consist of two parts:

One-dimensional flow:

$$c_v \frac{\partial^2 u_w}{\partial z^2} = \frac{\partial u_w}{\partial t}$$

Radial flow:

$$c_h \left(\frac{\partial^2 u_w}{\partial r^2} + \frac{1}{r}\frac{\partial u_w}{\partial r}\right) = \frac{\partial u_w}{\partial t} \tag{8.53}$$

A solution to Eq. 8.51 has been considered by Carilo (1942) as a combination of solutions of Eqs. 8.23 and 8.53, and accordingly we have

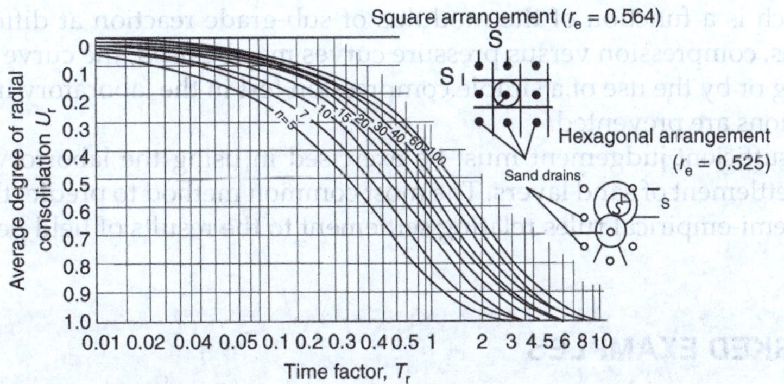


Fig. 8.22 Average degree of consolidation for radial flow

$$(1-U)=(1-U_z)(1-U_r) \tag{8.54}$$

where $U$ is the degree of consolidation for three-dimensional flow, $U_z$ the degree of consolidation for one-dimensional flow,and $U_r$ the degree of consolidation for radial flow

The time factor $T_r$ for radial flow is

$$T_r = \frac{C_h t}{(2r_e)^2} \tag{8.55}$$

where $r_e$ is the radius of influence (Fig. 8.21a). The time factor $T_v$ for one-dimensional consolidation is as given by Eq. 8.25.

The solution to Eq. 8.23 is already discussed, and the solution for Eq. 8.53 was given by Rendulic (1935) for free-strain condition and by Barron (1948) for equal-strain condition. It has been reported by Richart (1957) for values of $n(= r_e/r_w)$ greater than 5 that both the solutions give closer values. Hence, for all practical purposes, Fig. 8.22 (Barron, 1948) may be used regardless of the imposed boundary strain condition.

During drilling operation, a zone of soil surrounding the drain, called smear zone, is formed due to re-moulding of the clay, which causes a reduction in horizontal permeability. This factor has to be considered while selecting the diameter of the drain.

## 8.13 COMPRESSIBILITY OF SANDS

Sandy soils deform less under static loads than under dynamic loads. A static load does cause a reduction in the void space in the sand mass. In general, the settlement in non-cohesive soils occurs immediately after the application of load. This settlement is attributed to the volume changes caused by lateral yielding or shear strain that occurs in the soil. The settlement may be calculated considering the sand layer as an elastic medium (discussed in Chapter 15).

In uniform non-cohesive soil deposits, the soil rigidity (or modulus of elasticity) increases with depth because of increase in confining pressure, which is a function of overburden pressure. Further, if the loaded area is large, the settlement is governed by the contact pressure, which is a function of the modulus of sub-grade reaction at different points. In such situations, compression versus pressure curves may be used, the curve being obtained by field testing or by the use of a simple compression test in the laboratory in which the lateral deformations are prevented.

However, sufficient judgement must be exercised in using the labortory test results to predict field settlement of sand layers. The most common method to predict the settlement is by adopting semi-empirical rules relating settlement to the results of field penetration tests.

## WORKED EXAMPLES

**Example 8.1** The dial readings recorded during a consolidation test at a certain load are as follows:

| Time (minutes) | Dial reading ($mm \times 10^{-5}$) | Time (minutes) | Dial reading ($mm \times 10^{-5}$) |
|---|---|---|---|
| 0.0 | 240 | 15 | 622 |
| 0.10 | 318 | 30 | 738 |
| 0.25 | 340 | 60 | 842 |
| 0.50 | 360 | 140 | 930 |
| 1.00 | 385 | 240 | 975 |
| 2.00 | 415 | 1,200 | 1,070 |
| 4.00 | 464 | 1,620 | 1,090 |
| 8.00 | 530 | | |

The initial thickness of the sample is 25 mm. Compute the coefficient of consolidation by square root of time and logarithm of time-fitting methods.

***Solution***

| Dial reading ($mm \times 10^{-5}$) | $\sqrt{Time}$ ($\sqrt{min}$) | Dial reading ($mm \times 10^{-5}$) | $\sqrt{Time}$ ($\sqrt{min}$) |
|---|---|---|---|
| 240 | 0 | 622 | 3.873 |
| 318 | 0.316 | 738 | 5.477 |
| 340 | 0.500 | 842 | 7.746 |
| 360 | 0.707 | 930 | 11.832 |
| 385 | 1.000 | 975 | 15.492 |
| 415 | 1.414 | 1,070 | 34.641 |
| 464 | 2.000 | 1,090 | 40.249 |
| 530 | 2.828 | | |

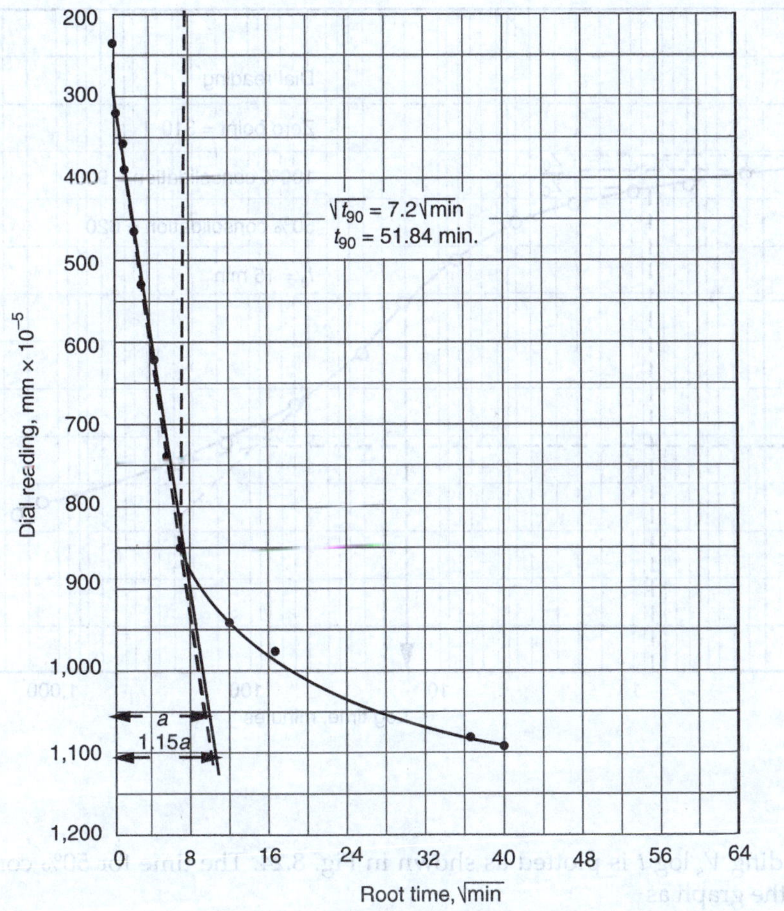


Fig. 8.23

Dial reading versus $\sqrt{\text{Time}}$ is plotted as shown in Fig. 8.23. The $\sqrt{\text{Time}}$ for 90% consolidation is read from the graph as

$$\sqrt{t_{90}} = 7.2\sqrt{\text{min}}$$

$$t_{90} = 51.84 \text{ min}$$

$$\text{Average thickness of the sample} = 25 - \frac{(1090 - 240) \times 10^{-5}}{2} = 24.96 \text{ mm}$$

$$\text{Length of drainage path } d = \frac{24.96}{2} = 12.48 \text{ mm}$$

$$c_v = \frac{(T_v)_{90} d^2}{t_{90}} = \frac{0.848 \times 12.48^2}{51.84} = 2.55 \text{ mm}^2/\text{min}$$

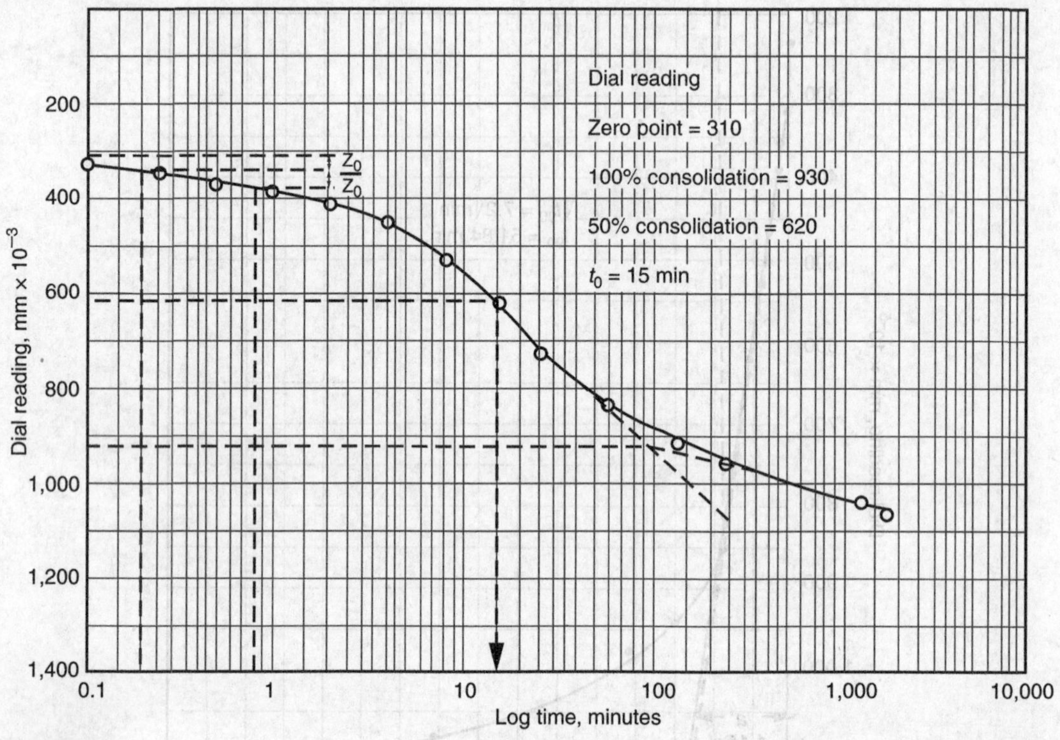


Fig. 8.24

Dial reading $V_s$ log $t$ is plotted as shown in Fig. 8.24. The time for 50% consolidation is read from the graph as

$$t_{50} = 15 \text{ min}$$

$$c_v = \frac{(T_v)_{50} d^2}{t_{50}} = \frac{0.196 \times 12.48^2}{15} = 2.04 \text{ mm}^2/\text{min}$$

**Example 8.2** A large highway interchange fill was built over a 15 m thick soil structure. The initial void ratio was 0.78. After 70% settlement, the void ratio was 0.69. Estimate the total anticipated settlement.

***Solution***

From Eq. 8.13,

$$U_z = \frac{e_0 - e_t}{e_0 - e_f}$$

That is,

$$0.70 = \frac{0.78 - 0.69}{0.78 - e_f}$$

Therefore,

$$e_f = 0.65$$

From Eq. 8.7,

$$\Delta H = H \frac{\Delta e}{1 + e_0}$$

Therefore,

$$\Delta H = 15 \times \frac{(0.78 - 0.65)}{1 + 0.78} = 1.096 \text{ m}$$

**Example 8.3** A layer of clay of 8 m thickness is located between two sand layers. The coefficient of consolidation of the clay is $5 \times 10^{-2}$ $m^2/s$. The fill material was placed over a very large area above the top sand layer. After 3 years, what percentage of the additional pressure due to the fill material will be carried by the soil grains at the middle of the clay layer?

***Solution***
From Eq. 8.26,

$$T_v = \frac{c_v t}{d^2}$$

That is,

$$T_v = \frac{5 \times 10^{-2}}{1000 \times 1000} \times \frac{3 \times 365 \times 24 \times 60 \times 60}{(8/2)^2} = 0.296$$

From Eq. 8.29,

$$T_v = \frac{\pi}{4} (U_z / 100)^2$$

That is,

$$U_v = \frac{\pi}{4} \left( \frac{U_z}{100} \right)^2$$

That is,

$$\frac{U_z}{100} = \sqrt{\frac{2 \times 0.296}{\pi}} = 0.434$$

Therefore,

$$U_z = 43.4\%$$

Additional pressure due to fill material carried by the soil grains is 43.4%.

**Example 8.4** A laboratory specimen of clay 20 mm thick, drained at the top and bottom, has taken 360 seconds to reach 50% consolidation when the pressure was increased from 75 to 150 kN/m$^2$. The initial void ratio was 0.80, and the final void ratio due to the increment of load was 0.45. Determine the coefficient of permeability of the clay.

***Solution***

For 50% consolidation, $T_v = 0.196$. From Eq. 8.26,

$$c_v = \frac{T_v d^2}{t}$$

That is,

$$c_v = \frac{0.196}{360} \times \left(\frac{20}{100 \times 2}\right)^2 = 5.44 \times 10^{-8} \text{ m}^2/\text{s}$$

From Eq. 8.24,

$$c_v = \frac{k}{\gamma_w}\left(\frac{1+e_0}{a_v}\right)$$

That is,

$$k = \frac{c_v a_v \gamma_w}{1+e_0} = \frac{5.44 \times 10^{-8} \times 4.67 \times 10^{-3} \times 9.81}{1+0.80}$$

Therefore,

$$k = 1.39 \times 10^{-9} \text{ m/s}$$

**Example 8.5** A soil has a compression index, $C_c$, of 0.28. At a stress of 120 kN/m$^2$, the void ratio was 1.02. Calculate (i) the void ratio if the stress on the soil is increased to 180 kN/m$^2$ and (ii) the total settlement of the stratum of 6 m thickness.

***Solution***

$$C_c = \frac{e_0 - e_1}{\log_{10}(p_1/p_0)}$$

or

$$0.28 = \frac{1.02 - e_1}{\log_{10}(180/120)}$$

or

$$e_1 = 1.02 - 0.28 \log_{10} \frac{180}{120} = 0.971$$

Now,

$$S_c = \frac{H}{1+e_0} C_c \log_{10} \frac{p_1}{p_2}$$

or

$$S_c = \frac{5 \times 1000}{1+1.02} \times 0.28 \times \log_{10} \frac{180}{120} = 122 \text{ mm}$$

**Example 8.6** A soil profile at a certain location is as shown in Fig. 8.25. A rigid circular foundation of 5 m diameter rests on the sand. The contact pressure at the underside of the foundation is 250 kN/m$^2$. The average coefficient of compressibility of the clay is $0.606 \times 10^{-3}$ m$^2$/kN for the stress range encountered. Estimate the ultimate settlement of the foundation assuming a 30° load distribution.

***Solution***

Consider that there is no capillary rise of water in sand. Thus,

Sand:

$$\gamma_d = \frac{G\gamma_w}{1+e} = \frac{2.65 \times 9.807}{1+0.54} = 16.9 \text{ kN/m}^3$$

$$\gamma_{sat} = \frac{G+e}{1+e}\gamma_w = \left(\frac{2.65+0.54}{1+0.54}\right) 9.807 = 20.3 \text{ kN/m}^3$$

Clay:

$$\gamma_{sat} = \left(\frac{2.7+1.2}{1+1.2}\right) 9.807 = 17.4 \text{ kN/m}^3$$

The initial effective stress at the centre line of the clay layer

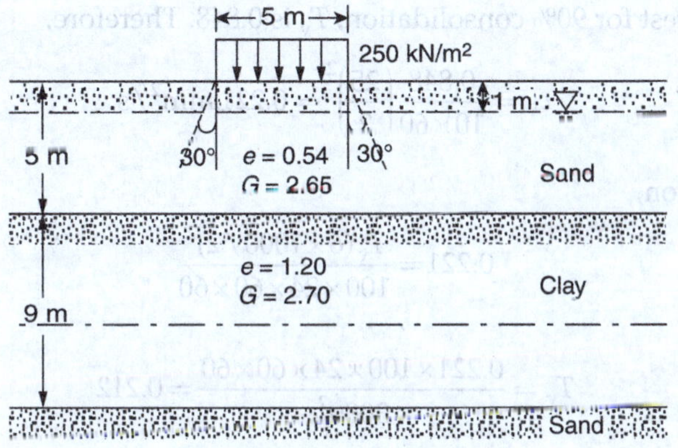


Fig. 8.25

$$p_0 = (1\times16.9 + 4\times20.3 + 4.5\times17.4 - 8.5\times9.807) = 93.0\ \text{kN/m}^2$$

The final effective stress at the centre line of the clay layer

$$p_1 = 93 + 250\times\frac{\pi\times5^2}{4}\times\frac{4}{\pi(5\times2\times4.75)^2}$$
$$= 122.7\ \text{kN/m}^2$$

Now,

$$m_v = \frac{a_v}{1+e_0} = \frac{0.606\times10^{-3}}{1+1.2} = 0.275\times10^{-3}\ \text{m}^2/\text{kN}$$

$$S_c = m_v H\Delta p$$
$$= 0.275\times10^{-3}\times9\times1000(122.7-93.0)$$
$$= 73.5\ \text{mm}$$

**Example 8.7** In an oedometer test, a clay specimen initially 25 mm thick attains 90% consolidation in 10 minutes. In the field, the clay stratum from which the specimen was obtained has a thickness of 6 m and is sandwiched between two sand layers. A structure constructed on this clay experienced an ultimate settlement of 200 mm. Estimate the settlement at the end of 100 days after construction.

***Solution***

$$c_v = \frac{T_v d^2}{t}$$

For the oedometer test for 90% consolidation, $T_v$ is 0.848. Therefore,

$$c_v = \frac{0.848}{10\times60}\left(\frac{25}{2}\right)^2 = 0.221\ \text{mm}^2/\text{s}$$

For the field condition,

$$0.221 = \frac{T_v(6\times1000/2)^2}{100\times24\times60\times60}$$

$$T_v = \frac{0.221\times100\times24\times60\times60}{3000^2} = 0.212$$

For $U_z \le 60\%$, the maximum $T_v$ value is 0.283. Therefore, the expression for $T_v$ to be used is

$$T_v = \left(\frac{\pi}{4}\right)\left(\frac{U_z}{100}\right)^2$$

$$0.212 = \left(\frac{\pi}{4}\right)\left(\frac{U_z}{100}\right)^2$$

$$U_z = \sqrt{\frac{0.212 \times 4 \times 1002}{\pi}} = 51.96\%$$

We know $U_z$ is also given as

$$U_z = \frac{S_t}{S_c} \times 100$$

$$S_t = \frac{U_z S_c}{100} = \frac{51.96 \times 200}{100} = 103.9 \text{ mm}$$

**Example 8.8** A 2 m thick layer of compressible soil with an average void ratio of 1.02 is subjected to a total overburden pressure of 150 kN/m$^2$. Laboratory consolidation tests show that the equation for the $e$–log $p$ curve is

$$e = e_0 - 0.125 \log_{10} \frac{p}{p_0}$$

A large structure at the ground surface increases the pressure in the compressible layer by 70 kN/m$^2$.

1. Find the change in volume per unit volume of the compressible layer.
2. If the soil is laterally confined so that all the volume change results in change in height, calculate the settlement that would result due to the increased load.

***Solution***

After an increased pressure of 70 kN/m$^2$,

$$e = 1.02 - 0.125 \log_{10} \frac{150 + 70}{150} = 0.999$$

1. From Eq. 8.11,

$$m_v = \frac{\Delta e / \Delta p}{1 + e_0}$$

$$m_v = \frac{1}{1 + 1.02} \frac{(1.02 - 0.999)}{70} = 1.49 \times 10^{-4} \text{ m}^2/\text{kN}$$

2. From Eq. 8.41, $S_c = m_v H \Delta p$. Therefore,

$$S_c = 1.49 \times 10^{-4} \times 2 \times 100 \times 70 = 2.09 \text{ cm}$$

**Example 8.9** A normally consolidated clay layer of 10 m thickness has a unit weight of 20 kN/m$^3$ and specific gravity of 2.72. The liquid limit of the clay is 58%. A structure constructed on the clay increases the overburden pressure by 10%. Estimate the ultimate consolidation settlement. There is no secondary compression.

***Solution***
For normally consolidated clays,

$$C_c = 0.009(w_L - 10\%)$$
$$= 0.009(58 - 10) = 0.432$$

Considering the unit weight as saturated unit weight, $e_0$ is calculated from

$$\gamma_{sat} = \frac{G + e_0}{1 + e_0}\gamma_w$$

That is,

$$20 = \frac{2.72 + e_0}{1 + e_0}9.81$$

Solving, $e_0 = 0.656$. From Eq. 8.39, the consolidation settlement

$$S_c = \frac{H}{1 + e_0}C_c \log\frac{p_1}{p_0}$$
$$= \frac{10}{1 + 0.656} \times 0.432 \times \log\frac{p_0 + 0.1p_0}{p_0} \times 100$$

That is,

$$S_c = 10.8 \text{ cm}$$

**Example 8.10** The overburden pressure on a normally consolidated 5 m thick clay structure is 250 kN/m$^2$. A laboratory consolidation test followed by a suitable correction gave two points on the field curve:

$e_1 = 1.12 \quad p_1 = 150 \text{ kN/m}^2$

$e_2 = 1.02 \quad p_2 = 450 \text{ kN/m}^2$

How much settlement does the clay layer experience due to an additional load intensity of 150 kN/m$^2$?

***Solution***
Overburden pressure = 250 kN/m$^2$

$e_1 = 1.12 \quad p_1 = 150 \text{ kN/m}^2$

$e_2 = 1.02 \quad p_2 = 450 \text{ kN/m}^2$

From Eq. 8.9,

$$C_c = \frac{e_0 - e_1}{\log(p_1/p_0)} = \frac{1.12 - 1.02}{\log(450/150)} = 0.209$$

From Eq. 8.39,

$$S_c = \frac{0.209}{1+1.12} \times 5 \times \log\frac{450}{150} \times 100$$

$$= 12.6 \text{ cm}$$

**Example 8.11** A homogeneous clay layer, 9 m thick, is expected to have an ultimate settlement of 308 mm. After a time span of 2 years, the average settlement was measured to be 108 mm. How much longer will it take for the average settlement to attain 220 mm?

***Solution***

Degree of consolidation $U_z = (S_t/S_c) \times 100\%$

$$U_z = \frac{108}{308} \times 100 = 35.07\%$$

For $U_z \leq 60\%$, $T_v$ is obtained from

$$T_v = \left(\frac{\pi}{4}\right)\left(\frac{U_z}{100}\right)^2$$

That is,

$$(T_v)_1 = \left(\frac{\pi}{4}\right)\left(\frac{35.07}{100}\right)^2 = 0.0966$$

Now, $c_v = (T_v d^2/t)$ from the first condition

$$c_v = \frac{(T_v)_1 d^2}{t}$$

Again,

$$U_z = \frac{220}{308} \times 100 = 71.4\%$$

Now, $T_v$ is obtained from

$$T_v = -0.933 \log_{10}\left|1 - \frac{U_z}{100}\right| - 0.0851$$

That is,

$$(T_v)_2 = -0.933 \log_{10}\left|1 - \frac{71.4}{100}\right| - 0.0851 = 0.422$$

$$t = \frac{(T_v)^2 d^2}{t} = \frac{(T_v)_2 d^2 t_1}{(T_v)_1 d^2} = \frac{(T_v)_2}{(T_v)_1} t_1$$

$$= \frac{0.422}{0.0966} \times 2 = 8.73 \text{ years}$$

The additional time for the average settlement to attain 220 mm is 8.73 years.

**Example 8.12** In a consolidation test, the pressure on a sample was increased from 150 to 300 kN/m$^2$. The void ratio after 100% consolidation under 150 kN/m$^2$ was 0.945, and that under 300 kN/m$^2$ was 0.812. The coefficient of permeability of the soil was $25 \times 10^{-6}$ mm/s, and the initial height of the sample was 20 mm. Determine (i) the coefficient of compressibility, (ii) the coefficient of volume compressibility, (iii) the coefficient of consolidation, and (iv) the time taken in days for 90% consolidation of the layer of this clay, 0.5 mm thick in the field and sandwiched between an impervious layer beneath and the pervious layer on top.

***Solution***

$$\Delta e = 0.945 - 0.812 = 0.133$$

$$\Delta p = 300 - 150 = 150 \text{ kN/m}^2$$

$$a_v = \frac{\Delta e}{\Delta p} = \frac{0.133}{150} = 8.87 \times 10^{-4} \text{ m}^2/\text{kN}$$

$$m_v = \frac{a_v}{1 + e_0} = \frac{8.87 \times 10^{-4}}{1 + 0.945} = 4.56 \times 10^{-4} \text{ m}^2/\text{kN}$$

$$c_v = \frac{k}{\gamma_w} = \frac{25 \times 10^{-6}}{1000 \times 9.807} \times \frac{1}{4.56 \times 10^{-4}} = 5.59 \times 10^{-6} \text{ m}^2/\text{s}$$

$$t = \frac{d^2}{c_v} T_v$$

$$t = \frac{5^2 \times 0.848}{5.59 \times 10^{-6} \times 60 \times 60 \times 24} = 43.9 \text{ days}$$

**Example 8.13** A clay deposit of 5 m thick is covered with coarse sand at top and bottom and has a permeability of 0.025 m/yr. It is additionally loaded by a pressure of 65 kpa. Laboratory tests on the field specimen indicate that the consolidation in the field will be only 50% in a six-month period. Find the settlement of the clay deposit in one year.

***Solution***

Coefficient of consolidation $C_v = \dfrac{T_r d^2}{t}$

$$= \frac{0.197 \times 2.5^2}{0.5} = 2.463\ w^2/y$$

$$\left.\begin{array}{l}\text{Coefficient of volume}\\ \text{compressibility}\end{array}\right\} m_v = \frac{k}{C_v\ \gamma_\omega} = \frac{0.025}{2.463 \times 9.8}$$

$$= 1.036 \times 10^{-3}\ \text{m}^2/\text{km}.$$

Settlement, $S_f = m_r H_0 \Delta\sigma'$

$$= 1.036 \times 10^{-3} \times 5.0 \times 65$$

$$= 0.337\ \text{m}.$$

As the consolidation will be 50% in six months, in one year the consolidation will be complete and the settlement will be 0.337 m.

**Example 8.14** Two clay specimens, *P* and *Q*, whose initial void ratios at equilibrium under a pressure of 1 km/m$^2$ are 0.56 and 0.65, respectively, decrease to 0.51 and 0.63, respectively, when the pressure is increased to 1.5 km/m$^2$. The thickness of specimens initially was 35 and 25 mm for *P* and *Q*, respectively. The time taken by specimen *P* to reach 50% consolidation is 30% of that required by specimen *Q* for reaching the same degree of consolidation of permeability of *P* and *Q*.

***Solution***

$$m_v = \frac{\Delta e}{1+e_0}\frac{1}{\Delta\sigma'}.$$

Hence for specimen *P*:

$$(m_v)_P = \frac{0.56-0.51}{1+0.56} \times \frac{1}{0.5} = \frac{0.05}{1.56 \times 0.5}$$

For specimen *Q*:

$$(m_v)_Q = \frac{0.65-0.63}{1+0.65} \times \frac{1}{0.5} = \frac{0.02}{1.65 \times 0.5}.$$

Hence, $\dfrac{(m_v)_P}{(m_v)_Q} = \dfrac{0.05}{1.56 \times 0.5} = \dfrac{1.65 \times 0.5}{0.02} = 2.64.$

Now,
$$\frac{t_P}{t_Q} = \frac{(h_P)^2}{(h_Q)^2}\frac{(C_v)_Q}{(C_v)_P}$$

i.e.,
$$\frac{300}{100} = \frac{\left(\frac{35}{2}\right)^2}{\left(\frac{25}{2}\right)^2} \times \frac{(C_v)_Q}{(C_v)_P}$$

$$\frac{(C_v)_Q}{(C_v)_P} = \frac{(17.5)^2}{(12.5)^2} \times \frac{1}{0.30} = 6.53.$$

We know,
$$C_v = \frac{k}{m_r \gamma_\omega}$$

$$\therefore \quad \frac{k_P}{k_Q} = \frac{(C_v)_P (m_v)_P \times \gamma_\omega}{(C_v)_Q (m_v)_Q \times \gamma_\omega}$$

$$= 6.53 \times 2.64$$

$$= 17.25.$$

## POINTS TO REMEMBER

8.1 A soil mass subjected to a change in the stress system resulting in a volume decrease is termed the compressibility or compression of soil. The compression may be immediate and/or time dependent.

8.2 The time-dependent volume decrease is attributed to compression of soil matter, compression of water and air within the voids, and escape of water or air from the voids.

8.3 One-dimensional consolidation is based on the following considerations: (i) the displacements are vertical, (ii) there are no lateral strains, (iii) flow of water from the soil layer is in a vertical direction only, and (iv) the change in void ratio is a direct function of the vertical component of effective stress.

8.4 The pressure ($\sigma'$)–void ratio ($e$) curve, obtained from a consolidation test, depicts the consolidation history of the soil. The initial flatter portion of the pressure–void ratio curve is called the recompression curve, and the steeper portion after the break in the curve is called the virgin compression curve.

8.5 The pre-consolidation pressure ($p_c$), the maximum stress ever experienced by the soil, is reflected by the point of transition in the first loading portion of the pressure–void ratio curve.

8.6 Compression index ($C_c$), a dimensionless parameter, is the slope of the linear portion of the pressure–void ratio curve on a semi-log plot.

8.7 Coefficient of compressibility ($a_v$) is the second slope for a given pressure increment of the pressure–void ratio curve.

8.8 Coefficient of volume compressibility ($m_v$) is the compression of a soil layer per unit of original thickness due to a given increase in pressure.

8.9 Degree of consolidation ($U_z$) is the ratio expressed as a percentage of the amount of consolidation at a given time within a soil mass to the total amount of consolidation obtainable under a given stress condition.

8.10 If the present effective overburden pressure in a deposit is the maximum pressure to which the deposit has ever been consolidated at any time in the past, such a deposit is called a normally consolidated clay deposit.

8.11 A clay deposit that has been fully consolidated under a pressure $p_c$, in the past, longer than the present overburden pressure $p_0$ is called an over-consolidated (pre-consolidated or pre-compressed) clay deposit.

8.12 Rapid natural deposition or deposits under recent fillings may not be fully consolidated under the present overburden pressure; such a clay deposit is called under-consolidated clay deposit.

8.13 Terzaghi's one-dimensional consolidation theory predicts the time rate of consolidation, i.e., it predicts the water pressure at any elapsed time and at any location in a soil deposit.

8.14 Time factor is a dimensionless factor containing the physical constants of a soil stratum influencing its time rate of consolidation.

8.15 Methods used to determine the coefficient of consolidation are called time-fitting methods. They are $\sqrt{\text{time}}$ -fitting and log(time)-fitting methods.

8.16 Consolidation does not cease even after dissipation of excess pore water pressure but continues. This continued time-dependent compression is referred to as the secondary compression.

8.17 Total settlement is a combination of immediate settlement ($S_i$), settlement due to consolidation ($S_c$), and settlement due to secondary compression ($S_s$).

8.18 Settlement in coarse-grained soils occurs immediately, which is attributed to the volume changes caused by lateral yielding or shear strain which occur in the soil.

## QUESTIONS

### *Objective Questions*

8.1 Identify the true and false statements from the following:

1. The coefficient of volume compressibility ($m_v$) of a soil is always less than its coefficient of compressibility ($a_v$).
2. For an intensive normally consolidated soil, the compression index can also be found from remoulded samples.

3. Settlement resulting from the dissipation of the excess pore water pressure is referred to as primary consolidation.
4. The rate of consolidation for a given soil deposit is a constant.
5. The time factor is a non-dimensional parameter and is independent of the degree of consolidation.

8.2 When consolidation of saturated soil sample takes place, the degree of saturation
(a) Decreases (b) Increases
(c) Remains constant (d) Decreases initially and then increases

8.3 The liquid limit of a saturated normally consolidated soil is 50%. The compression index of the soil for the virgin compression curve will be
(a) 0.36 (b) 0.505
(c) 0.605 (d) 0.705

8.4 Consolidation time
(a) Increases with increasing compressibility
(b) Increases rapidly with decreasing size of soil mass
(c) Decreases with increasing permeability
(d) Is dependent on the magnitude of the stress change

8.5 The value of volume compressibility is not constant but varies with the
(a) Thickness of the compressible layer (b) Level of effective areas
(c) Coefficient of permeability (d) Time factor

8.6 Which soil parameter is considered as a measure of the degree of over-consolidation?
(a) Pre-consolidation pressure (b) Compression index
(c) Over-consolidation ratio (d) Coefficient of consolidation

8.7 Compression in saturated soils may be caused by
1. Rearrangement of particles 2. Seepage of water
3. Crushing of particles 4. Elastic distortions
Of these statements,
(a) 1, 2, and 3 are correct (b) 2, 3, and 4 are correct
(c) All are correct (d) 3, 4, and 1 are correct

8.8 Indicate the incorrect statement. In a conventional consolidation test, the following factors do not significantly affect the result:
(a) Moderate variation in room temperature
(b) Variation in pore fluid properties
(c) Variation in specimen size
(d) Variation in duration of load increment

8.9 *Assertion A:* Both $m_v$ and $a_v$ decrease with increasing stress.
*Reason R:* Both $m_v$ and $a_v$ are influenced by the stress history.
*Choose the correct codes.*
(a) Both A and R are true, and R is the correct explanation of A.
(b) Both A and R are true, and R is not the correct explanation of A.
(c) A is true, but R is false.
(d) A is false, but R is true.

8.10 A sand exhibits considerable secondary compression due to
(a) High stresses (b) High pore water pressure
(c) Presence of less fines (d) Delayed consolidation

## Descriptive Questions

8.11 Explain why the measured values of time and rate of settlement do not agree with the predicted values.
8.12 How do you estimate the field $e$–$p$ curve of an over-consolidated clay ?
8.13 Briefly explain the physical meaning of the coefficient of consolidation $c_v$.
8.14 Settlements computed are normally higher than the actual settlements. Discuss this statement critically.
8.15 Explain why there is a significant time lag in the settlement of clay soils but not in sandy soils.
8.16 Why does it take infinite time for complete consolidation to occur?
8.17 Explain in sequence the occurrences that take place when a saturated clay deposit undergoes consolidation.
8.18 The water table in a lake has been lowered by 20 m. Will this cause consolidation settlement of the lake-bed sediments? Explain.

## EXERCISE PROBLEMS

8.1 An oedometer test was made on a sample of saturated marine clay. The dial readings corresponding to the final equilibrium condition under each load are given below:

| Pressure (kN/m$^2$) | 13.3 | 26.6 | 53.3 | 106.4 | 213.0 | 426.0 | 852.0 |
|---|---|---|---|---|---|---|---|
| Dial reading (mm$\times10^{-3}$) | 275 | 540 | 965 | 1,880 | 3,340 | 5,000 | 6,600 |

The diameter and initial thickness of the sample were 60 and 38 mm, respectively. The final water content of the sample was 35.2% and $G = 2.72$.
1. Compute the void ratio and plot the $e$–log $p$ curve.
2. Estimate the maximum overburden pressure by the Casagrande method.
3. Draw the field curve and determine the compression index.

8.2 Compute the compression index of the soil from the following experimental data:

| Load (kN/m$^2$) | 0 | 10 | 50 | 100 | 200 | 400 | 0 |
|---|---|---|---|---|---|---|---|
| $\Delta h$ (mm) | 0 | −0.7 | −2.18 | −0.70 | −0.72 | −0.85 | +0.75 |

The initial thickness of the sample was 0.20 mm, the specific gravity of soil particles 2.70, and the final water content of the soil 38.2%. Also determine the pre-consolidation pressure of the soil.

8.3 The thickness of a saturated specimen of clay under a consolidation pressure of 100 kN/m$^2$ is 22.12 mm, and its water content is 20%. On increase of the consolidation pressure to 200 kN/m$^2$, the specimen's thickness decreases by 1.28 mm. Determine the compression index for the soil. $G = 2.7$.

8.4 A clay stratum 2.5 m thick lies over a sandy stratum and has drainage on both sides. Calculate the values of the pore water pressure and the effective stress in the middle of the clay stratum when 60% of consolidation has taken place under an increment of load from 500 to 2,500 kN/m$^2$.

8.5 A layer of soft clay is 6 m thick and lies under a newly constructed building. The pressure due to the sand overlying the clay layer is 250 kN/m$^2$, and the new construction increases the overburden by 100 kN/m$^2$. If the compression index is 0.5, compute the settlement taking the natural water content of the clay layer to be 45% and the specific gravity of soil grains to be 2.7.

8.6 A clay layer whose settlement under a given loading is expected to be 16 cm settles 4 cm at the end of 2 months. How many months will be required to reach a settlement of 8 cm? How much settlement will occur in 10 months?

8.7 The following are the time dial readings in a consolidation test on a clay sample when the load is increased from 100 to 200 kN/m$^2$. The thickness of the sample at the start of the load increment is 23 mm, and the void ratio is 0.95.

| **Elapsed time (min)** | **Dial reading (division) 1 div. = 0.0025 mm** |
|---|---|
| 0.00 | 1,012 |
| 0.25 | 890 |
| 1.00 | 850 |
| 2.25 | 833 |
| 4.00 | 818 |
| 6.25 | 802 |
| 9.00 | 788 |
| 12.25 | 774 |
| 16.00 | 760 |
| 20.25 | 746 |
| 25.00 | 732 |
| 30.25 | 718 |
| 36.00 | 704 |
| 49.00 | 676 |
| 64.00 | 651 |
| 81.00 | 630 |
| 100.00 | 610 |
| 144.00 | 572 |
| 270.00 | 490 |
| 520.00 | 441 |
| 1250.00 | 408 |
| 1440.00 | 371 |

Determine the coefficient of consolidation using Taylor's method.

8.8 The following readings were obtained in a consolidation test:

| Pressure (kPa) | 0 | 50 | 100 | 200 | 400 | 800 |
|---|---|---|---|---|---|---|
| Thickness of sample after 24 hours (mm) | 19 | 18.98 | 18.95 | 18.80 | 18.31 | 17.76 |

The final water content was 30.1% and $G = 2.75$. Plot a graph of void ratio versus effective pressure, and from the curve find the coefficient of compressibility when the pressure is 300 kN/m$^2$. If the coefficient of permeability is $20\times10^{-6}$ mm/s, find the coefficient of consolidation of the soil.

8.9 A consolidation test on a sample of clay gave the following results:

| Effective pressure (kN/m$^2$) | 0 | 20 | 50 | 100 | 200 | 400 |
|---|---|---|---|---|---|---|
| Void ratio | 2.10 | 2.07 | 1.98 | 1.84 | 1.60 | 1.34 |

The layer of clay from which the consolidation sample was taken has a thickness of 1.2 m and is overlaid by 6 m of sand. The porosity and specific gravity of the sand layer are 40% and 2.65, respectively. There is another sand layer present beneath the clay layer. The groundwater table is 2 m above the top of the clay. The clay layer has a specific gravity surface, which is estimated to increase the pressure by 110 kN/m$^2$ at the top and by 75 kN/m$^2$ at the bottom of the clay layer. Compute the settlement of the structure due to the consolidation of the clay layer.

If the coefficient of permeability of the clay layer is $25 \times 10^{-6}$ mm/s, how long will it take for 50% of the settlement to take place?

8.10 In a consolidation test on a soil, the void ratio of the sample decreases from 1.24 to 1.12 when the pressure is increased from 200 to 400 kN/m$^2$. Calculate the coefficient of consolidation (in m$^2$/year) given that the coefficient of permeability of the soil during this pressure increment is $8.5\times10^{-8}$ cm/s.

8.11 The time to reach 60% consolidation is 35 seconds for a sample of 1.2 cm thickness tested in the laboratory under the conditions of double drainage. How long will the corresponding layer in nature require to reach the same degree of consolidation if it is 11 m thick and drained on one side only?

8.12 A clay layer has a drainage arrangement as shown in Fig. 8.26. The coefficient of consolidation of the soil is $10^{-1}$mm$^2$/s. A fill is placed on the soil, which induces a pressure of 50 kN/m$^2$ at the surface.

1. Make a neat drawing showing
   (a) The distribution of pore pressure before the fill is placed
   (b) The distribution of pore pressure immediately after the fill is placed
2. How long will it take for the average consolidation of the layer to exceed 90%?

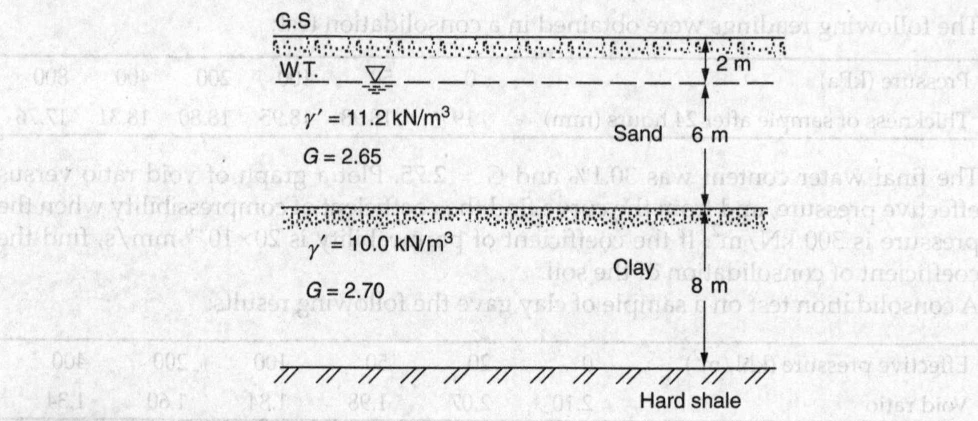
G.S
W.T.
2 m
γ′ = 11.2 kN/m³
G = 2.65
Sand
6 m
γ′ = 10.0 kN/m³
G = 2.70
Clay
8 m
Hard shale

**Fig. 8.26**

# 9

# Shear Strength of Soils

## CHAPTER HIGHLIGHTS

Stress–strain curve – Mohr–Coulomb failure criterion – Peak and residual shear strength – Strain- and stress-controlled tests – Laboratory measurement of shear strength: drainage condition, direct shear, triaxial shear, unconfined compression, ring shear tests, choice of rate of deformation – Field measurement of shear strength: vane shear test, bore hole shear device – Shear strength of saturated cohesive soil – Pore pressure coefficients – Sensitivity of cohesive soils – Shear strength of granular soils

## 9.1 INTRODUCTION

The capability of a soil to support a loading from a structure, or to support its own overburden, or to sustain a slope in equilibrium is governed by its shear strength.

The shear strength of a soil is of prime importance for foundation design, earth and rockfill dam design, highway and airfield design, stability of slopes and cuts, and lateral earth pressure problems. It is highly complex because of various factors involved in it, viz., the heterogeneous nature of the soil, the water table location, the drainage facility, the type and nature of construction, the stress history, time, chemical action, or environmental conditions.

The shearing strength and related deformations of a foundation or a construction soil can be obtained from carefully extracted field samples. Field situations may demand an *in situ* determination of shear strength for reasons of expediency and economy. This chapter encompasses both theoretical and experimental approaches for determination of shear strengths.

## 9.2 STRESS–STRAIN CURVE

Consider an element of soil subjected to a varying shear stress under a constant normal stress. Figure 9.1 shows a typical shear stress–shear strain curve. Initially, when the

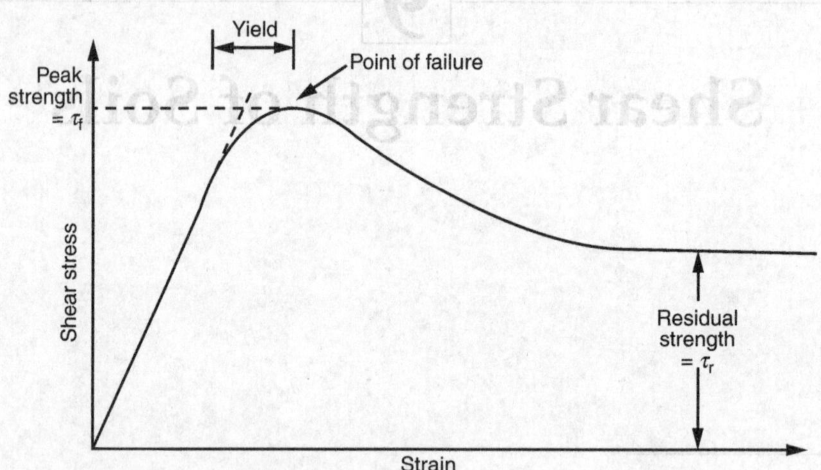


**Fig. 9.1** Typical shear stress–shear strain curve

shear stress is low, the soil behaves like an elastic material depicting a linear stress–strain curve. An increase in shear stress causes an increase in shear strain, and at a particular stress level, significant plastic shear strain starts to develop, and that point is referred to as *yield*. The shearing resistance of the soil increases with the plastic shear strain, and the material is said to *work hardening* or *strain hardening*. The strain hardening can only increase the resistance to a particular maximum shear stress, and that maximum resistance is called the *peak shear strength* or simply the *shear strength* of the soil ($\tau_f$), i.e., the shear strength of a soil is the maximum shear stress that can be resisted by the soil. This yield level is considered to be *unstable,* and the soil is said to *fail*. In some soils, the maximum shearing resistance decreases after this point, and the soil is said to be *strain softening* or *work softening*. After a continued large strain, the shearing resistance attains a constant level, and the corresponding shearing resistance is called the *residual shear strength* or simply the *residual strength* ($\tau_r$).

## 9.3 MOHR–COULOMB FAILURE CRITERION

It has been shown above that on any point within a soil mass, when the shear stress reaches the shear strength of the soil, the soil fails at that point. Coulomb, in 1773, suggested a simple relationship of shear strength of soil being a linear function of the normal stress as follows:

$$\tau_f = c' + \sigma_n' \tan \phi' \tag{9.1}$$

where $\tau_f$ is the shear strength of soil, $c$ the cohesion intercept, $\sigma_n$ the normal stress acting on the plane of failure, and $\phi$ the angle of shearing resistance.

Equation 9.1 is given in terms of the total stress. After Terzaghi's classical effective stress concept, the above equation is modified in terms of effective stresses as

$$\tau_f = c' + \sigma_n' \tan \phi' \tag{9.2}$$

where $c'$ and $\phi'$ are the shear strength parameters in terms of the effective stress and $\sigma'_n = \sigma_n - u_w$ is the effective normal stress. The shear strength parameters $c'$ and $\phi'$ are not

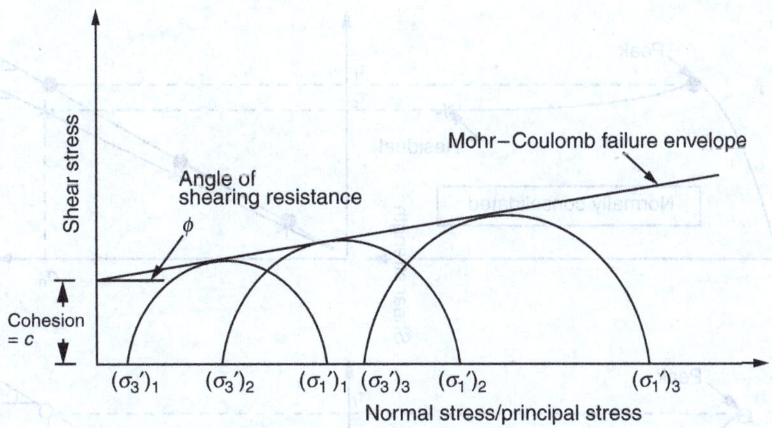


Fig. 9.2 Mohr–Coulomb failure envelope

constants for a given soil but depend on many factors such as void ratio, initial structure, and pore pressure, and hence these parameters are referred to the type of test and conditions under which they were measured. Equations 9.1 and 9.2 are referred to as Coulomb's failure envelopes or Coulomb's failure loci.

The state of stresses in a soil mass at the time of failure may be represented in terms of the effective major and minor principal stresses. Figure 9.2 represents three Mohr circles for three sets of stresses at the time of failure in a soil mass. A tangent drawn to these circles is termed Mohr's envelope, and this satisfies Eq. 9.1. Any combination of stresses that falls within this envelope represents a stable condition. A state of stresses plotting above the failure envelope cannot exist. Thus, another way of stating Coulomb's failure condition is that if the Mohr circle for a state of stresses at a point is tangential to Coulomb failure loci, then that point is said to be in a state of failure. This is generally known as the Mohr–Coulomb failure criterion, and the failure loci are called Mohr–Coulomb failure envelopes. Because of its simplicity, the Mohr–Coulomb failure criterion is widely used. In reality, the failure envelope may not be a straight line but for the range of normal stresses involved in soil, the straight line approximation is justified, and the shear strength parameters are determined for that stage.

## 9.4 PEAK AND RESIDUAL SHEAR STRENGTHS

A knowledge of shear stress at and after failure and the shape of the stress–strain curves are of paramount importance in studies of stability problems and, more particularly, in materials experiencing progressive failure* (Hvorslev, 1960).

Figure 9.3 represents typical stress–strain curves for normally consolidated (NC) and over-consolidated (OC) clays tested very slowly. At a given effective normal stress perpendicular to the plane of shear, the peak strength in both cases corresponds to the maximum shear stress the clay can resist. Continued straining after the strain at peak strength decreases the

*The processes that cause a gradual reduction in strength to the residual value are termed progressive failures.

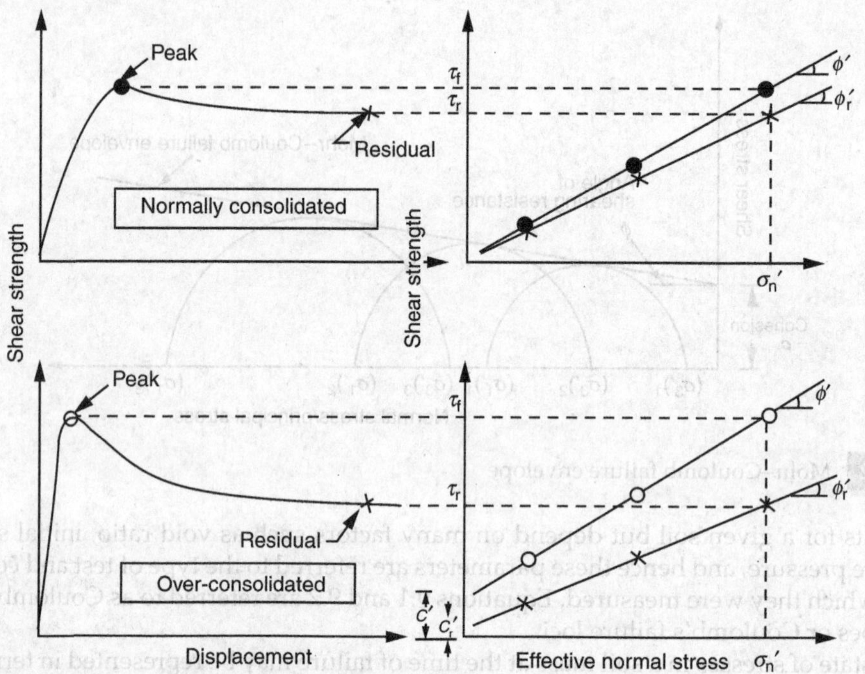


**Fig. 9.3** Simplified shear strength parameters of clay

resistance, and further, at large strains the residual strength stage is attained. Hence, peak and residual strength parameters can be expressed as

$$\tau_f = c' + \sigma_n' \tan \phi' \tag{9.3}$$

$$\tau_r = c_r' + \sigma_n' \tan \phi_r'$$

where $c'$ is the cohesion intercept corresponding to peak strength, $c'_r$ the cohesion intercept corresponding to residual strength, $\phi'$ the angle of shearing resistance corresponding to peak strength, and $\phi'_r$ the angle of shearing resistance corresponding to residual strength.

Some of the main features which are revealed from Fig. 9.3 (Skempton, 1964) are as follows:

1. The post-peak drop off in strength is more pronounced in over-consolidated clays (e.g., stiff clays) and less in normally consolidated clays (e.g., soft clays). With high clay contents, the drop in both cases is appreciably increased.
2. At the residual condition, the strength at a given effective stress is independent of past stress history.
3. The strength envelope for residual condition is generally located at a little lower than that for the peak strength of a normally consolidated clay but extremely lower than that of over-consolidated clay.

The residual cohesion intercept $c'_r$ is relatively low and can be taken to be zero. In general, the value of $\phi'_r$ decreases with increasing clay content. The complete shape of the stress–displacement curve up to displacement at residual condition clearly depicts the degree of brit-

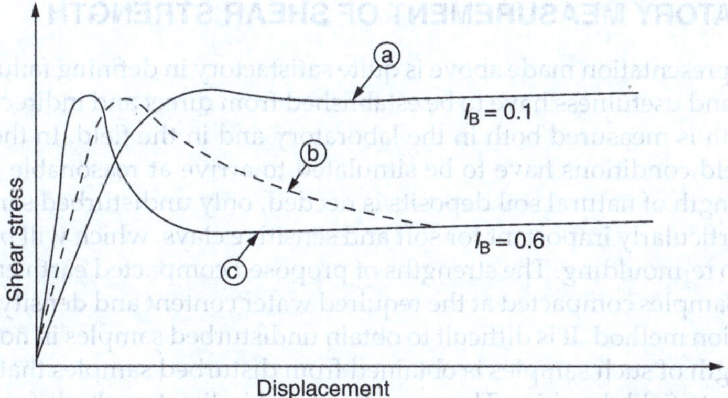


**Fig. 9.4** Stress–displacement curves showing different degrees of brittleness

tleness of the clay. Figure 9.4 represents three typical stress–displacement curves. Bishop (1967) introduced a factor called *brittleness index** $I_B$ relating the peak and residual strengths as

$$I_B = \frac{\tau_f - \tau_r}{\tau_f}$$

In Fig. 9.4, Curve *a* represents a plastic non-brittle clay, whereas Curve *b* represents a highly brittle clay. Skempton and Hutchinson (1969) showed that the shape of the stress–displacement curve is also significant since two different stress–displacement curves yield the same brittleness index (e.g., Curves *b* and *c*). This anomaly may be taken into account by considering the energy parameter introduced by Bishop (1967), which is expressed as the ratio of the work done in shearing from peak to residual to the work done if the clay continued shearing at peak strength.

In over-consolidated clay slopes, Skempton (1964) observed that field shear strength ($\overline{\tau}$), associated with a slip, lies somewhere between $\tau_f$ and $\tau_r$. Skempton (1964) related these strengths by a factor called *residual factor*, *R*:

$$R = \frac{\tau_f - \overline{\tau}}{\tau_f - \tau_r}$$

In physical terms, *R* is the proportion of the total slip surface in the clay along which its strength has fallen to the residual value. A value of $R = 0$ signifies no reduction in strength to have occurred, and $R = 1$ represents a complete reduction to residual strength. Although it is known that *R* increases with time, no method is yet available to determine the same.

The peak shear strength parameters depend on factors such as per cent clay content, drainage condition, type of loading, consolidation history, stress level, anisotropy, and other environmental factors. The residual shear strength is independent of many of the factors listed above, but $\phi'_r$ decreases with increasing clay content.

*This is equal to $(1 - \lambda R)$, where $\lambda R$ is Haefeli's residual coefficient (Haefeli, 1965).

## 9.5 LABORATORY MEASUREMENT OF SHEAR STRENGTH

The theoretical presentation made above is quite satisfactory in defining failure in a soil mass, but its validity and usefulness have to be established from direct and indirect measurements.

Shear strength is measured both in the laboratory and in the field. In the laboratory test samples, the field conditions have to be simulated to arrive at reasonable strength parameters. If the strength of natural soil deposits is needed, only undisturbed samples have to be used. This is particularly important for soft and sensitive clays, which will be disturbed substantially due to re-moulding. The strengths of proposed compacted earth embankments are obtained from samples compacted at the required water content and density, adopting a particular compaction method. It is difficult to obtain undisturbed samples in non-cohesive soils. The shear strength of such samples is obtained from disturbed samples that are re-moulded in the laboratory to field densities. There are direct and indirect methods for determining the shear strength in the field.

### 9.5.1 Drainage Condition

The drainage condition during the test significantly influences the measured values, and the shear strength parameters $c$ and $\phi$ are referred to test conditions. One of the following three basic sets of drainage conditions may be used.

**Unconsolidated–Undrained**. When no drainage of water is allowed during the entire test process, the test is referred to as the undrained test. Thus, an increase in the total stress produces a corresponding stress increase in pore water pressure. As no account of pore water pressure is taken, the shear strength parameters are called total shear strength parameters.

$$\tau_f = c_u + \sigma_n \tan \phi_u \tag{9.4a}$$

where $c_u$ and $\phi_u$ are called undrained shear strength parameters. For a saturated soil under undrained condition $\phi_u = 0°$, then

$$\tau_f = c_u \tag{9.4b}$$

**Consolidated–Undrained**. The specimen is tested in two stages. During the first stage, called the consolidation stage, the specimen is allowed to consolidate under conditions of constant applied stress and full drainage. During the second stage, called the shear stage, the specimen is subjected to a shear under no drainage condition with or without pore pressure measurements. Accordingly, the shear strength parameters are referred to as effective shear strength parameters, $c'_{cu}$ and $\phi'_{cu}$, and total shear strength parameters, $c_{cu}$ and $\phi_{cu}$.

$$\tau_f = c'_{cu} + \sigma'_n \tan \phi'_{cu} \tag{9.5a}$$

and

$$\tau_f = c_{cu} + \sigma_n \tan \phi_{cu} \tag{9.5b}$$

**Consolidated–Drained**. The first stage of the test is similar to that of the consolidated–undrained test. When the consolidation stage is complete, the specimen is sheared with full drainage so that the rate of loading does not increase the pore water pressure. Since there is no increase in pore water pressure, the increase in total stress is equal to the increase in effective stress. The shear strength parameters are referred to as $c'_d$ and $\phi'_d$.

$$\tau_f = c'_d + \sigma'_n \tan \phi'_d \tag{9.6}$$

### 9.5.2 Shear- and Strain-Controlled Tests

Shearing can be accomplished by controlling (Fig. 9.5) either the rate of strain or the rate of stress (Jumikis, 1962). In the *strain-controlled* apparatus, the shearing deformation is continuously applied at a constant speed, and shear force is measured by means of a proving ring or a load transducer. Such a test can be continued even after the failure of the specimen. In the *stress-controlled* apparatus, the magnitude of the shearing force is the controlled variable. The stress is increased uniformly or in increments. The applied shear load is continuously kept till the deformation ceases. This test cannot be continued after the failure of the specimen. Generally, the strain-controlled test is preferred, as a better control on the rate of loading can be had and the test can be run even after failure. The stress-and strain-controlled direct shear tests are shown in Fig. 9.5.

### 9.5.3 Direct Shear Test

The direct shear test is a simple and widely used test. The apparatus is a shear box with separated lower and upper halves. The lower half is fixed to the frame, and the upper half is capable of moving horizontally relative to the lower half. A rectangular prism of soil with a cross section 60 mm×60 mm is placed in the box with approximately half of the specimen within either section (Fig. 9.6). In case of a non-cohesive soil, the specimen has to be prepared in the box itself at the required void ratio. Suitable porous discs may be placed one on top and the other at the bottom of the specimen, depending on the drainage condition. A pressure pad is placed on the top and a vertical load applied.

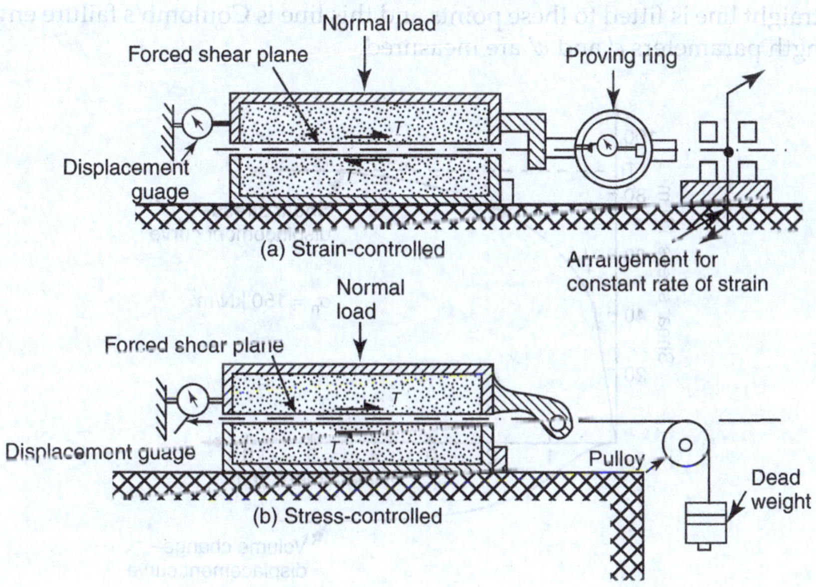


Fig. 9.5 Two types of direct shear testing devices

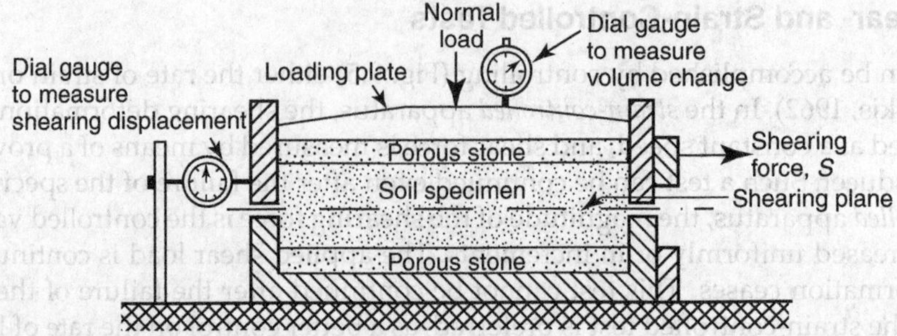


**Fig. 9.6** Schematic diagram of direct shear apparatus

The upper half of the apparatus is then moved laterally, forcing the sample to shear across the plane between the two halves of the apparatus. The magnitude of the shearing force is measured. Changes in sample thickness that occur during the shearing process are also recorded so that volume change versus shearing stress or shearing strain can be studied. The test is repeated on three or four specimens taken from the same soil sample. The detailed test procedure and typical results are given in Chapter 10.

**Test Curves**. Typical shear stress–displacement and volume change–displacement curves are shown in Fig. 9.7. The slope of the curve is the stress–strain modulus of the soil $E_s$. This modulus is an indication of the stiffness or resistance to deformation of the material. The higher the value of $E_s$, the stiffer or stronger is the soil. The failure stress is represented by the maximum shear stress value (i.e., peak value) if the stress–displacement curve is of work-softening type. Otherwise, the shear stress corresponding to 20% strain is taken as the shear strength.

**Failure Envelope**. Normal and shear stresses are calculated and plotted in Fig. 9.8. A best possible straight line is fitted to these points and this line is Coulomb's failure envelope. The shear strength parameters $c'$ and $\phi'$ are measured.

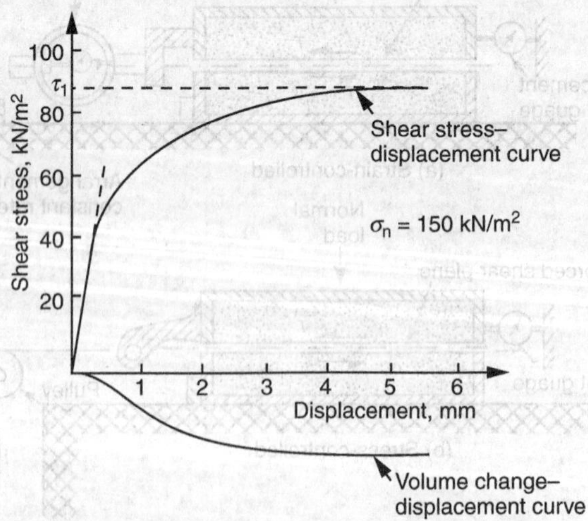


**Fig. 9.7** Typical shear stress and volume change curves for direct shear test

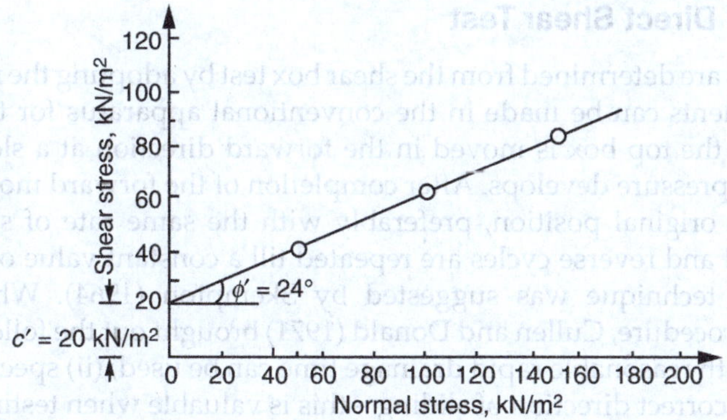


**Fig. 9.8** Coulomb's failure envelope from direct shear test results

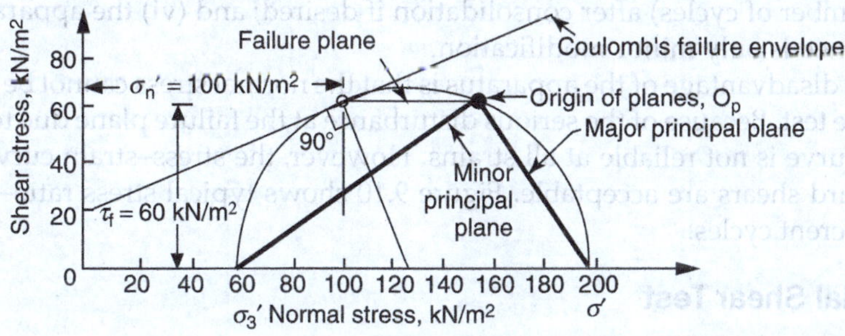


**Fig. 9.9** Mohr diagram for a direct shear test at failure

**Mohr Stress Circle**. For one normal stress, the Mohr circle is plotted in Fig. 9.9, and the pole, the major and minor principal stresses, and planes are shown.

**Merits**. The following are some of the merits of direct shear test: (i) as the draining time is relatively less, consolidated–undrained and consolidated–drained tests may be conducted even on soils of low permeability; (ii) this is the only test wherein both the shearing stress and the normal stress on the plane of failure are measured directly; (iii) volume changes during the test can be measured easily; (iv) the shear strength parameters $c$ and $\phi$ obtained from this test are as reliable as triaxial values; (v) using a reversing technique, the residual shear strength of the soil can be determined; (vi) the direct shear test apparatus is much more adaptable to electronic read-out equipments, and hence, a long duration test can be done conveniently; and (vii) as the displacements in a direct shear test take place in two directions, vertically and parallel to the direction of shear, this test is also referred to as plane strain test. This condition is similar to the strains developed in many field problems such as very long foundations or long embankments representing a two-dimensional case.

**Demerits**. The following are the demerits of direct shear test: (i) the shear–stress distribution over the plane of failure is non-uniform and (ii) the drainage cannot be controlled, and so the pore pressure behaviour cannot be obtained from the test.

### 9.5.4 Reversal Direct Shear Test

Residual strengths are determined from the shear box test by adopting the reversal technique. Suitable arrangements can be made in the conventional apparatus for the reversing technique. In general, the top box is moved in the forward direction at a slow rate so that no excess pore water pressure develops. After completion of the forward motion, the top box is pulled back to its original position, preferably with the same rate of strain, and the test repeated. Forward and reverse cycles are repeated till a constant value of residual strength is obtained. This technique was suggested by Skempton (1964). While attempting to standardize this procedure, Cullen and Donald (1971) brought out the following advantages: (i) thin samples with reasonable rapid drainage time can be used; (ii) specimens can be readily oriented in the correct direction of sliding – this is valuable when testing specimens with defects, such as faults, fissures, and pre-existing slip planes; (iii) large deformations can be obtained by continuous reversal of shear direction; (iv) reasonably large areas of failure surface can be tested; (v) specimens can be readily wire cut (sometimes samples are cut to reduce the number of cycles) after consolidation if desired; and (vi) the apparatus is widely available and needs only minor modification.

The serious disadvantage of the apparatus is that the normal stress cannot be kept constant throughout the test. Because of the serious disturbance at the failure plane due to reversal, the stress–strain curve is not reliable at all strains. However, the stress–strain curves in the first and last forward shears are acceptable. Figure 9.10 shows typical stress ratio–displacement curves for different cycles.

### 9.5.5 Triaxial Shear Test

The triaxial shear test is a versatile apparatus suitable for all types of soil and test conditions. Cylindrical soil specimens having a height–diameter ratio of 2:1 are used, and the conventional size of the test specimen is 76 mm height and 38 mm diameter. The general arrangement of the apparatus is shown in Fig. 9.11.

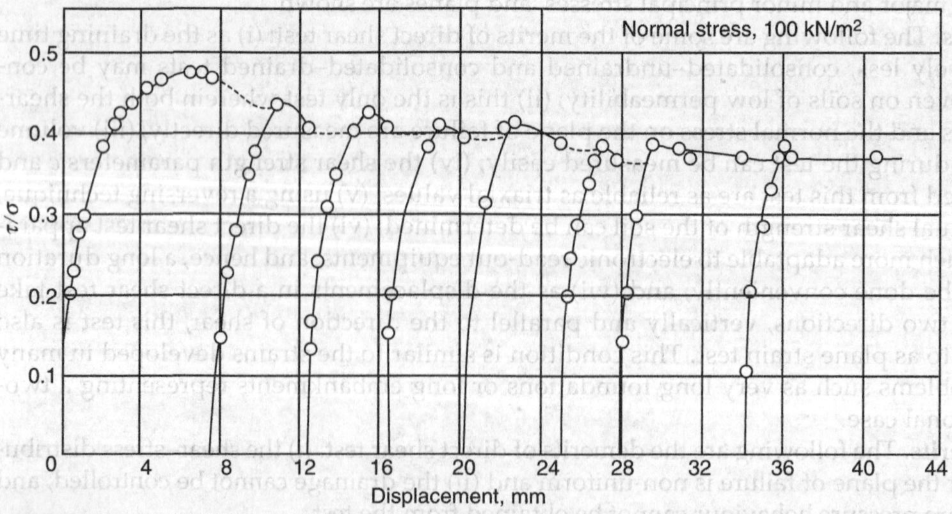


**Fig. 9.10** Stress ratio–displacement curves

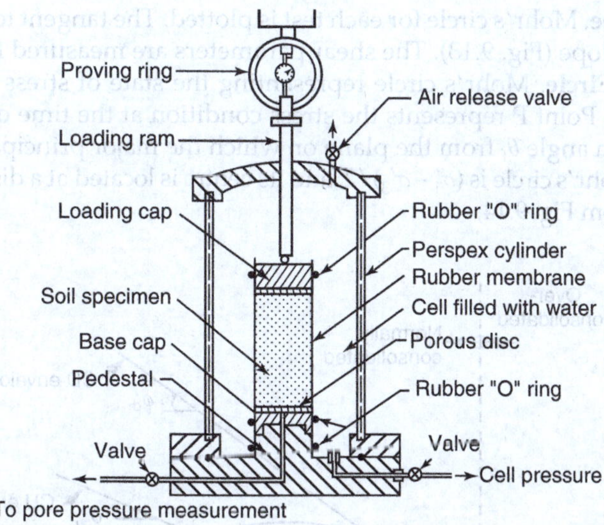


**Fig. 9.11** Schematic diagram of triaxial test

The soil specimen is covered by a thin membrane which extends over a top cap and a bottom pedestal. The perspex cell is filled with water, and the required cell or *confining pressure* ($\sigma_3$) is applied. This confining pressure acts horizontally on the cylindrical surface of the sample and vertically on the top of the specimen. An additional vertical stress called *deviator stress* ($\sigma_1 - \sigma_3$) is then applied and steadily increased until failure of the specimen occurs. The test procedure is repeated on different specimens taken from the same sample for different confining pressures and the corresponding deviator stresses calculated.

Detailed test procedures for undrained, consolidated–undrained, and drained tests are given in Chapter 10.

**Typical Stress–Strain Curve**. Figure 9.12 shows typical stress–strain curves for undrained, consolidated–undrained, and drained shear tests.

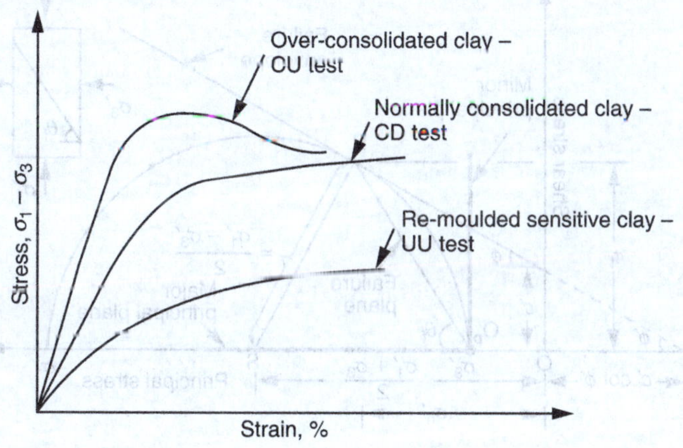


**Fig. 9.12** Typical stress–strain curves from triaxial shear tests

**Failure Envelope**. Mohr's circle for each test is plotted. The tangent to the resulting circles gives Mohr's envelope (Fig. 9.13). The shear parameters are measured from the plot.

**Mohr's Stress Circle**. Mohr's circle representing the state of stress for one specimen is shown in Fig. 9.14. Point P represents the stress condition at the time of failure on a failure plane inclined at an angle $\theta_f$ from the plane on which the major principal stress acts.

The radius of Mohr's circle is $(\sigma'_1 - \sigma'_3)/2$, and its centre is located at a distance of $(\sigma'_1 + \sigma'_3)/2$ from the origin. From Fig. 9.14,

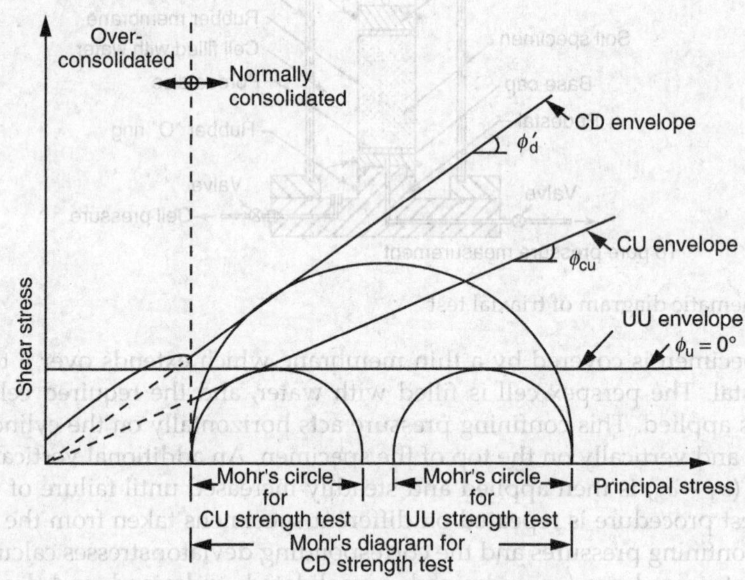


Fig. 9.13 Typical strength envelops for CD, CU, and UU tests on clay

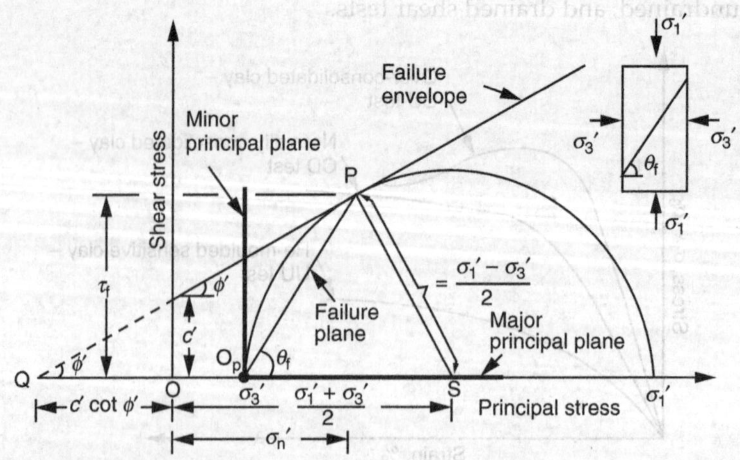


Fig. 9.14 Stress conditions at failure

Rearranging,
$$\sin\phi' = \frac{1/2(\sigma_1' - \sigma_3')}{c'\cot\phi' + 1/2(\sigma_1' + \sigma_3')} \tag{9.7}$$

or
$$\sigma_1' = \sigma_3' \tan^2(45° + \phi'/2) + 2c' \tan^2(45° + \phi'/2) \tag{9.8a}$$

$$\sigma_1' = \sigma_3' N_\phi + 2c'\sqrt{N_\phi} \tag{9.8b}$$

where
$$N_\phi = \tan^2(45° + \phi'/2) \tag{9.9}$$

The quantity $N_\phi$ is referred to as the *flow value.*

This is the relationship between the effective major and minor principal stresses in terms of effective shear strength parameters.

$$\theta_f = 45° + \phi'/2 \tag{9.10}$$

The coordinates of point P arc

$$\tau_f = \tfrac{1}{2}(\sigma_1' - \sigma_3')\sin 2\theta_f \tag{9.11}$$

and

$$\sigma_n' = \tfrac{1}{2}(\sigma_1' + \sigma_3') - \tfrac{1}{2}(\sigma_1' - \sigma_3')\cos 2\theta_f \tag{9.12}$$

The pole and the major and minor principal planes and stresses are shown in Fig. 9.14.

**Modified Failure Envelope.** Any state of stress can also be represented by a stress point instead of a Mohr circle, the point being represented by $\frac{1}{2}(\sigma_1' - \sigma_3')$ and $\frac{1}{2}(\sigma_1' + \sigma_3')$. Such a plot is shown in Fig. 9.15 for the Mohr circles represented in Fig. 9.2. A modified failure envelope is obtained with the equation of the form

$$\tfrac{1}{2}(\sigma_1' - \sigma_3') = c^* + \tfrac{1}{2}(\sigma_1' + \sigma_3')\tan\phi^* \tag{9.13}$$

where $c*$ and $\phi*$ are the modified shear strength parameters. The parameters $c'$ and $\phi'$ are then given by

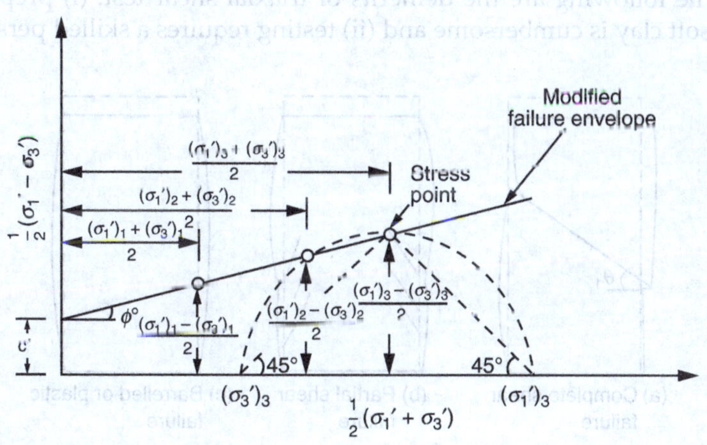


**Fig. 9.15** Alternative representation of stress conditions

$$\phi' = \sin^{-1}(\tan \phi^*) \tag{9.14}$$

$$c' = \frac{c^*}{\cos \phi'} \tag{9.15}$$

Lines drawn from the stress point at angles of 45° (as shown in Fig. 9.15) intersect the horizontal axis at $\sigma'_3$ and $\sigma'_1$.

**Types of Failure**. The diameter of the sample increases due to the shortening of the sample under load, and the middle of the sample bulges. Certain specimens fail with predominant failure planes, while in others, multiple parallel shear planes of the type shown in Fig. 9.16 are formed. In some saturated specimens of low shear strength, no shear planes are formed, but the sample shortens, bulges, and has the appearance of a barrel. In such cases, because of the continued increase of area, no definite pronounced peak will appear in the stress–strain curve, and the deviator stress corresponding to 20% strain is taken as the deviator stress at failure.

**Triaxial Test for Residual Strength**. In general, the triaxial test is to be preferred for measuring shear strength along discontinuities, as here the kinematic restraints are minimal (Skempton and Petley, 1967). But the limited maximum strain which can be applied to the specimen in the triaxial test is usually too small to permit the determination of residual strength in an originally intact clay. But where a pre-existing fissure surface is to be tested, or where an inclined plane has been made in the clay specimen, the triaxial test can be used, provided the effects of horizontal components of the load are taken into account.

**Merits**. The following are the merits of triaxial shear test: (i) pore pressure may be controlled and measured accurately in the test; (ii) suitable confining pressures can be applied (isotropic, anisotropic, or cyclic) to initially consolidate the sample to some pre-determined state; and (iii) special strain tests, such as increasing (compression) or decreasing (extension) axial load can be performed.

**Demerits**. The following are the demerits of triaxial shear test: (i) preparation of sand specimen and soft clay is cumbersome and (ii) testing requires a skilled person.

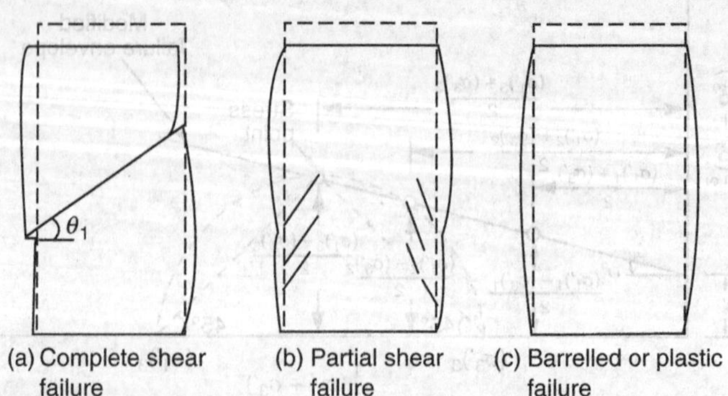


**Fig. 9.16** Types of failure in the triaxial test

### 9.5.6 Unconfined Compression Test

This is essentially a special case of the triaxial compression test where the minor principal stress is zero. The test may be carried out in the laboratory or in the field. In the laboratory, the standard triaxial apparatus can be used without a perspex cylinder. Generally, a membrane is not needed, but it is sometimes used to prevent the sides of the specimen from drying. A field apparatus is also available to obtain the unconfined compressive strength from calibrated charts. The apparatus is shown diagrammatically in Fig. 9.17.

The soil specimen is placed between the central fixed plate and the lower moving plate. The upper moving plate is raised by turning the handle. The movement of the upper plate causes an extension in the spring, and a force proportional to the extension is transmitted to the specimen. A chart is attached to the upper moving plate, and the movement is recorded by a pencil attached to the lower moving plate. As the vertical movement corresponds to the extension of the spring, the lateral movement of the pencil corresponds to the compression of the specimen.

Thus, the maximum extension of the spring corresponds to the maximum load on the sample ($P$), and the unconfined compressive strength is given as

$$\sigma_1 = \frac{P(1-\varepsilon)}{A_0} \tag{9.16}$$

where $\varepsilon$ is the longitudinal strain and $A_0$ the original cross-sectional area of the specimen.

This test is one of the easiest and simplest for determining the shear strength of cohesive soils. It can be used for both undisturbed and re-moulded samples. The detailed test procedure is given in Chapter 10.

**Test Curves.** The stress–strain curve for an undisturbed specimen and the curve after re-moulding are shown in Fig. 9.18. The ratio of undisturbed strength to re-moulded strength is called the sensitivity of the soil (discussed in Section 9.9).

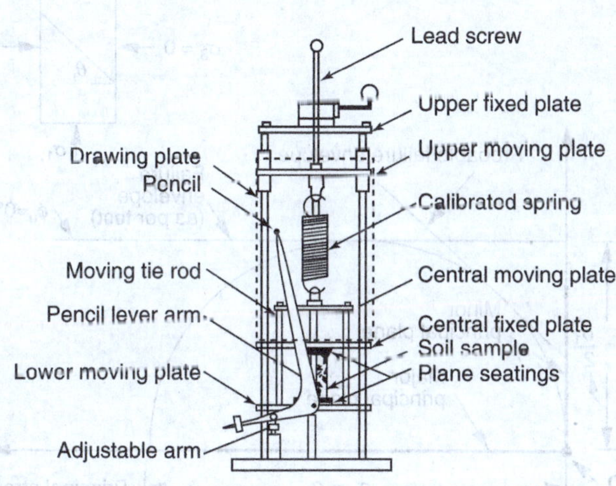


**Fig. 9.17** Unconfined compression test field apparatus

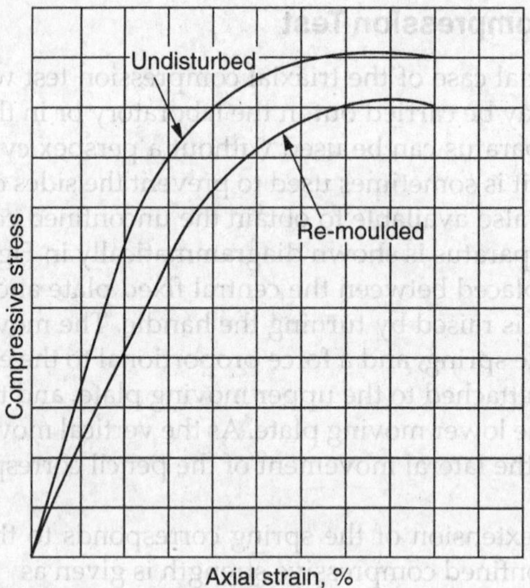


Fig. 9.18 Typical stress-strain curve for undisturbed and re-moulded insensitive clay

**Failure Envelope and Mohr Stress Circle**. The Mohr envelope is shown in Fig. 9.19. Since $\sigma_3 = 0$, only one circle can be drawn. The pole and the major and minor principal stresses and planes are also shown in Fig. 9.19. This test is applicable only for saturated unfissured clays. The shear strength $\tau_f = c_u$. If the failure occurs along a well-defined shear plane, the angle $\theta_f$ is measured, and then $\phi$ and $c$ are calculated as shown below.

$$\theta_f = 45° + \phi / 2$$

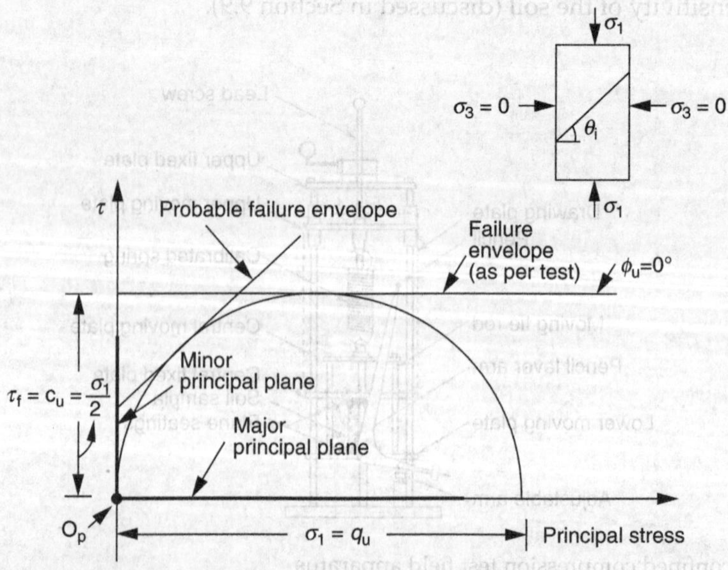


Fig. 9.19 Mohr diagram for unconfined compression test at failure

or

$$\phi = 2\theta_f - 90° \tag{9.17}$$

and

$$c = \sigma_1 / 2 \tan \theta_f$$

**Merits**. The following are the merits of the unconfined compression test: (i) this is the simplest test for determining the undrained shear strength of soil and (ii) a quick test and field apparatus gives fast determination of undrained strength.

**Demerits**. The following are the demerits of unconfined compression test: (i) fissured samples may not give reliable results because there is no confining pressure to close the fissures and (ii) samples with high sand content give unreliable results.

### 9.5.7 Ring Shear Test

The first determination of residual strength of an undisturbed clay under drained condition using a ring shear apparatus was proposed by Tiedemann (1937). Hvorslev (1960) also used a similar ring shear apparatus. In general, the two obvious advantages of any type of torsion or ring shear test are that there is no change in the area of the cross-section of the shear plane as the test proceeds and that the sample can be sheared through a large displacement of any magnitude.

A simple test of this type is one in which a laterally confined normally loaded annular specimen is twisted with controlled rate of displacement or torque (Fig. 9.20). The rate of rotation must be slow enough such that no excess pore water pressure develops in the specimen and the specimen remains under drained condition. The shear displacement may be continued to a desired level, and there is no restriction to the magnitude. Shear stress, which is calculated from the applied torque, is plotted against shear displacement, as shown in Fig. 9.21. For purposes other than the measurement of residual strength, it has all the disadvantages of the shear box, such as high local concentrations of strain, uncertainty about the direction of principal stresses as the test proceeds, and the enormous time taken for the test.

### 9.5.8 Choice of Rate of Deformation

The rate of deformation of a test is based on the drainage condition, the purpose of the test, and the convenience for accurate booking of readings. For the unconsolidated–undrained

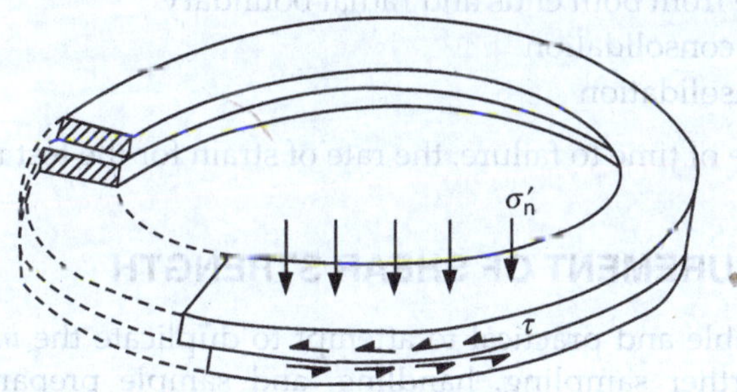


Fig. 9.20 Schematic representation of ring shear test

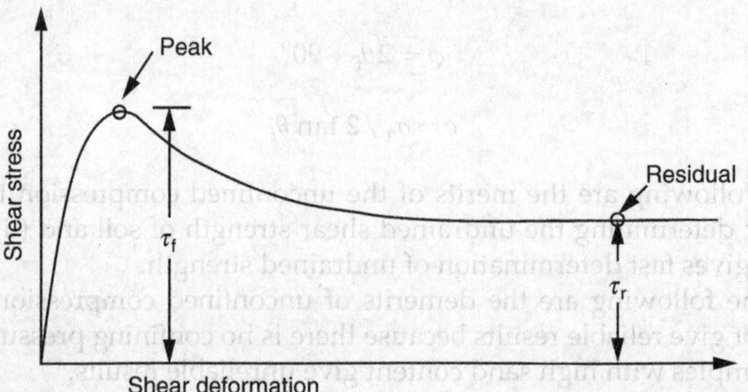


Fig. 9.21 Typical shear stress and shear deformation curve from ring shear test

test, the percentage of strain varies from 0.5 to 2 per minute in triaxial and unconfined compression tests, and the strain rate is about 1.5 mm/min in direct shear test. In the consolidated–undrained triaxial tests with pore water pressure measurements, the rate of strain ranges from 0.05 to 0.1% per minute. Drained tests have to be conducted still at a slow rate such that during the entire test, no excess pore water pressure develops. The conventional strain rate is about 0.01 to 0.2 mm/min.

Although suitable rates for carrying out drained tests may be found by a process of trial and error, they can be estimated from the consolidation data of the soil. The time required for failure ($t_f$) when 95% dissipation is ensured may be obtained from the expression

$$t_f = \frac{b^2}{n_d c_v (1 - U_z)} = \frac{20h^2}{n_d c_v}$$

where

$2\,h$ = initial thickness or height of the specimen

$n_d$ = a constant for drainage
 = 0.75 – drainage from one end only
 = 3.00 – drainage from both ends
 = 32 – drainage from radial boundary only
 = 35 – drainage from both ends and radial boundary

$c_v$ = coefficient of consolidation

$U_z$ = degree of consolidation

From a knowledge of time to failure, the rate of strain for the test may be calculated.

## 9.6 FIELD MEASUREMENT OF SHEAR STRENGTH

It is not always feasible and practical to attempt to duplicate the *in situ* conditions in the laboratory tests. Further sampling, handling, and sample preparation disturb the soil structure and hence the shear strength of the soil. It is thus desirable to test the soil in the *in situ* condition.

### 9.6.1 Vane Shear Test

Vane shear test is one field test that is coming into wider use for determining the *in situ* strength of cohesive soils. In particular, the test is very suitable for soft clays and may not give reliable results if the clay contains sand or silt.

The equipment consists of thin-bladed vanes that can be pushed into the soil with a minimum of disturbance (Fig. 9.22). The length of the vane is equal to twice its overall width. A torque is applied gradually to the upper end of the rod. Shear failure takes place over the surface and the ends. The shear strength is obtained from the expression

$$T = \pi c_u \left( \frac{d^2 b}{2} + \frac{a d^3}{4} \right) \tag{9.18}$$

where

$T$ = torque at failure and $a = 2/3$ for uniform end shear
$d$ = overall vane width = $3/5$ for parabolic end shear
$b$ = vane length = $1/2$ for triangular end shear
$c_u$ = shear strength (undrained)

It is assumed that the shear strength is isotropic. Re-moulding can be done by rotating several revolutions, and then the torque is measured, which corresponds to the re-moulded shear strength. A detailed description of the test is given in Chapter 10.

For measurements of low undrained shear strength of saturated clays, a laboratory vane shear apparatus can be used (IS: 2720 – Part 30, 1980). The apparatus may be hand operated or motorized with an arrangement to rotate the vane at the rate of 0.1°/s. The penetrating edge of the vane blade is sharpened to enable easy insertion into the soil without disturbance. A hand-operated vane is illustrated in Fig. 9.23.

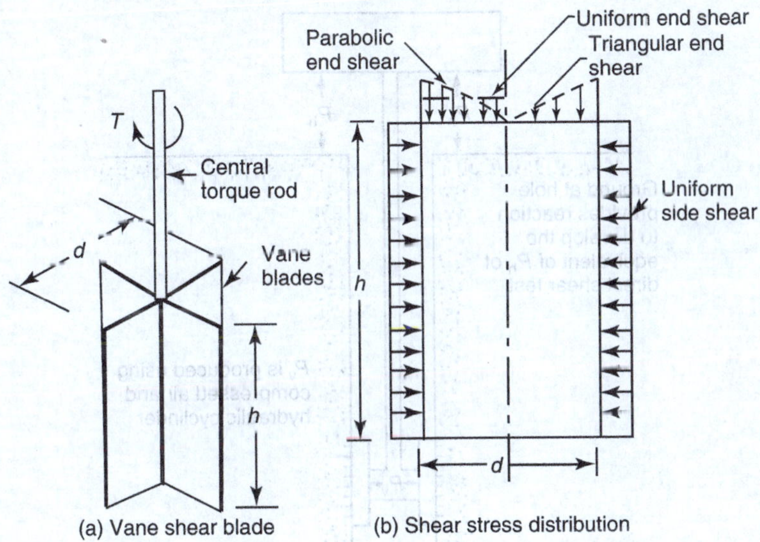


**Fig. 9.22** Scheme diagram of vane shear test

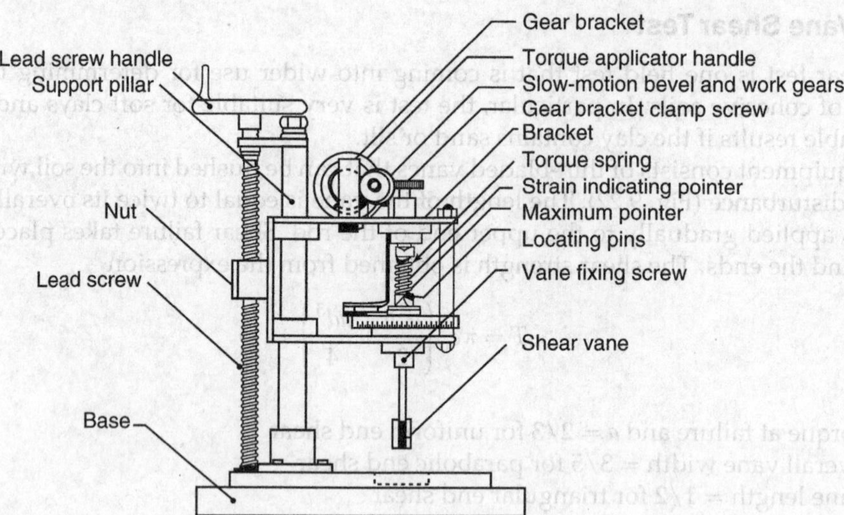


**Fig. 9.23** Laboratory vane shear apparatus (*Source:* IS: 2720 – Part 30, 1980)

### 9.6.2 Borehole Shear Device

A borehole shear device (Fig. 9.24) is a field method which has been developed by Wineland (1975). This device is inserted into the bore and expanded into the soil with a pressure $p_v$. The shear is caused by a pull $p_b$, as in a direct shear device. The test device operates by applying the direct shear device technique *in situ*. This test approximates a consolidated–undrained test. The reader may get additional details from Wineland (1975).

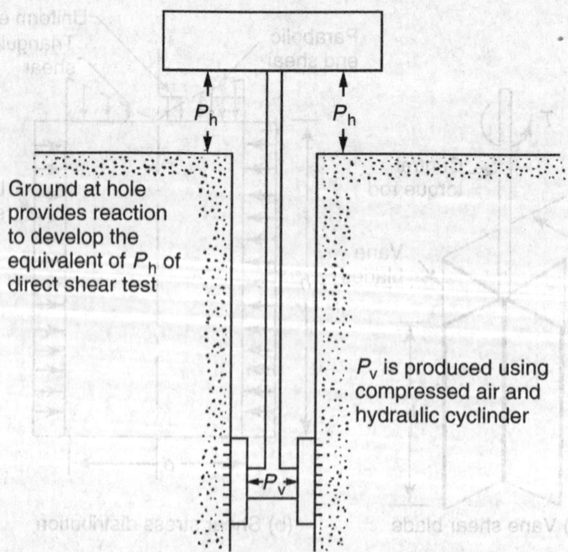


**Fig. 9.24** Borehole shear device

## 9.7 SHEAR STRENGTH OF SATURATED COHESIVE SOILS

The shear strength of cohesive soils depends on the degree of saturation, effective consolidation pressure (stress history), and loading and drainage conditions. In order to understand its behaviour, it is customary to treat cohesive soils in a saturated state, as this is the worst condition that can prevail in the field. Thus, the parameters estimated will be generally conservative. Because of the heterogenic nature of the soil, researchers concentrated many of their studies on re-moulded cohesive soils in evolving the basic concepts concerning the strength characteristics of saturated cohesive soils. From the results obtained, the strength behaviour of *in situ* soils has been analysed.

### 9.7.1 Undrained Strength

If a saturated sample of re-moulded soil is subjected to a consolidation pressure in a triaxial test (or a normal stress in a direct shear test), the effective stress remains unchanged for no drainage condition. Any subsequent change in all-round pressure (or normal stress) simply results in a corresponding change in pore water pressure. Similarly, the subsequent change in $\sigma_1$ will result in the same change in pore water pressure. Therefore, the undrained strength is constant and independent of all-round pressure (Fig. 9.25). Thus, in intact clays the undrained strength can be determined simply from an unconfined compression test. The undrained shear strength is represented by $c_u$, the undrained cohesion, and $\phi_u = 0$. If in a slip plane, $\theta_f > 45°$, then the undrained cohesion will be less than $c_u$ (for which $\phi > 0°$). As the initial condition is not changed, there will be only one effective stress circle, and the same can be obtained by measuring the pore water pressure at failure.

The undrained shear strength may be applied in field problems where the change in total stress is immediately compensated by a change in pore water pressure, because of non-availability of sufficient time for dissipation of pore water pressure. The typical examples in saturated clays are initial bearing capacity of footings or embankment foundations, initial stability of slopes or cuts, and initial stability of a braced excavation.

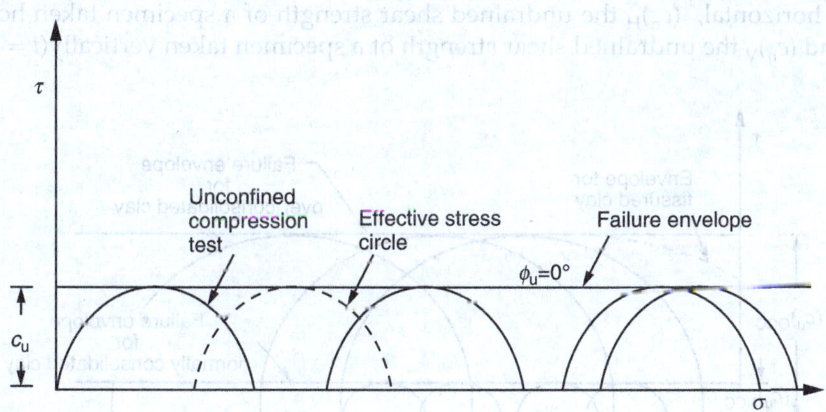


Fig. 9.25 UU strength of saturated clays

The undrained shear strength of *in situ* samples can be obtained from the unconfined compression test if the clay is intact and if fissured undrained triaxial test can be used. In such cases, the *in situ* samples should have the same void ratio. In fissured clays, the failure envelope is curved at low values of all-round pressure. However, at large pressures, the fissures close and behave like an intact sample. Proper judgement should be exercised in using such undrained strength results of fissured clays in practical problems. Figure 9.26 shows failure envelopes for normally consolidated and over-consolidated intact soils. For sampling in sensitive clays, thin-walled tube samplers can be used. For extra-sensitive and quick clays, *in situ* vane shear tests are highly desirable.

In uniform, normally consolidated clays, the undrained shear strength increases approximately linearly with depth. That is, the ratio of undrained strength, $c_u$, to the effective overburden pressure, $p_o$, is approximately constant. Skempton (1957) proposed a correlation between the ratio $c_u/p_o$ and the plasticity index as

$$\frac{c_u}{p_o} = 0.11 + 0.0037 I_p \quad (\text{for } I_P > 10\% \text{ and scatter of } \pm 0.05) \tag{9.19a}$$

and

$$\frac{c_u}{p_o} = 0.45 w_L \quad (\text{for } w_L > 0.4 \text{ and scatter of } \pm 0.10) \tag{9.19b}$$

Equations 9.19a and b may be used to estimate the value of $c_u$ for normally consolidated clays.

Based on the nature of deposition and the subsequent consolidation, clay particles in cohesive soils have a tendency to orient in a direction perpendicular to the major principal stress. Such orientations may cause the cohesive soil to show varied strength in different directions, or in other words, the clay may be anisotropic with respect to strength. A Casagrande and Carrillo (1944) proposed an expression (Eq. 9.20) for the directional variation of the undrained shear strength as

$$(c_u)_i = (c_u)_H + [(c_u)_V - (c_u)_H]\sin^2 i \tag{9.20}$$

where $(c_u)_i$ is the undrained shear strength of a specimen whose axis is inclined at an angle $i$ with the horizontal, $(c_u)_H$ the undrained shear strength of a specimen taken horizontally ($i = 0°$), and $(c_u)_V$ the undrained shear strength of a specimen taken vertically($i = 90°$).

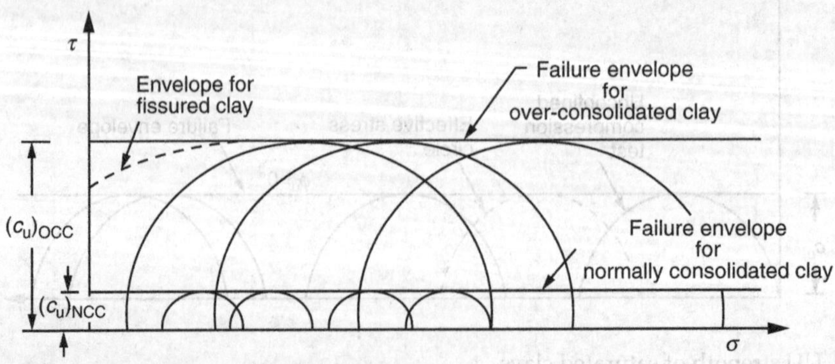


Fig. 9.26 Failure envelope for NC and OC intact clays

**Table 9.1** Typical shear strength values for clays

| Consistency of clay | Shear strength (N/mm²) |
|---|---|
| Very soft | <0.120 |
| Soft | 0.120–0.240 |
| Medium | 0.240–0.480 |
| Stiff | 0.480–0.960 |
| Very stiff | 0.960–1.920 |
| Hard | >1.920 |

*Source:* Lambe (1951).

The ratio of $(c_u)_V$ and $(c_u)_H$ is called coefficient of anisotropy. For natural deposits, this coefficient varies from 0.75 to 2.0.

A classification of saturated clays based on the undrained shear strength obtained from unconfined compressive strength is presented in Table 9.1 (Lambe, 1951).

### 9.7.2 Consolidated–Undrained Strength

The consolidated–undrained test in a triaxial or in a direct shear test consists of testing a soil under undrained condition after the initial value of void ratio has been changed by consolidation. Thus, the undrained strength forms a function of the confining pressure in triaxial test (or normal stress in direct shear test). If pore pressures are measured at the time of failure, then the effective stresses can be determined. Figure 9.27 shows the plots of both the total and the effective stress circles for a re-moulded soil from a consolidated–undrained triaxial test. The shear strength parameters $c_{cu}$ and $\phi_{cu}$ are obtained from total stress circles, and $c'$ or $c'_{cu}$ and $\phi'$ or $\phi'_{cu}$ are obtained from effective stress circles. The consolidated–undrained total stress parameters ($c_{cu}$ and $\phi_{cu}$) should be regarded as a rough guide to the undrained shear strength of the soil.

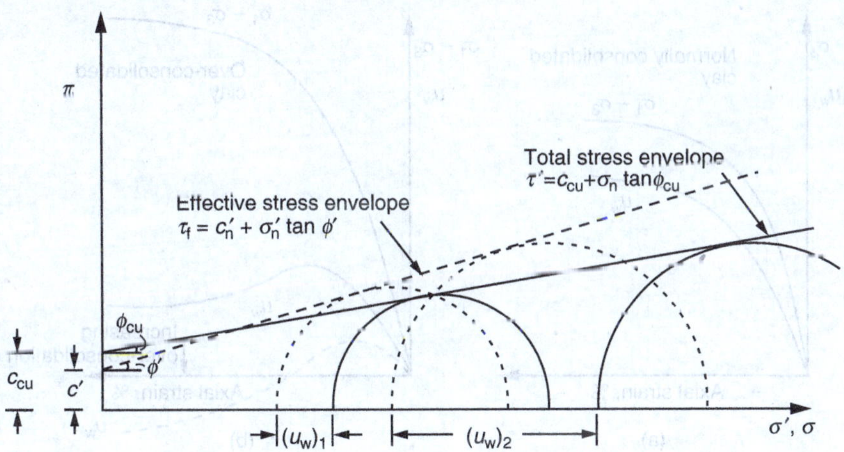


**Fig. 9.27** Effective and total CU shear test plots for saturated clays

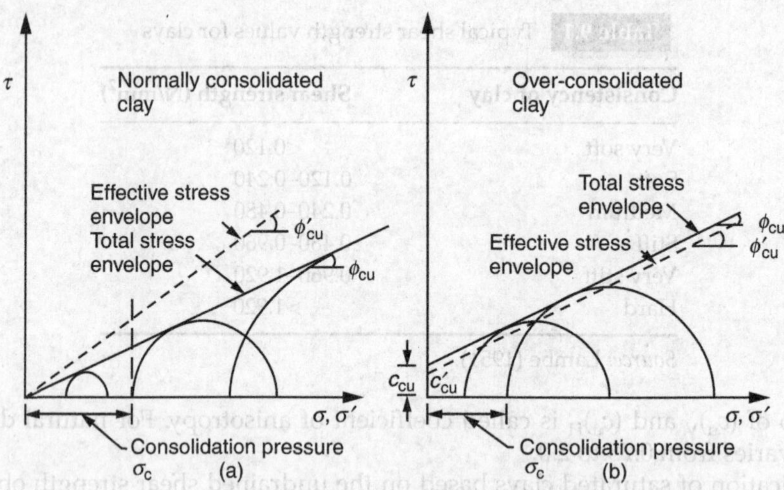


**Fig. 9.28** Effective and total stress CU test plots for NC and OC clays

The consolidated–undrained shear tests in terms of total and effective stresses for normally and over-consolidated clays are shown in Fig. 9.28, and the corresponding deviator stress and pore water pressure variations with strain are shown in Fig. 9.29. In normally consolidated clays, as a result of positive excess pore water pressure during shear (with no drainage), $\sigma_1$ and $\sigma_3$ are greater than $\sigma'_1$ and $\sigma'_3$, respectively, and hence, $\phi'_{cu} > \phi_{cu}$. However, $(\sigma_1 - \sigma_3)$ and $(\sigma'_1 - \sigma'_3)$ are equal; hence, the Mohr circles have the same diameter, but the effective stress circles are shifted to the left, reflecting higher $\phi'_{cu}$ than $\phi_{cu}$. In the over-consolidated case, because of negative pore water pressure, the effective stress circles are shifted to the right. In this case, $\phi_{cu}$ may be slightly greater or lesser than $\phi'_{cu}$ but $c_{cu}$, is always greater than $c'_{cu}$.

*In situ* clays have been consolidated anisotropically, i.e., the effective vertical and horizontal pressures are unequal. In the laboratory test, the consolidation is effected isotropically.

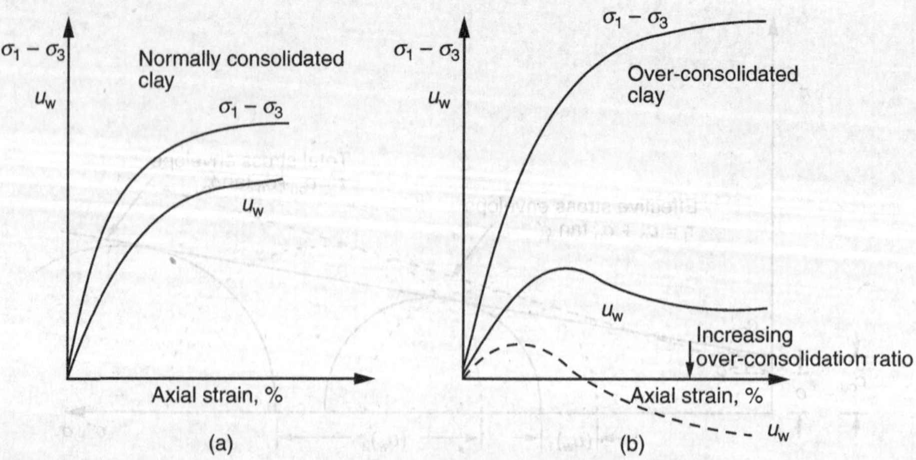


**Fig. 9.29** Deviator stress and pore water pressure variation with strain from CU test on NC and OC clays

This isotropic consolidation leads to a lower void ratio than the *in situ* one, and hence, the laboratory undrained strength overestimates the field value. As this is an unsafe situation, the specimen should be anisotropically reconsolidated in the laboratory.

### 9.7.3 Strength in Terms of Effective Stresses

The drained shear and the consolidated–undrained shear tests (in terms of effective stresses) should result in the same Mohr failure envelope for re-moulded samples. So parameters $c'$ and $\phi'$ can be obtained from drained triaxial tests (or direct shear tests). The strain rate adopted should be such that full dissipation of excess pore water pressure is maintained throughout the test. Thus, at any time of the test, the total and effective stresses are equal. The volume change during the application of principal stress difference is measured to correct the area of cross-section of the specimen. The strength in terms of effective stresses can also be obtained from the consolidated–undrained shear test with pore water pressure measurement at the time of failure. Typical results of failure envelopes for normally and over-consolidated samples are shown in Fig. 9.30, and typical results of deviator stress and volume change are given in Fig. 9.31. In drained tests, normally consolidated specimens fail at a low strain. A decrease in volume in normally consolidated clays occurs after failure, and in over-consolidated clays, an initial decrease in volume is followed by an increase at and after peak failure. With strain increase, a normally consolidated clay hardens, whereas an over-consolidated clay softens.

In saturated re-moulded clays, the difference between $\phi'_{cu}$ and $\phi'$ and $c'_{cu}$ and $c'$ is sufficiently small. It is necessary in practical problems to consider effective stress parameters whenever pore water pressure can change independently of the total stresses. Typical examples are earth dams under steady-state seepage conditions and natural earth slopes without excess pore water pressure but in equilibrium with the water table. The $c'$ and $\phi'$ parameters may be obtained from drained shear tests or from consolidated–undrained tests, with pore water pressure measurements. The values thus obtained are different, unlike in re-moulded clays. Natural slopes which are subjected to progressive failure have to be analysed with residual shear strength parameters, $c'_r$ ($\approx$0) and $\phi'_r$. The values of $c'_r$ and $\phi'_r$ may be obtained from a ring shear test or from a reversible direct shear box test.

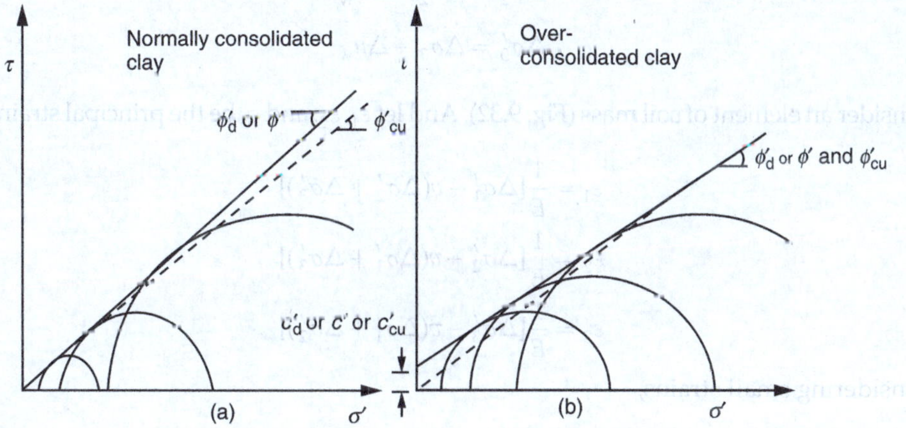


Fig. 9.30 Failure envelopes from drained tests on NC and OC clays

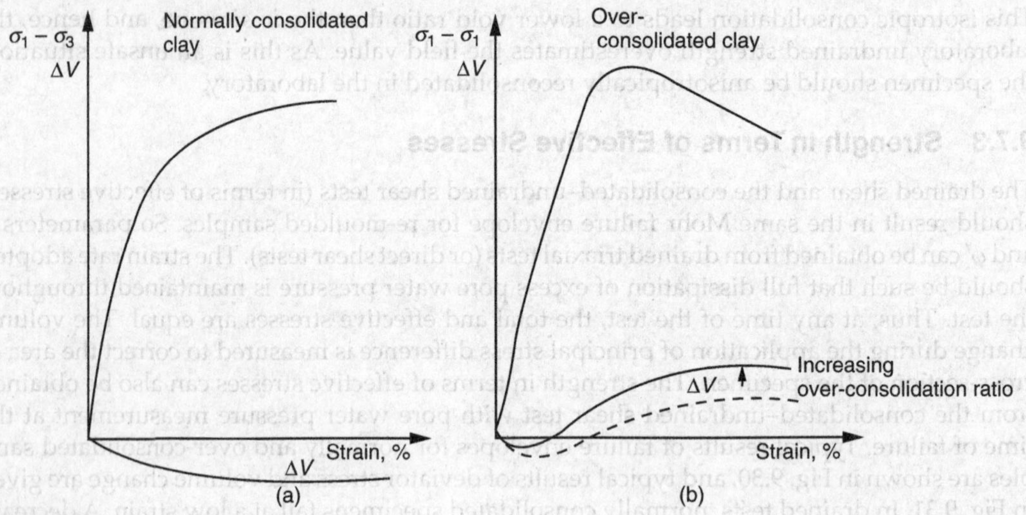


**Fig. 9.31** Deviator stress and volume change variation with strain from drained test on NC and OC clays

## 9.8 PORE PRESSURE COEFFICIENTS

### 9.8.1 Theory

In many practical problems involving deformation of soil masses, it is essential to estimate the magnitude of the changes in pore water pressure resulting from changes in the state of stress. In a saturated soil, changes in the principal stresses of $\sigma_1$, $\sigma_2$, and $\sigma_3$ result in a change in pore water pressure, $\Delta u_w$, for no drainage condition. Let the volume change be $\Delta V$. The changes in effective stresses will be

$$\Delta\sigma_1' = \Delta\sigma_1 - \Delta u_w$$
$$\Delta\sigma_2' = \Delta\sigma_2 - \Delta u_w \tag{9.21}$$
$$\Delta\sigma_3' = \Delta\sigma_3 - \Delta u_w$$

Consider an element of soil mass (Fig. 9.32). And let $\varepsilon_1$, $\varepsilon_2$, and $\varepsilon_3$ be the principal strains. Then,

$$\varepsilon_1 = \frac{1}{E}[\Delta\sigma_1' - v(\Delta\sigma_2' + \Delta\sigma_3')]$$
$$\varepsilon_2 = \frac{1}{E}[\Delta\sigma_2' - v(\Delta\sigma_1' + \Delta\sigma_3')] \tag{9.22}$$
$$\varepsilon_3 = \frac{1}{E}[\Delta\sigma_3' - v(\Delta\sigma_1' + \Delta\sigma_2')]$$

Considering small strains,

$$\frac{\Delta V}{V} = \varepsilon_1 + \varepsilon_2 + \varepsilon_3 \tag{9.23}$$

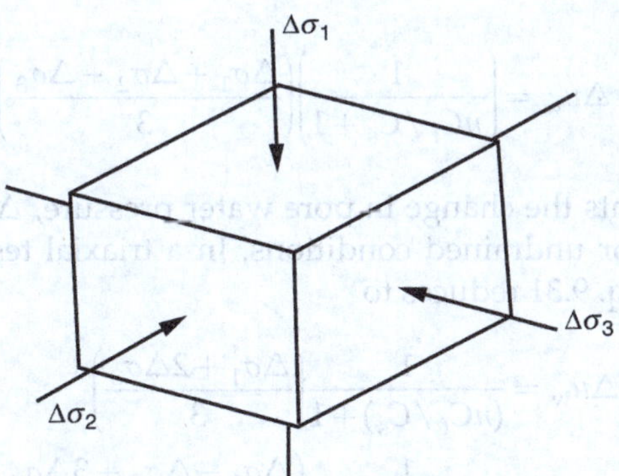


Fig. 9.32 Three-dimensional stress system on a soil element

Substituting for $\varepsilon_1$, $\varepsilon_2$, and $\varepsilon_3$ from Eq. 9.22, we have

$$\frac{\Delta V}{V} = \frac{1-2v}{E}(\Delta\sigma_1' + \Delta\sigma_2' + \Delta\sigma_3') \tag{9.24}$$

Let the compressibility of soil skeleton be $C_s$ and be represented as

$$C_s = \frac{3(1-2v)}{E} \tag{9.25}$$

Now representing the change in effective stresses in terms of the change in total stresses and pore water pressure, we have

$$\frac{\Delta V}{V} = C_s\left(\frac{\Delta\sigma_1 + \Delta\sigma_2 + \Delta\sigma_3}{3} - \Delta u_w\right) \tag{9.26}$$

As only change in volume causes consequent change in pore water pressure, the compressibility of the pore fluid $C_f$ is given as

$$C_f = \frac{1}{V_w}\frac{\Delta V_w}{\Delta u_w} \tag{9.27}$$

Since the soil is saturated, $V_w = nV$ and $\Delta V_w = \Delta V$. Therefore,

$$C_f = \frac{1}{nV}\frac{\Delta V}{\Delta u_w} \tag{9.28}$$

or

$$\frac{\Delta V}{V} = C_f n \Delta u_w \tag{9.29}$$

Equating Eq. 9.26 and Eq. 9.29, we have

$$C_s\left(\frac{\Delta\sigma_1 + \Delta\sigma_2 + \Delta\sigma_3}{3} - \Delta u_w\right) = C_f n \Delta u_w \tag{9.30}$$

Rearranging,

$$\Delta u_{\mathrm{w}} = \left(\frac{1}{nC_{\mathrm{f}}/C_{\mathrm{s}}+1}\right)\left(\frac{\Delta\sigma_1+\Delta\sigma_2+\Delta\sigma_3}{3}\right) \tag{9.31}$$

Equation 9.31 represents the change in pore water pressure, $\Delta u_{\mathrm{w}}$, due to changes in the total principal stresses for undrained conditions. In a triaxial test, it is customary to take $\Delta\sigma_2 = \Delta\sigma_3$ and, hence, Eq. 9.31 reduces to

$$\Delta u_{\mathrm{w}} = \frac{1}{(nC_{\mathrm{f}}/C_{\mathrm{s}})+1}\left(\frac{\Delta\sigma_1+2\Delta\sigma_3}{3}\right)$$

$$= \frac{1}{(nC_{\mathrm{f}}/C_{\mathrm{s}})+1}\left(\frac{\Delta\sigma_1-\Delta\sigma_3+3\Delta\sigma_3}{3}\right)$$

or

$$\Delta u_{\mathrm{w}} = \frac{1}{(nC_{\mathrm{f}}/C_{\mathrm{s}})+1}\left[\Delta\sigma_3+\frac{1}{3}(\Delta\sigma_1-\Delta\sigma_3)\right] \tag{9.32}$$

As the soil mass is not an elastic and isotropic material, the coefficients have been replaced by two parameters, *A* and *B*, referred to as pore pressure coefficients by Skempton (1954). Thus,

$$\Delta u_{\mathrm{w}} = B[\Delta\sigma_3 + A(\Delta\sigma_1-\Delta\sigma_3)] \tag{9.33}$$

The change in pore water pressure, $\Delta u_{\mathrm{w}}$, is due to an isotropic stress increase $\Delta\sigma_3$ together with an axial stress increase $(\Delta\sigma_1 - \Delta\sigma_3)$, as happens in a conventional triaxial test.

An overall coefficient $\bar{B}$ can be obtained by dividing Eq. 9.33 by $\Delta\sigma_1$. Thus,

$$\frac{\Delta u_{\mathrm{w}}}{\Delta\sigma_1} = B\left[\frac{\Delta\sigma_3}{\Delta\sigma_1}+A\left(1-\frac{\Delta\sigma_3}{\Delta\sigma_1}\right)\right]$$

$$= B\left[1-(1-A)\left(1-\frac{\Delta\sigma_3}{\Delta\sigma_1}\right)\right]$$

or

$$\frac{\Delta u_{\mathrm{w}}}{\Delta\sigma_1} = \bar{B}$$

where

$$\bar{B} = B\left[1-(1-A)\left(1-\frac{\Delta\sigma_3}{\Delta\sigma_1}\right)\right] \tag{9.34}$$

Now,

$$B = \frac{1}{(nC_{\mathrm{f}}/C_{\mathrm{s}})+1} \tag{9.35}$$

Water is an incompressible pore fluid, and hence, in a saturated soil $C_{\mathrm{f}} \ll C_{\mathrm{s}}$. Hence, $B \approx 1.0$. Thus, for a saturated soil, a uniform increase in total stress results in an equal rise in

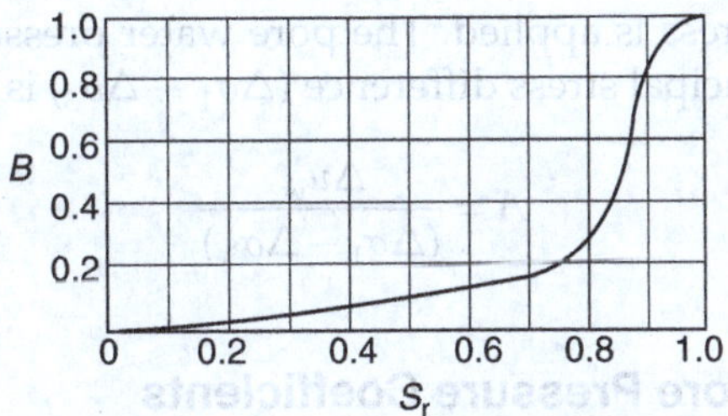


**Fig. 9.33** Typical Variation of $B$ with the degree of saturation $S_r$

pore water pressure, and the effective stress remains unchanged. That is, $\Delta u_w = \Delta\sigma_3$ during application of all-round pressure and $\Delta u_w = (\Delta\sigma_1 - \Delta\sigma_3)$ during application of deviator stress.

The presence of air in the voids increases the compressibility of the pore fluid in partially saturated soils. Thus, $C_f \gg C_s$, and hence, $B<1.0$. Figure 9.33 shows a typical variation of $B$ with the degree of saturation.

The pore pressure coefficient $A$ is not a constant but depends on the stress history, stress level, strain level, strain rate, and mode of application of stress (viz., increasing or decreasing). The value of $A$ at the time of shear failure is denoted as $A_f$ and is often quoted as a guide value. Values of $A_f$ for different soils are shown in Table 9.2.

### 9.8.2 Measurement of Pore Pressure Coefficients

The pore pressure coefficients, $A$ and $B$, may be measured in a consolidated–undrained test. The coefficient $B$ is determined during the consolidation stage and $A$ during the shear stage of the test.

Soon after the application of all-round pressure, and before permitting drainage, the change in pore water pressure due to change in $\sigma_3$ can be measured. Then, $B = \Delta u_w / \Delta\sigma_3$. After allowing for drainage and at the end of completion of consolidation, the drainage valve

**Table 9.2** Values of pore pressure coefficient $A_f$

| Soil type | $A_f$ |
|---|---|
| Dense fine sands | –0.3 to 0.0 |
| Medium fine sands | 0.0 to 1.0 |
| Very loose fine sands | 2.0 to 3.0 |
| Heavily pre-compressed clays | –0.5 to 0.0 |
| Lightly pre-compressed clays | 0.3 to 0.7 |
| Normally consolidated clays | 0.7 to 1.30 |
| Very sensitive to quick clays[a] | 1.2 to 2.5 |
| Compacted sandy clays | 0.25 to 0.75 |
| Compacted clay gravels | –0.25 to 0.25 |

[a]Refer Section 9.9.

is closed and a deviator stress is applied. The pore water pressure change ($\Delta u_w$) resulting from the change in the principal stress difference ($\Delta\sigma_1 - \Delta\sigma_3$) is measured; then,

$$A = \frac{\Delta u_w}{(\Delta\sigma_1 - \Delta\sigma_3)}$$

### 9.8.3 Application of Pore Pressure Coefficients

The knowledge of anticipated change in pore water pressure due to change in total stress provides a means whereby the margin of uncertainty in construction work can be greatly reduced. The following are some of the uses of pore pressure coefficients in practice (Bishop, 1954).

1. The initial pore water pressure caused due to the stressing of a natural stratum forming a foundation or an earth dam, which is necessary to estimate the stability during and after construction.
2. The pore water pressure induced due to the stressing of compacted impervious fill of an earth dam during construction, which is needed to control the stability and amount of settlement after construction.
3. The change in pore water pressure caused by the removal of water load on the impervious fill of an earth dam due to rapid draw-down.
4. The pressure changes resulting from the removal of the weight of soil in the formation of slopes and cuttings in natural strata.

## 9.9 SENSITIVITY OF COHESIVE SOILS

Changes in stress and environment with time may result in the soil having a higher strength in the undisturbed state than in the re-moulded state. The term *sensitivity* ($S_t$) is used to describe this difference in strength. This is given by the ratio of the undisturbed strength to the re-moulded strength, where the strength may be undrained, consolidated–undrained, or drained. The undrained strength value is most commonly used. The re-moulded strength is obtained by re-moulding the specimen used for the undisturbed strength, and both the tests should have the same water content. The sensitivity may vary from 1 to as high as 100, and thus, accordingly, they are classified as insensitive, medium sensitive, sensitive, extra sensitive, and quick. Table 9.3 gives the sensitivity classification. Figure 9.34 shows typical stress–strain curves before and after re-moulding, for sensitive clays.

Highly over-consolidated clays are insensitive. This is partly due to the low natural water content in the soil deposits. Glacial till clays are medium sensitive, whereas a few glacial clays and most freshwater deposits are very sensitive or extra sensitive.

A few of the freshwater deposits and many marine deposits are quick. The sensitivity of a majority of cohesive deposits ranges from 2 to 8.

A part of the disturbance caused due to re-moulding is attributed to the disturbance of adsorbed water in clay layers. If the re-moulded soil is left undisturbed at the same water content, the lost strength will be regained. This phenomenon is called *thixotrophy*.

**Table 9.3** Sensitivity classification

| Description | Sensitivity, $S_t$ |
|---|---|
| Insensitive | <2 |
| Medium sensitive | 2–4 |
| Sensitive | 4–8 |
| Very sensitive | 8–16 |
| Slightly quick | 16–32 |
| Medium quick | 32–64 |
| Quick | >64 |

*Source:* Bjerrum (1954).

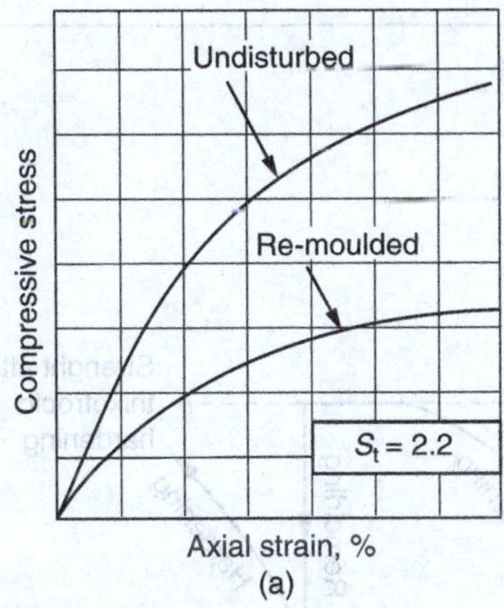


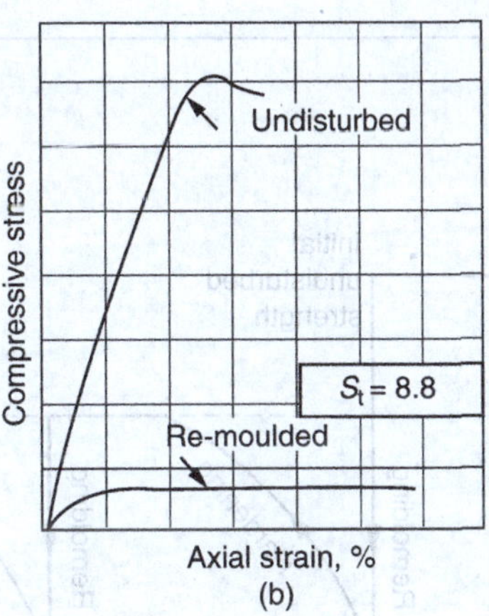


**Fig. 9.34** Typical stress–strain curves for sensitive clays

## 9.10 THIXOTROPHY OF CLAYS

Clays with an initial flocculant structure may lose strength due to disturbance or re-moulding. However, with passage of time, the clay may gain the original strength due to thixotropy.

Thixotropy is defined as an isothermal, reversible, time-dependent process which occurs under constant composition and volume. This phenomenon is attributed to a process of softening caused by re-moulding followed by a time-dependent regain of the original hard state.

The extent of strength gain by the thixotropic effect depends on the type of clay. The clay mineral that absorbs large quantities of water into their lattice structure, e.g., montmorillonite, may have high thixotropic gain. Figure 9.35 shows the gain in strength of a soil due to thixotropic effect.

Most of the soils partly regain the lost strength with time after remoulding. Such a material is referred to as partially thixotropic. The nature of the strength gain and the time for partially thixotropic material is shown in Fig. 9.35b. The difference in strength, that the undisturbed strength and the thixotropic strength gain, is attributed to the destruction of the clay-particle structure.

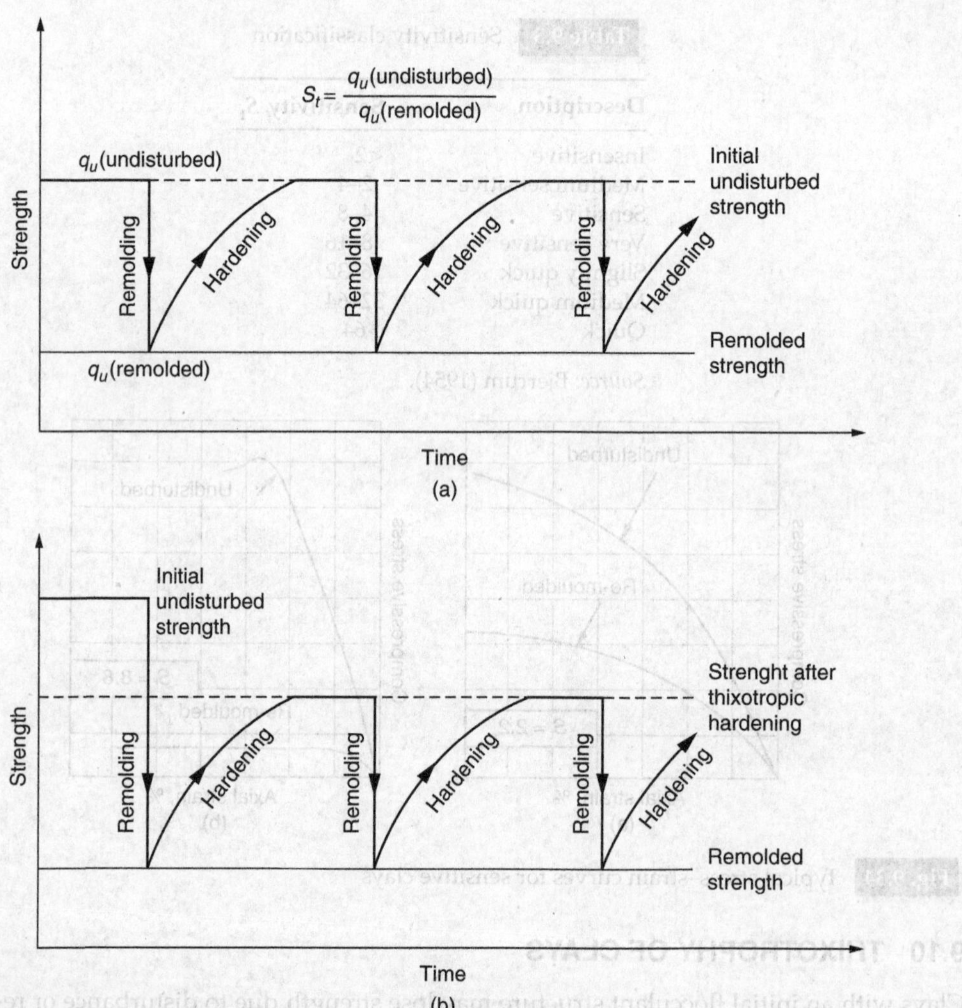


**Fig. 9.35** Behaviour of (a) thixotropic material; (b) partially thixotropic material

## 9.11 SHEAR STRENGTH OF GRANULAR SOILS

The shear strength of granular soils is a relatively simple phenomenon unlike that of clayey soils. Particle orientation and shape, surface roughness, grain-size distribution, and surface forces are significant factors. The strength of granular material is affected largely by the initial void ratio and the effective stresses. The $c'$ value of uncemented granular soil is zero. The parameters necessary to completely define the shear strength of a granular soil, tested under slowly applied shear deformation, are $\phi'$ and $\sigma'_n$.

Hence,

$$\tau_f = \sigma'_n \tan \phi' \tag{9.36}$$

and $\phi'$ is influenced slightly by the presence of water. Only the drained strength of a granular soil is useful in practice because of high permeability. Undrained strength is insignificant except during an earthquake.

The relevant shear strength parameter (Eq. 9.36) may be obtained from the results of direct shear tests or drained triaxial tests. The characteristics of dry and saturated granular soils are the same, provided all are under effective stress conditions and in the case of saturated soils no excess pore water pressure exists.

The principal difficulty is in the evaluation of $\sigma'_n$. Because of shear strains in granular soils, significant volume change occurs under drained and undrained shear conditions. The volume change mechanism is complex and is associated with random displacements of soil particles in the shear zone (Leonards, 1962).

Typical stress–strain curves and the respective volume change diagrams with different initial void ratios are shown in Fig. 9.36. Such relationships between shear stress and shear displacement under the same confining pressure can be obtained. The dense sample shows a definite peak in the stress–strain curve, followed by a decrease in the deviator stress with increase in strain until it levels out to a residual value. The loose sample does not have a peak point, and the failure criterion is governed by the strength at 20% strain, while at large strains the curve merges with the dense sample.

In dense samples there will be a considerable inter-locking between particles. Before complete shear failure can take place, the inter-locking must be overcome in addition to the frictional resistance at the points of contact. The degree of inter-locking decreases after the peak, and the shear stress necessary to continue shear displacement is correspondingly reduced. In a loose sample, there is no inter-locking, and hence, after the peak, the shear stress necessary to continue the shear displacement remains unchanged.

In a loose sample the void decreases rapidly at first and then attains a minimum value. It then increases slightly until the peak stress is reached and continues to show less or no change in volume. In a dense sample after an initial adjustment, there is a rapid increase in void ratio, which then gradually decreases and approaches a constant value. At sufficiently large strains, a void ratio is reached at which continued shear deformations will occur at constant volume. This void ratio has been referred to as *critical void ratio*, $e_{cr}$. This shows

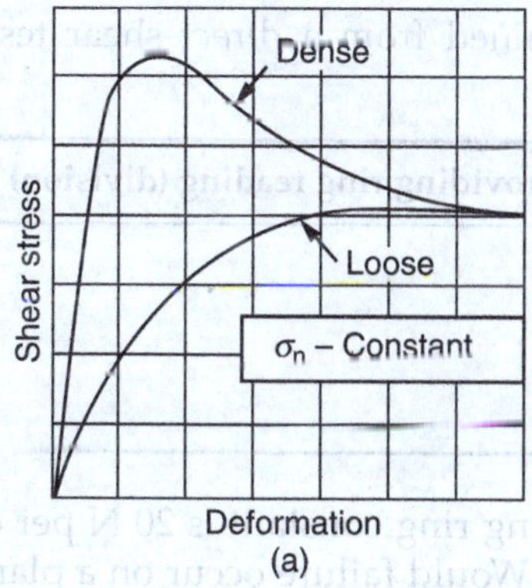


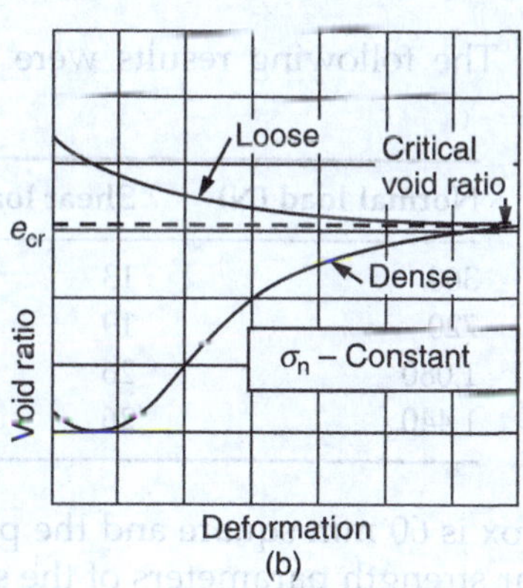


**Fig. 9.36** Typical stress–strain and volume change for granular soils

that if the initial void ratio ($e_0$) of a sample is at the critical void ratio, the shear deformation would take place at constant volume; if $e_0$ is higher than $e_{cr}$, then volume reduction would occur; and if $e_0$ is lower than $e_{cr}$, then volume increase would occur.

The appropriate strength envelope (peak or residual) for a given problem depends on the field conditions that are to be simulated. In most practical situations, large strains cannot be tolerated and the peak stress is used to define failure. Typical values of the effective angle of shearing resistance, $\phi'$, for granular soils are given in Table 9.4.

It is very rare to test undisturbed non-cohesive samples using direct or triaxial shear tests. As the direct shear test is simple and rapid, it is preferred for tests on sand. Further, it is believed that the direct shear test is a plane strain test. Both triaxial and direct shear tests give values of $\phi'$ as 2° to 8° smaller than those given by the plane strain test. Lade and Lee (1976) related the angles of shearing resistance of coarse-grained soils from the plane strain test, $\phi'_{ps}$, as

$$\phi'_{ps} = 1.5\phi'_{tr} - 17° \qquad (\phi'_{tr} > 34°)$$

$$\phi'_{ps} = \phi'_{tr} \qquad (\phi'_{tr} \leq 34°)$$

**Table 9.4** Typical effective angle of shearing resistance, $\phi'$, for coarse-grained soils

| Soil | $\phi'$° | |
|---|---|---|
| | **Loose sample** | **Dense sample** |
| Gravel | 34–40 | 40–50 |
| Uniform | 27 | 33 |
| Well-graded sand | 33 | 45 |
| Gravel | 35 | 50 |
| Silty sand | 25–35 | 30–36 |

## WORKED EXAMPLES

**Example 9.1** The following results were obtained from a direct shear test on a sandy clay sample:

| Normal load (N) | Shear load providing ring reading (division) |
|---|---|
| 360 | 13 |
| 720 | 19 |
| 1,080 | 26 |
| 1,440 | 26 |

If the shear box is 60 mm square and the proving ring constant is 20 N per division, estimate the shear strength parameters of the soil. Would failure occur on a plane within this soil at a point where the normal stress is 320 kN/m$^2$ and the corresponding shear stress is 138 kN/m$^2$?

***Solution***

| Normal load (N) | Normal stress (kN/m²) | PR dial reading | Shear stress (kN/m²) |
|---|---|---|---|
| 360 | $360/[(0.06)^2 \times 1000] = 100$ | 13 | $(13 \times 20)/[(0.06)^2 \times 1000] = 72.2$ |
| 720 | 200 | 19 | 105.6 |
| 1,080 | 300 | 26 | 144.4 |
| 1,440 | 400 | 32 | 177.7 |

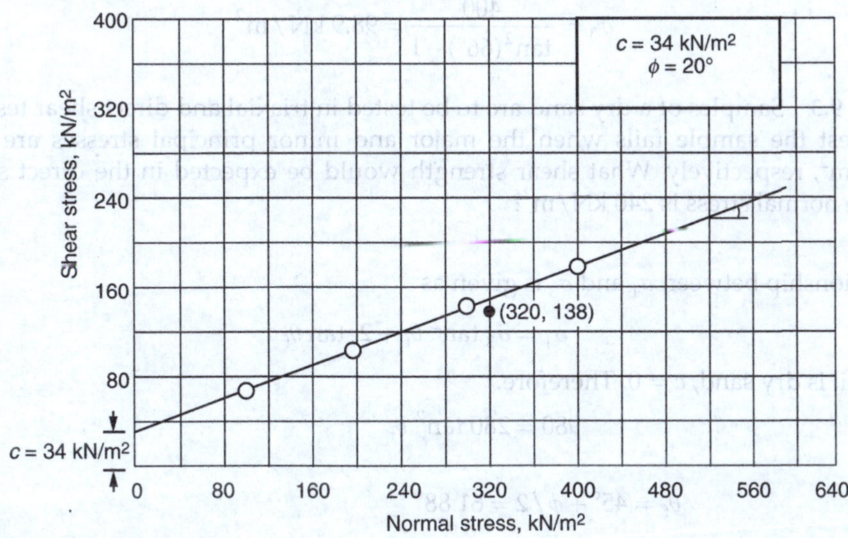


**Fig. 9.37**

The shear stresses are plotted against the corresponding normal stresses as shown in Fig. 9.37. The straight line having the best fit to the plotted points is drawn. The shear strength parameters taken from the plot are given as

$$c = 34 \text{ kN/m}^2 \quad \phi = 20°$$

The stress state $\tau_n = 138 \text{ kN/m}^2$ and $\sigma_n = 320 \text{ kN/m}^2$ plots below the failure envelope and therefore would not produce failure.

**Example 9.2** A specimen of fine dry sand, when subjected to a triaxial compression test, failed at a deviator stress of 400 kN/m². It failed with a pronounced failure plane with an angle of 24° to the axis of the sample. Compute the lateral pressure to which the specimen would have been subjected.

***Solution***

The failure angle $\theta_f = 90° - 24° = 66°$. From Eq. 9.8,

$$\sigma_1 = \sigma_3 \tan 2\theta_f + 2c \tan \theta_f$$

Since the soil is dry sand, $c = 0$.Therefore,

$$\sigma_1 = \sigma_3 \tan 2\theta_f$$

Deviator stress $\Delta\sigma = \sigma_1 - \sigma_3$

or

$$\sigma_1 = \Delta\sigma + \sigma_3 = 400 + \sigma_3$$

or

$$400 + \sigma_3 = \sigma_3 \tan^2(66°)$$

or

$$\sigma_3 = \frac{400}{\tan^2(66°) - 1} = 98.9 \text{ kN/m}^2$$

**Example 9.3** Samples of a dry sand are to be tested in triaxial and direct shear tests. In the triaxial test the sample fails when the major and minor principal stresses are 980 and 280 kN/m$^2$, respectively. What shear strength would be expected in the direct shear test when the normal stress is 240 kN/m$^2$?

***Solution***

The relationship between $\sigma_1$ and $\sigma_3$ is given as

$$\sigma_1 = \sigma_3 \tan^2 \theta_f + 2c \tan \theta_f$$

As the soil is dry sand, $c = 0$. Therefore,

$$980 = 280 \tan^2 \theta_f$$

or

$$\theta_f = 45° + \phi/2 = 61.88°$$

or

$$\phi = (61.88° - 45°) \times 2 = 33.83°$$

In the direct shear test, $\sigma_n = 240$ kN/m$^2$.

$$\tau_f = \sigma_n \tan\phi = 240 \tan 33.83° = 160.7 \text{ kN/m}^2$$

**Example 9.4** A boring log reveals that a thin layer of silty clay exists at a depth of 15 m below the natural ground surface. The soil above this layer is a silt having $\gamma_d = 15.5$ kN/m$^3$ and $w = 28\%$. The groundwater table is found to exist approximately near the ground surface. Triaxial shear tests on the undisturbed silty clay samples give the following results:

$$c_{cu} = 48.3 \text{ kN/m}^2, \phi_{cu} = 13° \text{ and } c'_d = 41.4 \text{ kN/m}^2, \phi'_d = 23°$$

Estimate the shearing resistance of the silty clay on a horizontal plane (i) when the shear stress builds up rapidly and (ii) when the shear stress builds up very slowly.

***Solution***

Total unit weight of silt $\gamma_{sat} = \gamma_d(1 + w)$

$$= 15.5\left(1 + \frac{28}{100}\right) = 19.84 \text{ kN/m}^3$$

Submerged unit weight $\gamma' = \gamma_{sat} - \gamma_w = 19.8 - 9.807 = 10.03\ \text{kN/m}^3$
Effective pressure at a depth of 15 m $\sigma'_n = 15 \times 10.03 = 150.45\ \text{kN/m}^2$
Total pressure at the depth of 15 m $= 15 \times 19.84 = 297.6\ \text{kN/m}^2$
For a rapid build-up of stresses there is no time for dissipation of pore water pressure, and the total stress parameters are used. Therefore,

$$\text{Shear strength } \tau_f = c_{cu} + \sigma_n \tan\phi_{cu}$$
$$= 48.3 + 297.6 \tan 13° = 117.0\ \text{kN/m}^2$$

For a slow build-up of stresses, there is no excess pore water pressure, and the effective stress parameters are used. Therefore,

$$\text{Shear strength } \tau_f = c'_d + \sigma'_n \tan\phi'_d$$
$$= 41.4 + 150.45 \tan 23° = 105.3\ \text{kN/m}^2$$

**Example 9.5** Triaxial compression tests were conducted on a specimen from a large sample of undisturbed clay. Tests 1 to 4 were run slowly, permitting complete drainage, and Tests 5 to 8 were run without permitting drainage. Plot Mohr's modified strength envelope, and determine the shear strength parameters for both kinds of tests.

| | **1** | **2** | **3** | **4** | **5** | **6** | **7** | **8** |
|---|---|---|---|---|---|---|---|---|
| $(\sigma_1 - \sigma_3)$ at failure (kN/m$^2$) | 447 | 167 | 95 | 37 | 331 | 155 | 133 | 119 |
| $\sigma_3$ (kN/m$^2$) | 246 | 89 | 36 | 6 | 481 | 231 | 131 | 53 |

***Solution***

Compute the points for plotting modified Mohr envelope as shown in the table below:

| **Test no.** | $(\sigma_1 - \sigma_3)$ | $\sigma'_3$ | $\sigma'_1$ | $(\sigma'_1 - \sigma'_3)/2$ | $(\sigma'_1 + \sigma'_3)/2$ |
|---|---|---|---|---|---|
| 1 | 447 | 246 | 693 | 223.5 | 469.5 |
| 2 | 167 | 89 | 256 | 83.5 | 172.5 |
| 3 | 95 | 36 | 131 | 47.5 | 83.5 |
| 4 | 37 | 6 | 43 | 18.5 | 24.5 |
| | $(\sigma_1 - \sigma_3)$ | $\sigma_3$ | $\sigma_1$ | $(\sigma_1 - \sigma_3)/2$ | $(\sigma_1 + \sigma_3)$ |
| 5 | 331 | 481 | 812 | 165.5 | 646.5 |
| 6 | 155 | 231 | 386 | 77.5 | 308.5 |
| 7 | 133 | 131 | 264 | 66.5 | 197.5 |
| 8 | 119 | 53 | 172 | 59.5 | 112.5 |

Mohr's modified strength envelopes are plotted as shown in Figs. 9.38 and 9.39 for Tests 1 to 4 and Tests 5 to 8, respectively.

From Fig. 9.38, the modified parameters are obtained as

$$c^* = 10\ \text{kN/m}^2 \quad \text{and} \quad \phi^* = 23°$$

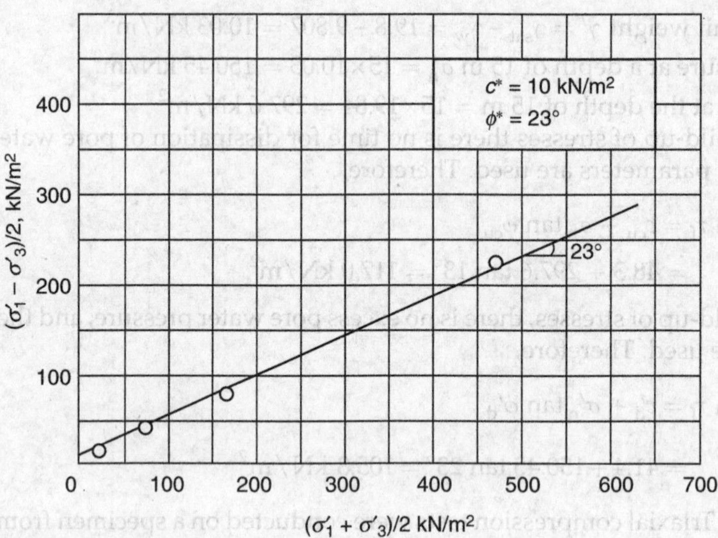


Fig. 9.38

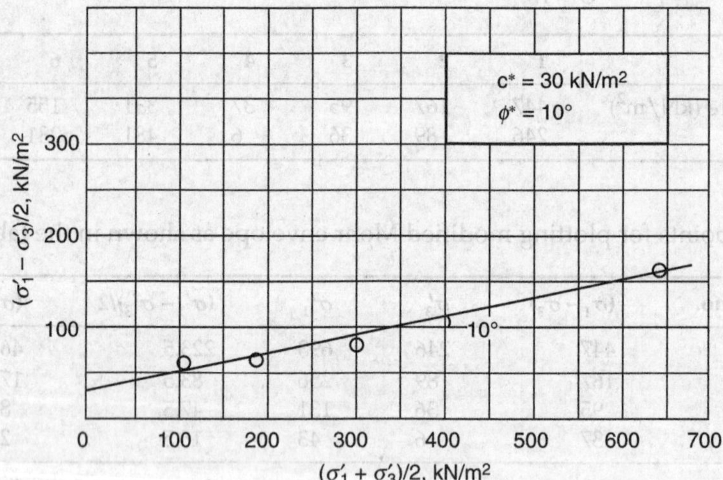


Fig. 9.39

Therefore, $\phi' = \sin^{-1}(\tan 23°) = 25.1°$
and

$$c' = \frac{10}{\cos(25.1°)} = 11.04 \text{ kN/m}^2$$

From Fig. 9.39, the modified parameters are obtained as

$$c^* = 30 \text{ kN / m}^2 \text{ and } \phi^* = 10°$$

$$\phi_u = \sin^{-1}(\tan 10°) = 10.16°$$

and

$$c_u = 30/\cos(10.16°) = 30.48 \text{ kN/m}^2$$

**Example 9.6** A vane of 80 mm diameter and 160 mm height has been pushed into an *in situ* soft clay at the bottom of a bore hole. The torque required to rotate the vanc was 76 N-m. Determine the undrained shcar strength of the clay. After the test, the vane was rotated several times, and the ultimate torque was found to be 50 N-m. Estimate the sensitivity of the clay.

***Solution***
Rearranging Eq. 9.18,

$$c_u = \frac{T}{\pi(1/2bd^2 + 1/6d^3)}$$

$$= \frac{76\times10^{-3}}{\pi\left[1/2\times0.160\times(0.08)^2 + 1/6(0.08)^3\right]} = 40.5 \text{ kN/m}^2$$

Therefore, the undisturbed undrained strength = 40.5 kN/m$^2$.
After re-moulding, the undrained shear strength is obtained as

$$c_u = \frac{50\times10^{-3}}{\pi\left[1/2\times0.160\times(0.08)^2 + 1/6(0.08)^3\right]} = 26.65 \text{ kN/m}^2$$

Therefore, the re-moulded undrained strength = 26.65 kN/m$^2$

$$\text{Sensitivity } S_t = \frac{\text{Undisturbed undrained strength}}{\text{Re-moulded undrained strength}} = \frac{40.50}{26.65} = 1.52$$

**Example 9.7** A normally loaded deposit of undisturbed clay extends to a depth of 15 m from the ground surface, with the groundwater level at 5 m depth from the ground surface. Laboratory test on the clay shows a plasticity index of 68% saturated and dry unit weights as 19.2 and 14.5 kN/m$^3$, respectively. An undisturbed specimen for unconfined compressive strength is taken at 10 m depth. What unconfined compressive strength is it likely to exhibit?

***Solution***
The effective overburden pressure at a depth of 10 m $p_0 = 5\times14.5 + (19.2-9.807)\times 5 = 119.45$ kN/m$^2$
Now,

$$c_u/p_0 = 0.11 + 0.0037\, I_p$$

or

$$c_u = p_0(0.11 + 0.0037\times I_p)$$

$$= 119.45(0.11 + 0.0037\times68) = 43.2 \text{ kN/m}^2$$

Unconfined compressive strength $= 43.2\times2 = 86.4$ kN/m$^2$

**Example 9.8** A triaxial sample was subjected to an ambient pressure of 200 kN/m$^2$, and the pore pressure recorded was 50 kN/m$^2$. In this state, the sample was found to be fully saturated. Then, the cell pressure was raised to 300 kN/m$^2$. What would be the value of pore pressure? Then, a deviator stress of 150 kN/m$^2$ was applied to the sample. Assuming the pore pressure parameter $A$ to be 0.50, determine the pore pressure value.

***Solution***

Change in cell pressure $\Delta\sigma_3 = 300 - 200 = 100$ kN/m$^2$

We know that the pore pressure parameter $B$ is given as

$$B = \frac{\Delta u_w}{\Delta\sigma_3}$$

As the soil is saturated, $B = 1$. Therefore,

$$u_w = \Delta\sigma_3 = 100 \text{ kN/m}^2$$

Pore pressure after increase of cell pressure $= u_w$

$$= (u_w)_0 + \Delta u_w$$

$$= 50 + 100 = 150 \text{ kN/m}^2$$

The pore pressure change due to deviator stress change is given as

$$\Delta u_w = A(\Delta\sigma_1 - \Delta\sigma_3)$$

$$= 0.5 \times 150 = 75 \text{ kN/m}^2$$

Pore pressure after application of deviator stress $= (u_w)_0 + \Delta u_w$

$$= 150 + 75 = 225 \text{ kN/m}^2$$

**Example 9.9** A sample of stiff clay was tested in a triaxial shear test and found to have a cohesion $c$ of 200 kN/m$^2$ and angle of shearing resistance of 37°. What will be its effective compressive strength if a horizontal hole is made with zero confining stress and a water pressure of 220 kN/m$^2$?

***Solution***

Plot the Coulomb envelope taking $c = 200$ kN/m$^2$ and $\phi = 37°$, as shown in Fig. 9.40.

From the origin, draw a line with angle $\theta_f = 45° + \phi/2(63.5°)$ to intersect the envelope at D.

At D, erect a perpendicular to cut the $x$-axis at E. With E as centre and ED as radius, draw a circle which will pass through the origin and intersect the $x$-axis at F. Point F represents the total compressive stress.

From the plot, $\sigma_1 = 760$ kN/m$^2$. The pore water pressure $u_w = 220$ kN/m$^2$. Therefore,

Effective compressive strength $= \sigma_1' = \sigma_1 - u_w$

$$= 760 - 220 = 540 \text{ kN/m}^2$$

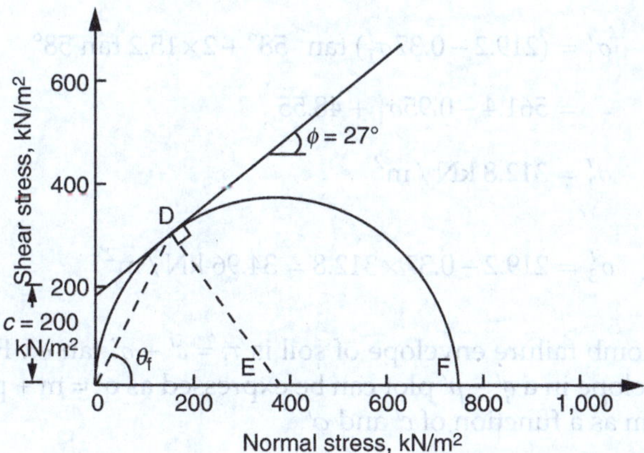


Fig. 9.40

**Example 9.10** In a triaxial test on a saturated clay, the sample was consolidated under a cell pressure of 160 kN/m$^2$. After consolidation, the cell pressure was increased to 350 kN/m$^2$, and the sample was then failed under undrained condition. If the shear strength parameters of the soil are $c' = 15.2$ kN/m$^2$, $\phi' = 26°$, $B = 1$, and $A_f = 0.27$, determine the effective major and minor principal stresses at the time of failure of the sample.

***Solution***

Pore pressure at the time of failure $(u_w)_f = (u_w)_0 + \Delta u_w$.

Pore pressure soon after increase in cell pressure

$$= B\Delta\sigma_3 = 1\times(350-160) = 190 \text{ kN/m}^2$$

Therefore,

$$(u_w)_f = 190 + A(\sigma_1 - \sigma_3)$$
$$= 190 + 0.27(\sigma_1 - \sigma_3)$$

Now,

$$\sigma_3' = \sigma_3 - (u_w)_f = 350 - 190 - 0.27(\sigma_1 - \sigma_3)$$

Since $\sigma_1 - \sigma_3 = \sigma_1' - \sigma'_3$

$$\sigma_3' = 160 - 0.27(\sigma_1' - \sigma_3') = 160 - 0.27\sigma_1' + 0.27\sigma_3'$$

or

$$\sigma_3' = \frac{160 - 0.27\sigma_1'}{1 - 0.27} = 219.2 - 0.37\sigma_1'$$

We know from Eq. 7.8 that

$$\sigma_1' = \sigma_3' \tan^2\theta_f + 2c'\tan\theta_f$$

or

$$\theta_f = 45 + \frac{\phi'}{2} = 45° + \frac{26°}{2} = 58°$$

Therefore,

$$\sigma_1' = (219.2 - 0.37\sigma_1')\tan^2 58° + 2\times 15.2 \tan 58°$$

$$= 561.4 - 0.95\sigma_1' + 48.55$$

$$\sigma_1' = 312.8 \text{ kN/m}^2$$

and

$$\sigma_3' = 219.2 - 0.37\times 312.8 = 34.96 \text{ kN/m}^2$$

**Example 9.11** Coulomb failure envelope of soil is $\tau_f = c' + \sigma' \tan\phi'$. For the same soil, the modified failure envelope in a $q' - p'$ plot can be expressed as $q' = m + p' \tan\alpha$. Express $\alpha$ as a function of $\phi'$ and m as a function of $c'$ and $\phi'$.

***Solution***

From Fig. 9.41

$$\sin\phi' = \frac{AB}{AC} = \frac{AB}{CO+OA} = \frac{\frac{(\sigma_1' - \sigma_3')}{2}}{c'\cot\phi' + \frac{(\sigma_1' + \sigma_3')}{2}}$$

Rearranging in a straight-line equation form, we get:

$$\frac{(\sigma_1' - \sigma_3')}{2} = c'\cos\phi' + \frac{(\sigma_1' - \sigma_3')}{2}\sin\phi'$$

This can be represented as:

$$q' = m + p' \tan\alpha$$

Then $m = c' \cos\phi'$
and $\tan\alpha = \sin\phi'$
i.e., $\alpha = \tan^{-1}(\sin\phi')$.

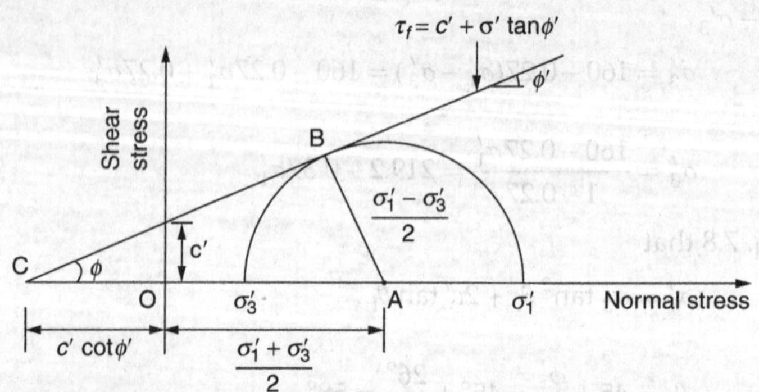


Fig. 9.41

## POINTS TO REMEMBER

9.1 Peak shear strength of a soil is the maximum shear stress that can be resisted by the soil.

9.2 In a strain-hardening soil the peak shear strength is referred to the point at which significant shear strain starts. In a strain-softening soil the peak shear strength is well defined and after a continued large strain, the shearing resistance attains a constant level, and the corresponding shearing resistance is called the residual shear strength.

9.3 Coulomb suggested a simple linear relationship of shear strength ($\tau_f = c' + \sigma' \tan \phi'$) controlled by the shear strength parameters $c'$ (cohesion intercept) and $\phi'$ (angle of shearing resistance). These parameters, $c'$ and $\phi'$, are not constants for a given soil but depend on factors like void ratio, initial stress, pore pressure, drainage conditions, and type of test.

9.4 Coulomb's failure condition is stated in another form – that if the Mohr's circle for a state of stress at a point is tangential to Coulomb's failure loci, then that point is said to be in a state of failure. This is known as Mohr–Coulomb failure criterion, and the failure loci is called Mohr–Coulomb failure envelope.

9.5 Peak shear strength parameters depend on per cent clay contact, drainage condition, type of loading, consolidation history, stress level, anisotropy, and other environmental factors. Residual shear strength is independent of many of the above factors, but $\phi'_r$ decreases with increasing clay content and $c'_r$ is almost zero.

9.6 Shear strength parameters are designated, based on the drainage conditions of the test: (i) when no drainage is allowed during both the stages (i.e., consolidation stage and shear stage) of the test, then the parametres are referred to as undrained shear strength parameters ($c_u$ and $\phi_u$); (ii) drainage is allowed during the consolidation stage and no drainage during the shear stage and the parameters are referred to as consolidated–undrained shear strength parameters $c_u$ and $\phi_u$; and (iii) drainage is allowed during both the stages, and the shear strength parameters are effective or drained shear strength parameters $c'_d$ and $\phi'_d$.

9.7 Undrained shear strength parameters may be applied in field problems where the change in total stress is immediately compensated by a change in pore water pressure. The consolidated–undrained total stress parameters may be taken as a rough guide to the undrained shear strength of the soil. Whenever pore water pressure can change independently of the total stresses, effective shear strength parameters should be used.

9.8 Change in pore water pressure resulting from changes in the state of stress (which occurs in many practical problems involving deformation of soil masses) are estimated using Skempton's pore pressure parameters $A$ and $B$.

9.9 Changes in stress and environment with time may result in cohesive soil having a higher strength in the undisturbed state than in the re-moulded state. The term *sensitivity of cohesive soils* is used to describe this difference in strength, which is given by the ratio of the undisturbed strength to the re-moulded strength.

9.10 Shear strength of granular soils depends on particle shape, orientation, surface roughness, grain-size distribution, initial void ratio, and effective stresses. Only the drained strength parameter ($\phi'_d$) is useful in practice because of high permeability. Undrained strength is insignificant except during an earthquake.

## QUESTIONS

### Objective Questions

9.1 State whether the following statements are true or false. Justify your choice.
1. Pore pressure parameter $B$ is a function of strain at failure.
2. In a partially saturated soil, the $\chi$-parameter is always greater than unity.
3. Mohr's failure theory does not consider the effect of intermediate principal stress.
4. The shear strength of granular material is affected largely by the initial void ratio.
5. Consolidated–undrained and drained tests on normally consolidated clays show zero cohesion.

9.2 When a saturated soil mass is loaded under undrained conditions, the load according to Terzaghi's concept is
(a) Borne entirely by water
(b) Borne entirely by soil solids
(c) Shared equally by soil solids
(d) Shared between soil solids and water proportional to their volumes

9.3 The unconsolidated–undrained strength of an intact saturated clay does not depend on
(a) Major principal stress (b) Maximum shear stress
(c) Minor principal stress (d) Maximum prinicipal stress ratio

9.4 Cohesionless soils whose natural void ratios are above the critical will ______ in volume during shear.
(a) Decrease (b) Remain constant
(c) Increase (d) Initially increase and then remain constant

9.5 For a very heavily over-consolidated clay sample, the probable value of pore pressure parameter $A$ at failure is likely to be
(a) 0.85 (b) 0.35
(c) 0.0 (d) 0.20

9.6 Consider the following statements.
1. Volume change is considered usually as three-dimensional effect.
2. Plastic flow is the mass movement of soil laterally.
3. Shear failure occurs where part of the soil mass moves as a single unit along a defined surface of rupture.

Of these statements,
(a) 1 and 2 are correct (b) 2 and 3 are correct
(c) 3 and 1 are correct (d) 1 alone is correct

9.7 The Mohr theory of rupture implies that there is no influence on failure by
(a) Minor principal stress ($\sigma_3$) (b) Intermediate principal stress ($\sigma_2$)
(c) Major principal stress ($\sigma_1$) (d) Principal stress difference ($\sigma_1 - \sigma_3$)

9.8 Identify the incorrect statement. Effective stress shear parameters of a clay can be obtained from
(a) Drained triaxial shear test
(b) Drained direct shear test
(c) Consolidated–undrained triaxial shear test with pore water pressure measurements
(d) Unconsolidated–undrained triaxial shear test with pore water pressure measurements

9.9 Both the shear stress and the normal stress on the plane of failure are measured directly in
(a) Triaxial shear test (b) Vane shear test
(c) Direct shear test (d) Unconfined compression test

9.10 The unconfined compression test is a special type of
(a) Vane shear test
(b) Unconsolidated–undrained triaxial test
(c) Unconsolidated–undrained direct shear test
(d) Drained triaxial test

## *Descriptive Questions*

9.11 Explain why the angle of shearing resistance of a soil is not always the same as the angle of internal friction.

9.12 Discuss the type of laboratory triaxial test you would recommend to be carried out for the following field problems:
1. The initial stability of a foundation on saturated clay
2. The stability of a clay foundation of an embankment, the rate of construction being such that some consolidation of the clay occurs
3. The long term-stability of a slope in stiff fissured clay

9.13 Explain why the angle of the failure plane observed in a shear test might differ more often from that predicted from a Mohr diagram at failure.

9.14 Define critical void ratio. Explain the shear behaviour of a soil whose void ratio is less than the critical void ratio.

9.15 An undrained triaxial shear test is conducted on a fully saturated cohesionless soil specimen. How does this shearing resistance compare with that from a drained test if the initial condition of the specimen was dense?

9.16 How are the drainage conditions adopted in a triaxial shear test realized in the field?

9.17 Explain how a negative pore water pressure develops in a consolidated–undrained test on a over-consolidated clay.

9.18 Discuss at least three factors which govern the shear strength of cohesionless soils.

## EXERCISE PROBLEMS

9.1 Describe the state of samples A to D when the Mohr circles describing their state of stresses are as follows: For A the Mohr circle is a dot on the normal stress axis, for B the Mohr circle is too small to touch the failure envelope, and for C the Mohr circle is so large that part of the circle is above the failure envelope.

9.2 The following observations were taken in a series of tests in a 60 mm×60 mm direct shear box.

| Normal load (N) | Max. shear load (N) |
|---|---|
| 100 | 150 |
| 200 | 230 |
| 300 | 308 |
| 400 | 380 |

Compute the values of $c$ and $\phi$ for the soil. Also, find the orientation of principal planes for Test 1.

9.3 A soil sample taken from a sand deposit is tested in a direct shear test and found to have an angle of shearing resistance of 32° at a unit weight of 19.8 kN/m$^3$. Estimate the shear strength of the soil in a horizontal plane, at a depth of 4.5 m below the ground surface. A structure proposed to be built on the site will cause the vertical and shear stresses to increase by 65 and 50 kN/m$^2$, respectively, at the same depth. Check whether the shearing stress exceeds the shear strength of the soil at that depth. Will the structure be stable if the groundwater rises to the ground surface?

9.4 A clay stratum of 10 m depth is just sheared due to an adjacent structure leaning against it. The lateral pressure at 10 m depth is estimated to be 115 kN/m$^2$. If the clay is completely saturated and the failure might be under undrained condition, what is the shear strength of the clay? The saturated unit weight of the clay is 22.5 kN/m$^3$.

9.5 A series of unconsolidated–undrained triaxial tests on saturated clay yielded the following results:

| Lateral stress (kN/m$^2$) | 20 | 40 | 60 |
|---|---|---|---|
| Deviator stress at failure (kN/m$^2$) | 22.2 | 22.0 | 22.2 |

Determine the shear strength parameters $c_u$ and $\phi_u$.

9.6 In an undrained triaxial test on a sample of saturated clay, the confining pressure is maintained at 100 kN/m$^2$. At what vertical applied pressure in addition to the confining pressure should the sample fail? Cohesion $c$ for the soil = 50 kN/m$^2$. What is the additional vertical pressure required for failure if the confining pressure is 200 kN/m$^2$ instead of 100 kN/m$^2$?

9.7 Two consolidated–undrained triaxial tests are performed on undisturbed silty clay samples. One sample is consolidated under an all-round pressure of 170 kN/m$^2$, and when subjected to shear, it fails at an added axial stress of 124 kN/m$^2$. The pore water pressure at the time of failure is found to have a positive value of 110 kN/m$^2$. The second sample is consolidated under a pressure of 430 kN/m$^2$ and fails at an added axial stress of 310 kN/m$^2$. The corresponding pore water pressure at the time of failure is 270 kN/m$^2$. Find the total and effective shear strength parameters of the soil. Also, compute the pore pressure parameter $A$ at the time of failure. Take $B = 1$.

9.8 The following results were obtained during a consolidated–undrained triaxial test with pore pressure measure:

| | Test no. 1 | Test no. 2 | Test no. 3 |
|---|---|---|---|
| Chamber pressure (kN/m$^2$) | 100 | 200 | 300 |
| Principal stress difference (kN/m$^2$) | 150 | 190 | 240 |
| Pore pressure at failure (kN/m$^2$) | 50 | 75 | 135 |

Estimate the effective shear strength parameters by plotting a modified Mohr–Coulomb plot.

9.9 A consolidated–undrained test yields shear strength parameters as $c' = 12.8$ kN/m$^2$ and $\phi' = 28°$. The pore pressure parameters at the time of failure were $B = 1.0$ and $A_f = -0.18$. If a specimen with identical initial condition fails at 165 kN/m$^2$ in an unconfined compression test, estimate the initial value of the suction pore pressure in the soil.

9.10 A direct shear test conducted on identical soil specimens gave the following results:

| Normal stress (kN/m$^2$) | Shear stress (kN/m$^2$) |
|---|---|
| 50 | 40 |
| 100 | 70 |

Determine the shear strength parameters. If an undrained triaxial test was conducted on the same soil and at the same density and water content with a cell pressure of 75 kN/m$^2$, estimate the deviator stress at failure.

9.11 An undisturbed sample with a unit weight of 16 kN/m$^3$ has been extracted from a depth of 7 m below the ground surface for shear testing in the laboratory. What stress condition would you apply to the specimen prepared from this sample in (i) the direct shear apparatus and (ii) the triaxial equipment before shearing the specimen to simulate the conditions in the ground? During sampling, no groundwater table is encountered but the groundwater rises to the ground surface during the rainy season.

9.12 A sample of dense sand is tested in the following tests:

1. Direct shear with a normal stress of 150 kN/m$^2$
2. Triaxial shear with a confining pressure of 150 kN/m$^2$

Find the maximum shear stress at failure in both the cases if the angle of internal friction of the sand is 36°. Explain your results with the Mohr–Coulomb envelope.

9.13 A 7 m high embankment is constructed with a soil whose effective shear strength parameters are $c' = 62$ kN/m$^2$, $\phi' = 22°$, and $\gamma = 15.8$ kN/m$^3$. The pore pressure parameters as determined from triaxial tests are $A = 0.39$ and $B = 0.94$. Find the shear strength of the soil at the base of the embankment just after the fill has been raised from 7 to 10 m. Assume that the dissipation of pore water pressure during this stage of construction is negligible and that the lateral pressure at any point is held at half the vertical pressure.

9.14 A soil specimen measuring 85 mm in length and 40 mm in diameter fails at a load of 90 N when subjected to the unconfined compression test. The axial deformation at the time of failure is found to be 6 mm. What is the shear strength of the soil sample?

9.15 In an unconfined compression test on a saturated clay, the unconfined compressive strength was found to be 160 kN/m$^2$, It is known that the same soil showed an angle of shearing resistance of 10° in a consolidated–undrained test. What is the percentage of error, and is it conservative or unconservative to use $c_u = q_u/2$? Give reasons.

9.16 A vane shear test was performed on a uniform normally consolidated clay at a certain depth and the following data obtained:

| | | |
|---|---|---|
| $T = 65$ N-m | $d = 65$ mm | $b = 110$ mm |
| $w_L = 68.4$ % | $w_p = 34.1$% | |

Estimate the effective overburden pressure present at the depth of testing.

9.17 An attempt has been made to measure the anisotropic undrained shear strength of a soft clay by conducting two vane shear tests with different sizes of vanes. The heights of the vanes were 150 and 300 mm, and they had the same diameter of 75 mm. The measured torques were 65 N-m and 150 N-m for 150 mm and 300 mm height vanes, respectively. Determine the undrained cohesion in the vertical and horizontal directions. Assume the shear stress distribution on the ends as parabolic.

# 10

# Laboratory Measurement of Soil Properties

## CHAPTER HIGHLIGHTS

Preparation of dry soil samples – Specific gravity of soil solids: density bottle method, pycnometer or gas jar method – Water content – In-place density: core-cutter method, sand replacement method – Grain-size distribution: sieve analysis, pipette method and hydrometer method – Liquid limit – Plastic limit – Shrinkage factors – Linear shrinkage – Permeability: constant head, falling head – Free swell – Proctor compaction – Density index – Consolidation – Unconfined compression – Direct shear – Triaxial shear – California Bearing Ratio (CBR)

## 10.1 INTRODUCTION

All geotechnical engineering problems in civil engineering are solved by a combination of theoretical knowledge and practical knowledge of the geology and history of the site under consideration and of the knowledge of geotechnical properties of the soil or rock obtained from laboratory and field tests. The problems associated with the construction of structures may have different aspects, such as settlement predictions, strength requirements, stability, and effects of groundwater. The data needed to evaluate the salient features are obtained from a site investigation and testing programme and are interrelated. Quite often pilot site investigations, involving sampling, are carried out to establish the type and characteristics of the soil to be studied. From such a pilot investigation, a laboratory or field testing programme is then decided upon, which considers the size of samples, the quality of samples, and the frequency of sampling with respect to variations in the soil. After deciding the appropriate laboratory or field tests, the major site investigation is undertaken with the distinct purpose of fulfilling a considered test programme. It is the responsibility of the engineer in charge to present the test results and findings in a report form with all relevant details in a sensible and concise way.

In this chapter, the essentials of laboratory testing of soils, including methods of sample collection, sample preparation, testing methods, data collection, and presentation of results, are given in a lucid form. The testing techniques explained in this chapter follow quite closely the Indian Standards for testing soils, and the relevant facilities are found in most institutions in India. Further, only testing methods which are relevant to an undergraduate course in geotechnical engineering are dealt with. However, reference has been made to certain tests which have some bearing on the main tests.

## 10.2 TEST NO.1: PREPARATION OF DRY SOIL SAMPLES FOR VARIOUS TESTS

**Scope**

To prepare dry soil samples from bulk soil samples received from the field

**Apparatus**

Wooden mallet
Trays
Pulverizing apparatus – mortar and rubber-covered pestle
Sampler – a suitable riffle-sampler
Sieves – sizes 19, 9.5, 4.75, and 2 mm and 425 μm
Oven with accurate temperature control in the range 105 to 110°C or other suitable apparatus
Balance of 10 kg capacity, 100 g sensitivity
Balance of 1 kg capacity, 1 g sensitivity
Balance of 250 g capacity, 0.01 g sensitivity

**Procedure**

1. Dry the bulk field soil samples in air or in the sun. In wet weather, use a drying apparatus with temperature not exceeding 60°C. Break the clods with a wooden mallet to hasten drying.
2. Remove organic matter, tree roots, pieces of bark, and shells, and make a note of them. But for the organic content or lime content test use the whole sample.
3. Dry in ovens with temperature not exceeding 10°C. Depending on the type of test (Table 10.1) the temperature of the oven is adjusted. Do not resort to chemical drying of samples.
4. Break the big clods with a wooden mallet, and do further pulverization in the mortar with the pestle.
5. Sieve the pulverized soil through a specified sieve depending on the type of test (Table 10.1). Repeat the pulverization till the required quantity is collected. Take care not to break up the individual soil particles.
6. Decide the actual quantity of the soil sample and the bulk field sample (Table 10.1) based on the type of test. When a small representative quantity is required from a bigger soil mass, obtain the same by quartering or riffling.

## 10.3 TEST NO.2: SPECIFIC GRAVITY OF SOIL SOLIDS

**Scope**

To determine the specific gravity of soil solids

**Apparatus**

Density bottle with stopper – 50 ml capacity
Gas jar with rubber bung – 1,000 ml capacity
Ground glass plate or a plastic slip cover
Mechanical shaking apparatus
Balances of 0.001 and 0.2 g accuracies
Hot water bath
Oven with accurate temperature control in the range 105 to 110°C
Thermometer – 0 to 50°C range with 0.1° graduation
Vacuum desiccator – 200 to 250 mm

**Table 10.1** Quantity of soil sample required for various tests

| Sl. no. | Name of test | Type of drying | Amount of soil sample required for test | Degree of pulverization (passing IS sieve size) |
|---|---|---|---|---|
| 1. | Specific gravity | Oven | 50 g for fine-grained soils, 400 g for others | 2 mm |
| 2. | Water content | Oven | As given in test | – |
| 3. | Grain size analysis | Air | As given in test | – |
| 4. | Liquid limit | Air | 270 g | 425 μm |
| 5. | Plastic limit | Air | 60 g | -do- |
| 6. | Shrinkage factors | Air | 100 g | -do- |
| 7. | Linear shrinkage | Air | 450 g | -do- |
| 8. | Permeability | Oven | 2,500 g for 100 mm dia mould<br>5,000 g for 200 mm dia mould | 9.5 mm |
| 9. | Free swell index | Oven | 20 g | 425 μm |
| 10. | Compaction | Air | 6,000 g | 19 mm |
| 11. | Consolidation | Air/oven | 500 g | – |
| 12. | Unconfined compressive strength | Oven | 1,000 g | – |
| 13. | Vane shear | Air/oven | 250 g | – |
| 14. | Direct shear | Air/oven | 120 g | 4.75 mm |
| 15. | Triaxial compression | Oven | 1,000 g to 5,000 g | – |
| 16. | Swelling pressure | Air/oven | 2 kg | 2 mm |

*Note:* All oven drying is done for 24 hours at 110 ± 5°C except for tests 1, 2, 8, 9, and 16, which are dried at 105°C to 110°C.
*Source:* IS: 2720 – Part 1, (1983).

Source of vacuum – good filter pump or vacuum pump
Spatula or glass rod
Wash bottle

**Procedure**

**(a)** ***Laboratory method: Density bottle method for fine-grained soils***

1. Weigh a clean and dry (dried at 105°C to 110°C and cooled in the desiccator) density bottle with the stopper to the nearest 0.001 g ($M_1$).
2. Transfer about 5 to 10 g of soil sample, oven dried at 105°C to 110°C and passing 2 mm IS test sieve, to the bottle and weigh to the nearest 0.001 g ($M_2$).
3. Add air-free distilled water into the bottle such that the soil in the bottle is fully covered.
4. Keep the bottle (without stopper) and contents in a vacuum desiccator and evacuate gradually till the pressure is reduced to 20 mm of mercury and allow it to remain for at least 1 hour until no further loss of air is apparent. Alternatively, remove the entrapped air by heating the bottle and contents in a water bath or sand bath.
5. Release the vacuum and remove the bottle from the desiccator.
6. Stir the soil in the bottle using a spatula or glass rod and carefully transfer the particles adhering to the blade or rod by washing off with a few drops of air-free water.
7. Put back the bottle and contents in the desiccator and once again evacuvate.
8. Repeat Steps 4 to 7 till no more air is released from the soil.
9. Remove the bottle and contents from the desiccator and add air-free water till the bottle is full. Insert the stopper and immerse the bottle up to the neck in a water bath till the contents of the bottle attain the constant temperature of the bath. If there is a decrease in the volume, add more water and keep it in the water bath again. Repeat the procedure till the volume remains constant.
10. Remove the stoppered bottle from the bath, wipe the outside, and weigh to the nearest 0.001 g ($M_3$).
11. Clean the bottle, top it with air-free water, fix the stopper, and immerse it in the water bath till it attains the constant temperature of the bath. Make good if there is any decrease in volume. Repeat the procedure till the volume remains constant.
12. Remove the stoppered bottle from the bath, wipe the outside, and weigh to the nearest 0.001 g ($M_4$).
13. Repeat Steps 2 to 12 for two more samples.

**(b)** ***Field method : Gas jar method for all soils***

1. Weigh a clean and dry gas jar and ground glass plate or plastic slip cover to the nearest 0.2 g ($M_1$).
2. Add 200 g of fine-grained soil or 400 g of medium- or coarse-grained soil into the glass jar and weigh the jar along with the slip cover to the nearest 0.2 g ($M_2$).
3. Add 500 ml of water at room temperature (±2°C) to the soil. Set aside the jar and its contents for 4 hours in case of medium- or coarse-grained soils.
4. Push a rubber stopper into the mouth of the jar and shake by hand until the particles are separated and are in suspension. Shake again in a shaking apparatus for a period of 20 to 30 minutes.
5. Remove the stopper and wash off the stopper and the top of the jar carefully into the jar. Also disperse any froth with a fine spray of water.

6. Add some water to the gas jar, allow the soil to settle, and then fill the jar up to the brim with more water.
7. Place the slip cover on the top, taking care not to trap any air under the plate.
8. Carefully dry the outside of the jar and weigh the jar and contents along with the slip cover to the nearest 0.2 g ($M_3$).
9. Clean the gas jar, fill completely with water up to the brim, place the slip cover, dry the outside, and weigh to the nearest 0.2 g ($M_4$).
10. Repeat Steps 2 to 9 for two more samples.

**Computations**

Specific gravity at $T$°C,

$$G = \frac{M_2 - M_1}{(M_4 - M_1) - (M_3 - M_2)}$$

If a liquid other than water is used, then the specific gravity is calculated as follows:

$$G = \frac{G_L(M_2 - M_1)}{(M_4 - M_1) - (M_3 - M_2)}$$

where $G_L$ is the specific gravity of the liquid used at $T$°C.

The specific gravity is usually reported at 27°C (unless otherwise specified) and is calculated thus:

$$(G)_{27°C} = K(G)_{T°C}$$

where

$$K = \frac{\text{Relative density of water at } T°\text{C}}{\text{Relative density of water at 27°C}}$$

The mean value based on three samples is reported to the nearest 0.01. Tests are repeated if the results differ by more than 0.03 from the mean value.

Typical observations of data and test results of specific gravity from the density bottle method are shown in Table 10.2.

**Discussion**

The major source of error is the complete removal of air from the sample. To ensure accurate results, the soil should be left in vacuum for several hours. Soils often contain a substantial

**Table 10.2** Data and results of specific gravity test from density bottle method

| | |
|---|---|
| 1. Mass of density bottle including stopper ($M_1$) g | 36.632 |
| 2. Mass of density bottle + stopper + oven-dried soil ($M_2$) g | 45.842 |
| 3. Mass of density bottle + stopper + oven-dried soil + air-free water ($M_3$) g | 72.013 |
| 4. Mass of density bottle + stopper + air-free water ($M_4$) g | 66.273 |
| 5. Specific gravity at 40°C | 2.654 |
| 6. Specific gravity at 27°C | 2.644 |

proportion of heavy or light particles. Such soils may give erratic values of specific gravity, and the tests have to be repeated a sufficient number of times to obtain a reasonable average. Presence of organic matter may decrease the specific gravity.

For soils containing soluble salts, kerosene or white spirit may be preferred in place of water. Conventionally, oven-dried soil is used. If there is a possibility of loss of water of hydration at the oven temperature, the soil may be dried at a temperature less than 80°C (IS: 2720 – Part 3/Sec 1, 2, 1980, 1981).

## 10.4 TEST NO.3: WATER CONTENT DETERMINATION BY OVEN-DRYING METHOD

**Scope**

To determine the water content of a given soil by the oven-drying method

**Apparatus**

Non-corrodible air-tight moisture cups
Balance of 0.01 g sensitivity
Oven with accurate temperature control of 110 ± 5°C
Desiccator with a suitable desiccating agent

**Procedure**

1. Weigh a dry and clean moisture cup with lid ($M_1$).
2. Take a representative sample of wet soil in the cup, replace the lid, and weigh ($M_2$).
3. Keep it in the temperature-controlled oven with the lid removed and allow it to dry for 24 hours.
4. When the sample has dried to a constant weight, replace the lid and cool the cup in a desiccator.
5. Weigh the moisture cup with the lid with dried soil ($M_3$).

**Computations**

The water content $w$ of a sample is given as

$$w = \frac{M_2 - M_3}{M_3 - M_1} \times 100\%$$

**Results**

The water content of the soil is represented as a percentage to two significant figures. Typical data and results of water content determination are given in Table 10.3.

**Table 10.3** Data and result of water content determination

| | |
|---|---|
| Moisture cup No. | 9 |
| Mass of cup ($M_1$) g | 19.99 |
| Mass of cup and wet soil ($M_2$) g | 52.31 |
| Mass of cup and dry soil ($M_2$) g | 49.33 |
| Mass of moisture ($M_2 - M_3$) g | 2.98 |
| Mass of dry soil ($M_3 - M_1$) g | 29.34 |
| Water content $w = [(M_2 - M_3)/(M_3 - M_1)] \times 100\%$ | 10.16 |

**Discussion**

The factors which are essential for accurate determination of water content are the mass of the wet representative sample, and the temperature and duration of the drying of sample.

As per IS recommendations (IS: 2720 – Part 2, 1973), the following masses of soil have to be used to provide reasonable results.

| Size of particles more than 90% passing | Minimum mass of soil specimen to be taken for test (g) |
|---|---|
| 425 μm IS sieve | 25 |
| 2 mm IS sieve | 50 |
| 4.75 mm IS sieve | 200 |
| 10 mm IS sieve | 300 |
| 20 mm IS sieve | 500 |
| 40 mm IS sieve | 1,000 |

The effects of temperature and duration are discussed in Chapter 2.

The oven-drying method is recommended by Indian Standards as the standard method. Other methods are the sand bath method, alcohol method, infrared lamp method, torsion balance method and calcium carbide method (IS: 2720 – Part 2, 1973). The latter two methods are rapid methods.

## 10.5 TEST NO. 4: IN-PLACE DRY DENSITY OF SOIL BY CORE-CUTTER METHOD

**Scope**

To determine the in-place dry density of soil by the core-cutter method

**Apparatus**

Cylindrical core-cutter of steel with steel dolly (Fig. 10.1a)
Steel rammer (Fig. 10.1b)
Balance of 1 g sensitivity
Steel rule
Palette knife
Straight edge
Apparatus for water-content determination
Apparatus for extracting samples from the cutter

**Procedure**

1. Measure inner dimensions (nearest to 0.25 mm) of the core-cutter and find its volume ($V_c$).
2. Weigh the cutter without the dolly ($M_c$).
3. Clear and level a small area of about 300 mm$^2$ where the in-place density is to be determined.
4. Place the cutter on the levelled surface. Keep the dolly on the cutter and advance the cutter into the subsoil, using the rammer until about 15 mm of the dolly protrudes above the surface.

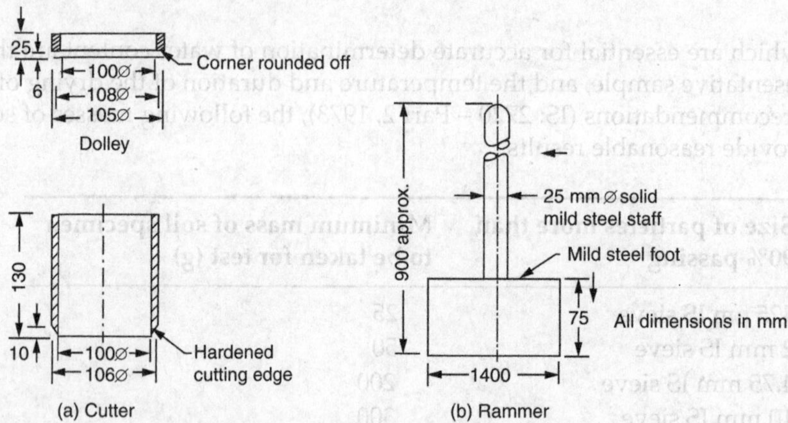


**Fig. 10.1** Apparatus for core-cutter method (*Source:* IS: 2720 – Part 29, 1975)

5. Dig the soil around the cutter using a spade or pickaxe and bodily remove the cutter with soil allowing some soil to project from the lower end of the cutter. Trim the top and bottom of the cutter by means of a palette knife and straight edge.
6. Weigh the cutter with soil, and without dolly ($M_{sc}$).
7. Remove the soil from the cutter and determine the water content of the soil.

**Computations**

The bulk density $\rho$ is given as

$$\rho = \frac{M_{sc} - M_c}{V_c} \text{ g/cc}$$

and the dry density $\rho_d$ is given as

$$\rho_d = \frac{\rho}{1 + \frac{w}{100}} \text{ g/cc}$$

The bulk unit weight $\gamma$ is given as

$$\gamma = 9.807\rho \text{ kN/m}^3$$

and the dry unit weight $\gamma_d$ is given as

$$\gamma_d = 9.807\rho_d \text{ kN/m}^3$$

**Results**

The dry density (in g/cc) and the dry unit weight (in kN/m$^3$) of the soil are reported to the second decimal place and the water content of the soil (per cent) to two significant figures. A knowledge of the specific gravity of the soil-solids of the soil will enable us to find the void ratio and the degree of saturation of the soil. A typical data sheet with relevant results is shown in Table 10.4.

**Table 10.4** Data and test results of in-place density

| | |
|---|---|
| Length of core-cutter | 130 mm |
| Diameter of core-cutter | 100 mm |
| Volume of core-cutter ($V_c$) | 1,021 ml |
| Mass of core-cutter ($M_c$) | 1,335 g |
| Mass of core-cutter with wet soil ($M_{sc}$) | 3,056 g |
| Mass of wet soil ($M_{sc} - M_c$) g | 1,721 |
| Bulk density [$\rho = (M_{sc} - M_c)/V_c$] | 1.69 g/cc |
| Bulk unit weight ($\gamma = 9.807\rho$) | 16.6 kN m$^2$ |
| Moisture cup | No. 306 |
| Mass of cup | 30.85 |
| Mass of cup and wet soil | 49.67 g |
| Mass of cup and dry soil | 46.15 g |
| Water content | 23.0% (wt.) |
| Dry density $\rho_d = \dfrac{\rho}{1 + (w/100)}$ | 1.37 g/cc |
| Dry unit weight ($\gamma_d = 9.807\rho_d$) | 13.48 kN/m$^3$ |

**Discussion**

The core-cutter method is convenient and quick; it works best on fine-grained soils but cannot be used on stony or non-cohesive soils. For the purpose of this test, a soil is a fine-grained soil if not less than 90% of it passes a 4.75 mm IS sieve. The Indian Standards (IS: 2720 – Part 29, 1975) recommend repeat determinations (at least three) and averaging out of results. Further, the number of determinations should be such that an additional test will not alter the average significantly. This method is less accurate than the sand replacement method.

For determination of the bearing capacity of soils, for calculation of the overburden pressure in settlement computations, and for stability analysis of natural slopes, the in- place density of natural soil is needed. In all earth dam and embankment projects, the in-place density is used to check the compaction criterion, and hence this test is usually referred to as the control test.

## 10.6 TEST NO. 5: IN-PLACE DRY DENSITY OF SOIL BY THE SAND REPLACEMENT METHOD

**Scope**

To determine the in-place dry density of soil by the sand replacement method.

**Apparatus**

Sand pouring cylinder (Fig. 10.2a)
Tools for excavating holes
Cylindrical calibration can (Fig. 10.2b)
Balance of 1 g sensitivity
Glass plate – 450 mm$^2$, 9 mm thick
Metal tray or container

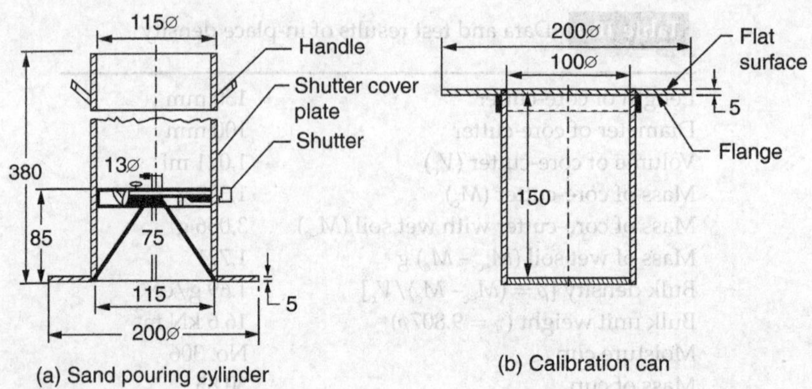


**Fig. 10.2** Apparatus for sand replacement method (*Source:* IS: 2720 – Part 28, 1974)

Metal tray with hole – 300 mm$^2$, 40 mm deep, with a 100 mm hole in the centre
Clean and closely graded natural sand passing the 1 mm IS sieve but retained on the 600 μm IS sieve
Apparatus for water content determination

**Procedure**

1. Fill the pouring cylinder with clean sand till the level of sand is about 10 mm from the top, and weigh ($M_1$). Maintain this mass constant throughout the test, for which the pouring cylinder has to be calibrated.
2. Place the pouring cylinder on a glass plate and close the tap when the conical portion has been filled.
3. Collect the sand on a glass plate carefully and weigh the sand. Repeat Steps 1 to 3 at least three times and take the average mass of sand filling the cone ($M_2$).
4. Measure the internal dimensions of the calibration can and find its volume. Fill the can with water up to the brim and find the mass. From this mass of water, find the volume and check the previous value obtained based on the measurement of internal dimensions. Let the volume be $V$.
5. Place the pouring cylinder concentrically on top of the calibration can with initial mass $M_1$. Open the shutter and allow the sand to fill it. Tap the cylinder to ensure that the can and the conical portion are completely filled with sand. Weigh the cylinder.
6. Repeat Step 5 at least thrice and record the average mass $M_3$ of the cylinder after filling the cone and the can.
7. Clean and level an area of 450 mm$^2$ of the soil to be tested.
8. Place the square tray with a central hole on the prepared surface, excavate a circular hole of 100 mm diameter and 150 mm depth. Carefully collect all the excavated soil, and find its mass ($M_s$). In fine-grained soils push a core-cutter into the soil until its edge is flush with the levelled surface. Remove the soil within the core-cutter approximately up to a depth of 100 mm and collect and weigh the soil ($M_s$). Keep the core-cutter in position during the rest of the test procedure.

9. Take some representative soil for water content determination.
10. Remove the tray and place the pouring cylinder concentrically on the hole with initial mass $M_1$. Open the shutter and allow the sand to fill it. Tap the cylinder to ensure that the hole and the conical portion are completely filled with sand. Weigh the cylinder ($M_4$).

**Computations**

Mass of sand filling calibration can $M_a = (M_1 - M_3 - M_2)$ g

Bulk density of sand $\rho_{sd} = \dfrac{M_a}{V}$ g/cc

Mass of sand required to fill the excavated hole $M_b = (M_1 - M_4 - M_2)$ g

Bulk density of soil $\rho = (M_s/M_b) \times \rho_{sd}$ g/cc

Bulk unit weight $\gamma = 9.807\rho$ kN/m$^3$

Dry density of soil $\rho_d = \left(\dfrac{\rho}{1+(w/100)}\right)$ g/cc

Dry unit weight, $\gamma_d = 9.807\rho_d$ kN/m$^3$

**Results**

The dry density (in g/cc) and unit weight (in kN/m$^3$) of the soil are reported to the second decimal place and the water content of the soil (per cent) to two significant figures. A typical data sheet with relevant results is shown in Table 10.5.

**Discussion**

As the in-place dry density of a soil varies from point to point, Indian Standards (IS: 2720 – Part 28, 1974) recommend repeat tests at different locations close to each other and taking the mean value. However, the number of determinations should be such that an additional test should not make a significant difference in the mean value.

In granular soils with little or no fines, there is a possibility of error because of the slumping of the sides of the excavated hole. It is customary to check the calibration bulk density during each day's work as there is a possibility of bulking of sand due to atmospheric humidity.

For fine- and medium-grained soils, a pouring cylinder of 3 litre capacity is used, and for fine-, medium-, and coarse-grained soils a greater capacity of 16.5 litres is used. Here soils with particles less than 2 mm in size are considered as fine-grained soils. Further, if the depth of excavation exceeds 150 mm for some reason or the other, a different calibration can with a depth comparable to the depth of the hole has to be chosen.

The sand replacement method is relatively slow but can be used on any type of soil.

Two more methods of determining the in-place density are available, viz., the ring and water replacement method (IS: 2720 – Part 3, 1971) and the rubber-balloon method (IS: 2720 – Part 34, 1972). The ring and water replacement method is suitable for coarse-grained soils, including gravels, cobbles, boulders, and rocks, while the rubber-balloon method is suitable for compacted or firmly bonded soils.

**Table 10.5** Data and test results of in-place density determination

| | | |
|---|---|---|
| (a) | Calibration | |
| | Mass of pouring cylinder with sand before pouring ($M_1$) | 11,686 g |
| | Average mass of sand filling cone only ($M_2$) | 1,093 g |
| | Volume of calibration can ($V$) | 1,178 ml |
| | Average mass of pouring cylinder and sand after filling can and cone ($M_3$) | 8,850 g |
| | Mass of sand filling calibration can ($M_a = M_1 - M_3 - M_2$) g | 1,743 g |
| | Bulk density of sand $\rho_{sd} = \dfrac{M_a}{V}$ | 1.48 g/cc |
| (b) | Measurement of soil density | |
| | Mass of excavated soil from the hole ($M_s$) | 2,807 g |
| | Mass of pouring cylinder and sand after filling hole and cone ($M_4$) | 8,245 g |
| | Mass of sand required to fill the excavated hole $M_b = (M_1 - M_4 - M_2)$ | 2,348 g |
| | Bulk density of soil $\rho = \dfrac{M_s}{M_b} \times \rho_{sd}$ | 1.77 g/cc |
| | Bulk unit weight, $\gamma = 9.807\rho$ | 17.35 kN/m$^3$ |
| | Moisture cup | No. 18 |
| | Mass of cup | 16.95 g |
| | Mass of cup and wet soil | 29.96 g |
| | Mass of cup and dry soil | 29.30 g |
| | Water content | 5.34% (wt.) |
| | Dry density of soil $\rho_d = \left(\dfrac{\rho}{1+(w+100)}\right)$ | 1.68 g/cc |
| | Dry unit weight $\gamma_d = 9.807\rho_d$ | 16.48 kN/m$^3$ |

## 10.7 TEST NO.6: GRAIN-SIZE DISTRIBUTION BY SIEVE ANALYSIS

**Scope**

To determine the grain-size distribution of a soil by sieve analysis

**Apparatus**

Balance of 0.1 g sensitivity

Sieves – 100 mm, 63 mm, 20 mm, 10 mm, 4.75 mm, 2.4 mm, 1.2 mm, 600 μm, 300 μm, 150 μm, and 75 μm IS sieves

Oven with accurate temperature control in the range from 105 to 110°C

Trays and buckets

Brushes for cleaning sieves

Mortar with a rubber-covered pestle

Mechanical sieve shaker

Reagents – sodium hexametaphosphate or a mixture of sodium hydroxide and sodium carbonate, or any other dispersing agent

**Procedure**

1. Prepare the soil sample received from the field, as suggested in Test No. 1.
2. Take a certain quantity of soil* (IS: 2720 – Part 4, 1975) and separate the soil fraction passing and retained on the 4.75 mm sieve.
3. Conduct a separate sieve analysis test for each fraction.
4. Sieve the soil retained on the 4.75 mm sieve by hand sieving through the following set of sieves: 100 mm, 63 mm, 20 mm, 10 mm, and 4.75 mm. Agitate the sieve while sieving such that the soil sample rolls in an irregular motion over the sieve. Rub the sample with a rubber pestle, if necessary, and re-sieve to ensure that only individual particles are retained.

| Maximum size of material present in substantial quantities (mm) | Mass of soil to be taken for test (kg) |
|---|---|
| 80 | 60 |
| 40 | 25 |
| 25 | 13 |
| 20 | 6.5 |
| 12.5 | 3.5 |
| 10 | 1.5 |
| 6.3 | 0.75 |
| 4.75 | 0.40 |

5. Record the mass of material retained on each sieve. If the soil contains more than about 20% of gravel particles with cohesive particles adhering to them, then wash the gravel on the 4.75 mm sieve with sodium hexametaphosphate solution and record the correct mass of soil retained on the 4.75 mm sieve, and thereby record the correct mass of soil passing the 4.75 mm sieve.
6. Sieve the soil passing the 4.75 mm sieve through the following sieves: 2 mm, 1 mm, 600 μm, 300 μm, 150 μm, and 75 μm. Arrange the sieves in descending order of sieve openings with the 2 mm sieve at the top. Place the cover and a receiver at the top and bottom of the sieves, respectively. Keep the entire set of sieves on a sieve shaker and allow the sample to be sieved for a minimum period of 10 minutes.
7. Record the mass of material retained on each sieve.

**Computations**

1. Percentage retained on any sieve $= \dfrac{\text{Mass of soil retained}}{\text{Total soil mass}} \times 100$
2. Cumulative percentage retained on any sieve = sum of percentages retained on all coarser sieves
3. Percentage finer $N = 100 -$ (cumulative percentage retained)

---

*If more than 500 g of soil passes the 4.75 mm sieve, take about 500 g of soil and calculate the combined percentage finer ($N$) for the second sieve analysis from the relation $N = N' \times (M_2/M_1)$, where $N'$ is the percentage finer for the second sieve analysis based on the soil taken for the second sieve test, $M_1$ is the mass of soil taken for the entire sieve analysis (as taken in Step 2), and $M_2$ is the mass of soil passing 4.75 mm sieve.

**Results**

The grain-size distribution curve is plotted by taking the percentage finer on the arithmetic scale and the sieve opening on the logarithmic scale. Typical test results are presented in Table 10.6, and the grain-size distribution curve is given in Fig. 10.3.

**Discussion**

A wet sieve analysis has to be performed if the material passing the 4.75 mm sieve contains more clay-size particles. For this, the soil is soaked in a dispersing agent. The dispersing agent is prepared by mixing 2 g of sodium hexametaphosphate or 1 g of sodium hydroxide with 1 g of sodium carbonate in 1 litre of water. The soaked soil specimen is washed through the nest of sieves. The collected material in each sieve is dried and weighed.

## 10.8 TEST NO. 7: GRAIN-SIZE DISTRIBUTION BY PIPETTE METHOD

**Scope**

To determine the grain-size distribution of a soil by the pipette method

**Apparatus**

Sampling pipette – as illustrated in Fig. 2.5 with a capacity of approximately 10 ml and fit enough to arrange to a required depth as shown in Fig. 10.4
Glass sedimentation tubes – 50 mm in diameter, 350 mm long, marked at 500 ml volume, with rubber bungs to fit a minimum of two numbers
Weighing bottles – fitted with round stoppers or crucibles with suitable lids, approximately 25 mm in diameter and 50 mm high. Mass of bottles is found to the nearest 0.001 g
Constant temperature bath – capable of being maintained at 27 ± 0.1°C with provision to immerse the tube up to the 500 ml mark
Stirring apparatus – mechanical stirrer with a speed of 8,000 to 10,000 rpm when loaded and with dispersion cups with baffle rod
Sieves – 2 mm, 425 μm, 75 μm IS sieves
Balance of 0.001 g sensitivity
Oven with an accurate temperature control in the range from 105 to 110°C
Stopwatch
Desiccator
Evaporating dish
Conical beaker – 650 ml or 1 litre capacity
Funnel – Buchner or Hirch about 70 mm in diameter
Filter flask – 500 ml capacity
Measuring cylinder – 100 ml capacity
Pipette – 25 ml capacity
Glass filter funnel – about 100 mm in diameter
Wash bottle
Filter paper
Blue litmus paper
Glass rod – 4 to 5 mm in diameter and 150 to 200 mm long
Thermometer – 0 to 50°C, accurate to 0.5°C
Reagents – hydrogen peroxide – 20 volume solution

**Table 10.6** Sieve analysis test results

| IS sieve no. | Particle size (mm) | Mass of soil retained (g) | Percentage retained | Cumulative percentage retained | Percentage finer |
|---|---|---|---|---|---|
| 100 mm | 100.0 | 0.0 | 0.0 | 0.0 | 100.00 |
| 63 mm | 6.3 | 5.5 | 1.1 | 1.1 | 98.90 |
| 20 mm | 20.0 | 4.5 | 0.9 | 2.0 | 98.00 |
| 10 mm | 10.0 | 5.8 | 1.16 | 3.16 | 96.84 |
| 4.75 mm | 4.75 | 29.0 | 5.80 | 8.96 | 91.04 |
| 2.0 mm | 2.0 | 70.2 | 14.04 | 23.0 | 77.00 |
| 1.0 mm | 1.0 | 124.6 | 24.92 | 47.92 | 52.08 |
| 600 μm | 0.6 | 69.1 | 13.82 | 61.74 | 38.26 |
| 300 μm | 0.3 | 58.2 | 11.64 | 73.38 | 26.62 |
| 150 μm | 0.15 | 82.0 | 16.40 | 89.78 | 10.22 |
| 75 μm | 0.075 | 19.3 | 3.86 | 93.64 | 6.36 |
| Pan | – | 31.8 | | | |

*Note*: Total mass of soil taken is 500 g.

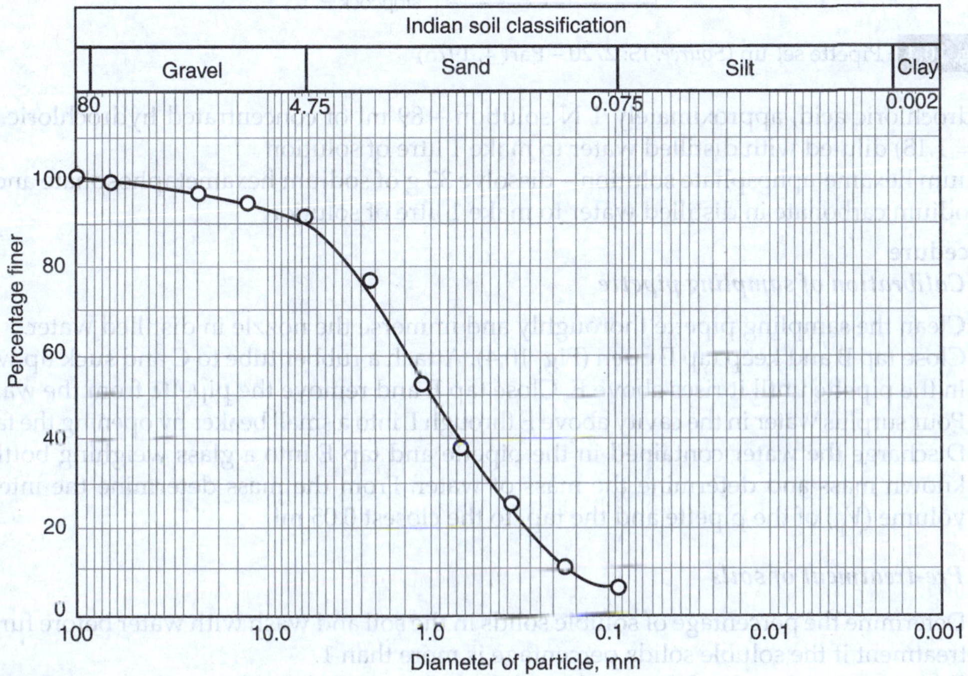


**Fig. 10.3** Grain size distribution curve from sieve anlaysis

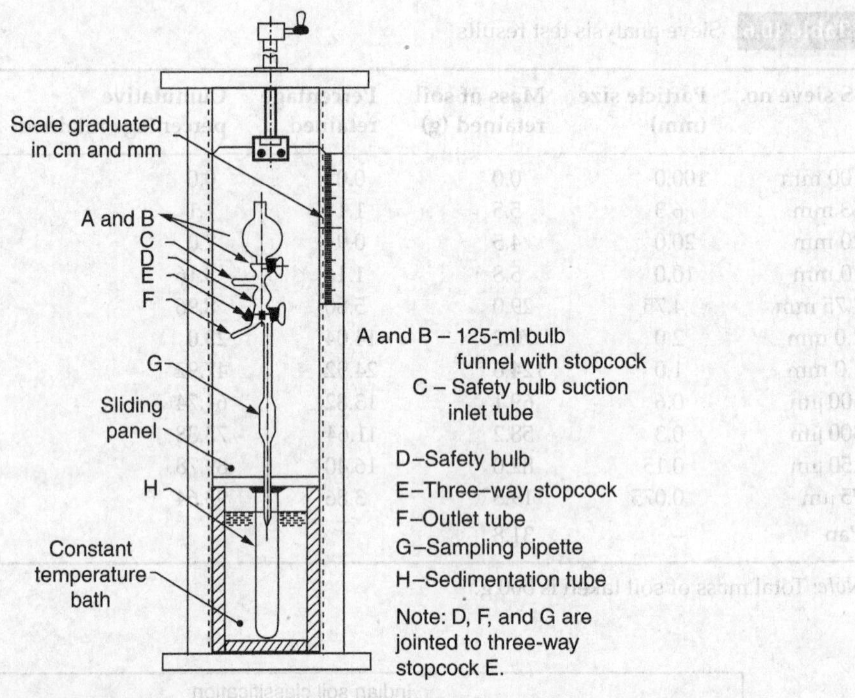


**Fig. 10.4** Pipette set-up (*Source:* IS: 2720 – Part 4, 1975)

Hydrochloric acid, approximately 1 N solution – 89 ml of concentrated hydrochloric acid ($G = 1.18$) diluted with distilled water to make 1 litre of solution

Sodium hexametaphosphate solution – dissolve 33 g of sodium hexametaphosphate and 7 g of sodium carbonate in distilled water to make 1 litre of solution

**Procedure**

(a) ***Calibration of sampling pipette***

1. Clean the sampling pipette thoroughly and immerse the nozzle in distilled water.
2. Close tap B and keep tap E open (Fig. 10.4). Attach a rubber tube to C and suck up water in the pipette until it rises above E. Close tap E and remove the pipette from the water.
3. Pour surplus water in the cavity above E through F into a small beaker by opening the tap E.
4. Discharge the water contained in the pipette and tap E into a glass weighing bottle of known mass and determine the mass of water. From the mass determine the internal volume ($V_p$) of the pipette and the tap, to the closest 0.05 ml.

(b) ***Pre-treatment of soils***

5. Determine the percentage of soluble solids in the soil and wash with water before further treatment if the soluble solids percentage is more than 1.
6. Take two samples of soil, 50 g each, and the soil passing the 4.75 mm IS sieve. Determine the water content of one of the samples.
7. Out of the other sample, use the entire 50 g if the soil contains more fine sand or use 20 g if the soil is clayey. Find the mass of soil accurately to 0.001 g ($M_a$) and place it in a 650 ml conical beaker.

8. Add 50 ml of distilled water and gently boil the soil suspension till the volume is reduced to about 40 ml. Add 75 ml of hydrogen peroxide and allow the sample to stand overnight covered with a cover glass.
9. Heat the sample gently, taking care to avoid frothing over. Agitate frequently either by stirring or by shaking the beaker. When vigorous frothing has subsided on addition of fresh hydrogen peroxide, reduce the volume to 30 ml by boiling.
10. If the soil contains calcium carbonate, add 10 ml of hydrochloric acid after cooling the solution obtained in Step 9. Stir the solution with a glass rod for a few minutes and allow it to stand for about 1 hour or for longer periods. Continue the treatment till the solution gives an acid reaction to litmus.
11. Filter the solution, pre-treated with peroxide and acid alone, using the Buchner of Hirch funnel and wash with warm water until the filtrate shows no acid reaction to litmus. Transfer the wet soil to an evaporating dish and wash the funnel and filter paper with minimum water. Dry the contents of the evaporating dish, cool in a desiccator, and weigh accurately. Record the mass of soil remaining after pre-treatment ($M_b$).
12. Omit pre-treatment of the soil if it does not contain calcium compounds or soluble solids and has a low (less than 2%) organic content.

### (c) *Dispersion of soil*

13. Add about 25 ml of sodium hexametaphosphate solution to the mixture, warm gently for about 10 minutes, and then transfer the mixture to the cup of a mechanical stirrer using a jet of water.
14. Stir the soil suspension for 15 minutes.
15. Transfer the suspension through a 75 μm IS sieve placed on a receiver and wash off all traces of suspension adhering to the dispersion cup.

### (d) *Sedimentation*

16. Transfer the suspension, that has passed through the sieve, to a sedimentation tube and make the volume to 500 ml by adding distilled water.
17. Add 25 ml of sodium hexametaphosphate solution in a 500 ml sedimentation tube (comparison tube), and add water to make the level exactly 500 ml.
18. Immerse the sedimentation tube with soil suspension in a constant temperature bath (if used) and note down the temperature of the bath. Fix a rubber bung on the mouth of the sedimentation tube and allow the suspension to attain the temperature of the bath.
19. Remove the sedimentation tube from the constant temperature bath and shake it thoroughly by inverting the tubes several times. Replace the tube in its position in the apparatus and remove the rubber bung carefully without disturbing the tubes.
20. Keep tap E closed and lower the pipette vertically into the suspension until the end is $100 \pm 1$ mm below the surface of the suspension. Take care to lower the pipette into the suspension about 15 seconds before collecting the sample.
21. Open tap E and draw up a sample ($V_p$ ml) into the pipette till the pipette and the bore in the tap E are filled and then close tap E. Complete this operation within a time of 10 seconds.

22. Withdraw the pipette and wash with distilled water the surplus suspension drawn above the bore of the tap E through the outlet tube F, by opening the tap E in such a way as to connect D and F. Allow distilled water to run from bulb funnel A into D and out through F until no solution remains in the suspension. Repeat this operation during each time of sampling.
23. Keep a tared weighing bottle under the end of the pipette and open the tap E so that the contents of the pipette are delivered into the bottle. Wash the inner walls by allowing distilled water to run from bulb A, through E, into the pipette.
24. Repeat Steps 17 to 20 after expiry of a particular time approximately corresponding to particle diameters 0.02 mm, 0.06 mm, 0.002 mm, and 0.001 mm. Take the time of settling to a depth of 100 mm of particles of various diameters for a given temperature from Table 10.7.
25. Place the weighing bottles along with the contents in the oven. After drying, cool in a desiccator and weigh to the nearest 0.001 g. Find the mass of the solid materials in the sample ($M_1$, $M_2$, $M_3$, and $M_4$ for each respective sampling time).
26. Also take a sample of volume $V_p$ from the comparison tube and find the mass of the solid material ($M_s$) in the sample tube.
27. Determine the specific gravity of soil solids from Test No. 2.

**Computations**

1. Loss in mass after pre-treatment is given as

$$P = 100 - \frac{M_b(100 + w)}{M_a}$$

where $P$ is the loss in mass in percentage, $M_b$ the mass of soil after pre-treatment, $w$ the air-dry moisture content of the soil taken for analysis, and $M_a$ the mass of air-dry soil.

2. The diameter of the particle is given as

$$D = \sqrt{\frac{30\eta_w}{980(\rho_s - \rho_w)}}\sqrt{\frac{H_e}{t}}$$

3. The mass of solid material in 500 ml of suspension for each sampling time is given as

$$M_i' \text{ or } M_s' = \frac{M_i \text{ or } M_s}{V_p} \times 500$$

where $M_i'$ is the mass of material in 500 ml from respective samplings (e.g., $M'_1$, $M'_2$ $M'_3$, etc.), $M'_s$ the mass of sodium hexametaphosphate in 500 ml of solution, $M_i$ the mass of material in $V_p$ ml from respective samplings (e.g., $M'_1$, $M_2'$, $M_3'$, etc.), and $M_s$ the mass of sodium hexametaphosphate in $V_p$ ml of suspension.

4. Percentage finer is given as $N' = (M_i' - M_s')/M_b \times 100$
5. Combined gradation may be calculated based on the total soil sample taken for analysis.

**Results**

Typical test results are presented in Table 10.8 and the gradation curve is shown in Fig. 10.5.

**Table 10.7** Rate of settling of particles at various temperatures

| Temp. (°C) Hours (1) | 0.001 (2) | 0.002 (3) | 0.003 (4) | 0.004 (5) | 0.005 (6) | 0.006 (7) | 0.007 (8) | 0.008 (9) | 0.009 (10) | 0.01 (11) | 0.02 (12) | 0.03 (13) | 0.04 (14) | 0.05 (15) | 0.06 (16) | 0.07 (17) | 0.08 (18) | 0.09 (19) | 0.01 (20) |
|---|---|---|---|---|---|---|---|---|---|---|---|---|---|---|---|---|---|---|---|
| | Diameter (mm) (Time for a fall of 100 mm) | | | | | | | | | | | | | | | | | | |
| | | | | | Minutes | | | | | | | Seconds | | | | | | | |
| 10 | 40.80 | 10.90 | 4.53 | 2.55 | 98.0 | 68.0 | 53.0 | 38.3 | 30.2 | 24.5 | 367 | 163 | 92 | 59 | 40.8 | 31.8 | 22.9 | 18.1 | 14.7 |
| 11 | 39.64 | 9.91 | 4.40 | 2.48 | 95.2 | 66.47 | 51.5 | 37.2 | 29.3 | 23.8 | 357 | 158 | 89 | 57 | 39.6 | 30.9 | 22.3 | 17.6 | 14.3 |
| 12 | 38.55 | 9.63 | 4.28 | 2.41 | 92.6 | 64.2 | 53.1 | 36.1 | 28.5 | 23.1 | 347 | 154 | 87 | 55 | 38.5 | 30.0 | 21.7 | 17.1 | 13.9 |
| 13 | 37.48 | 9.37 | 4.16 | 2.34 | 90.0 | 62.5 | 48.7 | 35.1 | 27.7 | 22.5 | 337 | 150 | 84 | 54 | 37.5 | 29.2 | 21.1 | 16.7 | 13.5 |
| 14 | 36.39 | 9.10 | 4.04 | 2.28 | 87.6 | 60.8 | 47.3 | 34.2 | 27.0 | 21.9 | 328 | 146 | 82 | 52 | 36.4 | 28.4 | 20.5 | 16.2 | 13.1 |
| 15 | 35.45 | 8.86 | 3.93 | 2.22 | 85.1 | 59.2 | 46.1 | 33.3 | 26.2 | 21.3 | 319 | 142 | 80 | 51 | 35.4 | 27.6 | 19.9 | 15.8 | 12.5 |
| 16 | 34.49 | 8.62 | 3.83 | 2.16 | 82.8 | 57.6 | 44.9 | 32.4 | 25.5 | 20.7 | 310 | 138 | 78 | 50 | 34.5 | 26.9 | 19.4 | 15.3 | 12.4 |
| 17 | 33.64 | 8.41 | 3.73 | 2.10 | 80.8 | 56.1 | 43.7 | 31.5 | 24.9 | 20.2 | 302 | 134 | 76 | 48 | 33.6 | 26.2 | 18.9 | 14.8 | 12.1 |
| 18 | 32.73 | 8.18 | 3.64 | 2.04 | 78.6 | 54.6 | 42.5 | 30.7 | 24.2 | 19.6 | 294 | 131 | 74 | 47 | 32.7 | 25.5 | 18.4 | 14.5 | 11.8 |
| 19 | 31.89 | 7.98 | 3.54 | 1.99 | 76.6 | 53.2 | 41.4 | 29.9 | 23.6 | 19.1 | 287 | 127 | 72 | 46 | 31.9 | 24.8 | 17.9 | 14.1 | 11.5 |
| 20 | 31.10 | 7.77 | 3.45 | 1.94 | 74.6 | 51.8 | 40.4 | 29.1 | 23.0 | 18.6 | 280 | 124 | 70 | 45 | 31.1 | 24.2 | 17.5 | 13.8 | 11.5 |
| 21 | 30.28 | 7.57 | 3.36 | 1.89 | 72.7 | 50.6 | 39.4 | 28.4 | 22.4 | 18.2 | 273 | 121 | 68 | 44 | 30.3 | 23.6 | 17.0 | 13.4 | 10.9 |
| 22 | 29.55 | 7.38 | 3.28 | 1.85 | 70.9 | 49.3 | 38.4 | 27.7 | 21.8 | 17.7 | 266 | 118 | 66 | 42 | 29.5 | 23.0 | 16.6 | 13.1 | 10.6 |
| 23 | 28.81 | 7.21 | 3.20 | 1.80 | 69.2 | 48.1 | 37.5 | 27.0 | 21.3 | 17.3 | 259 | 115 | 65 | 41 | 28.8 | 22.4 | 16.2 | 12.8 | 10.4 |
| 24 | 28.12 | 7.03 | 3.12 | 1.76 | 67.5 | 46.9 | 36.6 | 26.4 | 20.8 | 16.9 | 253 | 113 | 63 | 40 | 28.1 | 21.9 | 15.8 | 12.5 | 10.1 |
| 25 | 27.78 | 6.86 | 3.05 | 1.72 | 65.9 | 45.8 | 35.7 | 25.8 | 20.3 | 16.5 | 243 | 110 | 62 | 39 | 27.8 | 21.4 | 15.5 | 12.2 | 9.9 |
| 26 | 26.81 | 6.71 | 2.98 | 1.68 | 64.4 | 44.7 | 34.8 | 25.2 | 19.8 | 16.1 | 241 | 107 | 60 | 39 | 26.8 | 20.9 | 15.1 | 11.9 | 9.6 |
| 27 | 26.19 | 6.54 | 2.91 | 1.64 | 62.9 | 43.7 | 34.0 | 24.6 | 19.4 | 15.7 | 236 | 105 | 59 | 38 | 26.2 | 20.4 | 14.7 | 11.6 | 9.4 |
| 28 | 25.6 | 6.40 | 2.84 | 1.60 | 61.4 | 42.7 | 33.2 | 24.0 | 19.0 | 15.4 | 231 | 102 | 58 | 37 | 25.6 | 19.9 | 14.4 | 11.4 | 9.2 |
| 29 | 25.04 | 6.25 | 2.78 | 1.56 | 60.1 | 41.7 | 32.5 | 23.4 | 18.5 | 15.0 | 226 | 100 | 56 | 36 | 25.0 | 19.4 | 14.1 | 11.1 | 9.0 |
| 30 | 24.46 | 6.12 | 2.72 | 1.53 | 58.8 | 40.8 | 31.8 | 22.9 | 18.1 | 14.9 | 221 | 98 | 55 | 35 | 24.5 | 19.0 | 13.8 | 10.8 | 8.8 |
| 31 | 23.95 | 5.98 | 2.66 | 1.50 | 57.5 | 39.9 | 31.1 | 22.4 | 17.7 | 14.4 | 216 | 96 | 54 | 34 | 23.9 | 18.6 | 13.5 | 10.6 | 8.6 |
| 32 | 23.44 | 5.86 | 2.60 | 1.47 | 56.3 | 39.1 | 30.4 | 21.9 | 17.3 | 14.1 | 211 | 94 | 53 | 34 | 23.4 | 18.2 | 13.2 | 10.4 | 8.5 |
| 33 | 22.95 | 5.74 | 2.55 | 1.44 | 55.1 | 38.3 | 29.8 | 21.5 | 17.0 | 13.8 | 206 | 92 | 52 | 33 | 22.9 | 17.8 | 12.9 | 10.2 | 8.4 |
| 34 | 22.50 | 5.62 | 2.56 | 1.41 | 54.0 | 37.5 | 29.2 | 21.1 | 16.6 | 13.5 | 202 | 90 | 51 | 32 | 22.5 | 17.5 | 12.6 | 10.1 | 8.1 |
| 35 | 20.01 | 5.50 | 2.45 | 1.38 | 52.9 | 36.7 | 28.6 | 20.7 | 16.3 | 13.2 | 198 | 89 | 50 | 32 | 22.0 | 17.2 | 12.4 | 9.8 | 7.9 |

*Source:* IS: 2720 – Part 4 (1975).

**Table 10.8** Data and test results of the pipette method

| | |
|---|---|
| Specific gravity of solids | 2.66 |
| Mass of pre-treated soil passing 75-μm sieve taken for analysis ($M_b$) | 25 g |
| Volume of suspension | 500 ml |
| Mass of dispersing agent in 500 ml suspension ($M'_s$) | $\left(\frac{40}{1000}\right)25\times\frac{500}{500}=1$ g |
| Temperature | 20°C |

| Assumed diameter (mm) | Corresponding time from Table 10.7 | Actual diameter (mm) | $M'_i$ (g) | $N'=\frac{M'_i-M'_s}{M_b}\times 100\%$ |
|---|---|---|---|---|
| 0.020 | 280 seconds | 0.02010 | 18.5 | 70.0 |
| 0.006 | 51.8 minutes | 0.00604 | 16.2 | 60.8 |
| 0.002 | 7.77 hours | 0.00201 | 14.3 | 53.2 |
| 0.001 | 31.10 hours | 0.00101 | 13.1 | 48.4 |

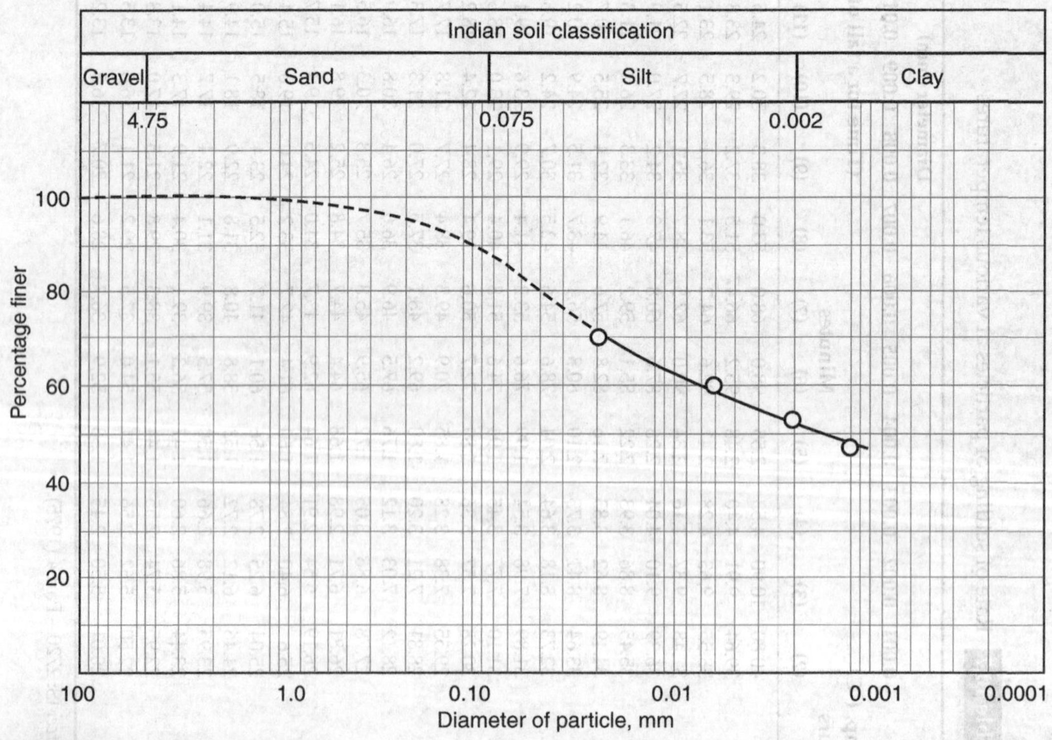


**Fig. 10.5** Grain-size distribution curve from pipette method

**Discussion**

This method is not recommended if less than 10% of the material passes the 75 μm IS sieve. If a constant temperature bath is not available, the test may be performed at room temperature. Note down the temperature and incorporate necessary corrections in the expression for determining the diameter of the particle.

## 10.9 TEST NO. 8: GRAIN-SIZE DISTRIBUTION BY THE HYDROMETER METHOD

**Scope**

To determine the grain-size distribution of soil by the hydrometer method

**Apparatus**

Hydrometer – range 0.995 to 1.03 with an accuracy of 0.0005
Glass measuring cylinders – two of 1,000 ml capacity, 70 mm diameter, and 330 mm height
Thermometer – 0 to 50°C with an accuracy of 5°C
Water bath – maintained at constant temperature
Stirring apparatus
Balance of 0.01 g sensitivity
Oven with accurate temperature control in the range from 105 to 110°C
Stopwatch
Desiccator
Evaporating dishes
Wide-mouth conical flask or conical beaker of 1,000 ml capacity
Buchner or Hirch funnel – 100 mm diameter filter flask
Measuring cylinder – 100 ml capacity wash bottle with distilled water
Filter papers
Reagents – hydrogen peroxide, hydrochloric acid, sodium hexametaphosphate
Blue litmus paper

**Procedure**

(a) *Calibration of hydrometer*

1. Immerse the hydrometer in a graduated jar and note down the increase in volume as read on the graduation, or weigh the hydrometer to the nearest 0.1 g and record the mass in grams as the volume of the hydrometer ($V_h$) in millilitres.
2. Obtain the cross-sectional area of the jar ($A_f$) by dividing the volume between two calibration marks by the distance between the same two marks.
3. Record the distance from the lowest calibration mark on the stem of the hydrometer to each of the major calibration marks [$R_h = 1000(r_h - 1)$, where $r_h$ is the actual reading on the hydrometer stem].
4. Record the distance from the neck of the bulb to the nearest calibration mark.
5. Compute the distance $H_1$ corresponding to a reading $R_h$ as the sum of the distances measured in Steps 3 and 4.
6. Record the distance ($h$) from the neck to the bottom of the bulb.
7. Compute the effective depth $H_e$ corresponding to each of the major calibration marks $R_h$ from the expression

$$H_e = H'_e - \frac{V_h}{2A_f}$$

where

$$H'_e = H_1 + \frac{h}{2}$$

8. Obtain a graphical relationship between $H_e$ and $R_h$ by plotting a smooth curve between them. Use this calibration curve for readings beyond 4 minutes. For readings at timings of 0.5, 1, 2, and 4 minutes, use the expression $H'_e = H_1 + h/2$ to obtain the calibration curve.
9. Insert the hydrometer in a 1,000 ml measuring cylinder with 700 ml water and note the readings corresponding to the upper and lower limits of the meniscus. Record the difference between the two readings as the meniscus correction $C_m$.

(b) ***Pre-treatment of soil***

10. Determine the percentage of soluble solids. If it is more than 1, wash the soil with water before further treatment.
11. From the air-dried sample passing the 4.75 mm sieve, obtain two samples of mass 50 or 100 g.
12. Use 50 g of soil for a clayey soil and 100 g for a soil with a little sand along with fines.
13. Determine the water content of one sample and place the other in a wide-mouth conical flask after finding the mass of the soil ($M_a$) accurately.
14. Add 150 ml of hydrogen peroxide and stir the mixture gently with a glass rod and allow it to stand overnight.
15. Now gently heat the conical flask carefully to avoid frothing over.
16. Reduce the volume to about 50 ml by boiling after subsidence of the frothing.
17. Cool the mixture and add about 50 ml of hydrochloric acid if the soil contains calcium compounds. Stir the solution for a few seconds and then allow it to stand for some more hours. Add more acid if the soil still contains a considerable amount of calcium. Check the solution for acid reaction to litmus.
18. Omit pre-treatment of the soil if it does not contain calcium compounds or soluble solids.
19. Filter the mixture prepared at Step 17 and wash with warm water until no acid reaction is noticed.
20. Transfer the damp soil to an evaporating dish of known mass, wash the filter paper and funnel, and transfer them to the dish.
21. Dry the evaporating dish with its contents in an oven at 105 to 110°C.
22. After drying, cool in a desiccator and find the mass of the soil. Record it as mass of soil after pre-treatment ($M_b$).

(c) ***Dispersion of soil***

Add 50 ml of sodium hexametaphosphate solution to the soil and follow the steps given in pipette method (Test no. 7) for dispersion.

**(d)** ***Sedimentation analysis***

23. Transfer to a 1,000 ml measuring cylinder the suspension that has passed through the sieve and add distilled water to make the volume of the suspension exactly 1,000 ml.
24. Add 50 ml of sodium hexametaphosphate solution in a 1,000 ml measuring cylinder, add distilled water to make 1,000 ml, and maintain it at the same temperature as that of the soil suspension. This cylinder has to be used when the hydrometer is not in use. The combined correction C of temperature and dispersing agent is taken from this cylinder.
25. Close the mouth of the measuring cylinder containing the suspension, shake vigorously, and finally invert end to end.
26. After shaking, immediately make the measuring cylinder stand in the constant temperature bath (if used) and start a stopwatch.
27. Insert the hydrometer, take readings after periods of 0.5, 1, 2, and 4 minutes, and thereafter rinse it and place it in the distilled water cylinder (prepared in Step 24).
28. Re-insert the hydrometer and take readings at intervals of 8, 15, and 30 minutes and 1, 2, and 4 hours after starting the test. Remove the hydrometer, rinse it and place it in the distilled water cylinder after every reading. After 4 hours take hydrometer readings once or twice within 24 hours. Take one reading at the end of 24 hours.
29. Record the temperature once during the first 15 minutes and then after every subsequent reading. Take hydrometer readings in the distilled water cylinder (prepared in Step 24) corresponding to these temperatures and calculate the combined correction, $C$. The correction for temperature is positive for temperatures greater than the calibrated temperature of the hydrometer. The dispersion agent correction is negative. The combined correction is negative for the range of laboratory temperatures.

**Computations**

1. Compute the loss in mass in pre-treatment (as in Test No. 7).
2. The diameter of the particles in the suspension, at any time $t$, is given as

$$D = \sqrt{\frac{30\eta_w}{980(\rho_s - \rho_w)}}\sqrt{\frac{H_e}{t}}$$

where $t$ is the time elapsed between the beginning of sedimentation and the taking of the hydrometer reading, in minutes.
The hydrometer reading corrected for the meniscus is given as

$$R_h = R_h' + C_m$$

where $R_h'$ is the hydrometer reading at the upper rim of the meniscus.
3. The percentage finer $N'$, based on the mass $M_b$, is given as

$$N' = \frac{100G}{M_b(G-1)}\bar{R}_h$$

where $\bar{R}_h = R_h + C$.

**Results**

The results of the grain-size analysis are presented in a graph by taking the diameter of the particle in a log scale and the percentage finer in an arithmetic scale. Details of calibration data and calibration curve are shown in Table 10.9 and Fig. 10.6, respectively. Test data and results are given in Table 10.10, and the graduation curve is presented in Fig. 10.7.

**Discussion**

As per Indian Standards, this method is considered as a subsidiary method. It is not recommended if less than 10% of the material passes the 75 μm IS sieve.

The hydrometer and pipette methods give fairly accurate results, but both are time-consuming. A new device called the plummet balance is in use in different laboratories (Malhotra and Chandra, 1982). A plummet balance is nothing but a specific gravity balance. It is based on the principle that the depth of immersion of a plummet in a suspension is

**Table 10.9** Data for calibration of hydrometer
Volume of hydrometer $(V_h) = 102.3$ ml
Area of cross-section of jar $(A_j) = 42.3$ cm$^2$
Height of bulb $(h) = 11.8$ cm

| Sl. no. | Hydrometer reading | $R_h = 1000(r_h - 1)$ | $H_1$ | $H'_e = H_1 + h/2$ | $H_e = H'_e - (V_h/2A_j)$ |
|---|---|---|---|---|---|
| 1. | 0.995 | –5.00 | 8.90 | 14.80 | 13.59 |
| 2. | 1.000 | 0.00 | 6.45 | 12.35 | 11.14 |
| 3. | 1.005 | 5.00 | 4.05 | 9.95 | 8.74 |
| 4. | 1.010 | 10.00 | 1.65 | 7.55 | 6.34 |

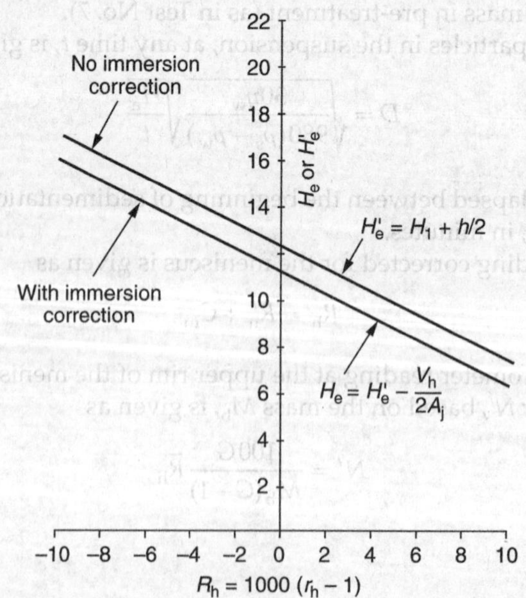


**Fig. 10.6** Hydrometer calibration curve

**Table 10.10** Data and test results from the hydrometer test
Mass of pre-treated soil passing 75 μm sieve taken for analysis ($m_b$) = 20 g
Specific gravity of soil particles of minus 75 μm ($G$) = 2.65
Meniscus correction ($C_m$) = +0.05

| Lapsed time (min) | Hydrometer reading ($r'_h$) | $R' = 1000\,(r'_h - 1)$ | Temperature (°C) | $R_h = R'_h + C_m$ | $C$ | $H'_e$ or $H_e$ (cm) | $D$ (mm) | $R_h = r_h + C$ | Percent finer, $N'$ |
|---|---|---|---|---|---|---|---|---|---|
| 0.25 | 1.00925 | 9.25 | 27.5 | 9.75 | –0.55 | 7.6 | 0.0686 | 9.20 | 73.9 |
| 0.50 | 1.00875 | 8.75 | 27.5 | 9.25 | –0.55 | 7.8 | 0.0491 | 8.70 | 69.9 |
| 1.00 | 1.00725 | 7.25 | 27.5 | 7.75 | –0.55 | 8.4 | 0.0361 | 7.20 | 57.8 |
| 2.00 | 1.00600 | 6.00 | 27.5 | 6.50 | –0.55 | 9.0 | 0.0264 | 5.95 | 47.8 |
| 4.00 | 1.00500 | 5.00 | 27.5 | 5.50 | –0.55 | 9.6 | 0.0193 | 4.95 | 39.8 |
| 8.00 | 1.00300 | 3.00 | 27.5 | 3.50 | –0.55 | 10.2 | 0.0103 | 2.95 | 23.7 |
| 15.00 | 1.00150 | 1.50 | 27.5 | 2.00 | –0.55 | 10.2 | 0.0103 | 1.45 | 10.0 |
| 30.00 | 1.00100 | 1.00 | 28.0 | 1.50 | –0.55 | 10.4 | 0.0073 | 0.95 | 7.6 |
| 60.00 | 1.00075 | 0.75 | 28.0 | 1.25 | –0.55 | 10.5 | 0.0052 | 0.70 | 5.6 |
| 75.00 | 1.00050 | 0.50 | 28.0 | 1.00 | –0.55 | 10.7 | 0.0018 | 0.45 | 3.6 |
| 140.00 | 1.00025 | 0.25 | 27.5 | 0.75 | –0.55 | 10.8 | 0.0011 | 0.20 | 1.6 |

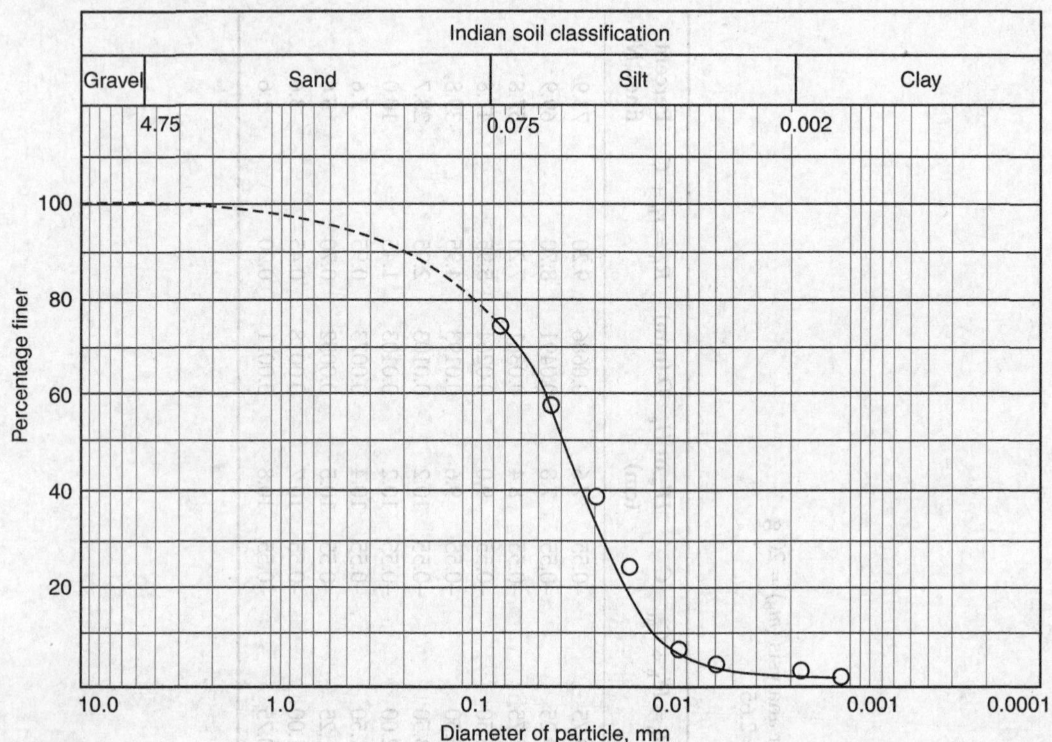


Fig. 10.7 Grain size distribution curve from hydrometer method

countered by the movement of a needle-shaped beam on a graduated scale, and the reading of the beam on the scale represents the percentage fraction of a particular size in a given time of fall (Marshall, 1956). Malhotra and Chandra (1982) used this apparatus on six different fine-grained soils and found it more suitable in clayey soils. This method has the advantage of being quick, but it is yet to be standardized.

## 10.10 TEST NO. 9: LIQUID LIMIT OF SOIL

**Scope**

To determine the liquid limit of a soil using a mechanical liquid limit device

**Apparatus**

Mechanical liquid limit device – as illustrated in Fig. 10.8a
Grooving tools – as illustrated in Fig. 10.8b
Evaporating dish
Spatula
Palette knives
Balance of 0.01 g sensitivity
Wash bottle or beaker
Apparatus for water content determination

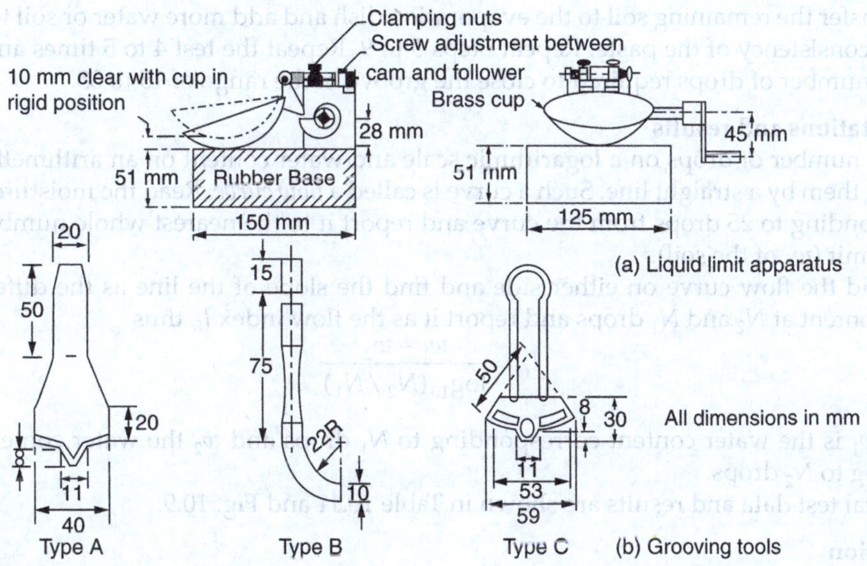


Fig. 10.8 Liquid limit device and tool (*Source:* IS: 2720 – Part 5, 1970)

**Procedure**

1. Clean and check the liquid limit device to see that it is in working order. Also clean the grooving tools.
2. Use a gauge or the handle of the grooving tool and the adjusting plate of the liquid limit device such that the cup falls exactly 10 mm for one revolution of the handle. After adjustment, secure the plate by tightening the screw.
3. Weigh about 120 g of the soil sample passing the 425 μm IS sieve and transfer it to an evaporating dish or on to the flat glass plate.
4. Mix the soil with distilled water to form an uniform paste.
5. Take a portion of the paste in the cup (of the liquid limit device) above the spot where the cup rests on the base, squeeze down, spread into position, and level to a depth of 10 mm at the point of maximum thickness.
6. Divide the soil in the cup by firmly running the grooving tool (Type A) diametrically such that a sharp groove is formed. Use Type B or C grooving tools for non-adhesive soils.
7. Turn the crank at the rate of two revolutions per second until the two parts of the soil come in contact with the bottom of the groove along a distance of about 12 mm and record the number of drops needed.
8. Add a small quantity of soil from the evaporating dish, mix it thoroughly, and repeat Steps 6 and 7 until two consecutive runs give the same number of drops for closure of the groove.
9. Take a representative slice of the soil sample, about the width of the spatula, at right angles to the groove, including that portion of the groove in which the soil flowed together, for water content determination.

10. Transfer the remaining soil to the evaporation dish and add more water or soil to change the consistency of the paste. Repeat Steps 5 to 9. Repeat the test 4 to 5 times and record the number of drops required to close the groove in the range of 15 to 35.

**Computations and results**

Plot the number of drops on a logarithmic scale and water content on an arithmetical scale and join them by a straight line. Such a curve is called a *flow curve*. Read the moisture content corresponding to 25 drops from the curve and report it to the nearest whole number as the liquid limit ($w_L$ of the soil).

Extend the flow curve on either side and find the slope of the line as the difference in water content at $N_2$ and $N_1$ drops and report it as the flow index $I_f$, thus

$$I_f = \frac{w_1 - w_2}{\log_{10}(N_2 / N_1)}$$

where $w_1$ is the water content corresponding to $N_1$ drops and $w_2$ the water content corresponding to $N_2$ drops.

Typical test data and results are shown in Table 10.11 and Fig. 10.9.

**Discussion**

In general, natural soils used for liquid and plastic limit tests (given elsewhere) should not be oven dried. Drying causes the particles to sub-divide and also causes the removal of absorbed water. It is reported (Lambe, 1951) that oven-dried organic soils tend to show a lower $w_L$ value than those of soils that have not been dried.

Thus, sometimes natural soils are directly used for a test without oven drying if all the particles are less than 425 μm in size. In case some stones are present, the wet soil is rubbed through the 425 μm IS sieve till a sufficient quantity of soil is collected to run the test.

A soil with a low clay content has to be tested immediately after thorough mixing with water. In case of tearing of the sides of the groove or slipping of the soil, the groove may be cut in stages. Instead of flowing, some soils tend to slide; in such cases discard the result and report that the liquid limit could not be obtained (IS: 2720 – Part 5, 1970).

Another method of finding the liquid limit is the use of the cone penetrometer method, which was discussed in Chapter 2. The mechanical liquid limit device has been recognized as a routine test. The cone penetrometer has been reported to have more advantages compared with the mechanical device (IS: 2720 – Part 5, 1970); yet it has not been accepted as a routine test.

**Table 10.11** Test data and results of liquid limit test

| Determination no. | 1 | 2 | 3 | 4 | 5 |
|---|---|---|---|---|---|
| Number of drops | 16 | 19 | 22 | 27 | 31 |
| Moisture cup no. | 36 | 63 | 54 | 90 | 81 |
| Mass of cup (g) | 17.33 | 17.82 | 15.06 | 17.40 | 15.71 |
| Mass of cup with wet soil (g) | 23.40 | 24.83 | 21.13 | 22.96 | 22.40 |
| Mass of cup with dry soil (g) | 21.45 | 22.67 | 19.29 | 21.38 | 20.62 |
| Water content (%) | 47.33 | 44.54 | 43.50 | 39.70 | 36.25 |
| Liquid limit from plot (%) | 4 | | | | |
| Flow index from plot | 32.6 | | | | |

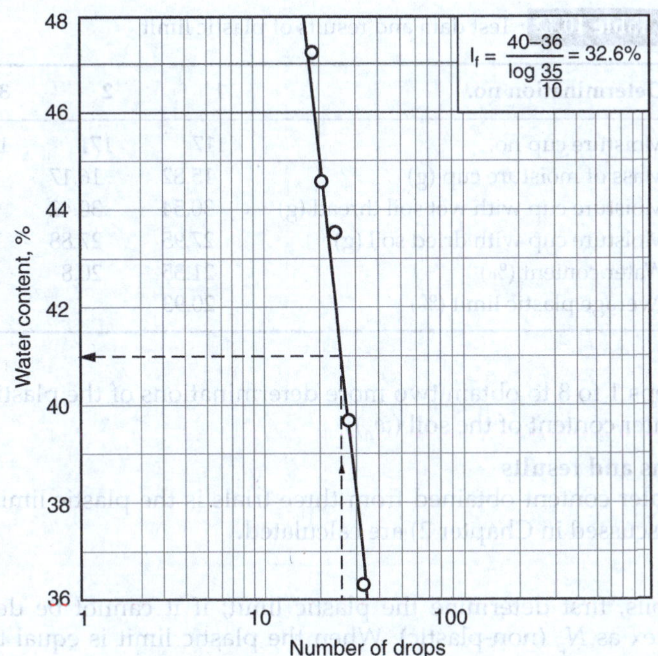


Fig. 10.9 Liquid limit flow curve

## 10.11 TEST NO. 10: PLASTIC LIMIT OF SOIL

**Scope**

To determine the plastic limit of a soil

**Apparatus**

Evaporation dish or flat glass plate
Palette knife or spatula
Surface for rolling-ground – glass plate about 200 mm × 150 mm
Balance of 0.01 g sensitivity
Rod – 3 mm in diameter and about 100 mm long
Apparatus for water content determination

**Procedure**

1. Take about 20 g of soil passing through the 425 μm IS sieve in an evaporating dish or glass plate. Add distilled water and thoroughly mix such that the soil mass becomes plastic enough to be easily moulded with the fingers. For clayey soils, allow sufficient time for moisture equilibrium.
2. Take about 8 g of this wet soil, make a ball out of it, and roll it on the glass plate with the palm of the hand to form a thread of uniform diameter. Continue the rolling till the thread is of 3 mm diameter. Knead the soil together to form a uniform mass and roll again. Continue the process of rolling and kneading until the thread just crumbles at 3 mm diameter.

**Table 10.12** Test data and results of plastic limit

| Determination no. | 1 | 2 | 3 |
|---|---|---|---|
| Moisture cup no. | 117 | 171 | 126 |
| Mass of moisture cup (g) | 15.82 | 16.17 | 16.26 |
| Moisture cup with wet soil thread (g) | 30.54 | 30.42 | 30.41 |
| Moisture cup with dried soil (g) | 27.95 | 27.88 | 27.99 |
| Water content (%) | 21.35 | 20.8 | 20.63 |
| Average plastic limit (%) | 20.93 | | |

3. Repeat Steps 1 to 3 to obtain two more determinations of the plastic limit. Also, find the natural water content of the soil ($w_n$).

**Computations and results**

The mean water content obtained from three trials is the plastic limit of the soil ($w_p$). The indices (as discussed in Chapter 2) are calculated.

**Discussion**

For sandy soils, first determine the plastic limit; if it cannot be determined, report the plasticity index as $N_p$ (non-plastic). When the plastic limit is equal to or greater than the liquid limit, the plasticity index is reported as zero.

Typical data and results of a plastic limit test are shown in Table 10.12.

## 10.12 TEST NO. 11: SHRINKAGE FACTORS OF SOIL

**Scope**

To determine the shrinkage limit, shrinkage ratio, shrinkage index, and volumetric shrinkage of soils

**Apparatus**

Evaporating dishes
Spatula
Shrinkage dish – 45 mm diameter and 15 mm height
Straight edge – 150 mm in length
Glass plates, plain and with metal prongs – 75 mm × 75 mm, 3 mm thick
Glass cup – 50 to 55 mm in diameter and 25 mm in height
Sieve – 425 μm IS sieve
Balance of 0.1 g sensitivity
Mercury
Desiccator – with any desiccating agent other than sulphuric acid

**Procedure**

1. Clean the shrinkage dish and weigh it ($M_1$). Fill the dish with mercury. Remove the excess mercury by pressing a glass plate over the top of the dish. Weigh the mercury, divide it by the density of mercury, and obtain the volume of the dish which is the volume of the wet soil ($V$).

2. Take about 30 g of soil, passing the 425 μm IS sieve, in an evaporating dish and thoroughly mix with water of an amount slightly greater than the liquid limit. The soil–water mixture should be capable of flowing if allowed to drop.
3. Coat the inside of the shrinkage dish with a thin layer of grease and fill one-third of the dish with soil–water mixture. Tap the dish on a firm surface, cushioned by several layers of blotting paper, rubber sheet, or similar material. In three operations, completely fill the dish. Strike off the excess soil paste with a straight edge, clean the outside surface, and weigh ($M_2$).
4. Dry the soil pat in air until the colour changes from dark to light, and then dry it in a temperature-controlled oven. After drying, cool it in air and weigh the shrinkage dish and dry the soil pat ($M_3$).
5. Fill the glass cup with mercury and remove the excess mercury by pressing the glass plate with three prongs firmly over the top of the cup.
6. Place the glass cup with mercury in a large evaporating dish and place the dry soil pat on the surface of the mercury.
7. Force the soil pat under the mercury carefully by means of the glass plate with the prongs, so that the soil pat is completely submerged in mercury (Fig. 10.10). Collect the displaced mercury, weigh it, and find its volume, which is the volume of the dry soil pat ($V_0$).

**Computations**

Moisture content of wet soil pat

$$w = \frac{M - M_0}{M_0} \times 100$$

where

$$M = M_2 - M_1$$

$$M_0 = M_3 - M_1$$

Shrinkage limit (re-moulded soil)

$$w_s = \left[ w - \left( \frac{V - V_0}{M_0} \right) \times 100 \right] \%$$

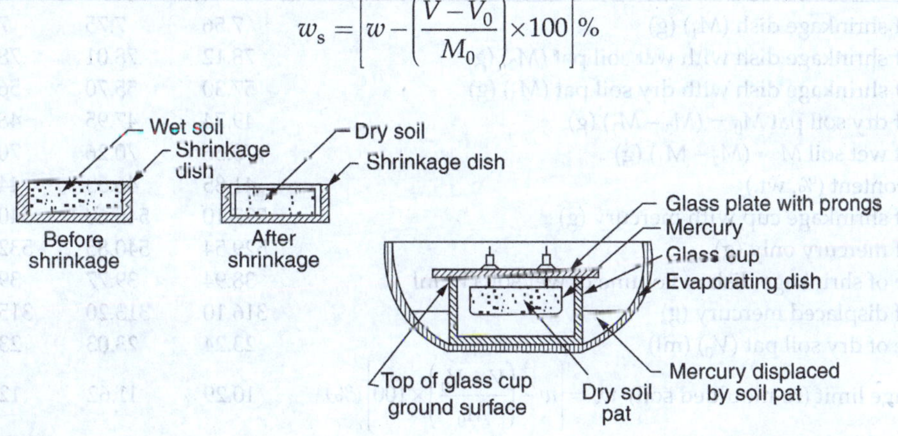


Fig. 10.10 Liner shrinkage mould (*Source*: IS: 2720 – Part 20, 1966)

Shrinkage index

$$I_S = I_p - w_s$$

Shrinkage ratio

$$R = \frac{M_0}{V_0}$$

Volumetric shrinkage (or volume change)

$$V_s = (w_1 - w_s)R$$

where $w_1$ is the given moisture content in percent.
Shrinkage limit of re-moulded soil when the specific gravity is known,

$$w_s = \left(\frac{1}{R} - \frac{1}{G}\right) \times 100$$

**Results**
Shrinkage limit (re-moulded soil) tests are repeated at least three times, and the average value is reported, and if any test shows a variation of 2% against the mean, the test is repeated. The shrinkage limit for a typical re-moulded soil is given in Table 10.13.

**Discussion**
In order to determine the shrinkage limit of undisturbed soils, prepare a wet soil pat of dimensions 45 mm diameter and 15 mm height and round off its edges to prevent the entrapment of air during mercury displacement (IS: 2720 – Part 6, 1972). Air-dry and then

**Table 10.13** Data and results of shrinkage limit test

| Determination no. | 1 | 2 | 3 |
|---|---|---|---|
| Mass of shrinkage dish ($M_1$) (g) | 7.56 | 7.75 | 7.67 |
| Mass of shrinkage dish with wet soil pat ($M_2$) (g) | 78.12 | 78.01 | 78.16 |
| Mass of shrinkage dish with dry soil pat ($M_3$) (g) | 57.30 | 55.70 | 56.35 |
| Mass of dry soil pat $M_0 = (M_3 - M_1)$ (g) | 49.74 | 47.95 | 48.68 |
| Mass of wet soil $M = (M_2 - M_1)$ (g) | 70.56 | 70.26 | 70.49 |
| Water content (%, wt.) | 41.85 | 46.53 | 44.80 |
| Mass of shrinkage cup with mercury (g) | 537.10 | 548.60 | 540.20 |
| Mass of mercury only (g) | 529.54 | 540.85 | 532.53 |
| Volume of shrinkage dish = volume of wet soil ($V$) ml | 38.94 | 39.77 | 39.16 |
| Mass of displaced mercury (g) | 316.10 | 313.20 | 315.50 |
| Volume of dry soil pat ($V_0$) (ml) | 23.24 | 23.03 | 23.20 |
| Shrinkage limit (re-moulded soil) $w_s = \left[w - \left(\frac{V - V_0}{M_0}\right) \times 100\right]$ (%) | 10.29 | 11.62 | 12.01 |

Average shrinkage limit = 11.31%

oven-dry the pat; cool it and weigh it ($M_{0s}$). Using the mercury displacement procedure, find the volume of the pat ($V_{0s}$). Then shrinkage limit of undisturbed soil,

$$w_{su} = \left(\frac{v_{0s}}{M_{0s}} - \frac{1}{G}\right) \times 100\%$$

The shrinkage limit test is useful in obtaining a quantitative indication of how much volume change can occur with changes in the water content.

## 10.13 TEST NO. 12: LINEAR SHRINKAGE OF SOIL

**Scope**

To determine the linear shrinkage of re-moulded soil

**Apparatus**

Two palette knives
Flat glass plate or evaporating dish
Cast iron or brass mould (as in Fig. 10.11)
Oven with accurate temperature control in the ranges 60 to 65°C and 105 to 110°C
Callipers
Silicone or any grease

**Procedure**

1. Thoroughly clean the mould and measure its length to get the initial length of the specimen ($L_1$).
2. Apply a thin layer of grease to the inner walls of the mould so as to prevent the soil from adhering to the sides of the mould.
3. Take 150 g of the soil sample passing through a 425 μm sieve in an evaporating dish.
4. Add sufficient quantity of distilled water (as a rough measure, this may be about 2% above the limit of the soil) and thoroughly mix. Set it aside for 24 hours for moisture equilibrium to be attained.
5. Place the thoroughly mixed soil in the mould such that it is slightly above the sides of the mould.
6. Remove the entrapped air bubbles by gently tapping the mould on a soft pad.

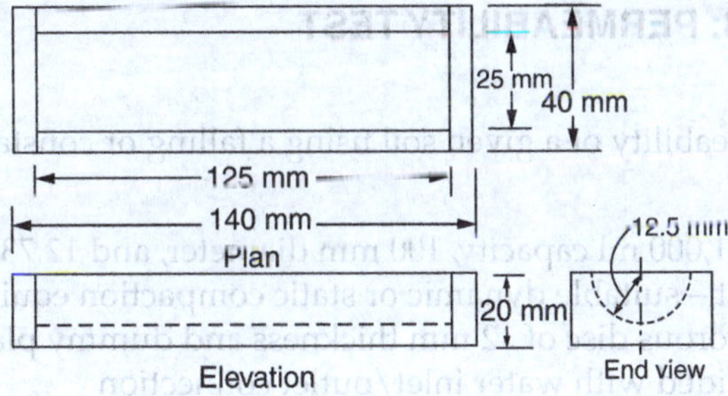


Fig. 10.11 Liner shrinkage mould (*Source*: IS: 2720 – Part 20, 1966)

**Table 10.14** Linear shrinkage and result of data

| Determination no. | 1 | 2 | 3 |
|---|---|---|---|
| Initial length of specimen ($L_i$) (mm) | 140 | 140 | 140 |
| Length of oven dry specimen ($L_d$) (mm) | 121.12 | 120.8 | 121.4 |
| Linear shrinkage $= [1-(L_d/L_i)]\times 100\%$ | 13.43 | 13.71 | 13.29 |
| Average linear shrinkage = 13.48% | | | |

7. Level the soil along the top mould with the palette knife.
8. Dry the mould in three stages, viz., in air, in the oven at a controlled temperature of 60 to 65°C and finally in the oven at a controlled temperature of 105 to 110°C. The time needed to dry during each stage depends on the type of soil. However, as a general guide, 24 hours may be allowed during each stage.
9. Remove the mould from the oven, cool, and measure the mean length of the soil bar ($L_d$); if the specimen has curved, measure along the mean arc.
10. Repeat the test for two more specimens.

**Computations**

The linear shrinkage of the soil is given as

$$\left(1-\frac{\text{Length of oven}-\text{dry specimen}}{\text{Initial length of the specimen}}\right)\times 100\%$$

**Results**

The liner shrinkage of the soil is represented as a percentage to the nearest whole number. Test data and results are shown in Table 10.14 for a typical case.

**Discussion**

Soil of low plasticity may not show cracks when subjected to rapid drying and in such soils the drying may be done directly at 110°C. For a highly colloidal clay, the drying process may have to be slowed down to prevent cracking. In soils of varying particles size, segregation of larger particles to the bottom of the mould may be avoided by reducing the soil–water wetness (IS: 2720 – Part 20, 1966).

## 10.14 TEST NO. 13: PERMEABILITY TEST

**Scope**

To determine the permeability of a given soil using a falling or constant head permeameter

**Apparatus**

Permeameter mould – 1,000 ml capacity, 100 mm diameter, and 12.73 mm height
Compaction equipment – suitable dynamic or static compaction equipment
Drainage base – with porous disc of 12 mm thickness and dummy plate of 12 mm thickness to suit the mould, provided with water inlet/outlet connection
Drainage cap – with porous disc, 12 mm thick, and water inlet/outlet connection to constant head tank

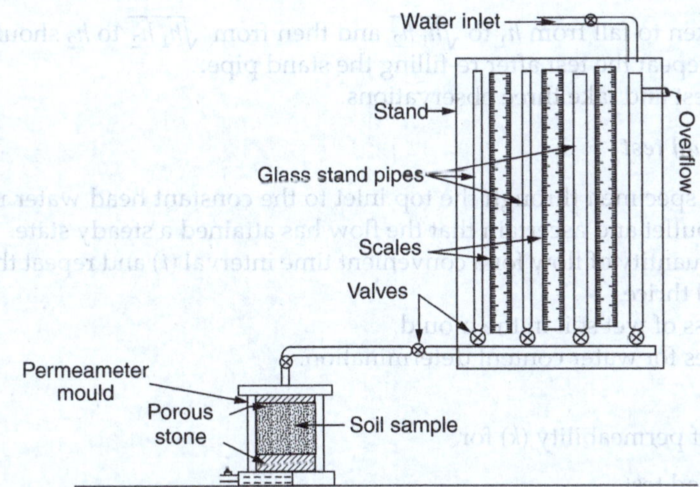


**Fig. 10.12** Permeability test set-up

Set of stand pipes – glass stand pipes of diameter 5 to 20 mm, suitably mounted on stand (Fig. 10.12)
Miscellaneous apparatus – IS sieves, mixing tray, graduated cylinder, metric scale, stopwatch, 75-gauge wire, thermometer, and source of water

**Procedure**

1. Take 2.5 kg of sample (as suggested in the standard compaction test) and the desired water content (may be field water content or optimum moisture content depending on the dry density requirement), spread uniformly, and allow moisture equilibrium to be attained.
2. Weigh the empty permeameter. Attach the extension collar, grease the inside of the mould and collar, and keep the assembly on a firm base.
3. Choose the type of compaction and compactive effort to suit the field condition, and complete the compaction process.
4. Remove the collar, level the soil, detach the base plate, and weigh.
5. Assemble the mould, drainage base, and cap along with porous discs (saturate the porous discs before use).
6. Saturate the specimen, by allowing water to flow with a sufficient head through it or by immersion for a high-permeability specimen and by subjecting it to a high head (for a day or two) for permeable specimen.

(a) ***Falling head test***

7. Connect the specimen through the top inlet to a selected stand pipe of inside area ($a$). Open the bottom outlet and note down the interval ($t$) required for the water level to fall from the initial head ($h_1$) to a known final head ($h_2$), the heads being measured above the centre of the outlet.
8. Fill the reservoir again to a higher $h_1$ and note the time taken for the water level to fall to $\sqrt{h_1 h_2}$ and then to $h_2$ again.

9. The time taken to fall from $h_1$ to $\sqrt{h_1 h_2}$ and then from $\sqrt{h_1 h_2}$ to $h_2$ should be the same. Otherwise, repeat the test after re-filling the stand pipe.
10. Report the test and take three observations.

**(b)** ***Constant head test***

11. Connect the specimen through the top inlet to the constant head water reservoir. Open the bottom outlet and ascertain that the flow has attained a steady state.
12. Collect the quantity of flow for a convenient time interval ($t$) and repeat this for the same time interval thrice.
13. Find the mass of wet soil in the mould.
14. Keep samples for water content determination.

**Computations**

The coefficient of permeability ($k$) for

1. The falling head test,

$$k = \frac{2.303\,aL}{At}\log_{10}\frac{h_1}{h_2}$$

2. The constant head test,

$$k = \frac{qL}{Ah}$$

The permeability at 27°C is given as

$$k_{27°C} = k_T\frac{\eta_{\omega T}}{\eta_{\omega 27°C}}$$

**Results**

The coefficient of permeability is reported in mm/s or m/s at 27°C. Typical test results are given in Tables 10.15 and 10.16 for falling and constant head permeabilities, respectively. The void ratio, degree of saturation, and dry density are presented in Table 10.17.

**Table 10.15** Data and test results of falling head test

Length of specimen ($L$) = 127 mm
Area of specimen ($A$) = 7,854 mm$^2$
Volume of specimen ($V$) = 9,97,458 mm$^3$
Area of stand pipe ($a$) = 113 mm$^2$
Specific gravity of soil ($G$) = 2.65
Temperature of water = 30°C

| Sl. no. | Initial head, $h_1$ (mm) | Final head, $h_2$ (mm) | Time, $t$ (seconds) | $\log_{10}(h_1/h_2)$ | $k_T$ (mm/s) | $k_{27}$ (mm/s) |
|---|---|---|---|---|---|---|
| 1. | 1,200 | 550 | 122 | 0.339 | 0.0117 | 0.0110 |
| 2. | 1,200 | 400 | 173 | 0.477 | 0.0116 | 0.0109 |
| 3. | 1,200 | 250 | 244 | 0.681 | 0.0118 | 0.0110 |
| | | | | Average: | 0.0117 | 0.0110 |

**Table 10.16** Data and test results of constant head test

Length of specimen ($L$) = 127 mm
Area of specimen ($A$) = 7,854 mm$^2$
Volume of specimen ($V$) = 9,97,458 mm$^3$
Specific gravity of soil ($G$) = 2.65
Temperature of water = 30°C

| Sl. no. | Time, $t$ (s) | Head, $h$ (mm) | Hydraulic gradient, $h/L$ | Quantity, $Q_3$ (mm$^3$) | $q = Q/t$ (mm$^3$/s) | $k_T$ (mm/s) | $k_{27}$ (mm/s) |
|---|---|---|---|---|---|---|---|
| 1. | 150 | 300 | 2.36 | 32,400 | 214.8 | 0.0116 | 0.0109 |
| 2. | 300 | 300 | 2.36 | 66,000 | 217.0 | 0.0117 | 0.0110 |
| 3. | 450 | 300 | 2.36 | 96,300 | 216.5 | 0.0117 | 0.0110 |
| | | | | | Average: | 0.0117 | 0.0110 |

**Table 10.17** Void ratio, degree of saturation, dry density (same for both tests)

Mass of saturated soil ($M$) = 2,087 g
Mass of moisture cup = 18.3 g
Mass of cup with wet soil = 37.09 g
Mass of cup with dry soil = 34.10 g
Water content = 18.92%

Mass of dry soil in the mould $\left(M_s \dfrac{M}{1+(w/100)}\right)$ g

$$\rho_d = \frac{M_s}{V} = 1.76 \text{ g/cc}$$

$$e = \frac{G\rho_w}{\rho d} - 1 = 0.51$$

$$S_r = \frac{\omega G}{e} = 98.3\%$$

**Discussion**

Permeability tests can also be conducted on undisturbed specimens. Prepare carefully a specimen 85 mm in diameter and 127 mm in height to suit the permeameter. Place the specimen centrally over the porous disc and fill the annular gap with a cement slurry or bentonite sand mix in the ratio 1:9. Fix the drainage cap. Now a falling or constant head test may be conducted, depending on the type of soil.

The constant head test is usually preferred for sandy soils and the variable head test for silty and clayey soils. A separate constant head method for granular soils has been recommended by Indian Standards (IS: 2720 – Part 36, 1975). This method is suitable for disturbed granular soils containing less than 10% soil passing 75 μm IS sieve. This range of particle sizes is used for construction of embankments and base courses under pavements. Granular soils with a particle size up to 20 mm can be tested using this method under laminar flow conditions.

Although these two laboratory methods are routinely used in various laboratories, they do not provide a reliable value for the following reasons:

1. A soil specimen in the laboratory is always disturbed to some extent and does not exist in the same state as in the field.
2. A laboratory specimen does not simulate the orientation of an *in situ* stratum to the flow of water.
3. Boundary conditions are not the same as simulated in the laboratory, e.g., smooth walls of the mould do not exist in the field.
4. There is a difference between the field and laboratory hydraulic gradients.
5. Complete saturation conditions are not possible in a laboratory sample, and the effect of entrap-ped air bubbles on the coefficient of permeability may be severe.
6. Prediction of the behaviour of a large formation *in situ* from the test results for a small sample is highly unreliable.

It is apparent that the laboratory determined $k$ is not representative and is hence not reliable. But tests on undisturbed samples might improve this situation.

## 10.15 TEST NO. 14: FREE SWELL INDEX OF SOILS

**Scope**
To determine the free swell index of a soil

**Apparatus**
Graduated glass cylinders of 100 ml capacity
Sieve – 425 μm IS sieve

**Procedure**

1. Take 10 g of oven-dry soil passing through the 425 μm sieve and pour it into a 100 ml graduated jar. Similarly, prepare another cylinder with the same weight of soil.
2. Fill one with kerosene oil and the other with distilled water up to the 100 ml mark.
3. Remove the entrapped air from both the cylinders by shaking and/or stirring with a glass rod.
4. Allow both the cylinders to settle down for 24 hours.
5. Read out the level of the soil in the kerosene-filled graduated jar ($V_k$). Kerosene, being a non-polar liquid, does not cause swelling of the soil.
6. Also read out the level of soil in the distilled water-filled graduated jar ($V_d$).

**Computations**
The free swell index of the soil can be calculated from the expression

$$\text{Free swell index} = \frac{V_d - V_k}{V_k} \times 100\%$$

**Results**
The free swell index is expressed as a percentage to two significant figures. Typical data and results are given in Table 10.18.

**Table 10.18** Data and results from a free swell index test

| | |
|---|---|
| Volume of soil sample read from kerosene-filled graduated cylinder ($V_k$) | 43 ml |
| Volume of soil sample read from distilled water–filled graduated jar ($V_d$) | 56 ml |
| Free swell index $\frac{V_d - V_k}{V_k} \times 100 = 30\%$ | |

**Discussion**

To get accurate results for highly swelling soils, the quantity of sample taken may be reduced to 5 g or the volume of cylinder may be increased to 250 ml (IS: 2720 – Part 40, 1977).

## 10.16 TEST NO. 15: MOISTURE CONTENT–DRY DENSITY RELATIONSHIP (STANDARD PROCTOR COMPACTION TEST)

**Scope**

To determine the relation between the moisture content and the dry density of a soil

**Apparatus**

Cylindrical metal mould – 1,000 ml diameter, with detachable base, and height of 127.3 mm with extension collar (Fig. 10.13)
Metal rammer – 2.6 kg with 310 mm fall (Fig. 10.14)
Sample extruder

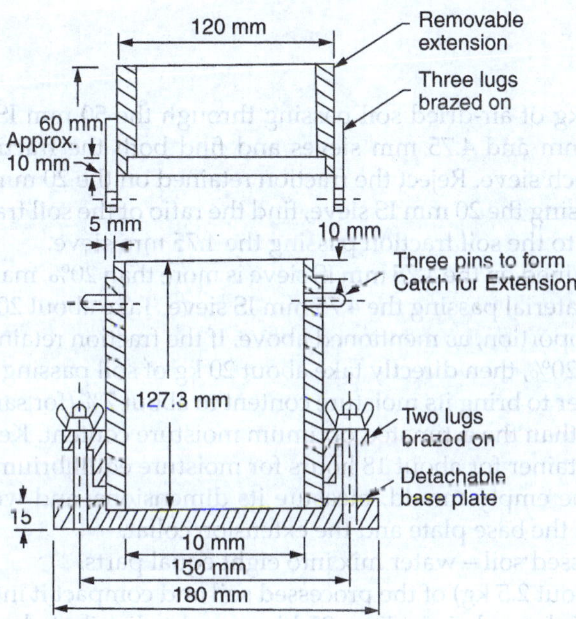


**Fig. 10.13** Mould for compaction (*Source:* IS: 2720 – Part 7, 1974)

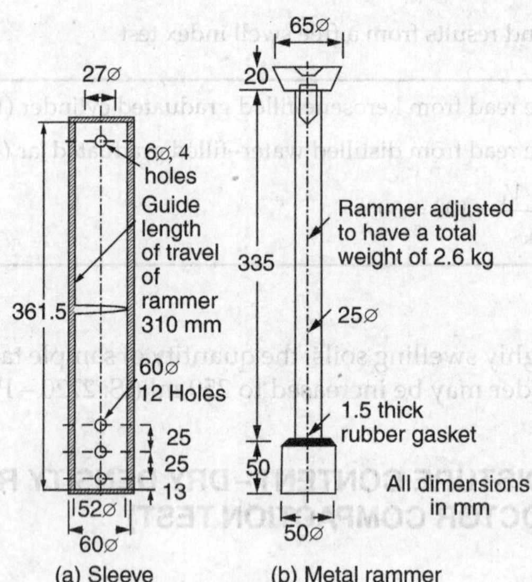


**Fig. 10.14** Metal rammer and sleeve (*Source:* IS: 2720 – Part 7, 1974;)

Apparatus for water content determination
Balance of 10 kg capacity with 1 g sensitivity
Steel straight edge
Sieves – 50 mm, 20 mm, and 4.75 mm IS sieves
Mixing tools

**Procedure**

1. Weigh about 25 kg of air-dried soil passing through the 50 mm IS sieve. Sieve the soil through the 20 mm and 4.75 mm sieves and find both the fraction passing and that retained in the each sieve. Reject the fraction retained on the 20 mm sieve.
2. From the soil passing the 20 mm IS sieve, find the ratio of the soil fraction retained on the 4.75 mm IS sieve to the soil fraction passing the 4.75 mm sieve.
3. If the fraction retained on the 4.75 mm IS sieve is more than 20%, maintain the ratio of this material to the material passing the 4.75 mm IS sieve. Take about 20 kg of the material in the calculated proportion, as mentioned above. If the fraction retained on the 4.75 mm IS sieve is less than 20%, then directly take about 20 kg of soil passing the 20 mm IS sieve.
4. Add enough water to bring its moisture content to about 7% (for sandy soils) or 10% (for clayey soils) less than the estimated optimum moisture content. Keep the processed soil in an airtight container for about 18 hours for moisture equilibrium.
5. Clean and dry the empty mould, measure its dimensions, and weigh it to the nearest gram ($M_m$). Fit in the base plate and the extension collar.
6. Divide the processed soil – water mix into eight equal parts.
7. Take one part (about 2.5 kg) of the processed soil and compact it into the mould in three equal layers, each layer, being given 25 blows to be distributed uniformly. Score each layer with a spatula before putting in the soil for the succeeding layer.

8. Remove the collar and carefully level off to the top of the mould by means of a straight edge. Weigh the mould ($M$).
9. Eject the soil from the mould, cut at the middle, and take representative samples for water content determination.
10. Repeat Steps 7 to 9 for 5 or 6 samples, using a fresh part of the soil specimen each time, after adding a higher water content than in the preceding specimen, so that at least two readings, one below and above the optimum moisture content, are available.

**Computations**

Compute the volume ($V_m$) of the mould from its height and diameter.

$$\text{Bulk density } \rho = \left(\frac{M - M_m}{V_m}\right) \text{ g/cc}$$

$$\text{Dry density } \rho_d = \left(\frac{\rho}{1 + w/100}\right) \text{ g/cc}$$

$$\text{Void ratio } e = \left(\frac{G\rho_w}{\rho_d} - 1\right)$$

$$\text{Porosity } n = \left(1 - \frac{\rho_d}{G\rho_w}\right) \times 100\%$$

**Results**

Plot a curve of water content versus dry density. The dry density (rounded to two decimal places) corresponding to the maximum point of the curve and the corresponding moisture content (rounded to the first decimal place) shall be reported as the maximum dry density ($\rho_{d\,max}$) and the optimum moisture content (OMC), respectively. A typical test results on a soil is presented in Table 10.19. The moisture content–dry density curve is plotted in Fig. 10.15.

**Discussion**

Instead of a 1,000 ml capacity mould, the Indian Standards (IS: 2720 – Part 7, 1974) also recommend a 2,250 ml mould to be used; in that case, for each layer 56 blows are given with the standard hammer. The Standard/Proctor Test is also termed a light compaction test.

As the material retained on the 20 mm IS sieve has been rejected for the test, a correction is applied to get the corrected maximum dry density and OMC.

$$\text{Corrected maximum dry density} = \frac{\rho_0 \times \rho_{d\,max}}{n_1\rho_{d\,max} + n_2\rho_0}$$

$$\text{Corrected } OMC = n_1 A_0 + n_2 w_0$$

where $\rho_0$ is the density of over-size particles (i.e., $G_0\rho_w$, where $G_0$ is the specific gravity of the over-size particles), $\rho_{d\,max}$ the maximum dry density obtained in the test, in g/cc, $n_1$ the fraction by weight of over-size particles in the total soil expressed as a ratio, $n_2$ the fraction by weight of portion passing 20 mm IS sieve (or 4.75 mm IS sieve) expressed as a fraction of the total soil, $A_0$ the water absorption capacity of over-size material, if any,

**Table 10.19** Data and test results from compaction test

Type of test: Standard/Proctor test

| | |
|---|---|
| Volume of mould ($V_m$) | = 1,000 ml |
| Mass of the rammer | = 2.6 kg |
| Fall of the rammer | = 310 mm |
| Specific gravity | = 2.65 |
| Percentage of material: | |
| (i) Retained on 20 mm IS sieve | = 5 |
| (ii) Passing 20 mm IS sieve and retained on 4.75 mm IS sieve | = 0 |
| (iii) Passing 4.75 mm IS sieve | = 90 |
| Ratio of (ii) to (iii) | = 1:9 |

| Determination no. | 1 | 2 | 3 | 4 | 5 | 6 |
|---|---|---|---|---|---|---|
| Mass of mould ($M$) (g) | 6,245 | 6,245 | 6,245 | 6,245 | 6,245 | 6,245 |
| Mass of mould with compacted soil ($M_m$) (g) | 8,130 | 8,211 | 8,260 | 8,279 | 8,268 | 8,220 |
| Mass of compacted soil ($M_m - M$) (g) | 1,885 | 1,966 | 2,015 | 2,034 | 2,023 | 1,955 |
| Wet density $\rho = (M_m - M)/V_m$ | 1.885 | 1.966 | 2.015 | 2.034 | 2.023 | 1.995 |
| Moisture cup no. | 9 | 18 | 27 | 45 | 54 | 36 |
| Mass of cup and wet soil (g) | 41.20 | 37.12 | 40.47 | 40.35 | 39.46 | 40.48 |
| Mass of cup and dry soil (g) | 39.35 | 35.18 | 38.18 | 37.89 | 36.82 | 37.71 |
| Mass of cup (g) | 22.67 | 20.74 | 22.84 | 22.65 | 21.84 | 22.37 |
| Dry density $\rho_d = \dfrac{\rho}{1+w/100}$ (g/cc) | 1.697 | 1.733 | 1.753 | 1.751 | 1.720 | 1.690 |
| Void ratio $e = \left(\dfrac{G\rho_w}{\rho_d} - 1\right)$ | 0.562 | 0.529 | 0.512 | 0.513 | 0.541 | 0.568 |
| Porosity $n = \left(1 - \dfrac{\rho_d}{G\rho_w}\right) \times 100$ | 36.0 | 34.6 | 33.9 | 33.9 | 35.1 | 36.2 |

From the plot $\rho_{d\,max} = 1.755$ g/cc and OMC = 15.45%.

expressed as the percentage of water absorbed, and $W_0$ the OMC obtained in the test in per cent.

This formula is based on the assumption that the volume of the compacted portion passing the 20 mm IS sieve (or 4.75 mm IS sieve) is sufficient to fill the voids between the over-size particles (IS: 2720 –Part 7, 1974).

With field compacting equipment becoming heavier and more efficient, it has become necessary to increase the amount of compacting energy in the laboratory test, and hence a standard test for heavier compaction (Modified Proctor Test) has been suggested (IS: 2720 – Part 8, 1983).

The procedure for conducting the test with heavier compaction is similar to that of light compaction, but with slightly modified equipment. In this case, a rammer with a mass of 4.89 kg and a fall of 450 mm is used.

The soil is compacted in five layers, with each layer being given 25 blows for a 1,000 ml mould or 56 blows for a 2,250 ml mould.

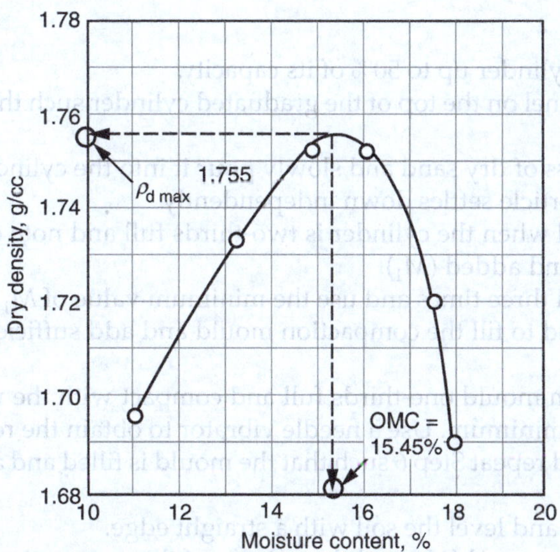


**Fig. 10.15** Moisture content–dry density curve

Another method which uses a constant weight of soil is available for determination of the moisture–density relation for the soil passing through the 4.75 mm IS sieve (IS: 2720 – Part 9, 1971). This is a rapid method which can be used as a field control method. It may also be used as a rapid laboratory test. However, it cannot be used as a substitute for the standard tests discussed earlier.

The compaction tests (both standard and modified) are satisfactory for cohesive soils. Clean sands and gravels which are displaced easily during the rammer blows do not indicate proper compaction characteristics. A knowledge of the maximum dry density and OMC obtained from this test suggests that the maximum density is obtainable in the field using a suitable roller and adopting a moulding water content almost equal to the OMC. A check can be made on the field-compacted soil by adopting field control tests.

## 10.17 TEST NO. 16: DENSITY INDEX OF NON-COHESIVE SOILS

**Scope**

To determine the density index (relative density) of non-cohesive free draining soils

**Apparatus**

Graduated cylinder – 1,000 ml
Large glass funnel and glass rod
Balance of 0.1 g sensitivity
Compaction mould
Wooden hammer
Needle vibrator
Straight edge
Hand scoop

**Procedure**

1. Fill the graduate cylinder up to 50% of its capacity.
2. Place the large funnel on the top of the graduated cylinder such that the tip of the funnel is in the water.
3. Take a known mass of dry sand and slowly pour it into the cylinder through the funnel such that every particle settles down independently.
4. Stop pouring sand when the cylinder is two-thirds full and note down the volume ($V_1$) and the mass of sand added ($M_1$).
5. Repeat Steps 1 to 4 three times and use the minimum value of $M_1/V_1$.
6. Take sufficient sand to fill the compaction mould and add sufficient water to saturate it completely.
7. Fill the compaction mould one-thirds full and compact with the wooden hammer such that the voids are minimum. Use a needle vibrator to obtain the required condition.
8. Place more soil and repeat Step 6 such that the mould is filled and about 50% of the collar is full.
9. Remove the collar and level the soil with a straight edge.
10. Find the mass of wet sand ($M_2$) and the volume of the compaction mould ($V_2$).
11. Keep a certain quantity of soil for water content determination.
12. Repeat Steps 6 to 10 three times and use the maximum value of $M_2/V_2$.
13. Find the filled density of the soil using the sand replacement method.
14. Find the field void ratio $e$ from the field density, knowing the specific gravity of soil solids.

**Computations**

Minimum dry density

$$\rho_{d\,min} = \frac{M_1}{V_1}$$

Maximum void ratio

$$e_{max} = \frac{G\rho_w}{\rho_{d\,min}} - 1$$

Maximum dry density

$$\rho_{d\,max} = \frac{M_2}{V_2}$$

Minimum void ratio

$$e_{min} = \frac{G\rho_w}{\rho_{d\,max}} - 1$$

Density index

$$D_r = \frac{e_{max} - e}{e_{max} - e_{min}} \times 100\%$$

**Results**

The density index is expressed as a percentage. Typical observations and test results are given in Table 10.20.

**Table 10.20** Density index test results

| Determination no. | 1 | 2 | 3 |
|---|---|---|---|
| Inital mass of sand (g) | 2,000 | 2,000 | 2,000 |
| Mass of sand after pouring in graduated jar (g) | 1070.5 | 1002.5 | 854.4 |
| Mass of sand ($M_1$) (g) | 929.5 | 997.5 | 1145.6 |
| Volume of sand ($V_1$) (ml) | 650 | 700 | 800 |
| Minimum dry density $\rho_{d\,min} = \dfrac{M_1}{V_1}$ (g / cc) | 1.43 | 1.425 | 1.432 |
| Average maximum void ratio ($G = 2.65$) $e_{max} = \dfrac{G\rho_w}{\rho_{d\,min}} - 1 = 0.86$ | | | |
| Minimum void ratio | | | |
| Mass of mould (g) | 6,245 | 6,245 | 6,245 |
| Mass of mould + wet soil (g) | 8,331 | 8,342 | 8,339 |
| Mass of mould ($V_2$) (ml) | 1,000 | 1,000 | 1,000 |
| Mass of wet soil ($M_2$) (g) | 2,086 | 2,097 | 2,094 |
| Mass of cup (g) | 22.67 | 2,0.74 | 22.84 |
| Mass of cup + wet soil (g) | 42.95 | 40.72 | |
| Mass of cup + dry soil (g) | 39.62 | 37.45 | |
| Water content (%) | 19.65 | 19.55 | |
| Maximum dry density $\rho_{d\,max} = \dfrac{M_2 / V_2}{1 + w / 100}$ (g / cc) | 1.743 | 1.754 | |
| Average minimum void ratio $e_{min} = \dfrac{G\rho_w}{\rho_{d\,max}} - 1 = 0.51$ | | | |
| Void ratio infield $e = 0.62$ | | | |
| Density index $D_r = \dfrac{e_{max} - e}{e_{max} - e_{min}} \times 100\% = 68.6\%$ | | | |

**Discussion**

This procedure is just sufficient for obtaining a fairly accurate value, provided the loosest density is obtained carefully (Prakash, 1969). A more comprehensive method has been given in IS: 2720 – Part 14 (1983).

## 10.18 TEST NO. 17: CONSOLIDATION TEST

**Scope**

To determine the consolidation properties of soil

**Apparatus**

Consolidation ring – a rigid ring with a smooth and polished inner surface and provided with a cutting edge to facilitate preparation of specimens. The minimum diameter of the ring should be 60 mm with a diameter–height ratio of 3.0 (Fig. 10.16)

Porous stone – shall be of silicon carbide, aluminium oxide, or other porous materials with high porosity such that free drainage is assured throughout the test. The diameter of the porous stone is 0.2 to 0.5 mm less than that of the inside diameter of the ring. The stone size varies depending on the type of ring (Fig. 10.16)

Consolidation cell – a container to house the consolidation ring has a provision to hold water and allow measurement of the change in height of the specimen at its central axis

Dial gauge – has a length of travel of 50% of the specimen height with an accuracy of at least 0.001% of the specimen height

Loading device – capable of taking axial loads in suitable increments with a suitable lever ratio and of maintaining this for a large duration of time with an admissible variation of ±1% of the applied load (Fig. 10.17). There should be no significant impact during load application. It should be located in an area free from vibrations

Sample extruder

Trimming equipment

Equipment for water content determination

Balance of 0.01 g sensitivity

Stopwatch with least count of 1 second

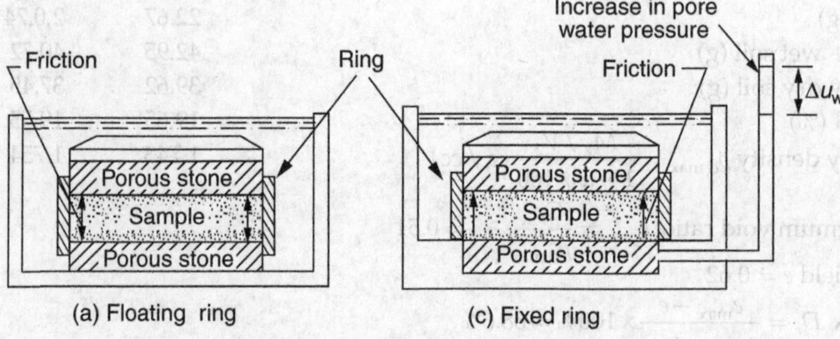


**Fig. 10.16** Types of consolidometers

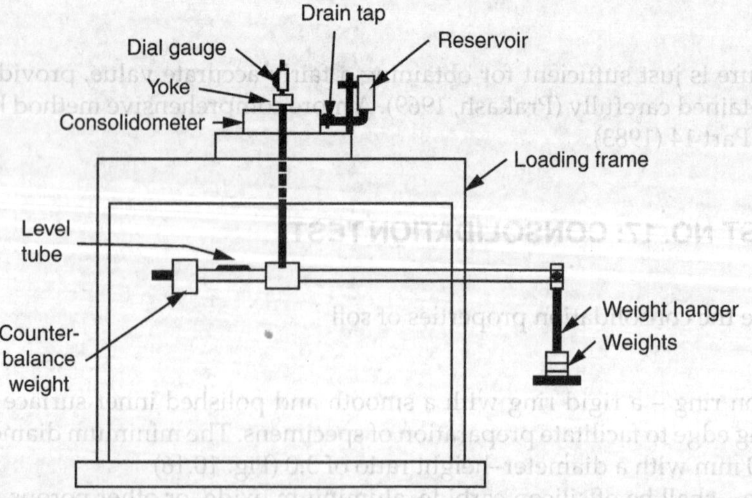


**Fig. 10.17** Consolidation test set-up

**Procedure**

1. Find the mass of the empty consolidation ring ($M_1$).
2. Coat the inside surface with silicone grease or oil. Trim a sample carefully to fit the consolidation ring and weigh the mass along with the ring ($M_2$). Keep a small quantity of the soil from the trimmings for water content determination.
3. Record the thickness of the specimen. In case of difficulties in measuring the thickness, take the thickness of the ring as the initial thickness.
4. Depending on the type of ring (fixed or floating), choose the correct size of the porous stone. Place the ring and the specimen centrally on the saturated bottom porous stone and place the upper saturated porous stone, followed by the loading cap.
5. Place the consolidometer in the loading device and attach the dial gauge. Fill the consolidometer with water, apply a seating load of 5 kN/m$^2$, and allow it to reach moisture equilibrium in 24 hours.
6. Apply the first load increment and simultaneously take deformation readings at elapsed times of 0.25, 0.50, 1, 2, 4, 8, 15, 30, and 60 minutes and 2, 4, 8, and 24 hours.
7. After 24 hours apply the increment load, keeping mind the fact that the applied pressure at any loading stage should be double that at the preceding stage. Apply the following loading sequence: 10, 20, 40, 80, 160, 320 kN/m$^2$. Each time repeat Step 6.
8. On completion of the final loading, unload the specimen with pressure decrements which decrease the load to one-fourth the previous load. Take dial gauge readings during each stage of unloading. If desired, the same time interval as adopted during loading may be adopted. Keep the last unloading at a pressure of 5 kN/m$^2$ for 24 hours to minimize the swelling during disassembly.
9. Remove the ring, wipe the water on the outside of the ring, and find the mass ($M_3$).
10. After drying, weigh the specimen with the ring and find the mass ($M_4$).

**Computations**

(a) ***Coefficient of consolidation***

Plot dial gauge reading versus $\sqrt{t}$ or versus log $t$ for each load increment, and find the coefficient of consolidation from the following expressions.

(i) Square root of time method:

$$C_v = \frac{0.848\,(H_{av}/2)^2}{t_{90}}$$

(ii) Logarithm of time method:

$$C_v = \frac{0.197\,(H_{av}/2)^2}{t_{50}}$$

where $H_{av}$ is the average thickness of the specimen for that load increment.

**(b) *Coefficient of compressibility***

Volume of solids, $V_s = \dfrac{M_s}{G\rho_w}$

where $M_s = M_4 - M_1$.

The equivalent height, $H_s = V_s / A$
where $A$ is the area of specimen.

The void ratio $e$, at the end of each pressure increment, is given as

$$e = \frac{H}{H_s} - 1$$

where $H$ is the height of specimen at the end of each pressure increment.

The coefficient of compressibility $a_v$, with units inverse of those for stress is given as

$$a_v = \frac{\Delta e}{\Delta p}$$

**(c)** ***Compression index $C_c$***

Plot the void ratio $e$ versus log $p$. The slope of the straight line portion of the curve is the compression index $C_c$ which is given as

$$C_c = \frac{\Delta e}{\log(p_2 / p_1)}$$

where $p_2$ and $p_1$ are the successive values of pressure and $\Delta e$ is the change in the void ratio over the above range of pressures.

**Results**

The consolidation test results are presented in the form of the following curves:

1. $e$ versus log $p$
2. Dial reading versus log $t$ for different stress ranges
3. Dial reading versus $\sqrt{t}$ for different stress ranges
4. $a_v$ versus log $p$
5. $c_v$ versus log $p$

Some typical test results are presented in Tables 10.21 to 10.23, and the corresponding plots are given in Figs. 10.18 and 10.19.

**Discussion**

As the effects of sample preparation are the same for any size of sample, larger samples provided more reliable results. The floating ring reduces the frictional loss along the sides of the sample between the soil and ring, and hence, the test rate is about four times faster. The fixed ring has the advantage of measuring the $k$ value of the sample as it is tested. The curve fitting methods are discussed in Chapter 6. Other details of the test can be obtained from IS: 2720 – Part 15 (1986).

**Table 10.21** Data and some details of a consolidation test

| | |
|---|---|
| Height of consolidation ring | 25 mm |
| Diameter of consolidation ring | 60 mm |
| Area of cross-section of consolidation ring ($A$) | 28.27 cm$^2$ |
| Mass of empty consolidation ring ($M_1$) | 162.50 g |
| Mass of wet soil + ring ($M_2$) | 182.70 g |
| Initial thickness of sample | 20 mm |
| Initial water content | 20.9% (wt.) |
| Specific gravity of soil solids ($G$) | 2.68 |
| Initial void ratio ($e_0 = wG$) | 0.560 |
| Mass of wet soil + ring (after completion of consolidation) ($M_3$) | 180.96 g |
| Mass of dry soil + ring ($M_4$) | 259.48 g |
| Mass of dry soil ($M_s = M_4 - M_1$) | 96.98 g |
| Volume of soil solids $\left(V_s = \dfrac{M_s}{G\rho_w}\right)$ | 36.19 ml |
| Final water content | 19.03% |

**Table 10.22** Dial gauge reading versus time for three loadings

| Elapsed time (minutes) | Time (minutes)$^{1/2}$ | Consolidation pressure (kN/m$^2$) (Dial gauge reading least count = 0.01 mm) | | |
|---|---|---|---|---|
| | | **10.0** | **20.0** | **40.0** |
| 0 | 0.00 | 195.0 | 175.40 | 150.60 |
| 0.25 | 0.50 | 191.0 | 173.25 | 148.00 |
| 0.50 | 0.71 | 190.25 | 172.00 | 147.00 |
| 1.00 | 1.00 | 189.75 | 170.75 | 145.50 |
| 2.00 | 1.41 | 189.00 | 169.00 | 143.25 |
| 4.00 | 2.00 | 187.75 | 166.50 | 140.25 |
| 8.00 | 2.83 | 186.50 | 163.75 | 137.25 |
| 15.00 | 3.87 | 185.25 | 161.75 | 134.75 |
| 30.00 | 5.48 | 184.50 | 159.50 | 133.00 |
| 60.00 | 7.75 | 183.25 | 158.25 | 131.75 |
| 120.00 | 10.96 | 182.75 | 157.75 | 129.25 |
| 180.00 | 13.42 | 182.50 | 157.25 | 129.00 |
| 1440.00 | 37.95 | 175.40 | 150.60 | 123.03 |

**Table 10.23** Final results from consolidation test

| Applied pressure (kN/m²) | Final dial reading | Compression, $\Delta H$ (mm) | $e = (H/H_s - 1)$ | $\Delta e$ | $\Delta p$ (kN/m²) | $a_v = \Delta_e/\Delta_p$ (m²/kN) | $H_{av}$ (mm) | $t_{90}$ (minutes) | $c_v$ (mm²/min) |
|---|---|---|---|---|---|---|---|---|---|
| 195.0 | 0 | 20 | 0.560 | 0 | 0 | 0 | 20 | – | – |
| 175.4 | 0.196 | 18.80 | 0.547 | 0.013 | 10 | $1.30 \times 10^{-3}$ | 19.90 | 17.64 | 4.76 |
| 150.6 | 0.248 | 19.55 | 0.548 | 0.019 | 10 | $1.90 \times 10^{-3}$ | 19.68 | 12.25 | 6.70 |
| 123.0 | 0.276 | 19.27 | 0.506 | 0.022 | 10 | $2.20 \times 10^{-3}$ | 19.41 | 10.24 | 7.80 |

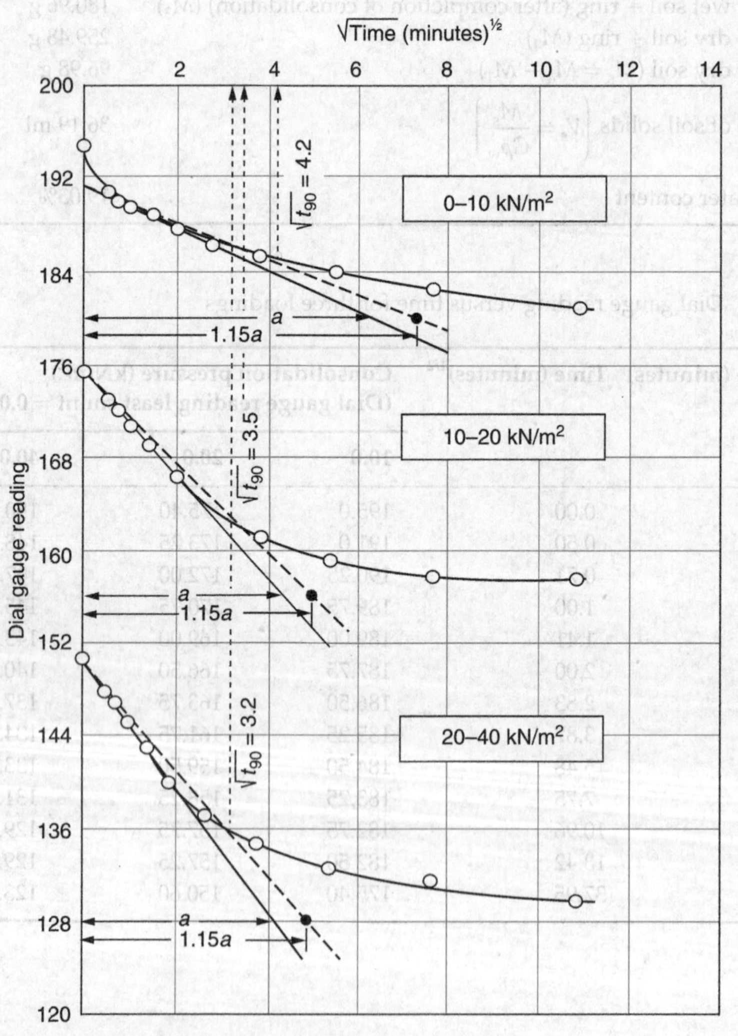


**Fig. 10.18** Curve fittings by $\sqrt{\text{time}}$ method

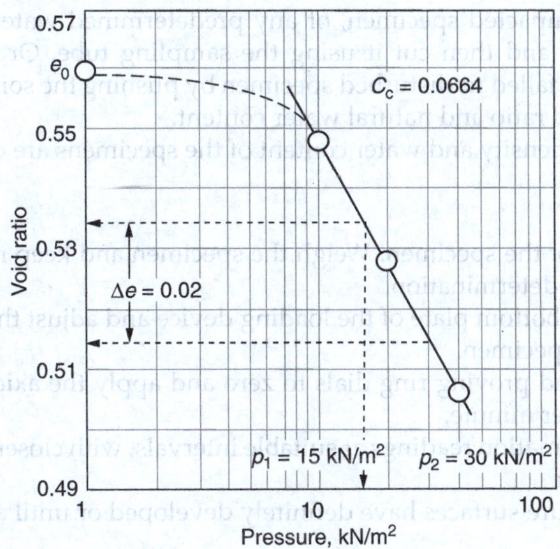


**Fig. 10.19** $e$–log $p$ plot

## 10.19 TEST NO. 18: UNCONFINED COMPRESSION TEST

**Scope**

To determine the compressive strength and sensitivity of a cylindrical sample of cohesive soil

**Apparatus**

Unconfined compression testing machine (strain controlled)
Sampling tube
Sample ejector
Deformation dial gauge – 0.01 mm graduations and specific travel to permit 20% axial strain
Vernier callipers – of least count 0.1 mm
Timer
Oven with accurate temperature control in the range 110 ± 5°C
Balance of 0.001 g sensitivity
Miscellaneous equipment, such as specimen trimmers, carving tools, re-moulding apparatus, moisture cups, etc.

**Procedure**

**(a) *Preparation of test specimen***

Undisturbed, compacted, or re-moulded specimens may be prepared, depending on the case.

1. Prepare undisturbed cylindrical specimens (38 mm diameter, 76 mm length) from large undisturbed field samples using a lathe or trimmer. Alternatively, directly obtain field samples in thin sampling tubes of the same diameter as that of the specimen. Obtain the required length by ejecting the sample through a split mould.

2. Prepare a re-moulded compacted specimen, of any predetermined water content and density, in a large mould and then cut it using the sampling tube. Or prepare a re-moulded specimen from a failed undisturbed specimen by pushing the soil inside a split mould, with the same void ratio and natural water content.
3. In both the cases, the wet density and water content of the specimens are determined.

**(b) *Compression test***

4. Measure the dimensions of the specimen. Weigh the specimen and keep representative samples for water content determination.
5. Place the specimen on the bottom plate of the loading device and adjust the upper plate to make contact with the specimen.
6. Adjust the deformation and proving ring dials to zero and apply the axial load with a strain rate of 0.5% to 2% per minute.
7. Record the force and deformation readings at suitable intervals, with closer spacing during initial stages of the test.
8. Apply the load till the failure surfaces have definitely developed or until an axial strain of 20% is reached.
9. Carefully sketch the failure pattern, and if the specimen has failed with a pronounced failure plane, measure the angle of the failure surface with the horizontal.
10. Take water content representative samples from the failure zone of the specimen.

**Computations**

Stress–strain values are calculated as

$$\text{Axial strain } \varepsilon = \frac{\Delta L}{L_0}$$

where $\Delta L$ is the the change in the specimen length (mm) and $L_0$ the initial length of the specimen (in mm).

The average cross-section area $A$ at a particular strain is given by

$$A = \frac{A_0}{1-\varepsilon}$$

where $A_0$ is the initial average area of cross-section of the specimen.

$$\text{Compressive stress } \sigma_1 = \frac{P}{A}$$

where $P$ is the compressive force.

Plot $\sigma_1$ versus $\varepsilon$ and obtain the maximum stress which gives the unconfined compressive strength $q_u$. In case no pronounced peak is observed, take the strength corresponding to 20% strain as the unconfined compressive strength.

For $\phi = 0$ condition, the shear strength or cohesion of the soil may be taken to be equal to half the unconfined compressive strength.

**Table 10.24** Data and test results from unconfined compression test

Type of specimen (undisturbed, compacted, or compacted and re-moulded)
Initial length ($L_0$) = 83 mm
Initial diameter ($D_0$) = 38 mm
Initial area ($A_0$) = 1,134 mm$^2$
Initial mass of specimen = 75.8 g
Initial density = 1.76 g/cc
Initial water content = 15.5%
Rate of strain adopted = 1.27 mm/min

| Sl. no. | Elapsed time (minutes) | Load (N) | Deformation (mm) | Strain (%) | Area $A = A_0/(1 - \varepsilon)$ (mm$^2$) | Stress (N/mm$^2$) |
|---|---|---|---|---|---|---|
| 1. | 0.50 | 30.97 | 0.10 | 0.1205 | 1135.4 | 0.027 |
| 2. | 1.00 | 61.94 | 0.30 | 0.3615 | 1138.1 | 0.054 |
| 3. | 1.50 | 89.65 | 0.60 | 0.7229 | 1142.3 | 0.079 |
| 4. | 2.00 | 118.90 | 1.00 | 1.2048 | 1147.8 | 0.104 |
| 5. | 3.00 | 128.77 | 1.20 | 1.4458 | 1150.6 | 0.112 |
| 6. | 4.00 | 140.18 | 1.40 | 1.6868 | 1153.5 | 0.122 |
| 7. | 5.00 | 146.70 | 1.50 | 1.8072 | 1154.9 | 0.127 |
| 8. | 6.00 | 148.33 | 1.60 | 1.9277 | 1156.3 | 0.128 |
| 9. | 8.00 | 144.71 | 1.70 | 2.0482 | 1157.7 | 0.125 |
| 10. | 10.00 | 143.44 | 1.80 | 2.1687 | 1159.1 | 0.124 |

Final water content = 15.5%
Unconfined compressive strength ($q_u$) = 0.128 N/mm$^2$
Undrained shear strength or cohesion $\left(c_u = \frac{q_u}{2}\right) = 0.064 \text{ N/mm}^2$

## Results

The observations made during the test are recorded as shown in the data sheet for a typical case (Table 10.24).

## Discussion

The strain-controlled unconfined compression test is used universally. The test is somewhat sensitive to the strain rate, and can be performed only in a strain-controlled machine. But a stress-controlled test may show an erratic strain response due to incremental changing of loads.

The shear strength obtained from the unconfined compressive strength is not very reliable for at least three reasons:

1. The lateral restraint present in the field is not properly simulated in the laboratory.
2. There is no control on the internal soil conditions (degree of saturation, pore water pressure, etc.).
3. The end platens because of lateral restraint alter the internal stresses.

These three factors have been properly taken care of in modern triaxial shear equipment (discussed elsewhere).

The unconfined compressive strength test is a quick test and gives the approximate shear strength of a cohesive soil. Another advantage of the unconfined compression test is that the failure occurs along the weakest portion of the clay and hence provides a conservative shear strength value. Keeping in view the deficiencies of the test, a reasonable interpretation has to be made. The unconfined compression test gives misleading results with heterogeneous soils because of the boundary condition (IS: 2720 – Part 10, 1973).

## 10.20 TEST NO. 19: DIRECT SHEAR TEST

**Scope**
To determine the shear strength of a soil using direct shear apparatus

**Apparatus**
Shear box – size 60 mm$^2$ and about 50 mm deep to suit particles with size less than 4.75 mm–grid plates, porous stones, base plate, loading pad
Container for shear box (Fig. 10.20)
Loading frame with proving ring/load cell
Micrometer dial gauges – 0.01 mm accuracy
Sample trimmer or specimen cutter
Stopclock
Balance of 1 kg capacity with 0.1 g sensitivity
Spatula and a straight edge

**Procedure**
**(a)** ***Preparation of specimen***

1. Prepare undisturbed specimens of the required size by trimming from suitable large undisturbed samples.
2. For re-moulded cohesive soils, compact the soil to the required density at the appropriate water content, extract the sample from the mould, and then trim to the required size or use a standard specimen cutter. Alternatively, compact the soil in the shear box itself to the desired density.

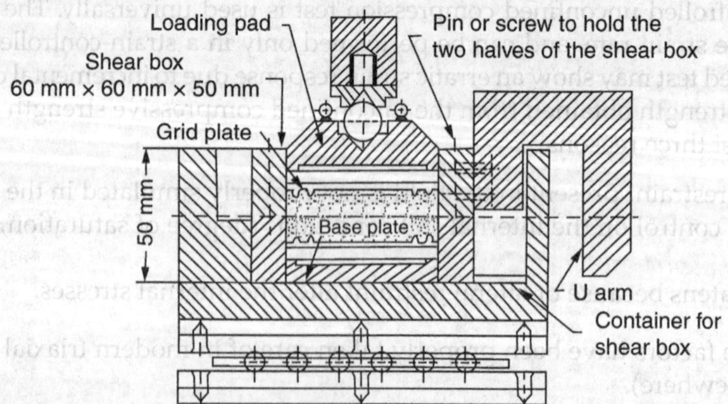


Fig. 10.20 Assembly of shear box and container (*Source:* IS: 2720 – Part 13, 1972)

3. For re-moulded non-cohesive soils, place the soil in the shear itself and tamp it till the required density is reached. Use a porous stone, if necessary, depending on the type of test.
4. Weigh the cut or trimmed specimen and record the mass of soil used in the case of a non-cohesive soil to find the bulk density of the specimen.

**(b)** ***Shear tests***

(i) *Undrained test*

5. Keep plain grid plates, one on either side of the specimen (with serrations of grid plates at right angles to the direction of shear), and place the specimen with grid plates on the base plate.
6. Place the loading pad on the top grid plate and add water in the shear box container so as to prevent drying of the specimen. Apply the required normal stress, depending on the field condition or design requirement.
7. Choose a suitable strain rate such that no drainage takes place during the test. Raise the upper part of shear box such that a gap of about 1 mm is left between the two parts of the box.
8. Apply the shear load at the chosen strain rate till failure or to 20% longitudinal displacement, whichever occurs first.
9. Record the shear load reading and longitudinal displacement. Ensure that no drainage has taken place by noting the vertical compression dial.
10. Remove the soil specimen and keep it in the oven for water content determination.
11. Repeat the test on three more separate specimens with the same initial conditions.

(ii) *Consolidated undrained test*

12. Follow Step 5 but use perforated grid plates and saturated porous stones at the top and bottom of the specimen.
13. Follow Step 6 and record the vertical compression (caused due to consolidation) and the time elapsed. Ensure that the consolidation is complete.
14. Follow Steps 7 to 11.

(iii) *Consolidated drained test*

15. Follow Steps 12 and 13.
16. Adopt a slow rate of strain during load application such that complete drainage occurs with 95% pore pressure dissipation.
17. Follow Steps 10 and 11.

**Computations**

Calculate the proving ring constant and hence the load at different displacements. Calculate the shear stress using the corrected area, which is given as

$$\text{Corrected area} = A_0\left(1-\frac{\delta}{3}\right)$$

where $A_0$ is the initial area of specimen and $\delta$ the displacement.

**Results**

Plot the shear stress versus the longitudinal displacement readings and note down the maximum shear stress and the corresponding longitudinal displacement for the particular normal stress. Plot the normal stress versus the maximum shear stress and obtain the shear strength parameters, $c$ and $\phi$.

For the consolidated undrained and consolidated drained tests, report the consolidation pressure and the consolidation characteristics. Typical data and results are shown in Tables 10.25 to 10.28. Figures 10.21 and 10.22 represent the shear stress–displacement and Coulomb's strength envelope, respectively.

**Discussion**

The procedure outlined above is for soils with particle size not greater than 4.75 mm (IS: 2720 – Part 13, 1972). For a test procedure for soils containing gravel, the reader may refer to IS: 2720 – Part 39/Sec. 1 (1977).

Failure in direct shear may be considered to occur at maximum shear stress or at maximum obliquity of the Mohr failure envelope. The angle of shearing resistance obtained considering the maximum shear stress is less than the other one, and the error is on the safe side. The error involved is much more important in sands than in clays.

**Table 10.25** Data of direct shear test

| | |
|---|---|
| Type of test | Consolidated undrained test |
| Rate of strain | 1.27 mm/min |
| Soil specimen | Undisturbed |
| Size = 60 mm × 60 mm | Area = 36 $cm^2$ |
| Height = 25 mm | Volume = 90 $cm^3$ |
| Initial wet weight = 186.07 g | Initial water content = 48.6% |
| Bulk density = 2.07 g/cc | |

**Table 10.26** Consolidation details
Normal stress = 0.01 $N/mm^2$

| Time (Hour) | Time (Minutes) | Vertical dial reading | Vertical dial difference | Thickness of specimen (mm) |
|---|---|---|---|---|
| | 0 | 0 | 0 | 25.0 |
| | 15 | 20 | 20 | 24.8 |
| | 30 | 23 | 3 | 24.77 |
| 1 | 00 | 24 | 1 | 24.76 |
| 1 | 30 | 27 | 3 | 24.73 |
| 2 | 00 | 30 | 3 | 24.70 |
| 2 | 30 | 32 | 2 | 24.68 |
| 3 | 00 | 36 | 4 | 24.64 |
| 3 | 30 | 37 | 1 | 24.63 |
| 4 | 00 | 38 | 1 | 24.62 |
| 5 | 00 | 38 | 0 | 24.62 |
| 6 | 00 | 38 | 0 | 24.62 |
| 7 | 00 | 38 | 0 | 24.62 |

**Table 10.27** Shearing details
Normal stress = 0.04 N/mm$^2$

| Displacement dial reading | Displacement $\delta$ (cm) | Area correction | Corrected area (mm$^2 \times 100$) | Stress dial reading | Shear force (5)×1.108 (N) | Shear stress (N/mm$^2$) |
|---|---|---|---|---|---|---|
| 20 | 0.02 | 0.993 | 35.75 | 29 | 32.14 | 0.0089 |
| 40 | 0.04 | 0.987 | 35.53 | 53 | 58.72 | 0.0165 |
| 60 | 0.06 | 0.980 | 35.28 | 71 | 78.67 | 0.0223 |
| 80 | 0.08 | 0.973 | 35.03 | 85 | 94.18 | 0.0269 |
| 100 | 0.10 | 0.967 | 34.81 | 97 | 107.48 | 0.0309 |
| 120 | 0.12 | 0.960 | 34.56 | 106 | 117.45 | 0.0340 |
| 140 | 0.14 | 0.953 | 34.31 | 114 | 126.31 | 0.0368 |
| 160 | 0.16 | 0.947 | 34.09 | 117 | 129.64 | 0.0380 |
| 180 | 0.18 | 0.940 | 33.84 | 119 | 131.85 | 0.0390 |
| 200 | 0.20 | 0.933 | 33.59 | 121 | 134.07 | 0.0399 |
| 220 | 0.22 | 0.927 | 33.37 | 122 | 135.18 | 0.0405 |
| 240 | 0.24 | 0.920 | 33.12 | 121.5 | 134.62 | 0.0407 |
| 260 | 0.26 | 0.913 | 32.87 | 120 | 132.93 | 0.0405 |
| 280 | 0.28 | 0.907 | 32.65 | 117 | 129.64 | 0.0397 |

**Table 10.28** Shear parameters

| Test no. | Normal stress (N/mm$^2$) | Shear stress (N/mm$^2$) | Cohesion (N/mm$^2$) | Angle of shearing resistance (°) |
|---|---|---|---|---|
| 1 | 0.04 | 0.0397 | | |
| 2 | 0.06 | 0.0510 | 0.018 | 28 |
| 3 | 0.08 | 0.0590 | | |
| 4 | 0.12 | 0.0810 | | |

## 10.21 TEST NO. 20: TRIAXIAL SHEAR TEST

**Scope**
To determine the shear strength of a soil using triaxial shear apparatus

**Apparatus**
Triaxial compression machine – 50 kN capacity (strain-controlled) with a strain rate range of 0.05 to 7.5 mm/min (Fig. 10.23)
Triaxial cell to accommodate 38 mm diameter sample
Constant cell pressure system with a capacity of 1,000 kN/m$^2$
Volume measuring device
Pore pressure apparatus
Rubber membrane
Membrane stretcher
Sample trimming device

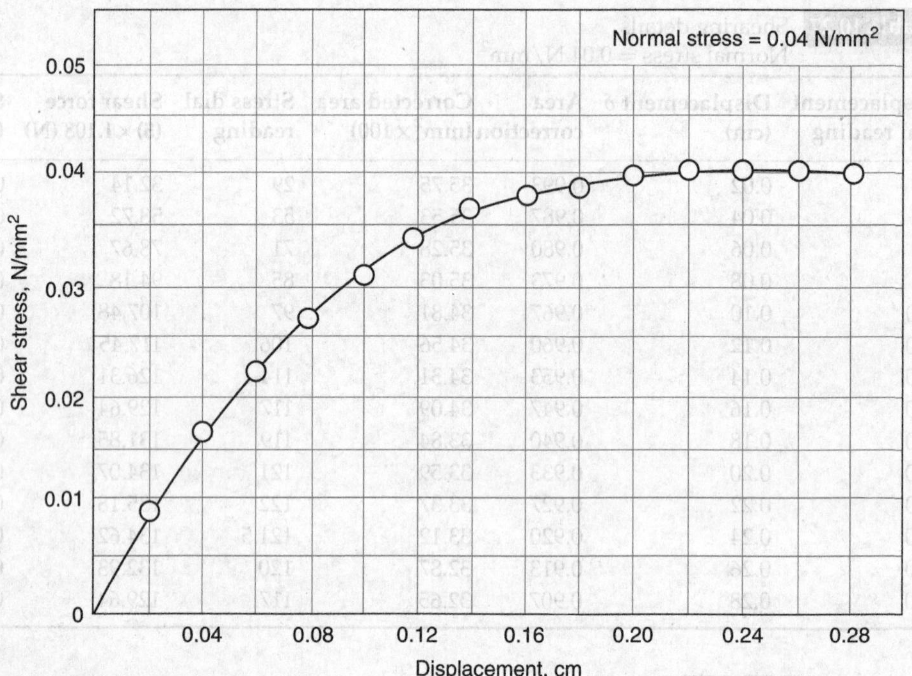


**Fig. 10.21** Shear stress versus displacement

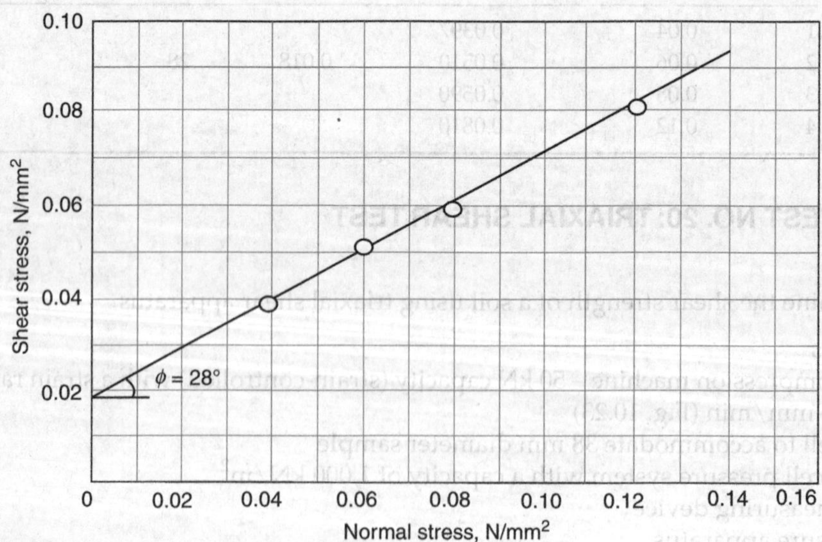


**Fig. 10.22** Coulomb's strength envelope

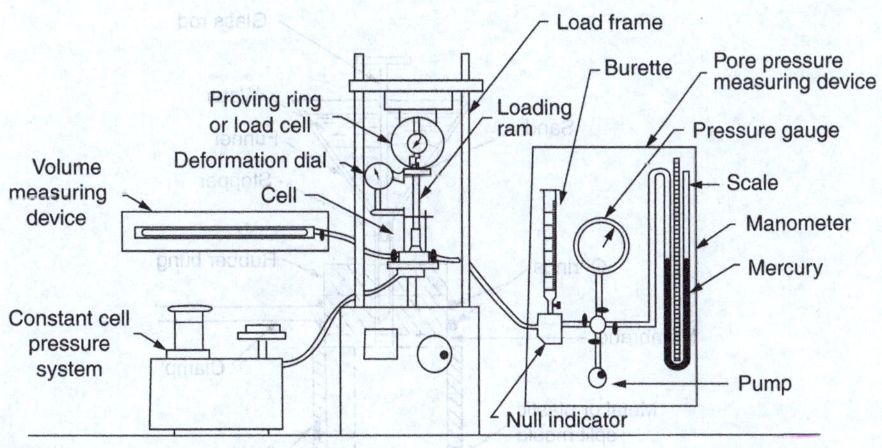


Fig. 10.23 Triaxial test assembly

Split mould
Trimming knife
Stopwatch
Apparatus for water content determination

**Procedure**

**(a)** ***Preparation of specimen***

(i) *Cohesive soil*

1. Undisturbed, compacted, or re-moulded cohesive soil specimens may be prepared as explained for the unconfined compression test (Test no. 18).

(ii) *Non-cohesive soil*

2. Attach a rubber membrane to the base platen (Fig. 10.24) using rubber O-rings. Place a porous stone on the base of the platen.
3. Take a known mass of dry sand so that the sample density can be obtained and approximately duplicated for successive tests.
4. Place a split mould around the membrane and fold the top portion of it over the mould.
5. Carefully transfer the sand to the membrane in two or three layers and tamp each layer with a glass rod to obtain the shape and density. If the test is to be conducted under saturated conditions, the sand may be placed in water and then transferred to the membrane.
6. Place a porous stone on the top of the sample and then place the top platen. Apply silicone grease to the sides of the platen to obtain a better leak-proof seal. Roll the membrane on to the top platen and seal it with rubber O-rings.
7. Attach a tube from the top platen to the vacuum outlet and apply a vacuum of 200 to 250 mm of mercury to the sample, or if the test is to be carried out in a saturated condition, attach a tube to the base, connect it to the U-tube manometer of the pore pressure apparatus, and apply a small negative pore water pressure to keep the specimen straight.
8. Remove the split mould and check for holes and leaks.
9. Take the average height and diameter to obtain the density.

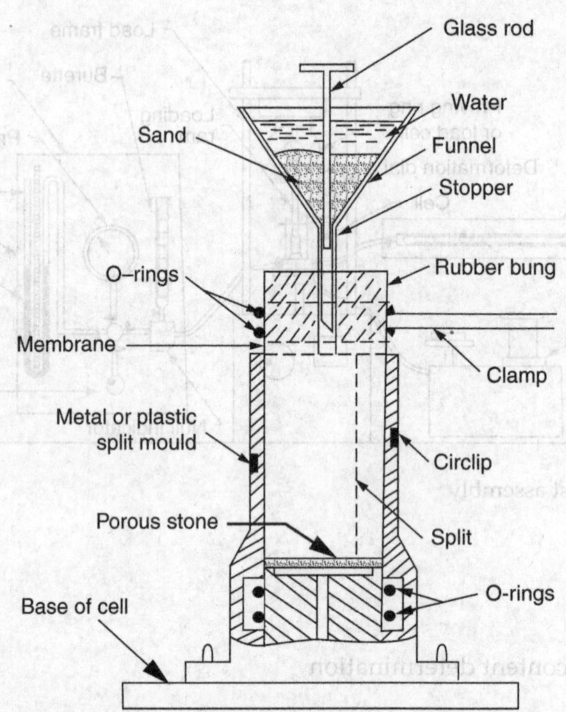


**Fig. 10.24** Preparation of saturated cohesionless soil specimen

**(b) Shear tests**

(i) *Undrained test*

10. Measure the dimensions of the specimen. Weigh the specimen and keep a representative sample for water content determination.
11. Place a solid Perspex platen over the specimen, which in turn is placed over another Perspex platen. Place the loading cap on the top platen.
12. Insert a rubber membrane using a membrane stretcher and fix two O-rings, one at the bottom and the other on the top of the platen or loading cap. (Steps 10 to 12 are not needed for non-cohesive soil specimens prepared following Steps 2 to 9.)
13. Place one cell on the triaxial cell base and transfer the same to the compression machine, and just make load contact of the loading ram.
14. Close the drainage valve, fill the cell with water, and apply the pre-determined chamber pressure.
15. Adjust the deformation and proving ring dials to zero, and apply the axial load with a strain rate of 0.5% to 2% per minute.
16. Record the force and deformation readings at suitable intervals with a closer spacing during the initial stages of the test.
17. Apply the load till the proving ring dial recedes backwards or until an axial strain of 20% is reached.

18. Unload the specimen and drain off the cell fluid. Dismantle the cell and carefully remove the membrane and note down the mode of failure.
19. Weigh the specimen and take representative water content samples from the failure zone of the specimen.
20. Repeat the test on three or more identical specimens under increased cell pressures.

(ii) *Consolidated undrained test*

21. Follow Steps 1 to 10. Place the specimen over a saturated porous stone which in turn is placed on top of the specimen, and then place a loading cap with a drainage outlet.
22. To quicken the process of consolidation of the specimen, place a series of small threads or a strip of filter paper around the surface of the specimen. Now insert the rubber membrane and fix the O-rings, one at the bottom and another at the top.
23. Follow Step 13. Connect the drainage valve to a volume-measuring device or to a burette. Fill the cell with water and apply the pre-determined chamber pressure.
24. Open the drainage valve and note the volume change during consolidation. Ascertain the completion of the volume change by noting down the constant water level in the burette or volume change device. Close the drainage valve.
25. If the pore water pressure is to be measured, connect a pressure transducer or a null pressure indicator device to the saturation line.
26. Adjust the null-indicator to the initial position or the transducer output to the initial reading. Adjust the deformation and proving ring dials to zero and apply the axial load at a slow rate such that the pore pressure readings can be taken conveniently.
27. Record the force, deformation and pore pressure readings at suitable intervals with a closer spacing during the initial stages of the test.
28. Follow Steps 17 to 20.

## Computations

1. The axial strain

$$\varepsilon = \frac{\Delta L}{L_0}$$

2. The average cross-sectional area $A$ at a particular strain is calculated as done in the unconfined compression test.
3. The deviator stress

$$\Delta\sigma = \frac{\text{PRR} \times \text{PRC}}{A}$$

Plot the deviator stress versus the strain and obtain the stress at the peak point unless the stress at 20% strain occurs first. Compute the major principal stress for each test as

$$\sigma_1 = \sigma_3 + \Delta\sigma$$

Also compute the pore water pressure corresponding to the maximum deviator stress. Compute the effective major and minor principal stresses as

$$\sigma_3' = \sigma_3 - u_w \quad \text{and} \quad \sigma_1' = \sigma_1 - u_w$$

**Table 10.29** Data and test results for dry non-cohesive soil from the triaxial test

Length of specimen = 76 mm
Diameter of specimen = 38 mm
Proving ring constant (PRC) 1 div = $3.13 \times 10^{-3}$ kN

| | Vertical deformation (mm × 0.01) | $\varepsilon = \Delta L/L_0$ | $1-\varepsilon$ | $A = A_0/(1-\varepsilon)$ ($m^3 \times 10^{-3}$) | Proving ring reading (PRR) | Deviator stress $\Delta\sigma = \dfrac{PRR \times PRC}{A}$ | $\sigma'_1 = \sigma'_3 + \Delta\sigma'$ ($kN/m^2$) |
|---|---|---|---|---|---|---|---|
| Sample 1 | 0 | 0 | 0.1 | 1.134 | 0 | 0 | 140.0 |
| $\sigma'_3 = 140\ kN/m^2$ | 38 | 0.005 | 0.995 | 1.140 | 116 | 318.5 | 458.5 |
| | 143 | 0.019 | 0.981 | 1.156 | 220 | 595.7 | 735.7 |
| | 287 | 0.038 | 0.962 | 1.179 | 247 | 655.7 | 795.7 |
| | 430 | 0.057 | 0.943 | 1.202 | 250 | 651.0 | 791.0 |
| | 776 | 0.102 | 0.898 | 1.263 | 258 | 639.4 | 779.4 |
| | 1,346 | 0.177 | 0.823 | 1.378 | 265 | 601.9 | 741.9 |
| | 1,742 | 0.229 | 0.771 | 1.471 | 272 | 578.8 | 718.8 |
| Sample 2 | 0 | 0 | 1 | 1.134 | 0 | 0 | 210.0 |
| $\sigma'_3 = 210\ kN/m^2$ | 73 | 0.010 | 0.990 | 1.145 | 199 | 544.0 | 754.0 |
| | 238 | 0.031 | 0.969 | 1.170 | 256 | 684.9 | 894.9 |
| | 412 | 0.054 | 0.946 | 1.200 | 333 | 868.6 | 1078.6 |
| | 500 | 0.066 | 0.934 | 1.214 | 342 | 881.8 | 1091.8 |
| | 685 | 0.090 | 0.910 | 1.246 | 352 | 884.2 | 1094.2 |
| | 1,082 | 0.142 | 0.858 | 1.322 | 355 | 840.5 | 1050.5 |
| | 1,515 | 0.199 | 0.801 | 1.416 | 360 | 795.8 | 1005.8 |
| Sample 3 | 0 | 0 | 1 | 1.134 | 0 | 0 | 280.0 |
| $\sigma'_3 = 280\ kN/m^2$ | 88 | 0.012 | 0.988 | 1.147 | 311 | 848.7 | 1128.7 |
| | 272 | 0.036 | 0.964 | 1.176 | 369 | 982.1 | 1262.1 |
| | 472 | 0.062 | 0.938 | 1.209 | 402 | 1040.7 | 1320.7 |
| | 572 | 0.075 | 0.925 | 1.226 | 410 | 1046.7 | 1326.7 |
| | 772 | 0.102 | 0.898 | 1.262 | 421 | 1044.2 | 1324.2 |
| | 1,215 | 0.160 | 0.840 | 1.350 | 425 | 985.4 | 1265.4 |
| | 1,447 | 0.190 | 0.810 | 1.401 | 433 | 967.4 | 1247.4 |

Plot Mohr's circles for both the total and effective principal stresses and obtain the shear strength parameters ($c$, $\phi$ and $c'_{cu}$, $\phi'_{cu}$).

**Results**
The results are presented in the form of a stress–strain curve, strain–volume change curve, stress–pore pressure curve, and Mohr–Coulomb plot. Some typical test results are presented in Tables 10.29 and 10.30 and in Figs. 10.25 to 10.27.

**Discussion**
For certain special field conditions, the samples have to be consolidated anisotropically. This is done using a dead-load frame and applying vertical pressure in conjunction with the cell pressure $\sigma_3$ to develop the desired stress ratio

$$K = \frac{\sigma'_h}{\sigma'_v}$$

Pore pressure and volume measuring devices should be perfectly de-aired before use to obtain accurate results. Pore pressure measurements can be performed using a pressure transducer. In such cases, connect the pressure transducer to the saturation line and in turn connect the output of the transducer to a voltmeter. Pressure transducers should be very

**Table 10.30** Data and test results for dry non-cohesive soil from the triaxial test
Length of specimen = 76 mm
Diameter of specimen = 38 mm
PRC 1 div = $3.13 \times 10^{-3}$ kN

| Vertical deformation (mm) | $\varepsilon = \Delta L/L_0$ | $1-\varepsilon$ | $A=A_0/(1-\varepsilon)$ ($m^2 \times 10^{-3}$) | PRR | Pore pressure $u_w$ (kN/m$^2$) | Deviator stress, $\Delta\sigma'$ = PRR×PRC/$A$ (kN/m$^2$) | $\sigma'_3=\sigma_3 - u_w$ (kN/m$^2$) | $\sigma'_1 = \sigma'_3 + \Delta\sigma'$ (kN/m$^2$) |
|---|---|---|---|---|---|---|---|---|
| 0 | 0 | 1 | 1.134 | 0 | 0 | 0 | 420 | 420 |
| 3.05 | 0.040 | 0.960 | 1.181 | 680 | 110 | 461 | 310 | 571 |
| 3.81 | 0.050 | 0.950 | 1.194 | 699 | 112 | 468 | 308 | 580 |
| 5.08 | 0.067 | 0.933 | 1.215 | 835 | 131 | 550 | 289 | 681 |
| 6.35 | 0.084 | 0.916 | 1.238 | 998 | 172 | 645 | 248 | 817 |
| 7.62 | 0.100 | 0.900 | 1.260 | 1,101 | 218 | 699 | 202 | 917 |
| 8.89 | 0.117 | 0.883 | 1.284 | 1,168 | 241 | 728 | 179 | 969 |
| 10.16 | 0.134 | 0.866 | 1.310 | 1,221 | 249 | 746 | 171 | 995 |
| 11.43 | 0.150 | 0.850 | 1.334 | 1,266 | 255 | 759 | 165 | 1,014 |
| 12.70 | 0.167 | 0.833 | 1.361 | 1,281 | 252 | 753 | 168 | 1,005 |
| 13.97 | 0.184 | 0.816 | 1.390 | 1,291 | 249 | 743 | 171 | 992 |
| 15.24 | 0.201 | 0.799 | 1.419 | 1,298 | 248 | 732 | 172 | 980 |
| 16.51 | 0.217 | 0.783 | 1.448 | 1,306 | 248 | 722 | 172 | 970 |

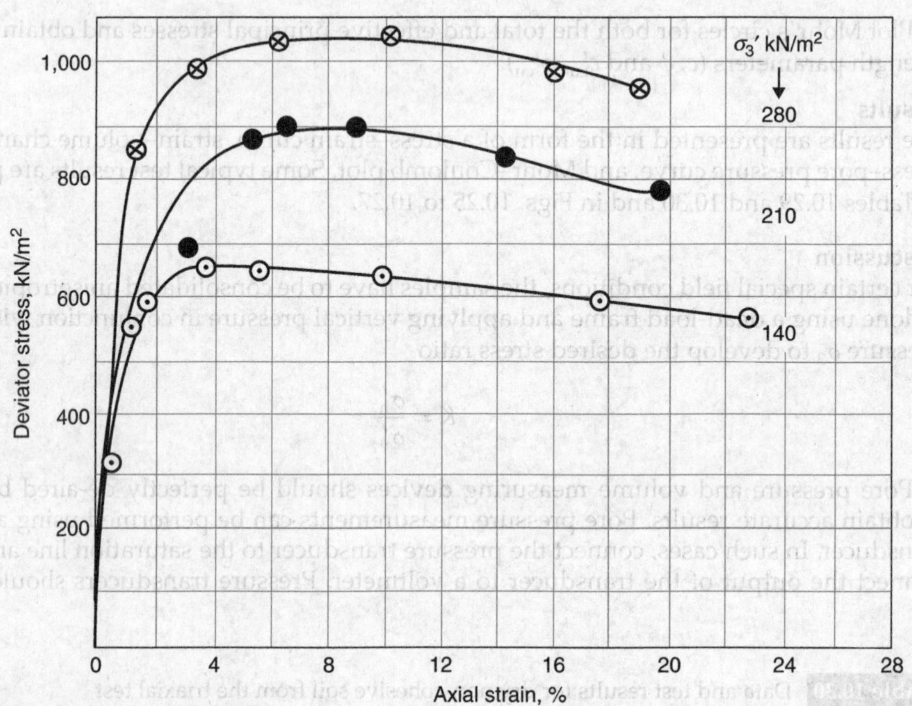


**Fig. 10.25** Deviator stress–strain curves

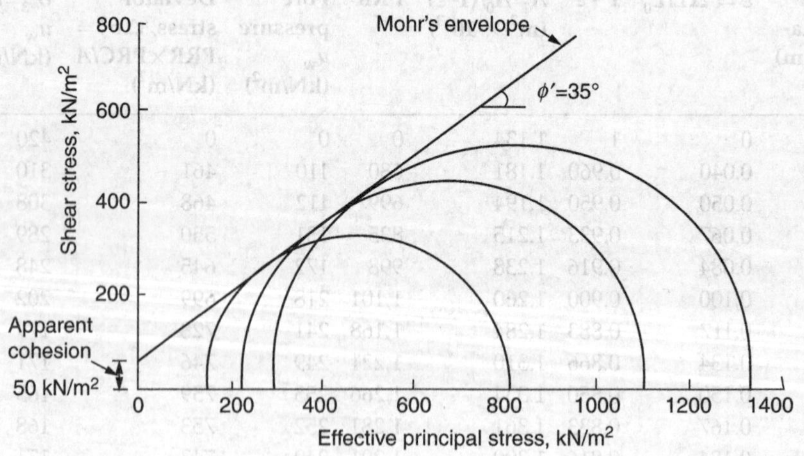


**Fig. 10.26** Mohr–Coulomb plot for dry cohesionless soil

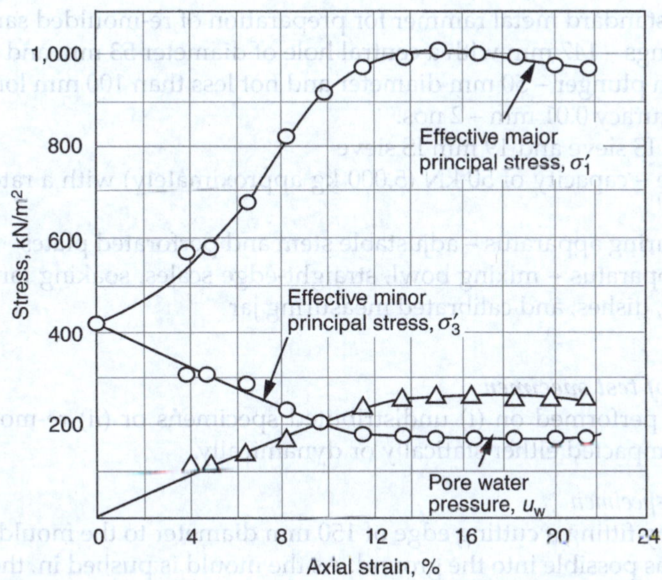


**Fig. 10.27** Stress plots with strain

sensitive, even to small volume displacements. Special-type loading rams have to be used for testing sensitive clays.

For more details of the triaxial equipment and procedure, the reader may refer to IS: 2720 – Part 11 (1971), Part 12 (1981), and Part 35 (1974).

Other special tests which can be performed using the triaxial apparatus are the extension test, the decreasing $\sigma_3$ test, the constant volume test, etc. (Bishop and Henkel, 1962).

## 10.22 TEST NO. 21: CALIFORNIA BEARING RATIO (CBR) TEST

**Scope**

To determine the California bearing ratio

**Definition**

The California bearing ratio (CBR) is expressed as the percentage of force per unit area required to penetrate a soil mass with a circular plunger of 50 mm diameter at a rate of 1.25 mm/min compared with that required for the corresponding penetration in a standard material. The ratio is usually determined for penetration values of 2.5 and 5 mm. In general the penetration value at 2.5 mm is greater than at 5 mm and the penetration value corresponding to 2.5 mm is taken as the design CBR value. However, if the ratio at 5 mm is consistently higher than that at 2.5 mm, the ratio at 5 mm is used.

**Apparatus**

Cylindrical mound – 150 mm inner diameter and 175 mm height
Collar – 50 mm height and 150 mm diameter
Base plate – 10 mm height
Metal spacer disc – 148 mm diameter and 47.7 mm in height

Metal rammer – standard metal rammer for preparation of re-moulded samples
Annular metal rings – 147 mm with a central hole of diameter 53 mm and weight 2.5 kg
Metal penetration plunger – 50 mm diameter and not less than 100 mm long
Dial gauges – accuracy 0.01 mm – 2 nos.
Sieves – 4.75 mm IS sieve and 19 mm IS sieve
Loading machine – capacity of 50 kN (5,000 kg approximately) with a rate of strain of 1.25 mm/min
Expansion measuring apparatus – adjustable stem and perforated plates
Miscellaneous apparatus – mixing bowl, straight-edge scales, soaking tank or pan, drying oven, filter paper, dishes, and calibrated measuring jar

**Procedure**

(a) ***Preparation of test specimen***

The test may be performed on (i) undistributed specimens or (ii) re-moulded specimens which may be compacted either statically or dynamically.

(i) *Undistributed specimen*

This is obtained by fitting a cutting edge of 150 mm diameter to the mould and pushing the mould as gently as possible into the ground. As the mould is pushed in, the soil is dug from the outside. When the mould is full of soil, it is removed by under-digging. Then the top and bottom surfaces are trimmed flat so as to get a specimen of the required length ready for testing. If the soil is hard and the mould cannot be pressed, then a large undisturbed lump of soil is cut out, from which the required specimen for the mould is made. If the specimen is loose in the mould, the annular cavity is filled with paraffin wax. The density of the soil and the water content are determined so as to determine the dry density.

(ii) *Re-moulded specimen*

The dry density to be determined for the re-moulded specimen may be the field density, maximum dry density, or any other density. The water content required for preparation of the specimen may be the OMC or the field moisture context.

The soil for the re-moulded specimen shall pass a 19 mm sieve. Allowance for larger size particles shall be made by replacing the soil by an equal amount of material which passes a 19 mm IS sieve but is retained in a 4.75 mm sieve.

The required quantity of wet soil is prepared and compacted in the mould either statically or dynamically as required.

In both cases of compaction, if the specimen is to be soaked, the water content of the soil before and after compaction is determined. If the specimen is not soaked, a representative sample of material from one of the pieces of the material cut after penetration shall be taken to determine the water content.

(b) ***Test for swelling***

(i) Place a filter paper over the specimen and the adjustable stem. Place the perforated plate on the compacted soil specimen in the mould.
(ii) Place weights on the compacted soil specimen to produce a surcharge equal to the weight of the base material and pavement to the nearest 2.5 kg.
(iii) Immerse the whole mould and weights in a tank of water allowing free access of water to the top and bottom of the specimen.

(iv) Mount the tripod of the expansion mounting device on the edge of the mould and record the initial dial gauge reading.
(v) Keep the set-up for 96 hours without disturbance and note the readings every day against the time of reading. Maintain a constant water level throughout the period.
(vi) At the end of the soaking period, note the change in the dial gauge reading and remove the tripod and the mould from the water tank.
(vii) Allow the specimen to drain for 15 minutes downwards.
(viii) Remove the weights, the perforated plate, and the filter plate.
(ix) Weigh the mould with the soaked soil specimen.

(c) ***Penetration test***

(i) Place the mould containing the specimen, with the base plate in position and the top face exposed, on the lower plate of the testing machine (Fig. 10.28).
(ii) Place on the specimen the required number of surcharge weights to simulate the intensity of loading equivalent to the base material and pavement.
(iii) In order to prevent upheaval of soil into the holes of the surcharge weights, place 2.5 kg of annular weights on the surface prior to seating the penetration plunger and then the balance surcharge weights.
(iv) Apply a seating load of 4 kg so that free contact is established between the surface of the specimen and the plunger.
(v) Set the load and deformation gauges to zero.
(vi) Apply the load on the plunger into the soil at a rate of 1.25 mm/min.
(vii) Note the readings of the load at penetrations of 0.5, 1.0,1 .5, 2.0, 2.5, 4.0, 5.0, 7.5, 10.0, and 12.5 mm.
(viii) Note the maximum load and penetration if the maximum load occurs at a penetration less than 12.5 mm.
(ix) Raise the plunger and detach the mould from the loading machine.
(x) Determine the water content of the soil sample taken from the top 30 mm layer of the specimen.
(xi) Find also the average water content of the specimen by taking samples from the entire depth of the specimen.
(xii) In case of undisturbed specimens from the field, carefully examine the presence of any oversize soil particles which may affect the results if they happen to be located directly below the penetration plunger.
(xiii) As a check, the penetration test may be repeated on the rear side of the specimen.

(*d*) ***Expansion ratio***

The expansion ratio is calculated from the expression

$$\text{Expansion ratio} = \frac{d_f - d_s}{h} \times 100$$

where $d_f$ is the final dial gauge reading in mm, $d_s$ the initial dial gauge reading in mm, and $h$ the initial height of the specimen in mm.

The expansion ratio is used to identify qualitatively the potential expansiveness of the soil.

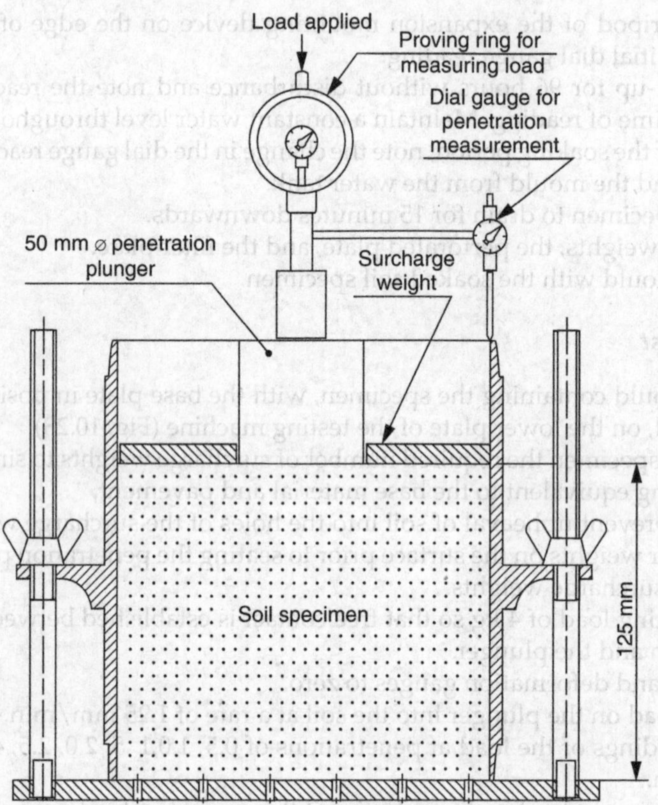


Fig. 10.28 Set-up for CBR test

### (e) *Load–penetration curve*

Plot the load–penetration curve. This curve is usually convex upwards although the initial portion of the curve may be convex downwards due to surface irregularities. A correction shall then be applied by drawing a tangent to the point of greatest slope and then transposing the axis of the load so that zero penetration is taken as the point where the tangent cuts the axis of penetration. The corrected load–penetration curve will then consist of the tangent from the new origin to the point of tangency on the re-shifted curve and then the curve itself.

### (f) *The CBR*

The CBR values are usually calculated for penetrations of 2.5 and 5.0 mm. Corresponding to the penetration value at which the CBR value is desired, the corrected load value should be taken from the load–penetration curve and the CBR calculated as follows:

$$\text{CBR} = \frac{P_\text{T}}{P_\text{s}} \times 100$$

where $P_\text{T}$ is the corrected unit (or total) test load corresponding to the chosen penetration from the load–penetration curve and $P_\text{s}$ the unit (or total) standard load for the soil depth of penetration as for $P_\text{T}$ taken from the table given in Fig. 10.29.

Generally, the CBR value at 2.5 mm penetration will be greater than that at 5.0 mm penetration and in such a case, the former shall be taken as the CBR value for design purposes. If the CBR value corresponding to a penetration of 5.0 mm exceeds that for 2.5 mm, the test shall be repeated. If identical results follow, the CBR corresponding to 5.0 mm penetration shall be taken for the design.

(g) ***Presentation of results***

(i) Weight of mould with base plate = 7,445 g
(ii) Weight of mould with base plate + wet soil = 12,495 g
(iii) Weight of wet soil = 5,050 g
(iv) Weight of mould + wet soil after soaking = 12,820 g
(v) Weight of water absorbed = 325 g
(vi) Percentage of water absorbed = 6.44%
(vii) Moisture content = 15.90%

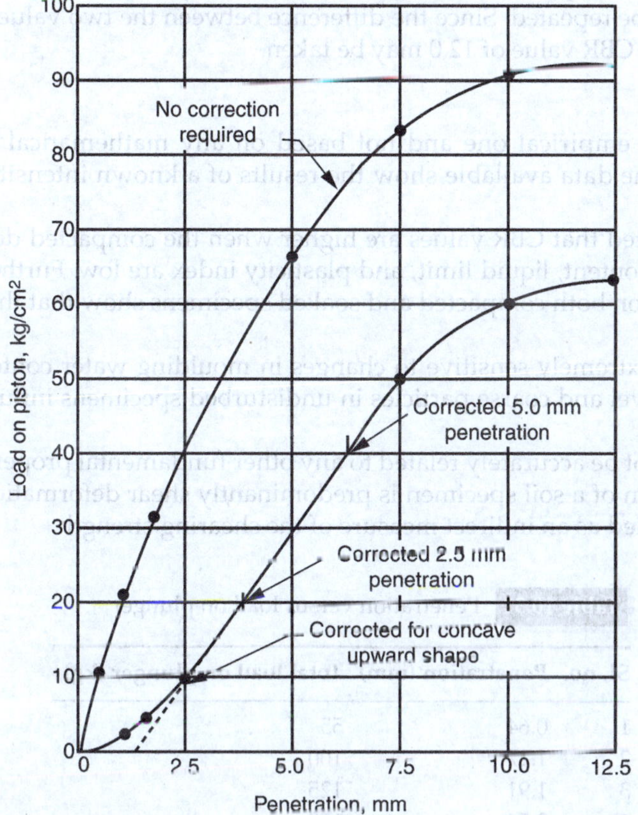


| Penetration depth (mm) | Unit standard pressure (kg/cm$^2$) | Total standard load (kgf) |
|---|---|---|
| 2.5 | 70 | 1,370 |
| 5.0 | 105 | 2,055 |

**Fig. 10.29** Correction load penetration curves

Penetration values versus load on plunger is given in Table 10.31 and presented in Fig. 10.30.

From Fig. 10.30, the load on plunger at 2.5 mm = 155 kg
5.0 mm = 255 kg

Then

$$\text{CBR value at 2.5 mm penetration} = \frac{155}{1370} \times 100 = 11.3$$

$$\text{CBR value at 5.0 mm penetration} = \frac{255}{2055} \times 100 = 12.4$$

**Result**

The CBR value at 5.0 mm penetration is greater than the CBR at 2.5 mm penetration. As per rule, the test has to be repeated. Since the difference between the two values is small for all practical purposes a CBR value of 12.0 may be taken.

**Discussion**

The CBR test is an empirical one and not based on any mathematical reasoning. It is only of use when the data available show the results of a known intensity of traffic on a pavement.

It has been reported that CBR values are higher when the compacted densities are high and when the clay content, liquid limit, and plasticity index are low. Further, the results of the penetration test on both compacted and soaked specimens show that the results are not reproducible.

CBR values are extremely sensitive to changes in moulding water content and density. The presence of gravel and coarse particles in undisturbed specimens influence greatly the CBR value.

CBR values cannot be accurately related to any other fundamental property of soil. However, the deformation of a soil specimen is predominantly shear deformation, and the CBR values can be regarded as an indirect measure of the shearing strength.

**Table 10.31** Penetration versus load on plunger

| Sl. no. | Penetration (mm) | Total load on plunger (kg) |
|---|---|---|
| 1 | 0.64 | 55 |
| 2 | 1.27 | 100 |
| 3 | 1.91 | 125 |
| 4 | 2.54 | 150 |
| 5 | 5.08 | 250 |
| 6 | 7.62 | 350 |
| 7 | 10.16 | 400 |
| 8 | 12.17 | 450 |

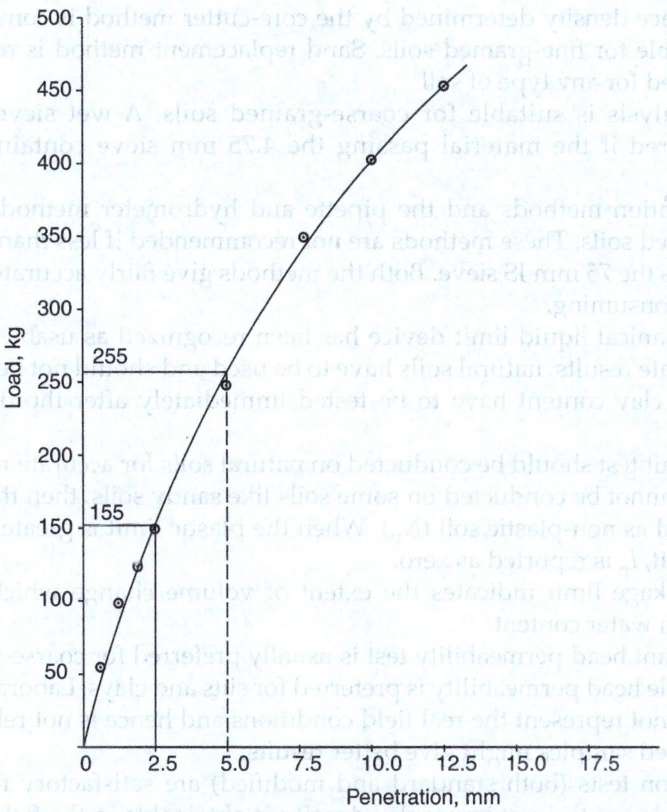


Fig. 10.30 Load – penetration curve

## POINTS TO REMEMBER

10.1 Dry soil samples have to be prepared for laboratory tests whenever needed. Drying may be done in air or oven as the case may be. Oven drying is generally done for 24 hours at 110 ⊥ 5°C.

10.2 The density bottle method is a laboratory method for determination of the *G* of fine-grained soils. The pycnometer or gas jar method is used for all soils. In the test the major source of error is the complete removal of air from the sample. For soils containing soluble salts, kerosene or white spirit may be preferred in place of water.

10.3 The factors which are essential for accurate determination of water content are the mass of the wet representative sample, the temperature, and the duration of drying of the sample. The ovendrying method of water content determination is recommended by Indian Standards as the standard method.

10.4 The in-place density determined by the core-cutter method is convenient and quick and suitable for fine-grained soils. Sand replacement method is relatively slow but can be used for any type of soil.

10.5 Sieve analysis is suitable for coarse-grained soils. A wet sieve analysis has to be preferred if the material passing the 4.75 mm sieve contains more clay-size particles.

10.6 Sedimentation methods and the pipette and hydrometer methods are suitable for fine-grained soils. These methods are not recommended if less than 10% of the material passes the 75 mm IS sieve. Both the methods give fairly accurate results, but both are time consuming.

10.7 The mechanical liquid limit device has been recognized as usable in a routine test. For accurate results, natural soils have to be used and should not be oven dried. Soils with low clay content have to be tested immediately after thorough mixing with water.

10.8 Plastic limit test should be conducted on natural soils for accurate results. If the plastic limit cannot be conducted on some soils like sandy soils, then the plasticity index is reported as non-plastic soil ($N_p$). When the plastic limit is greater than or equal to liquid limit, $I_p$ is reported as zero.

10.9 The shrinkage limit indicates the extent of volume change which can occur with changes in water content.

10.10 The constant head permeability test is usually preferred for coarse-grained soils, and the variable head permeability is preferred for silts and clays. Laboratory $k$ determination does not represent the real field conditions and hence is not reliable. But silts on undisturbed samples might give better results.

10.11 Compaction tests (both standard and modified) are satisfactory for cohesive soils. A knowledge of the maximum dry density is obtainable in the field using a suitable roller and adopting a moulding water content almost equal to the OMC.

10.12 In a consolidation test, a floating ring reduces the frictional loss along the sides and the test is faster. A fixed ring has the advantage of providing the $k$ value of the sample. Larger samples provide more reliable results.

10.13 The unconfined compressive strength test is a quick test and gives the approximate shear strength of a cohesive soil. Further, in the test, failure occurs along the weakest portion and hence provides a conservative shear strength value.

10.14 Failure in a direct shear test may be considered to occur at the maximum shear stress or at the maximum obliquity of the Mohr failure envelope. The $\phi$ angle obtained considering the maximum shear stress is less than the others, which is on the safe side.

10.15 In order to obtain accurate results from the triaxial test, the pore pressure and volume measuring devices should be perfectly air dried. A special type of loading ram has to be used for testing sensitive clays. For certain field conditions, the samples have to be consolidated anisotropically.

10.16 The CBR is expressed as the percentage of the force per unit area required to penetrate a soil mass with a standard plunger compared with that required for the corresponding penetration in a standard material.

## QUESTIONS

### Objective Questions

10.1 State whether the following statements are true or false:

(1) The size of particles smaller than 0.075 mm is generally obtained from a wet mechanical analysis.

(2) The rate of secondary compression is dependent on the specimen thickness.

(3) Changes in the laboratory temperature affects the permeability, which in turn affects the coefficient of consolidation.

(4) The vane shear test cannot be used where the apparent angle of internal friction ($\phi_u$) is not equal to zero.

(5) In an unconfined compression test, the inclination of the failure plane is always 45°.

10.2 Increase in permeability of a soil results due to change from

(a) Large to small size particles for the same void ratio

(b) High to a low viscous fluid at the same temperature

(c) Dry side of optimum to wet side in a compacted specimen at the same porosity

(d) Flocculated to dispersed structure at the same dry density

10.3 For a highly fissured clay the best method of finding the undrained shear strength is

(a) The direct shear test

(b) The triaxial shear test with $\sigma_3 = 0$

(c) The field vane shear test

(d) The unconfined compression test

10.4 The effective shear strength parameters of a sand can be obtained by conducting

(a) Consolidated undrained tests on saturated samples in triaxial shear

(b) Unconsolidated undrained tests on saturated specimens with pore water pressure measurement in triaxial shear

(c) The field vane shear test with the a low rate of loading

(d) Consolidated undrained tests on dry sand in direct shear

10.5 In a laboratory consolidation test, will the coefficient of consolidation alter (answer yes or no)

(a) If the pore fluid is replaced by salt water?

(b) If the rate of loading is changed?

(c) If the room temperature is increased?

(d) If the load duration is reduced?

### Descriptive Questions

10.6 State the maximum and minimum sizes of particles which may be determined by hydrometer analysis and give reasons for these limitations.

10.7 What are the inherent errors in using Stokes law to determine the grain-size distribution of fine-grained soils?

10.8 Explain the reasons for plotting a grain-size distribution curve on a semi-logarithmic plot rather than on an arithmetic scale.

10.9 Why should only distilled water be used in running the tests for limits?

10.10 In an Atterberg limit test, the drop of the cup was found to be 0.95 cm. If the liquid limit as obtained was 72%, is the true liquid limit is greater or lesser than 72%?

10.11 During the determination of the volume of a soil pat, a certain quantity of air was entrapped between the plate and the pat. How will this affect the shrinkage limit result?

10.12 Explain why the dry density is used instead of the wet density in describing the density of a soil mass.

10.13 How are we justified in using laboratory methods for determining the coefficient of permeability of soils?

10.14 How do you ensure that saturation is complete in a variable head permeability test specimen?

10.15 What are the effects of the friction on the loading ram in a triaxial test on the shear strength of a soil? How will you eliminate the friction on the loading ram?

10.16 How will you ensure when failure occurs in a soil specimen tested in the direct or triaxial shear apparatus?

10.17 Explain the reasons for the loss in strength of clay as a result of re-moulding.

10.18 Explain the influence of the end restraint on the triaxial shear test on saturated specimens.

10.19 It is always preferable to obtain samples for the consolidation test from strata underlying a building site. Why?

10.20 In conventional laboratory compression testing, what is the cause of soil volume decrease?

# 11

# Lateral Earth Pressure

**CHAPTER HIGHLIGHTS**

Limit analysis and limit equilibrium methods – Earth pressure at rest – Rankine's states of plastic equilibrium – Rankine's earth pressure theory – Coulomb's earth pressure theory – Culmann's graphical method – Poncelet's graphical method

## 11.1 INTRODUCTION

In the field of civil engineering there are many problems associated with lateral earth pressure. Some of the structures which require an estimation of lateral pressure for their design are retaining walls, sheet pile walls, buried pipes, basement walls, braced excavations, cofferdams, thrust blocks, and others.

Lateral pressures most typically develop against structures supporting soil or water. While designing retaining structures for waterfront facilities, such as cofferdams, quay walls, etc., one must consider the effects of both soil and water pressure. Lateral pressure depends on several factors, such as physical and time-dependent behaviour of soil, soil deformation, surface roughness of wall, and movement of retaining structure and imposed loading.

The state of stress in the backfill of a retaining structure depends on the movement of structure with reference to the backfill. The backfill material is said to be in a *state of elastic equilibrium* when the stress involved and the corresponding strain are within elastic limits. This generally occurs for no or very little movement of the wall. Further increase in stresses develops shear stresses at some point in the body, reaching the shear strength of the soil. Subsequent increase in stresses causes a substantial increase in strain, producing a condition known as *plastic flow*. The soil mass prior to the onset of the plastic flow condition is said to be in a *state of plastic equilibrium*, and the load or stress in this condition is referred to as the *collapse load*. The determination of the collapse load, adopting plasticity theory, is rather complex. However, plasticity theory also provides simplified analyses (as discussed below).

## 11.2 LIMIT ANALYSIS AND LIMIT EQUILIBRIUM METHODS

All the earth pressure problems, viz., earth-retaining structures, bearing capacity of foundations, and slope stability, may be solved by limit analysis or limit equilibrium methods.

The limit analysis method is based on a yield criterion and its associated flow rule, which considers the stress–strain relationship (Drucker and Prager, 1952). It can be used to calculate lower and upper bounds to the true collapse load. By a suitable choice of stress and velocity fields, the analysis may produce the same result, which would then be the exact value of the collapse load.

The lower-bound theorem states that if an equilibrium distribution of stress can be found which balances the applied loads and the boundary conditions and nowhere violates the yield criteria, which includes $c$ and $\phi$, the soil mass will not fail or will be just at the point of failure.

The upper-bound theorem states that the collapse will occur if, for a compatible plastic deformation, the rate at which the external forces do work on the body equals or exceeds the rate of internal dissipation of energy.

Limit equilibrium analysis considers a limiting value that can be reached when the forces acting to cause failure balance the forces resisting failure. This method adopts the following basic elements:

1. An assumed failure surface of a simple shape (e.g., planar, circular, or log-spiral) is considered.
2. A reasonable assumption about the stress distribution along the failure surface is made.
3. An estimation of mobilized shear strength is made, and the same is assumed to act simultaneously along the failure surface.

Based on the above basic elements, an overall equilibrium equation is developed, and the problem is solved by simple statics. Thus, the limiting values, viz., earth pressure on retaining structures, bearing capacity of foundations, and factor of safety of slopes, are computed. Most practical problems are statically indeterminate and need assumptions regarding force systems and directions of their applications.

The application of limit analysis to practical problems has not yet been completely successful because of difficulties in obtaining a proper stress–strain relationship. The limit equilibrium method has been in wide use because of its simplicity. However, sufficient judgement has to be exercised while making assumptions about the shape of the slip surface and stress distribution.

## 11.3 EARTH PRESSURE AT REST

During the formation of a soil deposit, the soil mass at a point is acted on by the vertical geostatic stress, $\sigma_v$, of the overburden. This vertical stress causes a vertical compression of the soil and at the same time produces a lateral strain. This lateral strain is completely restrained due to formation of all-round lateral stress of equal magnitude. With time, the vertical compression and lateral creep strains become zero, and a stable state of stress is created. Because of zero strain, a situation of effective vertical and horizontal stresses is attained. This state of equilibrium is called the *at-rest condition* or *$K_0$-condition.*

Consider an element of soil in such a homogeneous and isotropic soil bounded by a level ground surface. The effective horizontal and vertical stresses are shown in Fig. 11.1a. For the at-rest condition, the ratio of horizontal to vertical stress is called the *coefficient of lateral stress at rest* or *lateral stress ratio at rest* or *coefficient of earth pressure at rest*, $K_0$, that is,

$$K_0 = \frac{\sigma'_{h_0}}{\sigma'_v} \tag{11.1}$$

where

$$\sigma'_v = \gamma z \tag{11.2}$$

and $\sigma'_{h_0}$ is the effective lateral stress for the at-rest condition; that is,

$$\sigma'_{h_0} = K_0 \sigma'_v \tag{11.3}$$

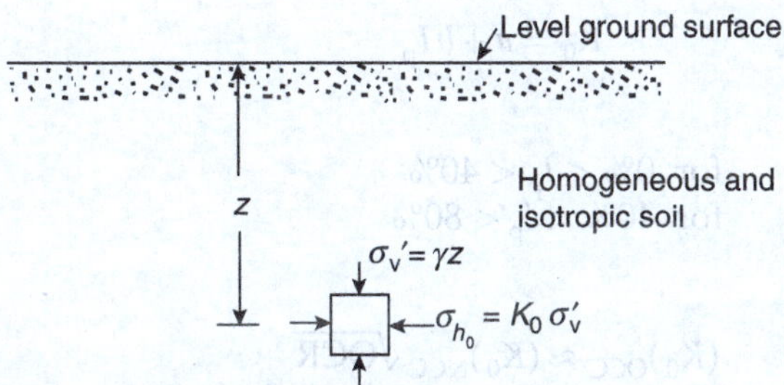


(a) Sub-surface stresses in the soil mass

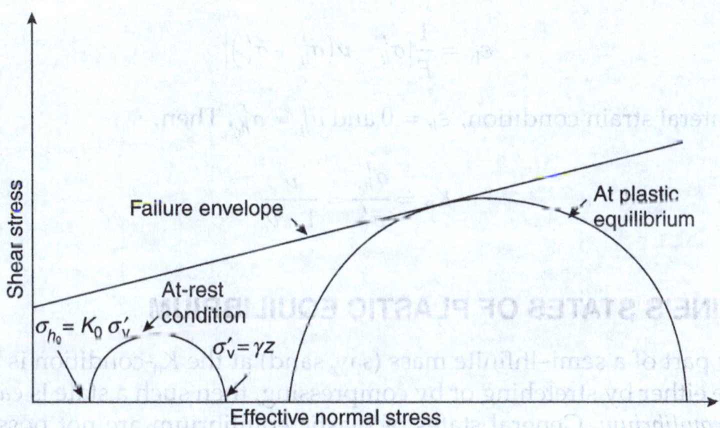


(b) Stress related to failure envelope for the at-rest condition

**Fig. 11.1** Subsurface stresses for the at-rest condition

These stresses are represented by a Mohr's circle along with the shear strength envelope in Fig. 11.1b. The location of Mohr's circle well below the failure envelope indicates a stable equilibrium condition. An increase in stresses would still keep the soil in an elastic equilibrium until the stresses are increased further to cause a failure, and the soil is then said to be at plastic or limiting equilibrium. Mohr circles for these two conditions are shown in Fig. 11.1b.

In general, for many situations, $K_0 < 1$, except in over-consolidated clays (OCC) where $K_0$ may be as high as 3.0. For normally consolidated clays (NCC), $K_0 < 1$, and for sand deposits, $K_0$ varies from 0.40 to 0.50.

It is impossible to determine $K_0$ by measuring $\sigma'_{h_0}$ *in situ*. Therefore, certain correlations have been suggested. Brooker and Ireland (1965) have suggested correlations (Eqs. 11.4 to 11.6) for $K_0$ in terms of the plasticity index $I_p$ and effective friction angle $\phi'$; that is,

$$K_0 = M - \sin\phi' \tag{11.4}$$

where

$M = 1$ for NCC and non-cohesive soils
$= 0.95$ for OCC for over-consolidation ratio (OCR) $> 2$

For NCC, $K_0$ is also given as

$$K_0 = a + b I_p \tag{11.5}$$

where

$a = 0.40$ and $b = 0.007$ for $0\% < I_p < 40\%$
$a = 0.64$ and $b = 0.001$ for $40\% < I_p < 80\%$

and for OCC,

$$(K_0)_{OCC} \approx (K_0)_{NCC}\sqrt{OCR} \tag{11.6}$$

The term $K_0$ may be related to Poisson's ratio, $\nu$, based on the theory of elasticity. The equation of lateral strain is given as

$$\varepsilon_h = \frac{1}{E}\left[\sigma'_h - \nu\left(\sigma'_h + \sigma'_v\right)\right] \tag{11.7}$$

For the no lateral strain condition, $\varepsilon_h = 0$ and $\sigma'_h = \sigma'_{h_0}$. Then,

$$K_0 = \frac{\sigma'_{h_0}}{\sigma'_v} = \frac{\nu}{1-\nu} \tag{11.8}$$

## 11.4 RANKINE'S STATES OF PLASTIC EQUILIBRIUM

Suppose every part of a semi–infinite mass (say, sand) at the $K_0$-condition is brought on the verge of failure either by stretching or by compressing, then such a state is called the *general state of plastic equilibrium*. General states of plastic equilibrium are not possible in normal practical problems except when a geological process is involved. Normal practical problems of interest cause deformations only to a limited extent (for example, yielding of a retaining wall), and hence a *local state of plastic equilibrium* is produced.

Suppose a rigid, frictionless, infinite wall, backfilled with a dry non-cohesive soil, is allowed to move a slight distance away from the retained soil mass (Fig. 11.2a). The soil starts to expand or stretch in the direction following the movement of the wall, resulting in the decrease of horizontal stress from the initial at-rest condition. When adequate lateral movement has occurred, the horizontal stress is decreased to a certain magnitude such that the full shear strength of the soil is mobilized. There is no possibility for a further reduction in the horizontal stress, and such a stress condition is called the *active stress*, $\sigma'_{h_a}$, and the ratio

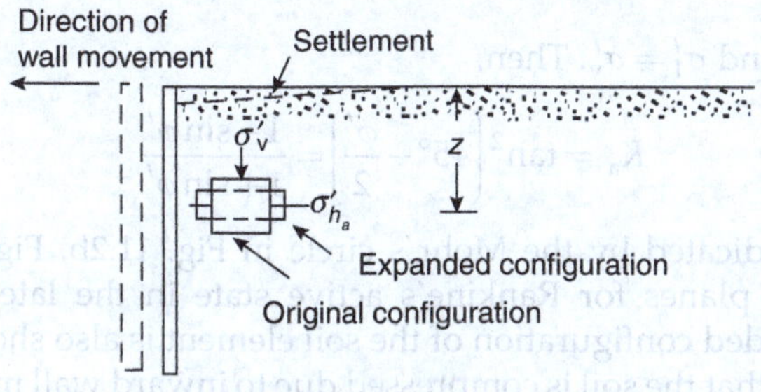


(a) Wall and soil movement for the active case

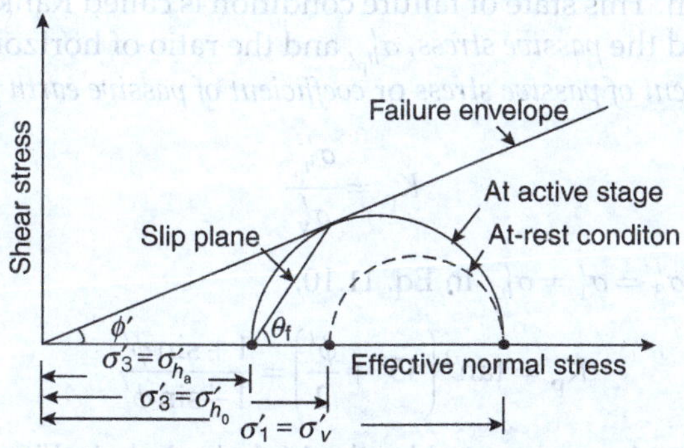


(b) Mohr's circle related to the active state

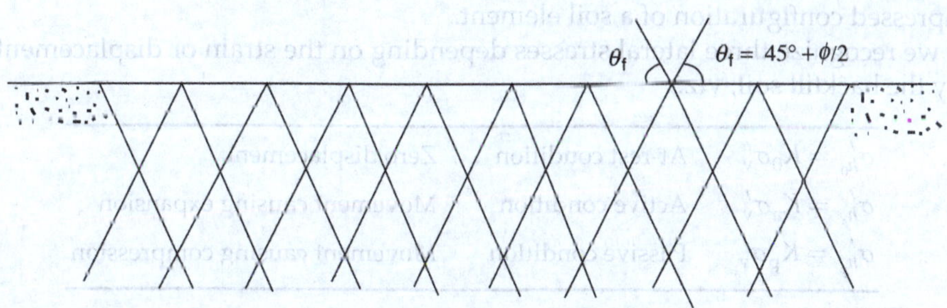


(c) Inclination of slip planes for active state

**Fig. 11.2** Rankine's state of plastic equilibrium – active state

of horizontal to vertical stress is referred to as the *coefficient of active stress* or *coefficient of active earth pressure*, $K_a$.

$$K_a = \frac{\sigma'_{h_a}}{\sigma'_v} \tag{11.9}$$

We know that

$$\sigma'_3 = \sigma'_1 \tan^2\left(45° - \frac{\phi'}{2}\right) - 2c' \tan\left(45° - \frac{\phi'}{2}\right) \tag{11.10}$$

Here, $c' = 0$, $\sigma'_3 = \sigma'_{h_a}$, and $\sigma'_1 = \sigma'_v$. Then,

$$K_a = \tan^2\left(45° - \frac{\phi'}{2}\right) = \frac{1 - \sin\phi'}{1 + \sin\phi'} \tag{11.11}$$

This condition is indicated by the Mohr's circle in Fig. 11.2b. Figure 11.2c also shows the inclination of slip planes for Rankine's active state in the laterally expanding soil. In Fig. 11.2a, the expanded configuration of the soil element is also shown.

Now let us suppose that the soil is compressed due to inward wall movement. This causes an increase in the horizontal stress from the at-rest condition (Fig. 11.3a). When sufficient lateral movement occurs, the maximum shear strength of the soil is mobilized and the horizontal stress is maximum. This state of failure condition is called Rankine's passive state, the horizontal stress is called the *passive stress*, $\sigma'_{h_p}$, and the ratio of horizontal to vertical stress is referred to as the *coefficient of passive stress* or *coefficient of passive earth pressure*, $K_p$:

$$K_p = \frac{\sigma'_{h_p}}{\sigma'_v} \tag{11.12}$$

Substituting $c' = 0$ and $\sigma'_3 = \sigma'_1 = \sigma'_{h_p}$ in Eq. 11.10,

$$K_p = \tan^2\left(45° + \frac{\phi'}{2}\right) = \frac{1 + \sin\phi'}{1 - \sin\phi'} \tag{11.13}$$

This condition of stress is represented by the Mohr's circle in Fig. 11.3b, and the inclination of slip lines for Rankine's passive state is shown in Fig. 11.3c. Figure 11.3a also shows the compressed configuration of a soil element.

Now we recognize three lateral stresses depending on the strain or displacement experienced by the backfill soil, viz.,

| | | |
|---|---|---|
| $\sigma'_{h_0} = K_0\sigma'_v$ | At-rest condition | Zero displacement |
| $\sigma'_{h_a} = K_a\sigma'_v$ | Active condition | Movement causing expansion |
| $\sigma'_{h_p} = K_p\sigma'_v$ | Passive condition | Movement causing compression |

The horizontal displacement required to attain the active state is substantially less than that required to obtain the passive state.

The active state is a condition of loosening strains, where the frictional resistance is mobilized to reduce the force necessary to hold the soil in position. As the soil cannot stretch

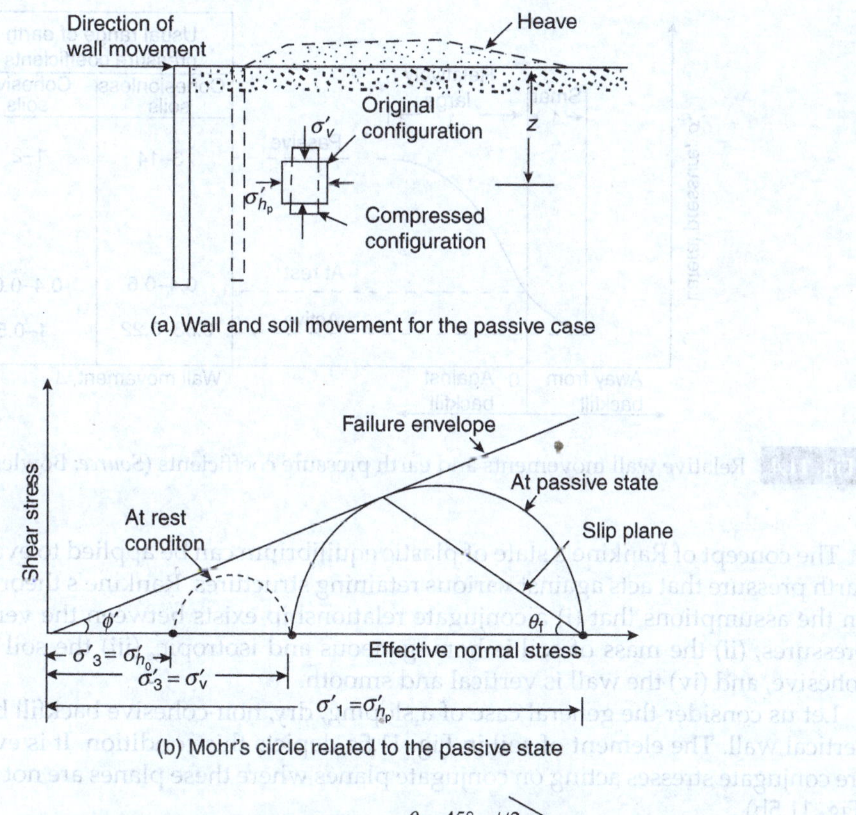


**Fig. 11.3** Rankine's state of plastic equilibrium – passive state

more, the magnitude of this strain is less. On the other hand, a passive state is a condition of densifying the soil by a lateral strain, where the frictional resistance is mobilized to increase the force to cause more strain. Figure 11.4 illustrates the relative movements and the order of magnitude of lateral earth pressure coefficients. For example, when $\phi' = 30°$, $K_a = 0.333$, and $K_p = 3.0$; then $K_p = 10K_a$.

## 11.5 RANKINE'S EARTH PRESSURE THEORY

In the course of various attempts at designing of earth-retaining structures, several earth pressure theories have been suggested since 1687. Coulomb's and Rankine's are perhaps the two best-known theories and are frequently referred to as *classical earth pressure theories*.

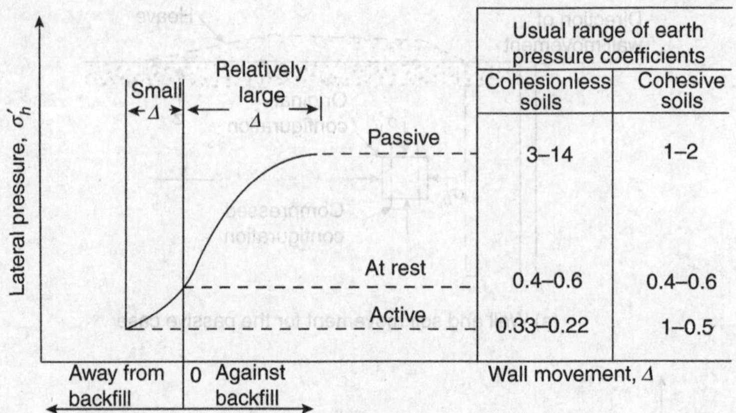


**Fig. 11.4** Relative wall movements and earth pressure coefficients (*Source:* Bowles, 1982)

The concept of Rankine's state of plastic equilibrium can be applied to evaluate the lateral earth pressure that acts against various retaining structures. Rankine's theory (1857) is based on the assumptions that (i) a conjugate relationship exists between the vertical and lateral pressures, (ii) the mass of soil is homogeneous and isotropic, (iii) the soil is dry and non-cohesive, and (iv) the wall is vertical and smooth.

Let us consider the general case of a sloping, dry, non-cohesive backfill behind a smooth vertical wall. The element of soil in Fig. 11.5a depicts this condition. It is evident that these are conjugate stresses acting on conjugate planes where these planes are not principal planes (Fig. 11.5b).

Consider Mohr's circle in Fig. 11.5c for the active condition. Draw a line passing through the origin with an inclination $i$, the slope angle, which cuts the Mohr circle at A and C. Now OC and OA represent the vertical and lateral stresses, respectively. Drop a perpendicular DB to the slope line from the centre D of the Mohr circle.

Now,

$$\frac{\text{OA}}{\text{OC}} = \frac{\text{OB} - \text{AB}}{\text{OB} + \text{AB}} \tag{11.14}$$

$$\text{OB} = \text{OD} \cos i$$

$$r = \text{OD} \sin \phi'$$

$$\text{BD} = \text{OD} \sin i$$

$$\text{AB} = \sqrt{r^2 - \text{BD}^2} = \sqrt{(\text{OD} \sin \phi')^2 - (\text{OD} \sin i)^2}$$

$$\text{OC} = \gamma z \cos i$$

Substituting the above in Eq. 11.14, we have

$$\frac{\text{OA}}{\text{OC}} = \left[\frac{(\text{OD})\cos i - (\text{OD})\sqrt{\sin^2 \phi' - \sin^2 i}}{(\text{OD})\cos i + (\text{OD})\sqrt{\sin^2 \phi' - \sin^2 i}}\right] \tag{11.15}$$

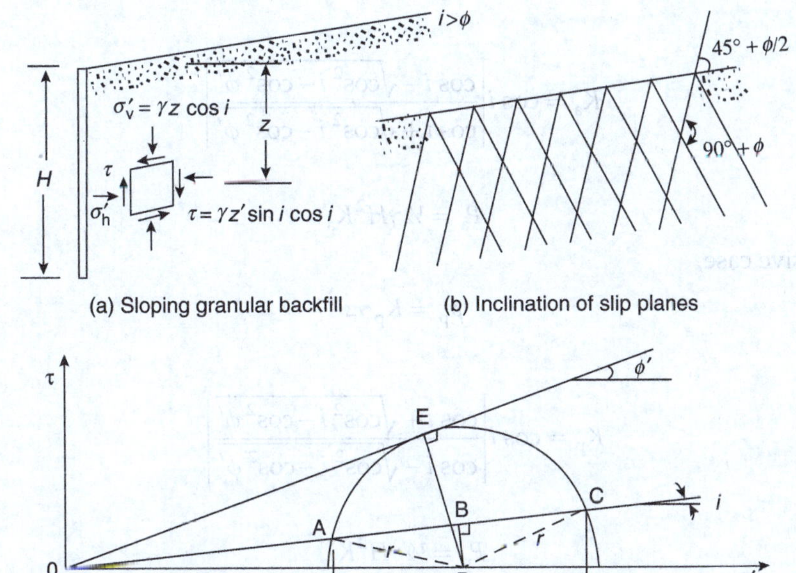


**Fig. 11.5** Lateral pressure and slip planes in granular sloping backfill

Reducing after substituting,

$$\sin^2 \phi' = 1 - \cos^2 \phi' \text{ and } \sin^2 i = 1 - \cos^2 i$$

we have

$$\frac{OA}{OC} = \left[ \frac{\cos i - \sqrt{1 - \cos^2 \phi' - 1 + \cos^2 i}}{\cos i + \sqrt{1 - \cos^2 \phi' - 1 + \cos^2 i}} \right]$$

or

$$\frac{OA}{OC} = \left[ \frac{\cos i - \sqrt{\cos^2 i - \cos^2 \phi'}}{\cos i + \sqrt{\cos^2 i - \cos^2 \phi'}} \right] \tag{11.16}$$

or

$$\sigma'_{h_a} = \left[ \frac{\cos i - \sqrt{\cos^2 i - \cos^2 \phi'}}{\cos i + \sqrt{\cos^2 i - \cos^2 \phi'}} \right] \gamma z \cos i \tag{11.17}$$

Let $p_a = \sigma'_{h_a}$ ; then,

$$p_a = K_a \gamma z \tag{11.18a}$$

where

$$K_a = \cos i \left[ \frac{\cos i - \sqrt{\cos^2 i - \cos^2 \phi'}}{\cos i + \sqrt{\cos^2 i - \cos^2 \phi'}} \right] \tag{11.18b}$$

Therefore,

$$P_a = \tfrac{1}{2}\gamma H^2 K_a \tag{11.19}$$

For the passive case,

$$p_p = K_p \gamma z \tag{11.20a}$$

where

$$K_p = \cos i \left[ \frac{\cos i + \sqrt{\cos^2 i - \cos^2 \phi'}}{\cos i - \sqrt{\cos^2 i - \cos^2 \phi'}} \right] \tag{11.20b}$$

Therefore,

$$P_p = \tfrac{1}{2}\gamma H^2 K_p \tag{11.21}$$

Equations 11.19 and 11.21 are Rankine's expressions for the lateral pressures for a wall of height $H$ with a backfill of unit weight $\gamma$ for the active and passive cases. These forces act at a height of $H/3$ from the base inclined at an angle $i$ to the normal of the wall.

### 11.5.1 Effect of Level Backfill Surface

Level ground surface is a simplified condition and is most often adopted in practice. Considering that all the conditions remain identical except for $i = 0$, the expressions $K_a$ and $K_p$ reduce to those for Rankine's fundamental states of plastic equilibrium. Setting $i = 0$ in Eqs. 11.18b and 11.20b, we have

$$K_a = \frac{1 - \sin \phi'}{1 + \sin \phi'} \tag{11.22}$$

and

$$K_p = \frac{1 + \sin \phi'}{1 - \sin \phi'} \tag{11.23}$$

For this situation, Mohr's circle is re-drawn as in Fig. 11.6. Thus, the total active thrust $P_a$ is given as

$$P_a = \tfrac{1}{2}\gamma H^2 K_a \tag{11.24}$$

$$P_p = \tfrac{1}{2}\gamma H^2 K_p \tag{11.25}$$

where $K_a$ and $K_p$ are taken from Eqs. 11.22 and 11.23, respectively. The pressure distribution and the point of application of the forces are shown in Fig. 11.7.

Further discussions are confined to a level backfill surface but with different backfill materials and loading conditions.

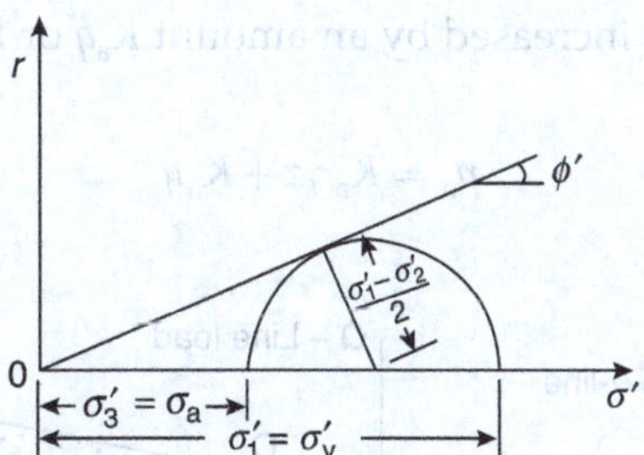


Fig. 11.6 Mohr's circle for active state – level backfill

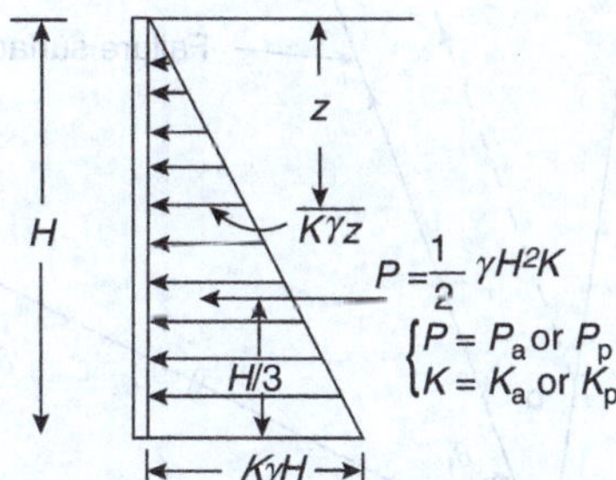


Fig. 11.7 Lateral pressure distribution

### 11.5.2 Effect of Surcharge Load on Backfill Surface

Consider a dry, non-cohesive level backfill (Fig. 11.8a) with a uniform surcharge load $q$ applied all over the surface. It may be assumed that the vertical effective stress is increased by the amount of surcharge. Then, at any depth $z$,

$$\sigma_v' = \gamma z + q \tag{11.26}$$

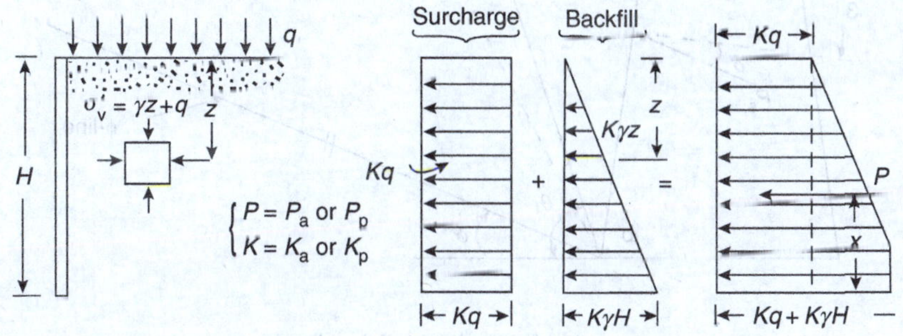


Fig. 11.8 Lateral pressure due to uniform surcharge of a level backfill

Thus the lateral pressure is increased by an amount $K_a q$ or $K_p q$ as the case may be. Therefore, for the active case,

$$p_a = K_a \gamma z + K_a q \tag{11.27}$$

and for passive case,

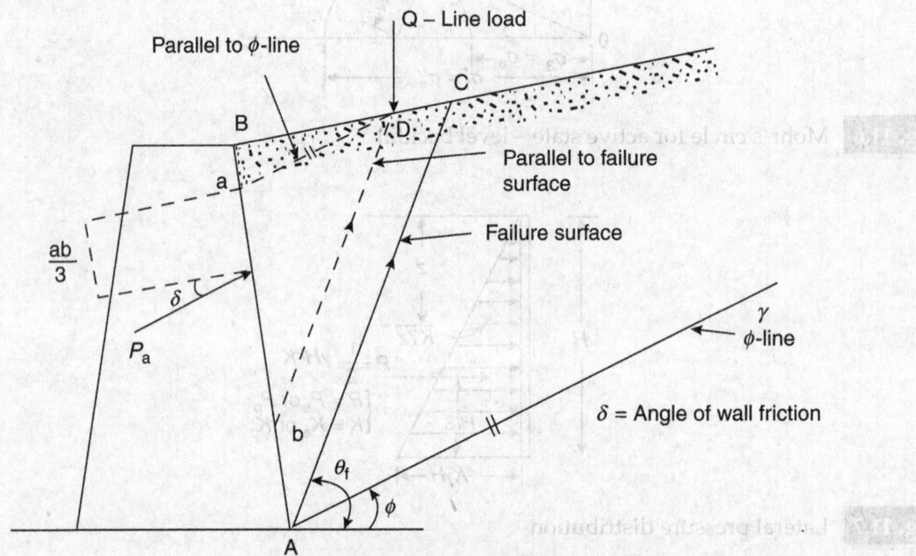


(a) Line load left of the slip plane

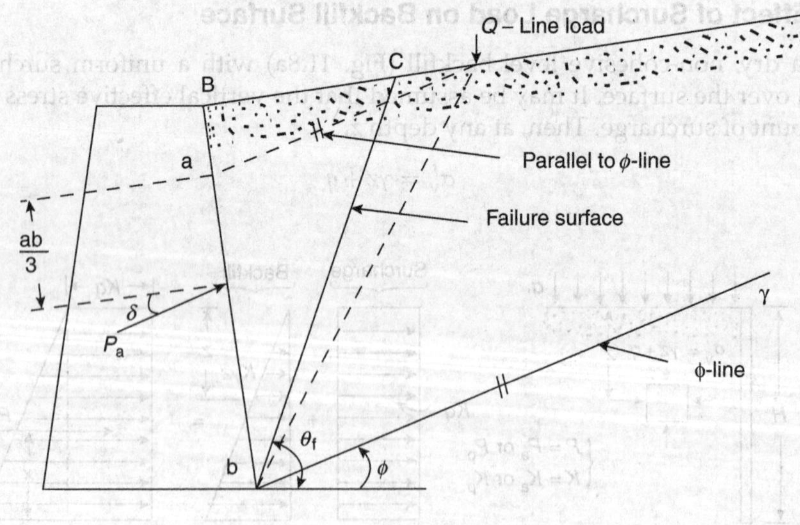


(b) Line load right of the slip plane

**Fig. 11.9** Procedure for estimating the line of action of the resultant active thrust $P_a$ caused by a line load (*Source:* Dunn et al., 1980)

$$p_p = K_p \gamma z + K_p q \tag{11.28}$$

This shows that the lateral pressure varies linearly with depth due to unit weight and remains constant with depth due to surcharge load. Figure 11.8b illustrates the pressure distribution. The area of the entire diagram gives the active thrust, $P_a$, or the passive resistance, $P_p$ on the wall. The line of action of $P_a$ and $P_p$ can be determined by considering the moments of the individual and total areas about the base.

Concentrated surcharge loads $Q$ (Fig. 11.9) running parallel to the wall may be induced on the backfill (e.g., continuous footing, railroad tracks, etc.). Increased stresses on the wall due to concentrated surcharge may be computed based on Boussinesq's equation. It is a laborious procedure and generally not recommended. However, graphical methods (discussed elsewhere) are more expedient for this purpose (refer Example 11.8).

The point of action of active thrust or passive resistance is obtained by following the procedure detailed below. The failure surface is located using any graphical method. The concentrated load may lie within or away from the failure wedge. If the concentrated load is within the failure wedge, then lines *Db* and *Da* are drawn parallel to the failure surface and the $\phi'$ line, respectively (Fig. 11.9a), and points *a* and *b* are located. If the concentrated load is away from the failure wedge (Fig. 11.9b), then the heel represents point *b*, and point *a* is located as explained earlier. Then, the point of application of the active thrust is at a distance of $ab/3$ from point *a*.

### 11.5.3 Effect of Water Table on a Backfill

Consider again a non-cohesive level backfill with the water table at the surface. The vertical stress at any depth $z$ can be split into two, viz., one due to soil grains and the other due to water; that is,

$$\sigma'_v = \gamma' z + \gamma_w z \tag{11.29}$$

Hence, the active and passive cases will be

$$P_a = K_a \gamma' z + K_w \gamma_w z$$

and

$$P_p = K_p \gamma' z + K_w \gamma_w z \tag{11.30}$$

where $K_w$ is the lateral coefficient for water, which is always 1. Thus,

$$P_a = K_a \gamma' z + \gamma_w z \tag{11.31}$$

and

$$P_p = K_p \gamma' z + \gamma_w z \tag{11.32}$$

The pressure distribution is shown in Fig. 11.10. As before, $P_a$ and $P_p$ can be determined from the area of the diagrams.

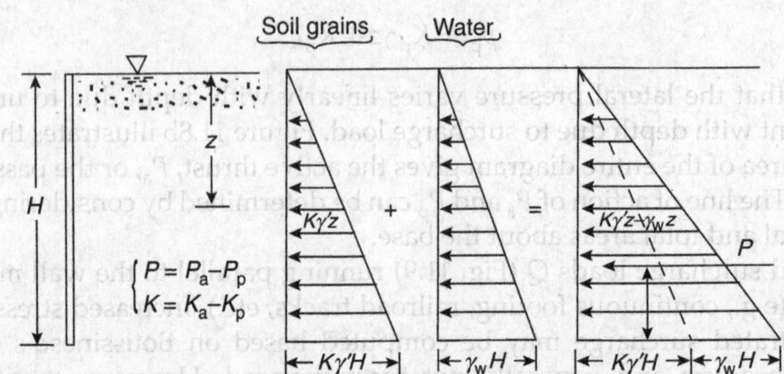


Fig. 11.10 Pressure distribution when the water table is at the surface

Let us consider a partial submergence now. Let the water table be at a depth of $H_1$ from the level surface (Fig. 11.11). Since the soil is non-cohesive, it is reasonable to assume that the pore pressure above the water table is everywhere zero. The dry unit weight is considered for soil above the water table and submerged unit weight below. Below the water table the unit weight of water is also taken into account. Thus,

$$\left.\begin{aligned} p_a &= K_a \gamma_d z \\ p_p &= K_p \gamma_d z \end{aligned}\right\} \quad \text{for } (0 < z \leq H_1) \tag{11.33}$$

and

$$\left.\begin{aligned} p_a &= K_a \gamma' z + \gamma_w z + K_a \gamma_d H_1 \\ p_p &= K_p \gamma' z + \gamma_w z + K_p \gamma_d H_1 \end{aligned}\right\} \quad \text{for } [H_1 \leq z \leq (H_1 + H_2)] \tag{11.34}$$

The problem may also be solved by assuming the first layer to act as a surcharge load, $q (q = \gamma_d H_1)$, on the second layer. The pressure diagram is drawn for the retaining wall with height $H_1$. The second layer is treated separately, as if loaded by a surcharge load $q$ and the water table at the surface, the height of the retaining wall being $H_2$. The pressure distribution is shown in Fig. 11.11. Values of $P_a$ and $P_p$ are obtained from the area of the diagrams.

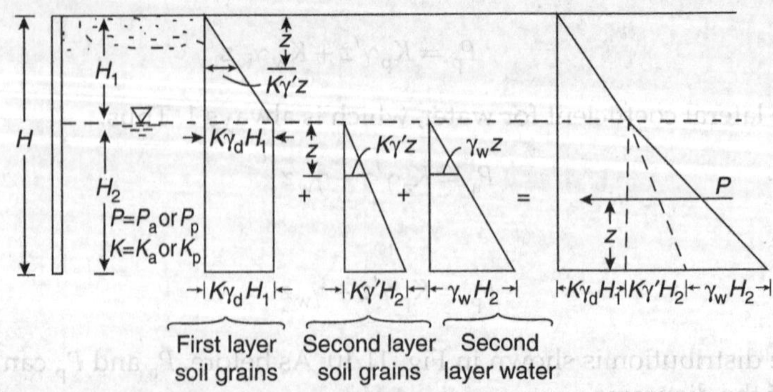


Fig. 11.11 Pressure distribution for partial submergence in a backfill

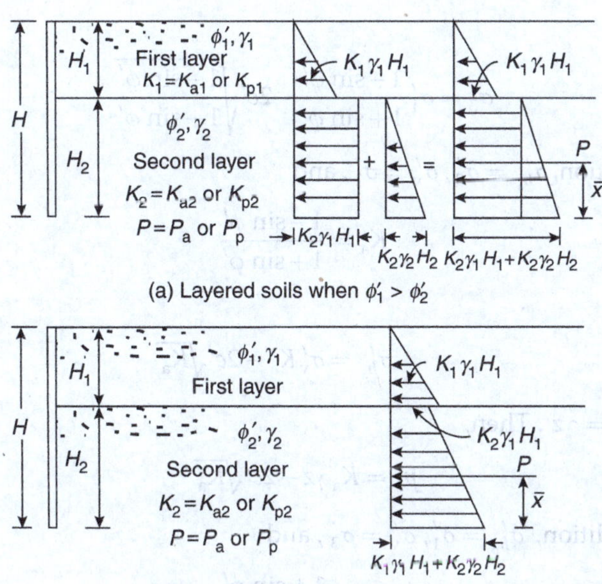


**Fig. 11.12** Pressure distribution for layered soils in a backfill

### 11.5.4 Effect of Stratified Soils in the Backfill

Consider two dry, non-cohesive soils in the level backfill. This is similar to the case of partial submergence. Here the angles of shearing resistances are different in the two layers, whereas in the previous case they were the same in both the layers. If $\phi'_1$ and $\phi'_2$ are the angles of shearing resistances, $\gamma_1$ and $\gamma_2$ the unit weights in the top and bottom layers of heights $H_1$ and $H_2$, and $K_{a1}$, $K_{p1}$ and $K_{a2}$, $K_{p2}$ the lateral coefficients for the respective layers, then

$$p_a = K_{a1}\gamma_1 z \text{ and } p_p = K_{p1}\gamma_1 z \text{ for } (0 \le z \le H_1) \tag{11.35}$$

$$p_a = K_{a2}\gamma_2 z + K_{a2}\gamma_1 H_1 \text{ and}$$
$$p_p = K_{p2}\gamma_2 z + K_{p2}\gamma_1 H_1 \text{ for } [H_1 \le z \le (H_1 + H_2)] \tag{11.36}$$

Here again, for the second layer, the first layer acts like a surcharge. Similar methods may be used if the number of layers is more than two. For a three-layer case, the top and middle layers will act as surcharges on the bottom layer.

The pressure distribution is shown in Fig. 11.12, and the values of $P_a$ and $P_p$ are calculated as usual.

### 11.5.5 Effect of $c$–$\phi$ Soils as Backfill

Rankine's theory was originally proposed only for non-cohesive soils. It was extended by Bell (1915) to include $c$–$\phi$ soils, and its application has also been widened. Consider a dry $c$–$\phi$ soil level backfill. We know that

$$\sigma'_1 = \sigma'_3 \frac{1+\sin\phi'}{1-\sin\phi'} + 2c'\sqrt{\frac{1+\sin\phi'}{1-\sin\phi'}} \tag{11.37}$$

or

$$\sigma_3' = \sigma_1' \frac{1-\sin\phi'}{1+\sin\phi'} - 2c'\sqrt{\frac{1-\sin\phi'}{1+\sin\phi'}} \tag{11.38}$$

For the active condition, $\sigma_{h_a}' = \sigma_3'$, $\sigma_v' = \sigma_1'$, and

$$K_a = \frac{1-\sin\phi'}{1+\sin\phi'}$$

Thus,

$$\sigma_{h_a}' = \sigma_v' K_a - 2c'\sqrt{K_a}$$

Let $p_a = \sigma_{h_a}'$ and $\sigma_v' = \gamma z$. Then,

$$p_a = K_a \gamma z - 2c'\sqrt{K_a} \tag{11.39}$$

For the passive condition, $\sigma_{h_p}' = \sigma_1'$, $\sigma_v' = \sigma_3'$, and

$$K_p = \frac{1+\sin\phi'}{1-\sin\phi'}$$

Substituting the values of $\sigma_{h_p}'$, $\sigma_v'$, and $K_p$ in Eq. 11.37, we have

$$\sigma_{h_p}' = \sigma_v' K_p + 2c'\sqrt{K_p}$$

Let $p_p = \sigma_{h_p}'$ and $\sigma_v' = \gamma z$. Then,

$$p_p = K_p \gamma z + 2c'\sqrt{K_p} \tag{11.40}$$

The pressure diagrams for these two cases are shown in Fig. 11.13a and b. In the active case, there exists a tension up to a depth of $z = z_0$, where the active pressure $p_a = 0$. Further, at the surface ($z = 0$), the active pressure has a value of $p_a = -2c'\sqrt{K_a}$.

The soil up to a depth of $z_0$ will be in a state of tension and will neither impart any pressure on the wall nor provide support. When the tension is released, tension cracks will develop from the surface up to a depth of $z_0$. From a practical point of view, the tension zone is ignored, and the active thrust is calculated only for the height ($H$–$z_0$) from the base (the pertinent area is shown shaded in Fig. 11.13a); that is, let $p_a = p_a$ at $z = H$; then,

$$p_a' = K_a \gamma H - 2c'\sqrt{K_a} \tag{11.41}$$

Thus,

$$P_a = \tfrac{1}{2} p_a' (H - z_0) \tag{11.42}$$

is acting at a height of $(H - z_0)/3$ from the base.

The depth of the tension zone, $z_0$, is obtained by setting $p_a = 0$ with $z = z_0$ in Eq. 11.39:

$$z_0 = \frac{2c'}{\gamma\sqrt{K_a}} \tag{11.43}$$

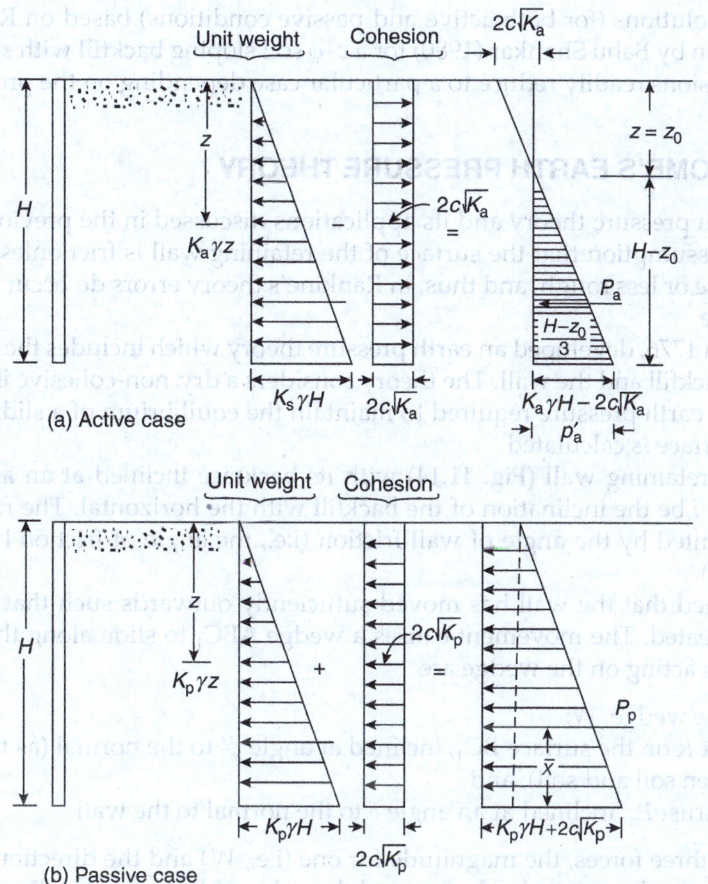


**Fig. 11.13** Active and passive pressure distributions: $c$–$\phi'$ soil as backfill

The existence of a tension zone in $c$–$\phi$ soils suggests that an unsupported excavation would be theoretically possible. The maximum unsupported depth of excavation, $H_c$, may be theoretically taken as 2 $z_0$, where the tensile stress is equal to the cohesive strength.

Hence,

$$H_c = 2z_0 = \frac{4c'}{\gamma\sqrt{K_a}} \tag{11.44}$$

The application of this theoretical depth in practice should be done more cautiously. Any factor which reduces the cohesion (e.g., the possibility of water entering the crack and causing a reduction in shear strength) may affect the estimation drastically. However, such depths may be adopted for short-duration works.

For many reasons, a cohesive backfill is not recommended in practice since changes in water content may significantly alter the performance of a retaining wall due to frequent swelling and shrinking of the soil. Further, placing, densifying, and maintaining a cohesive backfill is extremely difficult.

Analytical solutions (for both active and passive conditions) based on Rankine's theory have been given by Babu Shankar (1980) for a $c$–$\phi$ soil sloping backfill with surcharge. These general expressions readily reduce to a particular case depending on the problem.

## 11.6 COULOMB'S EARTH PRESSURE THEORY

Rankine's earth pressure theory and its applications discussed in the previous section were based on the assumption that the surface of the retaining wall is frictionless. But in reality, the wall is more or less rough, and thus, in Rankine's theory errors do occur which might be on the safe side.

Coulomb, in 1776, developed an earth pressure theory which includes the effect of friction between the backfill and the wall. The theory considers a dry, non-cohesive inclined backfill, and the lateral earth pressure required to maintain the equilibrium of a sliding wedge with a plane slip surface is calculated.

Consider a retaining wall (Fig. 11.14) with its backface inclined at an angle $\beta$ with the horizontal. Let $i$ be the inclination of the backfill with the horizontal. The roughness of the wall is represented by the angle of wall friction (i.e., the angle of friction between the soil and the wall, $\delta$).

It is presumed that the wall has moved sufficiently outwards such that the active-state condition is created. The movement causes a wedge $ABC_1$ to slide along the plane surface $BC_1$. The forces acting on the wedge are

1. weight of the wedge, $W$;
2. the resultant $R$ on the surface $BC_1$, inclined at angle $\phi'$ to the normal (as the sliding takes place between soil and soil); and
3. the active thrust $P_a$, inclined at an angle $\delta$ to the normal to the wall.

Among the three forces, the magnitude for one (i.e., $W$) and the direction of all three are known; the force polygon can be drawn, and the value of $P_a$, corresponding to the assumed wedge $ABC_1$, can be determined. Different trial wedges are taken and the corresponding $P_a$ values determined. The maximum value of $P_a$ may be mathematically expressed as

$$P_a = \tfrac{1}{2}\gamma H^2 K_a \tag{11.45}$$

where $K_a$ is Coulomb's active earth pressure coefficient

$$= \frac{\sin^2(\beta+\phi')}{\sin^2\beta\,\sin(\beta-\delta)\{1+[\sin(\phi'+\delta)\,\sin(\phi'-i)/\sin(\beta-\delta)\,\sin(i+\beta)]\}^2} \tag{11.46}$$

The mathematical treatment for Eq. 11.46 is beyond the scope of this book. The active thrust $P_a$ acts at a height $H/3$ from the base and is inclined at an angle $\delta$ to the normal of the back surface.

For computing Coulomb's passive resistance $P_p$, consider the retaining wall and other details shown in Fig. 11.14b. Again, taking into account the forces which contribute to the equilibrium, viz., $W$, $R$, and $P_p$, the force triangle is drawn and the passive resistance $P_p$ determined. The minimum value of $P_p$ from the plot of $P_p$ versus wedge locations represents Coulomb's passive resistance. Mathematically, this can be expressed as

$$P_p = \tfrac{1}{2}\gamma H^2 K_p \tag{11.47}$$

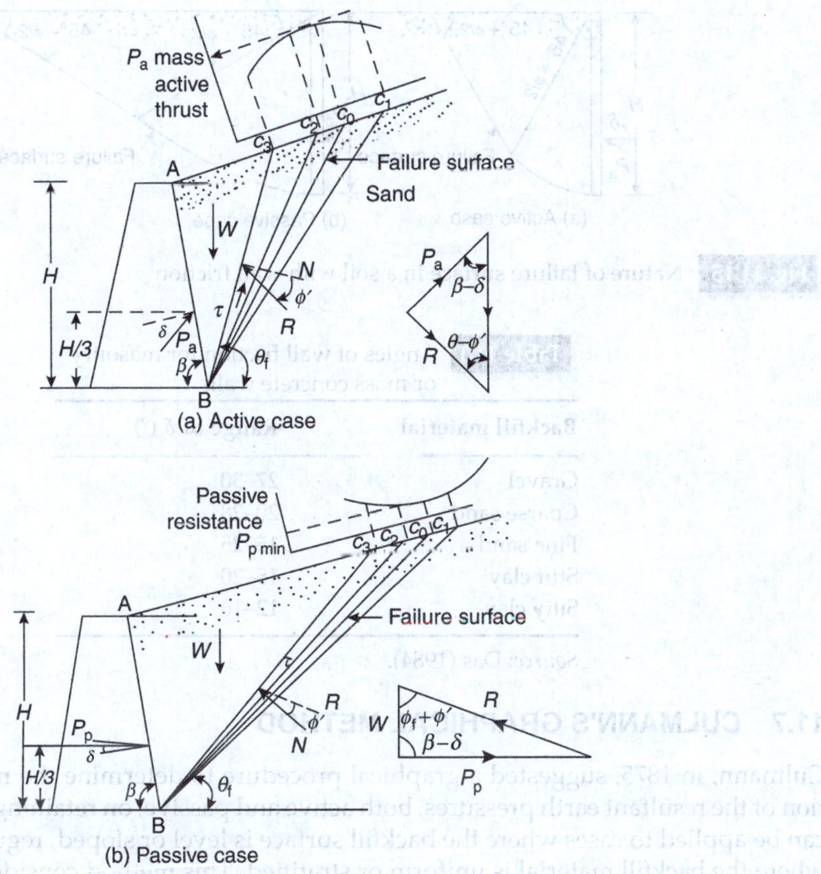


**Fig. 11.14** Coulomb's active and passive wedges

where $K_p$ is Coulomb's passive earth pressure coefficient

$$= \frac{\sin^2(\beta-\phi')}{\sin^2\beta\sin(\beta+\delta)\{1-[\sin(\phi'+\delta)\sin(\phi'+i)/\sin(\beta+\delta)\sin(\beta+i)]\}^2} \tag{11.48}$$

The passive resistance $P_p$ acts at a height of $H/3$ from the base at an angle $\delta$ to the normal.

Coulomb's plane failure surface assumption omits to take into account the actual or true nature of the failure surface (Fig. 11.15). Although the active pressure is not significantly affected by the plane surface assumption, this is not the case for passive pressure. The estimated value of $P_p$ is on the unsafe side and increases with increase in wall friction. Therefore, rigorous analyses have been carried out by researchers, assuming different shapes, such as circle, ellipse, and log-spiral, for the slip surface.

Retaining walls are generally constructed with mass concrete or masonry. The wall friction angle, $\delta$, depends on the type of backfill material and the type of wall, as shown in Table 11.1. The value of $\delta$ may be assumed to be between $\phi'/2$ and $2/3\phi'$.

The above treatment is applicable only for an unloaded backfill. But Babu Shankar (1981) has provided analytical solutions for the active case based on Coulomb's theory for inclined backfill of a $c'$–$\phi'$ soil with uniform surcharge load.

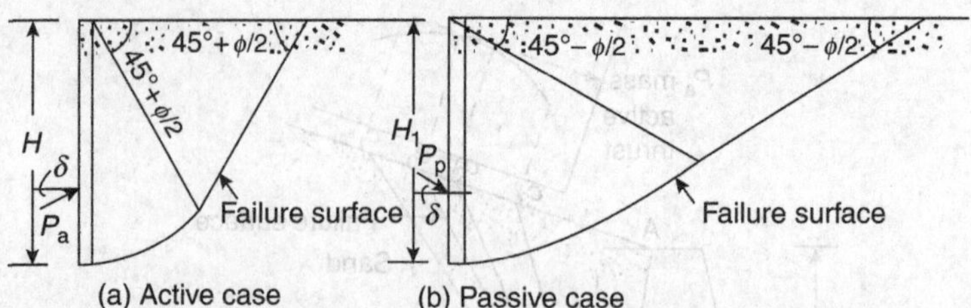


Fig. 11.15 Nature of failure surface in a soil with wall friction

Table 11.1 Angles of wall friction for masonry or mass concrete walls

| Backfill material | Range of $\delta$ (°) |
|---|---|
| Gravel | 27–30 |
| Coarse sand | 20–28 |
| Fine sand | 15–25 |
| Stiff clay | 15–20 |
| Silty clay | 12–16 |

*Source*: Das (1984).

## 11.7 CULMANN'S GRAPHICAL METHOD

Culmann, in 1875, suggested a graphical procedure to determine the magnitude and location of the resultant earth pressures, both active and passive, on retaining walls. This method can be applied to cases where the backfill surface is level or sloped, regular or irregular and where the backfill material is uniform or stratified. This method considers the effect of wall friction and varied surcharge load conditions. Although this method was basically proposed for non-cohesive soils, it can be extended to cohesive soils also. However, it demands a constant angle of shearing resistance in the backfill.

Consider a retaining wall with a sloping non-cohesive backfill in the active state (Fig. 11.16). The step-by-step graphical construction is as follows:

1. Choose a suitable scale, and draw the retaining wall along with the backfill.
2. Draw a line AD from the point A, making an angle $\phi$ above the horizontal.
3. Draw another line AE at an angle $\psi$ $(=\beta-\delta)$ from the line AD.
4. Consider a wedge $ABC_1$ with $AC_1$ as the slip surface.
5. Determine the weight $W_1$ of the wedge $ABC_1$.
6. Select a convenient force scale and represent $W_1$ on the line AD as $AW_1$.
7. From $W_1$, draw a line parallel to AE to meet the assumed slip surface $AC_1$ at $F_1$.
8. Choose another wedge $ABC_2$ and repeat Steps 5 to 7 and find point $F_2$.
9. Establish similar points, and connect these points of intersection with a smooth curve, called Culmann's curve.
10. Draw a tangent to Culmann's curve parallel to AD. Point F represents such a tangent point. An irregular curve may have more than one tangent.

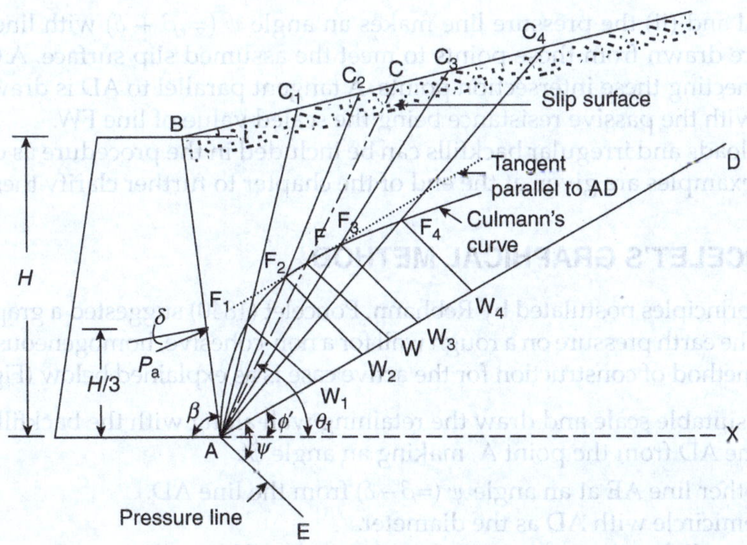


Fig. 11.16 Culmann's graphical method – active case

11. Draw FW parallel to AE. The magnitude of FW, based on the selected scale, represents the active thrust $P_a$. If several tangents to the curve are possible, the largest of them becomes the value of $P_a$.
12. The failure surface is AFC and is inclined at $\theta_f$ to the horizontal.

Culmann's procedure for the determination of passive resistance $P_p$ is similar to that for the active case, with some notable differences (Fig. 11.17): (i) line AD makes an angle $\phi$ below

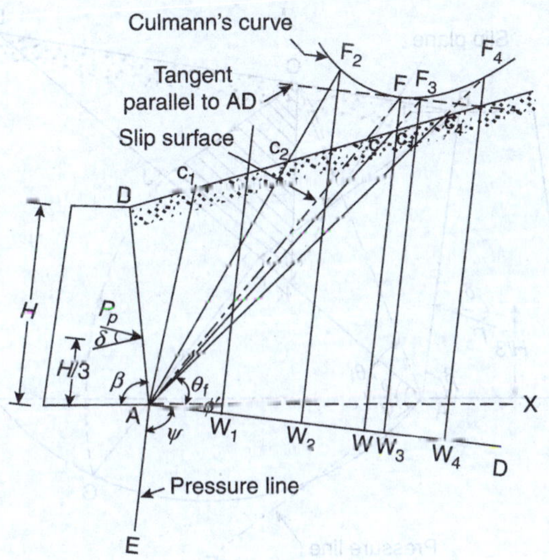


Fig. 11.17 Culmann's graphical method – passive case

the horizontal and (ii) the pressure line makes an angle $\psi$ $(= \beta + \delta)$ with line AD. Parallel lines to AE are drawn from these points to meet the assumed slip surface. A Culmann line is drawn connecting these intersection points. A tangent parallel to AD is drawn to the Culmann curve with the passive resistance being the scaled value of line FW.

Surcharge loads and irregular backfills can be included in the procedure as discussed earlier. Worked examples are given at the end of the chapter to further clarify these conditions.

## 11.8 PONCELET'S GRAPHICAL METHOD

Based on the principles postulated by Rebhann, Poncelet (1840) suggested a graphical method to determine the earth pressure on a rough wall for a non-cohesive, homogeneous, and inclined backfill. The method of construction for the active case is as explained below (Fig. 11.18).

1. Choose a suitable scale and draw the retaining wall along with the backfill.
2. Draw a line AD from the point A, making an angle $\phi'$.
3. Draw another line AE at an angle $\psi$ $(=\beta - \delta)$ from the line AD.
4. Draw a semicircle with AD as the diameter.
5. Draw BF parallel to the pressure line $\psi$ to meet the $\phi'$ line at F.
6. Draw a perpendicular FG at F to meet the semicircle at G.
7. With A as centre and AG as radius, draw an arc to meet AD at J.
8. From J, draw JC parallel to the pressure line to meet the backfill line at C.
9. With J as centre and JC as radius, draw an arc to meet AD at K.
10. Join AC and KC.
11. Find the area of the triangle JKC; then,

    $P_a =$ (Area of triangle) $\times \gamma$

    where $\gamma$ is the unit weight of the backfill. AC represents the failure surface.

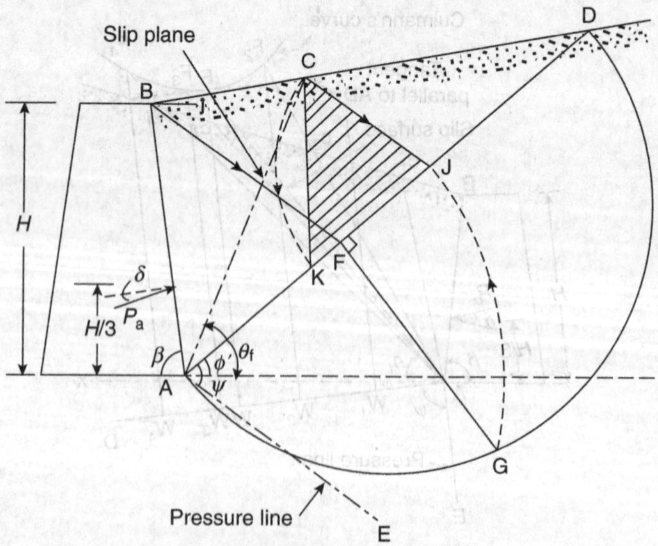


Fig. 11.18 Poncelet's graphical method – active case

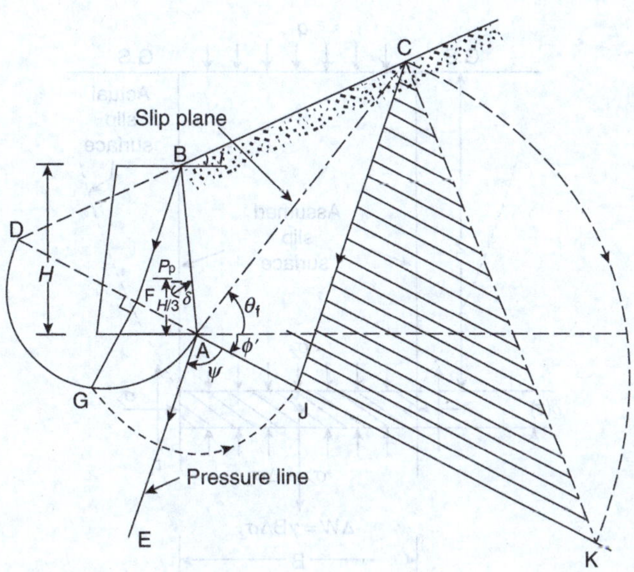


**Fig. 11.19** Poncelet's graphical method – passive case

When the slope of the backfill surface $i$ and the angle of shearing resistance $\phi'$ are equal or nearly equal, slight modifications are made in the procedure. The modifications are highlighted in the worked examples.

Poncelet's procedure for the determination of passive resistance $P_p$ is similar to that for the active condition, with some notable differences (Fig. 11.19): (i) line AD makes an angle $\phi'$ below the horizontal and is projected backwards to meet the extended backfill surface at D and (ii) the pressure line makes an angle $\psi$ $(= \beta + \delta)$ with line AD. Other steps are similar to those followed for the active case. For all angles of $\phi'$ and $i$, the procedure is the same, as these lines are the converging ones, and hence no modification is needed.

## 11.9 ARCHING OF SOILS

In a supported soil mass, when a certain part of the soil mass yields, then the soil adjoining the yielding part also gets displaced from its original position. The deformation of the parted soil is resisted by mobilization of shearing resistance along the zones of contact between the yielding and non-yielding portions of the soil. As the direction of mobilization of shear strength is opposed to the direction of deformation of the yielding soil, there is a reduction in pressure on the yield part of the support and a consequent increase in the pressure of the adjoining stationary parts. This phenomenon of the transfer of pressure from the yielding part of a soil mass to the non-yielding part of the mass is referred to as *arching*.

Consider the yielding of a horizontal strip (Fig. 11.20). The actual failure surfaces will be curved starting from the yielding point to the ground surface. Let the slip surfaces be assumed to be vertical rising from the yielding strip to the surface.

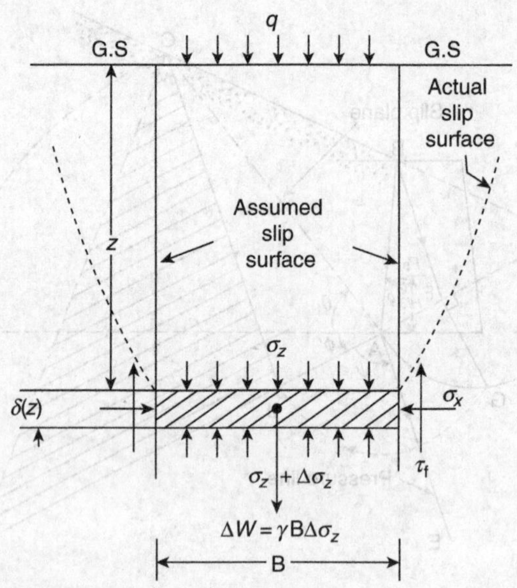


Fig. 11.20 Yielding of a horizontal strip

Consider the equilibrium of the yielding slice of width $B$ and unit length; then,

$$B[\sigma_z + \Delta\sigma_z] = B\sigma_z + \Delta W - 2\Delta z \tau_f$$

Substituting for

$$\Delta W = \gamma B \Delta z$$

and

$$\tau_f = c' + \sigma_x \tan\phi'$$

and

$$\sigma_x = K\sigma_z$$

where $K$ is an empirical constant, we have

$$B\Delta\sigma_z = \gamma B \Delta z - 2c'\Delta z - 2K\sigma_z \tan\phi' \Delta z$$

or

$$\frac{\Delta\sigma_z}{\Delta z} = \frac{\gamma - 2c'}{B} - \frac{2K \tan\phi'}{B}\sigma_z$$

Also, at $z = 0, \sigma_z = q$.

A solution of the above equation yields the following expression:

$$\sigma_z = \frac{B[\gamma - (2c'/B)]}{2K \tan\phi'}\left[1 - e^{(-2Kz/B)\tan\phi'}\right] + q e^{(-2Kz/B)\tan\phi'}$$

When $c' = q = 0$

$$\sigma_z = \frac{B\gamma}{2\,K\,\tan\phi'}\left[1 - e^{(-2Kz/B)\tan\phi'}\right]$$

For a cohesionless soil $c = 0$, and for the no surcharge condition $q = 0$; the above equation gives the intensity of vertical pressure on the yielding strip, considering that the mobilization of shearing resistance takes place along the full length of the surface $z$.

For the condition of $z = \infty$, $\sigma_z$ becomes constant and independent of $z$; that is,

$$\sigma_{z=\infty} = \frac{B\gamma}{2\,K'\tan\phi'}$$

It has been reported in the literature that at distances of more than 2.5 $B$ in sand, the yielding of the strip has no effect on the state of stress in sand.

## WORKED EXAMPLES

**Example 11.1** Compute the total lateral force acting against a 10 m high, vertical, smooth, unyielding wall, which retains a normally consolidated clay. The soil parameters are $\gamma = 21\text{kN/m}^3$, $I_p = 35\%$. The water table is at the ground surface.

***Solution***

Since the soil is normally consolidated, the coefficient of earth pressure at rest, $K_0$, is obtained from Eq. 11.5. Thus,

$$K_0 = a + bI_p$$

For $I_p = 35\%$, $a = 0.40$ and $b = 0.007$, $K_0 = 0.40 + 0.007 \times 35 = 0.645$.

Total vertical pressure $\sigma_v = \gamma H = 21 \times 10 = 210\ \text{kN/m}^2$

Therefore, total lateral pressure $\sigma_{h_0} = K_0\sigma'_v = 0.645 \times 210 = 135.5\ \text{kN/m}^2$

Total lateral force $P_{h_0} = ½(K_0\sigma'_v)H = ½(135.5) \times 10 \times 1$

$= 677.5\ \text{kN/m}$ length of wall

**Example 11.2** An 8 m high vertical, smooth retaining wall above the water table supports a 15° soil slope. The retained backfill has a unit weight of 18.6 kN/m$^3$, and the shear strength parameters are $c' = 0$ and $\phi' = 35°$. Compute the total active thrust on the wall, and also find the directions of the two sets of failure planes relative to the horizontal.

***Solution***

Draw the failure envelope and the backfill slope inclined at 15° to the origin (Fig. 11.21).

The vertical stress at a depth of 8 m $= \sigma'_z = \gamma z \cos i$

$$\sigma'_z = 18.6 \times 8 \times \cos 15°$$
$$= 143.7\ \text{kN/m}^2$$

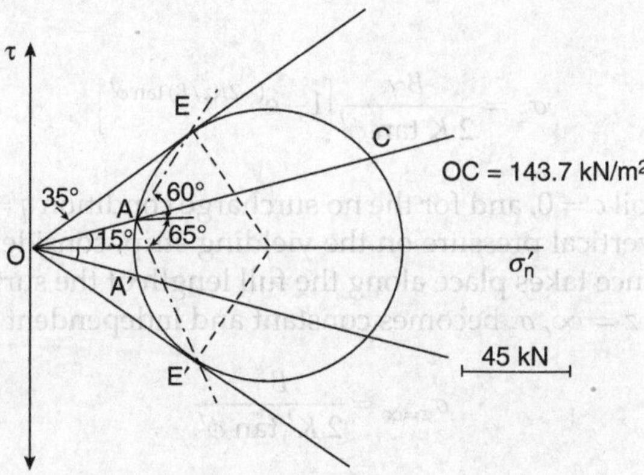


Fig. 11.21

Choose a stress of 1 mm = 1.5 kN and set off this stress (distance OC) along the 15° line. Draw a Mohr circle passing through this point C and tangential to the failure envelope. Then measure the distance OA or OA′, which represents the active pressure to scale. Therefore,

$$p_a = 45 \text{ kN/m}^2$$

Then,

$$P_a = \tfrac{1}{2} p_a H = \tfrac{1}{2} \times 45 \times 8 = 180 \text{ kN/m}$$

The failure planes are parallel to AE and AE′. The directions of these lines are measured as 60° and 65° (= 90° + $\phi'$ = 90° + 35° = 125°).

**Example 11.3** For the retaining wall shown in Fig. 11.22, make a sketch of the distribution of active pressure on the wall, giving the principal values. Compute the thrust per metre length of the wall neglecting cohesive and frictional forces on the back of the wall.

***Solution***

This is the same condition as given in Fig. 11.12b.

For the sand layer:

$$K_{a1} = \frac{1 - \sin 25^\circ}{1 + \sin 25^\circ} = 0.406$$

For the gravel layer:

$$K_{a2} = \frac{1 - \sin 33^\circ}{1 + \sin 33^\circ} = 0.353$$

Sand layer:

$$p_1 = K_{a1} \gamma_1 H_1 = 0.406 \times 18.2 \times 3 = 22.17 \text{ kN/m}^2$$

$$p_1' = K_{a2} \gamma_1 H_1 = 0.353 \times 18.2 \times 3 = 19.27 \text{ kN/m}^2$$

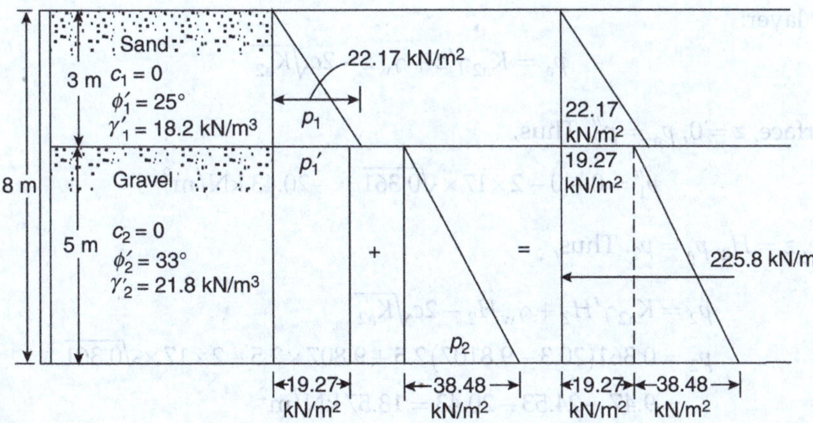


Fig. 11.22

Gravel layer:

$$p_2 = K_{a2}\gamma_2 H_2 = 0.353 \times 21.8 \times 5 = 38.48 \text{ kN/m}^2$$

The lateral pressure diagram is shown in Fig. 11.22.

$$P_a = ½ \times 22.17 \times 3 \times 1 + 5 \times 19.27 \times 1 + ½ \times 38.48 \times 5 \times 1 = 225.8 \text{ kN/m}$$

**Example 11.4** An 8 m high retaining wall supports a 5.5 m deep sand ($\gamma_d = 18.5 \text{ kN/m}^3$, $\phi = 34°$) overlying a saturated sandy clay ($\gamma_{sat} = 20.3 \text{ kN/m}^3, \phi = 28°, c = 17 \text{ kPa}$). The groundwater level is located at the interface of two layers. Sketch the lateral stress distribution up to a depth of 8 m for an active condition.

***Solution***

For the sand layer:

$$K_{a1} = \frac{1 - \sin 34°}{1 + \sin 34°} = 0.283$$

For the sandy clay layer:

$$K_{a2} = \frac{1 - \sin 28°}{1 + \sin 28°} = 0.361$$

Sand layer:

$$p_1 = K_{a1}\gamma_d H_1 - 2c\sqrt{K_{a1}}$$

$$p_1 = 0.283 \times 18.5 \times 5.5 - 0 = 28.8 \text{ kN/m}^2$$

$$p_1' = K_{a2}\gamma_d H_1 = 0.361 \times 18.5 \times 5.5 = 36.73 \text{ kN/m}^2$$

Sandy clay layer:

$$p_a = K_{a2}\gamma' z + \gamma_w z - 2c\sqrt{K_{a2}}$$

At the interface, $z = 0$, $p_a = p_1''$. Thus,

$$p_1'' = 0 + 0 - 2 \times 17 \times \sqrt{0.361} = -20.43 \text{ kN/m}^2$$

At the base, $z = H_2$, $p_a = p_2$. Thus,

$$p_2 = K_{a2}\gamma' H_2 + \gamma_w H_2 - 2c\sqrt{K_{a2}}$$

$$p_2 = 0.361(20.3 - 9.8107)2.5 + 9.807 \times 2.5 - 2 \times 17 \times \sqrt{0.361}$$

$$= 9.47 + 24.53 - 20.43 = 13.57 \text{ kN/m}^2$$

(a) (b) (c)

The components of $p_2$, i.e., (a), (b), and (c), are shown in Fig. 11.23.
Resultant pressure at the ground surface = 0 kN/m²
Resultant pressure at the bottom of the first layer = 28.8 kN/m²
Resultant pressure at the top of the second layer is

$$p_1' + p_1'' = 36.73 - 20.43 = 16.3 \text{ kN/m}^2$$

Resultant pressure at the bottom of the second layer is

$$p_1' + p_2 = 36.73 + 13.57 = 50.3 \text{ kN/m}^2$$

The lateral stress distribution is shown in Fig. 11.23.

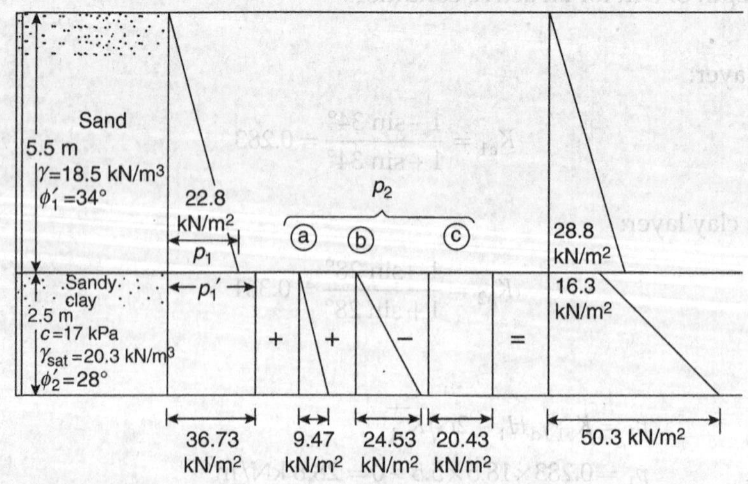


Fig. 11.23

**Example 11.5** A two-layer cohesive horizontal backfill is supported by a 10 m high vertical smooth wall. Determine the Rankine active force per unit length of the wall both before and after a tensile crack occurs in the top layer. Also, determine the line of action of the resultant in both cases. The soil layer parameters are given below:

0–5 m, Top layer: $c_u = 12\ \text{kN/m}^2, \phi_u = 0°, \gamma = 17\ \text{kN/m}^3$
5–10 m, Bottom layer: $c_u = 35\ \text{kN/m}^2, \phi_u = 10°, \gamma = 18\ \text{kN/m}^3$

***Solution***

Top layer

$$K_{a1} = \frac{1-\sin 0°}{1+\sin 0°} = 1.0$$

Bottom layer

$$K_{a2} = \frac{1-\sin 10°}{1+\sin 10°} = 0.704$$

$$p_a = K_a \gamma_z - 2c\sqrt{K_a}$$

*Top layer*. When $z = 0$,
Cohesion component $p_0 = -2c_1\sqrt{K_{a1}}$
$= -2\times12\times1 = -24\ \text{kN/m}^2$
Weight component $p_0' = 0$
When $z = H_1$,
Cohesion component $p_1 = -2\times12\times1 = -24\ \text{kN/m}^2$
Weight component $p_1' = K_{a1}\gamma_1 H_1 = 1\times17\times5 = 85\ \text{kN/m}^2$
*Bottom layer*. For the bottom layer, the weight of the first layer acts as a surcharge $q\ (= \gamma_1 H_1)$. Therefore, we consider the second layer separately, and the general equation is

$$p_a = K_{a2}\gamma_2 z - 2c_2\sqrt{K_{a2}} + K_{a2}\gamma_1 H_1$$

When $z = 0$ (at the interface),
Cohesion component $p_2 = 2c_2\sqrt{K_{a2}} = -2\times35\times\sqrt{0.704} = 58.73\ \text{kN/m}^2$
Weight component $p_2' = 0$
Surcharge component $p_2'' = K_{a2}\gamma_1 H_1 = 0.704\times17\times5 = 59.84\ \text{kN/m}^2$
When $z = H_2$ (at the bottom of the wall),
Cohesion component $p_3 = -2\times35\times\sqrt{0.704} = -58.73\ \text{kN/m}^2$
Weight component $p_3' = K_{a2}\gamma_2 H_2 - 0.704\times18\times5 = 63.36\ \text{kN/m}^2$
Surcharge component $p_3'' = 0.704\times17\times5 = 59.84\ \text{kN/m}^2$
The pressure distribution for the no-tension condition is shown in Fig. 11.24a.
The total thrust

$$P_{a1} = \tfrac{1}{2}\times85.0\times5 - 24\times5 + \tfrac{1}{2}\times63.36\times5 + 59.84\times5 - 58.73\times5$$
$$P_{a1} = 212.5 - 120 + 158.4 + 299.2 - 293.65 = 256.45\ \text{kN}$$

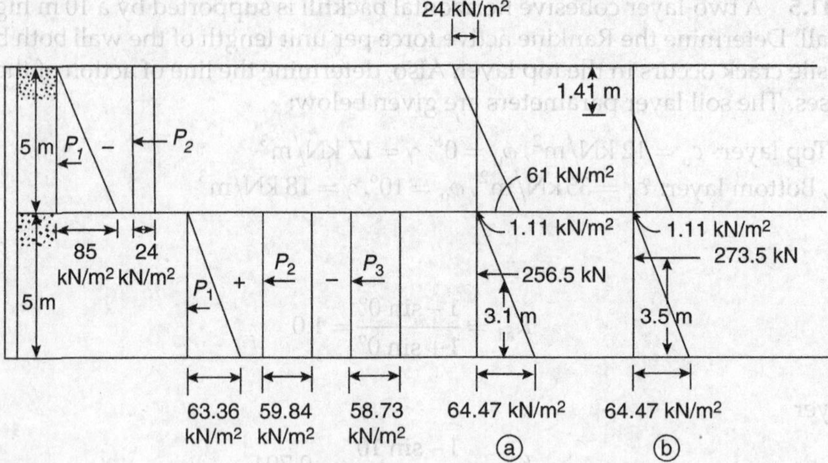


Fig. 11.24

The line of action,

$$\bar{x}_1 = \frac{212.5\times(5/3+5)-120(5/2+5)+158.4\times(5/3)+299.2\times(5/2)-293.65\times(5/2)}{256.45}$$

$$\bar{x}_1 = 3.1\ \text{m}$$

Pressure distribution after the development of the tension crack is shown in Fig. 11.24b. Here,

$$z_0 = \frac{2c_1}{\gamma\sqrt{K_{a1}}} = \frac{2\times 12}{17\times 1} = 1.41\ \text{m}$$

The total thrust

$$P_{a2} = \tfrac{1}{2}\times 61(5-1.41)+158.4+299.2-293.65$$

$$P_{a2} = 109.5+158.4+299.2-293.65 = 273.5\ \text{kN}$$

$$\bar{x}_2 = \frac{109.5\times[(3.59/3)+5]+158.4\times(5/3)+299.2\times(5/2)-293.65\times(5/2)}{273.5}$$

$$\bar{x}_2 = 3.5\ \text{m}$$

**Example 11.6** A vertical smooth-faced 8 m high retaining wall yields when rotated about the bottom. Estimate the movement at the top of the retaining wall required to establish an active case. The soil retained is a dry sand with angle of internal friction equal to 37°.

***Solution***

Figure 11.25 shows the active zone that would develop if the retaining wall AB yields by rotating about the bottom to a position A′B.

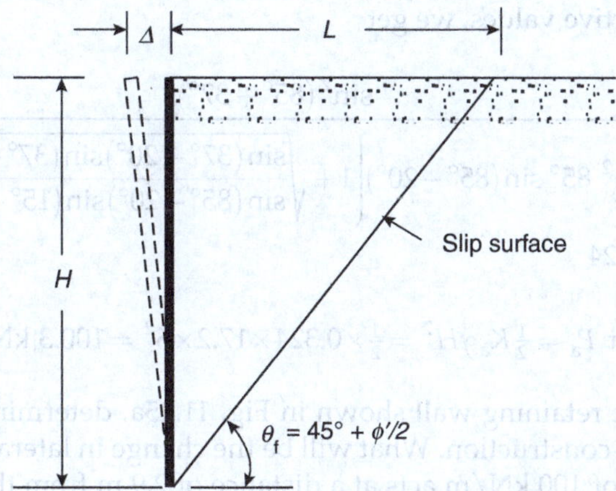


Fig. 11.25

The required yield (Δ) at the top of the wall is specified in terms of the width (*L*) of the active zone at the top of the wall. Therefore,

$$\text{Yield strain } \varepsilon = \frac{\Delta}{L}$$

Based on triaxial shear tests, the yield strain required to establish the active case is approximately 0.005. Now,

$$\theta_f = 45° + \frac{\phi'}{2} = 45° + \frac{37°}{2} = 63.5°$$

$$L = H\tan(90° - \theta_f^{\circ}) = 8 \times \tan(90° - 63.5°)$$

$$L = 3.99 \text{ m}$$

Therefore, movement, Δ, at the top of the wall = 0.005×3.99 = 0.0199 m

**Example 11.7** A retaining wall of 6 m vertical height has the pressure face inclined at 85° to the horizontal and has a 20° angle of wall friction. The backfill is sloping at 15° to the horizontal and has the following properties: $c = 0$, $\phi' = 37°$, $\gamma = 17.2 \text{ kN/m}^3$. Compute Coulomb's active force per unit length of the wall.

***Solution***

For the given problem, $\beta = 85°$, $\delta = 20°$, $\phi' = 37°$, and $\gamma = 17.2 \text{ kN/m}^3$ and $H = 6$ m. The active earth pressure coefficient is obtained from

$$K_a = \frac{\sin^2(\beta + \phi')}{\sin^2\beta \sin(\beta - \delta)\left(1 + \sqrt{\dfrac{\sin(\phi' + \delta)\sin(\phi' - 1)}{\sin(\beta - \delta)\sin(i + \beta)}}\right)^2}$$

Substituting the respective values, we get

$$K_a = \frac{\sin^2(85° + 37°)}{\sin^2 85° \sin(85° - 20°)\left(1 + \sqrt{\dfrac{\sin(37° + 20°)\sin(37° - 15°)}{\sin(85° - 20°)\sin(15° + 85°)}}\right)^2}$$

$$K_a = 0.324$$

Therefore, active thrust $P_a = \frac{1}{2}K_a\gamma H^2 = \frac{1}{2}\times 0.324 \times 17.2 \times 6^2 = 100.3$ kN.

**Example 11.8** For the retaining wall shown in Fig. 11.25a, determine the active thrust on the wall by Culmann's construction. What will be the change in lateral thrust and the line of action (i) if a line load of 100 kN/m acts at a distance of 2.9 m from the face of the wall and (ii) a uniform surcharge of 36 kN/m$^2$acts on the surface?

***Solution***

The retaining wall is drawn to a scale of 1mm = 100 mm.

*No surcharge load*. Different trial wedges are taken and their weights are computed and tabulated in Column 2 in the following table:

| Wedge no. | No surcharge load | Weight of wedge (kN) | |
|---|---|---|---|
| | | Line load (Col. 2 + 100) | Uniform surcharge |
| (1) | (2) | (3) | (4) |
| $W_1/W'_1$ | 241 | 341 | 336 |
| $W_2/W'_2$ | 359 | 459 | 550 |
| $W_3/W'_3$ | 511 | 611 | 781 |
| $W_4/W'_4$ | 654 | 754 | 999 |

Taking a force scale of 1 mm = 10 kN, the weights of the wedges are represented (as $AW_1$, $AW_2$, etc.) on the $\phi$-line. Establish points to plot the Culmann curve. This is shown in Fig. 11.26a by a solid line. Draw a tangent to the curve and measure the active thrust represented by FW.

Active thrust $P_{a1}$ = (FW × Force scale)

$$P_{a1} = 17 \times 10 = 170 \text{ kN/m}$$

*With line load*. Locate the line load position. The weight of the line load, $q$, will be acting on all the wedges, and the revised weights of the wedges are shown in Column 3 of the table. Represent the weights of the wedges (as $AW'_1$, $AW'_2$, etc.) with the same scale. Again, plot the Culmann curve (shown by broken lines), and find the active thrust represented by $F'W'$. Therefore,

Active thrust on the wall with a line load on the backfill

$$= P_{a2} = 23 \times 10 = 230 \text{ kN/m}$$

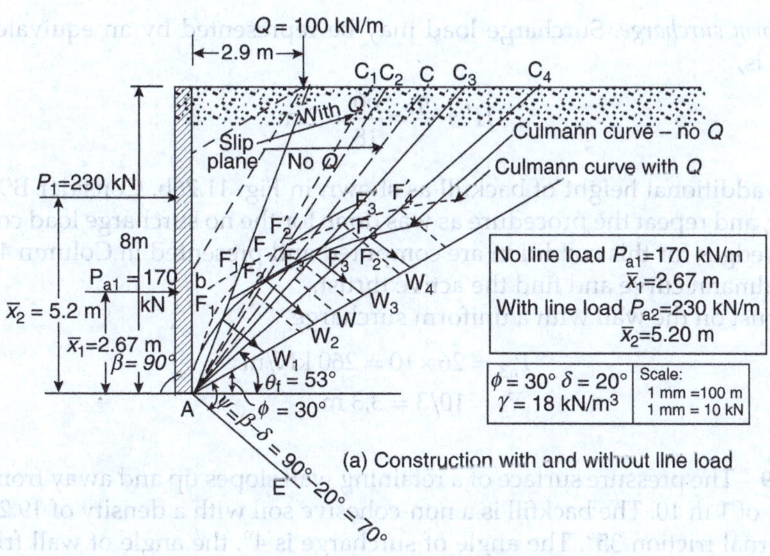


(a) Construction with and without line load

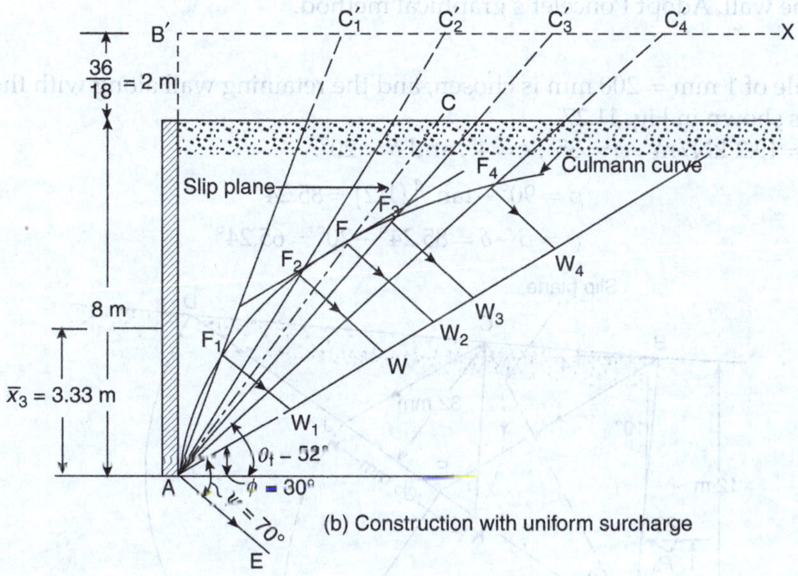


(b) Construction with uniform surcharge

Fig. 11.26

As the line load is left of the slip plane, establish the line of action following the procedure given in Fig. 11.9a. Thus, $\bar{x}_2$ is obtained as

$$\bar{x}_2 = 5.2 \text{ m}$$

*With uniform surcharge.* Surcharge load may be represented by an equivalent height of backfill; that is,

$$H' = \frac{q}{\gamma} = \frac{36}{18} = 2 \text{ m}$$

Draw the additional height of backfill as shown in Fig. 11.26b. Consider B′X as the new level surface, and repeat the procedure as was done for the no surcharge load condition. The weights of wedges for this condition are computed and presented in Column 4 of the table. Draw the Culmann curve and find the active thrust.

Active thrust on the wall with a uniform surcharge

$$P_{a3} = 26 \times 10 = 260 \text{ kN/m}$$
$$\bar{x}_3 = 10/3 = 3.3 \text{ m}$$

**Example 11.9** The pressure surface of a retaining wall slopes up and away from the backfill with a batter of 1 in 10. The backfill is a non-cohesive soil with a density of 19.2 kN/m$^3$ and angle of internal friction 35°. The angle of surcharge is 4°, the angle of wall friction is estimated to be 20°, and the vertical height of the wall is 12 m. Compute the maximum active thrust on the wall. Adopt Poncelet's graphical method.

***Solution***

A linear scale of 1 mm = 200 mm is chosen, and the retaining wall along with the surcharge is drawn, as shown in Fig. 11.27.

Here, $\gamma = 19.2$ kN/m$^3$, $\phi = 35°$, $i = 4°$, and $\delta = 20°$.

$$\beta = 90° - \tan^{-1}(1/2) = 85.24°$$
$$\psi = \beta - \delta = 85.24° - 20° = 65.24°$$

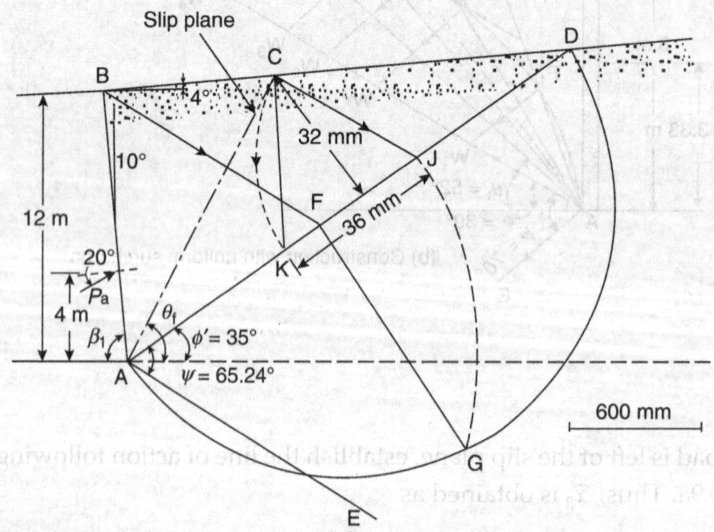


Fig. 11.27

Area of triangle JKC $= \frac{1}{2} \times 36 \times 32 \times \frac{200}{1000} \times \frac{200}{1000} = 23.04\ \text{m}^2$

Active thrust $P_a$ = (Area of ΔJKC) × $\gamma = 23.04 \times 19.2 = 442.4$ kN/m

Indication of failure plane $\theta_f = 62°$

**Example 11.10** Rework Example 11.9 for the case of a vertical wall with (i) $i = 25°$ and (ii) $i = 35°$.

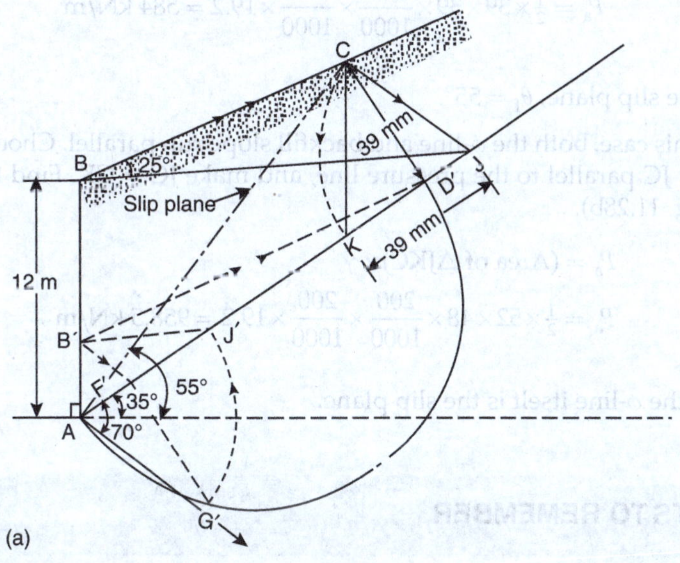


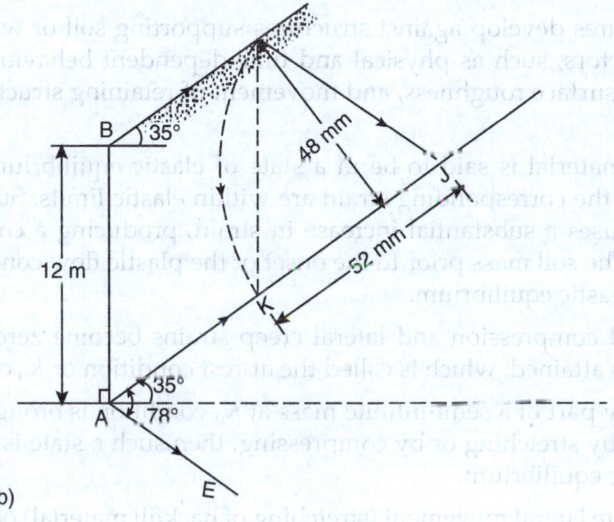


Fig. 11.28

*Solution*

Case I: For such cases, the backfill slope and $\phi$-line may not meet within the space available on the paper. In such cases, choose an arbitrary point B′ on the pressure face, considering AB′ as the wall, and proceed. Establish point J′ (similar to J in Example 11.9).

Join B′J′ and draw a line BJ parallel to B′J′. From J, draw JC parrallel to AE. Make JC = JK. Join C and K. Find the area of ΔJKC (as shown in Fig. 11.28a). Therefore,

$$P_a = (\text{Area of } \Delta\text{JKC}) \times \gamma$$

$$P_a = \tfrac{1}{2} \times 39 \times 39 \times \frac{200}{1000} \times \frac{200}{1000} \times 19.2 = 584 \text{ kN/m}$$

Inclination of the slip plane, $\theta_f = 55°$.

Case II: For this case, both the $\phi$-line and backfill slope run parallel. Choose any point J on the $\phi$-line, draw JC parallel to the pressure line, and make JC = CK. Find the area of ΔJKC (as shown in Fig. 11.28b).

$$P_a = (\text{Area of } \Delta\text{JKC}) \times \gamma$$

$$P_a = \tfrac{1}{2} \times 52 \times 48 \times \frac{200}{1000} \times \frac{200}{1000} \times 19.2 = 958.5 \text{ kN/m}$$

For such cases, the $\phi$-line itself is the slip plane.

## POINTS TO REMEMBER

11.1 Lateral pressures develop against structures supporting soil or water. They depend on several factors, such as physical and time-dependent behaviour of the soil, soil deformation, surface roughness, and movement of retaining structures and imposed loading.

11.2 The backfill material is said to be in a state of elastic equilibrium when the stress involved and the corresponding strain are within elastic limits. Subsequent increase in stresses causes a substantial increase in strain, producing a condition known as plastic flow. The soil mass prior to the onset of the plastic flow condition is said to be in a state of plastic equilibrium.

11.3 When vertical compression and lateral creep strains become zero, a state of stable equilibrium is attained, which is called the at-rest condition or $K_0$ condition.

11.4 Suppose every part of a semi-infinite mass at $K_0$-condition is brought on the verge of failure either by stretching or by compressing, then such a state is called the general state of plastic equilibrium.

11.5 When adequate lateral movement (stretching of backfill material) occurs, the horizontal stress decreases to a certain magnitude such that the full shear strength of the soil is mobilized. This horizontal stress condition is called Rankine's active state, and the stress is referred to as the active stress.

11.6 When sufficient lateral movement (compression of backfill material) occurs, then the maximum shear strength of the soil is mobilized and the horizontal stress is at a maximum. This state of failure is called Rankine's passive state, and the horizontal stress is called the passive stress.

11.7 The ratio of horizontal stress to vertical stress in the active state is referred to as the coefficient of active stress or coefficient of active earth pressure, $K_a$. The ratio of horizontal stress to vertical stress in the passive state is called the coefficient of passive stress or coefficient of passive earth pressure, $K_p$. The earth pressure coefficients may vary from 0.14 to 14 from active to passive condition in cohesionless soils and 0.5 to 2 in cohesive soils.

11.8 Rankine's theory assumes that (i) there is a conjugate relationship between vertical and lateral pressures, (ii) the soil is homogeneous and isotropic, (iii) the soil is dry and non-cohesive, and (iv) the wall is vertical and smooth.

11.9 The maximum unsupported depth of excavation, $H_c$, may be theoretically taken as twice the depth of the tension zone (i.e., $H_c = 2z_0$), where the tensile stresses equal the cohesive strength. The application of this depth in practice should be done very carefully.

11.10 Coulomb's earth pressure theory includes the effect of friction between the backfill and the wall, and a dry non-cohesive inclined backfill. The lateral earth pressure required to maintain the equilibrium of a sliding wedge with a plane slip surface is calculated.

## QUESTIONS

### Objective Questions

11.1 The lateral earth pressure coefficients $K_a$ and $K_p$ refer to
(a) Effective stresses (b) Total stresses
(c) Neutral stresses (d) None of the above

11.2 The active earth pressure caused by a cohesionless backfill on a smooth vertical surface may be reduced by
(a) Saturating the backfill soil with water
(b) Compacting the backfill soil
(c) Reducing the effective stress of the backfill
(d) Providing surcharge load on the backfill

11.3 If for an inclined backfill, with the angle of backfill inclination $i$ and angle of shearing resistance $\phi$ are equal, then for the Rankine condition the active earth pressure coefficient is

(a) $\cos^2 i$ (b) $\sin^2 i$
(c) $\cos i$ (d) $\sin i$

11.4 The presence of a water table in the backfill serves to increase the earth pressure due to
(a) Decrease in cohesion (b) Increase in surcharge
(c) Increase in the unit weight (d) Increase in wall friction

11.5 Identify the incorrect statement. Lateral pressure can be developed under the following conditions:
(a) Earthquake (b) Swelling pressure
(c) Ice formation (d) Over-consolidation

11.6 *Assertion A:* Earth pressure is not a unique property of a soil.

*Reason R:* Earth pressure is a function of backfill material, load on backfill, groundwater condition, and deflection of retaining structure.
*Select the correct code.*
(a) Both A and R are true, and R is the correct explanation of A
(b) Both A and R are true, and R is not the correct explanation of A
(c) A is true, but R is false
(d) A is false, but R is true

11.7 The state of shear failure accompanying a minimum earth pressure is called the
(a) At-rest state (b) Active state
(c) Passive state (d) None of the above

11.8 Identify the incorrect statement. Culmann's method is chiefly used under the following conditions:
(a) The wall has an inclined or broken back
(b) The backfill surface is irregular
(c) Backfill carries a surcharge
(d) High seepage pressure is exerted on the wall

11.9 The amount of translation needed to produce an active pressure condition in a dense cohesionless soil is
(a) $0.001H$ to $0.002H$ (b) $0.002H$ to $0.004H$
(c) $0.01H$ to $0.02H$ (d) $0.02H$ to $0.05H$

where $H$ is the height of the wall.

11.10 A sandy loam backfill has a cohesion of 14 kN/m$^2$, a friction angle of 18°, and unit weight of 16.5 kN/m$^3$. Then, the depth of the tension crack is
(a) 2.00 m (b) 3.33 m (c) 1.98m (d) 2.63 m

## Descriptive Questions

11.11 Explain with reasons the use of the at-rest lateral soil pressure condition for the design of basement walls.
11.12 Explain the possible consequences of the over-compaction of a backfill material.
11.13 Give a critical comparison of the Coulomb and Rankine earth pressure theories.

11.14 Comment on the influence of wall friction on the passive earth pressure using Coulomb's method for granular soils.

11.15 How do tension cracks influence the distribution of active earth pressure in pure cohesive soils?

11.16 State whether the following statements are true or false. Justify your choice with supporting arguments.

(i) Rankine's earth pressure analysis considers neither strains nor displacements.

(ii) "$K_0$-condition" is when no lateral deformations occur in the soil mass.

(iii) The critical height for open cuts for brittle clay soil is directly proportional to the unit weight and inversely proportional to the unconfined compressive strength.

(iv) The active earth pressure is decreased, while the passive earth pressure is increased, due to the application of a uniform surcharge load.

(v) The lateral stress under the passive condition is due to compression of the backfill.

## EXERCISE PROBLEMS

11.1 A retaining wall 6.5 m high supports an over-consolidated clay backfill with a plasticity index of 32% and an over-consolidation ratio of 2.3. Determine the lateral force per unit length of wall and the location if the yield of the wall is completely prevented. The unit weight of the soil is 17.6 kN/m$^3$.

11.2 A 4 m high smooth vertical wall retains a mass of dry loose sand. Compute the total lateral force per metre acting against the wall if the wall is prevented from yielding. The sand has a 30° angle of internal friction and unit weight of 14.8 kN/m$^3$. Also, estimate the lateral force per metre run of the wall if sufficient yield of the wall is permitted so as to develop the active Rankine state.

11.3 A vertical frictionless pressure face of an 8 m high retaining wall supports a non-cohesive 5° sloping backfill. The unit weight of the soil is 18 kN/m$^3$, and the angle of shearing resistance is 32°. Draw a Mohr circle representing the state of stresses, and hence, compute the lateral passive resistance per linear length of the wall.

11.4 A wall 15 m high has to be designed so as to retain dry sand. Under loose condition the sand has a void ratio of 0.65 and $\phi'$ of 32°, and under dense condition the void ratio and $\phi'$ are 0.41 and 43°, respectively, and $G = 2.65$. Compute the resultant lateral pressures for active and passive cases for both the density conditions. Recommend a suitable resultant lateral force if the wall has to be designed for (i) the active case and (ii) the passive case.

11.5 A dockside retaining wall 10 m high retains a non-cohesive backfill with a horizontal surface level with the top of the wall. The properties of the backfill material are, $G = 2.65$, $e = 0.55$, and $\phi = 32°$. An additional superimposed load of 20 kN/m$^2$ is induced at the surface of the backfill due to construction of warehouses and dockyard traffic. Compute the lateral thrust on the wall when the water table is (i) 2 m below the level surface, (ii) 5 m below the level surface, and (iii) at the bottom of the wall. Neglect wall friction.

11.6 A dry granular level backfill of a 6.3 m high retaining wall weighs 16.2 kN/m$^3$. The active thrust on the wall is believed to be 75 kN/m length of the wall. It is intended to increase the height of the wall and, at the same time, to keep the force on the wall within permissible limits. The backfill to a depth of 2.8 m from the top is removed. The removed portion is replaced by a material such as cinder with $\gamma = 8.2$ kN/m$^3$. If the portion of the additional height is also to be filled with cinder, estimate the additional height of the wall without increasing the initial active thrust. Neglect the wall friction, and assume that both the backfill soil and the cinder have the same friction angle.

11.7 Figure 11.29 represents a backfill behind a smooth vertical retaining wall. Estimate the magnitude and line of action of the lateral active force per metre length of the wall. What would be the reduction in the lateral force if drainage facility is provided to lower the water table to the base of the wall?

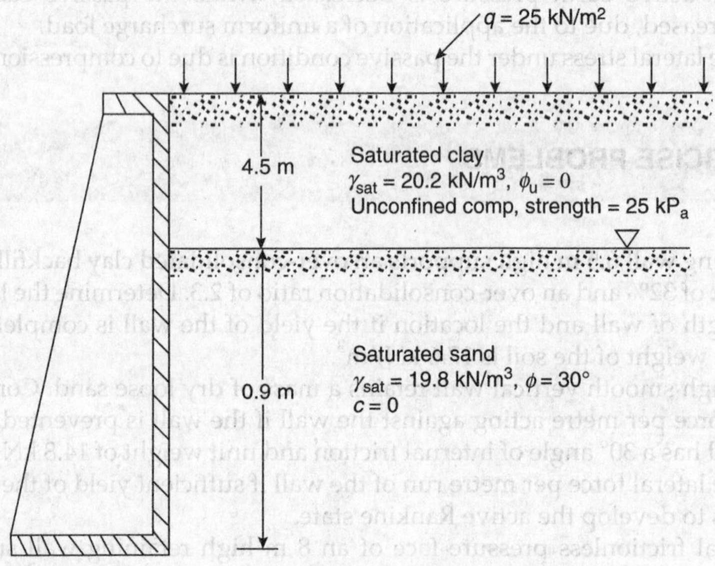


Fig. 11.29

11.8 A vertical wall 10 m high retains two horizontal layers of a saturated cohesive backfill with a level surface. The top 4 m of the backfill has an undrained cohesion of 18.2 kPa and a bulk unit weight of 18.6 kN/m$^3$. The bottom clay layer has a bulk unit weight and an undrained cohesion of 22 kN/m$^3$ and 23.6 kPa, respectively. Estimate the likely depth of the tension zone behind the wall. Compute the total active force if tension cracks develop, and also locate the application of the resultant lateral force.

11.9 A 9.5 m high vertical, smooth retaining wall is supporting three layers of soil with the following details:

| Layer no. | Depth (m) | Total unit weight (kN/m$^3$) | Cohesion (kPa) | Angle of shearing resistance (°) |
|---|---|---|---|---|
| Top | 3.5 | 19.2 | 17.5 | 0 |
| Middle | 3 | 20.2 | 0 | 30 |
| Bottom | 3 | 21.2 | 0 | 34 |

Compute the active thrust per metre run of the wall if the water table is located at the interface of the top and middle layers.

11.10 It is intended to excavate a vertical unsupported cut of depth 5 m. The natural soil has a unit weight of 17.5 kN/m$^3$ and the shear strength parameters are $c' = 30$ kPa and $\phi = 6°$. The groundwater table is deeper than the cut. Determine (i) the stress at the top and bottom of the cut, (ii) the maximum depth of the potential tension crack, and (iii) the maximum unsupported excavation depth.

11.11 A vertical retaining wall of height 6.5 m retains a non-cohesive level backfill weighing 19.2 kN/m$^3$, with the angle of friction being 18°. Compute the total thrust on the wall adopting Culmann's graphical method. Later, it is planned to place a piece of machinery weighing 30 kN on the surface, parallel to the crest of the wall. Find the minimum horizontal distance from the back of the wall at which the machinery could be placed without increasing the pressure on the wall. Take $\phi = 30°$.

11.12 The front of a retaining wall slopes at an angle of 80° to the horizontal. The depth of soil in front of the wall is 2.5 m. The soil surface is horizontal and the soil dry. The other properties of the soil are $c' = 0$, $\phi' = 32°$, $\delta = 18°$, and $\gamma = 18.2$ kN/m$^3$. Estimate the total passive resistance developed at the front of the wall.

11.13 For the retaining wall shown in Fig. 11.30, determine the active lateral force per metre length of the wall.

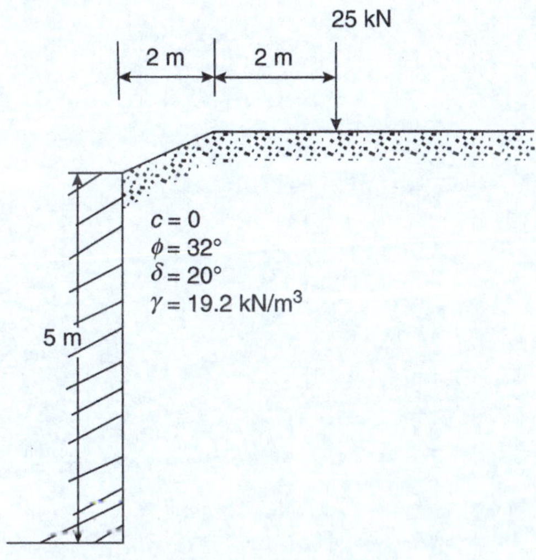


Fig. 11.30

11.14 For the retaining wall shown in Fig. 11.31, compute the lateral active force per metre length of the wall using Poncelet's graphical construction. Check the value using Culmann's graphical method.

Fig. 11.31

# 12

# Earth-Retaining Structures

**CHAPTER HIGHLIGHTS**

Gravity-type retaining walls: proportioning, earth pressure consideration, stability requirements, backfill materials and drainage, joints – Sheet pile walls: cantilever types, anchored types, wales, tie-rods and anchorages – Braced excavations: earth pressure distribution, heave and stability

## 12.1 INTRODUCTION

In Chapter 11, the two basic earth pressure theories were discussed at length. These theories are applied to the design of earth-retaining structures. In general, earth-retaining structures are constructed when abrupt changes in the ground surface elevation are needed to protect unstable slopes. The typical structures are various types of retaining walls, sheet piles, braced sheeting of excavations, bulkheads or abutments, basement or pit walls, etc. These may be self-supporting (e.g., gravity- or cantilever-type walls) or they may be laterally supported by means of bracing of anchorages. The retaining materials may be soil and water, coal or ore piles and water.

## 12.2 GRAVITY-TYPE RETAINING WALLS

Gravity-type walls provide slope and soil retention on account of their weight, which can consist of masonry, concrete mass, concrete in combination with soil weight, or the weight of earth mass alone. In addition to weight, they are aided by the passive resistance developed in front of the wall. They are all free to deflect at the top and thereby mobilize active earth pressure. Representative types of gravity-type walls are shown in Fig. 12.1. Massive walls are uneconomical because of the large wall material used for the dead weight. Reinforced concrete cantilever walls are more economical because the backfill itself is aimed to provide most of the required dead weight.

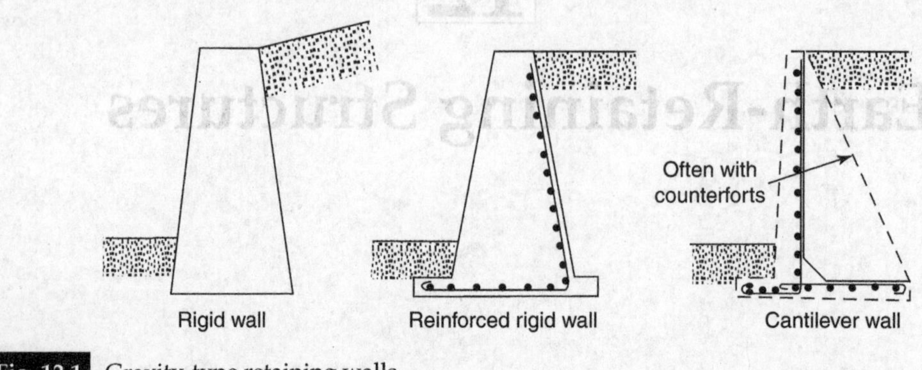


Fig. 12.1 Gravity-type retaining walls

### 12.2.1 Proportioning Retaining Walls

Simple gravity and cantilever walls are quite common. For the design of a retaining wall, a preliminary dimension has to be assumed. This is referred to as proportioning, which enables the engineer to decide the basic components of the wall for analysis. If unstatisfactory results are obtained after the required stability checks, the sections are modified and re-checked. Figure 12.2 represents the proportions of various wall components for preliminary selection (Das, 1984). The top of the stem of any retaining wall should not be less than 0.3 m for construction activities. The bottom of the base should be below the weathered soil due to seasonal variation, and in no case less than 0.6 m. The counterfort walls may have similar

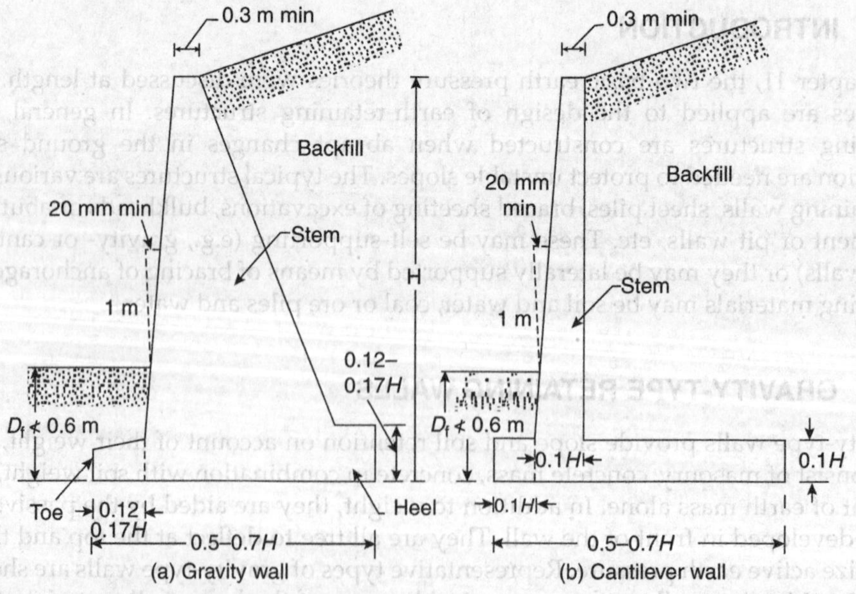


Fig. 12.2 Proportioning retaining wall

dimensions to those of cantilever walls. The counterforts are 0.3 m thick and are spaced at centre-to-centre distances of 0.3 to 0.7$H$, where $H$ is the height of the retaining wall.

### 12.2.2 Earth Pressure Consideration

Cantilever and gravity walls (Fig. 12.3) are both liable to rotational and translational movements, and hence Rankine or Coulomb theories may be used for the calculation of lateral pressure.

If Rankine's theory is to be applied, then it is assumed that the soil is retained by a vertical face (shown by broken lines in Fig. 12.3) extending upwards from the heel to the ground surface. Then, the active pressure is evaluated keeping AB as the face of the wall. While checking for stability, the active thrust ($P_a$), the weight of soil above the heel ($W_s$), and the weight of the concrete ($W_c$) should be taken into consideration. This assumption is theoretically correct as long as the zone bounded by the line BC (Fig. 12.3) is not obstructed by the stem of the wall. However, Coulomb's theory can be used directly on the real wall surface without any assumption.

### 12.2.3 Stability Requirements

The retaining wall as a whole must satisfy the following stability requirements:

1. Safety against overturning
2. Safety against sliding
3. Safety against bearing capacity failure
4. Safety against overall stability

**Check for Overturning.** Figure 12.3 represents the forces acting on a cantilever retaining wall for an active condition. The passive resistance ($P_p$) in front of the wall should not be relied on unless the soil is firm and undisturbed. The passive resistance, $P_p$, is given as

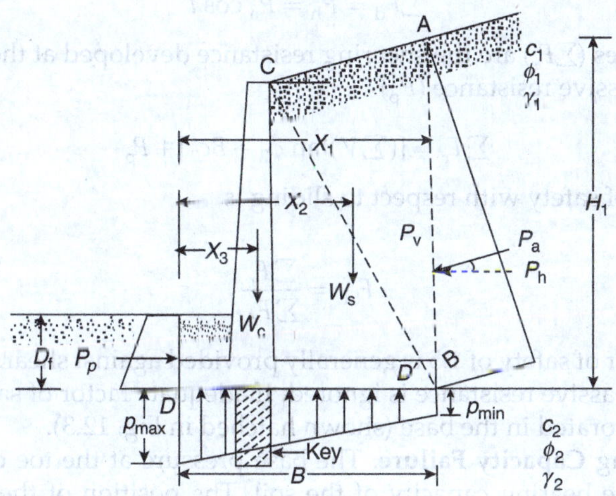


Fig. 12.3 Retaining wall with details of forces

$$P_p = \frac{1}{2}K_p\gamma_2 D_f^2 + 2c_2\sqrt{K_p}\,D_f$$

where $c_2$, $\phi_2$, $\gamma_2$, and $K_p$ are the parameters related to the soil in the wall and the foundation.

The active thrust, $P_a$, is determined by applying Rankine's theory on the vertical surface AB. The weight of the wall ($W_c$) and the weight of the soil above the heel ($W_s$) are calculated.

It is generally assumed that if overturning were to occur, it would do so about the toe of the wall. Thus, the factor of safety against overturning is defined as the ratio of resisting to disturbing moments about the toe.

Let $P_h$ and $P_v$ be the components of the active force $P_a$, then, the sum of disturbing moments ($\sum M_d$) is given as

$$\sum M_d = P_h\left(\frac{H_1}{3}\right) \tag{12.1}$$

where $P_h = P_a \cos i$.

The sum of stabilizing moments ($\sum M_r$) is given as

$$\sum M_r = W_c x_1 + W_s x_2 + P_v x_3 + P_p\left(\frac{D_f}{3}\right) \tag{12.2}$$

Therefore,

$$F_{OT} = \frac{\sum M_r}{\sum M_d} \tag{12.3}$$

where $F_{OT}$ is the factor of safety against overturning. This should not be less than 2.0.

**Check for Sliding**. The factor of safety against sliding along the base is defined as the ratio of the resisting forces to the sum of the horizontal disturbing forces.

The only horizontal force causing the sliding is $P_h$, hence, the sum of driving forces ($\sum F_d$) is given as

$$\sum F_d = P_h = P_a \cos i \tag{12.4}$$

The resisting forces ($\sum F_r$) are the shearing resistance developed at the base ($S_h = (\sum V)\tan \phi_2 + Bc_2$) and the passive resistance ($P_p$)

$$\sum F_r = [(\sum V)\tan \phi_2 + Bc_2] + P_p \tag{12.5}$$

Then, the factor of safety with respect to sliding is

$$F_{SL} = \frac{\sum F_r}{\sum F_d} \tag{12.6}$$

A minimum factor of safety of 1.5 is generally provided against shear. As discussed earlier in many cases, the passive resistance is ignored. If adequate factor of safety is not achieved, a key may be incorporated in the base (shown hatched in Fig. 12.3).

**Check for Bearing Capacity Failure**. The base pressure at the toe of the wall must not exceed the allowable bearing capacity of the soil. The position of the resultant force $R$ is determined by dividing the algebraic sum of the moments of all forces about any point on the base by the vertical component $\sum V$ (Fig. 12.4).

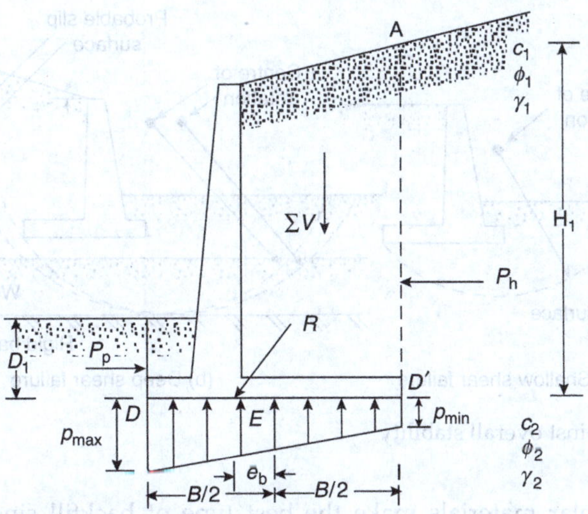


**Fig. 12.4** Forces to check for bearing capacity failure

In order to keep the base pressure compressive over the entire base width, the resultant $R$ must act within the middle third of the base, that is, the eccentricity ($e_b$) of the base resultant must not exceed $B/6$. Adequate safety against overturning of the wall will be ensured when the resultant falls within the middle third of the base. Considering a linear variation of the base pressure, the maximum and minimum pressures on the base are computed:

$$p_{max} = \frac{\sum V}{B}\left(1 + \frac{6e_b}{B}\right) \tag{12.7}$$

$$p_{min} = \frac{\sum V}{B}\left(1 - \frac{6e_b}{B}\right) \tag{12.8}$$

The value of $p_{max}$ should be less than the allowable bearing capacity of the soil. The allowable soil pressure considers both the safety against shear failure and the permissible settlement. Generally, a factor of 3 is provided against shear failure.

The value of $p_{min}$ becomes negative when eccentricity $e_b$ exceeds $B/6$. This should be completely avoided as the tensile strength of the soil is very small.

**Check Against Overall Stability**. Apart from safety of the retaining wall against the three factors explained above, the wall should also be safe against overall stability. Because of the presence of a weak layer at an immediate depth below the base in cohesive soils, there is a possibility of the entire soil slipping along with the wall due to inadequate strength, excess pore water pressure, removal of resistance near the toe, etc. Some of the possible failures are shown in Fig. 12.5. The factor of safety against the overall stability should not be less than 1.50.

### 12.2.4 Backfill Materials and Drainage

Backfill materials for retaining structures should be designed to minimize the lateral pressure. A good backfill material should satisfy two important requirements, viz., high long-term strength and free drainage.

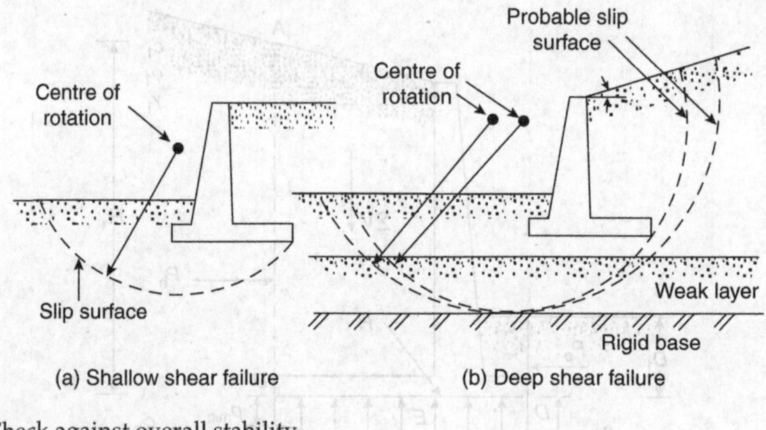


**Fig. 12.5** Check against overall stability

In general, granular materials make the best type of backfill since they maintain an indefinite active state of stress and have free drainage. Clay backfills tend to creep and have a very low permeability. They should be avoided as climatic changes are likely to cause successive swelling and shrinkage of the soil. Swelling imposes unpredictable pressures on the wall and its movements, and subsequent shrinkage may result in the formation of cracks in the backfill surface.

Poorly graded to well-graded sands and gravels form an excellent backfill because of their free-draining characteristics. Silty/clayey sand and gravels function as good backfill materials provided they are kept dry or are provided with adequate drainage arrangement. Low to high plastic clays and silts can be graded as poor backfill material. Organic silts and clays and peat should not be used as backfill because of the swelling and shrinking behaviour of such soils.

An important consideration is the control of the water table in the backfill. The easiest way to control groundwater is to provide a free-draining backfill. Further, as a result of rainfall, or other reasons, the backfill may get saturated and increase the pressure on the wall, creating an unstable condition. Weep holes and/or perforated drainage pipes are provided to drain away such water and reduce the development of pore water pressure. Weep holes should have a minimum diameter of 0.1 m and should be spaced adequately. Filter materials are provided behind the weep holes and around the drainage pipes to prevent the possible washing out of backfill materials into weep holes or drains (Fig. 12.6).

The backfill material has to be compacted to attain maximum strength and, hence, minimum active thrust on the wall. However, over-compaction has to be avoided and sufficient care taken not to disturb the wall while compacting the backfill.

### 12.2.5 Joints in Retaining Walls

A retaining wall is provided with construction, contraction, or expansion joints (Fig. 12.7).

**Construction Joints**. These are provided between two successive pours of concrete, and are vertical and horizontal joints. Either each surface of the concrete is cleaned and roughened before placing the next pour of concrete or a key is provided between the joints.

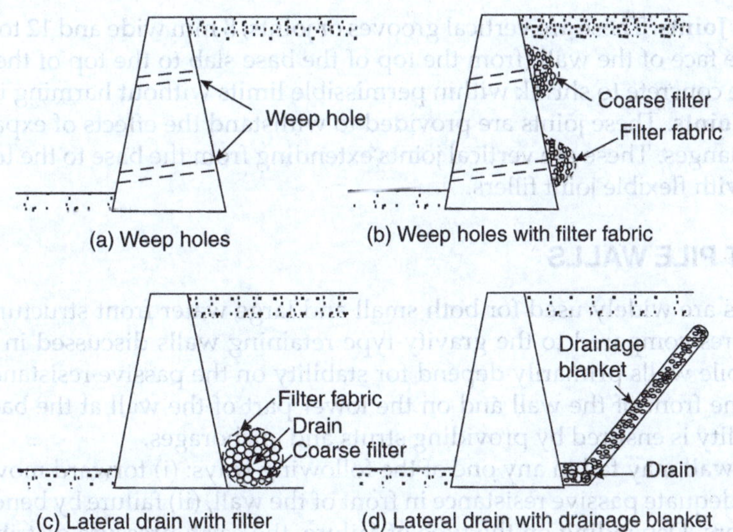


**Fig. 12.6** Drainage arrangements

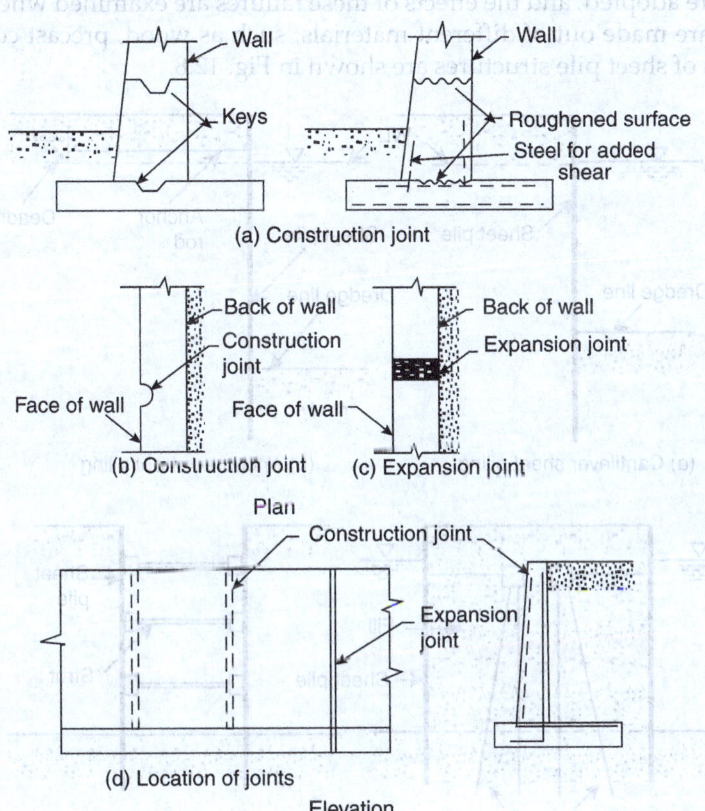


**Fig. 12.7** Joints in retaining walls

**Contraction Joints**. These are vertical grooves or joints, 8 mm wide and 12 to 16 mm deep, provided in the face of the wall (from the top of the base slab to the top of the wall). These joints allow the concrete to shrink within permissible limits without harming it.

**Expansion Joints**. These joints are provided to withstand the effects of expansion due to temperature changes. These are vertical joints extending from the base to the top of the wall and are filled with flexible joint fillers.

## 12.3 SHEET PILE WALLS

Sheet pile walls are widely used for both small and large water front structures. These are flexible structures compared to the gravity-type retaining walls discussed in the previous section. Sheet pile walls primarily depend for stability on the passive resistance developed by the soil in the front of the wall and on the lower part of the wall at the back. In certain types, the stability is ensured by providing struts and anchorages.

A sheet pile wall may fail in any one of the following ways: (i) forward movement of the base due to inadequate passive resistance in front of the wall, (ii) failure by bending, and (iii) failure of anchors. Depending on the type of failure, the earth pressure distribution varies, and it does not follow the conventional distribution adopted in rigid walls. However, simple distributions are adopted, and the effects of these failures are examined wherever necessary.

Sheet piles are made out of different materials, such as wood, precast concrete, or steel. Different types of sheet pile structures are shown in Fig. 12.8.

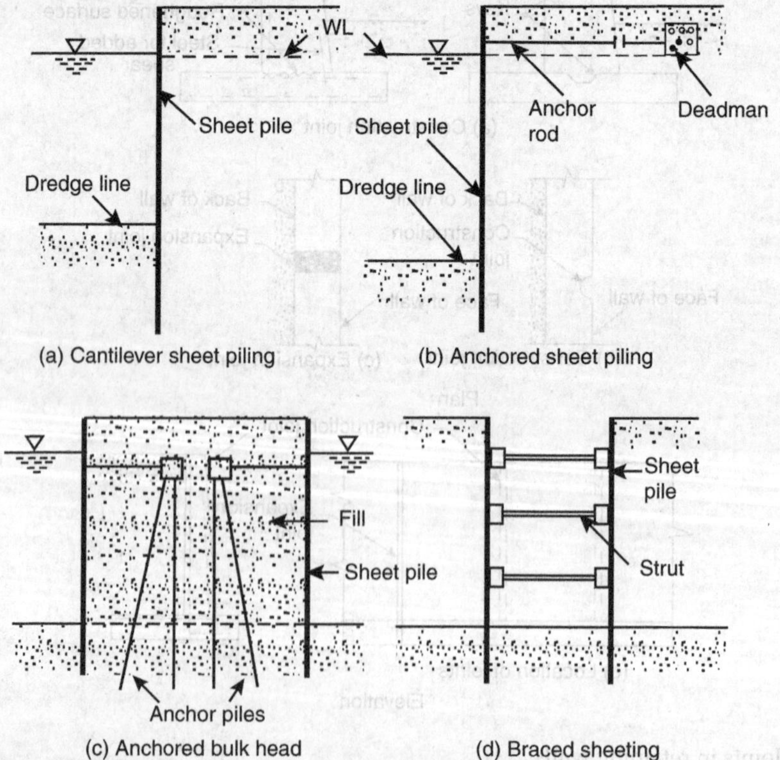


**Fig. 12.8** Sheet pile structures

### 12.3.1 Cantilever Sheet Pile Walls

Sheet piles of this type are constructed by driving the sheet pile to adequate depth into the soil below the dredge line so that they develop a cantilever beam-type reaction to resist the active pressures on the wall. These piles are economical only for a moderate height of up to about 10 m. Because of the cantilever action, the lateral deflection of this type of wall is more. As the stability of the wall depends primarily on the pressure developed in front of the wall, any action of lowering the dredge line (e.g., by erosion or scour) should be controlled. The use of such walls is primarily meant for temporary installations.

The embedment depth varies with different soils. Also, the pressure distribution varies with the type of soil and water level conditions. The wall rotates about the point O, and the development of active and passive conditions on either side of the wall is as shown in Fig. 12.9a. The probable actual pressure distribution is shown in Fig. 12.9b. However, for design purposes, the distribution is simplified, as shown in Fig. 12.9c.

**Cantilever Sheet Piling in Granular Soils.** In this case, both the retained soil and that below the dredge line are sands and are assumed to have the same properties. Appropriate values of $\gamma$ and $\phi$ should be used for a layered system. For non-level ground surfaces, Coulomb earth pressure theory may be applied and for the rest, Rankine's. In computing earth pressures on the wall, it is not wise to rely on the vertical shearing forces between the soil and the wall, and thus the application of Rankine's theory is justified.

The positions of ground surface, dredge line, water table, and pressure diagram are depicted in Fig. 12.10. Let $\phi$, $K_a$, $K_p$, $\gamma$, $\gamma_{sat}$, and $\gamma'$ be the angle of shearing resistance, the active and passive earth pressure coefficients, and the total, saturated, and submerged unit weights, respectively. Then, the pressures $p_1$ and $p_2$ are given as

$$p_1 = \gamma z_1 K_a \tag{12.9}$$

and

$$p_2 = (\gamma z_1 + \gamma' z_2) K_a \tag{12.10}$$

At the dredge line the hydrostatic pressures on either side of the wall cancel each other.

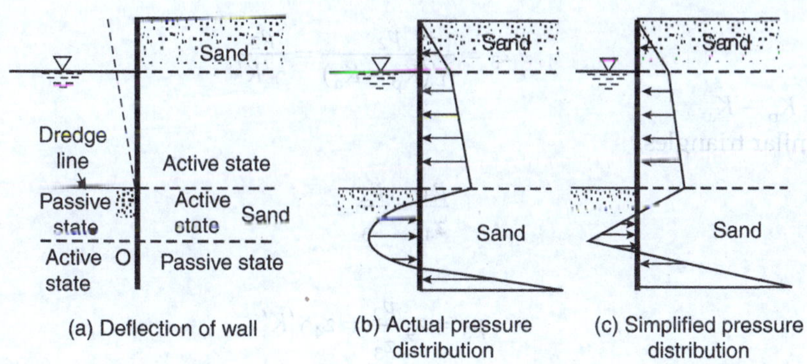


**Fig. 12.9** Cantilever sheet piling

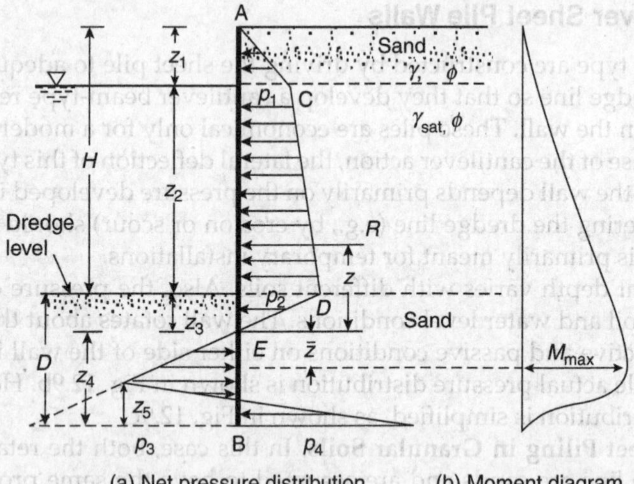

(a) Net pressure distribution (b) Moment diagram

**Fig. 12.10** Cantilever sheet piling in granular soils

In order to determine the net lateral pressure below the dredge line and at the point of zero pressure, consider any depth z from the dredge level. Again, hydrostatic pressures cancel each other, and the active passive pressure at depth z may be given as

$$p_a = p_2 + \gamma' z K_a \tag{12.11}$$

$$p_p = \gamma' z K_p \tag{12.12}$$

The net lateral pressure, $p_z$, is obtained as

$$p_z = p_a - p_p = p_2 + \gamma' z K_a - \gamma' z K_p$$

$$p_z = p_2 + \gamma' z (K_a - K_p)$$

At depth $z = z_3$, $p_z = 0$

$$p_2 + \gamma' z_3 (K_a - K_p) = 0$$

or

$$z_3 = \frac{p_2}{\gamma'(K_p - K_a)} = \frac{p_2}{\gamma' K'} \tag{12.13}$$

where $K' = K_p - K_a$.

From similar triangles,

$$\frac{p_3}{z_4} = \frac{p_2}{z_3}$$

or

$$p_3 = z_4 \frac{p_2}{z_3} = z_4 \gamma' K' \tag{12.14}$$

At the bottom of the sheet pile wall, the passive pressure acts from *right* to *left* and the active pressure from *left* to *right*; hence,

$$p_p = (\gamma z_1 + \gamma' z_2 + \gamma' D)K_p \tag{12.15}$$

and

$$p_a = \gamma' D K_a \tag{12.16}$$

Therefore,

$$p_p - p_a = p_4 = (\gamma z_1 + \gamma' z_2)K_p + \gamma' D K' \tag{12.17}$$

or

$$D = z_3 + z_4 \tag{12.18}$$

Considering the equilibrium of all horizontal forces,

$$R - \tfrac{1}{2} p_3 z_4 + \tfrac{1}{2} z_5 (p_3 + p_4) = 0 \tag{12.19}$$

where $R$ is the area of the pressure diagram ACDE.

Considering the equilibrium of all the moments about the point B,

$$R(z_4 + \bar{z}) - \left(\tfrac{1}{2} z_4 p_3\right)\left(\frac{z_4}{3}\right) + \tfrac{1}{2} z_5 (p_3 + p_4)\left(\frac{z_5}{3}\right) = 0 \tag{12.20}$$

From Eq. 12.19,

$$z_5 = \frac{p_3 z_4 - 2R}{p_3 + p_4} \tag{12.21}$$

On substituting Eq. 12.21 in Eq. 12.20 and rearranging, a fourth-order equation in $z_4$ is obtained:

$$z_4^4 + c_1 z_4^3 + c_2 z_4^2 + c_3 z_4 + c_4 = 0 \tag{12.22}$$

where

$$\left.\begin{aligned} c_1 &= \frac{p_0}{\gamma' K'} \\ c_2 &= \frac{8R}{\gamma' K'} \\ c_3 &= -\frac{6R[2\bar{z}\gamma' K' + p_0]}{(\gamma' K')^2} \\ c_4 &= -\frac{R(6\bar{z} p_0 + 4R)}{(\gamma' K')^2} \\ p_0 &= (\gamma z_1 + \gamma' z_2)K_p + \gamma' z_3 k' \end{aligned}\right\} \tag{12.23}$$

A trial-and-error solution may be adopted to solve the equation for $z_4$, whereby $D$ is obtained. The factor of safety is applied either by arbitrarily increasing the depth by 20 to

40% or reducing the $K_p$ term by a factor (about 1.5 to 2.0). Generally, the former method is preferred. Thus, the design depth

$$D_d = 1.2 \text{ to } 1.4D \tag{12.24}$$

The variation of the bending moment with depth is shown in Fig. 12.10 b. The point of zero shear corresponds to the point of maximum bending moment. Let $z'$ be the point of zero shear from the point E. Then,

$$R = \tfrac{1}{2}(z')^2 k'\gamma' \tag{12.25}$$

or

$$z' = \sqrt{\frac{2R}{K'\gamma'}} \tag{12.26}$$

The maximum bending moment is obtained as

$$M_{max} = R(\bar{z} + z') - \tfrac{1}{2}\gamma' K'(z')^2 \left(\frac{z'}{2}\right) \tag{12.27}$$

The section modulus $S$ of the sheet pile wall is

$$S = \frac{M_{max}}{\sigma_{fa}} \tag{12.28}$$

where $\sigma_{fa}$ is the allowable flexural stress of the sheet pile.

**Cantilever Sheet Piling in Cohesive Soils with Granular Backfill.** In certain cases, sheet piles have to be driven into cohesive soils with undrained cohesion (i.e., $\phi = 0°$ case). The pressure diagram with other details is shown in Fig. 12.11. The values of $p_1$ and $p_2$ are the

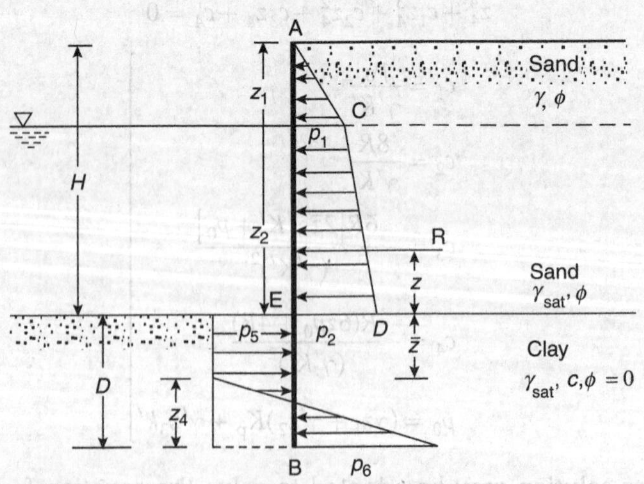


Fig. 12.11 Cantilever sheet piling in cohesive soil with granular backfill

same as those determined by Eqs. 12.9 and 12.10. The net pressure below the dredge line is calculated from $p_a$ and $p_p$. At a depth $z$ from the dredge line,

$$p_a = [\gamma z_1 + \gamma' z_2 + \gamma_{sc} z K_{ac} - 2c\sqrt{K_{ac}}] \tag{12.29}$$

where $\gamma_{sc}$ is the saturated unit weight of clay and $K_{ac}$ is the active earth pressure coefficient for clay (as $\phi = 0°$, $K_{ac} = 1$). Similarly,

$$p_p = \gamma_{sc} z K_{pc} + 2c\sqrt{K_{pc}} \tag{12.30}$$

where $K_{pc}$ is the passive earth pressure coefficient for clay (as $\phi = 0°$, $\gamma_{pc} = 1$). Thus, the net pressure, $p_5$, is given as

$$\begin{aligned} p_5 &= p_p - p_a = [\gamma_{sc} z + 2c] - [\gamma z_1 \gamma' z_2 + \gamma_{sc} z] + 2c \\ p_5 &= 4c - (\gamma z_1 + \gamma' z_2) \end{aligned} \tag{12.31}$$

At the bottom of the sheet pile, the passive resistance from right to left is

$$p_p = (\gamma z_1 + \gamma' z_2 + \gamma_{sc} D) + 2c \tag{12.32}$$

Similarly, the active pressure from left to right is

$$p_a = \gamma_{sc} D - 2c \tag{12.33}$$

Hence, the net pressure

$$p_6 = p_p - p_a = 4c + (\gamma z_1 + \gamma' z_2) \tag{12.34}$$

Considering the equilibrium of horizontal forces,

$$R - [4c - (\gamma z_1 + \gamma' z_2)]D + \tfrac{1}{2} z_4 [4c - (\gamma z_1 + \gamma' z_2) + 4c + (\gamma z_1 + \gamma' z_2)] = 0 \tag{12.35}$$

where $R$ is the area of the pressure diagram ACDE.

Simplifying,

$$z_4 = \frac{D[4c - (\gamma z_1 + \gamma' z_2)] - R}{4c} \tag{12.36}$$

Considering the equilibrium of moments about the point B,

$$R(D + \bar{z}) - [4c - (\gamma z_1 + z_2)]\frac{D}{2} + \frac{1}{2} z_4 (8c)\left(\frac{z_4}{3}\right) = 0 \tag{12.37}$$

Combining Eqs. 12.36 and 12.37, we get

$$D^2[4c - (\gamma z_1 + \gamma' z_2)] - 2DR - \frac{R(R + 12c\bar{z})}{(\gamma z_1 + \gamma' z_2) + 2c} = 0 \tag{12.38}$$

Equation 12.38 is solved for $D$. This is increased by 20 to 40%, hence

$$D_d = 1.2 \text{ to } 1.4D \tag{12.39}$$

The maximum bending moment occurs at a depth $z'$ from the dredge line, which is obtained as

$$R - p_5 z' = 0$$

or

$$z' = \frac{R}{p_5} \tag{12.40}$$

Therefore,

$$M_{max} = R(z' + \bar{z}) - \frac{p_5 z'^2}{2} \tag{12.41}$$

The section modulus is found from Eq. 12.28.

**Cantilever Sheet Piling in Cohesive Soil with Cohesive Backfill**. Sheet piling with cohesive backfill is treated in the same way as granular backfill. However, additional consideration with regard to consolidation of the clay layer, formation of tension crack, and the effect of shrinking on stability is required. Because of the uncertainty of clay backfill, granular backfills are generally preferred. Here, as both the soils are clay, $\phi = 0°$ and $K_{ac} = K_{pc} = 1$.

The pressure diagram is given in Fig. 12.12. Now,

$$z_0 = \frac{2c}{\gamma' \sqrt{K_{ac}}} \tag{12.42}$$

$$p_2 = K_{ac}\gamma z_1 + K_{ac}\gamma'_{sc} z_2 - 2c\sqrt{K_{ac}} = K_{ac}(\gamma_{sc} z_1 + \gamma'_{sc} z_2) - 2c\sqrt{K_{ac}}$$

or

$$p_2 = \gamma z_1 + \gamma'_{sc} z_2 - 2c \tag{12.43}$$

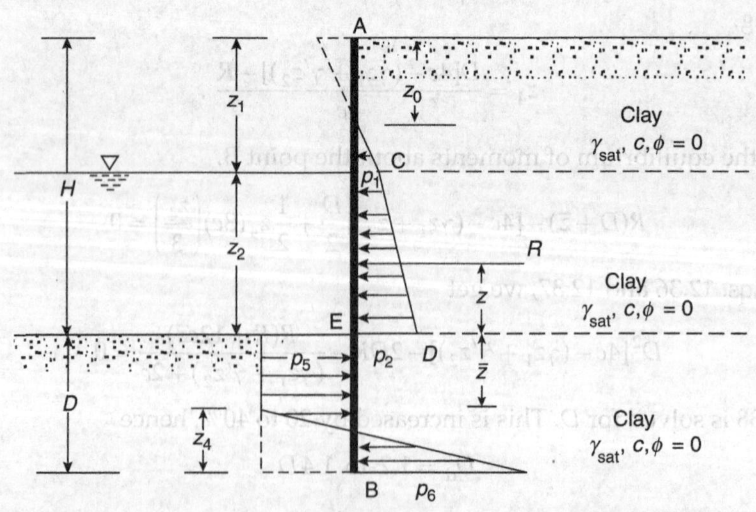


Fig. 12.12 Cantilever sheet piling in cohesive backfill

where $\gamma'_{sc}$ is the submerged unit weight of clay. Other treatments are similar to that given to granular soils, and $p_5$ is modified as

$$p_5 = (\gamma'_{sc} z + 2c) - (\gamma z_1 + \gamma'_{sc} z_2 - 2c) - \gamma_{sc} z$$

or

$$p_5 = 4c - (\gamma z_1 + \gamma'_{sc} z_2) \tag{12.44}$$

### 12.3.2 Anchored Sheet Pile Walls

The category of flexible structures called anchored sheet pile walls or anchored bulkheads is commonly used in water front construction. The construction of such walls consists of driving a sheet pile to the required depth, followed by dredging in front of the piling and backfilling behind the piling. The upper end of the sheet piling is attached to the anchor block through a tie rod. Such a provision of anchors reduces the depth of penetration and the cross-section area of the sheet pile. The use of more than one anchor may be necessary to reduce the lateral deflection and the bending moment. These walls achieve stability due to the passive resistance developed in front of the wall and the resistive force offered by the anchor force offered by the anchor system.

The behaviour of anchored bulk heads is highly complex, and therefore, considerable simplifications have to be made in their design. There are two basic methods of analysis of anchored bulkheads: (i) free earth support method and (ii) fixed earth support method.

The free earth support method assumes that the piling is rigid and may rotate at anchor rod level. As the depth of embedment is considered less in this analysis, the toe of the pile is not restricted and hence the bending moment near the toe is negligible (Fig. 12.13a). The forces acting only on the sheet pile are from lateral soil pressures and the anchor pull, and failure occurs by rotation about the anchor rod. The theoretical embedment depth is

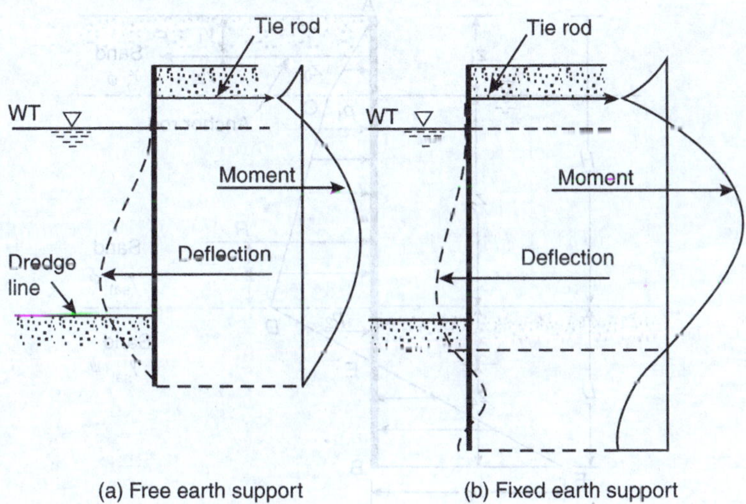


Fig. 12.13 Deflection and moment diagrams for anchored sheet piles

increased by 20% to 40% to arrive at the design depth or $K_p$ may be reduced prior to the computation of embedment depth.

The fixed earth support method considers the lowest section of the sheeting to be fixed in the earth. In this case, the depth of embedment is considered more, and hence, the base of the wall is assumed to be entirely restrained from rotation by the passive resistance of the soil behind it. This passive resistance is in addition to the pressures considered in the free earth support method (Fig. 12.13b). Since failure by forward movement of the toe is unlikely in a wall designed in this way, no factor of safety is applied to the passive resistance of the soil in front of the wall.

**Free Earth Support Method for Penetration of Sandy Soil.** The assumed pressure diagrams and details of other terms are illustrated in Fig. 12.14. The values of $p_1$ and $p_2$ are given as

$$p_1 = \gamma z_1 K_a \tag{12.45}$$

$$p_2 = (\gamma z_2 + \gamma' z_2) K_a \tag{12.46}$$

The value of $z_3$ is given by Eq. 12.13 as

$$z_3 = \frac{p_2}{\gamma' K'}$$

At the bottom of the wall, the net pressure can be given as

$$p_3 = \gamma' K' z_4 \tag{12.47}$$

Considering the equilibrium of the horizontal direction,

$$R - (\text{area of EBF}) - F_a = 0 \tag{12.48a}$$

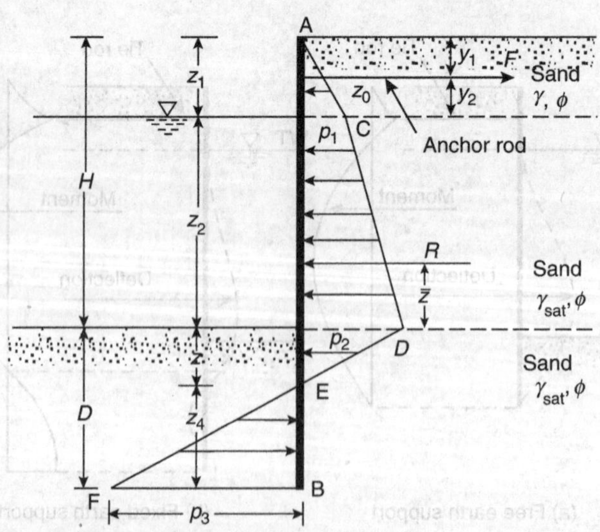


Fig. 12.14 Anchored sheet pile wall penetrating sand

where $R$ is the area of the pressure diagram ACDE and $F_a$ is the tension in the rod per unit length of the wall. Or

$$R - \tfrac{1}{2} p_3 z_4 - F_a = 0$$

Hence,

$$F_a = R - \tfrac{1}{2} p_3 z_4 \tag{12.48b}$$

Taking the moment about the anchor rod,

$$-R[(H + z_3) - (\bar{z} + y_1)] + \tfrac{1}{2} p_3 z_4^2 (y_2 + z_2 + z_3 + \tfrac{2}{3} z_4) = 0$$

or

$$z_4^3 + 1.5 z_4^2 (y_2 + z_2 + z_3) - \frac{3R[(H + z_3) - (\bar{z} + y_1)]}{\gamma' K'} = 0 \tag{12.49}$$

The solution for Eq. 12.49 is obtained by the trial-and-error method. The theoretical depth of penetration,

$$D = z_3 + z_4$$

The design depth

$$D_d = 1.2D \text{ to } 1.4D \tag{12.50}$$

The point of zero shear, $z$, from the ground surface is obtained from

$$\tfrac{1}{2} p_1 z_1 - F_a + p_1 (z - z_1) + \tfrac{1}{2} K_a \gamma' (z - z_1)^2 = 0 \tag{12.51}$$

From the knowledge of $\bar{z}$, the magnitude of bending moment is obtained.

**Free Earth Support Method for Penetration of Clay.** The assumed pressure distribution and other details are shown in Fig. 12.15. Here, the soil below the dredge line is saturated clay

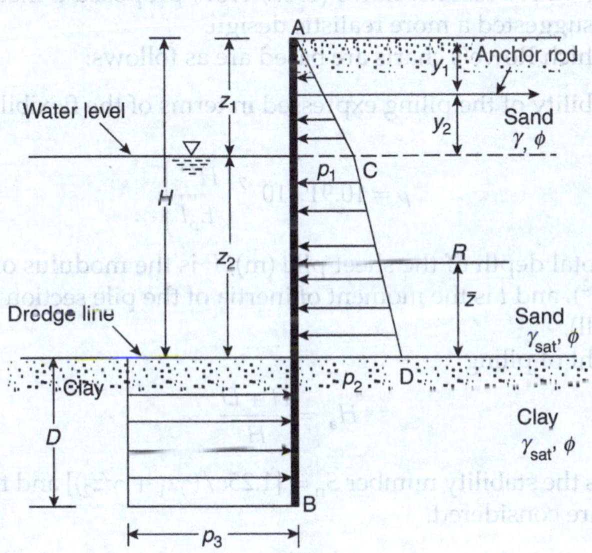


Fig. 12.15 Anchored sheet pile wall penetrating clay

under undrained condition (i.e., $\phi = 0$). The pressure distribution diagram and the values of $p_1$ and $p_2$ are the same as in the previous case. The net pressure distribution diagram below the dredge line (i.e., $z = H$ to $H + D$) can be given as

$$p_3 = 4c - (\gamma z_1 + \gamma' z_2) \tag{12.52}$$

Considering static equilibrium in the horizontal direction,

$$F_a = R - p_3 D \tag{12.53}$$

where $R$ is the area of the pressure diagram above the dredge line. Taking the moment about the anchor rod,

$$R(H - y_1 - \bar{z}) - p_3 D\left(y_2 + z_2 + \frac{D}{2}\right) = 0$$

Simplifying,

$$p_3 D_2 + 2p_3 D(H - y_1) - 2R(H - y_1 - \bar{z}) = 0 \tag{12.54}$$

From the above equation, the theoretical depth $D$ is determined. This depth is increased, and the design depth is determined as

$$D_d = 1.2 \text{ to } 1.4D \tag{12.55}$$

In this case, the maximum bending moment will occur at a depth of $z_1 < z < H$, and hence, the maximum bending moment is determined.

**Rowe's Moment Reduction Method**. The hydrostatic earth pressure distribution is valid only for rigid walls. As the sheet pile walls are flexible in nature, the conventional pressure distribution is affected, and hence, the bending moment differs. Generally, this reduces the bending moment. Thus the bending moment calculated based on the free earth support method gives conservative results. Rowe (1952, 1957) proposed a method for reducing the moments and thus suggested a more realistic design.

The factors on which Rowe's charts are based are as follows:

1. The relative flexibility of the piling expressed in terms of the flexibility number

$$\rho = 10.91 \times 10^{-7}\left(\frac{H'^4}{E_p l}\right) \tag{12.56}$$

where $H'$ is the total depth of the sheet pile (m), $E$ is the modulus of elasticity of the pile material (MN/m$^2$), and $I$ is the moment of inertia of the pile section per metre of the wall (m$^4$/m of the wall).

2. The relative height of piling

$$H_a = \frac{H + D}{H}$$

3. For cohesive soils the stability number $S_n = [1.25c/(\gamma z_1 + \gamma' z_2)]$ and for non-cohesive soils relative density are considered.

Figure 12.16 represents the moment reduction curves for non-cohesive soils. The reduced design moment $M_r$ is obtained by noting down the values corresponding to the particular

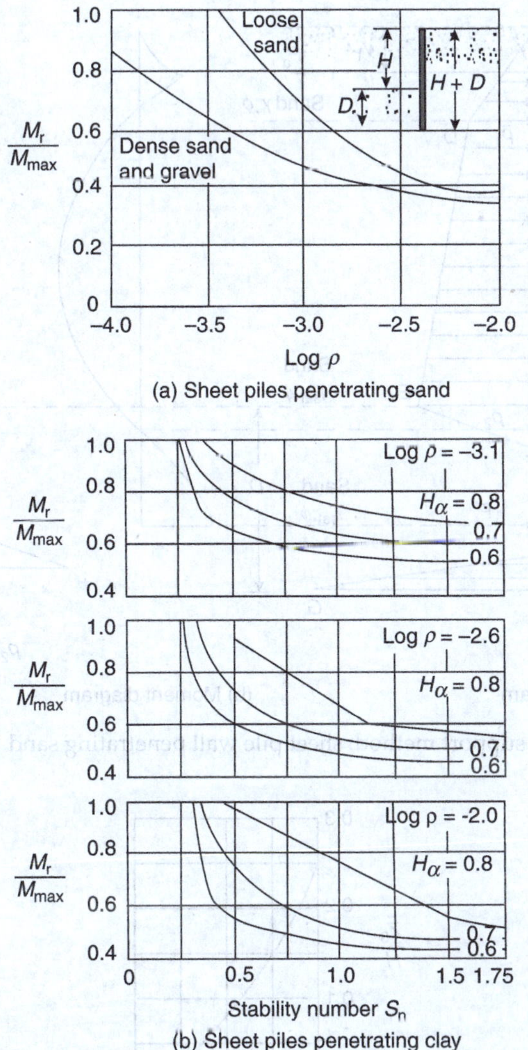


**Fig. 12.16** Moment reduction charts (*Source:* Rowe, 1957).

log $\rho$ and density for cohesionless soil and $H_a$, $S_n$, and log $\rho$ for cohesive soil. Suitable interpolations may be made wherever necessary.

**Fixed Earth Support Method for Penetration of Sandy Soil.** As discussed earlier, the toe of the pile is restrained (Fig. 12.13b). The assumed pressure distribution moment diagram and identification of terms are illustrated in Fig. 12.17. Point C in the moment diagram is the point of contraflexure. The pile may be assumed to be hinged at this point C. Thus, the portion of the piling above the point C can be considered as a beam resisting the net earth pressures through the anchor force and the shear force $P$. This is termed the *equivalent*

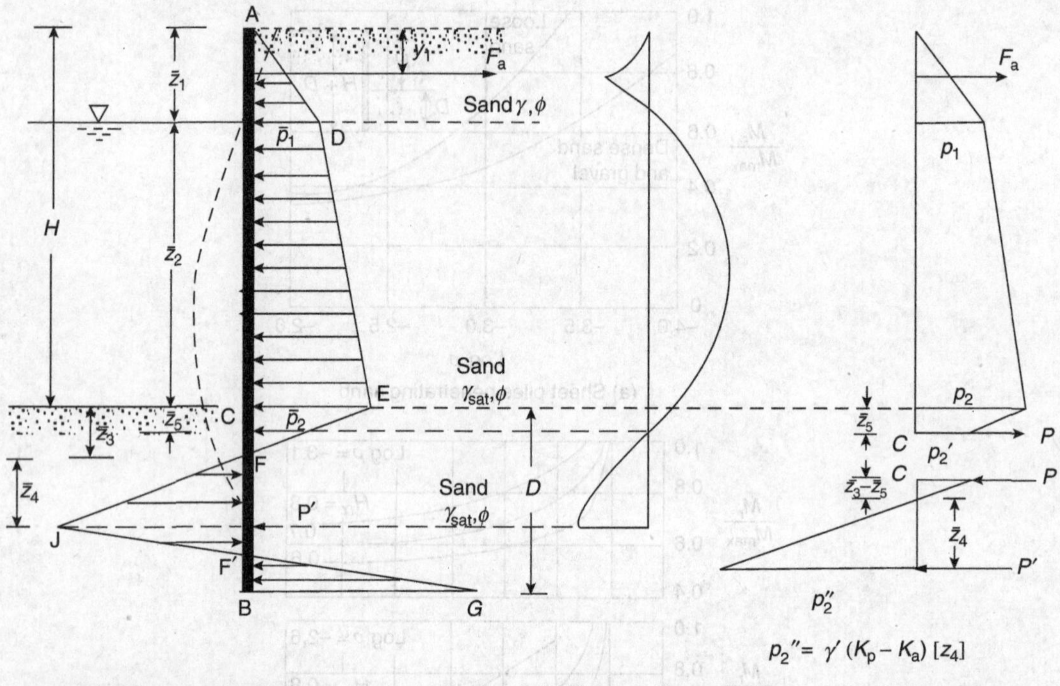

(a) Pressure diagram (b) Moment diagram (c) Determination of $z_5$

Fig. 12.17 Fixed earth support method: sheet pile wall penetrating sand

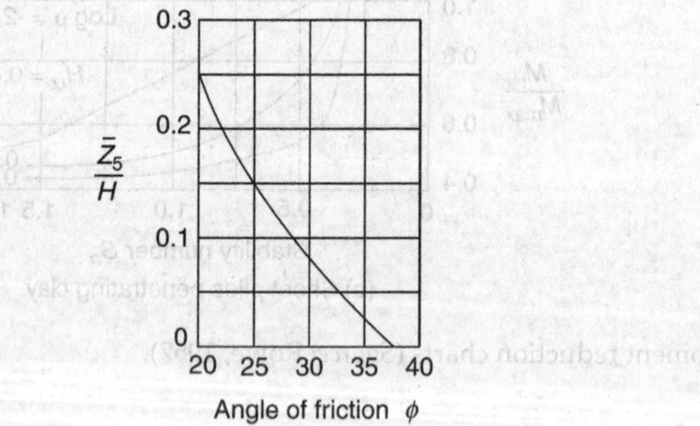


Fig. 12.18 Chart to find point of contraflexure (*Source:* Blum, 1931)

*beam method* (Blum, 1931). Blum provided a chart (Fig. 12.18) relating the angle of shearing resistance and the distance from the point of contraflexure to the dredge line, $z_5$, as shown in Fig. 12.17.

The knowledge of $\phi$ and $H$ enables the determination of $z_5$. Now, as discussed above, the portion above point C is treated as a beam, and the shear force $P$ is calculated using the moments about the anchor rod. With $P$ known, considering the moment equilibrium about

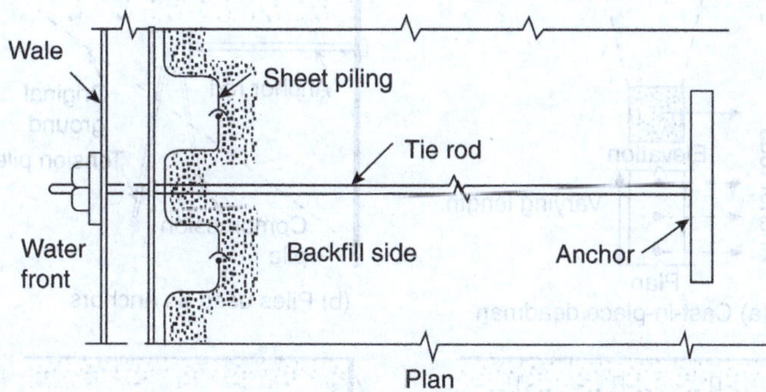


Fig. 12.19 Arrangement of wale, tie rod, and anchor

the base yields an expression where the only unknown, $D$, can be determined. The depth is increased by 20% to 40%, and hence the design depth $D_d = 1.2$ to $1.4D$. The force $F_a$ on the anchor rod may be determined using the moments about the point of contraflexure, C. With the value of design depth known, the calculation of the bending moment and, subsequently, the selection of the section can be done.

### 12.3.3 Wales, Tie rods, and Anchorages for Sheet Piling

Wales are longitudinal members of a rolled channel section usually provided back to back along the sheet pile length, as shown in Fig. 12.19. A wale is attached to the back of the wall if a flush-front face is required, otherwise it is placed in a horizontal position in front of the wall. Wales may be designed simply as supported beams with spans equal to the distance between tie rods. A cable or a steel bar, threaded to allow vertical alignment and tension adjustments, acts as a tie rod. Sufficient protection is provided against corrosion of tie rods by treating them with a coat of paint or asphaltic material. The spacing of the rods depends on the total anchor force to be provided and the capacity of each tie rod. Theoretical anchor forces are increased by 20 to 30% for design purposes, particularly in cohesive soils, with an allowable stress in steel approaching 80 to 90 % of the yield stress.

Anchors are basically classified as tie-back and deadman-type anchors. Figure 12.20 shows a variety of anchor schemes. The tie-back anchors are preferred in places where it is possible to encroach on the adjacent ground to instal the anchor. This type permits an unobstructed area in front of the wall for dredging and other installations. The only disadvantage is in encountering underground utilities. Deadman-type anchors are constructed in place by pouring concrete or by embedding a pre-cast beam. In order to develop sufficient passive resistance, the deadman is constructed at an adequate distance from the sheet pile wall.

## 12.4 BRACED EXCAVATIONS

Excavations of soils to significant depths are laterally supported temporarily by braced sheeting and permanently by retaining walls. Braced sheeting basically consists of a sheet piling to support the sides of the excavation, with stability being maintained by means of

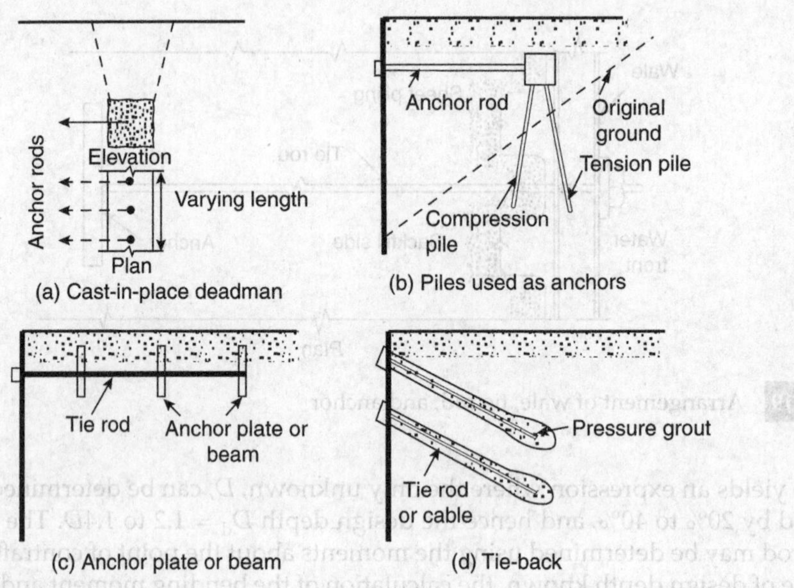


**Fig. 12.20** Types of anchors

struts across the excavation. A variety of materials, methods, and procedures have been in use. The choices are influenced by factors such as subsurface condition, excavation depth, working space, climate and season, equipment, and labour available.

Two common techniques adopted for lateral bracing are illustrated in Fig. 12.21. In the first method, generally referred to as *lagging*, wooden or steel *soldier beams* are driven into the ground before excavation. As the excavation progresses, horizontal wooden sheeting or steel plates (called laggings) are placed between the solider beams. When excavation reaches the desired depth, *wales* and *struts* are carefully installed.

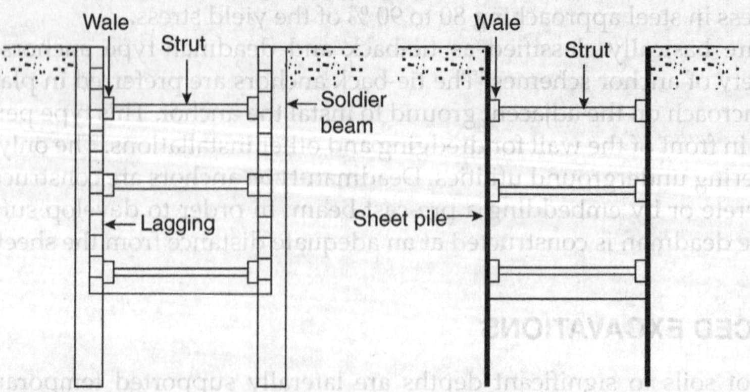


**Fig. 12.21** Techniques for lateral bracing

In the second method, *interlocking sheet piles* are driven to a depth greater than the anticipated depth of the excavation. As the excavation progresses, wales are installed horizontally along the excavation at intermittent depths. These wales are supported by struts placed horizontally at the required spacings.

### 12.4.1 Earth Pressure Distribution

The pressure distribution against bracing is different from the conventional distribution and depends on the deformation condition from top to bottom. Because of less yielding at the top, the lateral earth pressure will be close to earth pressure at rest, but the degree of yielding increases with depth and the earth pressure at the bottom may be different from that of Rankine's active pressure. Typical deflection patterns of these two types of walls are shown in Fig. 12.22. Thus, no theory can be directly applied; it depends on valid empirical methods. Such suggestions were given by Peck (1969) and Tschebotarioff (1949) based on the results from field tests. The pressure distributions on braced sheeting for sand and clay as recommended by them are shown in Fig. 12.23. Further, the point of application of the resultant pressure has been found at mid-height rather than at one-third height as in Rankine's case.

### 12.4.2 Failure of Braced Cuts

Because of deformation changes with depth, failure of the soil of a braced excavation takes the shape shown in Fig. 12.22b and c. This shows that the lower part of the soil is in a state of plastic equil-ibrium whereas the upper part is in a state of elastic equilibrium. Initial failure of one of the struts leads to a progressive failure of the entire system. Since it is essential that no individual strut should fail, the pressure distributions shown in Fig. 12.23 are representative of random distributions obtained from field measurements. For medium to dense sands, a uniform distribution of 0.65 times the Rankine active value has been found to be appropriate.

The behaviour of a braced cut in clay depends on the value of $\gamma H/c$. For $\gamma H/c < 4$, most of the clays retained will be in a state of elastic equilibrium, and for $\gamma H/c > 4$, plastic zones develop near the bottom of the cut. Hence, appropriate pressure diagrams for clay should be used in the design to avoid a failure.

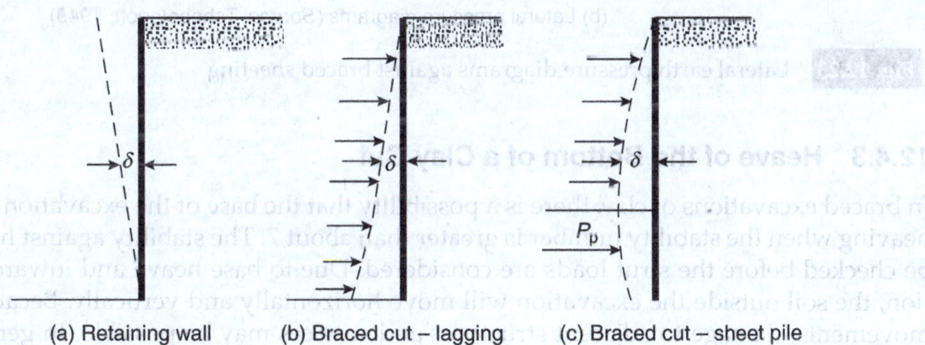


Fig. 12.22 Nature of yielding of walls

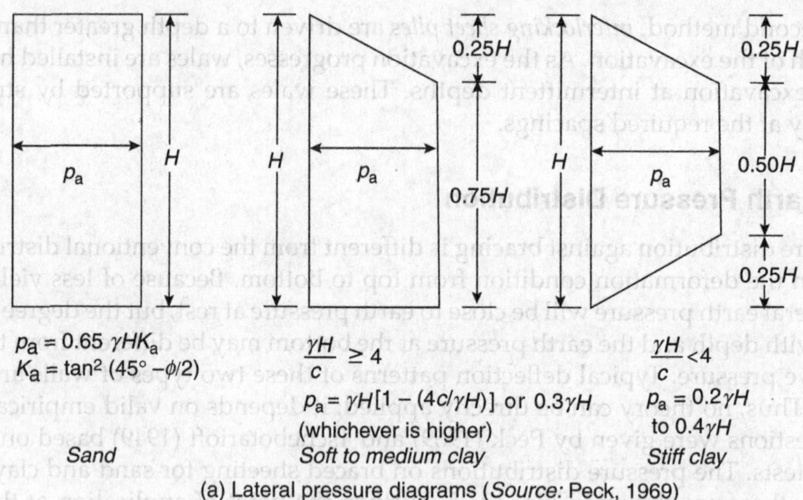

(a) Lateral pressure diagrams (*Source:* Peck, 1969)

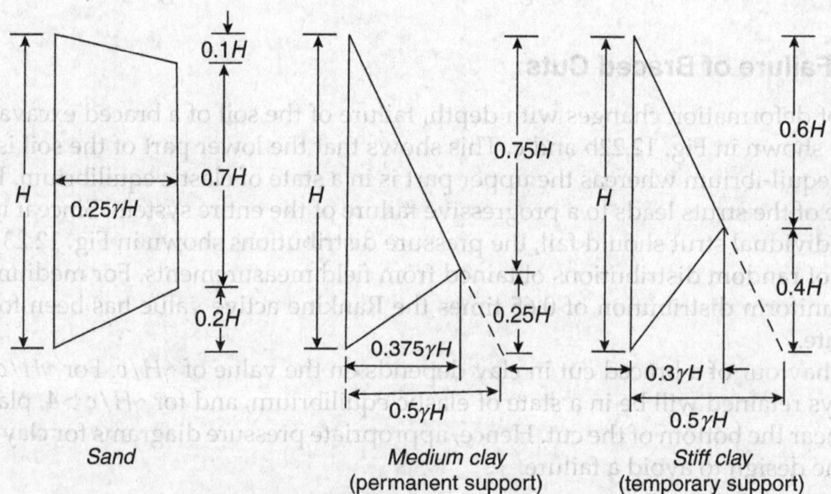

(b) Lateral pressure diagrams (*Source:* Tshebotarioff, 1949)

**Fig. 12.23** Lateral earth pressure diagrams against braced sheeting

### 12.4.3 Heave of the Bottom of a Clay Cut

In braced excavations of clay, there is a possibility that the base of the excavation will fail on heaving when the stability number is greater than about 7. The stability against heave has to be checked before the strut loads are considered. Due to base heave and inward deformation, the soil outside the excavation will move horizontally and vertically. Because of such movements, damage to adjacent structures and services may be possible. In general, if the flexibility of the wall system is large and the time for installation of struts or anchors is more, movement has to be anticipated.

This factor of safety of braced excavation against bottom heave can be computed based on Terzaghi's (1943) analysis. Figure 12.24 shows the failure pattern of the bottom of the excavation, in which surfaces CD and CF are arcs of the circles.

Consider the vertical load per unit length of the cut at the level of the bottom of the cut along the line BD and AF as

$$Q = \gamma HB' - c_u H \tag{12.57}$$

where $B' = 0.7B$ and $c_u$ = cohesion (for the $\phi = 0°$ condition).

Based on Terzaghi's bearing capacity theory (discussed in Chapter 14), treating $B'$ as the width of a continuous foundation, the net ultimate load per unit length of the foundation is given as

$$Q_{nu} = 5.7 c_u B' \tag{12.58}$$

Hence, the factor of safety against heave is

$$F_h = \frac{Q_{nu}}{Q} = \frac{5.7 c_u B'}{\gamma HB' - c_u H} = \frac{1}{H}\left(\frac{5.7 c_u}{\gamma - (c_u / 0.7B)}\right) \tag{12.59}$$

The above equation is based on the assumption that the clay layer is homogeneous and extends at least up to a depth of $0.7B$ below the bottom of the cut. If a hard rock or a hard material is located at a depth $D < 0.7B$, the failure surface will make a tangent with that surface. In such a case, Eq. 12.59 is modified as

$$F_h = \frac{1}{H}\left(\frac{5.7 c_u}{\gamma - (c_u / D)}\right) \tag{12.60}$$

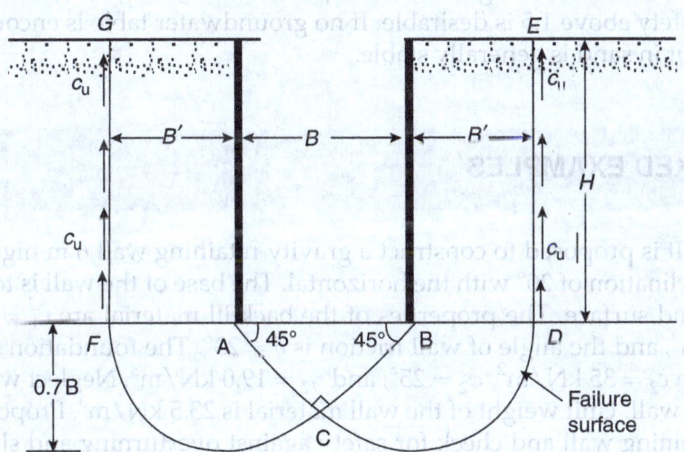


**Fig. 12.24** Failure mechanism of bottom heave

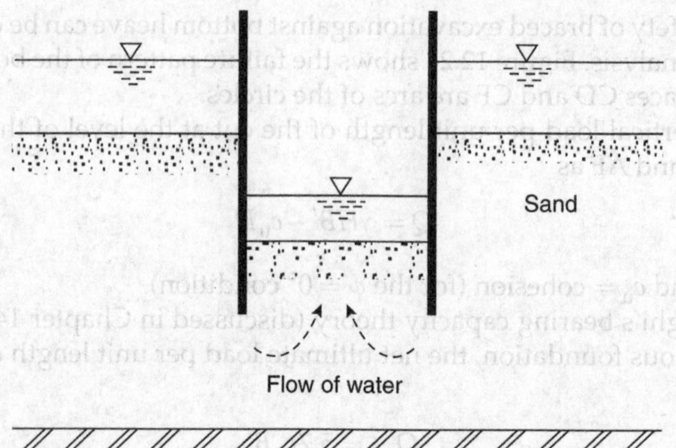


Fig. 12.25 Flow of water through the bottom of the cut in sand

### 12.4.4 Stability of the Bottom of the Cut in Sand

The bottom of the cut is stable as long as the water level inside the excavation is higher than that outside. If the water level inside the cut is lowered by pumping such that the water level inside is lower than that outside, instability may be created as a result of an upward seepage of water into the cut. Such a situation is shown in Fig. 12.25. Here, the factor of safety against piping has to be checked. After drawing the flow net, the maximum exit gradient is compared with the critical gradient. Thus, the factor of safety against piping is given as

$$F_p = \frac{i_c}{i_e} \tag{12.61}$$

where $i_c$ is the critical hydraulic gradient and $i_e$ is the maximum exit gradient.

A factor of safety above 1.5 is desirable. If no groundwater table is encountered, then the bottom of the cut in sand is generally stable.

## WORKED EXAMPLES

**Example 12.1** It is proposed to construct a gravity-retaining wall 6 m high, with a backfill sloping at an inclination of 20° with the horizontal. The base of the wall is to be placed 1.5 m below the ground surface. The properties of the backfill material are $c_1 = 0$, $\phi_1 = 36°$, and $\gamma_1 = 18.1$ kN/m$^3$, and the angle of wall friction is $\delta = 23°$. The foundation soil is a cohesive friction soil with $c_2 = 35$ kN/m$^2$, $\phi_2 = 25°$, and $\gamma_2 = 19.0$ kN/m$^3$. Neglect wall friction in the front face of the wall. Unit weight of the wall material is 23.5 kN/m$^3$. Proportion the dimensions of the retaining wall and check for safety against overturning and sliding. The water table is located at a greater depth.

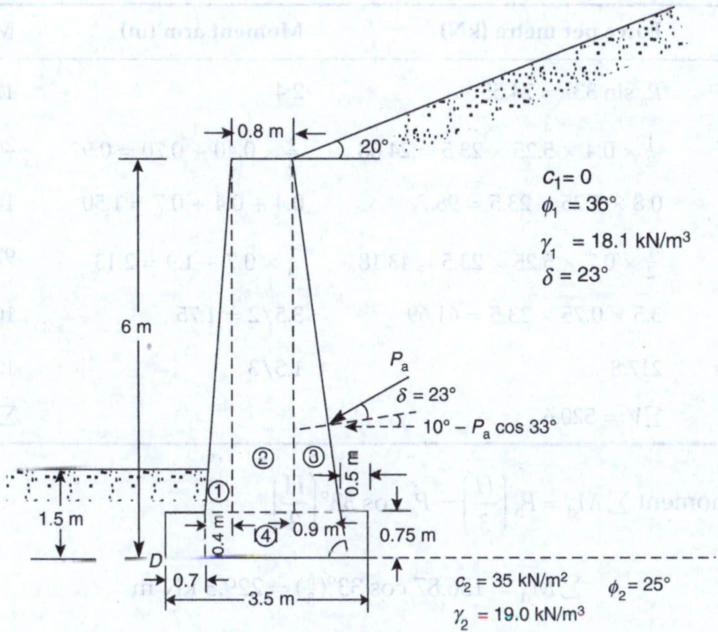


Fig. 12.26

***Solution***

Based on the details given in Fig. 12.2 for proportioning the wall, the dimensions of the retaining wall for a height of 6 m are shown in Fig. 12.26.

As the back of the wall and the soil surface are both inclined and there is wall friction, Coulomb's theory can be applied to get $K_a$ from Eq. 12.45. Here, $\beta = 80°$, $\delta = 23°$, $i = 20°$, and $\phi = 36°$. Thus,

$$K_a = \frac{\sin^2(80° + 36°)}{\sin^2 80° \; \sin(80° - 23°)\left[1 + \sqrt{\dfrac{\sin(36° + 23°)\; \sin(36° - 20°)}{\sin(80° - 23°)\; \sin(20° + 80°)}}\right]} = 0.42$$

Therefore,

$$P_a = \tfrac{1}{2} K_a \gamma_1 H^2 = \tfrac{1}{2} \times 0.42 \times 18.1 \times 6^2 = 13.87 \text{ kN/m}$$

Based on Rankine's theory, the passive resistance can be computed as the wall friction in the front face equals zero. Therefore,

$$K_p = \frac{1 + \sin 25°}{1 - \sin 25°} = 2.47$$

$$P_p = \tfrac{1}{2}(2.47)(19)(1.5)^2 + 2(35)(\sqrt{2.47})(1.5)$$

$$P_p = 52.80 + 165.02 = 217.8 \text{ kN/m}$$

Considering the moments about point D, the following table is prepared:

| Details | Force per metre (kN) | Moment arm (m) | Moment (kN-m) |
|---|---|---|---|
| $P_v$ | $P_a \sin 33° = 74.55$ | 2.4 | 178.8 |
| Wall (Section 1) | $\frac{1}{2} \times 0.4 \times 5.25 \times 23.5 = 24.68$ | $\frac{2}{3} \times 0.40 + 0.70 = 0.97$ | 24.0 |
| Wall (Section 2) | $0.8 \times 5.25 \times 23.5 = 98.7$ | $0.4 + 0.4 + 0.7 = 1.50$ | 148.1 |
| Wall (Section 3) | $\frac{1}{2} \times 0.7 \times 5.25 \times 23.5 = 43.18$ | $\frac{1}{3} \times 0.7 + 1.9 = 2.13$ | 92.1 |
| Wall (Section 4) | $3.5 \times 0.75 \times 23.5 = 61.69$ | $3.5/2 = 1.75$ | 108.0 |
| Passive pressure | 217.8 | 1.5/3 | 106.0 |
| | $\sum V = 520.6$ | | $\sum M_r = 657$ |

Overturning moment $\sum M_d = P_h\left(\frac{H}{3}\right) = P_a \cos 33°\left(\frac{H}{3}\right)$

$$\sum M_d = 136.87 \cos 33°(\tfrac{6}{3}) = 229.3 \text{ kN m}$$

$$F_{or} = \frac{\sum M_r}{\sum M_d} = \frac{657}{229.3} = 2.87 > 2$$

Hence, safe against overturning:

$$\sum F_r = [(\sum V) \tan \phi_2 + Bc_2] + P_p$$

$$\sum F_r = 514.8 \tan 25° + 3.5 \times 35 + 217.8 = 577.26 \text{ kN/m}$$

$$\sum F_d = P_a \cos a = 136.87 \cos 33° = 114.8 \text{ kN/m}$$

$$F_{SL} = \frac{\sum F_r}{\sum F_d} = \frac{577.26}{114.80} = 5.02 > 1.5$$

Hence, safe against sliding also.

**Example 12.2** For the cantilever retaining wall shown in Fig. 12.27, determine the maximum and minimum pressures under the base of the cantilever. The relevant shear strength parameters of the backfill and foundation soil are $c' = 0$, $\phi = 35°$, and unit weight of the soil $\gamma = 17.5$ kN/m$^3$. The unit weight of the wall material is 23.5 kN/m$^3$. Find also the factor of safety against sliding, considering the reduced value of base friction as $2/3\phi°$.

***Solution***

Considering the vertical face A′B′, Rankine's theory can be applied to determine the active earth pressure. Thus, $K_a$ is obtained from Eq. 11.18b:

$$K_a = \cos 10°\left[\frac{\cos 10° - \sqrt{\cos^2 10° - \cos^2 35°}}{\cos 10° - \sqrt{\cos^2 10° + \cos^2 35°}}\right] = 0.28$$

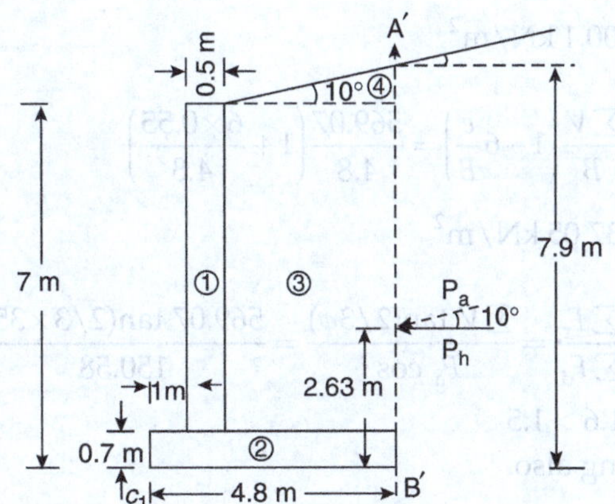


Fig. 12.27

$$P_a = \tfrac{1}{2} K_a \gamma H_1^2 = \tfrac{1}{2} \times 0.28 \times 17.5 \times 7.92 - 152.9 \text{ kN/m}$$

$$P_v = P_a \sin 10^\circ = 152.9 \sin 10^\circ = 26.55 \text{ kN/m}$$

$$P_b = P_a \cos 10^\circ = 152.9 \cos 10^\circ = 150.58 \text{ kN/m}$$

Considering the moments about point D, the following table is prepared:

| **Details** | **Force per metre (kN)** | **Moment arm (m)** | **Moment (kN-m)** |
|---|---|---|---|
| $P_v$ | $P_a \sin 10^\circ = 26.55$ | 4.8 | 127.4 |
| Wall (Section 1) | $0.5 \times 6.3 \times 23.5 = 74.03$ | $1 + 0.5/2 = 1.25$ | 92.5 |
| Wall (Section 2) | $0.7 \times 4.8 \times 23.5 = 78.96$ | $4.8/2 = 2.40$ | 189.5 |
| Soil (Section 3) | $3.3 \times 6.3 \times 17.5 = 363.83$ | $3.3/2 + 1.5 = 2.6$ | 946.0 |
| Soil (Section 4) | $\frac{1}{2} \times 3.3 \times 0.9 \times 17.5 = 25.99$ | $\frac{2}{3} \times 3.3 + 1.5 = 3.7$ | 95.8 |
| | $\Sigma V = 569.07$ | | $\Sigma M_r = 1451.2$ |

$$\text{Overturning moment } \Sigma M_d = P_h \left(\frac{H_1}{3}\right) = 150.58 \times \frac{7.9}{3} = 396.5 \text{ kN-m}$$

$$\text{Eccentricity } e = \frac{B}{2} - \frac{\Sigma M_r - \Sigma M_d}{\Sigma V} = \frac{4.8}{2} - \frac{1451.2 - 396.5}{569.07}$$

$$e = 2.4 - 1.85 = 0.55 < \frac{4.8}{6} = 0.80$$

$$\text{Pressure at toe } p_{max} = \frac{\Sigma V}{B}\left(1 + 6\frac{e}{B}\right) = \frac{569.07}{4.8}\left(1 + \frac{6 \times 0.55}{4.8}\right)$$

$$p_{max} = 200.1 \text{ kN/m}^2$$

Pressure at heel $p_{min} = \frac{\sum V}{B}\left(1-6\frac{e}{B}\right) = \frac{569.07}{4.8}\left(1+\frac{6\times 0.55}{4.8}\right)$

$$p_{min} = 37.05 \text{ kN/m}^2$$

$$F_{SL} = \frac{\sum F_r}{\sum F_d} = \frac{\sum V \tan(2/3\phi)}{P_a \cos i} = \frac{569.07 \tan(2/3\times 35)}{150.58}$$

$$F_{SL} = 1.6 > 1.5$$

Hence, safe against sliding also.

**Example 12.3** A cantilever sheet pile wall with a simplified pressure distribution is shown in Fig. 12.28. Determine the depth of penetration, considering a factor of safety of 2 against passive resistance.

***Solution***

Since wall friction is zero, Rankine's theory can be applied. Thus,

$$K_a = \frac{1-\sin 38°}{1+\sin 38°} = 0.24 \quad \text{and} \quad K_p = \frac{1}{K_a} = 4.2$$

Reduced coefficient $K'_p = \frac{K_p}{F} = \frac{4.2}{2} = 2.1$

$$P_a = \tfrac{1}{2}\times 0.24\times 19(D+2)(D+2) = 2.28(D+2)^2$$

$$P_p = (2.1\times 19\times D)\times \tfrac{1}{2}\times D = 19.95D^2$$

Taking moments about B,

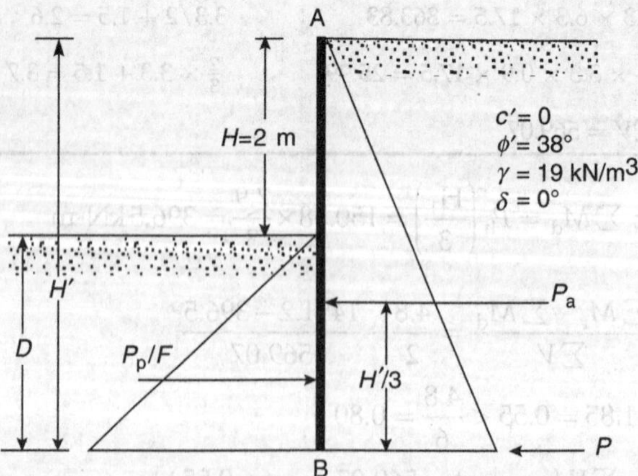


Fig. 12.28

$$19.95D^2 \times \frac{D}{3} = 2.28(D+2)^2\left(\frac{D+2}{3}\right)$$

$$D^3 = 0.114(D+2)^3$$

or

$$\left(\frac{D}{D+2}\right)^3 = 0.114$$

or

$$D = 1.18 \text{ m}$$

This simplified distribution does not give the exact distribution of pressure near the base, and the calculated driving depth may be increased by 20% in addition to the reduction made in the passive resistance. Therefore,

Embedment depth $D_d = 1.20 \times 1.18 = 2.26$ m

**Example 12.4** A cantilever sheet pile is to retain 3.5 m of sand. Water table is at 0.5 m from the top of the backfill. For the sand $\gamma = 19$ kN/m$^3$, $\gamma_1 = 12.2$ kN/m$^3$, $K_a = 0.2$ and $K_p = 5$. Find the depth of penetration for a factor of safety of 1.4.

***Solution***

Figure 12.10 is redrawn (Fig. 12.29) with the following data: $z_1 = 0.5$ m, $z_2 = 1.5$ m, $K_a = 0.2$, $K_p = 5$, $\gamma = 19$ kN/m$^3$ and $\gamma' = 12.2$ kN/m$^3$.

$$p_1 = K_a\gamma z_1 = 0.2 \times 19.0 \times 0.5 = 1.9 \text{ kN/m}^2$$

$$p_2 = K_a(\gamma z_1 + \gamma' z_2) = 0.2(19.0 \times 0.5 + 12.2 \times 3) = 9.22 \text{ kN/m}^2$$

From Eq. 12.12,

$$z_3 = \frac{p_2}{\gamma'(K_p - K_a)} = \frac{9.22}{12.2(5-0.2)} = 0.158 \text{ m}$$

$$R = \overset{(1)}{\tfrac{1}{2}p_1 \times z_1} + \overset{(2)}{p_1 \times z_2} + \overset{(3)}{\tfrac{1}{2}(p_2 - p_1)z_2} + \overset{(4)}{\tfrac{1}{2}p_2 \times z_3}$$

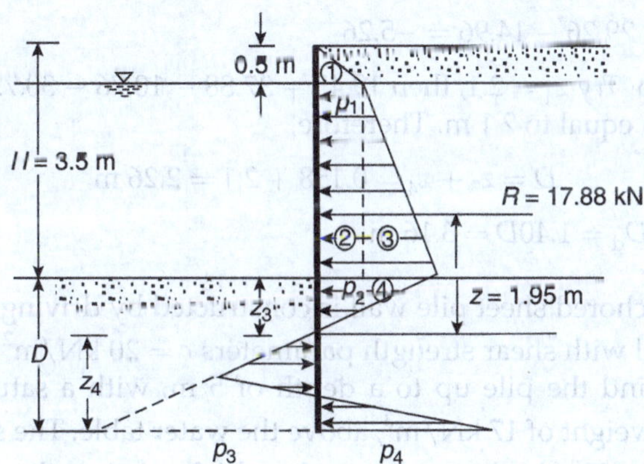


Fig. 12.29

$$= \tfrac{1}{2}\times 1.9\times 0.5 + 1.9\times 3 + \frac{(9.22-1.9)\times 3}{2} + \tfrac{1}{2}\times 9.22\times 0.158$$

$$= 0.475 + 5.7 + 10.98 + 0.728 = 17.88 \text{ kN}$$

$$\bar{z} = \frac{0.475\left(\frac{1}{3}\times 0.5 + 3.0 + 0.158\right) + \left\{5.7\times\left(\frac{3}{2}+0.158\right) + 10.98\times\left(\frac{2}{3}+0.158\right)\right\} + 0.728\times\frac{2}{3}\times 0.158}{17.88}$$

or

$$\bar{z} = \frac{1.58 + 33.15 + 0.077}{17.88} = 1.95 \text{ m}$$

$$K' = K_p - K_a = 5 - 0.2 = 4.8 \quad \text{and} \quad \gamma' K' = 58.56 \text{ kN/m}^3$$

$$p_0 = (\gamma z_1 + \gamma' z_2)K_p + K'\gamma' z_3$$

$$= (19\times 0.5\times 12.2\times 3)5 + 58.56\times 0.158 = 239.75 \text{ kN/m}^2$$

$$c_1 = \frac{p_0}{\gamma' K'} = \frac{239.75}{58.56} = 4.09$$

$$c_2 = \frac{-8R}{\gamma' K'} = \frac{-8\times 17.88}{58.56} = -2.44$$

$$c_3 = \frac{-6R(2\bar{z}\gamma' K' + p_0)}{(\gamma' K')^2} = \frac{-8\times 17.88(2\times 1.95\times 58.56 + 239.75)}{(58.56)^2}$$

$$= -14.65$$

$$c_4 = \frac{R(6\bar{z}p_0 + 4R)}{(\gamma' K')^2} = \frac{17.88(6\times 1.95\times 239.75 + 4\times 17.88)}{(58.56)^2}$$

$$= -14.96$$

$$z_4^4 + 4.09 z_4^3 - 2.44 z_4^2 - 14.63 z_4 - 14.96 = 0$$

For $z_4 = 2$, the above equation gives

16 + 32.72 – 9.76 – 29.26 – 14.96 = –5.26

which is less than zero. Try $z_4 = 2.1$, then 19.45 + 37.88 – 10.76 – 30.72 – 14.96 = 0.887 = 0. Hence, $z_4$ can be taken equal to 2.1 m. Therefore,

$$D = z_3 + z_4 = 0.158 + 2.1 = 2.26 \text{ m}$$

Hence, design depth, $D_d = 1.40D = 3.16$ m.

**Example 12.5** An anchored sheet pile wall is constructed by driving a line of piling into a saturated cohesive soil with shear strength parameters $c = 20$ kN/m$^2$ and $\phi = 0°$. Granular backfill is placed behind the pile up to a depth of 5 m, with a saturated unit weight of 20 kN/m$^3$ and a unit weight of 17 kN/m$^3$, above the water table. The shear strength parameters are $c' = 0$ and $\phi' = 33°$. Anchor rods are placed 1.0 m below the surface of the backfill. The water levels in front of the pile as well as behind it are both 3 m below the surface of the backfill. Determine the design depth of penetration of the piling. Also, find the tension in the anchor rod. Use the free earth support method.

***Solution***

Figure 12.15 is redrawn (Fig. 12.30) with the following given data: $H = 5$ m, $z_1 = 2$ m, $z_2 = 3$ m, $y_1 = 1$ m, and $y_2 = 1$ m.

$$K_a = \frac{1-\sin 33^\circ}{1+\sin 33^\circ} = 0.835$$

$$\gamma' = 20 - 9.81 = 10.19 \text{ kN/m}^3$$

$$p_1 = K_a \gamma z_1 = 0.835 \times 17 \times 2 = 28.39 \text{ kN/m}^2$$

$$p_2 = (\gamma z_1 + \gamma' z_2)K_a = (17 \times 2 + 10.19 \times 3)0.835$$

$$p_2 = 53.92 \text{ kN/m}^2$$

$$R = \underset{(1)}{\tfrac{1}{2}\times 28.39\times 2} + \underset{(2)}{\tfrac{1}{2}\times(53.92-28.39)\times 3} + \underset{(3)}{28.39\times 3}$$

$$R = 28.39 + 38.30 + 85.17 = 151.9 \text{ kN}$$

$$\bar{z} = \frac{28.39(1/3\times 2+3)+38.30\times 1/3\times 3+85.17\times(3/2)}{151.9} = 1.8 \text{ m}$$

$$p_3 = 4c - (\gamma z_1 + \gamma' z_2) = 4 \times 20 - (17 \times 2 + 10.19 \times 3) = 15.43 \text{ kN/m}^2$$

Taking the moments about the anchor rod,

$$R(H - y_1 - \bar{z}) - p_3 D[H + (D/2) - y_1] = 0$$

or

$$151.9\,(5 - 1 - 1.8) - 15.43D[5 + (D/2) - 1] = 0$$

or

$$334.148 - 7.78D^2 - 61.72D = 0$$

or

$$D = 3.7 \text{ m}$$

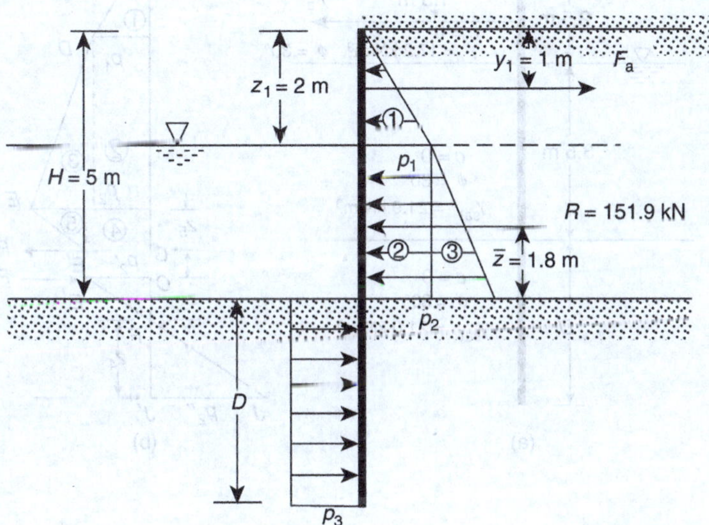


Fig. 12.30

The design depth $D_d = 1.2 \times 3.7 = 4.44$ m. Force on anchor rod $F_a = 151.9 - 15.43 \times 3.7 = 94.81$ kN.

**Example 12.6** The cross-section of an anchored sheet pile is shown in Fig. 12.31a. Determine the design depth of penetration. Use the fixed earth support method.

***Solution***

$$K_a = \frac{1-\sin 30^\circ}{1+\sin 30^\circ} = 0.33; \quad K_p = 3; \quad K' = K_p - K_a = 2.67$$

$$\gamma' = 21 - 9.81 = 11.2 \text{ kN/m}^3, \quad y_1 = 1.5 \text{ m}, \quad y_2 = 1 \text{ m}$$

$$p_1 = \gamma z_1 K_a = 17.2 \times 2.5 \times 0.33 = 14.19 \text{ kN/m}^2$$

$$p_2 = (\gamma z_1 + \gamma' z_2)K_a = (17.2 \times 2.5 + 11.20 \times 3.5)0.33 = 27.13 \text{ kN/m}^2$$

$$z_3 = \frac{p_2}{\gamma' K'} = \frac{27.13}{11.2 \times 2.67} = 0.91 \text{ m}$$

$z_5/H$ for $\phi = 30^\circ$ is obtained from Fig. 12.18 as $z_5/H = 0.08$. Or

$$z_5 = 0.080 \times 6.0 = 0.48 \text{ m}$$

$$p_2' = \frac{p_2(z_3 - z_5)}{z_3} = \frac{27.13(0.91 - 0.45)}{0.91} = 13.71 \text{ kN/m}^2$$

To determine the unknown force $P$, taking moment of the pressure diagram ADEE′C about the anchor rod,

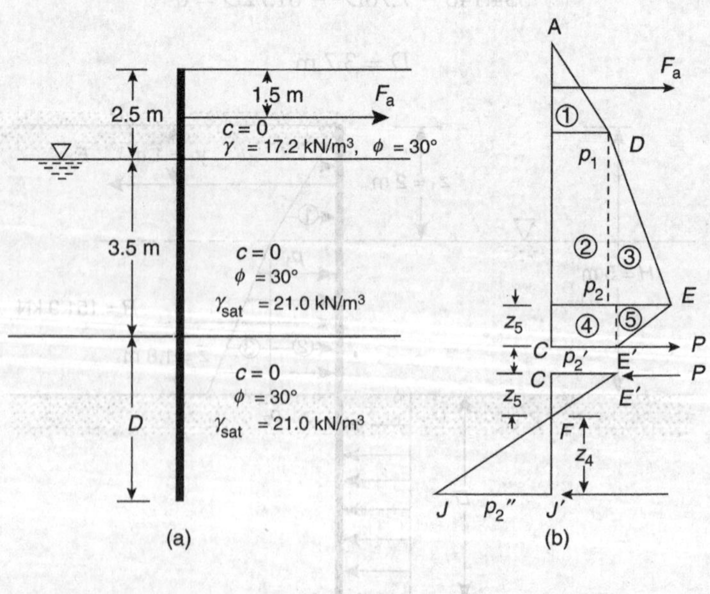


Fig. 12.31

$$\left(\frac{1}{2}\times p_1\times z_1\right)\times\left(y_2-\frac{z_1}{3}\right)+(p_1\times z_2)\times\left(\frac{z_2}{2}+y_2\right)+\left[\frac{1}{2}\times(p_2-p_1)\times z_2\right]$$
$$\times\left(z_2\times y_2-\frac{z_2}{3}\right)+(p_2'\times z_5)\times\left(\frac{z_5}{2}+z_2+y_2\right)+\left[\frac{1}{2}(p_2-p_2')z_5\right]$$
$$\times\left(\frac{z_5}{2}+z_2+y_2\right)-P\times(z_5+z_2+y_2)=0$$

or

$$\left(\frac{1}{2}\times 14.19\times 2.5\right)\times\left(1-\frac{2.5}{3}\right)+(14.19\times 3.5)\times\left(\frac{3.5}{2}+1\right)$$
$$+\left[\frac{1}{2}\times(27.13-14.19)\times 3.5\right]\times\left(3.5+1-\frac{3.5}{3}\right)+(13.71\times 0.48)\times\left(\frac{0.48}{2}+3.5+1\right)$$
$$+\left[\frac{1}{2}(27.13-13.71)\times 0.48\right]\times\left(\frac{0.48}{3}+3.5+1\right)-P(0.48+3.5+1)=0$$
$$2.96+136.58+75.48+31.19+15.01-4.98P=0$$

or

$$P=\frac{261.22}{4.98}=52.45\text{ kN}$$

$$p_2''=\gamma'K'z_4=11.2\times 2.67z=29.9\,z_4\text{ kN/m}^2$$

Taking the moment of the diagram CE′FJ′J about J′,

$$p(z_4+z_5)+\left(\frac{1}{2}p_2'\times z_5\right)\left(\frac{2z_5}{3}+z_4\right)-\left(\frac{1}{2}p_2''z_4\right)\frac{z_4}{3}=0$$

or

$$52.45(0.48+z_4)+\left(\frac{1}{2}13.71\times 0.48\right)\left(\frac{2}{3}\times 0.48+z_4\right)-\left(\frac{1}{2}\times 29.9\times z_4\times z_4\right)\frac{z_4}{3}=0$$

or

$$25.18+52.45z_4+1.05+3.29z_4-4.98z_1^3=0$$

or

$$z_4^3-11.149z_4-26.23=0$$

Take $z_4=4.0$; then,

$64-44.76-26.23=-6.99<0$

Now try $z_4=4.18$; then,

$73.03-46.77-26.33=-0.03\cong 0$

Therefore,

$D=z_3+z_4=0.91+4.18=5.09$ m

$D_\text{d}=1.2\times 5.09=6.1$ m

**Example 12.7** It is proposed to construct a 5 m deep trench in a stiff clay with $c = 40$ kN/m$^2$, $\phi = 0°$, and $\gamma = 18.5$ kN/m$^3$ and to timber it with horizontal struts at 1, 2.5, and 4 m below the top. Draw the earth pressure envelope and make reasonable assumptions to estimate the load that each strut can carry per metre run of excavation.

***Solution***

$$\frac{\gamma H}{c} = \frac{18.5 \times 5}{40} = 2.3 < 4$$

Hence, the earth pressure distribution suggested by Peck (1969) for stiff clay (Fig. 12.23a) is considered. The earth pressure envelope along with the strut positions are shown in Fig. 12.32a.

$$p_{\mathrm{a}} = 0.30\gamma H = 0.3 \times 18.5 \times 5 = 27.75 \text{ kN/m}^2$$

Assuming hinges at the reaction points, the entire pressure distribution and the same in split forms are shown in Fig. 12.32b and c. Taking the moments of the forces about $R'_2$

$$R_1 \times 1.5 = \left(\frac{1}{2} \times 1.25 \times 27.75\right) \times \left(\frac{1}{3} \times 1.25 + 1.25\right) + (1.25 \times 27.75) \times \frac{1.25}{2}$$

or

$$R_1 = 33.76 \text{ kN/m run on the top strut}$$
$$R'_2 = 17.34 + 34.69 - 33.76 = 18.27 \text{ kN}$$

$$R''_2 \times 1.5 = (1.25 \times 27.75) \times \left(\frac{1.25}{2} \times 1.25 - 1.0\right) + \left(\frac{27.75 + 22.9}{2} \times 0.25\right) \times \frac{0.25}{2}$$

or

$$R''_2 = 20.76 \text{ kN}$$
$$R'_3 = 34.69 + 6.33 - 20.76 = 20.26 \text{ kN}$$

Therefore, the force on the central strut $= R'_2 + R''_2 = 18.27 + 20.7 = 38.97$ kN / m run. Taking moment about $R_4$,

$$R''_3 \times 1 = \tfrac{1}{2} \times 1 \times 22.9 \times \tfrac{2}{3} \times 1$$

$$R''_3 = 7.63 \text{ kN}$$

The total force on the lower strut $= R'_3 + R''_3 = 20.26 + 7.63 = 27.89$ kN / m run.

$$\text{Reaction } R_4 = \tfrac{1}{2}(5 + 2.5) \times 27.75 - 33.76 - 38.97 - 27.89$$

or

$$R_4 = 3.44 \text{ kN/m}$$

The reaction $R_4$ is assumed to be provided by the soil.

**Example 12.8** A long 5 m wide and 8 m high vertical channel has to be constructed in a deep cohesive soil with $c = 36$ kN/m$^2$ and $\gamma = 18$ kN/m$^3$. Before protecting the sides using sheet piles, it is intended to check the safety of the bottom of the channel against heave. Consider the excavation to be completed rapidly and find the factor of safety against heave. What will be the change in the factor of safety if a hard material is present at 2.5 m from the bottom of the channel?

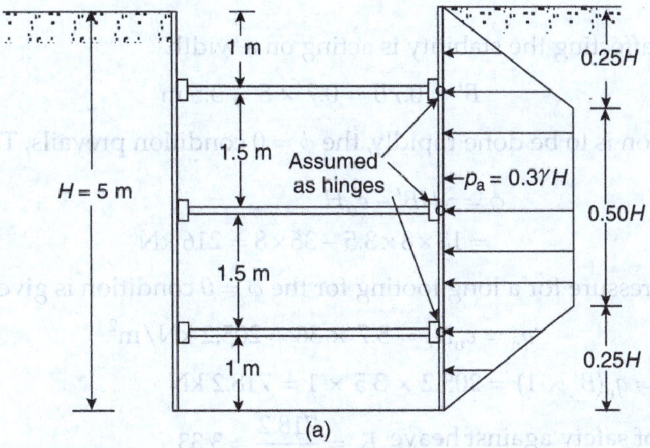
1 m
1.5 m
1.5 m
1 m
H = 5 m
Assumed
as hinges
$p_a = 0.3\gamma H$
0.25H
0.50H
0.25H
(a)

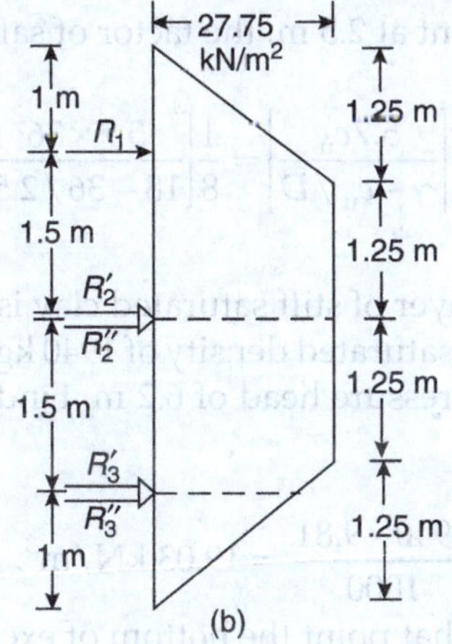
27.75
kN/m²
1 m
$R_1$
1.5 m
$R_2'$
$R_2''$
1.5 m
$R_3'$
$R_3''$
1 m
1.25 m
1.25 m
1.25 m
1.25 m
(b)

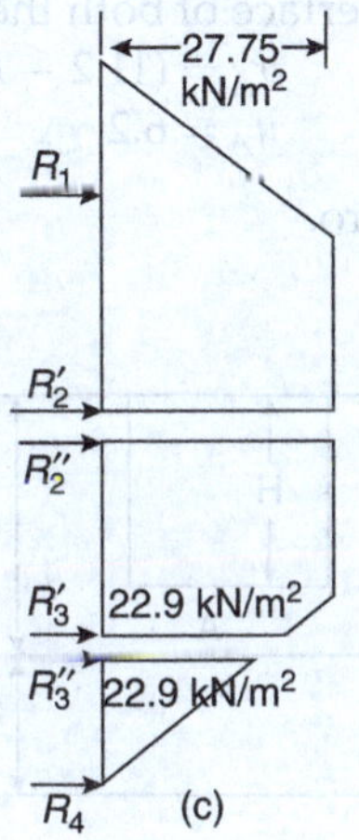
27.75
kN/m²
$R_1$
$R_2'$
$R_2''$
$R_3'$
22.9 kN/m²
$R_3''$
22.9 kN/m²
$R_4$
(c)

Fig. 12.32

***Solution***

The vertical load affecting the stability is acting on a width

$$B' = 0.7B = 0.7 \times 5 = 3.5 \text{ m}$$

Since the excavation is to be done rapidly, the $\phi = 0$ condition prevails. Therefore,

$$\phi = \gamma HB' - c_u H$$
$$= 18 \times 8 \times 3.5 - 36 \times 8 = 216 \text{ kN}$$

The net bearing pressure for a long footing for the $\phi = 0$ condition is given as

$$q_n = c_u N_c = 5.7 \times 36 = 205.2 \text{ kN/m}^2$$

Net bearing load $= q_n(B' \times 1) = 205.2 \times 3.5 \times 1 = 718.2$ kN

Therefore, factor of safety against heave $F_h = \dfrac{718.2}{216} = 3.33$

When the hard material is present at 2.5 m, the factor of safety is obtained by taking $B' = D$; then,

$$F_h = \frac{1}{H}\left[\frac{5.7c_u}{\gamma - c_u/D}\right] = \frac{1}{8}\left[\frac{5.7 \times 36}{18 - 36/2.5}\right] = 7.13$$

**Example 12.9** A 11.2 m thick layer of stiff saturated clay is underlain by a 2.3 m thick layer of sand. The saturated clay has a saturated density of 1940 kg/m$^3$ and the sand as 1825 kg/m$^3$. The sand layer is at a artesian pressure head of 6.2 m. Find the maximum depth of cut that can be made in the clay.

***Solution***

$\gamma_{sat}$ of clay = 1,940 kg/m$^3$ = $\dfrac{1940 \times 9.81}{1000}$ = 19.03 kN/m$^3$.

Let the depth of cut be $H$, at that point the bottom of excavation will heave.

The stability of a point $A$, as the interface of both the layers is considered.

From Fig. 12.33

$$\sigma_A = (11.2 - H)\gamma_{sat}$$
$$u_A = 6.2\,\gamma_\omega.$$

For heave to occur $\sigma'_A$ should be zero.

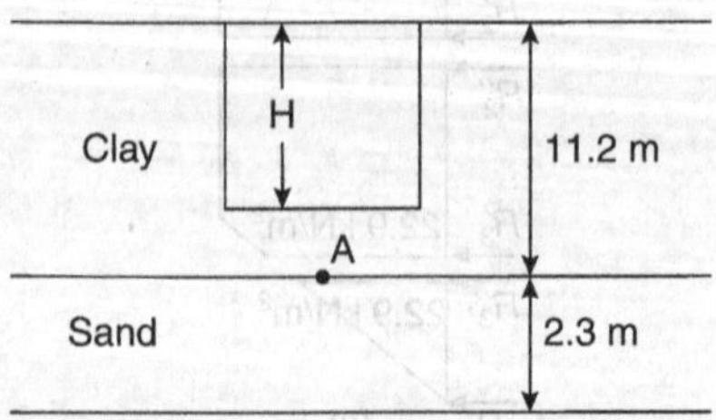


Fig. 12.33

Therefore, $\sigma_A - u_A = (11.2 - H)\, 19.03 - 6.2 \times 9.81 = 0$

i.e., $11.2 - H = \dfrac{6.2 \times 9.81}{19.03}$

$\therefore H = 112 - \dfrac{6.2 \times 9.81}{19.03} = 8.00$ m.

Maximum cut can be made is 8.00 m.

## POINTS TO REMEMBER

12.1 Gravity-retaining walls provide slope and soil retention by their weight, which may consist of masonry, concrete, concrete in combination with soil weight, or the weight of earth mass alone. Cantilever wall is a type of gravity wall which is economical as the backfill is designed to provide the most of the required dead weight.

12.2 Cantilever and gravity walls are both liable to rotational and translational movements, and hence, Rankine's and Coulomb's theories may be used for the calculation of lateral pressure.

12.3 Retaining walls have to satisfy the following stability requirements: (i) safety against overturning, (ii) safety against sliding, (iii) safety against bearing capacity failures, and (iv) safety against overall stability.

12.4 Backfill materials for retaining structures should have high long-term strength, free drainage and impact, and less lateral pressure. Poorly graded to well-graded sands and gravels form excellent backfill material.

12.5 Sheet pile walls are flexible structures compared to gravity-type retaining walls and are widely used for both small and large water front structures. Sheet pile walls are made out of wood, precast, concrete, or steel. The two types are cantilever and anchored sheet pile walls.

12.6 A sheet pile wall may fail in any one of the following ways: (i) forward movement of the base due to inadequate passive resistance in front of the wall, (ii) failure by bending, and (iii) failure of anchors. Depending on the type of failure, the earth pressure distribution varies, and it does not follow the conventional distribution adopted in rigid walls.

12.7 Wales, tie rods, and anchorages are provided to keep a sheet pile in the required position for the expected lifetime. Wales are longitudinal members of a rolled channel section usually provided back to back along the sheet pile length. A cable or a steel rod, threaded to allow vertical alignment and tension adjustments, acts as a tie rod. Anchors may be of tie-back or deadman type.

12.8 Sheeting for braced excavations basically consists of a sheet piling to support the sides of the excavation, with stability being maintained by means of strut across the excavation.

12.9 Failure of the soil of a braced excavation may occur due to deformation changes with depth or heave of the bottom of a cut or due to upward seepage of water.

## QUESTIONS

### *Objective Questions*

12.1 State whether the following are true or false.

1. Structures that are restrained from yielding should be designed to resist at-rest lateral pressures.
2. Stability analysis of retaining walls based on Rankine's theory results in unconservative wall design.
3. For free-standing retaining walls, active or passive pressures can develop only by translation of the wall.
4. The free earth support method considers the lowest section of sheeting to be fixed in the earth.
5. The spacing of the rods in anchored bulkheads depends on the forces taken by each rod.

12.2 Rowe's method for reducing the moment in anchored sheet piling basically depends on the
(a) Depth of fixity of wall at the base
(b) Modulus of rigidity of the wall material
(c) Relative flexibility of the piling
(d) Area of cross-section of the pile

12.3 For the design of braced excavation, the earth pressure distribution is based on
(a) Rankine's hydrostatic distribution
(b) Coulomb's distribution in the classical form
(c) Apparent pressure envelopes based on field studies
(d) None of the above

12.4 Failure of braced excavation in clay due to bottom heave may be avoided by
(a) Reducing the flexibility of the wall system
(b) Increasing the time for installation of struts or anchors
(c) Loading the ground surface with some surcharge
(d) Increasing the $\gamma H/c$ value to be >8

12.5 The qualities required for a material to cause minimum earth pressure with the least movement are
(a) Free draining, rigid, and light in weight
(b) Rigid, free draining, and with high angle of internal friction
(c) Free draining, light in weight, and with low angle of internal friction
(d) Free draining, loose, and light in weight

12.6 It is a general practice to provide the face of a cantilever retaining wall with a small batter to compensate for the
(a) Forward tilting
(b) Lateral sliding
(c) Overturning
(d) Forward sliding

12.7 The criterion for the design of a gravity-retaining wall is
(1) Safe against sliding
(2) Safe against overturning
(3) Safe against tensile stress
(4) Safe against bearing capacity failure

Of these statements,
(a) 1, 2, and 4 are correct
(b) All are correct
(c) 1, 3, and 4 are correct
(d) 2, 3, and 4 are correct

12.8 In the design of a cantilever sheet pile wall, the calculated depth is increased arbitrarily by 20% to allow for
(a) The development of passive resistance
(b) The reduction of active thrust
(c) Sufficient grip length
(d) Erosion

12.9 Identify the incorrect statement. A sheet pile wall may fail in any one of the following ways:
(a) Forward movement of the base
(b) Failure by bending
(c) Failure by shear
(d) Failure of anchors

12.10 Horizontal timber plants placed by hand as the excavation proceeds are referred to as
(a) Wales
(b) Compression members
(c) Lagging
(d) Sheeting

## *Descriptive Questions*

12.11 In a sheet pile wall, supporting and penetrating clay, how is the thrust likely to alter when the clay swells or consolidates?

12.12 Explain why the hydrostatic-type linear earth pressure distribution is not valid in a strutted excavation?

12.13 What are the possible signs of distress of masonry retaining walls? Suggest a few remedial measures.

12.14 Explain why weep holes are provided in retaining walls.

12.15 Explain why only granular materials are preferred for the backfill of a retaining wall.

12.16 What is the necessity to remove weak natural soil behind a bulkhead prior to placement of a granular fill? Explain.

12.17 How will you design a deadman-type anchor. What factors would you consider in deciding the location of a deadman?

12.18 What do you understand by repeated yielding? How can you counter this effect on a wall?

12.19 How will you decide the location of a railway line on top of a cohesive backfill of a rough-surfaced vertical retaining wall?

## EXERCISE PROBLEMS

12.1 Check the stability of the concrete retaining wall shown in Fig. 12.34. The backfill material is a mixture of sand and gravel with the following properties: $\gamma = 19.6$ kN/m$^3$ and $\phi = 33°$. The tangent of the coefficient of friction between the concrete and the soil is 0.48. The unit weight of concrete is 2.5 kN/m$^3$. The retaining wall is placed on a very dense gravelly bed with an allowable soil pressure of 380 kN/m$^2$.

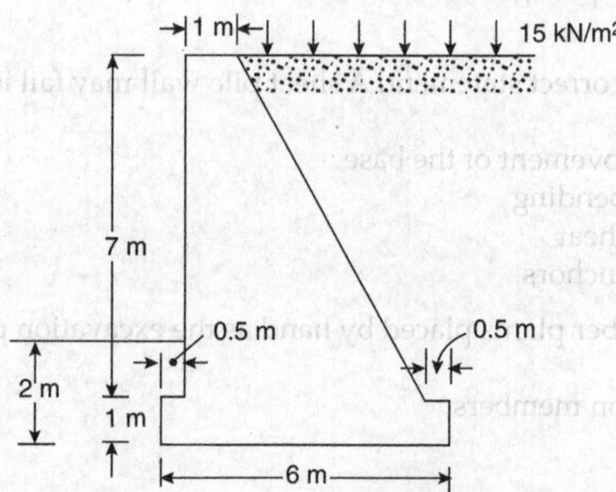


Fig. 12.34

12.2 Estimate the minimum and maximum pressures under the base of a cantilever retaining wall shown in Fig. 12.35. Also, check the stability against overturning and sliding. The properties of the backfill material are $\gamma = 18.2$ kN/m$^3$ and $\phi = 38°$. The friction angle at the base of the wall is given as 27°, and the unit weight of concrete is 23.5 kN/m$^3$.

12.3 Determine the minimum safe width of a gravity-retaining wall, supporting 5 m of a granular fill having a dry unit weight of 18.5 kN/m$^3$ and an angle of friction 32°. The pressure surface of the retaining wall has a batter of 1:6. The backfill is sloped with an angle of inclination of 15°. The base of the wall is located at a depth of 2 m from the ground surface. The properties of the foundation soil are $c = 15$ kN/m$^2$, $\phi = 25°$, and $\gamma = 19.0$ kN/m$^3$.

12.4 For the sheet pile wall system shown in Fig. 12.36, determine the depth of penetration considering the sheet pile as a cantilever type. What will be the percentage of reduction in the depth if tie rods are placed at 1.5 m from the top and 3 m from the centres?

12.5 It is intended to design a cantilever sheet pile wall to support a varved clay as detailed in Fig. 12.37. Compute the depth required considering a factor of safety of 2.5 against

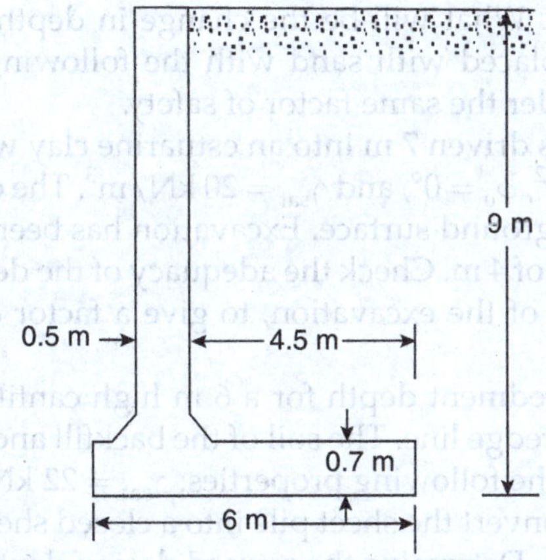


Fig. 12.35

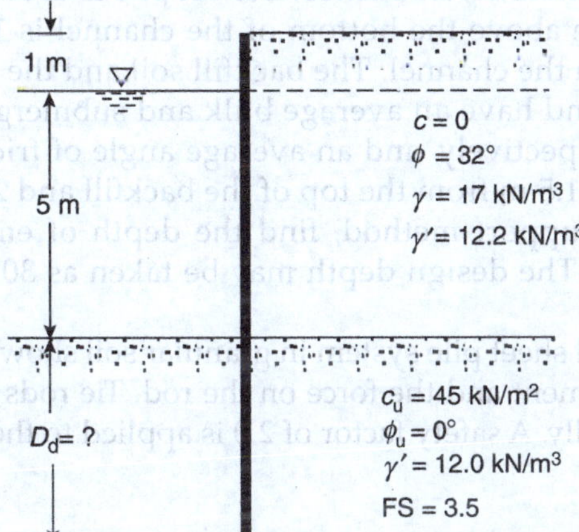


Fig. 12.36

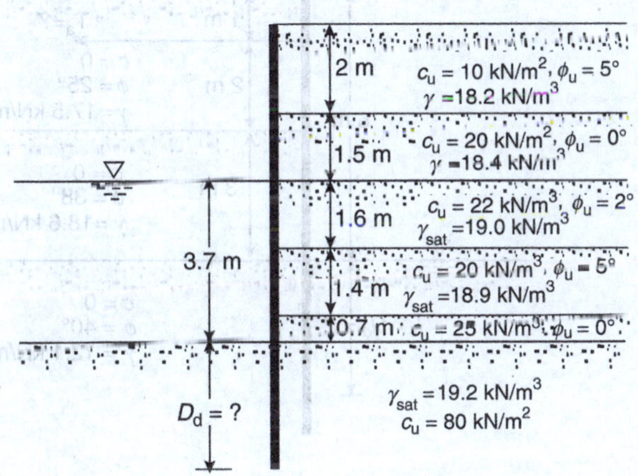


Fig. 12.37

passive resistance. What will be the change in depth of embedment if the top two soft layers are replaced with sand with the following properties: $\phi = 30°$ and $\gamma = 18\ \text{kN/m}^3$? Consider the same factor of safety.

12.6 A sheet pile wall is driven 7 m into an estuarine clay which has the following properties: $c_u = 18\ \text{kN/m}^2$, $\phi_u = 0°$, and $\gamma_{sat} = 20\ \text{kN/m}^3$. The original groundwater is located at 1.5 m from the ground surface. Excavation has been carried out on one side of the wall up to a depth of 4 m. Check the adequacy of the depth of penetration of the piling below the bottom of the excavation, to give a factor of safety of 2.0 with respect to passive resistance.

12.7 Compute the embedment depth for a 6 m high cantilever pile supporting 4 m high water above the dredge line. The soil of the backfill and that below the dredge line are the same, having the following properties: $\gamma_{sat} = 22\ \text{kN/m}^3$ and $\phi = 30°$. It is decided at a later date to convert the sheet pile into a closed sheet pile by providing a tie rod at 1.5 m from the top. Determine the revised design depth of embedment (with a safety factor of 1.4) and the force on the tie rod.

12.8 For a shipping channel, an anchored sheet pile is used to support a fill. The height of the sheet pile above the bottom of the channel is 10 m, and it supports a 8 m head of water in the channel. The backfill soil and the soil beneath the channel are both granular and have an average bulk and submerged unit weights of 18.6 and 12.8 $\text{kN/m}^3$, respectively, and an average angle of friction of 32°. The anchor rod is positioned at 1.5 m from the top of the backfill and 2.5 m centre to centre. Using the free earth support method, find the depth of embedment and the force on the anchor rod. The design depth may be taken as 30% more than the theoretical depth.

12.9 For the anchored sheet pile system in granular soil shown in Fig. 12.38, determine the depth of embedment and the force on the rod. Tie rods are placed at 3.5 m centre to centre horizontally. A safety factor of 2.0 is applied to the passive resistance.

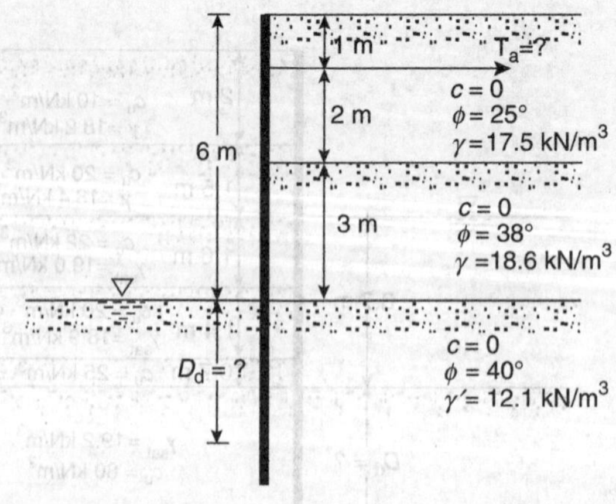


Fig. 12.38

12.10 An anchored sheet pile wall supports 5 m of fully saturated soil having the following relevant properties: $c = 5$ kN/m$^2$, $\phi = 32°$, and $\gamma_{sat} = 20$ kN/m$^3$. The groundwater is 0.5 m below the top of the wall. Horizontal anchors are installed at depths of 1.2 and 2.5 m from the centre. Use the free earth support method and determine the minimum safe driving depth, adopting a factor of safety of 1.50. Also, estimate the force transmitted by each anchor rod.

12.11 In a river bank protection scheme, an anchored sheet pile wall is driven to support sand up to a depth of 4.5 m. Anchor rods are provided at 1.0 m below the top and at 3 m centre to centre. The sand has a friction angle of 32°. The surface of the retained material is to be horizontal and level with the top of the wall. During heavy rains the water level rises to a level of 0.5 m below the top of the wall. Neglecting cohesion and friction on the surface of the piles, use the fixed earth support method to find the design depth of the pile. It is required to provide a factor of safety of 1.5 against the depth of penetration. Determine the force and diameter of the anchor rod if the tensile strength of the material of the rod is $90 \times 10^3$ kN/m$^2$.

12.12 An anchored sheet pile wall is constructed by driving a line of piling as shown in Fig. 12.39. The rods are spaced at 2.5 m centre to centre, 1.5 m below the surface of the backfill. The water level in front of the wall and the water table behind the wall are both 3 m below the surface of the backfill. Determine the design depth considering a factor of safety of 2.0 with respect to passive resistance. Also, find the force on the anchor rod. Adopt the fixed earth support method.

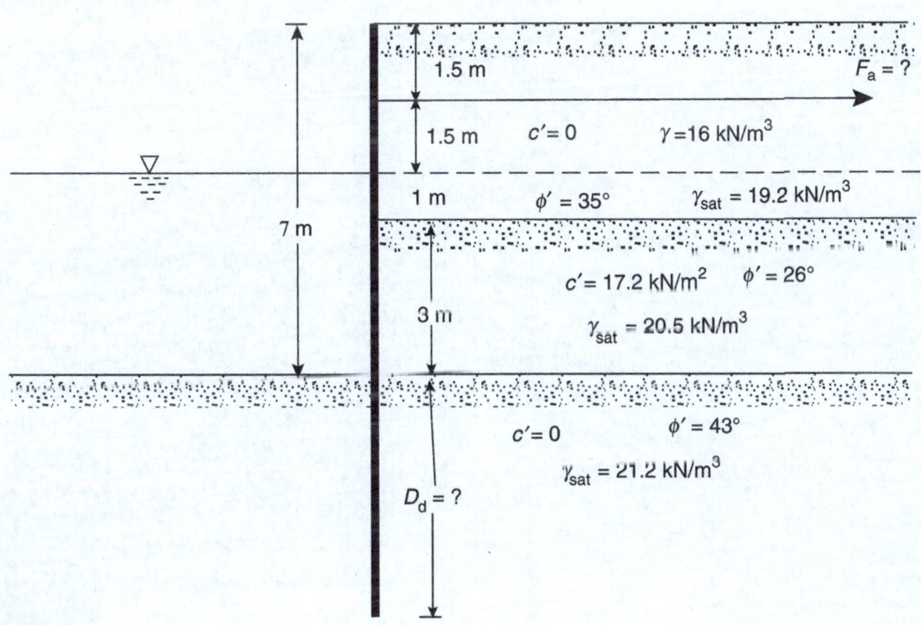


Fig. 12.39

12.13 A 3.2 m wide and 6.5 m deep cut is proposed to be made in a moist deposit of sand with shear strength parameters $c = 0$ and $\phi = 30°$. Find the total load on the timber sheeting if $\gamma = 19.5$ kN/m$^3$.

12.14 The sides of an excavation 5 m deep in stiff clay are to be supported temporarily by timber. The struts are placed at 1, 2.5, and 4 m below the top. Assuming a suitable pressure distribution, estimate the load that each strut can carry per metre run of excavation. The relevant properties of the soil are $\gamma_{sat} = 21$ kN/m$^3$, unconfined compressive strength $q_u = 200$ kN/m$^2$, and $\phi = 0°$.

12.15 A strutted excavation 1.5 m wide is executed in a saturated plastic clay with a unit weight of 18 kN/m$^3$. The bottom of the excavation yields when the height reaches 10 m. Estimate the approximate shear strength of the soil that prevailed during failure.

12.16 A strutted excavation 4 m $\times$ 8 m in plan is to be taken up for the installation of a machinery. The depth of the excavation is 5 m in a saturated stiff clay which has an undrained strength of 40 kN/m$^2$ and a unit weight of 19 kN/m$^3$. There is another supporting machinery to be placed on the surface of the ground in line with the vertical face of the excavation. This machinery will be inducing an overall surcharge of 15 kN/m$^2$. Estimate the factor of safety against base failure.

# 13
# Stability of Slopes

## CHAPTER HIGHLIGHTS

Causes of slope failures – Short- and long-term failures – Types of land slides and slope movements – Factor of safety – Infinite and finite slopes – Analysis of infinite slopes – Analysis of finite slopes: planar and circular failure surfaces: $\phi u = 0$ analysis, friction circle method, Fellenius method of slices, Bishop's simplified method – Taylor's stability chart – Location of critical circle

## 13.1 INTRODUCTION

Landslides are the downward and outward movements of slope materials because of exhaustion of required shear strength. The slope materials may be composed of natural rock and soil, artificial fills, or combinations thereof. Potential landslides in natural slopes may be identified either by aerial photographs or by ground reconnaissance. Slides also occur in man-made structures such as embankments and earth dams. Sufficient care has to be taken to choose the correct construction material and to adopt a suitable construction procedure to avoid sliding of the slope during or after construction. Further, the stability of foundations and earth-retaining walls against ground break or rupture of soil is also important. One of the causes of ground break is insufficient depth of the embedment of the foundation or the retaining wall combined with low shear strength. In principle, the analysis consists of determining the factor of the slopes against shear failure so as to ascertain the stability of natural slopes, cuts, embankments, earth dams, and ground break.

## 13.2 CAUSES OF SLOPE FAILURES

The causes of failure of slopes may be external or internal. External causes are those which produce an increase in the stress at unaltered shearing resistance of the material. They include steepening of the slope, deposition of material along the edge of slopes, and earthquake forces.

Internal causes are those that lead to a slide without any change in surface conditions which involve unaltered shearing stresses in the slope material. Some of these conditions are the decrease in shearing resistance brought about by excess pore water pressure, leaching of salts, softening, breakage of cementation bonds, and ion exchange. Intermediate between landslides due to external and internal causes are those due to rapid draw-down, to surface erosion, and to spontaneous liquefaction. Terzaghi (1950) reviewed the processes which cause landslides by several modes of action of agents and represented them in a lucid form, as shown in Fig. 13.1. As an example, some of the activities which may provoke or improve a landslide are shown in Fig. 13.2.

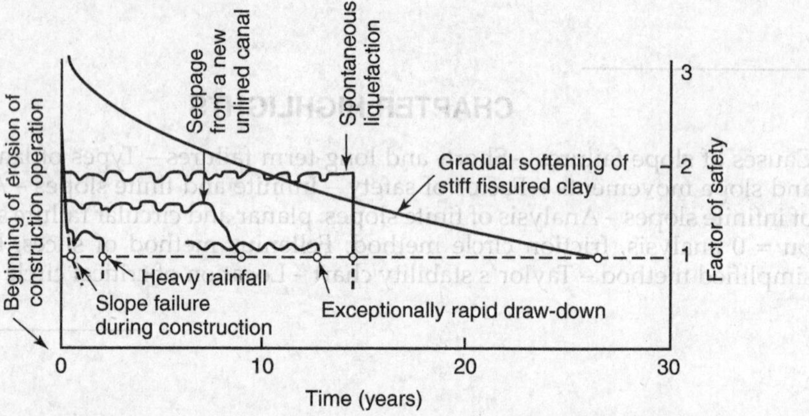


**Fig. 13.1** Variations in the factor of safety of different slopes of recent origin (*Source*: Terzaghi, 1950)

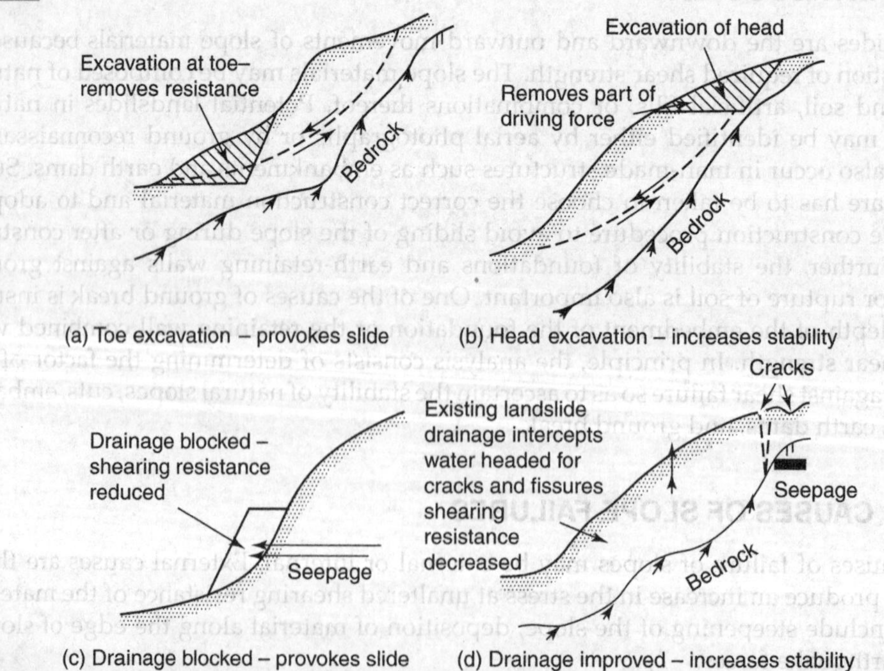


**Fig. 13.2** Activities that decrease or increase the probabilities of slides (*Source*: Woods, 1950)

## 13.3 SHORT- AND LONG-TERM FAILURES

Stability of natural slopes and cuts may be studied under two conditions, namely, short- and long-term conditions. The short-term instability is due to non-availability of sufficient time for the dissipation of pore water pressure. The long-term condition is one in which the pore water pressure gradually adjusts itself in the long run and shows values corresponding to a certain groundwater condition.

In the stability analysis of slopes, one may adopt effective or total stress analysis depending on the field situation. In the effective stress analysis, the proportion of the shear strength mobilized (actual stress) for limiting equilibrium is expressed (for a factor of safety of $F$) as

$$\tau = \left(\frac{c'}{F}\right) + (\sigma_n - u_w)\left(\frac{\tan\phi'}{F}\right) \tag{13.1}$$

Two factors that should be known for the application of the above equation are $\sigma_n$ and $u_w$. Generally, a suitable stress distribution is assumed, and an appropriate value of the pore water pressure, $u_w$, has to be used, which depends on the class of stability problem (Bishop and Henkel, 1962): (i) Class A problems, where the pore water pressure is an independent variable and the value of $u_w$ is obtained from the groundwater level if there is no flow net if a state of steady seepage exists and (ii) Class B problems, where the magnitude of pore water pressure depends on a change in stress.

In the total stress analysis, the proportion of the shear strength mobilized for the zero condition is expressed as

$$\tau = \frac{c_u}{F} \tag{13.2}$$

In natural and earth dam slopes (during steady seepage condition), pore water pressure is controlled by the prevailing groundwater conditions, and hence, they fall under Class A problems. On the other hand, in cuts and free-standing excavations in clay, pore water pressure changes because of stress release due to excavation, and hence, they fall under Class B problems.

## 13.4 TYPES OF LANDSLIDES AND SLOPE MOVEMENTS

A systematic classification of slides in clay and other mass movements was proposed by Skempton and Hutchinson (1969). This includes five basic types and six complex forms of movements (Fig. 13.3).

### 13.4.1 Basic Types of Landslides

**Falls**. The removal of earth support causes bulging at the toe and tension crack at the top. The development of cracks induces additional stresses on the separating mass and leads to an ultimate failure. Clay falls occur in steep slopes and are typical short-term failures. Such failures are found mostly in over-consolidated fissured clays.

**Rotational Slides (Slips, Slumps)**. These types of slides are common in fairly uniform clays or shales. The curved surface of failure, being concave upwards, imparts a backward tilt to the slipping mass, resulting in sinking at the rear and heaving at the toe. Such slides are deep-seated, and the failure surfaces may be circular or non-circular.

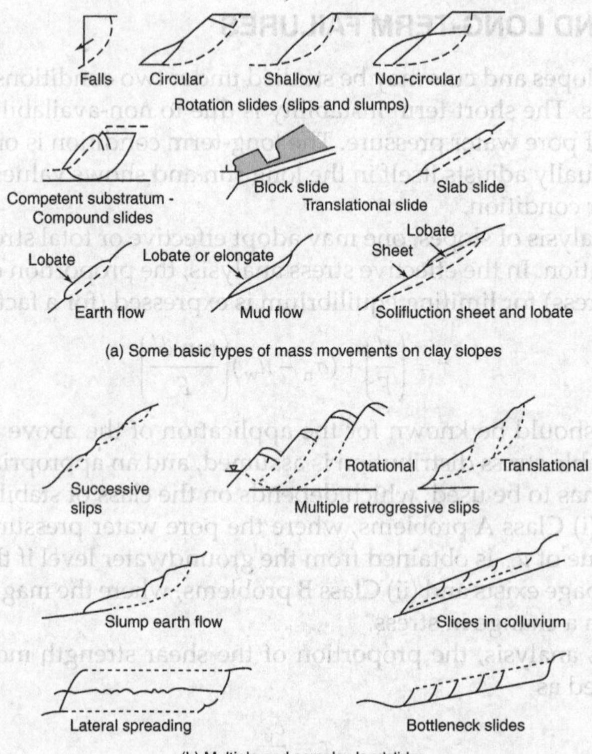


**Fig. 13.3** Types of mass movements on clay slopes (*Source*: Skempton and Hutchinson, 1969)

**Compound Slides**. The surface of failure is predetermined by the presence of heterogeneity within the slope material. Heterogeneity usually consists of a weak soil layer or a structural feature or a boundary between two materials, for example, clay and rock or weathered and unweathered material. Such heterogeneity prevents simple rotational slides but introduces a translational element in the movement in combination with or without rotational slide. Compound slides usually occur in soils with heterogeneity at moderate depth.

**Translational Slides**. These are planar and most commonly occur in a mantle of weathered material, the heterogeneity being at a shallow depth. Moreover, such slides occur as block or slab slides. Block slides are found in marls and sandstones, whereas slab slides are a type of translational failure in more weathered clay slopes.

**Flows**. These are mass movements which may be of either earth flow or mud flow. While earth flows are slow movements of softened weathered debris, mud flows are glacier-like in form and are often well developed below the bar in fissured clays.

### 13.4.2 Multiple and Complex Slides

A slide may include several types of basic movements within its various parts or at different stages in its development. These are referred to as multiple and complex slides.

**Successive Slips**. Successive rotational slips consist of an assembly of individual shallow rotational slips. Moreover, they are common in over-consolidated fissured clays at later stages of the free degradation process of the slopes.

**Multi Retrogressive Slips**. Multiple slides develop from single failures and are predominantly rotational but sometimes translational. The cause for more numerous individual retrogressive failures may be less cohesion of the sliding mass. These slides occur more frequently in actively eroding slopes of fairly high relief in which a thick stratum of over-consolidated fissured clay or clay–shale is overlain by a rock of considerable strength.

**Slump Earth Flows**. These are a common type of mass movement intermediate between rotational slides and mud flows. They develop typically in rotational slides of considerable displacement, where the toe of the slipping mass is much broken by over-riding, which in the presence of water softens and forms a mud flow.

**Slide in Colluvium**. Colluvium develops typically in the accumulation zones below freely degrading cliffs (Hutchinson, 1967). The sliding material is usually so shifted and weathered that individual slipped masses are no longer distinguishable. Other types of slides in colluvium involve the renewal of movements in debris, which is associated with individual old slides.

**Spreading Failures**. These are a particular type of retrogressive translational slides. The initial rapid movement reduces considerably and stops within a few minutes because of the gentle slopes involved.

**Quick Clay Slides**. Such slides generally begin with an initial rotational slip in the bank of a stream incised into quick clay deposits. The slipping mass is in part re-moulded to the consistency of a liquid, which runs out of the cavity carrying flakes of the stiff, weathered crust. In general, quick clays may fail in one of the above-mentioned ways.

### 13.4.3 Rates of Land Movement

Terzaghi (1950) gave a qualitative description of movements typically associated with a landslide. Excluding mud flow, four types of movements may be recognized (Skempton and Hutchinson, 1969; Terzaghi, 1950): (i) creep, (ii) pre-failure movements, (iii) movements during slide, and (iv) post-failure movements.

**Creep**. Invariably, all slopes are subject to creep, although at an imperceptible rate. Terzaghi (1950) distinguished between seasonal or mantle creep and continuous or mass creep (Fig. 13.4). Seasonal creep is confined within the zone of seasonal changes of moisture and temperature; at least part of the horizontal component of the ground movement is produced by thermal expansion and contraction, swelling and shrinkage, freezing and thawing, and other seasonal processes. Mantle creep may range from less than 1 to a few millimetres per year. But in moderate climates, significant movements may extend to a depth of as much as 25 mm (Terzaghi and Peck, 1967). The mass or continuous creep moves at a fairly constant rate but at a depth below the material subject to mantle creep. The load at which creep begins is much smaller than the shear stress at failure. Similar behaviour was reported by Bishop (1966) using laboratory-drained long-term creep tests on clays.

**Pre-failure Movements**. Pre-failure movements form a basis for the prediction of failure. The line *Oa* shows the movement which preceded the slide. The distance $OD_1$ depends primarily on the thickness of the zone within which the state of stress approaches the state

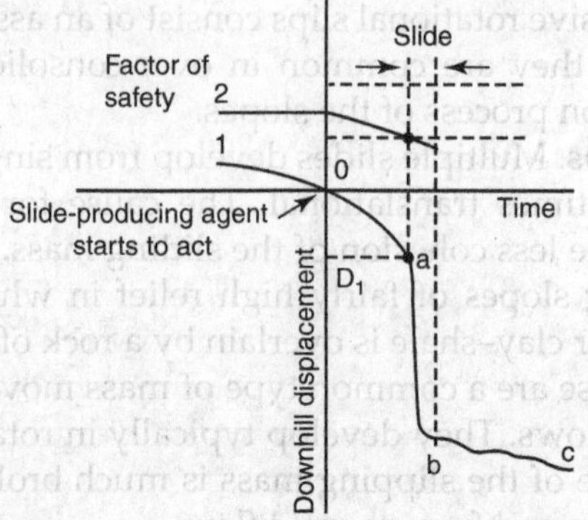


**Fig. 13.4** Landslide movement (*Source*: Terzaghi, 1950)

of failure and on the type of clay. Skempton and Hutchinson (1969) confirmed this from several field examples. Some techniques of measuring pre-failure movements in slopes are discussed by Terzaghi and Peck (1967).

**Movements During Slide**. During the first phase of the slide, the sliding mass advances at an accelerated rate, as shown by the upper part of the curve *ab*. The maximum velocity of the movement depends on the average slope angle of the surface of sliding, the resistance available, and the nature of stratification. For a clay with a perfectly plastic stress–strain curve after failure, the downslope movement is slow and attains a stable position with a factor of safety 1. In clays, such as over-consolidated fissured clays, which show a pronounced peak, the slide accelerates and is carried past the stable position by its own momentum, coming to rest with a factor of safety higher than 1 on the residual strength. But quick clays experience the fastest movement, and the decrease in shearing resistance may be of the order of 90%. In contrast, slides in more or less homogeneous masses of residual soil, or clay with low sensitivity, seldom attain a velocity of more than 0.3 m/min.

**Post-failure Movement**. After the descent (Fig. 13.4, Point *b*), the movement passes into a slow creep unless the slide has radically altered the physical properties of the sliding mass. In the majority of clays, the shear strength on the slip surface after failure may be at or very close to the residual strength. Heavily over-consolidated clays experience post-failure movements, and the speed of movement ranges from 0 to 6 m/year. Slides in normally consolidated or quick clays generally exhibit no post-failure movements. Steps in the trend of the line *bc* in Fig. 13.4 correspond to creep with seasonal effects.

## 13.5 FACTOR OF SAFETY

In any stability analysis, some measure of the degree of safety has to be provided. Such a measure of safety may be a factor like a limiting stress or strain or a comparative ratio of resistance. Working stresses in any earth structure are much less than the shear strength of the soil so as to ensure the safety of the structure. The working stress is the actual stress at a point or along a continuous surface and may be defined as *developed* or *mobilized strength*. In slope stability problems, shear strength is the governing factor for stability; hence, the mobilized or developed shear strength ($\tau$) is also important. If this mobilized strength is less than the available strength ($\tau_f$) of the soil, then the slope is said to be stable. Thus, the factor of safety may be defined, in a form most convenient and acceptable to practical engineers, as the ratio of the shearing resistance available along a slip surface to the total mobilized shearing resistance; that is,

$$F = \frac{\tau_f}{\tau}$$

In other words, $F$ measures the factor by which the shear strength will have to be reduced ($\tau = \tau_f/F$) to bring the structure to a state of imminent collapse.

If in the mobilization process, both cohesion and friction contribute in equal proportion, then the factor of safety is referred to as the factor of safety with respect to strength; that is,

$$F_s = \frac{\text{Available shear strength}}{\text{Mobilized shear strength}}$$

$$F_s = \frac{c' + \sigma'_n \tan\phi'}{\tau} \quad (13.3)$$

or

$$\tau = \frac{c'}{F_s} + \frac{\sigma'_n \tan\phi'}{F_s}$$

This may be written in a more general form as

$$\tau = \frac{c'}{F_c} + \frac{\sigma'_n \tan\phi'}{F_\phi} \quad (13.4)$$

where $F_c = c'/c_m$ is the average factor of safety for the cohesional component of strength and $F_\phi = \tan\phi'/\tan\phi_m$ is the average factor of safety for the frictional component of strength, where $c_m$ and $\phi_m$ are the mobilized cohesion and friction, respectively. Thus, Eq. 13.3 is the case for which $F_c = F\phi = F_s$.

In a non-cohesive soil, $c = 0$; hence,

$$\tau_f = \sigma'_n \tan\phi'$$

Hence, Eq. 13.4 reduces to

$$\tau = \frac{\sigma'_n \tan\phi'}{F_\phi} \quad (13.5)$$

where $F_\phi$ is the factor of safety with respect to friction.

If for a condition $F_\phi$ is unity (i.e., full friction has mobilized) or zero ($\phi_u - 0°$ condition), then the ratio of the actual cohesion to the cohesion required for stability is defined as the factor of safety with respect to cohesion ($F_c$).

The cohesion required for stability is directly proportional to the height of the slope. Hence, the factor of safety with respect to cohesion (when $F_\phi$ is unity) is sometimes referred to as the factor of safety with respect to height ($F_H$). This is nothing but the ratio of the critical height $H_c$ to the actual height $H_a$, the critical height being the maximum height at which the slope will be stable. For this case, $F_c = F_H$; hence,

$$F_{\mathrm{H}} = \frac{c'}{c_{\mathrm{m}}} \tag{13.6}$$

and

$$\tau = \frac{c'}{F_{\mathrm{H}}} + \sigma'_{\mathrm{n}} \tan \phi' \tag{13.7}$$

The safety factors defined above are simply standards of comparison and have no physical meaning beyond that given by their respective definitions.

## 13.6 BASIC CONCEPTS OF SLOPE STABILITY ANALYSIS

All sloping surfaces are subjected to shearing stresses on nearly all the internal surfaces. The shear strength available should overcome the same at all points. If the shear stresses are more and if these points are adjacent and continuous, then a surface of rupture is to be formed at the verge of failure.

Thus, the stability analysis of slopes is based on two aspects, viz.,

1. finding the most severely stressed internal surface and the associated shearing stress (mobilized shear strength) along the surface and
2. finding the shear strength along the above surface.

Finding the most severely stressed surface is possible by adopting a rigorous theory of elasticity or plasticity approach. In a routine analysis, such a rigorous approach is not warranted and is of only academic interest.

Thus, the problem is treated as a two-dimensional one which theoretically demands a long length of slope normal to the section. This situation exists in a majority of the cases. A reasonable shape of the failure surface can be assumed for the required ground condition, and the stability of the soil above such a surface is analysed using the limit equilibrium or limit analysis method.

All the methods of analysis are based on the following assumptions:

1. The shear stress (mobilized shear strength) along the assumed surface is the same at all points (this may not be true in practice since the point that first failed would have experienced large deformation and hence less mobilized shear strength).
2. Coulomb shear strength relationship is applicable (however, the correct shear strength parameters should be used depending on the field condition).
3. The seepage and water pressure are uniform and known all along the surface.
4. Depending on the method of analysis, an assumption regarding the distribution of stresses has to be made to make the problem a determinate one.

## 13.7 INFINITE AND FINITE SLOPES

The term *infinite slope* is given to any slope of great extent with uniform soil conditions at any given depth below the surface. This implies that the soil stratum is not necessarily homogeneous with depth but the strata of different soils are parallel to the surface of the slope.

Nature never provides such an idealized condition, but from a practical standpoint, such a simplification is enough. The usual plane of failure for such slopes is planar parallel to the surface and along a weak layer. Generally, a typical column is taken as representative of the soil mass, and the forces causing the flow are analysed (refer Section 13.8).

The term *finite slope* is given to any slope of finite extent (i.e., with limited height), e.g., slopes of embankments, dams, cuts, canals, etc. While analysing, the entire mass of soil above a slip surface is considered and analysed along with the forces causing the flow.

The stability of infinite and finite slopes is related to earth pressure problems. A small movement along the slope makes the upper portion of the slope to stretch. This movement is sufficient to bring in the active state and causes tension cracks (Fig. 13.5). The lower portion resists the movement and evidently attains the passive state. As the shear strength on the slope surface of the upper portion fully mobilizes, the lower portion is no longer in a position to support the weight of the material above it, and the passive state is fully reached, resulting in the failure of the material.

Earth structures always have their lengths parallel to the bases of the slope, much greater than their width or height. It is feasible to find the extent of soil mass parallel to the base. Although some resistance is available at the ends of the slide, it is not easy to evaluate the same. Hence, only the resistance available at the lower boundary of the slip is considered,

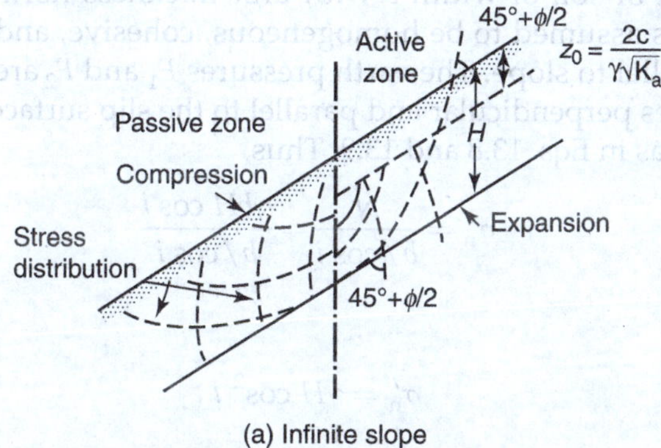


(a) Infinite slope

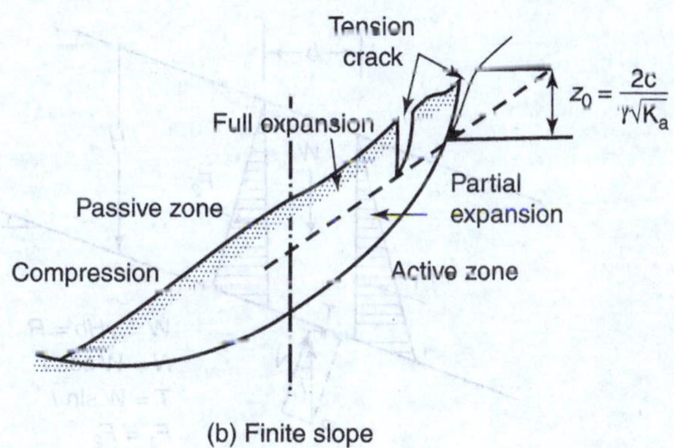


(b) Finite slope

**Fig. 13.5** Earth pressures acting in slopes (*Source*: Hunt, 1986)

ignoring the end effects. This allows the analysis to be treated as a plane strain problem. The analysis is made adopting either a limit equilibrium technique or a limit analysis technique.

The limit equilibrium method is used commonly in stability analysis. This method does not consider the stress–strain relationship of the soil but concentrates only on the equilibrium and yield conditions. Some researchers in geotechnical engineering have approached the problem by using the limit analysis method (e.g., Chen, 1969; Ramiah et al., 1972, etc.) The methods explained in the subsequent sections are based on limit equilibrium analysis.

## 13.8 ANALYSIS OF INFINITE SLOPES

Natural infinite slopes are of heterogeneous materials and are quite often subjected to seepage forces. It is extremely complicated to fit in a method to suit these requirements. For convenience, the slopes may be delineated as cohesive or non-cohesive soil with or without seepage. The engineer has to exercise his judgement and adopt a particular method where a specific condition may fit.

### 13.8.1 Infinite Dry or Moist Cohesive Slope

Consider an element of soil of width $b$ with unit thickness normal to the cross-section (Fig. 13.6). The soil is assumed to be homogeneous, cohesive, and without seepage, and the slip plane is parallel to slope. The earth pressures $F_1$ and $F_2$ are assumed to be equal.

Resolving the forces perpendicular and parallel to the slip surface, the normal and shear stresses are obtained as in Eqs. 13.8 and 13.9. Thus,

$$\sigma_n' = \frac{N}{b/\cos i} = \frac{\gamma bH \cos i}{b/\cos i}$$

That is,

$$\sigma_n' = \gamma H \cos^2 i \tag{13.8}$$

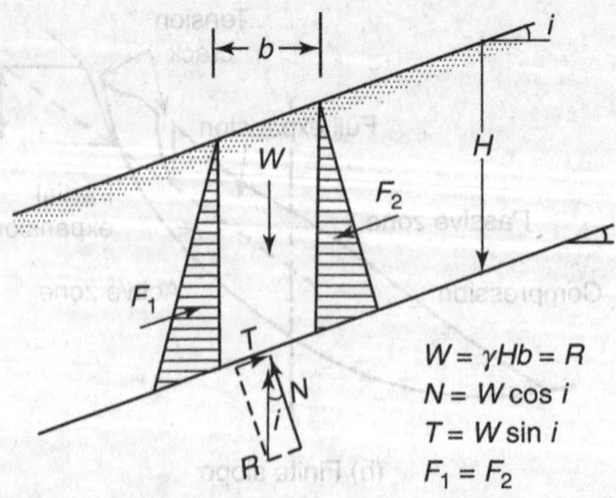


Fig. 13.6 Forces on an element of an infinite dry or moist cohesive slope

and

$$\tau = \frac{T}{b/\cos i} = \frac{\gamma bH \sin i}{b/\cos i}$$

or

$$\tau = \gamma H \sin i \cos i \tag{13.9}$$

Equating Eq. 13.9 to mobilized shear strength,

$$\tau = c_{\mathrm{m}} + \sigma_{\mathrm{n}}' \tan \phi_{\mathrm{m}}$$

$$c_{\mathrm{m}} + \gamma H \cos^2 i \tan \phi_{\mathrm{m}} = \gamma H \sin i \cos i \tag{13.10}$$

Rearranging Eq. 13.10, an expression for critical depth $H = H_c$ for clay stratum is given as

$$H_{\mathrm{c}} = \frac{c_{\mathrm{m}}}{\gamma}\left[\frac{\sec^2 i}{\tan i - \tan \phi_{\mathrm{m}}}\right] \tag{13.11}$$

The strength envelope for a cohesive soil is represented by the line ABC (Fig. 13.7), and line ODC is the line parallel to the slope. The shear strength corresponding to the normal stress OF is BF, which is larger than the mobilized shear strength on the slip represented by FD. Hence, under such stress conditions, no sliding occurs. But sliding would occur when the normal stress is OE, and under this condition there is an increase in shear stress and complete mobilization has taken place. The depth, $H$, at which the shear stress on the slip plane equals the shear strength of the soil is referred to as the *critical depth*, $H_c$. Any depth greater than this will not be stable, and sliding would occur. The factor of safety $F$ can be represented as

$$F = \frac{\tau_{\mathrm{f}}}{\tau}$$

$$F = \frac{c'}{\gamma H \sin i \cos i} + \frac{\tan \phi'}{\tan i} \tag{13.12}$$

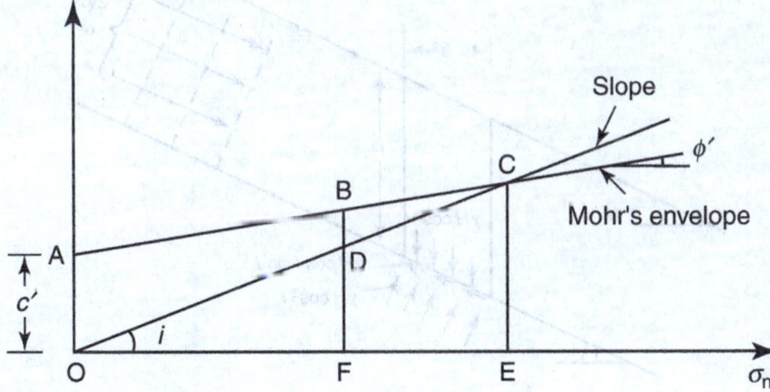


**Fig. 13.7** Limiting slope condition

### 13.8.2 Infinite Cohesive Slopes with Seepage

Let us consider a condition in which the water table is at the surface of the slope and seepage takes place (Fig. 13.8).

The pore water pressure at a depth $H$ is given as $\gamma_w H \cos^2 i$. Thus, the normal and shear stresses are given by Eqs. 13.13 and 13.14, respectively.

$$\sigma_n' = (\gamma - \gamma_w) H \cos^2 i$$

or

$$\sigma_n' = \gamma' H \cos^2 i \tag{13.13}$$

and

$$\tau = \gamma H \sin i \cos i \tag{13.14}$$

But the mobilized shear strength is

$$\tau = c_m + \sigma_n' \tan \phi_m$$

or

$$\tau = c_m + \gamma' H \cos^2 i \tan \phi_m \tag{13.15}$$

Equating Eqs. 13.14 and 13.15

$$c_m + \gamma' H \cos^2 i \tan \phi_m = \gamma H \sin i \cos i \tag{13.16}$$

Rearranging,

$$H_c = \frac{c_m \sec^2 i}{\gamma \tan i - \gamma' \tan \phi_m} \tag{13.17}$$

and

$$F = \frac{c'}{\gamma H \sin i \cos i} + \left(\frac{\gamma'}{\gamma}\right) \frac{\tan \phi'}{\tan i} \tag{13.18}$$

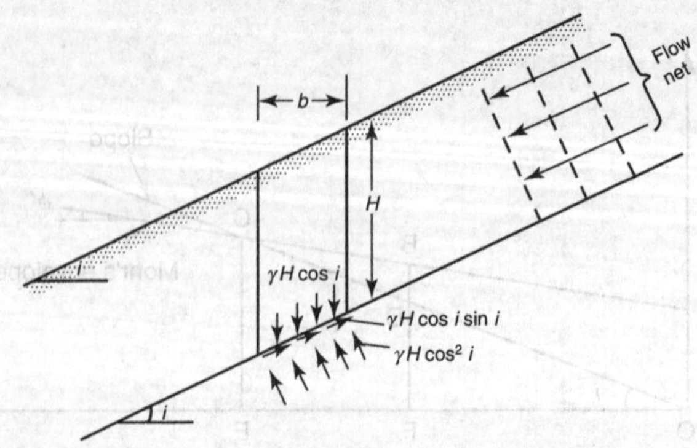


Fig. 13.8 Forces on an element of an infinite cohesive slopes with seepage

### 13.8.3 Infinite Non-cohesive Slopes with Seepage

Consider the same Fig. 13.8 with $c_m = 0$. Thus, the critical slope $i_c$ for this condition is obtained by letting $c_m$ be equal to zero in Eq. 13.16; that is,

$$i_c = \tan^{-1}\left(\frac{\gamma'}{\gamma}\right) \tan \phi_m \tag{13.19}$$

and

$$F = \frac{\tau_f}{\tau}$$

or

$$F = \frac{\gamma' H \cos^2 i \tan \phi'}{\gamma H \sin i \cos i}$$

or

$$F = \left(\frac{\gamma'}{\gamma}\right)\left(\frac{\tan \phi'}{\tan i}\right) \tag{13.20}$$

### 13.8.4 Infinite Dry or Moist Non-cohesive Slope

Considering Fig. 13.8 and letting $c_m$ be equal to zero in Eq. 13.10, we have

$$\gamma H \sin i \cos i = \gamma H \cos^2 i \tan \phi_m$$

or

$$\tan i = \tan \phi_m$$

$$i = \phi_m \tag{13.21}$$

So the maximum angle that could be maintained by a dry slope in a cohesionless soil is the angle of shearing resistance of the soil.

The factor of safety is

$$F = \frac{\gamma H \cos^2 i \tan \phi'}{\gamma H \sin i \cos i}$$

or

$$F = \frac{\tan \phi'}{\tan i} \tag{13.22}$$

## 13.9 ANALYSIS OF FINITE SLOPES

### 13.9.1 Planar Failure Surface

It is not uncommon to find a plane surface in a soil deposit or embankment with a specific plane of weakness. Excavation in a stratified deposit quite often leads to a planar failure along a plane parallel to the strata. In embankment dams with sloping cores, planes of weakness within the bank consist of two or three planar surfaces.

Culmann, in 1866, considered a simple failure mechanism of a slope of homogeneous soil with the plane failure surface passing through the toe of the slope.

Figure 13.9 shows a typical slope with a plane failure surface. The weight of the wedge is given as

$$W = \tfrac{1}{2} hL\gamma \tag{13.23}$$

An expression for $b$ can obtained from geometry as

$$AB = \frac{h}{\sin(i-\theta)} = \frac{H}{\sin i}$$

Thus,

$$h = \frac{H \sin(i-\theta)}{\sin i} \tag{13.24}$$

$$= \tfrac{1}{2} L\gamma H \frac{\sin(i-\theta)}{\sin i} \tag{13.25}$$

The force due to shear strength along plane AC is

$$S = c'L + W \cos\theta \tan\phi'$$

The weight component parallel to the plane AC is $W \sin\theta$. Thus, the factor of safety is

$$F = \frac{c'L + W \cos\theta \tan\phi'}{W \sin\theta}$$

That is,

$$F = \frac{c' + \frac{1}{2}\gamma H[\sin(i-\theta) \sin i] \cos\theta \tan\phi'}{\frac{1}{2}\gamma H[\sin(i-\theta)/\sin i] \sin\theta} \tag{13.26}$$

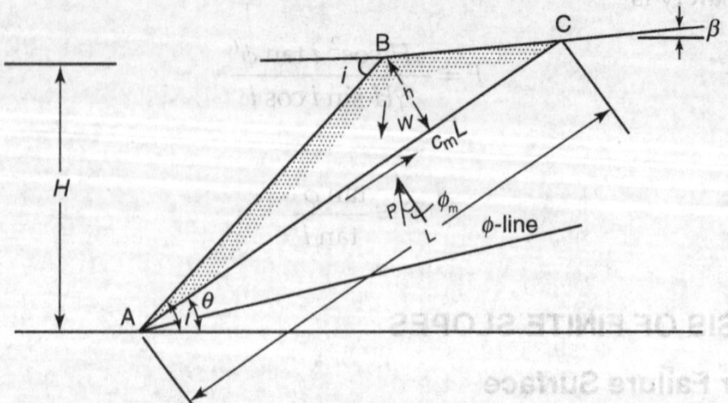


Fig. 13.9 Culmann's slip plane

Now, referring to the force polygon (Fig. 13.10) and using Fig. 13.10, the sine rule is

$$\frac{c_m L}{\sin(\theta-\phi_m)}=\frac{W}{\cos\phi_m}$$

Substituting for $W$, we have

$$\frac{c_m L}{\sin(\theta-\phi_m)}=\frac{\frac{1}{2}L\gamma H\sin(i-\theta)}{\sin i\cos\phi_m}$$

or

$$\frac{c_m}{\gamma H}=\frac{1}{2}\left[\frac{\sin(\theta-\phi_m)\sin(i-\theta)}{\sin i\cos\phi_m}\right] \tag{13.27}$$

where $c_m/\gamma$, known as the stability number. Thus for failure to occur the stability number has to bc at a maximum. Thus, differentiating Eq. 13.27 with respect to $\theta$, making $\phi_m=\phi'$ and equating it to zero, we get

$$\cos(\theta\;\;\phi')\sin(i-\phi)-\sin(\theta-\phi')\cos(i-\theta)=0$$

from which we have

$$\frac{\sin(\theta-\phi')}{\cos(\theta-\phi')}=\frac{\sin(i-\theta)}{\cos(i-\theta)}$$

or

$$\tan(\theta-\phi')=\tan(i-\theta)$$

Thus,

$$(\theta-\phi')=(i-\theta)$$

Representing $\theta=\theta_f$,

$$\theta_f=\frac{1}{2}(i+\phi') \tag{13.28}$$

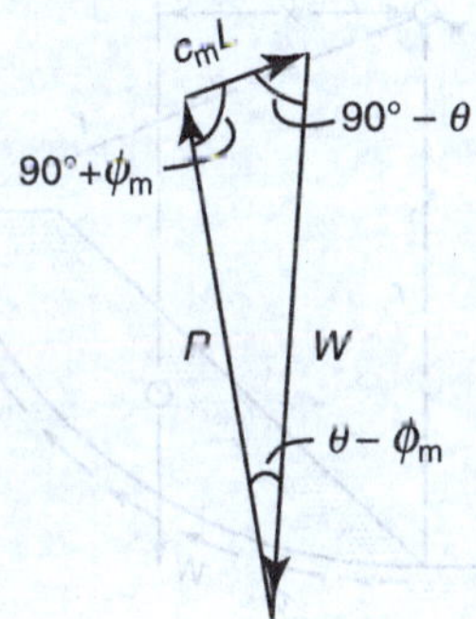


**Fig. 13.10** Forces acting on sliding mass

which is the expression for the angle of inclination of the critical slip plane. This method is suitable for very steep slopes.

This factor of safety is not comparable with the one defined previously.

### 13.9.2 Circular Failure Surfaces

The planar surface discussed above is not the one usually associated with most slope failures. But the actual failure surfaces are curved. The mode of rupture in clay masses is reported to be deep-seated with rotational movements over a curved rupture surface. It is reported that the rupture mass slides down a sliding surface in a definite pattern resembling that of a cycloid. Generally, the failure surfaces have arcs somewhat flatter at the ends and sharper at the centre. Due to large variations in the soil properties and slope characteristics, the failure surface with a general shape could be the best (discussed in the next section)

Based on the studies of failure of the quay wall in the harbour of Gothenburg, Sweden, in 1916, the circular rupture surface was first proposed by Petterson (1955). Further, field investigation by the Swedish Geotechnical Commission justified circular arcs as close approximations of actual slip surfaces in homogeneous and isotropic soil conditions. But some significant deviations may occur if discontinuities exist in the soil.

The methods described under this section consider the circular arc as the shape of the failure surface. In addition to the assumptions made in the limit equilibrium analysis, it is further assumed that the mass of soil above the rupture surface moves as a single rigid mass and the movement is similar to a rigid body motion.

The centre of rotation for a slip circle lies somewhere above the slope. For a given slope, a large number of potential slip circles exist with varying radii and different centres. Some circles may pass through the toe of the slope, while others may be deep-seated and cut the ground surface in front of the toe. A number of slip circles are chosen and safety factors calculated adopting a method of analysis. The slip circle giving the lowest factor of safety is referred to as the *critical slip circle*, along which failure is most likely to occur.

**The $\phi_u = 0$ Analysis.** This is a total stress analysis which may be applied to the case of a newly constructed slope or a cut in a fully saturated condition. As this method was first adopted in Sweden, it is also referred to as the Swedish circular arc method.

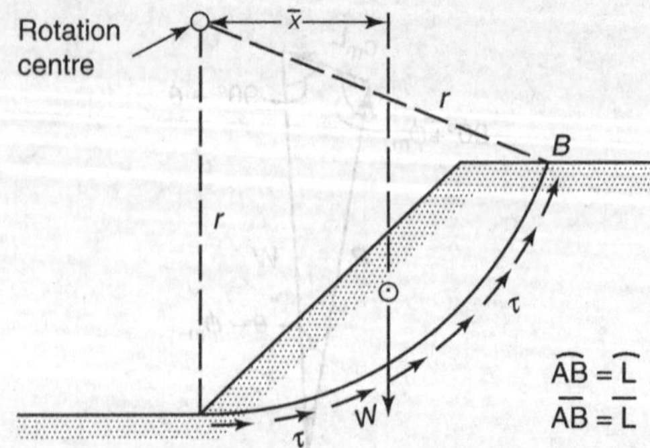


Fig. 13.11 The $\phi u = 0$ analysis

A trial slip circle with radius $r$ is shown in Fig. 13.11. The disturbing force is the weight ($W$) of the segment of soil within the arc (taking the full weight, both above the water level and that submerged below). This force causes an instability due to the moment of the weight ($W$); that is,
Disturbing moment $= W\bar{x}$
where $\bar{x}$ is the moment arm.
This force produces the resisting moment, which is the strength along the surface and is given as
Resisting moment $= c_u \hat{L} r$
The factor of safety is given as

$$F_c = \frac{\text{Resisting moment}}{\text{Disturbing moment}} \tag{13.29}$$

$$F_c = \frac{c_u}{W}\frac{\hat{L}r}{\bar{x}}$$

Alternatively, let $c_m$ be the mobilized shearing strength of the soil along the slip surface necessary for equilibrium; then,

$$W\bar{x} = c_m \hat{L} r$$

or

$$c_m = \frac{W\bar{x}}{\hat{L}r}$$

Therefore,

$$F_c = \frac{\text{Available cohesion}}{\text{Mobilized cohesion}} = \frac{c_u}{c_m} \tag{13.30}$$

or

$$F_c = \frac{c_u}{W}\frac{\hat{L}r}{\bar{x}}$$

Both definitions (Eqs 13.29 & 13.30) give to the formula as the centre of rotation is the same for the assumed slip surface.

If the minimum factor of safety is less than unity, the slope is considered unstable. The minimum factor of safety should be generally equal to or greater than 1.5.

In cohesive soils, due to stretching of the upper portion of the slope, tension cracks are formed, and the development of the slip circle is terminated at the tension crack depth (Fig. 13.12). The depth of the tension crack is given as $z_0 = 2c_u/\gamma$, where $\gamma$ is the unit weight of the soil. No shear strength mobilization is possible along this length; instead, if the crack is filled with water, the disturbing moment due to water pressure has to be taken into account.

If $P_w$ $(= \frac{1}{2}\gamma_w z_0^2)$ is the force due to water pressure, then the disturbing moment is $P_w\bar{y}$; thus,

$$F_c = \frac{c_u \hat{L}_{AG} r}{W\bar{x} + P_w\bar{y}} \tag{13.31}$$

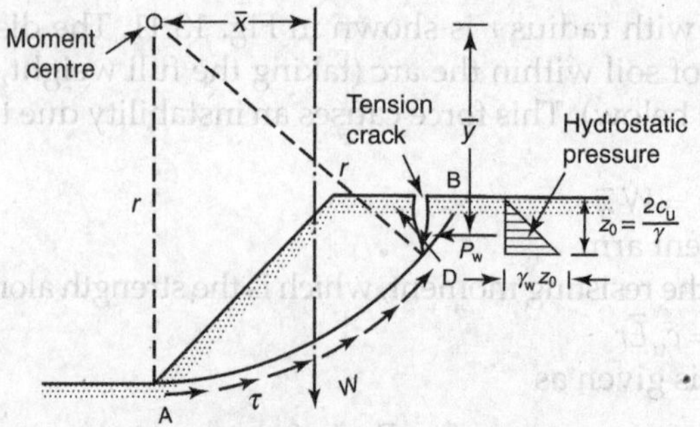


Fig. 13.12 Effect of tension crack on $\phi u = 0$ analysis

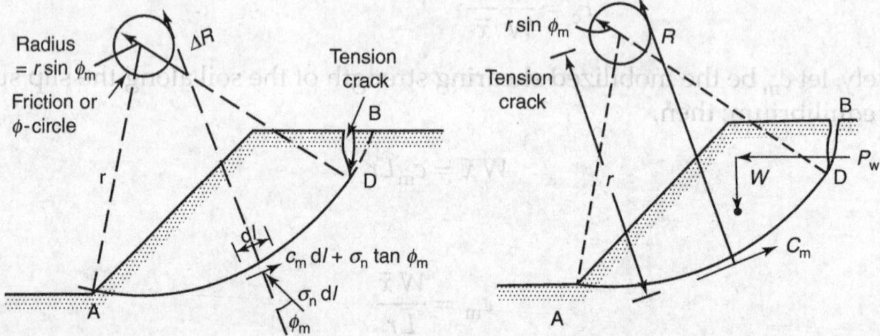


Fig. 13.13 The friction circle method

This method can be extended for multi-layered soils (under undrained condition) and for submerged slopes.

**Friction Circle Method**. In this method, which is based on total stress analysis, both cohesion and the angle of internal friction are taken into account. A circular failure arc from a trial centre is shown in Fig. 13.13.

Consider an element of length $dl$ on the trial slip circle ADB. The reaction $\phi R$ on the elemental length is directed against the direction of motion of the sliding wedge and inclined at an angle $\phi_m$ to the normal, at the point of $\Delta R$. Thus, $\Delta R$ is tangent to a concentric circle of radius $r \sin \phi_m$. This smaller circle of radius $r \sin \phi_m$ is known as the *friction circle* or *$\phi$-circle*.

The forces acting on the element $dl$ are

1. the shearing force due to cohesion, $c_m\, dl$;
2. the shearing force due to friction, $\sigma_n \tan \phi_m\, dl$; and
3. the normal force, $\sigma_n\, dl$.

The reaction $\Delta R$ is the vector sum of the forces $\sigma_n \tan \phi_m\, dl$ and $\sigma_n\, dl$. Consider the sum of all the forces $c_m\, dl$ along the arc AD. Resolve the forces parallel and perpendicular to chord AD. The sum of forces parallel to chord AD is given as

$$C_m = c_m \overline{L}$$

where $\bar{L}$ is the chord length of AD and the sum of components normal to AD is zero. Take the moment of the forces $c_m\,dl$ about the centre, and equating it to the moment due to resultant force $C_m$, we have

$$C_m r_1 = c_m r\hat{L}$$

where $\hat{L}$ is the arc length of AD, or

$$c_m \bar{L} r_1 = c_m r\hat{L}$$

or

$$r_1 = \frac{\hat{L}}{\bar{L}} \tag{13.32}$$

That is, the resultant mobilized cohesive force $C_m$ acts at a distance $r_1$ and parallel to the chord. The equilibrium of the wedge is analysed by considering the following four vectors (Fig. 13.14): the weight $W$, a resultant cohesive force $C_m$, the reaction $R$, and the force due to hydrostatic pressure of water in the tension crack $P_w$.

The weight vector is equal to the area of the wedge times the unit weight of the soil. It acts vertically downwards through the centroid of the wedge. This vector can be drawn to a suitable scale. The force $P_w = \frac{1}{2}\gamma_w z_0^2$. This acts horizontally at a height of $\frac{2}{3}z_0$ from the top ground surface.

Let $Q$ be the resultant of $W$ and $P_w$. Thus, the direction and magnitude of $Q$ are known. The direction of $C_m$ is known, but the magnitude and/or direction of $R$ should be fixed to draw the force polygon.

If $F_\phi$ is assumed, the friction circle for $\phi_m$ equal to $\tan^{-1}(\tan\phi/F_\phi)$ can be drawn. Hence, the direction of $R$ is fixed if we assume that the resultant reaction also makes a tangent with the $\phi$-circle. This is not strictly true, and the resultant $R$ actually makes a tangent with a friction circle with a slightly larger radius (say, $Kr\sin\phi$). The error involved in the assumption is only 20% for deep-seated circles. The value of $K$ may be read out from Fig. 13.15 for a particular central angle.

With the knowledge of the directions of the forces $C_m$ and $R$, the force polygon is completed and the required $C_m$ is measured (Fig. 13.14). The factor of safety with respect to cohesion, $F_c = C/C_m = c\bar{L}/C_m$, is then computed with the assumed, $F_\phi$. The value obtained for $F_c$ is compared with the assumed $F_\phi$. If $F_c \neq F_\phi$, the analysis is repeated until $F_c = F_\phi$.

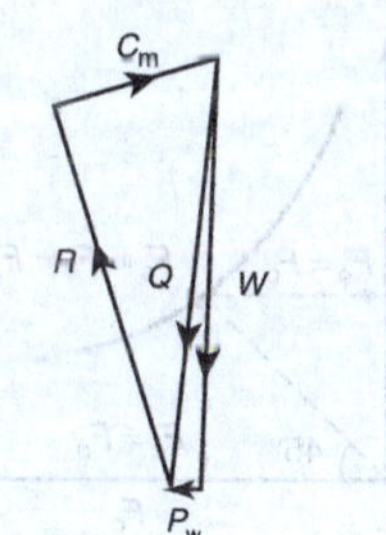


**Fig. 13.14** Force polygon for the $\phi$-circle method

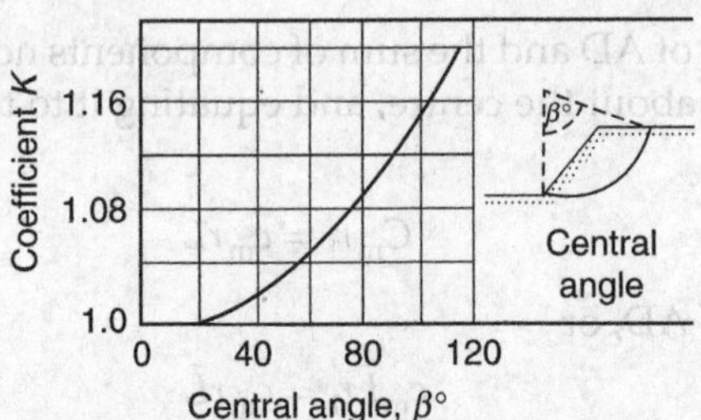


**Fig. 13.15** The coefficient K of the $\phi$-circle assumption

Alternatively, a series of $F_c$ values for assumed $F_\phi$ values are obtained, and the same are plotted versus assumed $F_\phi$ values. A 45° line drawn from the origin intersects the curve at a point whose projection on both the axes gives the value $F = F_c = F_\phi$ (Fig. 13.16). This value of $F$ is nothing but the factor of safety with respect to strength, $F_s$.

The friction circle method is limited to homogeneous soils and a total stress analysis. This can be extended for problems considering effective stress also; however, the method of slices (discussed in the next section) is more adaptable for such problems.

**Taylor's Stability Chart**. Taylor (1937, 1948) proposed stability coefficients for the analysis of homogeneous slopes in terms of total stress based on the friction circle method. Neglecting tension cracks, consider two slopes of different heights with similar slope and failure surfaces (Fig. 13.17). For such geometrically similar failure surfaces, the force diagrams are similar. This shows that the ratio $C_m/W$ is a constant.

But $C_m = c\bar{L}/F_c$ and $W = (\text{Area})\times\gamma$.

The factor $\bar{L}$ and area are functions of height of slope $H$. So

$$C_m \propto \frac{cH}{F_c} \quad \text{and} \quad W \propto H^2\gamma$$

Hence,

$$\frac{C_m}{W} = \frac{cH/F_c}{H^2\gamma}$$

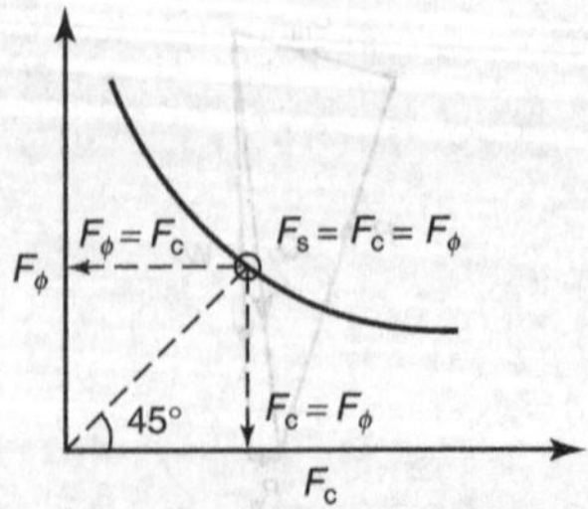


**Fig. 13.16** $F_c$–$F_\phi$ plot to find $F_s$

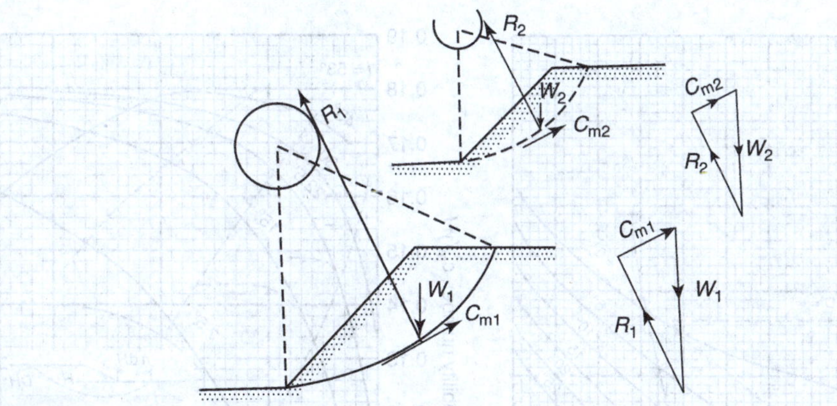


Fig. 13.17 Concept diagrams for Taylor's stability numbers

or

$$\frac{c}{F_c \gamma H} = \frac{c_m}{\gamma H} = S_n \tag{13.33}$$

This coefficient $S_n$, which is non-dimensional and depends only on the geometry of the embankment, is referred to as Taylor's *stability number*.

Values of $S_n$ and slope angle $i$ are related for different values of $\phi_m$ and the depth factor $D$, as shown in Fig. 13.18a and b. This chart can be used to find safety factors with respect to cohesion, friction, or strength.

As Taylor's stability numbers were determined from total stress analysis, the use of these charts for effective stress conditions may lead to a serious error. These charts are applicable at the end of construction and under short-term stability conditions. The curves are often utilized to determine the safe inclination for a given height or the maximum or critical height for a given inclination.

Thus, for the $F_\phi = 1$ condition, if $H_c$ is the critical height for the given slope and soil properties and $H_a$ is the actual height, then the factor of safety with respect to height may be calculated as

$$F_H = H_c / H_a$$

The critical height, $H_c$, of a slope in $c$–$\phi$ soils is expressed as

$$H_c = N_s \frac{c}{\gamma}$$

where $N_s$ is a *stability factor* depending on $\phi$ and $i$. Keeping in view the non-possibility of a base failure unless $\phi < 3°$, a chart (Fig. 13.19) is available (based on Taylor's data) to determine the critical height (Terzaghi and Peck, 1967). Evidently, all the points on the curve correspond to failure along the circles. The stability factor $N_s$ is the reciprocal of Taylor's stability number for a particular case of $F_c = F_\phi = 1.0$.

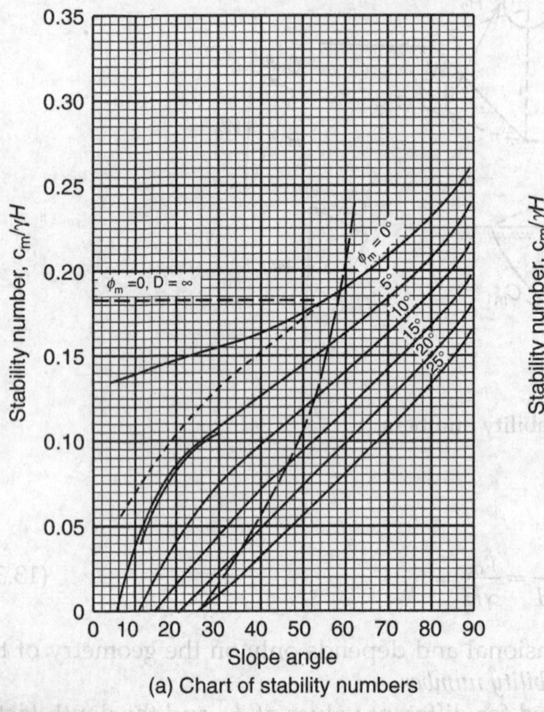

(a) Chart of stability numbers

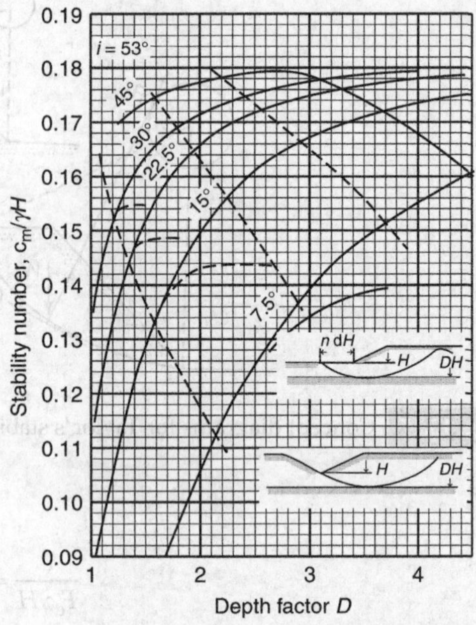

(b) Chart of stability numbers for the case of zero friction angle and limited depth

**Fig. 13.18** Stability charts (*Source:* Taylor, 1937, 1948)

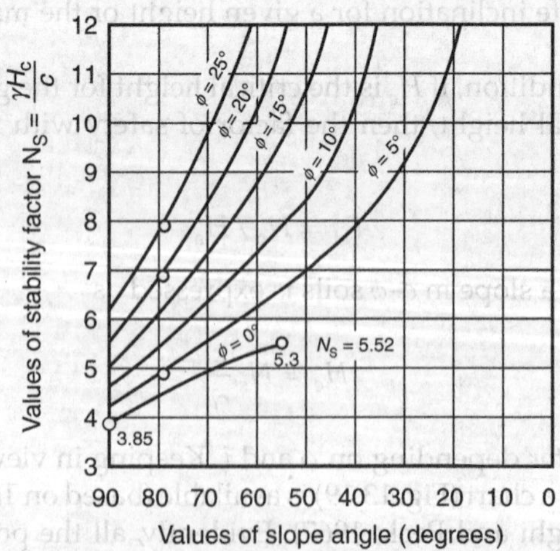


**Fig. 13.19** Relation between slope angle and stability factor (*Source:* Terzaghi and Peck, 1967)

**Location of Critical Circle for Total Stress Analysis**. Fellenius, in 1936, proposed a simple method of finding the centre P of a critical toe circle for a homogeneous slope with $\phi = 0°$ condition. This point P is located with the help of directional angles $\alpha_1$ and $\alpha_2$ as given in Table 13.1 (Fig. 13.20). Jumikis (1962) extended this for $c$–$\phi$ soils and gave a method of locating the locus on which the probable centre of a toe circle may lie. P is a point on the straight line PQ, the locus of the centre of critical slip circles. The point Q has its coordinates $H$ downwards from the toe and $4.5H$ horizontally away.

After obtaining the line PQ, trial centres are taken on PQ, and the factor of safety corresponding to each centre is calculated. These factors of safety are plotted as shown in Fig. 13.21. The point on the extended line PQ corresponding to the lowest factor of safety is thus the critical centre. This method is applicable only to homogeneous soils and provides an approximate location of the critical centre for use in an iterative method.

Taylor's analysis also provides the data necessary to locate the critical centre of the circle for relatively steep slopes (refer Taylor, 1948).

**Table 13.1** Location angles for critical circle (based on $\phi = 0$ analysis)

| Slope $H{:}V$ | Slope angle $i$ (°) | Directional angles (°) | |
|---|---|---|---|
| | | $\alpha_1$ | $\alpha_2$ |
| 0.58:1 | 60 | 29 | 40 |
| 1:1 | 45 | 28 | 37 |
| 1.5:1 | 30.8 | 26 | 35 |
| 2:1 | 26.6 | 25 | 35 |
| 3:1 | 18.4 | 25 | 35 |
| 5:1 | 11.3 | 25 | 35 |

*Source:* Fellenius (1936).

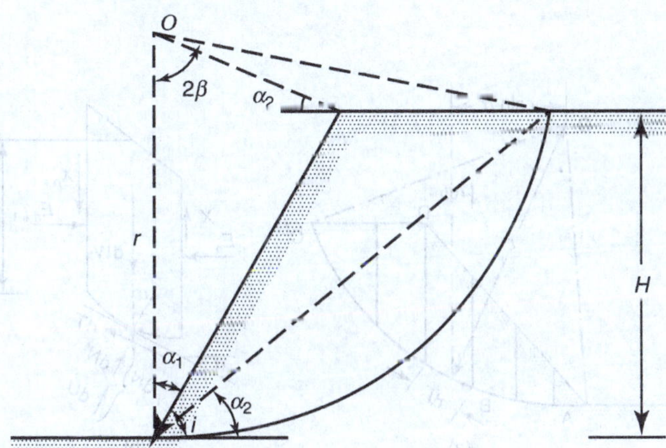


**Fig. 13.20** Location of critical centre for the $\phi = 0$ case (*Source:* Fellenius, 1936)

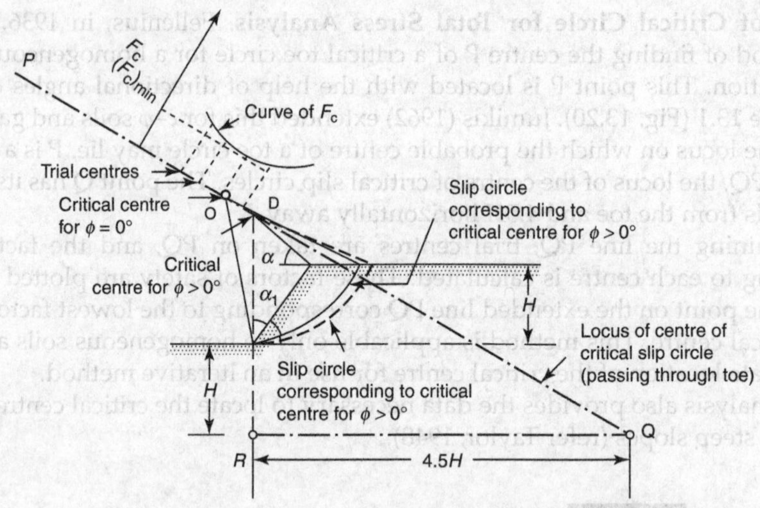


Fig. 13.21 Location of critical centre for the $\phi > 0°$ case

**Method of Slices – Fellenius Method**. The method of slices is a more generalized analysis suitable for different soils and pore water pressure conditions. This method is quite often referred to as effective stress analysis. This method was pioneered by Swedish engineers and more particularly by Fellenius (1936) and Petterson (1955).

The soil profile inside the assumed slip circle is divided into a convenient number of vertical strips or slices, as shown in Fig. 13.22. The base of each slice is assumed to be a planar surface, and other dimensions of a slice are shown in Fig. 13.22b.

The factor of safety is defined with respect to strength. Mutual support between slices comes by way of inter-slice forces.

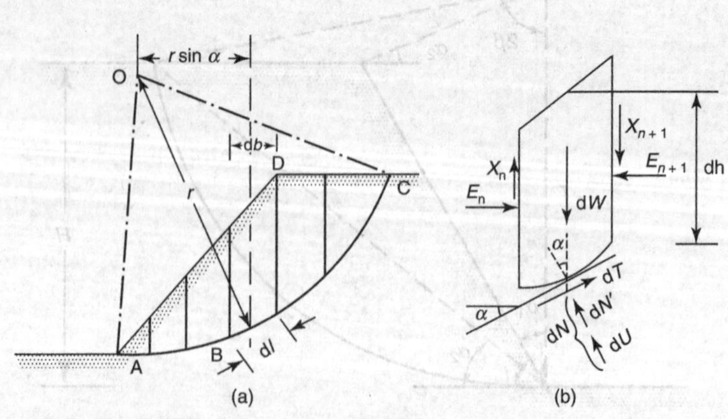


Fig. 13.22 The method of slices

Considering a unit dimension normal to the section, the forces acting on a slice will be as follows:

| | |
|---|---|
| d$W$ | Weight of each slice including any external boundary forces |
| d$N$ | Total normal force at the base of slice = $\sigma_n$ d$l$ |
| d$U$ | Force due to water pressure at the base of slice = $u_w$ d$l$ |
| d$N'$ | Effective normal force at the base of slice = d$N$ − d$U$ = $\sigma'_n$ d$l$ |
| d$T$ | Shear force induced along the base = d$W$ sin $\alpha$ = $\tau$ d$l$ |
| $\alpha$ | Angle of inclination of base of slice |

Also, $E_n$ and $E_n + {}_1$ are the inter-slice normal forces on the $n$th and ($n$ + 1)th faces, and $X_n$ and $X_n + {}_1$ are the inter-slice shear forces on the $n$th and ($n$ + 1)th faces. An assumption for $(E_n - E_n + {}_1)$ and $(X_n - X_n + {}_1)$ has to be made to remove the statical indeterminacy of the problem.

Considering the moment of forces d$T$ and d$W$ about the centre of rotation,

$$\sum \mathrm{d}T\, r = \sum \mathrm{d}W\, r \sin\alpha \tag{13.34}$$

or

$$\sum \tau\, \mathrm{d}l\, r = \sum \mathrm{d}W\, r \sin\alpha$$

but

$$\tau = \frac{\tau_f}{F_s}$$

Therefore,

$$\sum \frac{\tau_f\, \mathrm{d}l}{F_s} = \sum \mathrm{d}W \sin\alpha$$

or

$$F_s = \frac{\sum \tau_f\, \mathrm{d}l}{\sum \mathrm{d}W \sin\alpha}$$

$$F_s = \frac{\sum (c'\, \mathrm{d}l + \sigma'_n\, \mathrm{d}l \tan\phi')}{\sum \mathrm{d}W \sin\alpha}$$

Therefore,

$$F_s = \frac{\sum (c'\, \mathrm{d}l + \mathrm{d}N' \tan\phi')}{\sum \mathrm{d}W \sin\alpha} \tag{13.35}$$

A proper estimation of d$N'$ in each slice will yield the factor of safety $F_s$ for a given failure arc.

In the Fellenius method, the inter-slice forces are assumed to be equal and opposite, i.e., $(E_n - E_n + {}_1) = 0$ and $(X_n - X_n + {}_1) = 0$. An estimation of d$N'$ can be obtained by resolving the forces normal to the base; that is,

$$\mathrm{d}N' = \mathrm{d}W \cos\alpha - u_w\, \mathrm{d}l \tag{13.36}$$

Hence, the factor of safety in terms of effective stress is given as

$$F_s = \frac{\sum[c'\,dl + (dW\cos\alpha - u_w\,dl)\tan\phi']}{\sum dW\sin\alpha} \tag{13.37}$$

From the $\alpha$ values for each slice, $dW \sin\alpha$ and $dW\cos\alpha$ are determined. A minimum factor of safety is obtained by choosing different slip surfaces. This method in general gives conservative values with an error of about 5% to 20% in comparison with more exact methods. For $\phi_u = 0$, the factor of safety reduces to the same Eq. 13.31 for the $\phi = 0$ analysis.

**Method of Slices – Bishop's Simplified Method**. In the Fellenius approach, the omission of side forces violates the equilibrium requirements with respect to translation. Bishop (1955) suggested a method considering all the equilibrium equations. In the exact method, both the inter-slice forces were considered along with the moment equilibrium. He presented a simplified form of the exact method by assuming $(X_n - X_n +_1) = 0$ but $E_n \neq E_n +_1$.

Resolving the forces parallel to the base of the slice,

$$dT = \frac{1}{F_s}(c'\,dl + dN'\tan\phi')$$

Resolving the forces in the vertical direction,

$$dW = dN'\cos\alpha + u_w\,dl\cos\alpha + \frac{c'dl}{F_s}\sin\alpha + \frac{dN'}{F_s}\tan\phi'\sin\alpha$$

Therefore,

$$dN' = \left(dW - \frac{c'\,dl}{F_s}\sin\alpha - u_w\,dl\cos\alpha\right) \Big/ \left(\cos\alpha + \frac{\tan\phi'\sin\alpha}{F_s}\right) \tag{13.38}$$

Substituting for $dl = db\sec\alpha$ and substituting for $dN'$ from Eq. 13.38 in Eq. 13.35, we have

$$F_s = \frac{1}{\sum dW\sin\alpha}\sum\left[\{c'\,db + (dW - u_w\,db)\tan\phi'\}\frac{\sec\alpha}{1+\tan\phi'\tan\alpha/F_s}\right]$$

or

$$F_s = \frac{\sum[\{c'\,db + (dW - u_w\,db)\tan\phi'\}1/m_\alpha]}{\sum dW\sin\alpha} \tag{13.39}$$

where

$$m_\alpha = \cos\alpha\left[\frac{1+\tan\phi'\tan\alpha}{F_s}\right] \tag{13.40}$$

The pore pressure can be taken as a function of the overburden pressure at any point by means of a non-dimensional pore pressure ratio,

$$r_u = \frac{u_w}{\gamma\,dh} \tag{13.41}$$

Therefore, for any slice,

$$r_u = \frac{u_w}{dW/dh}$$

Hence, Eq. 13.39 can be rewritten as

$$F_s = \frac{\sum[\{c'\,db + dW(1-r_u)\tan\phi'\}1/m_\alpha]}{\sum dW \sin\alpha} \tag{13.42}$$

As the factor of safety appears on both sides of Eq. 13.42, an iterative procedure has to be adopted to arrive at its value. For manual use of the equation, the value of $m_\alpha$ may be read from Fig. 13.23 for an assumed $F_s$ value, and a new $F_s$ value can be obtained. Similarly, for a different $F_s$ value on the RHS, the corresponding $F_s$ value on the LHS can be found. From a plot of $(F_s)_{RHS}$ and $(F_s)_{LHS}$, the value of $F_s$ can be determined. This method does not satisfy the force equilibrium condition fully, and the error involved in $F_s$ is insignificant. The simplified method errs on the conservative side (about 3%) with reference to the exact method, and the two methods may not lead to the same critical circle. Because of the repetitive nature of the calculation, the method is more suitable for solution by a computer.

**Effective Stress Stability Charts.** Bishop and Morgenstern (1960) have presented stability coefficients similar to Taylor's coefficients in principle. These coefficients are based on effective stress and the pore pressure ratios. A simple expression for factor of safety, in terms of two stability coefficients, has been suggested; that is,

$$F_s = m - nr_u \tag{13.43}$$

where $m$ and $n$ are the stability coefficients and $r_u$ is the pore pressure ratio. The coefficients $m$ and $n$, and in turn $F_s$, depend on (i) the slope angle, $i$; (ii) the angle of shearing resistance, $\phi'$; (iii) the depth factor, $D$; and (iv) the non-dimensional parameter, $c'/\gamma H$. Charts are

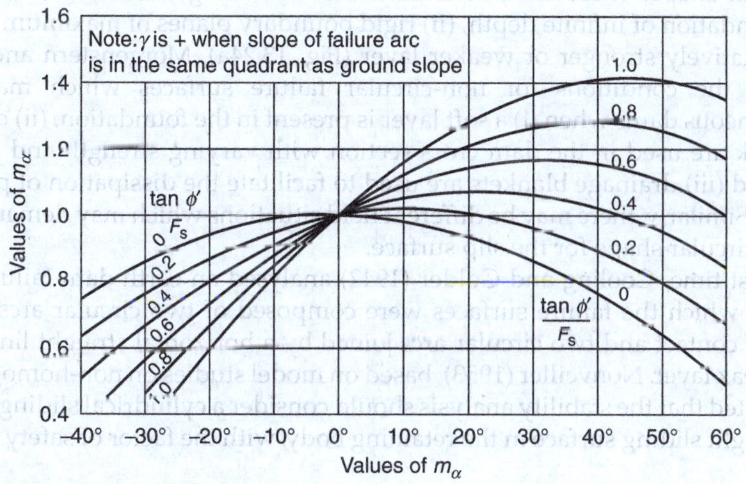


Fig. 13.23 Graph for determination of $m_\alpha$ (*Source:* Lambe and Whitman, 1978)

available for three depth factors, viz., $D = 1.0$, 1.25, and 1.5 (refer Bishop and Morgenstern, 1960). It has been reported that the factor of safety is not very sensitive to changes in the value of the depth factor.

Cousins (1978, 1980) developed stability charts considering homogeneous soil with constant pore pressure ratio, effective shear strength parameters, and tension crack. They are based on the friction circle method. Cousins' stability number is the reciprocal of Taylor's stability number, but in terms of effective cohesion,

$$N_F = \frac{\gamma H F_s}{c'} \tag{13.44}$$

Cousins defined another term, $\lambda_{c\phi}$, to group the soil properties and slope height, where

$$\lambda_{c\phi} = \frac{\gamma H \tan \phi'}{c'} \tag{13.45}$$

Charts are provided for different slope angles and pore pressure ratios, with and without tension cracks and water in tension cracks (refer Cousins, 1978, 1980).

Cousins also provided separate charts to locate the coordinates $X$ and $Y$ for the centre of the critical slip circle for different pore pressure ratios and $c$–$\phi$ values (refer Cousins, 1978, 1980).

It has been shown by Cousins (1980) that a tension crack tends to reduce the factor of safety by 8% to 10%, and the presence of water in the tension crack further reduces the factor of safety by 10%.

### 13.9.3 Non-circular Failure Surfaces

For simple idealized problems, the assumption of a circular failure surface is sufficiently accurate. However, there are many practical cases where the slip surface departs from the simple circular shape. These conditions may arise in homogeneous dams* (Bennett, 1951) with (i) a foundation of infinite depth, (ii) rigid boundary planes of maximum or zero shear, and (iii) a relatively stronger or weaker layer (Fig. 13.24a). Morgenstern and Price (1965) have shown the conditions for non-circular failure surfaces which may prevail in non-homogeneous dams when (i) a soft layer is present in the foundation, (ii) different types of soil or rock are used in the dam cross section with varying strength and pore pressure condition, and (iii) drainage blankets are used to facilitate the dissipation of pore pressures (Fig. 13.24b). Similarly, there may be different field situations which may demand on analysis using a non-circular shape for the slip surface.

For the first time, Cooling and Golder (1942) analysed an earth dam failure (for $\phi = 0°$ condition) in which the failure surfaces were composed of two circular arcs tangential at their point of contact and two circular arcs joined by a horizontal straight line through the centre of a weak layer. Nonveiller (1953), based on model studies on non-homogeneous earth dams, suggested that the stability analysis should consider a cylindrical sliding surface in the core and straight sliding surface in the retaining body, with the factor of safety being defined

*Discussed in Chapter 20.

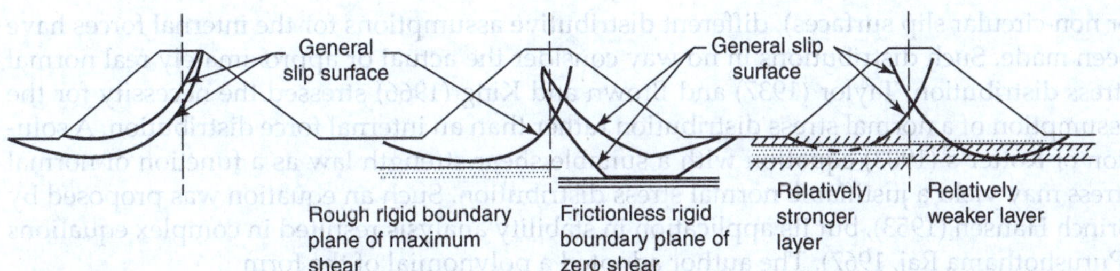


(a) Homogeneous dam and foundation (*Source:* Bennett, 1951)

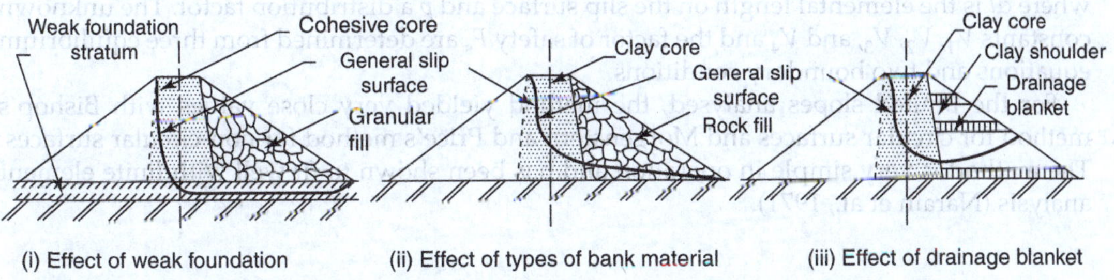


(b) Non-homogeneous dam and foundation (*Source:* Morgenstern and Price, 1965)

**Fig. 13.24** Practical cases for non-circular failure surfaces

as the ratio of the passive pressure of the retaining body necessary for the maintenance of equilibrium to the available passive pressure.

Janbu (1954) was the first to present a stability analysis with a general shape for the slip surface, adopting the requirement of the sum of horizontal forces to be equal to zero as the stability criterion in finding the factor of safety for a given surface. Janbu's solution may be applied safely to elongated shallow slip surfaces, but it errs when applied to deep slip surfaces.

Nonveiller's (1965) method is an extension of Bishop's exact method but with a general shape for the slip surface and an arbitrary point as the moment centre. This method needs a justifiable $X_n$ and $E_n$ distribution.

Morgenstern and Price (1965, 1967) presented a method of slice analysis with a general shape for the slip surface. Two different equations have been formed, one satisfying the no-rotation condition of the slice about its mid-point and the other satisfying the Coulomb–Mohr failure criterion for effective stress. The solutions of these equations have to be obtained by suitable inter-slice force distributions. It is mandatory to use computers to obtain the solution.

In general, a stability analysis problem can be made determinate only when the unknown normal stress is explored by suitable assumptions. Except for the Fellenius (1936) method, all methods were based on the equilibrium equation and thereby placed the burden of indeterminacy on the internal forces. In these methods (based on the method of slices with circular or non-circular slip surfaces), different distributive assumptions for the internal forces have been made. Such distributions in no way consider the actual or approximately real normal stress distribution. Taylor (1937) and Brown and King (1966) stressed the necessity for the assumption of a normal stress distribution rather than an internal force distribution. A solution of Kotter's (1903) equation with a suitable shear strength law as a function of normal stress may yield a justifiable normal stress distribution. Such an equation was proposed by Brinch Hansen (1953), but its application in stability analysis resulted in complex equations (Purushothama Raj, 1967). The author adopted a polynomial of the form

$$\sigma_n = (\mathrm{d}l)^p (V_1 + V_2\,\mathrm{d}l + V_3\,\mathrm{d}l_2)^2 + V_4 \tag{13.46}$$

where $\mathrm{d}l$ is the elemental length on the slip surface and $p$ a distribution factor. The unknown constants $V_1$, $V_2$, $V_3$, and $V_4$ and the factor of safety $F_s$ are determined from three equilibrium equations and two boundary conditions.

For the limited slopes analysed, the method yielded very close values with Bishop's method for circular surfaces and Morgenstern and Price's method for non-circular surfaces. The method is very simple in operation and has been shown to fit well with finite element analysis (Narain et al., 1971).

## 13.10 SELECTION OF SHEAR STRENGTH PARAMETERS AND STABILITY ANALYSIS

Construction of earth structures involves stability requirements in four cases: (i) during construction, (ii) at the end of construction, (iii) during the working stage (intermediate time), and (iv) under long-term condition. During and end of construction fall under the same category. Table 13.2 gives the four cases and the corresponding stability analysis.

**Table 13.2** Selection of strength parameter in slope stability

| Situation | Preferred method | Comment |
|---|---|---|
| 1. End of construction with saturated soil; construction period short compared to consolidation time | $\phi_u = 0°$ analysis | $c'-\phi'$ analysis permits check during construction using actual pore pressures |
| 2. Long-term stability | $c'-\phi'$ analysis with pore-pressures given by equilibrium groundwater conditions | |
| 3. End of construction with partially saturated soil; construction period short compared to consolidation time | Either method: $c_u$, $\phi_u$ from unconsolidated undrained tests or $c'$, $\phi'$ plus estimated pore pressures | $c'-\phi'$ analysis permits check during construction using actual pore pressures |
| 4. Stability at intermediate times | $c'-\phi'$ analysis with estimated pore pressures | Actual pore pressures must be checked in field |

*Source:* Lambe and Whitman (1979).

## 13.11 SLOPE PROTECTION MEASURES

Slopes that are susceptible to sliding should be protected so that the area will be safe. Slopes which have failed recently are likely to fail under long-term condition.

Slopes have been protected by adopting some successful techniques. In general, the corrective or protective measures involve (i) reducing the mass or loading which contributes to sliding (ii) improving the shearing strength along the anticipated zone of failure, and (iii) providing certain materials which will provide resistance to movement.

The protective measure to be adopted depends on different field conditions, viz., the type of soil in the slope, the volume or depth of the soil involving in sliding, the groundwater conditions, assessment of the complete area which may require stabilization, the space available to undertake corrective measures, topographical conditions prevailing in the area, and the possible changes that could occur due to the vibratory measure undertaken.

Some of the protective measures which could be adopted are given in Fig. 13.25 (McCarthy, 1982). Figure 13.25a shows different techniques for reducing the weight of the moving mass. When a base failure is anticipated, a term may be provided near the toe (Fig. 13.25b). If a zone near the toe is susceptible to erosion, a protective rock-fill blanket followed by a riprap can be provided (Fig. 13.25c). Shearing resistance of the soil is reduced due to high groundwater and excess pore-water pressure. This could be avoided by lowering the groundwater or intercepting the surface water. Figure 13.25d shows such a situation. Driven piles are sometimes used to keep the moving part intact with the original ground (Fig. 13.25e). Sometimes driven piles, sheet piling, and construction of retaining wall help by providing lateral support and increasing the resistance of slopes to sliding (Fig. 13.25f). If it is intended to construct a building in the vicinity of a slope, the procedure given in Fig. 13.25g may be adopted.

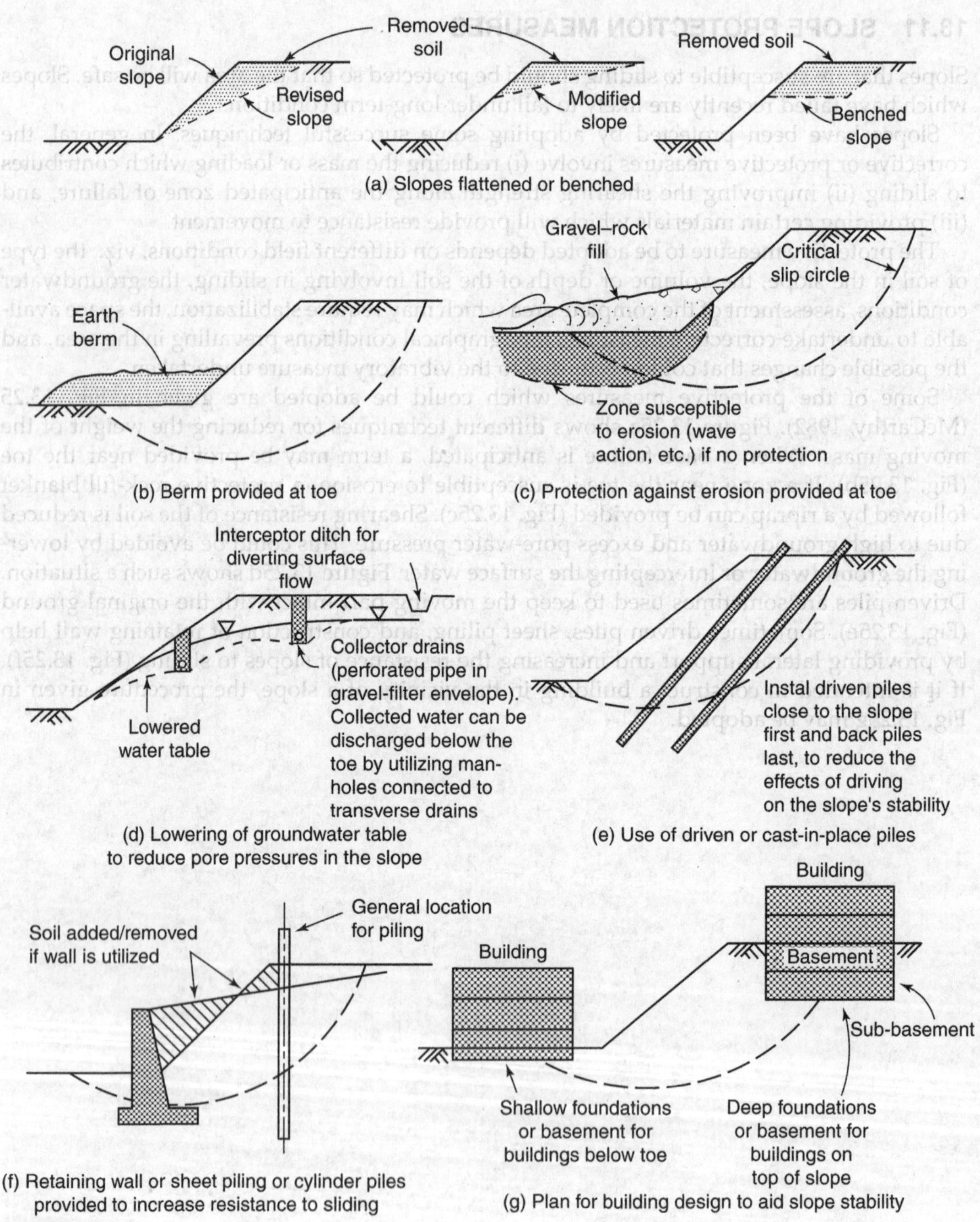


**Fig. 13.25** Methods to improve and protect slope stability (*Source:* McCarthy, 1982)

## WORKED EXAMPLES

**Example 13.1** An infinitely long slope having an inclination of 26° in an area is underlain by firm cohesive soil ($G = 2.72$ and $e = 0.50$). There is a thin, weak layer of soil 6 m below and parallel to the slope surface ($c = 25$ kN/m$^2$, $\phi' = 16°$). Compute the factor of safety when the slope is dry. If groundwater flow could occur parallel to the slope on the ground surface, what factor of safety would result?

***Solution***

When the slope is dry, the factor of safety can be obtained from Eq. 13.12; that is,

$$F = \frac{c'}{\gamma_d H \sin i \cos i} + \frac{\tan \phi'}{\tan i}$$

Here,

$$\gamma_d = \frac{G\gamma_w}{1+e} = \frac{2.72 \times 9.807}{1+0.5} = 17.8 \text{ kN/m}^3$$

Substituting

$$i = 26°,\ c' = 25 \text{ kN/m}^2,\ \phi' = 16°,\ \gamma_d = 17.8 \text{ kN/m}^3, \text{ and } H = 6 \text{ m}$$

$$F = \frac{25}{17.8 \times 6 \times \sin 26° \cos 26°} + \frac{\tan 16°}{\tan 26°} = 1.18$$

When there is seepage of water, the factor of safety can be obtained from Eq. 13.18; that is,

$$F = \frac{c'}{\gamma H \sin i \cos i} + \left(\frac{\gamma'}{\gamma}\right)\frac{\tan \phi'}{\tan i}$$

Here,

$$\gamma = \frac{G+e}{1+e}\gamma_w = \frac{2.72+0.5}{1+0.5} \times 9.807 = 21.05 \text{ kN/m}^3$$

or

$$\gamma' = 21.05 - 9.807 = 11.24 \text{ kN/m}^3$$

Hence,

$$F = \frac{25}{21.05 \times 6 \times \sin 26° \cos 26°} + \left(\frac{11.24}{21.05}\right)\frac{\tan 16°}{\tan 26°} = 0.816$$

**Example 13.2** A finite slope has an inclination of 48° with a horizontal ground surface. The height of the slope is 15 m, and the details of the soil are $c = 26$ kPa, $\phi = 18°$, and $\gamma = 17.2$ kN/m$^3$. Compute the factor of safety assuming a plane rupture surface. Adopt Culmann's method.

***Solution***

Here, $i = 0$ as the ground is horizontal (Fig. 13.9). The angle of the inclination of the critical slip surface is given as

$$\theta_f = \frac{1}{2}(i+\phi) = \frac{1}{2}(48^\circ + 18^\circ) = 33^\circ$$

$$L = \frac{H}{\sin \theta_f} = \frac{15}{\sin 33^\circ} = 27.54 \text{ m}$$

$$b = \frac{H \sin(i-\theta_f)}{\sin i} = \frac{15 \sin(48^\circ - 33^\circ)}{\sin 48^\circ} = 5.22 \text{ m}$$

$$W = \frac{1}{2}bL\gamma = \frac{1}{2} \times 5.22 \times 27.54 \times 17.2 = 1236.3 \text{ kN}$$

$$S = c'L + W \cos \theta_f \tan \phi' = 26 \times 27.54 + 1236.3 \times \cos 33^\circ \tan 18^\circ$$

$$S = 1052.93 \text{ kN}$$

$$T = W \sin \theta_f = 1236.3 \sin 33^\circ = 673.34 \text{ kN}$$

$$F = \frac{S}{T} = \frac{1052.93}{673.34} = 1.56$$

**Example 13.3** A 12 m deep cut with 1:1 slope is made in a layered clay deposit with the following details:

| Depth (m) | Soil | Cohesion (kPa) |
|---|---|---|
| 0–5 | Very soft clay | 10 |
| 5–8 | Medium stiff clay | 50 |
| 8–15 | Stiff clay | 100 |
| 15 | Rock | – |

Assume the average unit weight of the three layers to be 18 kN/m$^3$. Compute the factor of safety against sliding corresponding to the rotation centre shown in Fig. 13.26.

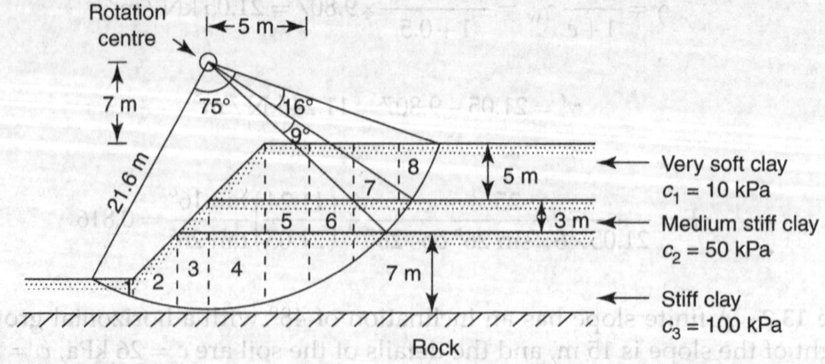


Fig. 13.26

***Solution***

As a hard surface is available near the toe of the slope, a base failure should be anticipated.

The slip surface is drawn tangential to the rock base.

In order to calculate the overall moment produced by the sliding mass, the mass of soil above the slip surface is divided into slices and the moments of individual slices is taken about the rotation centre.

| Slice No. | Weight of slice (Area × 1 × unit wt. of soil) (kN) | Lever arm (m) | Moment (kN-m) |
|---|---|---|---|
| 1 | 63.45 | 8.2 | –520.3 |
| 2 | 303.04 | 5.0 | –1515.2 |
| 3 | 627.65 | 1.0 | –627.7 |
| 4 | 896.36 | 3.0 | 2689.1 |
| 5 | 966.97 | 7.0 | 6768.8 |
| 6 | 833.60 | 11.0 | 9169.6 |
| 7 | 621.76 | 15.0 | 9326.5 |
| 8 | 234.39 | 18.4 | 4312.7 |

Driving moment = 32266.7 – 2663.2 = 29603.5 kN-m

Resisting moment = $r(c_1 l_1 + c_2 l_2 + c_3 l_3)$

Since,

$$l_1 = r\theta_1 = 21.6 \times 16° \times \frac{\pi}{180°} = 6.03 \text{ m}$$

$$l_2 = r\theta_2 = 21.6 \times 9° \times \frac{\pi}{180°} = 3.39 \text{ m}$$

$$l_3 = r\theta_3 = 21.6 \times 75° \times \frac{\pi}{180°} = 28.27 \text{ m}$$

Therefore, the resisting moment = 21.6 (10 × 6.03 + 50 × 3.39 + 100 × 28.27) = 66026.88 kN m

$$F = \frac{\text{Resisting moment}}{\text{Driving moment}}$$

$$F = \frac{66026.88}{29603.5} = 2.23$$

**Example 13.4** The bank of a canal is 9.4 m in height and has a face inclination of 30°. The material is homogeneous silty clay of unit weight 20 kN/m$^3$, cohesion 30 kPa, and angle of shearing resistance 20°. For the trial slip circle shown in Fig. 13.27, find the factor of safety with respect to cohesion by using the friction circle method, if $F_\phi = 1.50$.

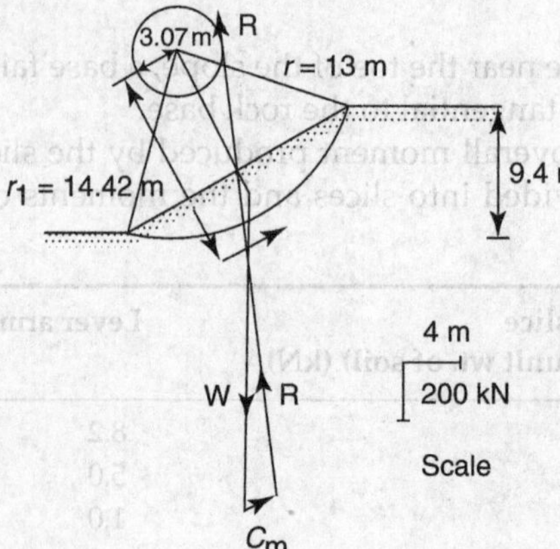


Fig. 13.27

***Solution***

$$\phi_m = \tan^{-1}\left(\frac{\tan\phi'}{F_\phi}\right) = \tan^{-1}\left(\frac{\tan 20°}{1.50}\right) = 13.64°$$

Radius of friction circle $r_0 = r \sin\phi_m = 13 \sin 13.64° = 3.07$ m

Area of sliding mass $= \frac{2}{3} \times 18 \times 3.8 = 45.6\ \text{m}^2$

Weight of driving mass = 45.6 × 1 × 20 = 912 kN
Central angle = 88°
Therefore,

$$\hat{L} = 13 \times 88° \times \frac{\pi}{180°} = 19.97\ \text{m}$$

$$r_1 = \frac{\hat{L}}{L} = 13 \times \frac{19.97}{18} = 14.42\ \text{m}$$

$C_m$ is read from the force polygon as 125 kN. Therefore,

$$c_m = \frac{C_m}{L \times 1} = \frac{125}{18 \times 1} = 6.94\ \text{kPa}$$

Therefore,

$$F_c = \frac{c}{c_m} = \frac{30}{6.94} = 4.32$$

**Example 13.5** A 60° sloping embankment has a height of 6.5 m. The embankment soil possesses the following properties: $\gamma = 18\ \text{kN/m}^3$, $\phi = 28°$, and $c = 20$ kPa. Determine the factor of safety with respect to strength. Use Taylor's chart.

***Solution***
Mobilized friction

$$\phi_m = \tan^{-1}\left(\frac{\tan\phi}{F_\phi}\right)$$

Assuming $F_\phi = 1.6$,

$$\phi_m = \tan^{-1}\left(\frac{\tan 28°}{1.6}\right) = 18.4°$$

From Fig. 13.18a, for $i = 60°$ and $\phi_m = 18.4°$, the value of $c_m/\gamma H$ is obtained by interpolation as 0.1007. Therefore,

$$c_m = 0.1007 \times 18 \times 6.5 = 11.78 \text{ kPa}$$

and

$$F_c = \frac{c}{c_m} = \frac{20}{11.78} = 1.698$$

Now take $F_\phi = 1.65$, then

$$\phi_m = \tan^{-1}\left(\frac{\tan 28°}{1.65}\right) = 17.86°$$

Again, after interpolation, $c_m/\gamma H$ for $i = 60°$ and $\phi_m = 17.86°$ is obtained as

$$\frac{c_m}{\gamma H} = 0.1018$$

Therefore,

$$c_m = 0.1018 \times 18 \times 6.5 = 11.91 \text{ kPa}$$

and

$$F_c = \frac{20}{11.9} = 1.68$$

$$F_\phi = 1.67, \quad \phi_m = 17.66°, \quad \text{and} \quad \frac{c_m}{\gamma H} = 0.1022$$

Trying again with $c_m = 0.1022 \times 18\ 6.5 = 11.96$ kPa,

$$F_c = \frac{20}{11.96} = 1.672$$

$$F_c = F_\phi = 1.671$$

Therefore, the factor of safety with respect to strength $F_s = 1.671$.

**Example 13.6** It is proposed to construct a 10 m high highway embankment with the following soil properties: $c = 18.8$ kN/m$^2$, $\gamma = 17$ kN/m$^3$, and $\phi = 10°$. What is the inclination required for the embankment if the design $F_c = 1.5$ and $F_\phi = 1.0$?

*Solution*

Stability number, $S_n = \dfrac{c_m}{\gamma H} = \dfrac{c}{F_c \gamma H}$

Substituting the values, we have

$$S_n = \frac{18.8}{1.5 \times 17 \times 10} = 0.0737$$

As $F_\phi = 1.0$, the mobilized friction angle $\phi_m = \phi = 10°$.

From Fig. 13.18a, for $S_n = 0.0737$ and $\phi_m = 10°$, the slope angle is read as 29°. Therefore, the required inclination of the embankment, $i = 29°$.

## POINTS TO REMEMBER

13.1 Causes of failure of slopes may be external or internal. External causes are those which produce an increase in the shearing stresses at unaltered shearing resistance of the material. Internal causes are those which lead to a slide without any change in surface conditions which involve unaltered shearing stresses in the slope material.

13.2 The short-term instability of a slope is due to non-availability of sufficient time for the dissipation of pore water pressure. The long-term condition of a slope is one in which the pore water pressure gradually adjusts itself in the long run and shows values corresponding to a certain groundwater condition.

13.3 Types of land slides may be falls, rotational slides, compound slides, translational slides, flows, and multiple and complex slides. Rates of landslides are recognized as creep, pre-failure movements, movements during slide, and post-failure movements.

13.4 Factor of safety of a slope is defined as the ratio of shearing strength available along a slip surface to the total mobilized shearing strength. Factor of safety is also defined in certain cases with respect to cohesion, friction, or height of a slope.

13.5 Any slope of great extent with uniform soil conditions at any given depth below the surface is termed as *infinite slope*. Any slope of finite extent, i.e., with limited height, is termed as *finite slope*. All natural slopes are infinite slopes: slopes of embankments, dams, cuts, canals, etc., are finite slopes.

13.6 Slip surfaces are generally curved, deep-seated, somewhat flatter at the ends, and sharper at the centre. Analysis with general shape as the slip surface is cumbersome. For all practical purposes, most of the analyses use a circular slip surface.

13.7 $\phi_u = 0$ analysis is a total stress analysis which may be applied to the case of a newly constructed slope or a cut in a fully saturated condition.

13.8 Friction circle method (assumes a circular slip surface) is based on total stress analysis; both cohesion and the angle of internal friction are considered with friction completely mobilized. Then, the factor of safety is defined with respect to cohesion.

13.9 Fellenius' method of slices (assumes a circular slip surface) is a more generalized analysis suitable for different soils and pore water pressure conditions and based on effective stresses.

13.10 Bishop's method of slices (assumes a circular slip surface) considers all equilibrium conditions, including side forces on slices.

13.11 Taylor's stability chart for homogeneous soil is based on the friction circle method and total stresses and provides a factor of safety for a given slope angle and mobilized friction angle.

## QUESTIONS

### Objective Questions

13.1 State whether the following are true or false:

1. The maximum possible slope angle in a granular soil is equal to the friction angle of the soil.
2. Gravitational forces tend to cause instability in natural slopes.
3. The term *infinite slopes* is given to earth masses of varying inclinations and non-uniform soil conditions of unlimited extent.
4. Tension cracks do not significantly affect the safety factor of a slope.
5. The most critical circle is the one along which failure is most likely.

13.2 Total stress method of stability analysis may be applied to find the factor of safety in the case of a newly cut slope in

(a) Fissured over-consolidated saturated clay
(b) Non-fissured over-consolidated saturated clay
(c) Normally consolidated saturated clay
(d) Partially saturated expansive clay

13.3 Total stress method of stability analysis may be applied to find the factor of safety of an embankment dam under end-of-construction condition

(a) Method of slices with $\phi > 0°$ condition
(b) $\phi_u = 0°$ analysis
(c) Friction circle method (ignoring the effect of tension cracks)
(d) Friction circle method with tension crack

13.4 In stability analysis, mobilized shear strength is referred to as

(a) Maximum shear stress
(b) Applied shear stress
(c) Developed cohesion only
(d) Developed friction only

13.5 Bishop's simplified method of slices satisfies

(a) All the statical equilibrium conditions
(b) Only the vertical force equilibrium condition
(c) Only the moment equilibrium condition
(d) All the conditions except the horizontal force equilibrium conditions

13.6 Inclination of a clay slope

(a) Can be greater than the angle of shearing resistance
(b) Cannot be greater than the angle of shearing resistance

(c) Cannot be greater than the angle of repose
(d) Cannot be greater than 45°

13.7 Identify the incorrect statement. The following factors cause instability of slopes because of increased stresses:
(a) Removal of part of the slope by excavation
(b) Shock caused by earthquake or blasting
(c) Water pressure in cracks
(d) Swelling of clays by adsorption of water

13.8 The effective stress method of stability analysis is used
(1) For analysing the long-term stability of slopes
(2) For analysing dense, moderately compressible soil material
(3) For analysing the stability of compressible soils where some drainage of water takes place when a load is applied

Of these statements,
(a) 1, 2, and 3 are correct (b) 1 and 2 are correct
(c) 2 and 3 are correct (d) 3 and 1 are correct

13.9 A base failure is likely to occur when
(a) $\phi_u > 0°$ and $\beta < 53°$ (b) $\phi_u = 0°$ and $\beta > 53°$
(c) $\phi_u = 0°$ and $\beta < 53°$ (d) $\phi_u > 0°$ and $\beta > 53°$
where $\beta$ is the slope angle and $\phi_u$ is the undrained friction angle

13.10 In order to use Taylor's stability chart for sudden draw-down condition, the weighted friction angle, $\phi_w$, should be equal to
(a) $(\gamma'/\gamma_w)\phi_u$ (b) $(\gamma_w/\gamma')\phi_u$ (c) $(\gamma'/\gamma_{sat})\phi_u$ (d) $(\gamma_w/\gamma')\phi_u$

## Descriptive Questions

13.11 Derive an equation for the factor of safety of an infinite slope in a cohesionless soil, assuming that seepage is
1. emerging from the slope at an angle $\alpha$, which is less than the slope angle $i$, or
2. flowing parallel to the slope at a certain depth from the surface.

13.12 Explain the various causes of the failure of earth slopes.

13.13 Explain why a high factor of safety of 2.5 to 3 for shallow foundations and a low factor of safety of 1.1 to 1.5 for stability of slopes are adopted.

13.14 It is often stated that refinements in stability analysis by using different methods is generally not as significant as the correct use of the shear parameters of the soil. Discuss the validity of this statement.

13.15 Discuss the different investigations needed to effect corrective measures in a landslide area.

13.16 Distinguish between the total and effective stress approaches of stability analysis. Indicate the advantages and shortcomings of the total stress approach.

13.17 Explain the various types of failures of finite slopes indicating the situations in which they are likely to occur.

## EXERCISE PROBLEMS

13.1 A 21° infinite slope consists of an uniform 5 m thick layer of sandy clay. At 5 m depth, a shale ledge runs parallel to the surface. A laboratory investigation on the sandy clay revealed the following properties: $c = 20$ kPa, $\phi = 15°$, $\gamma = 18$ kN/m$^3$. Compute the factor of safety against sliding on the shale and ledge if (i) no water exists at the top of the shale and (ii) the water level is at the surface of the slope.

13.2 A sub-surface investigation on a 12° natural slope revealed the presence of bedding planes dipping toward the slope at an angle of 40°. A 60° cut slope is to be excavated to a depth of 8 m as shown in Fig. 13.28. Estimate the factor of safety of the slope. The shear strength parameters of the soil in the bedding plane are, $c = 15$ kN/m$^2$ and $\phi = 28°$. The average unit weight of the soil, on the bedding plane and above, is 18.5 kN/m$^3$.

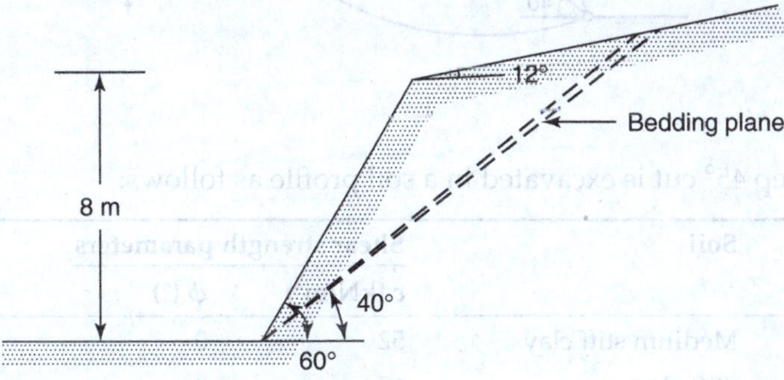


Fig. 13.28

13.3 A 45° cut was made in a clayey silt soil with $c' = 12$ kPa, $\phi = 30°$, and $\gamma = 19.5$ kN/m$^3$. A sub-surface exploration revealed the presence of a thin soft clay with $c = 13$ kPa and $\phi = 0°$, at a depth of 18 m from the ground surface. Estimate the factor of safety of the slope against sliding along the composite slip surface, as shown in Fig. 13.29.

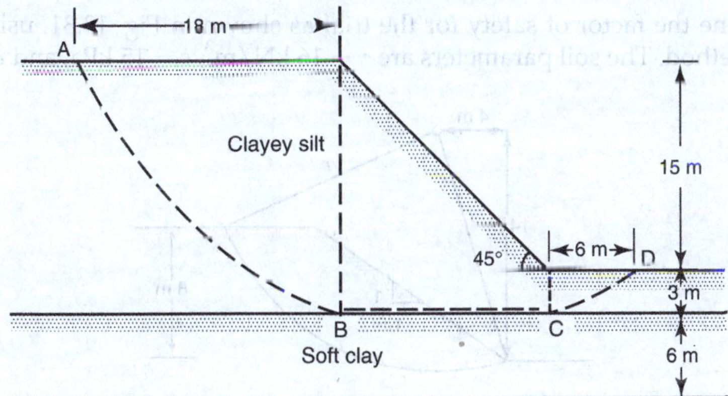


Fig. 13.29

13.4 A cutting in clayey soil is shown in Fig. 13.30. The undrained shear strength parameters are $c_u = 48\ \text{kN/m}^2$ and $\phi_u = 0°$. The unit weight of the soil is $20\ \text{kN/m}^3$. Compute the factor of safety against the slip surface shown when (i) no tension crack is formed, (ii) a tension crack exists with no water in the crack, and (iii) the tension crack is completely filled with water.

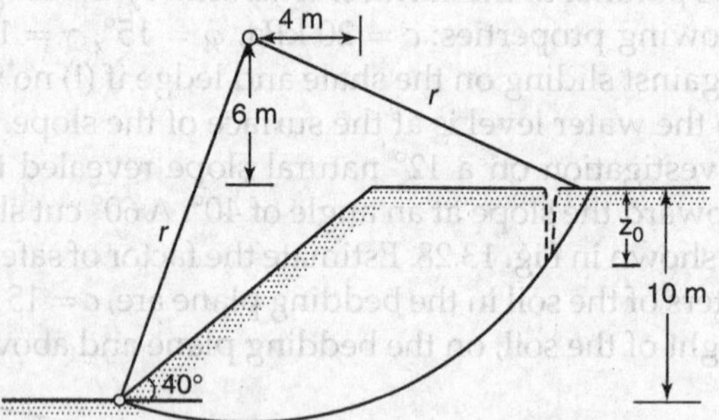


Fig. 13.30

13.5 A 15 m deep 45° cut is excavated in a soil profile as follows:

| Depth | Soil | Shear strength parameters | | Unit weight |
|---|---|---|---|---|
| | | c (kN/m²) | $\phi$ (°) | (kN/m³) |
| 0–2.5 | Medium stiff clay | 52 | 0 | 17.2 |
| 2.5–8.6 | Stiff clay | 60 | 0 | 18.5 |
| 8.6–18.2 | Very stiff (firm) clay | 73 | 0 | 19.1 |
| 18.2 | Shale (rock) | – | – | – |

1. Compute the factor of safety with respect to base failure assuming the centre of failure circle to be above the mid-point of the slope. Also, verify by various centres and radii.
2. Check the results, using Taylor's stability chart for an average cohesion.

13.6 Determine the factor of safety for the trial as shown in Fig. 13.31, using the friction circle method. The soil parameters are $\gamma = 16\ \text{kN/m}^3$, $c = 15$ kPa, and $\phi = 28°$.

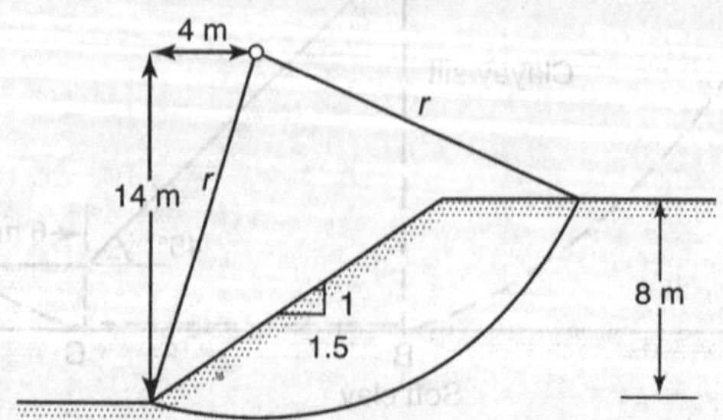


Fig. 13.31

13.7 An embankment is made of soil having a cohesion of 50 kPa, an angle of internal friction of 22°, and a unit weight of 19 kN/m$^3$. Locate the centre of rotation (for $\phi = 0$) by the Fellenius method and determine the factor of safety along a slip circle passing through the toe. Use the friction circle method.

13.8 A 15 m high clay embankment with a 45° slope has the following parameters: $c = 22$ kPa, $\phi = 0°$, and $\gamma = 18.2$ kN/m$^3$. What will be the factor of safety of this slope if a rock stratum exists 15 m beneath the toe elevation?

13.9 An excavation has to be made with an inclination of 35° in a soil with $c' = 28$ kPa, $\phi' = 26°$, and $\gamma = 18$ kN/m$^3$. What is the maximum height to which the excavation can be made if $F_c = 1.25$?

13.10 A canal is excavated to a depth of 5 m below the ground level through a soil stratum having the shear strength $\tau = c + \sigma_n \tan 15°$, $c = 16$ kN/m$^2$, void ratio $e_0 = 0.72$, and specific gravity $G = 2.70$. The bank of the canal has a slope of 1:1. Compute the factor of safety of the slope with respect to cohesion when the canal runs full. If it is suddenly and completely drawn down, what will be the change in the factor of safety?

13.11 A 10 m deep silty clay cut has an inclination of 45° and the following soil parameters: $c_u = 30$ kPa, $\phi_u = 10°$, and $\gamma = 18$ kN/m$^3$. Estimate the critical height of the slope in this soil.

13.12 A proposed cutting in a $c - \phi$ soil will be 15 m deep with a slope of 1*V*:2.5*H*. The soil has an average unit weight of 18.6 kN/m$^3$ and an average pore pressure ratio $r_u$ of 0.45. The shear strength parameters of the soil under different conditions are

$$c_u = 85 \text{ kN/m}^2, \quad \phi_u = 0°$$
$$c' = 12 \text{ kN/m}^2, \quad \phi' = 26°$$

Estimate the factor of safety against (i) immediate shear failure and (ii) long-term shear failure.

13.13 For the soil slope and trial slip surface shown in Fig. 13.32, estimate the factor of safety adopting Bishop's simplified method. A preliminary approximate calculation for the slip surface, based on the Fellenius method, gave a factor of safety of 2.

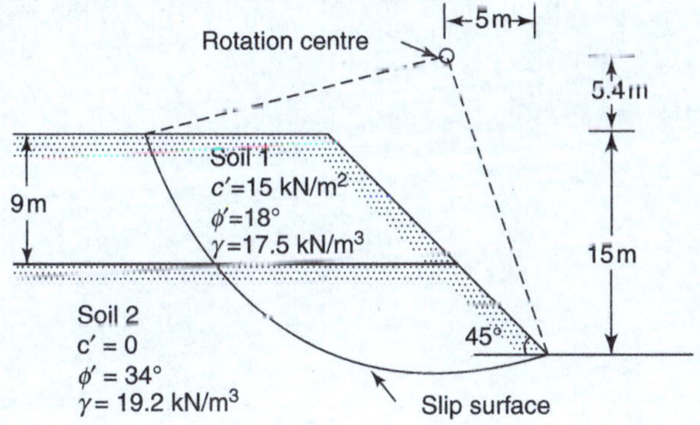


Fig. 13.32

13.7 An embankment is made of soil having a cohesion of 50 kPa, an angle of internal friction of 22°, and a unit weight of 19 kN/m³. Locate the centre of rotation (for $\phi = 0$) by the Fellenius method and determine the factor of safety along a slip circle passing through the toe. Use the friction circle method.

13.8 A 15 m high clay embankment with a 45° slope has the following parameters: $c = 22$ kPa, $\phi = 0°$ and $\gamma = 18.2$ kN/m³. What will be the factor of safety of this slope if a rock stratum exists 15 m beneath the toe elevation?

13.9 An excavation has to be made with an inclination of 35° in a soil with $c' = 25$ kPa, $\phi' = 26°$ and $\gamma = 18$ kN/m³. What is the maximum height to which the excavation can be made if $F_s = 1.25$?

13.10 A canal is excavated to a depth of 5 m below the ground level through a soil stratum having the shear strength $\tau = c + \sigma_n \tan 15°$; $c = 16$ kN/m², void ratio $e_0 = 0.72$, and specific gravity $G = 2.70$. The bank of the canal has a slope of 1:1. Compute the factor of safety of the slope with respect to cohesion when the canal runs full. If it is suddenly and completely drawn down, what will be the change in the factor of safety?

13.11 A 10 m deep silty clay cut has an inclination of 45° and the following soil parameters: $c_u = 30$ kPa, $\phi_u = 10°$, and $\gamma = 18$ kN/m³. Estimate the critical height of the slope in this soil.

13.12 A proposed cutting in a $c - \phi$ soil will be 15 m deep with a slope of 1V:2.5H. The soil has an average unit weight of 18.6 kN/m³ and an average pore pressure ratio $r_u$ of 0.45. The shear strength parameters of the soil under different conditions are

$$c_u = 35 \text{ kN/m}^2, \quad \phi_u = 0°$$

$$c' = 12 \text{ kN/m}^2, \quad \phi' = 26°$$

Estimate the factor of safety against (i) immediate shear failure and (ii) long-term shear failure

13.13 For the soil slope and trial slip surface shown in Fig. 13.32, estimate the factor of safety adopting Bishop's simplified method. A preliminary approximate calculation for the slip surface, based on the Fellenius method, gave a factor of safety of 2.

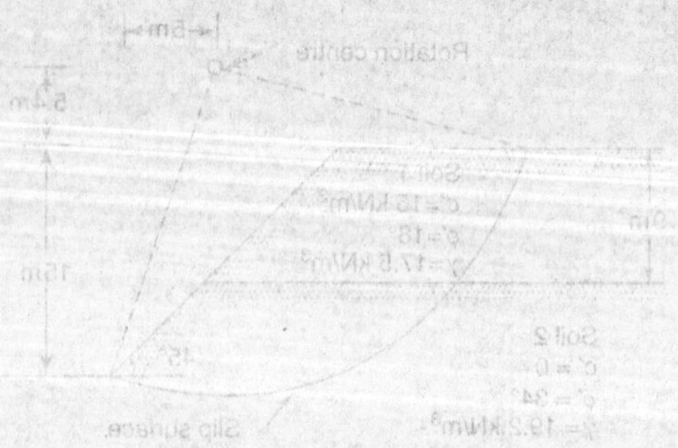

Fig. 13.32

# 14

# Bearing Capacity of Soils

## CHAPTER HIGHLIGHTS

Modes of failure – Bearing capacity theories: Terzaghi's bearing capacity theory – Effect of soil compressibility – Effect of water table – Foundation pressures – Special loading and ground conditions: eccentric load, inclined load, stratified soils, partially saturated soils and desiccated soils – Other bearing capacity theories: Modified bearing capacity formulae, Skempton's bearing capacity theory, Meyerhof's bearing capacity theory, Brinch Hansen's bearing capacity theory – Bearing capacity from building codes – Permissible settlements – Allowable bearing pressure – Bearing capacity from field tests – Bearing capacity from building codes – Factors affecting bearing capacity

## 14.1 INTRODUCTION

A foundation is that part of the structure which is in direct contact with the ground and transmits the load of the structure to the ground. It includes the soil or rock of the earth's crust or any special part of the structure which serves to transmit the loads into the soil or rock. The main purpose of the transmissions of load can be satisfied by a particular type of foundation that takes into account the properties of the supporting soil. A foundation functions properly only if the supporting soil performs properly. Consequently, the structural support is actually being provided by a soil–foundation system. This combination of soil and foundation (now referred to as soil–structure interaction) cannot be separated. Although engineers are aware of this relationship, it is common practice to consider the structure to be sound and to attribute the failure of the foundation to the failure of the supporting soil.

Foundations may be grouped as shallow or deep foundation depending on the depth of installation of foundation.

## 14.2 BEARING CAPACITY

The bearing capacity of a soil is the maximum intensity of loading which the soil can carry without being detrimental to the normal functioning of a foundation.

### 14.2.1 Bearing Capacity Criteria

The bearing capacity of a soil is based on the stability requirement of a foundation.

The two criteria on which the bearing capacity of a soil depends are shear strength and settlement.

1. The shear strength criterion is that the shear failure of the foundation or bearing capacity failure should not occur.
2. The settlement criterion is that the foundation shall not settle more than the safe or tolerable magnitude of settlement such that the anticipated settlement due to the applied pressure on the soil should not be detrimental to the stability of the foundation.

These two criteria are independent and have to be dealt with separately. The bearing capacity value to be decided for the design requirement of a foundation is the smaller of these two values based on the above two criteria. This smaller value of bearing capacity is referred to as allowable soil pressure (dealt in detail in Section 14.10).

### 14.2.2 Factors Affecting Bearing Capacity

Keeping in view the above two criteria, the following factors directly or indirectly affect the bearing capacity of a soil.

1. Type of soil, i.e., homogeneous, layered, expansive, etc., and its physical and engineering properties
2. Initial stress condition of the soil due to pre-history and due to the existing structure in and around the proposed foundation
3. Location of groundwater in the soil and its fluctuations with time
4. Type of foundation, i.e., shallow or deep, and other factors such as shape, size, and rigidity condition of the foundation
5. Depth and location of foundation
6. Allowable settlement of the foundation which shall not be detrimental to the functioning of the foundation
7. Natural calamities such as earth quake, floods, heavy wind, etc., of the region where the structure has to be located

## 14.3 MODES OF SHEAR FAILURE

The load–deformation relationship of a soil is not unique as it is in more homogeneous materials, like steel. Nevertheless, it can be generalized to a certain extent even in soils. A stratum subjected to loading through a footing will depict a reasonable elastic relationship up to a certain percentage of the ultimate strength. This phase of deformation is attributed to the densification of the stratum. Further increase in load causes a rapid increase in deformation. This increased rate of yielding is due to a combination of decrease in void ratio and the lateral displacement. Subsequent loading leads to excessive deformation and ultimate shear failure of the soil stratum. This pressure which has caused a shear failure of the supporting soil is usually referred to as the ultimate bearing capacity of the foundation. Figure 14.1 typifies a load–settlement relationship for the case of a footing on a hypothetical stratum. Different types of soils with varied conditions show wide variation in load–settlement relationships.

Three principal modes of shear failure have been identified, based on the model tests of strip footings on sand (Vesic, 1973).

### 14.3.1 General Shear Failure

General shear failure, usually associated with dense or stiff soils of relatively low compressibility, is said to occur when a continuously well-defined slip surface develops on one or both sides of the footing and extends from the edge of the footing to the soil surface. As the pressure is increased towards the ultimate value ($q_f$) the state of plastic equilibrium is attained, initially in the soil around the edges of the footing, which then gradually spreads downwards and outwards. Then, plastic equilibrium develops throughout the soil above the failure surface. The failure is reflected by the heave of the ground surface on both sides of the footing. The load–settlement curve is linear up to a substantial percentage of the ultimate load but thereafter shows a rapid yielding till the load intensity approaches the ultimate value (Fig. 14.2a).

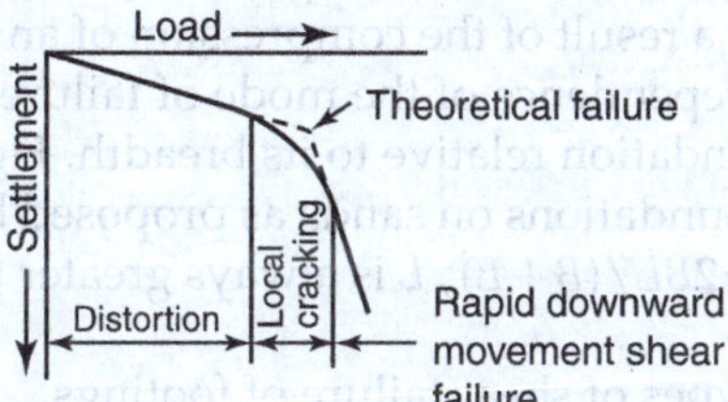


**Fig. 14.1** Load–settlement curve (*Source:* Vesic, 1973)

**Fig. 14.2** Modes of bearing failure (*Source:* Vesic, 1973)

### 14.3.2 Local Shear Failure

In case of local shear failure, usually associated with medium dense or medium stiff soils, the slip surface extends from the edges of the footing to a certain length (approximately up to the boundary of Rankine's passive state) but does not reach the ground surface. The soil undergoes a significant compression beneath the footing and only partial development of plastic equilibrium takes place. The heave of the ground is comparatively less and no tilting of foundation is expected (Fig. 14.2b).

The load–settlement curve displays a lesser degree of linearity. Because of high compressibility, a large settlement is characterized in the load–settlement curve, and no well-defined ultimate load is observed.

### 14.3.3 Punching Shear Failure

Punching shear failure, usually associated with loose or soft soils, is said to occur when there is compression beneath the footing accompanied by shearing in the vertical direction around the edges of footing. There is little horizontal strain and no apparent heave of soil around the footing (Fig. 14.2.c).

The load–settlement curve shows a relatively large settlement and the ultimate load is not well-defined. Although deformations are considerable, sudden collapse or tilting failures are not common. Punching shear failure is also possible in soils, if the foundation is located at a considerable depth or as a result of the compression of an underlying soft layer.

Vesic (1973) showed the dependence of the mode of failure on the compressibility of the soil and the depth of the foundation relative to its breadth. Figure 14.3 depicts the relationship for mode of failure of foundations on sands as proposed by Vesic (1973).

In the expression for $B^* = 2BL/(B+L)$, $L$ is always greater than $B$ for square and circular footings.

Table 14.1 illustrates the types of shear failure of footings.

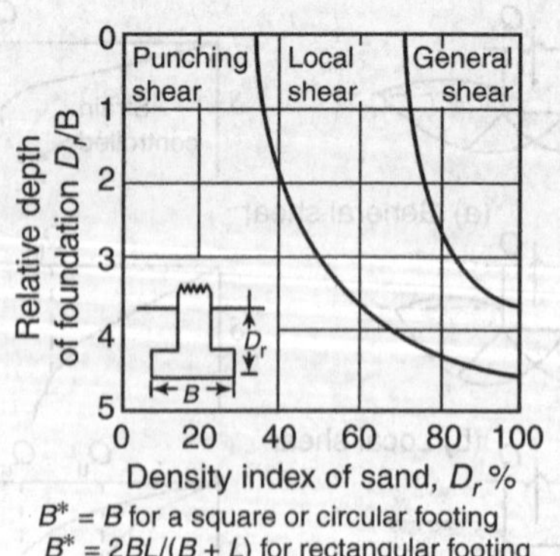


Fig. 14.3 Density index versus relative depth of foundation (*Source:* Vesic, 1973)

**Table 14.1a** Conditions for Modes of Failure (Kaviraj, 1988)

| Sl. no. | Foundation condition | Types of shear failure |
|---|---|---|
| 1. | Footings on the surface or at shallow depths in very dense sand. | General shear failure |
| 2. | Footings on saturated and normally consolidated clay under undrained loading. | General shear failure |
| 3. | Footings at deeper depth in dense sand. | Punching shear failure |
| 4. | Footings on the surface or at shallow depths in loose sand. | Punching shear failure |
| 5. | Footings on very dense sand loaded by transient dynamic load. | Punching shear failure |
| 6. | Footings on very dense sand underlain by loose sand or soft clay. | Punching shear failure |
| 7. | Footings on saturated and normally consolidated clay under drained loading. | Punching/local shear failure |

## 14.4 TERZAGHI'S BEARING CAPACITY THEORY

Most theories which are in wide use have emerged from Prandtl's theory of plastic equilibrium. Prandtl's theory considers the deformation or penetration effects of hard objects on soft materials. Based on this basic principle, the bearing capacity problem is considered as a rigid footing penetrating into a soft homogeneous material. The implied assumptions in adopting this theory are as follows: (i) the soil is isotropic and homogeneous, (ii) the soil is weightless, and (iii) the footing is long with a smooth base.

Figure 14.4 shows a generalized failure mechanism for a strip footing. This is the basic mechanism suggested by Prandtl. The condition of the footing (smooth or rough) and varied boundary wedge angles ($\alpha$ and $\beta$) have been adopted by researchers. Prandtl studied the effect of a long, narrow metal tool bearing against a smooth metal mass with $\alpha = 45° + \phi/2$ and $\beta = 45° - \phi/2$ and the curved portion of the slip surface is assumed as logarithmic spiral. Based on model studies, most researchers (e.g., Ko and Davidson, 1973) agreed that the curve fits to a logarithmic spiral for $\phi > 0°$ and a circle for $\phi = 0°$.

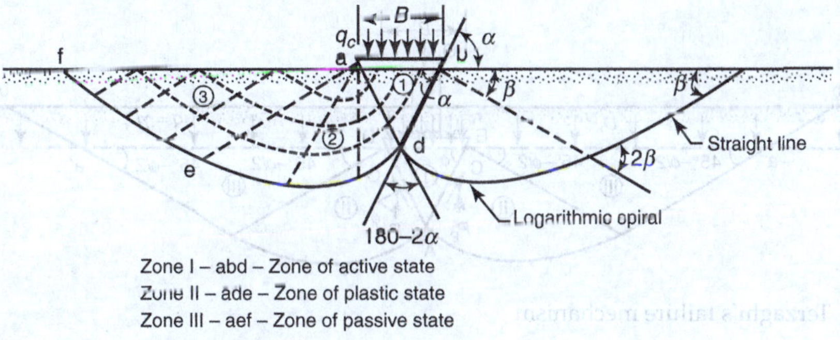


**Fig. 14.4** Generalized failure mechanism

Zone I is assumed to remain intact and at plastic state. The load is transmitted through this soil-wedge of Zone I. In Zone II, plastic flow develops with the formation of slip planes, as shown by broken lines in Fig. 14.4. Zone III is at passive state with plane slip surface.

The penetrating wedge (Zone I) pushes aside Zones II and III, and the shearing residence mobilizes along the logarithmic spiral and straight line segment. Based on this premise, the ultimate bearing capacity for a surface footing*

$$q_{nf} = c \cot \phi[\exp(\pi \tan \phi) \tan^2(45°+\phi/2)-1] \quad (14.1)$$

For the general case, it is necessary to consider the overburden pressure as surcharge $q$, otherwise when $c = 0$, the bearing capacity of the weightless soil would be zero. Ressner (1924) extended Prandtl's work by including the condition that the bearing area is located below the surface of the soil and the overburden is represented by a surcharge equation for ultimate bearing capacity given as

$$q_f = c \cot \phi[\exp(\pi \tan \phi) \tan^2(45°+\phi/2)-1]+q[\exp(\pi \tan \phi) \tan^2(45°+\phi/2)] \quad (14.2)$$

In order to consider the effect of the self-weight of the soil, an additional term must be added.

Terzaghi (1943) applied the developments of Prandtl–Ressner to soil foundation problems. He identified a foundation as shallow if the depth $D_f$ of the foundation is less than or equal to the width $B$ of the foundation. The assumed failure mechanism is shown in Fig. 14.5.

Terzaghi assumed a strip footing with a rough base placed at the depth $D_f$ on a homogeneous and isotropic soil medium. In the analysis, the shearing resistance of the soil above the base (*ab* and *a'b'* in Fig. 14.5) of the footing is not considered, but the effect of soil-weight above the base is considered by superimposing an equivalent surcharge intensity $q = \gamma D_f$. The development of the failure surface in the soil is governed by the general shear failure.

The soil immediately beneath the foundation forms a wedge (Zone I) which moves downwards (Fig. 14.5). The movement of the wedge forces the soil aside and produces two zones of shear (Zones II and III) consisting of radial shear zone (Zone II) immediately adjacent to

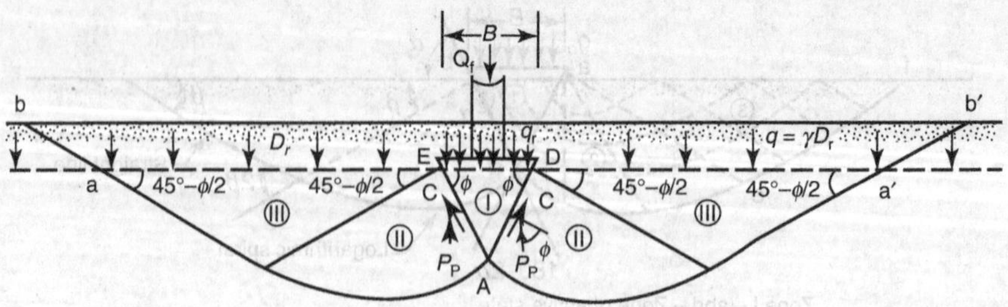


**Fig. 14.5** Terzaghi's failure mechanism

*This is defined as the ultimate beating capacity less overburden pressure, $\gamma D_f$, i.e., $q_f - \gamma D_f$, where $D_f$ is the depth of footing.

the wedge and linear shear (Zone III) beyond the radial. Here, Zone I is considered to be at Rankine active state, Zone II under radial shear, and Zone III at Rankine's passive state.

This situation can be compared with the passive case of a retaining wall. The penetrating wedge is in equilibrium when the downward load is resisted by forces on the inclined faces of the wedge. Cohesion and the resultant passive pressures contribute to the resistance along the inclined faces. For equilibrium in the vertical direction, at the verge of failure, $\sum V = 0$, thus,

$$q_f B = 2P_p + ADc \sin\phi$$

substituting $AD = B/2\cos\phi$,

$$q_f B = 2P_p + Bc \tan\phi \tag{14.3}$$

The value of $P_p$ has been represented as the vector sum of three components, viz., (i) cohesion, (ii) surcharge, and (iii) weight of the soil. Terzaghi assumed the method of superposition to be valid and presented the unit ultimate bearing capacity

$$q_f = cN_c + qN_q + \tfrac{1}{2}\gamma BN_\gamma \tag{14.4}$$

where $N_c$, $N_q$, and $N\gamma$ are non-dimensional bearing capacity factors and functions only for the angle of shearing resistance $\phi$,

$$N_c = \cot\phi[N_q - 1] \tag{14.5}$$

$$N_q = \frac{\exp[2(3\pi/4 - \phi/2)\tan\phi]}{2\cos^2[\pi/4 + (\phi/2)]} \tag{14.6}$$

$$N_\gamma = \frac{1}{2}\tan\phi\left(\frac{K_p}{\cos^2\phi} - 1\right) \tag{14.7}$$

where $K_p$ is the coefficient of passive pressure from Zones II and III.

The bearing capacity factors are given in Table 14.1.

Terzaghi (1943) stated that the value of the wedge angle $\alpha$ may lie between $\phi$ and $45° + \phi/2$. Purushothama Raj et al. (1972) using limit theorem (based on kinematical considerations) analysed the bearing capacity of shallow foundation varying the boundary wedge angles. The individual minimum values of $N_c$, $N_q$, and $N\gamma$ for critical wedge angles of $\alpha$ and $\beta$ based on the analysis is presented in Table 14.2. As it was generally observed that Terzaghi's values are high, the same behaviour has been highlighted by the limit analysis approach also. The critical wedge angles $\alpha$ and $\beta$ in Table 14.2 clearly show that $\beta$ angles were almost equal to $45° - \phi/2$, but $\alpha$ angles were found to vary significantly; however, their effect on bearing capacity factors was found to be negligible.

**Table 14.1b** Bearing capacity factors from Terzaghi's theory and the Bureau of Indian Standards

| $\phi$ (°) | $N_c$ | | $N_q$ | | $N_\gamma$ | |
|---|---|---|---|---|---|---|
| | Terzaghi | IS | Terzaghi | IS | Terzaghi | IS |
| 0 | 5.70 | 5.14 | 1.00 | 1.00 | 0.00 | 0.00 |
| 5 | 7.30 | 6.14 | 1.60 | 1.57 | 0.50 | 0.45 |
| 10 | 9.60 | 8.35 | 2.70 | 2.47 | 1.20 | 1.22 |
| 15 | 12.90 | 10.90 | 4.40 | 3.94 | 2.50 | 2.65 |
| 20 | 17.70 | 14.83 | 7.40 | 6.40 | 5.00 | 5.39 |
| 25 | 25.10 | 20.72 | 12.70 | 10.66 | 9.70 | 10.88 |
| 30 | 37.20 | 30.14 | 22.50 | 18.40 | 19.70 | 22.40 |
| 35 | 57.80 | 46.12 | 41.40 | 33.30 | 42.40 | 48.03 |
| 40 | 95.70 | 75.31 | 81.30 | 64.20 | 100.40 | 109.41 |
| 45 | 172.30 | 138.88 | 173.30 | 134.88 | 297.50 | 271.76 |
| 50 | 347.50 | 266.89 | 415.10 | 319.07 | 1153.20 | 762.89 |

**Table 14.2** Critical wedge angles and bearing capacity factors

| $\phi$ (°) | $N_c$ | | | $N_q$ | | | $N_\gamma$ | | |
|---|---|---|---|---|---|---|---|---|---|
| | $\alpha$ (°) | $\beta$ (°) | Value | $\alpha$ (°) | $\beta$ (°) | Value | $\alpha$ (°) | $\beta$ (°) | Value |
| 40 | 60.50 | 25.00 | 74.10 | 60.50 | 25.00 | 63.05 | 54.00 | 25.00 | 138.90 |
| 30 | 51.50 | 30.00 | 29.30 | 51.40 | 30.00 | 17.72 | 46.00 | 25.00 | 24.23 |
| 20 | 45.25 | 35.00 | 14.19 | 43.25 | 30.00 | 5.97 | 38.50 | 30.00 | 4.86 |
| 10 | 38.00 | 40.00 | 7.89 | 40.00 | 33.00 | 2.20 | 40.00 | 37.00 | 0.67 |
| 5 | 35.00 | 40.00 | 6.07 | 27.50 | 42.50 | 1.35 | – | – | 0 |

*Source:* Purushothama Raj et al. (1972).

Other notable contributions are made by Meyerhof (1951), Hansen (1970), and Vesic (1973). Meyerhof (1951) considered the effects of shearing resistance within the soil above foundation level, the shape and roughness of foundation. Hansen (1970) proposed a more generalized equation with shape and depth of foundation and the inclination of the load. Vesic (1973) reviewed different theories and showed that Meyerhof's and Hanzen's theories give almost same $N_c$ and $N_q$ values. Although $N_\gamma$ value of Meyerhof's has been in use, Vesic (1973) suggested that $N_\gamma$ is best represented by Eq. 14.8.

$$N_\gamma = 2(N_q + 1) \tan \phi \tag{14.8}$$

Based on these facts, Indian Standards recommended Vesic's values of bearing capacity factors, as given in Table 14.1. Plotted values are shown in Fig. 14.6.

Terzaghi's expression is valid for simplified conditions but can be modified, as discussed below, to adopt to different field conditions.

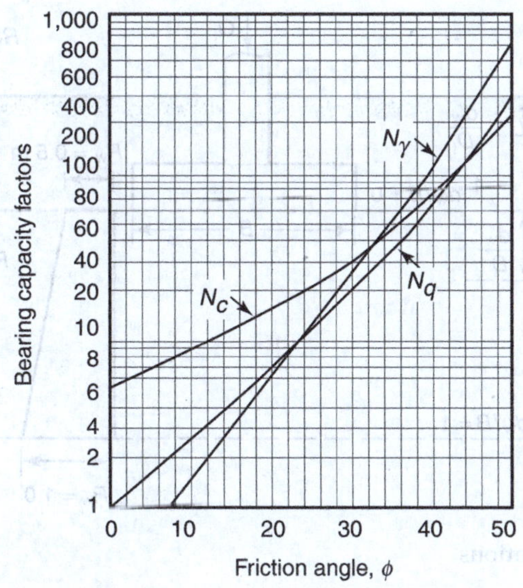


Fig. 14.6 Bearing capacity factors

**Effect of Soil Compressibility**. The general expression considers a relatively incompressible soil. This expression has to be modified in order to apply it to materials which undergo large vertical compression. Such a soil condition is identified as local shear failure and Terzaghi suggested modified shear strength parameters, $c_1 = (2/3)c$ and $\tan \phi_1 = (2/3) \tan \phi$. (The conventional notations for reduced shear parameters are $c'$ and $\phi'$, but they are referred to here as $c_1$ and $\phi_1$ to differentiate them from effective shear strength parameters.)

Accordingly, the ultimate bearing capacity for local shear failure is

$$q_f' = c_1 N_c + qN_q' + \tfrac{1}{2}\gamma B N_\gamma' \tag{14.9}$$

For obtaining values of $N_c'$, $N_q'$, and $N_\gamma'$, $\phi_1$ is calculated as $\phi_1 = \tan^{-1}$ (0.67 tan $\phi$). Then, $N_c$, $N_q$, and $N\gamma$ (IS recommended values) are read from Table 14.1 corresponding to the value of $\phi_1$ instead of $\phi$, which are values of $N_c'$, $N_q'$, and $N_\gamma'$, respectively (IS: 6403, 1981).

Vesic (1973) suggested that in sands the effect of relative density may be combined with the reduction factor $(\frac{2}{3} + D_r - \frac{3}{4}D_r^2)$ for the range $0 < D_r < 67\%$. But from a practical point of view foundations will never be laid on loose sand without proper densification.

**Effect of Water Table**. The general equation is based on the assumption that the water table is located well below the foundation. Some modifications are necessary depending on the location of the water table. In the general equation, there are two terms which are affected by water table movement: (i) the soil-weight component, $(½)\gamma B N\gamma$ and (ii) the surcharge component, $\gamma D_f N_q$.

Let us consider three locations of water table.

*Case I:* When the water table is well below the foundation, i.e., $d_w \geq B$. For this case, no correction is needed for both the components (Fig. 14.7).

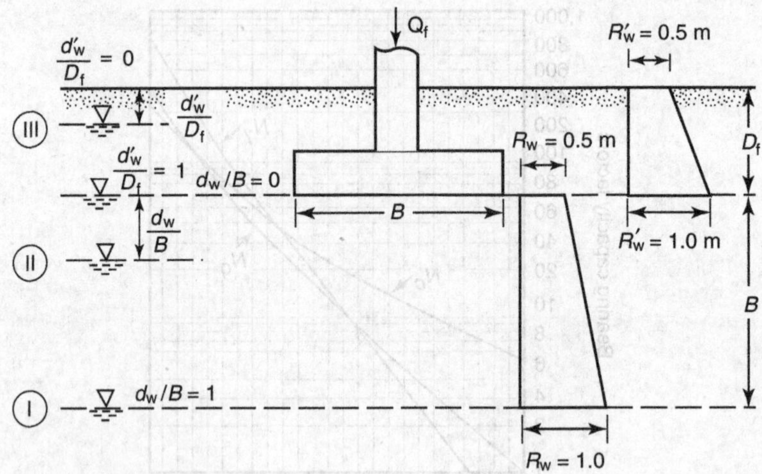


**Fig. 14.7** Water table locations

*Case II:* When the water table is anywhere from the base of the footing to a level well below the foundation, i.e., $0 \le d_w \le B$. In this case, only the soil-weight component is affected. This aspect can be considered by substituting an equivalent unit weight $\gamma_e$ in place of $\gamma$, i.e.,

$$\gamma_e = \frac{d_w \gamma + (B - d_w)\gamma'}{B}$$

or

$$\gamma_e = \gamma' + \frac{d_w}{B}(\gamma - \gamma') \tag{14.10}$$

The surcharge component is not affected.

*Case III:* When the water table is anywhere between the ground surface and the base of the footing, i.e., $0 \le d'_w \le D_f$. In this case, both the components are affected. For the surcharge component, the required substitution is

$$q = d'_w \gamma + (D_f - d'_w)\gamma' \tag{14.11}$$

For the soil-weight component, the required substitution is $\gamma'$ in place of $\gamma$ in the term $\frac{1}{2}\gamma B N_\gamma$.

Teng (1962) suggested water table correction factors, assuming the submerged unit weight of soil as 50% of the bulk unit weight of soil. Considering Case III, when the water table is at the ground surface, $d'_w / D_f = 0$ and $\gamma = \gamma'$, and at the base of the footing, $d'_w / D_f = 1$ and $\gamma = \gamma_{sat}$. This suggests a correction factor to have a value of 0.5 at $d'_w/D_f = 0$ and 1.0 at $d'_w/D_f = 1$, and such a factor may be $R'_w = (1/2)(1 + d'_w / D_f)$. For any intermediate point, a linear interpolation is made.

Considering Case II for $d_w/B = 0$, the correction factor should have a value of 0.5 when water table is at the base and for $d_w/B = 1.0$, the correction factor should be 1.0, such a factor may be $R_w = (1/2)(1 + d_w/B)$. For any intermediate point, a linear interpolation is made.

Thus, the general expression can be written with modification for water table as

$$q_f = cN_c + R'_w qN_q + \tfrac{1}{2} R_w \gamma BN_\gamma \tag{14.12}$$

The variations of $R_w$ and $R'_w$ are also shown in Fig. 14.7. For Case I, both $R_w$ and $R'_w$ have a value of 1. When the water table is at the base of the footing, $R'_w = 1.0$ and $R_w = 0.50$.

## 14.5 FOUNDATION PRESSURES

The total foundation pressure on the soil due to the weight of the structure is called the gross foundation pressure ($q_g$). The gross foundation pressure at the time of failure is nothing but the ultimate bearing capacity of the soil ($q_f$). The net foundation pressure ($q_n$) is the foundation pressure in excess of the pressure caused by the surrounding soil. So the net ultimate bearing capacity ($q_{nf}$) is the net foundation pressure at the time of failure. Thus,

$$q_n = q_g - \gamma D_f \tag{14.13a}$$

and

$$q_{nf} = q_f - \gamma D_f \tag{14.13b}$$

If $F$ is the factor of safety with respect to shear failure, then

$$F = \frac{q_{nf}}{q_n} \tag{14.14}$$

or

$$q_n = \frac{q_{nf}}{F} \tag{14.15}$$

Usually, a factor of safety of 3.0 is adopted against shear failure, and hence $q_n$ may be called the net bearing pressure $q_{ns}$ such that $q_{ns} = q_{nf}/F$. This leads to another definition of a term called gross safe pressure ($q_s$). That is,

$$q_s = q_{ns} + q$$

$$q_s = \frac{q_{nf}}{F} + q \tag{14.16}$$

$$q_s = \frac{1}{F}\left[cN_c + q(N_q - 1) + \frac{1}{2}\gamma BN_\gamma\right] + q$$

When the value of $\phi$ is relatively high there is no appreciable difference between the values of $F$ defined in terms of net and gross pressures. The safe pressure defined above is based on shear failure only and need not be minimum.*

## 14.6 SPECIAL LOADING AND GROUND CONDITIONS

The bearing capacity expressions given in the previous section deal with an idealized condition of loading and soil property. But in practice loading may be eccentric and inclined, and soil conditions may not be homogeneous but layered with varying shear parameters. Such special cases are briefly dealt with below.

### 14.6.1 Foundations with Eccentric Loading

A footing subjected to a concentric loading, with a moment or a load applied off the centre, causes the loading to be eccentric.

The analysis of an eccentrically loaded footing involves the evaluation of contact pressure beneath the footing and the ultimate bearing capacity. Let us consider the eccentricity in one direction only.

Eccentricity $e_y$ is given as (Fig. 14.8)

$$e_y = \frac{M}{Q} \tag{14.17}$$

The maximum and minimum pressures are given as

$$q_{\max} = \frac{Q}{BL}\left(1+\frac{6e_y}{B}\right) \tag{14.18}$$

and

$$q_{\min} = \frac{Q}{BL}\left(1-\frac{6e_y}{B}\right)$$

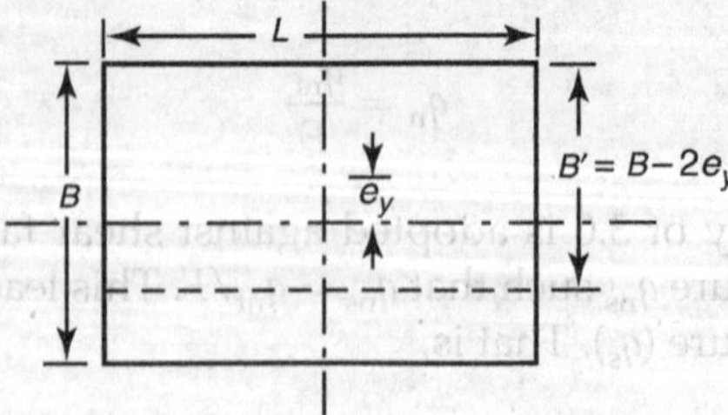


Fig. 14.8 One-directional eccentricity

*A minimum pressure is obtained after considering the settlement aspect also.

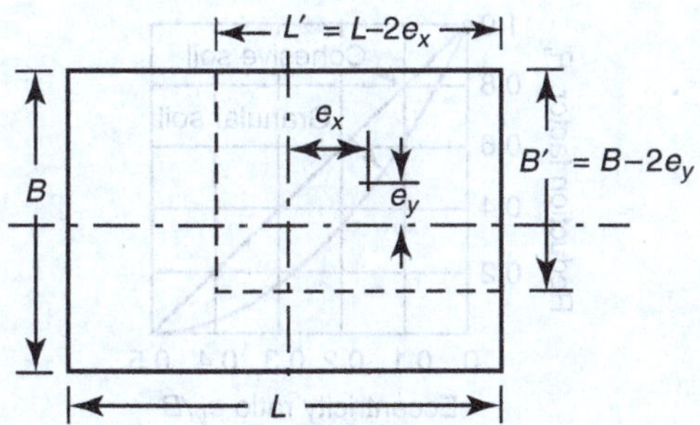


Fig. 14.9 Two-directional eccentricity

For $e_y = B/6$, $q_{\min}$ is zero, and for any condition where $e_y > B/6$, tension will develop. As soil cannot take tension, this has to be avoided; otherwise, there will be separation between the foundation and the soil.

If the eccentricity is in both the directions, $e_x$ and $e_y$ (Fig. 14.9) then the pressure is given as

$$q = \frac{Q}{BL}\left(1 \pm \frac{6e_x}{L} \pm \frac{6e_y}{B}\right) \tag{14.19}$$

Here again, a negative value of $q$ indicates tension between the soil and the bottom of footing. The footing has to be sufficiently weighted down by surcharge loads so as to rely on a proper bonding between the soil and the footing.

The concept of useful width was introduced by Meyerhof (1953) for the determination of the ultimate bearing capacity of eccentrically loaded footing and is also adopted in IS: 6403 (1981). The effective footing dimensions are

Effective length $L' = L - 2e_x$
Effective width $B' = B - 2e_y$
Effective footing area $= A' = B' \times L'$

By this concept, the area of the footing which is symmetrical about the load is taken as useful. The other portion is assumed to be excess. It is evident that the bearing capacity will decrease with increase in eccentricity. Meyerhof (1953) suggested a reduction factor to obtain the ultimate bearing capacity, as determined in the conventional way, considering the load is acting at the centroid of the footing. This bearing pressure is reduced by a reduction factor $R_e$. Thus, the reduced bearing pressure $(q_f)_e$ is

$$(q_f)_e = q_f R_e \tag{14.20}$$

where $R_e = 1 - 2(e_y/B)$ for the cohesive soil and $R_e = 1 - (e_y/B)^{1/2}$ for non-cohesive soils (for the range $0 < e_y/B < 0.3$). The reduction factor can also be read from Fig. 14.10.

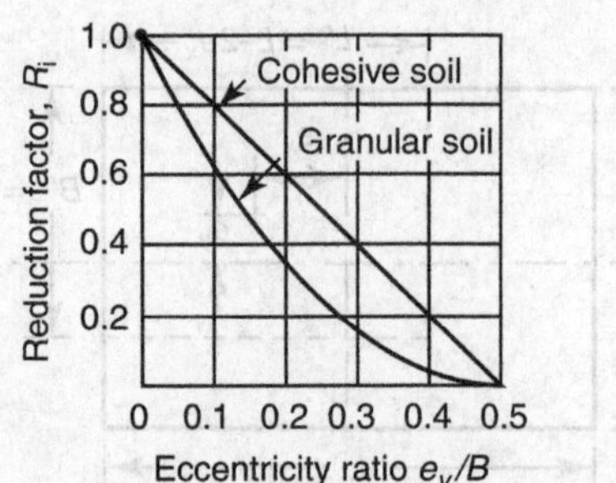


**Fig. 14.10** Bearing capacity of eccentrically loaded footing (*Source:* AREA, 1958)

### 14.6.2 Foundation Subjected to Inclined Load

The conventional method of analysis of footing subjected to inclined load is to resolve the force vertically and horizontally. The vertical component is used to determine the relevant bearing capacity and the horizontal force is ignored. However, the stability of footing against the horizontal force is analysed and a suitable factor of safety is adopted.

Meyerhof (1953) analysed the condition of inclined loads and presented a chart to find the reduction factor, $R_i$. According to his approach, the load is assumed vertical and the ultimate bearing capacity is determined. Then, it is corrected by a correction factor obtained from the chart (Fig. 14.11). Thus, the ultimate bearing capacity for a footing subjected to inclined load $(q_f)_i$ is

$$(q_f)_i = q_f R_i \tag{14.21}$$

where $R_i$ is the reduction factor.

Jambu (1957) extended Terzaghi's theory for inclined loads with due consideration of horizontal force and introduced an additional factor $N_h$ as

$$\frac{Q_u + N_h Q_b}{A} = cN_q + qN_q + \frac{1}{2}\gamma BN_\gamma \tag{14.22}$$

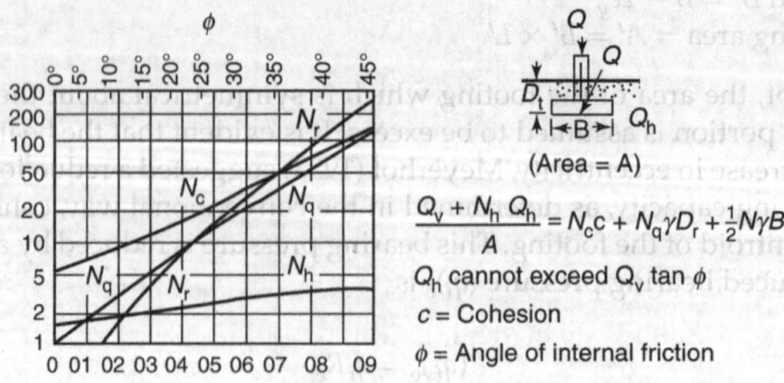


**Fig. 14.11** Bearing capacity charts for inclined loads (*Source:* Jambu, 1957)

Figure 14.11 shows the bearing capacity factors to use in the above equation. Modified bearing capacity formula (Eqs. 14.29 and 14.30) can also be used.

### 14.6.3 Foundations on Sloping Surface

Meyerhof (1957) proposed a method for determining the bearing capacity of footings on sloping ground surfaces. The bearing capacity equation for strip footing is given as

$$q_f = cN_{cq} + \frac{1}{2}\gamma BN_{\gamma q} \quad (14.23)$$

The bearing capacity factors $N_{cq}$ and $N\gamma_q$ depend on the slope of the ground and the relative position of the ground, in addition to the angle of shearing resistance of the soil.

Footing should not be placed on unsafe slopes. Before construction, the stability of the slope has to be checked. The construction of footing should not provoke a slide and the aspect has to be analysed. Further, if the slope material is under slow creep, the construction of footing on such slopes has to be avoided.

### 14.6.4 Foundations on Stratified Soil

Footings are sometimes constructed on soils where the rupture surface may not lie entirely on the first layer but may extend to the second layer. Further, depending on the strength of the soil in each layer, the rupture surface tends to increase in a weaker material and decrease in a stronger material.

A solution for the case of a footing at the surface of a two-layered saturated clay under undrained condition was first made by Button (1953). For $\phi = 0°$ condition, the curved sector becomes circular, and Button used a circular surface. The analysis was made for a two-layer system with cohesions $c_1$ in the top layer and $c_2$ in the bottom layer. The ultimate bearing capacity for surface footing is given as

$$q_f = c_1 N_c \quad (14.24)$$

where $N_c$ is the bearing capacity factor depending on $c_2/c_1$ ratio and the ratio of the thickness of the top layer to the width of footing. Figure 14.12 (Button, 1953) shows the variation of $N_c$ with $d/B$ and $c_2/c_1$. It is observed that when the upper layer is harder than the lower, the bearing capacity increases with the thickness of the top layer, and when the upper layer is softer, the bearing capacity decreases as its thickness increases. In this approach $c_1$ and $c_2$ are isotropic within their respective layers. This chart has been included in IS: 6403 (1981).

Button's solution was extended by Siva Reddy and Srinivasan (1967) for anisotropic soils, defined by a coefficient of anisotropy as

$$K = q_v / q_h \quad (14.25)$$

where $q_v$ is the vertical shear strength and $q_h$ the horizontal shear strength.

Charts for various $K$ values are available.

Different model analyses for two-layer systems were also attempted by Yamaguchi (1963), Meyerhof (1974), and Hanna and Meyerhof (1980). In these model analyses, some curves for estimating the capacity of sand overlying clay were presented.

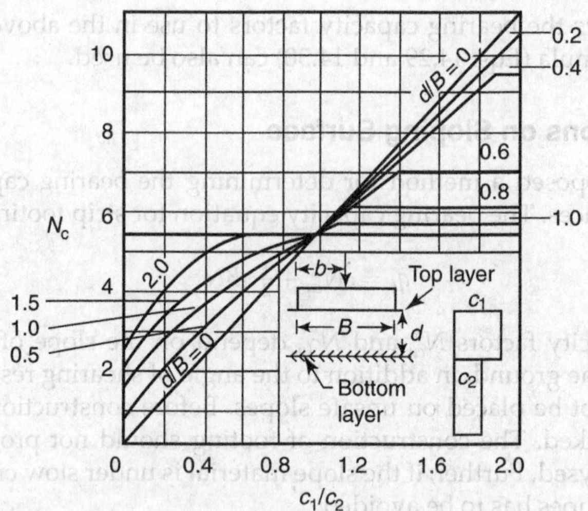


**Fig. 14.12** Bearing capacity factors for layered cohesive soil deposits (*Source:* IS: 6403, 1981)

A detailed bearing capacity for a more general case with both $c$ and $\phi$ was presented by Purushothama Raj et al. (1974) based on the upper bound limit theorem. Bearing capacity charts for varying cohesion with constant angle resistance were provided by them.

Purushothama Raj et al. (1975) also extended Button's solution for concentrically loaded footing to eccentrically loaded footing on an isotropic two-layer cohesive soil system based on upper bound limit theorem. Bearing capacity charts for different depth–width ratios $(d/B)$, cohesion ratios $(c_2/c_1)$, and eccentricity–width ratios $(e_y/B)$ for conventional and triangular foundation–soil contact conditions were presented.

### 14.6.5 Foundations on Partially Saturated Soil

Siva Reddy and Mogalaiah (1976) have presented a solution to the problem of ultimate bearing capacity of a strip footing on a partially saturated soil in the approach. Bishop's (1955) concept for effective stress of partially saturated soils and Skempton's (1954) pore pressure parameters have been used along with an initial negative pore water pressure. The solution is obtained by adopting the method of characteristics. Numerical results are given in the form of bearing capacity factors $N_c$, $N_q$, and $N\gamma$, for different values of pore pressure parameters $A$ and $B$, coefficient of earth pressure at rest $K_0$, ratio of initial negative pore water pressure $(u_w)_0$, and cohesion $c$.

It is inferred from their studies that the degree of saturation and initial negative pore water pressure have significant influence on the bearing capacity of partially saturated soils.

### 14.6.6 Foundations on Desiccated Soil

Desiccated cohesive soils show decrease in undrained cohesion with depth. Further, this decrease stabilizes at depths of about 3 to 4 m from the ground level. For this situation, a linear decrease of cohesion with depth may be considered. The gradient $\lambda$ and cohesion at the

**Table 14.3** Data for determining ultimate net bearing capacity

| $8\lambda B/q_{nf}$ | $q_{nf}/c_1$ |
|---|---|
| 0.0 | 5.7 |
| 0.2 | 5.0 |
| 0.3 | 4.5 |
| 0.6 | 4.0 |
| 0.7 | 3.6 |
| 1.0 | 3.2 |

*Source:* IS: 6403 (1981).

ground surface $c_1$ can be obtained from borehole data. A set of values relating $8\lambda B/q_{nf}$ and $q_{nf}/c_1$ are presented in Table 14.3. For a given footing width $B$, by trial and error $q_{nf}$ can be estimated by matching $q_{nf}/c_1$ and $8\lambda B/q_{nf}$ (IS: 6403, 1981), or a plot can be made between $(8\lambda B/c_1) = (8\lambda B/q_{nf} \times q_{nf}/c_1)$, and $(q_{nf}/c_1)$ using the values from Table 14.3, and with the knowledge of $(8\lambda B/c_1)$, $(q_{nf}/c_1)$ can be read from the plot.

Hence, from this value $q_{nf}$ is obtained.

### 14.6.7 Foundations on Rock

Sound rocks have strengths extremely higher than the pressure transferred by footings. Unfavourable rock conditions, heterogeneity, and overstressing of foundations may cause large differential settlement leading to failure. Particularly, porous limestones, volcanic rocks, some shales, and highly fractured rocks need special consideration.

In principle, Terzaghi's bearing capacity equation can be used provided strength parameters are obtained from triaxial shear tests on rocks. However, in rocks settlement criteria is more critical than shear failure.

The Bureau of Indian Standards (IS: 12070, 1987) suggests different methods for evaluating the bearing capacity of shallow foundation on rocks. They are explained below.

**Evaluation Based on Classification.** For preliminary design, the net safe bearing pressure based on classification (Table 14.4) has been suggested with a caution that it has to be

**Table 14.4** Net safe bearing pressure based on classification

| Material | $q_{ns}$ (kN/m$^2$) |
|---|---|
| Massive crystalline bed rock, including granite, diorite, gneiss, trap rock | 10,000 |
| Foliated rocks such as schist or slate in sound condition | 4,000 |
| Bedded limestone in sound condition | 4,000 |
| Sedimentary rock, including hard shales and sandstone | 2,500 |
| Soft or broken bed rock (excluding shale) and soft limestone | 1,000 |
| Soft shale | 400 |

*Source:* IS: 12070 (1987).

**Table 14.5** Net safe bearing pressure based on RMR

| Classification no. | I | II | III | IV | V |
|---|---|---|---|---|---|
| Description of rock | Very good | Good | Fair | Poor | Very poor |
| RMR | 100–81 | 80–61 | 60–41 | 40–21 | 20–0 |
| $q_{ns}$ (kN/m$^2$) | 6000–4480 | 4400–2880 | 2880–1510 | 1450–900–580 | 550–450–400 |

*Source:* IS: 12070 (1987).

checked before the final design. Further, net bearing pressure is also recommended based on rock mass rating (RMR) for different rocks (Table 14.5). The RMR values obtained up to a depth equal to width of the foundation should be used. The recommended values consider a limiting settlement of 12 mm.

**Evaluation Based on Core Strength.** For a rock mass with favourable bedding planes (i.e., rock surface parallel to the base of the foundation) and the walls of discontinuities closed, the safe bearing pressure is given as

$$q_s = q_0 N_j \tag{14.26}$$

where $q_0$ is the average uniaxial compressive strength of rock cores and $N_j$ the empirical coefficient depending on the spacing of discontinuities (Table 14.6) given as

$$N_j = \frac{3 + s/B_f}{10\sqrt{1 + 300\,\delta/s}}$$

where $\delta$ is the thickness of discontinuities (cm), $s$ the spacing of discontinuities (cm), and $B_f$ the footing width (cm).

The above relationship is valid for a rock mass with spacing greater than 0.3 m, opening of discontinuities less than 10 mm, and a foundation width greater than 0.3 m.

**Evaluation Based on Plate Load Test.** It is recommended that plate load test be conducted on poor rocks with safe bearing pressure less than 1,000 kN/m$^2$. From the plate load testing, the settlement of plate is computed from the formula as follows:

1. For massive or sound rock, $S_p = S_i \dfrac{B_p}{B_f}$ (14.27)

**Table 14.6** Value of $N_j$

| Spacing of discontinuities (cm) | $N_j$ |
|---|---|
| 300 | 0.40 |
| 100–300 | 0.25 |
| 30–100 | 0.10 |

*Source:* IS: 12070 (1987).

2. For laminated or poor rocks, $S_p = S_i \left[ \frac{B_p}{B_f} \left( \frac{B_f + 0.3}{B_p + 0.3} \right) \right]^2$ (14.28)

where $S_p$ is the settlement of plate (mm), $S_i$ the settlement of footing (mm), $B_p$ the width of plate (m), and $B_f$ the width of footing (m).

From the pressure–settlement curve, the safe bearing pressure is read for the calculated settlement of plate. It is further recommended to conduct at least three tests with different sizes of plates to check the results.

## 14.7 OTHER BEARING CAPACITY THEORIES

### 14.7.1 Modified Bearing Capacity Formulae (IS: 6403, 1981)

Because of mathematical complexity, no theoretical analysis is available for the shape of footing like square, rectangular, and circular. Empirical corrections based on model tests are recommended by Terzaghi (1943) and are subsequently simplified.

Terzaghi's expression (Eq. 14.4) is valid for $D_f/B \leq 1$. But in practice the depth factor has also to be considered. Further, the load may be inclined instead of vertical. Thus, modified ultimate net bearing capacity formulae taking into account the shape of the footing, depth of embedment, inclination of loading, and effect of water table has been recommended by the Bureau of Indian Standards (IS: 6403, 1981) as follows:

General shear failure:

$$q_{nf} = cN_c s_c d_c i_c + q(N_q - 1)s_q d_q i_q + \tfrac{1}{2}\gamma B N_\gamma s_\gamma d_\gamma i_\gamma R_w \quad (14.29)$$

Local shear failure:

$$q_{nf} = c_l N'_c s_c d_c i_c + q(N'_q - 1)s_q d_q i_q + \tfrac{1}{2}\gamma B N'_\gamma s_\gamma d_\gamma i_\gamma R_w \quad (14.30)$$

where $s_c$, $s_q$, and $s\gamma$ are shape factor corrections (values are given in Table 14.7), $d_c$, $d_q$, and $d\gamma$ are depth factor corrections (to be applied only when the backfilling is done with proper compaction) calculated as follows:

$$d_c = 1 + 0.2D_f / B\sqrt{N_\phi}$$

$$d_q = d_\gamma = 1 \quad \text{for } \phi < 10^\circ$$

$$d_q = d_\gamma = 1 + 0.1D_f / B\sqrt{N_\psi} \quad \text{for } \phi > 10^\circ$$

$$N_\phi = \tan^2(45^\circ + \phi/2)$$

and $i_c$, $i_q$, and $i_\gamma$ are load inclination factor corrections given as

$$i_c = i_q = \left(1 - \frac{\alpha}{90}\right)^2$$

$$i_\gamma = \left(1 - \frac{\alpha}{\phi}\right)^2$$

**Table 14.7** Shape factor corrections

| Shape of base of footing | Shape factor corrections | | |
|---|---|---|---|
| | $s_c$ | $s_q$ | $s_\gamma$ |
| Continuous strip | 1.00 | 1.00 | 1.00 |
| Rectangle (width $B$, length $L$) | $1 + 0.2B/L$ | $1 + 0.2B/L$ | $1 - 0.4B/L$ |
| Square (side $B$) | 1.30 | 1.20 | 0.80 |
| Circle (diameter $B$) | 1.30 | 1.20 | 0.60 |

*Source:* IS: 6403 (1981).

where $\alpha$ is the inclination of the load to the vertical in degrees.

$R_w = 1$, if the water table remains permanently at or below a depth $(D_f + B)$ beneath the ground level.

$R_w = 0.5$, if the water table is permanently located at a depth of $D_f$ likely to rise above the base of footing.

$R_w = \frac{1}{2}[1 + (d_w / B)]$, if the water table is likely to be permanently located at a depth $D_f < d_w < (D_f + B)$.

### 14.7.2 Skempton's Bearing Capacity Theory

Skempton (1951) proposed a simple expression for the ultimate bearing capacity of saturated clay under undrained condition ($\phi = 0$), for a rectangular footing of length $L$ and width $B$, and is given as

$$q_f = c_u N_c + \gamma D_f \tag{14.31}$$

Values of $N_c$ may be obtained from Fig. 14.13 or from the following expression:

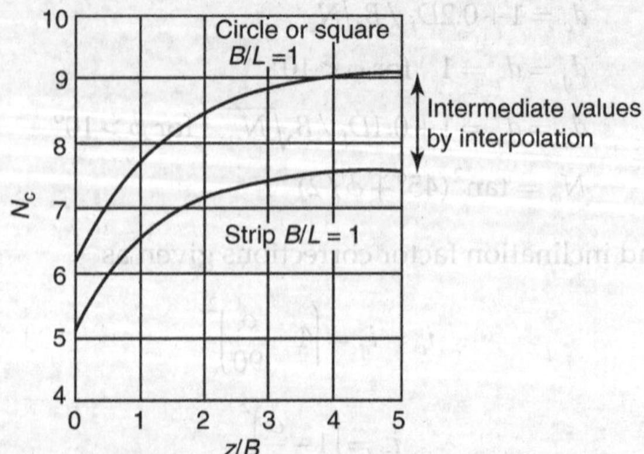


**Fig. 14.13** Skempton's values of $N_c$ when $\phi_u = 0°$ (*Source:* Skempton, 1951)

$$N_c = 5.14\left(1+0.2\frac{B}{L}\right)\left[1+\left(0.053\frac{D_f}{B}\right)^{1/2}\right] \tag{14.32}$$

The limiting maximum values ($D_f/B$) are

when $B/L = 0$, i.e., for strip footing, $N_c \ngtr 7$
when $B/L = 1$, i.e., for circular or square footing, $N_c \ngtr 9$

### 14.7.3 Meyerhof's Bearing Capacity Theory

Meyerhof (1963) presented a more comprehensive solution to solve the bearing capacity problem. This theory takes into account depths, shape, and inclination factors together in the bearing capacity equation. This theory is applicable for foundation of all shapes and loading conditions, except foundations on built up slopes.

The solution proposed by Meyerhof (1963) is given as

$$q_d = cN_c d_c s_c i_c + \bar{\gamma} D d_q s_q i_q + \tfrac{1}{2}\gamma B N_\gamma d_\gamma s_\gamma i_\gamma \tag{14.33}$$

where $c$ is the unit cohesion, $N_c$, $N_q$, $N_\gamma$ are the bearing capacity factors for a strip foundation, $d_c$, $d_q$, $d_\gamma$ the depth factors, $s_c$, $s_q$, $s_\gamma$ the shape factors, $i_c$, $i_q$, $i_\gamma$ the inclination factors for the load inclined at an angle $\alpha$ (degrees) to the vertical, $\bar{\gamma}$ the effective unit weight of soil above base level of foundation, $\gamma$ the effective unit weight of soil below foundation base, and $D$ the depth of foundation.

The depth, shape, and inclination factors are given in Table 14.8.

**Table 14.8** Depth, shape and inclination factors

| Factors | Equation | For |
|---|---|---|
| Depth | $d_c = 1+0.2\sqrt{N_\phi}\left(\frac{D}{B}\right)$ | any $\phi$ |
| | $d_q = d_\gamma = 1+0.1\sqrt{N_\phi}\left(\frac{D}{B}\right)$ | $\phi > 0$ |
| | $d_\gamma = d_q = 1$ | $\phi = 0$ |
| Shape | $s_c = 1+0.2N_\phi\left(\frac{B}{L}\right)$ | any $\phi$ |
| | $s_q = s_\gamma = 1+0.1N_\phi\left(\frac{B}{L}\right)$ | $\phi > 0$ |
| | $s_q = s_\gamma = 1$ | $\phi = 0$ |
| Inclination | $i_c = i_q = \left(1-\frac{\alpha^\circ}{90}\right)^2$ | any $\phi$ |
| | $i_\gamma = \left(1-\frac{\alpha^\circ}{\phi^\circ}\right)^2$ | $\phi > 0$ |
| | $i_\gamma = 0$ | $\phi = 0$ |

*Source:* Meyerhof (1963).

In Table 14.8, $B$ = width, $L$ = length of foundation, and $N_\phi = \tan(45°+\phi/2)$.

The theoretical bearing capacity factors for a shallow horizontal strip foundation are

$$\left.\begin{aligned} N_q &= e^{\pi \tan\phi} N_\phi \\ N_c &= (N_q - 1)\cot\phi \end{aligned}\right\} \tag{14.34a}$$

$$N_\gamma = (N_q - 1)\tan(1.4\phi) \tag{14.34b}$$

Equation 14.34a is the same as per Prandtl (1920), whereas Eq. 14.34b is proposed by Meyerhof (1961).

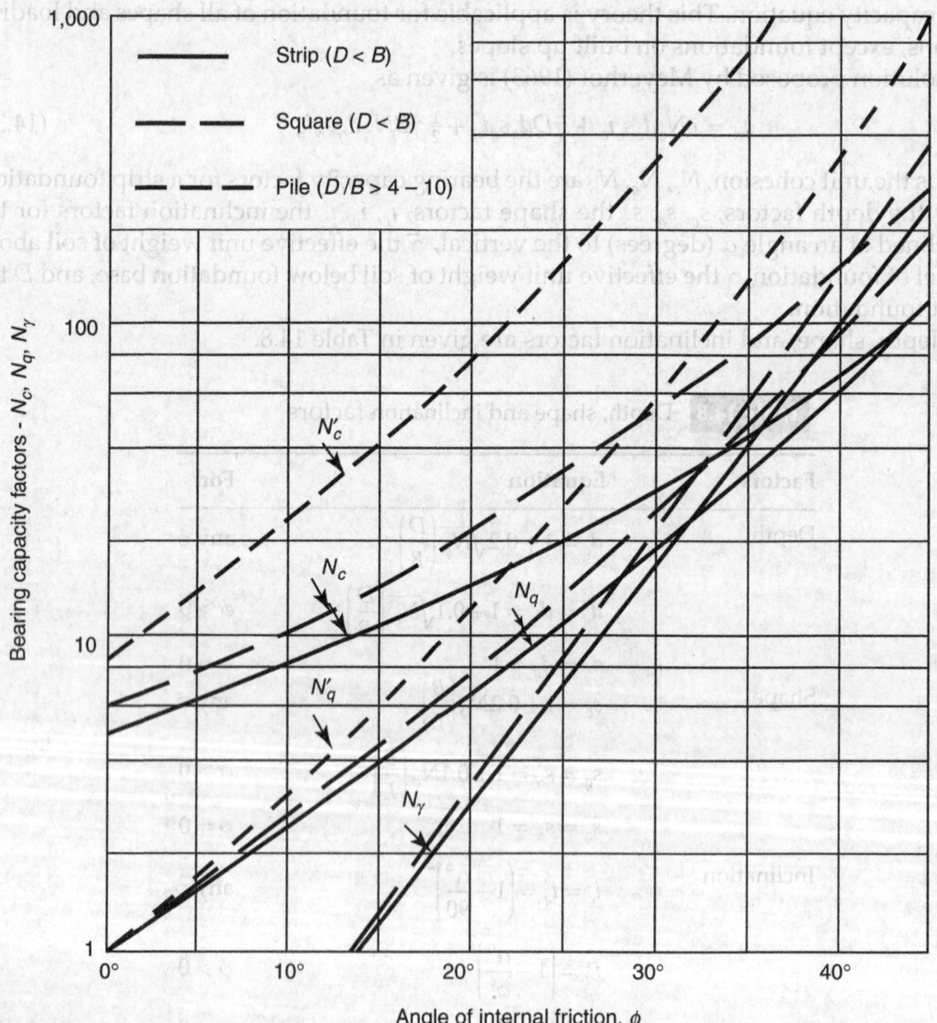


**Fig. 14.14** Bearing capacity factors for spread and pile foundations (*Source:* Meyerhof, 1963).

Figure 14.14 gives the $N$ factors for strip and square foundations. For rectangular foundations, the $N$ factors have to be interpolated. The $N'$ factors given in the same figure applies to point bearing capacity of pile foundation. At $\phi = 0°$, $N_c = 5.14$ for strip foundation and $N_c = 6.2$ for square foundation.

**Depth Factors**. The simple bearing capacity factors in Eq. 14.34 do not take into account the resistance of the soil above the foundation level which increases the bearing capacity. If the soil above the foundation level is quite compact, the effect of this on the bearing capacity may be considered by means of depth factors given in Table 14.8. The increase of bearing capacity can be estimated from depth factors by which the individual bearing capacity factors have to be multiplied. As the depth of the foundation increases, the depth factors increase at a decreasing rate and approach a maximum value which can be used for an estimate of the point resistance of piles.

**Shape Factors**. The bearing capacity factors given in Eq. 14.34 is for a strip foundation. The bearing capacity factors for rectangular foundations can be obtained by multiplying the individual $N$ factors in Eq. 14.34 of the corresponding shape factors given in Table 14.8.

**Eccentric Loading**. If the foundations are subjected to eccentric loads, vertical or inclined, the effective width $B'$ of the foundation has to be used in Eq. 14.34.

### 14.7.4 Brinch Hansen's Bearing Capacity Theory

Brinch Hansen's bearing capacity theory is in a way an extension of Meyerhof's work. In addition to the factors considered by Meyerhof, the foundation base tilt and foundations on slopes are included in Hansen's (1970) equation.

Hansen's equation can be put in a simple form as follows:

$$q_d = cN_cA_c + \bar{\gamma}DN_qA_q + \tfrac{1}{2}\gamma BN_\gamma A_\gamma \qquad (14.35)$$

where

$$A_c = d_c\, s_c\, i_c b_c g_c$$
$$A_q = d_q\, s_q\, i_q b_c g_q$$
$$A_\gamma = d_\gamma s_\gamma i_\gamma b_\gamma g_\gamma$$

$\bar{\gamma}$ is the effective unit weight of soil above the base, $\gamma$ the effective unit weight of soil below the base, $d_c$, $d_q$, $d_\gamma$ are depth factors, $s_c$, $s_q$, $s\gamma$ are shape factors, $i_c$, $i_q$, $i_\gamma$ are load inclination factors, $b_c$, $b_q$, $b_\gamma$ are base inclination factors, and $g_c$, $g_q$, $g_\gamma$ are ground surface inclination factors. The other factors are the same as given earlier. The equations for the various factors are given below:

1. $N$-factor

$$N_q = e^{\pi \tan\phi} N_\phi$$
$$N_c = (N_q - 1)\cot\phi \qquad \text{Meyerhof's factors} \qquad (14.36)$$
$$N_\gamma = 1.5(N_q - 1)\tan\phi$$

The bearing capacity factors are given in Fig. 14.15a.

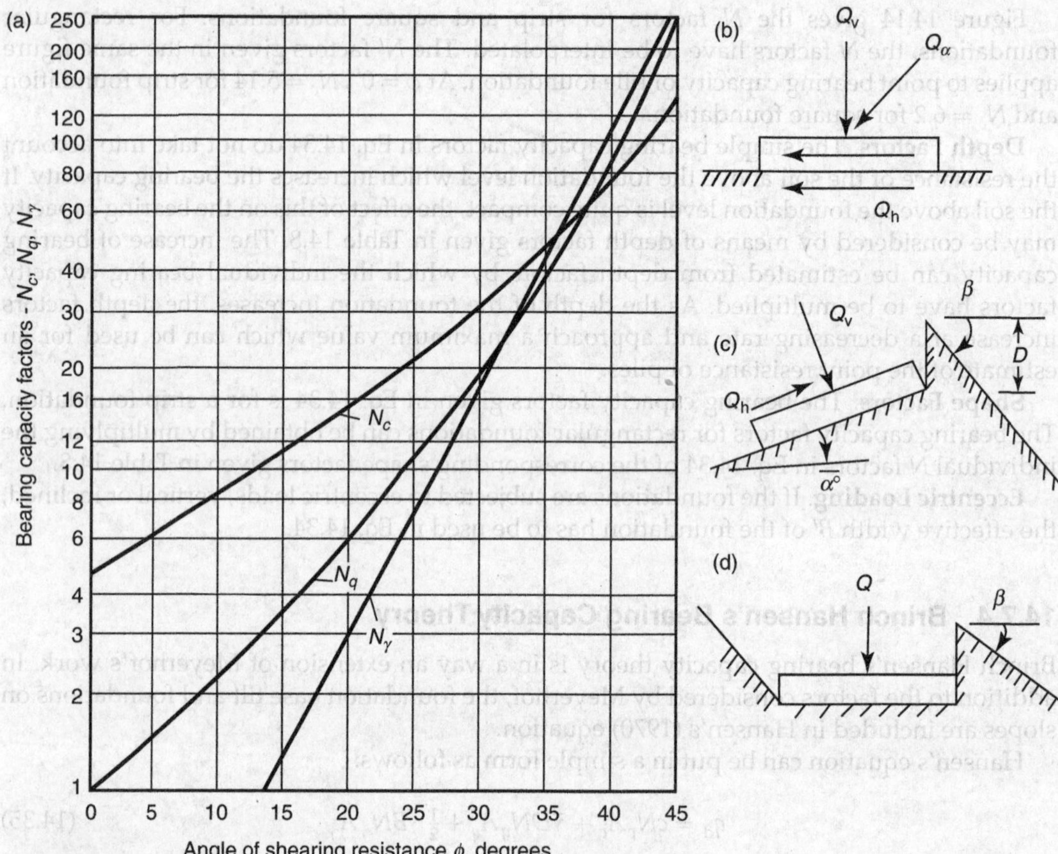


**Fig. 14.15** Bearing capacity factors and the definition of $Q_v$, $Q_h$, $\bar{\alpha}_0$, and $\beta$. (a) Bearing capacity factors, $N_c$, $N_q$, $N_\gamma$. (b) Definition of $Q_v$ and $Q_h$. (c) Definition of $\bar{\alpha}°$ and $\beta$. (d) Definition of ground inclination factor (Source: Hansen, 1970)

2. Depth factors

| Equation | Limiting value of $D/B$ | Limiting value of $\phi$ |
|---|---|---|
| $d_c = 1 + 0.4 \tan^{-1} D/B$ | >1 | >0 |
| $d_c = 1 + 0.4D/B$ | ≤1 | >0 |
| $d_c = 0.4D/B$ | ≤1 | $\phi = 0$ |
| $d_c = 0.4 \tan^{-1} D/B$ | ≥1 | $\phi = 0$ |
| $d_q = 1 + 2 \tan \phi(1 - \sin \phi)D/B$ | 1 | |
| $d_q = 1 + 2 \tan \phi(1 - \sin \phi) \times \tan^{-1}(D/B)$ | >1 | |
| $d\gamma = 1$ | | for all $\phi$ |

3. Shape factors

$$s_c = \left(1 + \frac{N_q}{N_c}\frac{B}{L}\right) \quad \text{for } \phi > 0$$

$$s_c = 0.2\frac{B}{L} \quad \text{for } \phi = 0$$

$$s_c = 1 \quad \text{for strip foundation}$$

$$s_q = 1 + \frac{B}{L}\tan\phi$$

$$s_\gamma = 1 - 0.4\frac{B}{L}$$

4. Load inclination factors for horizontal base

$$i_c = i_q - \frac{1 - i_q}{N_q - 1} \quad \text{for } \phi > 0°$$

$$i_c = 0.5 - 0.5\sqrt{1 - \frac{Q_h}{A_f c_a}} \quad \text{for } \phi = 0°$$

$$i_q = \left(1 - \frac{0.5Q_h}{Q_v + A_f c_a \cot\phi}\right)^5$$

$$i_\gamma = \left(1 - \frac{0.7Q_h}{Q_v + A_f c_a \cot\phi}\right)^5$$

The definitions of $Q_v$ and $Q_h$ are given in Fig. 14.15b.

5. Base inclination factors

$$b_c = \left(1 - \frac{\bar{\alpha}°}{147°}\right) \quad \text{for } \phi > 0°$$

$$b_c = \frac{\bar{\alpha}°}{147°} \quad \text{for } \phi > 0°$$

$$b_q = e^{-2\bar{\alpha}\tan\phi}$$

$$b_\gamma = e^{-2.7\bar{\alpha}\tan\phi}$$

The definitions of $\bar{\alpha}°$ and $\beta$ are shown in Fig. 14.15c.

6. Ground surface inclination factor

$$g_c = \left(1 - \frac{\beta°}{147°}\right) \quad \text{for } \phi > 0°$$

$$g_c = \left(\frac{\beta°}{147°}\right) \quad \text{for } \phi > 0°$$

$$g_q = g_\gamma = (1 - 0.5\tan\beta)^2$$

The definition of ground inclination is shown in Fig. 14.15d.

Wherever footings are subjected to eccentric loadings, the effective width $B'$ has to be used in Eq. 14.35 for evaluating the shape factors,

$$B' = B - 2e$$

If the eccentricity is in two directions for a rectangular foundation of width $B$ and length $L$, then the effective widths in each direction are

$$B' = B - 2e_x$$
$$L' = L - 2e_y$$
$$A_f' = B'L'$$

where $e_x$ and $e_y$ are the eccentricities in the $B$ and $L$ directions, respectively, and $A_f'$ the effective area. If $q_d$ is found out from Eq. 14.35 for an effective width of $B'$, then the ultimate load is

$$Q_d = A_f' \times q_d \tag{14.37}$$

For footings on a slope, $g$ factors are used to reduce the bearing capacity, however, these factors should be used cautiously as there is little experimental data available to confirm this factor.

## 14.8 BEARING CAPACITY OF SOILS FROM BUILDING CODE

Building codes are prepared in a traditional way based on the vast database on soils of different locations. These codes give a list of soil types and their safe bearing capacity. It is assumed that soil can sustain the pressure with respect to shear failure and without appreciable settlement.

As discussed in the previous section that bearing capacity of a soil depends on various factors, the bearing capacity given in a building code should be taken as a guiding value and not as one exact value (IS: 1904, 1986, revised).

The Bureau of Indian Standards has given presumptive bearing capacity values which are presented in Table 14.9. Following are the limitations of the bearing capacity values given in building codes:

1. The code values do not consider the effect of shape, size, depth of foundation, and the base condition (rough or smooth) of a foundation.
2. The code values do not consider the effect of water table and its fluctuation.
3. The code values assume that the soil is homogeneous in all directions.
4. The effect of settlement is not taken into account in the code values.
5. The code values are more simplified based on experience and have no theoretical base.
6. The code values used are to be updated.

**Table 14.9** Safe bearing capacity[a]

| S. no. | Type of rock or soil | Safe bearing capacity, kN/m$^2$ (t/m$^2$) | Remarks |
|---|---|---|---|
| I. ROCKS | | | |
| 1. | Rocks without lamination and defects, e.g., granite, trap, diorite | 3,240 (330) | |
| 2. | Laminated rocks, e.g., sandstone and limestone, in sound condition | 1,620 (165) | |

*Table 14.9 Contd.*

**Table 14.9** Contd.

| S. no. | Type of rock or soil | Safe bearing capacity, kN/m$^2$ (t/m$^2$) | Remarks |
|---|---|---|---|
| 3. | Residual deposits of shattered and broken bed rock and hard shale, cemented material | 880 (90) | |
| 4. | Soft rock | 440 (45) | |
| II. COHESIONLESS SOILS | | | |
| 5. | Gravel, sand and gravel, compact and offering high resistance to penetration when excavated by tools | 440 (45) | See note b |
| 6. | Coarse sand, compact and dry | 440 (45) | Dry means that the groundwater level is at a depth not less than width of the foundation below the base of the foundation |
| 7. | Medium sand, compact and dry | 245 (25) | |
| 8. | Fine sand, silt (dry lumps easily pulverized by fingers) | 150 (15) | |
| 9. | Loose gravel or sand–gravel mixture; loose coarse to medium sand, dry | 245 (25) | See note b Susceptible to long-term |
| 10. | Fine sand, loose and dry consolidation settlement | 100 (10) | Consolidation settlement |
| III. COHESIVE SOILS | | | |
| 11. | Soft shale, hard or stiff clay, dry | 440 (45) | Susceptible to long-term consolidation settlement |
| 12. | Medium clay, readily indented with a thumb nail | 245 (25) | |
| 13. | Moist clay, and sand–clay mixture which can be indented with strong thumb pressure | 150 (15) | |
| 14 | Soft clay indented with moderate thumb pressure | 100 (10) | |
| 15. | Very soft clay which can be penetrated easily with the thumb | 50 (5) | |
| 16. | Black cotton soil or other shrinkable or expansive clay in dry condition (50% saturation) | | See note c To be determined after investigation |
| IV. PEAT | | | |
| 17. | Peat | | See notes c and d To be determined after investigation |

*Table 14.9 Contd.*

**Table 14.9** Contd.

| S. no. | Type of rock or soil | Safe bearing capacity, kN/m$^2$ (t/m$^2$) | Remarks |
|---|---|---|---|
| V. MADE-UP GROUND | | | |
| 18. | Fills or made-up ground | | See notes b and d To be determined after investigation |

*Source:* IS: 1904 (1986, revised).

[a]Values listed in the table are from shear consideration only.

[b]Values are very much rough for the following reasons:

(i) Effect of characteristics of foundations (i.e., effect of depth, width, shape, roughness, etc.) has not been considered.

(ii) Effect of range of soil properties (i.e., angle of internal friction, cohesion, water table, density, etc.) has not been considered.

(iii) Effect of eccentricity and inclination of loads has not been considered.

[c]For non-cohesive soils, the values listed in the table shall be reduced by 50% if the water table is above or near the base of the footing.

[d]Compactness or looseness of non-cohesive soils may be determined by driving a cone of 65 mm dia and 60° apex angle by a hammer of 65 kg falling from 75 cm. If the corrected number of blows ($N$) for 30 cm penetration is less than 10, the soil is called loose; if $N$ lies between 10 and 30, it is medium; if more than 30, the soil is called dense.

## 14.9 PERMISSIBLE SETTLEMENTS

Settlement may be classified as uniform (or total) settlement, tilt, and non-uniform settlement. Structures on rigid foundations undergo uniform settlement (Fig. 14.16a). When the entire structure rotates, the structure is said to be under uniform tilt (Fig. 14.16b). If foundations of different elements of a structure undergo varied settlements, the foundation is said to be under non-uniform settlement (or differential settlement).

*Foundations may settle uniformly due to* (IS: 1904, 1986) (i) elastic, consolidation and secondary compression of soil, (ii) groundwater lowering, (iii) swelling and shrinkage of expansive soils caused by seasonal variations, (iv) surface erosion, creep, or landslides and effects of vegetation, and (v) mining subsidence, underground erosion by streams or floods, and adjacent excavation. Tilt of a structure generally occurs due to eccentric loading or sudden subsidence of a corner of a rigid foundation.

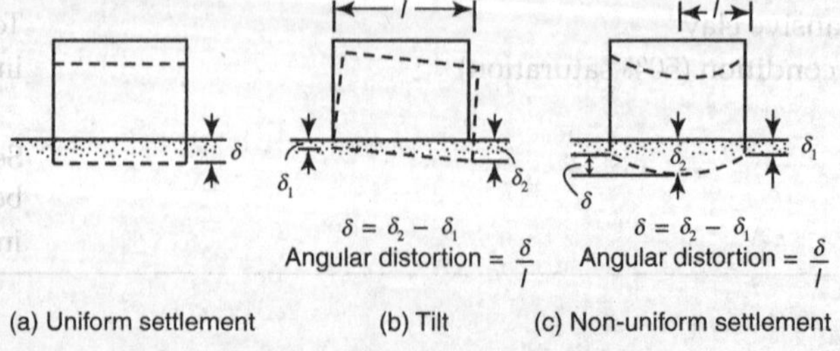


**Fig. 14.16** Types of settlement (*Source:* Lambe and Whitman, 1979)

**Table 10.10** Permissible maximum, differential settlements, and tilt (angular distortion) for shallow foundation in soils

| Sl. no. | Type of structure | Isolated foundation | | | | | | Raft foundations | | | | | |
|---|---|---|---|---|---|---|---|---|---|---|---|---|---|
| | | Sand and hard clay | | | Plastic clay | | | Sand and hard clay | | | Plastic clay | | |
| | | Maximum settlement (mm) | Differential settlement (mm) | Angular distortion (mm) | Maximum settlement (mm) | Differential settlement (mm) | Angular distortion (mm) | Maximum settlement (mm) | Differential settlement (mm) | Angular distortion (mm) | Maximum settlement (mm) | Differential settlement (mm) | Angular distortion (mm) |
| (1) | (2) | (3) | (4) | (5) | (6) | (7) | (8) | (9) | (10) | (11) | (12) | (13) | (14) |
| (i) | For steel structures | 50 | 0.0033$L$ | 1/300 | 50 | 0.0033$L$ | 1/300 | 75 | 0.0033$L$ | 1/300 | 100 | 0.0033$L$ | 1/300 |
| (ii) | For reinforced concrete structures | 50 | 0.0015$L$ | 1/666 | 75 | 0.0015$L$ | 1/666 | 75 | 0.0021$L$ | 1/500 | 100 | 0.002$L$ | 1/500 |
| (iii) | For multi-storeyed buildings | | | | | | | | | | | | |
| | (a) Reinforced concrete or steel framed buildings with panel walls | 60 | 0.002$L$ | 1/500 | 75 | 0.002$L$ | 1/500 | 75 | 0.0025$L$ | 1/400 | 125 | 0.0033$L$ | 1/300 |
| | (b) For loading bearing walls | | | | | | | | | | | | |
| | 1. $L/H = 2^a$ | 60 | 0.0002$L$ | 1/5,000 | 60 | 0.0002$L$ | 1/5,000 | Not likely to be encountered | | | | | |
| | 2. $L/H = 7^a$ | 60 | 0.0004$L$ | 1/2,500 | 60 | 0.0004$L$ | 1/2,500 | | | | | | |
| (iv) | For water towers and silos | 50 | 0.0015$L$ | 1/666 | 75 | 0.0015$L$ | 1/666 | 100 | 0.0025$L$ | 1/400 | 125 | 0.0025$L$ | 1/400 |

*Source:* IS: 1904 (1986).

The values given in the table may be taken only as a guide and the permissible total settlement/differential settlement and tilt (angular distortion) in each case should be decided as per requirements of the designer.

$L$ denotes the length of deflected part of wall/raft of centre-to-centre distance between columns.

$H$ denotes the height of wall from foundation footing.

[a]For intermediate ratios of $L/H$, the values can be interpolated.

*Non-uniform settlement can result from* (IS: 1904, 1986) (i) non-homogeneous subsoil condition, (ii) non-uniform pressure distribution on soil due to unequal loading, (iii) variation of water regime at the construction site, (iv) overstressing of adjacent site due to heavy structures and interference of pressure distribution, (v) unequal expansion of the soil due to excavation, (vi) non-uniform development of extrusion settlements, and (vii) non-uniform structural disruptions or volume changes due to freezing and thawing, shrinkage and swelling, etc.

Generally, the amount of uniform settlement is not a critical factor, but it is only a question of convenience. In practice, the settlement is often non-uniform and is of concern in the design of a foundation. Estimation of uniform settlement is much more simpler than that of differential settlement. On important jobs, it is essential to investigate and identify stronger and weaker subsoils and accordingly estimate the movements. On jobs of less importance, an empirical relationship between total and differential settlements is enough (e.g., 75% of total settlement may be taken as differential settlement).

The total and differential settlements should not exceed the permissible values. The permissible values of settlement for different types of structures are given in Table 14.10 (IS: 1904, 1986). The permissible differential settlement is obtained by taking the difference of maximum and minimum settlements. Tilt is computed by dividing the differential settlement by the distance between the points of related maximum and minimum settlements.

## 14.10 ALLOWABLE BEARING PRESSURE

A loaded foundation settles in direct proportion with increase in load. At higher load levels, the rate of increase of settlement is extremely large and the foundation is said to have broken into the ground or to have experienced a bearing capacity failure. It is evident that distinction between excessive settlement and failure by breaking into the ground is, in many instances, quite arbitrary.

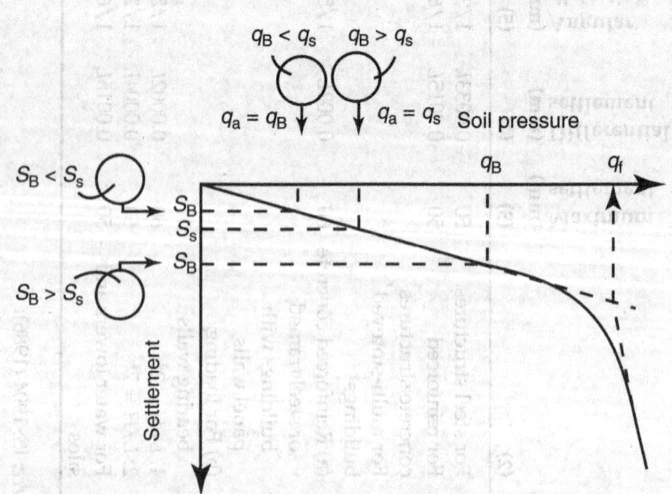


Fig. 14.17 Load–settlement curve of a foundation

Thus, every foundation has to satisfy two independent conditions. The first condition is that there should be adequate factor of safety against shear failure of foundation. Second, the settlement of the structure should not be great enough to damage the structure. Out of these two conditions, whichever gives a lower value of load intensity is referred to as the allowable soil pressure.

Let us consider the load–settlement curve of a foundation (Fig. 14.17). Let $q_f$ be the ultimate bearing capacity, $F$ the factor of safety, $q_s = q_f/F$ the safe soil pressure with respect to shear failure, $S_s$ the settlement corresponding to $q_s$, $S_B$ the permissible settlement of the foundation, $q_B$ the intensity of pressure corresponding to $S_B$, and $q_a$ the allowable soil pressure.

Now we can identify two conditions:

1. If the settlement $S_s$ corresponding to safe soil pressure is less than the permissible settlement $S_B$, the pressure $q_B$ corresponding to $S_B$ is greater than $q_s$. That is, the settlement criterion is satisfied but the shear failure criterion is violated. Hence, the allowable soil pressure is governed by the lesser pressure $q_s$. That is,

   $S_B > S_s$

   Therefore,

   $q_a = q_s \quad (\text{since } q_B > q_s)$

2. If the settlement $S_s$ is greater than $S_B$, the settlement criterion will be violated when $q_s$ is adopted. Hence, the allowable soil pressure is governed by the lesser pressure $q_B$. That is,

   $S_B < S_s$

   Therefore,

   $q_a = q_B \quad (\text{since } q_B < q_s)$

In general, the allowable soil pressures in sands, gravelly sands, and silty sands are governed only by the settlement considerations, except in narrow footings on loose sand. In many situations, the permissible settlement is reached at a pressure for which the factor of safety against shear failure is greater than 3.0. Settlement in sands occurs rapidly and about 80% to 90% of settlement takes place during construction.

The allowable soil pressure for clays, silty clays, and sandy clays is generally determined con-sidering a factor of safety of 3.0 with respect to shear failure. However, in certain cases the settlement criterion may predominate, for example, in normally consolidated clays. For homogeneous clays with less permeability, the factor of safety has to be checked immediately after construction, adopting the undrained shear strength. But in case of fissured clays, the permeability will be very high and the undrained shear strength condition may be very much on the conservative side.

The most important soils intermediate between sand and clay are silt and loess which have different characters. Loose silts behave worse than soft clays and are unsuitable for supporting footings. Medium or dense silts are those which have characteristics of a rock flour, or which have a certain plasticity. The allowable pressure on silts of the rock flour type may be computed roughly by adopting the procedure for sand, and that on plastic silts by the methods used for clay. No general rules can be established for silts in determining soil pressure.

## 14.11 ESTIMATION OF BEARING CAPACITY FROM FIELD TESTS

Over the years, various attempts were made for obtaining bearing capacity values by more direct approaches instead of using the more rigorous theoretical approaches. These methods are, in general, empirical in nature and warrants sufficient judgement from the user. There are three *in situ* tests used to estimate the bearing capacity of soils, viz., standard penetration test, cone penetration test, and plate load test.

### 14.11.1 Bearing Capacity Based on Standard Penetration Test

On the basis of the results of standard penetration tests (as explained in Chapter 11), Terzaghi and Peck in 1948 proposed correlations in the form of curves for estimating allowable soil pressure for footings on sand. The correlations represented in Fig. 14.18 allowable to situations in which the water table is at least 2*B* below the foundation level, where *B* is the width of the footing. The *N* values represented in Fig. 14.18 are the corrected values. If the water table is at or close to foundation level and the depth–width ratio of foundation is small, either the settlement is doubled or the allowable soil pressure is reduced by 50% for the same permissible settlement of 25 mm. If the depth–width ratio is close to unity, the values need to be reduced by only one-third.

Teng (1962) provided Eq. 14.38, which closely approximates the curves presented by Terzaghi and Peck (Fig. 14.18)

$$q_{na} = 34.5(N-3)\left(\frac{0.305B+1}{0.7B}\right)(\text{kN/m}^2) \tag{14.38}$$

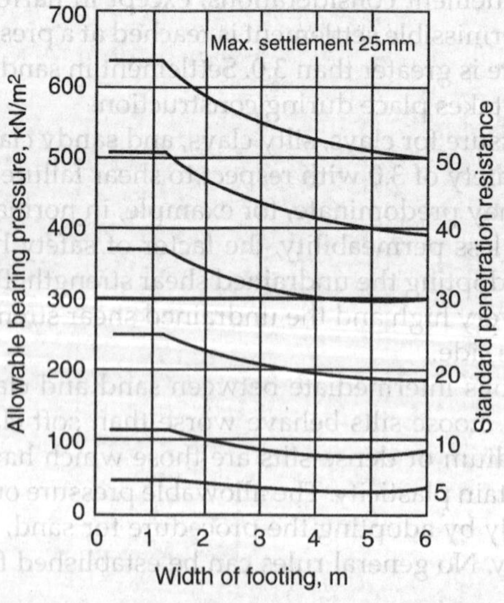


**Fig. 14.18** Allowable soil pressure based on *N* values (*Source:* Terzaghi and Peck, 1967)

where $B$ is given in metres. Meyerhof (1956) suggested a correlation as given in Eqs. 14.39 and 14.40

$$q_{na} = 12N\ (\text{kN/m}^2) \quad \text{for } B \le 1.22 \text{ m} \tag{14.39}$$

$$q_{na} = 8N\left(\frac{0.305B+1}{0.305B}\right)^2 (\text{kN/m}^2) \quad \text{for } B > 1.22 \text{ m} \tag{14.40}$$

where $q_{na}$ is the net allowable soil pressure for a permissible settlement of 25 mm.

The correlations given in the chart generally give conservative values. These values were so intended that the largest footing should not settle more than 25 mm even if it were situated on the most compressible pocket of sand. In general, the soil pressure for any settlement is

$$q'_{na} = \frac{S'}{S_a} q_{na}$$

where $q'_{na}$ is the net allowable soil pressure (in kPa) corresponding to the settlement $S'$ (in mm) and $S_a = 25$ mm.

Peck et al. (1974) revised the Terzaghi and Peck curves, accounting for water table location, based on research and observational data as

$$q'_{na} = C_w (0.41) N S' \tag{14.41}$$

where $C_w$ is the water table correction factor ($0.5 < C_w < 1.0$) and

$$C_w = 0.5 + 0.5 \frac{d'_w}{d_f + B}$$

Terzaghi and Peck curves and the correlations for $q_{na}$ are primarily intended for non-cohesive soils like sand and gravel and may be used for silts with judgements.

The Bureau of Indian Standards (IS: 6403, 1981) recommends to find the angle of shearing resistance $\phi$ from the corrected $N$ values and to compute $N_q$ and $N\gamma$ (from Table 14.1) to in turn evaluate the net ultimate bearing capacity.

### 14.11.2 Bearing Capacity Based on Cone Penetration Test

Meyerhof (1956, 1965) has suggested formulae for allowable soil pressures based on cone penetration test values restricting the settlement not to exceed 25 mm. His formulae are based on Terzaghi and Peck's curves for spread or strip footings on dry sands as

$$q_a = \frac{q_{cs}}{30} \ (\text{kPa}) \quad \text{and} \quad B \le 1.2 \text{ m} \tag{14.42}$$

and

$$q_a = \frac{q_{cs}}{50}\left(\frac{B+0.3}{B}\right) (\text{kPa}) \quad \text{and} \quad B > 1.2 \text{ m} \tag{14.43}$$

These formulae are based on the approximate rule that the $N$-value is one-quarter of the static cone resistance.

The value of $q_{cs}$ obtained using the above equations should be reduced by 50% if the sand within the stressed zone is submerged. Meyerhof further suggests that the values have to be doubled for raft or pier foundation.

Schmertmann (1975) gave a method of calculating the allowable soil pressure indirectly from the cone penetration test. He related $N\gamma$ with $q_{cs}$ as

$$N_\gamma = \frac{q_{cs}}{80} \text{ (kPa)} \tag{14.44}$$

With $N_\gamma$, one may work back to compute $\phi$ and then in turn $N_q$. Any standard theory may be used to calculate the bearing capacity with $N_q$ and $N\gamma$ values in silts and sands. Thornburn (1971) has presented a correlation between $q_{cs}$ and $N$ for particle range from 0.006 to 6 mm (Fig. 14.19). From a knowledge of $q_{cs}$ and the average particle size, we can find $N$ and hence the net allowable soil pressure. No standard correlation is yet available for clays using cone penetration test.

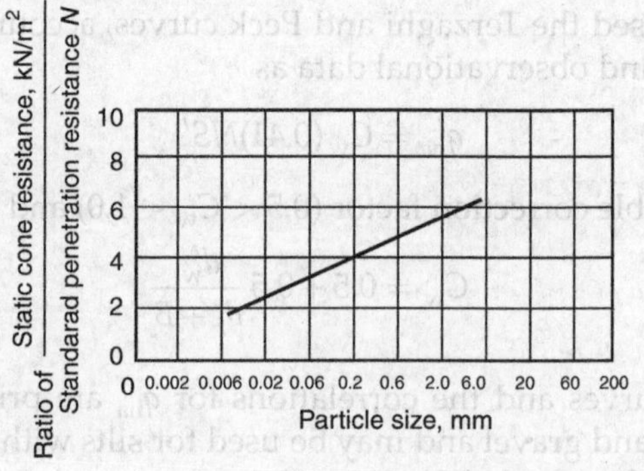


**Fig. 14.19** Relationship between static cone resistance and standard penetration resistance (*Source*: Tomlinson, 1986)

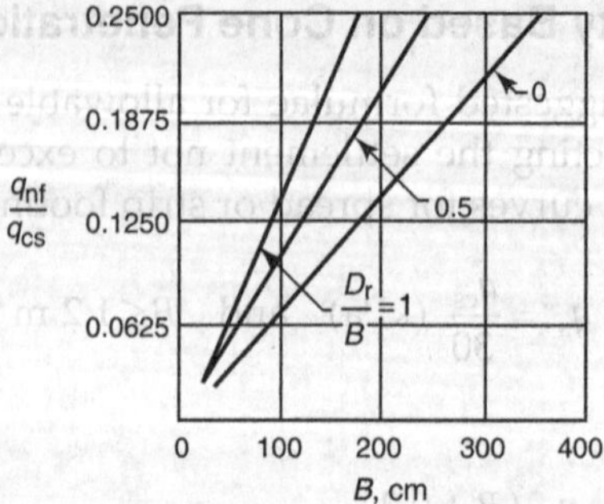


**Fig. 14.20** Chart for static cone test

For cohesionless soils, the Indian Standards code (IS: 6403, 1981) provides a chart relating $q_{nf}/q_{cs}$ and of footing $B$ (Fig. 14.20). Knowing the depth and width of footing, the value of $q_{nf}/q_{cs}$ is obtained and hence the net ultimate bearing capacity.

### 14.11.3 Bearing Capacity Based on Plate Load Test

The object of a plate load test is to obtain a load settlement curve of a soil at a particular depth. Such a curve is needed to estimate the ultimate bearing capacity, allowable soil pressure, and the settlement of footings.

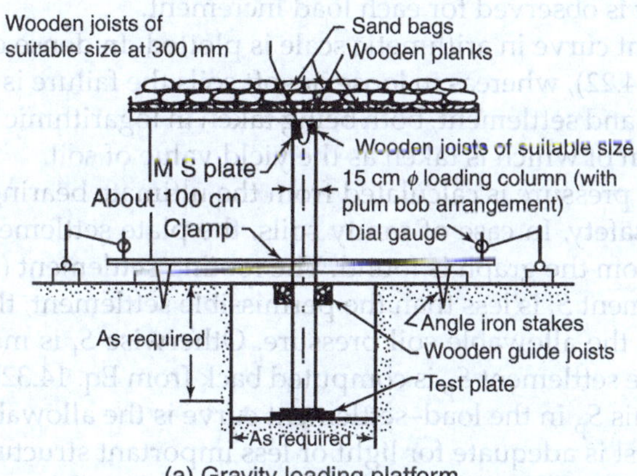


(a) Gravity loading platform

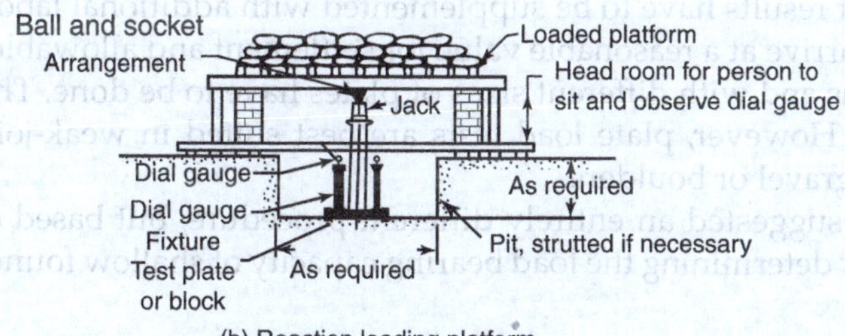


(b) Reaction loading platform

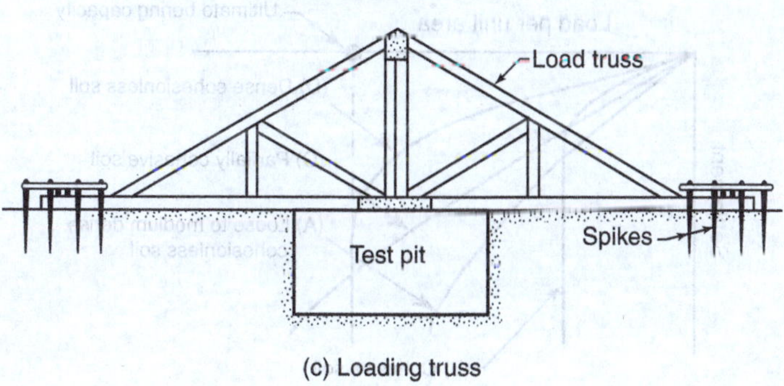


(c) Loading truss

**Fig. 14.21** Typical set-up for loading (*Source:* IS: 1888, 1982)

In order to load the soil to the required level a suitable loading device is needed. Three types of loading devices (Fig. 14.21) are recommended (IS: 1888, 1982), viz., gravity loading, reaction loading, and loading truss. The load is measured using a pressure gauge, proving ring or a cell. Settlements are recorded using dial gauges. Circular or square plates of size 300 to 750 mm are used. A Plate size of 300 mm is used for dense soils and 450 mm for loose soils.

Site for load test is located based on exploratory borings. A test pit of width equal to five times the size of plate is made at the proposed foundation level.

After placing the plate over a thin layer of sand at the bottom of the pit a seating pressure (70 g/cm$^2$) is applied. The load is applied in equal load increments with each increment not exceeding one-fifth of the estimated ultimate bearing capacity or 100 kN/m$^2$, whichever is less. The settlement is observed for each load increment.

A load–settlement curve in arithmetic scale is plotted. In dense or stiff soils, the failure is well-defined (Fig. 14.22), whereas in loose or soft soils the failure is not pronounced. In such cases, a plot of load and settlement, both being taken in logarithmic scales, gives two straight lines the intersection of which is taken as the yield value of soil.

The safe bearing pressure is calculated from the ultimate bearing capacity after allowing a certain factor of safety. In case of sandy soils, the plate settlement ($S_p$) corresponding to safe soil pressure from the graph is found. The footing settlement ($S_t$) is computed from Eq. 14.32b. If the settlement $S_t$ is less than the permissible settlement, then the safe soil pressure computed above is the allowable soil pressure. Otherwise $S_t$ is made equal to permissible settlement and plate settlement $S_p$ is computed back from Eq. 14.32b. Then, the soil pressure corresponding to this $S_p$ in the load–settlement curve is the allowable soil pressure.

The plate load test is adequate for light or less important structures under normal conditions. However, in the case of unusual soil types and for all heavy and special structures the plate load test results have to be supplemented with additional laboratory tests or field tests. In order to arrive at a reasonable value for settlement and allowable soil pressure, tests at different depths and with different sizes of plates have to be done. This is expensive and time-consuming. However, plate load tests are best suited in weak-jointed rocks or soils containing large gravel or boulders.

Housel (1929) suggested an entirely different procedure, but based on the results from plate load test, for determining the load bearing capacity of shallow foundations considering

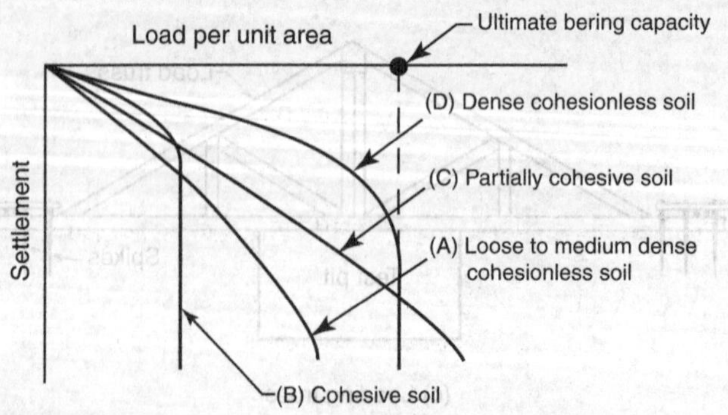


Fig. 14.22 Typical load–settlement curves (*Source:* IS: 1888, 1982)

settlement criterion. That is, it is required to find the dimensions of a foundation that has to carry a load $Q$ with a tolerable settlement $S_B$.

To obtain the relevant parameters, two plate load tests are conducted with different sizes of plates. From the load–settlement curves for equal settlement, the loads $Q_1$ and $Q_2$ are obtained. Then, the load is related to the area, perimeter, bearing pressure, and perimeter shear. That is,

$$Q_1 = A_1 m + P_1 n \tag{14.45}$$

$$Q_2 = A_2 m + P_2 n \tag{14.46}$$

where $A_1$ and $A_2$ are the areas of plates No. 1 and No. 2, respectively, $P_1$ and $P_2$ are the perimeters of plates No. 1 and No. 2, respectively, $m$ the constant corresponding to bearing pressure, and $n$ the constant corresponding to perimeter shear.

The constants $m$ and $n$ are obtained by solving the above equations. Then, for a given load $Q$ of the foundation, the area ($A$) and perimeter ($P$), and hence the dimensions are obtained from

$$Q = Am + Pn$$

## WORKED EXAMPLES

**Example 14.1** In a mass-housing complex scheme over a vast area, two types of soils were encountered. One of which is a partially saturated silty clay with $c_u = 5.8\ \text{kN/m}^2$, $\phi_u = 25°$, and $\gamma = 18.5\ \text{kN/m}^3$ and extends over most of the area. The other, predominantly clay having $c_u = 55\ \text{kN/m}^2$ spreads to a lesser extent. The water table is at a greater depth. As per the design, strip footings of the building have to be placed at 1 m depth. Compute the width of the footing required in each type of soil if the load intensity is 150 kN/m run. Adopt a factor of safety of 2.5 in both the soils, and only shear failure need to be considered. For $\phi = 25°$, take $N_c = 20.7$, $N_q = 10.7$, $N\gamma = 10.8$.

If there is a possibility of the water table rising to the ground surface, what should be the change in the width of footing in both areas. The submerged unit weight of the silty clay is $11.2\ \text{kN/m}^3$.

***Solution***

For partially saturated silty clay, the net safe bearing pressure

$$q_{ns} = \frac{1}{F}\left[cN_c + q(N_q - 1) + \frac{1}{2}\gamma B N_\gamma\right]$$

or

$$\frac{150}{B\times 1} = \frac{1}{2.5}\left[5.8\times 20.7 + 18.5\times 1(10.7-1) + \frac{1}{2}\times 18.5\times B\times 10.8\right]$$

or

$$\frac{150}{B} = 119.8 + 39.96B$$

$$39.96B^2 + 119.8B - 150 = 0$$

or

$$B^2 + 3.03B - 3.75 = 0$$

Therefore,

$$B = \frac{-3.03 \pm \sqrt{(3.03)^2 + 4\times1\times3.75}}{2\times1} = \frac{-3.03 \pm 4.92}{2}$$

Therefore, width of footing on partially saturated silty clay = 0.95 m.

For the clay soil with $\phi = 0°$, the values for bearing capacity factors are $N_c = 5.7$, $N_q = 1$, and $N\gamma = 0$. Hence, the net bearing pressure reduces to

$$q_{\text{ns}} = \frac{1}{F}[cN_c]$$

or

$$\frac{150}{B\times1} = \frac{1}{2.5}[55\times5.7]$$

or

$$B = \frac{150\times2.5}{55\times5.7} = 1.2\text{ m}$$

Therefore, the width of footing on clay = 1.2 m.

Because of submergence the unit weight of soil will be reduced to the submerged unit weight. Hence, the terms containing $\gamma$ should be replaced by $\gamma'$. Thus, footings only on silty clay will be affected. The footings on clay are independent of the unit weight. Hence, for footings on silty clay

$$q_{\text{ns}} = \frac{1}{F}\left[cN_c + \gamma' D_f(N_q - 1) + \frac{1}{2}\gamma' BN_\gamma\right]$$

or

$$\frac{150}{B\times1} = \frac{1}{2.5}\left[5.8\times20.7 + 11.21(10.7-1) + \frac{1}{2}\times11.2\times B\times10.8\right]$$

or

$$\frac{150}{B} = 91.48 + 24.19B$$

or

$$B^2 + 3.78B - 6.2 = 0$$

or

$$B = \frac{-3.78 \pm \sqrt{(3.78)^2 + 4\times1\times6.2}}{2\times1} = \frac{-3.78 \pm 6.25}{2} = 1.24$$

Hence a width of 1.24 m has to be provided for footings on silty clay.

**Example 14.2** A 1 m wide long footing is located at a depth of 1.5 m from the ground surface. The supporting soil is compressible and has shear strength parameters, $c'_{cu} = 30\ \text{kN/m}^2$ and $\phi'_{cu} = 25°$. The total unit weight of the soil, $\gamma = 18.3\ \text{kN/m}^3$. The water table is at a greater depth. Compute the safe load that can be carried by the long footing per metre length of the wall. Adopt a factor of safety of 3.0.

***Solution***

As the soil is compressible, the reduced shear strength parameters and bearing capacity factors corresponding to the local shear condition are used.

Therefore,

$$c_1 = \tfrac{2}{3} c'_{cu} = \tfrac{2}{3}\times 30 = 20\ \text{kN/m}^2$$

and

$$\tan\phi_1 = \tfrac{2}{3}\ \tan\phi'_{cu}$$

or

$$\phi_1 = \tan^{-1}[\ \tfrac{2}{3}\ \tan 25°] = 17.3°$$

For $\phi_1 = 17.3°$, the bearing capacity factors are taken as

$$N'_c = 13.91$$

$$N'_q = 5.17$$

$$N'_\gamma = 4.02$$

Based on Eq. 14.20,

$$q_s = \frac{1}{F}\left[c_1 N'_c + q(N'_q - 1) + \tfrac{1}{2}\gamma B N'_\gamma\right] + q$$

$$q_s = \tfrac{1}{3}[30\times13.91 + 18.3\times(5.17-1) + \tfrac{1}{2}\times18.3\times1\times4.02] + 18.3\times1.5$$

or

$$q_s = \tfrac{1}{3}[429.45] + 27.45 = 170.6\ \text{kN/m}^2$$

Therefore,

$Q_s = q_s \times B = 170.6 \times 1 = 170.6$ kN

Safe load that can be carried by the wall = 170.6 kN/m.

**Example 14.3** The construction of a strip footing is undertaken during a summer period and the water table is observed at 2.5 m from the ground surface (Fig. 14.23). During monsoon the water table rises to the ground surface. The relevant soil parameters are $\gamma = 19.2$ kN/m$^3$ and $\phi = 32°$. Determine the gross safe bearing capacity in both the cases for a factor of safety of 2.5. Use Teng's water table correction factors.

***Solution***

For the summer condition,

$$q_{ns} = \frac{1}{F}\left[R'_w q(N_q - 1) + \tfrac{1}{2} R_w \gamma B N_\gamma\right] + R'_w \gamma D_f$$

For $\phi = 32°$, the bearing capacity factors are $N_q = 23.2$ and $N\gamma = 30.2$.

$R_w$ is obtained from the expression for $d_w = 0.5$ m.

That is,

$$R_w = \tfrac{1}{2}\left(1 + \frac{d_w}{\text{B}}\right) = \tfrac{1}{2}\left(1 + \frac{0.5}{3}\right) = 0.58$$

and

$$R'_w = 1.0$$

Hence,

$$q_{ns} = \frac{1}{2.5}\left[1 \times 19.2 \times 2(23.2 - 1) + \tfrac{1}{2} \times 0.58 \times 19.2 \times 3 \times 30.2\right] + 1 \times 19.2 \times 2$$

$$= 542.8 + 38.4 = 581.2 \text{ kN/m}^2$$

For the monsoon condition,

$$R'_w = 0.50 \quad \text{and} \quad R_w = 0.5$$

Hence,

$$q_s = \frac{1}{2.5}\left[0.5 \times 19.2 \times 2(23.2 - 1) + \tfrac{1}{2} \times 0.5 \times 19.2 \times 3 \times 30.2\right] + 0.5 \times 19.2 \times 2$$

$$= 344.45 + 19.2 = 363.7 \text{ kN/m}^2$$

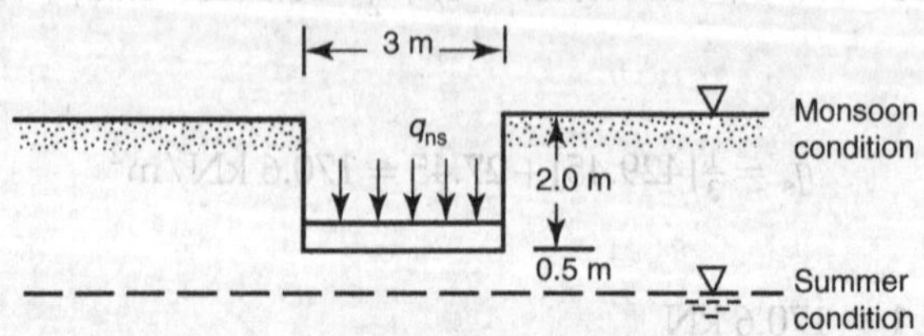


Fig. 14.23

**Example 14.4** In a warehouse building, two unequally loaded columns are combined by a rectangular combined footing. It is proposed to place the footings at a depth of 1.5 m on a saturated clay with the following soil properties: $c_u = 72$ kN/m$^2$, $\phi_u = 0°$, $\gamma = 17.8$ kN/m$^3$. The loads on the columns are 720 and 1,170 kN, with a spacing of 5 m, and the centre of the 720 kN column is placed at a distance of 0.8 m from the property line (Fig. 14.24). Neglecting the weight of the footing, estimate the dimension of the footing. Adopt a factor of safety of 3.

***Solution***

The general expression (Eq. 14.13) for net bearing capacity can be written as

$$q_{\text{nf}} = s_c c_u N_c + s_q q(N_q - 1) + \tfrac{1}{2} s_\gamma \gamma B N_\gamma$$

For a rectangular footing,

$$s_c = \left(1 + 0.2\frac{B}{L}\right)$$

$$s_\gamma = \left(1 - 0.4\frac{B}{L}\right)$$

and $s_q = s_c$.

Further, for $\phi_u - 0°$, adopting Terzaghi's values $N_c = 5.7$, $N_q = 1$, and $N\gamma = 0$,

$$q_{\text{nf}} = \left(1 + 0.2\frac{B}{L}\right) \times 5.7 c_u$$

$$= \left(1 + 0.2\frac{B}{L}\right) \times 5.7 \times 72$$

Assume $L/B = 4$, then

$$q_{\text{nf}} = \left(1 + \frac{0.2}{4.0}\right) \times 5.7 \times 72 = 430.92 \text{ kN/m}^2$$

$$q_{\text{ns}} = \frac{q_{\text{nf}}}{F} = \frac{430.92}{3} = 143.64 \text{ kN/m}^2$$

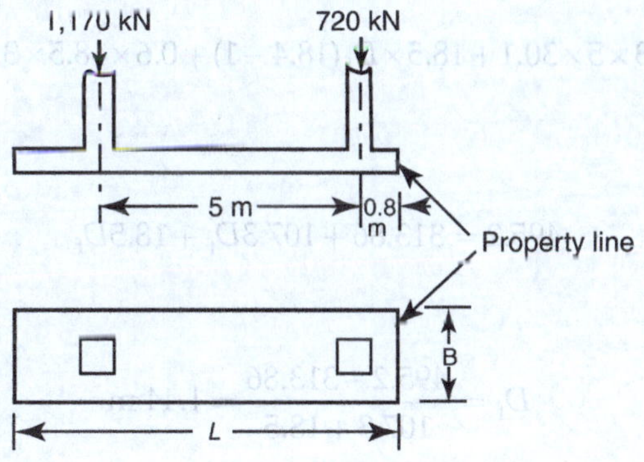


Fig. 14.24

The width $B$ and length $L$ can be obtained by satisfying $\sum V = 0$ and the centre of gravity requirement.

Equating the upward force with the downward force, or $q_{ns} \times$ footing area = column loads,

$$143.64 \times B \times L = 720 + 1170$$

$$B \times L = \frac{1890}{143.64} = 13.16 \text{ m}^2$$

The centres of gravity of the footing pressure and the loading should be in one line. Taking moment about the property line

$$720 \times 0.8 + 1170 \times 5.8 = 1890 \times \bar{x}$$

$$\bar{x} = \frac{7362}{1890} = 3.9 \text{ m}$$

Therefore, $L/2 = 3.9$ m or $L = 7.8$ m. That is,

$$B = \frac{13.16}{7.8} = 1.69 \text{ m} \quad \text{and} \quad \frac{L}{B} = 4.6$$

$L$ and $B$ are modified such that $L/B = 4$.

Hence, $L = 7.25$ m and $B = 1.82$ m.

**Example 14.5** A circular concrete pier of 3 m diameter carries a gross load of 3,500 kN. The supporting soil is a clayey sand having the following properties: $c = 5 \text{ kN/m}^2$, $\phi = 30°$, and $\gamma = 18.5 \text{ kN/m}^3$. Find the depth at which the pier is to be located such that a factor of safety of 3.0 is assured. The bearing capacity factors for $\phi = 30°$ are $N_c = 30.1$, $N_q = 18.4$, and $N\gamma = 22.4$.

***Solution***

The gross safe bearing pressure is given as

$$q_s = \frac{1}{F}\left[1.3N_c + q(N_q - 1) + 0.6\gamma BN_\gamma\right] + q$$

$$\frac{3500}{\frac{\pi \times (3.0)^2}{4}} = \tfrac{1}{3}[1.3 \times 5 \times 30.1 + 18.5 \times D_f(18.4 - 1) + 0.6 \times 18.5 \times 3.0 \times 22.4] + 18.5D_f$$

or

$$495.2 = 313.86 + 107.3D_f + 18.5D_f$$

or

$$D_f = \frac{495.2 - 313.86}{107.3 + 18.5} = 1.44 \text{ m}$$

Thus, depth of the location of pier = 1.44 m.

**Example 14.6** The weight of a heavy machinery is 7,600 kN and the base dimensions are 5.5 m × 3.5 m. The machinery has to be installed on a stiff clay soil with a cohesion of 150 kN/m$^2$, at a depth of 0.8 m below the ground surface. The total unit weight of the soil is 19.2 kN/m$^3$. Determine the size of the foundation required if the minimum factor of safety is 3.0. Assume the load to be rapidly applied so that undrained condition prevails ($\phi = 0$). Neglect the weight of the foundation.

***Solution***

Since the loading is made rapidly and the stratum is clay, Skempton's bearing capacity equation (Eq. 14.15) may be used. Thus,

$$q_f = C_u N_c + \gamma D_f$$

Provide an all-round clearance of 0.25 m then the length and width of footing may be taken as 6 and 4 m, respectively. For the condition, $N_c$ may be computed from Eq. 14.16. Thus,

$$N_c = 5.14\left(1+0.2\frac{B}{L}\right)\left[1+\left(\frac{0.053D_f}{B}\right)^{1/2}\right]$$

$$= 5.14\left(1+0.2\times\frac{4}{6}\right)\left[1+\left(0.053\times\frac{0.8}{4}\right)^{1/2}\right]$$

$$= 5.4\times1.25 = 6.43$$

Therefore,

$$q_{ns} = \frac{c_u N_c}{F} = \frac{150\times6.43}{3} = 321.5\ \text{kN/m}^2$$

Actual pressure transferred by the machine is $7600/(6\times4) = 316.67\ \text{kN/m}^2$

Hence, the recommended dimension of the foundation is 6 m × 4 m.

**Example 14.7** An eccentrically loaded rectangular footing of size 2.5 m × 3.5 m is placed at a depth of 1 m on a stiff saturated clay. The eccentricity is 0.2 m in each direction. The footing is loaded rapidly and the soil properties are $c = 105$ kN/m$^2$ and $\gamma = 17.8$ kN/m$^3$. Compute the safe net allowable bearing load on the footing if the factor of safety is 3.0 and the settlement is negligible.

***Solution***

Based on the useful width concept, the width and length of the footing are given as

$$L' = L - 2e_x = 3.5 - 2\times0.2 = 3.1\ \text{m}$$
$$B' = B - 2e_y = 2.5 - 2\times0.2 = 2.1\ \text{m}$$

The net safe soil pressure for strip footing is given by Eq. 14.19 as

$$q_{ns} = \frac{q_{nf}}{F} = \frac{1}{F}\left[c_u N_c + q(N_q - 1) + \tfrac{1}{2}\gamma B N_\gamma\right]$$

Modifying the above equation with shape factor correction and taking $N_c = 5.7$, $N_q = 1.0$, and $N\gamma = 0$, we have

$$q_{\text{ns}} = \frac{1}{F}\left[\left(1+0.2\frac{B'}{L'}\right)c_u N_c\right]$$

$$= \tfrac{1}{3}\left[\left(1+0.2\times\frac{2.1}{3.1}\right)\times 105\times 5.7\right] = 226.5\ \text{kN/m}^2$$

As the settlement is negligible, the net safe soil pressure with respect to shear strength is the net allowable soil pressure also.

Therefore, the net allowable load,

$$Q_{\text{na}} = (q_{\text{ns}})(\text{useful area}) = 226.5\times 2.1\times 3.1 = 1474.5\ \text{kN}$$

**Example 14.8** The corrected blow count from standard penetration test in a medium sand, observed at an average depth of 2.5 m was 22 blows per 305 mm. Laboratory tests conducted on the sample revealed the following physical properties: $c = 0$, $\phi = 30°$, and $\gamma = 18.5\ \text{kN/m}^3$. The water table was located at 4.5 m from the ground surface. It is planned to place a 2 m wide square footing at a depth of 2 m. Estimate the allowable gross bearing pressure for the soil if the factor of safety against shear failure is 2.5 and the limiting total settlement is 25 mm.

***Solution***

For limiting settlement, the net bearing pressure is given by Eq. 14.41 as

$$q'_{\text{na}} = C_{\text{w}}(0.41)NS'$$

Here,

$$C_{\text{w}} = 0.5 + 0.5\frac{d'_{\text{w}}}{D_{\text{f}}+B} = 0.5 + 0.5\times\frac{4.5}{2+2} = 1.063$$

Therefore,

$$q'_{\text{na}} = 1.063\times 0.41\times 22\times 25 = 239.7\ \text{kN/m}^2$$

The net bearing pressure based on the shear condition is given as

$$q_{\text{ns}} = \frac{1}{F}\left[1.2q(N_q - 1) + 0.8\times\gamma B N_\gamma\right]$$

For $\phi = 30°$, $N_q = 18.4$, and $N\gamma = 22.4$,

$$q_{\text{ns}} = \frac{1}{2.5}[1.2\times 18.5\times 2(18.4-1) + 0.8\times 18.5\times 2\times 22.4] = 574.2\ \text{kN/m}^2$$

The lower of the two net bearing pressures is taken as the net allowable bearing pressure.

Therefore, net allowable bearing pressure = 239.7 kN/m$^2$

Gross bearing pressure = 239.7 + 18.5 × 2 = 276.7 kN/m$^2$

**Example 14.9** Two plate load tests with square plates were performed on a soil deposit. For a 30 mm settlement, the following loads were obtained.

| Width of square plate (mm) | Load (kN) |
|---|---|
| 300 | 38.2 |
| 600 | 118.5 |

Determine the width of a square footing which would carry a net load of 1,500 kN for a limiting settlement of 30 mm.

***Solution***

Plate No. 1: $38.2 = (0.3)^2 m + 4 \times 0.3n$ (a)
Plate No. 2: $118.5 = (0.6)^2 m + 4 \times 0.6n$ (b)

Solving by elimination, we get,

(a) $\times 0.36 \rightarrow 0.032m + 0.43n = 13.75$
(b) $\times 0.09 \rightarrow 0.032m + 0.22n = 10.67$

Therefore,

$$n = \frac{13.75 - 10.67}{0.43 - 0.22} = 14.37 \text{ kN/m}$$

and

$$m = \frac{13.75 - 0.43 \times 14.67}{0.032} = 232.6 \text{ kN/m}^2$$

For the foundation,

$$Am + Pn = Q$$

Therefore,

$$B^2 \times 232.6 + 4B \times 14.67 = 1500$$

or

$$B^2 + 0.25B - 0.65 = 0$$

or

$$B^2 = \frac{-0.25 \pm \sqrt{[(0.25)^2 + 4 \times 10.65]}^{1/2}}{2} = \frac{-0.25 \pm 1.63}{2} = 0.69 \text{ m}$$

Therefore, the width of the square foundation = 0.69 m ≈ 0.70 m.

**Example 14.10** Plate load test data are given below. Plot the load–settlement curve and find the ultimate bearing capacity.

Width of plate = 300 mm
Least count of dial gauge = 0.01 mm

| Load intensity (kN/m²) | Dial gauge reading | | |
|---|---|---|---|
| | A | B | C |
| 0 | 0 | 0 | 0 |
| 55 | 186 | 192 | 192 |
| 110 | 362 | 365 | 353 |
| 165 | 766 | 758 | 756 |
| 220 | 1,886 | 1,889 | 1,865 |
| 280 | 4,810 | 4,806 | 4,784 |
| 335 | 14,006 | 14,010 | 13,984 |

***Solution***

Settlements are calculated by multiplying average values of dial gauge readings and the least count of dial gauges. The load–settlement curve is plotted as shown in Fig. 14.25. The ultimate bearing capacity is read from graph as 242 kN/m².

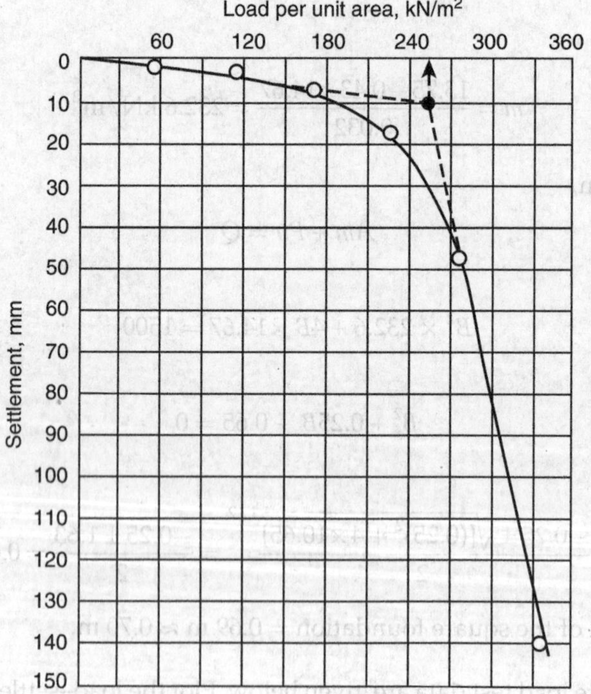


Fig. 14.25

## POINTS TO REMEMBER

14.1 General shear failure, usually associated with dense stiff soils of relatively low compress-ibility, is said to occur when a continuously well-defined slip surface develops on both sides of a footing and extends from the edge of the footing to the soil surface.

14.2 In case of local shear failure, usually associated with medium dense or medium stiff soils, the slip surfaces extend from the edges of the footings to a certain length only and do not reach the ground surface.

14.3 Punching shear failure, usually associated with loose or soft soils, is said to occur when there is compression beneath the footing accompanied by shearing in the vertical direction around the edges of the footing.

14.4 Prandtl's theory considers the deformation or penetration effects of hard objects on soft materials which is considered as the basic principle adopted in different bearing capacity theories.

14.5 $N_c$, $N_q$, and $N_\gamma$ are the bearing capacity factors which depend on angle of internal friction only.

14.6 Effect of soil compressibility is taken into account by considering the failure as a local shear failure and the corresponding bearing capacity factors, $N_c', N_q'$, and $N_\gamma'$, for the reduced friction angle $\phi_l = \tan-^1[2/3(\tan \phi)]$. The cohesion is also reduced as $c_l = 2/3c$.

14.7 Effect of water table is accounted by considering submerged unit weight in place of total unit weight depending on the location of water table.

14.8 Effect of different shapes of foundation are taken in to account by appropriate shape factors $s_c$, $s_q$, and $s_\gamma$.

14.9 The total pressure on the soil due to the weight of the structure is called the gross foundation pressure. The net foundation pressure is the foundation pressure in excess of the pressure caused by the surrounding soil. Thus, the net ultimate bearing capacity is the net foundation pressure at the time of failure.

14.10 Settlements may be classified as uniform (or total) settlement, tilt, and non-uniform settlement. The total and differential settlements should not exceed the permissible values which are denoted as permissible settlements. Permissible settlements depend on the type of structure and type of soil.

14.11 Allowable soil pressure is one which gives the lowest value based on the two conditions, viz., adequate factor of safety against shear failure and settlement should be less than permissible settlement.

## QUESTIONS

### *Objective Questions*

14.1 State whether the following statements are true or false:

(1) The bearing capacity factors for a clayey soil will depend on cohesion, shape, and size of the footing.

(2) Greater the width of foundation, greater is the settlement for the same pressure intensity.

(3) The safe bearing capacity of a surface strip footing on a saturated clay is approximately equal to the unconfined compressive strength.
(4) The correction factors to account for the effect of shapes are based on sound theoretical analysis.

14.2 The total settlement of a soil layer under any given loading is
(a) Proportional to the thickness of the layer
(b) Proportional to the square of the thickness of the layer
(c) Dependent on the length of the drainage path
(d) Dependent on factors other than the above

14.3 The ultimate bearing capacity of a footing on strip footing is reduced by 50% when the position of water table is at
(a) The base of the footing
(b) The ground surface
(c) A depth equal to 1.5 times the depth of foundation
(d) A depth equal to 0.5 times the depth of foundation

14.4 Two strip surface footings of equal lengths are placed on dry sand and the width of footing A is equal to half the width of footing B? Then the ratio of the load carrying capacities of A and B (i.e., $q_A/q_B$) is
(a) 1/2 (b) 1/4 (c) 2 (d) 1

14.5 Bearing capacity of a footing consists of the following components:
(1) The cohesion and friction of a weightless material carrying no surcharge
(2) The friction of a weightless material upon addition of a surcharge on the ground surface
(3) The friction of a material possessing weight and carrying no surcharge
Of these statements,
(a) 1, 2, and 3 are correct
(b) 1 and 2 are correct
(c) 2 and 3 are correct
(d) 3 and 1 are correct

14.6 Identify the incorrect statement. Bearing capacity of a footing on sand depends on
(a) Depth of footing
(b) Width of footing
(c) Position of water table
(d) Undrained shear strength

14.7 Plate load test results reflect only the character of the soil located within a depth ______ the width of the bearing plate. Choose the correct statement.
(a) Equal to
(b) Less than twice
(c) Equal to 2.5 times
(d) More than twice

14.8 The technique of reducing the net load on a soil by excavating soil up to a certain depth is called
(a) Load relief
(b) Buoyancy method

(c) Flotation
(d) Load reduction

14.9 Identify the incorrect statement. Meyerhof's bearing capacity equation includes
(a) Shape factor
(b) Width factor
(c) Depth factor
(d) Inclination factor

## *Descriptive Questions*

14.10 What factors determine whether a foundation type is shallow or deep?
14.11 Explain why bearing capacity equations for clay usually employ the undrained shear strength.
14.12 How will you proportion footings for equal settlements?

## EXERCISE PROBLEMS

14.1 On a cohesive friction soil, a square foundation of 2 m × 2 m is founded at 1 m depth. The soil has the following properties: $c_{cu} = 15.5$ kPa, $\phi_{cu} = 28°$, and $\gamma = 18.2$ kN/m$^3$. Determine (i) the net ultimate load, (ii) the gross ultimate load, (iii) the net safe load, and (iv) gross safe load on the footing if the factor of safety with respect to shear failure is 3.0. Assume that general shear failure occurs in the soil.
14.2 A strip footing of width 3.5 m is to be placed at a depth of 0.5 m below the ground surface on a compressible sandy silt having a bulk unit weight of 18.7 kN/m$^3$. The shear strength parameters of the soil are $c = 5$ kN/m$^2$ and $\phi = 22°$. Determine the net ultimate bearing pressure and the net safe load if the factor of safety against shear failure is 4.0.
14.3 The width of a square footing of an existing building is 2 m and is located at 1 m below the ground surface. It is proposed to add one additional floor which would make the total load on the column 1,800 kN Check whether the existing footing is adequate if it is intended to maintain a factor of safety of 3.0. The soil at the location has the following properties: $c = 20$ kPa, $\phi = 35°$, and $\gamma = 18.5$ kN/m$^3$.
14.4 Estimate the factor of safety of a 2 m square footing, located at a depth of 1.5 m and subjected to a 1,500 kN vertical load. A horizontal load of 300 kN is also applied at the base of the footing. The soil is dry sand with $\phi' = 33°$ and $\gamma = 18.2$ kN/m$^3$.
14.5 A long bridge pier, 3 m wide, carries a load of 1,640 kN per liner metre of its length. It is founded 5 m below the ground level on a soil whose angle of friction is 15° and unit cohesion 38.4 kN/m$^2$. The unit weight of the soil is 15.7 kN/m$^3$. Check for the safety of the pier if the factor of safety is 3.0.
14.6 A square footing of width 2.5 m is positioned on a medium dense sand at a depth of 2 m from the ground surface. The sand has a void ratio $e = 0.72$, specific gravity of soil solids $G = 2.65$, and the angle of shearing resistance $\phi = 35°$. Adopting a factor of safety of 2.5, find the safe load on the footing for the following water table positions:

(i) at 5 m from the ground surface, (ii) at 1.5 m from the base of the footing, and (iii) at 1.2 m from the ground surface.

14.7 The depth of a wall footing to be constructed on a saturated clay is 1 m. The soil parameters are $c_u = 65$ kPa, $\phi_u = 0°$, and $\gamma = 17.5$ kN/m$^3$. The wall imposes a load of 170 kN/m of wall length. Estimate the width of the footing to be provided so as to have a factor of safety of 3.0.

14.8 The load on a reinforced cement concrete column is 1,000 kN. The supporting soil is a dry, dense sand with the angle of friction of 41° and a unit weight of 18.2 kN/m$^3$. Find the size of the square footing for the following conditions considering a factor of safety of 3.0:

(a) if it is placed on the ground surface
(b) if it is placed at 1.5 m below the ground surface
(c) if water table rises to the ground surface for the case (b) above, the saturated unit weight is 21.3 kN/m$^3$.

14.9 A square column transfers a load of 1,650 kN on a $c$–$\phi$ soil and rests on a soil which weighs 19 kN/m$^3$ and has shear strength parameters as $c = 10$ kN/m$^2$ and $\phi = 36°$. Considering a factor of safety of 2.5, find the size of the footing if it is placed at the ground surface. Examine whether it would be cheaper to lower the footing if the column is 450 mm$^2$ and the footing is 500 mm thick than to place it at the ground surface. The cost of con-crete and the cost of excavation for hard soil at a site are Rs. 4,600/m$^3$ and Rs. 60/m$^3$, respectively.

14.10 Calculate the minimum depth of footing required below ground level in a clay stratum if the footing is to be safe

(i) For a continuous wall footing with a contact pressure of 65 kN/m$^2$ and width 1.6 m.
(ii) For a square footing with a contact pressure of 65 kN/m$^2$ and side width 1.6 m.

The undrained shear strength parameters are $c_u = 25$ kN/m$^2$, $\phi_u = 0°$, and $\gamma = 16$ kN/m$^3$. Adopt a factor of safety of 3.0. Discuss the effect of footing shape on the depth of footing.

14.11 During a sub-surface exploration programme, two cohesive layers are encountered. One forms the top layer of finite thickness 3 m, which is stiff clay and the bottom one is soft and showed undrained shear strengths of 135 and 50 kN/m$^2$, and the respective unit weights are 17.2 and 16.7 kN/m$^3$. It is intended to design a foundation 1.5 m from the ground surface. Compute the gross load for the foundation with a factor of safety of 2.5. If the layered system is assumed as homogeneous and isotropic with average values of cohesion and unit weights of both the layers, what is the percentage error involved?

14.12 Determine the safe load which can be imposed normal to the base of a strip footing which is 1.2 m wide and has its base inclined at 12° from the horizontal. One corner of the footing is located at 1.2 m from the ground surface. The footing rests on a saturated cohesive soil with a cohesion of 75 kN/m$^2$ and unit weight 18.2 kN/m$^3$.

14.13 A load bearing wall of an industrial building is to be located close to the edge of a slope as shown in Fig. 14.26. The shear strength parameters of the soil are $c_u = 45$ kN/m$^2$, $\phi_u = 0°$, and the unit weight of soil $\gamma = 18.2$ kN/m$^3$. Suggest a suitable width of strip footing for the given condition.

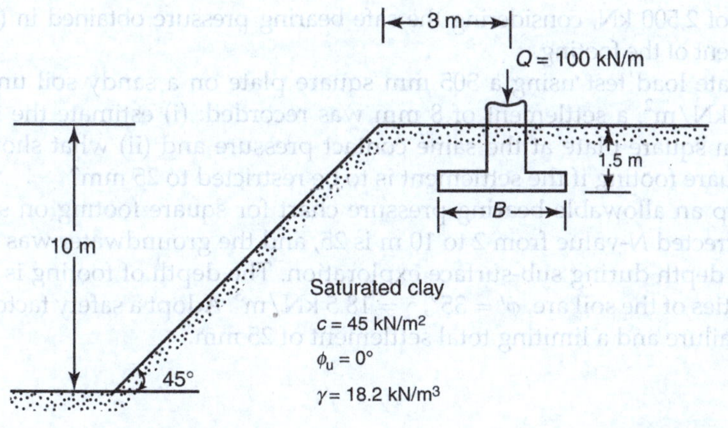


Fig. 14.26

14.14 A standard penetration test conducted at 2 m depth and a subsequent laboratory test revealed the soil at a location as medium dense sand with a blow count of 29 blows and a moist unit weight of 18 kN/m$^3$. It is planned to design a square footing on this sand to carry a load of 3,500 kN. As per the design requirement, the footing has to be designed for settlement criterion and the maximum total settlement should be limited to 25 mm. The water table is at 6 m from the ground surface. Find the size of the footing.

14.15 A square footing of 4 m width and 0.8 m thickness is supported by a sand having an average *N*-value of 30. The top of the footing is 1 m below the ground surface, and the water table is 1.2 m below the base of the footing. Determine the maximum load that the footing can carry if the settlement is not to exceed 15 mm.

14.16 The results of a plate load test conducted on a 300 mm square plate at a depth of 1 m on a dry sand is given below.

| Until applied pressure (kN/m$^2$) | Settlement (mm) |
|---|---|
| 50 | 3 |
| 100 | 5 |
| 150 | 9.8 |
| 200 | 13.0 |
| 250 | 19.0 |
| 275 | 22.0 |
| 300 | 28.0 |
| 325 | 39.0 |
| 350 | 65.0 |

Determine (i) the ultimate bearing pressure, (ii) the safe bearing pressure if the factor of safety is 3.0, (iii) the size of a square footing to be placed at the same depth and to carry a load of 2,500 kN, considering the safe bearing pressure obtained in (ii), and (iv) the settlement of the footing.

14.17 In a plate load test using a 305 mm square plate on a sandy soil under a pressure of 150 kN/m$^2$, a settlement of 8 mm was recorded: (i) estimate the settlement of a 600 mm square plate at the same contact pressure and (ii) what should be the size of a square footing if the settlement is to be restricted to 25 mm?

14.18 Develop an allowable bearing pressure chart for square footing on sand. The average corrected $N$-value from 2 to 10 m is 25, and the groundwater was encountered at a 15 m depth during sub-surface exploration. The depth of footing is 2 m. The other properties of the soil are $\phi' = 35°, \gamma = 18.5$ kN/m$^3$. Adopt a safety factor of 2.5 against shear failure and a limiting total settlement of 25 mm.

# 15

# Shallow Foundations

## CHAPTER HIGHLIGHTS

Design criteria – Types of shallow foundations – Selection of type of foundation – Location and depth of foundation Settlement of shallow foundation – Design considerations for a shallow foundation – Proportioning of combined footing – Mat foundation

## 15.1 INTRODUCTION

Structural foundations may be grouped under two broad categories – shallow foundations and deep foundations. This classification indicates the depth of foundation installation. A shallow foundation is one which is placed on a firm soil near the ground, and beneath the lowest part of the superstructure. A deep foundation is one which is placed on a soil that is not firm, and which is considerably below the lowest part of the superstructure. There is no exact definition which distinguishes one from the other.

## 15.2 DESIGN CRITERIA

While considering a shallow foundation for a given loading system, the foundation must meet certain design requirements. The three basic requirements are as follows:

1. Foundation placement, which involves the location and depth of foundation, requires a careful investigation of the past usage of the site and detailed information of the sub-surface stratum. The foundation placement should be such that any future influence should not affect its performance adversely.
2. Safety against bearing capacity is a requirement that involves suitable proportioning of footing to avoid a catastrophic collapse of the soil beneath the foundation. This

occurs if the shear strength of the soil is inadequate to support the applied load. This requirement makes it essential to have a complete knowledge of the geotechnical properties of the soils and rocks involved.

3. Tolerable foundation settlement involves keeping a check on the excessive settlement of a structure. Excessive settlement is caused due to the distortion of the soil mass as a result of the applied shear stresses and due to the consolidation of the supporting soil. This again requires a complete knowledge of the geotechnical properties of the soil to assess the anticipated settlement of the structure and the time required for completion of the same.

## 15.3 TYPES OF SHALLOW FOUNDATIONS

Shallow foundations are subdivided into a number of types according to their size, shape, and general configuration. They are described below.

### 15.3.1 Spread Footings

These footings are the most common of all types of footings involving minimum cost and complexity of construction (Fig. 15.1a). It necessarily provides the function of distributing the column load to a value compatible with the strength and deformation characteristics of the soil or rock on which the foundation is placed. These types of footings are also known as *pad footings*, *isolated footings*, and *square* or *rectangular footings* (for an $L/B$ ratio less than 5).

### 15.3.2 Combined Footings

These footings are formed by combining two or more columns (even with unequal loadings) into one footing. This arrangement averages out and provides a more or less uniform load distribution in the supporting soil or rock and, thus, prevents differential settlement. These footings are usually rectangular in shape but may be modified to a trapezoidal one to accommodate unequal column loadings (or columns close to property lines) and provided with a *strap* to accommodate wide column spacings or column close to property lines (Fig. 15.1b).

### 15.3.3 Continuous Footings

These footings carry closely spaced columns or a continuous wall so that the load intensity is low and uniform on the supporting soil or rock (Fig. 15.1c). In such footings, the load per unit length is considered accordingly. The load intensity is given in terms of force per unit length of the footing. These footings are also referred to as *strip footings* or *wall footings* (for an $L/B$ ratio greater than 5).

### 15.3.4 Mat Foundations or Footings

These are characterized by the feature that columns frame into the footing in two directions. Any number of columns can be accommodated, and the number can be as low as four (Fig. 15.1d). These are recommended for poor foundation soils and when the total area of footings exceeds 50% of the total plinth area.

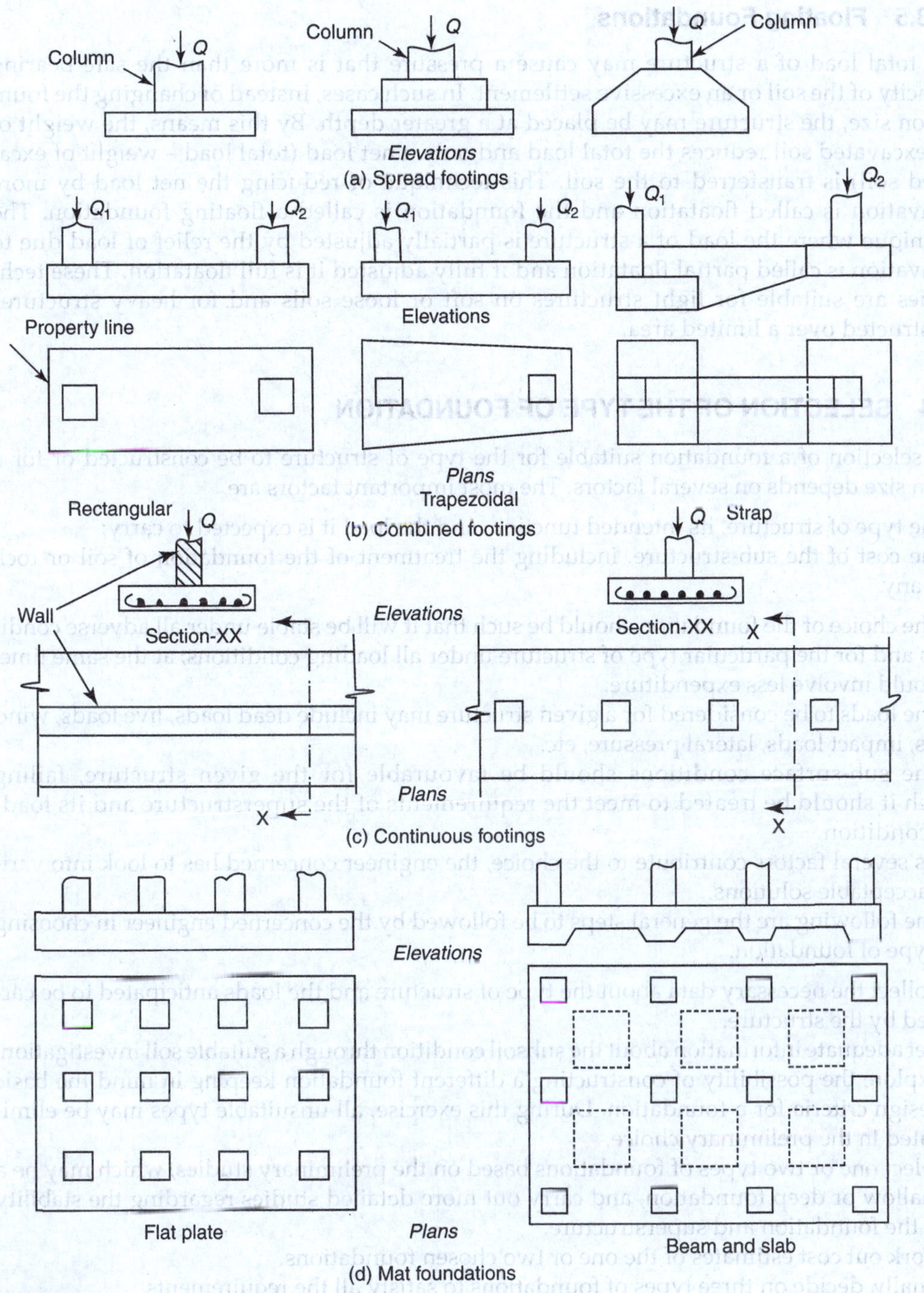


**Fig. 15.1** Types of shallow foundations

### 15.3.5 Floating Foundations

The total load of a structure may cause a pressure that is more than the safe bearing capacity of the soil or an excessive settlement. In such cases, instead of changing the foundation size, the structure may be placed at a greater depth. By this means, the weight of the excavated soil reduces the total load and only a net load (total load – weight of excavated soil) is transferred to the soil. This technique of reducing the net load by more excavation is called floatation and the foundation is called a floating foundation. The technique where the load of a structure is partially adjusted by the relief of load due to excavation is called partial floatation and if fully adjusted it is full floatation. These techniques are suitable for light structures on soft or loose soils and for heavy structures constructed over a limited area.

## 15.4 SELECTION OF THE TYPE OF FOUNDATION

The selection of a foundation suitable for the type of structure to be constructed or for a given size depends on several factors. The most important factors are

1. the type of structure, its intended function, and the load it is expected to carry;
2. the cost of the sub-structure, including the treatment of the foundation of soil or rock if any.

The choice of the foundation should be such that it will be stable under all adverse conditions and for the particular type of structure under all loading conditions; at the same time, it should involve less expenditure.

The loads to be considered for a given structure may include dead loads, live loads, wind loads, impact loads, lateral pressure, etc.

The sub-surface conditions should be favourable for the given structure, failing which it should be treated to meet the requirements of the superstructure and its loading condition.

As several factors contribute to the choice, the engineer concerned has to look into various acceptable solutions.

The following are the general steps to be followed by the concerned engineer in choosing the type of foundation:

1. Collect the necessary data about the type of structure and the loads anticipated to be carried by the structure.
2. Get adequate information about the subsoil condition through a suitable soil investigation.
3. Explore the possibility of constructing a different foundation keeping in mind the basic design criteria for a foundation. During this exercise, all unsuitable types may be eliminated in the preliminary choice.
4. Select one or two types of foundations based on the preliminary studies, which may be a shallow or deep foundation, and carry out more detailed studies regarding the stability of the foundation and superstructure.
5. Work out cost estimates of the one or two chosen foundations.
6. Finally, decide on three types of foundations to satisfy all the requirements.

## 15.5 LOCATION AND DEPTH OF THE FOUNDATION

The foundation must be located properly keeping in view both the horizontal and the vertical orientations, such that it is not affected by outside influences, apart from the general design criteria of bearing capacity and excessive settlement. Outside influences may include groundwater, volume changes, underground defects, adjacent structure, etc.

Thus, the depth and location of a foundation depend on the following data:

1. volume change of soil
2. adjacent structures
3. groundwater
4. underground defects

### 15.5.1 Volume Change of Soil

In general, the foundation must be placed below the zone of volume change. Highly plastic clays (called expansive soils) shrink significantly upon drying and swell significantly upon wetting. This volume change is greatest near the surface and decreases with increase in depth. So, during weather changes, a certain depth of the soil undergoes a volume change. This zone is called the zone of volume change.

Similarly, in areas where the air temperature falls below the freezing point, the moisture near the soil surface may freeze. The frozen moisture may melt due to increase in temperature. As the soil moisture freezes and melts, it alternately expands and contracts leading to volume changes of the soil.

Repeated expansion and shrinkage due to soil type or temperature change may cause the foundation to lift and drop. Such a sequence cannot be acceptable for the stability of a structure.

The IS code (IS: 1904, 1986) recommends that a foundation should be located at a minimum depth of 50 cm below the natural ground surface.

For expansive soils (black cotton soils of India) the zone of volume change varies from 1.5 to 3.5 m.

In order to avoid the above expansion and shrinkage, it is advisable to place the footing below the zone of volume change.

### 15.5.2 Adjacent Structures

The horizontal location of a footing is often affected by adjacent structures and property lines. The construction of a new structure may damage the existing adjacent structure by vibration and shock due to blasting, caving in due to nearby excavation, lowering of the water table or increasing the stress.

The Indian Standards code (IS: 1904, 1986) recommends the following for footings placed adjacent to a sloping ground or when the bases of footings are at different levels.

When the ground surface slopes downwards adjacent to a footing, the sloping surface should not encroach upon a frustum of bearing material under the footing, as shown in Fig. 15.2(a) and (b) for granular soils and clayey soils, respectively.

In order to avoid damage to an existing structure, the following norms may be followed (as shown in Fig.15.2c):

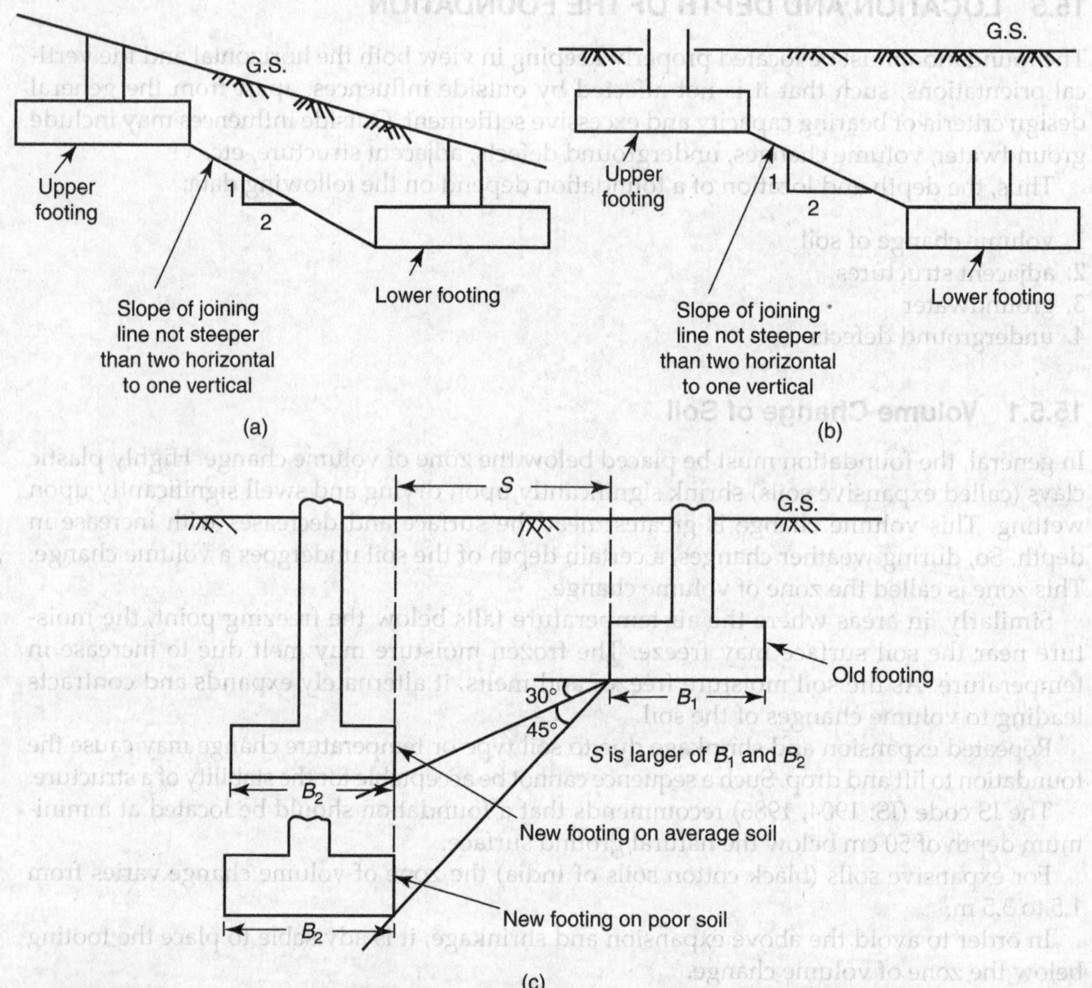


**Fig. 15.2** Footings at different levels for (a) granular soil, (b) clay soil, and (c) footings for old and new structures (*Source:* IS: 1904, 1980)

1. The footing should be placed at least at a distance *S* from the edge of the existing footing where *S* is the width of the larger footing.
2. The line from the edge of the new footing to the edge of the existing footing should make an angle of 45° or less.
3. When a new footing is placed lower than an old footing, the excavation for the foundation must be carefully done with a suitable bracing system so as to prevent damage to the existing structure.

Special care must be taken in placing a footing at or near a property line, so as to avoid encroachment of the footing into the adjacent property.

### 15.5.3 Groundwater

For several reasons, the presence of groundwater within the soil immediately around a footing is not desirable. The following points need consideration:

1. Construction of a footing below the groundwater level is difficult and expensive because the area must be dried prior to construction.
2. Existence of groundwater around a footing may reduce the bearing capacity of the soil, particularly in sands.
3. Excess groundwater around a footing may cause hydrostatic uplift problems.
4. In areas of sub-zero temperatures, frost action may predominate.
5. Existence of groundwater below a floor may add to waterproofing problems.

For the above reasons, as far as possible, footings should be placed above the groundwater level.

### 15.5.4 Underground Defects

The presence of underground defects may also affect the location of a footing. The underground defects may be faults, caves, mines, and man-made discontinuities, such as sewer lines, underground cables, and utilities. Construction of structures on or near tectonic faults should be avoided. Further, foundations should not be placed directly on caves mines or on man-made discontinuities.

## 15.6 CAUSES OF SETTLEMENT

Foundation settlement may occur due to the following reasons:

1. Elastic compression of the foundation and the underlying soil, also called immediate settlement, may be one cause. It is computed by idealizing the soil as an elastic material (dealt with in Section 15.7.1 in detail).
2. Plastic or inelastic compression of the underlying soil, called time-dependent settlement or settlement due to consolidation (both primary and secondary), which was dealt with in Chapter 8, could be another cause.
3. Groundwater lowering is another major cause for settlement to occur. Repeated raising and lowering of groundwater, particularly in granular soils, tends to reduce the void volume and causes settlement of the ground surface. Prolonged lowering of water table may cause settlement in fine-grained soils. Pumping of water or draining of water without proper filter material may also cause settlement.
4. Vibrations caused by pile driving, machinery, blasting, etc. may cause settlement, particularly in granular soils.
5. Other causes of settlement include volume change of soil, ground movement and excavation for adjacent structures, mining subsidence, etc.

## 15.7 SETTLEMENT OF SHALLOW FOUNDATIONS

Analytical methods are available for computing the settlement of shallow foundations under a symmetrical static vertical load only. Settlement due to other causes such as deterioration of the foundation, mine subsidence, and catastrophic settlement are not dealt with. The methods of estimating immediate settlement are discussed below.

### 15.7.1 Immediate Settlement

All highly permeable soils, including all non-cohesive soils, undergo immediate settlement. Immediate settlement also occurs in fine-grained soils. It is computationally convenient to idealize the soil as an elastic material, and the results from the mathematical theory of elasticity can be applied with full confidence to compute the settlement (Lambe and Whitman, 1979). It is generally accepted that immediate settlement predominates in non-cohesive soils and the estimation of settlement based on elastic theory is quite appropriate. In saturated clays, Leonards (1962) attributed the immediate settlement to shear strains caused by shear stresses. Further, as the shear stresses are small, the immediate settlement may be computed assuming that the soil mass behaves like an elastic solid (Leonards, 1962). Thus, immediate settlement, in metres, is calculated from

$$S_i = qB\frac{1-v^2}{E_u}I_f \tag{15.1}$$

where $q$ is the intensity of contact pressure (kN/m$^2$), $B$ the least lateral dimension of the loaded area (m), $E_u$ the undrained modulus of elasticity (kN/m$^2$), and $I_f$ the influence factor, which depends on rigidity and the shape of the loaded area (Table 15.1).

Equation 15.1 is basically given for a surface flexible loaded area. The same can be used for rigid footings by modifying the influence factor. As per IS: 8009 – Part 1 (1976), the total settlement of a rigid footing is taken to be 0.8 times the settlement at the centre of the flexible foundation. Hence, the influence factors for a rigid foundation are also given in Table 15.1.

Equation 15.1 is further based on the assumption that the elastic medium is a semi-infinite mass. For a compressible stratum of finite thickness ($H_t$), Steinbrenner's influence factor for settlement at the corner of the loaded area is recommended by Indian Standards (IS: 8009 – Part 1, 1976) (Fig. 15.3).

**Table 15.1** Influence factors for vertical displacement due to elastic compression

| Shape | Flexible | | | Rigid |
|---|---|---|---|---|
| | Centre | Corner | Average | |
| Circle | 1.00 | 0.64 | 0.85 | 0.80 |
| Rectangle | | | | |
| 1.0 | 1.12 | 0.56 | 0.95 | 0.90 |
| 1.5 | 1.36 | 0.68 | 1.20 | 1.09 |
| 2.0 | 1.53 | 0.77 | 1.31 | 1.22 |
| 5.0 | 2.10 | 1.05 | 1.83 | 1.68 |
| 10.0 | 2.52 | 1.26 | 2.25 | 2.02 |
| 100.0 | 3.38 | 1.69 | 2.96 | 2.70 |

*Source:* IS: 8009 – Part 1 (1976).

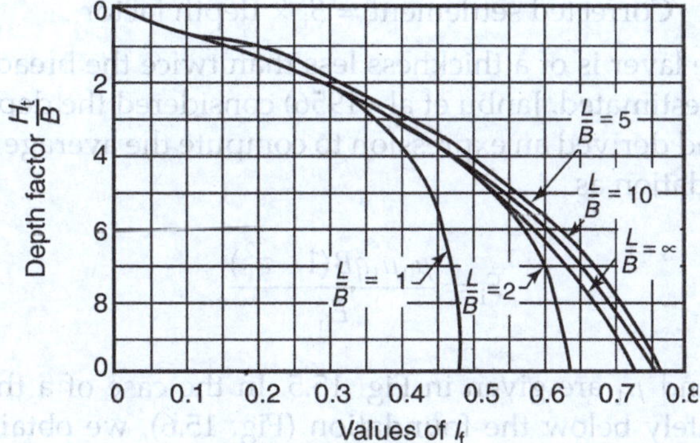


**Fig. 15.3** Steinbrenner's chart for the influence factor (*Source:* IS: 8009 – Part 1, 1976)

The above discussion has centred on foundations located at the ground surface. For foundations located at a certain depth, a depth factor correction has been suggested (IS: 8009 – Part 1, 1976). The depth factor can be read from Fig. 15.4 for different $L/B$ ratios. Hence,

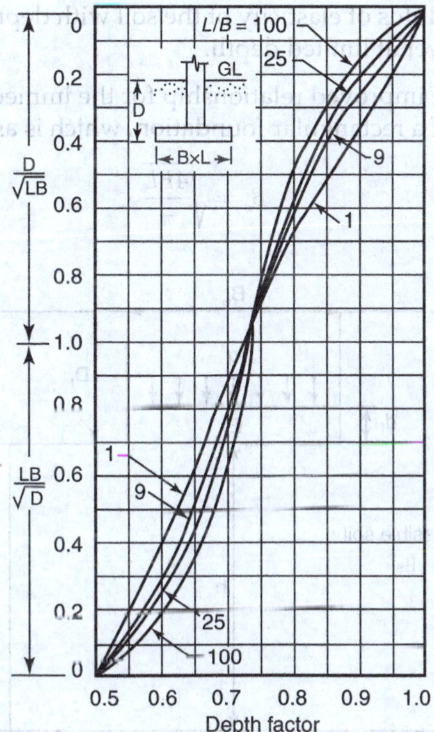


**Fig. 15.4** Fox's correction curves for settlements of flexible rectangular footings (*Source:* IS: 8009 – Part 1, 1976)

$$\text{Corrected settlement} = S_i \times \text{depth factor} \tag{15.2}$$

If the compressible layer is of a thickness less than twice the breadth, it is observed that the settlement is overestimated. Janbu et al. (1956) considered the depth as well as the finite compressible layer and derived an expression to compute the average immediate settlement under a flexible foundation as

$$S_i = \frac{\mu_0 \mu_1 q B (1 - v^2)}{E_u} \tag{15.3}$$

The values of $\mu_0$ and $\mu_1$ are given in Fig. 15.5. In the case of a thin layer (of thickness $H_t$) existing immediately below the foundation (Fig. 15.6), we obtain a value of $\mu_{1(t)}$ corresponding to the thickness $H_t$ and similarly find a value of $\mu_{1(b)}$ corresponding to the thickness $H_b$. Then, the immediate settlement due to the thin layer is computed by taking $\mu_1 = \mu_{1(b)} - \mu_{1(t)}$.

Improved relationship for estimating the immediate settlement of shallow foundation has been presented by Mayne and Poulos (1999). This improved relationship considers the following:

1. the rigidity of the foundation,
2. the depth of embedment of the foundation,
3. the increase in the modulus of elasticity of the soil with depth, and
4. the location of rigid layer at limited depth.

Figure 15.4a shows the improved relationship for the immediate settlement. Hence, $B_e$ is an equivalent diameter of a rectangular foundation, which is as follows:

$$B_e = \sqrt{\frac{4BL}{\pi}}.$$

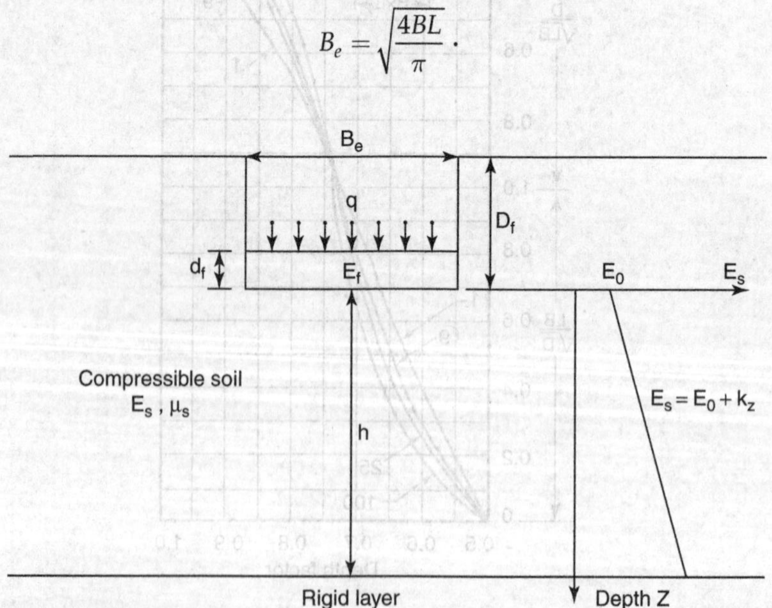


Fig. 15.4a Improved relationship for immediate settlement

In the above equation, $B$ and $L$ are the width and length of the foundation. For a circular foundation, $B_e$ is equal to the diameter $B$ of the foundation.

Hence $E_f$ is the modulus of elasticity of the foundation, $d_f$ is the thickness of the foundation, and $D_f$ is the depth of the foundation below the ground surface. A rigid layer is located at a depth $h$ below the bottom of the foundation.

Then, modulus of elasticity of the compressible soil, $E_s$, is given as:

$$E_s = E_o + k_z,$$

where, $E_o$ is the modulus of elasticity at the base of the foundation.

$z$ = depth.

$k$ = slope.

Now, the improved expression for the immediate settlement is given as follows:

$$(S_i)_i = q\, B_e \left( \frac{1 - v_s^2}{E_o} \right) I_G I_F I_E,$$

where, $I_G$ = influence factor for the variation of $E_s$ with the depth, which is a function of $E_o$, $k$, $B_e$, and $h$.

$I_F$ = foundation rigidity correction factor.

$I_E$ = foundation embedded correction factor.

Figure 15.4b shows the variation of $I_G$ with $\beta = E_o/k\,B_e$ and $h/B_e$.

The foundation rigidity correction factor $I_F$ is given as:

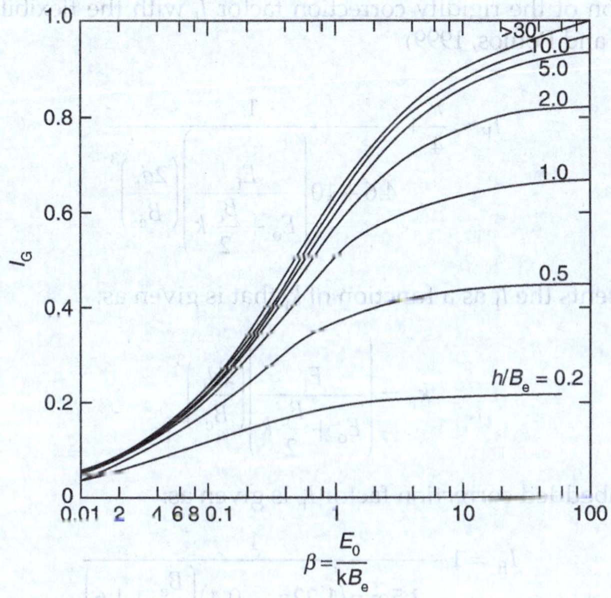


Fig. 15.4b Variation of $I_G$ with $\beta$ (*Source:* Mayne and Poulos, 1999)

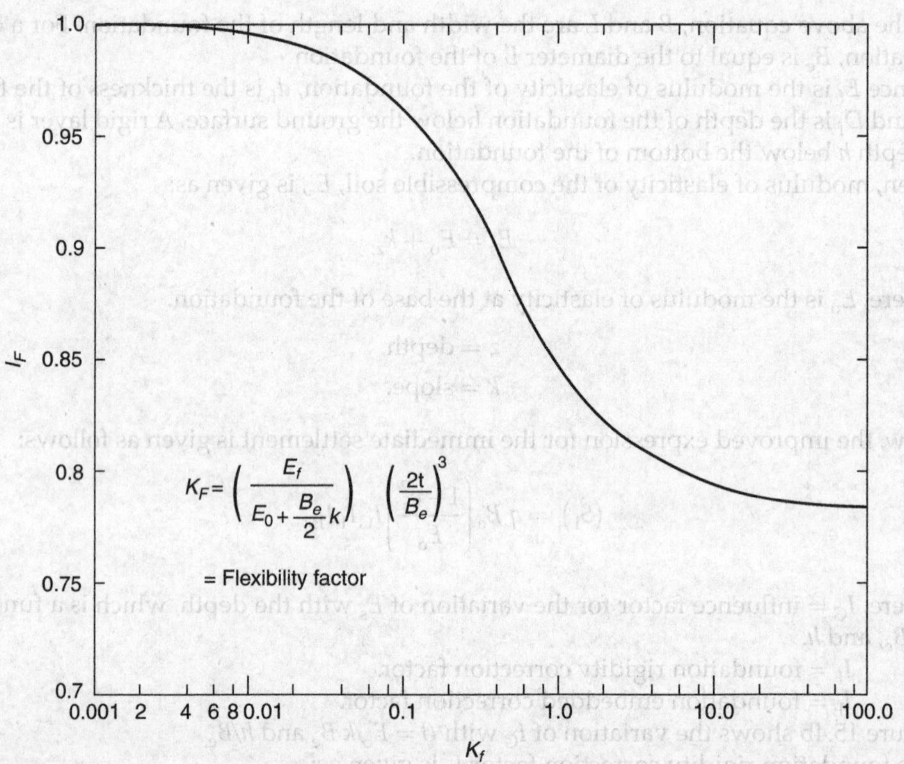


Fig. 15.4c Variation of the rigidity correction factor $I_F$ with the flexibility factor $K_f$ (*Source:* Mayne and Poulos, 1999)

$$I_F = \frac{\pi}{4} + \frac{1}{4.6 + 10\left(\dfrac{E_f}{E_o + \dfrac{B_e}{2}k}\right)\left(\dfrac{2d_f}{B_e}\right)^3}$$

Figure 15.4c presents the $I_F$ as a function of $k_f$ that is given as:

$$k_F = \left(\frac{E_f}{E_o + \dfrac{B_e}{2}k}\right)\left(\frac{2d_f}{B_e}\right)^3$$

Similarly, the embedded correction factor $I_E$ is given as:

$$I_E = 1 - \frac{1}{3.5\exp(1.22v_s - 0.4)\left(\dfrac{B_e}{D_f} + 1.6\right)}$$

Figure 15.4d shows the variation of $I_E$ with $D_f/B_e$.

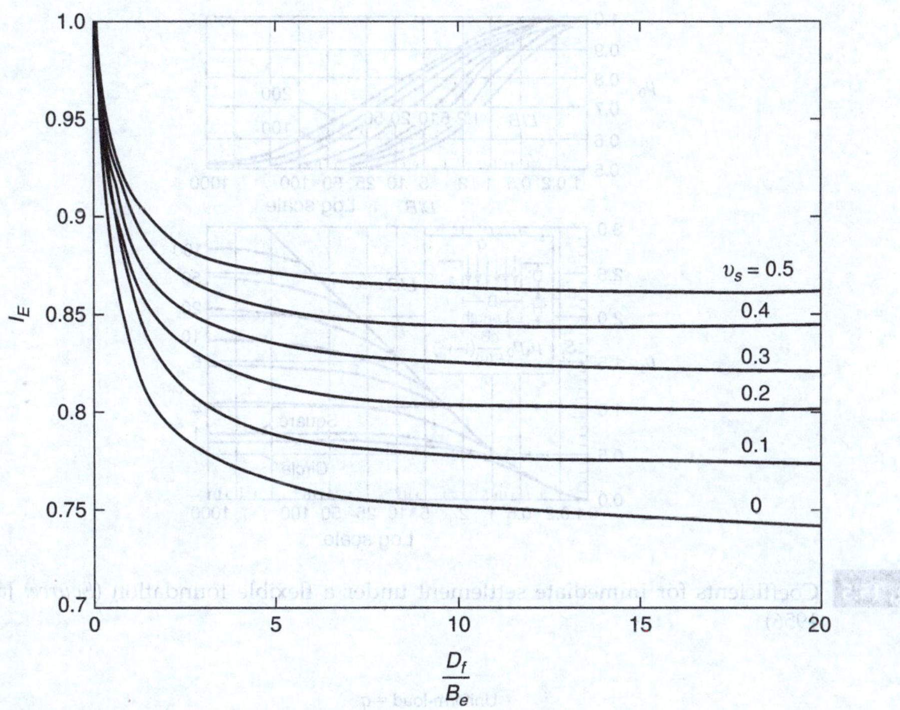


**Fig. 15.4d** Variation of the embedded correction factor $I_E$ (*Source:* Mayne and poulos, 1999)

## 15.7.2 Consolidation Settlement

As discussed in Chapter 8, settlements due to primary and secondary compression are important in fine-grained soils. Evaluation of settlement based on secondary compression has not yet been standardized. The methods of computing primary compression (already discussed) need certain modifications depending on the field situations. The usual field situation is that the clay layer is sandwiched between cohesionless soil layers or between a cohesionless soil layer at the top and rock at the bottom. These are conventional situations for which settlement equations, discussed in Chapter 8, can be used.

One major limitation of the estimation of consolidation settlement (based on oedometer results) is that the laboratory condition is one-dimensional whereas the actual field condition is different. Under one-dimensional conditions, the lateral strain is zero and the increase in pore water pressure is equal to the increase in total stress, i.e., $A = 1.0$. The field condition required for $A = 1$ is that the extent of the loaded area should be large compared with the thickness of the layer.

Under such conditions, the lateral strain is not zero and the total settlement comprises

1. immediate settlement for the undrained condition and
2. consolidation settlement due to drainage.

Skempton and Bjerrum (1957) have considered the thickness of the compressible layer, width of the loaded area, and pore pressure parameter $A$ and suggested a settlement coefficient $\mu_c$ to obtain the corrected consolidation settlement ($S_c'$).

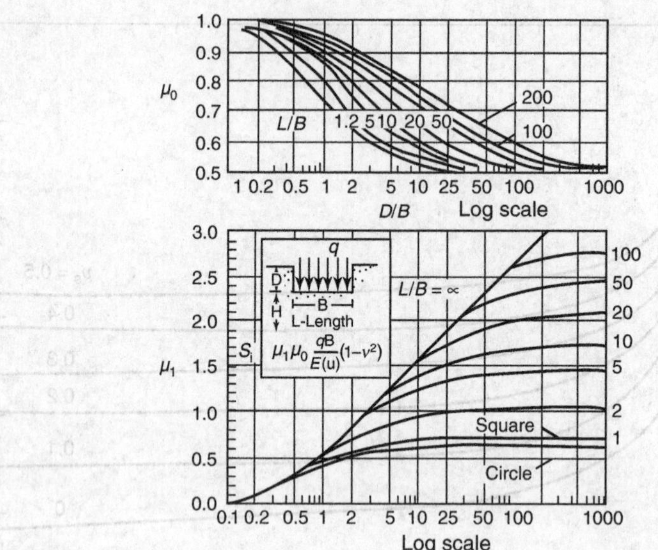


Fig. 15.5 Coefficients for immediate settlement under a flexible foundation (*Source:* Janbu et al., 1956)

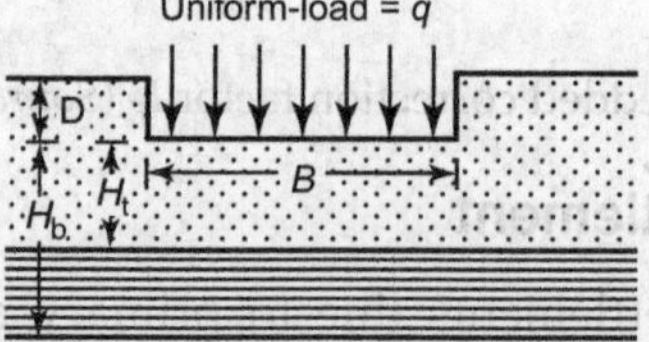


Fig. 15.6 Thin layer below foundation

$$S_c' = \mu_c S_c$$

where $\mu_c$ is a factor related to the pore pressure parameter $A$ and the ratio $H_t/B$.

This settlement coefficient chart (Fig. 15.7) suggested by Skempton and Bjerrum has been adopted in the Indian Standards (IS: 8009 – Part 1, 1976). In the absence of data for the parameter $A$, $\mu_c$ values from Table 15.2 may be used.

### 15.7.3 Evaluation of Settlement from Field Tests

There is much difficulty in the sampling of cohesionless soils and no standard test procedure is available for finding the compressibility characteristics of cohesionless soils. In such soils, the settlement may be estimated based on the data from *in situ* bearing tests such as the static cone penetration test, standard penetration test, and plate bearing test.

From a static cone penetration test (IS: 4968 – Part 3, 1976), a curve showing the depth and static cone penetration resistance ($q_{cs}$) is drawn (Fig. 15.8). The entire profile is divided

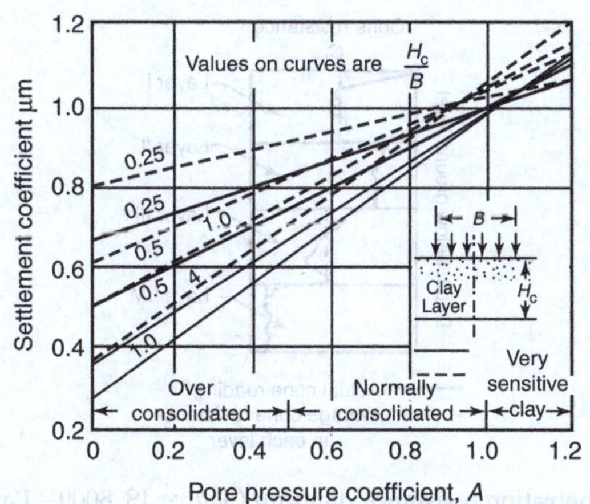


**Fig. 15.7** Settlement coefficients for circular and strip footings (*Source:* IS: 8009 – Part 1, 1976)

**Table 15.2** Values of $\mu_c$

| Type of clay | $\mu_c$ |
|---|---|
| Very sensitive clays (soft alluvial, estuarine, and marine clays) | 1.0–1.2 |
| Normally consolidated clays | 0.7–1.0 |
| Over-consolidated clays | 0.5–0.7 |
| Heavily over-consolidated clays | 0.2–0.5 |

into strata or layers with constant $q_{cs}$ for each layer. The settlement for each layer due to the foundation load is calculated from Eq. 15.4. Then, the settlements corresponding to all the layers are added to get the total settlement ($S$)

$$S_{cs} = 2.303\frac{H_t}{C}\log_{10}\left[\frac{p_0+\Delta p}{p_0}\right] \tag{15.4}$$

where

$$C = \frac{3}{2}\frac{q_{cs}}{p_0}$$

$H_t$ is the thickness of each layer and $p_0$ the initial effective pressure at mid-height of the layer. Hence, $S = \sum S_{cs}$.

Indian Standards (IS: 8009 – Part 1, 1976) also provides a chart for dry cohesionless soils relating settlement of a footing of width $B$ under unit intensity of pressure for different $N$ values (Fig. 15.9). For a given pressure on the footings, the settlement is proportional to the intensity of pressure. If the water table is located at a shallow depth from beneath the footing base the correction factor $R_w$ is found and the settlement obtained from Fig. 15.9 is multiplied by $R_w$.

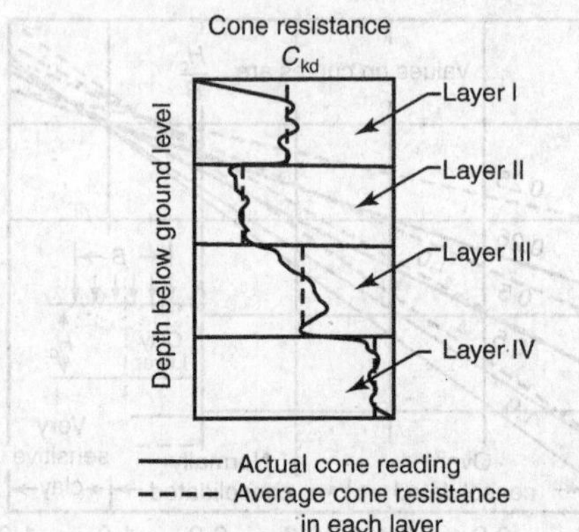


Fig. 15.8 Static cone penetration resistance diagram (*Source:* IS: 8009 – Part 1, 1976)

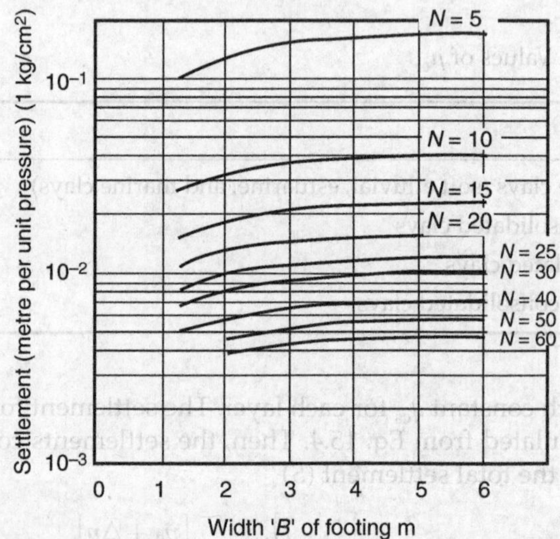


Fig. 15.9 Settlement per unit pressure from standard penetration resistance (*Source:* IS: 8009 – Part 1, 1976)

The settlement can also be estimated from the plate load test (IS: 8009 – Part 1, 1976) data. The test details are explained in Chapter 14. The plate load test (IS: 1888, 1982) is conducted at the required depth and the settlement ($S_t$) of the proposed foundation is found from Eq. 15.5.

$$S_t = S_p \left[ \frac{B_f(B_p + 0.3)}{B_p(B_f + 0.3)} \right]^2 \tag{15.5}$$

The water table correction factor can also be applied, if necessary, as explained in the previous paragraph.

### 15.7.4 Reliability of Settlement Computations

In general, the allowable bearing pressure in cohesionless soils is governed by settlement criteria rather than shear. Further, in such soils, the immediate settlement predominates with some creep effects. In sands, most of the settlement occurs during the initial stage of construction of the building due to the combination of construction vibrations and loads. The settlement predictions for cohesionless soils are based on conservative methods (Bowles, 1984). But there are many factors which affect the settlement of footings in sand, viz., relative density, amount of fines and gradation of sand, the size of the loaded area, the position of the water table, *in situ* stresses, and capillarity. A comprehensive critical review of the evaluation of settlement in sand was presented by Ramasamy (1984) and was modified by him later (Ramasamy, 1986). This modified method considers, among other routine factors, the effect of past loading history, non-linear load settlement behaviour of footings, and the effect of fines in sand. Further, the effects of embedment (e.g., Kaniraj, 1977; Ramasamy et al., 1982), capillary zone (Ramasamy et al., 1986), bearing area and plan dimension (Kaniraj, 1977), and stiffness of footing (Arora and Varadarajan, 1984) on settlement in sand have been reported in the literature.

In other soils, depending on the percentage clay fraction and organic matter, all the three components of settlement may be present. Consolidation settlement may predominate in inorganic silts and clays and in highly organic soils it could be secondary compression.

Consolidation settlement can be reasonably predicted provided test results from undisturbed samples are available (Mac Donald and Skempton, 1955; Skempton, 1955). The effect of depth of embedment change in bearing area and plan dimensions on the settlement of normally consolidated clay has been studied by Kaniraj and Ranganathan (1977). It has been shown that there is a decrease in settlement with increase in depth of embedment, bearing area, and length of footing.

## 15.8 DESIGN STEPS FOR A SHALLOW FOUNDATION

Generally, a footing may have to carry a load-bearing wall or a single column or more than one column to support the structure. Columns generally carry different loads depending on the location and based on the type of structure.

Each column has to carry a different type of load, and the major areas are dead and live loads. Further, the entire amount of live load is not borne by the column for the entire life period. Hence, it is generally considered a service load which will be applied on the column during the entire lifespan, which is taken as dead load plus 50% of the live load for ordinary buildings. A large percentage of live load should be used in warehouses and other storage floors. The design of column footings based on service loads is generally adequate.

Footings may be designed by adopting the following procedure (Teng, 1962):

1. Calculate the loads acting on the footing.
2. Obtain a soil profile or soil profiles showing the soil stratification at the site.
3. Establish the maximum water level.
4. Obtain the pertinent field and laboratory measurements and test results.
5. Determine the depth and location of the footing.

6. Determine the bearing capacity of the supporting stratum.
7. Proportion the footing sizes.
8. Check the footing contact pressure.
9. Check the stability of the footing against sliding, overturning, and uplift pressures.
10. Estimate the total and differential settlements.
11. Design the footing structure.
12. Assess the need for foundation drains, waterproofing or damp proofing.

## 15.9 PROPORTIONATING FOOTING SIZE

Footing sizes are basically designed for safe bearing and then checked for permissible total and differential settlements. The size is modified if the permissible settlements are not satisfied. While proportionating, the following procedure may be followed (Teng, 1962):

1. Calculate the load on the structure:

   $L_{l+d}$ = live load + dead load

   for the column which has the largest live load to dead load ratio.
2. Calculate the service load for the same column:

   $L_s$ = dead load + $C$ (live load)

   where $C$ is ½ for ordinary buildings and ¾ for warehouses and storage floors.
3. Decide the storage floor's safe bearing pressure:

   $Q_a$ = safe bearing pressure from theories or field tests
4. Compute the design pressure:

   $Q_d$ = design pressure

   for all footings except the one with the lowest live load to dead load ratio.

   $$q_d = L_c / A$$

   where $L_s$ is the service load. That is,

   $$\text{proportioned area of footing} = \frac{\text{service load}}{q_d}$$
5. Compute the area of footing supporting the column with the lowest live load to dead load ratio:

   $$A = (L_{1+d}) / q_s$$
6. Decide the length and width of footing and check for permissible settlement and alter width if needed.

## 15.10 DESIGN OF COMBINED FOOTINGS

As discussed earlier, a combined footing is used when equal or unequal columns are positioned so closely that individual footings are not practicable. These footings are usually rectangular in shape. In order to accommodate unequal column loads or columns close to property lines, the rectangular shape is modified to a trapezoidal shape. Sometimes, a strap is provided to combine two columns which have a wider spacing or if one of them is close to a property line.

The conventional methods of design of combined footings are based on the following considerations:

1. The footing is infinitely rigid and does not have any bearing on the pressure distribution.
2. The soil pressure beneath the footing is linearly distributed (or is distributed on a plane surface) such that the centre of the soil pressure coincides with the line of action of the resultant force of all the loads.

### 15.10.1 Rectangular Combined Footing

The following steps are adopted for the design of a rectangular combined footing (Fig. 15.10):

1. Find the total column loads which are to be positioned in the footing and determine the line of action of the resultant.
2. Obtain the soil pressure distribution (stress per unit length of the footing).
3. Find the width, $B$, of the footing.
4. Draw the shear force diagrams along the length of the footing.
5. Draw the bending moment diagram along the length of the footing.
6. Design the footing as a continuous beam to withstand the shear and bending moment.
7. Design the footings for transverse bending also in the same manner as for spread footings.

Then, the structural design details for the footing are worked out.

### 15.10.2 Trapezoidal Combined Footing

As discussed earlier, a trapezoidal combined footing is preferred (i) when two unequally loaded columns are encountered and (ii) when the property line is quite close to the exterior column. In the first case, the higher loaded column will have a larger width. The soil pressure distribution is linear or uniformly varying (not uniform) as shown in Fig. 15.11.

The width of the trapezoid is determined from the following equations:

$$\frac{L}{2}(B_1 + B_2)q_a = P_e + P_i$$

where $q_a$ is the allowable soil pressure. That is,

$$B_1 + B_2 = \frac{2(P_e + P_i)}{Lq_a}$$

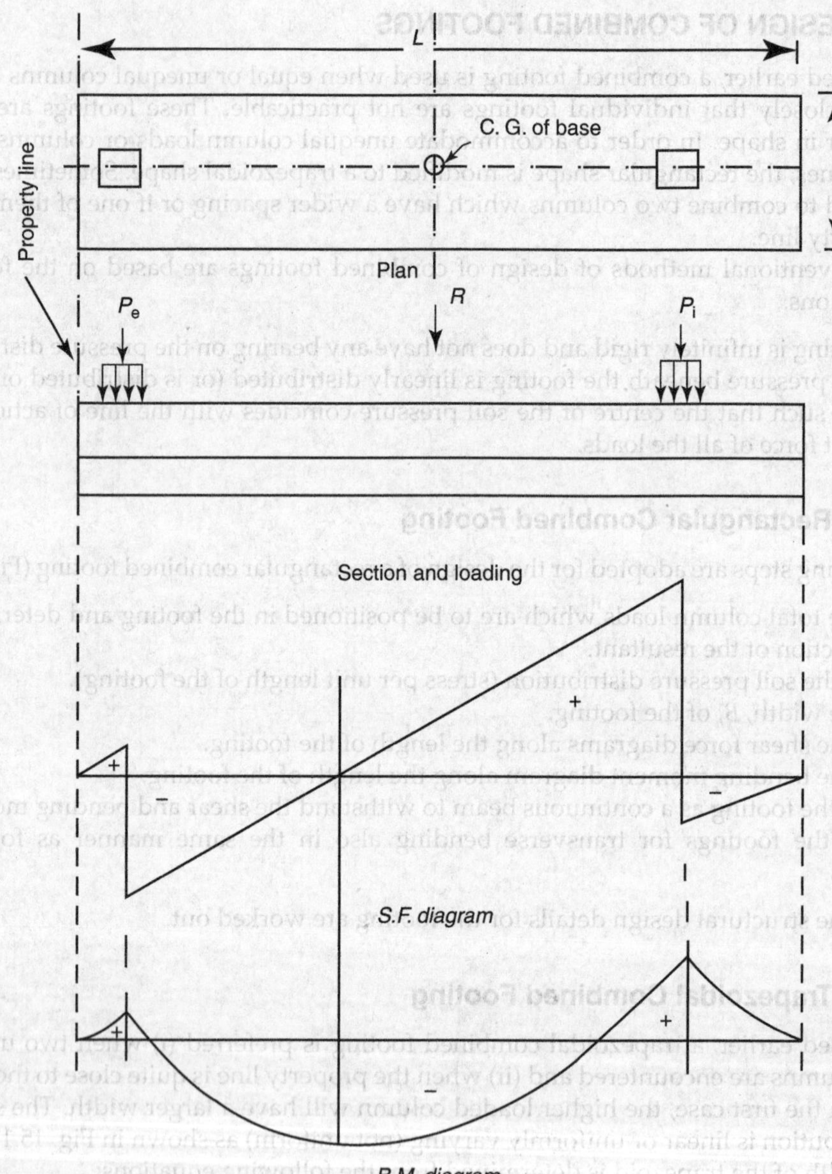


Fig. 15.10 Rectangular combined footing

By taking moments about the property line or left edge and simplifying,

$$\frac{2B_1 + B_2}{B_1 + B_2} = \frac{3}{L}\left[e_1 + \frac{2P_i L}{P_e + P_i}\right]$$

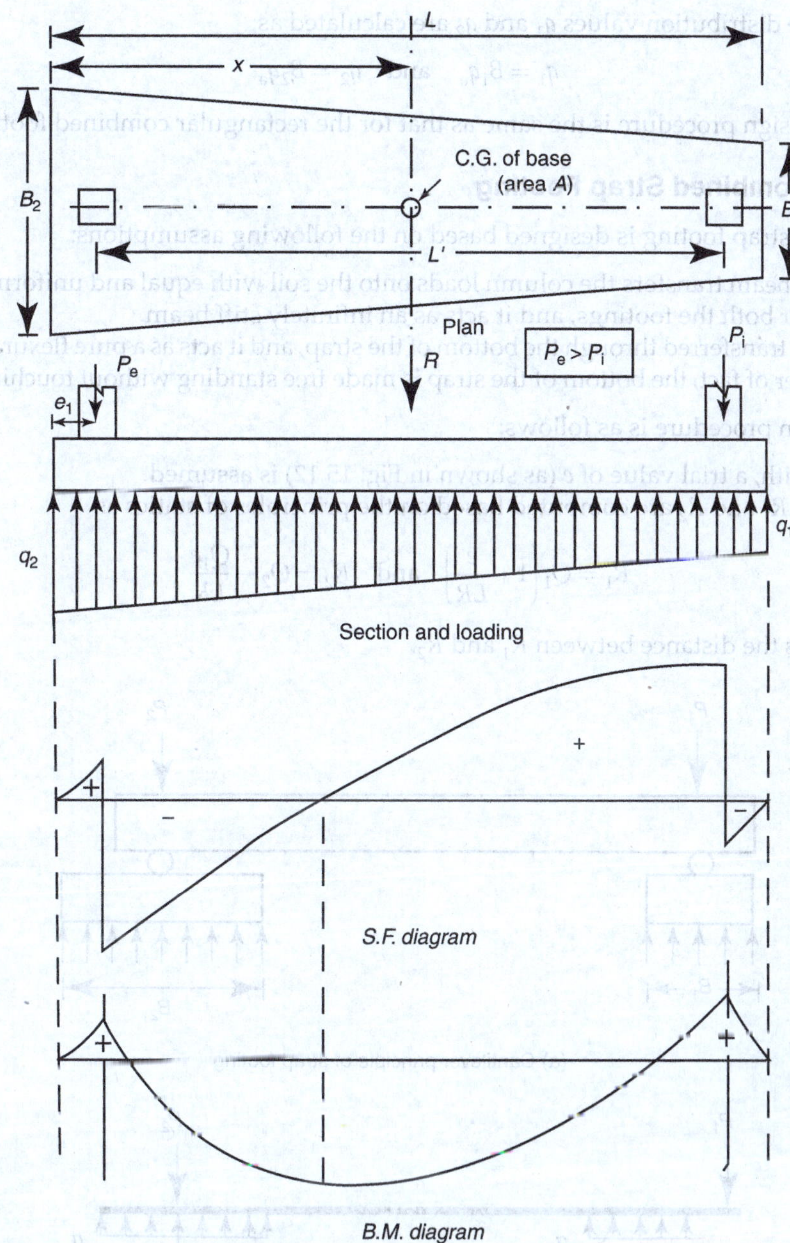


**Fig. 15.11** Trapezoidal combined footing

$B_1$ and $B_2$ may be obtained from the above two equations. That is,

$$B_1 = \frac{2A}{L}\left(\frac{3\bar{x}}{L} - 1\right) \quad \text{and} \quad B_2 = \frac{2A}{L} - B_1$$

The pressure distribution values $q_1$ and $q_2$ are calculated as

$$q_1 = B_1 q_a \quad \text{and} \quad q_2 = B_2 q_a$$

The other design procedure is the same as that for the rectangular combined footing.

### 15.10.3 Combined Strap Footing

A combined strap footing is designed based on the following assumptions:

1. The strap beam transfers the column loads onto the soil with equal and uniform soil pressure under both the footings, and it acts as an infinitely stiff beam.
2. No load is transferred through the bottom of the strap, and it acts as a pure flexural member. As a matter of fact, the bottom of the strap is made free standing without touching the soil.

The design procedure is as follows:

1. To start with, a trial value of $e$ (as shown in Fig. 15.12) is assumed.
2. Reactions $R_1$ and $R_2$ are computed based on the principles of statics as

$$R_1 = Q_1\left(1+\frac{e}{LR}\right) \quad \text{and} \quad R_2 = Q_2 - \frac{Q_1 e}{LR}$$

where $R$ is the distance between $R_1$ and $R_2$.

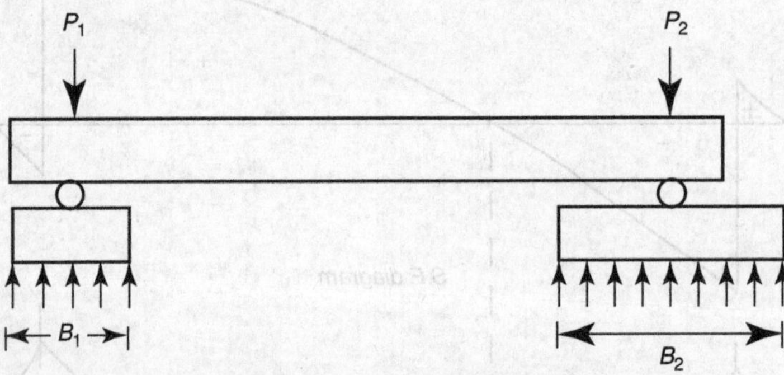

(a) Cantilever principle of strap footing

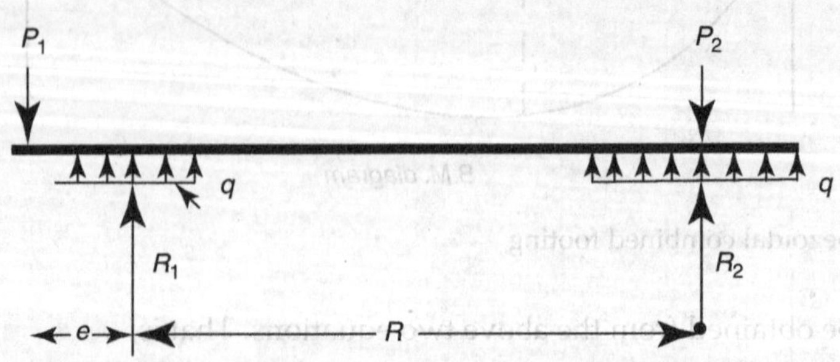

(b) Load–reaction details

Fig. 15.12 Strap footing

3. With a knowledge of the safe soil pressure, the tentative footing areas are computed.
4. Knowing the footing areas, $e$ is calculated.
5. Steps 1 to 4 are repeated till the $e$ value is identical to the final one.
6. The footings are designed as a simple spread footing subjected to uniform soil pressure.
7. Shear and bending moments are computed and the strap beam is designed.

Based on the above design procedure, the centre of gravity of the two footing areas will coincide with the resultant of the column loads $Q_1$ and $Q_2$.

## 15.11 MAT FOUNDATION

A mat or raft foundation is a large footing, usually supporting several columns in two or more rows. The choice between strap footing, combined footing, and raft foundation depends on the soil type and the relative cost. In a well-bearing ground, strap footing may be more economical than combined footing, whereas in a soil of low bearing capacity a large strap footing may be less preferred due to the cost.

As a general rule, mat foundations are used where the soil has low bearing capacity. Since the bearing capacity increases with increase in width and depth of the foundation, the mat foundation gains a two-fold advantage.

The mat foundation, in general, is a flat concrete slab having uniform thickness throughout the entire area. This is adopted where the column spacing is fairly small and uniform and the column loads relatively small. However, for large column loads, the slab has to be thicker.

### 15.11.1 Types of Mat Foundations

Some of the common types of mat foundations are shown in Fig. 15.13 (Das, 1984):

1. flat plate with a mat of uniform thickness (Fig. 15.13a);
2. flat plate thickened under columns (Fig. 15.13b);
3. beams and slab, wherein the beams run both ways and the columns are located at the intersection of the beams (Fig. 15.13c);
4. slab with basement walls as a part of the mat, with walls providing stiffness for the mat (Fig. 15.13d)

Mats may be placed directly on the soil or may be supported on piles. Piles are preferred particularly when the height of groundwater is large.

### 15.11.2 Bearing Capacity of Mat Foundations

The gross ultimate bearing capacity ($q_{ug}$) of a mat foundation can be determined in the same way as for a shallow foundation. The term $B$ to be used in the general equation is the smallest dimension of the mat. The net ultimate bearing capacity ($q_{un}$) is

$$q_{un} = q_{ug} - q$$

where $q$ is the surcharge.

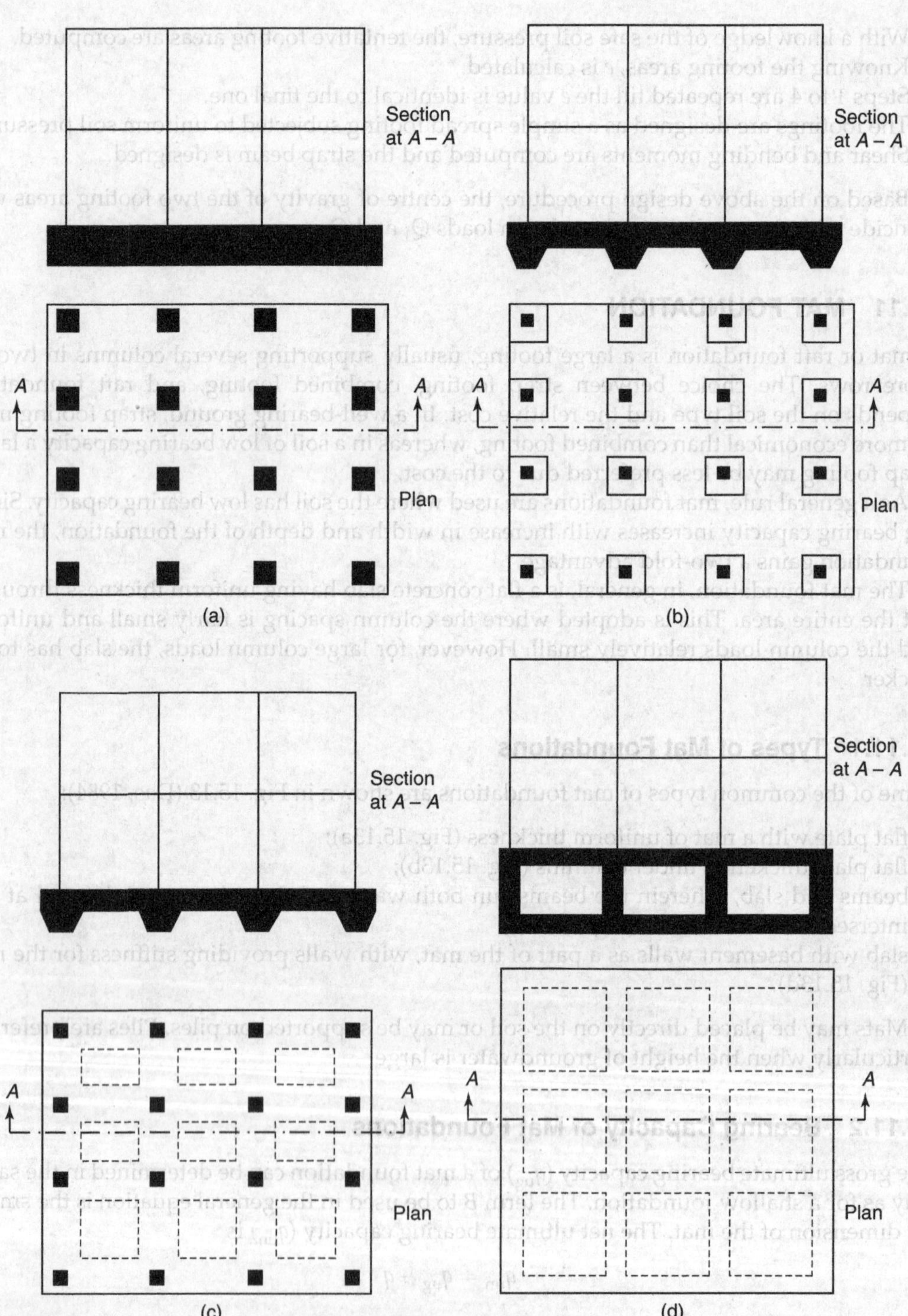


**Fig. 15.13** Types of mat foundations (*Source:* Das, 1984)

A suitable factor of safety, varying from 1.75 to 3, is used. For mats on clay, the factor of safety should not be less than 3 under dead load and maximum live load. Also, for mats on sand, again, the factor of safety should not be less than 3. On no account should the factor of safety be less than 1.75. In general, under normal working conditions, the factor of safety against bearing capacity failure of mats on sand is very large. On granular deposits, the net ultimate bearing capacity may be found based on Standard Penetration Resistance numbers.

Generally, the settlement permissible is about double that for spread footing, for the reason that the depth of the zone of influence is likely to be much larger than that of the spread footing. For this reason, the loose soil pockets under a mat may be evenly distributed, resulting in a smaller differential settlement. Generally, the permissible total settlement is 50 mm and differential settlement is 19 mm.

### 15.11.3 Design Methods

The design of mat foundations may be done using any one of the following three methods: conventional (rigid) method, simplified elastic (flexible) method, and truly elastic foundation method.

The conventional method of design assumes that (i) the mat is infinitely rigid; i.e., the deflection of the mat does not influence the pressure distribution and (ii) the soil pressure is a plane surface such that the centroid of the soil pressure coincides with the line of action of the resultant force of all the loads acting on the foundation.

The simplified flexible method assumes that the soil behaves like an infinite number of springs where none of the springs is affected by the others. The elastic constant of the springs is equal to the coefficient of sub-grade reaction, which is defined as the unit pressure required to produce a unit settlement. A number of procedures have been developed for the analysis of beams based on the simplified elastic foundation concept.

In the truly elastic foundation method, the soil is assumed to be a truly elastic material obeying Hooke's law in all directions. This method is very rarely used because of its complexity.

The conventional method of design is similar to that of the design of combined footings.

## WORKED EXAMPLES

**Example 15.1** The circular foundation of a ground-level oil tank of 20 m diameter, transmits to the soil a uniform contact pressure of 250 kN/m$^2$ at a 3 m depth. Determine the immediate settlement under the centre of the foundation. The properties of the soil are as follows: $E_u = 60$ MN/m$^2$, $v = 0.45$, and $\gamma = 22$ kN/m$^3$.

***Solution***

As the foundation is for an oil tank, consider it to be a flexible one. From Table 15.1 for circular flexible footings, the value of $I_f$ at the centre is 1.0

$$S_i = \frac{qB(1-v^2)}{E_u} I_f$$

Net contact pressure $q = 250 - (22 \times 3) = 184\ \text{kN/m}^2$. Therefore,

$$S_i = \frac{184 \times 20(1-0.45^2) \times 1 \times 10^3}{60 \times 10^3} = 48.9\ \text{mm}$$

**Example 15.2** A rectangular footing of dimensions 4 m × 2 m is founded at a depth of 2 m on a soil with $E_u = 48\ \text{MN/m}^2$ and $\nu = 0.50$. A rigid layer is laid under the soil at a depth of 10 m from the ground surface. The foundation transmits a uniform contact pressure of $200\ \text{kN/m}^2$. Estimate the average immediate settlement likely to occur.

***Solution***

Refer to Fig. 15.5

$$\frac{D}{B} = \frac{2}{2} = 1 \quad \text{and} \quad \frac{L}{B} = \frac{4}{2} = 2$$

For

$$\frac{D}{B} = 1 \quad \text{and} \quad \frac{L}{B} = 2,\ \mu_0 = 0.78$$

$$\text{and} \quad \frac{H}{B} = \frac{10}{2} = 5$$

For $H/B = 5$ and $L/B = 2.0$, $\mu_1 = 0.84$. Therefore,

$$S_i = \mu_0 \mu_1 \frac{qB}{E_u}(1-\upsilon^2)$$

$$= \frac{0.78 \times 0.84 \times 200 \times 2(1-0.5^2)10^3}{48 \times 10^3} = 4.1\ \text{mm}$$

**Example 15.3** A flexible rectangular foundation of dimensions 3 m × 1.5 m is placed at a depth of 2 m saturated in a clayey soil of infinite depth. The undrained modulus of clay, $E_u = 45\ \text{MN/m}^2$, $\nu = 0.42$, and the unit weight of clay $\gamma = 19.8\ \text{kN/m}^3$. The foundation transmits uniform contact pressure of $230\ \text{kN/m}^2$. Determine the average immediate settlement expected to occur under the foundation. If a hard stratum exists below the clay stratum at a depth of 6 m, what will be the change in the settlement?

***Solution***

From Eq. 15.1 the immediate settlement for flexible footing on an saturated clay of infinite depth is given as

$$S_i = qB\frac{1-\nu^2}{E_u} I_f$$

$$= 230 \times 1.5 \times \frac{(1-0.42^2)}{45 \times 10^3} \times 1.20 \times 10^3 = 7.6\ \text{mm}$$

From Eq. 15.3, the immediate settlement for flexible footing on saturated clay underlain by a hard stratum is given as

$$S_i = \mu_0 \mu_1 \frac{qB(1-\nu^2)}{E_u}$$

From Eq. 15.3 for $L/B = 2$ and $D/B = 2/1.5 = 1.33$, $\mu_0 = 0.85$. For $L/B = 2$ and $H/B = 6/1.5 = 4$, $\mu_1 = 0.80$. Therefore,

$$S_i = 0.85 \times 0.80 \times 230 \times \frac{1.5(1-0.42^2)}{45\times10^3} = 4.03 \text{ mm}$$

**Example 15.4** Settlement of a square footing of dimensions 1.2 m × 1.2 m carrying a load of 220 kN/m$^2$ is 30 mm. What would be the settlement of a footing measuring 3 m × 3 m carrying a load of 160 kN/m$^2$? The subsoil conditions are identical in both the footings.

***Solution***

Consider the immediate settlement equation

$$S_i = qB\left(\frac{1-\nu^2}{E_u}\right)I_f$$

For a given soil,

$$S_i = qBI_fX$$

where $X = \left(\frac{1-\nu^2}{E_u}\right)$

For the 1.2 m × 1.2 m footing, $30 = 220 \times 1.2 \times 0.62(X)$.

Therefore,

$$X = \left(\frac{30}{220\times1.2\times0.62}\right)$$

For the 3 m × 3 m footing,

$$S_i = 160\times3\times0.84\left(\frac{30}{220\times1.2\times0.62}\right) = 73.9 \text{ mm}$$

**Example 15.5** A combined footing has to be proportioned for the two columns detailed below.

| Column loads | Column $C_1$ | Column $C_2$ |
|---|---|---|
| Dead load | 550 kN | 700 kN |
| Live load | 400 kN | 800 kN |

The distance between the columns is 5.0 m. The footing should not be beyond 0.5 m from the face of the column.

The soil pressures to be considered are:

For dead load (DL) + reduced live load (LL) = 160 kN/m$^2$
For dead load + live load = 230 kN/m$^2$

***Solution***

| Total column loads | Column $C_1$ (kN) | Column $C_2$ (kN) | Total (kN) |
|---|---|---|---|
| DL + reduced LL | 550 + 200 = 750 | 700 + 400 = 1100 | 1860 |
| DL + LL | 550 + 400 = 950 | 700 + 800 = 1500 | 2450 |

1. Uniform soil pressure under DL + 50% LL condition.

Let $$\bar{x} = \frac{(1100 \times 5.0)}{1860} = 2.94 \text{ m}$$

Therefore,

$$\text{Length } L = 2(2.94 + 0.50) = 6.88 \text{ m say } 7.00 \text{ m}$$

$$\text{Width } B = \frac{1860}{160 \times 7.00} = 1.66 \text{ m say } 1.70 \text{ m}$$

2. Uniform soil pressure under DL + LL condition

Let $\bar{y}$ be the distance of a resultant from column $C_1$

$$\bar{y} = \frac{1500 \times 5.0}{2450} = 3.06 \text{ m}$$

From column $C_1$ to the centre of gravity of footing is the eccentricity, $e = 3.06 - 3.00 = 0.06$ m

$$q_{\max} = \frac{2450}{7.0 \times 1.70}\left(1 + \frac{6 \times 0.06}{7.0}\right) = 216.5 \text{ kN/m}^2 < 230 \text{ kN/m}^2$$

$$q_{\min} = \frac{2450}{7.0 \times 1.70}\left(1 - \frac{6 \times 0.06}{7.0}\right) = 195.3 \text{ kN/m}^2$$

The size of the footing is 7.0 × 1.70 mm.

**Example 15.6** Proportion a trapezoidal combined footing for two columns 300 mm × 300 mm carrying column loads of 800 kN and 1200 kN if the spacing between the columns is 4.5 m. Take the allowable soil pressure as 250 kN/m$^2$ and the length of the footing as 5 m.

***Solution***

Let A and B be columns of loads 800 kN and 1,200 kN, respectively.
Length of the footing $L = 5$ m
Let the projection of footing beyond the column face be 0.5 m
Then, distance from centre to centre of the column, $L' = 5 - 2 \times 0.5 = 4$ m.
Total area required

$$A = \frac{800 + 1200}{250} = 8 \text{ m}^2$$

Let the distance of column loads from column A be

$$\bar{x}_1 = \frac{1200 \times 4}{2000} = 2.4 \text{ m}$$

Let $\bar{x}$ be the distance of the resultant column from the left edge $= 2.4 + 0.5 = 2.9$ m. Let $B_1$ be the longer width and $B_2$ be the shorter width. $B_1$ near to load of 1,200 kN and $B_2$ near to 800 kN. Therefore,

$$B_1 = \frac{2A}{L}\left(\frac{3\bar{x}}{L} - 1\right) = \frac{2\times 8}{5}\left(\frac{3\times 2.9}{5} - 1\right) = 1.73 \text{ m}$$

$$B_2 = \frac{2A}{L} - B_1 = \frac{2\times 8}{5} - 1.73 = 3.2 - 1.73 = 1.47 \text{ m}$$

Total area provided is

$$\frac{1.73 + 1.47}{2} \times 5 = 8 \text{ m}^2$$

The trapezoidal footing will have widths of 1.73 m and 1.47 m with a length of 5 m and with the centre to centre of columns as 4 m.

## POINTS TO REMEMBER

15.1 Shallow foundations are those placed on a firm soil near the ground and beneath the lowest part of the superstructure.

15.2 A shallow foundation for a given loading system must meet three design requirements, regarding (i) foundation placement, (ii) safety against bearing capacity failure, and (iii) safety against permissible settlement.

15.3 The types of shallow foundations are spread footings, combined footings, continuous footings, and mat foundations.

15.4 Settlement of a shallow foundation could be of two types: immediate settlement and consolidation settlement. Immediate settlement is computed by idealizing the soil as an elastic material and using the results from the mathematical theory of elasticity. Consolidation settlement is due to drainage.

15.5 Combined footings are used for combining two or more columns into one footing. If the loading on the columns is not very much different, a rectangular footing may be used. If the difference in column loads is more, a trapezoidal footing is preferred.

15.6 The pressure transmitted from the base of a foundation to the soil is termed the contact pressure. This depends on the rigidity of the foundation structure and the nature of the soil.

## QUESTIONS

### *Objective Questions*

15.1 Ultimate settlement of footings on cohesive soils is best estimated from the data from
(a) Plate load test
(b) Consolidation test

(c) Cone penetration test
(d) Standard penetration test

15.2 The permissible settlement of a shallow foundation on rock for evaluation of safe bearing pressure from a plate load test is taken as
(a) 12 mm (b) 25 mm
(c) 40 mm (d) 50 mm

15.3 Identify the incorrect statement. Non-uniform settlement can result from
(a) Non-uniform bearing stress
(b) Non-homogeneous subsoil conditions
(c) Non-uniform stress acting upon a homogeneous soil
(d) Variation of water required at the construction site

15.4 Identify the incorrect statement. Settlement of a structure can be important for the following reasons:
(a) Imminent rupture of the structure
(b) Appearance of the structure
(c) Utility of the structure
(d) Damage to the structure

15.5 The dependence of the settlement of a footing in sand on the width of the footing is
(a) Directly proportional
(b) Indirectly proportional
(c) Logarithmically proportional
(d) None of the above

15.6 The major problem with settlement analysis is
(1) obtaining reliable values of the elastic parameters;
(2) obtaining a reliable stress profile from the applied load.
Of these statements
(a) 1 alone is correct
(b) 2 alone is correct
(c) 1 and 2 are correct
(d) None of them is correct

15.7 The vertical displacement of a rigid loaded area under a pressure $q$ is taken to be ______ times the displacement at the centre of the flexible area. Choose the correct factor from those listed below:
(a) 1.4 (b) 1.2
(c) 0.80 (d) 0.60

15.8 The influence factor for vertical displacement at the centre of a rectangular area varies from
(a) 1.52 to 2.10
(b) 1.12 to 1.52
(c) 1.12 to 1.00
(d) 1.00 to 0.76

15.9 The influence factor for vertical displacement at the centre of a flexible circular area is
(a) 1.0 (b) 1.5 (c) 2.0 (d) 0.80

### Descriptive Questions

15.10 Indicate the circumstances under which combined footings are adopted.

15.11 What precautions are to be taken while locating a footing (i) on a slope and (ii) adjacent to an existing structure?

15.12 What are the different types of settlements which are to be considered in the design of a shallow foundation?

## EXERCISE PROBLEMS

15.1 Determine the dimensions of a rectangular combined footing for the data given below:

| | Column A (kN) | Column B (kN) |
|---|---|---|
| Dead load | 480 | 680 |
| Live load | 350 | 450 |

Average allowable soil pressure is 250 kN/m$^2$. Distance between centre to centre of column is 4.6 m. The projection beyond column A should not exceed 0.4 m.

15.2 It is decided to provide a strap footing for two columns A and B as detailed below:

Column loads: load on A = 1,500 kN; load on B = 1,450 kN.
Size of column = 0.5 m.
Centre to centre of columns = 5.8 m.
Allowable soil pressure = 370 kN/m$^2$.

Determine the size of the footing for columns A and B.

15.3 Column loads on columns A and B are 1,920 kN and 1,500 kN, respectively. Column B is a boundary column. Proportion a trapezoidal footing. The allowable soil pressure is 200 kPa.

# 16

# Pile Foundations

## CHAPTER HIGHLIGHTS

Classification of piles: material composition, installation methods, ground effects, function as foundation – Pile-driving equipment – Bearing capacity of single pile: statistical methods, pile-driving formulae, wave equation, based on SPT values, pile load test – Negative skin friction – Under reamed piles – Pile groups: group capacity, group in filled ground – Group settlement – Pile cap

## 16.1 INTRODUCTION

The design and construction of deep foundations for transferring the weight of the superstructure through soft or weak soils, to deep load-bearing strata is a challenging job for a civil engineer. Piles, piers, and caissons are the most common types of deep foundations. The mechanism for deriving support from the soil or rock below and adjacent to the foundations is similar for any system. Each system differs in its method of construction. Piles are slender structural members normally installed by driving with hammer or by vibrating, and occasionally by auguering. Pre-drilling or other procedures may be necessary to permit penetration to the desired depth.

## 16.2 CLASSIFICATION OF PILES

Piles can be classified according to their material composition, installation method, ground effect, and their function as a foundation.

### 16.2.1 Material Composition

Based on material composition, the piles may be further classified as timber, steel, concrete, or composite piles. Timber piles are the oldest types of piles made from tree

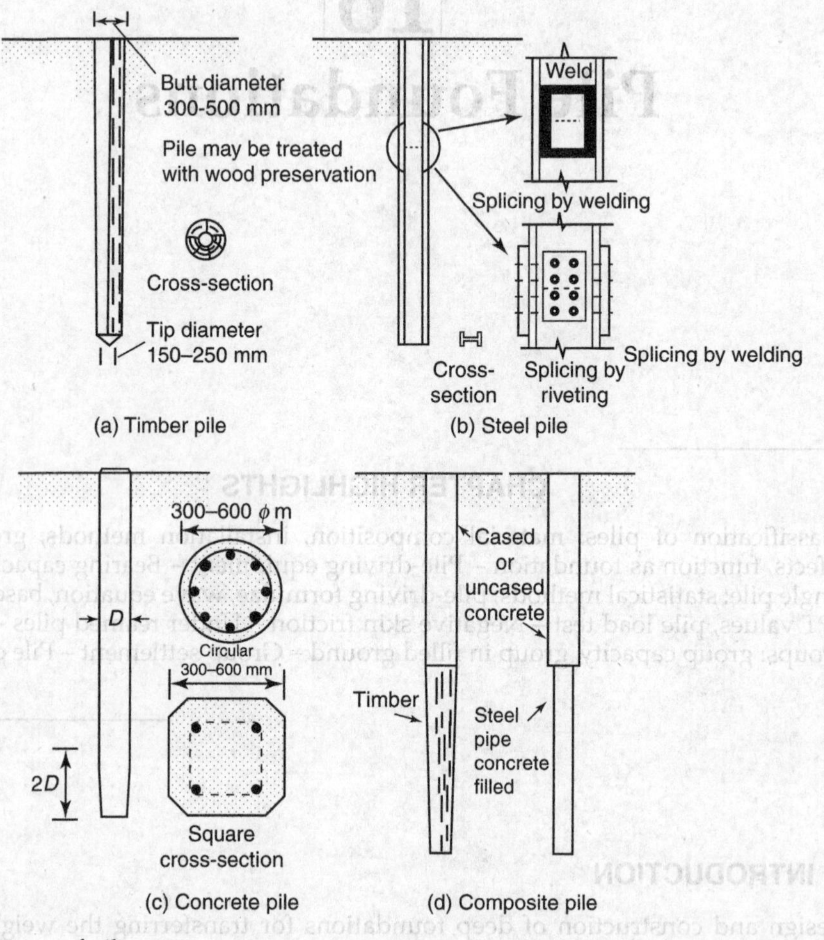


**Fig. 16.1** Types of piles

trunks that have had their branches carefully trimmed off. The maximum length of a pile is about 20 m. The timber should be straight, sound and without any defects (Fig. 16.1a). Timber piles are installed by driving. Overdriving of timber piles may result in splitting, crushing, and/or shearing of piles. Timber piles have a long life if prevented from alternative wetting and drying. A plain timber pile permanently below water will not decay. The life of timber piles may be increased by treating them with preservatives. Preservative protection treatment is required to protect timber piles from marine borers (if for a marine environment), or from wood-infesting insects, such as termites, or from decay (wet rot) if the pile is embedded in soil above water table. Timber piles find extensive use for compaction of soils, for supporting structures, and for protecting water-front structures. As per Indian Standards (IS: 2911, Part 2, 1980), piles are classified as Class A or Class B depending on the use. Piles used for railway and highway bridges, trestles, docks, and wharves are categorized as Class A. These piles shall have butt diameter or the sides of square not less than 30 cm. Piles used for foundation work and other temporary works are categorized as Class B. Such piles may have diameters less than 30 cm. For compaction piles, usually a 10 cm diameter is recommended.

Steel piles generally used are either pipe piles or rolled steel H-section piles. Pipe piles are normally filled with concrete. Steel piles are spliced by welding or riveting. The steel piles may be provided with driving points or shoes – for driving through dense materials. Because of the high strength of steel, these piles withstand driving pressures. Strength, relative ease of splicing, and economy are some of the advantages of steel piles (Fig. 16.1b). Steel piles are affected by corrosive agents such as salt, acid, moisture, and oxygen. To account for corrosion, an additional thickness of steel is generally recommended. Protection from corrosion is effected by applying epoxy coatings on the pile surface before driving, or by concrete encasement in most corrosive zones of the steel pile.

Concrete piles are cast to specified lengths and shapes of circular, square, or octagonal cross-sections with reinforcement. The reinforcement is provided to enable the pile to resist the bending moment developed during lifting and transportation (Fig. 16.1c).

Piles can also be pre-stressed using high tensile steel cables. The cables are pre-tensioned up to about 1,300 MN/m$^2$ before pouring concrete around them. After curing, the cables are cut, thus producing a compressive force on the pile section. Concrete piles are frequently used in marine environment. These piles are limited to a length of 25 m and the diameter is generally less than 0.5 m. Concrete piles may be of precast or in-place type. The installation process of cast-in-place piles are discussed in the next section.

For cast *in situ* concrete piles the reinforcing cage depends on the installation condition, the nature of the subsoil, and the nature of load transmission is 0.4% of the sectional area. The minimum cover for main reinforcement should not be less than 50 mm. The lateral reinforcement imparts adequate rigidity. The minimum diameter and spacing of links or spirals are 6 and 150 mm, respectively (IS: 2911, Sections 1 and 3, 1979). The area of main reinforcement for precast piles shall not be less than the following percentages of the cross-sectional area of the piles:

1. Pile length < 30 times the least width: 1.25%
2. Pile length 30 to 40 times the least width: 1.5%
3. Pile length > 40 times the least width: 2%

The lateral reinforcement resists the driving stresses induced in the piles and should be in the form of hoops or links of diameter not less than 6 mm. The volume of lateral reinforcement shall not be less than the following:

1. At each end of the pile for a distance of about three times the least width – not less than 0.6% of the gross volume of pile and
2. In the body of the pile – not less than 0.2% of the gross volume of piles.

Close spacing is provided near the ends and the maximum spacing over a length of three times the least width of the pile (IS: 2911 – Part 1/Section 3). The cover of concrete is not less than 40 mm, and in places where corrosion is anticipated, the cover should be at least 50 mm.

Materials and method of manufacture for cement concrete are based on IS: 456. Consistency of concrete for cast *in situ* piles shall be suitable to the method of installation of piles. The minimum grade of concrete to be used is M15. The grade of concrete for driven piles is as follows:

| | | |
|---|---|---|
| 1. | For hard driving (where driving stress is more than 1,000 kN/m$^2$) | Grade not less than M20 |
| 2. | For easy driving (where driving stress is less than 1,000 kN/m$^2$) | Grade not less than M15 |

Clean water, free from acids and other impurities, shall be used in the manufacture of concrete (IS: 2911 – Part 1, 1979).

Upper and lower portions of composite piles are made of different materials. Composite piles may be of steel and concrete or timber and concrete. Steel and concrete piles consist of a lower portion of steel and an upper portion of cast-in-place concrete (Fig. 16.1d). Timber and concrete piles usually have their lower portion as timber and their upper portion above water table as concrete. It is extremely difficult to give proper connection between two different materials. Hence, these types of piles are not widely used.

### 16.2.2 Installation Methods

Based on installation techniques, piles are classified as driven piles and cast *in situ* piles. Driven piles may be concrete, steel, or timber. Concrete piles are classified as driven precast concrete piles, driven cast *in situ* concrete piles, and bored cast *in situ* concrete piles (IS: 2911 – Part 1, 1979). Driven precast pile is the one constructed in concrete (reinforced or pre-stressed) in a casting yard and subsequently driven in the ground when it has attained sufficient strength. Driven cast *in situ* pile is formed within the ground by driving a casing (with bottom closed) of uniform diameter, permanent or temporary, and subsequently filling in the hole so formed with plain or reinforced concrete. When the casing is left permanently, it is termed as *cased pile* and when the casing is taken out it is referred to as *uncased pile*. Bored cast *in situ* pile is formed within the ground by excavation or boring a pile within it, with or without the use of a temporary casing and subsequently filling it with plain or reinforced concrete. Cased and uncased piles hold good here too. There is another type of bored cast *in situ* pile in which the compaction of surrounding ground and freshly filled concrete in pile bore is simultaneously achieved by suitable method, and such a pile is referred to as *bored compaction pile*.

Driven piles are installed by hammer impact or by a vibrating machine. The installation of any type of driven pile causes displacement and disturbance of the soil around the pile. However, in case of pipe piles without shoe and in H-piles, the displacement is negligible. Loose granular soils are densified because of driving, thereby increasing the frictional resistance. On the other hand, driving causes re-moulding in cohesive soils, and temporarily reduces the soil shear strength along the pile surface. The reduced strength is regained with time due to thixotropic effect.

For cast *in situ* piles after making the pre-excavated hole, required reinforcement is placed and concrete is poured around. These piles tend to relieve the lateral earth pressure, and hence, reduce the strength along the shaft. Such piles are designed taking into account only the end bearing. However, with time the resistance along the shaft improves substantially. Thus, driven piles show higher resistance than cast *in situ* piles immediately after placement. Figure 16.2 shows some types of cast *in situ* piles.

### 16.2.3 Ground Effects

Piles are sometimes employed to compact soils and such piles are referred to as *displacement* or *compaction piles*. These piles displace a substantial volume of soil during installation (Fig. 16.3a). In granular soils, there is a tendency for compaction, whereas in clays, heaving of the ground surface often results. The ratio of heave volume to pile volume is about 50% for clays and 30% for silty clays. Driven piles installed in pre-drilled holes are also called *non-displacement piles*. In foundations for bridge abutments and piers, to safeguard the foundation from damage due to scour, piles are used (Fig. 16.3b). Piles are also used to prevent the movement of earth slopes and earthen banks by driving piles through the moving mass into the firm ground (Fig. 16.3c).

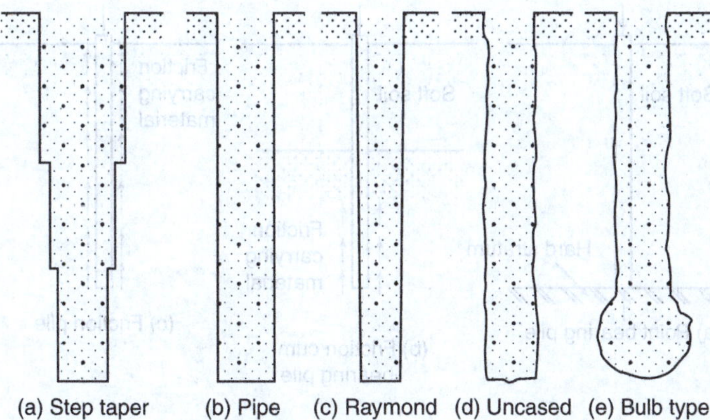


**Fig. 16.2** Some types of cast *in situ* piles

### 16.2.4 Function as Foundation

Piles function as foundation under different conditions and they are referred to accordingly.

When the topsoil is soft or too weak to support the superstructure, piles are employed to transmit the load to the underlying bedrock; such piles are called *end-bearing piles* or *point-bearing piles*. If the bedrock is not existing at a reasonable depth below the ground surface, the load is transferred through friction along the pile shaft. The frictional resistance developed at the soil–pile interface is utilized to support the structural load; such piles are called *friction piles* (Fig. 16.4a–c).

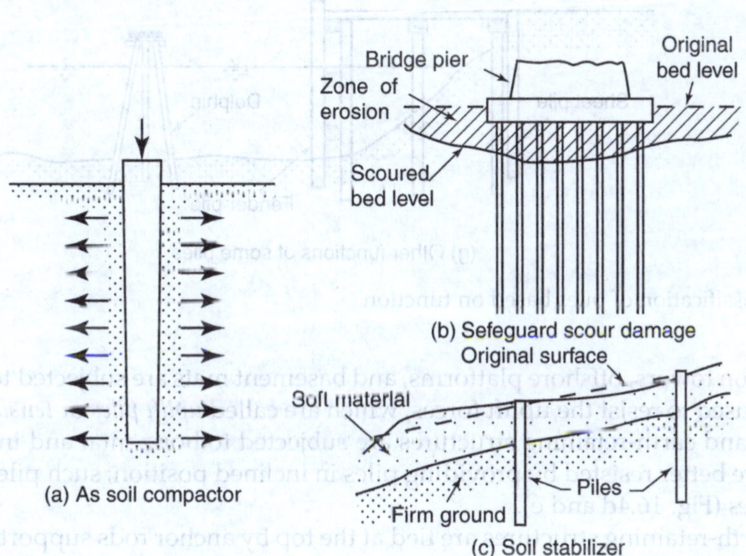


**Fig. 16.3** Piles for improving ground condition

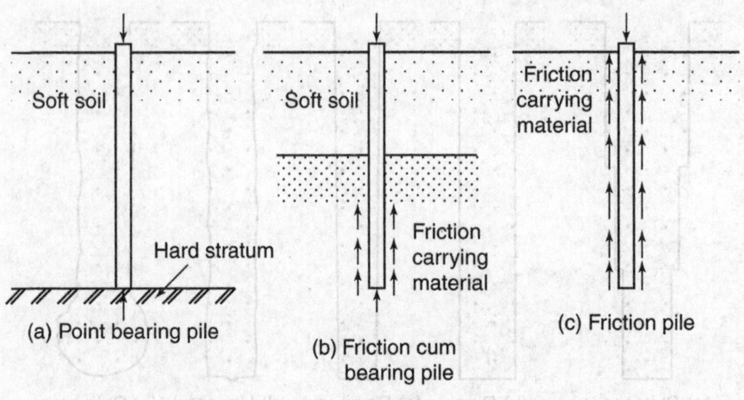


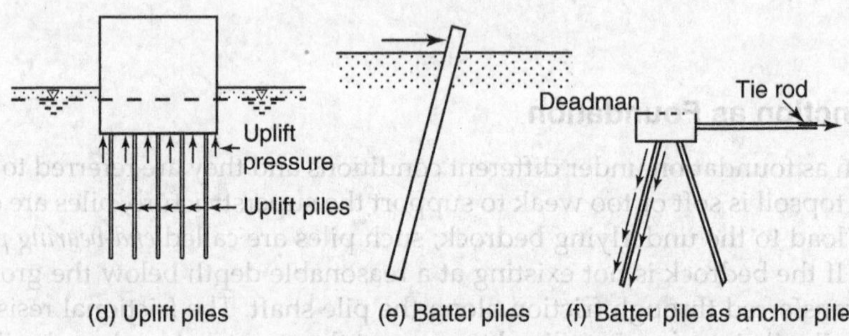


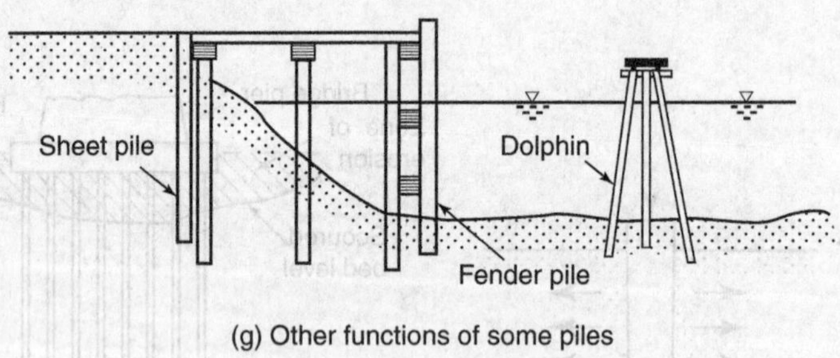


**Fig. 16.4** Classification of piles based on function

Transmission towers, offshore platforms, and basement mats are subjected to uplift forces and piles are used to resist the uplift forces, which are called *uplift piles* or *tension piles*. Some of the water- and earth-retaining structures are subjected to horizontal and inclined forces. Such forces are better resisted by providing piles in inclined position; such piles are referred to as *batter piles* (Fig. 16.4d and e).

Flexible earth-retaining structures are tied at the top by anchor rods supported by a deadman. The deadman is in turn supported by piles. These piles function as anchorage against horizontal pull from the sheet pile walls or other pulling forces; such piles are called *anchor piles*. In order to protect water front structures against impact from ships or other floating objects, fender piles and dolphins are used (Fig. 16.4f and g).

## 16.3 PILE-DRIVING EQUIPMENT

Pile-driving equipment mainly consists of pile frames, pile winches, and piling hammers.

Pile frames consist essentially of *leaders* which are a pair of steel members extending to full height of the frame (Fig. 16.5). They guide the hammer and the pile during the process of driving. Leaders can be extended at the top by telescopic boom when long piles have to be drawn. Pile frames are usually mounted on standard tracked crane base machines for mobility on land sites or on framed bases for mounting on stagings for marine construction. Pile frames have to remain in correct position throughout the driving of pile. Any possibility of a settlement of frame should be avoided, so that the weight of frame is not transmitted to a partially driven pile.

Piling winches may have single, double, and triple drums which can raise hammer and pile separately. Light winches have only one drum. Winches are provided with reversing facilities, such that in addition to their main purpose of lifting the hammer and pile, they can also be used for raking and rotating. Winches are powered by steam, diesel or petrol engines, or electric motors. Pile winches are mounted on the base of pile frames.

Most piles are installed from the ground surface by means of hammers or vibratory drivers. Piles can also be inserted by jetting or partial augering. General categories of pile hammers include drop hammer, single-acting hammer, double-acting hammer, and diesel hammers (Fig. 16.6).

Drop hammer is the simplest form of hammer used in conjunction with light frames and for piles driven for pile load test. Drop hammers are solid masses of steel, 10 to 50 kN in mass, fitted with a lifting eye and lugs for sliding in the leaders. The main disadvantages of a drop hammer are uncontrolled drop of fall and slow rate of hammering. Drop hammers have the advantage that they can be operated within a sound-proof box in places where noise abatement procedures are necessary.

Single-acting hammers use steam or compressed air to raise the hammer ram. At the ready-for-driving position, the steam is cut off and the cylinder falls freely on to the pile helmet. The maximum height of a drop is usually about 1.4 m and the rate of blow is about

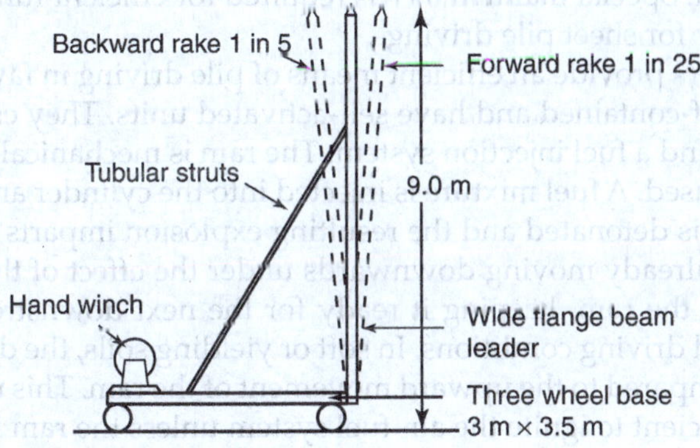


**Fig. 16.5** Schematic diagram of light pile frame

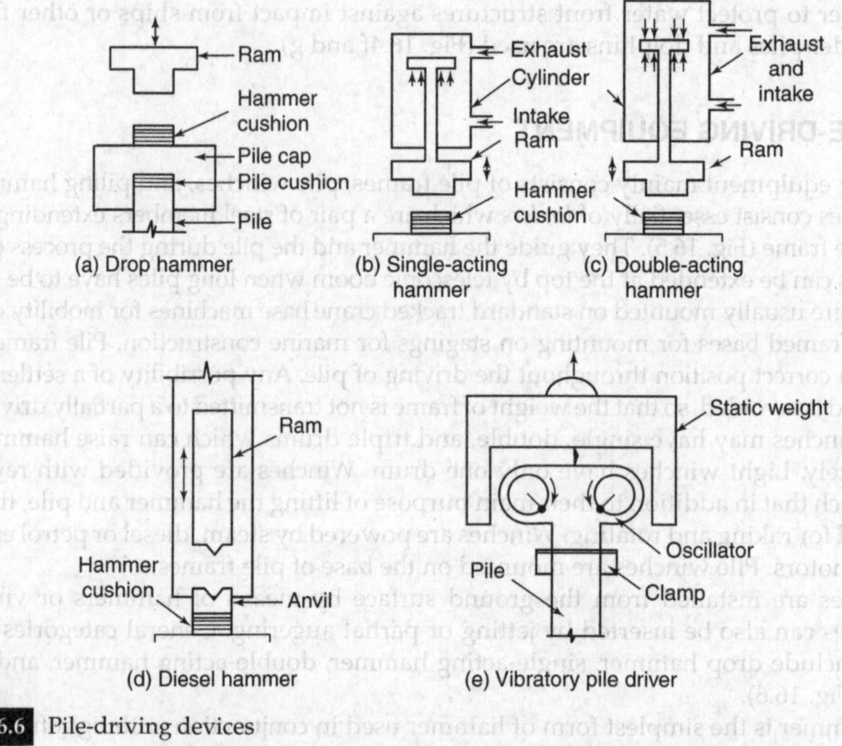


**Fig. 16.6** Pile-driving devices

60 strokes per minute. These hammers have to rely on the mass of ram. For effective driving, the mass of hammer ram should be equal or greater than the mass of pile.

Double-acting hammers use steam or compressed air to raise the ram to the driving position and also to accelerate the ram's downward thrust. This increases the impact velocity of the ram. The rate of driving varies from 300 blows per minute for light types to 100 blows per minute for heavier types. The weight of the hammer ram is in the range of 0.9 to 23 kN. The major advantage of the double-acting hammer over single-acting hammer is its tremendous operating speed. Special maintenance is required for efficient functioning. These hammers are used mainly for sheet pile driving.

Diesel pile hammers provide an efficient means of pile driving in favourable ground conditions. These are self-contained and have self-activated units. They essentially consist of a ram, an anvil block, and a fuel injection system. The ram is mechanically raised to the top of the cylinder and released. A fuel mixture is injected into the cylinder and compressed by the falling ram. The fuel is detonated and the resulting explosion imparts an additional impact to the pile, which is already moving downwards under the effect of the hammer blow. The explosion also raises the ram, keeping it ready for the next downstroke. Diesel hammers work well under hard driving conditions. In soft or yielding soils, the downward movement of the pile is large compared to the upward movement of the ram. This upward distance covered may not be sufficient to ignite the air–fuel system unless the ram is lifted up manually.

Vibratory methods of driving sheet piles or bearing piles are best suited to sandy or gravelly soils. Pile-driving vibrators consist of two counter-rotating masses, which produce a

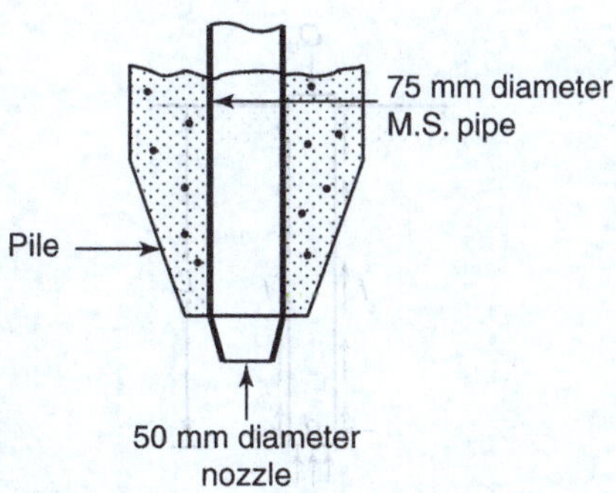


**Fig. 16.7** Pile with central jet pipe

dynamic vertical force on the pile and move the pile downwards. Driving steel piles in loose to medium dense sands is easy by this method but difficult in dense sands. Vibrators are rarely used in stiff clays. Vibrators are also used for extracting piles, and in large-diameter bored and cast *in situ* piling works, for sinking and extracting of pile casings.

Water jetting is a technique used to aid the penetration of pile into a sand or sandy gravel stratum. In this technique, water is discharged at the pile point to wash and loosen the sand and gravel. Jetting should be cut off at least 1 m above the required level and the pile is driven for the balance length. It is sometimes a difficult problem to dispose of the large quantity of water and sand flowing at ground level from around the piles (Fig. 16.7).

Each pile is provided with a cap or helmet at the top of the pile. The purpose of the helmet is to hold the resilient dolly and packing, which are provided between the hammer and the pile to prevent shattering of the pile head. A cushion may be used between the pile and the cap. This has the effect of evening out the hammer impulses. Longer dollies or followers are used when driving piles below the level of the bottom of the leaders.

## 16.4 BEARING CAPACITY OF SINGLE PILE

The bearing capacity of a single pile depends on the structural strength of the pile and the supporting strength of the soil, and the smaller of the two controls the design load. In order to satisfy the first criterion, the pile load is restricted so as to avoid damage from overdriving, to avoid overstressing of the pile under design load, and to avoid buckling failure of the pile. Generally, based on the material and dimension of the pile, allowable stress in piles is fixed by codes to account for the above-mentioned factors. However, the failure of a pile is controlled by the second criterion, i.e., the supporting strength of the soil, unless the pile material or the construction is below standard. Thus, further discussion is confined to the supporting strength of the soil.

Load on a pile is partly carried by skin friction and partly by the resistance offered by the soil at the base of the pile called the point resistance (Fig. 16.8). As observed in shallow foundations, an increase in load (within 20% of the failure load) on a pile is followed by a proportionate increase in settlement. But after a certain load level, the rate of increase of settlement is far out of proportion with the rate of increase of load. Thus, the ultimate load

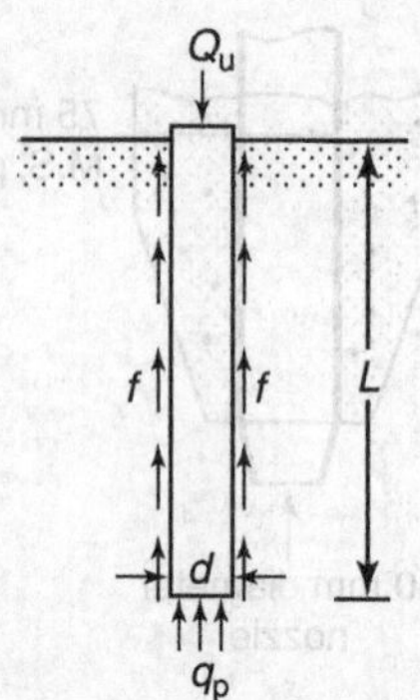


Fig. 16.8 Ultimate load of single pile

or bearing capacity of a single pile indicates the load at which the settlement of the pile increases continuously with no further increase in load.

Therefore, the ultimate bearing capacity of a pile $Q_u$ is given as

$$Q_u = Q_f + Q_p \tag{16.1}$$

where $Q_f$ is the resistance due to skin friction and $Q_p$ the resistance due to point bearing. That is,

$$Q_u = fA_s + q_p A_p \tag{16.2}$$

where $A_s$ is the effective surface area of the pile in contact with soil along the embedded shaft length, $A_p$ the bearing area of pile tip, $f$ the average unit skin friction or adhesion between soil and pile surface, and $q_p$ the bearing pressure of soil at the tip.

Depending on the type of soil penetrated, $Q_f$ and $Q_p$ are evaluated. For simplicity, the analysis for piles in sand and clay are discussed below. However, the analysis can be extended for layered soils by computing the skin friction for the appropriate layer.

### 16.4.1 Pile Capacity from Statical Methods for Driven Piles

For a driven pile in sand, the unit skin friction depends on the soil pressure acting normal to the pile surface and the coefficient of friction between the soil and pile material (Fig. 16.9). Thus, the unit skin friction acting at any depth in sand is

$$f = \sigma_h' \tan \delta$$

$$f = K\sigma_v' \tan \delta$$

where $\sigma_h'$ is the horizontal soil pressure acting at any depth $z$ in a soil mass, $\sigma_v'$ the effective over-burden pressure acting at the same depth within a soil mass, $K$ the lateral earth pressure coefficient, and $\tan \delta$ the coefficient of friction between soil and the pile surface.

Values of coefficient of friction depend on the type of soil and pile material. McCarthy (1982) reports values varying from 0.20 to 0.45 depending on the roughness of pile surface. Indian Standards (IS: 2911 – Part 1, 1979) recommends $\delta = \phi$.

The earth pressure is approximately equal to the passive condition at the top of the pile and may be near to the at-rest-condition at the pile tip. Further, the method of installation also influences the earth pressure coefficient. The following values of $K$ (Table 16.1) are recommended by Das (1990).

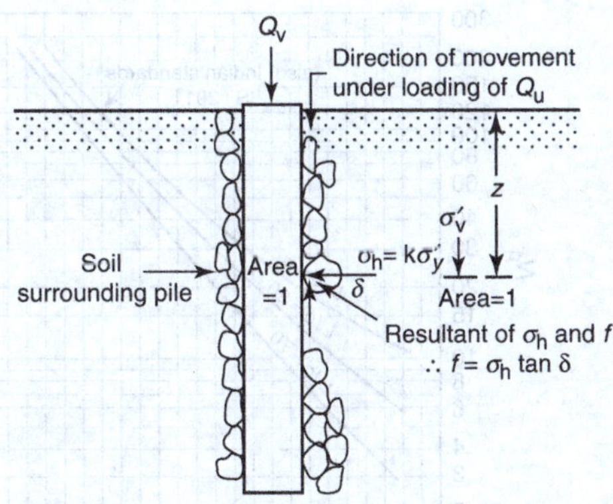


**Fig. 16.9** Soil–solid friction developing against pile shaft

Hence, the total skin friction acting along the embedded length of the pile is computed considering an average effective vertical overburden pressure. Thus,

$$Q_f = fA_s = [K(\sigma'_v)_a \tan\delta]A_s \tag{16.3}$$

where $(\sigma'_v)_a$ is the average effective vertical overburden pressure,

$$(\sigma'_v)_a = \frac{\gamma L}{2}$$

where $L$ is the length of the pile,

$$A_s = \pi dL$$

where $d$ is the average diameter of the pile.

The end-bearing component of pile capacity can be determined by a method similar to that of shallow foundations. This is given for deep foundations in non-cohesive soils as

$$Q_p = 0.5\gamma d_p N_\gamma + \sigma'_v N_q \tag{16.4}$$

where $N_q$ is the bearing capacity factor for pile foundation (Fig. 16.10), $N\gamma$ the bearing capacity factor for shallow foundations, and $d_p$ the pile tip diameter or width.

**Table 16.1** Coefficient of earth pressure

| Pile | K |
|---|---|
| Bored or jetted piles | $K_0 = 1 - \sin\phi$ |
| Low-displacement-driven piles | $K_0 - 1.4K_0$ |
| High-displacement-driven piles | $K_0 - 1.4K_0$ |

*Source*: Das (1990).

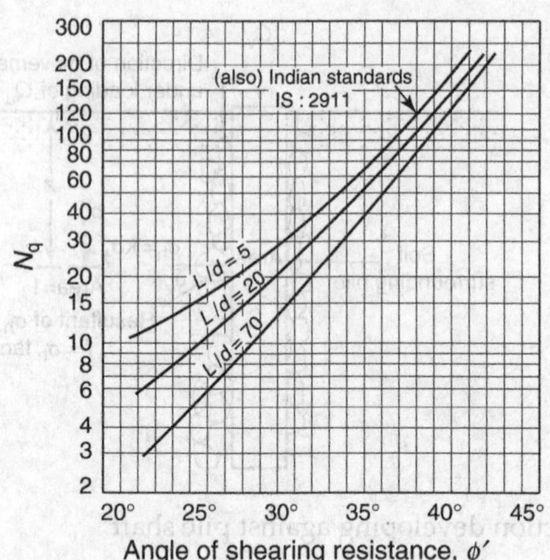


**Fig. 16.10** Values of $N_q$ for pile formulae (*Source:* Berezantzev et al., 1961)

In most of the driven piles, the first term of Eq. 16.4 is small compared to the second term because of limited $d_p$ dimensions. Thus, for practical considerations, the point-bearing resistance can be written as

$$Q_p = (\sigma'_v N_q) A_p \tag{16.5}$$

Various theoretical analyses for the point-bearing pressure have been attempted and among them the Berezantzev et al. (1961) value for $N_q$ have been in use for commonly encountered soil conditions (Fig. 16.10). $N_q$ curve as recommended by Indian Standards (IS: 2911 – Part 1, 1979) is also presented in Fig. 16.10. For design purposes, $\sigma'_v = \gamma L$ is considered. Thus, the ultimate bearing capacity of driven piles in sand is given as

$$Q_u = [K(\sigma'_v)_a \tan\delta] A_s + (\sigma'_v N_q) A_p \tag{16.6}$$

In saturated clays, pile driving causes the re-moulding of soil in the vicinity of the pile because of displacement and disturbance. But the strength is regained with time, and the rate of strength gain depends on the consolidation and thixotropic characteristics of the clay.

Presently, there are several approaches available for obtaining the unit skin friction. Two of the approaches adopt effective stresses in the analysis (Burland, 1973; Vijayvergiya and Focht, 1972) and the other one uses an empirical adhesion factor. Burland's method, commonly referred to as the $\beta$-method, is widely used.

The common approach is to use undrained cohesion on the presumption that the dissipation of pore water pressure takes much more time. But recent studies have shown that dissipation of excess pore pressure takes place fairly quickly, and at the time of the final loading of pile, the soil is almost at the drained condition. Thus, the unit skin friction based on Burland's approach is

$$f = \beta(\sigma'_v)_a \tag{16.7}$$

where $\beta = K \tan\phi'_{re}$ and $\phi'_{re}$ = drained friction angle of re-moulded clay.

The value of $K$ is conservatively taken as

$$K = K_0 = 1 - \sin \phi'_{re} \quad \text{for normally consolidated clays} \tag{16.8a}$$

and

$$K = K_0\sqrt{\text{OCR}} = (1 - \sin \phi'_{re})\sqrt{\text{OCR}} \quad \text{for over-consolidated clays} \tag{16.8b}$$

Thus,

$$Q_f = [\beta(\sigma'_v)_a]A_s \tag{16.9}$$

Indian Standards (IS: 2911 – Part 1, 1979) recommends Eq. 16.1 taking into account the adhesion factor

$$Q_f = \alpha c_s A_s \tag{16.10}$$

where $\alpha$ is the adhesion (or reduction) factor as given in Table 16.2 and $c_s$ the average undrained cohesion along pile shaft.

The end bearing is related only to the undrained strength of clay. For $\phi = 0°$ and $N_q = 1$, the contribution by end bearing is very small. So the point bearing resistance is expressed as a function of $c_p$ and $N_c$, i.e.,

$$Q_p = (c_p N_c)A_p \tag{16.11}$$

where $c_p$ is the average undrained cohesion at pile tip and $N_c$ is 9.0 for intact clays (Skempton, 1951 and IS: 2911 – Part 1, 1979) and $N_c$ is 6.75 for fissured clays (Skempton, 1951).

Thus, the ultimate bearing capacity of a pile in clay is

$$Q_u = [\beta(\sigma'_v)_a]A_s + [c_p N_c]A_p \tag{16.12a}$$

and

$$Q_u = \alpha c_s A_s + c_p N_c A_p \quad \text{(as per IS: 2911 – Part 1, 1979)} \tag{16.12b}$$

In both the cases, a factor of safety ($F$) of 2.5 is adopted to arrive at the allowable bearing capacity, $Q_a$, as

$$Q_a = \frac{Q_u}{F} \tag{16.13}$$

**Table 16.2** Value of $\alpha$

| Consistency | Value of $\alpha$ |
|---|---|
| Soft to very soft | 1.0 |
| Medium | 0.7 |
| Stiff | 0.4 |
| Stiff to hard | 0.3 |

*Source*: IS: 2911 – Part1 (1979).

### 16.4.2 Pile Capacity from Statical Methods for Cast *In Situ* or Bored Piles

For such piles in relatively homogeneous soils, the ultimate capacity is due to end-bearing and skin friction, as considered in driven piles. Because of the disturbance caused by construction, there will be a possibility of the loss of strength at the bottom of the pile for a depth of 2 m. Similarly, there is an equal possibility of disturbance and loss of strength occurring in the surface zone of soil for a depth of about 2 m. Thus, for computing skin friction, a depth of 4 m has to be omitted in the evaluation of skin friction. This is more important in clayey soils.

The ultimate capacity of the cast *in situ* pile in sand is similar to that of driven piles in sand (Eq. 16.6); that is,

$$Q_u = [k(\sigma'_v)_a \tan\delta]A_s + (\sigma'_v N_q)A_p$$

Here, $(\sigma'_v)_a$ is the effective average vertical pressure considering the limits imposed for effective depth $L'$ (i.e., $L' = L - 4$) where $L$ is in metres.

As no effective stress approach is made yet for bored piles, the ultimate pile capacity in clay is obtained from Eq. 16.12b; that is,

$$Q_u = \alpha c_s A_s + c_p N_c A_p$$

Sometimes, bored piles may be provided with an enlarged base, and in such cases the values of $f$ for belled foundation has to be used. The value of $A_p$ in such cases is evaluated for the foundation base. The values may be adopted as discussed for driven piles.

### 16.4.3 Pile Capacity from Pile-Driving Formulae

In qualitative terms, it may be considered that a pile is capable of sustaining a greater load if it exerts a greater resistance against driving. Based on this principle, many dynamic formulae were suggested to obtain pile capacity considering the energy needed for driving the piles. One of the earliest of these dynamic equations, commonly referred to as the Engineering News Record (ENR) formula, is derived on the basis of the work–energy theory.

The total driving energy caused by the hammer hitting the pile is equal to the weight of the hammer times the height of drop or stroke. This energy is consumed by the work done in penetrating the pile and by certain losses. This means that

$$\text{Driving energy} = (\text{work of pile penetration}) + (\text{loss of energy})$$

or

$$E = RS + E_L \tag{16.14}$$

where $E$ is the driving energy, $R$ the pile resistance, $S$ the pile penetration per blow, and $E_L$ the loss of energy, including loss in impact, in driving cap, in pile, and in soil. If $E_L$ is assumed to be proportionate to the pile resistance, it can be written as

$$E_L = RC$$

where $C$ is the empirical constant.

According to the ENR formula, the pile resistance is the ultimate load $Q_u$. Then

$$E = Q_u S + Q_u C \tag{16.15}$$

or

$$Q_u = \frac{E}{S+C} \tag{16.16}$$

The average value obtained from the last few driving blows is taken as the pile penetration $S$ (m). Recommended values of $C$ are

$$C = 2.54 \times 10^{-2} \text{ (m)} \quad \text{for drop hammers}$$

$$C = 0.254 \times 10^{-2} \text{ (m)} \quad \text{for steam hammers} \tag{16.17}$$

and

$$E = Wh$$

$$E = \eta H_E$$

where $W$ is the weight of ram or hammer (kN), $h$ the height of free fall of the ram (m), $\eta$ the hammer efficiency, and $H_E$ the rated energy of single or double acting hammer (kN-m).

A factor of safety, $F = 6.0$, has been used to estimate the allowable pile capacity. The ENR formula has been modified by Hiley and is given as

$$Q_u = \frac{\eta W h \eta_b}{S + C/2} \tag{16.18}$$

where $\eta$ is the hammer efficiency as given in Table 16.3, $S$ the final set or penetration per blow (m), $C$ the sum of the temporary elastic compressions of the pile (m) $= C_1 + C_2 + C_3$, where $C_1$ is the temporary compression of dolly and packing, $C_2$ the temporary compression of pile, $C_3$ the temporary compression of ground, and $\eta_b$ the efficiency of the blow, representing the ratio of energy after impact to the striking energy of ram

$$= \frac{W + Pe_r^2}{W + P}, \quad \text{where } W \geq Pe_r$$

$$= \frac{W + Pe_r^2}{W + P} - \frac{W - Pe_r^2}{W + P}, \quad \text{where } W < Pe_r$$

where $P$ is the weight of pile, anvil, helmet, and follower, if any (kN), and $e_r$ the coefficient of restitution of the materials under impact.

The factor $\eta_b$ is given for the condition that pile is driven into the penetrable ground. If the pile finds refusal in rock, $0.5P$ has to be adopted in place of $P$ in the expression for $\eta_b$.

**Table 16.3** Hammer efficiency

| Hammer type | Efficiency |
|---|---|
| Single- and double-acting hammers | 0.70–0.85 |
| Diesel hammer | 0.80–0.90 |
| Drop hammers | 0.70–0.90 |

*Source:* Das (1990).

**Table 16.4** Coefficient of restitution

| Pile material | Coefficient of restitution, $e_r$ |
|---|---|
| Cast-iron hammer and concrete piles (without cap) | 0.40–0.50 |
| Wood cushion on steel piles | 0.30–0.40 |
| Wooden piles | 0.25–0.30 |

*Source:* Das (1990).

The coefficient of restitution, $e_r$, of the material under impact are given in Table 16.4.

A factor of safety of 2 to 2.5 may be adopted to obtain the allowable load-bearing capacity of a pile.

The dynamic formulae discussed above are based on the premise that the soil resistance remains constant during and after the driving operation. In coarse-grained soils, this condition is fulfilled because of high permeability. In fine-grained soils, water cannot escape readily during driving, and thus, the excess water tends to reduce the frictional resistance along the periphery of the pile. Further, the driving operation reduces the shear strength of the surrounding soil. But because of thixotropic effect, the soil regains its strength rapidly. Thus, dynamic formulae are not dependable for the determination of pile capacity in soils containing more fines. Dynamic formulae give varying results even in coarse-grained soils. The dynamic formulae can be only applied to small jobs in granular soils, localities of known soil conditions, and to terminate the driving of a pile based on the data from a test pile.

## 16.4.4 Pile Capacity from Wave Equation

The major drawback with dynamic pile-driving formulae is that they do not stimulate soil–pile interaction or the time-dependent nature of the problem. The application of driving data to determine pile capacity requires knowledge of the effect that a hammer blow has on the pile and supporting soil. The impact of a pile-driving hammer causes stress waves to be transmitted through the length of the pile. These waves cause elastic compression and tension in the pile and interact with the soil both along the surface of the pile and at the pile tip. Apart from this, the pile-driving accessories, like hammer, anvil, cap block, pile cap, and cushion, influence the pile-driving behaviour. These variables were accounted for by Smith (1962) through an analogy, termed the wave equation, of the pile behaviour and a mathematical model.

The model presented by Smith was based on the propagation of an elastic wave through a long rod. The model considered the complete pile-driving operation, including pile-driving accessories, soil–pile interaction and the time-dependent nature of the elastic pile deformation. The model has been given in the form of a partial differential equation. The actual and the idealized systems are shown in Fig. 16.11. The springs simulate the axial resistance of the pile and the spring constant, $k$ relate to the elasticity of the pile. The spring damping $R$ represents the frictional resistance of the soil surrounding the shaft of the pile and the soil resistance at the pile tip. The various weight values $W$ correspond to the weight of the incremental sections of the pile, and the spring damping at the pile tip accounts for the point resistance.

The partial differential equation has been solved using computers and is now widely used by civil engineers. Bowles (1984) has presented a more detailed discussion on this approach.

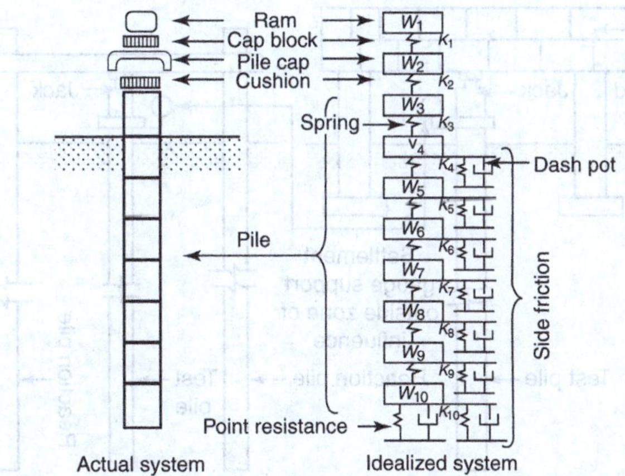


Fig. 16.11 Pile representation for wave equation analysis (*Source*: Smith, 1962)

### 16.4.5 Pile Capacity Based on SPT Values on Non-Cohesive Soils

The empirical correlation of pile capacity and SPT value has been widely used. The correlation between blow counts and pile capacity has not been accepted as a standard method. A careful judgement has been exercised by the engineers while adopting the correlation.

Meyerhof (1976) has suggested a formula that may be used for a non-cohesive material, particularly for sand deposits. It is

$$q = 40N\frac{L}{d} \le 400N\,(\text{kN/m}^2) \tag{16.19}$$

and

$$f = 2N\,(\text{kN/m}^2) \tag{16.20}$$

Hence, the pile capacity can be expressed as

$$Q_u = fA_s + qA_p \tag{16.21}$$

### 16.4.6 Pile Capacity from Pile Load Test

The pile load test is the most reliable method of determining the capacity of a pile. A test pile is installed adopting the same proposed procedure. It may be loaded to near-failure condition or up to the working load level. In the latter case, the pile shall form one of the permanent piles of the foundation. A careful record has to be maintained during installation and during the load test.

Three types of tests are conducted on piles, namely, vertical load test, lateral load test, and pull-out test. The results from vertical load test is used to estimate the vertical load-carrying capacity of the soil while the lateral and pull-out tests are used to estimate the lateral load-carrying capacity and the frictional resistance of the pile. Only the vertical load test is discussed below and the reader may refer to IS: 2911 – Part 4 (1985) for other tests.

The test procedure consists of applying a static load on the pile top in convenient load increments and recording the vertical deflections of the pile. Suitable reaction device is

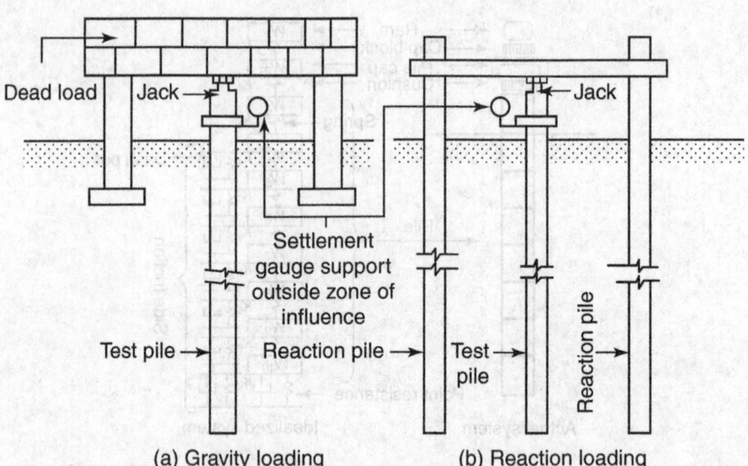


Fig. 16.12 Pile load test arrangements

adopted (Fig. 16.12). The reaction may be obtained from (i) a kentledge placed on a platform supported clear of the test pile with the centre of gravity of the kentledge passing through the axis of the pile or (ii) anchor piles installed at a distance not less than three times the test pile shaft diameter or 1.5 m, whichever is greater. The reaction for the test should be 25% more than the proposed final test load. Measurement of pile movements is related to a fixed reference mark. Reference marks would be supported on objects located outside the soil zone.

The pile head is made level by chipping off to natural horizontal plane and finished smooth and level with plaster of Paris. A bearing plate is placed before seating the hydraulic jack. Datum bar is set on immovable supports beyond a distance of 1.5 m from the edge of the pile. At least two dial gauges are fixed to the datum bar. A series of vertical downward increment of loads with intensity of 20% of the safe load on the pile are applied. Settlement readings are taken for each load increment till the rate of displacement is 0.003 mm/min. The test is continued till the maximum load is 1.5 times the working load or the maximum settlement of the test not exceeding 12 mm.

Pile load settlement curve is drawn and the safe load is obtained as the least of the following (IS: 2911 – Part 4, 1985):

1. The straight portions of the curve are extended and the ultimate load is found. A factor of safety is adopted and the safe load against shear failure is determined.
2. Two-thirds of the final load at which the total displacement attains a value of 12 mm.
3. Fifty percent of the load at which the total displacement equals 10% of the pile diameter in case of uniform diameter piles or 7.5% of bulb diameter in case of under-reamed piles (discussed in next section).

Typical test data and a load–settlement curve are given in Worked Example 16.5.

If soil conditions are uniform, a relatively low factor of safety of 1.5 to 2.0 may be enough. But in non-uniform deposits, being reflected by varying load test results, a higher factor of safety (about 3) is justified.

In addition to assessing the design load, the pile test may be used to establish the construction driving criteria. A comparison can be made with pile-driving records and the measured ultimate load, with dynamic pile-driving formulae, or with the wave equation to establish the driving specification. The skin friction and end-bearing components of the pile support capacity can be separated by running a tension test or a cyclic-load test on the pile (refer IS: 2911 – Part 4, 1985).

### 16.4.7 Negative Skin Friction

Negative skin friction is downward drag acting on the piles due to relative movement between the piles and the surrounding soil. This condition can develop where a soft or loose soil stratum located anywhere above the pile tip is subjected to compressive loading. The effect of negative skin friction is to increase the axial load in the pile and the pile settlement.

Negative skin friction can develop under different field conditions, such as the following (Fig. 16.13):

1. A cohesive fill is placed over a non-cohesive soil layer and a pile is driven into such a medium. The cohesive fill consolidates, and during the process of consolidation the fill imparts a downward drag on the pile.
2. A non-cohesive fill is placed over a soft cohesive layer and a pile installed in such a medium. Due to the overburden pressure, the cohesive fill consolidates and during the process imparts a downward drag on the pile.
3. In a saturated soil, lowering of the groundwater increases the vertical effective stresses in the soil medium. This induces settlement, and a pile driven in such a medium is subjected to a downward drag force.

The negative skin friction is computed for cohesive and non-cohesive fills as follows:

**Cohesive Fill Overlying Non-Cohesive Soil.** Based on $\beta$-method (as discussed earlier for piles in non-cohesive soils), the unit negative skin friction can be given as

$$f_n = K\sigma_v' \tan\delta \qquad (16.22)$$

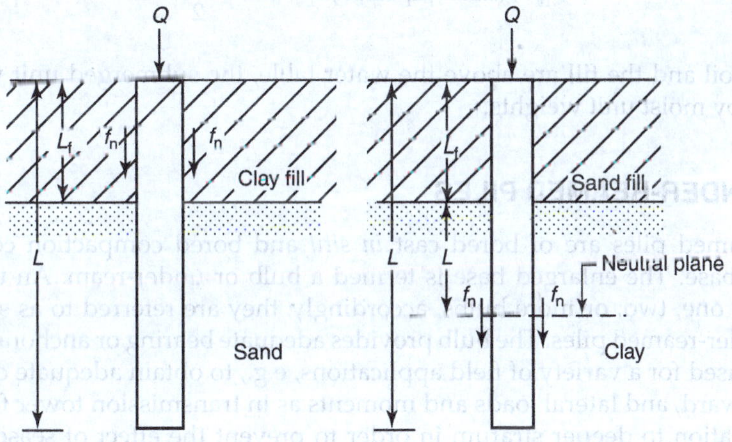


Fig. 16.13 Negative skin friction

where $K = K_0 = 1 - \sin P'_{re}$, $\sigma'_v = \gamma f z$ = vertical effective stress ($0 < z < L_f$), $\delta$ = soil–pile friction angle = $0.5 - 0.7 f'_{re}$, and $\gamma_f$ = unit weight of fill.

Hence, the total downward drag force, $Q_n$, is given as

$$Q_n = \int_0^{L_f} \pi d(K\gamma_f \tan\delta) z \, dz$$

$$Q_n = \frac{\pi d K \gamma_f L_f^2 \tan\delta}{2}$$

where $L_f$ is the depth of fill.

If the water table is at the ground surface $\gamma_f = \gamma'$ and if the fill is above the water table, $\gamma_f = \gamma$.

**Cohesive Soil Underlying Non-Cohesive Fill.** Vesic (1977) has shown that negative skin friction develops at the top portion of the pile in the compressing medium. That is, in the depth range $z = 0$ to $z = L_1$ (Fig. 16.13), and this depth is referred to as the *neutral depth*. The neutral depth may be given as (Bowles, 1982)

$$L_1 = \frac{(L - L_f)}{L_1}\left[\frac{L - L_f}{2} + \frac{\gamma'_f L_f}{\gamma'}\right] - \frac{2\gamma'_f L_f}{\gamma'} \tag{16.24}$$

where $\gamma'_f$ and $\gamma'$ are submerged unit weights of the fill and the underlying cohesive layer, respectively.

Hence, the total drag force

$$Q_n = \int_0^{L_1} \pi d f_n \, dz = \int_0^{L_1} \pi d K(\gamma'_f L_f + \gamma' z) \tan\delta \, dz$$

$$Q_n = (\pi d K \gamma'_f L_f \tan\delta) L_1 + \frac{L_1^2 \pi d K \gamma' \tan\delta}{2} \tag{16.25}$$

If the soil and the fill are above the water table, the submerged unit weights should be replaced by moist unit weights.

## 16.5 UNDER-REAMED PILES

Under-reamed piles are of bored cast *in situ* and bored compaction concrete piles with enlarged base. The enlarged base is termed a bulb or under-ream. An under-reamed pile may have one, two, or more bulbs, accordingly, they are referred to as single-, double-, or multi-under-reamed piles. The bulb provides adequate bearing or anchorage. Under-reamed piles are used for a variety of field applications, e.g., to obtain adequate capacity for downward, upward, and lateral loads and moments as in transmission tower foundation, to take the foundation to deeper stratum in order to prevent the effect of seasonal changes as in expansive soils, to reach firm strata and to take the foundations below scour level.

The under-reamed pile should satisfy all the design requirements for a conventional pile. In addition to these, in deep deposits of expansive soils, the minimum length of piles (irrespective of any other factors) should be 3 to 5 m below ground level. In poor bearing stratum or in recently filled-up grounds, the pile should pass through such soils and be seated in good bearing strata (IS: 2911 – Part 3, 1980).

The bulb diameter is taken two to three times the diameter of the shaft. For 30 cm diameter piles, the recommended spacing between the bulbs should not exceed 1.5 times the diameter of bulb and for greater diameter piles, the spacing is reduced to 1.25 times the stem diameter. The position of the top most bulb should be at a minimum depth of two times the bulb diameter, and for expansive soil, the minimum depth is 1.75 m from ground level. Further, the minimum clearance below the underside of pile cap and the bulb should be a minimum of 1.5 times the bulb diameter. Indian Standards (IS: 2911 – Part 3, 1980) recommends only two bulbs. Typical details of single and double under-reamed piles are shown in Fig. 16.14.

The bearing capacity of a single under-reamed pile (Fig. 16.15a) may be given based on static formula. Thus,

$$Q_u = Q_f + Q_p$$

or

$$Q_u = fA_s + q_b(A_b - A_{sh}) + q_{sh}A_{sh} \tag{16.26}$$

where $A_s$ is the surface area of the embedded shaft of the pile above and below the bulb, $A_b$ the cross-sectional area of the bulb $(= \pi d_u^2/4)$, $A_{sh}$ the cross-sectional area of the shaft

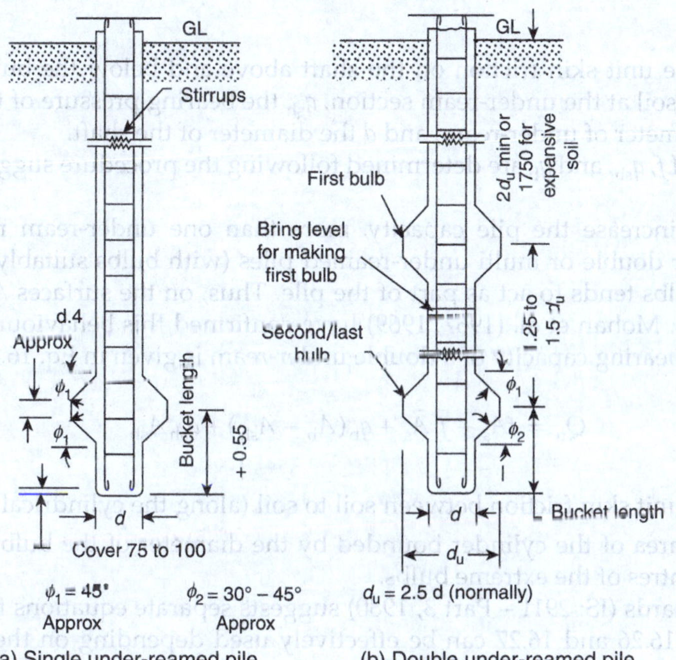


(a) Single under-reamed pile (b) Double under-reamed pile

**Fig. 16.14** Typical details of under-reamed piles (*Source*: IS: 2911 – Part 3, 1980)

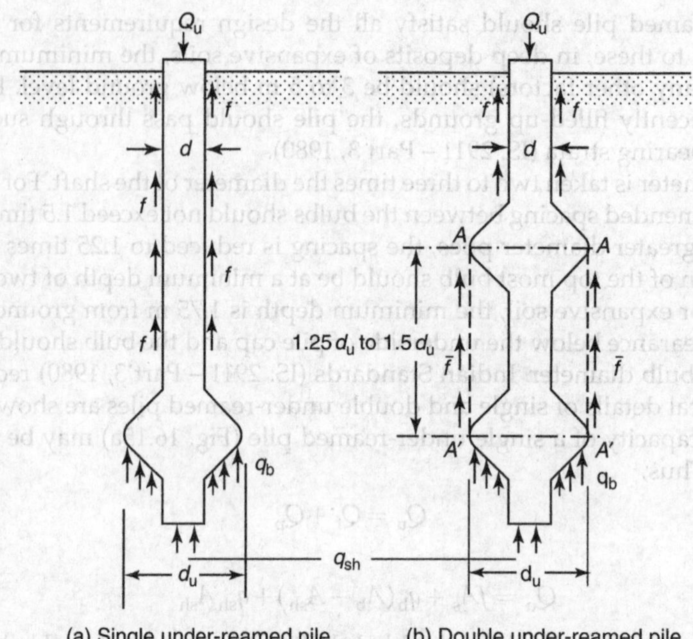


**Fig. 16.15** Details of forces acting on under-reamed piles

$(= \pi d^2/4)$, $f$ the unit skin friction on the shaft above and below the bulb, $q_b$ the bearing pressure of the soil at the under-ream section, $q_{sh}$ the bearing pressure of the soil at the pile base, $d_u$ the diameter of under-ream, and $d$ the diameter of the shaft.

The values of $f$, $q_{sb}$, and $q_b$ are determined following the procedure suggested under statistical methods.

In order to increase the pile capacity, more than one under-ream may be provided (Fig. 16.15b) for double or multi under-reamed piles (with bulbs suitably spaced), the soil between the bulbs tends to act as part of the pile. Thus, on the surfaces AA′ full soil resistance mobilizes. Mohan et al. (1967, 1969) have confirmed this behaviour from model and field tests. The bearing capacity of a double under-ream is given in Eq. 16.27:

$$Q_u = fA_s + \bar{f}\,\bar{A}_s + q_b(A_b - A_{sh}) + q_{sh}A_{sh} \tag{16.27}$$

where $\bar{f}$ is the unit skin friction between soil to soil (along the cylindrical surface AA′) and $\bar{A}_s$ the surface area of the cylinder bounded by the diameter of the bulb and the distance between the centres of the extreme bulbs.

Indian Standards (IS: 2911 – Part 3, 1980) suggests separate equations for clay and sand. However, Eqs. 16.26 and 16.27 can be effectively used depending on the type of soil and water table position as dealt for conventional piles using static methods. The bearing capacity of an under-reamed pile can also be determined by load test. Approximate safe loads on under-reamed piles are also provided in the code (IS: 2911 – Part 3, 1980).

## 16.6 PILE GROUPS

In the previous section, we discussed the pile capacity of single piles. When piles are used for foundation support, they are always used in a group. This is an important requirement to ensure that the imposed structural load lies within the support area provided by the foundation. Building codes never permit the use of less than three piles to support a major column, and less than two piles to support a foundation wall. Thus, the bearing capacity and settlement of pile groups are the end results needed for the design of the foundation.

### 16.6.1 Pile Group Capacity

The pile group capacity is not necessarily the individual pile capacity multiplied by the number of piles in the group. Soil disturbance caused by pile installation techniques and overlap of stresses between adjacent piles may reduce the group capacity from the sum of individual capacities. On the other hand, soil between individual piles might get densified or increased with adhesion, and the group may tend to behave as an equivalent single large pile, which may show higher group capacity.

Pile spacing (centre-to-centre) plays an important role in the group capacity of piles. Pile spacing is considered from two aspects of installing the piles and the nature of the load transfer to the soil. For end-bearing piles, the minimum spacing should be 2.5*d*, where *d* is the diameter. If the piles are resting on solid rock, the spacing of 2*d* is recommended. For friction piles, the spacing should be sufficient to avoid overlapping of stresses. In such cases, the recommended spacing is 3*d*. In case of loose sand or fill, a spacing of 2*d* may be adopted (IS: 2911 – Part 1, 1979).

Ideally, the piles in a group should be spaced in such a way that the group capacity is not less than the sum of individual pile capacity.

Vesic (1977) recommends an optimum spacing of 3 to 3.5*d*. Mohan (1981) suggests that the Indian Standards should follow Vesic's recommendations of 3 to 3.5*d*. As a general guide, 2*d* spacing is the minimum requirement and 3 to 3. 5*d* is the preferred spacing.

In order to understand the behaviour of pile groups, it is necessary to identify two types of them (Poulos and Davis, 1980), viz.,

1. A free-standing pile group, in which the pile cap is not in contact with the underlying soil.
2. A pile group in which the pile cap is in contact with the underlying soil (referred to as piled foundation).

In both the cases, an efficiency factor $\eta_g$ is defined as the ratio of the ultimate capacity of the group ($Q_{ug}$) to the sum of the individual pile capacity. It is expressed as

$$\eta_g = \frac{Q_{ug}}{n_p Q_u} \times 100\% \tag{16.28}$$

where $n_p$ is the number of piles in a group.

**Free-Standing Groups.** Several group efficiency formulae are in use relating pile spacing and number of piles. Converse–Labarre formula (Eq. 16.29) is one such formula quite often used by engineers

$$\eta_g = 1 - \left[\frac{(n_1 - 1)n_2 + (n_2 - 1)n_1}{90 n_1 n_2}\right]\theta \tag{16.29}$$

where $n_1$ is the number of rows, $n_2$ the number of piles in a row, $\theta = \arctan d/s$ (degrees), and $s$ the centre-to-centre spacing.

Chellis (1962) compared several formulae and found wide variation in $\eta_g$ values. The most widely used approach in clays is the one suggested by Terzaghi and Peck (1967). As per their approach, the group capacity is analysed based on the following two conditions and the lesser value is considered as the design load:

1. The sum of ultimate capacity of the individual piles in the group
2. The bearing capacity for block failure of the group (Fig. 16.16). That is,

$$Q_{us} = n_p Q_u \tag{16.30}$$

or

$$Q_{ub} = fA_{sg} + qA_{pg} \tag{16.31}$$

where $Q_{us}$ is the sum of individual pile capacity, $Q_{ub}$ the ultimate load capacity of block, and $A_{sg}$ the surface area of the group, that is,

$$A_{sg} = 2(B_1 + B_2)L$$

where

$$B_1 = (n_1 - 1)s + d$$

$$B_2 = (n_2 - 1)s + d$$

$$A_{pg} = B_1 \times B_2$$

For pile groups in clay

$$f = \alpha c_u$$

$$q = c_u N_c$$

and $N_c$ is Skempton's bearing capacity factor. Whitaker (1957) confirmed the existence of such failures from model studies in clays. Poulos and Davis (1980) observed that the values obtained from Eqs. 16.30 and 16.31 are not smooth transitions but abrupt and suggested the following empirical equation for clays:

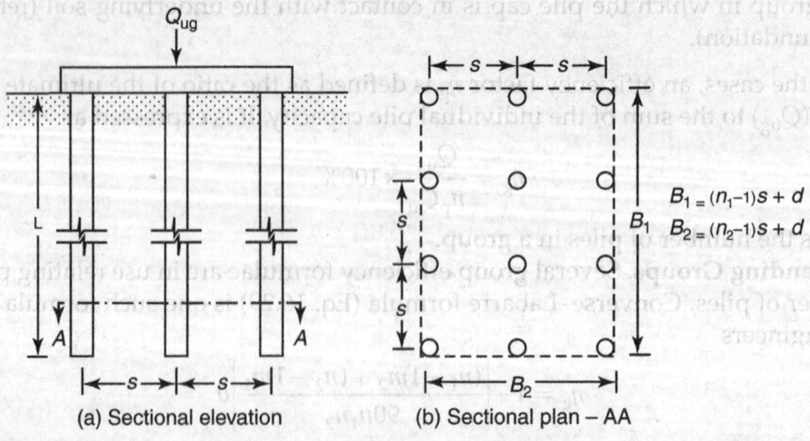


**Fig. 16.16** Pile group

$$\frac{1}{Q_{ug}^2} = \frac{1}{n_p^2 Q_u^2} + \frac{1}{Q_{ub}^2} \tag{16.32}$$

Equation 16.32 may be re-expressed as

$$\frac{1}{\eta_g^2} = 1 + \frac{n_p^2 Q_u^2}{Q_{ub}^2} \tag{16.33}$$

Comparative study of De Mello (1969) on efficiency of groups in clays showed that higher efficiency factors occur for

1. piles with smaller length–diameter ratios,
2. larger spacings, and
3. smaller number of piles in the group.

For spacing commonly used in practice, the efficiency factor is of the order of 0.70 to 0.85 in clays.

Only limited information is available for free-standing pile groups in sand. It has been established that the group efficiency in sands may often be greater than one (Poulos and Davis, 1980). But practical considerations limit the spacing to approximately 3*d* or the spacing corresponding to $\eta_g = 1$ (i.e., $Q_{ug} = nQ_u$).

**Piled Foundations**. In piled foundations, the pile group has the cap cast on or beneath the surface of the soil. For such cases in clays, Poulos and Davis (1980) suggested to take the lower value of ultimate load-bearing capacity from the following:

1. The ultimate load capacity based on block failure (Eq.16.31) plus the ultimate load capacity of that portion of the cap outside the perimeter of the block.
2. The individual pile capacity of cap and piles. That is,

$$Q_{ug} = n_p(\alpha c_u A_s + A_p C_p N_c) + N_{cc} C_c (B_1 \times B_2 - n_p \pi d^2 / 4) \tag{16.34}$$

where $c_p$ is the undrained cohesion at the level of pile tip, $c_c$ the undrained cohesion beneath pile cap, $N_{cc}$ the bearing capacity factor for rectangular cap $= 5.14(1 + 0.2B_1/B_2)$ for $B_1 > B_2$ (Skempton, 1951), and $N_c$ the bearing capacity factor (Skempton's values).

The first value applies to close pile spacing while the second for wider spacing (when individual action of piles occur). Model studies on piled foundations by Whitaker (1957) showed good agreement between the model test results and the predicted efficiency from the block-failure criterion.

In sands, it has been reported (Vesic, 1969) that the pile cap contributes significantly to the group capacity. For practical purposes, the contribution of the cap can be taken as equivalent to the bearing capacity of a strip footing with a half width equal to the distance from the edge of the cap to the outside of the pile (Poulos and Davis, 1980).

### 16.6.2 Pile Group in Filled Ground

In pile groups installed in a fill which is in the process of consolidation under its own weight or under the weight of the surface load, the weight of the whole mass of soil entrapped within the periphery of the group is transferred to the piles. For the first case (Fig. 16.17a), the additional load causing the drag down is due to the entire compressible soil located within the periphery. Thus, the total load on the pile group is

$$Q_{ug} = (\text{working load}) + B_1 \times B_2 \times \gamma_f' L_f' \qquad (16.35)$$

where $\gamma_f'$ is the submerged unit weight of fill and $L_f'$ the depth of fill over which the movement is sufficient to cause a drag down.

For the second case (Fig. 16.17b), the total load on the pile group at the level of bearing stratum is

$$Q_{ug} = (\text{working load}) + B_1 \times B_2 \times \gamma_f' L_f' + B_1 \times B_2 \times \gamma_n L_s' \qquad (16.36)$$

where $\gamma_f'$ is the submerged unit weight of soft stratum and $L_f'$ the depth of soft stratum over which movement is sufficient to cause a drag down.

However, the total load on the pile group will not exceed the ultimate skin friction on piles from fill and soft clay, that is,

$$Q_{ug} > (\text{working load}) + f'A_s' + f'A_s''$$

where $A_s'$ is the sum of surface areas of piles embedded in fill, $A_s''$ the sum of surface areas of piles embedded in soft clay, $f'$ the skin friction between fill and piles, and $f'$ the skin friction between soft clay and piles.

Group settlement is computed considering both the working load and the load transferred by the fill. Thus, there is no increase in the weight of soft stratum which should cause additional loading on the bearing stratum (Fig. 16.17b). Again, only the working load and the weight of the fill are causing settlement of the bearing stratum. The above approach assumes that the fill has been recently placed and has not had time to cause appreciable consolidation of the underlying strata.

The above-mentioned negative skin-friction problem will not occur if the piles are taken through the fill on to an incompressible stratum, such as bedrock or very compact sand and gravel.

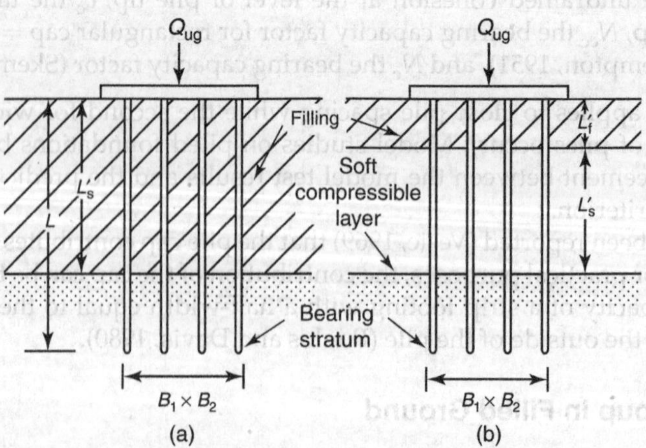


**Fig. 16.17** Pile groups in filled ground

### 16.6.3 Pile Group Settlement

The settlement observed at the pile cap level is due to compression within the pile from the loading plus the settlement occurring in the soil supporting the pile. The total settlement may be attributed to four separate causes: (i) the axial deformation of the pile, (ii) the deformation of the soil at the pile–soil interface, (iii) the compression of the soil between the piles, and (iv) the compression of the stratum below the tips of the pile. The settlement due to (i) and (ii) are very small and generally neglected, and settlement due to (iii) is difficult to evaluate. Hence, only the fourth factor is considered and expressed as the total settlement of a pile group.

A widely used procedure assumes that the pile group acts as a single large deep foundation, such as a pier or a raft. For pile groups that are essentially end bearing, the load is assumed to be applied at the pile tips as a uniform load over the area of the group. The stress distribution below the tips can be evaluated using the approximate 30° or 2:1 stress distribution (Fig. 16.18). For friction piles in clay, the load is assumed to be applied load at a depth of about two-thirds the length of the piles. The 30° or 2:1 distribution is adopted from that level.

For pile groups in sand and gravel, Meyerhof (1976) suggested an expression for the elastic settlement $S$ (mm) as

$$S = \frac{0.92q\sqrt{B_1 l}}{N'} \tag{16.37}$$

where $q = Q_g / (B_1 \times B_2)$ (kN/m$^2$), $N'$ is the corrected SPT value within the seat of settlement ($\approx B_1$ deep below the tip of the pile), $l$ the influence factor ($= 1 - L/B$, $B_1 \geq 0.5$), $Q_g$ the allowable load on the pile group (kN), and $L$ the length of the embedment of pile (m).

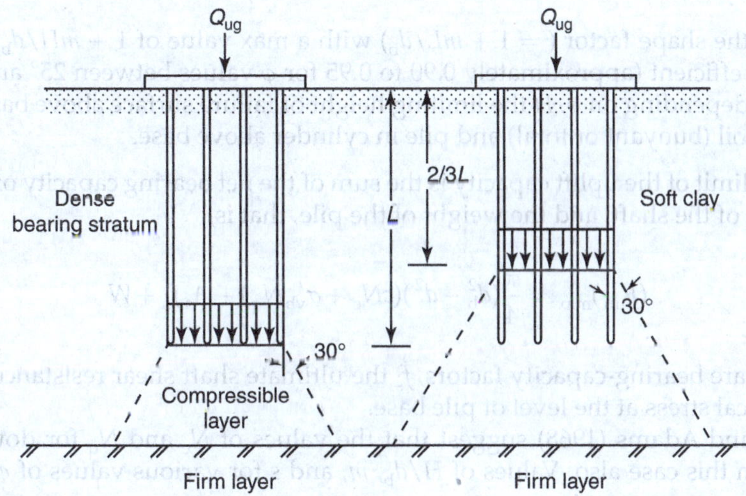


Fig. 16.18 Stress distribution for settlement analysis

## 16.7 UPLIFT RESISTANCE OF PILES

Under certain field conditions, such as tall chimneys, transmission towers, or jetty structures, foundation of structures are subjected to large overturning moments resulting in uplift forces. If foundations are on piles, they have to resist the upward forces. Methods of calculating the adhesion to resist uplift are the same as those used for bearing piles.

### 16.7.1 Uplift of Single Pile

For a uniform diameter in clay, the ultimate uplift resistance, $P_{uu}$, is given as

$$P_{uu} = \overline{c}_a A_s + W_p$$

where $W_p$ is the weight of pile and $\overline{c}_a$ the average adhesion along pile shaft.

For piles of uniform diameter in sand, the ultimate uplift capacity may be calculated as the sum of the shaft resistance plus the weight of the pile. Only a few data are available about the skin friction for upward loading.

For piles of uniform diameter in $c$–$\phi$ soils, Meyerhof and Adams (1968) suggest the following formulae for ultimate uplift load capacity.

1. Shallow depths ($L < d_b$)

$$P_{uu} = \pi c d_b L + s\frac{\pi}{2}\gamma d_b L^2 K_u \tan\phi + W$$

2. Great depths ($L > H$)

$$P_{uu} = \pi c d_b H + s\frac{\pi}{2}\gamma d_b (2L - H) K K_u \tan\phi + W$$

where $s$ is the shape factor ( $= 1 + mL/d_b$) with a max value of $1 + mH/d_b$, $K_u$ the earth pressure coefficient (approximately 0.90 to 0.95 for $\phi$ values between 25° and 40°), $m$ the coefficient depending on $\phi$, $H$ the limiting height of failure surface above base, and $W$ the weight of soil (buoyant or total) and pile in cylinder above base.

The upper limit of the uplift capacity is the sum of the net bearing capacity of the base, the side adhesion of the shaft, and the weight of the pile, that is,

$$(P_{uu})_{max} = \frac{\pi}{4}(d_b^2 - d^2)(cN_c + \sigma'_{vb} N_q) + A_s f_s + W$$

where $N_c$, $N_q$ are bearing-capacity factors, $f_s$ the ultimate shaft shear resistance, and $\sigma'_{vb}$ the effective vertical stress at the level of pile base.

Meyerhof and Adams (1968) suggest that the values of $N_c$ and $N_q$ for downward load can be used in this case also. Values of $H/d_b$, $m$, and $s$ for various values of $\phi$ are given in Table 16.5.

**Table 16.5** Factors for uplift analysis

| $\phi°$ | 20 | 25 | 30 | 35 | 40 | 45 | 48 |
|---|---|---|---|---|---|---|---|
| $H/d_b$ | 2.5 | 3.0 | 4.0 | 5.0 | 7.0 | 9.0 | 11.0 |
| $m$ | 0.05 | 0.10 | 0.15 | 0.25 | 0.35 | 0.50 | 0.60 |
| $s$ | 1.12 | 1.30 | 1.60 | 2.25 | 3.45 | 5.50 | 7.60 |

*Source:* Meyerhof and Adams (1968).

### 16.7.2 Uplift of Pile Group

Uplift of a group can be calculated as the lesser of the following (Meyerhof and Adams, 1968):

1. The sum of the uplift of the individual footings/piles.
2. The uplift load of an equivalent pier foundation consisting of the footings/piles and the enclosed soil mass.

Based on the uplift testing of footings in clays, Meyerhof and Adams (1968) have reported that the uplift efficiency increases with the spacing of footing or bases and as the depth of embedment decreases, but the efficiency decreases as the number of footings or bases in the group increases. These results are reported to be true for free-standing groups of piles in clays with downward loading (Whitaker, 1957).

In sands, for a given sand density, the uplift efficiency of the groups has been reported to increase roughly with the spacing of the footings or shafts and to increase when the depth of embedment becomes smaller (Meyerhof and Adams, 1968). Further, the uplift efficiency is reported to decrease as the number of footings or shafts in the group increases and as the sand density increases.

## 16.8 LATERAL RESISTANCE OF PILES

### 16.8.1 Causes of Lateral Forces

Sometimes, piles have to face lateral forces and moments; such a situation may arise in the following cases:

1. Quay walls and harbour structures where lateral forces are caused due to wave action and by the impact of ships during bettering.
2. Offshore structures subjected to wind and wave actions.
3. Earth-retaining structures supported by piles.
4. Transmission tower-foundation structures, that are to face a high wind force.
5. Lock-structure in dockyards.
6. Pile-supported structures constructed in earthquake areas.

Ultimate lateral resistance and deflection are the two main aspects of laterally loaded pile that are of interest. These problems are solved by simplified methods involving the concept of coefficient of sub-grade reaction. The ratio between the lateral pressure, $P$ and the deflection, $y$ produced at a particular point is called the coefficient of sub-grade reaction, a modulus of sub-grade reaction, or simply as the soil modulus, $k\ (= p/y)$.

### 16.8.2 Short and Long Piles

A vertical pile resists a lateral load by mobilization of passive pressure by the soil surrounding the pile. The distribution of the reaction of the soil depends on the following factors:

1. stiffness of the pile,
2. stiffness of the soil, and
3. fixity of the ends of the pile.

Thus the mode of failure, lateral resistance, and the deflection of a laterally loaded pile depend on the length of the pile and the surrounding soil. Accordingly, laterally loaded piles can be classified into two major categories, viz., short or rigid piles and long or elastic piles.

This behaviour of a pile is governed by the stiffness ($EI$) of the pile and the compressibility of the soil, the latter being a function of the soil modulus, $K$.

Pile stiffness and soil modulus are expressed as *stiffness factors*, which is explained as follows:

1. For a stiff preconsolidated clay with a constant soil modulus with depth, stiffness factor, $R = \left(\dfrac{EI}{KB}\right)^{1/4}$.
2. For cohesionless soils and most of the normally consolidated clays with soil modulus assumed to increase linearly with depth, then stiffness factor, $T = \left(\dfrac{EI}{n_h}\right)^{1/5}$.

In the above equation, where $E$ = modulus of elasticity of the pile material, $I$ = moment of inertia of the pile section, and $n_h$ = coefficient of modulus variation (force/$L^3$).

A laterally loaded pile is delineated as short or long based on the non-dimensional factor $L/R$ or $L/T$ as illustrated in Table 16.6.

References may be made to Kaviraj (1988) and Das (2002) for design charts for lateral resistance and deflection.

**Table 16.6** Criteria for short and long piles

| Pile type | Soil modulus | |
|---|---|---|
| | **Constant** | **Linearly increasing** |
| Short rigid | $L/R \leq 2$ | $L/T \leq 2$ |
| Long | $L/R \geq 3.5$ | $L/T \geq 4$ |

*Source:* Tomlinson, 1977.

## 16.9 INCLINED LOADING OF VERTICAL PILES

The ultimate load-carrying capacity of a vertical pile subjected to inclined loading is a function of both the lateral resistance and the vertical load capacity of the pile.

When there is only a small deviation from vertical loading the failure will occur essentially by axial slip or by bearing failure of the tip for downward loading. If the inclination of the applied load is large, lateral failure will occur due to load being applied perpendicular to the pile axis.

The above tow modes of failure will occur under the following conditions:

1. Axial failure is bound to occur when the ultimate lateral capacity exceeds the horizontal component of the ultimate inclined load, i.e., when

$$H_u > Q_u \sin\delta$$

or

$$H_u > P_u \tan\delta$$

where $Q_u$ is the ultimate inclined-load capacity of pile, $H_u$ the ultimate lateral-capacity of pile, $P_u$ the ultimate axial-load capacity of pile, and $\delta$ the angle of inclination of load from vertical.

2. Lateral failure is bound to occur when the ultimate lateral capacity is less than the horizontal component of the ultimate inclned load, i.e., when

$$H_u < P_u \tan\delta$$

For cohesive soils, it is generally assumed that the ultimate axial capacity of the pile is independent of the lateral component of the load and that the lateral load capacity is independent of the axial component of load. Then the inclined load capacity, $Q_u$, can be calculated as the lesser of the two values:

For axial failure

$$Q_u = P_u \sec\delta$$

For lateral failure

$$Q_u = H_u \operatorname{cosec}\delta$$

## 16.10 PILE CAP

The axial and lateral load carrying capacity of a pile group is significantly affected by a pile cap because of the resistance offered by the surrounding soil. This aspect is generally ignored in view of the potential danger of loss of soil support due to scour. However, in situations where the pile cap is fully embedded, it can contribute significantly to the group capacity. The pile group stiffness is increased by 5% to 15% when the pile cap rests in contact with the soil and also the settlement is reduced by 5% to 15% (Mohan, 1981).

Pile caps are almost invariably made of reinforced concrete. For designing pile caps, the load from the column at top of the cap and the reaction from the pile at the bottom of the cap are assumed to be distributed at 45° till the mid-depth of the cap. Based on this concept, the maximum bending moment and shear force are calculated at the critical sections (IS: 2911 – Part 1/Section 1). Other recommendations of the code are as follows:

1. The depth of pile cap should be sufficient enough to provide necessary anchorage of the column and pile reinforcement.

2. The pile cap should be rigid enough to distribute the imposed load on the piles in a group equitably.
3. The overhang of the pile cap should normally be 100 to 150 mm.
4. The pile should project 50 mm into the cap concrete.
5. The cap is cast over a levelling course of thickness of 75 mm.
6. In the case of large cap, where differential settlement may be imposed between piles under the same cap, due consideration for the consequential movement should be given.

## WORKED EXAMPLES

**Example 16.1** For a waterfront structure, round concrete piles are planned for use as foundation. The soil at the site is a medium-dense to dense sand having $\phi = 39°$, $\gamma = 19.7\ \text{kN/m}^3$, $\gamma' = 10.5\ \text{kN/m}^3$. The water table is located at a depth of 3.0 m from the ground surface. As per the design requirement, driven piles of 300 mm diameter and 8 m length have to be used. Determine the design load considering a factor of safety of 2.0. Assume $K = 3.0$, $\tan \delta = 0.45$.

***Solution***

Here, static formula can be used. Thus,

$$Q_u = Q_f + Q_p$$

$$f = K\sigma_v' \tan \delta$$

Because of the presence of water table, the effective overburden pressure should be separately calculated.

Hence, resistance due to skin friction

$$\begin{aligned} Q_f &= fA_s = (\sigma_v')K \tan \delta\ A_s \\ &= [(\sigma_v')_1 L_1 + (\sigma_v')_2 L_2] \tan \delta \pi d \\ &= \left[\frac{0 + 3\times 19.7}{2}\times 3 + \left(3\times 19.7 + \frac{3\times 19.7 + 5\times 10.5}{2}\right)\times 5\right]\times (3.0\times 0.45\times \pi \times 0.30) \\ &= (88.65 + 426.75)\times (3\times 0.45\times \pi\times 0.30) = 655.8\ \text{kN} \end{aligned}$$

Point bearing resistance

$$Q_p = qA_p = (\sigma_v' N_q)A_p = (3\times 19.7 + 5\times 10.5)N_q \times \pi \times \frac{0.3^2}{4}$$

From Fig. 16.10, for $\phi = 39°$, $N_q = 108$. Therefore,

$$Q_p = (3\times 19.7 + 5\times 10.5)\times 108\times \pi \times \frac{0.3^2}{4} = 852.0\ \text{kN}$$

$$Q_u = 655.8 + 852 = 1507.8\ \text{kN}$$

$$Q_a = \frac{Q_u}{F} = \frac{1507.8}{2.0} = 754\ \text{kN}$$

**Example 16.2** In a two-layered cohesive soil, bored piles of 400 mm are installed. The top layer has a thickness of 5 m and the bottom one is of considerable depth. The shear strength of the top clay layer is 45 kN/m$^2$ and that of the bottom is 100 kN/m$^2$. Determine the length of the bored pile required to carry a safe load of 380 kN, allowing a factor of safety of 2.0.

***Solution***

Consider an adhesion factor, $\alpha = 0.50$.

Let $L_1$ and $L_2$ be the depths of embedment of pile in top and bottom layers, respectively. Then,

$$Q_a = \frac{Q_u}{F}$$

$$Q_u = Q_a \times F = 380 \times 2 = 760 \text{ kN}$$

Also,

$$Q_a = \alpha (C_u)_1 \pi d L_1 + \alpha (C_u)_2 \pi d L_2 + (C_p)_2 N_c \frac{\pi d^2}{4}$$

Therefore,

$$760 = 0.5 \times 45 \times \pi \times 0.4 \times 5 + 0.5 \times 100 \times \pi \times 0.4 L_2 + 100 \times 9 \times \pi \frac{0.4^2}{4}$$

or

$$760 = 254.5 + 62.8 \times L_2$$

or

$$L_2 = \frac{760 - 254.5}{62.8} = 8 \text{ m}$$

Therefore, the length of the pipe is as given below:

$$L_1 + L_2 = 5 + 8 = 13 \text{ m}$$

**Example 16.3** A 12 m long 300 mm$^2$ square pre-cast concrete pile is driven into a sand stratum by a single-acting steam hammer. The weight of the CI hammer ram is 14 kN and the stroke is 750 mm. The pile showed a driving resistance of 5 blows/25 mm penetration. Estimate the ultimate bearing capacity of the pile based on the Hiley formula. Take $C = 0.00508$ m.

***Solution***

For a single-acting steam hammer, the hammer efficiency can be taken as 80% (from Table 16.3).

The coefficient of restitution for CI hammer and pile can be taken as 0.45 (from Table 16.4).

$$\text{Weight of pile, } W_p = 0.3 \times 0.3 \times 12 \times 23.5 = 25.38 \text{ kN}$$

$$\text{Penetration per blow} = \frac{25}{5} \times \frac{1}{1000} = 0.005 \text{ m}$$

$$Pe_r = 25.38 \times 0.45 = 11.4 \text{ kN}$$

That is, $W > Pe_r$. Therefore,

$$Q_u = \frac{\eta W h}{S + (C/2)} \times \frac{W + e_r^2 p}{W + p}$$

$$= \frac{0.80 \times 14 \times 0.75}{0.005 + 0.00254} \times \frac{14 + 0.45^2 \times 25.38}{14 + 25.38}$$

or

$$Q_u = 541.5 \text{ kN}$$

**Example 16.4** A 6 m thick layer of medium dense sand overlies a deep dense gravel. Series of standard penetration tests were undertaken and the sand stratum showed an average *N*-value of 21. From the tests on gravel, the *N*-value at the interface has been interpolated as 42. A round pile of 250 mm diameter is to be driven down through the sand gravel to give adequate end bearing. Taking a factor of safety of 3, determine the allowable load that the pile can carry.

***Solution***

As suggested by Meyerhof (1976), the ultimate bearing capacity is given as (Eq. 16.21)

$$Q_u = fA_s + qA_p$$

The lower value of *q* obtained from Eq. (a) or (b) is to be taken. Now,

$$q = 40N\frac{L}{d} \text{ kN/m}^2 \quad \text{(a)}$$

or

$$q = 400N \text{ kN/m}^2 \quad \text{(b)}$$

From (a)

$$q = 40 \times 42 \frac{6.0}{0.25}$$

$$= 40,320 \text{ kN/m}^2$$

From (b)

$$q = 400 \times 42 = 16,800 \text{ kN/m}^2$$

Therefore,

$$Q_u = 2 \times 21 \times \pi \times 0.25 \times 6 + 16,800 \times \frac{\pi(0.25)^2}{4}$$

$$= 197.8 + 824.3$$

$$= 1022.1 \text{ kN}$$

Allowable load,

$$Q_a = \frac{1022.1}{3} = 340.7 \text{ kN}$$

**Example 16.5** A pile load test is made on a 300 mm diameter test pile and the following data are obtained.

| Load (kN) | Settlement (mm) |
|---|---|
| 0 | 0 |
| 300 | 1.25 |
| 600 | 3.75 |
| 900 | 7.50 |
| 1,200 | 13.75 |
| 1,500 | 23.75 |
| 1,800 | 36.75 |

Determine the design load on the pile considering the settlement and shear criteria. Adopt a factor of safety of 2.0.

***Solution***

The load–settlement curve is plotted as shown in Fig. 16.19 and the tangents are drawn from the two straight portions. The ultimate load is read from the graph as

$$Q_u = 1050 \text{ kN}$$

Based on shear failure, the allowable load

$$(Q_a)_1 = \frac{1050}{3} = 350 \text{ kN}$$

Based on the settlement criterion, the safe load is least of the following (IS: 2911 – Part4, 1985):

1. 2/3 of the final load corresponding to 12 mm settlement

$$(Q_a)_2 = \tfrac{2}{3} \times 1125 = 750 \text{ kN}$$

2. 50% of the final load corresponding to 10% of the pile diameter (= 0.10 × 300 = 30 mm)

$$(Q_a)_3 = \tfrac{1}{2} \times 1650 = 825 \text{ kN}$$

Based on settlement, the allowable load = 750 kN.

Hence, the design load is least of the values obtained from shear and settlement criteria. Therefore, design load

$$Q_a = 350 \text{ kN}$$

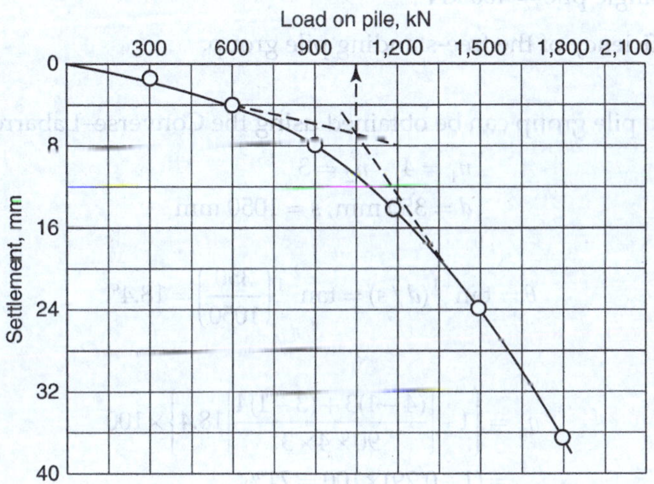


Fig. 16.19

**Example 16.6** A single under-reamed pile is to be installed in a hard clay deposit. The undrained strength obtained from a series of vane tests at different depths have shown a linear relationship, $c_u = 50 + 8D$ in kN/m$^2$, where $D$ is the depth in metres. The diameters of the pile shaft and the bulb are, respectively, 1 and 3 m. The centre of the under-ream is located at 16 m from the ground surface. Determine the allowable load on the pile to ensure an overall factor of safety of 2.0. Neglect the resistance offered by the shaft beneath the bulb.

***Solution***

At $D = 16$ m, $c_u = 178$ kN/m$^2$.

Therefore, average cohesion on shaft is

$$\frac{50+178}{2} = 114 \text{ kN/m}^2$$

From Table 16.2, $\alpha = 0.3$. Thus,

$$\begin{aligned} Q_u &= fA_s + qA_p \\ &= \alpha(c_u)_1 \pi d L + (c_u)_2 N_c \frac{\pi d_u^2}{4} \\ &= 0.3 \times 114 \times \pi \times 1 \times 16 + 178 \times 9 \times \pi \times \frac{3^2}{4} \\ &= 13{,}043 \text{ kN} \end{aligned}$$

$$Q_a = \frac{Q_u}{F} = \frac{13{,}043}{2} = 6{,}522 \text{ kN}$$

**Example 16.7** A 4 × 3 pile group has the following details:

Diameter of each pile, $d = 350$ mm
Centre-to-centre spacing of pile = 1,050 mm
Capacity of a single pile = 400 kN

Determine the efficiency of the free-standing pile group.

***Solution***

The efficiency of a pile group can be obtained using the Converse–Labarre formula. So,

$$n_1 = 4, \quad n_2 = 3$$

$$d = 350 \text{ mm}, s = 1050 \text{ mm}$$

Therefore,

$$\theta = \tan^{-1}(d/s) = \tan^{-1}\left(\frac{350}{1050}\right) = 18.4°$$

From Eq. 16.29,

$$\begin{aligned} \eta_g &= \left\{1 - \left[\frac{(4-1)3 + (3-1)4}{90 \times 4 \times 3}\right] 18.4\right\} \times 100 \\ &= (1 - 0.29) \times 100 = 71\% \\ &= \frac{Q_{ug}}{n_p Q_u} \end{aligned}$$

Therefore,

$$Q_{ug} = \eta_g \times n_p Q_u = 0.71 \times 12 \times 400 = 3408 \text{ kN}$$

**Example 16.8** A group of concrete piles is square in plan and consists of 9 piles each 12 m long and 500 mm diameter. The piles are bored piles and installed at a spacing of 3 $d$ in a deep clay deposit having an unconfined compressive strength of 62.4 kN/m$^2$.

At the tip of the pile and below, the undrained shear strength $c_u = 45$ kN/m$^2$. The average unit weight of the soil and concrete are 19.2 and 22.5 kN/m$^3$, respectively. Estimate the total ultimate load of the pile group.

***Solution***

Since the piles are bored piles, the difference between the weight of concrete and the excavated earth has to be considered.

The general equation for a single pile is written as

$$Q_u + (\text{weight of pile} - \text{weight of excavated soil}) = Q_f + Q_p$$

$$Q_u = fA_s + qA_p - (\text{weight of pile} - \text{weight of excavated soil})$$

Assuming the clay as medium stiff, $\alpha = 0.50$. Thus,

$$Q_u = 0.5 \times \left(\frac{62}{2}\right) \pi \times 0.50 \times 12 + 9 \times 45 \times \pi \times (0.5)^2 \times \frac{1}{4}\left[-\frac{\pi \times (0.5)^2}{4} \times 12(23.5 - 19.2)\right]$$

$$Q_u = 292.2 + 79.5 - 10.1 = 361.6 \text{ kN}$$

Considering the individual pile capacity,

$$n_p Q_u = 9 \times 361.6 = 325.4 \text{ kN}$$

Considering block failure of the group,

$$Q_{ug} = fA_{sg} + qA_{pg}$$

Now,

$$B_1 = B_2 = (n_1 - 1)s + d = (3 - 1) \times 3d + d = 7d = 3.5 \text{ m}$$

For a block failure, the skin friction development is predominantly between soil to soil, hence the adhesion factor can be taken as 1.0.

$$\begin{aligned} Q_{ug} &= 2\alpha(c_u)_1[B_1 + B_2]L + (c_u)_2 N_c [B_1 \times B_2] \\ &= 2 \times 1 \times (62/2)[3.5 + 3.5] \times 12 + 45 \times 9[3.5 \times 3.5] \\ &= 5208 + 2835 = 8043 \text{ kN} \end{aligned}$$

Here,

$$n_p Q_u < Q_{ug}$$

Therefore, group capacity ≈3,254 kN.

## POINTS TO REMEMBER

16.1 Pile is a type of deep foundation used for transferring the weight of the superstructure through soft of weak soils, to deep load-bearing strata. They are slender structural members normally installed by driving, by hammer or by vibrating, and occasionally by auguering.

16.2 Piles are classified based on material composition, installation methods, ground effects, and function as foundation.

16.3 Pile-driving equipment consist mainly of pile frames, pile winches, and pile hammers.

16.4 Bearing capacity of a single pile depends on the structural strength of the pile and the supporting strength of the soil, and the smaller of the two controls the design load.

16.5 Bearing capacity of a single pile may be found theoretically by statistical methods, pile-driving formulae, and wave equation.

16.6 Bearing capacity of a single pile may be found from field tests, viz., pile load test and SPT values.

16.7 Pile-driving formula is based on the qualitative principle that a pile is capable of sustaining a greater load if it exerts a greater resistance against driving.

16.8 Commonly used dynamic formula is the Engineering News Record (ENR) formula which is derived on the basis of the work–energy theory.

16.9 Pile capacity from wave equation is a model based on the propagation of an elastic wave through a long rod. The model considers the complete pile-driving operation, including pile-driving accessories, soil–pile interaction and the time-dependent nature of the elastic pile deformation.

16.10 Pile load test is the most reliable method of determining the capacity of a pile. The test necessitates a careful record to be maintained during installation and during load test.

16.11 Negative skin friction is a downward drag acting on the pile due to relative movement between the piles and the surrounding soil. The effect of negative skin friction is to increase the axial load in the pile and the pile settlement.

16.12 Piles are always used in a group. Bearing capacity and settlement of pile groups are needed for the design of a deep foundation.

## QUESTIONS

### *Objective Questions*

16.1 State whether the following statements are true or false:

1. The term composite pile is applied to a cross-section with more than one material.
2. Closely spaced piles embedded in clay often behaves like a group acting as a single large unit.
3. Dynamic formulae are generally used to establish pile-driving criteria for pile installation.
4. The development of skin friction along the shaft of a pile requires no shear strain in the soil adjacent to the pile.
5. In friction piles, the load is transmitted to the soil through the adhesion or frictional resistance along the shaft of the piles.

16.2 The negative skin friction or down-drag of a pile is a phenomenon which occurs when
(a) A compressible organic soil is found at the pile tip
(b) A soil layer surrounding a portion of the pile shaft settles more than the pile
(c) The groundwater table suddenly rises from the tip of the pile to ground surface
(d) A long pile is driven next to the short pile under consideration

16.3 Load tests on piles in clays are not carried out immediately but sufficient time is allowed for the soil to
(a) Come in contact with the pile surface
(b) Develop skin friction
(c) Regain its thixotropic strength
(d) Re-distribute the initial weight of pile

16.4 The action of driven piles in sands increases the relative density and thus the efficiency of a pile group may be
(a) Equal to 100% (b) Greater than 100%
(c) Approximately equal to 100% (d) Well below 100%

16.5 The point bearing and average skin frictional resistance of bored piles in cohesionless soil are ______ that for driven piles.
(a) Less than (b) Greater than
(c) Equal to (d) About 80% of

16.6 Indicate the incorrect statement. High group efficiency of pile can be obtained in a clayey stratum for
(a) Piles having smaller length-to-diameter ratio
(b) Larger spacing
(c) Small number of piles in a group
(d) Closer spacing

16.7 *Assertion A*: Longer pile groups should settle more than smaller groups for the same pile loads.
*Reason R:* This is due to the overlapping effect of stresses below the pile point from the additional piles.

*Choose the correct statement.*
(a) Both A and R are true, and R is the correct explanation of A.
(b) Both A and R are true, and R is not the correct explanation of A.
(c) A is true, but R is false.
(d) A is false, but R is true.

16.8 Consider the following statements:
1. The pile cap is in contact with the ground.
2. The piles are vertical.
3. Load is applied at the centre of the pile group.
4. The pile group is symmetrical and the cap is very thick.

The assumption that each pile in a group carries equal load may be correct when the criteria in statements
(a) 1 and 2 are met (b) 1, 2, and 3 are met
(c) 1, 3, and 4 are met (d) All are met

16.9 Settlement of a pile group predominantly depends on the
(a) Axial deformation of the pile
(b) Deformation of the soil at the pile–soil interface
(c) Compression of the soil between the piles
(d) Compression of the stratum below the tips of the pile

16.10 For settlement evaluation of friction pile groups in clay, the load is assumed to be applied as a uniform load at a depth of about ___ the length of the piles. Choose the appropriate statement
(a) One-thirds (b) Two-thirds
(c) One-fourths (d) One-half

## *Descriptive Questions*

16.11 Does the choice of a pile hammer have any relevance to the type of pile? If so, give reasons.
16.12 While driving large number of displacement piles for a foundation, how would you proceed: centre to out, outside to centre, or progressively from one side to the other? Explain.
16.13 In a pile group, what are the geometrical properties that are to be considered in bringing out a proper spacing of piles to ensure that they carry equal load?
16.14 List the circumstances under which a pile foundation becomes necessary.
16.15 What are the factors to be considered in the selection of pile hammer?
16.16 What type of piles would you recommend for the following types of soil and site conditions:
1. The subsoil is a weak underlain by a hard rock
2. For a multi-storeyed building in the central part of a city surrounded by existing buildings
3. For a harbour structure

16.17 Why do deep foundation units have typically long slender members?

## EXERCISE PROBLEMS

16.1 A tapered 10 m long wooden pile with a 200 m diameter at the tip and a 350 mm diameter at the butt is driven into a dry sand $\phi = 28°$, and dry unit weight, $\gamma = 17.2\ \text{kN/m}^3$. Compute the ultimate bearing capacity of the pile.

16.2 A single test pile of 0.25 m × 0.25 m square cross-section is driven through a stratum up to a depth of 10 m. It is observed that the undrained cohesive strength is varying from 12.4 $\text{kN/m}^2$ at its surface to 65.3 $\text{kN/m}^2$ at the depth of 10 m. Determine the safe load the pile can carry if the factor of safety is 2.5.

16.3 A pre-stressed concrete pile of 350 mm diameter is to be driven into a layered deposit of intact clay. The deposit consists of a normally consolidated clay up to a depth of 8 m from the ground surface, with $c_u = 30$ kPa, $\phi'_{re} = 15°$, and $\gamma = 16\ \text{kN/m}^3$ followed by 7 m of over-consolidated clay having $c_u = 100$ kPa, $\gamma = 19.6\ \text{kN/m}^3$, and an over-consolidation ratio of three. If the piles are driven to a depth of 12 m, estimate the total skin friction expected to develop.

16.4 A drilled, dry straight pile foundation is to be constructed in a layered clay stratum. The boring log and subsequent laboratory tests furnished the following details:

0–8 m depth $c_u = 45$ kPa $\gamma = 16.6$ kN/m$^3$
8–15 m depth $c_u = 80$ kPa $\gamma = 18.2$ kN/m$^3$

16.5 A sub-surface investigation at a site revealed the soil profile as shown in Fig. 16.20. A clock tower is to be constructed on the site over a pile foundation. As per the design requirement, piles of 300 mm diameter are to be driven to a depth of 14 m. Estimate the ultimate pile capacity of one single pile.

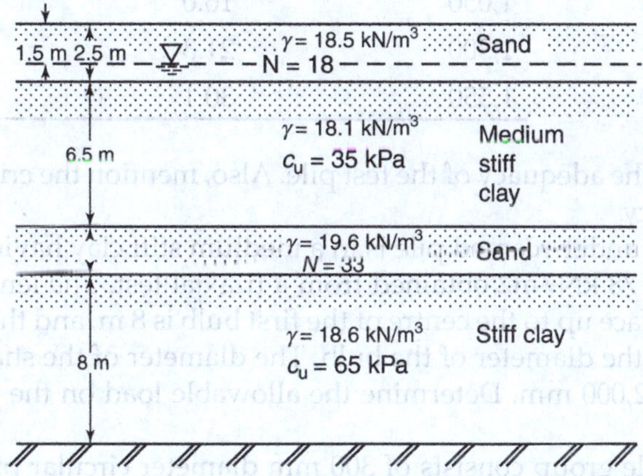


Fig. 16.20

16.6 Compute the safe bearing of a 500 mm diameter and 12.5 m long concrete pile driven into a granular medium by a hammer with a 30 kN ram and 900 mm stroke. The set is observed to be 12 mm per blow and the constant C = 25 mm. The hammer efficiency is 85% and coefficient of restitution is 0.50. Allow a factor of safety of 3.

16.7 In a multi-storeyed building site, soil boring revealed the presence of fine to coarse sand The standard penetration test information obtained from the boring log is as follows:

| Depth (m) | *N* value |
|---|---|
| 0.9 | 12 |
| 2.0 | 14 |
| 3.0 | 18 |
| 4.5 | 20 |
| 6.0 | 25 |
| 7.5 | 27 |

Compute the length and diameter of driven single pile to take a design load of 450 kN, considering a factor of safety of 2.0.

16.8 On a project, a pile load of 600 kN was desired. A pile load test was conducted and the results are given below.

| Load (kN) | Penetration (mm) |
|---|---|
| 0 | 0 |
| 300 | 4.5 |
| 550 | 5.5 |
| 800 | 9.5 |
| 1,050 | 16.0 |
| 1,300 | 21.3 |
| 1,550 | 30.1 |

Determine the adequacy of the test pile. Also, mention the criterion adopted to specify the adequacy.

16.9 A two-bulb under-reamed pile is in a medium stiff clay having an average undrained cohesion of 78 kN/m2 obtained from a triaxial test. The length of the shaft from the ground surface up to the centre of the first bulb is 8 m, and the centre-to-centre of bulb is 1.2 times the diameter of the bulb. The diameter of the shaft is 800 mm and that of the bulb is 2,000 mm. Determine the allowable load on the pile, allowing a factor of safety 2.0.

16.10 An eight-pile group consists of 300 mm diameter circular piles with centre-to-centre spacing of 900 mm both ways. The piles are driven to a depth of 20 m into a clay which has an unconfined compressive strength of 20 kPa and a unit weight of 17.5 kN/m$^3$. Also, the clay stratum is very deep. Determine the allowable load for the group, considering a factor of safety of 3.

16.11 It is proposed to transfer the total load of 3,000 kN of a structure through 10 m long bored piles in a deep deposit of clay having an average undrained shear strength of 90 kN/m2. The design diameter of the pile is 400 mm. Estimate the number of piles required, considering a factor of safety of 2.5. Also, suggest the arrangement of piles.

16.12 A nine-pile group is placed in a square pattern with centre-to-centre spacing of 0.9 m. The diameter and length of the pile are 300 mm and 11.5 m, respectively. The piles are driven into a sand deposit having $\phi = 32°$ and $\gamma = 18.5$ kN/m3. The sand deposit extends to a depth of 12.5 m followed by a 5 m depth of clay having $e_0 = 0.85$ and $C_c = 0.32$. Determine the pile group efficiency and the settlement of the group if the pile group carries a safe load of 4,500 kN. Allow a factor of safety of 2.0.

16.13 Forty RCC piles of 300 mm diameter are needed to carry a structural load of 1,650 kN from a column on a deep soft clay deposit. The undrained cohesion of the deposit is 22 kPa and the unit weight is 18.2 kN/m$^2$. Estimate the length and spacing of the piles such that the group has an efficiency of 100%. The adhesion factor can be taken as 0.75 and the factor of safety against shear is 2.5.

# 17

# Drilled Piers and Caisson Foundations

## CHAPTER HIGHLIGHTS

Drilled piers: uses and types of drilled piers, bearing capacity and settlement of drilled piers, construction methods – Caissons: types of caissons, bearing capacity and settlement of caissons – Construction procedures – Well foundations: types of wells, components of well foundations, design of wells, stability analysis, construction of well foundations – Shifts and tilts in well sinking

## 17.1 INTRODUCTION

Drilled piers and caissons provide a solid massive foundation for heavy loads and high horizontal thrusts. Drilled piers are structural members of relatively large-diameter massive struts constructed and placed in a pre-excavated hole. They are referred to variously by civil engineers as bored piles, large-diameter piles, foundation piers, sub-piers, and drilled caissons. The shafts can be enlarged at the base, resulting in belled or under-reamed piers. Caissons or well foundations are structural boxes or chambers that are sunk in place through the ground or water by systematically excavating below the bottom of the unit, which thereby descends to the final depth. These have a cross-sectional area and hence provide high bearing capacity, which is much larger than what may be offered by a cluster of piles.

## 17.2 DRILLED PIERS

### 17.2.1 Uses

Piers have distinct advantages compared to other types of deep foundations:

1. In areas where pile penetration is difficult, piers can be provided.
2. Vibration and heave of soil are not caused as in installation of a driven pile. This is a decisive factor when the adjacent structure is on spread footings or short piles.

3. Equipment used in the construction of drilled piers produces less noise and, hence, is quite suitable for areas near hospitals and similar institutions.
4. There is a possibility of inspection and physical testing of the soil or rock conditions at the bottom of the pier.
5. In the construction of piers, there is no displacement of volume of soil, and the problems of shifting and lifting are eliminated.
6. Drilled piers generally require light construction equipment.
7. They can resist high lateral stresses.

### 17.2.2 Types of Drilled Piers

The common type of drilled pier is the straight type (Fig. 17.1); such shafts are taken through the upper soil layers, and the end is placed on a firm soil layer or bed rock. In some instances, the shafts are provided with casing.

Sometimes a broad base called a bell is provided at the bottom of straight shafts; such piers are referred to as belled piers. The bell may be shaped like a dome or it may be angled. Under-reaming equipment is used to make the bells.

The third type is the extended straight shaft or socketed piers, in which the straight shafts are extended into the underlying rock layer.

### 17.2.3 Bearing Capacity of Drilled Piers

A drilled pier derives its supporting power from both skin friction and bottom bearing as in a pile. Generally, the skin friction developing along the shaft is less compared to the end bearing capacity. Since drilled piers are placed invariably on a hard stratum, which is less compressible, the maximum percentage of load is taken by the hard stratum only. The surface area of a drilled pier is less compared to the one available in a pile group. Thus, in many

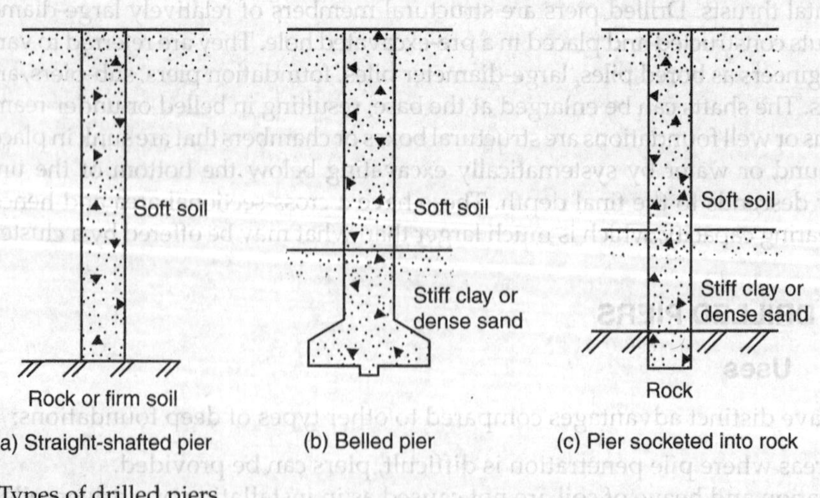


Fig. 17.1 Types of drilled piers

instances, the drilled piers are designed as a compression member subjected to a load on top and an equal reaction at the bottom, neglecting the skin friction.

The ultimate load-bearing capacity of drilled piers can be computed as for piles as

$$Q_u = Q_f + Q_p$$

**Piers in Cohesive Soils**. The skin resistance for the shaft in cohesive soils is computed as

$$Q_f = \alpha c A_s \tag{17.1}$$

where $A_s$ is the surface area of the shaft, $\alpha$ is the empirical adhesion factor (0.35 to 0.40), and the load-bearing capacity for a circular base is

$$Q_p = A_p(c\bar{N}_c + q'\bar{N}_q + 0.3\gamma D_b \bar{N}_\gamma) \tag{17.2}$$

where $\bar{N}_c$, $\bar{N}_q$, and $\bar{N}_\gamma$ are the bearing capacity factors for a deep foundation and $q'$ is the vertical effective stress at the level of the bottom of the pier. Also, $D_b$ is the diameter of the base and $A_p$ is the area of the base $= \pi D_b^2/4$. The last term is generally considered only for short piles and is neglected in other cases. Thus,

$$Q_p - A_p(c\bar{N}_c + q'\bar{N}_q) \tag{17.3}$$

The net ultimate bearing capacity $(Q_p)_n$ at the base can be approximated as

$$(Q_p)_n = A_p(c\bar{N}_c + q'\bar{N}_q - q')$$

$$(Q_p)_n = A_p[c\bar{N}_c + q'(\bar{N}_q - 1)] \tag{17.4}$$

It is customary to omit the surcharge term, for $\phi = 0°$, $\bar{N}_q = 1$; then,

$$(Q_p)_n = c_u \bar{N}_c A_p \tag{17.5}$$

Skempton's value of $\bar{N}_q = 9$ for $L/D_b \geq 4$ is widely used, where $L$ is the length of the pier. Thus,

$$(Q_p)_n = 9 c_u A_p \tag{17.6}$$

**Piers in Non-cohesive Soils**. For piers in non-cohesive soils, Berezantzev (1965) proposed the net allowable load $(Q_p)_a$ at the base limiting relative settlement $(\Delta H/D_b)$ to be about 0.20. Thus,

$$(Q_p)_a = [\gamma D_b(B_k)]A_p \tag{17.7}$$

Figure 17.2 presents values of $B_k$ for several selected $L/D_b$ ratios. For piers in non-cohesive soils, Eq. 17.4 may be written in terms of net base bearing pressure as

$$(Q_p)_n = A_p q'(\bar{N}_q - 1) \tag{17.8}$$

As the piers are invariably drilled and placed, compared to piles which are driven, a lower bound value of $\bar{N}_q$, as suggested by Vesic (1963), may be adopted (Fig. 17.3).

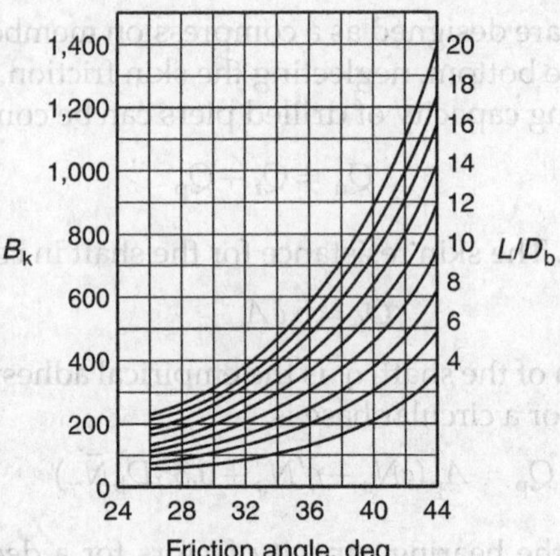


**Fig. 17.2** Coefficients Bk (*Source*: Berezantzev, 1965)

Allowing a suitable factor of safety to Eq. 17.8 and comparing it with Eq. 17.7, the lower value may be taken as the net safe base pressure, $(Q_p)_{ns}$.

The expression for skin resistance may be computed in the same manner as that for piles; that is,

$$Q_f = \int_0^L f \pi D_{sb} \, dz \tag{17.9}$$

where $f = K\sigma_v' \tan \delta$ and $D_{sb}$ is the diameter of the shaft.

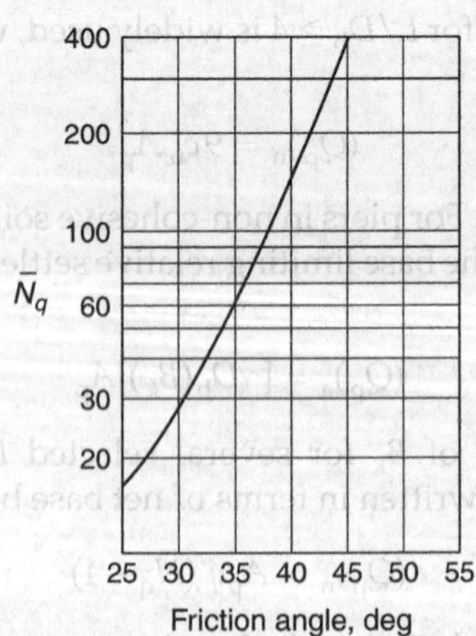


**Fig. 17.3** Bearing capacity factor $\bar{N}_q$ (*Source:* Vesic, 1963)

For piers with casing, $K < K_0$, and for uncased piers, $K = K_0$. The value of $\delta$ is taken as that of soil and casing in the first case and is equal to $\phi$ in the second case. For a uniform soil condition,

$$Q_f = (K\sigma_v' \tan\delta)\times(\pi D_{sb} L) \tag{17.10}$$

Applying the same factor of safety for skin friction, the net safe skin friction is given as

$$(Q_f)_s = \frac{Q_f}{F} \tag{17.11}$$

where $F$ is the factor of safety. Thus, the net allowable load on piers is

$$Q_{an} = (Q_p)_{ns} + (Q_f)_s \tag{17.12}$$

In granular soils, the value of $\phi$ depends on the confining pressure, and hence, a careful assessment of $\phi$ has to be made.

In some conditions, the negative skin friction can occur as discussed for piles. As many uncertain factors are involved, detailed estimates for skin friction are not very realistic. In many circumstances, it can be ignored. However, a rough estimate may be made based on the value of skin friction suggested by Terzaghi and Peck (1967), as given in Table 17.1.

When drilled caissons are sunk through sand, it is loosened by grabbing and surges into the dredging wells. This effect is not considered, as the reconsolidation of sand is rapid and the negative skin friction is again a relatively short-term effect.

### 17.2.4 Settlement of Drilled Piers

Drilled piers on sand or gravel will not be subjected to detrimental settlement as they are normally constructed on dense deposits. Settlement on sands may be computed, as done in shallow foundations, by obtaining or making an estimate of the stress–strain modulus $E$ and Poisson's ratio $\nu$ and using Eq. 14.33. Settlement may be computed as done for shallow foundations in clay. Since one of the purposes of using drilled piers is to place the large loads on

**Table 17.1** Observed values of skin friction for piers and caissons

| Type of soil | Skin friction (kN/m2) |
|---|---|
| Silt and soft clay | 7–29 |
| Very stiff clay | 48–192 |
| Loose sand | 12–34 |
| Dense sand | 34–67 |
| Dense gravel | 48–96 |

*Source:* Terzaghi and Peck (1967).

firm stratum, long-term settlement may not occur at all. However, immediate settlement of piers on saturated clays may be computed using the Burland et al. (1966) expression as

$$S = D_b(I_B)\frac{\bar{q}}{q_{ult}} \tag{17.13}$$

where

$$I_B = \varepsilon_1\left[0.295\left(9 + \frac{\gamma' L}{c_u}\right)\right] \tag{17.14}$$

where $\varepsilon_1$ is the axial strain at a deviator stress of one-third to half of the peak value and $\bar{q} = \gamma' D_b B_k$.

### 17.2.5 Construction Procedures of Drilled Piers

Construction procedures of drilled piers shafts are of three major types and they may be classified as (1) dry method, (2) casing method, and (3) wet method. Each procedure of construction is explained as follows:

**1. Dry method**

Dry method of construction is the simplest type wherein it is to remove all weak soils up to the foundation depth and construct the pier in the form of concrete column.

For construction of the shafts, excavations may be made by hand or machine. A hand-excavated pier is called as Chicago caisson or Gow caisson depending on the procedure adopted, and the machine-excavated piers so by the name of drilled caissons.

In the Chicago type, circular holes are excavated up to a certain depth by hand, and the sides are then lined with vertical boards called laggings (Fig. 17.4a(i)). They are secured in place by two circular steel rings. The excavation is continued up to the desired depths, and

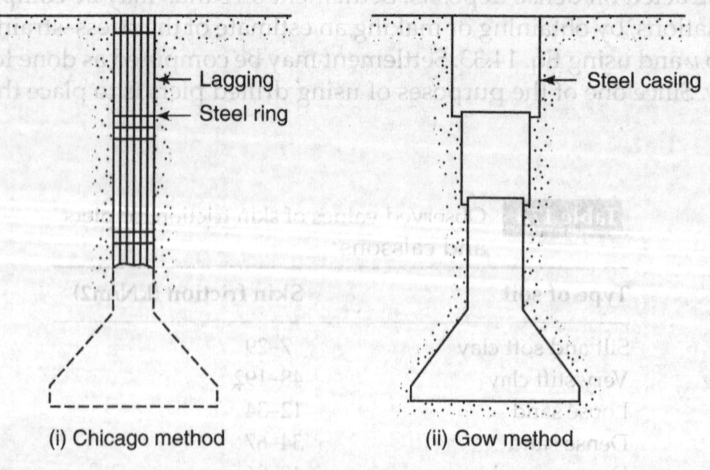


Fig. 17.4 (a) (i–ii) Methods of pier construction

then the bell of the pier is excavated. The hole is filled with concrete after completion of the excavation. In the Gow method (Fig. 17.4a(ii)) the excavation is done by hand, and the telescopic metal shells are used to maintain the shaft. The shells are removed, as concreting progresses piers up to a depth of 30 m can be constructed by this method.

Caisson shafts may be excavated more efficiently by utilizing mechanical devices such as an angle, a bucket, a chopping bit, or a chopping bucket. Open-helix auger is a common excavation tool that is attached to a shaft referred to as the kelly and pushed in to the soil and rotated. The augered soil is taken out, and an under-reaming tool is inserted into the hole to shape the bell. When boulders or hard stratum is encountered, the caisson is usually excavated by a chopping bucket or a chopping bit.

The sequence of operation of a vertical drilled shaft without a bell using mechanical device (O'Neill and Reese, 1999) is shown in Fig. 17.4 (a(iii)–(vi)).

The dry method is adopted in soils and rocks that are above the water table and will not cave in during the time of drilling the hole.

### 2. Casing method of construction

This method is adopted in situations wherein casing or excessive deformation is likely to occur during the time of excavation. The sequence of construction (O'Neill and Reese, 1999) is explained as follows (Fig. 17.4b):

(i) As in the case of dry method of construction, the excavation procedure is initiated (Fig. 17.4b(i)).
(ii) In the case of casing, if caving is met with bentonite slurry is pumped into the borehole (Fig. 17.4b(ii)). Drilling is continued until the excavation goes past the caving soil and a firm impermeable soil or rock is reached.
(iii) After this stage, a casing is introduced into the hole (Fig. 17.4b(iii)).
(iv) Using a submersible pump the slurry is bailed out of the casing (Fig. 17.4b(iv)).
(v) Then a small drill that can pass through the casing is introduced into the hole and the excavation is continued (Fig. 17.4b(v)).
(vi) If an enlarged base or bell is needed, an under-reamer is introduced (Fig. 17.4b(vi)).
(vii) If a reinforcing steel is needed, it is inserted up to the full length of the excavation. Concrete is then poured into the excavation and the casing is gradually pulled out (Fig. 17.4(vii)).
(viii) The completed view of the drilled pier is shown in Fig. 17.4 b(viii).

### 3. Slurry method of construction

In this method of construction, slurry is used to keep the borehole open during the full depth of excavation. The sequences of construction (O'Neill and Reese, 1999) are explained as follows:

(i) Along with the slurry the excavation is continued to the entire depth (Fig. 17.4 c(i)).
(ii) The reinforcement cage is placed, if needed (Fig. 17.4c(ii)).

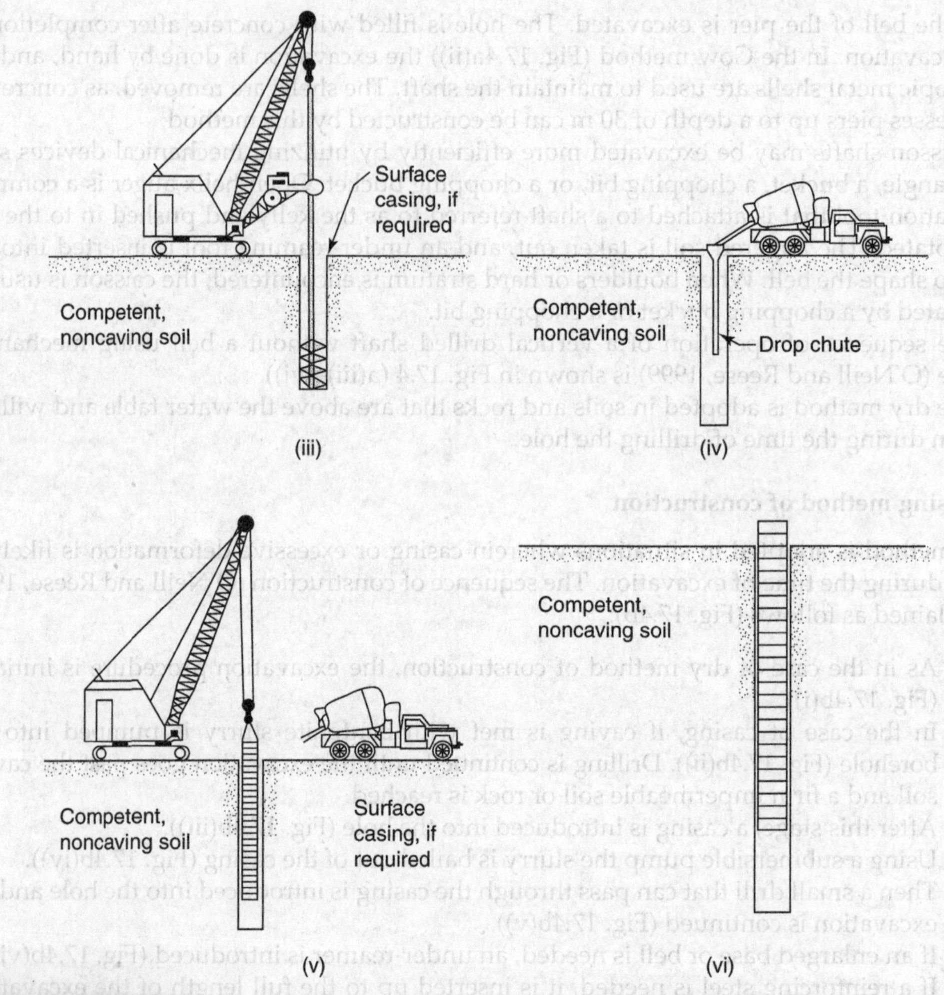


**Fig. 17.4 (a) (iii–vi)** Dry methods of construction (*Source:* O'Neill and Reese, 1999)

(iii) Then the concrete is placed in the drill hole, displacing the volume of slurry (Fig 17.4c(iii)).

(iv) The completed view of a shaft without a bell is shown in Fig. 17.4c(iv).

This method is also called as slurry displacement method.

A drilled caisson is usually inspected before the placement of concrete to ascertain

1. the correctness of alignment and dimensions,
2. the condition of the load-bearing stratum,
3. the risk of loss of ground and settlement of the adjacent area, and
4. the complete removal of the loose material.

Generally, the permissible misalignment is about 75 mm from the required location and 1% of the height or depth of the pier for verticality.

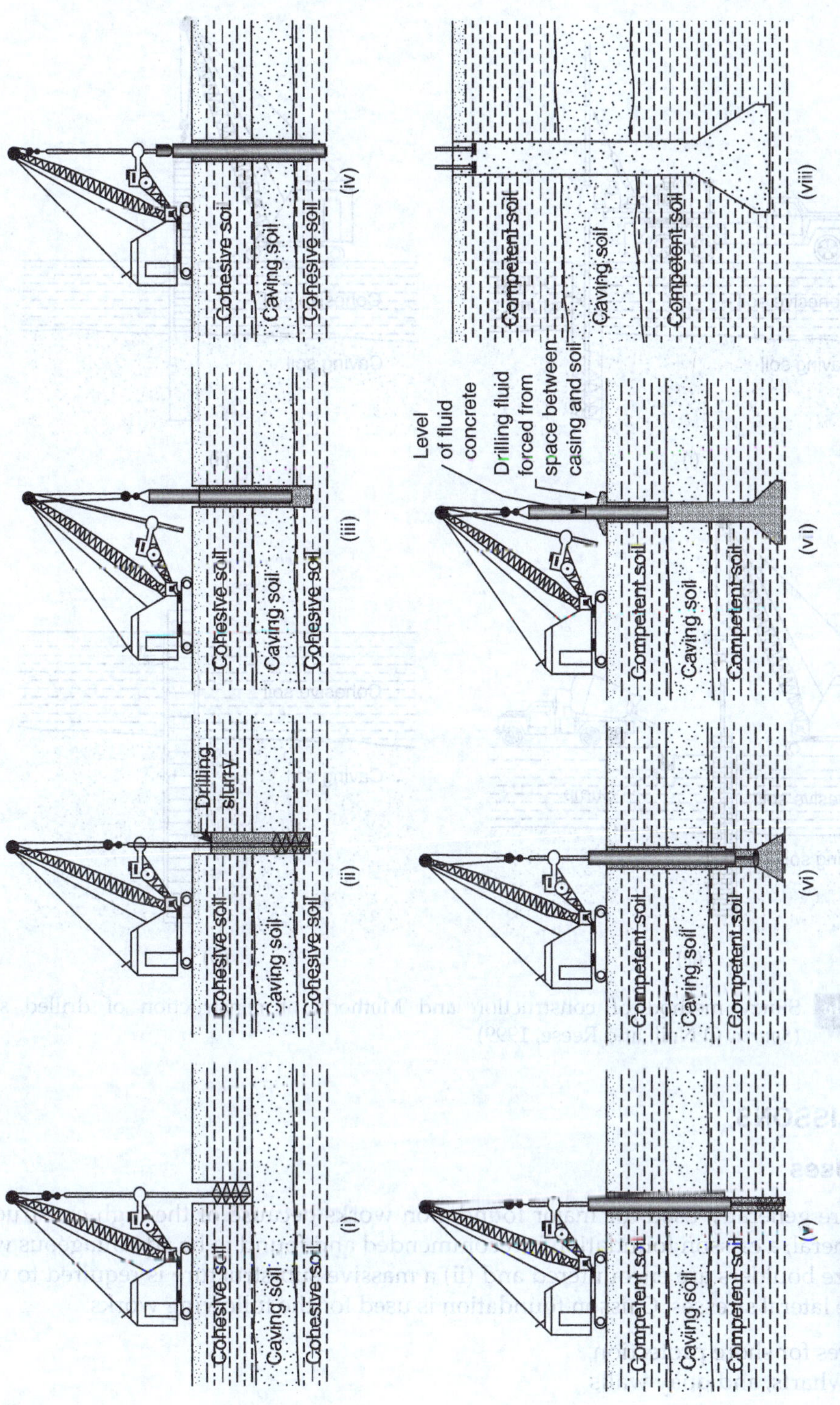


**Figure 17.4(b)** Casing method of construction (*Source:* O'Neill and Reese, 1999)

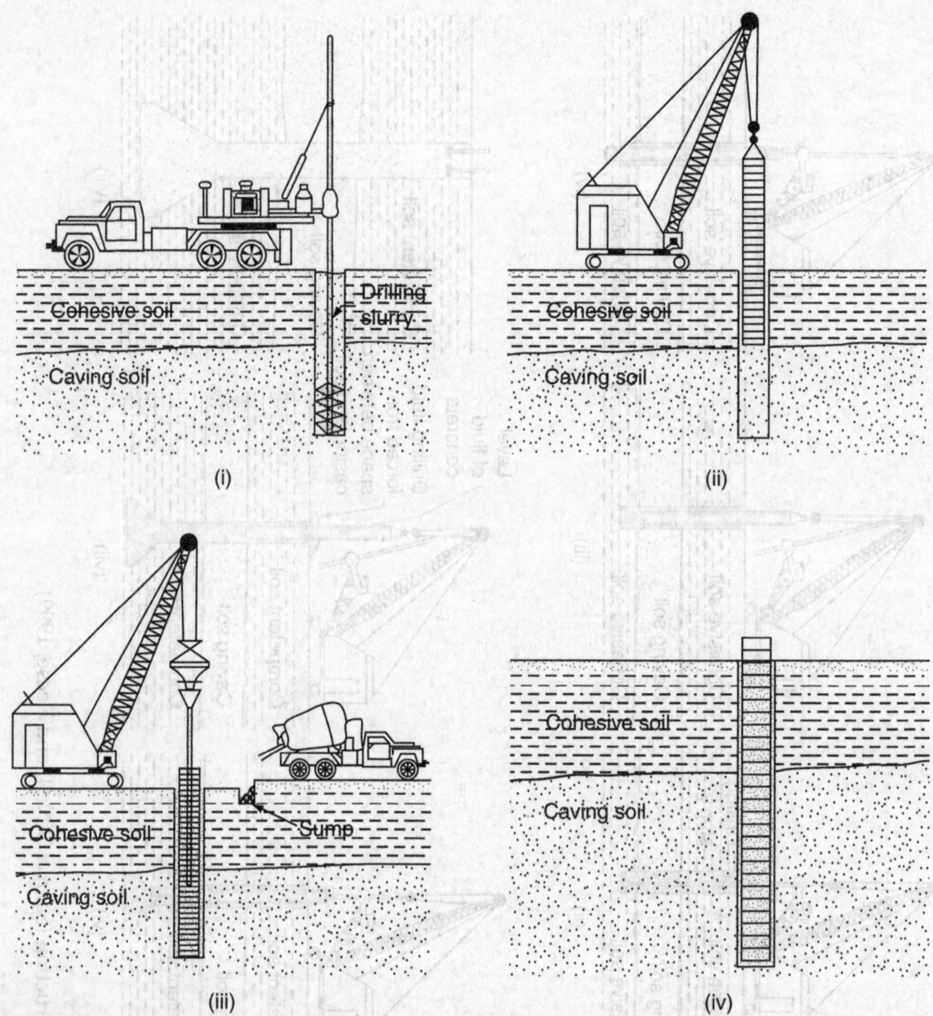


**Figure 17.4c** Slurry method of construction and Methods of construction of drilled shafts (*Source:* O'Neill and Reese, 1999)

## 17.3 CAISSONS

### 17.3.1 Uses

Caissons are generally used for major foundation works because of the high construction cost. In general, a caisson foundation is recommended and found to be advantageous when (i) large-size boulders are encountered and (ii) a massive sub-structure is required to withstand large lateral stresses. Caisson foundation is used for the following works:

1. Structures for shore protection
2. Docks, wharfs, and quay walls

3. Water front structures, such as large pump houses subjected to heavy vertical and horizontal loads
4. Bridge piers and abutments in rivers, lakes, etc.

Structures on land are generally founded on piles or drilled caissons, but for structures in water, in particular if the bearing stratum is 15 m below the water level, caissons are generally less expensive.

### 17.3.2 Types of Caissons

Caissons can be classified into three major types: (i) open caissons, (ii) box caissons, and (iii) pneumatic caissons.

**Open Caissons**. These are concrete or masonry shafts which remain open both at the top and at the bottom during construction (Fig. 17.5a). An open caisson of heavy mass concrete or masonry construction, containing one or more wells for excavations, is called a monolith (Fig. 17.5b). The conventional well foundation adopted in India is essentially an open caisson with minor changes in the construction procedure. The caisson is sunk into place, as the soil is removed from the inside, till the well sinks to the required depth. Then, a bottom concrete seal is made by depositing concrete. The well is pumped dry, after maturity of the bottom concrete seal, and filled with concrete or sand.

**Box Caissons**. These are structures with a closed bottom (Fig. 17.5c). They are constructed on land and transported and sunk in to the prepared foundations below the water level.

**Pneumatic Caissons**. These caissons have their top closed, and compressed air is used to stop the entry of water into the working chamber (Fig. 17.5d). Thus, excavation and concreting are done in a dry condition. The caisson is sunk as the excavation proceeds, and after reaching the required depth, the working chamber is filled with concrete.

### 17.3.3 Bearing Capacity and Settlement of Caissons

Computation of bearing capacity and settlement of caissons is similar to that of drilled piers.

### 17.3.4 Construction Procedure of Caissons

Construction of caissons requires heavy engineering equipment. A variety of usual and unusual problems have to be faced during the construction of caissons. Caissons may be constructed in slipways, on barrages, or on sand islands. Sometimes false bottoms are made to aid floating. For sinking the first few lifts of caissons, guide piles are commonly used. During the sinking process, if a hard or firm material is encountered under water, blasting may be necessary.

Open-well-type caissons have been used in India for many centuries for the foundations of river bridges. In the early days, the masonry of the wells was built on timber curbs and the caissons sunk by hand excavation from within the wells. Nowadays, the design and construction of well foundations in India have attained perfection. A detailed treatment of well foundations, as designed and constructed in India, is given in the next section.

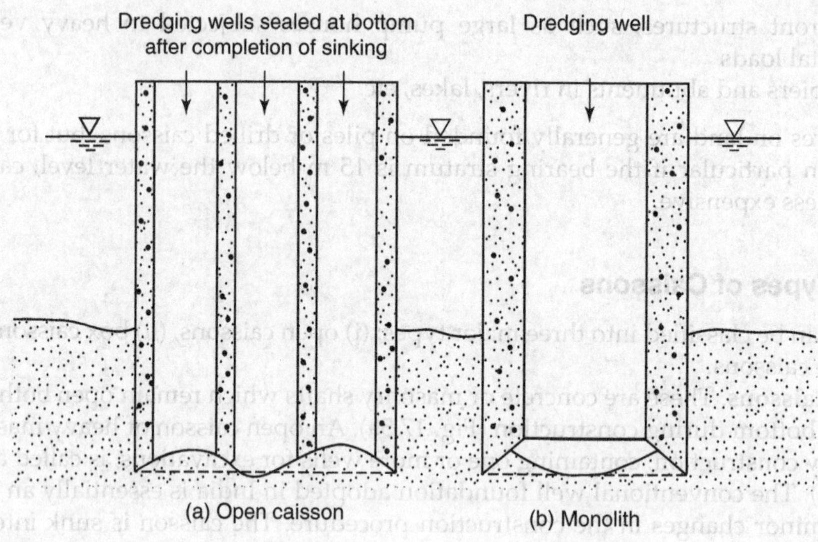


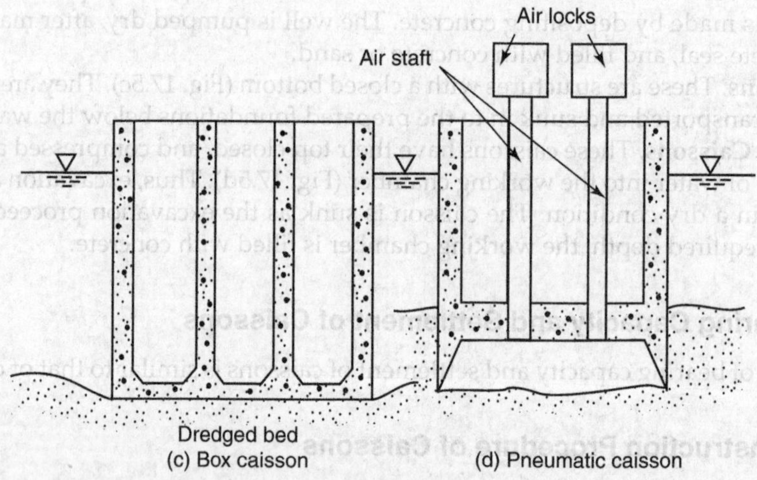


**Fig. 17.5** Types of caissons (*Source:* Thomlinson, 1986)

Open caissons and monoliths are provided with a cutting edge at the bottom, forming the lowermost portion of the shoe. The shoe has vertical outer steel skin plates (Fig. 17.6) and sloping inner steel haunch plates. Steel trusses are used in both directions for proper bracing. Trusses prevent the distortion of the shoe during fabrication, during towing to site, and during the early stages of sinking. After initial sinking, the space between the skin plates is filled with concrete (called steining). Over this rigid shoe, the steining is extended by placing reinforced concrete between form-work. Now the caisson sinks under its own weight while the soil is being excavated from the dredging wells. As the sinking operation progresses, additional lifts are successively installed. When the required foundation level is reached, the bottom of the caisson is plugged by a concrete seal by depositing under water a layer

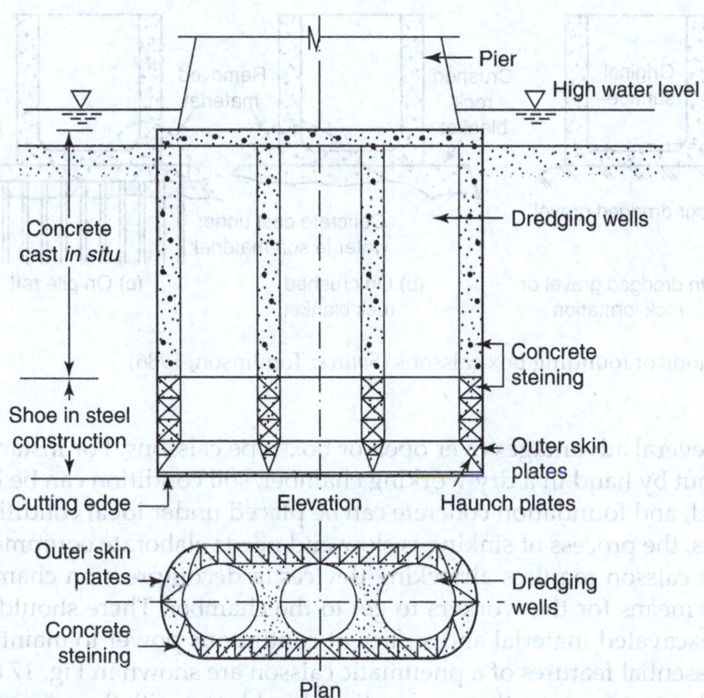


**Fig. 17.6** Details of an open caisson (*Source:* Tomlinson, 1986)

of concrete in the bottom of the wells. The wells are then pumped dry and more concrete is placed. Caissons can be further filled with concrete or sand depending on the condition followed, by a top concrete seal. During the process of grabbing under water in loose and soft materials, there is a possibility of surging and inflow of materials beneath the cutting edge. Further, as the sealing is done under water, open caissons have the disadvantage that the soil or rock at foundation level cannot usually be inspected before placing the sealing concrete. Open caissons are suitable in soft clays, silts, or gravels since excavation by grabbing will be easiest.

Monoliths are unsuitable for sinking in deep soft deposits because of their weight. They are usually used in quay walls where their heavy weight is favourable for resisting lateral and impact forces.

Reinforced cement concrete is used for making box caissons. They are constructed on land and floated to position after the concrete has strengthened. A box caisson during floating should be sufficiently safe against the danger of tipping or capsizing. The stability of a floating caisson can be analysed using the principles of hydrostatics. Box caissons can also be floated in horizontal or inverted positions and placed at the required location in a normal position. They can be founded on dredged gravel or rock formation, on crushed rock blanket over rock surface, or on piled raft. Box caissons are not suitable in sites where erosion can undermine the foundations. Figure 17.7 shows certain methods of founding box caissons.

Pneumatic caissons are used in situations where the soil flow into the excavated area is faster than it can be removed. They are also used in varying soil conditions. Pneumatic

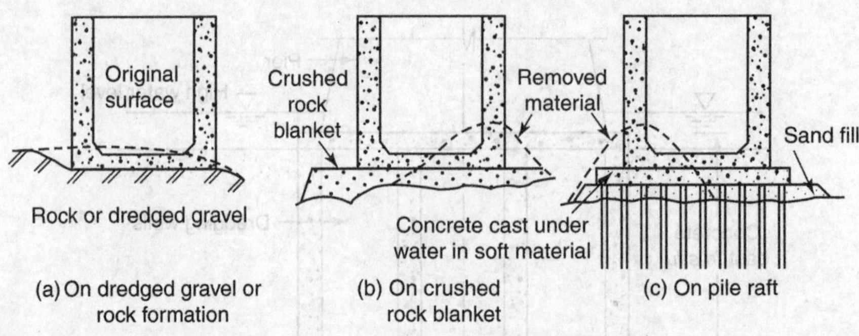


Fig. 17.7 Methods of founding box caissons (*Source:* Tomlinson, 1986)

caissons have several advantages over open or box-type caissons. For instance, excavations can be carried out by hand in a dry working chamber, soil condition can be inspected at the foundation level, and foundation concrete can be placed under ideal conditions. Compared to open caissons, the process of sinking is slow and needs elaborate equipment.

A pneumatic caisson requires airlocking devices, a decompression chamber, a working chamber, and a means for the workers to get to the chamber. There should be a provision to remove the excavated material and sufficient continuous power to maintain constant air pressure. The essential features of a pneumatic caisson are shown in Fig. 17.8. Since a pneumatic construction is expensive, the construction should start with the open method and continue as deep as practicable. When the condition requires a pneumatic operation, air shafts are introduced and the roof of the working chamber is formed either by concrete or by steel

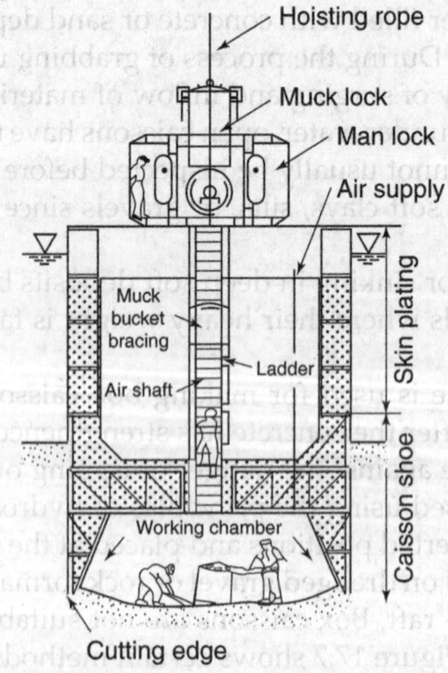


Fig. 17.8 General arrangement of a pneumatic caisson (*Source:* Tomlinson, 1986)

diaphragm. Then, water from the working chamber is removed by applying compressed air, followed by pumping if necessary. The excavation proceeds and the excavated material is removed from the air shaft by bucket devices or liquefied if feasible and pumped out, or if the soil is loose and sandy, it can be blown out using air pressure. As the excavation progresses, additional sections of shafts should be added, and the skin friction can be overcome by installing jets on the shaft walls. If sinking stops due to build-up of skin friction, it can be revoked by a process known as "blowing down". In this process, the air pressure is reduced to increase the effective weight of the caisson, thus increasing the sinking effect. When sinking reaches the required depth, concrete is placed to fill any base defects and also the remainder of the working chamber. The space between the roof of the working chamber and the concrete is filled with high-pressure grout. The rest of the procedure is similar to that for the open caisson. During the whole operation, the chamber pressure is maintained around 100 $kN/m^2$ above the atmospheric pressure. When chamber pressure of about 300 $kN/m^2$ is required, the workers should not be allowed to stay for more than 1½ to 2 hours.

## 17.4 WELL FOUNDATIONS

Well foundations are constructed either on dry ground or over an artificially formed island. The curbs are pitched in the correct position and then sunk into the ground to the desired level by grabbing the soil through the dredge holes formed by the masonry or concrete of steining. Caisson, in the Indian context, is a type of well foundation, and it is distinct owing to the method of commencing construction. That is, caissons are foundations for which the skin of the portion covering the curb and some length of steining of the well (to first cover the depth of the standing water) is fabricated or cast outside, floated to the final location, and lowered in position there. In India, this procedure of initial sinking is referred to as the caisson method.

Well foundations have all the advantages of open caissons. Additionally, in the Indian environment there is a lot of skill available. Further, under the present Indian conditions, well foundations are claimed to be the most economically viable method as cost of labour is cheaper compared to mechanical equipment (Ponnuswamy, 1986).

### 17.4.1 Types of Wells

Wells have different shapes, and accordingly, they are named as circular wells, twin circular wells, double-D wells, double octagonal wells, single and double rectangular wells, and multiple dredged holed wells. Among these, most common types used in India are circular, twin circular, and double D.

**Circular Wells**. These wells have uniform strength in all directions (Fig. 17.9a). As the weight per square of peripheral surface is highest, the sinking process is easy. They are simple in construction. Tilts and shifts can be controlled and corrected easily in this type.

**Twin Circular Wells**. In this type, two independent wells are placed close to each other and combined with a common well cap (Fig. 17.9b). These wells have all the advantages of circular wells. Further, the diameter and spacing of the two wells can be adjusted so as to accommodate the width and length of the pier. The minimum clear space between the well periphery is 0.6 to 1 m for a 6 to 7 m depth of well and 2 to 3 m for greater depth. This type

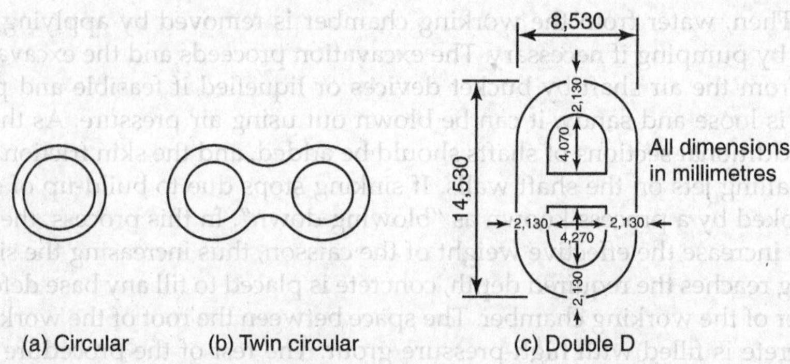


**Fig. 17.9** Types of wells (*Source:* Ponnuswamy, 1986)

of well is used with advantage when the sinking depth is less and hard foundation material is available.

**Double-D Wells**. This type is commonly adopted for deep foundations and major bridges with multiple land/line traffic (Fig. 17.9c). The length of the well is restricted to twice its width. Because of the presence of two wide dredge holes, casting and sinking are done more efficiently. In large-depth wells, the possibility of cracks is more due to the large bending moment. Further, dredging of the comers of the wells also poses some problem.

### 17.4.2 Components of Well Foundations

A typical sectional elevation of a well foundation is shown in Fig. 17.10. The bottom of a well structure is suitably tapered to end in a steel cutting edge. This tapered portion, called a well curb, is sufficiently strengthened with heavily reinforced concrete so as to take heavy loading. The main body of the well is called steining. The materials used for steining are brick or stone masonry, mass or reinforced cement concrete. After completion of the sinking at the stipulated level, the dredge hole is cleaned and filled with cement concrete. This bottom seal is called bottom plug. The remaining portion of the well may be filled partially or fully with saturated sand, water or left hollow. At the top finishing level, another plug is provided with plain or reinforced cement concrete; this is referred to as the top plug. This is provided to transmit the imposed load uniformly to the wall of the well. The topmost layer, called a well cap, is provided to accommodate the pier based on its shape and transmit the load to the steining. In certain cases, a lean top plug or no top plug is provided, but a well cap is invariably provided.

Brief design procedure of different components of well foundation is presented below based on the detailed treatment made by Ponnuswamy (1986).

### 17.4.3 Design of Wells

This basically involves finding the depth of the well, the size of the well, and the design of the other components.

**Depth of Scour**. As well foundations are constructed in river beds, they should be taken to a safe depth well below the anticipated scour level. Scour around piers depends on

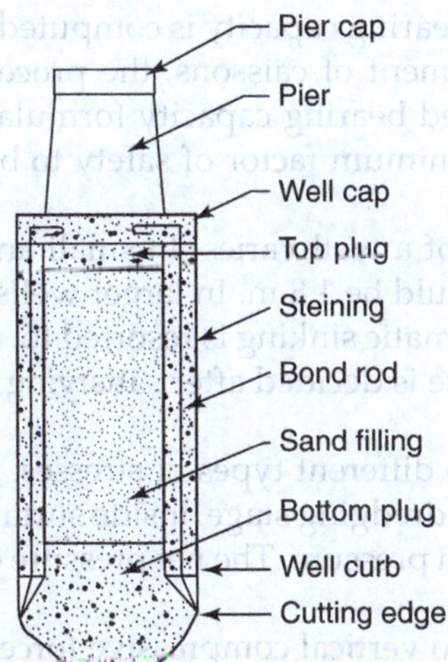


**Fig. 17.10** A typical section of a well foundation

several factors like flood discharge, the angle of attack of the flow, flow obstruction, etc. It is generally assumed that the flow at the high stage of the river is straight. The most common method is Lacey's formula, which is given as

$$D_s = 0.473\frac{q^{1/3}}{s_f} \tag{17.15}$$

where $D_s$ is the scour depth (m), $q$ the design discharge ($m^3/s$), $s_f$ the silt factor ($= 1.76D_m$), and $D_m$ the diameter of sand in the river bed (mm).

**Grip of Foundation**. The foundation should be taken well below the scour level to protect it from any movement due to the force of the stream flow and other external forces. The Indian Road Congress (IRC) method (reported by Ponnuswamy, 1986) recommends a grip of foundation (grip length) $D$ equal to $1/3D_s$ below the maximum scour level. However, the adequacy of the depth has to be checked and found stable against transverse/longitudinal forces by developing sufficient skin friction and passive earth resistance.

The depth of the foundation can be reduced if rock is met with. In a majority of the cases, wells are constructed on rocks and keyed to a minimum depth of 30 cm. If rock is available at a shallow depth and the mass of the well is not adequate to withstand the effect of longitudinal/transverse forces, the well is anchored down using mild steel (MS) rods or high-tensile wires. In soft rocks, the well is taken in the strata up to a sufficient depth.

**Size of the Well**. The practice of the Indian Railways is to design the wells based on point bearing resistance, whereas the IRC's practice is to consider both the skin friction and the point resistance. As the major portion of the load is transmitted by bearing, the size is decided based on point resistance only. Further, the well has to feasibly accommodate the pier with less cantilevering.

Safe bearing pressure for rocks can be found, as given in Chapters 15 and 23, based on either the field tests or the laboratory compression tests. A factor of safety of 6 to 8 is adopted. For disintegrated rocks and various soft varieties of rocks (where the recovery ratio is less than 35%), the safe bearing capacity is computed considering such rocks as soils. For bearing capacity and settlement of caissons, the procedure followed for piers can be adopted. Instead, the generalized bearing capacity formula (Eqs. 15.13 and 15.14) for base resistance may be used. The minimum factor of safety to be adopted for well foundations on soils is 2.0.

The size of the dredge hole of a well varies. In small and shallow wells, the minimum diameter of the dredge hole should be 1.8 m. In larger wells, where dredgers and chisels of large sizes are involved or pneumatic sinking is resorted to, the minimum size of the dredge hole should be 3 m. The final size is decided after satisfying the lateral stability condition of the wells.

**Steining**. This is subjected to different types of stresses. At the sinking it is subjected to water and earth pressure. At the dredging stage, inside surface is subjected to water pressure while outside surface to the earth pressure. The net pressure diagram for design requirement is shown in Fig. 17.11a.

Steining walls are subjected to vertical compressive forces. In situations where the steining is passing through a stiff clay followed by a soft stratum, there is a possibility of the bottom portion of the well getting dropped at the soft layer interface due to inadequate tensile strength of the well. Such dangers may be avoided by providing adequate bonding roads or, alternatively, the well should be reinforced. Some of the thumb rules recommended by the IRC for fixing the thickness of the steining are given below:

1. Cement concrete steining

   (i) For circular and dumbbell-shaped wells

$$T = k(0.01D_H + 0.1D_e) \tag{17.16}$$

where

$k = 1.1$ for sandy, silty, and soft clayey strata

$\quad = 1.25$ for hard strata including hard clay, boulders, kankar, shale, etc.

$D_H$ = height of the well

$D_e$ = external diameter of the well

   (ii) For rectangular and double-D wells

$$T = k(0.01D_H + 0.12) \tag{17.17}$$

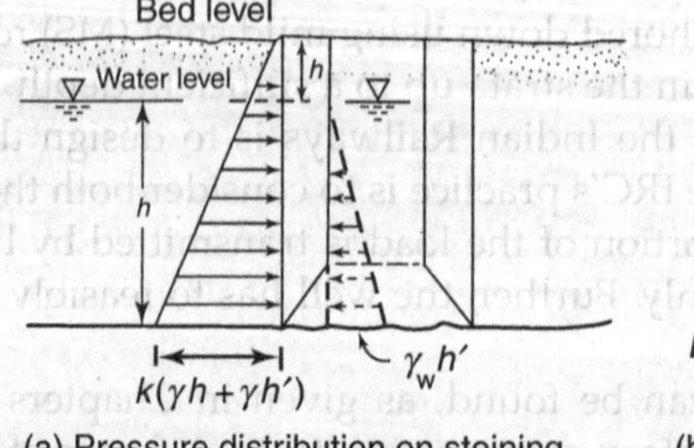


**Fig. 17.11** Pressure distribution on well steining and curb (*Source:* Ponnuswamy, 1986)

where

$k = 1.0$ for sandy strata
$= 1.1$ for soft clay
$= 1.15$ for clay
$= 1.20$ for boulders, shale, kankar, etc.

2. Brick steining

$$T = k\left(\frac{D_e}{8} + \frac{D_H}{40}\right) \tag{17.18}$$

where

$k = 1.0$ for sand
$= 1.1$ for soft clay
$= 1.25$ for hard clay

In order to keep on sinking continuously, the mass of the well should be greater than the skin friction. Otherwise, the well will reach a floating condition, and continued scooping may induce blowing of sand. This can be avoided by loading the wells with kentledges at the top to assist in sinking.

For a light structure, brick steining is used. In heavier and deeper structures, reinforced cement concrete or plain cement concrete with suitable bond rods is provided. It has been reported (Ponnuswamy, 1986) that M10 or M15 concrete wherever used is normally adequate.

**Curb**. The curb of a well transfers all the superimposed loads to the soil through the cutting edge while sinking. The material used for curbs may be timber or reinforced concrete. The present-day practice recommends heavy RCC well curbs with steel cutting edges for any type of well. Hoop tension is caused in the well curb due to bearing pressure, and suitable reinforcement has to be provided to withstand the same. The forces acting on the well are shown in Fig. 17.11b.

The total horizontal force on the well curb on both sides is

$$W \cot\theta \frac{D_e + D_i}{2}$$

where $D_i$ is the internal diameter of the well, $W$ the weight of the well and curb per unit length along the centre line of the steining, and $\theta$ the internal angle of the well. The force acting on the well curb on one side is half of the above.

Well curbs should be reinforced to resist these forces. A typical arrangement line of reinforcement in a well curb is shown in Fig. 17.12.

**Cutting Edge**. The cutting edge is provided at the bottom of the well below the curb to cut through the soil during sinking. It is generally made of steel and welded to an angle iron to fit the outer dimensions of the well steining.

During routine sinking in sand or clay, only very little stress is induced. But while resting on rock or pushing through boulders or cutting through logs of wood, high stresses are induced. The height $h_e$ of cutting edge is given as

$$h_e = \frac{q_u t}{f_c \tan\theta} \tag{17.19}$$

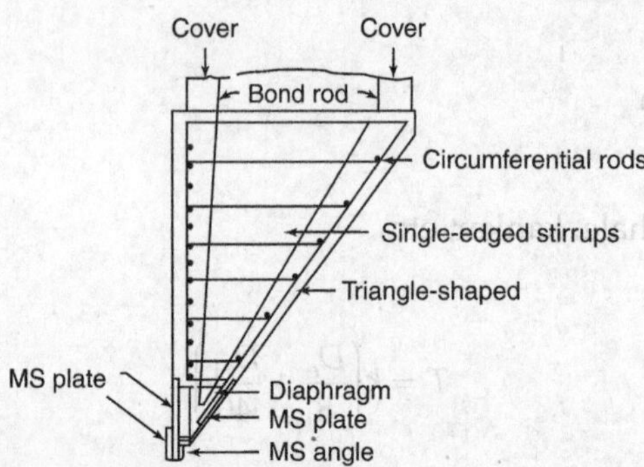


**Fig. 17.12** A typical well curb (*Source:* Ponnuswamy, 1986)

where $q_u$ is the crushing strength of rock, $t$ the thickness of the cutting edge, and $f_c$ the safe compressive stress of concrete.

The value of $\theta$ is usually taken as 30°. The choice of this angle has been proved to be suitable for easy access to the cutting edge. The thickness of the outer plate should not be less than 12 to 18 mm depending on the size of the well. Further, the unsupported plate height is limited to 7.5 to 10 cm.

**Bottom Plug**. After final grounding of the well to the required foundation level, a concrete plug is provided (Fig. 17.13). The bottom plug transfers the entire load, viz., the weight of the steining, the superimposed load, and the weight of saturated sand filled in the well, to the ground. The bottom plug functions as an inverted dome supported along the periphery

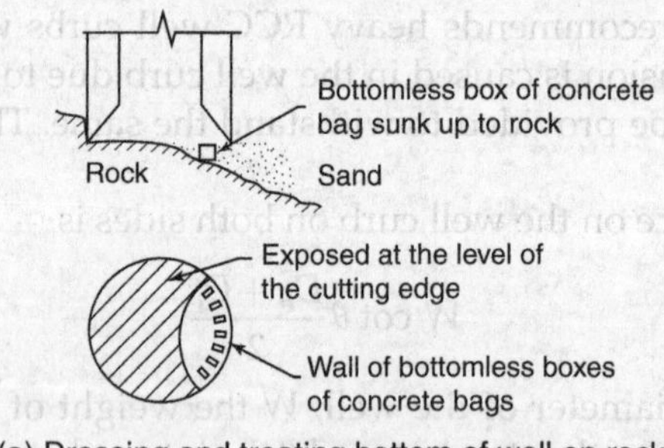


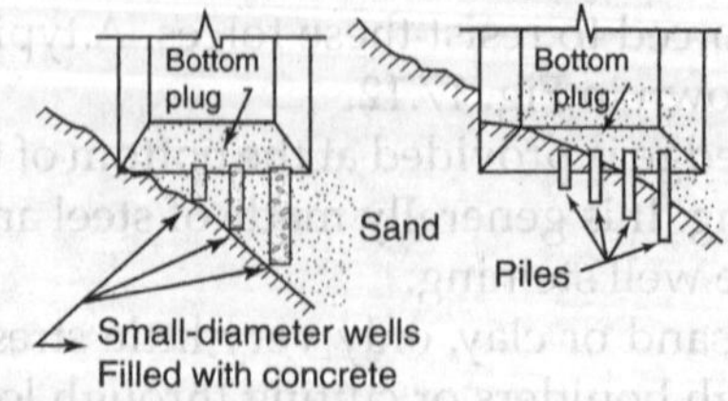


**Fig. 17.13** Bottom concrete plug

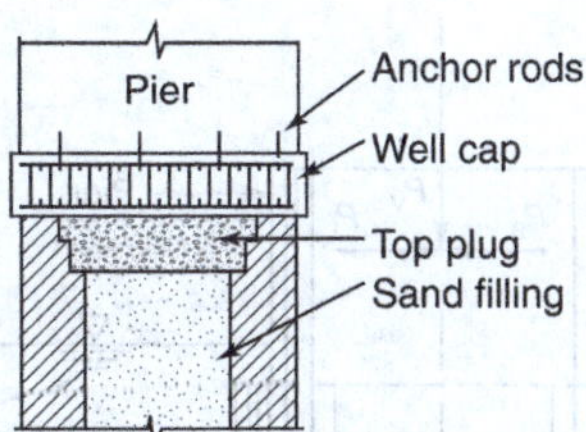


Fig. 17.14 Top plug and well cap

of the steining. As it is not feasible to provide reinforcement at the bottom, it is generally made thick, and a rich concrete mix (M20) is used.

**Sand Filling**. The bottom plug concrete is cured, and after curing, the well is filled with sand in a saturated condition. Sand filling, although not mandatory, provides stability to the bottom of the well, helps in the elimination of tensile forces at the base, and also cancels the hoop stresses induced in the steining. In earthquake-prone areas, it is preferable to make the well lighter by filling with saturated sand up to the top of well and then casting the top plug. But on the road bridges in India, either the well is kept empty or the filling is done only up to scour level.

**Top Plug**. A top plug is provided after the filling is completed. This enables transfer of the load of the pier and the superimposed load to the well steining (Fig. 17.14). The thickness of the top plug is generally kept greater than 50% of the smaller dimension of the dredge hole. If a sand filling is provided, no reinforcement is needed, and a plain cement concrete (PCC) mix of 1:2:4 is enough. In the case of larger dredger holes, it has been observed that the top plug provided is not sufficient but needs a reinforced well cap to transfer the superimposed load. In such cases, a lean concrete (1:3:6) top plug is provided, which forms an even base for the well cap.

**Well Cap**. As the shapes of the pier and the well are different, the well cap (entablature) forms an interim transition layer to accommodate the pier. The well cap is so designed that the base of the pier is provided with a minimum all-round offset. Further, the centre of the cap is made to coincide with that of the pier and not with that of the well. Such positioning nullifies the effect of the minor shifts which might have occurred during well sinking.

### 17.4.4 Stability Analysis of Well Foundations (Approximate Solution)

A well foundation supporting a bridge pier is subjected to vertical and horizontal forces. The forces causing instability (disturbing forces) of a well are self-weight of the well and its superstructure, live loads, water currents and buoyancy, forces due to environmental factors such as temperature, wind, and earthquake, breaking and tracking forces, forces on account of resistance of the well walls, and base and skin friction.

The stability requirement is to find the depth of the well below the maximum scour level (i.e., the grip length or grip of foundation) after allowing a suitable factor of safety after satisfying the static conditions.

Terzaghi (1943) gave an approximate solution based on the analysis of the free rigid bulk head. For the analysis, the critical combination of forces is considered.

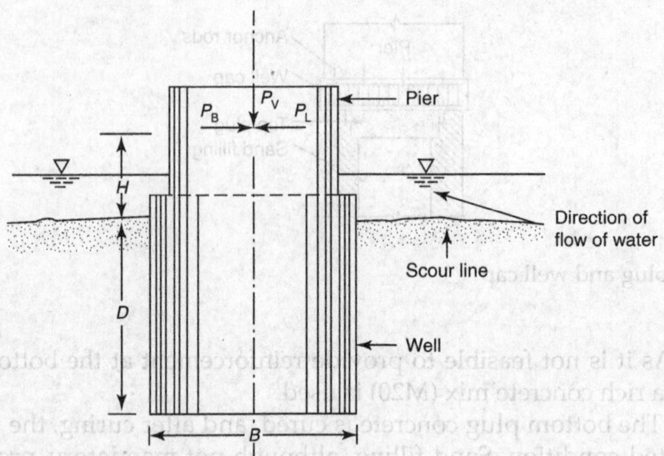


**Fig. 17.15** Forces acting on a well

Resolving all the forces in the vertical direction, one resultant force is obtained, $P_V$. Similarly, resolving the forces in two horizontal directions across and along the pier, two resultant horizontal forces, $P_B$ and $P_L$, respectively, are obtained (Fig. 17.15).

The critical force system to be considered for stability analysis is the one in the direction perpendicular to the water flow (i.e., in the direction of transverse axis of the pier). Thus, the resultant vertical force $P_V$ and the resultant horizontal force $P_B$ are considered for analysis.

Considering a state of plastic equilibrium, the forces and the earth pressure distributions acting on the well are shown in Fig. 17.16.

Pressure at any depth *z* below the scour level is

$$p = \gamma z(K_p - K_a) = \gamma z K' \tag{17.20}$$

$$z = DP_D = \gamma D K' \tag{17.21}$$

The well is assumed to fail as soon as the soil reaction at the bottom is equal to $P_D$. For equilibrium at that instant,

$$\begin{aligned}(P_B)_{max} &= \text{resultant of total pressure per unit length}\\ &= \text{area of } \Delta\text{AEF} - \text{area of } \Delta\text{BCF}\\ &= \tfrac{1}{2}\gamma D^2 K' - \tfrac{1}{2}2\gamma D K' D_1\end{aligned}$$

Therefore,

$$(P_B)_{max} = \tfrac{1}{2}\gamma D K'(D - 2D_1) \tag{17.22}$$

Taking moment about *E*,

$$(P_B)_{max} \times H_1 = \frac{1}{2}\gamma D^2 K' \frac{D}{3} - \frac{1}{2} \times 2D\gamma K' D_1 \times \frac{D_1}{3}$$

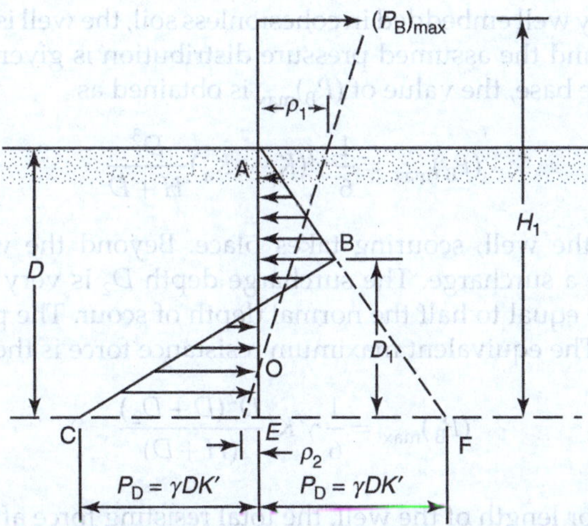


Fig. 17.16 Forces on well considered for analysis

Solving for $D_1$,

$$2D_1 = 3H_1 + \sqrt{9H_1^2 - 2D(3H_1 - D)} \tag{17.23}$$

Substituting Eq. 17.23 in Eq 17.22, an expression in depth $D$ is obtained. This is the grip length required to sustain the maximum horizontal force.

A safe depth can be obtained by reducing $P_D$ by a factor of safety, $F$, i.e., considering $P_D/F$. This theory is based on the further considerations that (i) the well is treated as a light bulk head, (ii) $K_p$ and $K_a$ are Rankine's earth pressure coefficients, and (iii) there is no friction at base and wall. Omission of these frictional forces yields a conservative $(P_B)_{max}$.

If $\rho_1$ and $\rho_2$ are the horizontal displacements, then the angular deflection of the centre line of the well, $\delta$, is given as

$$\tan \delta = \frac{1}{D}(\rho_1 - \rho_2) \tag{17.24}$$

The above analysis may be applied to the force system of a well provided the movements due to side friction and the resultant base reaction are ignored. This omission is on the safe side. The possible vibrations due to traffic over the bridge and water currents may minimize the side friction. Further, the deflection of a heavy well may be small, and thus the passive resistance will also be less. Thus, Rankine's coefficient of earth pressure is justifiable as it gives a conservative result. Equation 17.22 represents the maximum equivalent resisting force per unit length of well due to earth pressure.

### 17.4.5 Stability Analysis of a Heavy Well

In the Terzaghi approximate analysis, it is assumed that the bulkhead tends to rotate about some point O above the lower edge and tends to transfer the soil from elastic to plastic equilibrium.

In the case of a heavy well embedded in cohesionless soil, the well is assumed to invariably rotate about its base, and the assumed pressure distribution is given in Fig. 17.17a. Taking the moments about the base, the value of $(P_B)_{max}$ is obtained as

$$(P_B)_{max} = \frac{1}{6}\gamma'(K_p - K_a)\frac{D^3}{H+D}$$

Normally, around the well, scouring takes place. Beyond the well surroundings, the uncovered soil acts as a surcharge. The surcharge depth $D_2$ is very difficult to assess and may be assumed to be equal to half the normal depth of scour. The pressure distribution is shown in Fig. 17.17b. The equivalent maximum resistance force is then given as

$$(P_B)_{max} = \frac{1}{6}\gamma' K' \frac{D^2(D+D_2)}{(H+D)}$$

If $d$ is the diameter or length of the well, the total resisting force after allowing a factor of safety, $F$, is given as

$$P_a = \frac{(P_B)_{max}\, d}{F}$$

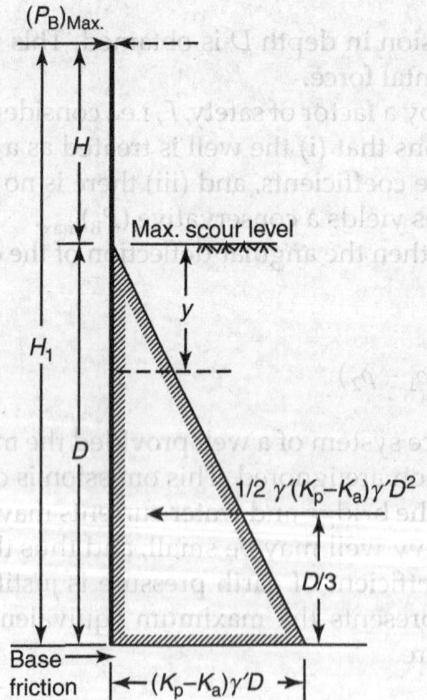


(a) Tiliting of a heavy well about base in cohesionless soil

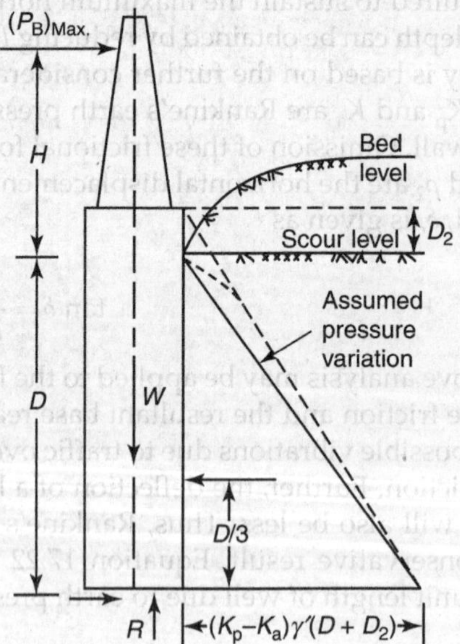


(b) Effect of surcharge on a well tilting about its base

**Fig. 17.17** Stability analysis of a heavy well

The factor of safety should not be less than 2.

The maximum pressure $f$ at the base of the well for the no overturning moment condition is

$$f = \frac{W}{A}$$

where $W$ is the net direct load on the well base, after making allowance for buoyancy and skin friction; $A$ is the area of the well base; and $z$ is the section modulus of the well base.

This maximum foundation pressure should be kept within the safe bearing capacity of the soil, assuming no tension occurs at the base.

The maximum moment on the steining occurs where the resultant shear force is zero. If the shear force is zero at a depth $y$ below the maximum level, then

$$P_a = \frac{\gamma' K' y^2 d}{2F}$$

That is,

$$y = \sqrt{\frac{2FP_a}{\gamma' K' d}}$$

### 17.4.6 Construction of Well Foundations

If the well has to be laid on a dry ground, the soil is excavated and levelled 15 to 30 cm above the surface water level. The line of the cutting edge is marked with reference to the centre of the pier and the axis. In case the well has to be in water, an artificial island is formed and compacted. The marking is made over the compacted and levelled layer. Wooden blocks of sand bags are placed tangentially or along the cutting line to prevent unnecessary settlement of failure. The cutting edge is assembled on the wooden or sand bags which will fall inside the dredge hole when grabbing starts.

The inner conical wooden shuttering is placed, followed by curb reinforcement. The cutter steel shuttering is positioned, and the curb is concreted. The concrete mix should not be leaner than M15. The inner conical shuttering is removed after 72 hours and the others after 48 hours.

The steining reinforcement or bond rods are placed, and additional shuttering is positioned and the next stage of steining concreted with a 1:3:6 mix; for brick or stone masonry 1:2 or 1:3 cement mortar is used. Two-metre height of steining or curb has to be built before open sinking.

The sinking may be done by sending men down and excavating the soil with shovels. The excavated soil is removed using baskets, and sinking is continued till the curb sinks to a depth of 0.5 to 1.5 m. Additional steining can be raised in heights of 1.2 to 1.5 m. At least 24 hours' timing is given before each stage of addition if ordinary cement is used, otherwise a longer setting time has to be allowed. The steining heights can be increased to 3 m after attaining a grip depth of about 6 m. The verticality of the well has to be ascertained at each stage.

Manual grabbing and removal of earth can be resorted to only for small depths. For greater depths and where subsoil water is encountered, dredgers have to be resorted to. The most commonly used dredger is the bells dredger. The dredgers are lowered into the well from overhead derricks or winches through pulley blocks are to be adopted. For the purpose of hoisting the grab and removing the material, different types of derricks are used. For minor jobs, timber Scotch Derrick is used. The dredging operation is stopped when the required height less the minimum free board is attained. The steining height is raised up to the free board level and then concreted.

The sinking process may be hastened by temporarily reducing the skin friction. The outer surface of the steining is finished smooth. The surface may also be coated with coal tar or bentonite solution. Alternatively, air or water jets may be used, which is an effective method of reducing skin friction.

The rate of sinking has to be monitored, and an advance work chart should be available for this purpose. The following rates of sinking (cm/day) have been recommended by Ponnuswamy (1986):

| | |
|---|---|
| Medium-sized well through sandy strata: | 60 to 90 cm/day |
| Medium-sized well through clayey strata: | 40 to 50 cm/day |
| Large-sized well through sandy strata: | 50 to 60 cm/day |
| Large-sized well through clayey strata: | 30 to 40 cm/day |
| Large-sized well through rocky/hard strata | |
| by driving: | 10 to 15 cm/day |
| by pneumatic sinking: | 15 to 25 cm/day |

After the required foundation level is reached, the bottom of the well has to be dressed and treated before plugging. Figure 17.13a shows the condition of the well sunk in sandy strata. When sinking is done by de-watering or by a pneumatic process, the bottom can be dressed by sending men down to the bottom. Shallow wells and those just resting on rock may be anchored with the help of mild steel dowels fixed to the rock before plugging. In case there is difficulty in keying in the rock, the arrangement shown in Fig. 17.13b and c may be adopted.

After curing of the bottom plug, the well is filled with sand or water; the top plug is provided followed by the well cap as discussed earlier.

### 17.4.7 Shifts and Tilts

In the process of sinking a well, it may get shifted from the true alignment or get tilted due to non-uniform sinking. Shift and tilt of each well have to be measured regularly during the entire operation. At each stage of casting of steining, observations have to be taken to this effect.

As per the IRC (1986) norms, the tilt should not exceed 1 in 80 and the shift should not be more than 5% of the maximum outside dimension of a well subject to a maximum of 150 mm.

Necessary corrective measures have to be taken while sinking proceeds to keep the tilts and shifts within permissible limits. The following measures can be employed to contain the tilts and shifts within the limits:

1. *Regulation of dredging:* The dredging is made at the higher side of the well. This may be done manually after de-watering. If the well has sunk to a great depth, this method will not be effective.
2. *Eccentric loading:* In this case, a bracket is provided at the top of the well on the higher side, and sand bags or other heavy materials are placed on the bracket. This causes eccentric loading, and minor tilts are rectified.
3. *Blocking and hooking:* Sometimes a wooden piece is placed temporarily under the cutting edge of the lower side to avoid further tilt while other methods of rectification are attempted at the higher side. Instead of a timber piece, the cutting edge on the lower edge may be hooked, and the hook is pulled of with the help of a winch if necessary.
4. *Pulling the well:* This technique is effective only during the initial stages of sinking. This is done by putting one or more steel ropes round the well and pulling it towards the higher side. Alternatively, the well may be pushed from the lower side using hydraulic jacks.
5. *Water jetting and excavation:* This method may be resorted to at the high side to reduce skin friction. This technique alone can be effective unless supported by other correction methods. Skin friction can be reduced by releasing the earth pressure on the higher side by making an open excavation.

## POINTS TO REMEMBER

17.1 Drilled piers are structural members of relatively large-diameter massive struts constructed of concrete placed in a pre-excavated hole.

17.2 Caissons or well foundations are structural boxes or chambers that are sunk in place through ground or water by simultaneously excavating below the bottom of the unit, which thereby descends to the final depth.

17.3 Types of drilled piers are straight shafted pier, belled pier, and pier socked into rock.

17.4 Drilled piers and caissons derive supporting power from both skin friction and bottom bearing as in a pile. The contribution by end bearing capacity is much more than by skin friction.

17.5 Settlement of piers and caissons in sands and gravels is mostly immediate, but in clays the settlement may be immediate or long term.

17.6 Types of caissons are open caisson, box caisson, and pneumatic caisson.

17.7 Caisson, in the Indian context, is a type of well foundation and it is distinct owing to the method of commencing construction. Well foundations have all the advantages of open caissons.

17.8 Types of wells are circular wells, twin circular wells, and double-D wells.

17.9 Components of a well foundation are cutting edge, well curb, bottom plug, staining, top plug, and well cap.

17.10 Design of a well foundation involves the following factors: (i) depth of scour, (ii) grip of foundation, (iii) size of the well, (iv) steining, (v) curb, (vi) cutting edge, (vii) bottom plug, (viii) sand filling, (ix) top plug, and (x) well cap.

## QUESTIONS

### Objective Questions

17.1 State whether the following statements are true or false:

1. The drilled pier derives its maximum supporting power from the end bearing.
2. Monolith is a type of open caisson with less mass concrete or masonry.
3. Sinking of a pneumatic caisson is slow and requires elaborate equipment.
4. Steining is subjected to water and earth pressure during the sinking stage.
5. Wells located in earthquake areas are completely filled with sand and compacted so that the well is made heavy.

17.2 Machine-excavated piers are designated as
(a) Chicago caisson (b) Gow caisson
(c) Drilled caisson (d) Open caisson

17.3 The critical force system to be considered for stability of a well foundation is the one in the
(a) Vertical direction
(b) Direction perpendicular to the water flow
(c) Direction parallel to the water flow
(d) Horizontal direction

17.4 The stability of a floating caisson can be analysed by the principle of
(a) Hydrodynamics
(b) Hydrolysis
(c) Hydrostatics
(d) Hydromechanics

17.5 Pneumatic caissons are preferred in situations where the soil flow into the excavated area is ______ than it can be removed.
(a) Faster
(b) Slower
(c) Initially faster
(d) Initially slower

17.6 Identify the wrong choice:
A cast-in-place broad pile with diameter greater than 75 cm is called a
(a) Caisson (b) Drilled shaft (c) Drilled pier (d) Short column

17.7 *Assertion A:* Unlike a driven pile, a drilled caisson does not compact the surrounding area.
*Reason R:* Skin friction along the shaft of a drilled caisson is relatively small.
*Select the correct code:*
(a) Both A and R are true, and R is the correct explanation of A.
(b) Both A and R are true, and R is not the correct explanation of A.
(c) A is true, but R is false.
(d) A is false, but R is true.

17.8 Identify the incorrect statement:
(a) Settlement analysis is necessary for drilled caissons resting on medium and soft clays.
(b) Drilled caissons carried to materials other than hard rock are often enlarged at the bottom.
(c) The drilled caisson shaft is designed in accordance with the principles of the long column.
(d) Drilled caissons carried to bedrock are generally safe.

## *Descriptive Questions*

17.9 Describe the method of analysis of a heavy well as right bulk head.
17.10 Briefly explain the procedure adopted in well sinking and bring out the problems that are encountered in open sinking.
17.11 Discuss in detail the methods to correct tilts and shifts of wells during sinking.
17.12 Explain briefly the method of assessing the load-carrying capacity of a well foundation in a saturated stiff clay.

## EXERCISE PROBLEMS

17.1 A bridge pier is to be founded at 5 m below the bed level of a river. The 8.5 m diameter pier has a gross load of 28,000 kN. The mean high and low water levels are 8.5 and 3.2 m above the bed level, respectively. The saturated unit weight of the clay is 21.2 kN/m$^3$. Adhesion between the clay and the surface of the pier is estimated as 36 kN/m$^2$, and this is effective for a depth of 3.5 m from the base of the pier. Compute the factor of safety against general shear failure at low and high water levels of the river. Use a value of $N_c = 7.5$.
17.2 A reinforced concrete floating caisson has outside dimensions of 5.5 m × 5.5 m × 5.5 m. The thickness of the well and the bottom is 400 mm. Examine whether this can provide stable flotation in fresh water.
17.3 An open caisson has an inside diameter of 3.5 m and a length of 15 m. The caisson is made of concrete, and it is intended to sink it by its own weight. Determine the steining thickness necessary.

# 18

# Ground Investigation

**CHAPTER HIGHLIGHTS**

Planning the ground investigation programme – Types of samples – Indirect methods: geophysical methods, sounding methods Semi-direct methods: wash boring, rotary drilling, auger boring – Direct methods: sampling process, sample disturbance, types of samples, accessible explorations, undisturbed sampling – Routine field tests – Recording of field data – Location, spacing, and depth of borings

## 18.1 INTRODUCTION

Before undertaking any civil engineering work, a thorough investigation of the ground is essential. Ground investigation refers to the methodology of determining surface and sub-surface features in the proposed construction area.

Information on surface conditions is necessary for planning the accessibility of the site, for deciding the disposal of removed material (particularly in urban areas), for removal of surface water in water-logged areas, for movement of construction equipment, and other factors that could affect construction procedures.

Information on sub-surface conditions is a more critical requirement in planning and designing the foundations of structures, de-watering systems, shoring or bracing of excavations, the materials to be used in construction, and site improvement methods.

The above information will enable a civil engineer to plan, design, and execute a construction project. In a nutshell, the purpose of a ground investigation is to

1. determine the geological conditions of rock and soil formation,
2. establish groundwater level,
3. select the type and depth of foundation,
4. determine the bearing capacity of the site,

5. evaluate the anticipated settlement of the structure,
6. locate and select the materials of construction, and
7. locate suitable transportation routes.

## 18.2 PLANNING THE GROUND INVESTIGATION PROGRAMME

Basically, in a ground investigation, one is interested in finding out details about sub-surface strata and their engineering properties, such as strength, deformation, and hydraulic characteristics. The programme should aim at obtaining the maximum information with minimum investment. For successful operation of the programme, the civil engineer must have an adequate knowledge of the current exploration techniques and laboratory tests.

The ground investigation, irrespective of the magnitude of the project, consists of four phases, which are discussed below.

1. *Available Information*. This is the first phase in which collection of published geological and topographical information of the area, hydrological data, details of existing or historic development, local regulations for construction activity, etc. are made.

2. *Reconnaissance*. This is the phase during which a first examination of the area is made by the engineer along with other specialists, such as the geologist, land surveyor, geotechnical engineer, etc. At this stage, a thorough study of the existing structures for the type of construction and defects, such as cracks and settlement, soil profiles in highway or railroad cuts and quarries, erosion in existing cuts, high-water marks on bridge abutments, rock outcrops, and history of flood and scour levels from the local people are collected.

3. *Preliminary Investigation*. This is an important phase of the entire programme. In this stage, the engineer planning the investigation programme should undertake two obligatory steps (Terzaghi and Peck, 1967). As the subsoils were formed under different geological environments, the first step towards a ground investigation is a thorough understanding of the geology of the site, which enables an efficient working out of the investigation programme. The second step is to obtain more details about the subsoil strata (e.g., thickness of individual strata) from one or two exploratory drill holes. All further steps depend on the magnitude of the job and the character of the soil profile.

4 *Detailed Investigation*. Additional borings are planned from the data obtained from the preliminary borings. If the subsoil is uniform in stratification, an orderly spacing may be planned. Often, additional borings are made to locate weak soil or rock zones, outcrops, etc., which may influence the design and construction of the project. Necessary *in situ* tests should also be performed. Sufficient samples should be procured to obtain relevant parameters for design and construction. Certain additional samples should be recovered to redefine the design or construction procedure.

## 18.3 TYPES OF SOIL AND ROCK SAMPLES

Soil and rock samples may be classified into the following three groups, based on the condition of the material.

**Non-Representative Samples**. These samples comprise mixtures of materials from various soil or rock layers. Here, there are possibilities of removal or exchange of some mineral constituents by washing and sedimentation. Such samples, also called wash or wet samples, are not a true representation of the material found at the bottom of the borehole. These samples are unsuitable for identification and laboratory tests.

**Representative Samples**. These samples consist of constituent minerals from each layer and are not mixed with material from other layers. The structure and water content of the sample are seriously altered. The soil samples are suitable for identification and classification purposes but not suitable for major laboratory tests. These samples are also referred to as dry samples but the samples do contain some moisture.

**Undisturbed Samples**. These are the types of samples in which the material has experienced so little disturbance that it is suitable for all laboratory tests. Thus, these samples may be used for permeability, consolidation, and strength tests. The term undisturbed is misleading to some extent, and hence, it is appropriate to replace it with undistorted (Hvorselv, 1984).

## 18.4 INDIRECT METHODS OF SUB-SURFACE EXPLORATION

Indirect methods consist of geophysical and sounding methods. In these methods, depths to the principal strata are established, based on some physical properties of the material, and the measurements are made on the ground surface. No samples are obtained in the geophysical method, but in some sounding methods, representative samples are obtained.

### 18.4.1 Geophysical Methods

This method involves the technique of determining underground materials by measuring some physical property of the material. Geophysical methods can be used rapidly and economically for large areas of great linear extent. These methods are more suitable for reconnaissance exploration, for location of general deposits of construction materials, and in identifying the depth of ground-water. The results obtained are generally satisfactory.

Two types of geophysical investigations have been found to be useful for civil engineering works. They are electrical resistivity and seismic refraction methods.

**Electrical Resistivity Method**. In this method the differences in the electrical resistivities of the various strata are detected. In general, both soil and rock are poor conductors of electricity, but because of the dissolved salts in the pore water, the conductivity of the ground is improved. An electric field is produced in the ground by means of two current electrodes, and measurements are made of the current ($I$) flowing between these electrodes and of the potential drop (voltage, $V$) between a pair of intermediate electrodes. A schematic diagram of the equipment (Wenner configuration) is shown in Fig. 18.1.

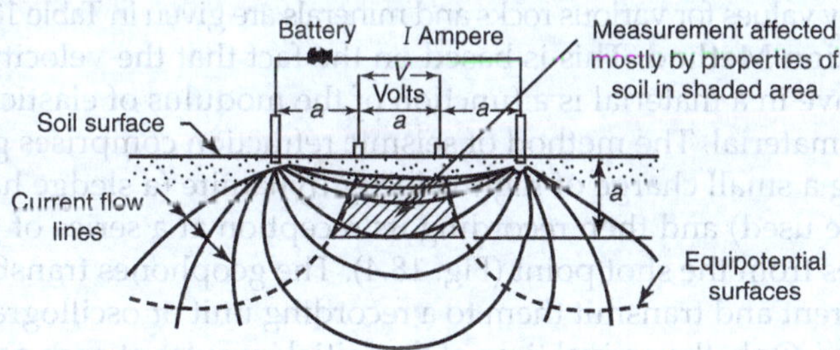


**Fig. 18.1** Schematic diagram of the resistivity method – Wenner configuration

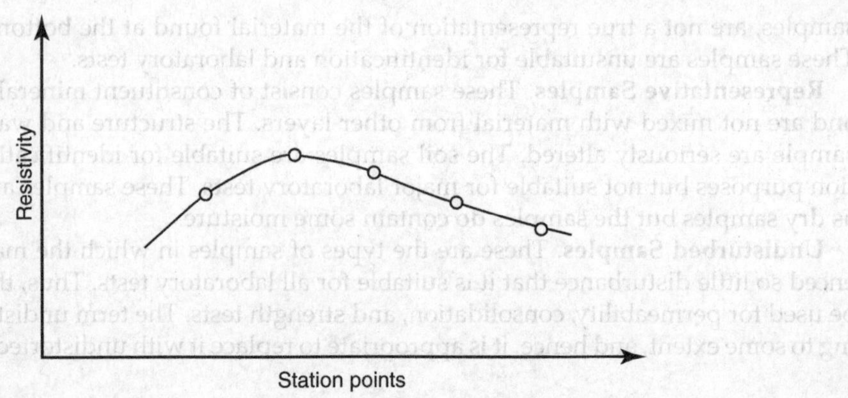


Fig. 18.2 Resistivity traverse plot

The resistivity ($\rho$) of the soil or rock between the two electrodes with spacing $a$ is given as

$$\rho = 2\pi aV / I \qquad (18.1)$$

Two different field procedures, viz., *electrical profiling* and *electrical sounding*, are in use for obtaining information about surface conditions. The first method is suitable for establishing boundaries between different underground materials, and has practical applications in the study of variation of sub-surface condition with depth and in detecting layered deposits and water-bearing strata.

In the electrical profiling method, four electrodes are kept at a constant spacing and moved across the area, and resistivity measurements are made. The information resulting from a profile line can be plotted with station points as the horizontal axis and resistivity along the vertical axis. A change in the plotted curve indicates a change in the underground materials (Fig. 18.2). From the series of profile lines, boundaries of areas underlain by different materials can be established on a map of the area.

In the electrical sounding method, the electrode spacing $a$ is progressively increased to pick up changes in resistivity with depth. For homogeneous materials, half the electric current flows to a depth of $a$ to $1.5a$ and half to a greater depth. On the basis of field measurements, the resistivity depth curves are plotted with resistivity along the horizontal axis and $a$ along the vertical axis (Fig. 18.3). Since the depth is directly related to electrode spacing, the series of resistivity data obtained will indicate changes of resistivity with depth, and hence provide information about the layering of material.

Average resistivity values for various rocks and minerals are given in Table 18.1 (IS: 1892, 1979).

**Seismic Refraction Method**. This is based on the fact that the velocity of a longitudinal or compression wave in a material is a function of the modulus of elasticity, Poisson's ratio, and density of the material. The method of seismic refraction comprises generating a sound wave by exploding a small charge of high velocity dynamite (a sledge hammer or a falling weight may also be used) and then recording its reception at a series of geophones located at various distances from the shot point (Fig. 18.4). The geophones transform the vibrations into an electric current and transmit them to a recording unit or oscillograph equipped with a timing mechanism. Only the arrival time of the initial impulse at each geophone is utilized.

The direct waves travelling with velocity, $V_T$, in the top layer arrive first at the geophones nearer to the shot point. The waves with the higher velocity, $V_B$, reach the farther detectors

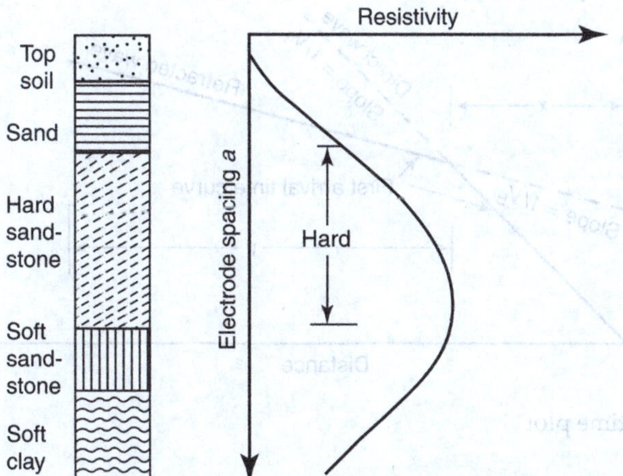


Fig. 18.3 Resistivity–depth curve

Table 18.1 Resistivity values for rocks and minerals

| Material | Mean resistivity (ohm m) |
|---|---|
| Limestone (marble) | $10^{12}$ |
| Quartz | $10^{10}$ |
| Rock salt | $10^{6}$–$10^{7}$ |
| Granite | 5,000–$10^{6}$ |
| Sandstone | 35–4,000 |
| Moraines | 8–4,000 |
| Limestones | 120–400 |
| Clays | 1–120 |

*Source:* IS: 1892 (1979).

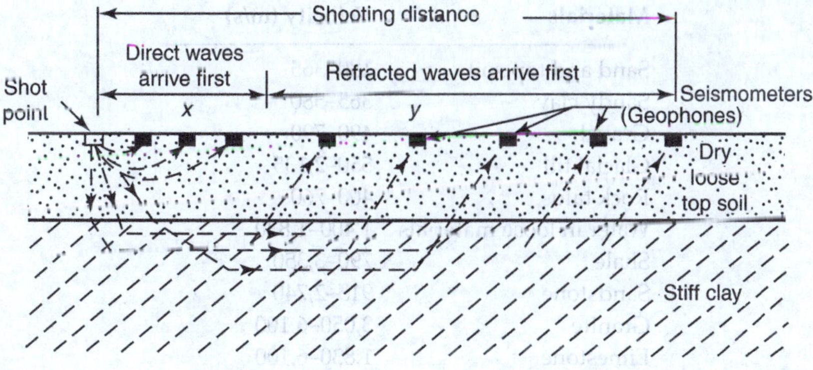


Fig. 18.4 Seismic refraction method

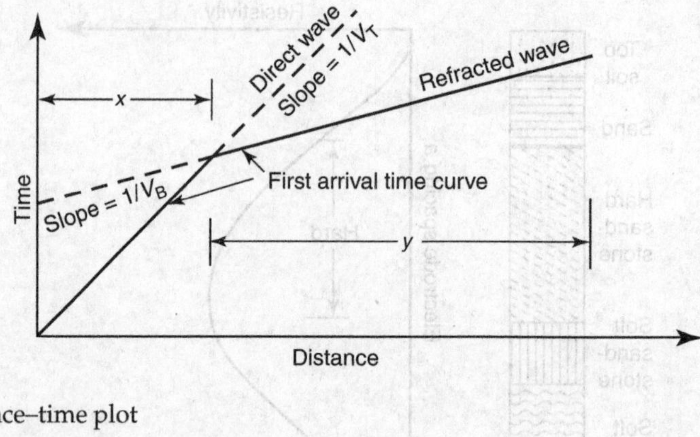


Fig. 18.5 Distance–time plot

first by travelling downwards in the lower velocity material (say top layer), horizontally in the higher velocity material, and then return to the surface.

The first arrival times are plotted against distance from the shot point (Fig. 18.5). This gives two time/distance straight lines with slopes $1/V_T$ and $1/V_B$. Let the distance of the intersection point from the shot point be $x$; then, the thickness of the top layer is given by

$$h_T = \frac{1}{2}(x)\left[\frac{V_B - V_T}{V_B + V_T}\right]^{1/2} \tag{18.2}$$

This method can only be used when the wave velocity is successively higher in each layer. Complications may be encountered when there is no sharp contrast in the velocities of the layers. In such cases, the travel–time diagram will be a curve. When inclined strata are encountered, only the average depth can be determined. However, if the positions of shot point and detectors are reserved, the actual depth and dip of the strata can be found. Further, in a multilayered series of strata a blind layer may occur when its wave velocity is less than that in the overlying layer. The range of velocities for different materials is given in Table 18.2 (IS: 1892, 1979).

Table 18.2 Range of shock wave velocities for different materials

| Materials | Velocity (m/s) |
|---|---|
| Sand and top soil | 180–365 |
| Sandy clay | 365–580 |
| Gravel | 490–790 |
| Glacial till | 550–2,135 |
| Rock talus | 400–760 |
| Water in loose materials | 1,400–1,830 |
| Shale | 790–3,350 |
| Sandstone | 915–2,740 |
| Granite | 3,050–6,100 |
| Limestone | 1,830–6,100 |

*Source:* IS: 1892 (1979).

### 18.4.2 Sounding Methods

Soil sounding or probing consists in forcing a rod, a rod enclosed in a sleeve pipe or a cone or a sample into the soil and observing the penetration or withdrawal resistance. Variation in this resistance indicates the existence of different soil strata, and the numerical values of the resistance permit an estimate of some of the physical properties of the strata.

The oldest and simplest form of soil sounding consists in driving a rod into the ground by repeated blows of a hammer. The penetration resistance of the rod may be directly correlated with some physical properties of the soil based on local experience. Variations in the diameter of the rod, in the driving force, and in the method of driving may affect the penetration resistance, and hence, the assessed physical property. Since the friction acting on the rod is cumulative with depth, the penetration resistance does not directly represent the strength or density of strata. Methods are available to separate the resistance offered by skin friction and point bearing. Comparisons of penetration resistance by rotation with the properties of strata are also being used.

Standardized methods of sounding are the standard penetration and cone penetration tests and are explained at the end of this chapter.

## 18.5 SEMI-DIRECT METHODS OF SUB-SURFACE EXPLORATION

Semi-direct methods are common boring and drilling methods combined with intermittent sampling. The depths of different layers are ascertained by the rate of advancement of boring tools or by means of non-representative samples obtained in the course of boring operations. Borings provide access to a particular layer for sampling. In general, only major changes in the character of the sub-surface materials can be detected by this method. The common boring methods are wash boring, rotary drilling, and auger boring.

### 18.5.1 Wash Boring

This method requires equipment to lift, rotate, and drop a bit, and a water supply unit. Due to the up and down movement of the chisel, the soil is loosened and broken up by the water jets. Water is pumped through a hollow drilling rod and released under pressure. It is continuously pumped till the soil particles are washed to the surface through the annular space between the rod and side of the borehole. The water–soil suspension is collected in a sump. The settled materials may be used for visual inspection. Wash boring can be used in most types of soils. An advantage of the method is that it produces a clear and fairly undisturbed environment at the bottom of the borehole for obtaining undisturbed samples using samplers. The details of wash boring are illustrated in Fig. 18.6.

### 18.5.2 Rotary Drilling

Although this method is primarily meant for rocks, it is also used in soils. In rotary drilling, the borehole is advanced by rapid rotation of the drilling bit, which cuts, chips, and grinds the material at the bottom of the holes into particles. Water or drilling mud is allowed

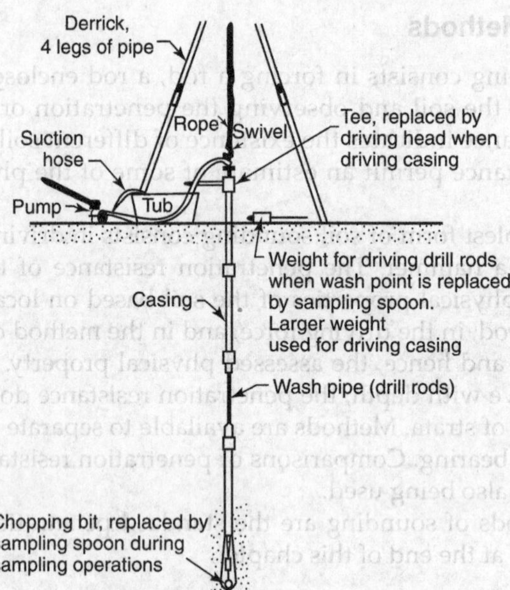


**Fig. 18.6** Typical arrangement for wash boring (*Source:* IS: 1892, 1979)

through the hollow rod, which then passes through narrow holes in the bit (Fig. 18.7). The drilling fluid cools and lubricates the drilling tool, and, as in wash boring, the loose debris are carried to the surface. The drilling mud also stabilizes the sides of the uncased borehole.

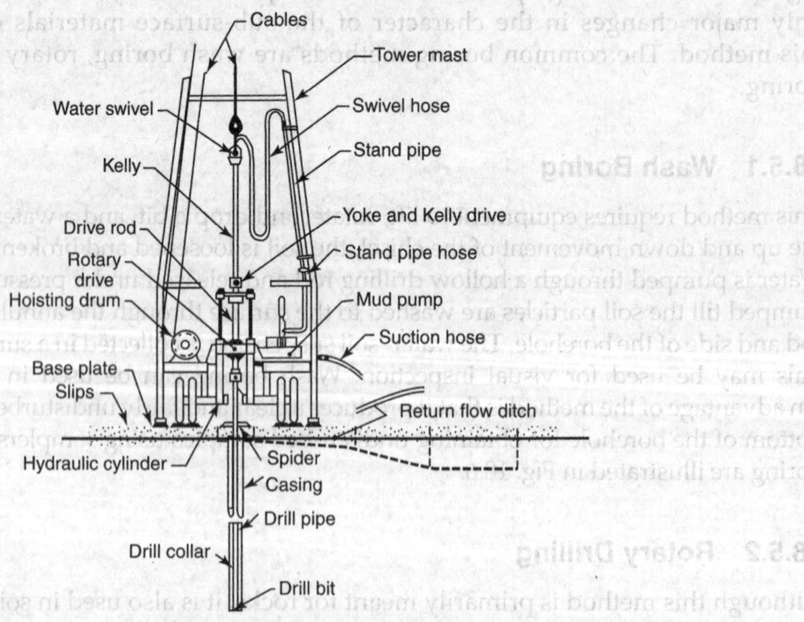


**Fig. 18.7** Typical arrangement for rotary drilling (*Source:* IS: 1892, 1979)

Open-hole drilling and core drilling are the two types of rotary drilling. Open-hole drilling is used in soils and weak rocks, and in other circumstances, as explained above. In core drilling, which is used in rocks and stiff clays, the bit cuts an annular hole in the material and an intact core is removed as a sample.

Rotary drilling is best suited for boring of diameter more than 100 mm. The rate of progress is greater in soils and rocks, and a uniform, clean hole with less disturbance of the soil is generally produced. This method is not suitable for coarse gravel, boulders, and badly fissured rocks.

### 18.5.3 Auger Boring

Hand-operated auger methods are suitable for identifying various types of soils with depth and also for getting information about the depth to the groundwater table. For deeper boring, solid or hollow-stem, continuous-flight augers (rotary augers) are frequently used. Such auger drilling requires rotary drill equipment to provide both a push against the drill head and a rotation. As the drill advances, additional auger flights are added and soil is brought to the surface in a disturbed form. Figure 18.8 shows various augers.

In auger boring it is possible to identify even disturbed soils. Since the borehole is kept dry, auger boring is particularly suitable for advancing borings above the water table to obtain undisturbed, partially saturated samples. It further facilitates the determination of free groundwater level.

## 18.6 DIRECT METHODS OF SUB-SURFACE EXPLORATION

Direct methods are boring and sampling methods which continuously provide representative or undisturbed samples. All accessible exploration, such as test pits, trenches, large diameter

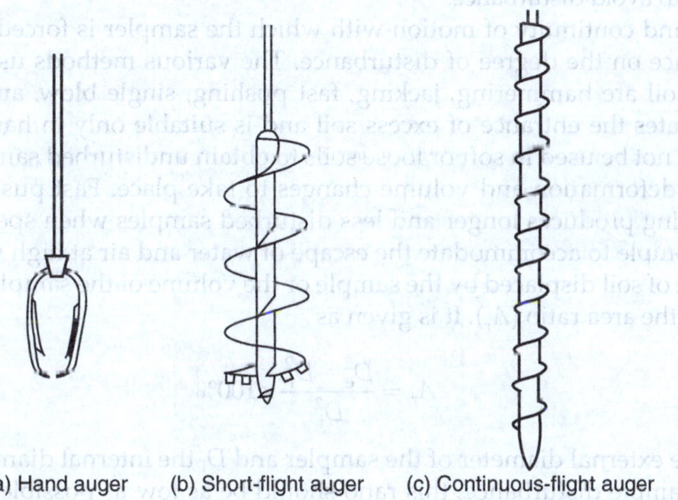


**Fig. 18.8** Types of augers

boring, shafts, drifts, etc., are grouped under direct methods. These accessible explorations allow direct examination of strata *in situ*.

### 18.6.1 Sampling Process

Soil samples are recovered from the ground by utilizing special techniques and sampling equipment. Typically, in a sampling process, a borehole is first advanced to the required depth by using a suitable boring method and the bottom of the borehole is then cleaned. Further, a soil sampler is advanced by driving it with a drop hammer or by pushing it with a hydraulic piston or jack, and then the sampler is brought to the surface.

Thereafter, some soil is removed from each end of the sampler, a thin layer of molten wax is applied to it and the ends of the tube are then covered by protective caps. All samples (both disturbed and undisturbed) should be clearly labelled with details about project name, date of sampling, borehole number, depth, and method of sampling. Extra care has to be exercised in handling, transportation, and storage of samples prior to testing. The use of a proper sampler and driving technique suited to the particular type of soil will reduce the sample disturbance.

### 18.6.2 Sample Disturbance

It is impracticable to obtain a sample that is totally undisturbed, especially from deep holes. During the boring and sampling processes, disturbances are bound to take place despite the use of elaborate and careful sampling techniques. In the case of clayey soils, there is a possibility of swelling due to stress relief. Soft clays are more susceptible to sample disturbances and the effect is more in low plastic clays. The disturbance caused in driving the sampling tool depends on the thickness of the tube and the manner in which it is driven. The rotation of the sample to free it from the ground causes disturbance. Thus, extension collars are provided at the top and bottom of the sampler to avoid disturbance.

The speed and continuity of motion with which the sampler is forced into the soil have a great influence on the degree of disturbance. The various methods used to force a sampler into the soil are hammering, jacking, fast pushing, single blow, and shooting. Hammering eliminates the entrance of excess soil and is suitable only in hard or dense coarse soils. It should not be used in soft or loose soils to obtain undisturbed samples. Slow jacking allows plastic deformation and volume changes to take place. Fast pushing, single heavy blow, or shooting produces longer and less disturbed samples when special vents are provided in the sample to accommodate the escape of water and air at high velocities.

The volume of soil displaced by the sample or the volume of the sample is represented by a factor called the area ratio ($A_r$). It is given as

$$A_r = \frac{D_e^2 - D_i^2}{D_i^2} \times 100\% \tag{18.3}$$

where $D_e$ is the external diameter of the sampler and $D_i$ the internal diameter.

To reduce sample disturbance, this ratio should be as low as possible (generally, $<10\%$) but, at the same time, maintaining sufficient strength in the wall of the tube. Further friction between the soil and inside tube wall would compress the sample. On the other hand, the presence of large stones, roots, and other materials may tear the sample, causing it to lengthen. A measure of such a disturbance is the recovery ratio ($L_r$), which is given as

$$L_r = \frac{\text{Length recovered}}{\text{Length pushed}} \tag{18.4}$$

$L_r < 1.0$ indicates compression and $L_r > 1$ shows expansion. Samples with a recovery ratio much different from 1.0 are substantially disturbed. Thus, the use of an appropriate type of sampler may reduce the sample disturbance to a great extent.

### 18.6.3 Types of Samplers

**Open-Drive Sampler.** It consists of a steel tube with a screw thread at each end. The lower end is generally fitted with a cutting shoe but sometimes with an extension piece. The upper end is fitted with a sampler head which incorporates a non-return valve. The non-return valve allows air and water to escape while the sample enters the sampler and closes as the sampler is raised to the surface, thus retaining the sample within the tube (Fig. 18.9). This is the simplest and most common type of sampler.

**Thin-Walled Sampler.** To obtain undisturbed samples in soft to firm clays and plastic silts, thin-walled samplers may be used (IS: 11594, 1985). No separate cutting shoe is attached to the lower end, but the sampler's lower end is itself machined to serve as a cutting edge (Fig. 18.10). Good-quality undisturbed samples are possible if $A_r < 10\%$ and the soil is not disturbed during the boring operation. This sampler can be used more conveniently in trial pits and shallow boreholes.

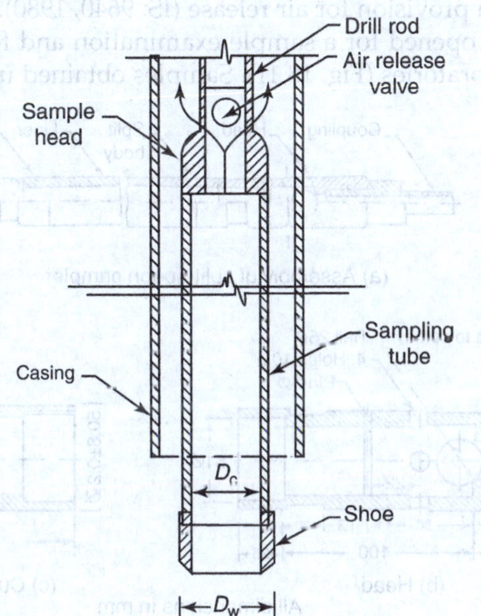


Fig. 18.9 Open-drive sampler

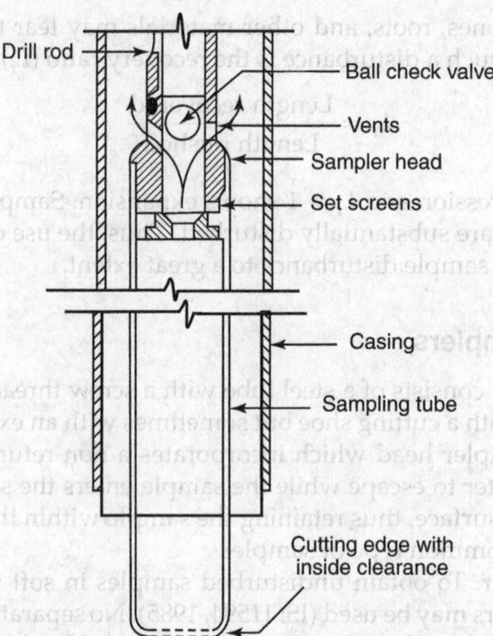


**Fig. 18.10** Thin-walled sampler

**Split-Spoon Sampler**. It consists of a longitudinally split tube or barrel fitted with a shoe and a sampler head with provision for air release (IS: 9640, 1980). The splitting aspect of the sampler permits it to be opened for a sample examination and for onward transmission in sample containers to laboratories (Fig. 18.11). Samples obtained using this sampler are rated

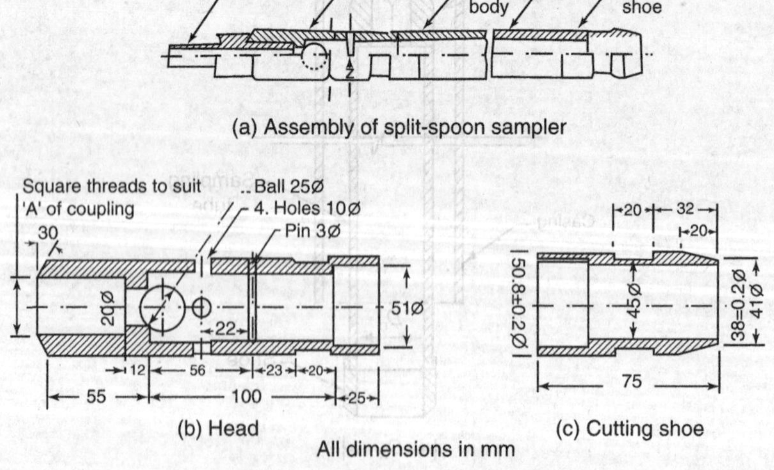


**Fig. 18.11** Details of split-spoon sampler (*Source:* IS: 9640, 1980)

as representative. This sampler is suited for sands and is used only in the standard penetration test (SPT). Split-soon samplers may be provided with a liner, which is a thin metal or plastic tube fitted within the split spoon, in which case it is called a composite sampler. The purpose of the liner is to protect the sampler during handling, shipping, and storage.

**Piston Sampler**. For very soft alluvial silts and clays, piston samplers are quite useful. These consist of a thin-walled tube which includes a piston device that serves to push the thin-walled tube into the undisturbed soil from the bottom of the boring (IS: 10108, 1982). The piston is locked in the lower position and the sampler is lowered to the bottom of the borehole. The piston is provided with a seal which prevents the entry of water and debris. When it is unlocked, the tube is driven down into the soil to the full length of travel of the piston. The whole assembly is withdrawn to the surface after locking the piston at the top of the tube (Fig. 18.12). The sampler is separated from the sample head and the piston. It is then sealed at both ends.

### 18.6.4 Accessible Explorations

Accessible exploration permits a direct visual examination of the subsoil and affords the most complete information about the ground. For major works, the locations of these should be decided in consultation with an engineering geologist. These are useful techniques, provided the overburden and groundwater conditions permit, and are universally adopted to obtain the required information for prospecting of materials for dam and embankment construction or concrete aggregates.

**Trial Pits**. Another method of sub-surface exploration is excavation of trial pits or test pits. It is one of the most dependable and informative methods of investigation. It is limited to a depth of 4 to 5 m. Relatively undisturbed samples can be obtained from walls or the bottom of pits by pushing a thin-walled steel tube. In cohesive soils, block samples can be cut by

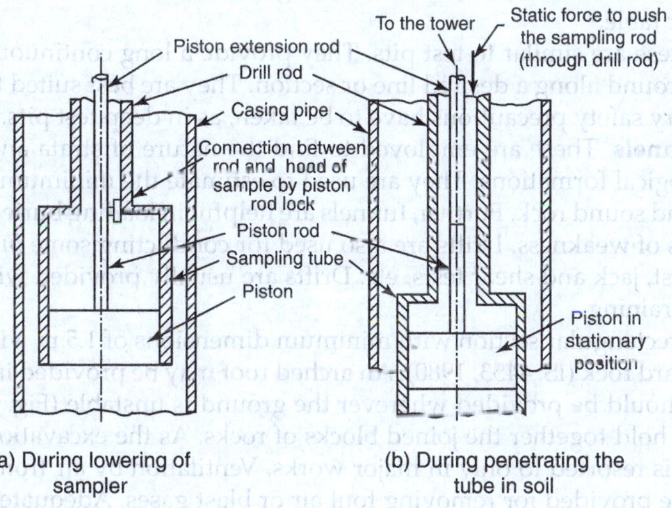


**Fig. 18.12** Schematic diagram explaining principle of operation of piston sampler (*Source:* IS: 10108, 1982)

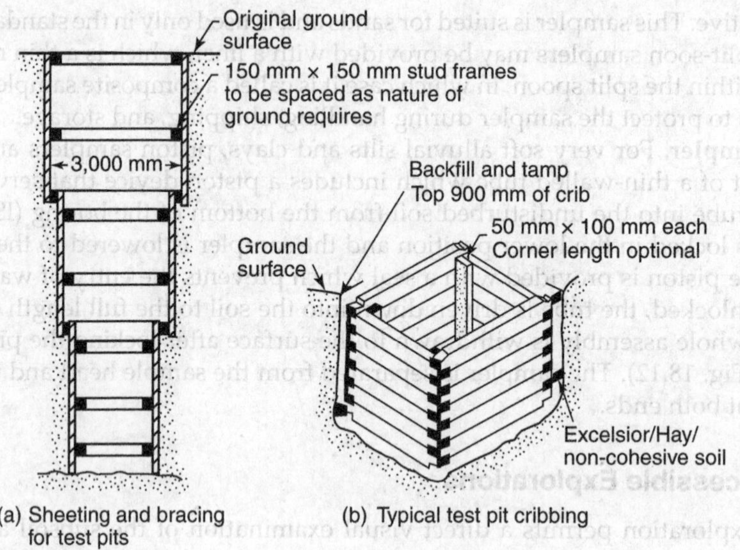


**Fig. 18.13** Arrangement for protecting test pits (*Source:* IS: 4453, 1980)

hand from the sides or bottom. Trial pits are suitable for all types of soils, and permit a most detailed examination of the soil formation for the entire depth.

Deeper pits have to be sheeted and braced (Fig. 18.13) or cribbed to prevent collapses (IS: 4453, 1980). Ventilation of deep test pits is necessary to prevent accumulation of dead air. This is done by providing pipes starting slightly above the floor and extending about 1 m above the top of the pit. Special precautions have to be exercised if the presence of obnoxious gases is anticipated (IS: 3764, 1966). A de-watering system has to be used if pits are to extend below the water table.

**Trenches**. These are similar to test pits. They provide a long continuous exposure of the surface of the ground along a desired line or section. They are best suited for exploration on slopes. Necessary safety precautions have to be taken, as in deep test pits.

**Drifts or Tunnels**. These are employed to find the nature of strata and the structure of particular geological formations. They are used to estimate the minimum excavation limits to reach fresh and sound rock. Further, tunnels are helpful in loading buried channels, faults, and other zones of weakness. Drifts are also used for conducting some *in situ* tests, such as plate bearing test, jack and shear tests, etc. Drifts are usually provided with a low outward slope for easy draining.

Generally, a rectangular section with minimum dimensions of 1.5 m width by 2 m height is followed in hard rock (IS: 4453, 1980). An arched roof may be provided in soft rock. Necessary supports should be provided wherever the ground is unstable (Fig. 18.14). Rock bolts may be used to hold together the joined blocks of rocks. As the excavation in rocks is slow and costly, this is resorted to only in major works. Ventilation by air from a compressor or a blower may be provided for removing foul air or blast gases. Adequate lighting arrangements are also provided for examination of the stratum.

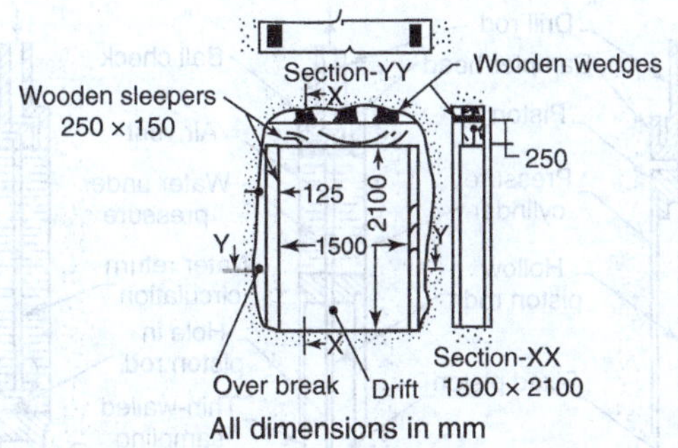


**Fig. 18.14** Typical method of supporting weak zones in drift (*Source:* IS: 4453, 1980)

**Shafts and Headings**. Shafts or deep pits are advanced by hand excavation with suitable sheeting. Headings or adits are horizontally excavated from the bottom of shafts, from the surface of steeply sloped grounds, or from quarry faces. Shafts and heading are not excavated from below the water table. These are very costly and used only in special investigations, such as pilot tunnels, mineral exploration surveys, etc.

Shafts may be circular (or rectangular) in section with minimum dimensions of 2.4 m × 2.4 m diameter, so as to provide ample room for movement of men and machinery (IS: 4453, 1980). As in deep open pits, dead air or blast gases may be removed by stove pipes starting above the floor and extending 1 m into open air above the mouth of the shaft. Air from a compressor or blower may also be used. A pumping system should be used when water is encountered.

### 18.6.5 Undisturbed Sampling of Soils

Undisturbed samples of soils are required for a number of tests. As discussed earlier, it is impossible to get a truly undisturbed sample. But a minimally disturbed sample is possible if certain procedures and precautions are observed. Although undisturbed sampling is somewhat expensive and time consuming, these samples are more valuable. Sampling techniques differ for fine and coarse-grained soils and are discussed below

**Undisturbed Sampling of Fine-Grained Soil.** A thin-walled sampler with stationary piston is recommended for clay and silt formations (IS: 10108, 1982). A thin-walled sampler with stationary piston, operated hydraulically is illustrated in Fig. 18.15. In this sampler, the piston is positioned at the required depth and the sampler is pushed down – hence the name stationary piston. The sampler head is connected tightly with a drill rod. The thin-walled sampler is connected to another piston head which moves in the pressure cylinder. The piston rod is hollow and a hole is provided in it which helps to release the pressure and thereby avoids over-driving. Operation of the system hydraulically minimizes the disturbance to the soil.

A borehole is made to the desired depth and cleaned. The borehole walls may be protected against cave-in by casing or bentonite mud. The fixed piston is kept at its lowest point, thus closing the lower end and preventing the entry of any foreign matter into the sampler. The different stages of sampling are illustrated in Fig. 18.15. The sample is teared at

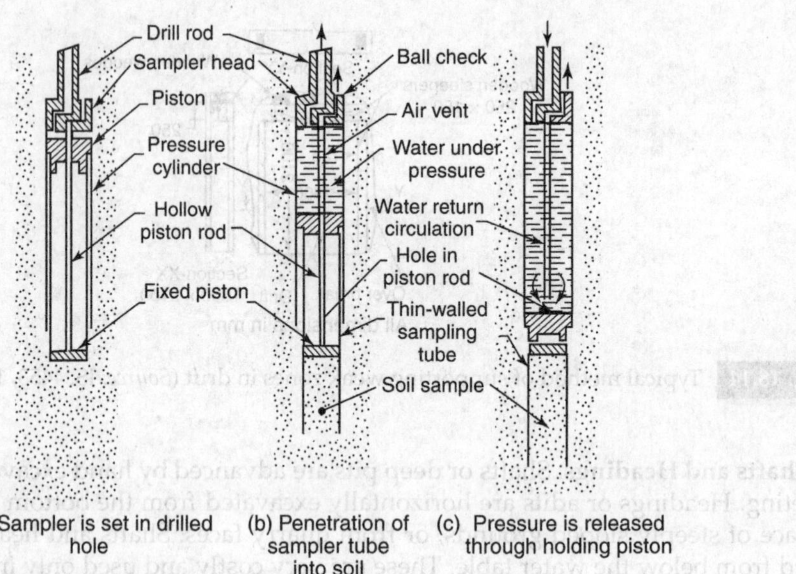


**Fig. 18.15** Diagrammatic sketch of hydraulically operated piston (*Source:* IS: 10108, 1982)

its bottom without causing any shock to the sample. In very loose sandy silt and silty sands, particularly below the water table, a core catcher (Fig. 18.16) may be used to avoid loss of samples while lifting.

**Undisturbed Sampling of Sands**. Cohesionless soils are always problematic as far as undisturbed sampling is concerned. Thus, *in situ* tests are preferred in cohesionless soils. Under favourable conditions, freezing is supposed to be the best method. By freezing, the lower part of the sample is solidified, which makes it easier to retain in the sampler. This is a very expensive technique. Two other methods of undisturbed sampling in uncemented sands are stationary piston sampling with drilling fluid circulation and compressed

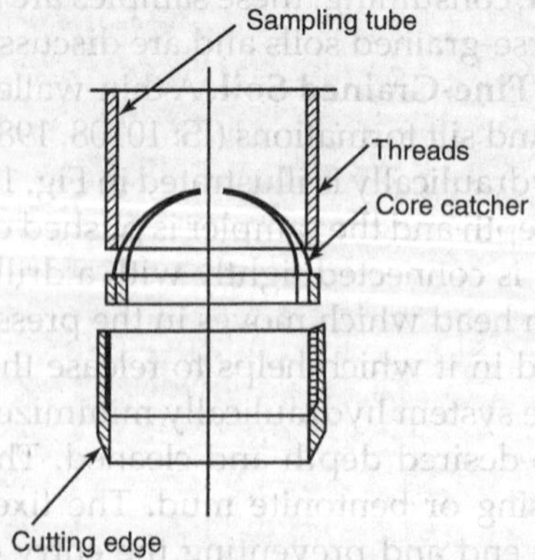


**Fig. 18.16** Core catcher fixed inside the cutting edge of the sampler (*Source:* IS: 10108, 1982)

air technique (IS: 8763, 1978). However, the sample obtained may be considered only as relatively undisturbed. These samples are normally used to determine the *in situ* density.

The first method works on the same principle as explained in the previous section but with an additional provision for circulation of the drilling fluid. In this method, a partial vacuum is created above the sample while withdrawing. The reader may refer to IS: 8763 (1978) for details.

The second method is important as it is suitable for sampling of sand below the water table. Compressed air is used in this technique and keeps the groundwater separated from the sample. This is necessary to avoid dispersion of sampled sand (Fig. 18.17). In a borehole, the sampler is pushed into the soil at the required depth with the help of a drill rod, a spacer block, and a shackle arrangement. The spacer block located above the bell prevents over-driving and allows the correct sample length. Now compressed air is forced into the bell, which in turn closes the diaphragm check valve and hence an excess pressure of 140 kN/m$^2$ is maintained inside the bell. This enables the water to be expelled from the bell. After the complete expulsion of water, the sampler along with the soil is withdrawn into the bell. The complete assembly is raised to the surface by means of a cable. During the process of withdrawal, water is continuously poured to keep the drill hole full and the air pump is also kept working. The spacer block is removed, the sampler is pushed out of the bell, and

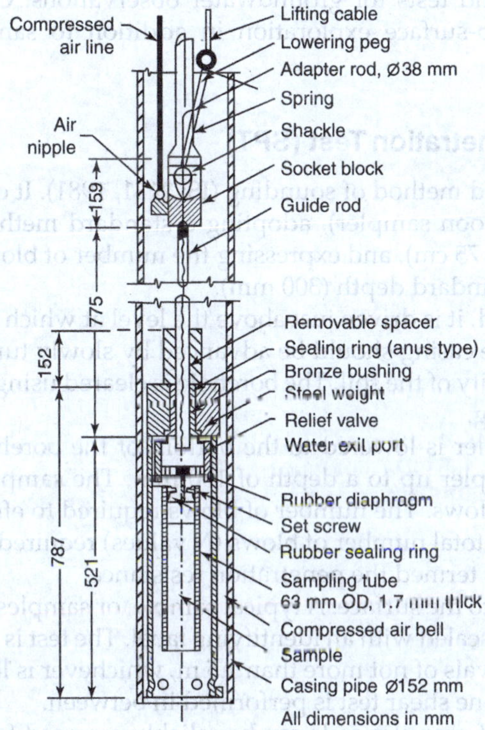


Fig. 18.17 General layout of sand sampler with auxiliary bell for compressed air (*Source:* IS: 8763, 1978)

the sampling tube is disconnected. A filter plug is placed at the lower end, the suction is released, and the undisturbed sample obtained.

**Undisturbed Sampling from Accessible Explorations**. Undisturbed samples may be obtained from accessible explorations, particularly from open pits and trenches. For this purpose, a pillar of dimensions 40 cm × 40 cm may be left if the sample is strong and undisturbed at the centre of the pit to extract the undisturbed sample of required size. If the sample is weak or to be transported to a far-off place, additional protection is required. A box with open ends is placed around the sample with a gap of about 25 mm all around, and the gap is filled with paraffin wax. The sample and the box are removed and additional wax is poured on the top and bottom, which prevents evaporation of moisture from the sample.

## 18.7 ROUTINE FIELD TESTS

Reliable results can be obtained from a carefully obtained sample and appropriate testing techniques in the laboratory. Even then, all the environmental conditions (e.g., in-place stress condition, moisture, etc.) cannot be simulated in the laboratory, and evidently, the soil samples undergo some disturbance during sampling and handling, which may greatly affect the test results. In addition, some valuable data can be obtained only by field testing. Three routine field tests are identified, viz., penetration test (standard penetration or cone penetration), vane shear test, and tests for groundwater observations. One or more of these are performed in every sub-surface exploration in addition to sampling. They are briefly explained below.

### 18.7.1 Standard Penetration Test (SPT)

This test is a standardized method of sounding (IS: 2131, 1981). It consists of driving a standard sampler (a split-spoon sampler), adopting a standard method of driving (a 63.5 kg weight with a free fall of 75 cm), and expressing the number of blows (as-value) required to push the sampler to a standard depth (300 mm).

Where a casing is used, it is driven just above the level at which the test is made. In cohesionless soils, the sample casing should be advanced by slowly turning or pushing so as to avoid change in the density of the soil. The borehole is cleared using augers, bailers with flap valves, or by wash boring.

The split-spoon sampler is lowered to the bottom of the borehole and driven with the hammer to seat the sampler up to a depth of 150 mm. The sampler is further driven to a depth of 300 mm or 50 blows. The number of blows required to effect each 150 mm of penetration is recorded. The total number of blows ($N$ values) required for the second and third 150 mm of penetration is termed the penetration resistance.

The sampler is raised to the surface. A typical sample, or samples, from the opened split is placed in jars, which are sealed with an identifying label. The test is repeated at every change in the stratum or at intervals of not more than 1.5 m, whichever is less. The intervals may be increased to 3 m if the vane shear test is performed in between.

The relative density of granular soils can be reliably assessed from SPT values. The correlation of $N$ values with the shear strength of cohesive soils is not very reliable. Several factors affect the blow count, and necessary judgement has to be exercised while adopting

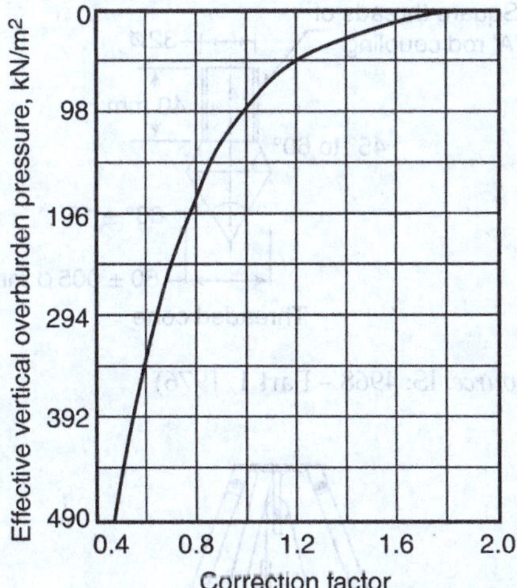


Fig. 18.18 Correction chart for $N$-value in cohesionless soil for overburden (*Source:* IS: 2131, 1981)

the results. In cohesionless soils, an increase in depth increases the overburden pressure; consequently, the number of blows needed to drive the sampler increases for a cohesionless soil of the same relative density. Hence, the $N$ value has to be corrected for overburden pressure in cohesionless soils. The corrected $N$ value ($N'$) is given as

$$N' = (N_{\text{observed}})(\text{correction factor}) \tag{18.5}$$

The correction factor is given in Fig. 18.18, which is as per IS: 2131 (1981). The value obtained has to be further corrected for dilatancy. If the stratum consists of fine sand and silt below the water table, the revised $N'$ value ($N''$) is given as

$$N'' = 15 + \tfrac{1}{2}(N' - 15) \tag{18.6}$$

### 18.7.2 Cone Penetration Test (CPT)

The CPT is another important sounding method. Two penetration tests have been standardized, viz., dynamic CPT (IS: 4968 – Parts 1 and 2, 1976) and static CPT (IS: 4968 – Part 3, 1976).

The dynamic cone penetration value ($q_{cat}$) is the resistance offered by a 60° cone for a 300 mm penetration when driven by a 65 kg weight with a 750 mm free fall.

The 50 mm diameter or 60° cone is shown in Fig. 18.19. The cone is threaded to the driving rod. The hammer head is joined to the other end of the rod and a guide rod, 150 cm long, is connected to the hammer head. The assembly is kept vertical as illustrated in Fig. 18.20. The cone is then driven into the soil using the 65 kg hammer falling through the free fall of 750 mm. The number of blows required for every 100 mm penetration is recorded. The process is repeated till the cone is driven to a depth of 300 mm.

The dynamic cone is a simple device by which the continuous soil resistance of a location can be obtained. Further, another advantage is that no borehole is needed. In the attempt to

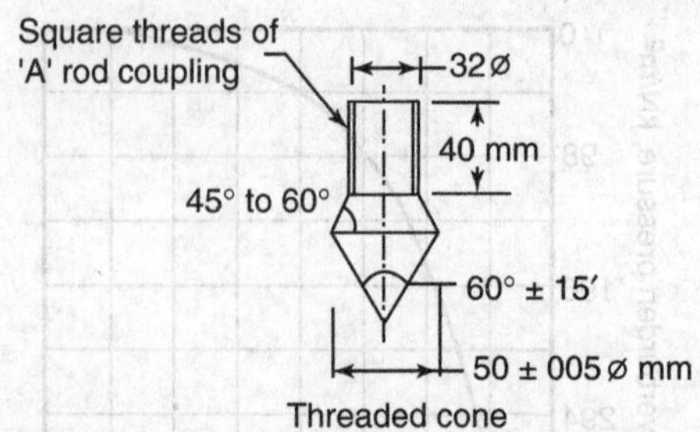


**Fig. 18.19** Dynamic cone (*Source:* IS: 4968 – Part 1, 1976)

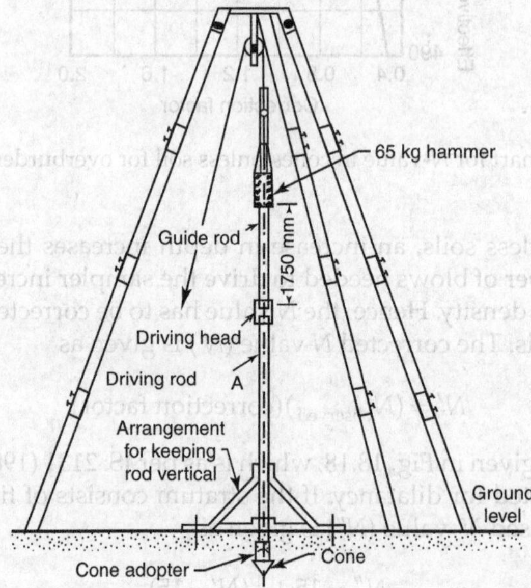


**Fig. 18.20** Typical assembly of equipment for cone penetration test (*Source:* IS: 4968 – Part 1, 1976)

arrive at a correlation between dynamic cone penetration and standard penetration, a wider cone of 62.5 mm with use of bentonite slurry has been suggested (IS: 4968 – Part 2, 1976).

Among the sounding tests, the static cone test is the best and can give more reliable values in locations below the water table and where SPT fails. Basically, the test consists of pushing the cone first and then the cone and friction jacket, thus finding the frictional resistance. Tolia (1978) reviewed the factors affecting the static cone and the influence of overburden pressure on static CPT values.

### 18.7.3 Vane Shear Test

The principle of the vane shear test and the laboratory method of determining the same have been discussed in Chapter 7. The vane shear test is suitable for saturated clays of soft medium consistency. This is more suitable for soils which are fissured or highly susceptible to sampling disturbance.

The height of the vane is twice its diameter. The overall diameter of the vane should be 37.5, 50, 65, 75, or 100 mm. The area ratio ($A_r$) of the vane is given as

$$A_r = \frac{8t(D-d)+\pi d^2}{\pi D^2} \times 100\% \tag{18.7}$$

where $t$ is the thickness of the vane blade (mm), $D$ the overall diameter of the vane (mm), and $d$ the diameter of central vane rod (mm).

As per Indian Standards (IS: 4434, 1978) the area ratio shall not exceed 18% for the 37.5 mm vane and 12% for the 50, 65, 75, and 100 mm diameter vanes.

The instrument should be capable of applying a torque to the vane and measuring the same. The torque applicator should be capable of controlling the speed at the rate of 0.1° per second.

There are two methods of testing, viz., testing from the bottom of a borehole and direct penetration from the ground surface. The first test method is explained below.

An arrangement (diagrammatic) for testing from the bottom of the borehole is shown in Fig. 18.21. About 5 minutes after insertion of vane, the torque is applied. The maximum torque applied is noted. Just after this, the vane is rotated through a minimum of 10 revolutions. After about a minute the remoulded strength is determined. The shear strength is computed from Eq. 9.18.

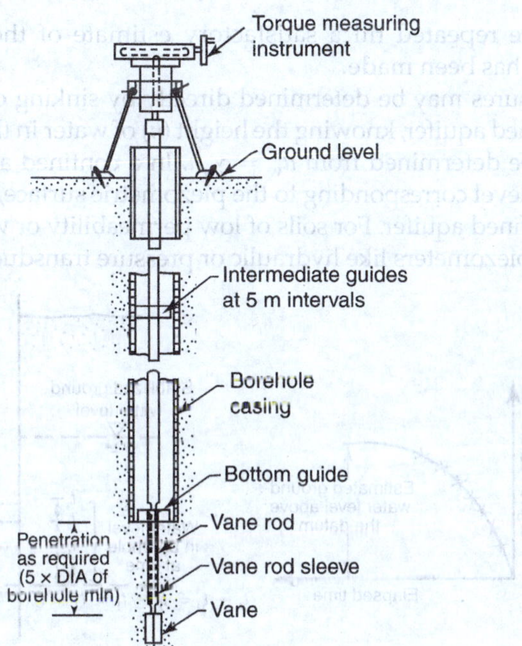


**Fig. 18.21** Diagrammatic vane test arrangement (for test from bottom of borehole) (*Source:* IS: 4434, 1978)

### 18.7.4 Groundwater Observations

There are two aspects to groundwater observations: viz., determination of levels and pore water pressure and permeability tests. Field permeability tests have been discussed in Chapter 5. Observations of water table and piezometric surface levels are discussed below.

Borehole observation is the simplest technique. Boreholes drilled for a sub-surface investigation can be kept open for 24 hours. The level of water is normally determined by lowering a chalked tape or a tape with a float or by an electrical switching device which is actuated on contact with water.

In a cohesive soil stratum, the stabilization of the water table may take time. In such situations, the location may be ascertained by adopting the extrapolation method. In this case, a plot of water level versus time is made and the groundwater level is estimated by extrapolating the curve until it becomes parallel to the time axis (Fig. 18.22). If several levels are noted at equal time intervals the following computational method is used.

Let the rise in water level from time $t_0$ to $t_1$ be $h_1$
Let the rise in water level from time $t_1$ to $t_2$ be $h_2$
Let the rise in water level from time $t_2$ to $t_3$ be $h_3$
And let $t_1 - t_0 = t_2 - t_1 = t_3 - t_2$ (Fig. 18.22). The depths of the observed water level below the stabilized groundwater level are

$$D_0 = \frac{h_1^2}{h_1 - h_2}, \quad D_1 = \frac{h_2^2}{h_1 - h_2}, \quad D_2 = \frac{h_3^2}{h_2 - h_3}$$

Observations are repeated till a satisfactory estimate of the location of the stabilized groundwater level has been made.

Pore water pressures may be determined directly by sinking open piezometers or a stand pipe. In an unconfined aquifer, knowing the height ($h$) of water in the stand pipe, the pore water pressure ($u_w$) can be determined from $u_w = \gamma_w h$. In a confined aquifer, the water level automatically rises to a level corresponding to the piezometric surface, representing the pore water pressure in the confined aquifer. For soils of low permeability or where a more rapid response is required, closed piezometers like hydraulic or pressure transducer piezometers are used.

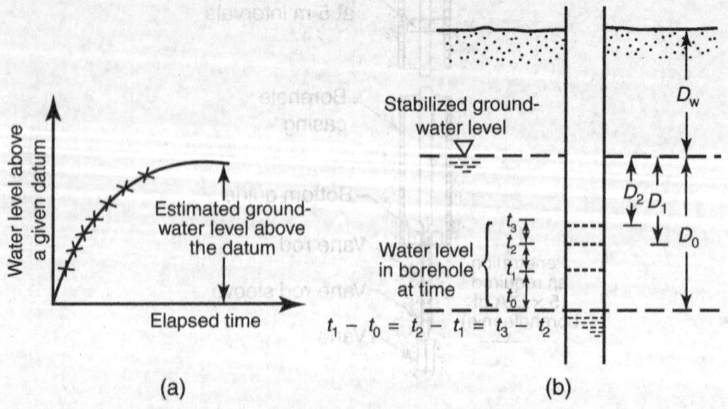


Fig. 18.22 Location of groundwater level (*Source:* Whitlow, 1983)

## 18.8 RECORDING OF FIELD DATA

Data obtained from exploratory work must be recorded accurately to derive maximum benefit from the effort. The following information should be routinely recorded in the field notes:

1. name and address of the project,
2. date of work and names of personnel involved,
3. location of boring or test pits with respect to an established coordinate system,
4. elevation of ground surface,
5. water table elevation,
6. elevation of upper level of each soil layer,
7. elevation at which soil samples were taken,
8. record of SPT or CPT results,
9. a field classification of each layer, and
10. description of drilling equipment and changes encountered in drilling.

After the completion of laboratory tests, the information obtained from the field is assembled and summarized to obtain the details of a soil profile, which is termed a boring log. A typical boring log of a deposit in Calcutta (Som, 1975) is shown in Fig. 18.23.

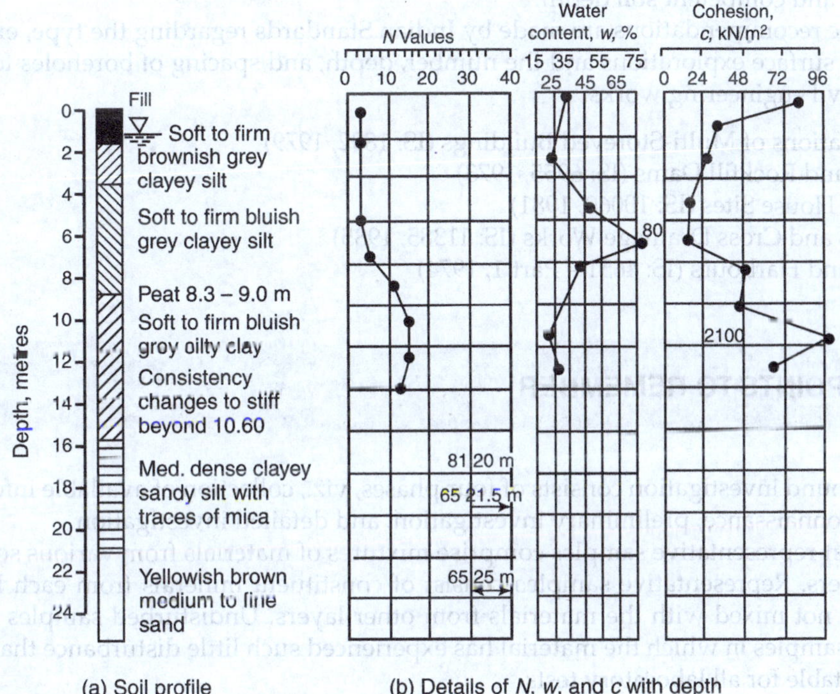


**Fig. 18.23** Boring log for a deposit in Calcutta (*Source:* Som, 1975)

## 18.9 LOCATION, SPACING, AND DEPTH OF BORINGS

There is no hard and fast rule to decide the boring location or spacing. The number of borings, or the spacing between borings for a project is related to the type, size, and weight of the structure planned. In general, borings should be located to obtain maximum information from the minimum number of boreholes. Primarily, as a few as two or three boreholes may be made to get a confident assessment of sub-surface conditions.

The type of structure often decides the number and the depth of borings. Usually, for a building project, boreholes are made at the building corners, at the centre of the site and locations where heavily loaded columns or machinery pads are proposed. Generally, at least one boring should be taken to a deeper stratum, preferably up to the bedrock if practicable. Other borings may be taken at least to the significant stress level. This significant depth corresponding to the associated stress is taken to be approximately two times the least width. It is not advisable to terminate the borings on soft or organic soils, if they are encountered. For highways and runways, sewer and water lines, borings are taken along the centre line. For airports, additional borings are needed for parking stands, taxiways, and terminal buildings. Boreholes for highways and runways are taken up to the bedrock to determine the soil stratification, so that some of the soils can be used for filling and construction. Water and drainage works also consider the water table location and rock line. Generally, for a preliminary survey of long-distance projects, such as highways and dams, the spacing of boring required is from 150 to 300 m. The final programme may eventually require borings of 30 to 70 m spacing. Boring in river beds for bridge piers and abutments, should ascertain the scour and competent soil depth.

Specific recommendations are made by Indian Standards regarding the type, extent, and details of surface explorations and the number, depth, and spacing of boreholes for the following civil engineering works:

1. Foundations of Multi-Storeyed Buildings (IS: 1892, 1979)
2. Earth and Rockfill Dams (IS: 6955, 1973)
3. Power House Sites (IS: 10060, 1981)
4. Canals and Cross Drainage Works (IS: 11385, 1985)
5. Ports and Harbours (IS: 4651 – Part 1, 1974)

## POINTS TO REMEMBER

18.1 Ground investigation consists of four phases, viz., collection of available information, reconnaissance, preliminary investigation, and detailed investigation.

18.2 Non-representative samples comprise mixtures of materials from various soil or rock layers. Representative samples consist of constituent minerals from each layer and are not mixed with the materials from other layers. Undisturbed samples are types of samples in which the material has experienced such little disturbance that they are suitable for all laboratory tests.

18.3 In indirect methods (geophysical and sounding methods) depths to principal strata are established, based on some physical properties of the material, and the measurements are made on the ground surface.

18.4 In semi-direct methods (borings and rotary drilling) the depths of different layers are ascertained by the rate of advancement of boring tools or by means of non-representative samples obtained in the course of boring operations.

18.5 Direct methods include boring and sampling methods, which continuously provide representative or undisturbed samples.

18.6 Sample disturbance is caused due to the sample thickness, method of driving the samples, and rotation of samples during removal.

18.7 Accessible explorations permit a direct visual examination of the subsoil and afford the most complete information of the ground.

18.8 Routine field tests include the standard penetration test, cone penetration test, vane shear test, and groundwater observations.

## QUESTIONS

### *Objective Questions*

18.1 State whether the following are true or false:

1. In the electrical resistivity method, differences in the electrical potential of the strata are detected.
2. Non-representative samples comprise mixtures of materials from various soil or rock layers.
3. A value of recovery ratio greater than one indicates compression and less than one, expansion.
4. The cone penetration test is another important sampling method.
5. The field vane shear test is most suitable for saturated soft clays.

18.2 The actual planning of a sub-surface exploration programme includes

(1) Collection of all available information
(2) Reconnaissance of the area
(3) Preliminary site investigation
(4) Detailed site investigation

Of these statements,

(a) 1, 2, 3, and 4, are correct
(b) 1 and 2 are correct
(c) 3 and 4 are correct
(d) 2, 3, and 4 are correct

18.3 The basic requirement in a seismic refraction method is that the wave velocity in the upper layer must be ______ that in the lower layer.

(a) Less than
(b) Greater than

(c) Equal to
(d) Twice

18.4 The standard penetration test is most frequently used to measure the
(a) Shear strength of soft clays
(b) Undrained strength of fissured clays
(c) Relative density of granular soils
(d) Consistency of clays

18.5 Which of the following pairs is correctly matched?

| | |
|---|---|
| 1. Wash boring | Can be conveniently used even below the water table |
| 2. Percussion boring | Only method suitable for drilling base holes in bouldery and gravelly strata |
| 3. Auger boring | Samples recovered have high value |
| 4. Rotary boring | Useful only for sands and clays |

Select the correct answer using the codes given below:

*Codes:*
(a) 1 and 2 (b) 2 and 3 (c) 3 and 4 (d) 4 and 1

18.6 The degree of disturbance for a soil sample is usually expressed by
(a) Void ratio (b) Area ratio
(c) Recovery ratio (d) Consolidation ratio

18.7 Identify the incorrect statement. Undisturbed samples are obtained from
(a) Thin-walled tube samplers (b) Piston samplers
(c) Split-spoon samplers (d) Hand-trimmed samplers

18.8 Samples of highly fissured soils can be obtained from
(a) Accessible explorations (b) Open-drive sampling
(c) Split-spoon sampling (d) Thin-walled sampling

## *Descriptive Questions*

18.9 What sort of expertise would be needed to choose the borehole depth and number in a ground investigation?
18.10 How do you analyse the samples obtained after sampling for strength if they are fine sand, over-consolidated clay, and soft sandstone?
18.11 Explain the basic differences in explorations for foundations (including abutments) and construction materials.
18.12 Discuss the various stages of sample disturbance.
18.13 Discuss the factors which are relevant to the planning of a well-balanced exploration programme.
18.14 Compare soil boring and sampling methods with test pits.
18.15 Explain the following terms which are used in subsoil exploration.
1. Area ratio
2. Recovery ratio
3. Non-representative sample

4. Representative sample
5. Undisturbed sample

18.16 Briefly describe the types of sub-surface information that seismic refraction studies can provide.

18.17 If you are in charge of subsoil exploration of important structures, how do you decide the depth of exploration. List the factors you will consider and their importance.

18.18 Discuss briefly the methods of taking undisturbed samples in (i) non-cohesive soils and (ii) cohesive soils.

18.19 What steps will you take to economically investigate a site which is suspected to be erratic in nature?

18.20 If a thin-walled sampler is pushed fast into the following soils, what will be the effect of the disturbance: (i) sensitive clay, (ii) sandy silt, and (iii) sandy clay.

## EXERCISE PROBLEMS

18.1 An open drive sampler has an outside diameter of 76 mm and an inside diameter of 72 mm. What is the area ratio of the sampler?

18.2 During a sampling operation, the open-drive sampler is advanced 600 mm and the length of the recovered sample is 525 mm. What is the recovery ratio of the sample?

18.3 In a standard penetration test, the following observations were taken at a depth of 4 m below the ground level:

| | |
|---|---|
| First 15 cm | 31 blows |
| Second 15 cm | 32 blows |
| Third 15 cm | 23 blows |
| Fourth 15 cm | 36 blows |

Estimate the corrected SPT value for overburden if the average unit weight of the soil at 4 m depth is 20 kN/m$^2$.

# 19

# Soil Improvement

## CHAPTER HIGHLIGHTS

Soil improvement techniques – Surface compaction – Drainage methods: well-point systems, deep-well drainage, vacuum de-watering system, de-watering by electro-osmosis – Vibration methods: vibro-compaction, vibro-displacement compaction – Pre-compression and consolidation – Grouting and injection – Chemical stabilization – Soil reinforcement – Geotextiles and geomembranes – Other methods

## 19.1 INTRODUCTION

*In situ* soil characteristics of a construction site are different from those desired and, almost always, far from ideal for a designated need. With increased urban development, sites with favourable foundation conditions became rare. At times, the civil engineer has been forced to construct structures at sites selected for reasons other than soil conditions. Thus, it is increasingly important for the engineer to know the degree to which soil properties may be improved or if there is an alternative that can be thought of for the construction of an intended structure at the stipulated site.

If unsuitable soil conditions are encountered at the site of a proposed structure, one of the following four procedures may be adopted to ensure satisfactory performance of the structure (Mitchell, 1976):

1. Bypass the unsuitable soil by means of deep foundations extending to a suitable bearing material.
2. Redesign the structure and its foundations for support by the poor soil, a procedure that may not be either feasible or economical.
3. Remove the poor material and either treat it to improve and replace it, or substitute it by a suitable material.
4. Treat the soil in place to improve its properties.

Further, in the case of existing structures exhibiting foundation distress, in-place foundation treatment may be used as corrective measures. Nowadays, various processes are available by which the characteristics of the construction site can be improved to facilitate construction operation, to allow increased bearing pressures, or to reduce settlements. The techniques involved in the attainment of the required improvement facilities are referred to as geotechnical processes.

## 19.2 IMPROVEMENT TECHNIQUES

Soil improvement in its broadest sense is the alteration of any property of a soil to improve its engineering performance. This may be either a temporary process to improve the construction of a facility or may be a permanent measure to improve the performance of the completed facility. The result of an application of a technique may be increased strength, reduced compressibility, reduced permeability, or improved groundwater condition. Soil improvement techniques may be classified based on the nature of the process involved, material used, the desired result, etc. The various techniques discussed are surface compaction, drainage methods, grouting and injection, chemical stabilization, thermal stabilization, soil reinforcement, and application of geotextiles and geomembranes.

The factors that must be considered in the selection of the best technique in any case include the following (Mitchell, 1976):

1. Soil type – sand, clay, organic, etc.
2. Area and depth of treatment required depend on the geometric characteristics of the soil deposit and load distribution
3. Type of structure and load distribution
4. Soil properties – strength, compressibility, permeability, etc.
5. Permissible total and differential settlements
6. Material availability – stone, sand, water, admixture, stabilizers, etc.
7. Availability of skills and equipment
8. Environmental considerations – waste disposal, erosion, water pollution, etc.
9. Local experience and preferences
10. Economics

The discussion on different improvement techniques in the following sections is directed primarily at their use as a permanent measure of soil improvement. However, some of them are well suited to expedite or facilitate construction as well, for example, pumping to control water flow into excavations, the use of grounting to reduce liquefaction or settlement, the use of electro-osmosis for seepage control, and temporary slope stabilization by freezing.

## 19.3 SURFACE COMPACTION

One of the most widely used and the oldest techniques of soil densification is compaction. Construction of a new road, a runway, an embankment, or any soft or loose foundation site needs a compacted base for laying the structure. If the depth to be densified is less, then compaction on the surface of the soil alone may solve the problem. Surface compaction needs less skilled labour and is usually the most economical of the techniques available.

The field engineer can vary the content, amount of compaction, and type of compaction to attain the required density. The usual surface compaction devices are rollers, tampers, and rammers. All conventional rollers like smooth wheel, rubber-tyred, sheeps-foot, vibratory, and grid rollers can be used (as discussed in Chapter 4). A particular type of roller has to be chosen to suit a particular job. In order to achieve the required density, only the required number of passes should be allowed. Mere increase in the number of passes will not increase the density beyond a certain depth. For all practical purposes, granular soils can be surface compacted using vibratory rollers up to a depth of about 2 m.

In cohesive soils, the required percentage compaction can be obtained using any of the rollers and tampers, but vibrations are not effective. In the cases of sub-grades and base courses for heavy duty roads and airfields, the heavy rubber-tyred rollers may be used with advantage. In order to efficiently bond each layer for water-retaining structures a sheeps-foot roller may be used.

## 19.4 DRAINAGE METHODS

Groundwater causes some of the most difficult problems in excavation work. The presence of water increases the pore water pressure and decreases the shear strength. Further, heavy inflow of water to the excavations is liable to cause erosion or collapse of the sides of open excavations. Sometimes, there can be instability of the base due to upward seepage. However, from a knowledge of the soil and groundwater conditions, it is feasible to adopt certain methods to control the groundwater and ensure a safe and economical construction scheme.

### 19.4.1 Well-Point Systems

Well points are small well screens of sizes 50 to 80 mm in diameter and 0.3 to 1 m in length. The well-point system of ground water lowering comprises the installation of well points around the excavation. The well points are attached to a header pipe and in turn connected to a well-point pump. The water flowing by gravity to the fiter well is drawn by the vacuum up to the header main and discharged through the pump. A typical installation of a well-point system is shown in Fig. 19.1.

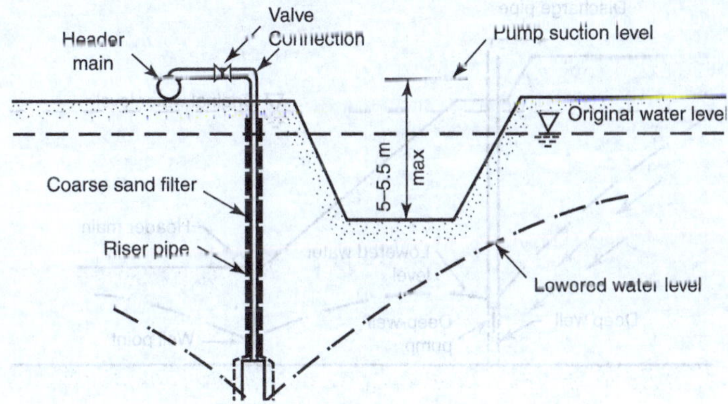


Fig. 19.1 Single-stage well-point installation

If the site is accessible and the water-bearing strata to be drained are not too deep, the well-point system will be the most suitable method. The equipment is reasonably simple and cheap and can be installed rapidly. Well points are generally used where the water table does not have to be lowered too much. In the case of large excavations or where the depth of excavation below the water table is more than 10 m or there is artesian pressure, deep wells and turbine pumps have to be adopted. The added advantage with the well-point system is that the water is filtered as it is removed from the ground and carries few little or no particles with it.

A single well point of 50 mm has a capacity of about 10 $l$/min. Permeability of the soil and the time available to effect the drawdown are the factors that govern the spacing of well points. In a highly permeable gravel medium, the spacing is about 0.3 m, whereas in fine to coarse sands, a spacing of about 0.75 to 1 m is satisfactory. In low permeability mediums like sandy silts, the spacing may be of the order of 1.5 m. A well-point equipment of normal size comprises 50 to 60 points to a single 150 or 200 mm pump with a separate 100 mm jetting pump.

### 19.4.2 Deep-Well Drainage

Where the soil formation becomes more pervious with depth, large diameter deep wells are suitable for lowering the groundwater table. In contrast to a well-point system, deep-well de-watering can be installed outside the zone of construction operation, and drainage effected to the depth of excavation required. Deep wells may be combined with the well-point system on certain occasions for lowering the groundwater table (Fig. 19.2). Deep wells are also suitable for relief of the artesian pressure in some field conditions. The cost of installation of a deep-well system is relatively high. Therefore, the process is generally restricted to jobs which have a long construction period, such as dry docks.

The procedure in installing a deep well is to sink a cased borehole having a diameter that is about 300 mm larger than the well casing. The diameter of the latter depends on the size of the submersible pumps. The inner well casing is inserted after the completion of the borehole. A perforated screen is provided over the length where de-watering of the soil is

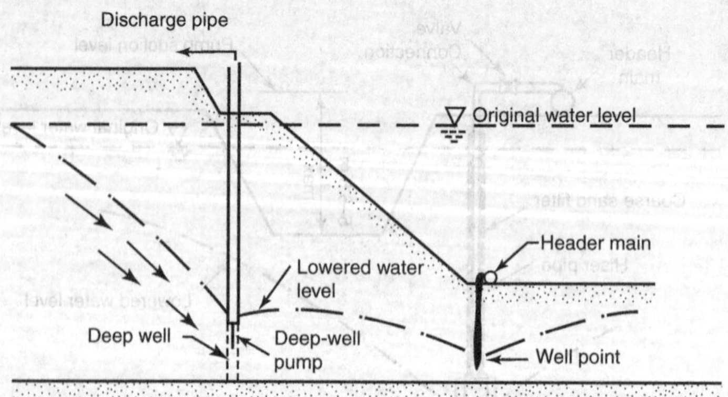


Fig. 19.2 Deep-well de-watering

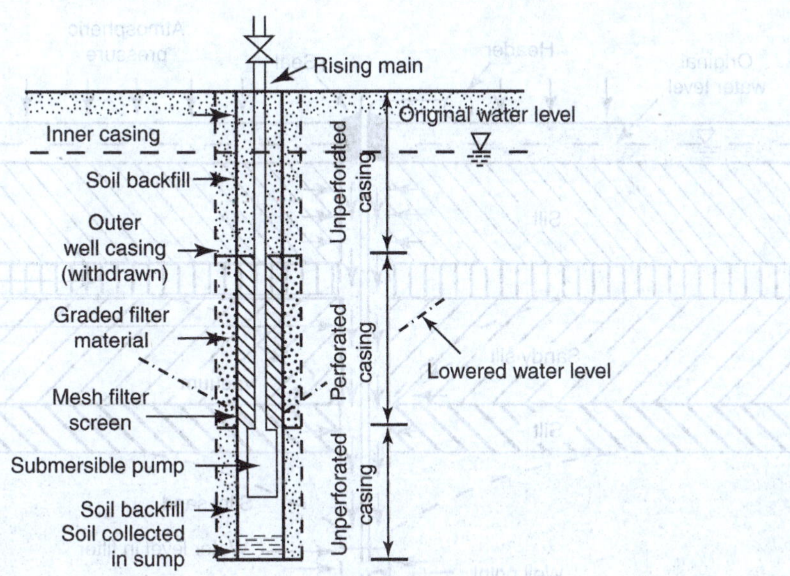


**Fig. 19.3** Details of deep-well installation (*Source:* Tomlinson, 1986)

required. Graded gravel filter material is placed between the well casing and the outer bore-hole casing over the length to be de-watered. The outer casing is withdrawn in stages as the filter material is placed. The space above the screen is backfilled with any available material. The details of the completed installation are shown in Fig. 19.3. Deep wells are spaced at 10 to 100 m intervals depending on the situation.

### 19.4.3 Vacuum De-watering System

Gravity methods are not very effective in fine-grained soils with permeability in the range of 0.1 to $10 \times 10^{-3}$ mm/s. Such soils can be stabilized by means of a vacuum well or well-point system. A vacuum de-watering system consists of wells or well points with the screen and riser pipes surrounded with a free drainage sand filter extending to within a few metres of the surface. A bentonite or impervious soil seal is provided at the remaining portion of the hole. By maintaining a vacuum in the well screen and the sand filter, the hydraulic gradient producing flow towards the well or well point is increased. For proper de-watering, closer spacing of wells is essential (Fig. 19.4).

### 19.4.4 De-watering by Electro-osmosis

Fine-grained soils are more troublesome to drain because of capillary forces acting on the pore water. If the vacuum well-point system is ineffective, application of an electrical gradient may be made. When a direct electric current is passed through a saturated soil, water moves towards the cathode. If the water is removed at the cathode, the soil decreases in volume resulting in increased shear strength. This process is called de-watering by electro-osmosis. Casagrande (1952) has shown that the electro-osmotic flow is dependent on the

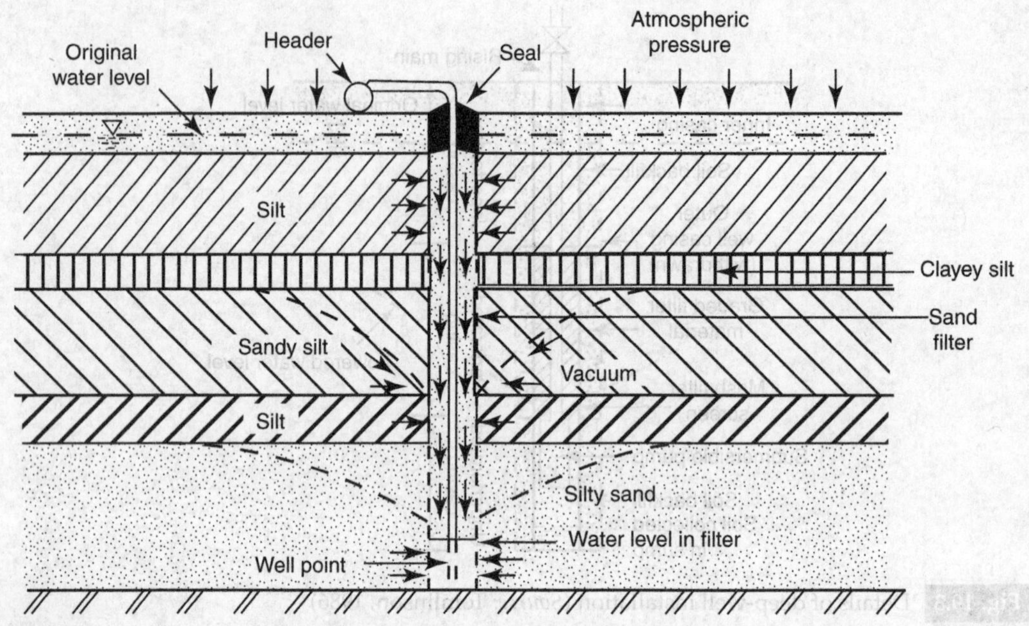


**Fig. 19.4** Vacuum de-watering system (*Source:* Tomlinson, 1986)

porosity of the soil and the electric potential. A comparison of electro-osmotic flow with hydraulic flow through a single capillary is shown in Fig. 19.5.

The general layout of the electrodes depends upon the purpose for which they are intended. Figure 19.6 shows electrode arrangements for two field situations. Sheet piles of any shape and old pipes of 25 to 50 mm diameter can be used as anodes. Since the

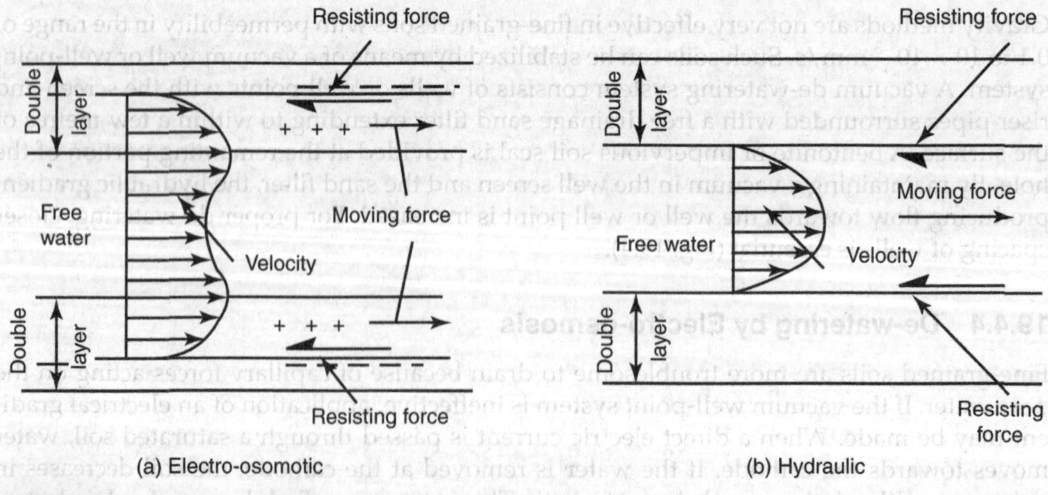


**Fig. 19.5** Comparison of electro-osmotic and hydraulic flows

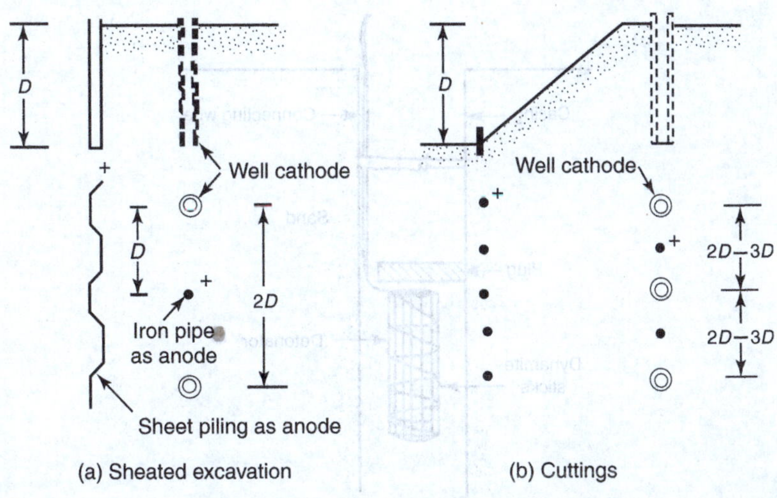


**Fig. 19.6** Electrode arrangements

anodes corrode considerably in the course of a few weeks of electro-osmotic treatment, they should be replaced as soon as the current drops to less than 30% of the initial consumption. Perforated tubes form cathodes and the cathode wells are connected to a pumping system.

Electro-osmosis is resorted to only to remedy a difficult situation where other methods have failed. This method has a high cost of installation and initial running cost. But the power consumption, and hence running cost, decreases considerably after the ground is stabilized.

## 19.5 VIBRATION METHODS

### 19.5.1 Vibro-Compaction

Vibro-compaction methods can be effectively used for rapid densification of saturated non-cohesive soils. Vibrations and shock waves in loose deposits of such materials cause liquefaction followed by densification and settlement accompanying the dissipation of excess pore water pressures. The effectiveness of these methods decrease with increase in the percentage of fines in the soil, since the fines reduce the permeability of the material, which is a prime factor for liquefaction. Similarly, the effect is less in partially saturated soils because of surface-tension forces.

**Blasting.** This technique consists of detonating a certain amount of explosive charge at a certain depth of the soil required to be compacted. Localized spontaneous liquefaction develops in saturated deposits due to sudden shock waves and soil grains get displaced. In dry soils too the shocks place the particles in a more irreversible way and the new density attained is permanent.

Pipes of about 100 mm in diameter are driven into the soil strata to the required depth. After removing the soil from inside the pipe, sticks of dynamite and an electric detonator are wrapped in water-proof bundles and lowered through the casing (Fig. 19.7). A wad of paper

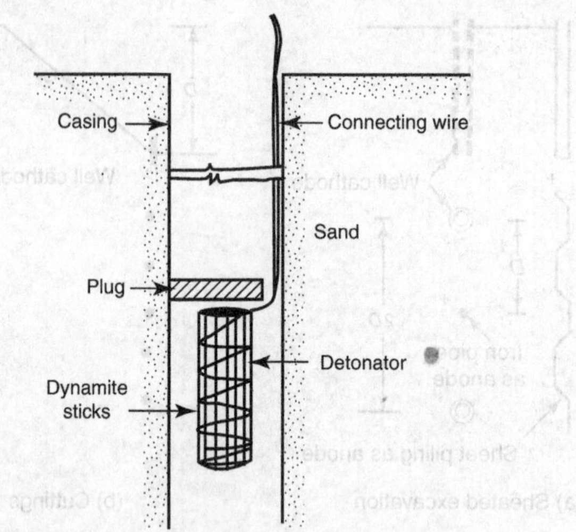


Fig. 19.7 Installation of explosives (*Source*: Mitchell, 1970)

or wood is placed against the charge of the explosives to protect it from misfire. The casing is withdrawn, and in order to obtain the full force of the blast, the hole is backfilled with sand. The electrical circuit is closed and the charge is fired. The surface settlements are measured by taking levels or from screw plates embedded at certain depths below the ground surface.

The correct amount of charge has to be used so that it is just enough to shatter the soil particles uniformly and at the same time prevent the formation of craters. Layman (1942) has suggested an empirical formula which can be taken as a rough guide.

$$W \approx 164CR^3$$

where $W$ is the weight of explosive ($N$), $C$ the coefficient (=0.0025 for 60% dynamite), and $R$ the radius of the sphere of influence (m).

To avoid cratering, the minimum depth of charge should be greater than $R$. Charge spacings less than 3 m should be avoided; values of 3 to 8 m are typical. The centre of charges should be located at a depth of two-thirds the thickness of the layer to be densified. It is found that repeated blasts of small charges are more efficient than a single large blast (Hall, 1962). If the depth of the deposits to be densified is 10 m or less, compaction is carried out in single tier only; for deeper deposits, more tiers need to be used. Charges should be exploded from the bottom-most tier in an upward direction in a uniform manner. The uppermost portion of the stratum is always loosened and can be compacted by any surface compaction device. The typical grid spacing and firing pattern as suggested by Mitchell (1970) is given in Fig. 19.8.

The blasting technique is less expensive and involves less time, labour, and equipment. Although blasting is one of the most economical stabilization methods, it suffers from the disadvantages of non-uniformity, potential adverse effects on adjacent structures, and the dangers associated with the use of explosives in populated areas.

**Vibrating Probe**. Vibrating probe, also known as Terra-probe, is a patented process used to densify loose sands. A 760 mm open-ended tubular probe is vibrated into the ground from a vibratory pile driver operating at 15 Hz. Vibrations of 10 to 25 mm amplitude are made in a

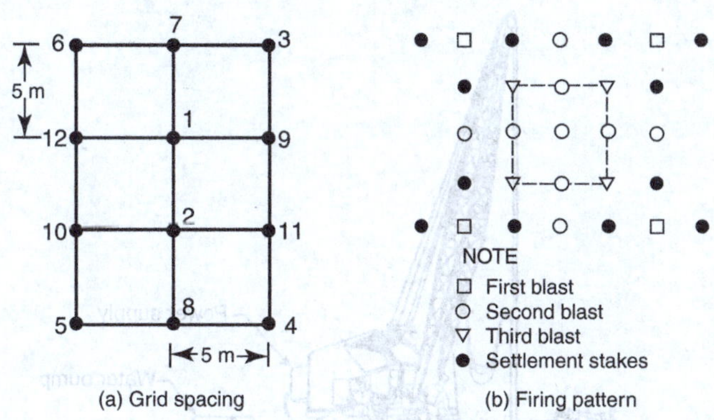


**Fig. 19.8** Typical charge spacing and firing (*Source*: Mitchell, 1970)

vertical mode. After reaching the planned penetration depth, the probe is withdrawn slowly while vibrations continue. Effective treatment can be obtained between depths of 4 m below the ground surface and about 20 m. The operation is most efficient where groundwater is within 2 to 3 m of the surface. Water jets can be attached to the probe, or ponding of the surface can be done to assist the penetration and densification.

Test sections of the order of 10 to 20 m on a side are desirable to evaluate the effectiveness and to determine the required spacing in any given case. A square pattern is often used, with a fifth probe at the centre of each square giving more effectively increased densification than a reduced spacing. The density achieved by this process is generally lower. Since the procedure does not require a sand fill, it can be applied effectively for offshore sites.

It is essential that lift thickness, soil type, and roller type be matched. If lift thickness is too great, then low-density layers will form alternately between high-density layers. If lift thickness is too small, then much of the effort is lost through repeated over-compaction of near-surface layers.

## 19.5.2 Vibro-Displacement Compaction

The methods described in this section are similar to those dealt with in the preceding section except that the vibrations are supplemented by active displacement of the soil and, in the case of compaction piles and vibro-flotation, by backfilling the zones from which the soil has been displaced.

**Compaction Piles.** Driving displacement piles at close spacings can densify highly permeable and partially saturated soils. One effective procedure is to drive a pipe pile with a false bottom to the desired depth. The pipe is then filled with sand or other backfill materials in lifts. Each lift of materials is compacted concurrently with withdrawal of the pipe pile. The backfill not only gets compacted but also expands laterally at the bottom, forming a broadened base which is also referred to as a sand pile. Compaction piles are economical for smaller sites of moderate depths up to 15 m.

In another rapidly developing technique, soft cohesive strata or loose non-cohesive deposits are replaced by granular material which is compacted by ramming or vibration. Such compacted piles are referred to as granular piles or stone columns. Different installation

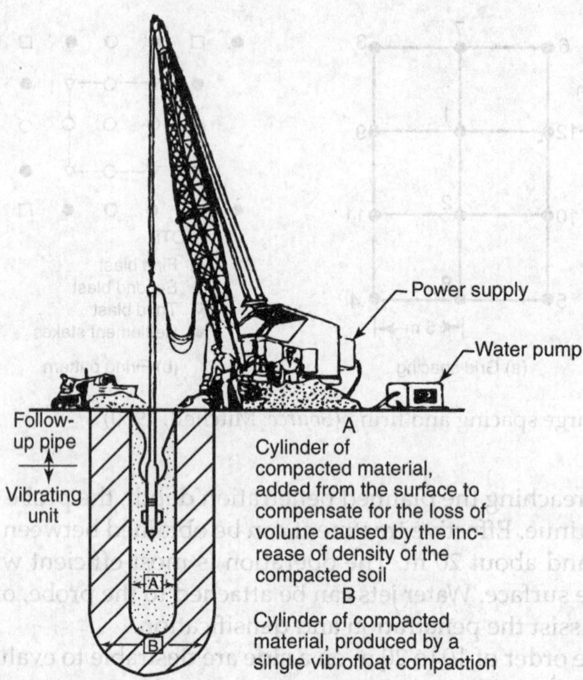


**Fig. 19.9** Vibro-flotation equipment (*Source:* Brown, 1976)

techniques are being adopted (Baumann and Bauer, 1974; Datye and Nagaraju, 1977; Ranjan and Rao, 1983). But most of the granular piles are installed adopting the vibration technique through a vibroflot (discussed in the next section). A comprehensive review and adoptability of stone columns in ground improvement are presented by Ranjan (1989).

**Vibroflotation**. Vibroflotation is a technique for densifying *in situ* non-cohesive soils with simultaneous vibration and saturation. This principle of densification was first published in 1936 by Steuerman in a Russian journal and later applied in Germany during 1939 for improvement of foundation soils for buildings.

The equipment required for vibroflotation involves a vibroflot probe, accompanying power supply, water pump, crane, and front-end loader (Fig. 19.9). The vibroflot probe is an essential piece of equipment consisting of a cylindrical penetrator, about 0.38 m in diameter and about 2 m in length, with an eccentric weight inside the cylinder developing a horizontal centrifugal force of about 100 kN at 1,800 rpm. A typical vibroflot consists of two parts. The lower part is the horizontal vibrating unit which connects to the upper part of the follow up pipe, the length of which can be varied depending on the compaction depth (Fig. 19.10). The water pump provides water to jet the vibroflot into the ground as the vibroflot is lowered with the crane. The front-end loader is used to supply the backfill material even as the *in situ* soils are densified.

The probe is freely suspended from a crane. Each compaction sequence has four basic steps (as suggested by Brown, 1976, and Vibroflotation Foundation Co., USA) (Fig. 19.11). They are as follows:

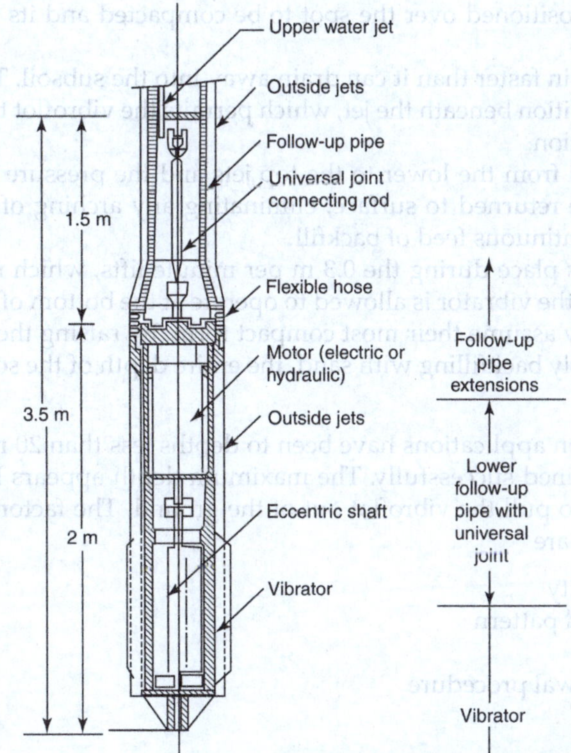


**Fig. 19.10** 100 HP vibroflot (*Source*: Brown, 1976)

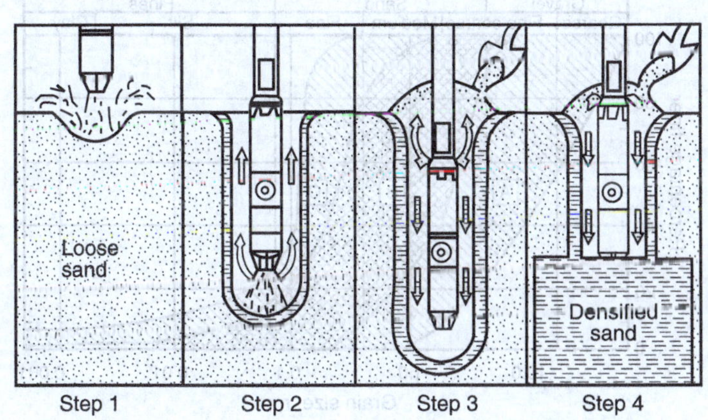


**Fig. 19.11** Vibroflotation compaction process (*Source*: Brown, 1976)

1. The vibroflot is positioned over the spot to be compacted and its lower jet is then fully opened.
2. Water is pumped in faster than it can drain away into the subsoil. This creates a momentary "quick" condition beneath the jet, which permits the vibroflot to settle due to its own weight and vibration.
3. Water is switched from the lower to the top jets and the pressure is reduced enough to allow water to be returned to surface, eliminating any arching of backfill material and facilitating the continuous feed of backfill.
4. Compaction takes place during the 0.3 m per minute lifts, which return the vibroflot to the surface. First, the vibrator is allowed to operate at the bottom of the crater. As the particles densify, they assume their most compact form. By raising the vibrator step by step and simultaneously backfilling with sand, the entire depth of the soil is compacted into a hard core.

Most vibroflotation applications have been to depths less than 20 m, although depths of 30 m have been attained successfully. The maximum depth appears limited mainly by the ability of the crane to pull the vibroflot out of the ground. The factors contributing to successful densification are

1. Equipment capacity
2. Probe spacing and pattern
3. *In situ* soil
4. Vibroflot withdrawal procedure
5. Backfill material
6. Workmanship

The two most important factors are the grain-size distribution of the soil and the nature of backfill material used. The range of the grain-size distribution of *in situ* soils suitable for vibroflotation (Brown, 1976) is shown in Fig. 19.12. The technique is best suited for densifying very loose sands below the water table that have grain-size distributions falling

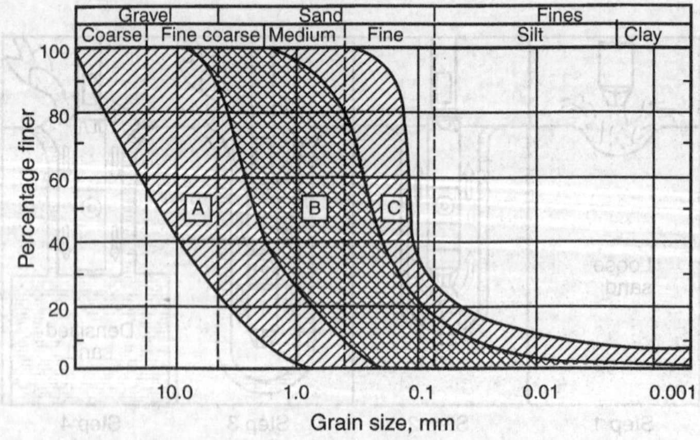


Fig. 19.12 Soils suitable for vibroflotation (*Source*: Brown, 1976)

entirely within Zone B. Layered clays, fine particles, cementation, and organics in the *in situ* soil pose a variety of difficulties for compaction by this technique. Soils with the grain-size distribution entirely within Zone C are very difficult to compact by vibroflotation. In general, the densities achieved and the zone of compaction decrease with increasing silt and clay contents. The fines and organics apparently damp out vibrations, stick the sand particles together between particles, and thus restrict the relative movement of particles necessary for densification.

Clay layers present in *in situ* soil also reduce the zone of compaction. Gravel, dense sand, and cemented sands are represented by Zone A. These soils have a reduced rate of probe penetration and the effect becomes still less when the water table is located at a greater depth. Hence, under these conditions, vibroflotation might prove to be uneconomical in the long run.

The suitability of backfill material depends on the gradation. Brown (1976) has developed a rating system to judge the suitability of the backfill material. The rating system is based on a suitability number defined as

$$\text{Suitability number} = 1.7\sqrt{\frac{3}{(D_{50})^2}+\frac{1}{(D_{20})^2}+\frac{1}{(D_{10})^2}}$$

where $D_{10}$, $D_{20}$, and $D_{50}$ are the particle sizes corresponding to 10%, 20%, and 50% finer. Table 19.1 gives the rating description. The withdrawal of the probe is also affected by the quality of the backfill material.

With conventional vibroflotation equipment, minimum relative densities in excess of 70% can be obtained for spacings up to about 2 m. Continuous square or triangular patterns are often used. Typical patterns for spread footings for allowable soil pressures up to 300 kN/m$^2$ are shown in Fig. 19.13.

**Heavy Tamping**. The most basic and simplest way of compacting loose soil is by repeated dropping of a weight on to the ground. Although this technique was used long ago, it has undergone rapid development after 1975 (Menard and Broise, 1975). The method, also known as deep dynamic compaction or deep dynamic consolidation, consists of allowing a very heavy weight (up to 400 kN) to fall freely on the ground surface from a height of 15 to 40 m. This leaves an impression on the ground. The tamping is then repeated either at the same location or over other parts of the area to be stabilized. In the case of non-cohesive

**Table 19.1** Backfill evaluation criteria

| Suitability number | Description of rating |
|---|---|
| 0–10 | Excellent |
| 10–20 | Good |
| 20–30 | Fair |
| 30–50 | Poor |
| >50 | Unsuitable |

*Source:* Brown (1976).

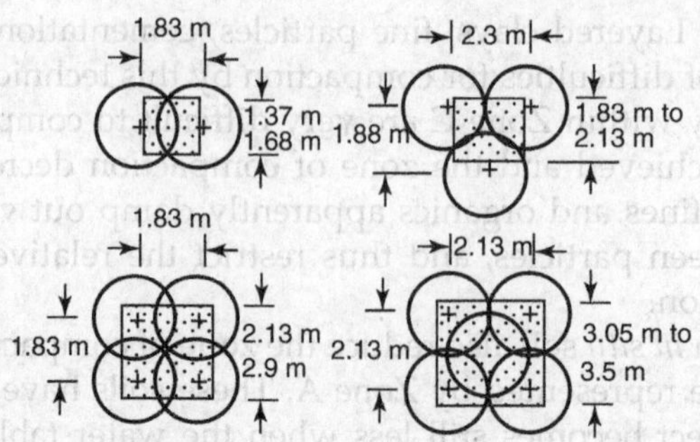


**Fig. 19.13** Typical vibroflotation patterns for footings (*Source*: Brown, 1976)

soils, the impact energy causes liquefaction, followed by settlement as water drains. Fissures formed around the impact points sometimes facilitate drainage in some soils. This method has been successfully used to treat various types of soils and fill deposits up to 20 m thick. This method can be adopted for densifying soils both above and below the water table. This technique produces equal settlements faster than if a static load is applied.

Since the variables involved are many, no satisfactory ideal rigorous model is available yet to understand the behaviour of dynamic consolidation. Only empirical information is available with regard to the depth of penetration of the compaction. Dobson and Slocombe (1982) have given an expression for the range of the effective depth in terms of the energy of impact as

$$1.26\sqrt{wh} < D < 3.16\sqrt{wh}$$

where $D$ is the effective depth (m), $w$ the weight being dropped (kN), and $h$ the height of drop (m).

As the analytical understanding of this process is in the formative stage, it is recommended to have a small test section at the site under consideration for necessary preliminary field evaluation (Koerner, 1985). Because of high-amplitude, low-frequency vibrations, a sufficient minimum distance should be maintained depending on the type of structure or facility.

## 19.6 PRE-COMPRESSION AND CONSOLIDATION

### 19.6.1 Pre-loading and Surcharge Fills

In this process, an earthfill or some other material is placed over the selected site. The amount of fill is sufficient enough to produce a stress in the soil equal to the one anticipated from the final structure. The soft soil is allowed to consolidate prior to construction. Since the consolidation takes a very long time, this method is suitable only for stabilization of thin layers.

If the thickness of fill placed for pre-loading is greater than that required for the final ground surface elevation, the excess fill is termed *surcharge fill*. Thus, both permanent fill and surcharge fill together cause a given amount of settlement in a shorter time than can the permanent fill alone, and hence the time required for stabilization is drastically reduced. Conventional consolidation theories can be followed to estimate the time required for consolidation. Both the primary consolidation and most of the secondary compression settlements can be taken out in advance by surcharge fills. Secondary compression settlements may be the major part of the total settlement of highly organic deposits or old sanitary landfill sites.

The rate of pre-load and surcharge fill placement has to be controlled depending on the bearing capacity of the soil. If the bearing capacity of the soil is inadequate, layers of fill can be placed only after a sufficient gain in shear strength is obtained. Geotechnical analyses can be carried out for predictions of the rates of consolidation, strength, and strength gain. The predicted values have to be checked from field measurements like piezometer readings and *in situ* strength tests.

The two main requirements for pre-loading are enough space and availability of fill material. Heaping of fill is the most common method of pre-loading although pre-loading can be successfully effected by the weight of water or by lowering the water table. Among the fill materials, granular soil is the most desirable because it does not turn into mud during rains. Ores and industrial products are generally satisfactory, but clayey soils are less desirable.

Pre-loading has been used successfully on virtually every type of naturally laid or man-made soil. Natural soils include loose sands and silts, soft silty clays, organic silts, and erratic alluvial deposits, whereas man-made fills may be miscellaneous depositions such as uncompacted dredged materials, industrial wastes, and rubbish sites previously used as urban dumps. Deposits to be handled with care are thick homogeneous layers of plastic clay and sanitary land fills.

The main advantage in the pre-loading is that the construction equipment needed is the same as that for simple earth-moving jobs. The additional equipment required for follow-up of pre-loading are relatively simple and inexpensive. By measuring ground movements, the effect of pre-loading can be immediately assessed. Pre-loading ensures uniformity of improvement because it eliminates local inhomogenities and it reduces considerably the danger of liquefaction by earthquake in sands. Compared with other methods of improving ground support, pre-loading costs much less (about 10% to 20% and about 20% to 40% with vertical drains). A detailed treatment of this subject can be found in Stamatopoulos and Kotzias (1985) where several field examples are given.

### 19.6.2 Vertical Drains

For deep clay deposits, pre-loading alone will take more time because of the long drainage path available for consolidation. An efficient way to do this is by providing vertical drains. Vertical drains are continuous vertical columns of pervious material installed in clayey soil for the purpose of collecting and discharging the water expelled during consolidation. Vertical drains with pre-loading will rapidly accelerate consolidation. A further aspect of the drains is that they reinforce the ground in which they are installed.

Vertical drains are mainly of two types: (i) sand drains, made by filling a cylindrical hole with sand and (ii) pre-fabricated drains, also known as "wickdrains" or simply "wicks." These drains can also be of flexible corrugated plastic pipe, wrapped inside a filter. For want of modern installation machines, these techniques are not yet available in India. However, one particular type of wick called "sandwick" (sand packed in filter stocking) has been used successfully in several installations in India (Dastidar et al., 1969; Som, 1975). Another type of drain called rope drain has been developed and used in several projects by the Central Buildings Research Institute (Mohan et al., 1977; Sengupta et al., 1980). The rope material consists of natural fibres such as coir. In such drains, the drainage capacity is the major constraint.

Until about 1950, most installed vertical drains were sand drains. Since 1980, fabricated drains have become popular because they cost less and can be installed quickly. In the technically advanced countries, it has been reported that the cost of pre-fabricated drains is about one-third the cost of sand drains. However, in India, only sand drains have been widely used (Datye and Nagaraju, 1975, 1976). Figure 19.14 shows a typical arrangement of vertical drains with drainage blanket and surcharge fill.

The holes required for installing sand drains are guarded against collapse by pipes or mandrels which are inserted by jetting, driving, rotating, or vibrating. The soil that originally occupied the space where the hole is being made is excavated by washing or augering (referred to as *non-displacement drains*) or, alternatively displaced downwards or sideways by driving closed bottom pipes (referred to as *displacement drains*). Under Indian conditions, displacement drains up to 400 mm in diameter can be installed by the equipment commonly used for driven cast in place of concrete piles (Datye, 1982). Adopting a modified technique, Datye (1982) reported that over 10,000 drains 200 mm in diameter have been successfully installed. The usual installation consists of 200 to 450 mm diameter sand drains installed at spacings of 2 to 5 m. Displacement drains are less expensive than augered or bored non-displacement drains.

In the installation of sand drains, the soil adjacent to the well is disturbed, causing a reduction in permeability. This effect is referred to as *smear*. Methods are available to account for smear effects based on the permeability and thickness of the smear zone.

Vertical drains are ineffective in fibrous organic deposits and in clay deposits with abundant pervious inclusions. Also, in sensitive clays, especially when drain installation is by the displacement methods, soil disturbance may result in high initial pore water pressures

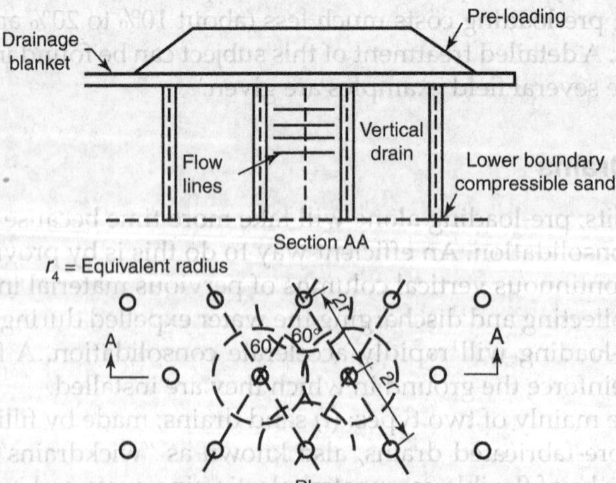


**Fig. 19.14** Arrangement of sand drains

and a zone of low permeability around the drain. However, it is certainly true that in many instances vertical drains have shortened the time required for soil stabilization.

### 19.6.3 Dynamic Consolidation

Heavy tamping, also referred to as dynamic consolidation, has also been used for cohesive soils. Pre construction settlements are usually of the order of two or three times the settlement predicted for the construction itself. This process also reduces the secondary compression. Treatment time is much less compared to surcharge loading with sand drains and the bearing capacity of clay increases by about 100% to 150%.

Because of the low permeability of cohesive soils, this technique takes more time than for non-cohesive soils. Each location is subjected to several blows and a rest period of 1 to 4 weeks is given before repeating the process. Drainage is facilitated by the radial fissures that form around impact points and by the use of horizontal and peripheral drains. Generally, the settlement is immediate. Because of time lapse, to be provided between successive cycles, only a limited area can be treated economically.

### 19.6.4 Electro-osmotic Consolidation

The application of electro-osmosis for de-watering purposes has been discussed earlier. The removal of water is an effective means of consolidation in fine-grained soils. Similar to a one-dimensional hydraulic field, a one-dimensional direct current field is formed due to the application of a current. The water flow rate is high in the initial stages, but decreases with time. It ceases when a hydraulic gradient, induced by water content variation, tending to cause a flow from the cathode towards the anode exactly balances the electricity-induced hydraulic gradient causing a flow from the anode towards the cathode. When this condition is satisfied, there is an increase in the effective stress. The amount of consolidation associated with this effective stress increase is obtained from a void ratio versus pressure relationship for the soil, determined in the usual manner. The rate of consolidation is governed by the same relationships that apply to consolidation under directly applied loading.

The efficiency of the method is decreased by gas generation and drying and fissuring at the electrode. The treated soils will have non-uniform changes in properties between electrodes because the induced consolidation depends on the voltage, and the voltage varies between the anode and cathode. In order to achieve a more uniform stress condition, reversal of the polarity of the electrodes may be desirable. Pre-loading or surcharge fill techniques may be combined with electro-osmosis to accelerate the consolidation.

## 19.7 GROUTING AND INJECTION

Grouting is a process whereby stabilizers, either in the form of a suspension or solution, are injected into sub-surface soil or rock for one or more of the following applications:

1. control of groundwater during construction;
2. void filling to prevent excessive settlement;
3. strengthening adjacent foundation soils to protect them against damage during excavation, pile driving, etc.;

4. soil strengthening to reduce lateral support requirements;
5. stabilization of loose sands against liquefaction;
6. foundation underpinning; and
7. reduction of machine foundation vibrations.

That the basic function of grouting is to serve should always be kept in mind when considering different aspects of the subject. The three basic functions are the following (Koerner, 1985):

1. *Permeation* or *penetration:* The grout flows freely with minimal effect into the soil voids or rock seams.
2. *Compaction* or *controlled displacement*: In this condition the grout remains more or less intact as a mass and exerts pressure on the soil or rock.
3. *Hydraulic fracturing:* Hydraulic fracturing or uncontrolled displacement occurs when the grouting pressures are greater than the tensile strength of the soil or rock being grouted; then, the latter material fails, and the grout rapidly penetrates into the fracture zone.

### 19.7.1 Suspension Grouts

Suspension grouts consist of solid particles like soil, cement, lime, asphalt emulsion, etc. carried in water. The solution grouts are numerous, viz., aqueous, non-aqueous, colloidal, etc. Particles in a suspension grout are of silt size, and hence these materials cannot be injected into the pores of soils finer than medium to coarse sand sizes. For successful grouting of soils, it has been established (Kravetz, 1958) that

$$\mathrm{GR} = \frac{(D_{15})\ \text{formation}}{(D_{85})\ \text{grout}} > 20$$

where GR is the groutability ratio, $D_{15}$ the particle size corresponding to 15% finer of the formation, and $D_{85}$ the particle size corresponding to 85% finer of the grout.

The criterion basically limits the use of suspension grouting to permeation of sands and gravels. Other considerations that must be taken into account in grouting design are the grout's setting time and its stability.

**Grouting with Soil**. Soil itself can be used to fill up some of the volumes in coarse-grained deposits. Even fine sands and silts used for this purpose settle down quite quickly after injection. The soil to be used as a grout should be a very fine-grained soil. Bentonite clay is the commonly used material. Viscosity, strength, and flow properties of a bentonite clay can be adjusted to suit the situation.

No flow of soil-grout occurs when the water-to-soil ratio is kept very low. Pressure is then exerted by the grout against the soil mass from the outlet of the grout pipe. This causes the densification and movement of the grout to adjacent areas. This technique was originally known as mudjacking and quite often used to raise pavement slabs or to underpin shallow building foundations.

To prevent a blow-out during grouting, the grouting pressure is generally limited to about 20 kN/m$^2$ per metre of depth. Higher pressures are used when grouting under heavy structures or in other situations where greater confinement exists. Sometimes, higher grouting pressure can be used by working from the outside of the area to be treated. In certain cases, high pressures are deliberately applied to widen or increase the fracture, thus providing added channels through which the grout can flow.

**Grouting with Soil–Cement Mixes**. Soil in combination with a stabilizing material, e.g., cement, would do better than soil alone. Grouts may have different properties depending on the amount and type of soil, cement, and water they contain. The viscosity of a grout depends on solid-to-water ratios and different cement-to-soil ratios.

In soil–cement systems, volumes of soil between four and six times the loose volume of cement are common. The volume of mixing water varies from about three-fourth to twice the volume of clay per bag of cement in cement–clay grouts, and from about one-third the loose volume of sand per bag of cement to an equal volume of the same in cement–sand grouts. Water–cement ratios in the range of 0.5:1 to 5:1 have been in use. The lower this ratio, the less likely cement segregation and filtering will be, but injection will be more difficult and the friction losses in the pumping system will be greater.

High viscosity systems with very low water content can be used as displacement grouts. The advantage of these soil-cement mixes over soil alone is the permanence of the grout but their dis-advantage is an increase in cost. The other factors, such as equipment, pressure, and pumping rates, however, are roughly the same.

**Grouting with Cement**. Cement grouting has been widely used, more often in seepage cut off beneath dams, but also in groundwater control in certain cases. Cement grouts are usually made from Portland cement and water. Sometimes the cement particles come out of the suspension before complete curing of the grout has occurred. This phenomenon is referred to as bleeding. This behaviour is more pronounced when water content and fluidity of the grout are greater.

When selecting a particular cement grout for use, one would obviously like to know its final strength, flow rate, set time, shrinkage, permeability, and durability. Other ingredients that are sometimes used in cement mixes are fine sand, clay (as discussed in the previous section), fly ash, fluidizers, accelerators, and retarders or expansion additives.

**Grouting with Lime**. Grouting with lime is a special form of grouting – pressure injected lime has been increasingly applied, especially for the stabilization of expansive soils for foundations of light structures. In this process, a lime slurry, containing 3 to 4 N weight of lime per litre of water plus a surfactant, is injected under high pressure (350 to 1,400 kN/m$^2$). Treatment locations may be spaced at 1 to 2 m laterally and 0.3 to 0.5 m vertically up to the depth of seasonal moisture variation. Pumping of the grout is continued at each depth until refusal, or until the slurry runs out at the ground surface. About 120 litres of grout per metre depth is quite usual.

The method is best suited for expansive soils with cracks, fissures, slickensides, fractures, and root holes. These passages, in conjunction with channels formed by hydraulic fracturing under higher injection pressure, provide channels for the slurry throughout the soil. Lime reacts with the soil adjacent to the cracks forming moisture barriers to protect the unreacted soil blocks against volume change. Free penetration of grout into soil pores is difficult because of the large lime particle size and small bore size.

**Displacement Grouting**. Displacement or compaction grouting is a specialized technique used for controlled densification of *in situ* soils at depth. This technique is not the same as conventional grout filling within a soil mass by penetrating it with a cementing material. The basic concept of compaction grouting is that of injecting a growing "bulb" of grout that

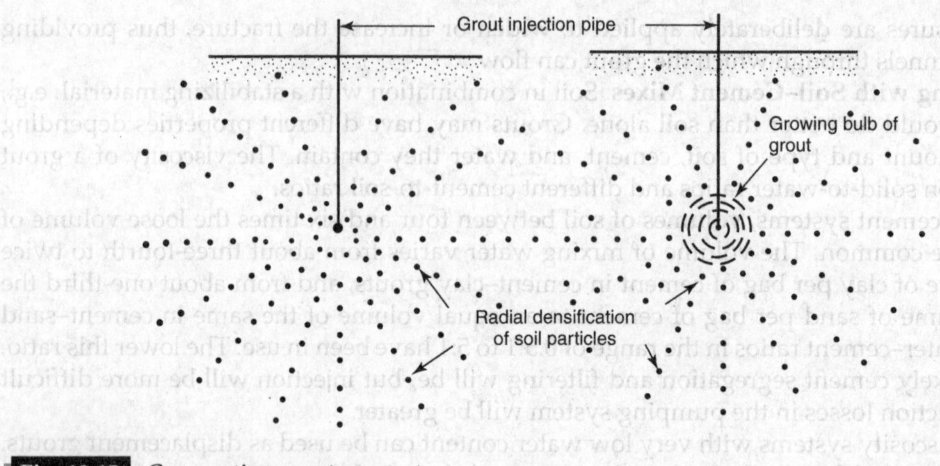


Fig. 19.15 Compaction grout concept

acts as a radial hydraulic jack, displacing the surrounding soil particles and thus radially compacting the soil from the point of injection (Fig. 19.15).

Most displacement grouts are composed of a cement–sandy-loam mixture containing three to five sacks of cement per cubic metre of soil. After thorough mixing, the materials are pumped using a mud jack. Equipments are available to pump a zero slump grout up to 30 m depth under a high pressure of 2,700 kN/m$^2$. Most applications of a displacement grout are for correction of differential settlements. This technique is more suitable in partially saturated cohesive or organic soil masses, silts, sands, and soils containing void pockets (Mitchell, 1970).

## 19.7.2 Solution Grouts

The advantages of solution grouting that serve to offset its high cost are (i) absence of particulate material, (ii) low viscosity, and (iii) control over setting time. Solution grouting is done using "one-shot" or "two-shot" systems. In the one-shot system, all required chemicals are injected together after re-mixing. Setting times are controlled by varying the catalyst concentration according to the grout concentration, water composition, and temperature. The two-shot system, wherein one chemical is injected followed by injection of a second chemical which reacts with the first to produce a precipitate in the soil pores, may also be used. Two-shot systems are slower and require higher injection pressures and more closely spaced grout holes.

The electro-osmotic driving force was used by Karpoff in 1953 as a means for injecting a stabilizer into a soil. In this method, the stabilizer is introduced at the anode and carried towards the cathode by electro-osmosis. For this direct current, electrical gradients of the order of 50 to 100 V/m were required. Such techniques may be attractive in cases in which the permeability of the soil is so low as preclude injection under high pressures or in situations where large pressures cannot be used. The method is likely to be expensive and effective only in special cases (Mitchell, 1970).

## 19.8 CHEMICAL STABILIZATION

Chemical stabilization in the form of lime, cement, fly ash, and a combination of the above is widely used in soil stabilization to (i) reduce the permeability of the soil, (ii) improve shear strength, (iii) increase bearing capacity, (iv) decrease settlement, and (v) expedite construction. Many cases of successful chemical stabilization attempts have been described in the literature. Chemical stabilization may be used for surface soils more successfully. Surface treatments are common in connection with the sub-grades or bases for pavement construction. Mixture of soil and chemicals are mixed either mechanically in place or by batch process. The optimum benefit of using these agents in stabilization must be determined by laboratory testing. The general principles of these admixtures as stabilizers are discussed below.

### 19.8.1 Lime Stabilization

Lime is an effective agent to be mixed with fine-grained soils with high plasticity. Hydrated high calcium lime [$Ca(OH)_2$], calcitic quick lime [CaO], monohydrated dolomitic lime [$Ca(OH)_2$, MgO], and dolomitic quick lime are the commonly used limes for stabilization. Addition of lime to soil causes chemical reactions, such as cation exchange and flocculation–agglomeration. In these reactions, the monovalent cations in clays are replaced by divalent calcium ions. Further, these reactions make highly active clays inactive, changing the texture of the clay soil. In general the reactions reduce the plasticity of the soil and thus improve the strength and deformation properties of the soil.

The percentage of lime for effective action is in the range of 5% to 10%. The first 2% to 3% lime by weight has a substantial influence in improving the workability and property of the soil. Lime stabilization can be applied in the field by any one of the following processes: (i) the *in situ* material (and borrowed material, if necessary) is mixed with the proper amount of lime at the site and then compacted after addition of moisture, (ii) the soil and the amount of lime with water are mixed at a plant and then hauled back to the site for compaction, and (iii) lime slurry can be poured through small drill holes or pressure injected to a depth of 3 m at a spacing of 2 m depending on the situation. Both hydrated lime and quick lime can be used. Quick lime has to be used cautiously as it may affect the tools and working personnel. The curing period has been reported to affect the strength of lime-stabilized soils considerably (Ramana Sastry, 1989).

### 19.8.2 Cement Stabilization

Cement has been widely used to stabilize sandy and clayey soils. Similar to lime, cement has the effect of decreasing the plasticity and increasing the strength of the soil. Stabilization is most effective in soil with less than about 40% of it has particle size of 2 μm. The percentage of cement (by volume) required for the stabilization of sandy soil is in the range of 6 to 10, for clays and silts of low plasticity it is 8 to 12, and for highly plastic clays it is about 10 to 14. Clays with calcium ions are easily stabilized by cement, whereas clays with sodium are better stabilized by lime stabilization.

Cement stabilization as a surface treatment is used in highway and runway sub-grades. For field compaction, the mixing may be done at the site or in a mixing plant and then transferred to the site. The mixed soil is then compacted to the required density with a predetermined quantity of water.

### 19.8.3 Fly Ash Stabilization

Fly ash is a by-product of the pulverized coal combustion process and has silica, alumina, and various oxides and alkalis as its constituents. It is fine-grained and pozzolanic in nature. Fly ash reacts actively with hydrated lime and hence is used in combination with lime as a stabilizer. A mixture of about 10% to 35% of fly ash and 2% to 10% of lime forms an effective stabilizer for the stabilization of highway bases and sub-bases. Soil–lime–fly ash mixes are compacted under controlled conditions with an adequate quantity of water.

## 19.9 SOIL REINFORCEMENT

Soil reinforcement is not a new concept but the principles already exist in nature and are demonstrated by animals, birds, and plants. The scientific basis for the modern concept of soil reinforcement lies in the idealization of the problem of soil reinforcement, wherein a weak soil is reinforced by high-strength thin horizontal membranes (Westergaard, 1938). The modern form of soil reinforcement was first applied by Vidal (1969). According to Vidal's concept, the interaction between the soil and the reinforcing horizontal membranes is solely due to friction generated by gravity. The first major retaining walls using the Vidal concept wre built in France in 1968. The new technology has been widely used in Europe and the USA. However, this technique has not yet become popular in India, except in a few cases (e.g., Nagaraj et al., 1982), the constraining factor being the non-availability and cost of reinforcing materials (Datye, 1982).

Reinforced soil is somewhat analogous to reinforced concrete. But a direct comparison between the functions of reinforcement in the two cases is not valid. The mode of action of reinforcement in soils is not through carrying developed tensile stresses, as in reinforced concrete, but rather of anisotropic reduction of the normal strain rate (Jones, 1985).

Jones (1985) identifies several soil reinforcement field applications, viz., in bridge works, dams, embankments, foundations, highways, housing, industries, military applications, railways, root pile systems, pipe works, and waterway structures. Some of the applications are shown in Fig. 19.16.

A wide variety of materials are in use as reinforcing materials. Early structures were formed using organic materials such as timber, straw, or reed for reinforcement. As these materials are less durable, new materials such as steel, concrete, glass fibre, rubber, aluminium, and thermo plastics have been used successfully.

Durability of reinforced soil is the basic requirement; hence sufficient judgement has to be exercised while selecting the reinforced material. Resistance to corrosion is the most significant factor affecting the durability. High alloy steel, aluminium, glass-fibre reinforced plastics (GRP), and geosynthetics (discussed in the next section) are non-corrosive and in general have long lives.

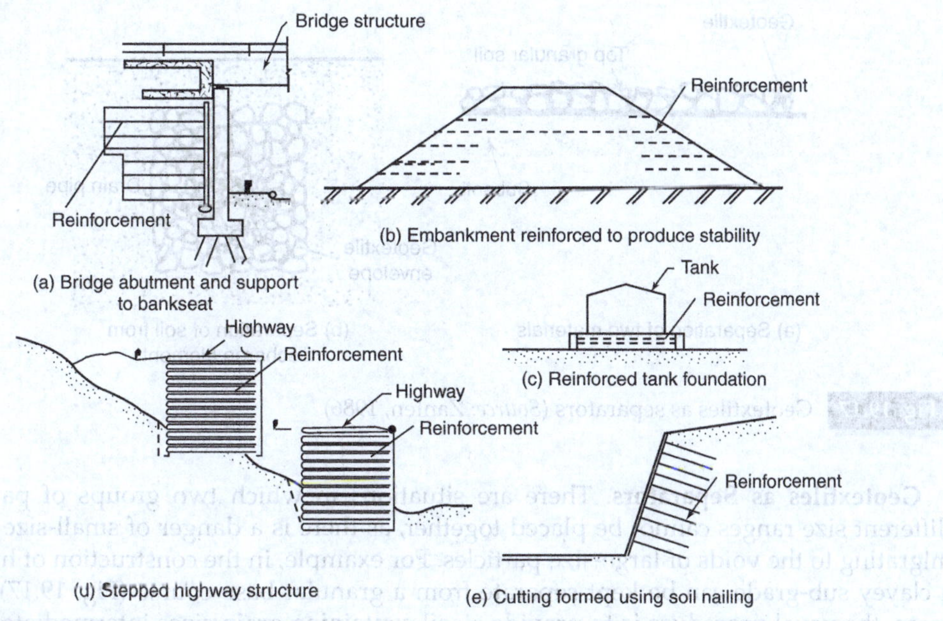


**Fig. 19.16** Some applications of soil reinforcement (*Source:* Jones, 1985)

## 19.10 GEOTEXTILES AND GEOMEMBRANES

### 19.10.1 Geotextiles

Geotextiles are porous fabrics manufactured from synthetic materials that are primarily petroleum products and others, such as polyester, polyethylene, polypropylene and polyvinyl chloride, nylon, fibre glass, and various mixtures of these. They are manufactured in thicknesses ranging from 10 to 300 mils (1 mil = 0.0254 mm), widths up to 10 m, and roll lengths up to about 600 m. Geotextiles are termed filter fabrics and have permeabilities comparable in range from coarse gravel to fine sand.

Geotextiles are manufactured in a variety of patterns, the most common methods being (Koerner, 1985)

- woven-made from continuous mono-filament or slit-film fibres;
- non-woven-made from continuous or staple fibres joined at the fibre cross-over by mechanical, thermal, or chemical bonding;
- grid-made from a sheet of polymer, punched and then elongated in at least one direction; and
- hybrid combinations of any of the above materials and techniques.

Geotextiles have been used in a variety of civil engineering works. Thus, in the selection of a proper geotextile, due importance has to be given to the major function that the geotextile is intended to perform. The four major functions are soil separation, filtration, drainage, and reinforcement. They are explained briefly below.

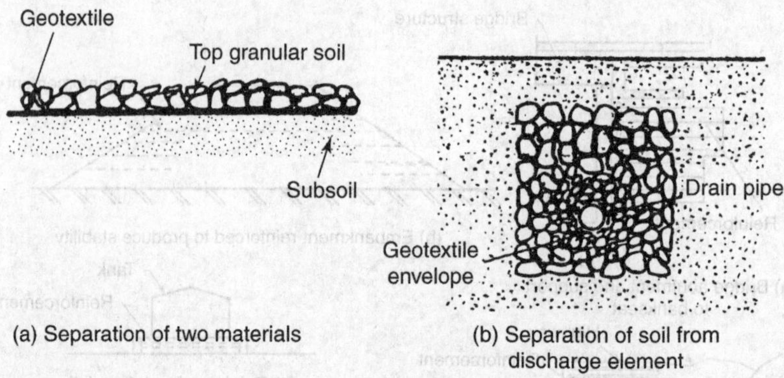


**Fig. 19.17** Geotextiles as separators (*Source:* Zanten, 1986)

**Geotextiles as Separators**. There are situations in which two groups of particles of different size ranges cannot be placed together, as there is a danger of small-size particles migrating to the voids of large-size particles. For example, in the construction of highways, a clayey sub-grade can be kept separate from a granular base course (Fig. 19.17). In such cases, the usual procedure is to provide a soil containing grain sizes intermediate between the soils. If there is a wide variation of grain sizes between the two soils, more than one intermediate grain-size soil has be provided. The obvious alternative in such a situation is the use of a geotextile. The advantages of using geotextiles as separators over single or multiple soil layers are the following: (i) simplicity in the construction, (ii) less excavation required, (iii) less weight placed, and (iv) short time required for construction.

**Geotextiles as Filters**. Geotextiles can be used more effectively as filters. Geotextile filtration occurs in fabrics where the flow of water transports some of the fine-grained particles of the protected soil onto the surface of the geotextiles. This modification in the soil and void of the fabric attains equilibrium after sometime. Only then does clear water start passing through the fabric. For relatively thin geotextiles, most of the filtering takes place within the soil, upstream from the fabric. This application of geotextiles has been widely used. Figure 19.18 signifies some of the applications of geotextiles as filters.

**Geotextiles as Drains**. Geotextiles themselves function as drains because they have a higher water transporting capacity than that of the surrounding material. Geotextile drainage occurs

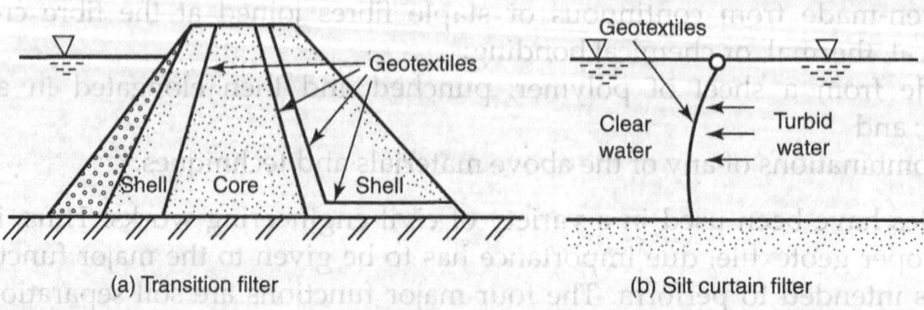


**Fig. 19.18** Geotextiles as filters (*Source:* Zanten, 1986)

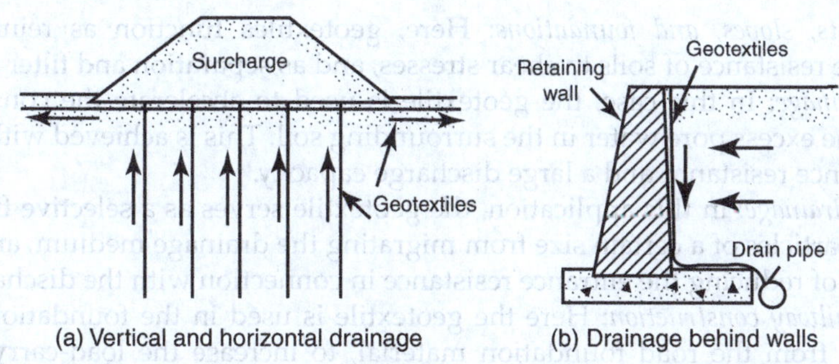


Fig. 19.19 Geotextiles as drains (*Source:* Zanten, 1986)

either cross-plane, when functioning primarily as a filter, or in-plane, when water is transmitted within the geotextile structure itself. In the latter case, a bulky geotextile or a composite system is needed. Figure 19.19 represents typical cases of application of geotextiles as drain.

Both the filtration and drainage systems have the following advantages: (i) faster installation, (ii) less soil to excavate and dispose of, (iii) less load, and (iv) greater system stability.

**Geotextiles as Reinforcement**. Since the tensile strength of soil is less, geotextiles, which have high tensile strengths, can contribute to the load-bearing capacity of the soil. Thus, geotextiles perform the function of reinforcement in soils. This application has solved many construction problems on soft and compressible soils. The most important use has been in road construction on soft sub-grades. Geotextiles are also used to reinforce walls and embankments. Figure 19.20 represents some of the applications of geotextiles as reinforcement.

**General Applications of Geotextiles**. Geotextiles can also be used in the following situations (Zanten, 1986):

1. *Bank and bed protection*: In this application, the geotextile protects the underlying material against excess pore water pressure. The filtering material is water permeable during its life time. This construction functions in a subsidiary way as a separating sheet and reinforcement.

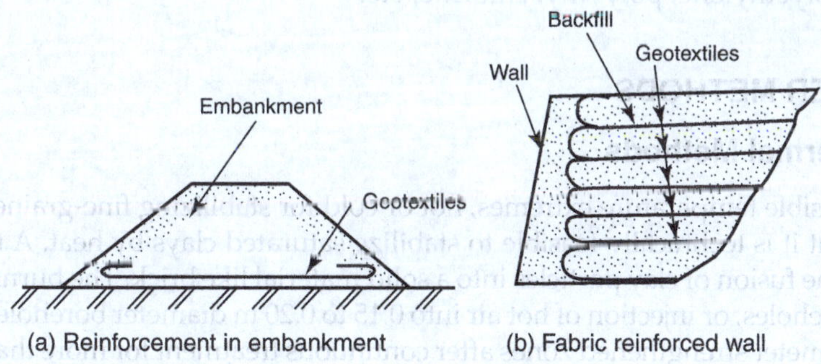


Fig. 19.20 Geotextiles as reinforcement (*Source:* Zanten, 1986)

2. *Embankments, slopes, and foundations*: Here, geotextiles function as reinforcement to increase the resistance of soils to shear stresses, and as separation and filter media.
3. *Vertical drainage*: In this case, the geotextile is used to accelerate the consolidation by reducing the excess pore water in the surrounding soil. This is achieved with drains with a low entrance resistance and a large discharge capacity.
4. *Horizontal drainage:* In this application, the geotextile serves as a selective filter that prevents soil particles of a certain size from migrating the drainage medium, and it also acts as a means of reducing the entrance resistance in connection with the discharge of water.
5. *Road and railway construction*: Here the geotextile is used in the foundation to separate the subsoil from the road foundation material, to increase the load-carrying capacity of the subsoil, and to drain surface water into the subsoil.

### 19.10.2 Geomembranes

Geomembranes are thin materials with very low permeability. They are flexible and are manufactured from synthetic or bituminous products. They may be strengthened, if necessary, with a fabric or film. Geomembranes differ from geotextiles with reference to the rate of permeability. The permeability is high (as in sand) in geotextiles and very low in geomembranes (as in bentonite or colloidal clay). For all practical purposes, geomembranes may be considered to be impermeable to both gases and fluids. This makes them ideal for constructing waterproof or gasproof barriers between adjacent bodies of soil, or soil and fluid.

Geomembranes are used in the following situations (Zanten, 1986):

1. Sealing against fluid percolation – e.g., on sea coasts, river banks, shipping canals and locks, reservoirs, terrain bunding, etc. The function of geomembranes in these cases is to form a barrier between the water and the surroundings and to ensure that water transport is reduced to a minimum.
2. Buffers against pollutants – e.g., permanent or temporary storage of waste products, waste-water treatment plants, basins for use in emergencies, roads in areas used for extraction of groundwater, etc. In these cases, the function of the geomembrane is to create a barrier between two media and prevent any mixing of these media.

As stated above, geomembranes are manufactured from synthetic (thermoplastic) or bituminous products. Some of the synthetic materials used are high-density polyethylene, low-density polyethylene, polyvinyl chloride, etc.

## 19.11 OTHER METHODS

### 19.11.1 Thermal Methods

Of the two possible temperature extremes, hot or cold for stabilizing fine-grained soils, heat is rarely used. But it is technically feasible to stabilize saturated clays by heat. A temperature of 100°C causes the fusion of clay particles into a solid material like brick. The burning of liquid or gas fuels in boreholes, or injection of hot air into 0.15 to 0.20 m diameter boreholes, can produce 1.3 to 2.5 m diameter strengthened zones after continuous treatment for more than 10 days. Dry or partly saturated weak clayey soils and loess are well-suited for this type of treatment. The economics of using heat, however, precludes its use in most construction projects.

Ground freezing appears to be gaining popularity in recent years. Frozen soil is far stronger and less pervious than unfrozen ground. Stabilizing soils by freezing has the following advantages: (i) high strength of stabilized soil, (ii) low permeability in the stabilized soil, (iii) barrier to seepage flow, (iv) protection from soil deformation, (v) noiseless operation, and (vi) applicability to a wide range of soils. Ground freezing is accomplished by bringing a refrigerant into the proximity of soil pore water. The pore water may be stationary or moving with a velocity of 2 m per day. The pore water around the refrigerant pipes freeze, and continued freezing yields a continuous wall of ice.

Ground freezing has applications in temporary underpinning and excavation stabilization. Other applications are in backfreezing of soil around pile foundations in permafrost and in maintenance of frozen soil under heated buildings on permafrost. Design on frozen ground involves properties of the frozen ground, heat flow, transfer of water to ice, and design of a refrigeration system.

### 19.11.2 Moisture Barriers

Expansive clays and partly saturated, lightly cemented soils, such as loess, can provide excellent foundation support for light structures, provided they are maintained relatively dry. Moisture barriers are sometimes used to keep water from critical zones under foundations.

Geomembranes discussed earlier can act as moisture barriers more effectively than sand or gravel blankets, and trenches may also act as effective moisture barriers, provided they can be maintained at a low degree of saturation. The hydraulic conductivity of such unsaturated coarse layers is so low as to effectively prevent significant water transfer across the layer.

### 19.11.3 Pre-wetting

One technique for the stabilization of expansive soils that can be effective under light structures, such as dwellings, is to flood the area prior to construction. Successful ponding is facilitated initially by the natural soil. After successful pre-wetting, the soil has a water content closer to that to be attained after construction; hence, subsequent volume changes are small.

Lime treatment of the surface layer to a depth of 0.3 to 0.5 m after ponding may be beneficial. This treatment provides a working platform for construction and an impermeable moisture barrier to retard subsequent desiccation of the pre-wetted soil.

### 19.11.4 Addition or Removal

The engineering property of a soil can be significantly changed by adding some selected soil or by removing some selected fraction of the soil. Sometimes poor soil may be removed and replaced with the same soil treated by compaction, with or without admixture. In general, the addition–removal technique of soil stabilization may work out economical in the absence of groundwater. Invariably, all the inorganic soils can be processed and treated to form an acceptable construction medium. This technique is not suitable for highly organic soils, peats, and sanitary fills. No generalized procedure can be given for this technique as it depends entirely on the problem and the site condition.

## POINTS TO REMEMBER

19.1 Soil improvement in its broadest sense is the alteration of any property of a soil to improve its engineering performance. This may be either a temporary process or a permanent measure to improve the performance of the completed facility.

19.2 Surface compaction is one of the most widely used techniques and is also one of the oldest techniques of soil densification.

19.3 Drainage methods like well-point systems, deep-well drainage, etc. are adopted to control the groundwater entry into the construction site, as a temporary or permanent measure, thereby ensuring a safe and economical construction scheme.

19.4 Vibration techniques produce shock waves which cause liquefaction followed by densification and settlement accompanying the dissipation of excess pore water pressure. These methods will be effective only in coarse-grained soils with less fines.

19.5 Compression of fine-grained soils is effectively done by pre-loading and surcharge fills, by installing vertical drains, and by dynamic consolidation.

19.6 Grouting is a process whereby stabilizers, either in the form of suspension or solution, are injected into sub-surface soil or rock fissures to control groundwater intrusion, prevent settlement, increase the strength of soil, etc.

19.7 Chemical stabilization uses lime, cement, fly ash, and a combination of the above for soil stabilization.

19.8 Geotextiles are porous fabrics manufactured from synthetic materials which are nowadays used for four major functions, viz., soil separation, filtration, drainage, and reinforcement.

19.9 Geomembranes are flexible materials with very low permeability, which are manufactured from synthetic or bituminous products. They are used for sealing against fluid percolation and as buffers against pollutants.

## QUESTIONS

### Objective Questions

19.1 State whether the following statements are true or false:

1. Well points are generally recommended where the water table needs to be lowered only by a small depth.
2. When an electrical gradient is applied in a saturated soil, water moves towards the anode.
3. Fine particles reduce the permeability of a material which is a prime factor for liquefaction.
4. Vibratory rollers are the best and most economical for attaining high density in cohesionless soils.
5. Suitability of a backfill material in vibroflotation is independent of the gradation of the material.

19.2 Fine-grained soils are more troublesome to drain because of
1. Small-sized grains which form small voids
2. High surface area which attracts more water
3. Presence of clay minerals
4. Capillary forces acting on the pore water

19.3 Indicate the incorrect statement. Injection method is used to
(a) Increase the strength
(b) Reduce the permeability
(c) Increase the grain size
(d) Reduce the compressibility of the soil

19.4 Generally, grouting can be used if the permeability, *k*, of the deposit is
(a) $<10^{-3}$ m/s
(b) $>10^{-3}$ m/s
(c) $<10^{-5}$ m/s
(d) $>10^{-5}$ m/s

19.5 Stone columns are not applicable in deposits of
(a) Highly organic silts or clays
(b) Loose sands
(c) Inorganic clays
(d) Inorganic silts

19.6 To avoid cratering in a blasting technique the minimum depth of charge should be ______ the radius of the sphere of influence.
(a) Equal to
(b) Greater than
(c) Less than
(d) Twice

19.7 The vibro-flotation technique is best suited for densifying
(a) Very loose sands below the water table
(b) Layered clays and fine cemented particles
(c) Organic clays with fine silt
(d) Coarse sands and gravels

19.8 The installation of sand drains causes the soil adjacent to the well to
(a) Reduce in shear strength
(b) Increase in porosity
(c) Reduce in permeability
(d) Increase in compressibility

## *Descriptive Questions*

19.9 Compare the use of sheep's foot and vibratory rollers in the surface compaction of granular soils.

19.10 In electro-osmotic stabilization, what are the different types of anodes used? What types of electrodes are used in the marine environment?

19.11 Why is cratering undesirable in the process of densifying granular soils by blasting with explosives?

19.12 How does deep dynamic compaction densify granular soils? Describe the influence of water content in the process.

19.13 Evaluate the technique of vibroflotation, compaction piles, and terraprobe with reference to equipment, time for compaction, and the maximum density achievable.

19.14 Considering the groutability of various types of soils, what method do you recommend to grout fine-grained soils?

19.15 Discuss the advantages of using fly ash in cement grouting over naturally available soils.

19.16 Describe with illustrations the differences between geotextiles and geomembranes.

19.17 Discuss the various methods used to control groundwater in excavations of soft clays. Indicate their relative suitability.

19.18 Describe a method suitable to stabilize a highway foundation in hilly terrain with high rainfall.

19.19 Two earth dams, each 100 m high, are to be constructed, on foundations whose soil properties are shown below. Discuss and suggest a foundation stabilizing technique in each case.

| Dam | A | B |
|---|---|---|
| Coefficient of permeability | 120 m/year | 0.6 m/year |
| Nature of soil | Medium sandy soil $D_r = 58\%$<br>$w_L = 60\%$ | Soft clay |
| Shear parameters | $c = 0, \phi = 32°$ | $c = 0.04\ \text{N/mm}^2, \phi = 2°$ |

19.20 How can the horizontal spacing of reinforcing strips be designed for the material in a retaining wall?

19.21 A highway alignment passes through a region where the subsoil is a highly compressible clay. Describe any one technique by which the consolidation of this clay can be hastened so that the construction of the road can be completed early.

## EXERCISE PROBLEMS

19.1 A grain-size analysis of a backfill material for a vibroflotation work yielded the following characteristic grain sizes: $D_{10} = 0.10$ mm, $D_{20} = 0.18$ mm, and $D_{50} = 0.55$ mm. Find the suitability number and give the rating of the material.

19.2 A surcharge fill has a volume of 6,000 m$^3$ and is placed at a dry unit weight of 20.0 kN/m$^3$. The borrow source for the fill has a dry unit weight of 15.6 kN/m$^3$ and $G = 2.68$. Estimate the volume of material required from the borrow to make the surcharge fill.

# 20

# Embankment Dams

**CHAPTER HIGHLIGHTS**

Types of embankment dams – Components of embankment dams – Other embankment details – Flow nets for earth dams – Design criteria for earth dams: safety against over-topping – Control of seepage and pressure in earth dams – Stability of upstream and downstream slopes – Selection of dam section: Earth dams – Rockfill dams

## 20.1 INTRODUCTION

Embankment dams are water impounding structures. These are flexible structures which can deform slightly to conform to the deflection of the foundation. Embankment dams are primarily made out of earth and rock fragments. Thus, the term embankment encompasses both earth and rockfill dams. These dams are also referred to as dykes or simply embankments or banks. Earth dams that are used to confine flood waters are called levees or guide banks. Dams that are both safe and economical can be constructed at a given site with the available materials.

Since the dawn of history, man has built dams of earth and rocks. Numerous tanks of reservoirs found in South India are over 2,000 years old. With modern technology earth dams of varying heights have been constructed in different parts of India, such as Ram Ganga Dam, Kishan Dam, Kothar Dam, Nagarjuna Sagar Dam. It is often said that the largest structure ever built by man is an earth dam.

## 20.2 TYPES OF EMBANKMENT DAMS

Based on the type of construction, embankment dams may be classified as rolled fill or placed earthfill dams and hydraulic fill dams. In the rolled fill type of construction, the major portion of the embankment is constructed in successive layers which are mechanically compacted. Materials needed for construction are invariably procured from the construction sites, known as borrow areas. Materials from borrow areas are transferred to the embankment location and spread to the required thickness; the proper amount of water is added and then the material is compacted by power-operated rollers. The materials used are unprocessed natural materials.

Some small quantities of specific materials required for drains, filters, etc. are transported to the site or processed by screening or otherwise at the site.

In hydraulic fill dams the embankment materials are suspended in water. The soil–water suspension (generally with about 85% water) is pumped to the required site and allowed to settle. With proper control of the suspension and the settling process, a fairly uniform construction can be achieved. However, because of segregation problems for coarse materials this method may not be suitable. The placed earthfill type is widely used.

Embankment dams are of three types, viz., diaphragm, homogeneous, and zoned.

### 20.2.1 Diaphragm Type

In this type of dam, a major portion of the embankment is constructed from pervious materials. A water barrier is formed using a thin diaphragm of impermeable material. The diaphragm may be located in the centre as a vertical core or placed as a blanket on the upstream face. The material used for the diaphragm may be earth, cement concrete, bitumen, etc. If an earth core is provided to serve as a diaphragm, the thickness should be less than 3 m or less than the height of the embankment.

### 20.2.2 Homogeneous Type

A purely homogeneous dam is constructed using a single kind of material excluding the material used for slope protection. Many of the low- to moderate-height dams are essentially homogeneous. The material selected for such dams should be sufficiently impervious and for stability requirements the slope should be relatively flat. Homogeneous dams of 6 to 8 m height have to be provided with some type of downstream drain, which helps to reduce the pore water pressure in the downstream portion of the dam and control any seepage. The types of drains provided in homogeneous dams are toe drains, horizontal blanket drains, and chimney drains.

### 20.2.3 Zoned Type

This is a more common type of dam constructed using, basically, pervious and impervious materials. The impervious material, called the *core*, is placed at the centre and is flanked by zones of pervious materials called *shells* or *casings* (discussed in the next section).

The central core is supported and protected by the shells. The upstream shell affords stability against sudden drawdown and the downstream one acts as a drain to control the line of seepage. The materials for the pervious zones may be sands, gravels, cobbles, or rocks or mixtures of these materials. The width of the core is controlled by the availability of material and design requirements, such as stability and seepage.

If a variety of soils is available in a location, there will be an ample range of material available to choose from for different sections of embankment. In such situations, there will be no constraint on the height of the dam and the zoned type will invariably yield a more economical section. If the major part of the dam is composed of rock, it is classified as a rockfill dam.

## 20.3 COMPONENTS OF EMBANKMENT DAMS

Each embankment dam consists of three basic components, viz., foundation, shell, and core (Fig. 20.1). Depending on the type of dam, additional appurtenances are added to enable the basic components to function efficiently.

### 20.3.1 Foundation

The foundation of a dam is the sole supporting medium resisting the vertical and horizontal forces. Depending on the foundation material, i.e., whether it is soil or rock, the foundation may allow or resist the flow of water. Rocks form the best foundation material provided they are free from faults, joints, or seams of soft shale or clay, etc. Sands and gravels also provide good support for dams, but adequate steps should be taken to control the seepage. Fine sands with relative densities less than about 65% should be compacted by vibration so as to avoid liquefaction. Clay foundations pose serious stability and settlement problems unless measures are taken to accelerate the consolidation. Because of their low shear strength they require a flat slope.

### 20.3.2 Casing

As pointed out earlier, the shell imparts stability and protects the core. All relatively pervious materials which are not prone to cracking under normal atmospheric conditions are suitable for casing. Table 20.1 shows (IS: 8826, 1978; 1498, 1970) the suitability of different soils for use as shell and core.

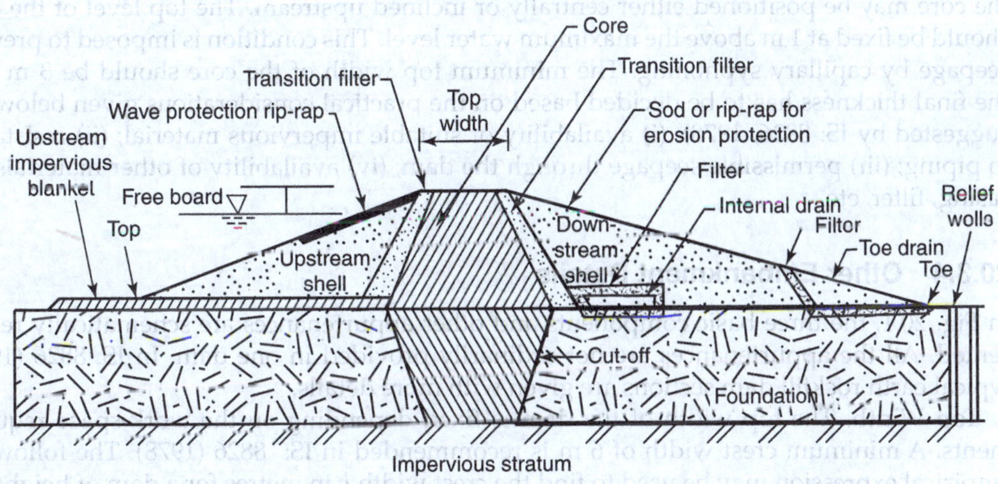


Fig. 20.1 Parts of an earth dam

**Table 20.1** Suitability of soils for construction of earth dams

| Relative suitability | Homogeneous dams | Zoned earth dams | | Impervious blanket |
|---|---|---|---|---|
| | | Core | Casing | |
| Very suitable | GC | GC | SW, GW | GC |
| Suitable | CL, CI | CL, CI | GM | CL, CI |
| Fairly suitable | SP, SM, CH | GM, SM, SC, CH | SP, GP | CH, SM, SC |
| Poor | – | ML, MI, MH | – | – |
| Not suitable | – | OL, OI, OH, Pt | – | – |

*Source*: IS: 8826 (1978); IS: 1498 (1970).

The upstream and downstream slopes of a casing have to be decided based on availability of material, foundation condition, height, and type of dam. The upstream ranges from 2:1 to 4:1 for stability requirements, and a flat one is chosen for material with low permeability. The usual downstream slope varies from 2:1 to 2.5:1. Goel et al. (1980), based on statistical shear strength data, have recommended design slopes for dams (up to 15 m high) which confirm the slope ranges given above.

### 20.3.3 Core

The core acts as an impermeable barrier and prevents the free seepage of water through the body of the dam. Soils with high degrees of compressibility, swelling, shrinking, and organic content are not suitable for cores. The Indian Standards' recommendation for cores are listed in Table 20.1.

Based on the availability of material, topography of the site, and diversion considerations, the core may be positioned either centrally or inclined upstream. The top level of the core should be fixed at 1 m above the maximum water level. This condition is imposed to prevent seepage by capillary syphoning. The minimum top width of the core should be 3 m and the final thickness has to be decided based on the practical considerations given below (as suggested by IS: 8826, 1978): (i) availability of suitable impervious material; (ii) resistance to piping; (iii) permissible seepage through the dam; (iv) availability of other materials for casing, filter, etc.

### 20.3.4 Other Embankment Details

In Fig. 20.1, the three basic components and other appurtenances are schematically represented. All the appurtenances are not ordinarily provided in one dam. In IS: 8826 (1978) typical earth rockfill dam sections are given with more details.

**Top Width**. The top width of the dam is fixed depending on the work-space requirements. A minimum crest width of 6 m is recommended in IS: 8826 (1978). The following empirical expression may be used to find the crest width $b$ in metres for a dam of height $H$:

$$b = \frac{H}{5} + 3 \quad \text{for very low dams} \tag{20.1}$$

$$b = 0.55\sqrt{H} + 0.2H \quad \text{for dams with } H < 30 \text{ m} \tag{20.2}$$

$$b = 1.65(H + 1.5)^{1/3} \quad \text{for dams with } H > 30 \text{ m} \tag{20.3}$$

**Free-Board.** This is the vertical distance between the reservoir level and the crest level. The free-board is said to be *minimum* for the maximum reservoir level and *normal* for the normal reservoir level. The primary function of a free-board is to save the embankment from over-topping by severe wave action and large inflow flood. For a rational determination of free-board, details of height and action of waves are required. Arthur (1973) recommends normal and minimum free-boards based on fetch* (Table 20.2).

**Cut-off.** On a permeable foundation, if the under seepage is very high it is necessary to extend the core down into the foundation up to the impervious layer. Such a provision not only reduces the loss of stored water but also prevents sub-surface erosion by piping. The cut-off should be positioned such that the centre line is in the upstream of the impervious core.

While deciding the width, the following Indian Standards recommendations may be considered: (i) provide sufficient working space for compaction equipment and to carry out certain grouting and (ii) provide safety against piping. To satisfy the second condition, a bottom width of 10% to 30% of hydraulic head has to be provided. A minimum width of 4 m is recommended. Side slopes of at least 1:1 or flatter may be provided.

**Internal Drain**. This is essential primarily to carry away the seepage passing through the core or the cut-off. Secondly, this is necessary to prevent undue saturation of the upper part of the downstream shell by rain or spray falling on the dam. An internal drainage system consists of inclined and horizontal filters, rock toe, toe drain, etc. For design details of internal drainage systems refer to IS: 9429 (1980).

**Transition Filter**. This is needed between a coarse and fine material to act as a transition medium to prevent the migration of fine material into coarse material. In zoned dams, this is provided between the core and the shell. The transition filter may be omitted when the seepage gradient is low and no wide variation exists in the grain sizes of core and shell.

For the reasons mentioned above, all the internal and toe drains require *protective filters.*

**Slope Protection**. Upstream slope protection is ensured by providing *rip rap*. This prevents erosion or wash by waves of upstream materials. Usually, it starts from above the maximum water level to just below the minimum.

**Table 20.2** Normal and minimum free-boards

| Fetch (km) | Normal free-board (m) | Minimum free-board (m) |
|---|---|---|
| <1.60 | 1.22 | 0.91 |
| 1.60 | 1.52 | 1.22 |
| 4.03 | 1.83 | 1.52 |
| 8.07 | 2.44 | 1.83 |
| 16.13 | 3.05 | 2.13 |

*Source:* Arthur (1973).

*Fetch is the distance over which the wind can act on a body of water.

The downstream slope is protected by providing rip rap or sod (turfing). The conventional practice is to provide the entire downstream slope with turfing, which prevents the slopes from rain cuts. In certain cases, even paving is resorted to, but it is more expensive. Sometimes *berms* are employed on the downstream face to accommodate access for maintenance. This also forms a break in the slope to reduce the rain wash. For design details of slope protection refer to IS: 8237 (1976).

**Upstream Blanket**. The horizontal upstream impervious blanket is provided to increase the path of seepage when full cut-off is not practicable on pervious foundations. The blanket is connected to the core of the dam and may or may not be provided with a partial cut-off. The permeability of the blanket material should be far less than that of the foundation material. Table 20.1 gives Indian Standards recommendations. For design details refer to IS: 8414 (1977).

**Relief Wells**. These are vertical or nearly vertical holes arranged in one or more lines or in a pattern so as to reduce the internal pressure and collect the seepage. Relief wells are also used as a corrective device in completed dams that later develop unforeseen seepage pressures. These wells may be installed either downstream or inside the structure.

## 20.4 DESIGN CRITERIA FOR EARTH DAMS

More than 200 earth dams have been studied by Middle Brook and the causes of failure have been attributed to the following:

| Types of failure | Percentage |
|---|---|
| Over-topping | 30 |
| Seepage through foundations | 25 |
| Upstream and downstream slope slides | 15 |
| Conduit leakage | 13 |
| Slope paving | 5 |
| Miscellaneous | 7 |
| Unknown | 5 |

In a broader sense, the above failures may be grouped under three main classes, viz., hydraulic failures (40%), seepage failures (30%), and structural failures (30%).

Based on the knowledge gained from the studies of failures of old and new dams, the following design criteria are laid out:

1. There should be no danger of over-topping of the dam.
2. The seepage flow through the dam and foundations should be controlled.
3. There should be no free passage of water from the upstream to the downstream face.
4. The upstream and the downstream slopes should be safe against sliding.
5. The crest and upstream and downstream faces should be protected from wave action, rain, and frost action.

### 20.4.1 Safety Against Over-Topping

The causes of over-topping may be either inadequate provision of spillway or the free-board. The provision of spillway capacity is based on the design flood, which is estimated from the principles of hydrology. Generally, for a 30 m dam a 100-year flood and for dams higher than 30 m a 250-year flood may be considered. The free-board requirements have already been discussed.

### 20.4.2 Control of Seepage and Pressure in Earth Dams

Uncontrolled seepage flow affects the earth dams broadly in two ways, viz., excess leakage and excess pore pressure or gradients. Seepage control is the correction of the conditions which lead to these troubles. While applying corrective measures care should be taken that a correction at one point does not bring in or increase the trouble in some other point. Data needed for analysis for these aspects are obtained from flow nets.

**Flow Nets for Earth Dams.** In Chapter 6, we discussed seepage under impermeable structures, such as flow beneath concrete dams, sheet pile walls, etc. These are termed confined flow conditions whereas the flow through a permeable structure, such as an earth dam, is an example of unconfined seepage.

In an earth dam, the phreatic surface constitutes the top flow line, and its position must be located first. The different boundary conditions for an earth dam are shown in Fig. 20.2. Seepage through the dam takes place below the phreatic line. As the pressure on the phreatic line is equal to the atmospheric pressure, the total head is equal to the elevation head. The soil is homogeneous and isotropic except at the rock filter; hence, the conventional properties of flow net are satisfied. The phreatic line takes different shapes depending on the materials used in the dams. Figure 20.3 shows different phreatic lines for various earth dam cross sections.

It has been mathematically shown that the basic shape of the phreatic surface is that of a parabola and it deviates only at the upstream and downstream faces, depending on the cross section of the dam. Thus, the entry and exit conditions are discussed first.

As the upstream entry surface is an equipotential line with maximum pressure head, the flow lines must intersect it at right angles as long as the equipotential line with maximum head (i.e., the upstream face of the dam) is less than 90° (Fig. 20.4a). When an upstream coarse filter is provided, the equipotential line with maximum head may be greater than or equal to 90°. In such cases, the phreatic surface's entry is horizontal since water at zero pressure cannot flow upwards (Fig. 20.4b).

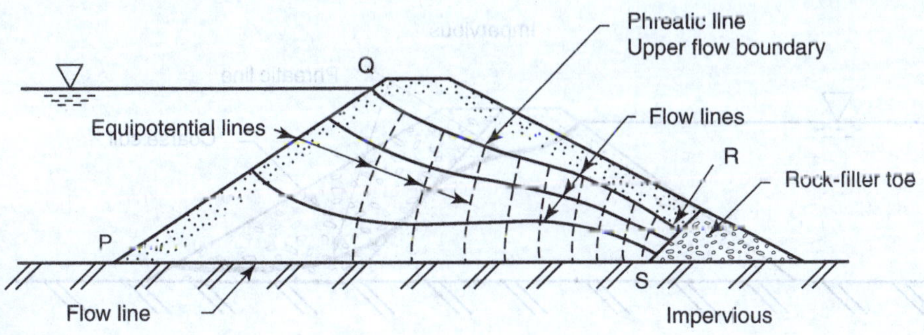


Fig. 20.2 Boundary conditions and flow net for a homogeneous dam

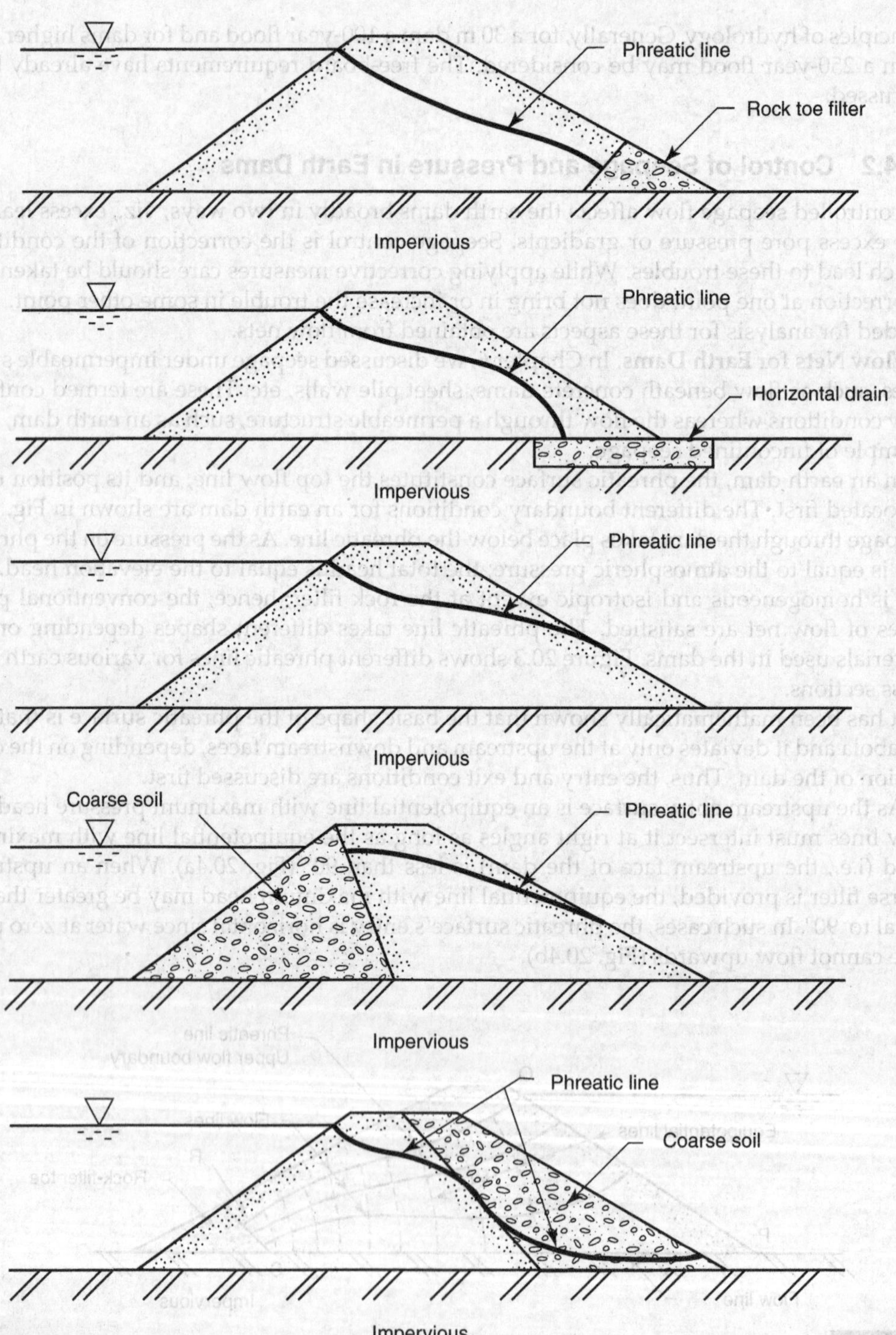


**Fig. 20.3** Phreatic lines for various dam cross sections

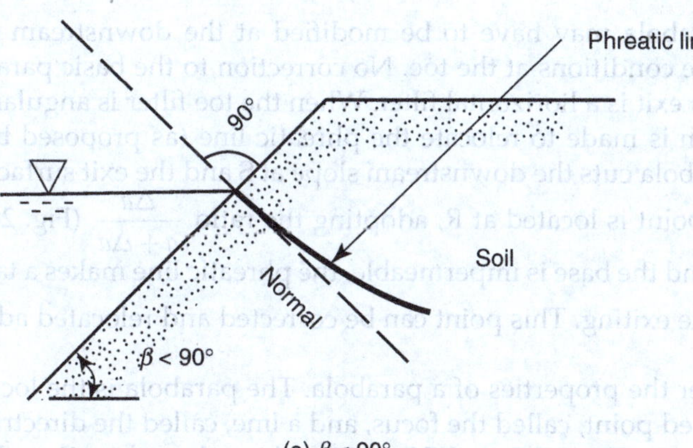

(a) $\beta < 90°$

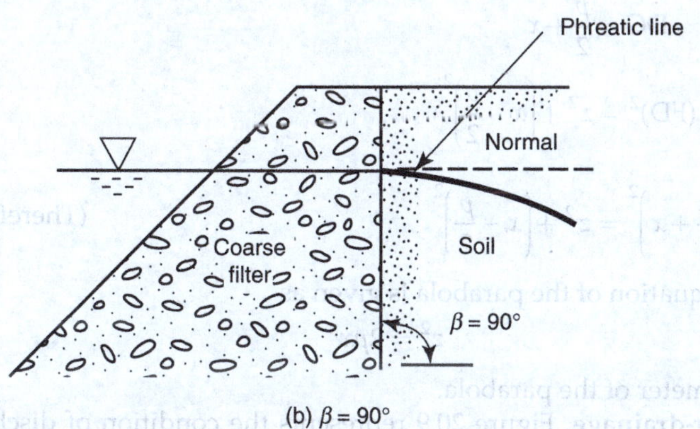

(b) $\beta = 90°$

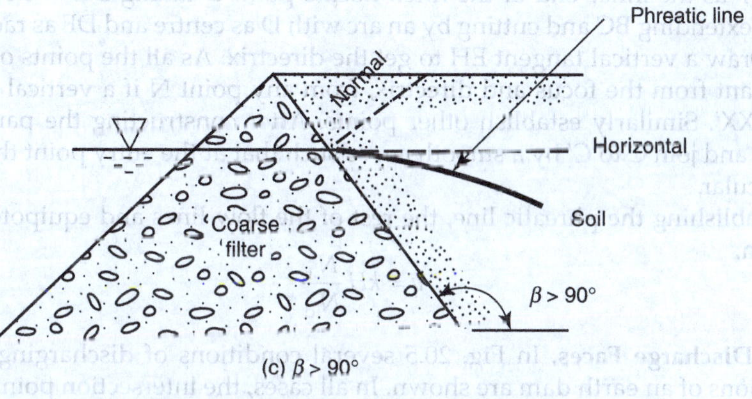

(c) $\beta > 90°$

**Fig. 20.4** Phreatic line details at entry to seepage zone (Source: Whitlow, 1983)

The theoretical parabola may have to be modified at the downstream or exit surface also, depending on the conditions at the toe. No correction to the basic parabola is needed when the downstream exit is a horizontal filter. When the toe filter is angular with $\alpha < 180°$ (Fig. 20.5), a correction is made to relocate the phreatic line (as proposed by Casagrande, 1937). If the basic parabola cuts the downstream slope at S and the exit surface intersects the base at F, the correct point is located at R, adopting the ratio $\frac{\Delta a}{a+\Delta a}$ (Fig. 20.6). In case no toe filter is provided and the base is impermeable, the phreatic line makes a tangent with the downstream face while exiting. This point can be corrected and relocated adopting $\frac{\Delta a}{a+\Delta a}$ ratio (Fig. 20.7).

Now, let us consider the properties of a parabola. The parabola is the locus of all points equi-distant from a fixed point, called the focus, and a line, called the directrix (Fig. 20.8).

By definition, FA = AB, FD = DC, and FO = OE, taking the vertex O as the origin and $x$ and $z$ as coordinates:

$$\text{DC} = \frac{p}{2} + x$$

or

$$(\text{FD})^2 = z^2 + \left(x - \frac{p}{2}\right)^2$$

or

$$\left(\frac{p}{2} + x\right)^2 = z^2 + \left(x - \frac{p}{2}\right)^2 \qquad \text{(Therefore, DC = FD)}$$

Simplifying, the equation of the parabola is given as

$$z^2 = 2px \tag{20.4}$$

where $2p$ is the parameter of the parabola.

**Horizontal Under-drainage**. Figure 20.9 represents the condition of discharging into a horizontal toe filter. Draw the dam and its appurtenances to a suitable scale. Assume the focus point F as the inner end of the filter. Locate point D taking DC = 0.3BC. Locate the directrix by extending BC and cutting by an arc with D as centre and DF as radius; i.e., make DF = DE. Draw a vertical tangent EH to get the directrix. As all the points on the parabola are equidistant from the focus and directrix, from any point N if a vertical is drawn at X, then FX = XX′. Similarly establish other points. After constructing the parabola, draw a normal at C and join C to C′ by a smooth curve such that at the entry point the phreatic line is perpendicular.

After establishing the phreatic line, the rest of the flow lines and equipotential lines are drawn. Then,

$$q = kH\frac{N_{\text{f}}}{N_{\text{d}}}$$

**Sloping Discharge Faces**. In Fig. 20.5 several conditions of discharging at the downstream portions of an earth dam are shown. In all cases, the intersection point of the bottom flow line with the discharge face is considered as the focus F.

After deciding the focus point F, the parabola as described earlier is developed. Considering Fig. 20.4a, point D and the point on the directrix E are found. The parabola is drawn as before. Let S be the point of intersection of the parabola with the downstream face. Now FS,

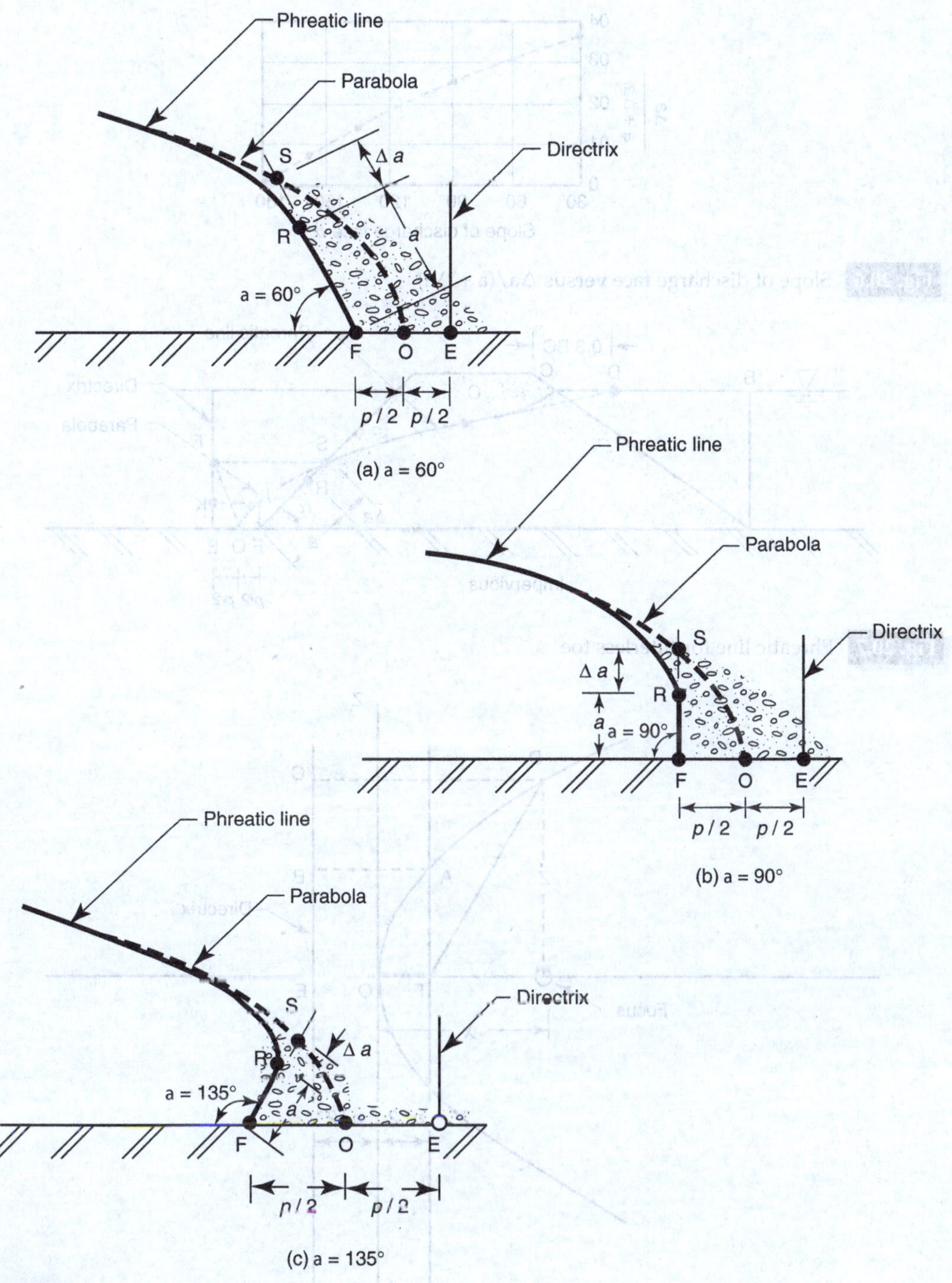


Fig. 20.5 Phreatic line details at exit from seepage zone (Source: Cernica, 1982)

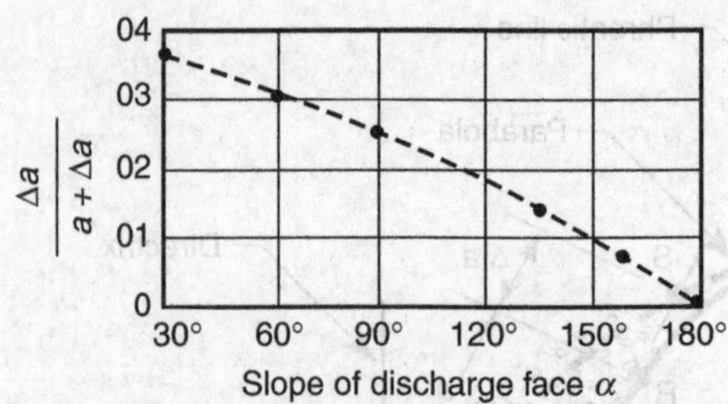


Fig. 20.6 Slope of discharge face versus $\Delta a/(a + \Delta a)$

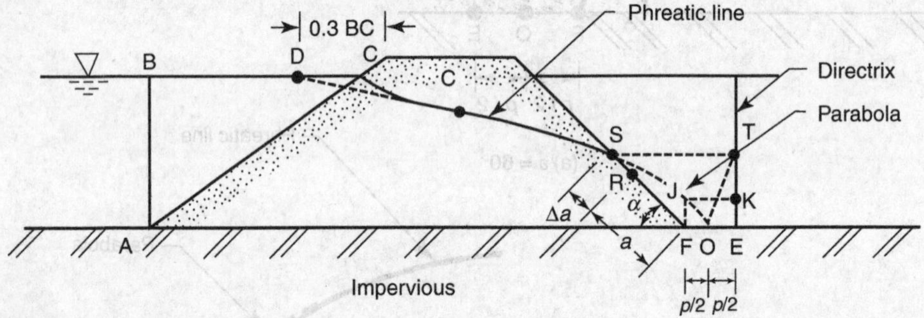


Fig. 20.7 Phreatic line for filterless toe

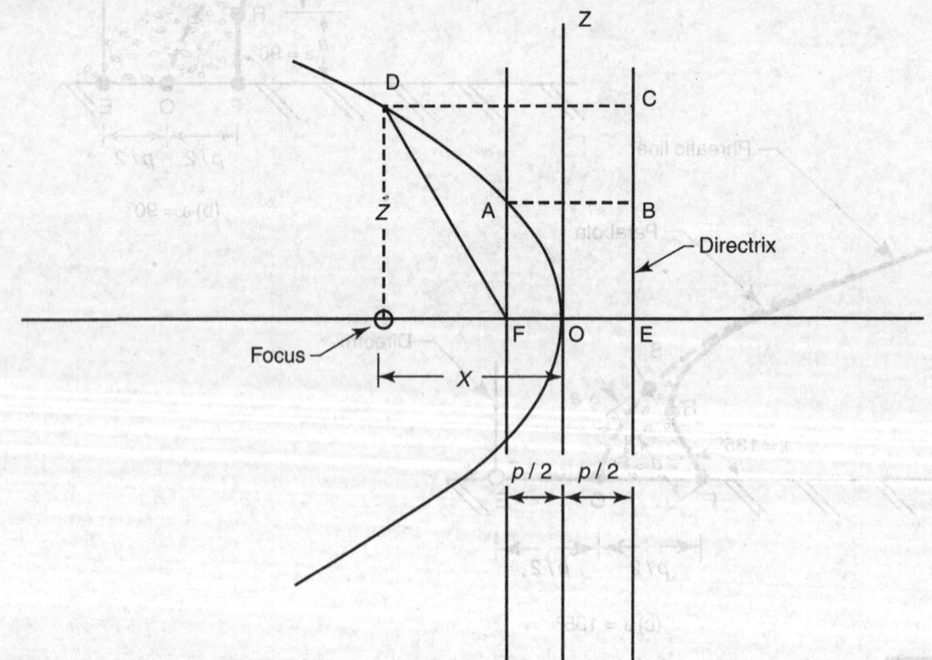


Fig. 20.8 The basic parabola

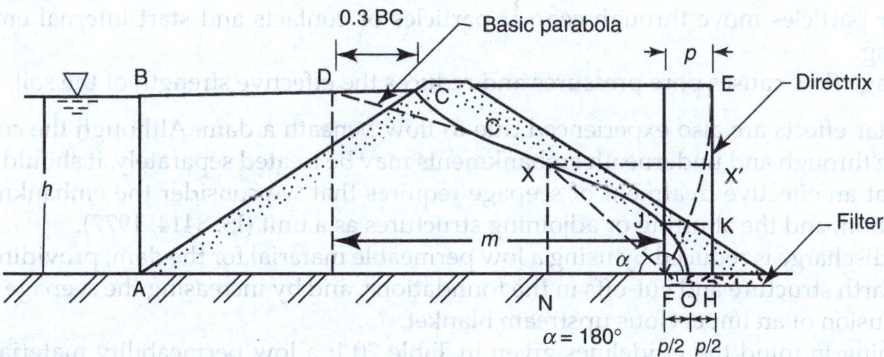


**Fig. 20.9** Construction of parabola for dam with horizontal drain

which is equal to $a + \Delta a$ is measured. $\alpha$ is found and from Fig. 20.6 $\Delta a/(a+\Delta a)$ is read. Knowing $a + \Delta a$ $a$ and hence the break out point R are found. The transition section between R and the parabola can be sketched by eye.

After drawing the phreatic line, the flow net can be completed and the discharge calculated.

For the condition $\alpha > 30°$, if the constructed flow net looks like confocal parabolae, the discharge may be calculated theoretically. We know that

$$q = kiA = k\frac{dz}{dx}A$$

Considering unit thickness of the dam, $A = z \times 1 = z$.

From Eq. 20.4, $z^2 = 2px$, or $z = \sqrt{2px}$, or $A = \sqrt{2px}$.

Therefore,

$$q = k\left(\frac{p}{\sqrt{2px}}\right)(\sqrt{2px}) = kp$$

From the properties of a parabola,

$$FD = DC = m + p$$

Also, $(FD)^2 = h^2 + m^2$, where $m = x - p/2$ and $h = z$. That is, $m + p = \sqrt{h^2 + m^2}$.

Solving for $p$ we get $p = \sqrt{h^2 + m^2} - m$. Therefore,

$$q = k\sqrt{h^2 + m^2} - m \tag{20.5}$$

**Control of Quantity of Seepage.** Excess seepage may be caused due to highly permeable dam material, short seepage paths, and defects, such as fissures and cracks brought in by uneven settlements. The seepage of reservoir water through the body of the dam, apart from causing excessive water loss (IS: 9429, 1980), creates the following problems:

1. Finer particles migrate and clog soil structure, thus preventing seepage and causing sloughing and weakening of soil strength.

2. Finer particles move through coarser particles or contacts and start internal erosion or piping.
3. Seepage flow causes pore pressures and reduces the effective strength of the soil.

Similar effects are also experienced due to flow beneath a dam. Although the control of seepage through and underneath embankments may be treated separately, it should be realized that an effective treatment of seepage requires that we consider the embankment, its foundation, and the abutting or adjoining structures as a unit (IS: 8414, 1977).

The discharge is reduced by using a low permeable material for the dam, providing a core in the earth structure and cut-offs in the foundations, and by increasing the seepage path by the inclusion of an impervious upstream blanket.

Keeping in mind the guidelines given in Table 20.1, a low permeability material has to be chosen depending on the availability of material in the field. If a suitable material is not available, then by proper selection and by mixing different soils available at the site, it is often possible to obtain a composite soil with sufficiently low permeability.

An effective way of reducing leakage through an embankment is by providing a relatively impervious core (Fig. 20.10a). The impervious core may be made of wood, steel, concrete, masonry, or soil and should remain intact and impervious throughout the life of the structure. Generally, clay cores are preferred. A core may be placed near the upstream (called the sloping core) or at the centre. An upstream core reduces the pore pressure in the downstream part of the embankment and increases its safety. As far as stability is concerned, the upstream core is less stable, particularly during a sudden drawdown, and more volume of soil is required. The central core requires minimum core material and is more stable during a sudden drawdown.

As discussed earlier, the provision of a cut-off in the foundation not only reduces the loss of water but also controls piping. A cut-off may be partial or complete, depending on whether its depth of penetration is partial or full (Fig. 20.10b and c). A properly constructed complete cut-off can reduce the seepage to a negligible amount, whereas a partial cut-off is less effective. All the materials used for cores may also be used for cut-offs.

Another efficient way of controlling seepage and gradient is by providing an impervious upstream blanket (Fig. 20.10d). By this provision, the length of the seepage path is increased, thus reducing the gradient and seepage. The length of blanket required is estimated using a flow net. Blankets are particularly useful when fissures and cracks exist in the foundation. However, the blanket itself can crack due to settlement of the foundation (Fig. 20.10e). Ordinarily, the blanket is effective in reducing the seepage only by 50%. Blankets are of compacted impervious soil, as suggested by Indian Standards (Table 20.1). Sometimes, various chemical additives are also employed to reduce the permeabilites.

Grouting is also an effective way of controlling excessive seepage flow. Foundation grouting is resorted to in specially drilled holes for the purpose of sealing off or filling joints, seams, fissures, or other openings encountered.

*Control of Pressures and Gradients.* Excess hydrostatic pressures cause boiling and piping, especially at points where there is less weight of structures to resist them. The methods of reducing excess pore pressures and gradients are (i) changing the direction of the seepage,

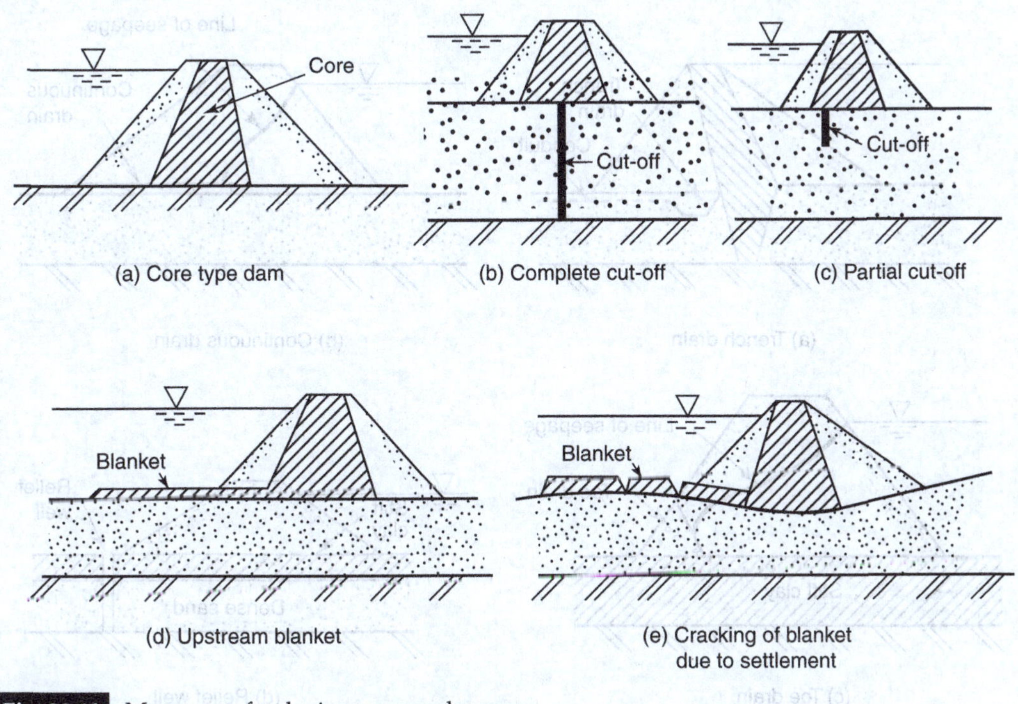


**Fig. 20.10** Measures of reducing seepage losses

(ii) installation of internal drains, (iii) incorporating relief wells, (iv) increasing the external load, and (v) providing cut-off walls and an upstream blanket.

Among the above-mentioned methods, internal drainage is the most effective. In principle, such a provision short circuits the seepage, reduces the excess pressures, changes the direction of movement, and hence shifts the point of high gradients to a safer place inside the structure. In Fig. 20.11, different measures of correcting excessive uplift pressures and gradients are given.

The trench drain (Fig. 20.11a) lowers the line of seepage in the homogeneous dam and prevents pore pressures and loss of strength in the casing of a zoned dam. In a foundation severely damaged by cracks and fissures, a continuous drain, as shown in Fig. 20.11b, is provided. In a highly pervious downstream shell and in small dams, toe drains (Fig. 20.11c) are provided. Apart from reducing the pressures at the downstream slope, it prevents saturation of the soil at the toe of the dam due to rainfall.

Relief wells are an important adjunct to most of the preceding basic schemes for seepage control and pressure relief. They are provided in earth dams where there are seams or pockets of pervious water-bearing strata at great depth which cannot be intercepted by other means (Fig. 20.11d). Usually, they are used in nearly all cases with upstream impervious blankets. Besides, they are also used along with other schemes to provide additional assurance that excess hydrostatic pressures do not develop. Continuous observation and maintenance of relief wells is essential so as to ensure the satisfactory performance of the overall system as regards seepage and pressure control.

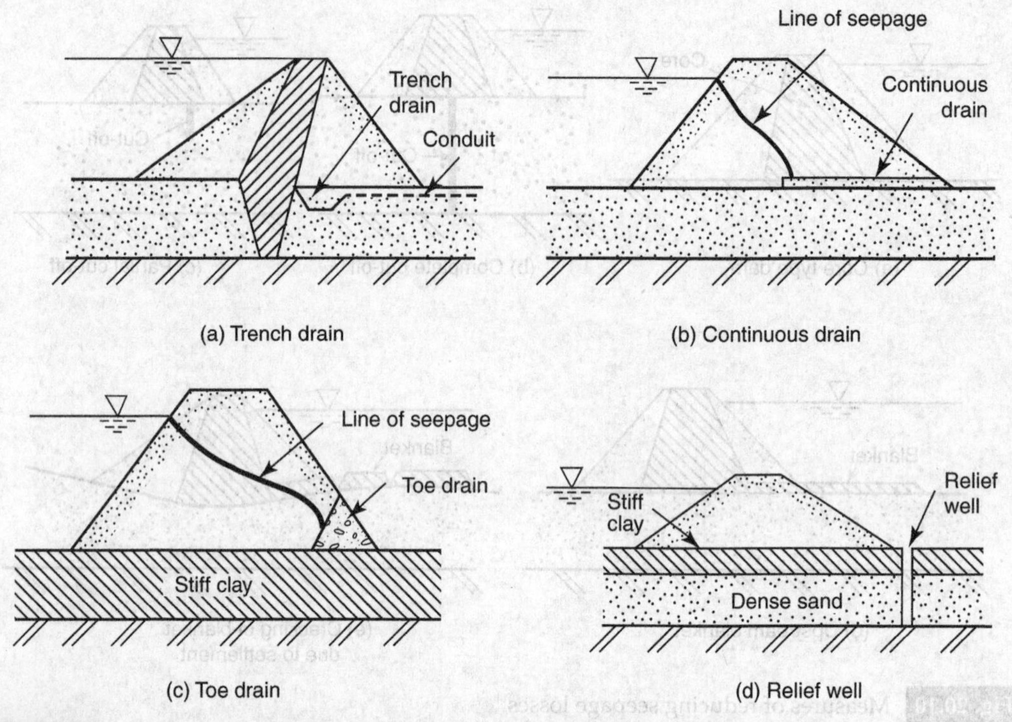


**Fig. 20.11** Measures for reducing excess uplift pressures and gradients

Another method for correcting pressures is to increase the downward load at the downstream side. Blocks of concrete have been used on the downstream top of a pervious stratum. In earth dams, a large over-sized toe drain can serve the same purpose.

All the drainage systems require protective filters to prevent the movement or erosion of the soil.

### 20.4.3 Protection Against Free Passage of Water Through Dams

The free passage of water from the upstream to the downstream will cause serious danger to the stability of a dam. Such free passage may be caused due to the following:

1. water moving along the exterior surfaces of pipes or conduits which are embedded in the body of the dam,
2. failure to form sufficiently strong bonds between successive layers of the dam,
3. failure to bind the lower layers of the dam properly to the foundation,
4. wrong placing of a pervious material in an otherwise impervious section in such a manner as to make a blind drain from the upstream to the downstream,
5. water following the smooth surfaces of concrete abutments or other concrete structures, and
6. burrowing animals.

Of the above-mentioned causes, the first four can be effectively controlled by strictly following the specifications during construction.

Water has a tendency to follow the smooth surface of abutments, pipes, conduits, etc. The best practice is to place pipes and conduits in trenches excavated in the original foundation material. Rocky abutments should be suitably shaped and prepared such that a perfect bond is obtained at the contact between the impervious core and the rock. Overhangs should be removed and the vertical surface excavated to form moderate slopes. Indian Standards (IS: 8826, 1978) recommends a wider impervious zone and thicker transitions at the abutment contacts to increase the length of the path of seepage.

Burrowing animals, such as musk rats and land squirrels, are responsible for piping failures in small dams. However, in modern dams, this danger is quite remote because (i) the core material is so densely compacted that it may be difficult to burrow into it and (ii) the crest width and the free-board are generally ample and animal holes do not penetrate to a great depth.

Apart from the important design criteria discussed so far, there are some special design requirements, such as control of cracking and stability in earthquake regions and at junctions. A detailed discussion of these factors is beyond the scope of this book. The reader may refer to some advanced book on earth dams and to IS: 8826 (1978).

### 20.4.4 Stability of Earth Dam Slopes

In the design of earth dams, both safety and economy call for thorough soil studies of the foundation and of the materials of construction, combined with stability computations.

The present-day stability analyses are based on the results of studies of actual slides in old dams. Stability analysis has to be performed for ascertaining the factor of safety for a new structure or for the redesign of an old structure. An effective stress method of analysis is recommended by Indian Standards (IS: 7894, 1975). Depending on the type of failure surface, either the circular arc method or sliding wedge method is adopted.

The stability of an earth dam with respect to slope or foundation failure depends on the magnitude and distribution of pore water pressure at all times. While designing an earth dam, both the upstream and downstream slopes have to be checked for adequate stability. For design purposes, the distribution of pore water pressure during construction, after the reservoir has been filled, during steady state of seepage, and during sudden drawdown are critical and have to be considered in the slope stability analysis.

The following are the conditions identified by Indian Standards (IS: 7894, 1975) as critical for the stability of an earth dam:

Case I – Construction condition with or without partial pool (for upstream and downstream slopes)
Case II – Reservoir partial pool (for upstream slope)
Case III – Steady seepage (for downstream slope)
Case IV – Sudden drawdown (for downstream slope)
Case V – Steady seepage with sustained rainfall (for downstream slope)
Case VI – Earthquake condition (for upstream and downstream slopes).

The methods of analysis and the above-mentioned conditions are discussed below.

**Methods of Analysis**. The circular arc method adopted for the analysis is the method of slices. This method was discussed in Chapter 13. The second method, the sliding wedge method, is most suitable for earth dams and is discussed below.

The sliding wedge method divides the sliding mass into two or three wedges. In the three-wedge system, the upper and lower wedges are called the driving or active wedge and resisting or passive wedge, respectively, and the middle segment is referred to as the sliding block. In a two-wedge system there is no middle segment. For earth dam this method is most frequently applied under two circumstances, as shown in Fig. 20.12.

The factor of safety may be computed by adopting the methods of slices with side forces (discussed in Chapter 13), the only difference being that the number of slices, instead of being large, is only two or three.

Now, an alternative method, more suitable in practice, as mentioned in Fig. 20.12, is considered (Sherard et al., 1972). Let us assume that sufficient movement has taken place to keep the active and passive wedges at failure. The active ($P_a$) and passive ($P_p$) forces acting on the planes *bc* and *de* are computed assuming no shear forces act on the planes. Now, the factor of safety is defined with reference to the stability of the central block. The unbalanced force acting on the central wedge is

$$P_1 = P_a - P_p \tag{20.6}$$

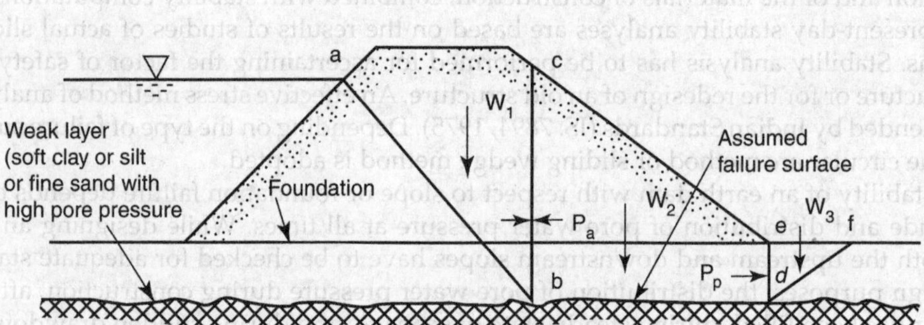


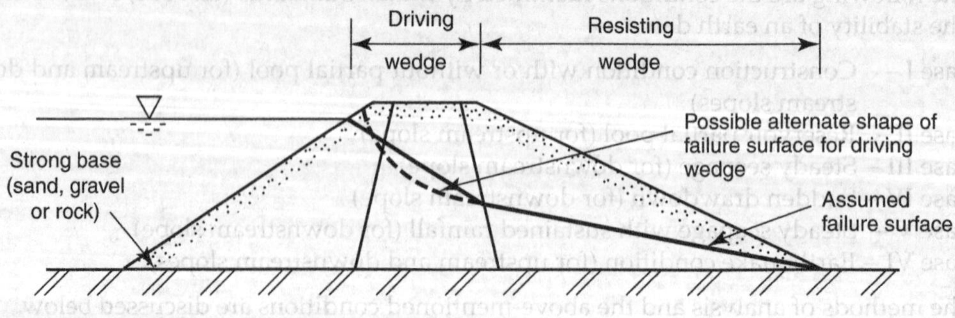


**Fig. 20.12** Sliding wedge method

The force available to resist the movement of the central block is

$$P_2 = C_{bd} + (W_2 - U_{bd}) \tan \phi'_{bd} \tag{20.7}$$

Therefore, the factor of safety

$$F = \frac{P_2}{P_1} \tag{20.8}$$

**Construction Condition or End of Construction**. This is the condition in which there is development of pore pressure due to overlying fill. Due to non-availability of sufficient time, the dissipation of pore pressure is partial and its estimation involves many uncertainties. Both upstream and downstream slopes have to be analysed.

In principle, total and effective stress methods of analysis can be applied, provided a correct estimation of undrained strength (with its associated pore pressure for the total stress analysis) and pore pressure (only in the effective stress analysis) can be made. In the effective stress analysis, a pore pressure value (in terms of pore pressure ratio, $r_u$) may be determined in the laboratory or assumed based on experience (Bishop and Bjerrum, 1960). Such an assumed pore pressure in the design can be measured during construction and modified accordingly. This is the advantage of the effective stress method over the total stress method (Bishop, 1957).

The pore water pressure at any point is

$$u_w = (u_w)_0 + \Delta u_w \tag{20.9}$$

where $(u_w)_0$ is the initial pore water pressure and $\Delta u_w$ the pore water pressure in the undrained condition. This may be represented in terms of the overall pore pressure coefficient $\bar{B}$ as

$$u_w = (u_w)_0 + \bar{B}\Delta\sigma_1 \tag{20.10}$$

Dividing by $\gamma h$ and letting $r_u = u_w / \gamma h$, we have

$$r_u = \frac{(u_w)_0}{\gamma h} + \frac{\bar{B}\Delta\sigma_1}{\gamma h} \tag{20.11}$$

For the no surcharge condition, $\Delta\sigma_1 = \gamma h$. Therefore,

$$r_u = \frac{(u_w)_0}{\gamma h} + \bar{B} \tag{20.12}$$

For higher initial water content, $(u_w)_0 = 0$ and $\bar{B}$ is high, then,

$$r_u \approx \bar{B} \tag{20.13}$$

The value of $\bar{B}$ is obtained from a triaxial test corresponding to the stress condition in the field. For the end of construction condition, a factor of safety of 1.3 may be allowed.

**Reservoir Partial Pool Condition**. This condition corresponds to the initial partial pool filling. It has been presumed that a condition of steady seepage has developed at some intermediate stages. For this condition, the upstream slope is critical. Minimum safety factors are computed for three reservoir level conditions, viz., one thirds, two-thirds, and full reservoir levels.

It has been suggested (IS: 7894, 1975) that all zones above the phreatic line are to be taken as moist and those below are to be taken as submerged, while calculating the resisting and driving forces. The partial condition is critical for high dams where the range of drawdown is small compared to the height of the dam.

**Steady-state Seepage Condition**. Conditions of steady seepage are established after a sufficient lapse of time from the time of filling of the reservoir. For steady seepage, the stability of the downstream slope is always analysed by effective stress analysis and the pore pressure is estimated from a flow net (Fig. 20.13a). Values of $r_u$ up to 0.45 are used in homogeneous dams. A least factor of safety of 1.5 is adopted for this condition.

**Sudden Drawdown Condition**. Upstream slopes of earth dams and natural slopes adjacent to the reservoir can undergo rapid drawdown if there is a sudden fall in the reservoir level (Fig. 20.13b). Generally, high pore pressures result during drawdown, unless there is an equal adjustment of pore pressure with the drawdown level.

An effective stress analysis can be adopted by calculating pore water pressure from a flow net or from a reasonable estimation depending on the time availability for consolidation. If the consolidation time is much less than the drawdown time, a flow net may be used to compute the pore water pressure. As the flow net pattern changes with dissipation, it is sufficient to construct the flow net immediately after the drawdown (Lambe and Whitman, 1979).

If the consolidation time is much longer than the drawdown time, the change in pore water pressure due to the change in water load may be assumed to take place under undrained conditions. The flow net for the steady seepage condition is considered (Fig. 20.13b). The pore water pressure at a point P on the trial slip circle is given as (Bishop and Bjerrum, 1960).

$$(u_w)_0 = \gamma_w (h + h_w - h') \tag{20.14}$$

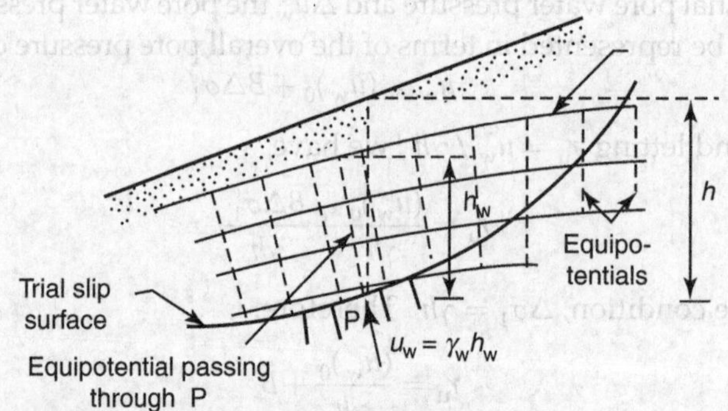


(a) At steady seepage

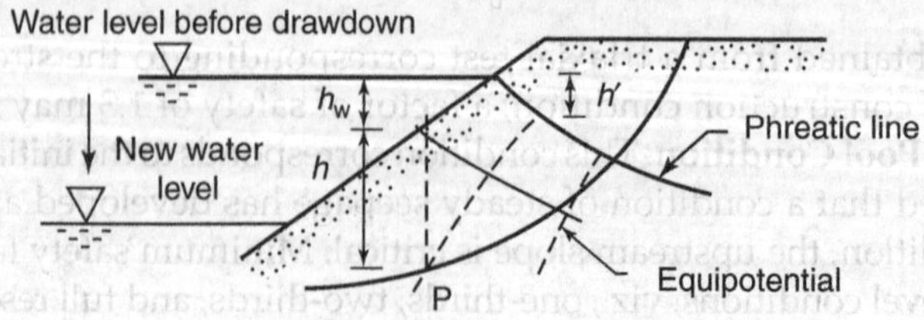


(b) After rapid drawdown

**Fig. 20.13** Pore pressures in steady seepage and drawdown conditions

Let $\sigma_1$ be $\gamma h$. Drawdown causes the piezometric level to fall to $h_w$, and thus, the change in total major principal stress is given by

$$\Delta\sigma_1 = -\gamma_w h_w$$

and

$$\Delta u_w = \bar{B}\Delta\sigma_1 = -\bar{B}\gamma_w h_w$$

$$u_w = (u_w)_0 + \Delta u_w$$

where $u_w$ is the pore water pressure at P immediately after drawdown; that is,

$$u_w = \gamma_w[h + h_w(1-\bar{B}) - h']$$

Dividing by the overburden pressure, $\gamma_{sat}h$, we get

$$\frac{u_w}{\gamma_{sat}} = \frac{\gamma_w}{\gamma_{sat}}\left[1 + \frac{h_w}{h}(1-\bar{B}) - \frac{h'}{h}\right] \tag{20.15}$$

When $\bar{B} = 1$ and $h' \approx 0$, a conservative value for $r_u$ is obtained. A pore pressure ratio of 0.3 to 0.4 is the typical value for a rapid drawdown condition. A factor of safety of 1.2 is acceptable for this condition. Morgenstern (1963) presented stability coefficients for a rapid drawdown condition.

Depending on the value of the coefficient of permeability of the shell material, Indian Standards (IS: 7894, 1975) has recommended that the pore water pressure in the casing may be allowed in the analysis in the following manner:

1. Full pore water pressures shall be considered if the coefficient of permeability is less than $10^{-4}$ cm/s.
2. No pore water pressures shall be considered if the coefficient of permeability is more than $10^{-2}$ cm/s.
3. A linear variation from full to zero pore water pressure shall be considered for the coefficients of permeability lying between $10^{-4}$ cm/s to $10^{-2}$ cm/s.

For the core material, the recommendation is to allow full pore water pressures for the core zone lying in the drawdown range.

For a zoned dam, the pore water pressure can be determined (IS: 7894, 1975) from the formula (Fig. 20.14)

$$u_w = \gamma_w[h_c + h_r(1-m)h] \tag{20.16}$$

where $u_w$ is the drawdown pore water pressure at any point, $h_c$ the height of core material at the point, $h_r$ the height of shell material at the point, $m$ the volume of water draining out from the shell per unit volume, and $h$ the drop in the head under steady seepage condition at the point.

As in the case of a homogeneous dam, the drawdown pore water pressures can be determined from the flow net.

**Steady Seepage with Sustained Raintall Condition.** This condition is critical again for the downstream slope. For this, a partial saturation of shell material due to rainfall is arbitrarily assumed. Accordingly, during analysis, the shell and other material lying above the phreatic line shall be considered as moist for computing driving forces and buoyant for resisting forces.

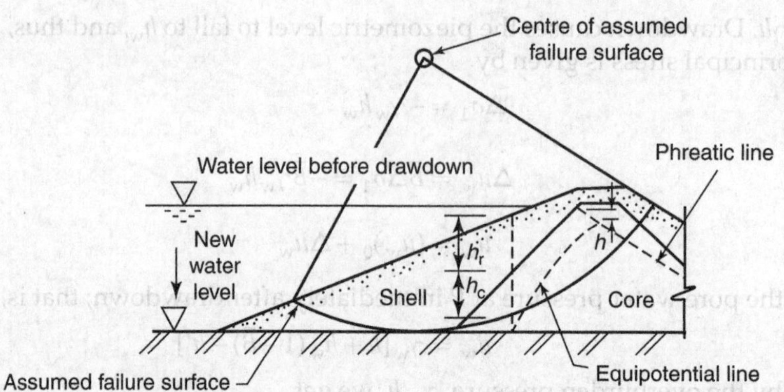


**Fig. 20.14** Criterion for drawdown pore pressure in compressible core

The saturation for the downstream shell material shall be assumed, based on Indian Standards recommendations (IS: 7894, 1975), as

1. 50%, if the coefficient of permeability is $10^{-4}$ cm/s or less,
2. 0%, if the coefficient of permeability is $10^{-2}$ cm/s or more, and
3. the percentage shall vary linearly from 50% to 0% for the coefficients of permeability lying between $10^{-4}$ cm/s and $10^{-2}$ cm/s.

**Earthquake Conditions**. For this case, both the upstream and downstream slopes are critical. The reader may refer to IS: 1893 (1975) for other details.

### 20.4.5 Protection of Crest, Upstream, and Downstream Faces

Some type of surfacing should be provided on the crest to protect it against damage by wave splash and spray, rainfall, and wind and traffic wear and tear. The conventional method is to place a 30 cm thick layer of selected fine rock or gravelly material with a suitable crown for surface drainage. A parapet wall is provided on the upstream side and the downstream edge is protected by a curb.

The upstream slope should be protected from destructive wave action. The different types of surface protection of the upstream slope include stone rip-rap (either dry-dumped or hand placed), concrete pavement, and steel facing. Sometimes sacked concrete or willow mattresses are used for relatively small and unimportant dams. The rock for rip-rap should consist of hard, dense, durable boulders or rock fragments from the quarries.

The downstream slope is protected by providing a heavy layer of coarse gravelly material, or it may be surfaced with top soil and planted with native vines, shrubs, or grass. If sufficient rock or cobble is available, it is preferable to provide a downstream rock or cobble fill. This, in addition to its primary function of providing a stabilizing weight, also furnishes a protective covering for the underlying earth slope. On high dams, the effects of surface run-off may be minimized by the use of berms or shoulders at intervals on the slope to collect and dispose off the run-off water.

## 20.5 SELECTION OF DAM SECTION

For the selection of an earth or rockfill dam section, Indian Standards (IS: 8826, 1978) has provided some recommendations which are discussed below.

### 20.5.1 Earth Dam

For all site conditions, no single type of cross section of an earth dam section is suitable. The following factors have to be considered while deciding the earth section:

1. availability of construction materials, their quantity, and nearness to site;
2. condition of foundation and cut-off requirements;
3. types of construction machinery;
4. construction schedule and diversion considerations;
5. climatic conditions and their interference in placement water content;
6. safety requirements as regards to stability and seepage.

A homogeneous section is generally preferred in a location where adequate construction material is available. If different types of soils are available in a location, a zoned dam is preferred. In zoned earth dams, the weaker materials are often utilized for random zones. Random zones are those that are generally provided below the minimum drawdown level on the upstream side and on the downstream of the inclined filter. The planned section is often altered or modified due to one reason or another.

In order to accommodate for post-construction vertical deformation resulting from compression of the embankment and foundation settlement, an extra height of the dam is provided. This extra height is provided as a longitudinal camber with zero at the abutments and maximum at the centre of the gorge. Generally, a provision of increase in height of 1% to 2% of the embankment height is made to account for this.

### 20.5.2 Rockfill Dams

Rockfill dams primarily consist of an impervious membrane for water tightness and supporting rockfill. The merit of each rockfill dam section has to be analysed for the intended site. The separation of the rockfill zone into several different zones and specification of a different gradation for each zone is considerably expensive. It is generally satisfactory to use quarry run rock in the entire rockfill zone. This ensures a reasonably uniform rock embankment with free drainage characteristics.

## WORKED EXAMPLES

**Example 20.1** Compute the factor of safety with respect to effective stress for the slip surface shown in Fig. 20.15. The shear strength parameters of the soil are $c' = 15$ kPa, $\phi' = 30°$, and $\gamma = 20$ kN/m$^3$. The groundwater level is also shown in the figure.

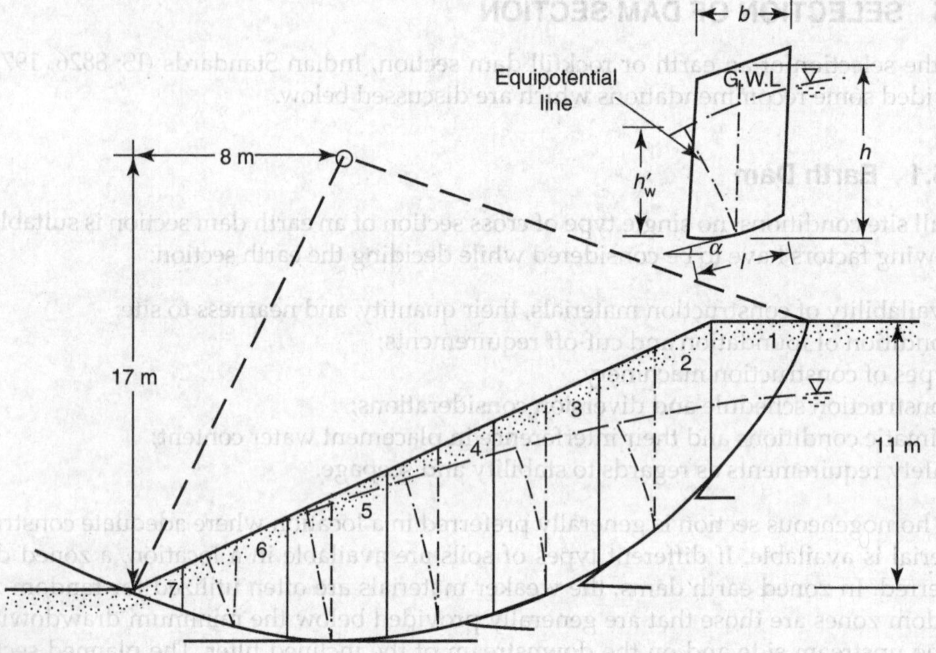


Fig. 20.15

*Solution*

For this problem, the Fellenius method of slices may be adopted. The sliding mass is divided into seven slices, as shown in Fig. 20.15. The width, height, length, and angle of inclination of the base are determined and tabulated as below:

| Slice no. | $b$ (m) | $h$ (m) | $l$ (m) | $\alpha$ (°) | $W = bh\gamma$ (kN) | $N = W \cos \alpha$ (kN) | $h_w$ (m) | $U = \gamma w h w l$ (kN) | $N' = N - U$ (kN) | $N' \tan \phi'$ (kN) | $T = W \sin \alpha$ (kN) | $c'l$ (kN) |
|---|---|---|---|---|---|---|---|---|---|---|---|---|
| 1. | 3.6 | 4.0 | 6.8 | 58 | 288.0 | 152.6 | 1.2 | 80.0 | 72.6 | 41.9 | 244.2 | 102.0 |
| 2. | 4.4 | 7.2 | 5.8 | 41 | 633.6 | 478.2 | 4.0 | 227.5 | 250.7 | 144.7 | 415.7 | 87.0 |
| 3. | 4.0 | 8.0 | 4.4 | 24 | 640.0 | 584.7 | 5.8 | 250.3 | 334.4 | 190.1 | 260.3 | 66.0 |
| 4. | 4.0 | 7.4 | 4.0 | 13 | 592.0 | 576.8 | 5.8 | 227.5 | 349.3 | 201.7 | 183.2 | 60.00 |
| 5. | 4.0 | 5.8 | 4.0 | 0 | 464.0 | 464.0 | 5.0 | 196.1 | 267.9 | 154.7 | 0 | 60.00 |
| 6. | 4.0 | 3.4 | 4.2 | −13 | 272.0 | 265.0 | 2.4 | 98.9 | 166.1 | 95.9 | −61.2 | 63.00 |
| 7. | 2.0 | 1.0 | 1.0 | −22 | 60.0 | 55.6 | 0.5 | 34.4 | 20.9 | 12.1 | −22.5 | 15.0 |
| | | | | | | | | | Σ | 841.1 | 969.7 | 453.0 |

$$F = \frac{\sum c'l + \sum N' \tan \phi'}{\sum W \sin \alpha}$$

$$F = \frac{453 + 841.1}{969.7} = 1.34$$

**Example 20.2** The natural moisture content of a borrow pit soil is 12.5% with a unit weight of 18.6 kN/m$^3$ with a specific gravity of 2.68. Compute the compacted volume of an embankment per 1.0 m$^3$ of a borrow pit soil. The void ratio of the proposed embankment is 0.42.

***Solution***

$$\left.\begin{array}{l}\text{Dry unit weight of}\\ \text{borrow pit soil, } \gamma_d\end{array}\right\} = \frac{\gamma}{1+\omega} = \frac{18.6}{1+0.125}$$

$$= 16.53 \text{ kN/m}^3.$$

$$\text{Void ratio of borrow pit } e_b = \frac{C\gamma_w}{\gamma_d} - 1$$

$$= \frac{2.68 \times 10}{16.53} - 1 = 0.62.$$

Volume of soil solids is same both in borrow pit and embankment, i.e.,

$$v_s = \frac{v_b}{1+e_b} = \frac{v_e}{1+e_e}.$$

$$\therefore \text{Volume of embankment} = (1+e_e)\frac{v_b}{1+e_b}$$

$$= (1+0.42) \times \frac{1}{(1+0.62)}$$

$$= 0.877 \text{ m}^3.$$

Hence for 1 cu m of borrow pit soil, 0.877 cu m of embankment can be constructed with a void ratio of 0.42.

## POINTS TO REMEMBER

20.1 Embankment dams are flexible structures primarily made out of earth and rock fragments and used to impound water.

20.2 Embankment dams are classified as rolled fill or placed earthfill dams and hydraulic fill dams. Rolled fill dams are constructed by compacting soils in successive layers with the available materials in and around the construction site. Hydraulic fill dams are constructed by pumping the soil water suspension to the required site and allowing it to settle.

20.3 Embankment dams are of three types, viz., diaphragm type, homogeneous type, and zoned type. In the diaphragm type, the embankment is constructed of pervious materials with a water barrier on the upstream side. Homogeneous dams are constructed from a simple kind of material. Zoned dams are more common types of dams with pervious materials as shells and impervious materials as cores.

20.4 The components of an earth dam are the foundation, casing, core, top width, freeboard, cut-off, internal drain, transition filter, slope protection, and relief wells. All the components are not provided in a single dam.

20.5 The salient features in the design of earth dams are (i) there should be no danger to the dam by over-topping, (ii) seepage through the body of the dam and foundation should be controlled, (iii) upstream and downstream slopes should be protected against sliding, and (iv) crest, upstream, and downstream faces should be protected.

20.6 In earth dams, the phreatic surface constitutes the top flow line which has to be located based on the boundary condition.

20.7 It has been mathematically shown that the basic shape of the phreatic surface is that of a parabola, and it deviates only at the upstream and downstream faces depending on the cross section of the dam.

20.8 Excess seepage may be caused due to highly permeable dam material, short seepage paths, and defects such as fissures and cracks caused by uneven settlement.

20.9 Excess seepage through the body or foundation of a dam is controlled by providing (i) an impervious core, (ii) partial or free cut-off in the foundation, or (iii) upstream blanket.

20.10 Excess uplift pressures and gradients are reduced by providing (i) trench drains, (ii) continuous downstream drains, (iii) toe drains, or (iv) relief wells.

20.11 Stability of the slopes of an earth dam are critical under the following conditions: (i) construction condition with or without partial pool (for upstream and downstream slopes), (ii) reservoir partial pool (for upstream slopes), (iii) steady seepage (for downstream slopes), (iv) sudden drawdown (for upstream slopes), (v) steady seepage with sustained rainfall (for downstream slopes), and (vi) earthquake conditions (for upstream and downstream slopes).

20.12 The circular arc method or sliding wedge method is adopted for stability analysis of slopes.

20.13 Crest, upstream, and downstream slopes should be protected from destructive wave action. The crest is protected by surfacing. The upstream slope is protected by rip-rap, concrete pavement, or steel facing. The downstream slope is protected by a heavy layer of coarse gravelly material or surfaced with top soil and planting shrubs or grass.

20.14 The factors to be considered while selecting an earth or rockfill dam are (i) availability of construction materials, their quantity, and proximity to the site; (ii) condition of the foundation and cut-off requirements; (iii) types of construction machinery; (iv) construction schedule and diversion considerations; (v) climatic conditions and their interference in placement water content; and (vi) safety requirements as regards stability and seepage.

## QUESTIONS

### *Objective Questions*

20.1 State whether the following statements are true or false:

1. In an earth dam, the phreatic surface constitutes the bottom flow line.
2. The upstream slope of an earth dam during steady seepage conditions is an equipotential line.
3. The shell imparts stability and protects the core.

4. Cut-off is a barrier to reduce seepage of water through the core and casing.
5. The material used for an impervious blanket should have far less permeability than the foundation soil.

20.2 Select the type of soil most suitable for the core of a zoned dam from the following:
(a) SC (b) GC (c) CH (d) CI

20.3 A transition filter is provided between a coarse and fine material to
(a) Give sufficient support for the core and shell.
(b) Prevent the seepage of water completely into the core.
(c) Prevent the migration of fine material to coarse material.
(d) Reduce the excess hydrostatic pressure.

20.4 In a homogeneous dam sudden drawdown of a reservoir level causes instability to the
(a) Downstream slope
(b) Upstream slope
(c) Both upstream and downstream slopes
(d) None of the slopes

20.5 The extra height provided in the crest of the dam is to allow for
(a) Compression of fill material
(b) Settlement of foundation
(c) Compression of fill material and settlement of foundation
(d) Extra safety against wave action

20.6 In an earth dam the critical condition(s) for which the stability has to be checked during construction with or without partial pool is/are
(a) Downstream slope
(b) Upstream slope
(c) Upstream and downstream slopes
(d) None

20.7 In the stability analysis of an upstream slope for the sudden drawdown condition, generally, no pore pressure is considered in the shell when the coefficient of permeability is greater than
(a) $10^{-3}$ cm/s (b) $10^{-4}$ cm/s (c) $10^{-1}$ cm/s (d) $10^{-2}$ cm/s

20.8 The sliding wedge method of analysis is generally applicable in the circumstances where
(a) One or more horizontal layers of weak soil exist in the upper part of the foundation.
(b) More than one type of material is used in the body of the earth dam.
(c) The dam material is a homogeneous one and resting on a pervious foundation.
(d) The dam material is a homogeneous one and resting on a very stiff soil.

## *Descriptive Questions*

20.9 Explain the design considerations for upstream and downstream slopes of an earth dam.

20.10 What is a rockfill dam? How is it different from an earth dam?

20.11 Explain why a thin core dam may be economical even when ample supplies of both pervious and impervious soils are available.

20.12 Under what conditions would it be advisable to place a blanket layer of coarse-grained material between an embankment and its foundation?

20.13 Define seepage force, piping, and roofing as understood in the study of stability of an earth dam resting on a relatively porous medium.

20.14 What is meant by construction pore pressure in earth dams? What are the methods adopted to reduce the same?

20.15 An earth dam foundation consists of clay having very low shear strength. It is proposed to construct a dam of 10 m height. Suggest a method of foundation treatment and design considerations so as to avoid a base failure.

20.16 What is the significance of a filter in an earth dam? Discuss the design criteria of filters.

20.17 State some general principles of soil selection in the design and construction of embankments.

20.18 Bring out the role of the nature of embankment soil on the cracking phenomena of embankments.

20.19 Write a detailed note on the construction of rockfill dams.

20.20 List the field tests you would conduct on a composite rolled fill construction. Explain the importance of each test.

## EXERCISE PROBLEMS

20.1 Construct a flow net and estimate the seepage of the homogeneous earth dam sketched in Fig. 20.16. The average coefficient of permeability is $4 \times 10^{-7}$ m/s.

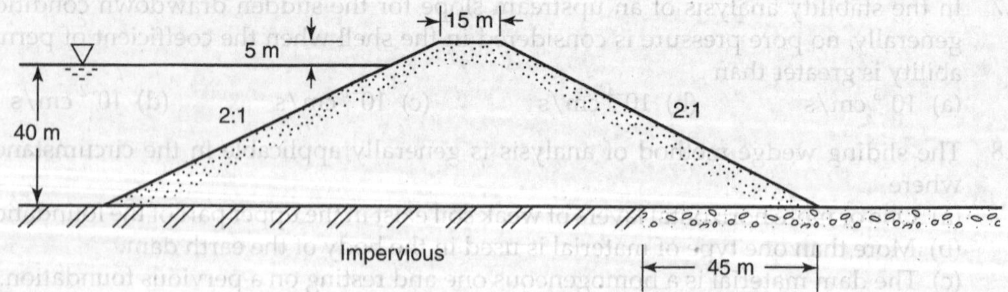


Fig. 20.16

20.2 Make a flow net for the earth dam given in Fig. 20.17 and estimate the seepage loss.

20.3 Figure 20.18 shows a section through an earth–rock dam. The clay is isotropic with a permeability of $10^{-8}$ m/s. Assuming the rockfill permeability to be infinite, determine the quantity of seepage through the core per metre width of the dam.

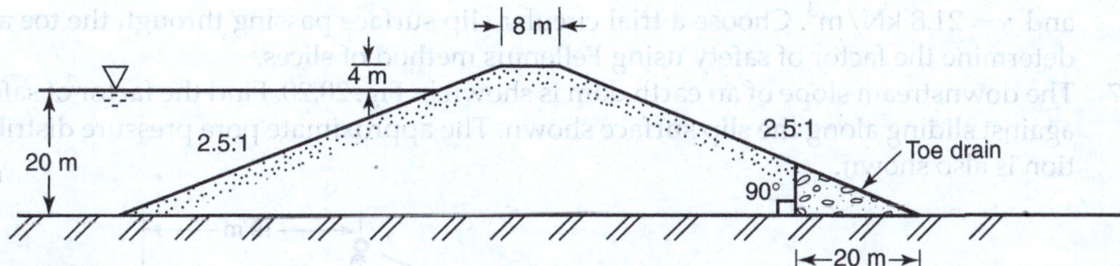


Fig. 20.17

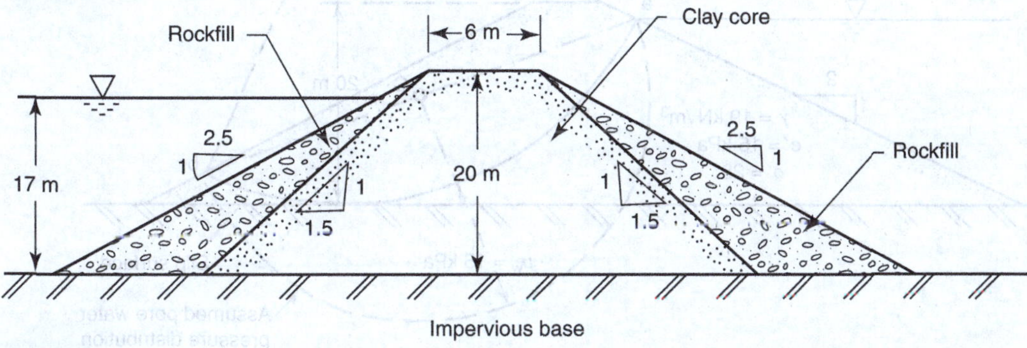


Fig. 20.18

20.4 Calculate the minimum length of under drain needed for the cross section of an earth dam shown in Fig. 20.19, such that the top seepage line shall not be nearer than 7.2 m from the downstream surface of the dam.

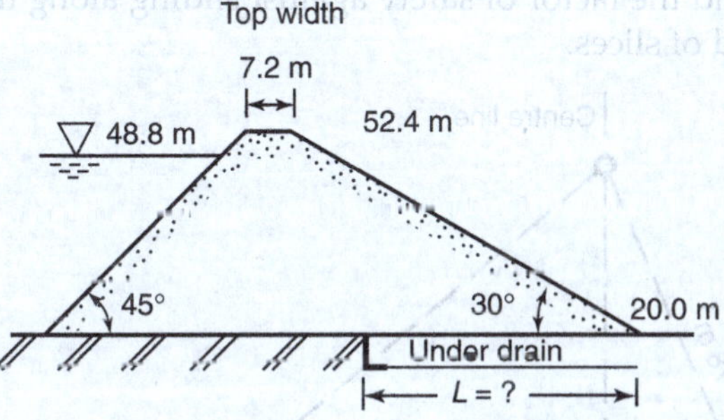


Fig. 20.19

20.5 A reservoir embankment 30 m in height consists of compacted earth with $c = 35\ \text{kN/m}^2$, $\phi = 15°$, and $\gamma = 19.6\ \text{kN/m}^3$. The embankment has a slope of 20°. Compute the factor of safety if a sudden drawdown of the reservoir occurs. Use Taylor's stability charts.

20.6 A 50 m high homogeneous earth dam with an upstream slope of 2.5 to 1 is founded on a rock base. The estimated pore pressure coefficient $r_u = 0.30$ at the end of construction. The soil parameters of the earth dam material are $c' = 22\ \text{kN/m}^2$, $\phi' = 27°$, and $\gamma = 21.8\ \text{kN/m}^3$. Choose a trial circular slip surface passing through the toe and determine the factor of safety using Fellenius method of slices.

20.7 The downstream slope of an earth dam is shown in Fig. 20.20. Find the factor of safety against sliding along the slip surface shown. The approximate pore pressure distribution is also shown.

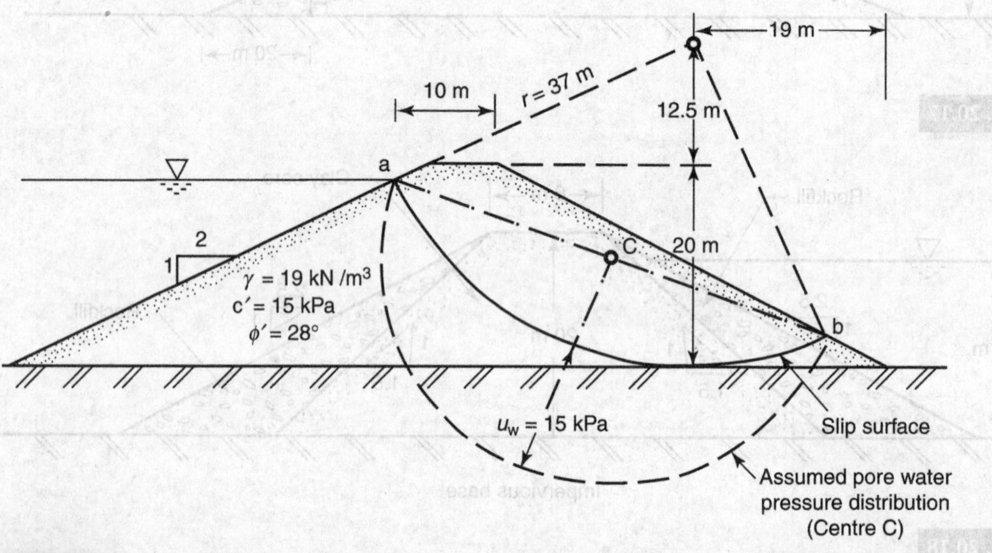


Fig. 20.20

20.8 The upstream slope of an earth dam under steady seepage conditions is shown in Fig. 20.21. The relevant parameters of the dam are $e = 0.60$, $G = 2.68$, $c' = 17\ \text{kN/m}^2$, and $\phi' = 26°$. Find the factor of safety against sliding along the slip surface using the ordinary method of slices.

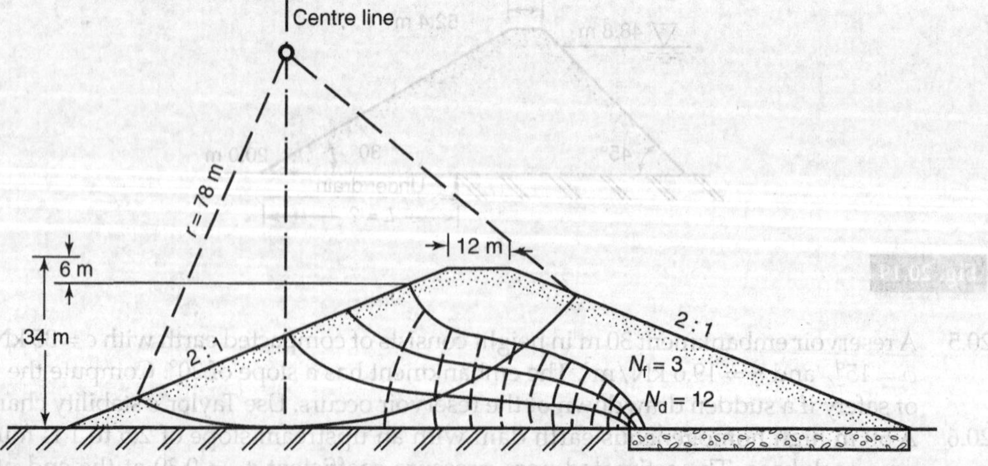


Fig. 20.21

# 21 Dynamic Loading of Soil

**CHAPTER HIGHLIGHTS**

Earthquake: seismic waves, magnitude – Other dynamic loads – Theory of vibration: free and forced vibrations with and without damping – Types of machines and machine foundations – Dynamic bearing capacity of shallow foundations – Design requirements – Methods of analysis for block foundation – Liquefaction of soils

## 21.1 INTRODUCTION

Elastic waves moving through a soil mass produce ground motion, which is transmitted through foundations to a structure as vibrations. The vibrations are caused by earthquakes or from construction activity such as rock blasting, pile driving, etc. On the other hand, vibrations also result from operating machinery and are transmitted through the foundations to the soil. Vibrations developed by operating machinery produce several effects which must be considered in the design of foundations. As large machines are usually supported on the soil and the impulses are directly transmitted to the soil, the design of machine foundations involves the problem of soil dynamics. A detailed review of the effects of dynamic loading on soil properties is given by Prakash (1981). Liquefaction of soils is an important phenomenon that occurs due to large vibrating forces.

## 21.2 EARTHQUAKES

When friction between rocks on either side of a fault is adequate to prevent the rocks from slipping easily or when the stressed rock is not already fractured, some elastic deformation occurs before failure. When the stress exceeds the rupture strength of the rock (or the friction between rocks along an existing fault), sudden movement occurs along the fault, resulting in an *earthquake*. After the rupture, the rocks snap back to the previous dimensions due to elasticity, which is referred to as *elastic rebound*. The stress release and the movement are reflected as the relative displacement of the rocks on either side of the fault following the earthquake. This released energy is propagated in the form of seismic waves, which pass on the energy through the earth media. During this process, all the structures erected on the earth's surface are subjected to vibrations.

The point inside the earth's surface at which the first movement or break occurs during an earthquake is called the *focus* or *biocentre* of the earthquake. The point on the earth's surface directly above the focus is the *epicentre*. The position of the focus is determined from a seismograph record.

### 21.2.1 Seismic Waves

The major seismic waves are body waves and surface waves. Body waves consist of *P*- and *S*-waves. *P*-waves are compressional waves travelling through the earth, similar to sound waves, which travel through the air. *S*-waves are shear waves involving a side-to-side sliding motion of the material. Surface waves travel along the surface, which is somewhat analogous to the movement of surface waves on water. Surface waves cause rocks and soils to be displaced in such a way that the ground surface undulates. Vertical ground motions are caused by Rayleigh waves, while Love waves cause horizontal motions. Most of the damages due to earthquake are caused by surface waves.

### 21.2.2 Magnitude

Magnitude of an earthquake is a measure of the amount of ground shaking (i.e., the amount of vertical motion) based on the amplitude of elastic wave it generates. Richter's magnitude scale, named after Prof. Charles Richter, a geologist, is most often used. The Richter scale starts from 2, and there is no upper limit. Table 21.1 gives the description of an earthquake in relation to its magnitude on the Richter scale.

The Richter scale is a logarithmic one; that is, an earthquake of magnitude 4 causes 10 times as much ground movement as one of magnitude 3, one hundred times as much as one of magnitude 2, and so on.

**Table 21.1** Magnitude of an earthquake

| Magnitude | Description |
|---|---|
| >8.0 | Great earthquake |
| 7.0–7.9 | Major earthquake |
| 6.0–6.9 | Destructive earthquake |
| 5.0–5.9 | Damaging earthquake |
| 4.0–4.9 | Minor earthquake |
| 3.0–3.9 | Small earthquake, usually felt |
| 2.0–2.9 | Detected but not felt |

*Source:* Montgomery (1990).

## 21.3 OTHER DYNAMIC LOADS

Apart from earthquakes, there are other sources of dynamic loads. Soils may also be subjected to vibrations due to bomb blasts, driving of piles in the vicinity, wind and water action on structures, landing of aircraft, etc. Some of the above sources are natural, and some are man-made.

The other major man-made sources of dynamic loads are machines, machine tools, engines, generators, turbines, forging and foundry machinery, vehicle and conveyers, mining, and transmission equipment, which cause different vibration problems. These dynamic forces and moments should be contained by proper means, otherwise it may lead to undesirable noise, discomfort, instability of structures, etc.

## 21.4 THEORY OF VIBRATIONS

The behaviour of a structure subjected to dynamic load is better understood by studying the mechanisms of vibrations caused by the dynamic load. The pattern of variation of a dynamic load with respect to time may be either transient or periodic. Transient vibrations may have non-periodic time history, e.g., vibrations caused by earthquakes, bomb blasts, quarry blasts, etc. The periodical motions can be resolved into sinusoidally varying components, e.g., vibrations due to reciprocating machine foundation.

A structure subjected to a dynamic load vibrates in one of the following four ways of deformation or a combination of one or more:

1. Extensional (Fig. 21.1a)
2. Shearing (Fig. 21.1b)

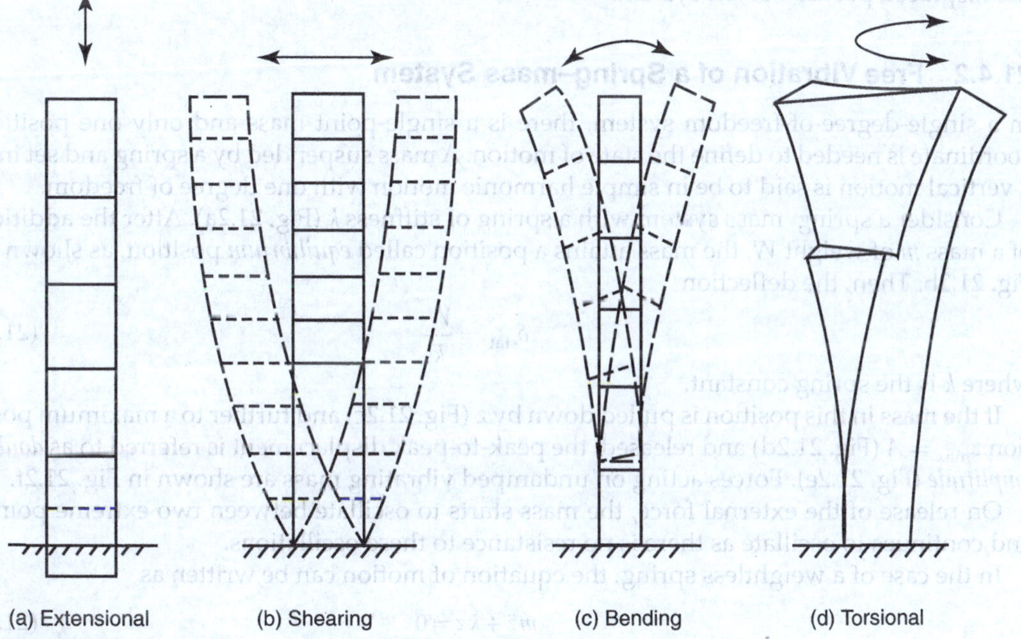


**Fig. 21.1** Different types of vibrations (*Source:* Swami Saran, 1999)

3. Bending (Fig. 21.1c)
4. Torsional (Fig. 21.1d)

### 21.4.1 Harmonic Vibrations

In order to understand the reasons for the additional precautions needed for oscillating machinery, it is essential to have a clear concept of the basic theory of harmonic vibrations.

Many of the foundation problems arising from oscillating machinery can be analysed by considering a simple harmonic application of force. Some of the terms relevant to machine foundation vibrations are discussed below.

The motion of a point in a straight line is said to be in *simple harmonic motion* such that the acceleration of the point is proportional to the distance of the point from some fixed origin and is always directed towards the origin.

A foundation system is said to be in *free vibration* when it is disturbed and then left free to vibrate about some mean position. Thus, if an elastic system vibrates under the action of inherent forces in the system and in the absence of any externally applied force, the frequency of vibration is termed *natural frequency*. Vibrations that result from exciting agencies are referred to as *forced vibrations*.

The foundation system is said to be at *resonance* when the frequency of the exciting agency (operating frequency) is close to the natural frequency. Under the condition of resonance, large forces and amplitudes of motion (the distance that a body moves from its position of rest when subjected to vibration) can be generated.

*Damping* is associated with energy dissipation and is the internal resistance offered by a foundation system to the vibration of a machine.

*Degrees of freedom of a system* is defined as the number of coordinates required to describe the displaced position of the system.

### 21.4.2 Free Vibration of a Spring–mass System

In a single-degree-of-freedom system, there is a single-point mass and only one position coordinate is needed to define the state of motion. A mass suspended by a spring and set into a vertical motion is said to be in simple harmonic motion with one degree of freedom.

Consider a spring–mass system with a spring of stiffness $k$ (Fig. 21.2a). After the addition of a mass $m$ of weight $W$, the mass attains a position called *equilibrium* position, as shown in Fig. 21.2b. Then, the deflection

$$\delta_{\text{stat}} = \frac{W}{k} \tag{21.1}$$

where $k$ is the spring constant.

If the mass in this position is pulled down by $z$ (Fig. 21.2c) and further to a maximum position $z_{\max} = A$ (Fig. 21.2d) and released, the peak-to-peak displacement is referred to as *double amplitude* (Fig. 21.2e). Forces acting on undamped vibrating mass are shown in Fig. 21.2f.

On release of the external force, the mass starts to oscillate between two extreme points and continues to oscillate as there is no resistance to these oscillations.

In the case of a weightless spring, the equation of motion can be written as

$$m\ddot{z} + kz = 0 \tag{21.2}$$

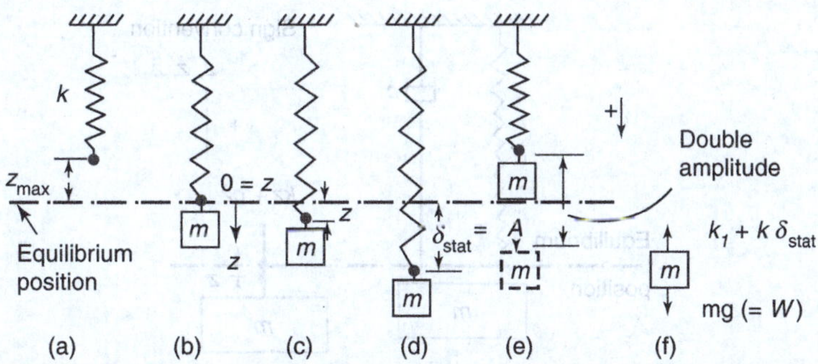


**Fig. 21.2** Spring–mass system: (a) unstretched spring; (b) equilibrium position; (c) mass in oscillating position; (d) mass in maximum downward position; (e) mass in upward position; (f) free-body diagram of mass corresponding to (c)

A solution to this equation is obtained (for the solution, refer Converse, 1962; Prakash, 1981), and after satisfying Eq. 21.1, the expression for $\omega_n$, the circular natural frequency of the system, is given as

$$\omega_n = \sqrt{\frac{k}{m}} \tag{21.3}$$

One cycle of motion is completed when $\omega_n T = 2\pi$, where $T$ is the period. Then,

$$T = \frac{2\pi}{\omega_n} = 2\pi\sqrt{\frac{m}{k}} \tag{21.4}$$

The natural frequency, $f_n$, the number of cycles executed in unit time, is given as

$$f_n = \frac{1}{T} = \frac{1}{2\pi}\sqrt{\frac{m}{k}} \tag{21.5}$$

The period $T_n$ is the free period of the spring–mass system and depends on both the spring constant and the mass. With reference to foundations on soil, the foundation and attached masses are taken as $m$ and the earth as the spring. In the above treatment, the mass of the spring (earth) is neglected, and hence this simple theory represents approximately the behaviour of foundation on soil.

### 21.4.3 Free Vibration with Viscous Damping

The governing equations for the free-vibration condition are based on the assumption that there was no loss of energy in the oscillating system due to internal friction or external forces opposing the motion. In practical problems, there is loss of energy and decrease in amplitude due to damping. In one type of damping, the force $F_d$ is taken directly proportional to the velocity of the oscillating mass, which is defined as *viscous damping*. That is,

$$F_d = -c\dot{z} \tag{21.6}$$

where $c$ is the coefficient of viscous damping.

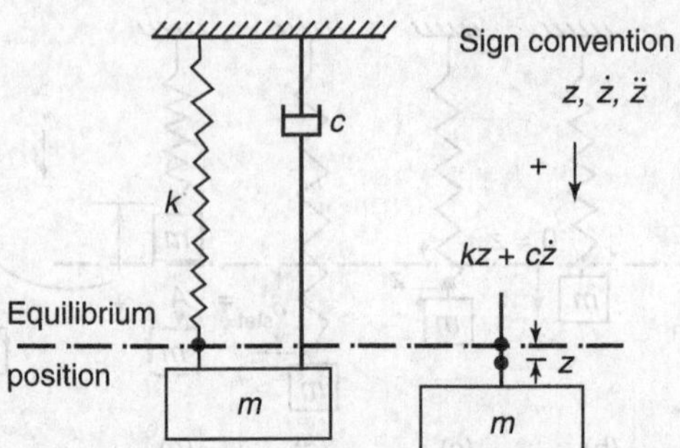


Fig. 21.3 Spring–mass–dashpot system

Consider a spring–mass–dashpot system (Fig. 21.3). The free-body diagram of a displaced mass under damped vibration condition is also shown in Fig. 21.3. The equation of motion may be written as

$$m\ddot{z} + c\dot{z} + kz = 0 \tag{21.7}$$

A solution to this equation (for the solution, refer Converse, 1962; Prakash, 1981) is based on two conditions. For one condition, the value

$$\left(\frac{c}{2m}\right)^2 = \frac{k}{m} = \omega_n^2 \tag{21.8a}$$

This condition is referred to as critical damping condition and Eq. 21.8a is rewritten with $c = c_c$ as

$$c_c = 2m\,\omega_n \tag{21.8b}$$

The ratio of actual damping ($c$) to critical damping ($c_c$) is referred to as the damping factor, $\zeta$. That is,

$$\zeta = \frac{c}{c_c} \tag{21.8c}$$

For the second condition, $k/m < (c/2m)^2$, the solution yields an expression for $\omega_{nd}$, the circular natural frequency for the damped free vibration, as

$$\omega_{nd} = \omega_n\sqrt{1-\zeta^2} \tag{21.8d}$$

Then, the period is given by

$$T = \frac{2\pi}{\omega_n\sqrt{1-\zeta^2}} \tag{21.9}$$

Figure 21.4 represents a typical damped oscillation for $\zeta < 1.0$.

### 21.4.4 Forced Vibrations with Viscous Damping

Consider a spring–mass–dashpot system subjected to a force $F$ (Fig. 21.5a), where

$$F = F_0 \sin \omega t \tag{21.10}$$

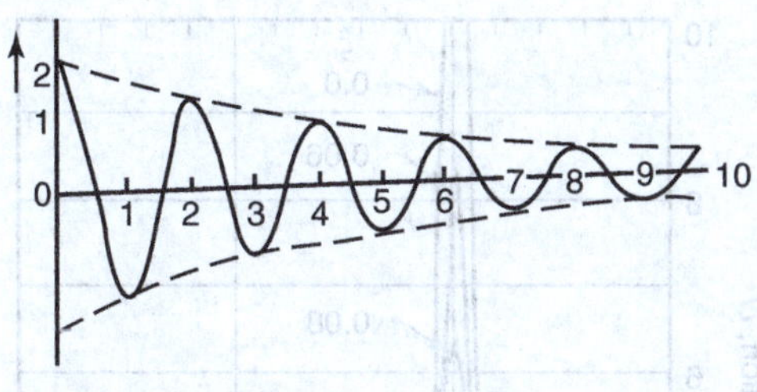


**Fig. 21.4** Typical damped oscillation

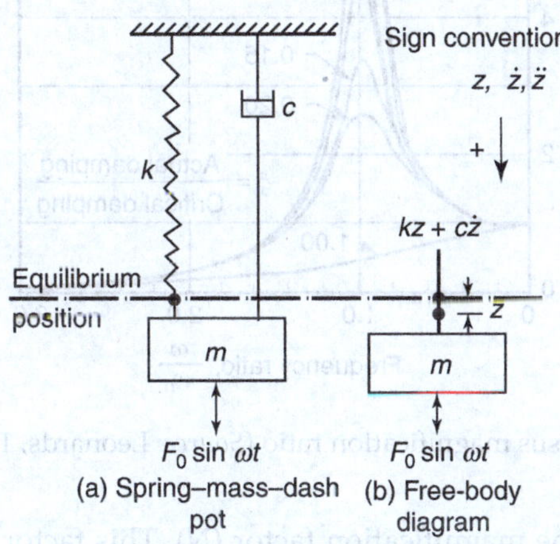


**Fig. 21.5** Forced vibrations with viscous damping

where $\omega$ is the frequency of the force of excitation. The free-body diagram is shown in Fig. 21.5b. The equation of motion is

$$m\ddot{z} + c\dot{z} + kz = F_0 \sin \omega t \tag{21.11}$$

A solution to this equation (for the solution, refer Converse, 1962; Prakash, 1981) gives the following expressions:

$$z_0 = \frac{F_0 / k}{\sqrt{(1-r^2) + (2\zeta r)^2}} \tag{21.12}$$

$$\theta = \tan^{-1}\left[\frac{2\zeta r}{1-r^2}\right] \tag{21.13}$$

where $z_0$ is the maximum displacement of the forced vibration, $\theta$ the phase angle between the applied force and the displacement, and $r = \omega / \omega_n$, the frequency ratio.

The factor $\Gamma_0/k$ represents the static deflection of the system ($\delta_{stat}$) under a static load equal to the dynamic force $F_0$. Thus,

$$\frac{z_0}{\delta_{stat}} = \frac{1}{\sqrt{(1-r^2)^2 + (2\zeta r)^2}} \tag{21.14}$$

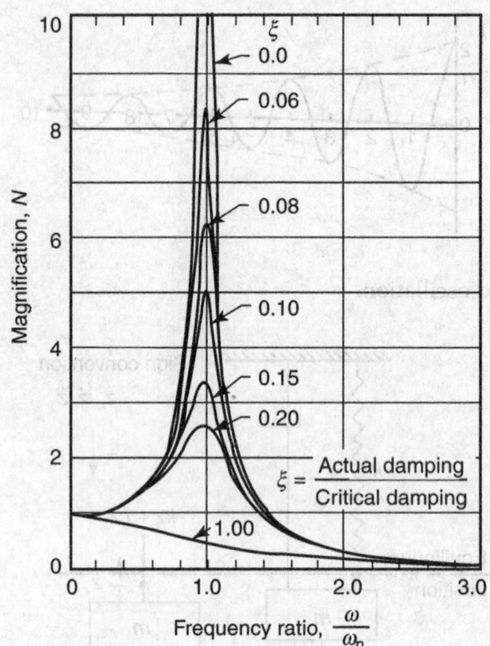


**Fig. 21.6** Frequency ratio versus magnification ratio (*Source:* Leonards, 1962)

Equation 21.14 is called the magnification factor ($N$). This factor signifies the increase in the maximum deformation of a forced vibration from that caused by a static load equal in magnitude to that of the dynamic force.

For the undamped vibration condition, $c = 0$, and $N$ becomes

$$\frac{1}{(1-r^2)^2} \tag{21.15}$$

For low values of $\omega$, the increase in amplitude is less, as $N$ is close to unity. For $\omega = \omega_n$, $N$ is infinite and the amplitude is infinite if sufficient damping is not available in the system. For $\omega > \omega_n$, the amplification gradually decreases. By providing adequate damping, the amplitude decreases depending on the magnitude of the damping ratio.

Figure 21.6 shows the magnification factor $N$ versus frequency ratio $r$. To avoid large amplitudes, $r$ should be kept less than 0.5 or greater than 1.5.

## 21.5 TYPES OF MACHINES AND MACHINE FOUNDATIONS

Machines generate different periodic forces. They can be broadly categorized under three heads (Prakash, 1981):

1. *Reciprocating machines:* These machines produce unbalanced force and work with low operation frequency* (speed) in the order of 600 revolutions (rev)/min. While analysing

*The rotating speed of the main drive in rpm or the frequency of the periodic force acting on the system.

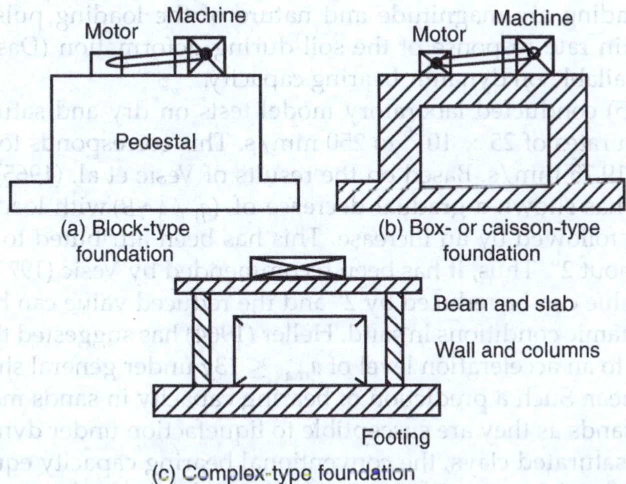


**Fig. 21.7** Typical machine foundations (*Source:* Prakash, 1981)

the foundation, the unbalanced force is considered to be sinusoidal. Compressors and reciprocating engines fall under this category.

2. *Impact machines:* These machines produce impact loads and operate with a speed range of 60 to 150 blows per minute. In these machines, dynamic load reaches a maximum in a very short time and attenuates immediately. Machines like forging hammers, stamping press, etc., are included in this category.
3. *Rotary machines:* These machines operate at frequencies ranging from very low to high. They are classified (IS: 2974 – Part 3, 1975) as low (frequency up to 1,500 rpm), medium (frequency 1,500 to 3,000 rpm), and high (frequency greater than 3,000 rpm). Machines which are included in this category are crushing mills, pumps, motor generators , turbo generators, etc.

Machine foundations can be broadly classified into three types: (i) block foundations, (ii) box or caisson foundations, and (iii) complex foundations (Fig. 21.7). Block foundation is the simplest type, consisting of a pedestal resting on a footing. Block foundations have a large mass and, hence, less natural frequency (defined in the next section). This type of foundation is suitable for compressors and reciprocating foundations. If more machines of a similar type are to be arranged in a machine shop, it can be done economically by providing a single, continuous mat.

In the box or caisson type of foundation, the mass is reduced, and by virtue of this, the natural frequency is increased. Such foundations can be used for medium-speed machines like forging hammers.

High-speed machines like steam turbines require a system of wall columns and beam slabs. Here, the elements of the systems are relatively flexible compared to block or box-type foundations.

## 21.6 DYNAMIC BEARING CAPACITY OF SHALLOW FOUNDATIONS

Dynamic loads may act in different directions. The dynamic loading in the vertical direction only is considered here. Such loadings may induce large, permanent deformations in foundations. The most important factors that need consideration while analysing a foundation under dynamic loading are magnitude and nature of the loading pulse, duration of the pulse, and the strain rate response of the soil during deformation (Das, 1983). Only very limited data are available on dynamic bearing capacity.

Vesic et al. (1965) conducted laboratory model tests on dry and saturated dense sands under varied strain rates of $25 \times 10^{-5}$ to 250 mm/s. This corresponds to a varying loading velocity of 14.4 to 19.75 mm/s. Based on the results of Vesic et al. (1965) and other similar works, Das (1983) has shown a gradual decrease of $(q_f / \frac{1}{2}\gamma B)$ with loading velocity up to a minimum value, followed by an increase. This has been attributed to a reduction of the friction angle by about 2°. Thus, it has been recommended by Vesic (1973) that for dynamic conditions the $\phi$ value can be reduced by 2° and the reduced value can be used for bearing capacity under dynamic conditions in sand. Heller (1964) has suggested that foundations on sand are subjected to an acceleration level of $a_{max} \leq 13g$ under general shear and $a_{max} > 13g$ under punching shear. Such a prediction of bearing capacity in sands may be unreliable in loose, submerged sands as they are susceptible to liquefaction under dynamic loads.

For footings on saturated clays, the conventional bearing capacity equation can be used for the $\phi = 0$ condition, but an appropriate cohesion value has to be introduced. Generally, undrained cohesion increases with strain rate. Carroll (1963) suggested that undrained cohesion under dynamic condition can be taken as 1.5 times the undrained cohesion under static condition. This modified cohesion can be used in the bearing capacity equations to obtain the dynamic bearing capacity in clays.

## 21.7 DESIGN REQUIREMENTS

The conventional considerations of safe bearing capacity against shear failure and allowable settlement are insufficient to ensure a satisfactory machine foundation design. In the earlier designs, the static load was increased by multiplying by a dynamic factor, and conventional static foundation analysis was done for the increased static load without any knowledge of the safety factor. Because of this uncertainty about the magnitude of the dynamic factor, erroneous behaviour of foundations was observed in practice. The other approach is to increase the mass of the foundation so as to absorb the vibrations. Cozens (1938) suggested desirable ratios of foundation mass to mass of the engine (Table 21.2). Evidently, this approach does not consider the type of soil and its environment, the frequency level or the amplitude of the machine. However, the data available in Table 21.2 can be used to obtain an initial rough estimate of the size of the foundation.

Thus, the design of a machine foundation has to satisfy the following additional criteria (Prakash, 1981):

1. Resonance should be avoided; that is, the natural frequency of the foundation–soil system should not be equal to the operating frequency. This can be achieved if the natural frequency of the system lies outside the zone of resonance.
2. The amplitude of vibration at the operating frequency should be less than the limiting amplitude. The limiting amplitudes are usually specified by the manufacturers or in Codes of Standards. Frequency and amplitude requirements for machine foundations are given in Table 21.3, based on data from the Indian Standards (IS: 2974, 1979).
3. The vibration of the machine should neither cause damage to machines and structures nor harm the health and comfort of the people. The degree of severity of vibration depends on the operating frequency and amplitude of motion.

**Table 21.2** Foundation mass required to absorb vibrations

| Type of engine | Ratio of foundation mass to mass engine |
|---|---|
| Steam engine (single cylinder) | Not less than 4.0:1 |
| Steam engine (compound) | 3.75:1 |
| Steam engine (triple expansion) | 2.50:1 |
| Gas engine (single cylinder) | 3.00:1 |
| Gas engine (two cylinders) | 3.00:1 |
| Gas engine (four cylinders) | 2.75:1 |
| Gas engine (six cylinders) | 2.25:1 |
| Gas engine (eight cylinders) | 2.00:1 |
| Diesel engine (two cylinders) | 2.75:1 |
| Diesel engine (four cylinders) | 2.40:1 |
| Diesel engine (six cylinders) | 2.10:1 |
| Diesel engine (eight cylinders) | 1.90:1 |
| Vertical steam engine (compound) coupled to generator | 3.80:1 |
| Vertical steam engine (triple expansion) coupled to generator | 3.50:1 |
| Horizontal steam engine (cross-compound) coupled to generator | 3.25:1 |
| Horizontal steam turbine coupled to generator | 3:1 to 4:1 (if of small output) |
| Vertical gas engine coupled to generator | 3.50:1 |
| Vertical diesel engine coupled to generator | 2.60:1 |

*Source:* Couzens (1938).

**Table 21.3** Foundation frequency and amplitude requirements for various machines

*Reciprocating machines*

$f_{op}$ should be greater than $2f_n$ or less than $0.5f_n$ in important machines; otherwise, $f_{op}$ should be greater than $1.5f_n$ or less than $0.6f_n$

*Drop and forge hammers*

$f_n$ should be greater than $2.5f_i$ or less than $0.7f_i$

Peak vertical amplitude of the foundation should not exceed 1.2 mm. Peak vertical amplitude of the foundation block should not exceed 0.8 mm if the foundation is on sand below the water table

With important structures nearby, maximum peak velocity of the foundation is 3 mm/s

*Rotary-type machines*

For high-speed machines, $f_n$ should be no closer than 20% to $f_{op}$. Permissible amplitudes at beating level:

1. For $f_{op} < 3,000$ rpm
   Vertical, 0.04–0.06 mm
   Horizontal, 0.07–0.09 mm
2. For $f_{op} > 3,000$ rpm
   Vertical, 0.02–0.03 mm
   Horizontal, 0.04–0.05 mm

Foundation vibrations should not exceed 50% of the above figures. For low-frequency machines ($f_{op} <$ 1,500 rpm), $f_n$ must be considerably greater than $f_{op}$, and foundation peak amplitude must not exceed 0.3 mm

*Note:* $f_{op}$ is the operating frequency of the machine, $f_i$ is the frequency of impact, and $f_n$ is the natural frequency of the machine–foundation system.

*Source:* IS: 2974 (1979); compiled by Moore (1985).

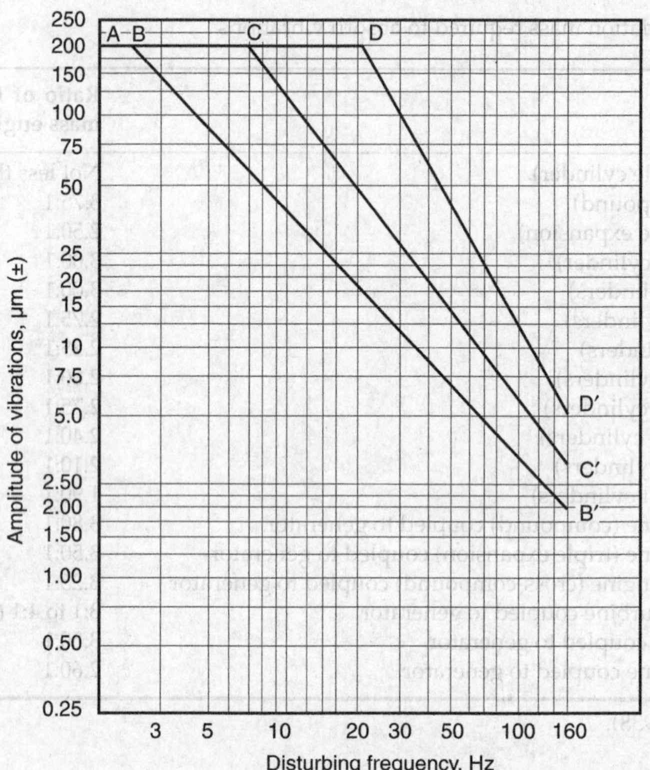


Line ADD′ limit to avoid damage to buildings
Line ACC′ limit to avoid service discomfort to persons
Line ABB′ limit to ensure reasonable comfort to persons
Note: These limits do not include any factor of safety

**Fig. 21.8** Amplitude limits of a foundation block (*Source:* IS: 2974 – Part 1, 1982)

A chart has been provided by the Indian Standards (IS: 2974 – Part 1, 1982), which gives various limits of frequency and amplitude for different conditions (Fig. 21.8).

The main steps to be followed in the design of a satisfactory machine foundation are as follows:

1. The design criteria to be adopted are finalized (as discussed above).
2. The dynamic loads which are to be transmitted to the foundation are computed.
3. The appropriate soil parameters under static and dynamic conditions are estimated from field and laboratory tests.
4. An appropriate method of analysis is chosen to compute the natural frequency and the vibration amplitude at operating frequency for each of the relevant modes of vibration. This is accomplished for an assumed foundation size, which is modified if necessary to satisfy the design criteria.

Moore (1985) suggests the following factors to be considered while proportioning the foundation and finalizing the details:

1. The centre of gravity of the machine and foundation should be as close as possible to the centroid of the foundation base area.
2. For low-speed machines ($f_{op} < 5$ Hz), the foundation has to be designed such that the natural frequency is much higher than the operating frequency. In case of high-speed machines ($f_{op} > 8$ Hz), the natural frequency should be maintained at a lower level.
3. In order to reduce or isolate transmission and vibrations from the machine foundation to other foundations and parts of the building, joints or vibration isolators are provided.
4. The foundation has to be designed such that it can be modified at a later date if some unforeseen vibration problems develop.

## 21.8 METHODS OF ANALYSIS FOR BLOCK FOUNDATION

In general, a machine foundation may undergo six independent displacement rotations, which are referred to as six degrees of freedom. They are translations in the $x$ (lateral), $y$ (longitudinal), and $z$ (vertical) directions (Fig. 21.9) and rotations about the $x$-axis (pitching), the $y$-axis (rocking), and the $z$-axis (yawing) (Fig. 21.9).

Of these six types of motion, some are independent and some are coupled. Translation along the $z$-axis and rotation about the $z$-axis can occur independently. But translation about the $x$-axis (or $y$-axis) and rotation about the $y$-axis (or $x$-axis) are coupled motions. Hence, in the analysis of a block foundation, one should consider four types of motions.

The theory of vibrations, based on one degree of freedom, is sufficient for the solution of many problems (Converse, 1962). Hence, analysis pertaining to vertical vibration only is considered.

A variety of theoretical approaches are available in the literature for determining the dynamic response of machine foundations. Two principal methods are common. The first considers the soil as an elastic half-space with the foundation on the surface. The second

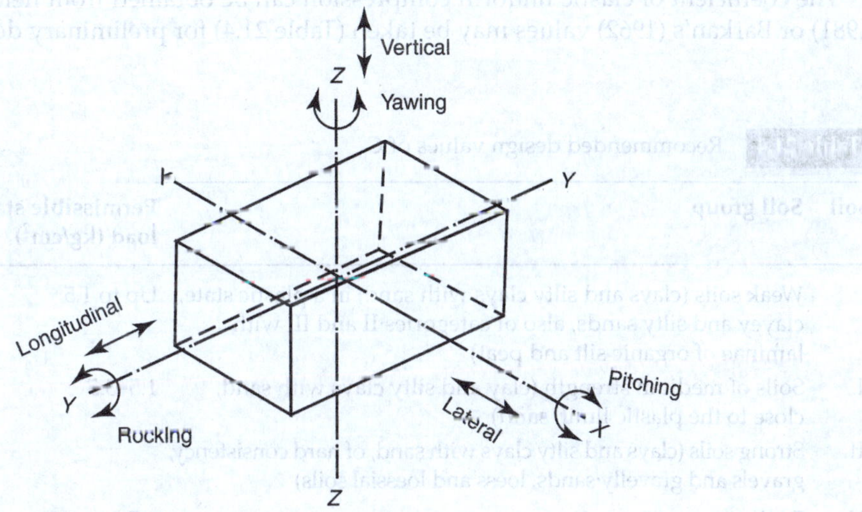


Fig. 21.9 Modes of vibration (*Source:* IS: 2974 – Part 1, 1982)

considers the foundation–soil system as a damped or undamped vibrating spring–mass system. The elastic half-space concept is more rational but complicated and beyond the scope of this book. In the spring–mass system, the soil behaviour is considered to be linear and elastic. Damping does not affect the resonant frequency, but it has a considerable effect on resonant amplitudes. As a machine foundation is designed avoiding the zone of resonance, the effect of damping on amplitudes computed at operating frequency is also small compared to that at resonance. Hence, for no damping condition, the relevant expressions for natural frequency and amplitude are given below.

For defining the soil constant, for use in the analysis, the soil is idealized as linear and elastic. Further, the soil constant varies with the stress produced and the elastic deformation formed below the block. This depends on the mode of vibration.

The soil below the base of the foundation block undergoes uniform compression when the block is subjected to vertical oscillations. Accordingly, the coefficient of elastic uniform compression ($C_u$) is defined as the ratio of uniform compression ($p$) and elastic settlement ($S_e$); that is,

$$C_u = \frac{p}{S_e} \tag{21.16}$$

But by definition the spring constant in the vertical direction, $k_z$, is given as

$$k_z = \frac{\text{Load}}{\text{Elastic deformation}} = \frac{pA}{S_e} \tag{21.17}$$

The equivalent spring constant of soil, $k_z$, is given as

$$k_z = C_u A \tag{21.18}$$

where $A$ is the area of the test plate.

The coefficient of elastic uniform compression can be obtained from field tests (Prakash, 1981) or Barkan's (1962) values may be taken (Table 21.4) for preliminary designs.

**Table 21.4** Recommended design values of $C_u$

| Soil | Soil group | Permissible static load (kg/cm$^2$) | $C_u$ (kg/cm$^3$) |
|---|---|---|---|
| I | Weak soils (clays and silty clays with sand, in a plastic state, clayey and silty sands, also of categories II and III with laminae of organic silt and peat) | Up to 1.5 | Up to 3.0 |
| II | Soils of medium strength (clay and silty clays with sand, close to the plastic limit, sand) | 1.5–3.5 | 3–5 |
| III | Strong soils (clays and silty clays with sand, of hard consistency, gravels and gravelly sands, loess and loessial soils) | | |
| IV | Rocks | >5.0 | >10.0 |

*Source:* Barkan (1962).

Let us consider a machine foundation of area $A$ subjected to an unbalanced vertical force of $F_0 \sin wt$ and placed at a depth of $D_f$. For the sake of analysis, $D_f = 0$ (this is a conservative assumption) is taken, and the equation of motion of this system is given as

$$m\ddot{z} + k_z z = F_0 \sin \omega t \tag{21.19}$$

The circular natural frequency, $\omega_{nz}$, of the system is

$$\omega_{nz} = \sqrt{\frac{C_u A}{m}} \tag{21.20}$$

The amplitude of motion $A_z$ is given as

$$A_z = \frac{F_0 \sin \omega_{op} t}{m(\omega_{nz}^2 - \omega_{op}^2)} \tag{21.21}$$

where $\omega_{op}$ is the circular operating frequency.

Equations 21.20 and 21.21 assume that the foundation rests on the surface. But under field conditions, foundations are always embedded in the soil partially or fully. The effect of embedment influences both the resonant amplitude and the frequency. In general, embedment causes a reduction in the amplitude at the resonant peak followed by a marginal increase in the resonant frequency. The two factors which contribute to this behaviour are the development of frictional force on the vertical sides of the foundation and the surcharge. Several analytical methods have been reported in the literature to study the effect of embedment with particular reference to vertical vibrations (e.g., Anandakrishnan and Krishnaswamy, 1969, 1973; Baranov, 1967; Ramiah et al., 1977; Krishnaswamy (1975), based on field vibratory tests, has shown that the amplitude of motion at resonance can be reduced by suitably adjusting the intensity of surcharge around the footing. The other notable factor which influences the vertical vibrations is soil moisture. It has been reported that soil moisture has a tendency to reduce the spring constant (Krishnaswamy and Anandakrishnan, 1975).

## 21.9 LIQUEFACTION OF SOILS

Liquefaction denotes a condition where a soil will undergo continued deformation at a constant low residual stress or with no residual resistance. Liquefaction occurs mostly due to earthquake forces which induce high pore water pressure resulting in low confining pressure. Liquefaction often appears in the form of sand foundations. When soil fails due to liquefaction, the structures founded on such soil sink. Liquefaction is more pronounced in sandy soils.

The shear strength of sand depends solely on internal friction. The effective shear strength for sands is

$$\tau = \sigma_n' \tan \phi' \tag{21.22}$$

where $\tau$ is the shear strength of sand, $\sigma_n'$ is the effective normal stress on any plane, and $\phi'$ is the angle of internal friction.

Ground vibrations in sand tend to reduce the volume; if there is no room for reduction in volume, the pore water pressure increases dramatically and is expressed as

$$\tau_{dyn} = (\sigma'_n - u_{dyn}) \tan \phi'_{dyn} \tag{21.23}$$

where $\tau_{dyn}$ is the shear strength of soil under dynamic condition, $u_{dyn}$ is the excess pore water pressure due to dynamic loading, and $\phi_{dyn}$ is the angle of internal friction under dynamic conditions.

It can be observed that because of increase of pore water pressure, the effective normal stress decreases, resulting in shear strength reduction. However, in sands under dynamic conditions, $\phi'_{dyn}$ is almost equal to $\phi'$.

For a zero shear strength condition, that is,

$$\tau_{dyn} = 0; \ \sigma'_n = u_{dyn}$$

or

$$\frac{u_{dyn}}{\sigma'_n} = 1 \tag{21.24}$$

Increase in pore water pressure results in reduction in shear strength. Complete transfer of inter-granular stress from soil grains to water is known as complete liquefaction. If the transfer of stress is partial, then it is called partial liquefaction.

In case of complete liquefaction, the effective stress reduces to almost zero, and the sand–water mixture behaves as a viscous material followed by starting of consolidation resulting in surface settlement.

Liquefaction of sand can develop at any zone of a deposit. Liquefaction of the upper layers may occur not as a direct result of the ground motion but because of development of the liquefaction condition in an underlying zone of the deposit. When liquefaction develops at a deeper level, the excess pore water pressure in the liquefied zone will dissipate due to flow of water in an upward direction. If the hydraulic gradient is greater than the critical gradient, then the upward flow of water induces a quick or liquefied condition in the surface layers of the deposit. The onset of liquefaction at one zone of a deposit may lead to liquefaction of other zones.

## POINTS TO REMEMBER

21.1 When stress along a plane (or fault) in the ground exceeds the rupture strength of rock, a sudden movement occurs along that plane of fault resulting in an earthquake.

21.2 After rupture, due to the elastic property of rocks, the ruptured portion pushes back to the original position, which is referred to as elastic rebound.

21.3 The point inside the earth's surface at which the first movement or break occurs during an earthquake is called the focus or biocentre. The point on the earth surface directly above the focus is the epicentre.

21.4 The major seismic waves are body waves and surface waves.

21.5 Magnitude of an earthquake is a measure of the amount of ground shaking based on the amplitude of elastic waves. Richter's scale represents the magnitude of an earthquake.

21.6 Other dynamic forces are due to bomb blasts, machineries, pile driving, landing of aircraft, etc.

21.7 Many of the foundation problems arising from oscillating machinery can be analysed by considering the simple harmonic application of force.

21.8 A foundation is said to be in free vibration when it is disturbed and then kept free to vibrate about some mean position. Vibrations that result from exciting agencies are referred to as forced vibration.

21.9 In an elastic system, vibrations which are taking place due to inherent forces in the system and are free from any external forces are termed natural frequency.

21.10 The foundation system is said to be at resonance when the frequency of the exciting agency (operating frequency) is close to the natural frequency.

21.11 Damping is associated with energy dissipation and is the internal resistance offered by a foundation system to the vibration of a machine.

21.12 Machines are broadly classified as (i) reciprocating machines, (ii) impact machines, and (iii) rotary machines.

21.13 The design of a machine foundation has to satisfy the following conditions: (i) resonance should be avoided, (ii) the amplitude of vibration at the operating frequency should be less than the limiting amplitude, and (iii) vibrations of the machine should neither cause damage to the machine and structures nor harm the health and comfort of the people.

21.14 Liquefaction denotes a condition where a soil will undergo continued deformation at a constant low residual stress or with no residual resistance.

## QUESTIONS

### *Objective Questions*

21.1 The major seismic waves are
(a) Body waves and surface waves
(b) Tension waves and translation waves
(c) Compound waves
(d) Rayleigh waves

21.2 An earthquake is said to be damaging when the Richter number is
(a) >8 (b) 6 to 6.9 (c) 5 to 5.9 (d) <4

21.3 When the frequency of the exciting agency (operating frequency) is close to the natural frequency, then the foundation system is said to be at
(a) The free vibrations stage (b) The forced vibrations stage
(c) Resonance (d) Low amplitude

21.4 Indicate the wrong statement.

1. In viscous damping, the damping force is taken directly proportional to the velocity of the oscillating mass.
2. The ratio of the actual damping to the critical damping is referred to as damping factor.
3. Magnification factor is the ratio of maximum displacement and static deflection of the system.
4. In order to avoid large amplitudes, the frequency ratio should be unity.

21.5 A rigid-body displacement of a block foundation produces two independent motions and ______ coupled motions.

(a) Four (b) Three (c) Two (d) Six

21.6 Ground vibration in saturated sand causes the pore water pressure to

(a) Remain constant (b) Decrease
(c) Increase steadily (d) Increase suddenly

21.7 In case of complete liquefaction, the effective stress

(a) Reduces to almost zero (b) Increases suddenly
(c) Remains constant (d) None of the above

## *Descriptive Questions*

21.8 How are earthquakes caused?
21.9 What is an elastic rebound?
21.10 Name the two seismic waves. Explain briefly the importance of Rayleigh waves.
21.11 How is the magnitude of an earthquake measured?
21.12 Discuss the different types of vibrations.
21.13 Bring out the difference between free and forced vibrations.
21.14 What is a spring–mass system? How is damping represented?
21.15 Briefly explain the method of analysis of a block foundation.
21.16 What are the limits of amplitude of vibration with reference to human beings?
21.17 Explain liquefaction of sands during an earthquake.

# 22

# Environmental Geotechnology

## CHAPTER HIGHLIGHTS

Environmental cycles – Natural cycles: oxygen cycle, nitrogen cycle, carbon cycle – Environmental imbalance: solid and liquid wastes – Petroleum contamination – Acid rain and acid drainage – Scrap wastes – Arid lands and deserts Wetlands – Coastal margins – Soil erosion – Land subsidence – Birth of environmental geotechnology – Contaminated soils: identification, classification, permeability, electrical properties, shear strength, consolidation – Applications – Load–environment design criteria

## 22.1 INTRODUCTION

Environmental geotechnology is an interdisciplinary science which covers soil and rock and their interactions with various environmental cycles. This field has grown rapidly since 1986.

Air, water, and soil are in eternal interaction with each other. Therefore, the phases of the pollution process should not be viewed in isolation but should be taken in integration with the other phases (Fang, 1997). Hence, designs that are based only on the physico-mechanical behaviour of the soil are not adequate, and the role of the environmental aspects has to be included in the design.

## 22.2 ENVIRONMENTAL CYCLES

Environmental cycles consist of atmosphere, biosphere, hydrosphere, lithosphere, and geo-microbiosphere.

The *atmosphere* is the layer of air surrounding the earth. The *biosphere* represents all plant and animal life. The *hydrosphere* deals with water in oceans. The *lithosphere* refers to soil and rock which contain combined silicon, oxygen, aluminium, and other elements.

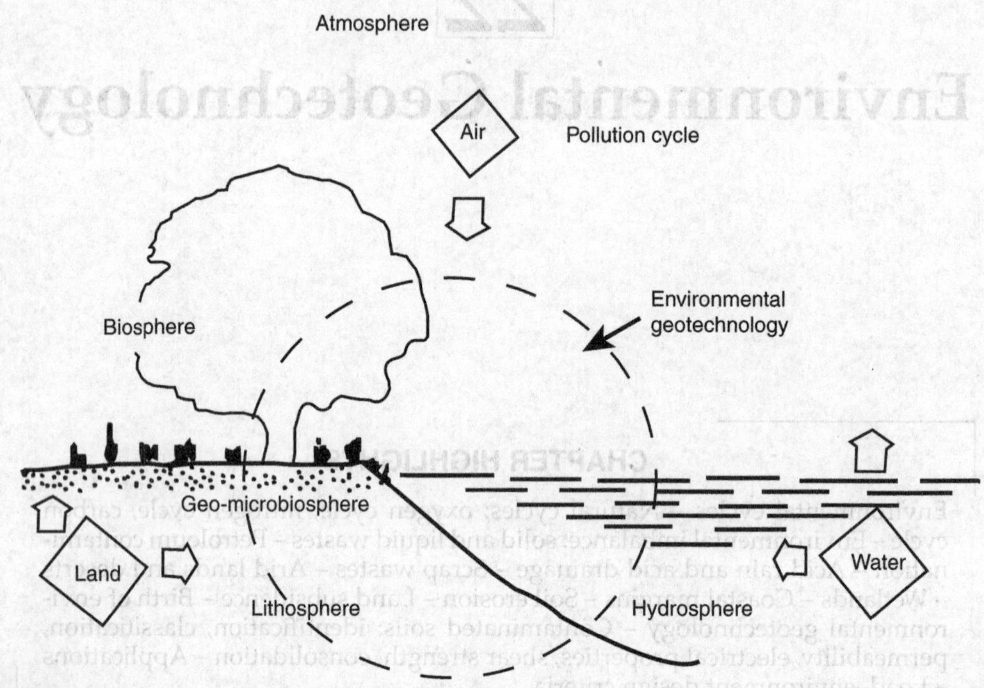


Fig. 22.1 Environmental cycles and their interactions (*Source:* Fang, 1997)

*Geo-microbiosphere* is a new term which does not come under the ecosphere. It is the study of the characteristics of tree and vegetation roots and bacterial activities in the groundwater and soil and subsequent response to the engineering behaviour of the soil–water system. Figure 22.1 shows the environmental cycles and their interactions.

## 22.3 NATURAL CYCLES

Many elements in the ecosphere are exchanged. These exchanges are caused from physical or physico-chemical interactions among the atoms and molecules that exist in each sub-sphere. Such cyclic exchanges are called natural cycles, viz., oxygen cycle, nitrogen cycle, and carbon cycle.

### 22.3.1 Oxygen Cycle

In the earth, oxygen is available along with silicon (Si) and aluminium (Al) as aluminium silicates (rocks). It may also be found in combination with metals, viz., metallic oxides, carbonates, sulphates, nitrates, and phosphates. The three substances, viz., water, carbon dioxide, and molecular oxygen, in combination with molecules of the living systems form the oxygen cycle.

### 22.3.2 Nitrogen Cycle

Nitrogen is the fourth abundant element next to carbon, hydrogen, and oxygen. Nitrogen cycle involves the transfer of nitrogen between atmosphere, biosphere, lithosphere, and hydrosphere in various chemical forms.

### 22.3.3 Carbon Cycle

In the biosphere, the carbon cycle consists of essential elements. The main function of the cycle is the oxidation of carbon-containing materials to atmospheric carbon dioxide, with photosynthesis returning it to the biosphere. Three important features of the cycle are as follows: (i) Fixed carbon content in plants is oxidized back to carbon dioxide, (ii) There is a large availability of carbon as carbonate in the sedimentary rocks and (iii) Carbon gets modified by fossil fuel combustion.

## 22.4 ENVIRONMENTAL IMBALANCE

Man-made environments and natural causes bring in an environmental imbalance. Man-made environments include solid and liquid wastes, petroleum contamination, acid rain, rubber tyres, etc. Natural causes include erosion, subsidence, landslides, land usage, etc. (Fang, 1997).

### 22.4.1 Solid and Liquid Wastes

Wastes may be of two forms: solid wastes and liquid wastes. Urban refuse is the major solid waste. With time, the solid wastes generate gases and liquid wastes, which are referred to as *leachates*. In solid and liquid wastes, hazardous and toxic wastes constitute 16%, sludge 14%, and nuclear wastes 12%.

### 22.4.2 Petroleum Contamination

Oil spills occur accidentally during transportation, both on land and on sea from storage tanks or during oil drilling. Underground storage tanks of petroleum products are not leak-proof, but 25% of them leak petroleum liquids into the ground. When a leak or oil spill occurs, soil around the point of leakage gets contaminated and requires special remediation and reclamation.

### 22.4.3 Acid Rain and Acid Drainage

Acid rain is formed by the emission of various pollutants into the air. The major sources are fossil fuel power plants, automotive exhausts, and industrial facilities. These emissions contain primarily sulphur dioxide, nitrogen, and other gases, which combine with oxygen and water vapour in air to form sulphuric acid, nitric acid, hydrochloric acid, and other acids. As acids are suspended high above the earth, they are carried far off by wind and are dropped or washed out from the air thus causing acid rain depositions. Also, acid rains are not uniformly distributed.

Acid drainage is due to water draining from old abandoned coal mines. The iron sulphides or pyrites in the mines get oxidized to sulphuric acid, which dissolves in water and drains through the mines. This acid water then enters streams, lakes, or ground soil–water systems, thereby resulting in pollution.

### 22.4.4 Scrap Wastes

The major scrap waste is from rubber tyres. Only a small amount is used for recycling. In most occasions, tyres pile up, particularly in developed countries, and cause fire hazard and health hazard.

### 22.4.5 Arid Lands and Deserts

It is estimated that deterioration of productive land to desert-like conditions is increasing and is at a rate of 14.8 million acres per year. The deserts thus formed constitute about 35% of the earth's surface, i.e., about 45 million square miles. Because of the loss of productive land due to desertification, more people are likely to be underfed and die of starvation. This should be avoided.

### 22.4.6 Wetlands

Urban development invariably uses large areas of wetlands for residential, industrial, and commercial purposes. Although additional useful lands are created for urban development, there are many detrimental effects. For example, various ecological effects may be caused by the construction of highway earth fills on wetlands, viz., inhibition of storm water and tidal distribution, increased water turbidity, alteration of water circulation patterns, removal of natural filtration systems, etc. Thus, construction of any structure on a specific wetland or on a problematic soil deposit is a complex problem.

### 22.4.7 Coastal Margins

Construction of structures on coastal lands and other activities on the area may lead to many changes in the coastal environments. Some of the factors are given below:

1. Construction of waterworks for irrigation, storm protection, and power may reduce the coastal land area, which may cause interception of water and sediments and shifting of channel locations.
2. Construction of industries on coastal land may result in the production of effluents which are let into coastal land or sea causing increased pollution of near-shore environments.
3. Construction and activities on coastal land leads to loss of wildlife habitat.
4. Further constructions lead to increase of rate of sea level.

### 22.4.8 Soil Erosion

Soil erosion is caused by the drag action of wind, rainfall, or wave action on the surface of an unprotected soil surface. Soil erosion causes soil particle detachment and transport. The sediments formed due to soil erosion (i) restrict the efficiency of drainage and irrigation ditches, fills, and reservoirs, (ii) pollute surface water, (iii) damage agricultural lands, and (iv) upset the natural ecology and environment.

### 22.4.9 Land Subsidence

Application of a load on a soil mass causes deformation due to (i) immediate deformation of soil structure, (ii) escape of pore water due to consolidation, (iii) time-dependent deformation, and (iv) a combination of all the above. Thus, subsidence depends on the applied load, soil properties, drainage condition, stress history, and environmental factors. Ground subsidence may also be caused due to natural causes such as tectonic movements, which include tectonic creep and tectonic earthquake. Ground surface movement may be caused by tree roots, floods, de-watering, etc.

## 22.5 BIRTH OF ENVIRONMENTAL GEOTECHNOLOGY

Environmental geotechnology has come into being because of two major reasons, viz., population growth and rising living standards.

An increase in population demands a need for more land. Many soil deposits which were claimed to be unfit for residential housing or for other construction projects have been compelled to be used now because of increase in population. Such soil deposits consist of landfills, wetlands, collapsible soil regions, mining subsidence areas, etc. Construction in such problematic soils requires additional techniques and use of new materials along with the conventional construction techniques and materials.

Rising living standards in a rapidly growing urban area causes hazardous pollution of air, water, and land and urban refuse, which collectively endanger the global environment.

Both problematic soil deposits and pollution problems have challenged the conventional soil mechanics concepts and methods of analysing soil behaviour under varied environmental conditions. Thus, environmental geotechnology has come into being to tackle the problems both on sound theoretical grounds and on practical considerations.

The birth of environmental geotechnology is based on practical and logical reasons and stems to provide (i) better productivity, (ii) healthier economy, and (iii) safe and secure environment. Environmental geotechnology is an emerging discipline comprising geotechnical engineering, environmental science and engineering, and soil agricultural science.

## 22.6 CONTAMINATED SOILS

Soils polluted due to various environmental reasons have to be identified before taking up construction activity. Further, apart from conventional tests, special tests have to be conducted so as to have a thorough knowledge of the behaviour of the soil.

### 22.6.1 Identification

During reconnaissance and field investigation, colour, odour, and presence of organic matter and other debris help in the proper identification of contaminated soils. Sometimes, water bubbles are formed by certain polluted waters. Cracking patterns and soil colour are affected by contaminated soil water.

### 22.6.2 Classification

The basic concept of classification of contaminated soil is based on the particle size. Even a small percentage of clay and colloids may contribute to the entire solid surface. Thus, the interaction of soil surface and the pore fluid depends on the pore fluid characteristics and the physico-chemical properties of the particle surface. The controlling parameters for proper classification, considering the environmental factors, are 2 μm fraction, specific surface, pH in pore fluid, ion-exchange capacity, absorption and adsorption, and dielectric constant.

### 22.6.3 Permeability

Permeability of a contaminated soil is significantly affected by pore fluid characteristics. Fang (1997) has reported that permeability increases when (i) the percentages of multivalent cations in the pore fluid are increased, (ii) the organic contents are increased, and (iii) the salt concentration of a pore fluid is increased.

### 22.6.4 Electrical Properties

Engineering properties of a contaminated soil–water system are closely related to the electrical properties of the system. In general, they depend on the electric surface charge, dielectric constant, and redox reaction. The pH in contaminated pore fluids will change the ion-exchange reactions, which in turn influences the electrical properties of the soil–water system.

### 22.6.5 Shear Strength

Temperature and electrolyte concentration have significant effect on the shear strength of contaminated soils. Soils formed from clay–water slurries, with a high salt concentration and subsequently leached, may show high sensitivity. Contaminated pore fluids due to industrial and hazardous/toxic waste may have adverse effects on shear strength.

### 22.6.6 Consolidation

As for shear strength, the effect of pore fluid and temperature may show different settlement characteristics in contaminated soil. More recently, Terzaghi's concept of consolidation has been extended to contaminated soils taking into account the electrokinetic, thermal, and electromagnetic effects.

## 22.7 APPLICATIONS

All civil engineering structures are founded directly or indirectly on ground soil or rock. The stability of the structure, both sub-structure and superstructure, depends on the stability characteristics of the soil, which is influenced to a great extent by local environmental factors. Foundation is the most important component of a structure whose performance in turn depends on the nature of ground or its improvement.

The current practice of foundation is based mostly on the load and settlement criterion. Now, the practice has to be modified to take into account the environmental factors into design. Thus, the modified design criteria should have a safe load–environment factor design.

If the existing ground does not satisfy the load–settlement–environmental factors, measures must be taken to improve the factor that will make the ground suitable for use. As discussed in Chapter 19, ground improvement comprises improving mechanical, hydrological, physico-chemical, biological, or a combination of these properties or any other properties of nature or man-made soil deposits. The purposes of ground improvement are as follows:

1. To strengthen ground soil before failure occurs,
2. To strengthen ground soil for the full life period,
3. To strengthen ground soil after premature or unexpected failure, and
4. To provide temporary ground improvement.

For deciding an effective ground improvement system, the following factors need to be considered: (i) sensitivity of soil to environment, (ii) ground soil–water pollution, (iii) adverse ground conditions, and (iv) selection of construction materials.

## 22.8 LOAD–ENVIRONMENT DESIGN CRITERIA

Conventional design approach is based on a safe load which could avoid a collapse, and the structure will be allowed to settle within the permissible limit. But the ground soil is extremely sensitive to the local environments such as temperature, moisture content, the characteristics of the pore fluid, weather conditions, groundwater fluctuations, and degree of groundwater and air pollution. Thus, civil engineering design of structures is not simple but interdisciplinary in nature and demands the help of all related fields so as to provide a safe and economical structure to have a long life.

Loads to be considered for a load–environment design criteria are, static loads, dynamic loads, and environmental loads.

Static load, also called as dead load, is the weight of the structure and all materials permanently attached to it. Dynamic loads are live load and impact load. Major live loads are human occupancy, partition walls, furniture, warehouse goods, and mechanical equipment. Local Building Code provides the live load requirements depending on the type of structure.

Environmental loads include earthquake, wind, snow, water pressure, and others. Forces due to water flow, ice flow, and wave forces have to be considered in the design of waterfront structures.

## POINTS TO REMEMBER

22.1 Environmental geotechnology is an interdisciplinary science which covers soil and rock and their interactions with environmental cycles.

22.2 Environmental cycles consist of atmosphere, biosphere, hydrosphere, lithosphere, and geo-microbiosphere.

22.3 Natural cycles are oxygen cycle, nitrogen cycle, and carbon cycle.

22.4 Man-made environments and natural causes bring in an environmental imbalance.

22.5 Man-made environments include solid and liquid wastes, petroleum contamination, acid rain, acid drainage, scrap wastes, etc.

22.6 Natural causes include acid lands and deserts, wetlands, coastal margins, soil erosion, land subsidence, etc.

22.7 Two major reasons, viz., population growth and rising living standards, have shown the way for the development of the field of environmental geotechnology.

22.8 Soils polluted due to various environmental reasons have to be identified before taking up any construction activity.

22.9 In addition to the conventional design requirements such as a safe load to avoid collapse and not to exceed permissible settlement, environmental effects on the soil have to be considered.

22.10 If the existing ground is not in a position to satisfy the load–settlement–environmental factors, the ground should be improved by any ground improvement techniques.

## QUESTIONS

### Objective Questions

22.1 Many elements in the ________ are exchanged.
(a) Hydrosphere (b) Ecosphere
(c) Microsphere (d) Lithosphere

22.2 Indicate the wrong statement. Man-made environments include
(a) Solid wastes (b) Liquid wastes
(c) Petroleum contamination (d) Land subsidence

22.3 *Assertion A:* Nitrogen is the fourth abundant element next to carbon, hydrogen, and oxygen.
*Choose the correct statement:*
*Reason R:* In the biosphere, carbon cycle, like nitrogen cycle, is of essential elements.
(a) A and R are true, and R is the explanation of A.
(b) A and R are true, and R is not the explanation of A.
(c) A is true, but R is false.
(d) A is false, but R is true.

22.4 Identify the wrong statement. Sediments formed due to erosion
(a) Resist the efficiency of drainage systems
(b) Damage agricultural lands
(c) Upset the natural ecology and environment
(d) Improve surface water quality

22.5 Choose the wrong term. For design of water-front structures, the following forces should be considered:
(a) Water flow (b) Ice flow
(c) Atmospheric flow (d) Wave force

22.6 Select the wrong statement.
(a) Dynamic loads are live load and impact load.
(b) Environmental design criteria consider safe load, dynamic load, and environmental load.
(c) Ground improvement may be for permanent or temporary purposes.
(d) Contaminated soils generally have high shear strength.

## Descriptive Questions

22.7 Explain clearly the natural cycles.
22.8 What are the factors involved in environmental imbalance?
22.9 Briefly explain the importance of land subsidence.
22.10 How are contaminated soils identified?
22.11 Briefly explain the need for consideration of environmental factors.
22.12 What are the additional factors to be considered in the load–environment design criteria?

# 23

# Introductory Rock Mechanics

**CHAPTER HIGHLIGHTS**

Index properties of rocks – Classification of rocks – *In situ* state of stress – Mechanical properties of rocks

## 23.1 INTRODUCTION

Rock mechanics is an inter-disciplinary subject covering various fields, such as engineering geology, mining, petroleum, and civil engineering. The common man believes that rocks are strong materials with reliability and permanence and need no treatment if they are used as foundations. But rocks present in nature have several geological discontinuities, such as faults, joints, fissures, etc. The behaviour of a rock mass subjected to a change in stress is governed by the mechanical properties of the rock material and the geological discontinuities.

Several major failures of dams and rock slopes have occurred in the past due to inadequate analysis and understanding of rock behaviour. Nowadays, tall structures, high dams, lengthy tunnels, and deep mines are constructed, which induce a large stress change. Thus, for an exact analysis and proper prediction of the behaviour of civil engineering structures on rock and the use of rock as a construction material, a complete understanding of the engineering properties and behaviour of rock masses is essential. The basic knowledge of soil behaviour, gained from the preceding chapters, forms a sound foundation for understanding the rock behaviour dealt in this chapter.

## 23.2 INDEX PROPERTIES OF ROCKS

Rocks occur with a variety of structures, fabrics, and components. In order to describe a rock quantitatively, we need to know about certain properties of the rocks which may be designated as index properties of rocks. Such properties are porosity, density or unit weight, permeability, strength, durability, and sonic velocity (Goodman, 1980). These properties are explained in the following paragraphs.

### 23.2.1 Porosity

Porosity of a rock, as in soils, is defined as the fraction expressing the proportion of void space to total space in the rock and is represented as a percentage. It identifies the relative proportion of solids and voids. The porosity of a rock depends on the particles-size distribution, sorting, grain shape, fabric, degree of compaction and cementation, solution effects, and mineralogical composition. The highest porosity is commonly attained when all the grains are of the same size. Large ranges of porosity can be obtained with irregularly shaped grains.

In sedimentary rocks, the average porosity is about 15%. In these rocks, there is a decrease in porosity with age and depth. Among the rocks, chalk is the most porous with a porosity of 50%. In unweathered igneous rocks, the porosity is usually less than 1% or 2% which, with the progress of weathering, increases to about 20%. Thus, the measurement of porosity in such rocks serves as an accurate index to rock quality.

Porosities of rocks are determined by adopting a variety of techniques, such as measured density, measured water content after saturation in water, mercury content after saturation with mercury using a pressure injector, and measured solid volume and pore air volume using Boyle's law. The standard saturation method (Franklin, 1970) and the air porosimeter (Ramana and Venkatanarayana, 1971) are the conventional methods used for determining porosity.

### 23.2.2 Density

The dry density or unit weight of a rock is its specific weight (kg/m$^3$ or kN/m$^3$ as in soil). From a knowledge of specific gravity of rock grains, the rock density or unit weight can be approximately found. For example, a rock with a specific gravity of 2.8 has a unit weight of 28 kN/m$^3$. The grain specific gravity can be determined by grinding the rock and adopting methods used for soils. Rocks exhibit a large range of density values compared to soils. The density of a rock is easily measured from the volume and mass of a carefully drilled core specimen. Table 23.1 lists the densities of some common rocks.

A knowledge of rock density has wide application in civil engineering and mining practices, e.g., in the selection of aggregates for roads, slabs and beams, and gravity dams and in the analysis of underground openings; besides, in oil shale deposits, it directly correlates to the oil yield and in coal deposits, it directly reflects the ash content.

### 23.2.3 Permeability

Water is always present and stored in the rock masses. It percolates through pores when subjected to hydraulic gradient. As in soils, permeability in rocks is defined as the ability of a rock to allow the passage of fluids into or through it, without impairing its structure. Permeability is selected as an index property of rock since it gives information about the interconnection between the pores or fissures.

**Table 23.1** Dry densities of some typical rocks

| Rock name | Dry density (g/cm$^3$) |
|---|---|
| Coal | 0.7–2.0 |
| Oil shale | 1.6–2.7 |
| Rock salt | 2.1 |
| Shale | 2.25–2.62 |
| Gypsum | 2.30 |
| Rhyolite | 2.37 |
| Syenite | 2.60–2.70 |
| Granite | 2.65 |
| Basalt | 2.77 |
| Quartz, mica schist | 2.82 |
| Diorite | 2.85 |
| Amphibolite | 2.99 |
| Gabbro | 3.00 |

*Source:* Goodman (1980).

Darcy's law is valid in most of the rocks. If the permeant is other than water and the temperature varies considerably, a more useful form of Darcy's law is

$$q = \frac{K}{\eta_w}\frac{dp}{dx}A \tag{23.1}$$

where $q$ is the discharge, $K$ the permeability coefficient independent of permeant (length$^2$), $p$ the fluid pressure (equal to $\gamma h$), $\eta_w$ the viscosity of the permeant, and $A$ the area of cross section.

As in soils, permeability of rocks can be assessed in the laboratory from a constant or falling head test. An alternative approach is to allow a radial flow in a hollow specimen (Bernaix, 1969) and measure the discharge either from inner surface to outer surface or from outer surface to inner surface. For radial flow in an unfractured rock with inter-connected pores, the permeability is given by

$$k = \frac{q\gamma_w}{2\pi L \Delta p}\log\left(\frac{R_2}{R_1}\right) \tag{23.2}$$

where $R_1$ and $R_2$ are the radii of inner and outer surfaces, $L$ the length of cylinder, and $\Delta p$ the difference between external and internal water pressure.

Berniax (1969) showed that porous rocks are unaffected by pressure changes. Further, it was shown that fissured rocks exhibit higher permeability in divergent flow than in convergent flow. It is presumed that some amount of hydraulic fracturing* would have occurred in divergent flow. Goodman and Sundaram (1980) reported similar behaviour based on the tests on tuff, schist, sandstone, and limestone. Some typical values of permeability of rocks are presented (Serafim, 1968) in Table 23.2.

*It is the process of initiation of a crack caused by water pressure.

**Table 23.2** Coefficients of permeability (to water) of various rock materials

| Rock materials | k (cm/s) |
|---|---|
| Sandstone | $1.6\times10^{-7}\simeq10^{-10}$ |
| Siltstone | $10^{-8}\simeq10^{-9}$ |
| Granite | $2\times10^{-10}\simeq5\times10^{-11}$ |
| Slate | $1.6\times10^{-10}\simeq7.0\times10^{-11}$ |
| Calcite | $9.3\times10^{-8}\simeq7.0\times10^{-10}$ |
| Limestone | $1.2\times10^{-7}\simeq7.0\times10^{-10}$ |
| Dolomite | $1.2\times10^{-8}\simeq4.6\times10^{-9}$ |
| Hard mudstone | $2\times10^{-6}\simeq6\times10^{-7}$ |

*Source:* Stagg and Zienkinwicz (1968).

The measurement of permeability in rocks has a significant bearing on many engineering and mining problems, viz., pumping of water, oil, or gas into or out of a porous formation, storing fluids in mined caverns for energy conversion, disposing brine waste in porous formation, assessing the water tightness of a reservoir, and predicting water inflows in a tunnel (Goodman, 1980).

### 23.2.4 Strength

A strength index is useful only if the results from tests on rocks are reproducible and inexpensive. The point load test described by Franklin and Broch (1972) is identified as a strength index test. In this test, a specimen is subjected to compression between two cone-shaped platens (Fig. 23.1), which causes tensile stresses normal to the axis of loading and the *point load strength* (tensile strength) $\sigma_{pl}$ is given as

$$\sigma_{pl} = \frac{P}{D^2} \tag{23.3}$$

where $P$ is the load at rupture and $D$ the distance between the point loads.

The *point load strength index* is reported as the point load strength of a 50 mm diameter core.

The point load strength test is simple and quick and can be carried out in the field itself. Rocks showing strong anisotropy, such as slates, schists, laminated sandstones, etc., have

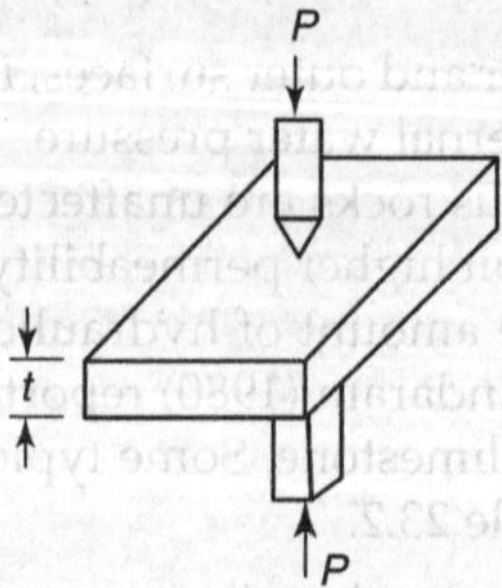


**Fig. 23.1** Point load test on rock prism (*Source:* IS: 10082, 1981)

**Table 23.3** Point load index

| Description | Point load strength index (MPa) | Equivalent uniaxial compressive strength (MPa) |
|---|---|---|
| Extremely high strength | >10 | >160 |
| Very high strength | 3–10 | 50–160 |
| High strength | 1–3 | 15–60 |
| Medium strength | 0.3–1 | 5–16 |
| Low strength | 0.1–0.3 | 1.6–5.0 |
| Very low strength | 0.03–0.1 | 0.5–1.6 |
| Extremely low strength | <0.03 | <0.5 |

*Source:* Franklin and Broch (1972).

to be tested both along and parallel to bedding planes. Franklin and Broch (1972) have provided a point load strength index for easy identification of rocks (Table 23.3). The unconfined compressive strength of rocks ($q_u$) is correlated to the point load strength test as

$$q_u = 24\sigma_{pl(50)} \tag{23.4}$$

where $q_{pl(50)}$ is the point load strength corresponding to a 50 mm diameter core.

### 23.2.5 Slaking and Durability

Hydration, slaking, solution, oxidation, and other processes produce changes in the properties of rocks. Such changes affect the durability of rocks. As there are many paths of deterioration, some index to represent the degree of durability of rocks is required.

One such index test is the slake durability test suggested by Franklin and Chandra (1972). The test essentially consists of rotating about 500 g of rock broken into 10 lumps and loaded in a drum which is turned at 20 rpm in a water bath for 10 minutes. The drum is 140 mm in diameter and 100 mm long with a 2 mm opening sieve mesh forming the cylindrical walls. After 10 minutes, the percentage of rock retained inside the drum, on a dry weight basis, is reported as the *slake durability index* ($I_d$). Gamble (1971) recommended a second 10 minute cycle and, based on the test results, proposed a Slake Durability Scale (Table 23.4).

**Table 23.4** Gamble's Slake Durability Scale

| Group name | Percentage retained after one 10-minute cycle (dry weight basis) | Percentage retained after two 10-minute cycles (dry weight basis) |
|---|---|---|
| Very high durability | >99 | >98 |
| High durability | 98–99 | 95–98 |
| Medium high durability | 95–98 | 85–95 |
| Medium durability | 85–95 | 60–85 |
| Low durability | 60–85 | 30–60 |
| Very low durability | <60 | <30 |

*Source:* Goodman (1980).

**Table 23.5** Description of rate and amount of slaking

| Amount of slaking | Liquid limit (%) |
|---|---|
| Very low | <20 |
| Low | 20–50 |
| Medium | 50–90 |
| High | 90–140 |
| Very high | >140 |
| *Rate of slaking* | *Change in liquidity index* |
| Slow | <0.75 |
| Fast | 0.75–1.25 |
| Very fast | >1.25 |

*Source:* Goodman (1980).

Morgenstern and Eigenbrod (1974) used a water absorption test to assess the amount and rate of slaking of argillaceous materials. They found that non-cemented claystone or shale absorbed water faster and that all materials eventually reached a final water content equal to their liquid limits. Materials with high liquid limits exhibited a high degree of slaking than those with low liquid limits. They have also shown that the rate of slaking is reflected by the rate of water content change following soaking. The rate of slaking is given by the *change in liquidity index* ($\Delta I_L$) following 2 hours of immersion in water; that is,

$$T\Delta I_L = \frac{\Delta w}{w_L - w_p} \tag{23.5}$$

where $\Delta w$ is the change in water content of the rock after soaking for 2 hours.

Table 23.5 provides the rate and amount of slaking as suggested by Morgenstern and Eigenbrod (1974), reported by Goodman (1980).

### 23.2.6 Sonic Velocity

The effect of discontinuities, such as fissures, in a rock mass can be estimated by comparing the *in situ* wave velocity with the laboratory sonic velocity of an intact core obtained from the same rock mass. The movement of stress waves in a rock mass depends on the elastic property and density of rocks.

In practice, a network of fissures in the specimen superimposes an overriding effect. Under these conditions, the sonic velocity can be utilized to index the degree of fissuring with rock specimens (Goodman, 1980).

Fourmaintraux (1976), reported by Goodman (1980), proposed a classification scheme based on longitudinal wave velocity for fissuring in rock specimens. Assuming the specimen has no pores or fissures, the longitudinal wave velocity ($V_l^*$) is calculated from the known mineral composition. Thus,

$$\frac{1}{V_l} = \sum_i \frac{C_i}{V_{l,i}} \tag{23.6}$$

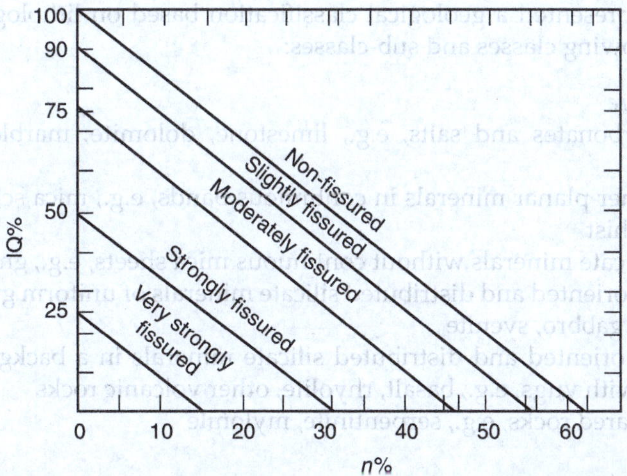


**Fig. 23.2** Classification scheme for fissuring in rock specimens (*Source:* Goodman, 1980)

where $V_{1,i}$ is the longitudinal wave velocity in the $i$th mineral constituent and $C_i$ the volume proportion.

Now the actual velocity of longitudinal waves ($V_1$) is measured. A quality index (IQ) is defined as

$$IQ = \left(\frac{V_1}{V_1^*}\right) \times 100\% \tag{23.7}$$

Further, Fourmaintraux (1976) showed the dependence of IQ on the porosity of the rock. Thus,

$$IQ = (100 - 1.6n_p)\% \tag{23.8}$$

where $n_p$ is the porosity of the non-fissured rock expressed in percentage.

Fourmaintraux (1976) presented a plot of IQ versus porosity (Fig 23.2) as a basis for describing the degree of fissuring of a rock specimen.

## 23.3 CLASSIFICATION OF ROCKS

In general, rock classification is required to help in making an initial assessment of a problem and to examine the areas where additional information has to be supplemented in order to obtain the required answer. Many civil engineering works require rock classification, which includes the assessment of slope stability, open and sub-surface excavations, foundation stability, and the selection of rocks as construction materials.

Geological classifications of rocks are based on a genetic point of view, viz., the conventional geological classification: igneous, metamorphic, and sedimentary. Deere (1979) and Goodman (1980) emphasized that the geological classification should also consider the lithology of rocks, which refers to mineralogy, texture, fabric, and structural anisotropy.

Goodman (1980) presented a geological classification based on lithology and divided the rocks into the following classes and sub-classes:

1. *Crystalline texture*
   (i) Soluble carbonates and salts, e.g., limestone, dolomite, marble, rock salt, trona, gypsum
   (ii) Mica or other planar minerals in continuous bands, e.g., mica schist, chlorite schist, graphite schist
   (iii) Banded silicate minerals without continuous mica sheets, e.g., gneiss
   (iv) Randomly oriented and distributed silicate minerals of uniform grain size, e.g., granite, diorite, gabbro, syenite
   (v) Randomly oriented and distributed silicate minerals in a background of very fine grain and with vugs, e.g., basalt, rhyolite, other volcanic rocks
   (vi) Highly sheared rocks, e.g., serpentinite, mylonile
2. *Clastic texture*
   (i) Stably cemented, e.g., silica-cemented sandstone and limonite sandstones
   (ii) With slightly soluble cement, e.g., calcite-cemented sandstone and conglomerates
   (iii) With highly soluble cement, e.g., gypsum-cemented sandstone and conglomerates
   (iv) Incompletely or weakly cemented, e.g., friable sandstone, tuff
   (v) Uncemented, e.g., clay-bond sandstones.
3. *Very fine grained rocks*
   (i) Isotropic, hard rocks, e.g., hornfels, some basalts
   (ii) Anisotropic on a macro-scale but microscopically isotropic hard rocks, e.g., cemented shales, flagstones
   (iii) Microscopically anisotropic hard rocks, e.g., slate, phyllite
   (iv) Soft, soil-like rocks, e.g., compaction shale, chalk, marl
4. *Organic rocks*
   (i) Soft coal, e.g., lignite and bituminous coal
   (ii) Hard coal
   (iii) Oil shale
   (iv) Bituminous shale
   (v) Tar sand

For engineering purposes, mere geological names are insufficient and should be accompanied by an engineering classification. Such a classification may be concerned with the physical properties of rocks and the nature of discontinuities within the rock masses such that systematic predictions can be made about their engineering behaviour. One such classification has been suggested by Deere and Miller (1966) for intact rocks. This classification is based on two important engineering properties of intact rocks, viz., the uniaxial compressive strength and the modulus of elasticity. Here the tangent modulus is used which is being computed at a stress level equal to one-half the ultimate strength of the rock. The uniaxial compressive strength is that determined on specimens with a length diameter ratio of at least two.

Based on uniaxial compressive strength, rocks are divided into five categories, and designated as *A*, *B*, *C*, *D*, and *E*, as shown in Table 23.6. The second element modulus of elasticity has been used as a ratio of the modulus to the uniaxial compressive strength,

**Table 23.6** Strength classification of intact rocks

| Class | Description | Uniaxial compressive strength (MPa) |
|---|---|---|
| A | Very high strength | >224 |
| B | High strength | 112–224 |
| C | Medium strength | 56–112 |
| D | Low strength | 28–56 |
| E | Very low strength | <28 |

*Source:* Deere and Miller (1966).

**Table 23.7** Modulus ratio classification of intact rocks

| Class | Description | Modulus ratio |
|---|---|---|
| H | High modulus ratio | >500 |
| M | Medium modulus ratio | 200–500 |
| L | Low modulus ratio | <200 |

*Source:* Deere and Miller (1966).

called the *modulus ratios* designated as *H*, *M*, and *L* (Table 23.7). The strength categories follow a geometric progression. Combining these two letters a rock may be classified as *AM*, *CH*, *BH*, *DL*, etc.

A classification chart with the value of compressive strength and modulus in logarithmic scale has been provided by Deere and Miller (Fig. 23.3). The strength classification is shown across the top of the chart and the modulus ratio is obtained from the plotted position with respect to the diagonal lines. The zone above the upper line with a modulus ratio of 500:1 represents the zone of *high* modulus ratio (*H*), the zone below the lower line with a modulus ratio of 200:1 shows a zone of *low* modulus ratio (*L*), and the stippled zone is referred to as *M* – the zone of *medium* modulus ratio. It is observed that different rock types occupy different unique positions in the chart. Deere and Miller (1966) attributed this characteristic variation in properties to the sensitiveness of rocks to mineralogy, fabric, and direction of anisotropy. This engineering classification for intact rock is considered as a workable and useful system.

Engineers are concerned with a classification system which considers various aspects of both the rock material and the rock masses. Such a classification system is referred to as the geomechanics classification system. Terzaghi (1946) was one of the earlier researchers who recognized the significance of discontinuities, their spacing, and the filling materials. The basic idea of classification of rock masses was given by John (1962), who considered the compressive strength, degree of jointing, and amount of alteration. The geomechanics classification system of Wickham et al. (1968), Bieniawski (1973, 1974a, b, 1975, 1976), and Barton et al. (1975) are very well received. Bieniawski's system may be rated high.

Bieniawski's system suggests a general rock rating (*RMR*) increasing with rock quality from 0 to 100. This system is based on five universal parameters: strength of rock (15), drill core quality (20), groundwater condition (10), joint and fracture spacing (30), and

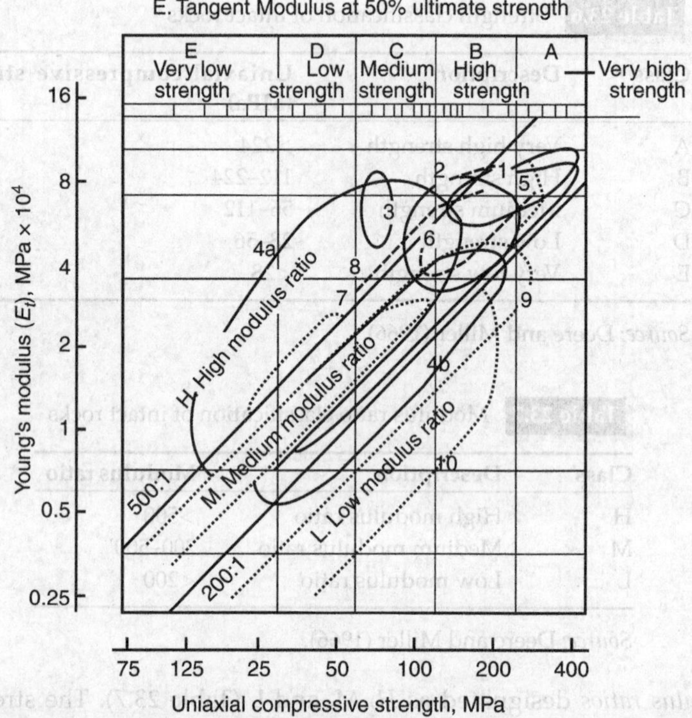


Metamorphic: 1. Quartzite, 2. Gneiss, 3. Marble, 4a. Schist, Steep foliation, 4b. Schist, flat foliation Igneous: 5. Diabase, 6. Granite, 7. Basalt and other flow rocks
Sedimentary: 8. Limestone and dolomite, 9. Sandstone, 10. Shale

**Fig. 23.3** Engineering classification of intact rock based on uniaxial compressive strength and modulus ratio (*Source:* Deere and Miller, 1966)

joint characteristics (25). The figures in parentheses refer to the maximum rating value of each parameter. In order to determine the RMR, the individual rating of the parameters are summed. The *RMR* values recommended by Bieniawski (1974a, b) are presented in Table 23.8.

**Table 23.8** Geomechanics classification of rock masses

| Class | Description of rock mass | Rock mass rating (*RMR*) |
|---|---|---|
| I | Very good rock | 81–100 |
| II | Good rock | 61–80 |
| III | Fair rock | 41–60 |
| IV | Poor rock | 21–40 |
| V | Very poor rock | 0–20 |

*Source:* Goodman (1980).

## 23.4 IN SITU STATE OF STRESS

The natural state of a stress at a point within a rock mass depends on the geologic processes that have acted on the mass. It is impossible to delineate with sufficient accuracy the various events which have brought about the *in situ* state of stress. However, it is often possible to estimate the stresses and their directions. But the margin of error cannot be ensured unless substantiated by field measurements.

A knowledge of rock stresses has been helpful in several civil engineering works, e.g., in choosing the orientation for a cavern, in deciding the layout of complex underground works, in fixing the location of pressure tunnels and penstocks, and in the economical planning for a large surface excavation.

### 23.4.1 Vertical Stress

The vertical normal stress can be generally taken as equal to the weight of the overlying rock and on an average can be taken as 0.027 MPa/m depth (Goodman, 1980). In the near-horizontal profiles, the principal stress directions are vertical and horizontal and are generally assumed to be true with depth also. This has been widely accepted in practice. However, this is not true beneath a valley side wherein one principal stress is normal to the slope and equals zero while the other two principal stresses lie in the plane of the slope. *In situ* stresses approach the strength of rock beneath V-shaped valleys.

The conventional vertical total stress equation

$$\sigma_v = \gamma z$$

has been supported by Brown and Hoek (1978) based on stress measurements. Effects of geological structure could alter the vertical stress and the direction of principal stresses. In major works, wherever geological heterogeneities can be expected, a thorough investigation has to be made using numerical methods like finite element method.

### 23.4.2 Horizontal Stress

The magnitude of horizontal stress ($\sigma_h$) is estimated (as done in soils) using the lateral stress coefficient ($K$)

$$\sigma_h = K\sigma_v$$

The expression is based on one-dimensional loading of an elastic half space with zero horizontal strain. Such a relationship has no bearing in a rock mass since it has experienced cycles of loading and unloading. However, the horizontal stress may lie between $K_a\sigma_v$ and $K_p\sigma_v$ (Fig. 23.4) where $K_a$ and $K_p$ correspond to conditions for normal and reverse faulting. For no fault condition, the assessment is difficult.

Based on published data on *in situ* stresses, Brown and Hoek (1978) suggested a hyperbolic relation for the limits of $\bar{K}(z)$ as follows:

$$0.3 + \frac{100}{z} < \bar{K} < 0.5 + \frac{1500}{z} \tag{23.9}$$

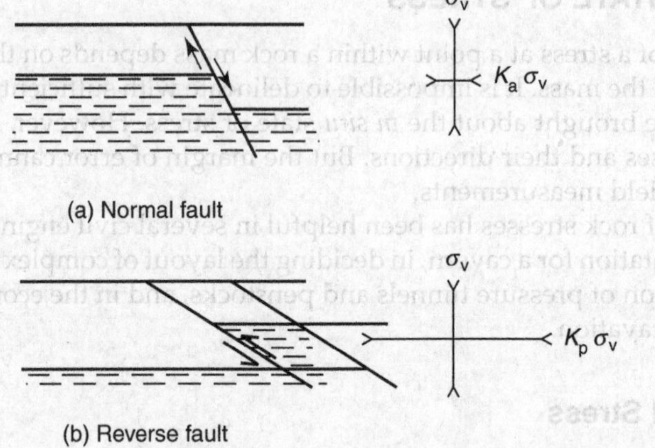


**Fig. 23.4** Stresses required to initiate normal and reverse faults (*Source:* Goodman, 1980)

where $\bar{K}$ is the ratio of average horizontal stress to vertical stress. It has been observed that the actual measured data are consistently found to be inverse with depth (Goodman, 1980). In a broader sense, one can estimate the variation of horizontal stress with depth without any field measurements.

### 23.4.3 Effective Stress in Rock Masses

Terzaghi's effective stress equation for saturated soils is an excellent approximation for soils although it is unlikely to be an exact statement. Skempton (1961) remarked that it is not necessarily true for rocks which are commonly subjected to very high stresses. Skempton suggested generalized effective stress expressions for saturated rocks in the form

1. For shear strength:

$$\sigma' = \sigma - \left(1 - a_c \frac{\tan \psi'}{\tan \phi'}\right) u_w \tag{23.10}$$

2. For volume change:

$$\sigma' = \sigma - \left(1 - \frac{C_s}{C}\right) u_w \tag{23.11}$$

where $\psi'$ is the angle of intrinsic friction of solids, $\phi'$ the angle of shearing resistance, $C_s$ the compressibility of solid particles, $C$ the volume compressibility, and $a_c$ the mineral contact area ratio.

For saturated rocks, tan $\psi'$/tan $\phi'$ is in the order of 0.1 to 0.3. This means that even if $a_c$ approaches 0.5 (rocks with very low porosity), Eq. 23.10 reduces to

$$\sigma' = \sigma - (0.85 \text{ to } 0.95) u_w \tag{23.12}$$

This shows that Terzaghi's effective stress equation used in soils is not an unacceptably poor approximation for rocks. Further, if the compressibility of grains is low, Eq. 23.11 reduces to Terzaghi's equation. Thus, Skempton (1961) concluded that Terzaghi's effective stress equation is valid in rocks too if the grains are incompressible and the yield stress of grains is independent of pressure.

If this is not true, the effective stress can be obtained from Eq. 23.13 (a generalized form of Eqs. 23.10 and 23.11):

$$\sigma' = \sigma - k u_w \tag{23.13}$$

where $k$ is a constant depending on the area of contact between grains and their mechanical properties (Jaeger and Cook, 1969). Terzaghi (1945) suggested a similar expression for rocks where $k$ is assigned a value lower than unity for rock material and unity for rock masses (when the surfaces of joints are considered).

However, Terzaghi's effective stress law used for soils has been confirmed to be valid for rocks also [see, e.g., Brace and Byerlee, 1966 (reported by Serafim, 1968); Goodman, 1980; Hubert and Rubey, 1959].

Based on Bishop's (1955) concept for effective stress in partially saturated soils, Skempton (1961) expanded Eqs. 23.10 and 23.11 for partially saturated materials as

1. For shear strength:

$$\sigma' = \sigma - \left(1 - a_c \frac{\tan \psi'}{\tan \phi'}\right) S_\chi u_w \tag{23.14}$$

2. For volume change:

$$\sigma' = \sigma - \left(1 - \frac{C_s}{C}\right) S_\chi u_w \tag{23.15}$$

where

$$S_\chi = 1 + (1 - \chi) \frac{u_a - u_w}{u_w} \tag{23.16}$$

As the parameter $\chi$ has serious limitations, the application of Eqs. 23.14 and 23.15 for rocks has not yet been explored.

### 23.4.4 Measurement of In Situ Stresses

Among several techniques available for measurement of *in situ* stresses, three techniques, viz., hydraulic fracturing, flat jack method, and overcoring method, are the best known. All the techniques involve perturbing the rock in order to create a response that can then be measured and analysed. In the hydraulic fracturing method, water pressure is used to create a crack in a borehole. From the knowledge of the tensile strength of the rock and the inferred concentration of stress, the initial stress in the plane normal to the bore hole is assessed. In the flat jack method, a slot is cut in the rock and then reloaded with the help of a jack. The *in situ* stress is related to the pressure needed to balance the deformation which occurs as a result of slot cutting. In the overcoring method, a large core sample is drilled, whereby a condition of stress release is created. The radial displacement or surface strains are monitored in a central parallel borehole. An analytical treatment based on an unloaded thick-walled cylinder model gives the required *in situ* stress. The reader may refer to Goodman (1980) for the details of the techniques.

## 23.5 MECHANICAL PROPERTIES OF ROCKS

Engineering properties of rocks and rock masses are needed in the analysis of every rock mechanics problem. This necessitates a knowledge of in situ stresses, modes of failure of rocks, the stress–strain behaviour of rocks, and the time-dependent behaviour of rocks. With this knowledge (based on a particular criterion of failure), a solution to the problem can be arrived at. Methods of assessing in situ stresses have been discussed in the previous chapter. The other aspects shall be discussed in this chapter.

### 23.5.1 Modes of Failure of Rock

As the *in situ* load configuration is complex, no single mode of rock failure can be identified. Depending on the field situation, the failures may be due to flexure, shear, compression, and tension. A flexure failure refers to failure by bending followed by tensile cracks. Such a failure may occur in the layers of a mine roof (Fig. 23.5a). Shear failure refers to rupture along a surface when the shear stresses have become critical. Slopes cut in weak material, such as weathered clay, shales, and crushed rock of fault zones exhibit shear failure (Fig. 23.5b). Crushing or compression failure occurs when high stresses are transferred through shortened rocks with less bearing area. A typical example is the failure of a mine pillar due to over extraction of ore (Fig. 23.5c). Failure due to direct tension occurs in sedimentary rocks on the flank of an anticline and sometimes in rock layers resting on convex upward slope surfaces (Fig. 23.5d and e).

### 23.5.2 Behaviour of Rocks in Uniaxial Compression

Uniaxial or unconfined compression test is the oldest, simplest, and most frequently used strength test for rocks. In this test, right circular cylinders or prisms of rocks are compressed parallel to their longitudinal axis. The specimen is compressed between two hard discs or end platens. The rigidity of the platens prevents the lateral expansion at the ends. The length to diameter ratio of 2 to 3 (IS: 9221, 1979) and preferably 2.5 is recommended (Dhir and Sangha, 1973). The diameter of the specimen shall not be less than 10 times the largest mineral grain in rock. As per Indian Standards recommendation, a 45 mm diameter specimen is preferred and in no case the diameter should be less than 35 mm. The load on the specimen is applied continuously with a stress rate of 0.50 to 1.0 MPa/s (IS: 9143, 1979). The uniaxial compression ($q_u$) is given as the ratio of peak load ($P$) to the initial cross-sectional area ($A$), that is, $q_u = P/A$. Table 23.9 gives some representative values of $q_u$ of some rocks (Jaegar and Cook, 1969). The stress–strain behaviour of rocks under uniaxial compression is discussed below.

Wawersik and Fairhurst (1970) explained the stress–strain behaviour of rocks tested under an unconfined condition. They divided the stress–strain paths between zero stress and the maximum compressive stress into three regions (Fig. 23.6). Region $A$ shows a steadily

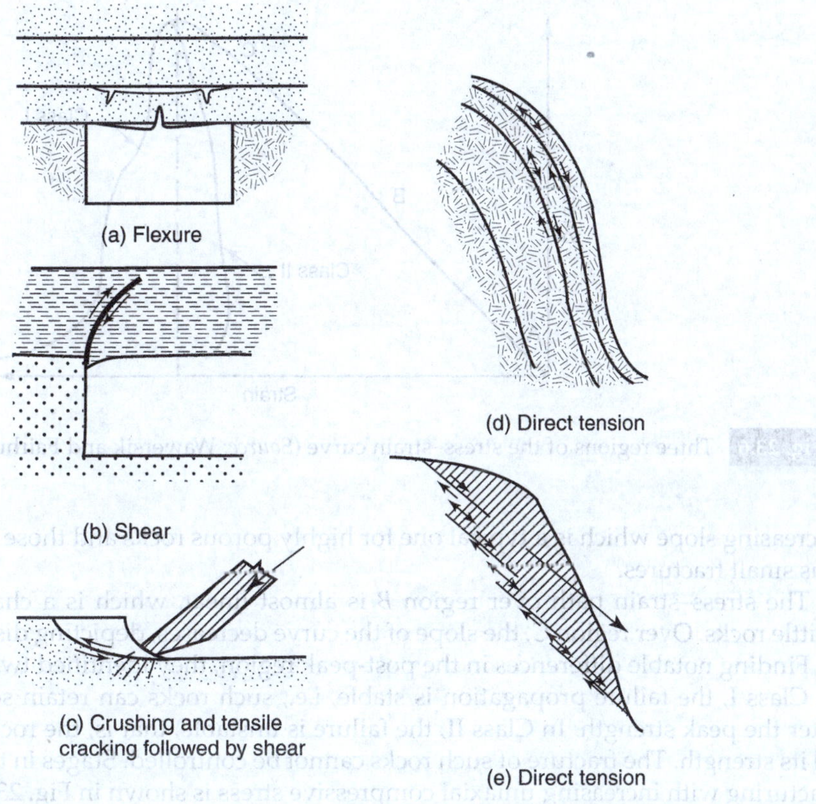


**Fig. 23.5** Examples of failure models involving breakage of rock (*Source:* Goodman, 1980)

**Table 23.9** Typical values of uniaxial compression, modulus of elasticity, and Poisson's ratio for different rocks

| Rock type | $q_u$ (MPa) | $E$ (MPa) | $\nu$ |
|---|---|---|---|
| Granite | 229.1[a] | 55,890[a] | 0.11[a] |
| Quartzite | 460.2[a]–200.1[d] | 78,660[a]–77,970[d] | – |
| Diabase | 487.1[a] | 99,360[a] | – |
| Marble | 152.9[b]–77.3[d] | 47,610[a]–64,860[d] | – |
| Granite | 173.2[b]–587.9[c] | 44,160[b]–82,800[c] | 0.20[c] |
| Shale | 171.8[c] | 67,620[c] | 0.23[c] |
| Dolerite | 331.2[d] | 84,180[d] | – |
| Sandstone | 37.0[d] | 9,660[d] | – |
| Limestone | 224.3[d] | 53,130[d] | – |

[a]Brace (1964) and Walsh (1965).
[b]Cook (1965).
[c]Cook et al. (1966).
[d]Wiebols et al. (1968).
*Source:* Jaeger and Cook (1969).

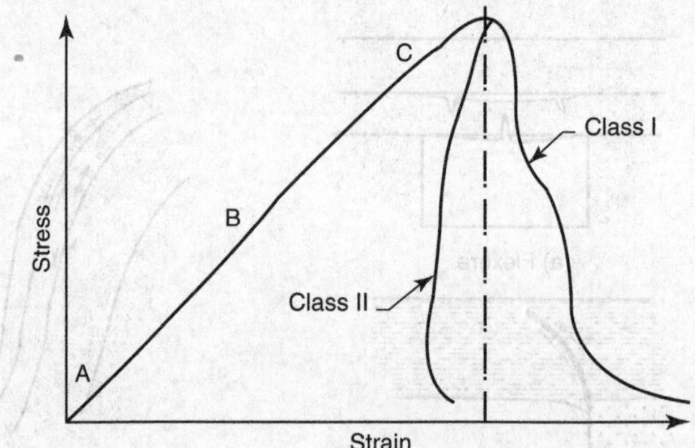


Fig. 23.6 Three regions of the stress–strain curve (*Source:* Wawersik and Fairhurst, 1970)

increasing slope which is a typical one for highly porous rocks and those containing numerous small fractures.

The stress–strain path over region *B* is almost linear, which is a characteristic one for brittle rocks. Over region C, the slope of the curve decreases, depicting dissipation of energy.

Finding notable differences in the post-peak region, they identified two classes of failure. In Class I, the failure propagation is stable, i.e., such rocks can retain some strength even after the peak strength. In Class II, the failure is unstable, that is, the rock has lost virtually all its strength. The fracture of such rocks cannot be controlled. Stages in the development of fracturing with increasing uniaxial compressive stress is shown in Fig. 23.7.

Based on a comprehensive study on 28 rock types, Deere and Miller (1966) classified the uniaxial stress–strain curves into six types (Fig. 23.8). The Type I curve represents a nearly straight-line behaviour of a brittle material until a sudden explosive failure occurs. Such failures are typical of basalts, dolerites, quartzites, and strong limestones. Softer limestones, siltstones, and tuff exhibit a continuous increase in elastic yield, which is characterized by a Type II curve. The stress–strain curve represented as Type III is typical of sandstone, granite, some dolostones and dolerites, and schist scored parallel to foliation. The S-shaped curve with an extremely stiff central portion is identified as Type IV which is typical of metamorphic rocks such as marbles gneiss. Schist scored perpendicular to foliation has the long

Fig. 23.7 Stages in the development of fracturing with increasing unconfined compressive stress (*Source:* Wawersik and Fairhurst, 1970)

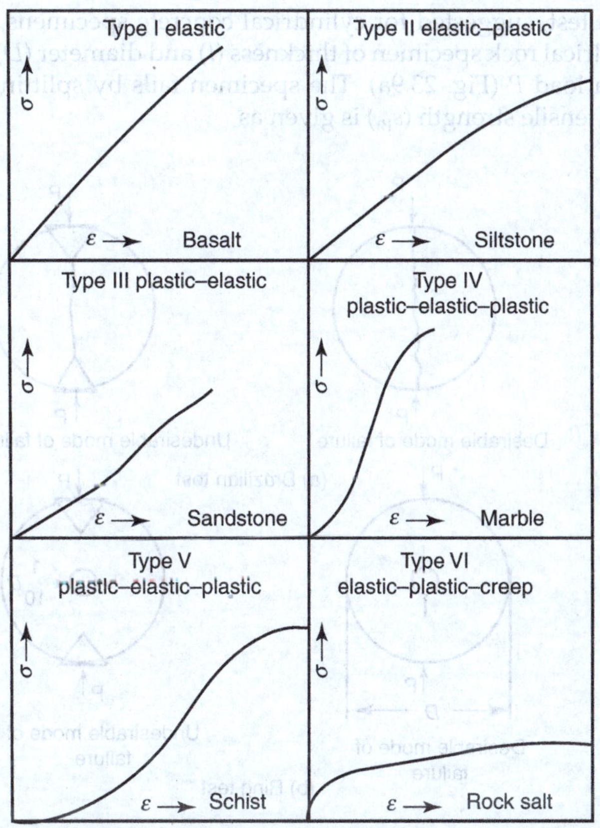


**Fig. 23.8** Typical stress–strain curves for rock in uniaxial compression loaded to failure (*Source:* Stagg and Zeinkiewicz, 1968)

sweeping S-shaped curve of Type V. The Type VI curve, which is characteristic of rock salt, has an initial straight-line portion, followed by an inelastic deformation and continuous creep. Types III, IV, and V curves are all characterized by initial concave upward portion followed by a steeper linear section as cracks close up. The upper part of these curves shows varying degrees of inelastic yielding as failure is approached.

### 23.5.3 Tensile Strength of Rocks

Direct measurement of tensile strength has to be obtained by subjecting a cylindrical specimen to direct extension. Suitable devices to hold the specimen without causing stress concentrations or bending stresses are yet to be developed. Alternatively, the ends of the specimen may be cemented with epoxy resin direct to steel end-platens (Fairhurst, 1961). Specially shaped specimens have also been used (Brace, 1964).

The tensile strength of rocks is lower than their compressive strength. Further, an accurate determination of tensile strength is not generally warranted. Thus, the tensile strength of a rock is usually measured by rapid indirect means (IS: 10082, 1981).

The Brazilian test, suggested for cylindrical concrete specimens, is a convenient one. In this test, a cylindrical rock specimen of thickness ($t$) and diameter ($D$) is loaded in a diametrical plane with a load $P$ (Fig. 23.9a). The specimen fails by splitting along the diametrical loading and the tensile strength ($s_{Bt}$) is given as

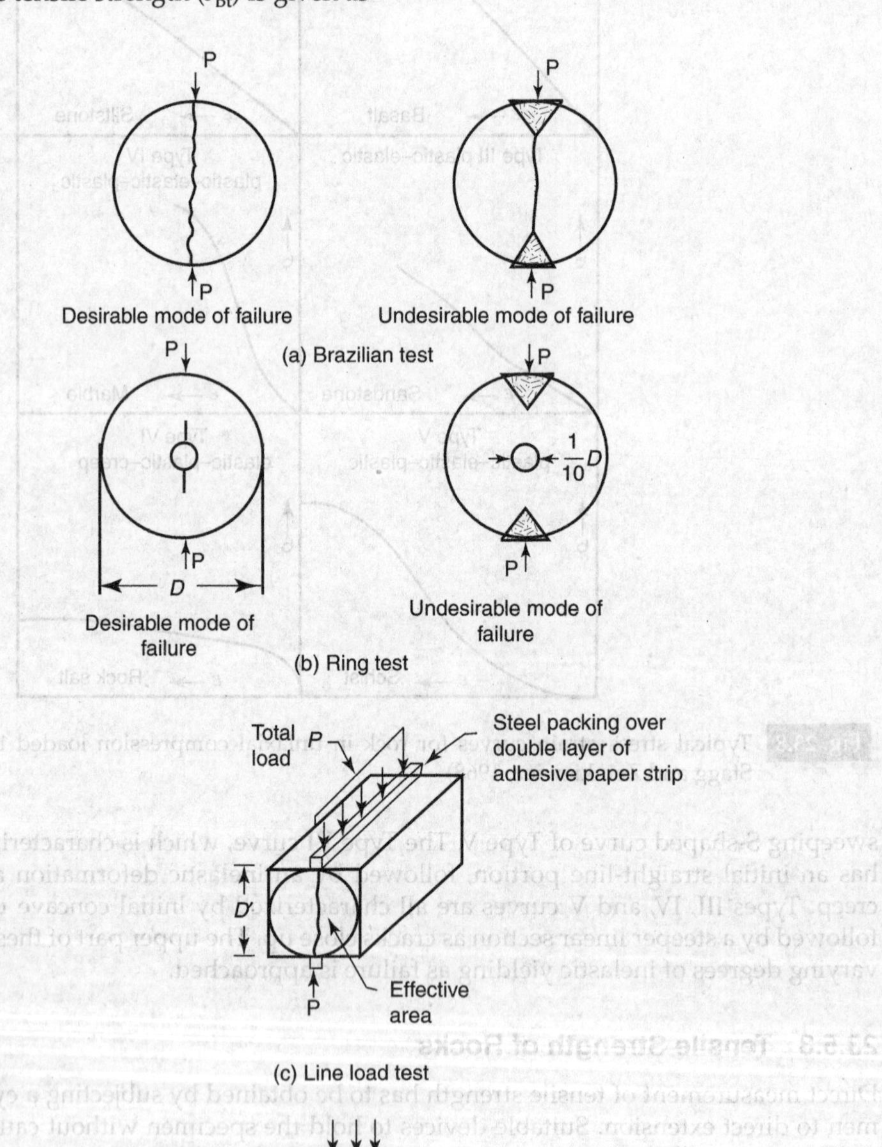


**Fig. 23.9** Tests for tensile strength of rocks [*Source (for a, b, and c):* IS: 10082, 1981]

$$\sigma_{Bt} = \frac{2P}{\pi D t} \tag{23.17}$$

As per Indian Standards (IS: 10082, 1981), the specimen diameter shall not be less than 45 mm and the thickness shall be approximately equal to half the diameter. A loading rate of 200 N/s is recommended, and the load has to be measured with 1% accuracy.

It has been reported that the Brazilian test yields higher tensile strength than the direct tension tests. The extra strength has been attributed to the effect of fissures (Tourenq and Denis, 1970, reported by Goodman, 1980).

In the ring test, which has come into vogue recently (IS: 10082, 1981), a hollow specimen is tested (Fig. 23.9b). The testing apparatus and procedure are same as in the Brazilian test. As the central hole acts as a stress raiser, the vertical splitting increases considerably. Further, the cracks simultaneously initiate from the top and bottom of the hole and propagate towards loading points. The tensile strength is obtained from the expression

$$\sigma_{ri} = \frac{2P[(6 + 38\,(D_0 / D_i)^2]}{\pi D_i t} \tag{23.18}$$

The ring test is considered to be an appropriate test as it is based on fracture mechanics consideration.

Another indirect test recommended by Indian Standards is the line load test. In this test, a rock prism is loaded along a line with point loads (Fig. 23.9c). The rate of loading is 200 N/s and the tensile strength is given as

$$\sigma_{lt} = \frac{P}{\pi D' t} \tag{23.19}$$

where $D'$ is the square base in millimetres. Equation 23.19 is applicable when the specimen rests directly on the lower platen. This value is increased by 1.96 times if the specimen is supported on a packing of steel bar.

The flexural test is another indirect tensile strength test. In this test, a cylindrical specimen of rock is loaded at three points with the bottom of the core supported on points near the ends (Fig. 23.9d). If $P$ is the failure load, $L$ the length between the supports, and $D$ the diameter of the specimen, then the flexural tensile strength $\sigma_{ft}$ is given as

$$\sigma_{ft} = \frac{8PL}{\pi D^3} \tag{23.20}$$

The flexural test also gives higher tensile strength than does the direct tension test.

The tensile strength test from point load has been discussed earlier. The point load test is the only test suitable for field testing of cores (IS: 8764, 1978). As discussed earlier, the test result can be used for evaluating the uniaxial compressive strength of unweathered rock from the expression

$$q_u = 22\,\sigma_{(pl)50} \tag{23.21}$$

Based on the brittle failure theory, the ratio of compressive strength to tensile strength is taken as 8:1, but in practice the ratio is found to be 15:1 to 25:1 (Bell, 1983). Hendron (1968) recommends a tensile strength of 5% to 10% of uniaxial compression strength for most rock mechanics problems.

### 23.5.4 Behaviour of Rocks in Triaxial Compression

The mechanical properties of rocks can be better understood from the results of a triaxial compression. In principle, the triaxial apparatus for a rock specimen is similar to that used for soil specimen. The major difference is the magnitude of lateral pressure and the application of deviatoric stress which demand special type of equipment. The rock specimen is enclosed in an impervious jacket before applying the confining pressure. Hydraulic oil is used as the confining fluid and the jacket is an oil-resistant rubber (e.g., polyurethane). For a short duration test, a bicycle inner tube may be used. Many varieties of triaxial cells are available in different rock mechanics laboratories. The reader may refer to Hoek and Franklin (1968) and Vogler and Kovari (1978) to know the details of equipments and testing.

At the maximum load stage, the stress conditions are $\sigma_1 = P/A$ and $\sigma_3 = p$, where $P$ is the maximum load along the axis of the specimen and $p$ is the pressure in the confining medium.

Most rocks are appreciably strengthened due to confinement. This is especially pronounced in highly fissured rocks. The effect of confining pressure on Berea sandstone and gabbro (after Serdengecti and Boozer, 1961; reported by Hendron, 1968) is shown in Fig. 23.10. An increase in the confining pressure causes an increase in the peak and residual strength and strain to failure. Further, the inelastic strain before failure increases with confining pressure in sandstones (Fig. 23.10a).

With increase in the confining pressure, the initial tangent modulus also increases in weaker rocks but the effect has been reported to be less in strong hard rocks (Hendron, 1968). But in gabbro, the strength increases and the initial elastic properties are not changed by the increase in the confining pressure (Fig. 23.10b). Further, in gabbro the increase in the confining pressure did not improve the inelastic action and ductile behaviour before ultimate failure. As for soils, the triaxial compression results are represented by Mohr's circles.

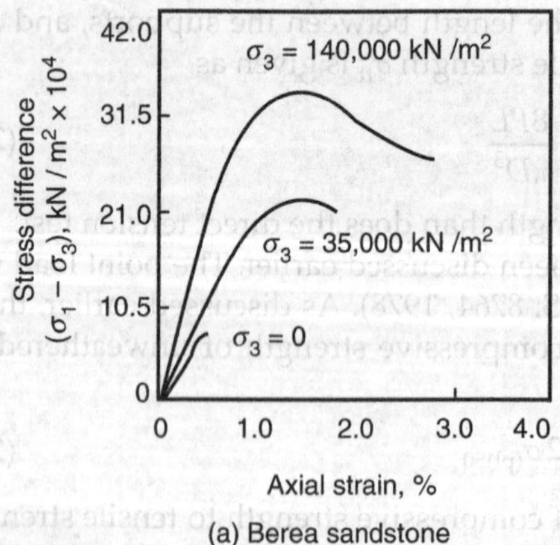

(a) Berea sandstone

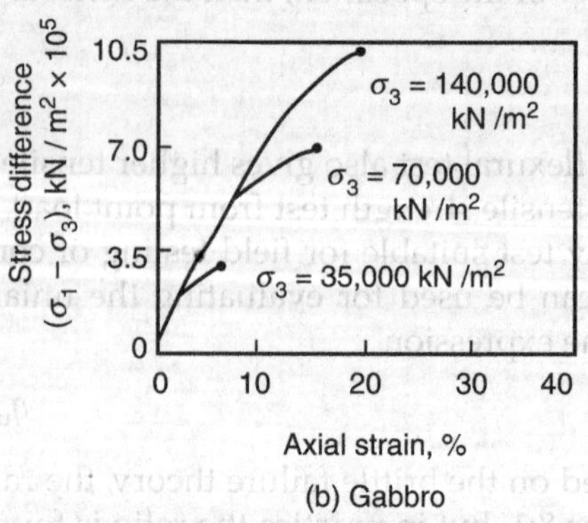

(b) Gabbro

**Fig. 23.10** Variation of stress–strain and strength properties (*Source:* Hendron, 1968)

### 23.5.5 Failure Theory Applicable to Rocks

The Mohr–Coulomb failure criterion is the simplest well-known criterion of failure for rocks (Goodman, 1980). This criterion is applicable for both peak and residual strength of rocks. For peak shear stresses, this is shown in Fig. 23.11 and stated in terms of normal and shear stresses as

$$\tau_p = s_i + \sigma_n \tan\phi \tag{23.22}$$

where $\tau_p$ is the peak shear stress or shear strength of rock, $\phi$ the angle of internal friction, $\sigma_n$ the normal stress on the shear plane, and $s_i$ the shear strength intercept.

The physical interpretation for Eq. 23.22 is that when the difference between applied shear stress ($\tau_p$) and the frictional resistance associated with the normal stress on the failure plane ($\sigma_n \tan\phi$) becomes equal to a constant $s_i$, failure is said to occur. But for a tensile normal stress, it is not justifiable to accept frictional resistance. However, the minimum $\sigma_3$ may be tensile as long as $\sigma_n$ is compressive. Hence, it is justifiable to extend the failure envelope up to the tensile region where $\sigma_3$ becomes equal to the uniaxial tensile strength $-\sigma_t$ and the minor principal stress can never become less than $-\sigma_t$.

Although the Mohr–Coulomb criterion is easy to work, a more accurate failure criterion can be obtained for rocks by fitting in an envelope to Mohr circles corresponding to stresses at failure (Fig. 23.12). Such an envelope is generally curved downwards. In most of the rocks, the failure envelope lies between a straight line and a parabola (Hoek, 1968; Jaegar and Cook, 1969). In practice, depending on the rock type, an empirical curve fitting may be resorted to without substantially violating the Mohr–Coulomb criterion (Eq. 23.25). For a straight-line failure condition, $\sigma_1$ and $\sigma_3$ are related as

$$\sigma_1 = 2s_i \tan\left(45° + \frac{\phi}{2}\right) + \sigma_3 \tan^2\left(45° + \frac{\phi}{2}\right) \tag{23.23}$$

For an uniaxial compression, $\sigma_3 = 0$ and $\sigma_1 = q_u$, then

$$q_u = 2s_i \tan\left(45° + \frac{\phi}{2}\right) \tag{23.24}$$

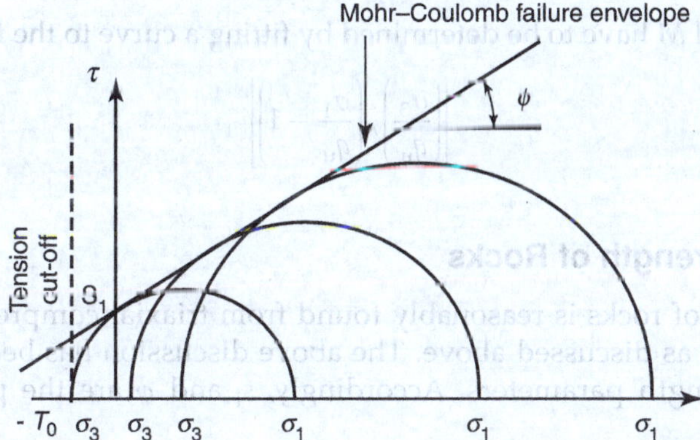


Fig. 23.11 Mohr–Coulomb failure criterion with a tension cut-off

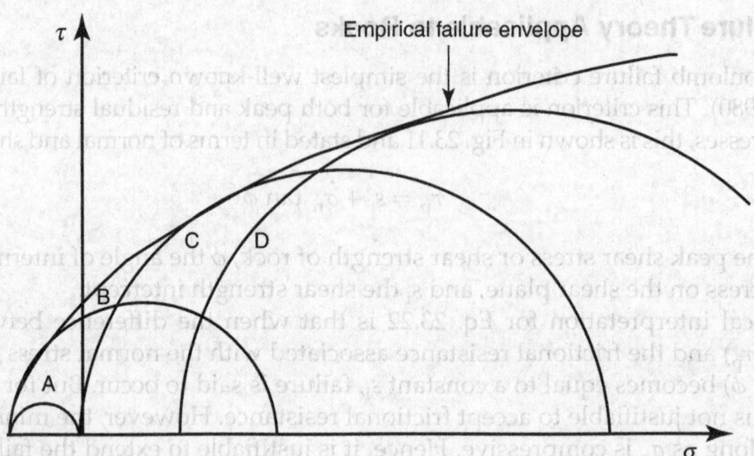


**Fig. 23.12** An empirical criterion of failure: A, direct tension; B, Brazilian; C, unconfined; D, triaxial compression

Hence, Eq. 23.23 reduces to

$$\sigma_1 = q_u + \sigma_3 \tan^2\left(45° + \frac{\phi}{2}\right) \tag{23.24a}$$

Dividing both sides of Eq. (23.24a) by $q_u$, we have

$$\frac{\sigma_1}{q_u} = 1 + \tan^2\left(45° + \frac{\phi}{2}\right)\left(\frac{\sigma_3}{q_u}\right) \tag{23.24b}$$

Equation 23.24b is a non-dimensional one and has been generalized by Bieniawski (1974a, b), by a power law to include the tension cut-off $\sigma_3 = \sigma_t$, as

$$\frac{\sigma_1}{q_u} = 1 + N\left(\frac{\sigma_3}{q_u}\right)^M \tag{23.25}$$

The constants $N$ and $M$ have to be determined by fitting a curve to the family of points

$$\left[\left(\frac{\sigma_3}{q_u}\right), \left(\frac{\sigma_1}{q_u} - 1\right)\right]$$

### 23.5.6 Shear Strength of Rocks

The shear strength of rocks is reasonably found from triaxial compression and adopting the failure criterion as discussed above. The above discussion has been made keeping in view the peak strength parameters. Accordingly, $s_i$ and $\phi$ are the peak shear strength parameters.

It has been reported that the Mohr–Coulomb envelope is also applicable for residual condition. As discussed in Chapter 7 for soils, the shear strength intercept for rocks at residual strength may also approach zero (i.e., $s_i = 0$) and the residual friction angle ($\phi_r$) may lie between zero and peak friction angle.

The presence of water in rocks reduces their strength. Some rocks on saturation weaken due to chemical deterioration of cement or clay binder. But the major reduction in strength is due to the water pressure in pores and fissures. In such situations, effective stresses have to be considered while defining the shear strength parameters.

Several unconfined shear test procedures, such as single, double, punch, and torsional shear tests, are available for rocks. In all these shear tests, the results depend on the testing technique and the rock material. A schematic testing arrangement and the relevant formula are presented by Bell (1983). It is recommended by Goodman (1980) that the shear strength of rocks should be assessed only from the triaxial compression test.

### 23.5.7 Elastic Properties of Rocks

The elastic behaviour of a material depends on its homogeneity, isotropy, and continuity. Homogeneity in rocks means the uniform distribution of constituent particles and isotropy, in terms of a rock, can be defined as a measure of its directional properties. Continuity refers to the amount of fracture and pore space in a particular mass of rock. Thus elastic properties of rocks depend on the size of the rock mass, its constituent minerals, and the degree of fracture.

Modulus of elasticity is the most important elastic property of a rock. Crystalline rocks have an S-shaped stress–strain curve (Fig. 23.13), and at low stress level the modulus of elasticity increases with the increase in stress. The initial tangent modulus, the tangent modulus and the secant modulus for a rock are shown in Fig. 23.13. Dhir and Sangha (1973) recommended the 50% tangent modulus (the tangent modulus at half the failure load) provides the best value.

Farmer (1968) distinguished three types of elastic behaviour in rocks, viz., quasi-elastic, semi-elastic, and non-elastic. The initial tangent modulus ranges from $6 \times 10^4$ to $11 \times 10^4$ MPa in quasi-elastic rocks, between $4 \times 10^4$ and $7 \times 10^4$ MPa in semi-elastic rocks, and it is less than $5 \times 10^4$ MPa in non-elastic rocks. The quasi-elastic rocks are fine-grained, compact, and massive and show a near linear stress–strain curve up to failure (Fig. 23.13a). Semi-elastic rocks are coarse-grained igneous rocks and fine-grained compacted sediments, and the stress–strain

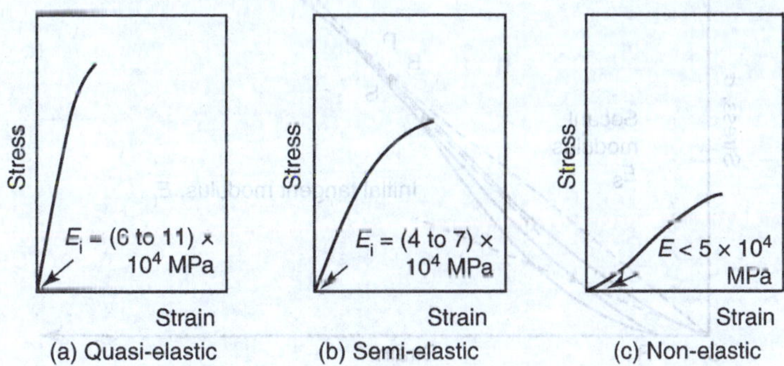


**Fig. 23.13** Typical stress–strain relationships for rocks (*Source:* Farmer, 1968)

curve shows a slope decrease with increasing stress (Fig. 23.13b). Less cohesive rocks with large pore space fall under the third category, whose stress–strain curve is shown in Fig. 23.13c.

Another important elastic property of rocks is the Poisson's ratio ($\nu$). Deere and Miller (1966), based on the studies on rocks, showed that $\nu$ varied between 0.125 and 0.341. In rocks subjected to uniaxial compression, $\nu$ remains more or less constant with increase in stress, and $\nu$ reaches the theoretical maximum of 0.50 at the failure range. The trend is opposite in rocks subjected to uniaxial tension. The $\nu$ values of some rocks are given in Table 23.9.

The laboratory uniaxial compression test is itself used to obtain circumferential and axial deformations or strains. The data needed to evaluate them are obtained by electrical resistance strain gauges, compressometers, optical devices, or other suitable means. While choosing the measuring devices, it should be noted that at least two circumferential and two axial strain measurements can be found for each increment of load (IS: 9221, 1979). The gauge length for strain measurements should be at least five times the grain-size diameter. The axial strain ($\varepsilon_a$) and the diametric strain ($\varepsilon_d$) are then calculated. Thus, Poisson's ratio is

$$\nu = \frac{\varepsilon_d}{\varepsilon_a}$$

The modulus of elasticity $E$ may be calculated from a stress–strain curve using any one of the methods explained earlier (Fig. 23.14). It is determined in the field by adopting uniaxial jacking test (IS: 7317, 1974).

### 23.5.8 Hardness

Hardness of a mineral is defined based on its capability to resist scratching and relative hardness has been used as the index for identification. For this, Mohr's hardness scale has been in use. But in rocks, the condition is much more complex and no clear definition is yet available. Deere and Miller (1966) defined rock hardness as a combined resistance to displacement and penetration of particles. Further, they noted that both hardness and toughness depend on the same factors.

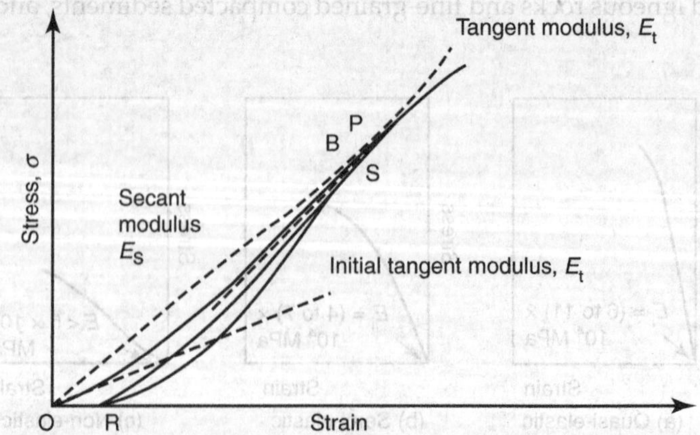


Fig. 23.14 Representative stress–strain curve for rock in uniaxial compression

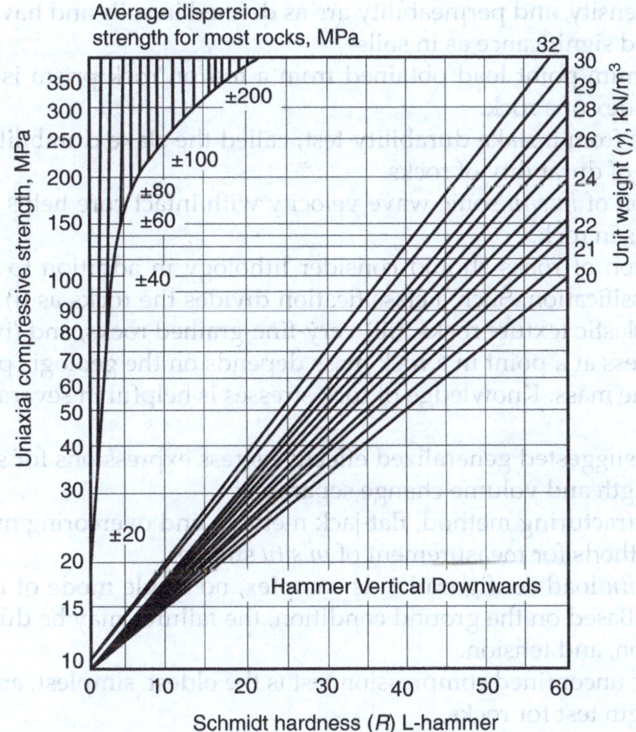


**Fig. 23.15** Correlation chart for Schmidt (*L*) hammer relating unit weight of rock, compressive strength, and rebound number (*Source:* Deere and Miller, 1966)

Thus attempts were made to determine the hardness of rock either from the measured indentation (Brinell, Rockwell, or Vickers) or from the rebound tests (Shores or Schmidt). Deere and Miller (1966) used both Shores's scleroscope and Schmidt rebound hammer and presented correlation charts with uniaxial compressive strength. Figure 23.15 shows a correlation chart for Schmidt (*L*) hammer. It is possible to use this hardness number and the dry unit weight for predicting the uniaxial compressive strength within 75% confidence limits (Hendron, 1968).

## POINTS TO REMEMBER

23.1 Rock mechanics is an interdisciplinary subject which covers various fields like geology, mining, petroleum, and civil engineering.

23.2 In order to identify the quality of a rock, certain properties like porosity, density, permeability, strength, durability, and sonic velocity have been grouped and designated as index properties.

23.3 Porosity, density, and permeability are as defined in soils and have the same characteristics and significance as in soils.
23.4 The maximum point load obtained from a test on rock prism is designated as the strength index of a rock.
23.5 The results from a slake durability test, called the slake durability index, represent the degree of durability of rocks.
23.6 Comparison of *in situ* sonic wave velocity with intact core helps in identifying discontinuities in rock.
23.7 Classification of rocks should consider lithology in addition to conventional geological classification. Such a classification divides the rocks as (i) crystalline texture rocks, (ii) clastic texture rocks, (iii) very-fine-grained rocks, and (iv) organic rocks.
23.8 State of stress at a point in a rock mass depends on the geologic processes that have acted on the mass. Knowledge of rock stresses is helpful in several civil engineering works.
23.9 Skempton suggested generalized effective stress expressions for saturated rocks for shear strength and volume change separately.
23.10 Hydraulic fracturing method, flat-jack method, and overcoring method are the best known methods for measurement of *in situ* stresses.
23.11 As the *in situ* load configuration is complex, no single mode of rock failure can be identified. Based on the ground condition, the failures may be due to flexure, shear, compression, and tension.
23.12 Uniaxial or unconfined compression test is the oldest, simplest, and most frequently used strength test for rocks.
23.13 The Brazilian test is the convenient test to determine the tensile strength of rocks.
23.14 The Mohr–Coulomb failure criterion is the simplest well-known criterion of failure for rocks and applicable both for peak and residual strengths.
23.15 Shear strength of rocks is reasonably found from triaxial compression and adopting the Mohr–Coulomb failure criterion.
23.16 Modulus of elasticity is the most important elastic property of rock. The tangent modulus determined at half of failure load provides the best value.
23.17 Rock hardness is a combination of resistance to displacement and penetration of particles.

## QUESTIONS

### *Objectives Questions*

23.1 State whether the following statements are true or false:

1. In sedimentary rocks, the porosity varies from close to zero to as much as 90%.
2. Uniaxial compressive strength is identified as a strength index test.
3. Most rocks are significantly strengthened by confinement.
4. Engineering classification of intact rocks is based on the modulus of elasticity and Poisson's ratio of the rocks.
5. Terzaghi's effective stress law used for soils is not valid for rocks.

23.2 Pororsity in a rock is found to be maximum when the grains are
(a) Of the same size (b) Of varying sizes
(c) In a specified size range (d) Arranged in a parallel array

23.3 The rate of slaking of a rock is governed by the change in
(a) Plasticity index (b) Flow index
(c) Liquidity index (d) Toughness index

23.4 The average uniaxial compressive strength of an intact rock is found to be 85 MPa. Based on this value, the rock may be classified as
(a) Low strength (b) High strength
(c) Very high strength (d) Medium strength

23.5 In a ring test on a hollow specimen for tensile strength determination, cracks initiate
(a) From the top and bottom and propagate towards the centre
(b) From the top only immediately below the loading
(c) From the top and bottom of the hole and propagate towards loading points
(d) From the bottom only and propagate towards the hole

### *Descriptive Questions*

23.6 The pores of a rock are filled with mercury by injecting it at high pressure. Derive an expression for porosity in terms of the volume of mercury filling the pores, specific gravity of component grains, and density of water.

23.7 Explain the factors which affect the shear strength of a rock mass? Why cannot the strength of a core sample be used for the mass?

23.8 How do modes of failure of rock specimens vary with effective confining pressure? Illustrate your answer by stress–strain curves.

23.9 For a three-point bending test of a core sample with circular cross section, derive an expression for the modulus of rupture.

23.10 In a homogeneous rock mass, how does erosion of top layers affect the ratio of horizontal stress to vertical stress?

## EXERCISE PROBLEMS

23.1 A sandstone is composed of a mixture of 73% quartz and 27% pyrite grains with a porosity of 16%. Determine the dry density of the sandstone if the specific gravity of quartz and pyrite grains are 2.65 and 5.0, respectively (Hint: Specific gravity, $G$, of the solid part of a rock is given as

$$G = \sum_{j=1}^{m} G_j V_j$$

where $G_j$ is the specific gravity and $V_j$ is its volume percentage of component $j$).

23.2 A sandstone core, 85 mm in diameter and 172 mm long, consists of quartz and feldspar grains. The sample was saturated with water and found to weigh 22.3 N when wet and 21.1 N after oven drying. Determine the wet and dry unit weights and its porosity.

23.3 Triaxial shear tests conducted on samples of rock revealed the shear strength parameters as $s_i = 12{,}600\ \text{kN/m}^2$ and $\phi = 38°$. An unlined tunnel is made in the rock where the confining pressure was zero and the water pressure was $850\ \text{kN/m}^2$. Under such a condition, what will be the compressive strength of the rock?

23.4 The triaxial strength of the rock specimen may be approximated by the following expression:

$$\frac{q_f}{q_c} = 0.1 + k\left(\frac{\sigma'_m}{q_c}\right)^n$$

where $q_f$ is the maximum shear stress, $\sigma'_m$ the effective mean normal stress, $q_c$ the uniaxial compressive strength, $k = 0.78$, and $n = 0.90$.
Plot Mohr's failure envelope and estimate shear strength parameters for a range of effective normal stresses of 250 to $500\ \text{kN/m}^2$. The uniaxial compressive strength may be taken as $2{,}000\ \text{kN/m}^2$.

23.5 A porous rock under triaxial test conditions yielded the shear strength intercept as 1.2 MPa and the angle of internal friction as 33°. Compute the uniaxial compressive strength and estimate the tensile strength of this rock.

# 24

# Pavements

**CHAPTER HIGHLIGHTS**

Components of pavements; Types and requirements; subgrade; pavement design – Wheel load – Flexible and rigid pavement; Methods – Group Index – CBR – Mc Leod – Burmister – Westergaard's – IRC

## 24.1 INTRODUCTION

A pavement is the main structure of a road or a runway, which is relatively a stable layer or a crust constructed over the natural soil. The natural soil is called a subgrade. The main function of a pavement is to support and distribute the wheel loads of vehicles through the underlying layer to the subgrade.

## 24.2 COMPONENTS OF PAVEMENT

Subgrade is the bottom-most level of a pavement, which is termed as the formation level over which other layer are laid, namely, sub-base, base and surface courses (Figure 24.1).

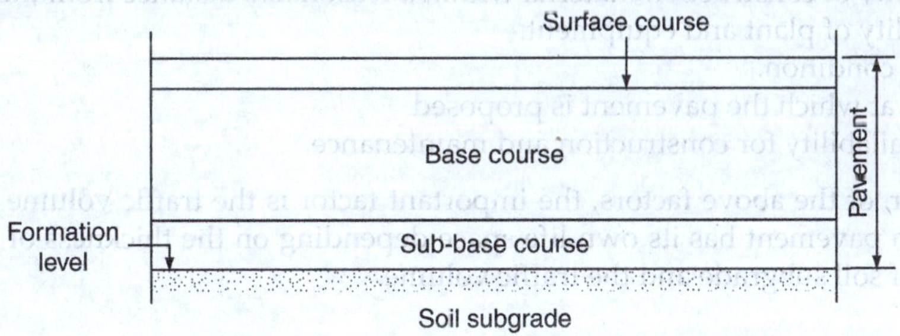


**Fig. 24.1** Components of a pavement

As heavy load of vehicles has to be carried, surface layer is the strongest layer provided at the top. The courses of decreasing strength are provided in order, such that the load is distributed smoothly on to the subgrade.

Sub-base, base and surfacing courses are also called soling, metalling and wearing courses, respectively. Sometimes a strong subgrade is available the sub-base may be omitted.

## 24.3 TYPES OF PAVEMENT

Pavements may be classified into two major categories, namely, flexible pavement and rigid pavement. The basic difference between these two types is the way by which they distribute the load over their subgrade.

The design of flexible pavement is based on the principle that the wheel loads of vehicles are distributed to the natural soil through the successive layers of granular materials. The intensity of load decreases with depth as the area of dissipation is increased. The strength of the subgrade decides the thickness of the flexible pavement. Water-bound Macadam (WBM) roads, gravel roads, earth roads, soil-stabilized roads, bitumen roads etc., consisting of layers of road-making materials compacted to form an elastic bed, are grouped under flexible pavements.

The design of rigid pavement is based on the strength of the structural slab, which tends to distribute the load over a wide area of the soil. The pavement slab is of Portland cement concrete, which has high rigidity that resists the deformation of the surface. For this reason, minor variation in subgrade strength does not affect the structural capacity and the performance of the rigid pavement.

A third category of pavement in between these two may be termed as semi-rigid. Cement grouted, lean cement concrete, soil–cement pavement etc. may fall under this group.

## 24.4 REQUIREMENTS OF PAVEMENT COMPONENTS

Selection of pavement components for the selected pavement such as type of pavement, base course, sub-base and surface course is governed by the following factors:

1. Type of subgrade soil.
2. Drainage conditions.
3. Type and volume of traffic.
4. Availability of construction material within a reasonable distance from the site.
5. Availability of plant and equipment.
6. Climatic condition.
7. Altitude at which the pavement is proposed.
8. Fund availability for construction and maintenance.

However, of the above factors, the important factor is the traffic volume or intensity of traffic. Each pavement has its own life space depending on the thickness of pavement section, type of soil subgrade and the traffic volume.

## 24.5 SUBGRADE

### 24.5.1 Preparation

Preparation of subgrade includes all operations before the construction of a pavement structure. Depending on the topography, the subgrade may be situated on the existing ground level or as embankment or as cutting.

In all cases, the side clearance is the first aspect. The site should be cleared off of the top soil, grass, roots, rubbish and other organic matters.

Next the grading operation is reworked so as to bring the vertical profile of the subgrade to the designed grade and camber. Necessary equipment may be used for removing the boulder. Levelling and finishing it is essential that the subgrade in compact to the standard density at an appropriate moisture content before the construction of a pavement structure.

### 24.5.2 Embankment and Cutting

Embankment is constructed when the grade line of a highway is required to be raised above the existing ground level. Necessity of raising the ground level for the construction of highway may aid due to the following reasons:

1. In order to make the subgrade level high above the ground-water level.
2. In order to prevent the capillary water to enter the subgrade by capillary action or surface water through permeation.
3. In order to satisfy the vertical alignment of the designed highway.

The design and construction of embankment comprises of:

1. deciding the height of the embankment,
2. finalizing side slopes and their short- and long-term stability,
3. embankment material,
4. compaction requirement and compacting with suitable equipment and
5. check for settlement and overall stability

Figure 24.2 shows the cross-section of a typical embankment

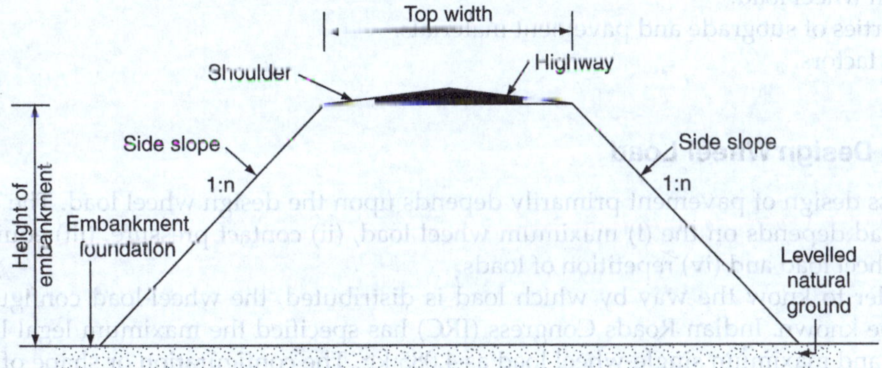


Fig. 24.2 Details of highway embankment

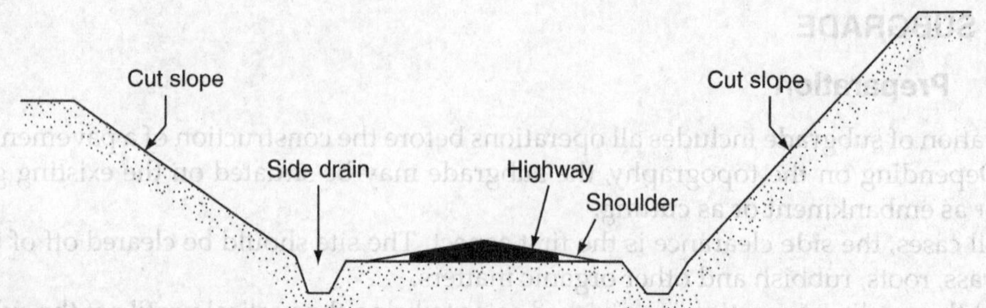


Fig. 24.3 Details of highway in cutting

Sometimes the grade of natural ground may not be as per design requirement, then it has to be properly graded by cutting. In this case, apart from the preparation of subgrade, the stability of side slopes and drainage arrangements are to be made.

Figure 24.3 shows a highway in cutting. Stability of cuts may be made by adopting the following measures:

1. By maintaining adequate flat slopes.
2. By consolidating the cut slopes.
3. By providing steps at suitable intervals as the sloping cut edge.
4. By constructing breast walls against cut slopes.
5. By developing grass turf.

## 24.6 PAVEMENT DESIGN

Pavement design comprises of two parts, namely, (i) mix design of materials to be used in each component layer and (ii) thickness design of the pavement and the component layer. Depending on the type of materials to be used, the load and other conditions, the mix design is made.

Various factors to be considered for the design of pavements are:

1. Design wheel load.
2. Properties of subgrade and pavement materials.
3. Other factors.

### 24.6.1 Design Wheel Load

Thickness design of pavement primarily depends upon the design wheel load. The design wheel load depends on the (i) maximum wheel load, (ii) contact pressure, (iii) equivalent single-wheel load and (iv) repetition of loads.

In order to know the way by which load is distributed, the wheel-load configuration should be known. Indian Roads Congress (IRC) has specified the maximum legal load as 8,170 kg and maximum single-wheel load as 4,085 kg. The configuration or shape of wheel load gives an idea how the load of a vehicle acts on the surface of the pavement. Figure 24.4 shows the typical wheel-load configuration of a tractor-trailer unit.

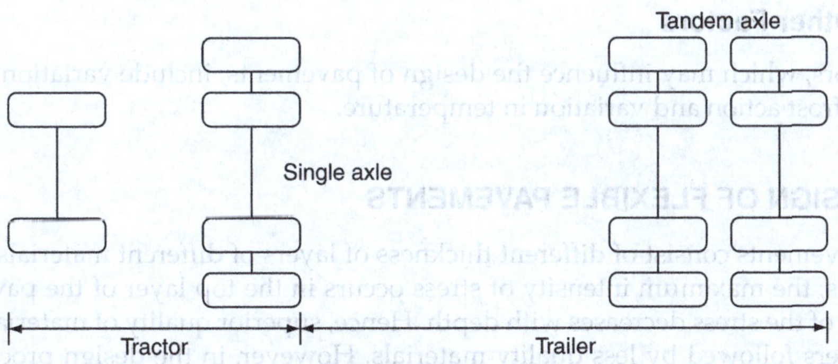


**Fig. 24.4** Wheel configuration of tractor trailer unit

While total load influencing the thickness of the pavement, the tyre pressure influences the quality of surface course. The wheel load is assumed to be distributed over a circular area. This distributed load is the tyre pressure, which is differently referred to as *inflation pressure* or *contact pressure*. Contract pressure, $p$, is given by the relationship

Contact pressure, $p = \text{h}A$
where $h$ = load on wheel and
$A$ = contact area or area of imprint.

The contact pressure is focused to be more than the tyre pressure when the tyre pressure is less than 7 kg/cm$^2$.

In order to have a maximum wheel load, dual wheel assembly is provided to the rear axles of the load vehicles. Because of this, the load die to both wheels are not to be transfer led to the pavement. However, there will be overlap pressure after a certain depth. The actual effect is in between a single-wheel load and a double-wheel load, which is called the equivalent single-wheel load (ESWL).

The deformation of a pavement or subgrade may be small due to the application of a single-wheel load. As the pavement experiences repeated loads, the accumulated deformation at any time may be freely elastic or partially plastic. This plastic deformation may increase with more repetition and ultimately may result in pavement failure. For design purposes, it is necessary to convert various wheel loads to a single-wheel load. Equivalent wheel loads accounting for repetition of loads are those which require save thickness and strength of pavement.

## 24.6.2 Properties of Subgrade and Pavement Components

The two strength characteristics of subgrade or pavement components are California bearing ratio (CBR) and elastic modulus.

CBR has been used for the design of flexible and rigid pavements. It is an *ad hoc* property of a subgrade and other materials, which shows relative significance and do not provide absolute measure.

Elastic modulus of subgrade soil are determined using the results from a plate-bearing test. Burmister's elastic two-layer systems can be used to compute elastic modulus of pavement materials for maximum deformation.

### 24.6.3 Other Factors

Other factors, which may influence the design of pavements, include variation in moisture condition, frost action and variation in temperature.

## 24.7 DESIGN OF FLEXIBLE PAVEMENTS

Flexible pavements consist of different thickness of layers of different materials. Due to the wheel loads, the maximum intensity of stress occurs in the top layer of the pavement. The magnitude of the stress decreases with depth. Hence, superior quality of materials is used in the top layers followed by less quality materials. However, in the design process, it is the basic requirement that under the application of the load none of the layer is overstressed. That is, no section of the pavement undergoes excessive deformation, which may result in localized depression or settlement.

There is no rational design method of flexible pavement. In recent years, there has been a considerable shift in the design concepts of flexible pavements. The flexible pavement design methods may be classified under three distinct groups.

**Empirical Methods:** These are based on soil classification, physical or strength parameters of soils and other factors such as climate and moisture.

**Sub-empirical Methods:** These are based on stress–strain functions, performance and serviceability concepts.

**Theoretical Methods:** These are based on theoretical analysis and mathematical computations, considering stress–strain behaviour of soils.

Some of the methods, which are in use, are discussed below.

### 24.7.1 Group Index Method

This method of flexible pavement design is based on the soil properties such as (i) particle-size distribution, (ii) liquid limit and (iii) plasticity index.

Group index is claimed to be an inverse measure of the thickness of pavement. The higher is the group index, the lower is the strength and greater is the thickness requirement for a given traffic volume. According to this method, subgrades have been divided into five categories based on group index (Table 24.1).

**Table 24.1** Subgrade quality and group index

| Subgrade quality | Value of group index |
|---|---|
| Excellent | 0 |
| Good | 0–1 |
| Fair | 2–4 |
| Poor | 5–9 |
| Very poor | 10–20 |

**Table 24.2** Classification of traffic

| Total volume of traffic (commercial vehicles) | Number of vehicles per day |
|---|---|
| Height | Less than 50 |
| Medium | 50–300 |
| Heavy | Over 300 |

Traffic is divided into three groups as given in Table 24.2.

In order to design the pavement thickness by this method, the following steps are followed:

Step I: The value of the Group Index is found using the details of the soil.

Step II: The anticipated traffic is estimated and classified as light, medium or heavy as indicated in Fig. 24.5 (a).

Step III: The appropriate design curve is chosen from Fig. 24.5 (b) and the total thickness of pavement (surface, base and sub-base courses) is found for the corresponding Group Index.

Step IV: The thickness of sub-base or base only is found from appropriate curve to find the thickness.

As discussed earlier, the method is only empirical, which considers only soil properties and not the strength and compressibility characteristics of the soil. These curves were presented as only approximate and may be modified by the engineer concerned to suit the local conditions based on his/her experience.

### 24.7.2 California Bearing Ratio (CBR) Method

California Bearing Ratio method was made during the Second World War on the basis of survey of the existing methods of pavement design performed by Corps of Engineers, USA. The method was adopted for the design of military airport pavement. Later, California State Highway Department carried out CBR tests on the there existing pavement layers including subgrade, sub-base and base courses. Based on the exhaustive study on the pavements, which have performed well and those which have failed, an empirical design chart was developed correlating the thickness of the pavement and the CBR value.

This design chart simply specifies that a material with a particular CBR requires a certain thickness of pavement layer as a cover for a given traffic load. It is evident that a higher load needs a thicker pavement layer.

Primarily, the design charts (Figure 24.6) were prepared by California State Highway Department for wheel loads of 3,175 kg and 5,443 kg, representing light and heavy traffic. Subsequently, the design curve for medium traffic of 4,082 kg wheel load was obtained by interpolation.

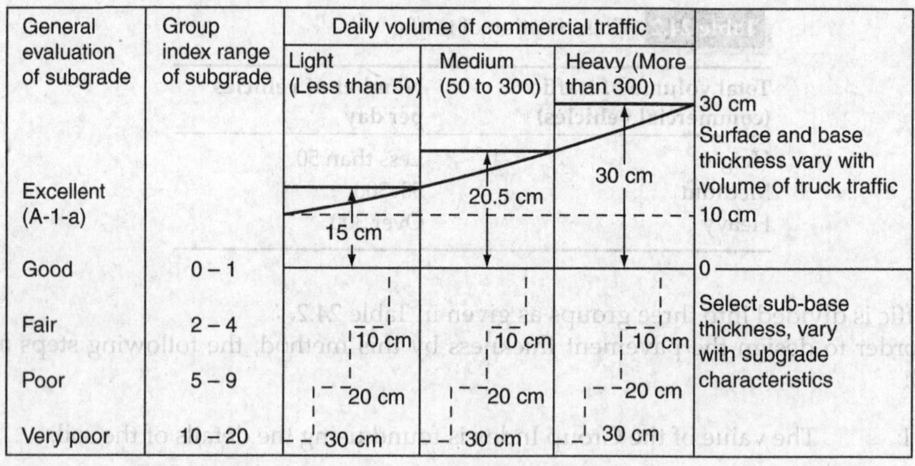


(a)

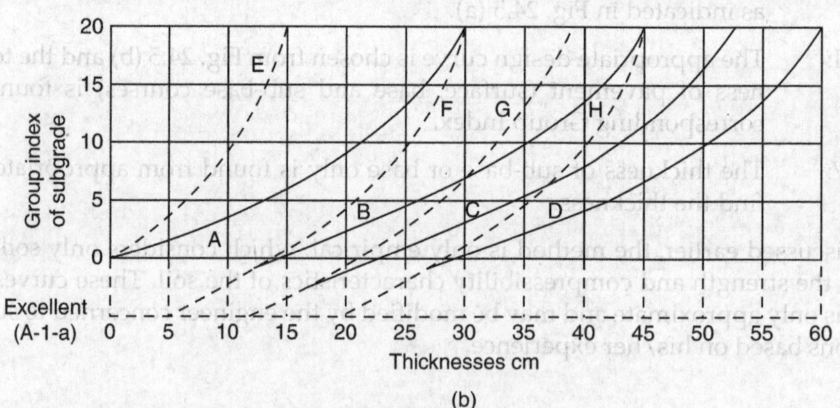


(b)

—— *Combined thickness of surface, base and sub-base*

– – – *Thickness of surface and base*

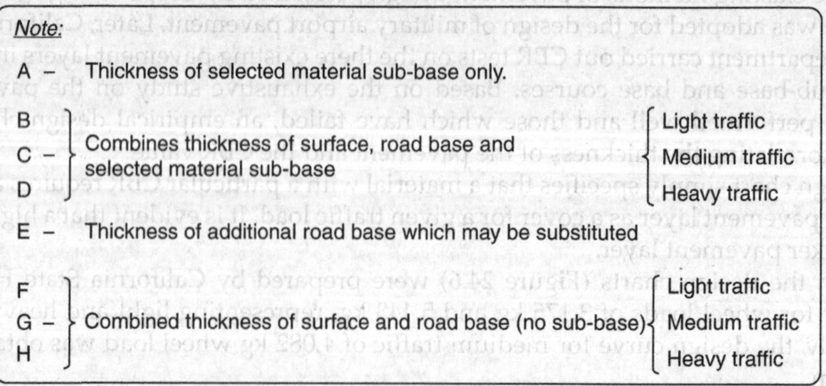


**Fig. 24.5** Design chart by Group Index value

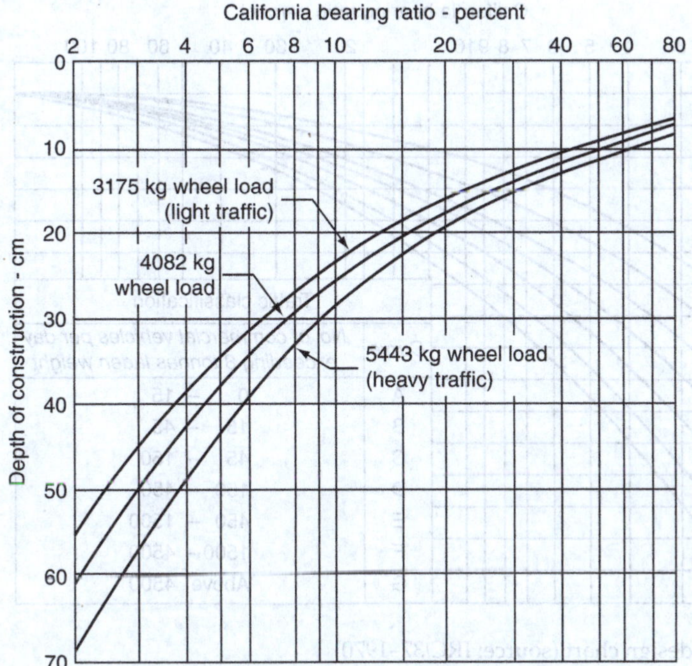


Fig. 24.6 CBR design chart (California State Highway Department)

On similar grounds, Indian Roads Congress (IRC: 37–1970) provided design charts similar to the one followed in UK, for different traffic volumes from 15 to over 4,500 commercial vehicles per day exceeding 8 tons loader weight (Figure 24.7).

In order to use the chart, the following steps are followed:

1. CBR value should be found for a soaked specimen of the subgrade soil.
2. Considering the design wheel load or the anticipated traffic, the appropriate design curve is chosen.
3. Corresponding to the CBR value and the chosen curve, the total thickness of flexible pavement needed to cover the subgrade is found.
4. If a superior material is to be used as a sub-base, whose CBR value is found corresponding to this CBR value of sub-base material, the required thickness of construction (over this material) on the already chosen traffic curve is used.
5. Then the thickness of the sub-base course is the total thickness minus the thickness over the sub-base.

The CBR method of flexible pavement design can be used to use materials of different CBR values as layers with appropriate thickness.

### 24.7.3 Mc Leod Method

The Canadian Department of Transport under the direction of Mc Leod conducted extensive plate-bearing tests to investigate the stability of air fields and highway pavements.

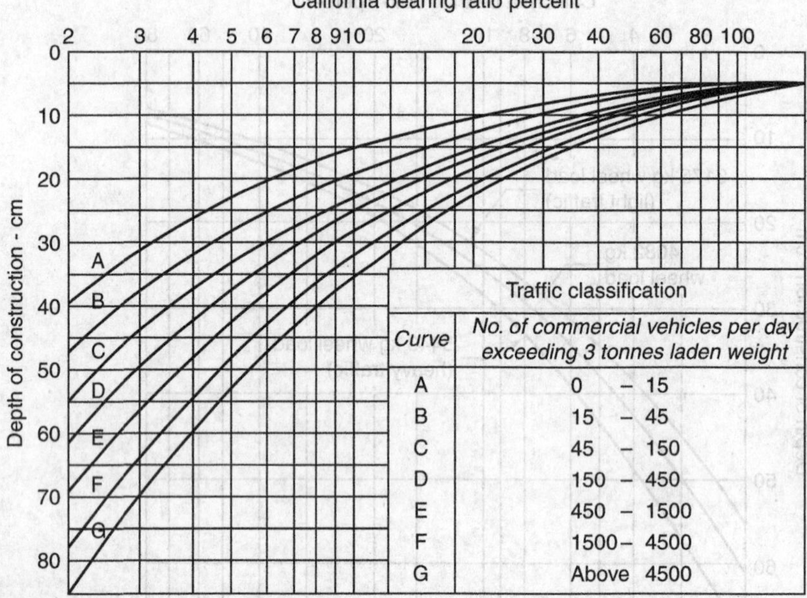


| Traffic classification | |
|---|---|
| Curve | No. of commercial vehicles per day exceeding 3 tonnes laden weight |
| A | 0 – 15 |
| B | 15 – 45 |
| C | 45 – 150 |
| D | 150 – 450 |
| E | 450 – 1500 |
| F | 1500– 4500 |
| G | Above 4500 |

**Fig. 24.7** CBR design chart (source: IRC 37–1970)

The plate-bearing tests were repetitive and were conducted on surface, base course, sub-base and subgrade. Based on these results Mc Leod developed a design procedure, which goes by his nature as Mc Leod method.

Mc Leod's empirical design formula is given as:

$$T = K \log_{10} \frac{P}{S} \tag{24.1}$$

where $T$ = required thickness of gravel base
$P$ = gross wheel load, kg.
$S$ = total subgrade support, kg (for the same contact area, deflection and number of repetitions of load, $P$)
$K$ = base course constant.

It has been found that the base course constant, $K$, is a function of the loaded area. Figure 24.8 shows the relationship between the plate diameter and base course constant.

The thickness may be calculated from the above equation for a given wheel load $P$ and the subgrade support determined from a plate-bearing test.

The subgrade support, $S$, is computed from the support corresponding to 0.5 cm deflection after ten repetitions using a 30 cm diameter plate.

Figure 24.9 represents the perimeter of the area of the plate to the ratio of subgrade support (kg/cm$^2$) at a particular deflection for any size of the plate to the subgrade support (in kg/cm$^2$) at 0.50 cm deflection on a 30 cm diameter plate.

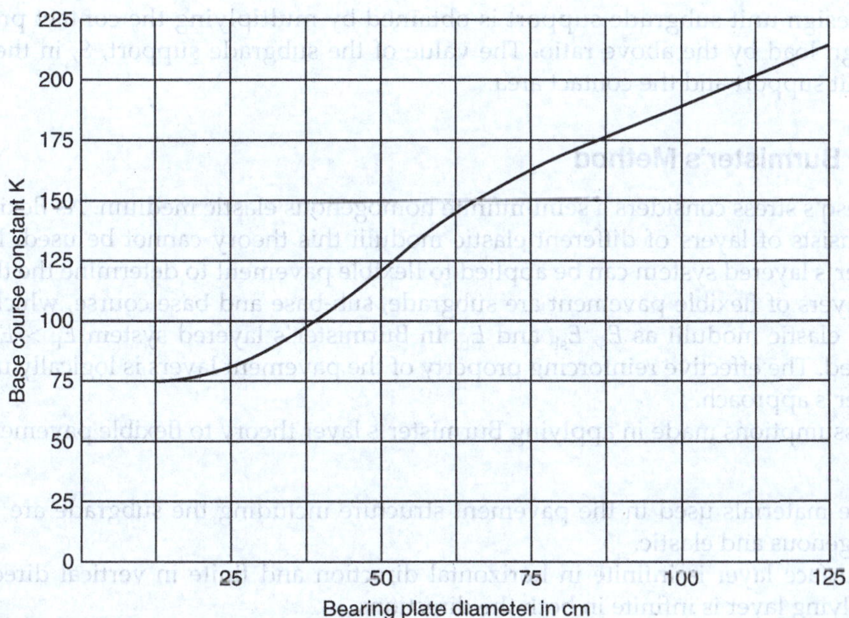


Fig. 24.8 Relation between plate diameter and base course constant

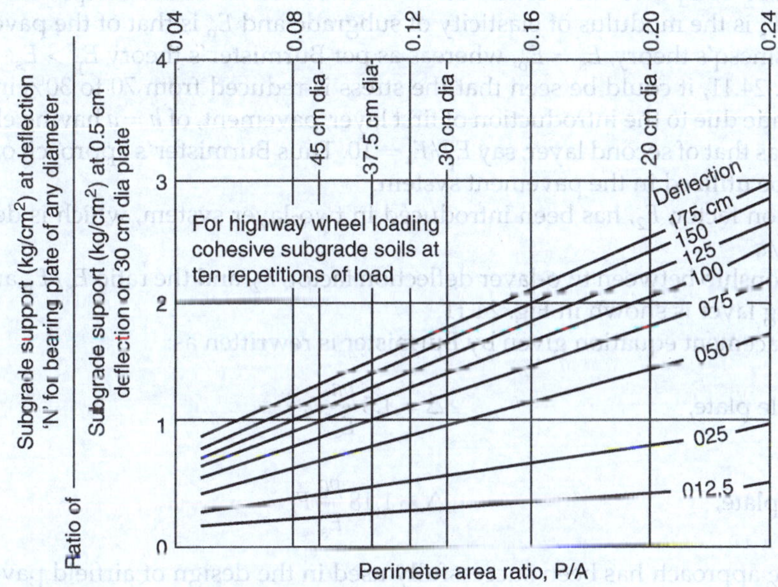


Fig. 24.9 Relationship of subgrade support with *P/A* ratio

The design unit subgrade support is obtained by multiplying the contact pressure of the design load by the above ratio. The value of the subgrade support, $S$, in the product of the unit support and the contact area.

### 24.7.4 Burmister's Method

Boussinesq's stress considers a semi-infinite homogenous elastic medium. As flexible pavement consists of layers of different elastic modulii this theory cannot be used. However, Burmister's layered system can be applied to flexible pavement to determine the thickness.

The layers of flexible pavement are subgrade, sub-base and base course, which may be assigned elastic modulii as $E_s$, $E_{sb}$ and $E_b$. In Burmister's layered system $E_b > E_{sb} > E_s$ is considered. The effective reinforcing property of the pavement layers is logically utilized in Burmister's approach.

The assumptions made in applying Burmister's layer theory to flexible pavements are as follows:

1. All the materials used in the pavement structure including the subgrade are isotropic, homogenous and elastic.
2. The surface layer is infinite in horizontal direction and finite in vertical direction; the underlying layer is infinite in both the directions.
3. The layers have continuous perfect contact.
4. The top layer is free of shearing and normal stresses outside the loaded area.
5. Poisson's ratio, $v$, is 0.5 in both the layers.

Figure 24.10 compares the vertical stress distribution of Boussinesq's and Burmister's analyses. If $E_s$ is the modulus of elasticity of subgrade and $E_p$ is that of the pavement; then, as per Boussinesq's theory, $E_s = E_p$, whereas as per Burmister's theory $E_p > E_s$.

From Fig. 24.11, it could be seen that the stress is reduced from 70 to 30% in the second layer, subgrade due to the introduction of first layer, pavement, of $h = a$ having elastic modulus 10 times as that of second layer, say $E_p/E_s = 10$. Thus Burmister's approach of reinforcing action may be utilized in the pavement system.

A deflection factor, $F_2$, has been introduced in two-layer system, which is dependent on $E_s/E_p$ and $h/a$.

The relationship between two-layer deflection factor, $F_2$, and the ratio $E_s/E_p$ and thickness of reinforcing layer is shown in Fig. 24.11.

The displacement equation given by Burmister is rewritten as:

For flexible plate,
$$\Delta = 1.5\frac{pa}{E_s}F_2 \tag{24.2}$$

For rigid plate,
$$\Delta = 1.18\frac{pa}{E_s}F_2 \tag{24.3}$$

The above approach has been successfully used in the design of airfield pavements. The layered system approach can also be utilized in the design of highway pavement. The diameter of the loaded plate may be taken as 30 cm and the design deflection may be considered as 0.50 or 0.25 cm.

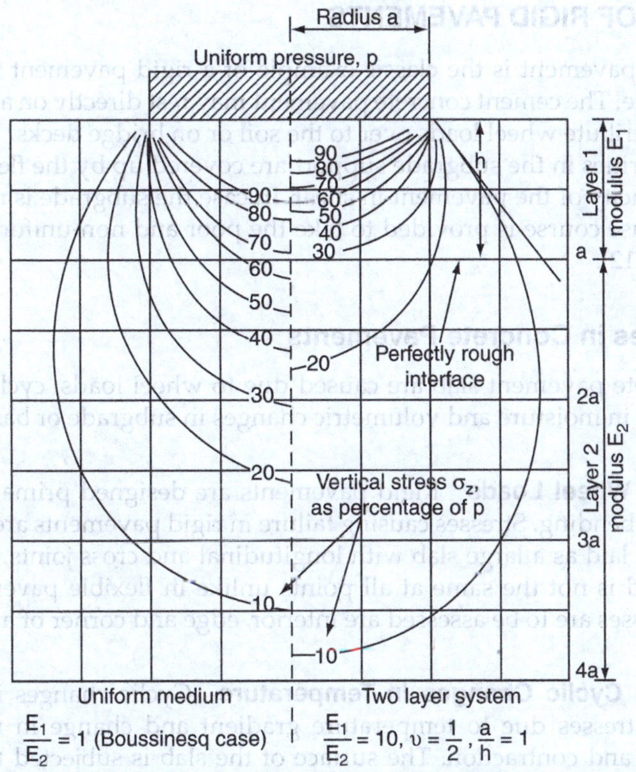


**Fig. 24.10** Comparison of vertical stress distribution by Boussinesq's and Burmister's approaches

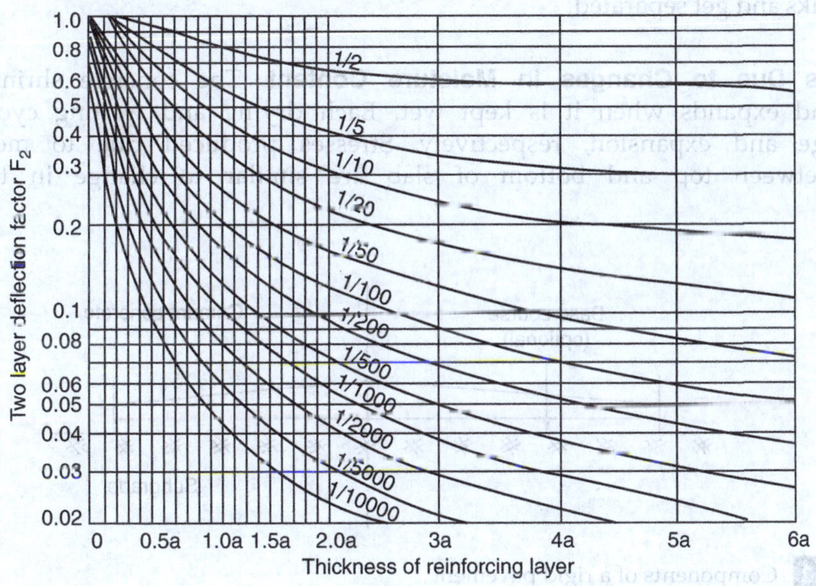


**Fig. 24.11** Relationship of $F_2$ and $h$ in a two-layer system (Burmister's method)

## 24.8 DESIGN OF RIGID PAVEMENTS

Cement concrete pavement is the classic example of a rigid pavement which provides an ideal traffic surface. The cement concrete pavement may rest directly on a properly prepared subgrade and distribute wheel loads over to the soil or on bridge decks.

Minor irregularities in the subgrade support are covered up by the flexural strength and load-transfer capacity of the pavement in shear. In case the subgrade is not proper and satisfactory, a sub-base course is provided to take the poor and non-uniform condition of the subgrade (Fig. 24.12).

### 24.8.1 Stresses in Concrete Pavements

Stresses in concrete pavement slab are caused due to wheel loads, cyclic changes in temperature, changes in moisture and volumetric changes in subgrade or base course.

**Stresses Due to Wheel Loads** Rigid pavements are designed primarily as the basis of their resistance to bending. Stresses causing failure in rigid pavements are in general tensile. As the concrete is laid as a large slab with longitudinal and cross joints, the induced stress due to wheel load is not the same at all points unlike in flexible pavements. The critical points where stresses are to be assessed are interior, edge and corner of a slab.

**Stresses Due to Cyclic Changes in Temperature** Cyclic changes in air temperature causes warping stresses due to temperature gradient and change in mean temperature causes expansion and contraction. The surface of the slab is subjected to a wide range of temperatures, whereas the subgrade or base remains relatively at constant temperature. Repeated change of temperature in night and day cause cracks on the slab and ultimately the slab breaks and get separated.

**Stresses Due to Changes in Moisture Content** The concrete shrinks when it dries and expands when it is kept wet. Each drying and wetting cycle causes a shrinkage and expansion, respectively. Stresses produced due to moisture gradient between top and bottom of slab are similar to change in temperature.

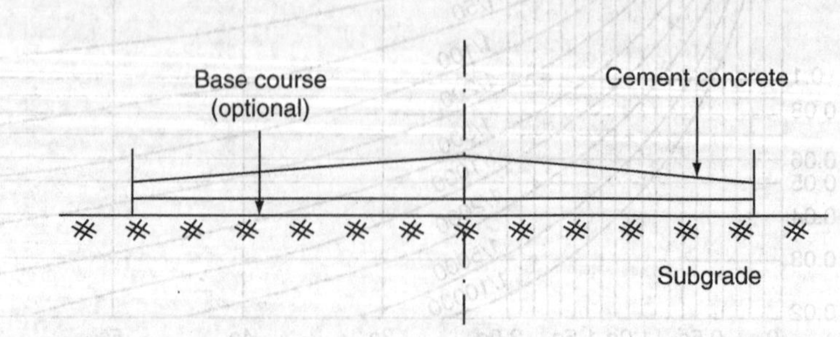


Fig. 24.12 Components of a rigid pavement

As the magnitude of stress produced due to moisture change is less, compared to other stresses, it is ignored.

**Stresses Due to Volumetric Changes of Subgrade** Due to changes in moisture content, because of capillary waters or otherwise, subgrade soils are subjected to volume changes. Clayey soils, particularly swelling soils, are much more affected than other soils.

Such variation in moisture in the subgrade causes non-uniform swelling and shrinking resulting in non-uniform support under the slab. Further, their non-uniform subgrade support may give rise to high stress concentration at certain parts of the pavements.

Sometimes, non-uniformity of subgrade may lead to subsidence causing lateral displacement of pavement slab. The movement may be accelerated due to further shrinking and swelling.

**Resultant Critical Stress and Location** Stresses, discussed above, in a combined form may cause distress on the pavement at a particular location as a resultant stress. The worst condition depends on the time of variation of temperature and moisture as loading is a continuous process.

The three critical locations of stresses in rigid pavement are loading at corner, loading at edge and loading at interior. The resultant stress varies on the atmospheric condition, day and night times, moisture variation in subgrade etc. In total, the three critical locations in the degree of decrease in intensity of stress are corner, edge and interior.

### 24.8.2 Westergaard's Stress Equations

Westergaard has postulated a theory based on the condition that the subgrade reaction is proportional to the deflection and the cement concrete slab is assumed to be homogenous and to have uniform elastic properties.

He has given expressions for interior, edge and corner loadings.

Stress, $S_i$, for the interior loading is given as:

$$S_i = \frac{0.316\,P}{h^2}\left[4\log_{10}\left(\frac{l}{b}\right)+1.069\right] \tag{24.4}$$

Stress, $S_e$, for edge loading is given as:

$$S_e = \frac{0.572\,P}{h^2}\left[4\log_{10}\left(\frac{l}{b}\right)+0.359\right] \tag{24.5}$$

Stress, $S_c$, for corner loading is given as:

$$S_c = \frac{3P}{h^2}\left[1-\left(\frac{a\sqrt{2}}{l}\right)^{0.6}\right] \tag{24.6}$$

where $P$ = wheel load, kg
$h$ = slab thickness, cm
$l$ = radius of relative stiffness, cm

$$l = \left[\frac{Eh^2}{12(1-v^2)k}\right]^{1/4} \tag{24.7}$$

$E$ = modulus of elasticity of cement concrete (kg/cm$^2$)
$\mu$ = Poisson's ratio of cement concrete slab
$k$ = modulus of subgrade reaction (kg/cm$^2$ per cm)
$b$ = equivalent radius of resisting section, cm

$$b = \sqrt{1.6a^2 + h^2} - 0.675h \tag{24.8}$$

$a$ = radius of wheel-load distribution, cm
When $a > 1.724h$, the value of $b = a$.

In order to compute the slab thickness, $h$, for a given maximum stress of $S_i$, $S_e$, or $S_c$, trial and error approach has to be resorted to.

### 24.8.3 Indian Road Congress (IRC) Formula (IRC 15, 1981)

Westergaard's stress equations have been modified by different people. Among the three locations, the edge and corner loading are generally critical.

Thus, Indian Road Congress has recommended a formula for edge and corner loading.

**IRC Edge-loading Formula** This equation is based on Westergaard's equation, but modified by Teller and Sutherland as:

$$S_e = 0.529\frac{P}{h^2}(1+0.54\mu)\left(4\log_{10}\frac{l}{b} + \log_{10} b - 0.4048\right) \tag{24.9}$$

**IRC Corner Load Formula** This equation is based on Westergaard's equation, but modified by Kelley as:

$$S_c = \frac{3P}{h^2}\left[1 - \left(\frac{a\sqrt{2}}{l}\right)^{1/2}\right] \tag{24.10}$$

The above equations and for finding load stresses at the edge and corner region are presented in the form of charts by IRC in Figure 24.13.

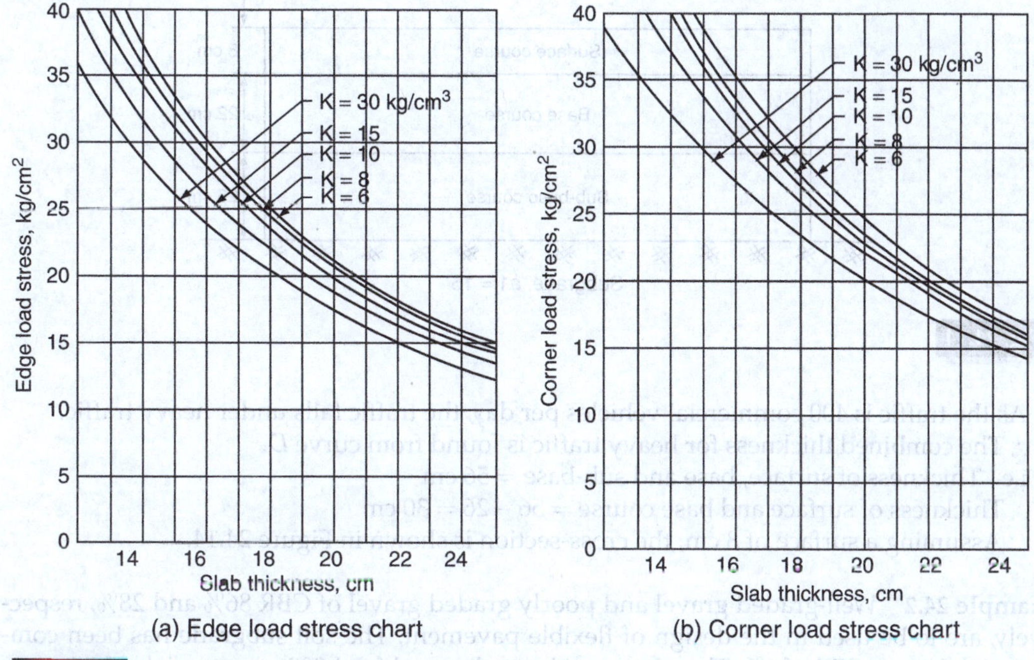

(a) Edge load stress chart (b) Corner load stress chart

**Fig. 24.13** IRC edge- and corner-load charts

These charts are applicable for a particular set of design parameters, viz.,

$$P = 4{,}100 \text{ kg}$$

$$E = 3 \times 10^5 \text{ kg/cm}^2$$

$$v = 0.15$$

$$a = 15 \text{ cm}$$

and curves for different $K$ values are provided. The design curves are applicable for slab thickness $h$, 15 to 25 cm.

## WORKED EXAMPLES

**Example 24.1** In a highway project, it is decided to provide flexible pavement on the basis of thickness obtained from group index method. The average group index in that area is 15. Design the pavement thickness if the anticipated traffic is over 400 commercial vehicles per day.

***Solution***

As the average group index is 15, the subgrade may be grouped under very poor category. From Figure 24.14 (b), the thickness of sub-base only from curve $A = 26$ cm.

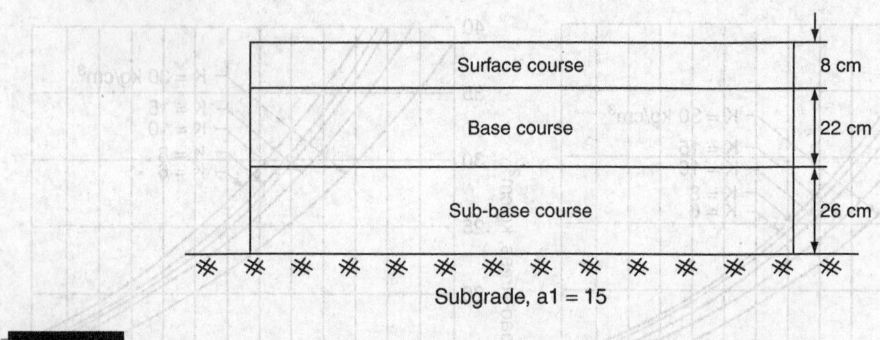


Fig. 24.14

As the traffic is 400 commercial vehicles per day, the traffic falls under heavy traffic.
∴ The combined thickness for heavy traffic is found from curve *D*.
i.e., Thickness of surface, base and sub-base = 56 cm
∴ Thickness of surface and base course = 56 − 26 = 30 cm
Assuming a surface of 8 cm, the cross-section is shown in Figure 24.14.

**Example 24.2** Well-graded gravel and poorly graded gravel of CBR 86% and 28%, respectively, are to be used in the design of flexible pavement. The soil subgrade has been compacted to have a CBR of 6%. The pavement has to be used for 4,000 commercial vehicles per day. Design the pavement using IRC curves. The bituminous surfacing is to be of 8 cm thick. The natural soil has a CBR of 4%.

***Solution***
As the number of commercial vehicles for which the road is to be used is 4,000, the design curve (Figure 24.5b) *F* is to be used.

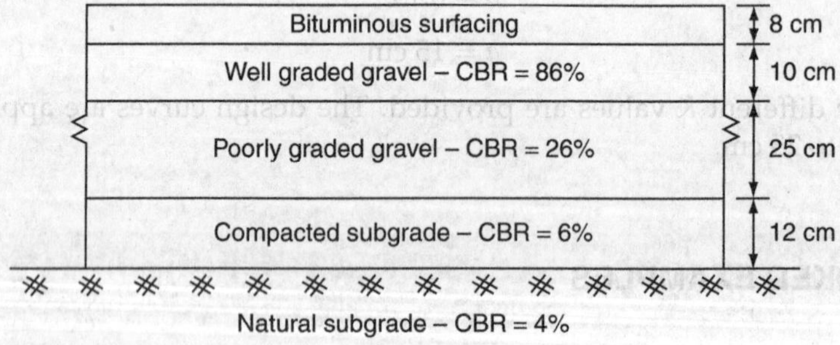


Fig. 24.15

$$\left.\begin{array}{l}\text{Total thickness needed for the natural}\\ \text{soil subgrade with a CBR of 4\%}\end{array}\right\} = 55 \text{ cm}$$

$$\left.\begin{array}{l}\text{Thickness needed to cover the compacted}\\ \text{subgrade of with CBR 26\%}\end{array}\right\} = 43 \text{ cm}$$

$$\left.\begin{array}{r}\text{Thickness needed to cover the poorly}\\ \text{graded gravel with CBR} = 28\%\end{array}\right\} = 8 \text{ cm}$$

$$\left.\begin{array}{r}\text{Total thickness over natural soil}\\ \text{subgrade}\end{array}\right\} = 55 \text{ cm}$$

Thickness of compacted soil subgrade = 55 – 43 = 12 cm
Thickness of poorly graded gravel = 43 – 18 = 25 cm
Thickness of well-graded gravel =18 – 8 = 10 cm
Bituminous surfacing = 8 cm
Thicknesses of different layers are shown in Fig. 24.15.

**Example 24.3** The flexural strength of concrete is 48 kg/cm$^2$ and modulus of elasticity is $2.8 \times 10^5$ kg/cm$^2$.The maximum wheel load is 4,200 kg with an impact factor of 5%. The tyre pressure is 5.2 kg/cm$^2$ and the modulus of subgrade reaction is 6.0 kg/cm$^2$. Assuming a factor of safety of 2.5, find the thickness of cement concrete pavement. Use Westergaard's corner-stress equation.

***Solution***

$$\left.\begin{array}{r}\text{Allowable flexural}\\ \text{strength}\end{array}\right\} = \frac{\text{Flexural strength}}{\text{Factor of safety}}$$

$$= \frac{48}{2.5} = 19.2 \text{ kg/cm}^2$$

Maximum wheel load = 4,200 kg

Impact factor = 5%

$$\text{Design wheel load} = 4,200 + \frac{5}{100} \times 4,200$$

$$= 4,410 \text{ kg.}$$

A pavement thickness of 20 cm is assumed and to calculate the radius of relative stiffness. Thus,

$$\text{Radius of contact area, } a = \sqrt{\frac{P}{pu}}$$

$$\therefore \quad a = \sqrt{\frac{4,410}{5.2 \times u}} = 16.43 \text{ cm}$$

$$\text{Radius of relative stiffness,} \quad l = \sqrt[4]{\frac{Eh^3}{12(1-v^2)K}}$$

$$= \sqrt[4]{\frac{2.8 \times 10^5 \times 20^3}{12(1-0.15^2)^6}}$$

$$= 75.11 \text{ cm}$$

Corner tensile stress, $$S_c = \frac{3P}{h^2}\left[1-\left(\frac{a\sqrt{2}}{l}\right)^{0.6}\right]$$

$$= \frac{3\times 4,410}{20^2}\left[1-\left(\frac{16.43\sqrt{2}}{75.11}\right)^{0.6}\right]$$

$$= 33.075\left[1-(0.309)^{0.6}\right].$$

$$= 16.74 \text{ kg/cm}^2$$

## POINTS TO REMEMBER

24.1 A pavement is a main structure of a road or a runway, which is relatively a stable layer or a crust constructed over the natural soil or subgrade. The main function of a pavement is to support and distribute the wheel loads of vehicles through the underlying layers to the subgrade.

24.2 Pavements may be classified into two major categories, namely, flexible pavement and rigid pavement.

24.3 Flexible pavement is the one which distributes the wheel load stresses to the subgrade through successive layers of granular materials.

24.4 Cement concrete pavement is the classic example of a rigid pavement, which provides an ideal traffic service. Cement concrete pavements may be directly placed on a properly prepared subgrade or a base.

24.5 The design wheel load depends on the (i) maximum wheel load, (ii) contact pressure, (iii) equivalent single-wheel load and (iv) repetition of loads.

24.6 The components of a flexible pavement are a sub-base course, base course and a surface course.

24.7 Stress in the concrete pavement slabs are caused due to wheel loads, cyclic changes in temperature, changes in moisture and volumetric changes in subgrade or sub-base.

24.8 Stresses in combined form may cause distress at a particular location of a pavement due to the resultant stress. The worst condition depends on the time of variation of temperature and moisture, as loading is a continuous process.

24.9 Critical positions of load are corner, edge and interior location of a pavement.

24.10 Westergaard's concept of temperature stresses has shown that temperature tends to produce two types of stresses in a concrete pavement, namely, warping stresses and frictional stresses.

## QUESTIONS

### Objective Questions

24.1 Bottom-most layer of pavement is known as
(a) Base course. (b) Sub-base course.
(c) Wearing course. (d) Subgrade.

24.2 Equivalent wheel load of loads of intensity '$p$' each is
(a) $2p$. (b) Greater than $2p$.
(c) In between $2p$. (d) Less than $2p$.

24.3 CBR method of design of flexible pavement is
(a) A theoretical method.
(b) An empirical method based on physical properties of subgrade soil.
(c) A semi-empirical method.
(d) An empirical method based on strength characteristics of subgrade soil.

24.4 The IRC recommended flexible pavement design method is
(a) CBR method. (b) Group index method.
(c) Westergaard's method. (d) Benkelman Beams method.

24.5 If the Group Index value of subgrade is between 6 and 10, then the subgrade is treated as
(a) Good. (b) Fair.
(c) Poor. (d) Very poor.

24.6 Flexible pavement distributes the wheel load
(a) Directly to subgrade.
(b) Through base course only.
(c) Through base and sub-base courses.
(d) Through sub-base course only.

24.7 Rigidity factor for a tyre pressure greater than 7 kg/cm$^2$ is
(a) Zero. (b) Greater than 1.
(c) Less than 1. (d) Equal to 1.

24.8 Inflation pressure influences the
(a) Total depth of pavement. (b) Quality of surface course.
(c) Both (a) and (b). (d) None of the above.

24.9 Critical location of wheel load in a rigid pavement design is taken as
(a) Interior-loading. (b) Edge-loading.
(c) Corner-loading. (d) Edge and corner loading.

24.10 Radius of relative stiffness is a function of
(a) $E$. (b) $h$ and $K$.
(c) $v$. (d) All the above.

### Descriptive Questions

24.11 Differentiate between rigid and flexible pavements.
24.12 What is equivalent single-wheel load? Explain briefly.
24.13 State how variations in climatic conditions affect the design of pavements.
24.14 Explain the Group Index method.
24.15 What are the methods of flexible-pavement design? Write the IRC recommendations for CBR method.
24.16 Critically appraise the CBR method of flexible-pavement design.
24.17 How Burmister's two-layer theory is used for the design of highway flexible pavement?
24.18 Bring out the merits and demerits of cement concrete pavement.
24.19 Describe briefly the factors influencing the design of rigid pavements.
24.20 Discuss the stresses caused in cement concrete pavements.
24.21 What are the critical resultant stresses and their locations?
24.22 Explain Westergaard's stress equation for cement concrete pavement.
24.23 Explain temperature stresses that develop in cement concrete pavement.

## EXERCISE PROBLEMS

24.1 Soil subgrade on a certain area showed the following properties:
Soil passing through 75μ sieve = 78%
Liquid limit = 65%
Plastic limit = 40%
Design the pavement section by Group Index method for heavy traffic with over 3500 commercial vehicles per day.

24.2 Using the following data, design the flexible pavement layers:
CBR of the subgrade soil = 5%.
CBR of poorly graded gravel sub-base = 15%.
CBR of well-graded gravel base = 80%.
Traffic of 3,200 vehicles per day may be considered.
Assume any other data found needed.

24.3 A plate-beating test was conducted with a 30 cm diameter plate on a subgrade soil, which showed a pressure of 4.2 kg/cm$^2$ at 0.50 cm deflection. Similar test conducted on a base course of 22 cm thick showed a pressure of 9.4 kg/cm$^2$ at 0.25 cm deflection. Design the pavement thickness for a wheel load of 5,800 kg with tyre pressure of 6.8 kg/cm$^2$. Use Burmister's approach.

24.4 A cement concrete pavement of 20 cm rests over a base course with a modulus of reaction of 15 kg/cm$^2$. Find the load stresses at the edges and corner region under a wheel load of 5,100 kg. By using IRC stress charts, assume $A$ = 15 cm, $E = 3 \times 10^5$ kg/cm$^2$ and $v$ = 0.15.

# List of Symbols

| Symbol | Definition |
|---|---|
| $A$ | Activity of clay |
| $A_a$ | Cross-section area |
| $A, \overline{A}$ | Pore pressure coefficient |
| $A'$ | Total inter-particle attraction |
| $A_r$ | Area ratio |
| $A_v$ | Air content or air void ratio |
| $a$ | Radius |
| $a_v$ | Coefficient of compressibility |
| $B$ | Buoyancy force |
| $B, \overline{B}$ | Pore pressure coefficient |
| $B, B_f$ | Width of footing |
| $b$ | Width |
| C | Clay fines |
| C | Inorganic clay |
| $C_c$ | Compression index |
| $C_\alpha$ | Coefficient of secondary compression |
| $C_e$ | Expansion index |
| $C_f$ | Compressibility of pore fluid |
| $C_s$ | Composite shape factor |
| $C_s$ | Compressibility of soil skeleton |
| $C_u$ | Uniformity coefficient |
| $C_z$ | Coefficient of curvature |
| $c$ | Cohesion intercept |
| $c'$ | Effective cohesion |
| $c^*$ | Modified cohesion intercept |
| $c_h$ | Coefficient of consolidation (horizontal drainage) |
| $c_m$ | Mobilized cohesion |
| $c_u$ | Undrained cohesion |
| $c_v$ | Coefficient of consolidation (vertical drainage) |
| $c_{cu}$ | Consolidated undrained cohesion |
| $c'_{cu}$ | Effective consolidated undrained cohesion |
| $c'_d$ | Effective drained cohesion |
| $c'_r$ | Effective cohesion for residual strength |
| $\overline{c}$ | Effective contact stress |
| $D, d$ | Diameter |
| $D$ | Particle size |
| $D$ | Depth |
| $D$ | Depth factor |
| $D$ | Length of drainage path |

| | |
|---|---|
| $D_d$ | Design depth of sheet pile |
| $D_s$ | Depth of scour |
| $D_f$ | Depth of foundation |
| $D_H$ | Height table position |
| $d_w$, $d'_w$ | Water table position |
| $E$ | Modulus of elasticity/stress-strain modulus |
| $e$ | Void ratio |
| $e_x$, $e_y$ | Eccentricity |
| $e_{cr}$ | Critical void ratio |
| $e_{max}$ | Maximum void ratio |
| $e_{min}$ | Minimum void ratio |
| $e_o$ | Original void ratio |
| $F$ | Factor of safety |
| $F_B$ | Factor of safety against overturning |
| $F_C$ | Factor of safety with respect to cohesion |
| $F_H$ | Factor of safety against heave |
| $F_{OT}$ | Factor of safety against overturning |
| $F_{SL}$ | Factor of safety against sliding |
| $F_a$ | Force on anchor rod |
| $F_b$ | Factor of safety with respect to height |
| $F_d$ | Driving forces |
| $F_p$ | Factor of safety against piping |
| $F_r$ | Resisting forces |
| $F_s$ | Factor of safety with respect to strength |
| $F_\phi$ | Factor of safety with respect to friction |
| $F'_p$ | Increase in factor of safety against piping |
| $f$ | Frictional resistance |
| $f$ | Frictional resistance between soil to soil |
| $G$ | Gravel |
| $G$ | Specific gravity of soil solids |
| $G$ | Shear modulus |
| GR | Groutability ratio |
| $g$ | Gravitational acceleration |
| $H$ | High compressibility |
| $H$, h | Head loss |
| $H$ | Height of dam |
| $H_c$ | Unsupported depth of excavation/critical height |
| $H_\alpha$ | Relative height of piling |
| $h_c$ | Capillary head |
| $h_c$ | Capillary rise |
| $h_u$ | Piezometric head |
| $I$ | Current |
| $I$ | Medium compressibility |
| $I$ | Moment of inertia |
| $I_B$ | Brittleness index |
| $I_D$ | Density index |
| $I_L$ | Liquidity index |
| $I_Q$ | Quality index |
| $I_c$ | Consistency index |
| $I_f$ | Flow index |

| | |
|---|---|
| $I_f$ | Influence factor |
| $I_p$ | Plasticity index |
| $I_t$ | Toughness index |
| $i$ | Hydraulic gradient |
| $i$ | Inclination of slope |
| $i_c$ | Critical hydraulic gradient |
| $i_e$ | Exit gradient |
| $K$ | Absolute permeability |
| $K$ | Coefficient of lateral pressure |
| $K_a$ | Coefficient of active earth pressure |
| $K_0$ | Earth pressure at rest |
| $K_p$ | Coefficient of passive earth pressure |
| $k$ | Coefficient of permeability |
| $k_s$ | Spring constant |
| $k_E$ | Coefficient of electrical conductivity |
| $k_e$ | Effective coefficient of permeability |
| $k_H$ | Effective coefficient of permeability in the horizontal direction |
| $k_V$ | Effective coefficient of permeability in the vertical direction |
| $L$ | Length of soil specimen/pile |
| $L$ | Low compressibility |
| $L_r$ | Recovery ratio |
| $M$ | Silt fraction fines |
| $M$ | Total mass |
| $M$ | A coefficient |
| $M_a$ | Mass of air |
| $M_d$ | Disturbing moment |
| $M_r$ | Resisting moment |
| $M_s$ | Mass of soil solids |
| $M_w$ | Mass of water |
| $m_v$ | Coefficient of volume compressibility |
| $N$ | Percentage finer |
| $N$ | Standard penetration value |
| $N_{cd}$ | Dynamic cone penetration value |
| $N_B$ | Boussinesq's vertical stress coefficient |
| $N_{CA}$ | Stress coefficient for circular loaded area – stress at any point |
| $N_{CC}$ | Stress coefficient for circular central load |
| $N_E$ | Stress coefficient for embankment loading |
| $N_\Gamma$ | Cousins stability number |
| $N_N$ | New Mark's stress coefficient |
| $N_R$ | Stress coefficient for rectangular loaded area |
| $N_S$ | Stability factor |
| $N_W$ | Westergaard's vertical stress coefficient |
| $N_c', N_q', N_\gamma'$ | Bearing capacity factors – deep foundation |
| $N_c, N_q, N_\gamma$ | Bearing capacity factors – general shear |
| $N_c', N_q', N_\gamma'$ | Bearing capacity factors – local shear |
| $N_d$ | Number of potential drops |
| $N_f$ | Number of flow channels |
| $N_\phi$ | Flow value |
| $n$ | Porosity |
| $n_d$ | Number of potential drops lost |

| | |
|---|---|
| $O$ | Organic clays and silts |
| $OCR$ | Over-consolidation ratio |
| $O_P$ | Origin of planes |
| $P$ | Poorly graded |
| $P$ | Maximum load on specimen/load at rupture |
| $P_a$ | Active thrust |
| $P_P$ | Passive resistance |
| $p_a$ | Unit active pressure |
| $p_c$ | Pre-consolidation pressure |
| $p_o$ | Overburden pressure |
| $p_P$ | Unit passive pressure |
| $p_s$ | Seepage pressure |
| $Q$ | Vertical load |
| $Q, q$ | Discharge |
| $Q_a$ | Allowable load |
| $Q_f$ | Resistance due to skin friction |
| $Q_n$ | Net ultimate load |
| $Q_p$ | Resistance due to point bearing |
| $Q_u$ | Ultimate load |
| $Q_n$ | Total drag force |
| $Q_{ug}$ | Ultimate group capacity |
| $q$ | Stress due to surcharge |
| $q$ | Uniformly distributed load/contact pressure |
| $q_a$ | Allowable soil pressure |
| $q_c$ | Cone penetration resistance |
| $q_f$ | Ultimate bearing capacity |
| $q_g$ | Gross foundation pressure |
| $q_n$ | Net foundation pressure |
| $q_{nf}$ | Net ultimate bearing capacity |
| $q_{ns}$ | Net safe bearing pressure |
| $q_s$ | Gross safe bearing pressure |
| $R$ | Radius of influence |
| $R$ | Residual factor |
| $\overline{R}_b, \overline{r}_b$ | Hydrometer reading |
| $R'$ | Total inter-particle repulsion |
| $R_n$ | Reynold's number |
| $R_w, R'_w$ | Water table correction factors |
| $r$ | Radius |
| $r_u$ | Pore pressure ratio |
| $S$ | Total settlement |
| $S$ | Section modulus of sheet pile wall |
| $S_B$ | Settlement of footing |
| $S_F$ | Seepage forces on particles |
| $S_c$ | Consolidation settlement |
| $S_i$ | Immediate settlement |
| $S_n$ | Stability number |
| $S_p$ | Settlement of plate |
| $S_r$ | Degree of saturation |
| $S_s$ | Settlement due to secondary compression |
| $S_t$ | Sensitivity of clay |

| | |
|---|---|
| $s_c, s_q, s_\gamma$ | Shape factor corrections |
| $T$ | Coefficient of transmissivity |
| $T$ | Torque |
| $T_r$ | Time factor (radial drainage) |
| $T_v$ | Time factor (vertical drainage) |
| $T_s$ | Surface tension |
| $t$ | Time |
| $U_r$ | Degree of consolidation (radial drainage) |
| $U_v$ | Degree of consolidation (vertical drainage) |
| $u_a, \overline{u}_a$ | Pore air pressure |
| $u^*$ | Equivalent pore pressure |
| $V$ | Volume |
| $V$ | Voltage |
| $V_a$ | Volume of air |
| $V_s$ | Volume of soil solids |
| $V_v$ | Volume of voids |
| $V_w$ | Volume of water |
| $v$ | Velocity |
| $v_s$ | Seepage velocity |
| W | Well graded |
| $W$ | Weight |
| $W_s$ | Weight of soil particles |
| $w$ | Water content |
| $w_L$ | Liquid limit |
| $w_P$ | Plastic limit |
| $w_s, w_{su}$ | Shrinkage limit of soil/undisturbed soil |
| $Z$ | Depth |
| $Z$ | Position head |
| $Z_0$ | Depth of tension zone |

*Greeks*

| | |
|---|---|
| $\alpha$ | Contact angle |
| $\beta$ | Inclination of retaining wall |
| $\gamma$ | Total unit weight of soil |
| $\gamma'$ | Submerged unit weight of soil |
| $\gamma_d$ | Dry unit weight of soil |
| $\gamma_s$ | Unit weight of soil solids |
| $\gamma_{sat}$ | Saturated unit weight of soil |
| $\gamma_w$ | Unit weight of water |
| $\Delta e$ | Change in void ratio |
| $\Delta H$ | Change in height |
| $\Delta u_w$ | Change in pore water pressure |
| $\Delta V$ | Change in volume |
| $\Delta\sigma$ | Change in total stress |
| $\Delta\sigma'$ | Change in effective stress |
| $\delta$ | Angle of wall friction |
| $\varepsilon$ | Strain |
| $\theta_f$ | Inclination of failure plane |
| $\mu$ | Micron |

| | |
|---|---|
| $\nu$ | Poisson's ratio |
| $\chi$ | A parameter |
| $\rho$ | Total density of soil |
| $\rho$ | Resistivity |
| $\rho$ | Flexibility number |
| $\rho'$ | Submerged density of soil |
| $\rho_d$ | Dry density of soil |
| $\rho_s$ | Density of soil solids |
| $\rho_{sat}$ | Saturated density of soil |
| $\rho_{sd}$ | Bulk density of soil |
| $\rho_w$ | Density of water |
| $\sigma$ | Total stress |
| $\sigma'$ | Effective stress |
| $\sigma''$ | Intrinsic effective stress |
| $\bar{\bar{\sigma}}$ | Mineral–mineral contact stress |
| $\sigma_{BT}$ | Brazilian tensile strength |
| $\sigma_{fa}$ | Allowable flexural stress |
| $\sigma_{ft}$ | Flexural tensile strength |
| $\sigma_h$ | Lateral stress |
| $\sigma'_h$ | Effective lateral stress |
| $\sigma_n$ | Total normal stress |
| $\sigma'_n$ | Effective normal stress |
| $\sigma_{pt}$ | Point load tensile strength |
| $\sigma_r$ | Radial stress or radial stress due to surface load |
| $\sigma_t$ | Tensile strength |
| $\sigma_v$ | Total vertical stress |
| $\sigma'_v$ | Effective vertical stress |
| $\sigma_z$ | Total vertical stress due to surface load |
| $\sigma_\theta$ | Circumferential stress due to surface load |
| $\sigma_1, \sigma_2, \sigma_3$ | Total major, intermediate and minor principal stresses |
| $\sigma'_1, \sigma'_2, \sigma'_3$ | Effective major, intermediate and minor principal stresses |
| $\tau_f$ | Peak shear strength |
| $\tau_r$ | Residual shear strength |
| $\tau_{xy}, \tau_{yz}, \tau_{zx}$ | Complementary shear stresses |
| $\phi$ | Angle of shearing resistance |
| $\phi$ | Potential function |
| $\phi'$ | Angle of shearing resistance for effective stress condition |
| $\phi^*$ | Modified friction angle |
| $\phi_{cu}$ | Consolidated undrained angle of shearing resistance |
| $\phi'_{cu}$ | Effective consolidated undrained angle of shearing resistance |
| $\phi'_d$ | Drained angle of shearing resistance |
| $\phi'_{ps}$ | Angle of shearing resistance for plain strain |
| $\phi_r$ | Residual angle of shearing resistance |
| $\phi'_{tr}$ | Angle of shearing resistance for triaxial test |
| $\phi_u$ | Undrained angle of shearing resistance |
| $\psi$ | Flow function |
| $\eta$ | Efficiency |
| $\eta$ | Viscosity of fluid |
| $\eta_g$ | Efficiency of pile group |
| $\eta_w$ | Viscosity of water |

# Bibliography

## Abbreviations in References

| | |
|---|---|
| ASCE | American Society of Civil Engineers, USA |
| ASTM | American Society for Testing and Materials, USA |
| BIS | Bureau of Indian Standards, India |
| ENR | Engineering News Record, USA |
| CGJ | Canadian Geotechnical Journal, Canada |
| GTECH | Geotechnique, Institution of Civil Engineers, London, UK |
| HRB | Highway Research Board, USA |
| ICE | Institution of Civil Engineers, London, UK |
| ICSMEF | Proceedings of International Conference on Soil Mechanics and Foundation Engineering |
| IGJ | Indian Geotechnical Journal, India |
| JGED | Journal of Geotechnical Engineering Division, ASCE (1974–2000) |
| JSMFD | Journal of Soil Mechanics and Foundation Division, ASCE (1955–1973, Incl.) |
| PICE | Proceedings of Institution of Civil Engineers, London, UK |
| STP | Special Technical Publication, ASTM |


Alam Singh and G. G. Chowdhary, *Soil Engineering,* Part 1, 3rd Ed., CBS Publishers and Distributors, New Delhi, 969 pp., 1973.

Anandakrishnan, M. and N. R. Krishnaswamy, "Response of soil-foundation systems," *Proc. 2nd South East Asian Conference on SMFE,* Vol. 1, pp. 107–119, 1969.

Anandakrishnan, M. and N. R. Krishnaswamy, "Response of embedded footings to vertical vibrations," ASCE, *JSMFED,* Vol. 99, p. 863, 1973.

AREA, *Manual of Recommended Practice,* Chicago, Construction and Maintenance Section, Engineering Division, Association of American Railroads, 1958.

Arthur, H. G., "Selection of type of dam," In: *Design of Small Dams by USBR,* Oxford and IBH Publishing Co., New Delhi, pp. 97–105, 1973.

Atterberg, A. "Uber die Physikalische Bodenuntersucshung and uber die plastiziat der Tone," *Int. Mitt. fur Bodenkund,* Vol. 1, pp. 10–43, 1911.

Babu Shankar, N., "Rankine's earth pressure theory for inclined backfills," *IGJ,* Vol. 10, pp. 361–367, 1980.

Babu Shankar, N., "Coulomb's theory of earth pressure for $c$–$\phi$ soils," *IGJ,* Vol. 1, pp. 99–104, 1981.

Baranov, V. A., "On the calculation of excited vibrations of an embedded foundation," (in Russian), *Voprosy Dynamiki Prochnocti,* No. 14, Polytechnical Institute of Riga, pp. 195–209, 1967.

Barkan, D. D., *Dynamics of Bases and Foundations,* McGraw-Hill, New York, 434 pp., 1962.

Barron, R. A., "Consolidation of fine grained soils by drain wells," *Trans. ASCE,* Vol. 113, pp. 718–742, 1948.

Barton, N. R., Lien, and J. Lunde, "Engineering classification of rock masses for design of tunnel support," *Rock Mech.*, Vol. 6, pp. 189–236, 1975.

Baumann, V. and G. E. A. Bauer, "The performance of foundation on various soils stabilised by vibro-compaction method," *CGJ*, Vol. 11, pp. 509–530, 1974.

Bell, A. L., "The lateral pressure and resistance of clay and the supporting power of clay foundations," In: *A Century of Soil Mechanics*, ICE, London, pp. 93–134, 1915.

Bell, F. G., *Fundamentals of Engineering Geology*, Butterworth, London, 648 pp., 1983.

Bennett, O. T., "Notes on embankment design," *Fourth Congress on Large Dams*, Vol. 1, p. 223, 1951.

Berezantzev, V. G. et al., "Load bearing capacity and deformation of piled foundations," *5th INCSMFE*, Vol. 2, pp. 11–15, 1961.

Berezantzev, V. G., "Design of deep foundations," *6th INCSMFE*, Vol. 2, pp. 234–237, 1965.

Bernaix, J., "New laboratory methods of studying the mechanical properties of rocks," *Int. J. Rock Mech., Min. Sci.*, Vol. 6, pp. 43–90, 1969.

Bertram, G. E., *An Experimental Investigation of Protective Filters*, Publication of the Graduate School of Engineering, Harvard University, No. 267, 1940.

Bieniawski, Z. T., "Engineering classification of jointed rock masses," *Trans. S. Afr. Inst. Civil Engineers*, Vol. 15, pp. 335–343, 1973.

Bieniawski, Z. T., "Geomechanics classification of rock masses and its application in tunnelling," *Proc. 3rd Int. Conf. Rock Mech.*, Denver, Vol. 2, 27–32, 1974a.

Bieniawski, Z. T., "Estimating the strength of rock materials," *Inst. Min. Metall.*, Vol. 74, pp. 312–320, 1974b.

Bieniawski, Z. T. (ed.), "Engineering properties of rocks with reference to tunnelling," In: *Tunnelling in Rock*, S. Afr. Inst. Civil Engineers/S. Afr. Nat. Gr. Rock Mech, CSIR, Pretoria, pp. 105–123, 1975.

Bieniawski, Z. T. (ed.), "Rock mass classification in rock engineering," *Proc. Symp. Exploration for Rock Engineering*, A. A. Balkema, Cape Town, Vol. 1, pp. 97–106, 1976.

Blum, H. Einspannungsverhtialtnisse bei Bohlwerken, Diss. Tech. Wilhelm Ernst und Sohn, Berlin, 1931.

Bjerrum, L., "Geotechnical properties of Norwegian marine clays," *GTECH*, Vol. 4, pp. 49–69, 1954.

Bjerrum, L., "Mechanism of progressive failure in slopes of over-consolidated plastic clays and clay shales," *Third Terzaghi Lecture*, ASCE, 1966.

Bishop, A. W., "The use of pore-pressure coefficients in practice," *GTECH*, Vol. 4, pp. 148–152, 1954.

Bishop, A. W., "The use of the slip circle in the stability analysis of slopes," *GTECH*, Vol. 5, pp. 1–5, 1955.

Bishop, A. W., "Some factors controlling the pore pressure set up during the construction of earth dams," *Proc. 4th INCSMFE*, Vol. 2, pp. 294–300, 1957.

Bishop, A. W., "The principle of effective stress," *Teknisk Ukeblad*, Vol. 39, p. 859, 1959.

Bishop, A. W., "The strength of soil as engineering materials," *Sixth Rankine Lecture*, Vol. 16, pp. 91–128, 1966.

Bishop, A. W., "Discussion on shear strength of stiff clays," *Proc. Geotechnical Conf.*, Oslo, Vol. 2, p. 142, 1967.

Bishop, A. W., "The influence of progressive failure on the method of stability analysis," *GTECH*, Vol. 2, p. 168, 1971.

Bishop, A. W., I. Alpan, E. E. Blight, and I. B. Donald, "Factors controlling the strength of partially saturated cohesive soils," *Proc. Research Conf. Shear Strength of Cohesive Soils*, ASCE, Boulder, pp. 503–532ff, 1960.

Bishop, A. W. and L. Bjerrum, "The relevance of the triaxial test to the solution of stability problems," *Proc. Research Conf. Shear Strength of Cohesive Soils*, ASCE, Boulder, pp. 437–501, 1960.

Bishop, A. W., G. E. Green, V. K. Garga, A. Andresen, and J. D. Brown, "A new ring shear apparatus and its application to the measurement of residual strength," *GTECH*, Vol. 21, pp. 273–328, 1971.

Bishop, A. W. and D. J. Henkel, *The Measurement of Soil Properties in the Triaxial Test*, 2nd Ed., Edward Arnold, London, 228 pp., 1962.

Bishop, A. W. and N. R. Morgenstern, "Stability coefficients for earth slopes," *GTECH*, Vol. 10, pp. 129–150, 1960.

Boussinesq, J., *Application des Potentials a L' Etude de L' Equilibre et du Movement des solides Elastiques*, Ganthier-Villars, Paris, 1885.

Bowles, J. E., *Engineering Properties of Soils and Their Measurement*, 2nd Ed., McGraw-Hill, New York, 213 pp., 1978.

Bowles, J. E., *Foundation Analysis and Design*, 3rd Ed., McGraw-Hill, New York, 816 pp., 1982.

Bowles, J. E., *Physical and Geotechnical Properties of Soils*, 2nd Ed., McGraw-Hill, New York, 478 pp., 1984.

Brace, W. F., "Brittle fracture of rocks," In: *Symp. State of Stress in Earth's Crest*, edited by W. R. Judd, Santa Monica, Elsevier, pp. 111–180, 1964.

Brace, W. F. and D. Byerlee, *Recent Experimental Studies of Brittle Fracture of Rocks*, Unpublished Report, MIT, Cambridge, MA, 1966.

Brooker, E. W. and H. O. Ireland, "Earth pressure at rest related to stress history," *CGJ*, Vol. 2, pp. 1–15, 1965.

Brown, R. E., *Vibration Compaction of Granular Hydraulic Fills*, Preprint 2657, ASCE National Water Resources and Ocean Engineering Convention, pp. 1–30, 1976.

Brown, C. B. and I. P. King, "Automatic embankment analysis," *GTECH*, Vol. 16, p. 252, 1966.

Brown, E. T. and E. Hoek, "Trends in relationships between measured *in situ* stresses and depth," *Int. J. Rock Mech. Min. Sci.*, Vol. 15, pp. 211–215, 1978.

Burland, J. B., "Shaft friction of piles in clay – A simple fundamental approach," *Ground Engineering*, Vol. 6, pp. 30–42, 1973.

Burland, J. B., et al., "The behaviour and design of large diameter bored piles in stiff clay," *Proc. Conf. on Large Bored Piles*, ICE, London, pp. 51–71, 1966.

Burmister, D. M., "The application of controlled test methods in consolidation testing," *Symposium on Consolidation Testing of Soils*, ASTM, STP No. 126, p. 83, 1951.

Burmister, D. M., "Principles of permeability testing of soils," *Symposium on Permeability of Soils*, ASTM, STP, No. 163, 1954.

Button, S. J., "The bearing capacity of footings on a two-layer cohesive subsoil," *3rd INCSMFE*, Vol. 1, pp. 332–335, 1953.

Carilo, N., "Simple two and three dimensional cases in the theory of consolidation of soil," *J. Math. Phys.*, Vol. 21, pp. 1–5, 1942.

Carroll, W. F., "Dynamic bearing capacity of soils, vertical displacements of spread footings on clay: Static and impulsive loadings," Tech. Report No. 3-599, U.S. Army Corps. Engineers, Waterways Experiment Station, Vicksburg, 1963.

Casagrande, A., "Research on Atterberg limits of soils," *Public Roads*, Vol. 13, pp. 121–136, 1932a.

Casagrande, A., "The structure of clay and its importance in foundation engineering," *Contributions to Soil Mechanics*, Boston Society of Civil Engineers, 1925–1940, Boston, pp. 72–112, 1932b.

Casagrande, A., "Characteristics of cohesionless soils affecting the stability of slopes and earthfills," *Journal of Boston Society of Civil Engineers*, Vol. 23, pp. 13–32, 1936a.

Casagrande, A., "The determination of the pre-consolidation load and its practical significance," *1st ICSMFE*, Vol. 3, pp. 60–64, 1936b.

Casagrande, A., "Seepage through dams," *Journal of New England Water Works Association*, Vol. 51, pp. 295–336, 1937.

Casagrande, A. and N. Carrillo, "Shear failure of anisotropic materials," In: *Contributions to Soil Mechanics*, Boston Society and Civil Engineers, 1940–1953, Boston, 1944.

Casagrande, A., "Classification and indentification of soils," *Trans. ASCE*, Vol. 113, pp. 901–930, 1948.

Casagrande, L., "Electro-osmotic stabilization of soils," *J. Boston Soc. Civil Engrs.*, Vol. 39, pp. 51–83, 1952.

Cernica, J. N., *Geotechnical Engineering*, Holt, Reinhart and Winston, New York, 488 pp., 1982.

Chellis, R. D., "Pile foundations," In: *Foundation Engineering,* edited by G. A. Leonards, McGraw-Hill, New York, 1962.

Chen, W. F., "Soil mechanics and theorems of limit plasticity," ASCE, *JSMFD,* Vol. 95, pp. 493–518, 1969.

Converse, F. J., "Foundations subjected to dynamic forces," In: *Foundation Engineering,* edited by G. A. Leonards, McGraw-Hill, New York, pp. 769–825, 1962.

Cook, N. G. W., "The failure of rock," *Int. J. Rock Mech. Min. Sci.,* Vol. 2, pp. 389–403, 1965.

Cook, N. G. W., E. Hoek, J. P. G. Pretorius, W. D. Ortlepp, and M. D. G. Salamon, "Rock mechanics applied to the study of rockbursts," *J. S. Afr. Inst. Min. Metall.,* Vol. 66, pp. 435–528, 1966.

Cooling, L. F. and H. O. Golder, "The analysis of the failure of an earth dam during construction," *J. ICE,* Vol. 19, pp. 38–42, 1942.

Coulomb, C. A., "Essai sur une Application des Regles des Maximis et Minimis a quelques Problems de Statique Relatifs a l' Architecture," *Mem. Acad. Roy. Des. Sci.,* Paris, Vol. 3, p. 38, 1976.

Cousins, B. F. "Stability charts for simple earth slope allowing for tension cracks," *Third Australian–New Zealand Conf. on Geomechanics,* Wellington, Australia, Vol. 2, pp. 1–5, 1980.

Cousins, B. F., "Stability charts for simple earth slopes," ASCE, *JGED,* Vol. 104, pp. 267–279, 1978.

Cozens, W. J., "Machinery foundations," *JIEE,* Vol. 82, pp. 327–328, 1938.

Craig, R. F., *Soil Mechanics,* English Language Book Society, Berkshire, 419 pp., 1986.

Cullen, R. M. and I. B. Donald, "Residual strength determination in direct shear," *Proc. First Australia–New Zealand Conf. on Geomechanics,* Vol. 1, p. 1, 1971.

Culmann, C. A., *Theorie der Stutz-und Futtermanern.* Section 8 in *Die Graphische Statik,* Meyer und Zeller, Zurich, 1866.

Culmann, C. A., *Die Graphische Statik,* Meyer und Zeller, Zurich, 644 pp., 1875.

Darcy, H., *Les Fountaines Publiques de la Ville de Dijon,* Dalmont, Paris, 674 pp., 1856.

Das, B. M., *Fundamentals of Soil Dynamics,* Elsevier, Amsterdam, 399 pp., 1990.

Das, B. M., *Principles of Foundation Engineering,* Cole Engineering Division, California, 595 pp., 1984.

Dastidar, A. G., S. Gupta, and J. K. Gosh, "Applications of sand wick in a housing project," *7th INCSMFE,* Vol. 2, pp. 59–64, 1969.

Datye, K. R., "Simpler techniques for ground improvements," *Fourth IGS Annual Lecture, IGH,* Vol. 12, pp. 1–82, 1982.

Datye, K. R. and S. S. Nagaraju, "Observation based design approach for preloading," *Proc. 5th Asian Regional Conference on SMFE,* Bangalore, pp. 19–27, 1975.

Datye, K. R. and S. S. Nagaraju, "Practical and economic aspects of the design of ground improvement system," *IGS Conference on Foundations and Excavations in Weak Soil,* Calcutta, p. c-9, 1976.

Datye, K. R. and S. S. Nagaraju, "Design approach and field control for stone columns," *10th INCSMFE,* Stockholm, Sweden, Vol. 3, pp. 637–644, 1977.

Deere, D. U., "Geological considerations," In: *Rock Mechanics,* edited by K. G. Stagg and O. C. Zienkiewic, John Wiley & Sons, New York, pp. 1–20, 1979.

Deere, D. U. and R. P. Miller, "Engineering classification and index properties for intact rock," Tech. Rep. No. AFWL-TR 65-116, Air Force Weapons Lab., Kirtland Air Base, New Mexico, 1966.

De Mello, V. F. B., "Foundations of buildings on clay, state-of-the-art report," *Proc. 7th INCSMFE,* Mexico City, Mexico, pp. 49–136, 1969.

Dhir, R. K. and C. M. Sangha, 'Relationships between size, deformation and strength for cylindrical specimens loaded in uniaxial compression," *Int. J. Rock Mech. Min. Sci. Geomech. Abstr.,* Vol. 10, pp. 699–712, 1973.

Dobson, T. and B. Slocombe, "Deep densification of granular fills," *Proc. 2nd Geotechnical Conference on Design and Construction,* Las Vegas, 21 pp., 1982.

Drucker, D. C. and W. Prager, "Soil mechanics and plastic analysis or limit design," *Q. Appl. Math.,* Vol. 10, pp. 157–165, 1952.

Dunn, I. S., L. R. Anderson, and F. W. Kiefer, *Fundamentals of Geotechnical Analysis,* John Wiley & Sons, New York, 414 pp., 1980.

Dupuit, J., *Etudes Theoriques et Practques sur le Mouvement des eaux dans les Canaux Decouverts et a travers les Terrains Permeables*, Dunod, Paris, 1863.

Fadum, R. E., "Influence values for estimating stresses in elastic foundation," *2nd ICSMFE*, Vol. 3, pp. 77–84, 1948.

Fairhurst, C., "Laboratory measurements of some physical properties of rock," *Proc. 4th Symp. on Rock Mech.*, Pennsylvania State University, pp. 105–118, 1961.

Farmer, I. W., *Engineering Properties of Rocks*, Spon, London, 1968.

Fang, H. Y., *Introduction to Environmental Geotechnology*, CRC Press, Boca Raton, New York, 652 pp., 1997.

Fellenius, W., "Calculation of the stability of earth dams," *Trans. 2nd Cong. Large Dams*, Washington, D.C., Vol. 4, p. 445, 1936.

Fellenius, W., *Erdatatische Berechnungen Mit Reibung und Kohasion (Adhasion) und Unter Annabme kreiszylindrischer Gleitflachen*, Wilhelm Ernst und Sohn, Berlin, 1947.

Foster, C. R., "Field problems: compaction," In: *Foundation Engineering*, edited by G. A. Leonards, McGraw-Hill, New York, pp. 1001–1002, 1962.

Foster, C. R. and R. G. Ahlvin, "Stresses and deflections induced by a uniform circular load," *HRB*, Vol. 33, pp. 467–470, 1954.

Fourmaintraux, D., "Characterization of rocks: Laboratory tests," In: *La Meecanique desroches appliqué aux auvrages du genie civil*, edited by Marc Panet et al., Ecole Nationale des Ponts et Chausses, Paris, 1976.

Franklin, J. A. and R. Chandra, "The Slake durability index," *Int. J. Rock Mech. Min. Sci.*, Vol. 9, pp. 325–342, 1972.

Franklin, J. E. and E. Broch., "The point load strength test," *Int. J. Rock Mech. Mining Sci.*, Vol. 9, pp. 669–697, 1972.

Franklin, J. L., *Index Properties of Rocks: Part I – Suggested Methods for Determination of Water Content, Porosity, Density and Related Properties*, Imperial College, London, 1970.

Gamble, J. C., Durability–Plasticity Classification of Shales and Other Argillaceous Rocks, Ph.D. thesis, University of Illinois, 1971.

Goel, M. C., R. P. Singh, and G. Mekonnen, "Design slopes for small homogeneous earth dams," *IGJ*, Vol. 17, pp. 485–498, 1980.

Goodman, R. E., *Introduction to Rock Mechanics*, John Wiley and Sons, New York, 478 pp., 1980.

Haefeli, R., "Creep and progressive failure in snow, soil, rock, and ice," *5th INCSMFE*, Vol. 3, p. 134, 1965.

Hall, C. E., "Compacting a dam foundation by blasting," ASCE, *JSMFD*, Vol. 88, pp. 33–51, 1962.

Hanna, A. M. and G. G. Meyerhof, "Design charts for ultimate bearing capacity for sands," *Overlying Clays, CGJ*, Vol. 17, pp. 300–303, 1980.

Hansen, J. B., *Earth Pressure Calculation*, Danish Technical Press, Copenhagen, 1953.

Hansen, J. B., "A revised and extended formula for bearing capacity," *Danish Geotechnical Institute, Bulletin*, Vol. 28, Copenhagen, 21 pp., 1970.

Hazen, A., "Physical properties of sand and gravels with reference to their use in filtration," *Rep. Mass. State Board of Health*, p. 539, 1892.

Hazen, A., *Discussion of "Dams on Sand Foundations,"* edited by A. C. Koenig, *Trans. ASCE*, Vol. 73, p. 199, 1911.

Heller, L. W., "Failure modes of impact-loaded footings on dense sand," Technical Report R-281, U.S. Naval Civil Eng. Laboratory, Port Hueneme, California, USA, 1964.

Hendron, A. J., "Mechanical properties of rock," In: *Rock Mechanics*, edited by K. G. Stagg and O. C. Zienkiewicz, John Wiley and Sons, New York, pp. 21–53, 1968.

Hoek, E., "Brittle failure of rocks," In: *Rock Mechanics*, edited by K. G. Stagg and O. C. Zienkiewicz, John Wiley and Sons, New York, pp. 99–124, 1968.

Hoek, E. and J. A. Franklin, "Sample triaxial cell for field or laboratory testing of rock," *Trans. Section A, Inst. Min. Metal.*, Vol. 77, pp. A22–A26, 1968.

Holtz, W. G. and H. J. Gibbs, "Engineering properties of expansive clays," *Trans. ASCE*, Vol. 121, pp. 641–677, 1956.

Hough, B. K., *Basic Soil Engineering*, Ronald Press Company, New York, 1969.

Housel, W. S., "A practical method for the selection of foundations based on fundamental research in soil mechanics," *Univ. Michigan Eng. Res. Bull.*, Vol. 13, 1929.

Hubert, M. K. and W. Rubey, "Mechanics of fluid-filled porous solids and its application to overthrust faulting," *Bull. Geol. Soc. Am.*, Vol. 70, p. 115, 1959.

Hunt, R. E., *Geotechnical Engineering Techniques and Practices*, McGraw-Hill, New York, 729 pp., 1986.

Hutchinson, J. N., "The free degradation of London clay cliffs," *Proc. Geotechnical Conf.*, Oslo, Norway, Vol. 1, p. 113, 1967.

Hvorslev, M. J., "Subsurface exploration and sampling of soils for civil engineering purposes," *Waterways Exp. Sta.*, Vicksburg, Mississippi, 521 pp., 1948.

Hvorslev, M. J., "Physical components of the shear strength on saturated clays," *Proc., Res. Conf. on Shear Strength of Cohesive Soils*, ASCE, Boulder, pp. 169–273, 1960.

*IRC: Specification for Road and Bridge Works*, Indian Roads Congress, New Delhi, pp. 167–181, 1986.

*IRC: Standard Specification and Code of Practice for Road Bridges*, Section VII, Indian Roads Congress, New Delhi, 1987.

IS: 401, *Code of Practice for Preservation of Timber*, BIS, 1967.

IS: 456, *Code of Practice for Plain and Reinforced Concrete*, BIS, 1978.

IS: 1080, *Design and Construction of Shallow Foundations in Soils (Other than Raft, Ring and Shell)*, BIS, 1985.

IS: 1498, *Classification and Identification of Soils for General Engineering Purposes*, BIS, 1970.

IS: 1888, *Method of Load Test on Soils*, BIS, 1982.

IS: 1892, *Code of Practice for Subsurface Investigation for Foundations*, BIS, 1979.

IS: 1893, *Criteria for Earthquake Resistant Design of Structures*, BIS, 1975.

IS: 1904, *Design and Construction of Foundations in Soils, General Requirements*, BIS, 1986.

IS: 2131, *Method for Standard Penetration Test for Soils*, BIS, 1981.

IS: 2132, *Code of Practice for Thin-Walled Tube Sampling of Soils*, BIS, 1986.

IS: 2720, *Methods of Test for Soils*, BIS.

IS: 2720 – Part 1, *Preparation of Dry Soil Samples for Various Tests*, BIS, 1983.

IS: 2720 – Part 2, *Determination of Water Content*, BIS, 1973.

IS: 2720 – Part 3/Sec. 1, *Determination of Specific Gravity – Fine Grained Soils*, BIS, 1980.

IS: 2720 – Part 3/Sec. 2, *Determination of Specific Gravity – Fine, Medium and Coarse Grained Soils*, BIS, 1981.

IS: 2720 – Part 4, *Grain Size Analysis*, BIS, 1975.

IS: 2720 – Part 5, *Determination of Liquid and Plastic Limits*, BIS, 1970.

IS: 2720 – Part 6, *Determination of Shrinkage Factors*, BIS, 1972.

IS: 2720 – Part 7, *Determination of Water Content – Dry Density Relation Using Light Compaction*, BIS, 1974.

IS: 2720 – Part 8, *Determination of Water Content – Dry Density Relation Using Heavy Compaction*, BIS, 1983.

IS: 2720 – Part 9, *Determination of Dry Density – Moisture Content Relation by Constant Weight of Soil Method*, BIS, 1971.

IS: 2720 – Part 10, *Determination of Unconfined Compressive Strength*, BIS, 1973.

IS: 2720 – Part 11, *Determination of the Shear Strength Parameters of a Specimen Tested in Unconsolidated Undrained Triaxial Compression Without the Measurement of Pore Water Pressure*, BIS, 1971.

IS: 2720 – Part 12, *Determination of Shear Strength Parameters of Soil from Consolidated Undrained Triaxial Compression Test with Measurement of Pore Water Pressure*, BIS, 1981.

IS: 2720 – Part 13, *Direct Shear Test*, BIS, 1972.

IS: 2720 – Part 14, *Determination of Density Index (Relative Density) of Cohesionless Soils*, BIS, 1983.

IS: 2720 – Part 15, *Determination of Consolidation Properties*, BIS, 1986.

IS: 2720 – Part 16, *Laboratory Determination of CBR*, BIS, 1987.

IS: 2720 – Part 17, *Laboratory Determination of Permeability*, BIS, 1986.

IS: 2720 – Part 20, *Determination of Linear Shrinkage*, BIS, 1966.

IS: 2720 – Part 21, *Determination of Total Soluble Solids*, BIS, 1977.
IS: 2720 – Part 22, *Determination of Organic Matter*, BIS, 1972.
IS: 2720 – Part 23, *Determination of Calcium Carbonate*, BIS, 1976.
IS: 2720 – Part 28, *Determination of Dry Density of Soils, In-place, by the Sand Replacement Method*, BIS, 1974.
IS: 2720 – Part 29, *Determination of Dry Density of Soils, In-place by the Core-Cutter Method*, BIS, 1975.
IS: 2720 – Part 30, *Laboratory Vane Shear Test*, BIS, 1980.
IS: 2720 – Part 31, *Field Determination of CBR*, BIS, 1990.
IS: 2720 – Part 33, *Determination of the Density of Soil, In-place, by the Ring and Water Replacement Method*, BIS, 1971.
IS: 2720 – Part 34, *Determination of the Density of Soil, In-place, by Rubber-Balloon Method*, BIS, 1972.
IS: 2720 – Part 35, *Measurement of Negative Pore Water Pressure*, BIS, 1974.
IS: 2720 – Part 36, *Laboratory Determination of Permeability of Granular Soils (Constant Head)*, BIS, 1975.
IS: 2720 – Part 38, *Compaction Control Test (Hilf Method)*, BIS, 1977.
IS: 2720 – Part 39/Sec. 1, *Direct Shear Test for Soils Containing Gravel-Laboratory Test*, BIS, 1977.
IS: 2720 – Part 40, *Determination of Free Swell Index of Soils*, BIS, 1977.
IS: 2911 – Part 1/Sec. 1, *Design and Construction of Pile Foundations – Driven Cast In Situ Concrete Piles*, BIS, 1979.
IS: 2911 – Part 1/Sec. 3, *Design and Construction of Pile Foundations – Driven Pre-cast Piles*, BIS, 1979.
IS: 2911 – Part 2, *Code of Practice for Design and Construction of Pile Foundations – Timber Piles*, BIS, 1980.
IS: 2911 – Part 3, *Code of Practice for Design and Construction of Pile Foundations – Under Reamed Piles*, BIS, 1980.
IS: 2911 – Part 4, *Load Test on Piles*, BIS, 1985.
IS: 2968 – Part 1, *Dynamic Method Using 50 mm Cone Without Bentonite Slurry*, BIS, 1976.
IS: 2974, *Code of Practice for Design and Construction of Machine Foundations*, BIS.
IS: 2974 – Part 1, *Foundations for Reciprocating Type Machines*, BIS, 1982.
IS: 2974 – Part 2, *Foundations for Impact Type Machines (Bammer Foundations)*, BIS, 1980.
IS: 2974 – Part 3, *Foundations for Rotary Machines (Medium and High Frequency)*, BIS, 1975.
IS: 2974 – Part 4, *Foundations for Rotary Type Machines for Low Frequency*, BIS, 1979.
IS: 2974 – Part 5, *Foundations for Impact Type Machines other than Hammers (Forging and Stamping Press, Pig Breaker Elevator and Hoist Tower)*, BIS, 1970.
IS: 3764, *Safety Codes for Excavation Work*, BIS, 1966.
IS: 4434, *Code for Practice for In Situ Vane Shear Test for Soils*, BIS, 1978.
IS: 4453, *Code of Practice for Subsurface Exploration by Pits, Trenches, Drifts and Shafts*, BIS, 1980.
IS: 4651 – Part 1, *Planning and Design of Ports and Harbours, Code of Practice – Site Investigation*, BIS, 1974.
IS: 4968 – Part 2, *Dynamic Method Using Cone and Bentonite Slurry*, BIS, 1976.
IS: 4968 – Part 3, *Static Cone Penetration Test*, BIS, 1976
IS: 5529, *In Situ Permeability Tests*, BIS.
IS: 5529 – Part 1, *Code of Practice for In Situ Permeability Test – Tests in Overburden*, BIS, 1985.
IS: 6403, *Code of Practice for Determination of Bearing Capacity of Shallow Foundations*, BIS, 1981.
IS: 6955, *Code of Practice for Subsurface Exploration for Earth and Rockfill Dams*, BIS, 1973.
IS: 7317, *Code of Practice for Uniaxial Jacking Test for Deformation Modulus of Rock*, BIS, 1974.
IS: 7894, *Code of Practice of Stability Analysis of Earth Dams*, BIS, 1975.
IS: 8009 Part 1, *Shallow Foundations subjected to Symmetrical Static Vertical Loads*, BIS, 1976.
IS: 8237, *Code of Practice for Protection of Slope for Reservoir Embankments*, BIS, 1976.
IS: 8414, *Guidelines for Design of Under-Seepage Control Measures for Earth and Rockfill Dams*, BIS, 1977.
IS: 8763, *Guide for Undisturbed Sampling of Sands*, BIS, 1978.
IS: 8764, *Method for Determination of Point Load Strength Index of Rocks*, BIS, 1978.
IS: 8826, *Guidelines for Large Earth and Rockfill Dams*, BIS, 1978.
IS: 9143, *Method for the Determination of Unconfined Compressive Strength of Rock Materials*, BIS, 1979.

IS: 9221, *Method of Determination of Modulus of Elasticity and Poisson's Ratio of Rock Materials in "Uniaxial Compression,"* BIS, 1979.
IS: 9429, *Code of Practice for Drainage System for Earth and Rockfill Dams*, BIS, 1980.
IS: 9640, *Specification for Split Spoon Sampler*, BIS, 1980.
IS: 9759, *Guidelines for Dewatering During Construction*, BIS, 1981.
IS: 10060, *Code of Practice for Subsurface Investigation for Power Houses Sites*, BIS, 1981.
IS: 10082, *Method of Test for Determination of Tensile Strength by Indirect Tests on Rock Specimens*, BIS, 1981.
IS: 10108, *Code of Practice for Sampling for Soils by Thin Wall Sampler with Stationary Piston*, BIS, 1982.
IS: 10379, *Code of Practice for Field Control of Moisture and Compaction of Soils for Embankment and Sub-grade*, BIS 1982.
IS: 11385, *Code of Practice for Subsurface Exploration for Canals and Cross Drainage Works*, BIS, 1985.
IS: 11594, *Specifications for Mild Steel Thin Walled Sampling Tubes and Sampler Heads*, BIS, 1985.
IS: 12070, *Code of Practice for Design and Construction of Shallow Foundations on Rocks*, BIS, 1987.
Iyer, T. S. R., "Marine deposits," *Proc. 5th Asian Regional Conference*, Vol. 1, pp. 35–69, 1975.
Iyer, T. S. R. and S. Padmanabha Pillai, "A study of the *in situ* stresses in lateritic profile," *Proc. Symposium on Strength and Deformation Behaviour of Soils*, Bangalore, Vol. 1, pp. 159–163, 1972.
Jaeger, J. C. and N. G. W. Cook, *Fundamentals of Rock Mechanics*, Chapman & Hall, 515 pp., 1969.
Janbu, N., "Application of composite slip surface for stability analysis," *European Conference on Stability of Earth Slope*, Stockholm, Discussion, Vol. 3, pp. 43–49, 1954.
Janbu, N. "Earth pressure and bearing capacity, calculations by generalized procedure of slices," *4th INCSMFE*, Vol. 2, pp. 207–212, 1957.
Janbu, N., L. Bjerrum, and B. Kjaernsli, *Veildring ved losing av fundamentering soppgaver*, Norwegian Geotechnical Institute, Publ. No. 16, Oslo, 93 pp., 1956.
John, K. W., "An approach to rock mechanics," ASCE, *ISMFD*, Vol. 88, pp. 1–30, 1962.
Jones, C. J. F. P., *Earth Reinforcement and Soil Structures*, Butterworth, London, 183 pp., 1985.
Jumikis, A. R., *Soil Mechanics*, Van Nostrand, New York, 576 pp., 1962.
Kaniraj, S. R., "Limiting of immediate settlement of shallow foundations," *IGJ*, Vol. 7, pp. 235–247, 1977.
Kaniraj, S. R. and B. V. Ranganathan, "Limiting of settlement of shallow foundations in normally consolidated clay, *IGJ*, Vol. 7, pp. 159–177, 1977.
Katti, R., K. Kulkarni, V. S. Chandrasekaran, G. Venkatachalan, and D. M. Dewaikar, "Regional soil deposits of India, state-of-the-art report," *Proc. 5th Asian Regional Conference*, Bangalore, Vol. 2, pp. 35–52, 1975.
Ko, H. Y. and L. W. Davidson, "Bearing capacity of footings in plane strain," ASCE, *JSMFD*, Vol. 99, pp. 1–23, 1973.
Koerner, R. M., *Construction and Geotechnical Methods in Foundation Engineering*, McGraw-Hill, New York, 496 pp., 1985.
Kotter, F., *Die Bestimmung des bruckes an gekrummten Geeitflachen*, Sitzugsber, kgl, Preuss, Akad. der Wiss, Berlin, 1903.
Kravetz, G. A., "Cement and clay grouting of foundations, the use of clay in pressure grouting," ASCE, *JSMFD*, Vol. 84, pp. 1546-1-30, 1958.
Krishnaswamy, N. R., "Study of surcharge effects on footings vibrations," *Geotechnical Engineering*, Vol. 16, pp. 61–64, 1975.
Krishnaswamy, N. R. and M. Anandakrishnan, "Influence on soil moisture on footing vibrations," *Proc. 5th Asian Regional Conference*, Bangalore, Vol. 1, pp. 295–298, 1975.
Ladd, C. C., *Strength and Compressibility of Saturated Clays*, Pan American Soils Course, Universidad Catolica Andres Bello, Caracas, Venezuela, 1967.
Lade, P. J. and K. L. Lee, *Engineering Properties of Soils*, Soil Mech. Laboratory, UCLA-Eng. 7652, 1976.
Lambe, T. W., "How dry is a dry soil," *HRB*, p. 491, 1949.
Lambe, T. W., *Soil Testing for Engineers*, Wiley Eastern, New Delhi, 165 pp., 1951.
Lambe, T. W., *The Structure of Inorganic Soils*, ASCE, STP No. 315, Vol. 79, 1953.

Lambe, T. W., *The Permeability of Compacted Fine Grained Soils*, ASTM, STP No. 163, 1955.

Lambe, T. W., *The Engineering Behaviour of Compacted Clay*, ASCE, *JSMFD*, Vol. 84, pp. 1665-1-35, 1958.

Lambe, T. W., "A mechanistic picture of shear strength in clay," *Proc. Research Conf. on Shear Strength of Cohesive Soils*, ASCE, Boulder, p. 437, 1960.

Lambe, T. W., "Soil stabilization," In: *Foundation Engineering*, Edited by G. A. Leonards, McGraw-Hill, New York, pp. 351–437, 1962.

Lambe, T. W., "Stress path method," ASCE, *JSMFD*, Vol. 93, pp. 309–331, 1967.

Lambe, T. W. and R. V. Whitman, *Soil Mechanics*, Spl. version, Wiley Eastern, New Delhi, 553 pp., 1979.

Lane, K. S. and D. E. Washburn, "Capillarity tests by capillarimeter and by soil filled tubes," *HRB*, p. 1, 1946.

Layman, A. K. B., "Compaction of cohesionless foundation soils by explosives," *Trans. ASCE*, Vol. 107, pp. 1330–1348, 1942.

Leonards, G. A., "Engineering properties of soils," In: *Foundation Engineering*, edited by G. A. Leonards, McGraw-Hill, New York, pp. 66–240, 1962.

Leonards, G. A. and B. K. Ramiah, "Time effects in the consolidation of clay," *Symposium on Time Rates of Loading in Testing Soils*, ASTM, STP No. 254, p. 116, 1959.

MacDonald, D. H. and A. W. Skempton, "A survey of comparisons between calculated and observed stresses and displacements, *ICE*, London, pp. 318–337, 1955.

Malhotra, B. R. and D. Chandra, "Sub sieve particle size analysis by different methods," *ICJ*, Vol. 12, pp. 271–280, 1982.

Mansur, C. I. and R. I., Kaufman, "Dewatering," In: *Foundation Engineering*, edited by G. A. Leonards, McGraw-Hill, New York, pp. 241–350, 1962.

Marshall, T. J., *A Plummet Balance for Measuring the size Distribution of Soil Particles*, CSIRO Publication, pp. 142–147, 1956.

McCarthy, D. F., *Essentials of Soil Mechanics and Foundations*, Reston Publishing Company, Virginia, 632 pp., 1982.

Menard, L. and Y. Broise, "Theoretical and practical aspects of dynamic consolidation," *GTECH*, Vol. 25, pp. 3–18, 1975.

Meyerhof, G. G., "The ultimate bearing capacity of foundations," *GTECH*, Vol. 2, pp. 301–331, 1951.

Meyerhof, G. G., "The bearing capacity of foundations under eccentric and inclined loads,", *3rd ICSMFE*, Vol. 1, pp. 440–445, 1953.

Meyerhof, G. G., "Penetration tests and bearing capacity of cohesionless soils," ASCE, *JSMFD*, Vol. 82, pp. 1–19, 1956.

Meyerhof, G. G., "The ultimate bearing capacity of foundations on slopes," 4th *IMCSMFE*, Vol. 1, pp. 384–386, 1957.

Meyerhof, G. G., "Some recent research on the bearing capacity of foundations," *CGJ*, Vol. 1, pp. 16–26, 1963.

Meyerhof, G. G., "Shallow foundations," ASCE, *JSMFD*, Vol. 91, pp. 21–31, 1965.

Meyerhof, G. G., "Ultimate bearing capacity of footings on sand layer overlaying clay," *CGJ*, Vol. 11, pp. 223–229, 1974.

Meyerhof, G. G., "Bearing capacity and settlement of pile foundations," ASCE, *JGED*, Vol. 102, pp. 195–228, 1976.

Meyerhof, G. G. and Adams, J. I., "The ultimate uplift of capacity of foundations," *JSMFD*, ASCE, Vol. 58, SM 6, pp. 1–29, 1968.

Michaels, A. S. and C. S. Lin, "The permeability of kaolinite," *Ind. Eng. Chem.*, Vol. 46, pp. 1239–1246, 1954.

Mitchell, J. K., "In-place treatment of foundation soils," ASCE, *JSMFD*, Vol. 96, pp. 73–110, 1970.

Mitchell, J. K., "Stabilization of soils for foundation of structures," Report prepared for U.S. Army Waterways Experiment Station, pp. 1–34, 1976.

Mohan, D., "A close look at problems of research and its application to pile foundations," Third IGS Annual Lecture, *IGJ*, Vol. 11, pp. 1–41, 1981.

Mohan, D., G. S. Jain, and D. Sharma, "Bearing capacity of multiple under-reamed bored piles," *Proc. 3rd Asian Conf. on SM & FE*, Haifa, Vol. 1, pp. 103–106, 1967.

Mohan, D., G. S. Jain, D. P. Sen Gupta, and Devendra Sharma, "Consolidation of ground by vertical rope drains," *IGJ*, Vol. 7, pp. 106–115, 1977.

Mohan, D., V. N. S. Murthy, and G. S. Jain, "Design and construction of multi under-reamed piles," *Proc. 7th INCSMFE*, Vol. 2, pp, 183–186, 1969.

Mohr, O., *Uber die Darstellung des spannun-gazustandes und des Deformation-zustandes eines korper-elements*, Zivilingenieur, 1882.

Montgomery, C. W., *Physical Geology, WMC Brown Publishers*, New York, 2nd Ed., 555 pp., 1990.

Moore, P. J., "Vibration criteria," In: *Analysis and Design of Foundations for Vibrations*, edited by P. J. Moore, Oxford and IBH Publishing Co., New Delhi, pp. 1–20, 1985.

Moore, P. J., "Design of shallow foundations," In: *Analysis and Design of Foundations for Vibrations*, edited by P. J. Moore, Oxford, IBH Publishing Co., New Delhi, pp. 1–20, 1985.

Morgenstern, N. R., "Stability charts for earth slopes during rapid drawdown," *GTECH*, Vol. 3, pp. 121–131, 1963.

Morgenstern, N. R. and V. E. Price, "The analysis of the stability of general slip surface," *GTECH*, Vol. 15, pp. 79–93, 1965.

Morgenstern, N. R. and K. D. Eigenbrod, "Classification of argillaceous soils and rocks," ASCE, *JGED*, Vol. 100, pp. 1137–1158, 1974.

Morgenstern, N. R. and V. E., Price, "A numerical method for solving the equations of stability of general slip surface," *Comp. J.*, Vol. 9, p. 388, 1967.

Nagaraj, T. S., A. Sridharan, and M. V. Paul Alexander, "*In situ* reinforced earth – An approach for deep excavation," *IGJ*, Vol. 12, pp. 101–111, 1982.

Narain, J., B. Singh, and P. Purushothama Raj, "The strain energy method of stability analysis for general slip surface," *Proc. 4th Asian Regional Conference of SMFE*, Bangkok, Vol. 1, pp. 151–157, 1971.

Newmark, N. M., *Influence Charts for Computation of Stresses in Elastic Foundations*, University of Illinois, Bull. No. 338, 1942.

Nonveiller, E., "The stability of slopes of dams composed of heterogeneous material," *3rd ICSMFE*, Vol. 2, p. 268, 1953.

Nonveiller, E., "The stability analysis of slopes with a slip surface of general shape," *6th ICSMFE*, Vol. 2, pp. 522–525, 1965.

Osterberg, J. O., "New piston type soil sampler," *ENR*, Vol. 148, pp. 77–78, 1952.

Osterberg, J. O., "Influence values for vertical stresses in a semi-infinite mass due to an embankment loading," *4th INCSMFE*, Vol. 1, pp. 393–394, 1957.

Peck, R. B., "Deep excavations and tunneling in soft ground, state-of-the art report, *7th INCSMFE*, Vol., pp. 225–290, 1969.

Peck, R. B., W. E. Hanson, and T. H. Thoruburn, *Foundation Engineering*, 2nd Ed., John Wiley & Sons, New York, 514 pp., 1947.

Petterson, K. E., "The early history of circular sliding surfaces," *GTECH*, Vol. 5, pp. 275, 1955.

Poncelet, J. V., *Memoire sur la stabilite des revetments er de loure foundations. Note addition elle sur les relatins analytiques qui lient entre elles la poussee et la butee de la terre.* Memorial de l'officiere de genie, Paris, Vol. 13, 1840.

Ponnuswamy, N., *Bridge Engineering*, Tata McGraw-Hill, New Delhi, 544 pp., 1986.

Poulos, H. G. and E. H. Davis, *Pile Foundation Analysis and Design*, John Wiley & Sons, 397 pp., 1980.

Proctor, R. R., "Design and construction of rolled earth dams (four articles)," *ENR*, Vol. 3, pp. 245–248, 286–289, 348–351, 372–376, 1933.

Prakash, S., *Introductory Soil Testing*, Asia Publishing House, New Delhi, 103 pp., 1969.

Prakash, S., *Soil Dynamics*, McGraw-Hill, New York, 426 pp. 1981.

Prandtl., "Uber die Hartc Plastischer Kirper," Nachr. Kgl. Gass Wiss. Gottingen Math, Ply, K1, 1920.

Purushothama Raj, P., "Stability analysis of slopes with general slip surface," M.E. Dissertation submitted to University of Roorkee, Roorkee, 77 pp., 1967.

Purushothama Raj, P., B. K. Ramiah, and K. Basavarajappa, "Bearing capacity of eccentrically loaded strip footing on two-layer cohesive soils," *Proc. 4th Southeast Asian Conference on Soil Engineering*, Malaysia, Vol. 1, pp. 63–75, 1975.

Purushothama Raj, P., B. K. Ramiah, and K. N. Venkatakrishna Rao, "Limit analysis for bearing capacity of shallow foundations," *Proc. Symp. on Strength and Deformation Behaviour of soils*, Vol. 1, pp. 191–196, 1972.

Purushothama Raj, P., B. K. Ramiah, and K. N. Venkatakrishna Rao, "Bearing capacity of strip footings in two-layered cohesive friction soils," *CGJ*, Vol. 11, pp. 32–45, 1974.

Ramana, Y. V. and B. Venkatanarayana, "An air porosimeter for the porosity of rocks," *Int. J. Rock Mech. Min. Sci.*, Vol. 8, pp. 29–53, 1971.

Ramana Sastry, M. V. B., S. G. Kumar, "Strength characteristics of lime stabilised clay–sand mixes," *IGJ*, Vol. 19, pp. 87–103, 1989.

Ramasamy, G., "Estimation of settlement of footings of sand – A critical reappraisal," *IGJ*, Vol. 14, pp. 112–126, 1984.

Ramasamy, G., "Settlement of footings on normally loaded and preloaded sand beds," *IGJ*, Vol. 16, pp. 364–382, 1986.

Ramasamy, G., A. S. R. Rao, and C. Prakash, "Effect of embedment on settlement of footings on sand," *IGJ*, Vol. 12, pp. 112–131, 1982.

Ramasamy, G., A. S. R. Rao, and P. S. Singh, "Influence of capillary zone on settlement of footings on sand," *IGJ*, Vol. 16, pp. 383–393, 1986.

Ramiah, B. K., L. S. Chickanagappa, and T. N. Ramamurthy, "Vertical vibrations of embedded footings," *Proc. 9th INCSMFE*, Tokyo, pp. 343–346, 1977.

Ramiah, B. K., Purushothama Raj, and V. A. Kumaresan, "Stability analysis of slopes subjected to progressive failure," *Proc. 3rd Southeast Asian Conference on soil Engineering*, Hong Kong, Vol. 1, pp. 171–177, 1972.

Ranjan, G., "Ground treated with granular piles and its response under load," *11th IGS Annual Lecture*, *IGJ*, Vol. 19, pp. 1–86, 1989.

Ranjan, G. and B. G. Rao, "Skirted granular piles for ground improvement," *Proc. 8th European Conference on Soil Mechanics and Foundation Engineering*, Helsinki, 1983.

Rankine, W. J. M., "On the stability of loose earth," *Philos. Trans. Royal Soc., London*, Vol. 147, pp. 9–27, 1857.

Rao, K. S. and B. C. Raymahashay, "Influence of clay minerals and iron oxides on selected properties of two lateritic soils," *IGJ*, Vol. 11, pp. 25–266, 1981.

Rassner, H., "Zum Erddruckproblem," *Proc. 1st Int. Conf. of Applied Mechanics*, Delft, pp. 295–311, 1924.

Rendulic, L., "Der hydrodynamische Spannungsausgleich in Zentral entwasserten Tonzylinder," *Wasserwirtschaft und Technik*, Vol. 2, pp. 250–253, 269–273, 1935.

Resenqvist, I. Th., "Physico-chemical properties of Soil–water systems," ASCE, *JSMFD*, Vol. 85, p. 31, 1959.

Richart, F. E., "Review of the theories for sand drains," *Trans. ASCE*, Vol. 124, pp. 709–736, 1957.

Richart, F. E., Jr., "Foundation vibrations," *Trans. ASCE*, Vol. 127, pp. 863–898, 1962.

Rowe, P. W., "Anchored sheet pile walls," *PICE*, Vol. 1, pp. 27–70, 1952.

Rowe, P. W., "Sheet pile walls in clay," *PICE*, Vol. 7, pp. 629–654, 1957.

Schmertmann, J. M., "The undisturbed consolidation of clay," *Trans. ASCE*, Vol. 120, pp. 1201–1233, 1955.

Schmertmann, J. M., "The measurement of *in situ* shear strength," *Spec. Conf. on In Situ Measurement of Soil Properties*, ASCE, Raleigh, Vol. 2, pp. 57–138, 1975.

Serafim, J. L., "Influence of interstitial water on the behaviour of rock masses," In: *Rock Mechanics*, edited by K. G. Stagg and O. C. Zienkiewicz, John Wiley & Sons, New York, pp. 55–97, 1968.

Serdengecti, S. and G. D. Boozer, "The effects of strain rate and temperature on the behaviour of rocks subjected to triaxial compression," *Proc. Symp. Rock. Mech. 4th Bull., Mineral Ind. Expr. Sta.*, Pennsylvania State University, No. 76, 1961.

Sen Gupta, D. P., B. Dey, and D. Rey, "Substrata treatment for a building using rope drains," *IGJ*, Vol. 10, pp. 322–331, 1980.

Sherard, J. L., R. J. Woodward, S. P. Gizienzki, and W. A. Clevenger, *Earth and Earth-Rock Dams*, John Wiley & Sons, New York, 725 pp., 1963.

Siva Reddy, A. and G. Mogaliah, "Bearing capacity of strip footings on partially saturated soils," *IGJ*, Vol. 6, pp. 220–245, 1976.

Siva Reddy, A. and R. J. Srinivasan, "Bearing capacity of footings on layered soils," ASCE, *JSMFED*, Vol. 93, pp. 83–99, 1967.

Skempton, A. W., "The bearing capacity of clays," *Proc. Building Research Congress*, Vol. 1, pp. 180–189, 1951.

Skempton, A. W., "The colloidal activity of clays," *3rd ICSMPE*, Vol. 1, pp. 57–61, 1953.

Skempton, A. W., "The pore pressure coefficients A and B," *GTECH*, Vol. 4, pp. 143–147, 1954.

Skempton, A. W., "Settlement analysis of six structures in Chicago and London," *PICE*, Vol. 4, pp. 525–544, 1955.

Skempton, A. W., "Discussion of the planning and design of the new Hong Kong airport," *PICE*, Vol. 7, pp. 305–307, 1957.

Skempton, A. W., *Effective Stress in Soils, Concrete and Rocks, Pore Pressure and Suction in Soils*, Butterworth, London, p. 4, 1961.

Skempton, A. W., "Long-term stability of clay slopes," *GTECH*, Vol. 14, pp. 77–101, 1964.

Skempton, A. W., "Some observation on tectonic shear zones," *Proc. 1st Int. Conf. on Rock Mechanics*, Vol. 1, p. 329, 1966.

Skempton, A. W. and L. Bjerrum, "A contribution to the settlement analysis of foundations on clay," *GTECH*, Vol. 7, pp. 168–178, 1957.

Skempton, A. W. and J. Hutchinson, "Stability of natural slopes and embankment foundation," *7th INCSMFE*, State of Art Volume, pp. 291–340, 1969.

Skempton, A. W. and D. J. Petley, "The strength along structural discontinuities in striff clays," *Geotechnical Conf.*, Oslo, Vol. 2, pp. 29–46, 1967.

Smith, E. A., "Pile driving analysis by the wave equation," *Trans. ASCE*, Vol. 127, pp. 1145–1193, 1962.

Som, N., "Regional deposits," *Proc. 5th Asian Regional Conference*, Bangalore, Vol. 2, pp. 70–81, 1975.

Sowers, G. F., "Engineering properties of residual soils derived from igneous and metamorphic rocks," *Proc. 2nd Pan. Am. Conf. on Soil Mechanics and Foundation Engineering*, Brazil, Vol. p. 39, 1963.

Sowers, S. F., *Introductory Soil Mechanics and Foundations: Geotechnical Engineering*, 4th Ed., Macmillan, New York, 621 pp., 1979.

Sridharan, A., "Some studies on the strength of partially saturated clays," Ph.D. Thesis, Purdue University, Indiana, 1968.

Sridharan, A. and G. Venkatappa Rao, "Mechanics controlling volume change of saturated clays and the role of the effective stress concept," *GTECH*, Vol. 23, pp. 359–382, 1973.

Sridharan, A. and G. Venkatappa Rao, "Shear strength behaviour of saturated clays and the role of the effective stress concept," *GTECH*, Vol. 29, pp. 177–193, 1979.

Stagg, K. G. and O. C. Zienkiweicz, *Rock Mechanics in Engineering Practice*, 12 contributing authors, John Wiley & Sons, New York, 442 pp., 1968.

Stamatopoulos, A. S. and P. C. Kotzias, *Soil Improvement by Preloadings*, John Wiley & Sons, New York, 261 pp., 1985.

Swami Saran, *Soil Dynamics and Machine Foundations*, Galgotia Publ., New Delhi, 486 pp., 1999.

Taylor D. W., "Stability of earth slopes," *J. Boston Soc. Civil Eng.*, Vol. 24, pp. 197–246, 1937.

Taylor D. W., "Review of pressure distribution theories, earth pressure cell investigations and pressure distribution data," Report to U.S. Army Waterways Experimental Station, 1945.

Taylor, D. W., *Fundamentals of Soil Mechanics*, Asia Publishing House, New Delhi, 700 pp., 1948.

Teng, W. C., *Foundation Design*, Prentice Hall, New Jersey, 466 pp., 1962.

Terzaghi, K., *Erdaumecbanik*, Franz Deuticke, Vienna, 1925.

Terzaghi, K., *Theoretical Soil Mechanics*, John Wiley & Sons, New York, 510 pp., 1943.

Terzaghi, K., "Conditions for the failure of concrete and rock," *Proc. ASTM*, Vol. 45, p. 777, 1945.

Terzaghi, K., "Introduction to tunnel geology," In: *Rock Tunneling with Steel Supports*, edited by R. Proctor and T. White, Commercial Shearing and Stamping Co., Youngstown, Ohio, pp. 17–99, 1946.

Terzaghi, K., "Mechanism of landslides," In: *Application of Geology in Engineering Practice*, Berkeley Vol., Geological Society of America, pp. 83–123, 1950.

Terzaghi, K. and R. B. Peck, *Soil Mechanics in Engineering Practice*, 2nd Ed., John Wiley & Sons, New York, 729 pp., 1967.

Tiedemann, B., "Ueber die schabfestigkeit, Bindiger Boden," *Die Bautechnik*, Vol. 15, pp. 400, 403, 1937.

Tolia, D. S., "The interpretation of static cone penetration tests," *IGJ*, Vol. 8, pp. 152–168, 1978.

Tomlinson, M. J., *Foundation Design and Construction*, 5th Ed., English Language Book Society, Essex, 842 pp., 1986.

Thorburn, S., "Discussion," In: *Proceedings of the Conference on the Behaviour of Piles*, Institution of Civil Engineers, London, p. 54, 1971.

Tourenq, C. and A. Denis, *The Tensile Strength of Rocks (in French)*, Lab de ponts et Chausses Paris, Research Rep. No. 4, 1970.

Tschebotarioff, G. P., "Large scale earth pressure tests with model flexible bulkheads," *GTECH*, Vol. 1, pp. 98–111, 1949.

Turnbull, W. J., "Compaction and strength tests on soil," Annual Meeting, ASCE, January, 1950.

Turnbull, W. J., A. A. Maxwell, and R. G. Ahlvin, "Stresses and deflections on homogeneous soil masses," *5th INCSMFE*, Vol. 2, 1961.

U.S. Department of Interior, *Earth Manual*, Bureau of Reclamation, Denver, 783 pp., 1963.

Van Olphen, H., *An Introduction to Clay Colloid Chemistry*, Interscience Publishers, New York, 301 pp., 1963.

Vesic, A. S., "Bearing capacity of deep foundations in sand," Highway Research Record No. 39, HRB, pp. 112–153, 1963.

Vesic, A. S., "Experiments with instruments pile groups in sand," *ASTM*, STP No. 444, pp. 177–222, 1969.

Vesic, A. S., "Analysis of ultimate loads of shallow foundations," ASCE, *JSMFD*, Vol. 99, pp. 45–73, 1973.

Vesic, A. S., *Design of Pile Foundations, National Cooperative Highway Research Program (NCHRP), Synthesis of Practice*, No. 42, Transportation Research Board, Washington, D.C., 68 pp., 1977.

Vesic, A. S., D. C. Bands, and J. M. Woodard, "An experimental study of dynamic bearing capacity of footings on sand," *Proc. 6th INCSMFE*, Montreal, Vol. 2, pp., 209–213, 1965.

Vidal, H., *The Principle of Reinforced Earth*, Highway Research Record No. 282, pp. 1–16, 1969.

Vijayvergiya, V. N. and J. A. Focht, Jr., "A new way to predict capacity of piles in clay," *4th Offshore Technology Conference*, OTC Paper No. 1718, Houston, 1972.

Vogler, U. V. and K. Kovari, "Suggested methods for determining the strength of rock materials in triaxial compression, ISRM Commission on Standardization of Laboratory and Field Tests," *Int. J. Rock Mech. Min. Sci. Geomech. Abstr.*, Vol. 15, pp. 47–51, 1978.

Walsh, J. B., "The effect of cracks in rocks on Poisson's ratio," *J. Geophys. Res.*, Vol. 70, pp. 5249–5257, 1965.

Wawersik, W. R. and C. Fairhurst, "A study of brittle rock fracture in laboratory compression experiments," *Int. J. Rock Mech. Min. Sci.*, Vol. 7, pp. 561–575, 1970.

Westergaard, H. M., "A problem of elasticity suggested by a problem in soil mechanics: Soft material reinforced by numerous strong horizontal sheets," In: *Contributions to the Mechanics of Solids*, Stephen Timoshenko 60th Anniversary Volume, New York, 1938.

Whitaker, T., "Experiments with model piles in groups," *GTECH*, Vol. 7, pp. 147–167, 1957.
Whitlow, R., *Basic Soil Mechanics*, Construction Press, London, 439 pp., 1983.
Wickham, G. E., H. R. Tiedemann, and E. H. Skinner, "Support determination based on geologic predictions," *Proc. 1st N. Am. Tunneling Conf.*, AIME, New York, pp. 43–64, 1968.
Wiebols, G. A., J. C. Jaeger, and N. G. W. Cook, "Rock property tests in a stiff testing machine," *10th Rock Mech. Symp.*, Rice University, Houston, 1968.
Wineland, J. D., "Borehole shear device," *Special Conf. on In Situ Measurement of Soil Properties*, ASCE, Raleigh, Vol. 1, pp. 511–522, 1975.
Woods, K. B. (ed.), *Highway Engineering Handbook*, McGraw-Hill, New York, pp. 11-3 to 11-80, 1959.
Wuerker, R. G., "Influence of stress rate on the strength and elasticity of rocks," *Q. Colorado School Mines*, Vol. 54, 1959.
Yamaguchi, H., "Practical formula of bearing value for two-layered ground," *Proc. 2nd Asian Regional Conference on SMFE*, Tokyo, Vol. 1, pp. 176–180, 1963.
Zanten, V. V., *Geotextiles and Geomembranes in Civil Engineering*, A. A. Balkema, Rotterdam, 658 pp., 1986.

# Index